BIOLOGY
CONCEPTS & CONNECTIONS

Campbell • Mitchell • Reece

BIOLOGY
CONCEPTS & CONNECTIONS

Neil A. Campbell

Lawrence G. Mitchell

Jane B. Reece

The Benjamin/Cummings Publishing Company, Inc.
Redwood City, California ▪ Menlo Park, California ▪
Reading, Massachusetts ▪ New York ▪ Don Mills, Ontario ▪ Wokingham, U.K. ▪
Amsterdam ▪ Bonn ▪ Paris ▪ Milan ▪ Madrid ▪ Sydney ▪ Singapore ▪ Tokyo ▪
Seoul ▪ Taipei ▪ Mexico City ▪ San Juan

Sponsoring Editor:
Edith Beard Brady (1990–1993),
Don O'Neal (1993–)

Science Executive Editor:
Robin J. Heyden

Senior Developmental Editor:
Susan Weisberg

Consulting Senior Developmental Editor:
Pat Burner

Final Reader/Reviewer:
Margot Otway

Editorial Assistants:
Natasha Banta, Hilair Chism,
Christine Ruotolo, Kimberly Viano

Executive Managing Editor:
Glenda Miles

Senior Production Editor:
Anne Friedman

Art and Design Manager:
Michele Carter

Senior Art Supervisor:
Kelly Hall

Art Supervisor:
M. Elizabeth Williamson

Art Coordinators:
Carol Ann Smallwood,
Karl Miyajima

Developmental Artists:
Carla Simmons, Barbara Cousins, Georg Klatt,
Mary Bryson, Laurie O'Keefe, Kevin Somerville

Artists:
Nea Bisek, Mary Bryson, Barbara Cousins,
Tom Dallman, Bill Glass, Illustrious, Inc.,
JAK Graphics, LTD, Georg Klatt, Laurie
O'Keefe, Carla Simmons, Kevin Somerville,
Terry Toyama, Pamela Drury-Wattenmaker.
See also Appendix 5.

Production Artists:
Steph Bradshaw, Liz Chiarolla, Val Felts,
Kelly Hall, Mark Konrad, Marshall Krasser,
Steve McGuire, Abigail Rudner,
Denise Schmidt

Production Assistants:
Val Felts, Shirley Bortoli

Senior Photo Editor:
Cecilia Mills

Photo Researchers:
Darcy Lanham Wilding, Amy Howorth

Text Design:
Gary Head, John Martucci

Cover Design:
John Martucci

Layout Artist:
Martucci Studio

Copyeditor:
Betsy Dilernia

Proofreader:
Anita Wagner

Indexer:
Katherine Pitcoff

Composition and Film Manager:
Lillian Hom

Digital Prepress:
PC&F, Inc.

Film Preparation:
Colotone Graphics

Manufacturing Supervisor:
Casimira Kostecki

Senior Manufacturing Coordinator:
Merry Free Osborn

Executive Marketing Manager:
Anne Emerson

Cover photo: © Art Wolfe
A luna moth on a lichen-covered tree. The
luna moth (*Actias luna*) belongs to the family
of giant silkworm moths. Found exclusively
in North America, the luna moth occurs
most frequently in the eastern half of the
continent, from southern Canada into
Mexico.

© 1994 by The Benjamin/Cummings Publishing Company, Inc.

Library of Congress Cataloging-in-Publication Data
Campbell, Neil A., 1946–
    Biology: concepts and connections / Neil A. Campbell, Lawrence G.
Mitchell, Jane B. Reece.
      p.   cm.
    Includes index.
    ISBN 0-8053-30920-9
    1. Biology.   I. Mitchell, Lawrence G.   II. Reece, Jane B.
III. Title.
QH308.2.C34   1993b
574--dc20   93-30071
                                                     CIP

2 3 4 5 6 7 8 9 10-DO-97 96 95 94

To Rochelle and Allison, with love

N.A.C.

To Mary, Roberta, Wesley, and Ben

L.G.M.

To my parents, George M. Reece and the late
Rose Long Reece, who showed me by their example
that learning is a lifelong endeavor

J.B.R.

About the Authors

Neil Campbell has taught general biology for 25 years and is the author of *Biology,* the most widely used text for biology majors. His enthusiasm for sharing the fun of science with students stems from his own undergraduate experience. He began at Long Beach State College as a history major, but switched to zoology after general education requirements "forced" him to take a science course. Following a B.S. from Long Beach, he earned an M.A. in Zoology from UCLA and a Ph.D. in Plant Biology from the University of California, Riverside. He has published numerous research articles on how certain desert plants survive in salty soil and how the sensitive plant (*Mimosa*) and other legumes move their leaves. His diverse teaching experiences include courses for non-biology majors at Cornell University, Pomona College, and San Bernardino Valley College, where he received the college's first Outstanding Professor Award in 1986. Dr. Campbell is currently a visiting scholar in the Department of Botany and Plant Sciences at UC Riverside.

Larry Mitchell is an affiliate professor of biological sciences at the University of Montana. He holds a B.S. in Zoology from the Pennsylvania State University and a Ph.D. in Zoology and Microbiology from the University of Montana. Dr. Mitchell has 21 years of experience teaching biology and zoology at both undergraduate and graduate levels, mostly at Iowa State University. While a professor at Iowa State, he helped develop self-paced instruction and videotaped lectures in environmental biology, as part of a science education project sponsored by the National Science

Foundation. Dr. Mitchell received the Outstanding Teacher Award at Iowa State in 1982. His research has focused on the ecology and systematics of parasitic protists. In addition to numerous research publications, he has coauthored the textbook *Zoology,* a laboratory manual, and a study guide for introductory biology. He has also developed television courses in general biology and has written, produced, and narrated programs on wildlife biology for public television. Since 1989, Dr. Mitchell has devoted most of his time to writing.

Jane Reece, like her coauthors, has had a career in biology divided among research, teaching, and textbook publishing—but in different proportions. After receiving an A.B. from Harvard University, she did graduate work at Rutgers University and at the University of California, Berkeley, where she earned a Ph.D. in Bacteriology. At Berkeley and later as a post-doctoral fellow at Stanford University, her research focused on genetic recombination in bacteria. She has taught introductory biology to a wide spectrum of students at Middlesex County College (New Jersey) and Queensborough Community College (New York). Since 1978, Dr. Reece has worked in the editorial department at Benjamin/Cummings. She has contributed to all three editions of Neil Campbell's *Biology* and is the sponsoring editor for *Molecular Biology of the Gene,* by James D. Watson et al.

Preface

So much biology, so little time! It's the problem all biology teachers face. How do we share the exciting progress in biology without crushing our students' enthusiasm and curiosity under an overload of information and vocabulary? How do we interest students in biology's connections to their own lives without neglecting basic biological concepts every educated person should understand? The subtitle of this textbook, *Concepts and Connections*, reflects the authors' convictions as teachers and scientists. We believe that students benefit more from developing and applying a framework of general biological concepts than from memorizing specific biological facts. We also believe that a conceptual framework will mean more to students if they understand its connections—how the ideas fit together and how they relate to students' lives. Many colleagues who share these teaching values encouraged us to produce a new kind of textbook; *Biology: Concepts and Connections* is our response.

Concept Modules Focus the Student on Each Chapter's Main Ideas

We created *Biology: Concepts and Connections* primarily to serve college students who are not biology majors, although it is also appropriate for courses that enroll both majors and nonmajors. We have made difficult decisions in balancing breadth and depth—hard choices about which topics to explain in greatest detail, and even harder choices about what to leave out of the book. One trait of the best classroom teachers is their skill at maintaining a focus on the big ideas of biology even as they enrich their students' understanding of these key concepts by judiciously applying a layer of details and useful terminology. It is that teaching model we adopted for *Biology: Concepts and Connections.*

Each chapter of this textbook consists of a series of **concept modules**. The heading of a module—for example, "Chemical cycles in our environment depend on bacteria"—announces a concept we think students should understand. The module's text and illustrations support the concept with explanation and evidence. The illustrations were developed along with the text to teach the concept in an integrated way. For example, if understanding the concept involves learning about some sequential process, such as how a cell builds a protein, then numbered steps in the text correspond to numbers in the illustration. In class-testing, students who described themselves as visual learners said that after they read a module heading, they next studied the figure and then read the text to learn more about the concept. Students who said they were verbal learners read the text right after reading the module heading, and then used the figure to reinforce what they had read. No matter what the student's learning style, the close integration of text and art within a module will keep the focus on the main concept.

We realize from our classroom experience that "getting lost in the forest because of the trees" is the major reason so many biology students fail to develop a conceptual understanding of life. To the beginning student, each bit of information seems equally important, and rote memorization is the unfortunate result. This book's concept modules will help students structure what they learn into a hierarchical scheme, where details and terms are clearly subordinate to the main ideas they support. Moreover, each chapter is structured to help students fit these ideas into an even broader conceptual framework. The concept modules are not isolated chunks of information, but are organized to help students structure their study of biology. Transitional statements and many transitional modules link the modules into a unified lesson that builds on the chapter introduction. In fact, a good way for a student to preview a chapter is to read the introduction and then page through the rest of the chapter to read the module headings. And after reading the chapter, reviewing the module headings is a good way to reinforce the main points.

At the end of each chapter, students will find a summary keyed to the modules, rephrasing the major ideas and directing students to the appropriate places in the chapter where they can work more on any concepts that are still giving them trouble. In crafting chapters that keep students focused on the main ideas instead of "getting lost in the forest," we hope to help students develop study skills that will serve them throughout their education.

Connections Provide Context

A conceptual framework is only as strong as its connections; without a context, even concepts are reduced to facts to be memorized for the next exam. *Biology: Concepts and Connections* helps students integrate what they learn into a context synthesized from content connections (relationships between different biological topics), evolutionary connections, ecological connections, human connections, interdisciplinary connections, and connections to science as a process.

Content Connections   A biology course can start with either molecules or ecosystems (or anything in between) and cover the same major concepts by the end of the course.

Biology is cyclical. A textbook, however, is linear, its chapters physically bound into a particular sequence. But in this text, professors will find it remarkably easy to rearrange chapter assignments to fit a variety of syllabi, including an "ecology first" course. The book's versatility is based on content connections that place each field of biology into broader context.

Because the biological level that is often most appealing to students is the organism, we link all other levels, from molecules to ecosystems, to whole organisms. To keep students connected to organisms, each chapter begins with a natural scene painted with words and pictures, an illustrated essay generally concentrating on a particular organism and something interesting about the way it lives. This opening essay sets a context for the chapter by using the organism to introduce the topic at hand, as when the endothermic great white shark introduces "Control of the Internal Environment" (Chapter 25) or the garden spider with its web of silk protein introduces "The Molecules of Cells" (Chapter 3). These chapter introductions are among the places where the book makes content connections that help students integrate the biology they learn, regardless of the order in which they study the chapters.

Evolutionary Connections   The book's strong organismal context is complemented by evolutionary connections that help students frame a historical view of biological diversity. For example, we trace the structure and physiology of plants in the evolutionary context of adaptation to the problems of living on land. The evolutionary theme is a unifying thread that runs through the entire book.

Ecological Connections   Considering organisms as members of populations, communities, and ecosystems makes their evolutionary adaptations more meaningful for students. For example, the diverse shapes, sizes, and colors of flowers begin to make sense in the context of symbiotic relationships with insects and other pollinators. *Biology: Concepts and Connections* places organisms in their environmental context and builds the interdependence of life into the student's conceptual framework.

Human Connections   Students are most readily motivated to learn biological concepts with which they can personally identify—concepts related to issues of health, economic survival, environmental quality, ethical values, and social responsibility. This book connects biology to students' lives with over eighty application modules that apply the basic concepts of chapters to health, social, and environmental concerns. In addition, the text highlights biology's relevance in many of the chapter introductions (hereditary deafness on Martha's Vineyard in Chapter 13, for example) and concept modules. And the learning aids at the end of each chapter include "Science, Technology, and Society" questions that encourage students to integrate biology and its implications into their world view. *Biology: Concepts and Connections* is not a "human biology" text, but we believe that linking concepts to human concerns makes biology more interesting and significant to our students.

Interdisciplinary Connections   *Biology: Concepts and Connections* views science as integral to a liberal education and fits biology into the context of a student's other courses. Throughout human history, science and its applications have evolved along with the rest of culture, its priorities shaped in part by the changing interests and needs of society. For example, the exploits of European explorers during the sixteenth and seventeenth centuries helped make the description and classification of biological diversity fashionable. Conversely, science and technology have had a profound impact on every other aspect of culture. The development of agriculture, for instance, was a breakthrough stage in the evolution of civilizations. Many of this book's chapter introductions, modules, and chapter-end questions relate biology to the social sciences and humanities.

Connections to Science as a Process   One of the most valuable contributions a biology course can make to students' general education is to give them experience with the power and limitations of the scientific process. *Biology: Concepts and Connections* introduces science as a process in Chapter 1 and illustrates how science works, with examples of key experiments and the historical development of theories in many other chapters. The book also personalizes science with "Talking About Science" modules, short profiles that feature influential scientists and reveal science as a social activity of creative men and women rather than an impersonal collection of facts. And many of the questions at the ends of chapters enable students to practice critical thinking by applying some elements of the scientific process: the development of testable hypotheses and the critical evaluation of evidence for and against those hypotheses. Other chapter-end questions give students a chance to explain biological concepts in their own words. Writing about biology is a great way to learn the subject. In fact, *Biology: Concepts and Connections* will work best for students who participate actively in learning how biology is connected to their lives.

* * * *

Long after students forget most of the specific facts and terms of biology, they will be left with general impressions and attitudes that formed during their semester or two in a general biology course. We hope that *Biology: Concepts and Connections* supports the instructor's goals for sharing the fun of biology. To help us do this even better in the next edition, we invite correspondence from students and instructors to:

Neil Campbell, Larry Mitchell, or Jane Reece

c/o Neil Campbell
Department of Botany and Plant Sciences
University of California
Riverside, CA 92521

Supplements

Study Guide
Richard Liebaert, Linn-Benton Community College
Written by the author of *Biology: Concepts and Connections* end-of-chapter study questions, this study guide offers a variety of interactive exercises.

Instructor's Guide
Fred Rhoades, Western Washington University
This Instructor's Guide contains lecture outlines and instructional activities. It also includes course outlines for instructors who use alternative syllabi. Also available on disk.

Test Bank
Contributions from Deborah Langsam, University of North Carolina-Charlotte, Linda Simpson, University of North Carolina-Charlotte, Lisa Shimeld, Crafton Hills College, David Tauck, Santa Clara University, and Marshall Sundberg, Louisiana State University.
The test bank includes questions at three levels: factual, conceptual, and applications-based. Also available on both IBM and Macintosh test-generating programs. (Available to qualified college adopters.)

Transparency Acetates, Masters, and Slides
All art from *Biology: Concepts and Connections* is available on either full-color acetates or black-line masters, with enlarged labels for effective classroom use. In addition, 35 mm slides are available for the same illustrations that appear as acetates. (Available to qualified college adopters.)

BioShow: The Videodisc
This is a videodisc of text art, original animations, and motion sequences to accompany *Biology: Concepts and Connections*. Art conversion from Neil Campbell's *Biology* to still figures, stepped figures, and animations was developed and executed by Tom Dallman, Ph.D. Art conversion from *Biology: Concepts and Connections* to still figures, stepped figures, and animations was developed by Iain Miller, Ph.D. BioShow is available to qualified college adopters and is accompanied by a barcode manual.

B/C Tutor Animated Tutorial Software
Donald Keefer, Loyola College
This animated tutorial program, keyed to the text, can be used by students for review or for quizzes. Available for both IBM and Macintosh.

Related titles coming in 1995
- **Laboratory Manual with Annotated Instructor's Edition and Preparatory Guide** by Jean L. Dickey, Clemson University
- **The Diversity of Life**, a complement to the diversity coverage in the text, by Lawrence G. Mitchell

Acknowledgments

One of life's great pleasures is working as part of a team. This book has been a team effort at every stage. The nucleus of our team was formed when biology editor Robin Heyden, now executive editor for science at Benjamin/Cummings, brought the three of us together as coauthors. Under Robin's leadership, the project was launched in a novel way, with the immediate formation of a book team representing the full range of publishing functions.

To our delight, our author trio has worked out even better than hoped. Not only have we had fun, but we believe the resulting book is better than any one of us could have produced individually. Senior developmental editor Susan Weisberg was our partner from the book's inception, and she has the unique distinction of being the only person other than ourselves to have read every word of our manuscript at least three times. Susan brought her uncompromisingly high standards, remarkable organizational skills, and extraordinary patience to bear on every module. Susan is a top-notch editor: She helped us keep in mind the overarching goals of the book as she helped us fine-tune the manuscript right down to the level of choosing the best color for the potatoes in a carbohydrate diagram. At the same time, Susan was a continual source of emotional support and realistic perspective.

The book has greatly benefited from the contributions of Richard Liebaert of Linn-Benton Community College, who created the Chapter Review sections. We are grateful to Richard for his patience with our constantly evolving ideas for this material and his incisive suggestions for improving the main text. Richard is also the author of the excellent study guide for the book. We thank Steve Lebsack of Linn-Benton for checking the Answers section in Appendix Three.

Another biologist who contributed substantially to the manuscript is John A. Mutchmor of Iowa State University. We thank John for creating a scientifically sound and well-designed first-draft figure manuscript for Unit Three.

In 1990, Robin Heyden was succeeded as sponsoring editor by Edith Beard Brady, another accomplished and creative editor. Edith brought new enthusiasm and a host of fresh ideas. She became our main "ear" to the needs and trends in introductory biology courses and obtained a wealth of valuable feedback on our manuscript from many dedicated instructors and students. Both Robin and Edith played major roles in shaping the book's vision and in supporting us during its creation. Don O'Neal became sponsoring editor in the spring of 1993 and has been instrumental in pulling together the large supplements package.

We are deeply indebted to all the instructors, students, and others who provided valuable criticisms and suggestions at various stages of the project. We benefited from discussions with numerous colleagues about specific topics. Charles D. Drewes, Eugenia S. Farrar, Richard J. Hoffman, and Edwin C. Powell of Iowa State University and Ira Herskowitz of UC San Francisco were especially helpful. Other colleagues and friends kindly provided photographs for use in the book (see the Photograph Credits in Appendix Four).

Over 100 instructors contributed their scientific and teaching expertise in the form of written reviews of the manuscript. In addition, 52 instructors and 15 students participated in group discussions focusing on various aspects of the book. We are particularly grateful to Mary Harris, Marshall Sundberg, and Kathy Thompson and their students at Louisiana State University for class-testing the second draft of Chapters 2–4 and 14–16. All of these people were crucial members of our team, and we gave serious attention to every one of their comments and suggestions. Nonetheless, we are fully responsible for any errors that remain. We hope readers who find problems will help us improve future editions by telling us about them.

Natasha Banta, Sissy Lemon, Christine Ruotolo, Kimberly Viano, and Thomas Viano provided editorial assistance—mocking up chapters so that reviewers could see, even in draft stages, how the text and art would actually work together; trafficking the many reviews; helping prepare the art manuscript; researching information for last-minute updating; and preparing the manuscript for production. Hilair Chism applied her biological knowledge and eagle eye to checking of the final art. We are especially grateful to Mary Mitchell, who researched new literature and suggested several of the topics for our introductory essays. Thanks also to Daniel Terdiman, Dan Gillen, Trudy Reece, and Marty Granahan for their help.

As the manuscript entered production, we were lucky to have senior developmental editor Pat Burner join the team. Benjamin/Cummings' reigning expert on art manuscript development, Pat played a major role in setting art styles that would give us figures both attractive and pedagogically effective. She also scrutinized the initial page layouts for every chapter, fine-tuning them for optimal clarity and teaching effectiveness. Another top publishing professional, Betsy Dilernia, copyedited our manuscript. Betsy's commitment to excellence and consistency is a model to us all.

Benjamin/Cummings broke new ground in using computer technology in developing and producing this book.

We are grateful to Gary Head of Publishing Principals for guiding Jane through the early months of working in QuarkXPress, for the use of his office and computer during that period, and for instructing her on the basic principles of good page layout. But most of all, we thank him for the clean, functional, inviting chapter design he created; it plays an important part in making our modular approach work. Also contributing significantly to the design and overall look of the book were Benjamin/Cummings art and design manager Michele Carter and designer John Martucci of Martucci Studio, who did most of the final page design. In this area, we also thank Bonnie Grover, who got the project off to a strong start before leaving to teach desktop publishing in Cairo, and Betty Gee, our first art supervisor.

During the production of the book, art supervisor M. Elizabeth Williamson and senior art supervisor Kelly Hall organized and trafficked hundreds of pieces of art. We appreciate their high standards and cooperative attitude when we wanted to make "one more change." We are particularly grateful to Kelly for her herculean efforts during the last two months of production. The book benefited greatly from her extensive production experience, knowledge of biology, professionalism, and creativity. Also working on the illustration program and/or layout were Shirley Bortoli, Val Felts, Denise Schmidt, Karl Miyajima, and Carol Ann Smallwood. It was a pleasure working with all of them, and they contributed to the book in no small way.

We can't talk about the look of the book—or even about how well it presents concepts and connections—without mention of our excellent artists. First, we want to thank all the artists who created art for Neil Campbell's *Biology*, because we made use of many of those pieces for this book. For new and modified figures, we were privileged to have biological artists Carla Simmons, Georg Klatt, Laurie O'Keefe, Barbara Cousins, Mary Bryson, and Kevin Somerville on the team. Carla Simmons, the principal artist on Campbell's *Biology,* also served as our art consultant. We benefited from the experience, knowledge, and talents of all these outstanding artists, both in their own drawings and paintings and in much of the artwork rendered in final form on the computer. Our hardworking computer artists were Terry Toyama, Nea Bisek, Pamela Drury-Wattenmaker, Bill Glass, Tom Dallman, Illustrious, Inc., and JAK Graphics.

Senior photo editor Cecilia Mills and photo researchers Darcy Lanham Wilding and Amy Howorth did a fine job on the challenging photo program. We appreciate their willingness to search far and wide for just the right photograph. (We could illustrate an entire chapter with the out-takes for the sloth photo in Chapter 1!)

The production of the book could not have come together as well as it did without the leadership of senior production editor Anne Friedman. Unflappably professional, Anne kept everything running smoothly, while simultaneously managing to make a number of creative contributions to the book, and we are grateful to her. We also thank executive managing editor Glenda Miles for administrative direction and advice. In addition, we thank final reader Margot Otway and indexer Kathy Pitcoff, who are both unusual in bringing a knowledge of biology to their jobs, and Benjamin/Cummings computer gurus Guy Mills and Ari Davidow. We gratefully acknowledge the important, behind-the-scenes work of composition and film buyer Lillian Hom, manufacturing supervisor Casi Kostecki, and senior manufacturing coordinator Merry Free Osborn in making the book itself an object of beauty and utility.

We also owe thanks to the Benjamin/Cummings marketing department, in particular executive marketing manager Anne Emerson. Anne has been a member of our team from the start. Through mail surveys, campus interviews, focus groups, and additional class-testing, Anne has helped us stay in touch with what professors and students want and need in an introductory biology textbook. Both the look and the content of the book reflect her input. We have enjoyed working with Anne and her staff, especially Nathalie Mainland, David Harris, Bob Ting, Rosemarie Forrest, and freelancer Karryl Nason. And throughout our years of work we have been bolstered by the enthusiastic encouragement of the sales groups.

None of this would have been possible without the confidence and encouragement of Jim Behnke (now president of the Addison-Wesley higher education group), Benjamin/Cummings president Sally Elliott, and editorial director Barbara Piercecchi, who have been supportive of the project and the authors throughout the years of the book's creation. We appreciate their willingness to make a major investment of human and material resources in the book.

Finally, we thank the members of our families who have shared with us on a daily basis the joys, frustrations, and long-term commitment of creating this book: Rochelle and Allison Campbell, Mary Mitchell, Bob Floyd, and Dan Gillen.

Neil Campbell
Larry Mitchell
Jane Reece

Reviewers

Focus Group Participants

Dan Alex
Chabot College
Ken Allen
Schoolcraft College
Dick Botwell
Missouri Western State College
Charles Brown
Santa Rosa Junior College
Virginia Buckner
Johnson County Community College
Christine Case
Skyline College
John Caruso
University of Cincinnati
Mary Colavito-Shepanski
Santa Monica College
Cindy Erwin
City College of San Francisco
Judy Daniels
Eastern Michigan University
Jean DeSaix
University of North Carolina, Chapel Hill
Eugene Fenster
Longview Community College
Michael Gaines
University of Kansas
George Garcia
University of Texas, Austin
Bill Glider
University of Nebraska, Lincoln
Richard Haas
California State University, Fresno
Margaret Heimbrook
University of Northern Colorado
Jean Helgeson
Collin County Community College

Kenneth Hutton
San Jose State University
Janis Jackson
McLennan Community College
Steve Jensen
Southwest Missouri State University
Carole Kelley
Cabrillo College
Robert Kitchin
University of Wyoming
Robert Krasner
Providence College
Paul Kugrens
Colorado State University
Jean Lagowski
University of Texas, Austin
Kenneth Laser
Salem State College
Elmo Law
University of Missouri, Kansas City
Diane Mack
William Rainey Harper College
Mike Martin
University of Michigan, Ann Arbor
Robert McGuire
University of Montevallo
Jan Mercer
Tarrant Community College–NE
Debra Meuler
Cardinal Stritch College
James Mickle
North Carolina State University
Henry Mulcahy
Suffolk University
Steve Muzos
Austin Community College–Rio Grande Campus

Gaylen Neufeld
Emporia State University
Paul Niehaus
Washtenaw Community College
Catherine O'Brien
San Jacinto College–South
James Schwarz
McLennan Community College
David Senseman
University of Texas, San Antonio
Ted Sherrill
Eastfield College
Jane Shoup
Purdue University Calumet
Gary Smith
Tarrant Community College–NE
Gil Starks
Central Michigan University
Brad Stith
University of Colorado, Denver
Richard Storey
Colorado College
Gerald Summers
University of Missouri, Columbia
Corrine Thomas
Saint Louis Community College, Meramec
Mark Wallert
Moorhead State University
Barry Welch
San Antonio College
Edwin Wodehouse
Skyline College

San Francisco City College Student Focus Group Participants

Cindy Erwin and Vic Chow, Faculty Sponsors
Sam Ang
Eric Johnson
Mildred Thomas
Quinton Tieu
Mary Yan

University of North Carolina at Chapel Hill Student Focus Group Participants

Jean DeSaix, Faculty Sponsor
Caroline Ames
Steven Austin
William Graham IV
Eugena Maria Harrington
Meredith Key

North Carolina State University Student Focus Group Participants

James Mickle, Faculty Sponsor
Susan Everett
Patrick Nolan
Dennis Rogers
Gabriel Smith
Mary Zadigian

Manuscript Reviewers

Daryl Adams
Mankato State University

Dawn Adrian Adams
Baylor University

Olushola Adeyeye
Duquesne University

Dan Alex
Chabot College

Sylvester Allred
Northern Arizona University

Jane Aloi
Saddleback College

Marjay Anderson
Howard University

Chris Barnhart
University of San Diego

William Barstow
University of Georgia

Ernest Benfield
Virginia Polytechnic Institute

Harry Bernheim
Tufts University

Richard Bliss
Yuba College

Lawrence Blumer
Morehouse College

William Bowen
University of Arkansas, Little Rock

Chris Brinegar
San Jose State University

Virginia Buckner
Johnson County Community College

Jerry Button
Portland Community College

James Cappuccino
Rockland Community College

Cathryn Cates
Tyler Junior College

Vic Chow
San Francisco City College

Mary Colavito-Shepanski
Santa Monica College

Robert Creek
Western Kentucky University

Judy Daniels
Monroe Community College

Lawrence DeFilippi
Lurleen B. Wallace College

James Dekloe
Solano Community College

Loren Denney
Southwest Missouri State University

Jean DeSaix
University of North Carolina at Chapel Hill

Jean Dickey
Clemson University

Stephen Dina
St. Louis University

Gary Donnermeyer
Iowa Central Community College

Charles Duggin
University of South Carolina

Betty Eidemiller
Lamar University

Cindy Erwin
City College of San Francisco

Nancy Eyster-Smith
Bentley College

Terence Farrell
Stetson University

Jerry Feldman
University of California, Santa Cruz

Dino Fiabane
Community College of Philadelphia

Kathleen Fisher
San Diego State University

Robert Frankis
College of Charleston

James French
Rutgers University

Shelley Gaudia
Lane Community College

Robert Gendron
Indiana University of Pennsylvania

Dana Griffin
University of Florida

Richard Haas
California State University, Fresno

Martin Hahn
William Paterson College

Leah Haimo
University of California, Riverside

Laszlo Hanzely
Northern Illinois University

Jim Harris
Utah Valley Community College

Mary Harris
Lousiana State University

Jean Helgeson
Collin County Community College

Ira Herskowitz
University of California, San Francisco

Paul Hertz
Barnard College

Margaret Hicks
David Lipscomb University

Laura Hoopes
Occidental College

Robert Howe
Suffolk University

George Hudock
Indiana University

Kris Hueftle
Pensacola Junior College

Charles Ide
Tulane University

Ursula Jander
Washburn University

Alan Jaworski
University of Georgia

Florence Juillerat
Indiana University-Purdue University, Indianapolis

Lee Kirkpatrick
Glendale Community College

Mary Rose Lamb
University of Puget Sound

Carmine Lanciani
University of Florida

Deborah Langsam
University of North Carolina, Charlotte

Steven Lebsack
Linn-Benton Community College

Richard Liebaert
Linn-Benton Community College

Ivo Lindauer
University of Northern Colorado

William Lindsay
Monterey Peninsula College

Melanie Loo
California State University, Sacramento

Joseph Marshall
West Virginia University

Presley Martin
Drexel University

William McComas
University of Iowa

Steven McCullagh
Kennesaw State College

James McGivern
Gannon University

Henry Mulcahy
Suffolk University

James Nivison
Mid Michigan Community College

Brief Contents

Contents

UNIT 2 Genetics

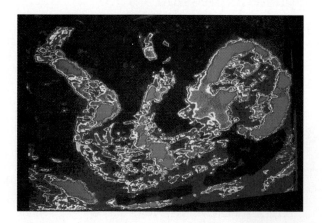

UNIT 3 Evolution and the Diversity of Life

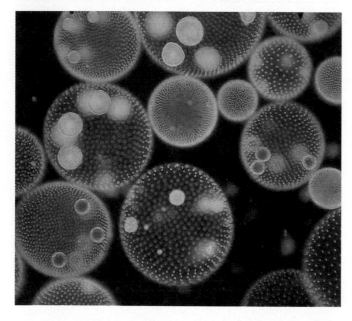

UNIT 4 Animals: Form and Function

UNIT 6 Ecology

How to Use This Book

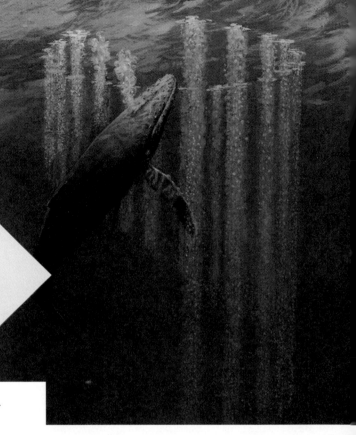

To the Student:

The following four pages show you how to use *Biology: Concepts and Connections* in the most effective way—so you can make studying easier and understand how biology connects with your own life.

Chapter openers are your entry into the chapter

■ Most chapters begin with an organism that is representative of the chapter's main subject.

■ The illustrated introductory essay tells you about the way the organism lives, its unique characteristics, and how it interacts with its environment.

■ You will find references to the organism helping to illustrate concepts later in the chapter.

19.10 Most animals have segmented bodies

The phyla we will now discuss illustrate another body feature that had profound influence on animal history—body **segmentation**, the subdivision of the body along its length into a series of repeated parts (segments). This feature played a central role in the evolution of many complex animals.

Segmentation is an obvious feature of an animal like an earthworm (Figure A), in which the segments are marked off externally by grooved rings. Internally, the coelom is partitioned by walls (only two are fully shown here). The nervous system (yellow) includes a ventral nerve cord with a dense cluster of nerve cells in each segment. Excretory organs (green), which dispose of fluid wastes, are also repeated in each segment. The digestive tract, however, is not segmented; it penetrates the segment walls and runs the length of the animal. The main channels of the circulatory system—a dorsal blood vessel and a ventral blood vessel—are also unsegmented. But they are connected by segmental vessels, including five pairs of accessory hearts near the anterior end. The main heart is simply the anterior region of the dorsal blood vessel.

The dragonfly (Figure B) is also segmented, though less uniformly than the earthworm. Its segments are most pronounced in its abdomen; its head and mid-region (thorax) are each formed from several fused segments. Each pair of its six walking legs and each pair of its four wings emerge from a body segment in the thorax.

Segmentation also occurs in the human body (Figure C). We have a backbone formed of a repeated series of bones called vertebrae, and muscles associated with our vertebrae are segmented. We also have segmented abdominal muscles, clearly visible in body builders as "stomach ripples."

A segmented body is advantageous in many ways. It allows greater body flexibility and mobility, and it probably evolved as an adaptation that aids in movement. The earthworm uses its flexible, segmented body to crawl over wet surfaces and burrow rapidly into the soil. In the dragonfly, segmentation occurs in both its tough external skeleton and the muscles that move its head, thorax, and abdomen. Also, the segments provide flexibility for flying, perching, mating, and laying eggs.

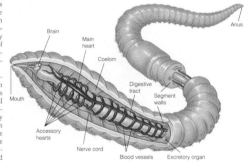

A. Segmentation in an earthworm

B. Dragonfly segmentation

C. Segmentation in the human body

Modules feature concepts

■ Each module focuses on a single concept which is stated in the heading.

■ Use the concept headings as your "road map" through the chapter.

■ To preview the chapter, read through the headings before you start studying individual modules.

■ Modules are numbered for easy reference and are never longer than two facing pages.

Whales are the largest animals in the world. Few other species, living or extinct, even approach their great size. The humpback whale, shown in the pictures here, is a medium-sized member of the whale clan. It can be 16 meters (53 ft) long and weigh up to 65,000 kg (72 tons), about as much as 70 midsize cars.

It takes an enormous amount of food to support a 72-ton animal. Humpback whales eat small fishes and crustaceans called krill. The painting at the left shows a remarkable technique they often use to corral food organisms before gulping them. Beginning about 20 meters below the ocean surface, a humpback swims slowly in an upward spiral, blowing air bubbles as it goes. The rising bubbles form a cylindrical screen, or "bubble net." Krill and fish inside the bubble net swim away from the bubbles and become concentrated in the center of the cylinder. The whale then surges up through the center of the net with its mouth open, harvesting the catch in one giant gulp.

Humpback whales are filter feeders, meaning they strain their food from seawater. Instead of teeth, these giants have an array of brushlike plates called baleen on each side of their upper jaw. You can see the white, comblike baleen in the open mouth of the whale in the photograph at the right. The baleen is used to sift food from the ocean. To start feeding, a humpback whale opens its mouth, expands its throat, and takes a huge gulp of seawater. When its mouth closes, the water squeezes out through spaces in the baleen, and a mass of food is trapped in the mouth. The food is then swallowed whole, passing into the stomach, where digestion begins. The humpback's stomach can hold about half a ton of food at a time, and in a typical day, the animal's digestive system will process as much as 2 tons of krill and fish.

The humpback and most other large whales are endangered species, having been hunted almost to extinction for meat and whale oil by the 1960s. Today, most nations honor an international ban on whaling, and some species are showing signs of recovery. Humpbacks still roam the Atlantic and Pacific oceans. They feed in polar regions during summer months and migrate to warmer oceans to breed when temperatures begin to fall. The photograph below was taken during summer in Glacier Bay, Alaska. Food is so abundant there that humpbacks harvest much more energy than they burn each day. Much of the excess is stored as a thick layer of fat, or blubber, just under their skin. After a summer of feasting, humpback whales leave Glacier Bay and head south to breeding and calving grounds off the Hawaiian Islands, some 6000 km (3600 mi) away. Living off body fat, they eat little, if at all, until they return to Alaskan waters eight months later.

In about four months, a humpback whale eats, digests, and stores as fat enough food to keep its 72-ton body active for an entire year—a remarkable feat, and a fitting introduction to this chapter on animal nutrition and digestion. We will return to the whale as we examine the diverse ways that animals obtain and process nutrients.

36.11 An energy pyramid explains why meat is a luxury for humans

The dynamics of energy flow apply to the human population as much as to other organisms. Like other consumers, we depend entirely on productivity by plants for our food. As omnivores, we eat both plant material and meat. When we eat grain or fruits, we are primary consumers; when we eat beef or other meat from herbivores, we are secondary consumers. When we eat fish like trout and salmon (which eat insects and other small animals), we are tertiary or quaternary consumers.

The energy pyramid on the left below indicates energy flow from primary producers to humans as vegetarians. The energy in the producer trophic level comes from a corn crop. The pyramid on the right illustrates energy flow from the same corn crop, with humans as secondary consumers, eating cattle. These pyramids are generalized models, based on the rough estimate that about 10% of the energy available in a trophic level appears at the next higher trophic level. Thus, the pyramids indicate that the human population has about ten times more energy available to it when people eat grain than when they process the same amount of grain through another trophic level and eat grain-fed beef. Put another way, the pyramids indicate that it takes about ten times more energy to feed the human population when we eat meat than when we eat plants directly.

Actually, the 10% figure is high for energy flow involving cattle and humans. As endotherms, cattle expend a great deal of the energy they take in on heat production—much more than do ectotherms, such as grasshoppers (see Module 25.4). Accounting for the energy loss in heat production, it may actually take closer to 100 times more energy to feed us on cattle (and other mammals and birds) than on plants directly.

Eating meat of any kind is an expensive luxury, both economically and environmentally. In many countries, people cannot afford to buy much meat or their country cannot afford to produce it, and people are vegetarians by necessity. Whenever meat is eaten, producing it requires that more land be cultivated, more water be used for irrigation, and more chemical fertilizers and pesticides be applied to croplands used for growing grain. It is likely that, as the human population expands, meat consumption will become even more of a luxury than it is today.

The laws of thermodynamics and the fact that energy does not cycle within ecosystems explain why the human population has a limited supply of energy available to it. We turn next to the subject of chemical nutrients, all of which differ from energy in that they follow cyclic pathways within ecosystems.

TROPHIC LEVEL

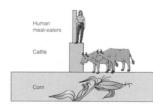

Food energy available to the human population at different trophic levels

36.12 Chemicals are recycled between organic matter and abiotic reservoirs

The sun keeps most ecosystems supplied with energy, but there are no extraterrestrial sources of water or the other chemical nutrients essential to life. Life, therefore, depends on the recycling of chemicals. In the next four modules, we look at the cyclic movement of four substances within the biosphere: water, carbon, nitrogen, and phosphorus. In each case, we see that chemicals pass back and forth between organic matter and the abiotic components of ecosystems. We call the part of the ecosystem where a chemical accumulates or is stockpiled outside of living organisms an abiotic reservoir. The main abiotic reservoirs are highlighted in white boxes in the figures. Let's begin with the cycling of water.

There are two kinds of modules

Blue-tabbed numbers identify modules that present basic scientific concepts.

Red-tabbed numbers identify modules that apply concepts to human issues. These include "Talking About Science" modules—profiles of men and women working in the world of science. (You can see an example on p. 288.)

vests energy stored in a glucose molecule by oxidizing the sugar and reducing O_2 to H_2O. This process involves a number of energy-releasing redox reactions, with electrons losing potential energy as they travel down an energy hill from sugar to O_2. Along the way, the mitochondrion uses some of the energy to synthesize ATP, as we saw in Chapter 6.

In contrast, the food-producing redox reactions of photosynthesis involve an uphill climb. As water is oxidized and CO_2 is reduced during photosynthesis, electrons gain energy by being boosted up an energy hill. The light energy captured by chlorophyll molecules in the chloroplast provides the boost for the electrons. Photosynthesis converts the light energy to chemical energy and stores it in sugar molecules.

Photosynthesis occurs in two stages linked by ATP and NADPH   **7.5**

The equation for photosynthesis is a simple summary of a very complex process. Actually, photosynthesis is not a single process, but has two stages, each with multiple steps. The steps of the first stage are known as the **light reactions**; these are the reactions that convert light energy to chemical energy and produce O_2 gas as a waste product. The steps of the second stage are known as the **Calvin cycle**; this is a cyclic series of reactions that assemble sugar molecules using CO_2 and the energy-containing products of the light reactions. The second stage of photosynthesis is named for American biochemist and Nobel laureate Melvin Calvin. In the 1940s, Calvin and his colleagues traced the path of carbon in the cycle, using the radioactive isotope ^{14}C to label the carbon from CO_2. The word "photosynthesis" capsulizes the two stages. *Photo-*, from the Greek word for light, refers to the light reactions; *synthesis*, meaning "putting together," refers to sugar construction by the Calvin cycle.

As indicated in the diagram here, the light reactions of photosynthesis occur in the thylakoid membranes of the chloroplast's grana. Light absorbed by chlorophyll in the thylakoid membranes furnishes the energy that eventually powers the food-making machinery of photosynthesis. Light energy is used to make ATP from ADP and phosphate. It is also used to drive a transfer of electrons from water to $NADP^+$, a hydrogen carrier similar to the NAD^+ that carries hydrogens in cellular respiration. Enzymes reduce $NADP^+$ to NADPH by adding a pair of light-excited electrons along with an H^+. This reaction temporarily stores the energized electrons. As $NADP^+$ is reduced to NADPH, water is split (oxidized), giving off O_2.

In summary, the light reactions of photosynthesis are the steps that absorb solar energy and convert it

into chemical energy stored in ATP and NADPH. Notice that these reactions produce no sugar; sugar is not made until the Calvin cycle, the second stage of photosynthesis.

The Calvin cycle occurs in the stroma of the chloroplast. The incorporation of carbon from CO_2 into organic compounds, shown in the figure as CO_2 entering the Calvin cycle, is called **carbon fixation**. After carbon fixation, enzymes of the cycle make sugars by further reducing the fixed carbon—by adding high-energy electrons to it, along with H^+.

As the figure suggests, it is NADPH produced by the light reactions that provides the high-energy electrons for reduction in the Calvin cycle. And ATP from the light reactions provides chemical energy that powers several of the steps of the Calvin cycle. The Calvin cycle does not require light directly. However, in most plants, the Calvin cycle runs during daytime, when the light reactions power the cycle's sugar assembly line by supplying it with NADPH and ATP.

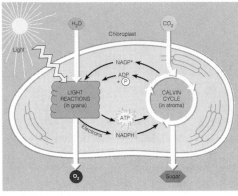

An overview of photosynthesis

Chapter 7   Photosynthesis: Using Light to Make Food   **113**

There are several ways to review a chapter

■ Start by rereading the module headings and looking over all the illustrations.

■ The Chapter Summary rephrases the main points of the chapter. Module numbers before each paragraph of the summary refer you back to the place in the chapter where you can review further if you need to.

Illustrations and text work together

■ Use both as you study.

■ The number in the colored tab stands for both the module number and the figure number.

Many modules have overview or review figures

■ Overview figures, such as Figure 7.5, present an overarching concept.

■ In later modules, these concepts are broken down into smaller units, each with a more detailed illustration.

■ Use the overview and review figures to remind you of how all the details fit together.

Genetic engineering could greatly increase crop yields

...cian readies a .22-caliber gun for shooting foreign plant pellets. The gun fires plastic bullets containing ...etal pellets coated with DNA. As indicated in the ... the bullet stays in the gun, but the particles are ... the plant cells. They travel through the cell ... the cytoplasm, and the foreign DNA becomes integrated into plant cell DNA. An engineered cell can be grown into a whole new plant that will produce proteins encoded by the foreign DNA, along with the proteins of the original parent cell.

Many new varieties of crop plants have already been produced with the gene gun and by an older technique that uses bacterial plasmids for gene transfer (see Module 12.14). Cotton and tobacco plants have been engineered that are resistant to viral attack. Potato plants have been engineered to synthesize their own insecticide, making them resistant to attack by beetles that can destroy whole crops. Tomato plants have been engineered to produce fruit that is

slow to spoil. In the future, genetic engineering may als... produce crop plants that can synthesize pharmaceuticals... industrial oils, and other useful chemicals.

A major goal of genetic engineering is to create crop plant... that provide more nutritious food—for instance, corr... wheat, and other grains that have a full complement of th... amino acids humans need to make proteins. Another goal ... to engineer varieties of nitrogen-fixing bacteria that are mor... efficient than naturally occurring varieties at making NH_4^+... Eventually, it may be possible to transplant genes for nitrogen... fixation directly into the DNA of nonlegume crop plants.

Genetic engineering holds great potential for increasin... agricultural production. There are potential problems... however. Gene-spliced crop plants, containing genes tha... resist natural diseases, might, for example, escape into th... wild and overgrow native species. Engineered crop plant... might also hybridize with their wild relatives, creatin... weeds that grow out of control. Also, there is a concern tha... new proteins in foods produced b... gene-spliced plants could be toxi... or cause serious allergies in som... people. Governments throughou... the world are grappling with ho... to proceed—whether to promot... the agricultural revolution offere... by gene splicing, or slow its pro... gress until more information ... available about the potential haz... ards. We will touch on the subjec... of agricultural productivity agai... in Chapter 33, which discusse... plant hormones.

Gunpowder

Gun

"Bullet"

Plant cells

DNA-coated pellets

Using a gene gun

Chapter Review

Begin your review by rereading the module headings and scanning the figures before proceeding to the Chapter Summary and questions.

Chapter Summary

Introduction–32.1 As a plant grows, its roots absorb water, inorganic nutrients, and oxygen from the soil. Its leaves take carbon dioxide from the air. Xylem and phloem transport water, nutrients, and the products of photosynthesis throughout the plant.

32.2 Root hairs greatly increase a root's absorptive surface. Water and solutes move freely through or between cells of the epidermis and cortex toward the center of the root. All water and solutes must pass through the endodermis before entering the xylem for transport upward. The endodermal cells admit only certain solutes. In some plants, solute transport may raise water pressure in the xylem. This root pressure can push water a short way up the stem.

32.3–32.4 Most of the force that moves water and solutes upward in the xylem comes from transpiration, the evaporation of water

from the leaves. Cohesion causes water molecules to stick togethe... relaying the pull of transpiration along a string of water molecules a... the way to the roots. The adhesion of water molecules to xylem ce... walls helps counter gravity. These processes are capable of movin... xylem sap, consisting of water and dissolved inorganic nutrients, t... the top of the tallest tree. Guard cells surrounding stomata in th... leaves control transpiration.

32.5 Phloem transports food molecules made by photosynthesi... by a pressure-flow mechanism. At a sugar source, such as a leaf, suga... is loaded into a phloem tube by photosynthetic cells. This raises th... solute concentration in the tube, and water follows by osmosis, rais... ing the pressure in the tube. As sugar is removed and stored in a suga... sink, such as the root, water follows. The increase in pressure at th... sugar source and decrease at the sugar sink causes phloem sap to flo... from source to sink. In the same way, sugar stored in roots may b... moved upward to developing leaves.

32.6–32.7 A plant's ability to make food depends on the nutrients ... obtains from its surroundings. Macronutrients, such as carbon, oxygen... nitrogen, and phosphorus, are required in large amounts, mostly to buil... organic molecules. Micronutrients, including iron, copper, and zinc, ac...

Consistent symbols and colors help you understand concepts

■ A number of symbols are used consistently throughout the book—the light energy arrow, the ATP starburst, and the shapes and colors of molecules, such as the purple protein molecules in Figure 28.7, are some examples.

■ The symbols provide an immediate visual clue to the meaning of the figure and how it relates to other figures you've seen.

Some figures take you through a sequence of levels or steps

■ Arrows often indicate the progression from large- to small-scale details.

■ Red-circled numbers identify the steps of a process. These numbers also appear in the text, keyed to the explanation of the steps in the process.

One of the key components of the nervous system is the **synapse**, the junction, or relay point, between two neurons. When an action potential arrives at the end of one neuron's axon, the information coded in the action potential passes to a receiving neuron across the synapse.

Synapses are either electrical or chemical. In an electrical synapse, action potentials themselves pass from one neuron to the next. The receiving neuron is stimulated quickly and always at the same level (same frequency of action potentials) as the transmitting neuron. Lobsters, crayfish, and many fishes can flip their tails with lightning speed because the neurons that carry signals for these movements communicate by electrical synapses. In the human body, electrical synapses are common in the heart and digestive tract, where nerve signals maintain steady, rhythmic muscle contractions. In contrast, chemical synapses are prevalent in most other organs, including skeletal muscles, and in the central nervous system, where signaling among neurons is complex and varied.

Unlike an electrical synapse, a chemical synapse has a narrow gap, called the **synaptic cleft**, separating a synaptic knob of the transmitting neuron from the receiving neuron. The cleft prevents the action potential in the transmitting neuron from spreading directly to the receiving neuron. Instead, the action potential is first converted to a chemical signal at the synapse. The chemical signal, consisting of molecules of **neurotransmitter** (see Module 26.1), then may generate an action potential in the receiving cell.

Now let's follow the sequence of events that occur at a chemical synapse in the figure to the right. The neurotransmitter is contained in vesicles in the synaptic knob of the transmitting neuron. Following the numbered sequence, ① an action potential (red arrow) arrives at the synaptic knob. ② The action potential triggers chemical changes that make neurotransmitter vesicles fuse with the plasma membrane of the transmitting cell. ③ The fused vesicles release their neurotransmitter molecules (green) into the synaptic cleft. ④ The released neurotransmitter molecules diffuse across the cleft and bind to receptor molecules on the receiving cell's plasma membrane. ⑤ The binding of neurotransmitter to receptor opens chemical-sensitive ion channels in the receiving cell's membrane. With the channels open, ions can diffuse into the receiving cell and trigger new action potentials. ⑥ The neurotransmitter is broken down by an enzyme, and the ion channels close. Step 6 ensures that the neurotransmitter's effect on the receiving cell is brief and precise.

You can review the aspects of the nervous system we have covered so far by thinking about the cellular events occurring right now in your own nervous system. Action potentials carrying coded information about the words on this page are streaming along sensory neurons from your eyes to your brain. Arriving at synapses with receiving cells (interneurons in the brain), the action potentials are triggering the release of neurotransmitters at the ends of the sensory neurons. The neurotransmitters are diffusing across synaptic clefts and triggering changes in some of your interneurons—changes that lead to integration of the signals and ultimately to a determination of what the signals mean (in this case, the meaning of words and sentences). When you finish this module, motor neurons in your brain will send out action potentials to muscle cells in your fingers, telling them to contract in just the right way to turn the page.

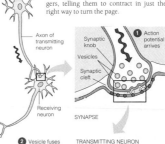

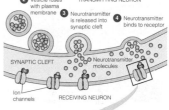

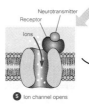

---

mainly as cofactors of enzymes. Growing plants in solutions of known composition enables researchers to determine nutrient requirements. Stunting, wilting, and color changes indicate nutrient deficiencies.

32.8–32.10 Soil characteristics determine whether a plant will be able to obtain the nutrients it needs to grow. Fertile soil contains a mixture of small rock and clay particles that hold water and ions. Humus—decaying organic material—holds nutrients, air spaces, and water and supports the growth of organisms that enhance soil fertility. Water-conserving irrigation, erosion control, and the prudent use of herbicides and fertilizers are aspects of good soil management. Organic farming protects the environment, and organically grown foods have become increasingly popular.

32.11–32.12 Relationships with other organisms aid plants in obtaining nutrients. Many plants form mycorrhizae, mutually beneficial associations with fungi. A network of fungal threads increases a plant's absorption of nutrients and water, and the fungus receives some nutrients from the plant. Parasitic plants such as mistletoe siphon sap from host plants. Carnivorous plants obtain some of their nitrogen by digesting insects.

32.13–32.16 Bacteria in the soil recycle nitrogen by decomposing organic matter. Other soil bacteria fix nitrogen from the air, and still others convert it to a form used by plants. Legume plants have a built-in source of nitrogen; nodules in their roots house nitrogen-fixing bacteria. Plants use nitrogen to make proteins, which are important in the human diet, as well as other important organic molecules. Plant scientists are using genetic engineering to develop new food crops for a growing world population.

Testing Your Knowledge

Multiple Choice

1. Houseplants require the smallest amount of which of the following nutrients?

 a. oxygen   d. iron
 b. phosphorus   e. hydrogen
 c. carbon

2. The clay particles in soil are important because they

 a. are composed of nitrogen needed by plants
 b. allow spaces for air and drainage
 c. fill spaces and keep oxygen out of the soil
 d. are charged and hold ions needed by plants
 e. supply humus needed by plants

3. By trapping insects, carnivorous plants obtain _____, which they need

 a. water . . . because they live in dry soil
 b. nitrogen . . . to make sugar
 c. phosphorus . . . to make protein
 d. sugars . . . because they can't make enough in photosynthesis
 e. nitrogen . . . to make protein

4. A major long-term problem resulting from flood irrigation is the

 a. drowning of crop plants
 b. accumulation of salts in the soil
 c. erosion of fine soil particles
 d. encroachment of water-consuming weeds
 e. excessive cooling of the soil

True/False *(Change false statements to make them true.)*

1. If a plant gets too hot, guard cells change shape and open the stomata.

2. Lower air pressure in the leaves "sucks" water to the tops of tall trees.

3. Potassium is carried from the roots to the leaves in the xylem.

4. Most of the organic material produced by a plant as it grows comes from materials obtained from the air.

5. Transpiration moves sugar from leaves to roots.

6. Carbon, nitrogen, oxygen, and chlorine are macronutrien

7. Negatively charged ions such as NO_3^- (nitrate) are easily soil.

Describing, Comparing, and Explaining

1. Explain how guard cells control the rate of water loss from hot, dry day. Why is this both helpful and harmful to the plan

2. Write a short paragraph describing the three ways in whi pend on bacteria for their supply of nitrogen.

3. Describe the characteristics of good topsoil. What are the roles of fine rock and clay particles, humus, and living organisms in the soil?

Thinking Critically

1. Acid rain is acidic because it contains an excess of hydrogen ions (H^+). One effect of acid rain is to deplete the soil of nutrients such as calcium (Ca^{2+}), potassium (K^+), and magnesium (Mg^+). Why do you think acid rain washes these nutrients from the soil?

2. Researchers have found that, in some situations, the application of nitrogen fertilizer to crops may have to be increased each year. Fertilizer decreases the rate of natural nitrogen fixation that occurs in the soil, and more fertilizer is needed to make up the difference. Explain how this might occur.

3. A tip for making cut flowers last longer without wilting is to cut off the cut ends of the stems under water and then keep them wet, so no air bubbles get into the xylem. Explain why this works.

Science, Technology, and Society

1. About 10% of U.S. cropland is irrigated. Agriculture is by far the biggest user of water in arid western states, including Colorado, Arizona, and California. The populations of these states are growing, and there is an ongoing conflict between cities and farm regions over water. To ensure water supplies for urban growth, cities are purchasing water rights from farmers. This is often the least expensive way for a city to obtain more water, and it is possible for some farmers to make more money selling water than growing crops. Discuss the possible consequences of this trend. Is this the best way to allocate water for all concerned? Why or why not?

2. The first genetically engineered food plant—a tomato that resists spoiling—is now in food stores. Citing health and safety concerns, several prominent chefs have announced they will boycott genetically altered foods. What might be some hazards of eating genetically engineered foods? Should the public be informed that the food they purchase has been genetically altered? Would you have any reservations about eating a tomato genetically engineered to resist spoiling? Why or why not?

3. This chapter discusses several trends in modern agriculture, such as organic farming and the use of genetic engineering to develop improved crops. How might organic farming benefit from advances in genetic engineering, and vice versa? How might organic farming and genetic engineering undermine each other?

Questions help you make connections

■ "Testing Your Knowledge" questions let you check what you've learned through multiple-choice, true/false, and matching exercises.

■ "Describing, Comparing, and Explaining" questions ask you to restate concepts in your own words.

■ "Thinking Critically" questions give you a chance to analyze biological information.

■ "Science, Technology, and Society" questions involve you in thinking about the ways the concepts connect to your own life.

■ The Answers section in Appendix Three provides answers and explanations for the Chapter Review questions.

Introduction 1

A hot-air dirigible carries a giant rubber raft over a thick cloud cover. We are in the tropics, just north of the equator over French Guiana, in eastern South America. Through breaks in the clouds you can just make out the tops of trees—the canopy of a dense, unbroken tropical rain forest. The dirigible's pilot skims lower and lets his cargo settle gently onto the trees. Soon the raft comes alive, as scientists climb up from the forest floor and step out onto the fiber netting connecting the raft's inflated rubber pontoons.

Over the next several months, research teams from around the world will live and work on the raft, studying life in the canopy of the rain forest. In the photograph on this page, French biologist Pierre Girard, armed with a laptop computer, identifies plants. He uses the hunting horn to herald the dirigible's visits. Meanwhile, one of his colleagues catches and classifies beetles, including many that scientists have never seen before. Another researcher monitors how fast plants are producing sugar by photosynthesis. The scientists say that life on the raft is like being at sea. When the tropical wind blows, the raft waves with the trees, and walking on it is like trying to keep your balance on a boat in a storm. It is wet much of the time in the canopy, and buggy; biting flies sing in your ears, as if to warn of tropical diseases they may carry.

Tropical rain forests cover equatorial lands in Central and South America, Africa, and Southeast Asia. Rainfall in these areas is at least 200 centimeters (80 inches) per year. Rain forests occupy only about 6% of Earth's land surface, but nowhere else is there as much life. Biologists are fond of saying that tropical rain forests are "megadiverse"—indeed, they are home to at least half the planet's species of organisms. Insects and flowering plants are the most prevalent forms of life here.

Despite their rich diversity, rain forests have not been studied extensively. Most rain forest animals live in the canopy, where the foliage is extremely dense and hard for humans to navigate. The height of the trees—often 30 meters (100 feet) or more—is also a problem. Some researchers are using a gigantic, 50-meter-high crane to sample the top of the canopy. The crane works well for long-term studies in one place, but moving from place to place is a major undertaking. A big advantage of the tree raft is mobility; the dirigible can relocate the raft quickly, enabling researchers to sample numerous sites in a short period of time.

Researchers have been using the raft since 1989, mostly to catalog the diverse life of the tropical forest. Because rain forests are being destroyed at a staggering rate, cataloguing what is there is an urgent task. Some researchers estimate that as many as 6000 tropical species are being exterminated each year as the human population grows and increases its demand for agricultural lands, wood, and living space.

Worldwide, biologists have identified and named about 1.5 million different species to date, including some 250,000 plants and about 1 million animals. Although these are large numbers, they probably represent only a fraction of all species on Earth. Indeed, scientists now estimate that between 5 million and 30 million different species exist on Earth. One of nature's great showplaces of different kinds of species, the rain forest forms an appropriate backdrop for this chapter's introduction to **biology**, the science of life.

Biological diversity can be arranged into five kingdoms

The canopy of a tropical rain forest is alive with sights, sounds, and scents, and a treetop raft exposes you to many of them. The sweet fragrance of showy orchids hangs in the heavy air. The loud calls of parrots, toucans, and other colorful birds compete with the hoots and howls of monkeys. A scarlet and blue tree frog about the size of your thumb sits motionless, then suddenly snaps up an ant that gets too close. Ants, mosquitoes, beetles, and other insects are literally everywhere—flying, crawling, jumping—and you hear them day and night.

The richness of life in a rain forest—the vast diversity of species—can be almost overwhelming. To make diversity here and elsewhere somewhat more comprehensible, biologists have devised ways of grouping species. The broadest groups are called the **kingdoms** of life. According to the most widely used scheme, there are five kingdoms: Monera, Protista, Plantae, Fungi, and Animalia. Every living individual (organism) belongs in one of these broad groups.

The organisms in these photographs, representing the five kingdoms, could all be found in one small area of a rain forest. The culture dish in Figure A shows growing bacteria, microscopic organisms that make up the kingdom Monera. The spots on the growth medium are colonies of bacteria, each color representing a different species. Found literally everywhere there is life, from rain forests and polar oceans

to your own skin and intestines, bacteria are the most widespread of all living organisms. Bacteria are distinguished from all other forms of life by their structure. A single bacterium is a unit called a cell. A **cell** is a unit of living matter separated from its environment by a boundary called a membrane. Every living being is composed of cells, but only bacteria have cells without a nucleus, a discrete internal structure that controls cellular activities.

Any pool of water containing bacteria would also support members of a second kingdom, the Protista, or protists. Protists include algae and protozoans. Like plants, algae make their own food by the process of photosynthesis. Protozoans, which are single-celled, are animal-like in that they eat other organisms, including algae and bacteria. Figure B shows a number of protists in a drop of water viewed with a microscope. The large, irregular, bluish cell is an amoeba, and the smaller cells are mostly other protozoans and single-celled algae (the greenish cells). Also present are multicellular algae, which are considered protists because of their similarities to single-celled algae.

Organisms in the remaining three kingdoms are multicellular. Kingdom Plantae, the plants, are photosynthetic and consist of cells with strong walls made of cellulose. The flowers of over 2000 species of bromeliads (the pineapple family), represented by the plant in Figure C, brighten the forests of Central America. Some bromeliads grow on the forest floor, and many others grow on tree surfaces high in the canopy.

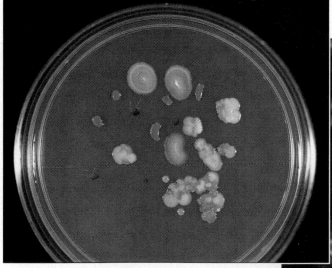

A. Kingdom Monera (bacteria)

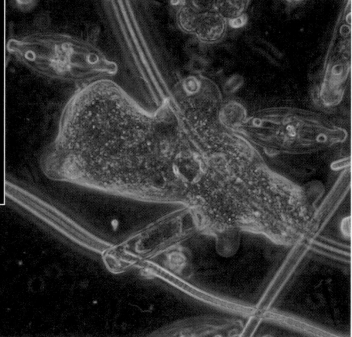

292×

B. Kingdom Protista

Life's diversity is evident almost everywhere. You can even find representatives of all five kingdoms on many city streets. There the most obvious examples of Animalia are likely to be people, with trees, shrubs, and grass representing Plantae. With the help of a microscope, you can find bacteria, fungi, and protists in any puddle of water or pile of moist soil. Less obvious, but just as significant, are signs of the basic similarities shared by all organisms. All organisms are made of similar molecules and of cells bounded by a membrane. In fact, the millions of species of organisms are variations on a set of basic features that we will explore throughout the book.

D. Kingdom Fungi

C. Kingdom Plantae

Kingdom Fungi, represented in Figure D by the colorful fly mushroom, is a large and diverse group that includes the molds, yeasts, and mushrooms. Fungi decompose the remains of dead organisms and absorb nutrients from the leftovers. Found throughout the world, the fly mushroom is highly toxic to animals. A powder made from this species was formerly used as an insecticide.

Representing the kingdom Animalia (animals), the sloth in Figure E resides in the rainforest canopy. Animals eat other organisms and are made of cells that lack rigid walls. Most animals are motile, moving about actively in search of food. The sloth is a slow-moving animal that spends most of its time hanging upside down eating leaves. Females even give birth to their young in this position.

There are actually members of three kingdoms in Figure E. The sloth is hanging from one of the largest members of the kingdom Plantae, one of the 30-meter-high trees dominating the rain forest. And the greenish tinge in the animal's hair is a luxuriant growth of photosynthetic bacteria, which belong to the kingdom Monera. This photograph emphasizes a theme reflected in our book's title: connections among living things. The sloth depends on trees for food and shelter, the bacteria gain access to sunlight necessary for photosynthesis by living on the sloth, and even the tree depends on animals and on bacteria far below to supply its roots with chemical nutrients. We will see many examples of biological connections as we explore life and its diversity in later chapters.

E. Kingdom Animalia (with Plantae and Monera)

1.2 Common threads connect all of life

A. Three species of orchids from the rain forest

The three species of rainforest plants in Figure A look quite different, but they all are variations on a common theme. For instance, their showy flowers all have a large liplike petal that attracts pollinating insects and guides them to the male and female parts of the flower. All three plants also produce tiny seeds made of only a few cells, and their seeds will start growing in nature only in the presence of specific kinds of fungi. Plants with this set of basic features are closely related; in fact, all three shown here are members of the orchid family.

Flower structure, type of seeds, and associations with other organisms provide important clues about relatedness within a group of plants. On a broader scale, fossils (the remains of extinct organisms) provide evidence of species relatedness over long periods of time. You might be able to trace your family history back for a century or two,

but fossils let us trace the ancestry of many species back thousands or millions of years. The skeleton in Figure B is reconstructed from the fossilized remains of a mammal, sometimes called the dawn horse, that lived in North America some 55 million years ago. The shapes of its skull, teeth, and leg bones suggest that it was one of the earliest members of the horse family. (The hand shows the scale.)

Fossils do more than provide clues to the ancestral relationships of species. They also show us that a changing cast of organisms has populated the Earth and that the history of life parallels the geological history of our planet. Environmental change has been one of Earth's hallmarks since it solidified from cosmic gas and dust about 4.5 billion years ago. The changes in species, which we call **evolution,** have been a central feature of life since it arose about 4 billion years ago.

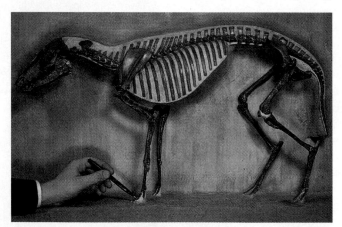

B. Reconstruction of "dawn horse" skeleton from fossils

1.3 Evolution is the core theme of biology

In November 1859, British biologist Charles Robert Darwin published one of the most important and controversial books ever written. Entitled *On the Origin of Species by Means of Natural Selection,* Darwin's book was an immediate best-seller and soon made his name almost synonymous with the concept of evolution.

Darwin showed how evolution could explain the common threads underlying life's diversity. Applied to the or-

chid example in Module 1.2, for instance, his theory contends that members of the orchid family are fundamentally similar because they have evolved from a common ancestor. Darwin

A. Charles Darwin in 1859

perceived that the fossil record chronicles the evolution of species. In the case of the horse, for example, the fossil record reveals the origin of the modern species (horses living today) from a succession of ancestors. In Darwin's words, species arise through a process of "descent with modification."

Most importantly, Darwin's *Origin of Species* proposed a mechanism to explain how evolution occurs. Darwin called the mechanism **natural selection,** and Figure B shows how it works. In part 1, we see a group of imaginary rainforest beetles of the same species. The group is a discrete unit called a population, with individuals breeding among themselves. The individuals exhibit varied traits that are inherited—in this case, three different body colors. As Darwin realized, heritable variation must be present in the population for natural selection to operate.

In part 2, predatory birds have arrived on the scene and are eating the yellow beetles, the ones they can see most easily. The yellow beetles are eliminated before they have a chance to reproduce and pass on the gene for yellow color. In part 3, the surviving, dark-colored beetles have reproduced. The group is now quite different from the original one; natural selection has produced a change in the proportions of the colors in the population. Here we see that natural selection is not a creative process, but an editing mechanism. In fact, *natural selection occurs as heritable variations are exposed to environmental factors that favor the reproductive success of some individuals over others.* In our example, birds are the relevant environmental factor.

Darwin's overall conclusion was that numerous small changes in populations caused by natural selection eventually result in big changes that alter species. He proposed that the evolution of species into new species results from an accumulation of minute changes resulting from natural selection over time. We elaborate on the theory of evolution in Chapters 14 and 15. We will see that while there have been some refinements of Darwin's ideas, his theory of

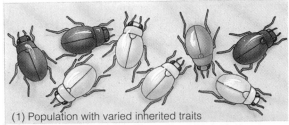

(1) Population with varied inherited traits

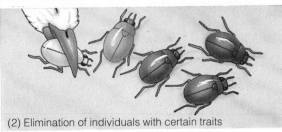

(2) Elimination of individuals with certain traits

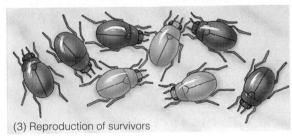

(3) Reproduction of survivors

B. Natural selection

evolution through natural selection has stood the test of time.

We see the exquisite results of natural selection in every kind of organism. Each species has its own special set of adaptations, features that evolved by means of natural selection. On the left in Figure C, we see the pangolin, a mammal that lives in East African rain forests. One of its main adaptations, its tough body armor of overlapping scales, protects it from most predators. The pangolin also has an unusually long, straplike tongue, which it uses to prod termites and ants out of their nests. Another mammal, the killer whale (orca), is adapted for life at sea (Figure C, right). It breathes air through nostrils on the top of its head and keeps in touch with its companions by emitting clicking sounds into the water. Orcas also use sound echoes to detect obstacles and to locate schools of fish or other prey, such as seals and sea lions.

The pangolin's armor and the orca's echolocating ability did not result from changes in individuals during their lifetimes. These traits arose over many, many generations, as individuals with heritable traits that made them best adapted to their environment had the greatest reproductive success.

Understanding how adaptations evolve by natural selection is key to the study of life. Indeed, evolution is biology's central theme—the one idea that makes sense of all we know about life.

Pangolin

Killer whale

C. Adaptations to the environment

1.4   Science is a powerful approach for understanding life

Charles Darwin stands out in history with Galileo, Newton, Einstein, and other great scientists who synthesized important theories about nature. A **theory** in science is an explanatory idea that is broad in scope and supported by a large body of evidence. A theory is different from a **hypothesis,** which is an educated guess a scientist proposes as a tentative explanation for a specific phenomenon. All ideas in science are open to question and refinement, but a theory is widely accepted. Thus, the theory of evolution is the central organizing principle of biology. Aspects of the theory may be refined and updated as we learn more, but the basic idea is sound.

Science has emerged from human curiosity about ourselves, our world, and the universe. Scientists are obsessively curious. Convinced that natural phenomena have natural causes, they ask questions about nature, pose hypotheses to answer the questions, and test the hypotheses. Scientists are also skeptics, suspicious of poorly documented or contrived answers to their questions.

Science is a powerful means of gaining information and understanding, but there are many questions that lie outside its realm. It is important to recognize the limitations of science. As a scientist, a biologist would not attempt to answer such questions as, What is the purpose of life? or What is the nature of the human spirit? These questions are unapproachable by science because they are not concerned with the natural world and its laws. The quest for their answers lies in the realms of philosophy and religion.

Striving to understand seems to be a basic human drive. As scientists, biologists employ the methods of science in seeking knowledge about life. Let's examine how scientists go about their work.

1.5   Biologists use the scientific method

The large, hawklike bird in Figure A lives in large caves in South American rain forests. It is called an oilbird, because native peoples rendered a fine cooking oil from the fat of its nestlings. In Figure B we see Roberto Roca, a Venezuelan biologist who heads the zoology section of the Nature Conservancy's Latin American Science Program. Roca recently completed a study of the oilbird's feeding habits in northern Venezuela. His research won the oilbird a breeding sanctuary in the rain forest.

Roca employed the **scientific method,** a process that scientists use to answer questions about nature. The method's key ingredients are observations, questions, hypotheses, predictions, and tests. Let's see how Roca's work illustrates the scientific method.

Observations set the stage. Before Roca began his studies, others had observed that oilbirds nest in large colonies in caves and eat the fruits of rainforest trees. At dusk, the adults fly out of the caves, returning later with fruit for their nestlings. The fruits eaten by oilbirds contain large seeds that the birds regurgitate (Figure C). At least some of the seeds dropped by oilbirds pile up in the caves.

These observations triggered *questions:* How far do oilbirds fly in search of food? Do they travel so far that they must eat fruits to sustain themselves before carrying any back to their young? If so, wouldn't the birds have to regurgitate some seeds in the forest? It occurred to Roca that if the adults drop seeds in the forest, they might play a role in

C. An oilbird regurgitating a seed

reforesting areas where trees had been destroyed. This idea became Roca's central *hypothesis:* Oilbirds help reseed the rain forest.

Unfortunately there was no way for Roca to test this hypothesis directly—that is, go out and see whether oilbirds drop seeds in remote areas at night. Instead he used his hypothesis and the known facts to deduce a testable *prediction.*

Roca surmised that adult oilbirds would gradually use up the fruit supply near the cave where they fed their nestlings and would have to travel farther for food as the nesting season wore on. He reasoned that the adults would eventually be flying so far that they would have to eat fruit to supply their own energy needs. If this were true, the adults would have to regurgitate seeds in the forest before returning to the cave.

A. An oilbird

B. Roberto Roca

6

Roca's first testable prediction was that adult oilbirds fly far from their cave late in the nesting season. To *test* this prediction, he captured 11 adult oilbirds and attached tiny radio transmitters to their backs. After releasing the birds, Roca climbed a nearby mountain, set up a receiver, and tracked the birds by radio. As weeks passed, the distances the birds flew increased. By the time the nestlings were four months old, the adults were traveling 100 kilometers (60 miles) from the cave to collect fruit each night. These observations supported Roca's first prediction and led to another one, that most of the seeds piled up in the cave were dropped there by nestlings, not by adults.

To test this second prediction, Roca performed an experiment in an oilbird cave. He suspended wire trays under rock ledges where there were solitary nests, placing some trays under fully occupied nests and some under nests with adults but no nestlings (Figure D). Seeds accumulated rapidly under nests with nestlings, but none appeared under nests without nestlings. These results supported Roca's central hypothesis by making it seem even more likely that the adults drop seeds in the forest.

At this point, Roca made some calculations. He determined the energy (caloric) content of the fruits the oilbird eats. A comparison of the fruits' energy content with the bird's energy needs told him that a typical adult would have to eat at least 50 fruits a day just to sustain itself during the breeding season, suggesting that the bird would drop many seeds in the forest. Roca estimated that 10,000 adults (the population of only one of 20 caves he studied) disperse at least 9 million seeds (almost 13 tons) each month during the breeding season—enough to reseed thousands of square kilometers of nearby rain forest destroyed by burning and logging. In 1989, the Venezuelan government, impressed by Roca's findings, established a 400-km² (250-mi²) preserve for oilbird nesting grounds.

Roca's work illustrates several important points about science and the scientific method. It shows us the logical sequence of the scientific method: Prior observations lead to questions and hypotheses; hypotheses lead to predictions and tests. Roca's study also demonstrates the use of a **controlled experiment,** another key ingredient in the scientific method. In performing a controlled experiment, a scientist actually carries out two parallel tests, one called the experimental, the other the control. Ideally, the experimental test differs from the control test by only a single factor, called the variable. In Roca's experiment, the variable was the presence of nestlings in the nest. For his experimental tests, he placed seed trays under nests with nestlings. For his controls, he placed trays under nests with adults only. Controls make it possible to draw clear-cut conclusions from the results of experiments. Without controls, Roca would not have been able to tell whether adults, nestlings, or both were producing the seed piles in the caves.

Another aspect of the scientific method is the need for *repetition* of tests. What if Roca had placed seed trays under only two nests, one with nestlings and one without? It would be risky to generalize from the results of such an experiment. For example, the two adults in one of these nests might have had different habits from most oilbirds. We can draw firm conclusions only after repeating experiments and obtaining consistent results. In Roca's case, he obtained consistent results from seed trays under ten nests.

Finally, it is important to recognize that the scientific method is an idealized model of how science works, and it is not always possible to adhere rigidly to the model. Many biological studies have to be performed where clear-cut (single-variable) controls are not practical. Such situations often arise when experiments must be conducted in nature rather than in a laboratory. In Roca's seed-collection experiment, for instance, there were actually three variables—three differences between the experimental nests and the control nests. In addition to the presence or absence of nestlings, the nests differed in age and in the presence or absence of eggs. The nests without nestlings were newer and contained eggs. Multiple variables like these can make results difficult to interpret, but they do not necessarily invalidate scientific tests. In Roca's case, the fact that no seeds were dropped from nests without nestlings did show that nestlings are the chief, if not the only, source of seeds in the cave.

It is characteristic of scientific work that hypotheses are not proved. Roca did not prove that oilbirds drop seeds in rain forests; he amassed evidence to support his hypothesis. Now Roca or another scientist can build on the hypothesis and make additional scientific discoveries. For Roca, it was also possible to take practical action—that is, use his evidence to save the oilbird's breeding territory and help conserve a rain forest.

D. An experiment to determine the source of seed piles in a cave

1.6 Life is organized at different levels

Nowhere on Earth is the saying, "It's hard to see the forest for the trees," more fitting than in a tropical rain forest. There is so much life here that it takes special effort to take in the whole, grand expanse of the forest.

In studying oilbirds, Roberto Roca worked at the metaphorical tree level. Likewise, most of the biologists aboard the big rainforest raft in the chapter's opening are studying individual plants and animals. As a logical and necessary starting place for studies of unexplored environments, they are naming and describing the different kinds of organisms in the rain forest.

An **organism** is an individual living thing—a bacterium, fungus, protist, plant, or animal. As an organism, the oilbird in the figure at the right represents one of several structural levels into which life is organized. Together the levels form a hierarchy, with each level building on the ones below it.

Above the organism level, a group of interbreeding individuals of one species, such as a flock of oilbirds living in a cave, is called a population. Above the population level, all the organisms in the rain forest are collectively called a community. Above the community is an ecosystem—the entire rain forest, in our example. The ecosystem is the most complex level of all. In addition to the organisms in the forest, the ecosystem includes the nonliving, physical features that affect the organisms, such as climate, soil, and sunlight. A biologist working in a rain forest might change the forest-tree metaphor to, "It's hard to see the ecosystem for the organisms."

Below the organism level, life's hierarchy unfolds within the individual organism. The oilbird's body consists of several organ systems, such as the circulatory system, the excretory system, and the nervous system, shown here. Each organ system consists of organs. For instance, the main organs of the nervous system are the brain, the spinal cord, and the nerves, which transmit messages between the spinal cord and other parts of the body.

As we continue downward through the hierarchy, each organ is made up of several different tissues, each of which consists of a group of cells with a specific function. The brain, for example, consists mainly of nervous tissue, in turn made up of nerve cells. The nervous tissue in the bird's brain has millions of microscopic nerve cells organized into a communication network of spectacular complexity. The nerve cells transmit signals that coordinate the oilbird's activities.

Finally, we reach the chemical level in the hierarchy. We use as our example a molecule of a neurotransmitter, a chemical that carries nerve signals from one nerve cell to another. A molecule is a cluster of atoms (the smallest particles of ordinary matter) held together by bonds. Each of the black, red, or white spheres in the molecule at the right represents an atom.

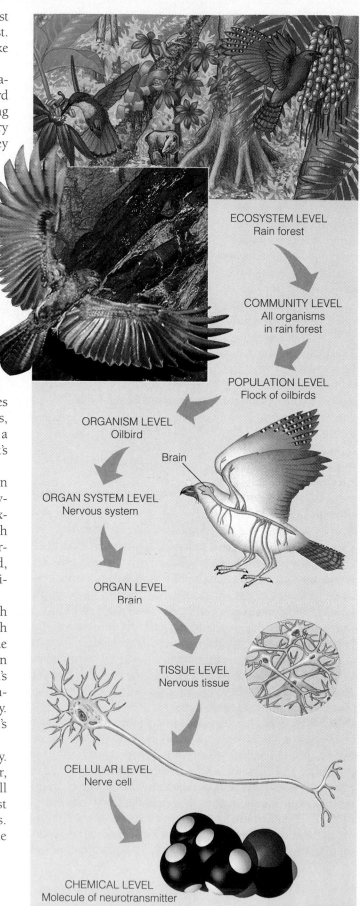

ECOSYSTEM LEVEL
Rain forest

COMMUNITY LEVEL
All organisms
in rain forest

POPULATION LEVEL
Flock of oilbirds

ORGANISM LEVEL
Oilbird

Brain

ORGAN SYSTEM LEVEL
Nervous system

ORGAN LEVEL
Brain

TISSUE LEVEL
Nervous tissue

CELLULAR LEVEL
Nerve cell

CHEMICAL LEVEL
Molecule of neurotransmitter

As we will discuss in later chapters, life's hierarchy builds from molecules to ecosystems. It takes many molecules to make a cell, many cells to make a tissue, several kinds of tissues to make an organ, and so on. And, as we see in the next module, it takes more than just the presence of the right parts to give each level the unique properties that define it.

At every level in life's hierarchy, the whole is greater than the sum of its parts

An organism—an oilbird, a tropical vine, a tree—has the unique property of being an independent living entity. At another level in life's hierarchy, a nerve cell has the unique ability to transmit signals rapidly over long distances, a capability far beyond those of any of the individual molecules that constitute it. And, at the organ level, a brain can experience a wide range of sensations, direct complicated behaviors, and modify behavior through learning—special abilities that extend way beyond those of the individual nerve cells that make it up.

Where do the special properties of a cell, an organ, an organism, or any structure in life's hierarchy come from? They result from the precise organization of component parts and the interactions among them. All the capabilities of a brain, for instance, result from the precise arrangement of nerve cells and the communications among them. If the arrangement is disrupted by a serious head injury, the brain loses its special properties even though all its components may still be present. Thus, the special properties of the whole—in this case, the brain—are far more than the sum of its parts.

Special features that result from a system's particular organization, and that do not exist without this organization, are called **emergent properties**.

A precise structural order and a resulting set of emergent properties define life

If you have ever tried to define life, you know that it resists a simple, one-sentence definition. Nonetheless, even a very young child perceives that a dog, a bug, or a tree is alive and a rock is not. Recognizing life is easier than defining it; we recognize living organisms by their general appearance and by what they do.

The phrases that are italicized in this module name the main properties we associate with the state of being alive. First of all, any living organism has a *precise structural organization*. The outward appearance of an organism—the oilbird in the photograph here, for instance—suggests a high degree of structural order. And the obvious order in the bird's beak, feathers, and eyes results from an underlying order in molecules and cells.

Precise structural organization is one of several features that, taken together, define life. Other defining features, describing what organisms do, emerge from structural order. For example, the *ability to take in energy and use it* to perform many kinds of work—flapping wings or walking, for instance—emerges from the precise arrangement of cells and molecules in a bird's digestive, circulatory, nervous, and muscular systems. Other properties of life, including the *ability to respond to stimuli* from the environment, the *capacity for growth and development,* and the *ability to reproduce,* all emerge from the precise arrangement of cells and body parts.

All of life's basic properties except one are apparent at the organism level in life's hierarchy. The *ability to evolve* is a property of the population level. As a reproducing adult, an organism is a member of a population, a group whose structure as a discrete unit in the environment is capable of evolving. As we discussed in Module 1.3, a population can develop heritable changes over generations as a result of natural selection.

Emergent properties are not unique to life. A diamond and the graphite in a pencil, for instance, both consist of carbon, but they have different properties because their carbon atoms are arranged differently. Thus, life is characterized by a *particular set of emergent properties* that arise from the special structural order inherent in organisms and populations.

Just as the concept of life is associated with a set of emergent properties, so is every one of the levels in life's hierarchy. We pursue this idea further in the next five modules.

1.9 Life's properties have a chemical basis

The precise arrangement of parts that makes an oilbird different from an orchid or a human has a chemical basis. Take one of your hairs for example. It was produced by living cells in the skin, but it consists entirely of nonliving material. The special properties of a hair emerge from a chemical order, a specific combination and arrangement of molecules. For instance, hair color comes from a particular assemblage of pigment molecules.

The figure here shows the structure of a hair in detail. A hair is made of many parallel fibers, and each of these consists of three smaller fibers, wound together like a rope. The smallest fibers are each a molecule of a protein, alpha-keratin. The ribbon on the far right of the figure shows the general shape of an alpha-keratin molecule. It is a long chain coiled into a structure called an alpha helix. Over 90% of a hair is alpha-keratin. The hair's shape (whether it is wavy, curly, or straight), its flexibility, and its insulating power all result from the structure and arrangement of its alpha-keratin molecules.

Proteins called beta-keratins, which exist in sheets rather than coils, form the feathers, beaks, and claws of birds. In all cases, the types and arrangement of the molecules determine the properties of the structure. Chemical structure and arrangement underlie all the properties of life. In the next module, we examine the chemical basis of reproduction and inheritance.

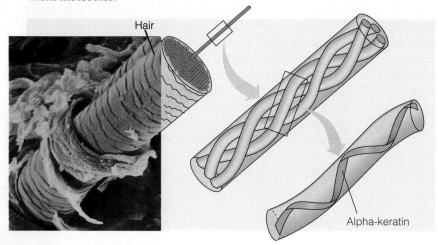

Hair

Alpha-keratin

Alpha-keratin, the main ingredient of hair

1.10 The continuity of life depends on DNA

There is a chemical connection between the larger, adult chameleon in Figure A and the baby chameleon perched on its horn. The baby has inherited molecules of deoxyribonucleic acid, or DNA, from its horned father (as well as from its mother). DNA passes from one generation to the next—for animals, plants, and all other organisms.

The computer graphic in Figure B (top of next page) shows part of a DNA molecule. Each DNA molecule consists of two very long helices, chains coiled around each other. Each helix is made up of four kinds of chemical building blocks called nucleotides (indicated by the four colors). DNA contains genetic information, a chemical blueprint for constructing other molecules that make up an organism. The information is encoded in sequences of nucleotides along the length of DNA.

The way DNA encodes information is analogous to the way we arrange the letters of the alphabet into sequences with specific meanings. The word TREES, for example, conjures up tall plants with woody trunks and green leaves; but STEER and TERSE, other five-letter words containing the same five letters, have quite different meanings. We can think of the four different nucleotides of DNA as the alphabet of inheritance. Specific sequential arrangements of these four chemical letters encode heritable information in genes.

A typical gene consists of hundreds or thousands of nucleotides, and each gene has its own specific nucleotide sequence. The information in an organism's DNA is vast. If the entire library of genes stored within a single one of your cells were written in letters the size of those you are now reading,

A. Horned chameleon and offspring

the information would fill more than a thousand books, each larger than this one.

The discovery of DNA and its role in inheritance has given us a deeper understanding of life. We know that all organisms use the same DNA code. A particular sequence of nucleotides means the same thing for one organism as it does for another. Thus, the differences among organisms reflect different nucleotide sequences in DNA. Put another way, different expressions of a common language determine the differences among organisms.

Studies of DNA from different kinds of organisms have provided strong evidence of the evolutionary connections among species. Species that appear closely related, such as the gorilla, chimpanzee, and human, turn out to have very similar DNA (DNA with similar nucleotide sequences). But human and ape DNA molecules are quite different from the DNA of any bird

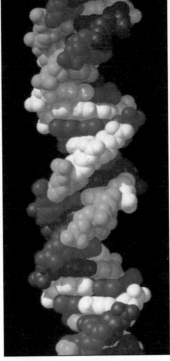

B. DNA

or reptile and even more different from that of a frog or fish. In fact, the degree of similarity between the nucleotide sequences in the DNA of two organisms is one measure of how closely they are related through evolution. Therefore DNA studies can actually help us trace the course of evolution.

The information contained in DNA underlies all the emergent properties that characterize life. DNA provides the blueprint for the growth and development of an organism's body form. It also contains the information cells need to make molecules required for energy use and for responding to environmental stimuli. Most importantly, DNA is capable of replication—that is, precise duplication of its nucleotide chains. DNA replication, as we will see in later chapters, underlies an organism's ability to reproduce.

All organisms are composed of cells

Life has a molecular basis, and molecules underlie life's special properties. But no molecule, including DNA, is itself alive. The lowest level in life's structural hierarchy where we actually see life is the cellular level. The cell is life's most basic unit of structure and function. All organisms are composed of cells, and some organisms, such as most bacteria and protists, consist of single cells. A single-celled organism performs all the activities of life and is a member of a population that can evolve. Figure A shows an example of such an organism, an amoeba, which is a protist.

In contrast to the cell of an amoeba, most cells of multicellular organisms have highly specialized roles. For instance, nerve cells serve as regulators in the animal body, controlling the activity of other cells. Figure B shows part of a nerve cell that controls several muscle cells. Structure clearly correlates with function here. The nerve cell has long branching extensions that make direct contact with the muscle cells. The contact points are specialized structures where signals transmitted along the nerve cell from the brain or spinal cord make the muscle cells contract. Long nerve cells give the brain precise, split-second control over distant parts of the body.

The structure of muscle cells also correlates with their function. Like nerve cells, they are long and can extend between two body parts. Their function is to contract, pulling body parts closer together, as happens when you make a fist or bend your knee. When we see an animal move its body in any way, we are seeing the result of cellular action in the nervous and muscular systems. Structure correlates with function in all the cells making up multicellular organisms.

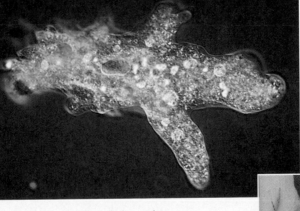

357×

A. An amoeba, a single-celled organism

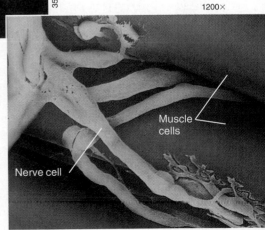

1200×

Muscle cells

Nerve cell

B. A nerve cell connecting with muscle cells

Every organism exhibits clear connections between form and function

Given a choice of tools, you wouldn't try to loosen a screw with a hammer or pound a nail with a screwdriver. The structure, or form, of a device is correlated with how it works, and this is as true for living organisms as it is for tools.

As the oilbird evolved in South American rain forests, it became superbly adapted through natural selection to feeding on the fruits of tall trees and raising its young in dark caves. For instance, the shape of the oilbird's beak—a structural feature—makes the beak an effective tool for plucking fruits from tropical trees. And the oilbird's strong wings, spanning more than a meter, correlate with its ability to fly great distances in search of food. We see the results of natural selection in the structural features that enable any organism to perform specific functions in its environment.

The beautiful double-collared sunbird in Figure A evolved elsewhere, in the rain forests of East Africa. Feeding on nectar and small insects, it is an African counterpart of the hummingbirds of the Americas. Many plants in tropical African forests depend on sunbirds for pollination. While feeding on nectar, the birds carry pollen (which contains cells that produce the plant's sperm) to the female parts of flowers. The plant nectar attracts sunbirds, and the flowers have special structures that promote pollination.

Illustrating the connection between structure and function, the sunbird's thin, curved beak is an effective probe for reaching nectar inside flowers. It also has a long tongue used for drinking nectar and a set of nose flaps that keep pollen out of its nostrils. The sunbird's short, rounded wings enable it to flit rapidly from flower to flower, and its clawed feet allow it to clasp branches tightly so it can perch and feed at the same time.

We can also see correlations between structure and function in the plants that sunbirds pollinate. The mistletoe in Figure B, for instance, produces flowers with a springlike device that releases when touched. When a sunbird probes a flower for nectar, the device releases a cloud of pollen onto the bird's head. Then, when the bird visits another flower, some of the pollen brushes off and pollinates the flower.

The connection between form and function, resulting from natural selection, is a central theme in biology—a theme that will appear often as we discuss such diverse topics as the chemistry of life, organ systems in animals and plants, and environmental biology.

A. Double-collared sunbird

B. An African mistletoe, a food source for sunbirds

Living organisms and their environments form interconnecting webs

Sunbirds and mistletoe are interdependent members of a rainforest ecosystem. The sunbird depends heavily on nectar to feed itself and its young. In turn, mistletoe and other flowering plants rely on sunbirds for pollination. This interdependency between birds and plants is part of a complex web of relationships among the organisms that live in the ecosystem. The web includes nonliving components of the environment such as air, water, soil, and sunlight.

The drawing at the right is a simplified view of some of the relationships among organisms in an African rainforest ecosystem. The arrows indicate the directions in which en-

ergy and nutrients pass. Plants dominate the scene and provide much of the food that supports the ecosystem. Plants, as well as certain bacteria and some protists (not shown), trap energy from sunlight and use carbon dioxide (CO_2) from the air, along with water (H_2O), to make food molecules by photosynthesis. Plants also absorb mineral nutrients from the soil. Animals like the sunbird, the gorilla, and many insects eat plants or plant parts. Mice and parrots eat mainly plant material but also some insects. Other animals, such as the leopard, many kinds of snakes, and meat-eating insects, prey on animals. In all cases, however, plants and photosynthetic bacteria and protists are the ultimate sources of food. Comprising another vital part of the ecosystem are the bacteria, fungi, and small animals in the soil that decompose the remains of dead organisms. These decomposers act as recyclers, changing the complex dead matter into simple mineral nutrients that plants can use. Thus, the web of relationships among plants, animals, microorganisms, and the physical environment give an ecosystem its structure. The weblike pattern of arrows in an ecosystem diagram indicates this basic structure visually.

The ecosystem is the highest level in the hierarchy of life. Like all the other levels, the structure and interactions of its parts give rise to unique functional properties. One of the emergent properties of an ecosystem is the ability to cycle chemical nutrients, moving them around from the air and soil to plants, to animals, and back to the air and soil. Another property that emerges from the unique interconnections of the ecosystem's web is the passage of energy from one organism to another. Plants and other photosynthesizers trap solar energy and convert it to chemical energy, which is then shuttled through the ecosystem's web, powering each organism. All life on Earth depends on these special properties of ecosystems.

A web of interactions in a rainforest ecosystem

Biology is connected to our lives in many ways

Global warming, air and water pollution, endangered species, genetic engineering, test-tube babies, nutrition, aerobic exercise and weight control, medical advances, AIDS and the immune system—is there ever a day that we don't see several of these issues featured in the news? Every one of these topics and many more have biological underpinnings. Biology, the science of life, has an enormous impact on our everyday life, and it is impossible to take an informed stand on many important issues without a basic understanding of life science.

Dangers To Forests Seen From Warming

GENEVA, Aug. 16 (Reuter) — Global warming poses just as much of a threat to tropical rain forests, which are already shrinking because the land is being cleared for farms and timber, as it does to the polar ice caps, the World Wide Fund for Nature said today.

The group, based in Switzerland, issued its opinion in a report as offi-

A. Biology in the news

Much of biology's impact on modern society stems from its contributions to technology and medicine. Many people tend to equate science and technology, but technology is actually the *application* of scientific knowledge. Many discoveries in biology have practical applications. The technology of modern birth control, for instance, grew out of an understanding of the structure and function of the human reproductive system.

One of the newest applications of biology is the development of techniques for artificially transferring genes (DNA) from one organism to another. Popularly called genetic engineering, this technology is being used today for producing certain medicinal drugs. It also has potential use in curing certain human diseases and in increasing crop productivity.

Perhaps the most important application of biology to our lives today is in helping us understand and respond to the environmental problems we currently face. One of our biggest environmental challenges is the possibility of global changes in weather and climate.

Tropical rain forests, which we have featured in this chapter, have a major effect on climate. In this capacity, they are vital to life as far away as Siberia and Antarctica. Consider the photographs in Figure B, which contrast an intact Costa Rican rain forest and an area being illegally burned to prepare it for farming. Every year vast areas of tropical rain forest are destroyed for agriculture or mining, as the human demands for farmland and mineral resources increase. Burning these forests kills off untold numbers of species. It also produces large amounts of carbon dioxide (CO_2) gas. The CO_2 traps heat from sunlight and can warm the atmosphere. Using computer models, some scientists predict that destruction of rain forests at the current rate, coupled with CO_2 increases from other sources (such as industrial pollution), could raise the global temperature. Higher temperatures might melt glacial and polar ice, cause worldwide flooding, and alter the world's climates even more drastically.

Interpreting news reports on problems of this magnitude requires some familiarity with many aspects of biology. These include plant, cell, and molecular biology as these subjects relate to the processes of photosynthesis and energy transformation; carbon and water cycling in ecosystems; the growth patterns of the human population; and the effect of climate and soil conditions on the distribution of life on Earth.

Biology—from the molecular level to the ecosystem level—is directly connected to our everyday lives. It may also help us find solutions to the many environmental problems that confront us. Biology offers us a deeper understanding of ourselves and our planet, and a chance to more fully appreciate life in all its diversity.

B. Tropical rain forest: intact (left) and being burned for farmland (right)

Unit 1
The Life
of the Cell

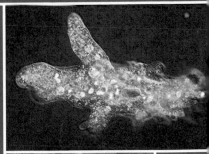

The Chemical Basis of Life 2

In the autumn, groves of quaking aspen trees turn Colorado mountainsides a rich gold. Up close you can hear the soft rustling of the leaves. It's almost constant, and in a gust of wind, whole trees seem to tremble, or quake. The trembling is an emergent property of aspen leaves, one that results from the unique combination of the parts of the leaves (Figure A). The stalk of each aspen leaf is flattened and acts like a hinge, letting the small, light leaf blade flutter back and forth in the slightest breeze. The fluttering seems to help the leaves breathe—that is, take up carbon dioxide and give off oxygen to the air. Plants use carbon dioxide and solar energy to make food (sugar molecules) by photosynthesis. They give off oxygen as a by-product of photosynthesis.

As we saw in Chapter 1, life exhibits emergent properties at several levels. Underlying the emergent property of fluttering aspen leaves, for instance, is a specific combination and arrangement of cells making up the stalks and leaves. And a particular arrangement of molecules makes up the cells.

In a sense, the brilliant gold of fall aspen leaves is also an emergent property. It arises from molecules of yellow and orange pigments, colored substances contained in cells near the leaf surface. These pigments are present in the leaf even when it is green during the summer months. The yellow and orange are unmasked when leaf cells die and lose their green pigment.

The green pigment in aspens and other plants is chlorophyll, a substance that absorbs energy from sunlight and uses it to drive photosynthesis. The computer model in Figure B shows a single molecule of chlorophyll. The spheres represent atoms, the tiny units of matter that make up all molecules. The function of a molecule such as chlorophyll depends on its atoms: their kind, number, and arrangement. Ulti-mately, all the emergent properties of aspens—everything the plants do, from quaking in a breeze to absorbing solar energy and manufacturing food—depends on how atoms are arranged in molecules and how molecules are arranged in the plant's leaves, stems, and roots.

Much of what we know about the functions of plants and other organisms comes from what is called a reductionist approach—a scientific approach based on the idea that the whole can best be understood by studying its parts. In fact, only by taking plants apart and studying the structure and function of their cells and molecules have biologists been able to determine how plants trap and use sunlight, take up carbon dioxide from the air, and absorb water from the soil. Nevertheless, a full understanding of any organism requires that the reductionist approach be combined with the study of the functioning whole. An aspen tree clearly has properties beyond those of its individual leaves, roots, trunk, and molecules. And at the chemical level, an intact chlorophyll molecule has properties lacking in an unorganized pile of its atoms, such as the property of absorbing light that can be used for photosynthesis.

In this unit of chapters, we take a reductionist approach and focus on the chemical components of life. Keep in mind, however, that the properties of life emerge from the organization of these chemical parts into cells and organisms that we will study throughout the book.

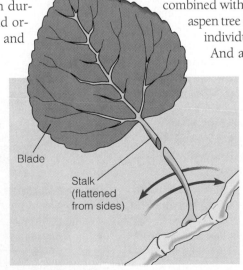

B. Chlorophyll molecule

A. Aspen leaf

Blade

Stalk (flattened from sides)

2.1 Biological function starts at the chemical level

Taking any biological system apart, we eventually end up at the chemical level. Let's now *begin* at the chemical level, with the chlorophyll molecule, and travel to the level of the plant leaf. Throughout this hierarchy, you will see the relationship between structure and the function of photosynthesis, and the emergence of new properties.

Figure A is a chemical diagram of chlorophyll. It shows every atom making up the molecule. There are five kinds of atoms in chlorophyll: C = a carbon atom, H = hydrogen, O = oxygen, N = nitrogen, and Mg = magnesium. Both this diagram and the computer model on the previous page give information about the specific arrangement of atoms in the molecule that enables chlorophyll to absorb light.

Chlorophyll by itself performs only the very first step in photosynthesis. The whole process requires an array of many molecules working together with chlorophyll. These molecules are all housed in chloroplasts, tiny structures inside plant cells. Figure B is a micrograph—a photograph taken with a microscope—of a chloroplast. The ability of a chloroplast to carry out photosynthesis results from the specific arrangement of all the molecules in the chloroplast.

Illustrating the next level in this hierarchy of emergent properties, the micrograph in Figure C shows about 10 cells from a plant leaf. Each of the multisided structures is a plant cell, and within the cells you can see the chloroplasts, which look like small spheres. Chloroplasts and other structures that perform specific functions inside cells are called organelles. Each of these plant cells functions as a living unit because of cooperation between its chloroplasts and many other organelles. There is also cooperation among the cells themselves. Together, they function as a tissue, a group of similar cells performing a specific function. The main function of this particular tissue is photosynthesis.

Finally, we come to the leaves (Figure D), which are plant organs. A leaf is composed of several different tissues, including photosynthetic tissue, epidermal tissue forming the leaf's protective covering, and vascular tissue. Among other functions, vascular tissue transports water needed for photosynthesis from the roots to the leaf. The leaf cannot carry out photosynthesis for long without help from the roots and stems, which are also organs. All the organs work together as parts of the whole plant. We could extend the figure to the level of the whole organism (the aspen tree) and beyond. In fact, we could continue all the way to the involvement of photosynthesis at the ecosystem level, where, for example, beavers and many other animals use the aspens for food.

We will now look more closely at life's most fundamental level of organization, beginning with the chemical elements.

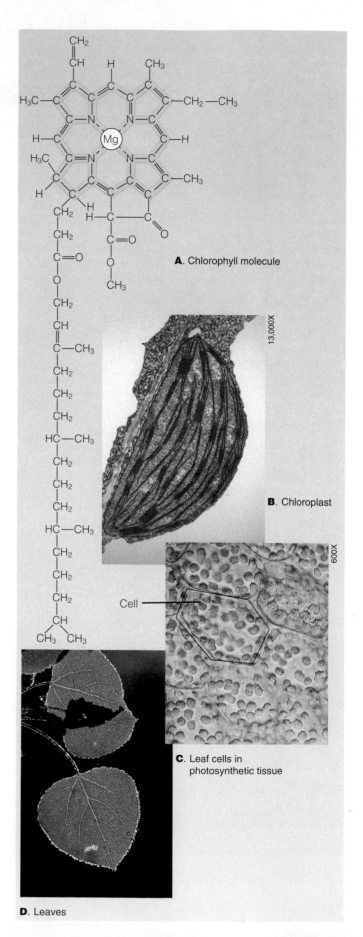

A. Chlorophyll molecule

13,000X

B. Chloroplast

600X

Cell

C. Leaf cells in photosynthetic tissue

D. Leaves

Life is composed of **matter,** which is anything that occupies space and has mass. Matter, in forms as diverse as rock, wood, metal, and air, is composed of chemical elements. A **chemical element** is a substance that cannot be broken down to other substances by ordinary chemical means. Today chemists recognize 92 elements occurring in nature; gold, copper, carbon, and oxygen are some examples. About a dozen more elements have been made in the laboratory. Each element has a symbol, the first letter or two of its English, Latin, or German name. For instance, the symbol for gold, Au, is from the Latin word *aurum;* the symbol O stands for the English word oxygen.

About 25 of the 92 natural elements are essential to life. As you can see in the table, four of these—oxygen (O), carbon (C), hydrogen (H), and nitrogen (N)—make up about

Healthy plant   Magnesium-deficient plant

Magnesium requirement by a plant

Naturally Occurring Elements in the Human Body		
Symbol	**Element**	**Wet Weight Percentage***
O	Oxygen	65.0 ⎫
C	Carbon	18.5 ⎬ 96.3
H	Hydrogen	9.5
N	Nitrogen	3.3 ⎭
Ca	Calcium	1.5
P	Phosphorus	1.0
K	Potassium	0.4
S	Sulfur	0.3
Na	Sodium	0.2
Cl	Chlorine	0.2
Mg	Magnesium	0.1

Trace elements (less than 0.01%): boron (B), Chromium (Cr), cobalt (Co), copper (Cu), fluorine (F), iodine (I), iron (Fe), manganese (Mn), molybdenum (Mo), selenium (Se), silicon (Si), tin (Sn), vanadium (V), and zinc (Zn).

*Includes water.

96% of the human body, which is typical of living matter. Calcium (Ca), phosphorus (P), potassium (K), sulfur (S), and a few other elements account for most of the remaining 4%. The **trace elements** listed at the bottom of the table are essential to at least some organisms, but only in minute quantities. Some trace elements, such as iron, are needed by all forms of life. Others are required only by certain species. The average human, for example, needs about 0.15 milligram (mg) of the trace element iodine each day. A deficiency of iodine prevents normal functioning of the thyroid gland and results in an abnormal enlargement called a goiter. Too much iodine can also cause a goiter.

The figure above demonstrates the importance of one essential element in plants. The plant on the left is a normal, well-nourished tobacco plant. The other is a tobacco plant grown in soil deficient in magnesium (Mg). Forming the core of the chlorophyll molecule (see Figure 2.1A), magnesium is required for photosynthesis. A plant that does not obtain an adequate supply of this essential element cannot produce enough food to survive in nature.

Elements can combine to form compounds 2.3

Chlorophyll is a **compound,** a substance containing two or more elements in a fixed ratio. Compounds are much more common than pure elements. In fact, few elements exist in a pure state in nature. Many compounds consist of only two elements—for instance, table salt (sodium chloride, NaCl) has equal parts of the elements sodium (Na)

and chlorine (Cl). In contrast, most of the compounds in living organisms contain at least three or four different elements, mainly carbon, hydrogen, oxygen, and nitrogen. Proteins, for example, are all formed mainly of these four elements, as is chlorophyll. Different arrangements of the atoms determine unique properties for each compound.

2.4 Atoms consist of protons, neutrons, and electrons

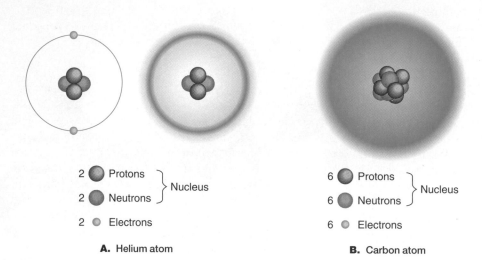

2 ● Protons ⎫
2 ● Neutrons ⎬ Nucleus
2 ○ Electrons

A. Helium atom

6 ● Protons ⎫
6 ● Neutrons ⎬ Nucleus
6 ○ Electrons

B. Carbon atom

Each element consists of one kind of atom, which is different from the atoms of other elements. An **atom,** named from a Greek word meaning indivisible, is the smallest unit of matter that still retains the properties of an element. It would take about a million atoms to stretch across the period printed at the end of this sentence.

Physicists have split the atom into more than a hundred types of subatomic particles. However, to understand basic atomic structure, we only have to consider three. A **proton** is a subatomic particle with a single positive electric charge (+). An **electron** is a subatomic particle with a single negative electric charge (−). The positive charge of a proton is equal in magnitude to the negative charge of an electron. Opposite charges (+ and −) attract each other. A third type of subatomic particle, the **neutron,** is electrically neutral (has no electric charge).

Figure A shows two models of an atom of the element helium (He). (Chemical symbols such as He stand for an atom of an element as well as for the element in general.) Both models in Figure A are highly diagrammatic, and we use them only to show the basic properties of atoms. Notice that two neutrons and two protons are tightly packed in the atom's central core, or **nucleus.** Two electrons, much smaller than the nuclear particles, orbit the nucleus at nearly the speed of light. The attraction between the negatively charged electrons and the positively charged protons keeps the electrons near the nucleus. The left-hand model shows the number of electrons in the atom. The right-hand model, slightly more realistic, shows the electrons as a spherical cloud of negative charge surrounding the nucleus. Neither model is drawn to scale. In real atoms, the electron cloud is much bigger compared to the nucleus.

Elements differ in the number of subatomic particles in their atoms. All atoms of a particular element have the same unique number of protons. This is the element's **atomic number.** Thus an atom of helium, with 2 protons, has an atomic number of 2. Carbon, with 6 protons, has an atomic number of 6 (Figure B). Note that in these atoms, the atomic number is also the number of electrons. When an atom has an equal number of protons and electrons, its net electric charge is 0 (zero).

In contrast to its atomic number, an atom's **mass number** is the sum of the numbers of protons and neutrons in its nucleus. For helium the mass number is 4; for the carbon atom shown here, it is 12. Mass is a measure of the amount of matter in an object. A proton and a neutron are almost identical in mass, and for convenience physicists use the mass of a single proton or neutron as a unit of mass. An electron has very little mass—only about $1/2000$ the mass of a proton. So an atom's mass is essentially equal to its mass number. An element's atomic mass is commonly referred to as its **atomic weight,** and this is given as a whole number—for example, 4 for helium.

Some elements have variant forms called **isotopes.** The different isotopes of an element have the same numbers of protons and electrons but different numbers of neutrons. The table shows the numbers of subatomic particles in the three isotopes of carbon. Carbon-12 (usually written ^{12}C), with 6 neutrons, makes up about 99% of all naturally occurring carbon. Most of the other 1% consists of ^{13}C, with 7 neutrons. A third isotope, ^{14}C, with 8 neutrons, occurs in minute quantities. Notice that all three isotopes have 6 protons—otherwise, they would not be carbon. Both ^{12}C and ^{13}C are stable isotopes, meaning their nuclei remain permanently intact. The isotope ^{14}C, on the other hand, is unstable, or radioactive. A **radioactive isotope** is one in which the nucleus decays spontaneously, giving off particles and energy. Radioactive isotopes pose serious risks to living organisms, but they also have many uses in biological research and medicine, as we see next.

Isotopes of Carbon			
	Carbon-12	Carbon-13	Carbon-14
Protons	6	6	6
Neutrons	6	7	8
Electrons	6	6	6

Living cells cannot distinguish radioactive isotopes from nonradioactive atoms of the same element. Consequently, organisms take up and use compounds containing radioactive isotopes in the usual way. Because radioactivity is easily detected, radioactive isotopes are useful as tracers—biological spies, in effect—for monitoring the fate of atoms in living organisms. To detect radioactivity, scientists use photographic film or instruments such as Geiger counters.

Biologists often use radioactive tracers to follow molecules as they undergo chemical changes in an organism. For example, plant researchers have used them to study photosynthesis. Knowing that plants take in carbon dioxide (CO_2) from the atmosphere and use it to make sugar molecules, they used the radioactive isotope ^{14}C to find out what particular molecules plants make in the process. When researchers provided plants with $^{14}CO_2$, the radioactive carbon appeared first in a compound called 3-PGA and then in several other compounds before showing up in the sugar glucose. These results showed the route by which plants make glucose from CO_2.

Medical diagnosis also makes use of radioactive tracers. Certain kidney disorders, for example, can be diagnosed by injecting tiny amounts of radioactive isotopes into a patient's blood and then measuring the amount of radioactive tracer passed in the urine. In this and most other diagnostic uses of radioactive tracers, the patient receives only a tiny amount of an isotope that decays completely in minutes or hours.

In recent years, radioactive tracers have been used for diagnosis in combination with sophisticated imaging instruments. Figure A shows a patient being examined by a powerful technique called PET (positron-emission tomography), which can monitor chemical processes as they actually occur in the body. The patient is first injected with an isotope that emits subatomic particles called positrons. A PET scanner then detects energy released by the positrons as the isotope moves through the body. Linked to the scanner is a computer that translates the energy data into an anatomical image. The color of the image varies with the amount of the isotope present in an area.

Figure B shows a PET image of the brain (top view) of a 2-year-old boy who had been suffering from seizures. The image shows normal activity in the brain's right half (red, yellow, and green colors) but very little activity in the left half (purple). With this information, surgeons removed the part of the left half that was producing seizures, and saved the boy's life. PET is also useful for diagnosing certain heart disorders and cancers.

Though radioactive isotopes have many beneficial uses, uncontrolled exposure to them can harm living organisms by damaging molecules, especially DNA, in cells. The particles and energy thrown off by radioactive atoms can break apart the atoms of molecules and may cause abnormal connections to form between atoms. The severity of the damage depends on the type and amount of radiation an organism absorbs. The explosion of a nuclear reactor at Chernobyl, Ukraine, in 1986 released large amounts of radioactive isotopes into the environment, killing 30 people within a few weeks. Thousands of others who were exposed to milder doses of the isotopes are expected to develop cancers from radiation damage in the future. Many survivors of the atomic bombs exploded over Japan in World War II are now, 50 years later, fighting radiation-induced cancers.

Natural sources of radiation can also pose a threat. Radon, a radioactive gas, is the second-leading cause of lung cancer (after smoking) in the United States. Radon is a common household contaminant in regions underlain by uranium-bearing rocks, and houses sealed up to prevent heat loss in winter are the ones most likely to contain hazardous levels of radon. Homeowners can buy a radon detector or hire a company to test their homes (state departments of environmental protection will provide references), in order to ensure that radon concentrations are at safe levels.

A. PET scanner

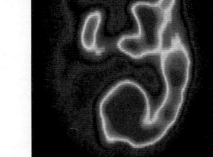

B. PET image of a brain

2.6 Electron arrangement determines the chemical properties of an atom

Of the subatomic particles we have discussed, it is mainly electrons that determine how an atom behaves when it encounters other atoms. Electrons vary in the amount of energy they possess. The farther an electron is from the nucleus, the greater its energy. Electrons in an atom occur only at certain energy levels, called **electron shells.** Depending on their atomic number, atoms may have one, two, or more electron shells, with electrons in the outermost shell having the highest energy. Each shell can accommodate up to a specific number of electrons. The innermost shell is full with only 2 electrons. So in atoms with more than 2 electrons, the remainder are found in shells farther from the nucleus. The outermost (highest-energy) shell can hold up to 8 electrons. It is the number of electrons in the outermost shell that determines the chemical properties of an atom. Atoms whose outer shells are not full tend to interact with other atoms—that is, to participate in chemical reactions.

The diagram below shows the electron shells of four biologically important elements. Small dotted circles represent the unfilled "spaces" in the outer electron shells. Because the outer shells of all four atoms are incomplete, all these atoms react readily with other atoms. The hydrogen atom is highly reactive because it has only 1 electron in its single electron shell, which can accommodate 2 electrons. Atoms of carbon, nitrogen, and oxygen are also highly reactive because their outer shells, which can hold 8 electrons, are incomplete. In contrast, the helium atom in Figure 2.4A has a single, first-level shell that is full with 2 electrons. As a result, helium is chemically inert (unreactive).

How does a chemical reaction enable an atom to fill its outer electron shell? When two atoms with incomplete outer shells react, each atom gives up or acquires electrons so that both partners end up with completed outer shells. Atoms do this by either sharing or transferring outer electrons. These interactions usually result in atoms staying close together, held by attractions called **chemical bonds.** Two or more atoms held together by chemical bonds form a **molecule.** In the next two modules, we look at the two strongest types of chemical bonds.

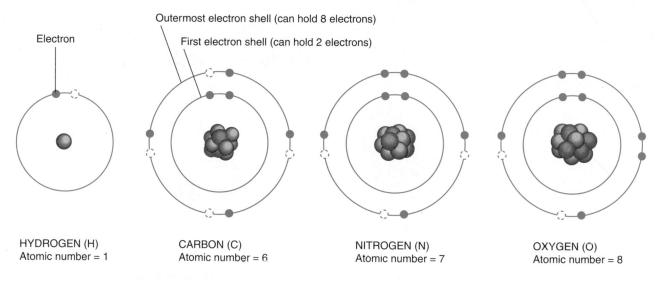

Electron

Outermost electron shell (can hold 8 electrons)

First electron shell (can hold 2 electrons)

HYDROGEN (H)
Atomic number = 1

CARBON (C)
Atomic number = 6

NITROGEN (N)
Atomic number = 7

OXYGEN (O)
Atomic number = 8

Atoms of four elements that are important for life

2.7 Ionic bonds are attractions between ions of opposite charge

At the top of the next page, Figure A shows how a sodium atom and a chlorine atom form the compound sodium chloride (NaCl). Notice that sodium has only 1 electron in its outer shell, whereas chlorine has 7. When these atoms collide, the chlorine atom strips sodium's outer electron away. In doing so, chlorine fills its outer shell with 8 elec-trons. Sodium, in losing 1 electron, ends up with only two shells, the outer shell having a full set of 8 electrons.

Since electrons are negatively charged particles, the electron transfer between the two atoms moves one unit of negative charge from sodium to chlorine. Sodium, with 11 protons but now only 10 electrons, has acquired a net electric

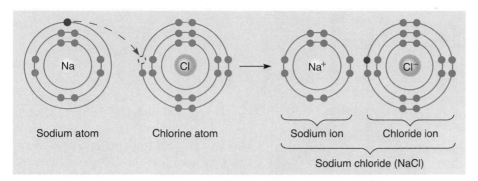

A. Formation of an ionic bond, producing sodium chloride

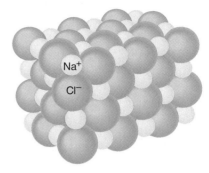

B. A crystal of sodium chloride

charge of +1. Chlorine, having gained an extra electron, now has a net electric charge of −1. In each case, an atom has become what is called an ion. An **ion** is an atom or a molecule with an electric charge resulting from a gain or loss of one or more electrons. As you can see in Figure A, the ion formed from chlorine is called a chloride ion. Two ions with opposite charges attract each other; when the attraction holds them together, it is called an **ionic bond.** The resulting compound, in this case NaCl, is electrically neutral.

Sodium chloride is a type of salt. As a group, salts are ionic compounds that often exist as crystals in nature. Figure B shows the atoms in a piece of crystalline sodium chloride. A crystal of NaCl can be of any size (there is no fixed number of ions), but sodium and chloride ions are always present in a 1:1 ratio. The ratio may be different in crystals of other salts.

Covalent bonds involve electron sharing   **2.8**

The second kind of strong chemical bond is the **covalent bond,** in which two atoms share one or more pairs of outer-shell electrons. For example, a covalent bond connects the two hydrogen atoms in the molecule H_2, a common gas in the atmosphere. The figure shows three ways to represent this molecule. The symbol H_2, called the molecular formula, merely tells you that the molecule consists of two atoms of hydrogen. The middle diagram shows that the atoms share 2 electrons and in doing so both fill their outer (only) shells. At the right, you see a structural formula. The line between the hydrogen atoms stands for the single covalent bond formed by the sharing of one pair of electrons.

The number of single covalent bonds an atom can form is equal to the number of additional electrons needed to fill its outer shell. Looking back at Figure 2.6, we can see that H can form one covalent bond; O can form two; N, three; and C, four. However, as you can see in the figure here, in an O_2 molecule each O atom does not form two single bonds. Instead, the two O atoms share two pairs of electrons, forming a **double bond.** In the structural formula, the double bond is indicated by a pair of lines between the O atoms.

H_2 and O_2 are molecules, but, because they are composed of only one element, they are not compounds. An example of a molecule that is a compound is methane (CH_4), a common gas produced by certain bacteria. As you can see in the figure, each of the four hydrogen atoms in this molecule shares one pair of electrons with the single carbon atom. The same type of bonding occurs in molecules of water, a compound so important to life that we devote the next five modules to it.

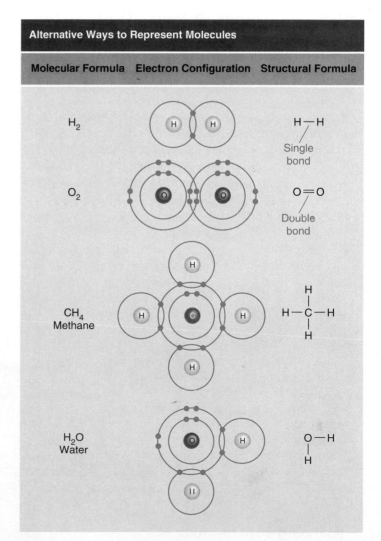

Molecular Formula	Electron Configuration	Structural Formula
H_2		H—H Single bond
O_2		O=O Double bond
CH_4 Methane		H—C—H
H_2O Water		O—H

Alternative Ways to Represent Molecules

2.9 Water is a polar molecule

A water molecule (H$_2$O), shown in the figure here, consists of two hydrogen atoms covalently bonded to a single oxygen atom. Atoms in a covalently bonded molecule are in a constant tug-of-war for the electrons of their covalent bonds. An atom's attraction for the shared electrons of the bond is called its **electronegativity.** The more electronegative an atom, the more strongly it pulls shared electrons toward itself. In molecules formed of only one element, such as O$_2$ and H$_2$, the two identical atoms exert an equal pull on the electrons. The covalent bonds in such molecules are said to be **nonpolar** because the electrons are shared equally between the atoms. Some compounds—for instance, methane (CH$_4$)—also have nonpolar bonds. The atoms in such compounds are not substantially different in electronegativity.

In contrast to O$_2$, H$_2$, and CH$_4$, water is composed of atoms

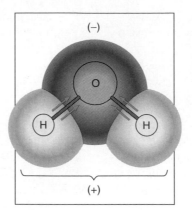

with very different electronegativities. Oxygen is one of the most electronegative of all the elements. As indicated by the arrows in the figure, O attracts the shared electrons in H$_2$O much more strongly than does H, so that the shared electrons are actually closer to the O atom than to the H atoms. This unequal sharing of electrons produces what is called a polar covalent bond between the oxygen atom and each of the hydrogen atoms in the water molecule. A **polar covalent bond** is a chemical bond in which shared electrons are pulled closer to the more electronegative atom, making it partially negative and the other atom partially positive. Thus, in H$_2$O, the O atom actually has a slight negative charge and each H atom a slight positive charge, even though H$_2$O as a whole is neutral. Because of its polar covalent bonds, water is a **polar molecule**—that is, it has a slightly negative pole and two slightly positive ones.

2.10 Water's polarity leads to hydrogen bonding and other unusual properties

The polarity of water molecules makes them interact with each other as shown in Figure A. The charged regions on each molecule are attracted to oppositely charged regions on neighboring molecules, forming weak bonds. Since the positively charged region in this special type of bond is always an H atom, the bond is called a **hydrogen bond.** As you can see in Figure A, the negative (O) pole on each water molecule can form hydrogen bonds to two H atoms; each H$_2$O molecule can hydrogen-bond to as many as four partners.

Water's polarity and hydrogen bonds give it unusual properties. Like no other common substance on Earth, water exists in nature in all three physical states: solid, liquid, and gas (Figure B). Moreover, water's properties and its abundance are major reasons life thrives on Earth.

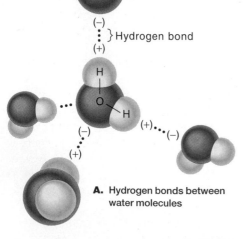

A. Hydrogen bonds between water molecules

B. The three states of water: ice, liquid water, and water vapor

Hydrogen bonds make liquid water cohesive

Hydrogen bonds between molecules of liquid water last for only a few trillionths of a second, yet at any instant, most of the molecules are hydrogen-bonded to others. This tendency of molecules to stick together, called **cohesion**, is much stronger for water than for most other liquids. The cohesion of water is important in the living world. Trees, for example, depend on cohesion to help transport water from their roots to their leaves. The evaporation of water from a leaf exerts a pulling force on water within the veins of the leaf. Because of cohesion, the force is relayed through the veins all the way down to the roots. As a result, water rises against the force of gravity.

Related to cohesion is **surface tension,** a measure of how difficult it is to stretch or break the surface of a liquid. Hydrogen bonds give water unusually high surface tension, making it behave as though it were coated with an invisible film. The insect in the photograph, called a water strider, takes advantage of the high surface tension of water. The insect is denser than water, yet it walks on ponds without breaking the surface.

Surface tension of water exploited by a water strider

Water's hydrogen bonds moderate temperature

If you have ever burned your finger on a metal pot while waiting for the water in it to boil, you know that water heats up much more slowly than metal. In fact, because of hydrogen bonding, water has a better ability to resist temperature change than most other substances. Because of this property, Earth's giant water supply moderates temperatures, keeping them within limits that permit life.

Temperature and heat are related, but different. A swimmer crossing San Francisco Bay has a higher temperature than the water, but the bay contains far more heat because of its immense volume. **Heat** is the amount of energy due to the movement of molecules in a body of matter. **Temperature** measures the intensity of heat—that is, the *average* speed of molecules rather than the *total* amount of heat energy in a body of matter.

When water is heated, the heat energy first disrupts hydrogen bonds and then makes water molecules move faster. Because heat is absorbed as the bonds break, water absorbs and stores a large amount of heat while warming up only a few degrees. Conversely, when water is cooled, more hydrogen bonds form. Heat energy is released when the bonds form, slowing the cooling process.

A large body of water can store a huge amount of heat from the sun during warm periods. At cooler times, heat given off from the gradually cooling water can warm the air. That's why coastal areas generally have milder climates than inland regions. Water's resistance to temperature change also stabilizes ocean temperatures, creating a favorable environment for marine life.

Hydrogen bonds also decrease water's tendency to evaporate, or vaporize. Liquids vaporize when some of their molecules move fast enough to overcome the attractions that keep the molecules close together. Heating a liquid increases vaporization by increasing the energy of the molecules. Water must absorb an unusually large amount of heat in order to vaporize because its hydrogen bonds tend to hold the molecules in place. So hydrogen bonds slow vaporization and give water a high boiling point (100°C). Liquid water is abundant on Earth largely because temperatures near its boiling point are rare on the planet's surface.

Another way water moderates temperatures is by evaporative cooling. When a substance evaporates, the surface of the liquid remaining behind cools down. This occurs because the molecules with the greatest energy (the "hottest" ones) tend to vaporize first. It is as if the 100 best runners at your college left school, lowering the average speed of the remaining students. Evaporative cooling prevents some land-dwelling organisms from overheating; sweating performs that function, for example. On a much larger scale, surface evaporation cools tropical oceans.

Evaporative cooling

2.13 Ice is less dense than liquid water

In contrast to the hydrogen bonds in liquid water, the hydrogen bonds in ice are stable, with each molecule bonded to four neighbors, forming a three-dimensional crystal. In the diagram below, compare the spaciously arranged molecules in the ice crystal to the tightly packed molecules in the liquid water. The ice crystal has fewer molecules than an equal volume of liquid water. In other words, ice is less dense than liquid water. Therefore ice floats, rather than sinking to the bottom of a pond or lake. Being less dense as a solid than as a liquid is one of water's unusual properties that help maintain life as we know it. If ice sank, it would seldom have a chance to thaw, and eventually all ponds, lakes, and even oceans would freeze solid.

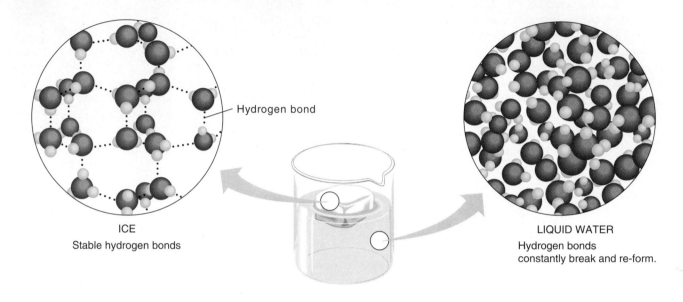

Hydrogen bond

ICE
Stable hydrogen bonds

LIQUID WATER
Hydrogen bonds
constantly break and re-form.

2.14 Water is a versatile solvent

A solution is a liquid consisting of a homogeneous mixture of two or more substances. The dissolving agent is called the **solvent,** and a substance that is dissolved is a **solute.** When water is the solvent, the result is called an aqueous solution (Latin *aqua,* water). As the solvent inside all cells, in blood, and in plant sap, water dissolves an enormous variety of solutes necessary for life.

Water's versatility as a solvent results from the polarity of its molecules. The figure shows how a crystal of table salt dissolves in water. The sodium and chloride ions at the surface of the crystal have affinities for different parts of the water molecules. The positive Na$^+$ ions (yellow) attract the electronegative regions (oxygen, red) of the water molecules. The negative Cl$^-$ ions (green) attract the positively charged hydrogen regions (gray). As a result, H$_2$O molecules surround and separate individual Na$^+$ and Cl$^-$ ions, dissolving the crystal in the process.

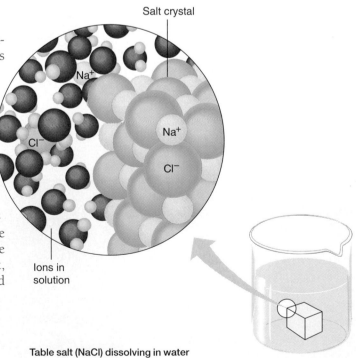

Salt crystal

Na$^+$

Na$^+$

Cl$^-$

Cl$^-$

Ions in
solution

Table salt (NaCl) dissolving in water

The chemistry of life is sensitive to acidic and basic conditions

In the aqueous solutions within organisms, most of the water molecules are intact. However, some of the water molecules actually break apart (dissociate) into ions. The ions formed are called hydrogen ions (H^+) and hydroxide ions (OH^-). For the proper functioning of chemical processes within organisms, the right balance of H^+ ions and negatively charged ions, such as OH^-, is critical.

A chemical compound that donates H^+ ions to solutions is called an **acid.** One example of a strong acid is hydrochloric acid (HCl), the acid in your stomach. In solution, HCl dissociates completely into H^+ and Cl^- ions. The more acidic a solution, the higher its concentration of H^+ ions.

A **base** (or alkali) is a compound that accepts H^+ ions and removes them from solution. Some bases, such as sodium hydroxide (NaOH), do this by donating OH^- ions; these combine with H^+ to form H_2O. The more basic (alkaline) a solution, the lower its H^+ concentration (and, for bases like NaOH, the higher its OH^- concentration).

To describe the acidity of a solution, we use the **pH scale** (pH stands for potential hydrogen). As shown in the figure, the scale ranges from 0 (most acidic) to 14 (most basic). Each pH unit represents a tenfold change in the concentration of H^+. For example, lemon juice at pH 2 has 100 times more H^+ than an equal amount of tomato juice at pH 4.

Pure water and aqueous solutions that are neither acidic nor basic are said to be neutral; they have a pH of 7. They do contain some H^+ and OH^- ions, but the concentrations of the two kinds of ions are equal. The pH of the solution inside most living cells is close to 7. Even a slight change in pH can be harmful, because the molecules in cells are very sensitive to H^+ and OH^- concentrations. However, biological fluids contain **buffers,** substances that resist changes in pH by accepting H^+ ions when they are in excess and donating H^+ ions when they are depleted. Buffers resist some changes in pH, but they are not foolproof. As we discuss in the next module, increasing acidity is a current environmental threat.

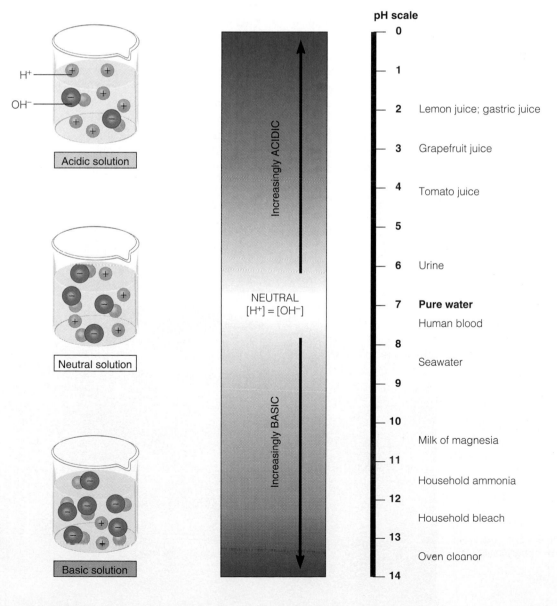

pH scale

pH	
0	
1	
2	Lemon juice; gastric juice
3	Grapefruit juice
4	Tomato juice
5	
6	Urine
7	**Pure water**
	Human blood
8	
	Seawater
9	
10	
	Milk of magnesia
11	
	Household ammonia
12	
	Household bleach
13	
	Oven cleaner
14	

Increasingly ACIDIC

NEUTRAL
$[H^+] = [OH^-]$

Increasingly BASIC

H^+
OH^-

Acidic solution

Neutral solution

Basic solution

Acid precipitation threatens the environment

Imagine arriving for a long-awaited vacation at a mountain lake only to discover that, since your last visit a few years ago, all fish and other forms of life in the lake have perished because of increased acidity of the water. Over the past two decades, thousands of lakes in North America, Europe, and Asia have suffered that fate, primarily as a result of **acid precipitation,** usually defined as rain or snow with a pH below 5.6. About 4% of the lakes in the U.S. are now dangerously acidic, with the number close to 10% in the eastern part of the country.

Effects of acid precipitation are also felt on land. The photograph below shows dead spruce and fir trees on Mount Mitchell in North Carolina, where acid precipitation and acid fog have greatly reduced the numbers of these high-mountain trees. In cities, acid in the air eats away the surfaces of buildings and contributes to smog.

Acid precipitation results mainly from the presence in the air of sulfur oxides and nitrogen oxides, air-polluting compounds composed of oxygen combined with sulfur or nitrogen. These oxides react with water vapor in the air to form sulfuric and nitric acids, which fall to the earth in rain or snow. Rain with a pH between 2 and 3—more acidic than vinegar—has been recorded in the eastern U.S. Acid fog of pH 1.7, approaching that of the digestive juices in our stomachs, has been recorded downwind from Los Angeles.

Sulfur and nitrogen oxides arise mostly from the burning of fossil fuels (coal, oil, and gas) in factories and automobiles. Electric power plants that burn coal produce more of these pollutants than any other single source. Ironically, the tall smokestacks built to reduce local pollution by dispersing factory exhaust help spread airborne acids. Winds carry the pollutants away, and acid rain may fall thousands of miles away from industrial centers.

The effect of acid in lakes and streams is most pronounced in the spring, as snow begins to melt. The surface snow melts first, drains down, and sends much of the acid that has accumulated over the winter into lakes and streams all at once. Early meltwater often has a pH as low as 3, and this acid surge hits when fish and other aquatic life are producing eggs and young, which are especially vulnerable to acidic conditions.

Strong acidity can break down the molecules of living organisms, and even if the molecules remain intact, they may not be able to carry out the essential chemical processes of life at very low pH.

While acid precipitation can clearly damage lakes and streams, its effects on forests and other land life are controversial. The damage to the forest on Mount Mitchell, for example, almost certainly results from acid fog and precipitation. The acid apparently causes changes in the soil that lead to mineral imbalances, lowered tolerance to cold, and general weakness in the trees. Nevertheless, careful studies over the past decade seem to show that the vast majority of North American forests are not suffering substantially from acid precipitation.

Many questions remain. We do not know for sure what the long-term effects of acid precipitation may be on plants and soils. Nor do we know much about the effects of airborne acids on terrestrial animals, including humans. Perhaps most important, we do not know how much we must reduce fossil-fuel emissions in order to prevent more damage.

As with most environmental issues, there are no easy solutions to the acid precipitation problem. There is some hopeful news, however. Between 1970 and 1985, emissions from such sources as automobiles and fossil-fuel-burning power plants in the U.S. dropped 30% for sulfur oxides and about 10% for nitrogen oxides. Laws that require reductions in emissions can go a long way toward alleviating the problem. But just as important is energy conservation. We all need to realize that unless we decrease our consumption of electricity and our dependence on gasoline-powered automobiles, we will continue to contribute to acid precipitation and other threats to the environment.

Spruce and fir trees killed by acid precipitation

The basic chemistry of life has an overriding theme: The structure of atoms and molecules determines the way they interact and behave. As we have seen in this chapter, the chemical properties of an atom are determined by the number and arrangement of its subatomic particles, particularly its electrons. Other unique properties emerge when atoms combine to form molecules and when molecules combine to form more complex substances, such as liquid water. Water is a particularly good example, because its unusual properties form the foundation of life itself.

Water can be made from hydrogen and oxygen molecules as follows:

$$2 H_2 + O_2 \longrightarrow 2 H_2O$$

This is an example of a **chemical reaction,** a process leading to chemical changes in matter. In this particular case, two molecules of hydrogen (H_2) react with one molecule of oxygen (O_2) to give two molecules of water (H_2O). The arrow indicates the conversion of the starting materials, called **reactants** (H_2 and O_2), to the resulting **product** (H_2O). Notice that the same *numbers* of hydrogen and oxygen atoms appear on the right and left sides of the arrow, although they are grouped differently. This shows that chemical reactions do not create or destroy matter; they only rearrange it in various ways. Chemical reactions involve the making and breaking of chemical bonds. In the example above, the bonds holding hydrogen atoms together in H_2 and those holding oxygen atoms together in O_2 are broken, and new bonds are formed to yield the H_2O product molecules (Figure A).

Organisms cannot make water from molecules of hydrogen and oxygen, but they do carry out a great number of chemical reactions that rearrange matter in significant ways. Let's examine one that relates to the chapter's opening essay. Much of the yellow-orange color in autumn leaves is from molecules of pigments called carotenoids. These organic molecules are important in photosynthesis, absorbing solar energy, as chlorophyll does. One of the most common carotenoids is beta-carotene. Shown in Figure B, beta-carotene is also abundant in carrots and many green vegetables and is sold in concentrated form as a food supplement. This substance can be important in our diet as a source of vitamin A, which is essential for normal vision and healthy skin. Our cells can make vitamin A from beta-carotene in the following way:

$$\underset{\text{Beta-carotene}}{C_{40}H_{56}} + O_2 + 4 H \longrightarrow \underset{\text{Vitamin A}}{2\ C_{20}H_{30}O}$$

Although there are many atoms in beta-carotene and vitamin A, the chemical reaction that converts one to the other is essentially a simple one. Two molecules of vitamin A are made from each beta-carotene molecule by splitting the beta-carotene molecule in half. Notice that beta-carotene has 40 carbon (C) atoms, whereas each vitamin A molecule has 20 carbons. The red arrow in Figure B indicates where beta-carotene is split. The other reactants are a molecule of O_2 and 4 H atoms contributed by other molecules in the cell. If you count up all the atoms, you will see that the same number of each type appears on each side of the reaction.

The conversion of beta-carotene to vitamin A is one example of thousands of chemical reactions routinely carried out in living cells. Like most of these reactions, our example involved compounds of the element carbon. We look at the carbon compounds of cells in more detail in Chapter 3.

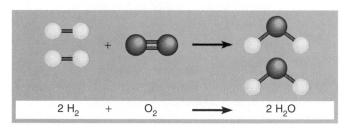

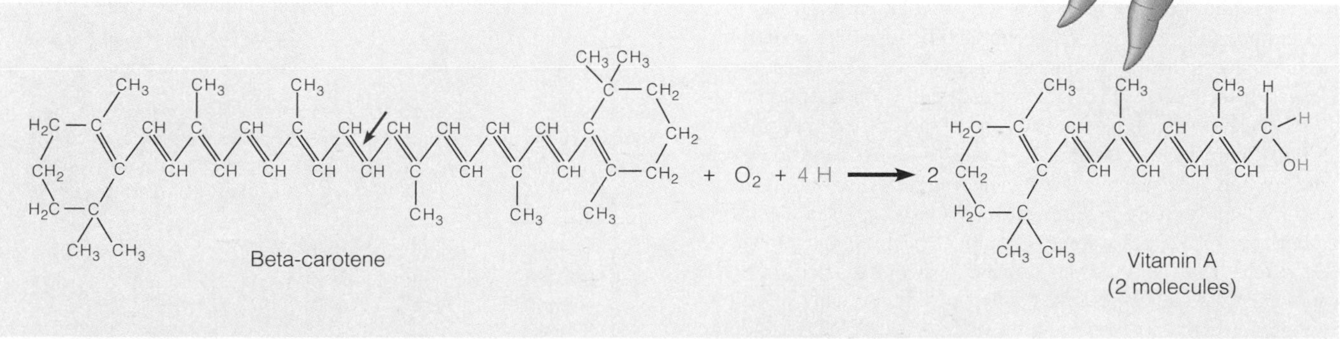

A. Breaking and making of bonds in a chemical reaction

B. Chemical reaction converting beta-carotene to vitamin A

Begin your review by rereading the module headings and scanning the figures, before proceeding to the Chapter Summary and questions.

Chapter Summary

Introduction–2.1 Everything an organism is and does depends on how atoms are arranged in molecules. Much of what we know about life has come from taking living things apart, all the way down to atoms. Studying the parts in order to understand the whole is called the reductionist approach. But this approach must be combined with study of the functioning whole; each level of organization—a molecule, cell, tissue, organ, or organism—has unique emergent properties.

2.2–2.3 About 25 chemical elements are essential to life. Carbon, hydrogen, oxygen, and nitrogen make up the bulk of living matter, but life also requires such trace elements as iron and zinc. Chemical elements combine in fixed ratios to form compounds, such as sodium chloride and chlorophyll.

2.4–2.5 The smallest particle of an element is an atom. An atom consists of protons and neutrons in a central nucleus, surrounded by electrons. Each atom is held together by attraction between its positively charged protons and negatively charged electrons; neutrons are neutral. Atoms of each element are characterized by a certain number of protons; the number of neutrons may vary. Variant forms of an atom are called isotopes. Some isotopes are radioactive. Though their radioactivity is sometimes harmful to life, isotopes can be useful tracers for studying biological processes.

2.6–2.8 Electrons are arranged in electron shells. The outermost shell determines the chemical properties of an atom. In most atoms, a complete outer shell holds 8 electrons. Atoms whose shells are not full tend to interact with other atoms and gain, lose, or share electrons. Electron gain and loss creates charged atoms, called ions, whose electrical attraction results in ionic bonds between them. Sodium loses an electron and chlorine gains one to become the ions that form sodium chloride, NaCl. Other atoms share pairs of electrons, binding the atoms together with covalent bonds. Atoms joined by covalent bonds form molecules, such as water, H_2O.

2.9–2.13 Atoms in a covalently bonded molecule may share electrons equally, in which case the molecule is said to be nonpolar. In a water molecule, the oxygen atom exerts a stronger pull on the electrons than hydrogen does, making the oxygen slightly negative and the hydrogens slightly positive. Water is therefore a polar molecule. The charged regions on water molecules are attracted to oppositely charged regions on nearby molecules, forming weak bonds called hydrogen bonds. Polarity and hydrogen bonds give water some unique properties. Like no other common substance, it exists in nature in solid, liquid, and gaseous states. Hydrogen bonds make water molecules cohesive, allowing for the movement of water from plant roots to leaves. Insects can walk on water because of surface tension created by cohesive water molecules. Water's ability to store heat moderates body temperature and climate. It takes a lot of energy to disrupt hydrogen bonds, so water is able to absorb a great deal of heat energy without a large increase in temperature. As water cools, a slight drop in temperature releases a large amount of heat. A water molecule takes much energy with it when it breaks away from its neighbors and evaporates, making evaporative cooling possible. Finally, molecules in ice are farther apart than in liquid water. Because ice is less dense than water, it floats, and lakes do not freeze solid.

2.14–2.15 Solutes whose charges or polarity allow them to stick to water molecules will dissolve in water, forming aqueous solutions. A compound that releases H^+ ions in solution is an acid, and one that accepts H^+ ions is a base. Acidity is measured on the pH scale, from 0 (most acidic) to 14 (most basic). Pure water and solutions that are neither acidic nor basic are neutral, having a pH of 7. Cells are close to pH 7 and kept that way by buffers—substances that resist pH change.

2.16 Some ecosystems are threatened by acid precipitation, which is formed when air pollutants from burning fossil fuels combine with water vapor in the air to form sulfuric and nitric acids. These acids can kill fish and damage buildings, and may injure trees. Regulations, new technology, and energy conservation may help us reduce acid precipitation.

2.17 In a chemical reaction, reactants interact, atoms rearrange, and products result. Living cells carry out thousands of chemical reactions that rearrange matter in significant ways.

Testing Your Knowledge

Multiple Choice

1. Your body contains the smallest amount of which of the following?
   - **a.** nitrogen
   - **b.** phosphorus
   - **c.** carbon
   - **d.** oxygen
   - **e.** hydrogen

2. Changing the _____ would change it into an atom of a different element.
   - **a.** number of electrons surrounding the nucleus of an atom
   - **b.** number of bonds formed by an atom
   - **c.** number of protons in the nucleus of an atom
   - **d.** electrical charge of an atom
   - **e.** number of neutrons in the nucleus of an atom

3. A solution at pH 6 contains _____ than the same amount of solution at pH 8.
   - **a.** 2 times more H^+
   - **b.** 4 times more H^+
   - **c.** 100 times more H^+
   - **d.** 4 times less H^+
   - **e.** 100 times less H^+

4. Most of the unique properties of water result from the fact that water molecules
   - **a.** are very small
   - **b.** are held together by covalent bonds
   - **c.** easily separate from one another
   - **d.** are constantly in motion
   - **e.** tend to stick together

5. A silicon atom has 6 electrons in its outer shell. As a result, it forms _____ covalent bonds with other atoms. *(Explain your answer.)*
   - **a.** 2
   - **b.** 3
   - **c.** 4
   - **d.** 6
   - **e.** 8

6. A can of cola consists mostly of sugar dissolved in water, with some carbon dioxide gas that makes it fizzy and makes the pH less than 7. In chemical terms, you could say that cola is an aqueous solution, where water is the _____, sugar is a _____, and carbon dioxide makes the solution _____.
   - **a.** solvent . . . solute . . . basic
   - **b.** solute . . . solvent . . . basic
   - **c.** solvent . . . solute . . . acidic
   - **d.** solute . . . solvent . . . acidic
   - **e.** not enough information to say

7. Which of the following is *not* a chemical reaction?

   a. Sugar and oxygen gas combine to form carbon dioxide and water.
   b. Sodium metal and chlorine gas unite to form sodium chloride.
   c. Hydrogen gas combines with oxygen gas to form liquid water.
   d. Solid ice melts to form liquid water.
   e. Sulfur dioxide and water vapor join to form sulfuric acid.

True/False *(Change false statements to make them true)*

1. Salt, water, and carbon are compounds.

2. Atoms in a water molecule are held together by the sharing of electrons.

3. A bathtub full of lukewarm water may hold more heat than a teakettle full of boiling water.

4. If the atoms in a molecule share electrons equally, the molecule is said to be nonpolar.

5. Ice floats because water molecules in ice are more tightly packed than in liquid water.

6. The smallest particle of an element is a molecule.

7. The pH scale is set up so that pure water has a pH of 1.

8. Most acid precipitation results from the presence of pollutants from aerosol cans and air conditioners.

9. An atom that has gained or lost electrons is called an ion.

10. Reactants are the substances produced by a chemical reaction.

Describing, Comparing, and Explaining

1. Make a sketch that shows how water molecules hydrogen-bond with one another. Why do water molecules form hydrogen bonds? What unique properties of water result from the tendency of water molecules to form hydrogen bonds?

2. Describe two ways in which the water in your body helps stabilize your body temperature.

3. Compare covalent and ionic bonds.

4. What is an acid? A base? How is the acidity of a solution described?

Thinking Critically

1. Animals obtain energy through a series of chemical reactions in which sugar ($C_6H_{12}O_6$) and oxygen gas (O_2) are reactants. This process produces water (H_2O) and carbon dioxide (CO_2) as waste products. Researchers want to know whether the oxygen in CO_2 comes from sugar or oxygen gas. How might they use a radioactive isotope to find out?

2. The diagram below shows the arrangement of electrons around the nucleus of a fluorine atom (left) and a potassium atom (right). Predict what would happen if a fluorine atom and a potassium atom came into contact. What kind of bond do you think they would form?

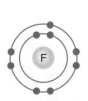
Fluorine atom

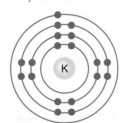
Potassium atom

3. A student performed the following experiment: She heated two identical glass marbles to the same temperature in an oven. Then she dropped one marble into a beaker containing 100 milliliters (ml) of alcohol. She dropped the other marble into an identical beaker containing 100 ml of water. The temperature of the alcohol rose 12°C, but the temperature of the water only went up 5°C. Explain the difference in temperature increase in the two liquids.

Science, Technology, and Society

1. The 1990 U.S. Clean Air Act includes a provision for "acid rain pollution credits." Under this plan, the government sets limits on the sulfur and nitrogen oxide emissions of power plants that burn fossil fuels. If an efficient, modern plant emits fewer pollutants than the standard, it may sell the unused portion of its pollution limit to another power plant elsewhere in the country, allowing that plant to emit more pollutants than would otherwise be allowed. For example, a low-emission power plant in Wisconsin recently sold its excess pollution credits to a utility in Pennsylvania. It was more economical for the Pennsylvania power plant to buy the pollution credits than to modernize its equipment. What advantages do you see in pollution credits? For whom? What are some drawbacks of this plan? For whom? What might happen if citizens, environmental groups, or speculators decided to buy pollution credits?

2. One solution to the problem of acid precipitation is to use nuclear energy to produce electricity. Development of nuclear power in the United States virtually stopped after the accident at the Three Mile Island power plant in Pennsylvania in 1979. But proponents of nuclear power contend that accelerated development of nuclear energy is the answer to U.S. energy needs and tightened pollution standards. Besides reduction of acid precipitation, what are other potential benefits of nuclear power? What are its possible costs and dangers? Do you think we ought to pursue development of nuclear power? Why or why not? If a new power plant were to be built near where you live, would you prefer it to be a nuclear or coal-fired plant? Why?

3. While waiting at an airport, one of the authors of this book once overheard this claim: "It's paranoid and ignorant to worry about industry or agriculture contaminating the environment with their chemical wastes. After all, this stuff is just made of the same atoms that were already present in our environment anyway." How might you counter this argument?

The Molecules of Cells 3

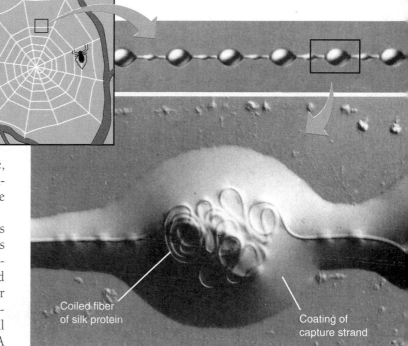

Coiled fiber
of silk protein

Coating of
capture strand

Part of a spider's capture strand

Ever since ancient times, people have been fascinated with spiders and the webs they spin. An old Navajo legend tells of a Spider Woman who invented weaving and taught all the textile arts to the Native Americans of the Southwest. According to Greek mythology, spiders originated when the maiden Arachne was transformed into one, as a punishment for boasting of her weaving skills. Our technical name for spider is arachnid, which is derived from the Greek word.

The ancient legends allude to special qualities of spiders and their webs—qualities that arise from unique properties of molecules. Web builders, like the spider in the photograph at the left, spin their webs with remarkable speed and agility. This web probably took no longer than an hour to construct, yet its symmetry and efficacy in capturing insects are without parallel. The spider's web-building skill depends on its genetic programming, built into the DNA molecules it inherited from its parents. The special qualities of the web, which is made of silk produced by glands in the spider's abdomen, result from unique properties of silk proteins associated with a mix of other molecules.

Silk proteins make a spiderweb remarkably strong and resilient, able to withstand a struggling insect's attempts to escape. The type of web in the photograph is known as an orb web for its roughly circular shape. The strands extending straight out from the center of the orb—the radial strands—are composed of dry, relatively inelastic proteins; they maintain the web's position and overall shape. In contrast, the orb's spiraling strand, which actually captures insects, is wet, sticky, and highly elastic. In fact, it can stretch up to four times its original length and then recover with little sag. How does a capture strand extend and contract in response to thrashing insects, to the weight of rainwater and dew, or to the tearing force of a strong wind? The web's elasticity apparently results from the coiling and uncoiling

of silk proteins. The photographs above show a single fiber of a capture strand at two magnifications. The coiled thread in the middle of the bottom photo is made of silk protein. It can unwind as needed and then recoil rapidly. Droplets of a watery glue coat the coiled protein thread. The glue helps trap prey and helps wind up the thread when prey no longer presses against it.

Spiders and their webs illustrate life's molecules in action. A spider's DNA and the proteins in its silk represent two major classes of molecules in living organisms. Life's diversity results from a seemingly endless variety of these and other molecules. However, a relatively small number of basic structural patterns underlies this diversity and helps us make sense of it. In this chapter, we describe the major molecules of living cells and how they are put together.

3.1 Life's diversity is based on the properties of carbon

Almost all the molecules a cell makes are composed of carbon atoms bonded to one another and to atoms of other elements. Carbon is unparalleled in its ability to form large, diverse molecules. Next to water, compounds containing carbon are the most common substances in living organisms.

Compounds containing carbon are known as **organic compounds.** Well over two million organic compounds are known, and chemists identify more each day. As we discussed in Chapter 2, an element's chemical properties are determined by the electrons in the outermost shell of its atoms. Thus, the enormous variety of carbon-based molecules derives from the tendency of carbon's outer-shell electrons to form chemical bonds, specifically covalent bonds (see Module 2.8). A carbon atom has 4 outer electrons in a shell that holds 8. Consequently, it has a strong tendency to complete its outer shell by sharing electrons with other atoms in four covalent bonds.

At the top of the figure here are three illustrations of methane, one of the simplest organic molecules. The structural formula simply shows that covalent bonds link four hydrogen atoms to the carbon atom. Each of the four lines in the formula represents a pair of shared electrons. The two models show that methane is three-dimensional, with the space-filling version portraying its overall shape more accurately. The ball-and-stick model indicates the angles of the molecule's covalent bonds (the white "sticks"). The four hydrogen atoms of methane are at the corners of an imaginary tetrahedron (an object with four triangular sides); the same tetrahedral shape occurs wherever a carbon atom has four single bonds. Different bond angles occur where carbon atoms form double bonds. The shape of a molecule, which depends partly on its bond angles, usually helps determine its function.

Methane and other compounds composed of only carbon and hydrogen are called **hydrocarbons.** The figure illustrates some variations in hydrocarbon structure. Carbon atoms, with attached hydrogens, can bond together in chains of various lengths to form compounds such as ethane or propane, a gas used as household fuel. The chain of carbon atoms in organic molecules like these is called a **carbon skeleton.** Carbon skeletons can be unbranched, as in butane, or branched, as in isobutane. Carbon skeletons may also include double bonds, as in 1-butene and 2-butene. These two compounds have the same molecular formula, C_4H_8, but differ in the position of their double bond. Compounds like these with the same molecular formula but different structure are called **isomers.** (Butane and isobutane are also isomers.) Each isomer has unique properties. Cyclohexane and benzene, the only hydrocarbons in the figure that are liquids rather than gases, are examples of carbon skeletons arranged in rings.

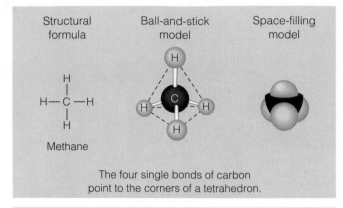

Structural formula   Ball-and-stick model   Space-filling model

Methane

The four single bonds of carbon point to the corners of a tetrahedron.

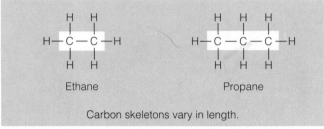

Ethane   Propane

Carbon skeletons vary in length.

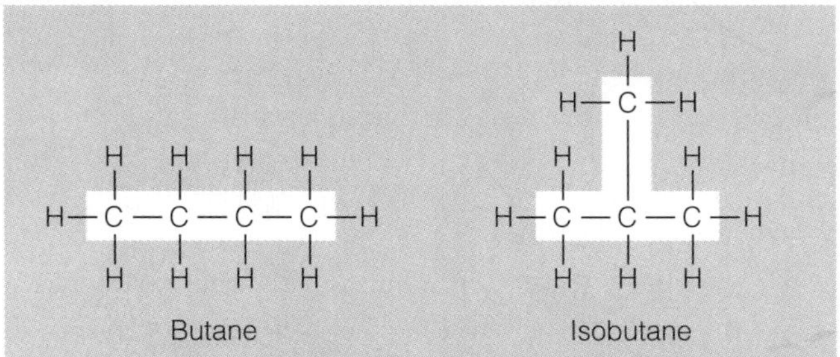

Butane   Isobutane

Skeletons may be unbranched or branched.

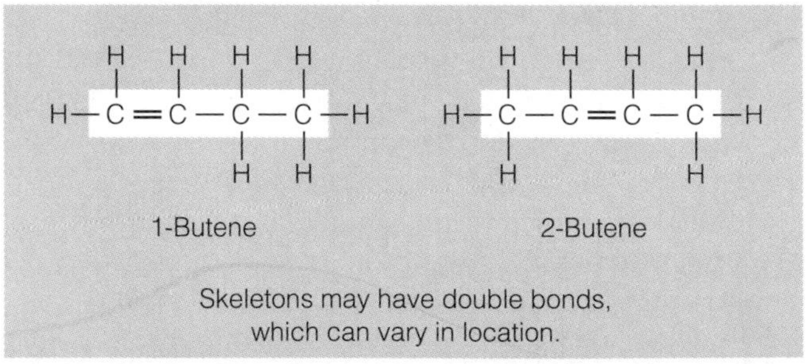

1-Butene   2-Butene

Skeletons may have double bonds, which can vary in location.

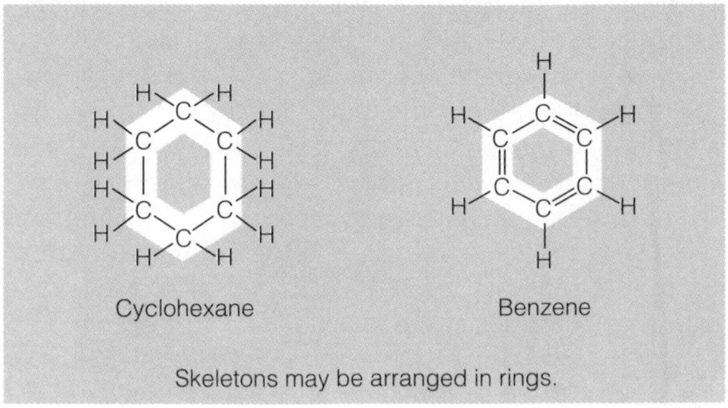

Cyclohexane   Benzene

Skeletons may be arranged in rings.

Functional groups help determine the properties of organic compounds

The unique properties of an organic compound depend not only on its carbon skeleton but also on certain groups of atoms that are attached to the skeleton. These assemblages of atoms are called **functional groups** because they are the parts of organic molecules usually involved in chemical reactions. The table below shows four functional groups important in the chemistry of life. All these functional groups are polar, because their oxygen or nitrogen atoms exert a strong pull on shared electrons. The polarity tends to make compounds containing these groups **hydrophilic** (water-loving) and therefore soluble in water—a necessary condition for their roles in water-based life.

A **hydroxyl group** consists of a hydrogen atom bonded to an oxygen atom. The oxygen is bonded to the carbon skeleton of an organic molecule. Ethanol and many other organic compounds containing hydroxyl groups are called alcohols.

A **carbonyl group** is a carbon atom linked by a double bond to an oxygen atom. If the carbon atom of the carbonyl group is at the end of a carbon skeleton, the compound is called an aldehyde. Compounds whose carbonyl groups are within a carbon chain are called ketones.

A **carboxyl group** consists of an oxygen atom double-bonded to a carbon atom that is also bonded to a hydroxyl group. The carboxyl group acts as an acid by contributing a hydrogen ion to a solution (see Module 2.15), and compounds containing this functional group are called **carboxylic acids.** Acetic acid, shown in the figure, gives vinegar its sour taste.

An **amino group** is composed of a nitrogen atom bonded to two hydrogen atoms. It acts as a base by picking up a hydrogen ion from a solution. Organic compounds with an amino group are called **amines.**

Though all the examples illustrated here contain only one functional group each, many biological molecules have two or more. For example, sugars have both hydroxyl and carbonyl groups, and thus can be classified as both alcohols and aldehydes (or ketones). Compounds called amino acids have carboxyl as well as amino groups. Amino acids are the building blocks of proteins, as we will see in a later module.

Functional Group	General Formula	Name of Compounds	Example	Where Else Found
Hydroxyl —OH (or HO—)	—O—H	Alcohols	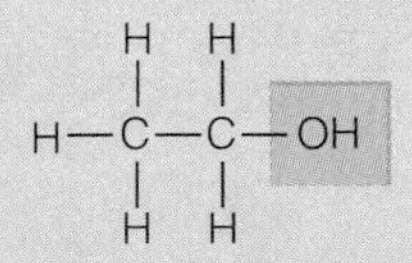 Ethanol	Sugars; water-soluble vitamins
Carbonyl \CO/	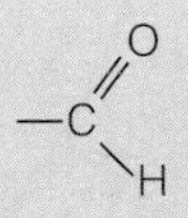 (aldehyde general formula)	Aldehydes	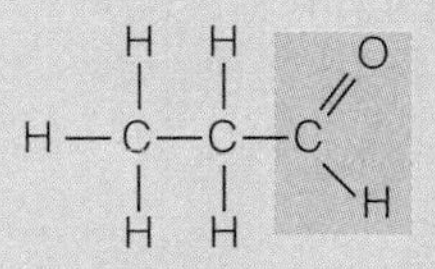 Propanal	Some sugars; formaldehyde (a preservative)
	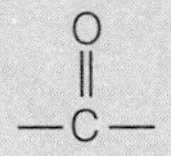 (ketone general formula)	Ketones	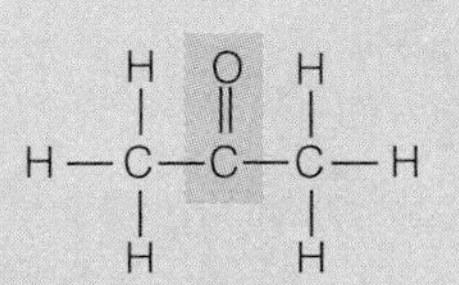 Acetone	Some sugars; "ketone bodies" in urine (from fat breakdown)
Carboxyl —COOH	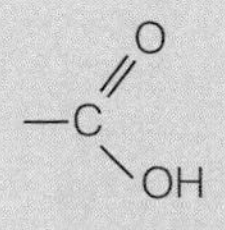 (carboxyl general formula)	Carboxylic acids	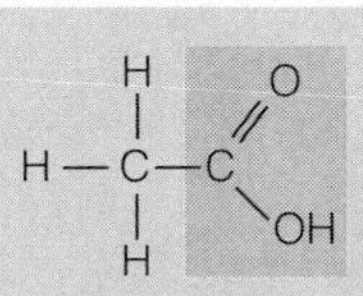 Acetic acid	Amino acids; proteins; some vitamins; fatty acids
Amino —NH₂ (or H₂N—)	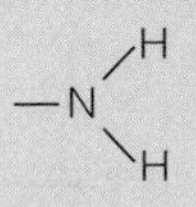 (amino general formula)	Amines	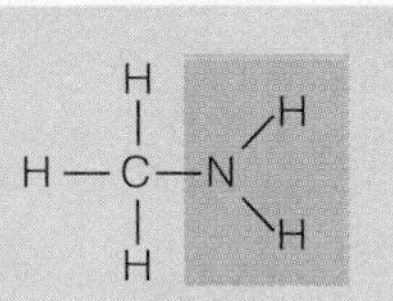 Methylamine	Amino acids; proteins; urea in urine (from protein breakdown)

3.3 Cells make a huge number of large molecules from a small set of small molecules

On a molecular scale, many of life's molecules are gigantic; in fact, biologists call them **macromolecules.** Proteins, one class of macromolecules, may consist of thousands of covalently connected atoms. A second class is nucleic acids, one of which is DNA. A third class of macromolecules is made up of carbohydrates—specifically, a group of large carbohydrates called polysaccharides. Lipids, including the fats, comprise a diverse fourth group of large organic molecules important in cells.

Cells make most of their large molecules by joining smaller organic molecules into chains called **polymers** (from the Greek *polys,* many, and *meros,* part). A polymer is a large molecule consisting of many identical or similar molecular units strung together, much as a train consists of many individual cars. The units that serve as the building blocks of polymers are called **monomers.**

Living cells make a vast number of different polymers. For proteins alone, there are about a trillion different kinds in nature, and the variety is potentially endless. One of life's most remarkable features is that a cell makes all of its diverse macromolecules from a small list of ingredients—about 40 to 50 common monomers plus a few others that are rare. Proteins, for example, are built from only 20 kinds of amino acids, arranged in chains typically about 100 amino acids long. DNA is built from just four kinds of monomers called nucleotides. As we will see, the key to the great diversity of protein and DNA molecules is arrangement—variation in the sequence in which the monomers are strung together.

The variety in polymers accounts for the uniqueness of each organism. The monomers used to make polymers, however, are essentially universal. Your proteins and those of a tree or spider are assembled from the same 20 amino acids; the amino acids are just arranged in different sequences. Life has a simple yet elegant molecular logic: Small molecules common to all or-ganisms are ordered into macromolecules, which vary from species to species and even from individual to individual.

Cells link monomers together to form polymers by a process called **dehydration synthesis** (Figure A). All unlinked monomers have hydrogen atoms (H) and also hydroxyl groups (—OH). For each monomer added to a chain, a water molecule (H_2O) is removed. Notice that two monomers contribute to the H_2O molecule, one monomer (the one at the right end of the short polymer in this example) losing a hydrogen atom, and the other monomer losing a hydroxyl group. As this occurs, a new covalent bond forms, linking the two monomers. The same process occurs regardless of the specific monomers.

Cells not only make macromolecules, they also have to break them down. For example, food that an organism ingests is often in the form of macromolecules. To digest these substances and make their monomers usable, a cell carries out **hydrolysis,** essentially the reverse of dehydration synthesis. Hydrolysis means to break (lyse) with water (hydro-), and cells break bonds between monomers by adding water to them, as Figure B shows. In the process, a hydroxyl group from a water molecule joins to one monomer, and a hydrogen joins to the adjacent monomer.

In later modules, we will examine each of the four classes of macromolecules in more detail. We will see that proteins, nucleic acids, and polysaccharides are all polymers assembled by dehydration synthesis from respective monomers: proteins from amino acids, nucleic acids from nucleotides, and polysaccharides from monosaccharides. Lipids are not truly polymers. However, one of the largest types of lipid, the fats, are formed by dehydration synthesis from several smaller molecules.

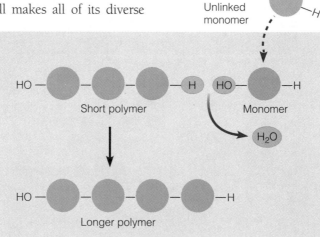

A. Dehydration synthesis of a polymer

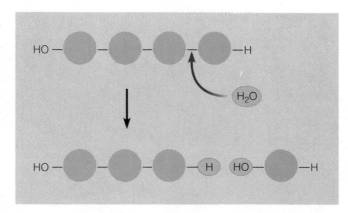

B. Hydrolysis of a polymer

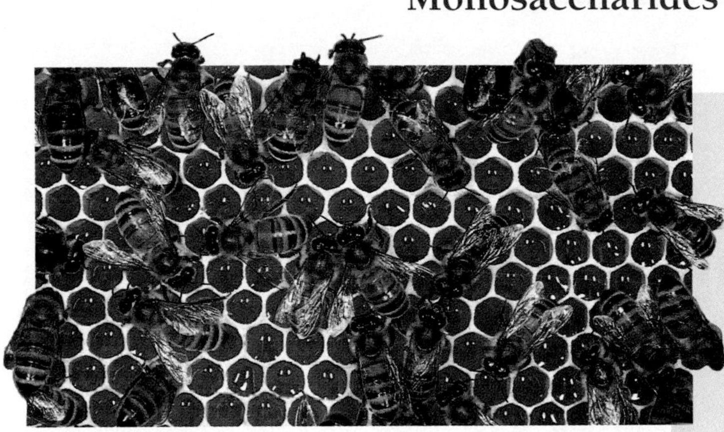

A. Bees with honey, a mixture of two monosaccharides

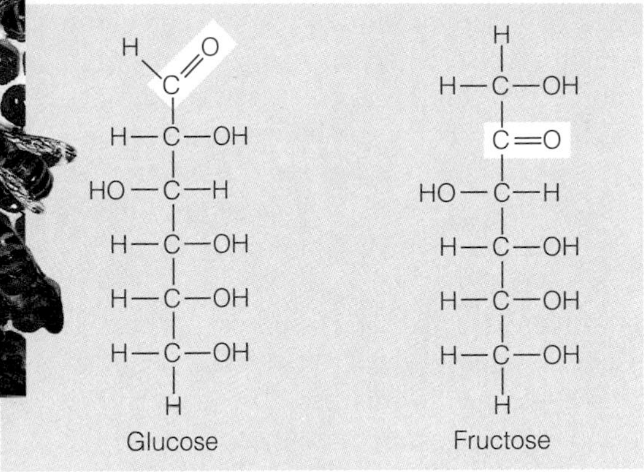

Glucose   Fructose

B. Structures of glucose and fructose

The name **carbohydrate** refers to a class of molecules ranging from small sugar molecules to large polysaccharides, which are long polymers of sugar monomers. The carbohydrate monomers (single-unit sugars) are **monosaccharides,** from the Greek *mono-*, single, and *sacchar,* sugar. The honey shown in Figure A consists mainly of monosaccharides called glucose and fructose. These and other single-unit sugars can be hooked together by dehydration synthesis to form more complex sugars and polysaccharides.

Monosaccharides generally have molecular formulas that are some multiple of CH_2O. For example, the formula for glucose, a common monosaccharide of central importance in the chemistry of life, is $C_6H_{12}O_6$. Its molecular structure, shown in Figure B, has the two trademarks of a sugar: a number of hydroxyl groups ($-OH$) and a carbonyl group ($>C=O$). The hydroxyl groups make a sugar an alcohol, and the carbonyl group, depending on its location, makes it either an aldehyde or a ketone as well. Glucose is an aldehyde, and fructose is a ketone.

If you count the numbers of different atoms in the fructose molecule in Figure B, you will find that its molecular formula is $C_6H_{12}O_6$, identical to that of glucose. Thus, glucose and fructose are isomers; they differ only in the arrangement of their atoms (in this case, the positions of the carbonyl groups, highlighted in white). Seemingly minor differences like this give isomers different properties, such as the ability to react with other molecules. In this case, the differences also make fructose taste considerably sweeter than glucose.

While the carbon skeletons of both glucose and fructose are six carbon atoms long, other monosaccharides have three to seven carbon atoms. The five-carbon sugars, called pentoses, and the six-carbon sugars, called hex-

oses, are among the most common. Note that many names for sugars have the suffix "-ose."

It is convenient to draw sugars as if their carbon skeletons were linear, but this does not always create an accurate picture. In aqueous solutions, many monosaccharides form rings, as shown for glucose in Figure C. (The small identifying numbers on the carbon atoms and the red and blue type should help you follow along.) The arrows indicate that glucose switches back and forth between the linear and ring forms when in solution. As shown at the right, the ring diagram of glucose and other sugars may be abbreviated by not showing the carbon atoms at the corners of the ring. Also, as shown in the abbreviated structure, the bonds in the ring are often drawn with varied thickness, indicating that the ring is a relatively flat structure with atoms and functional groups, such as $-OH$, extending above and below it.

Monosaccharides, particularly glucose, are the main fuel molecules for cellular work. Cells also use the carbon skeletons of monosaccharides as raw material for manufacturing other kinds of organic molecules, including amino acids. Monosaccharides that cells do not use immediately are usually incorporated as monomers into disaccharides and polysaccharides, as described next.

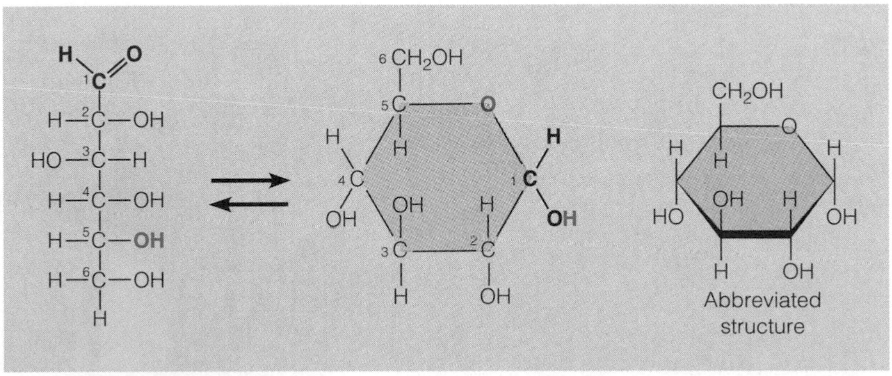

C. Linear and ring forms of glucose

3.5 Cells link single sugars to form disaccharides

Cells construct a **disaccharide**, or double sugar, from two monosaccharides by dehydration synthesis. The figure here shows how maltose, also called malt sugar, forms from two glucose monomers. A linkage forms when one monomer gives up a hydrogen atom from a hydroxyl group and the other gives up an entire hydroxyl group. As H_2O forms, an oxygen atom is left between two covalent bonds, linking the two monomers. Maltose, which is common in germinating seeds, is used in making beer.

The most common disaccharide is sucrose, which is made of a glucose monomer linked to a fructose monomer in a reaction just like the one for maltose. The main carbohydrate in plant sap, sucrose nourishes all the parts of the plant. We extract it from the stems of sugarcane or the roots of sugar beets to use as table sugar.

3.6 "How sweet it is . . ."

We are born preferring sweet-tasting foods, and although this preference lessens after childhood, it continues to influence our diet. Not only may we prefer a peach to a serving of squash, we also add concentrated sugar to many prepared foods. In addition to sucrose, the main sugars we consume are fructose, glucose, maltose, and lactose (milk sugar, a disaccharide); the first three are the ones added to foods as sweeteners. The United States is the world's leading market for such sweeteners, with the average American consuming about 125 pounds a year.

People have known how to concentrate sucrose from sugarcane for at least 5000 years, but concentrated sugars did not become cheap and widely available until the industrialization of sugar refining in the nineteenth century. A worldwide sugar industry soon arose. By 1950, it was split about 60-40 between cane sugar from the tropics and beet sugar from temperate regions. Occupying only a small part of the sweetener market was corn syrup, which was made by breaking down cornstarch, a (nonsweet) polysaccharide, to its glucose monomers. Because glucose is only half as sweet as sucrose, corn syrup was not a serious rival to sucrose.

The balance among sweeteners was drastically upset in the 1980s, when corn syrup producers developed a commercial method for converting much of the glucose in corn syrup to fructose, an isomer of glucose even sweeter than sucrose. The resulting high-fructose corn syrup (HFCS), which is about half fructose, is inexpensive and has replaced sucrose in many prepared foods. Coca-Cola and PepsiCo, once the largest commercial users of sucrose in the world, have almost completely replaced sucrose with HFCS. The changeover shook up the sucrose industry and hurt the developing economies of tropical countries where sugarcane is a major crop. Corn syrup is produced mainly in developed countries.

While many people continue to eat lots of sugar, a growing number are concerned about the possible health effects. Scientists have shown that sugars are the major cause of tooth decay: Bacteria living in the mouth convert sugars to acids, which eat away tooth enamel. However, studies investigating whether sugar consumption causes such diseases as diabetes, cancer, and heart disease have not revealed any direct connections. Similar efforts have failed to show any link between sugar intake and behavioral problems in children.

Besides tooth decay, high sugar consumption seems to be hazardous only to the extent that sugars replace other, more varied foods in our diet. The description of sugars as "empty calories" is accurate in the sense that even the less-refined sweeteners such as brown sugar and honey contain only negligible amounts of nutrients other than carbohydrates. For good health, we also require protein, fats, vitamins, and minerals. Evidence is also mounting that we need to include substantial amounts of "complex carbohydrates"—that is, polysaccharides—in our diet.

Polysaccharides are polymers of a few hundred to a few thousand monosaccharides linked together by dehydration synthesis. Some polysaccharides are storage molecules, which cells break down as needed to obtain sugar. **Starch,** a storage polysaccharide in plant roots and other tissues, consists entirely of glucose monomers, as shown in the figure. (The functional groups that extend from the glucose rings are omitted.) Starch molecules coil into a helical shape because of the angles of the bonds joining their glucose units. A starch helix may be unbranched (as shown below) or branched.

Plant cells often contain starch granules—actually masses of coiled starch molecules—that serve as sugar stockpiles. Plant cells need sugar for energy and as a raw material for building other molecules. They break starch down into glucose by hydrolyzing the bonds between the glucose monomers. Humans and most other animals are able to use plant starch as food by hydrolyzing it within their digestive systems. Potatoes and grains, such as wheat, corn, and rice, are the major sources of starch in the human diet.

Animals store excess sugar in the form of a polysaccharide called **glycogen.** Glycogen is identical to starch except that it is more extensively branched. Most of our glycogen is stored as granules in our liver and muscle cells, which

hydrolyze the glycogen to release glucose when it is needed. Also, our digestive system can hydrolyze glycogen in the meat we eat.

Many polysaccharides serve as building material for structures that protect cells and support whole organisms. **Cellulose,** the most abundant organic compound on Earth, forms cablelike fibrils in the tough walls that enclose plant cells and is a major component of wood. Cellulose resembles starch and glycogen in being a polymer of glucose, but its glucose monomers are linked together in a different orientation; they form an unbranched rod, rather than a coil. Arranged parallel to each other, several thousand cellulose molecules are joined by hydrogen bonds, forming part of a fibril. In wood, layers of cellulose fibrils combine with other polymers, making a material strong enough to support trees hundreds of feet high.

Unlike the glucose linkages in starch and glycogen, those in cellulose cannot be hydrolyzed by most animals. The cellulose in plant foods that passes unchanged through our digestive tract is commonly known as "fiber." It may help keep our digestive system healthy, but it does not serve as a nutrient. Most animals that do derive nutrition from cellulose, such as cows and termites, have cellulose-hydrolyzing microorganisms inhabiting their digestive tracts.

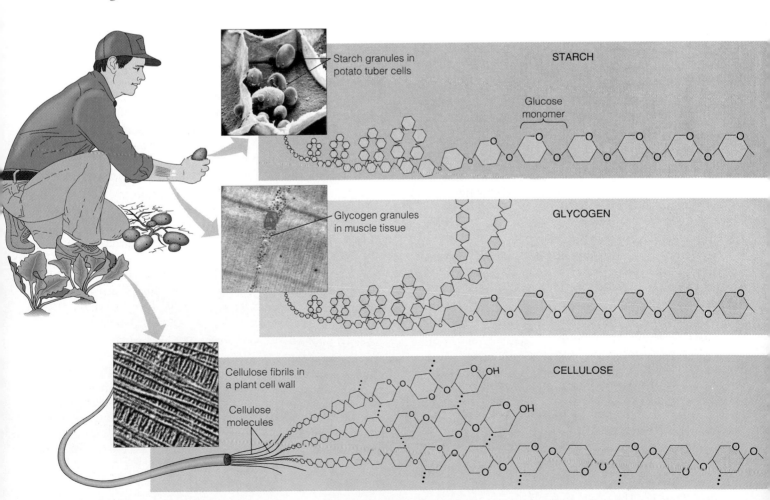

Starch granules in potato tuber cells

STARCH

Glucose monomer

Glycogen granules in muscle tissue

GLYCOGEN

Cellulose fibrils in a plant cell wall

Cellulose molecules

CELLULOSE

Lipids include fats, which are mostly energy-storage molecules

Lipids are diverse compounds that consist mainly of carbon and hydrogen atoms linked by nonpolar covalent bonds. Being mostly nonpolar, lipid molecules are not attracted to water molecules, which are polar. You can see the effect of this chemical difference in a bottle of salad dressing: The oil (a type of lipid) separates from the vinegar (which is mostly water). Other oils make the feathers in Figure A repel water, which you see as beads on the surface. By keeping feathers from absorbing water, oils help waterfowl stay afloat. Because they do not mix with water, lipids are said to be **hydrophobic** (literally "water-fearing").

Oils are a type of fat. A **fat** is a large lipid made from two kinds of smaller molecules: glycerol and fatty acids. Shown on the left in Figure B, glycerol is an alcohol with three carbons, each bearing a hydroxyl group. A fatty acid consists of a carboxyl group and a hydrocarbon chain with about 15 other carbon atoms. The carbons in the chain are linked to hydrogen atoms by nonpolar covalent bonds, making the hydrocarbon chain hydrophobic. The main function of fats is energy storage. A gram of fat stores more than twice as much energy as a gram of a polysaccharide such as starch.

Figure B shows how one fatty acid molecule can link to a glycerol molecule by dehydration synthesis. Figure C shows an actual fat molecule. It consists of three fatty acids hooked to one glycerol molecule as a result of dehydration synthesis occurring at all three hydroxyl sites on the glycerol. A synonym for "fat" is thus **triglyceride**, a term you may see on food labels or on medical tests for fat in the blood. The three fatty acids in a fat are often of different kinds, as in our example.

As shown by the third fatty acid in Figure C, some fatty acids contain double bonds, which cause kinks in the carbon chain. Double bonds prevent a carbon skeleton from bonding to the maximum number of hydrogen atoms. Fatty acids and fats with double bonds are said to be **unsaturated**—that is, having less than the maximum number of hydrogens. Fats with the maximum number of hydrogens are said to be **saturated**. The kinks in unsaturated fats prevent the molecules from packing tightly together and solidifying at room temperature.

A. Water beading on feathers

Corn oil, olive oil, and other vegetable oils are unsaturated fats. When you see "hydrogenated vegetable oils" on a margarine label, it means that unsaturated fats have been converted to saturated fats by adding hydrogen. This addition gives the lipids the consistency of margarine.

Most plant fats are unsaturated oils, whereas most animal fats are saturated. This is why butter and lard are solids at room temperature. Diets rich in saturated fats may contribute to cardiovascular disease by promoting a condition called atherosclerosis. In this condition, lipid-containing deposits called plaques build up on the inside surfaces of blood vessels, reducing blood flow.

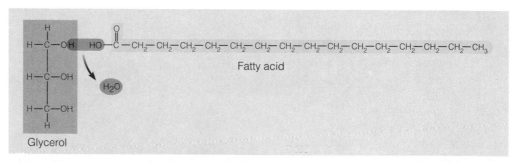

B. Dehydration synthesis linking a fatty acid to glycerol

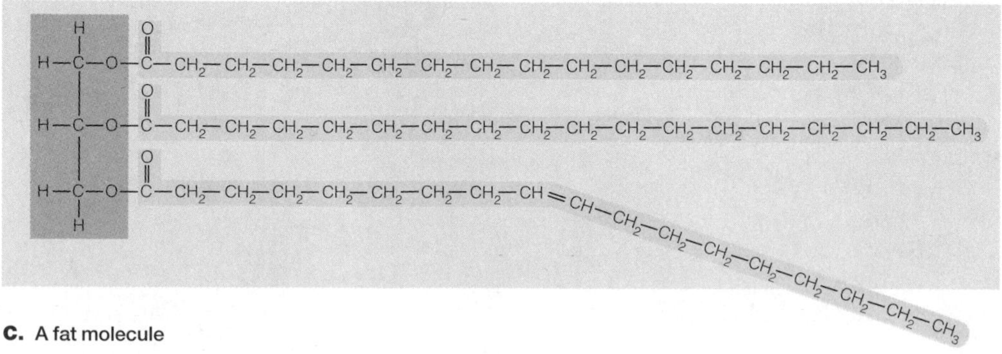

C. A fat molecule

Phospholipids, waxes, and steroids are lipids with a variety of functions

Fats are only one type of lipid important in living organisms. Other lipids are major constituents of cell membranes and perform such vital functions as protecting body surfaces and regulating cellular and body functions.

Phospholipids, a major component of cell membranes, are structurally similar to fats, but they contain the element phosphorus and have only two fatty acids instead of three. Phospholipids are very important biological molecules, as we will see when we discuss cell membranes in Chapter 5.

Waxes consist of one fatty acid linked to an alcohol. They are more hydrophobic than fats, and this characteristic makes waxes effective natural coatings for fruits such as apples and pears. Many animals, especially insects, also have waxy coats that help keep them from drying out.

Steroids are lipids whose carbon skeleton is bent to form four fused rings, as shown in this structural diagram of cholesterol. All steroids have the same ring pattern: three 6-sided rings and one 5-sided ring. (The diagram omits the carbons in the rings and the hydrogens attached to them.)

Cholesterol, a steroid

Cholesterol is a common substance in animal cell membranes, and animal cells also use it as a starting material for making other steroids, including the female and male sex hormones. Too much cholesterol in the blood may contribute to atherosclerosis.

Anabolic steroids make big bodies and big problems

Anabolic steroids are synthetic variants of the male hormone testosterone. Testosterone causes a general buildup in muscle and bone mass during puberty in males, and maintains masculine traits throughout life. Because anabolic steroids structurally resemble testosterone, they also mimic some of its effects.

Pharmaceutical companies first produced and marketed anabolic steroids in the early 1950s as a treatment for general anemia and for certain diseases that destroy body muscle. About a decade later, some athletes began using anabolic steroids to build up their muscles quickly and enhance their performance.

Today, anabolic steroids, along with other drugs, are banned by most athletic organizations. Nonetheless, many professional athletes admit to using them heavily, citing such benefits as increased muscle mass, strength, stamina, and aggressiveness. It is not surprising that some of the heaviest users are weight lifters, football players, and body builders.

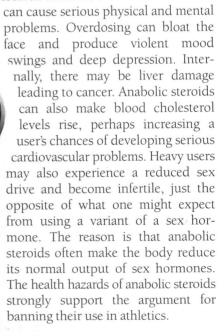

Using anabolic steroids is a fast way to increase general body size. They enable an athlete who is willing to cheat at a sport to increase body mass with less hard work. But at what cost? Medical research indicates that these substances can cause serious physical and mental problems. Overdosing can bloat the face and produce violent mood swings and deep depression. Internally, there may be liver damage leading to cancer. Anabolic steroids can also make blood cholesterol levels rise, perhaps increasing a user's chances of developing serious cardiovascular problems. Heavy users may also experience a reduced sex drive and become infertile, just the opposite of what one might expect from using a variant of a sex hormone. The reason is that anabolic steroids often make the body reduce its normal output of sex hormones. The health hazards of anabolic steroids strongly support the argument for banning their use in athletics.

3.11 Proteins are essential to the structures and activities of life

The name protein, from the Greek word *proteios,* meaning "first place," suggests the importance of this class of macromolecules. A **protein** is a biological polymer constructed from amino acid monomers. Every one of us has tens of thousands of different kinds of proteins, each with a unique, three-dimensional structure that corresponds to a specific function. Proteins are important to the structures of cells and organisms and participate in everything they do.

There are seven major classes of proteins. One class, called structural proteins, includes the silk of spiders, the hair of mammals (including our own), and fibers that make up our tendons and ligaments. Working together with such structural elements is a second class, called contractile proteins; the proteins that provide muscular movement are one example.

A third class of proteins is storage proteins, such as ovalbumin, the main substance of egg white. Ovalbumin serves as a source of amino acids for developing embryos. A fourth class, the defensive proteins, includes the antibodies carried in the blood, which fight infections. Transport proteins, a fifth class, include hemoglobin, the iron-containing protein in blood that conveys oxygen from our lungs to other parts of the body. Certain hormones, which help coordinate body activities by serving as messages from one cell to another, are examples of signal proteins, a sixth class.

Perhaps the most important class of proteins is the enzymes. An **enzyme** is a protein that serves as a chemical catalyst, an agent that changes the rate of a chemical reaction without itself being changed into a different molecule in the process. Enzymes promote and regulate virtually all chemical reactions in cells. We will have a lot to say about enzymes as we pursue the chemistry of life in later chapters.

Hair, made of structural proteins

3.12 Proteins are made from just 20 kinds of amino acids

Of all of life's molecules, proteins are the most diverse in structure and function. Protein diversity is based on different arrangements of a universal set of amino acids. **Amino acids** all have an amino group and a carboxyl group, as you can see in the general structure shown in Figure A. Both of these functional groups are covalently bonded to a central carbon atom called the alpha carbon (red). Also bonded to the alpha carbon is a hydrogen atom and a chemical group symbolized by the letter R. The R group is the variable part of an amino acid. In the simplest amino acid (glycine), the R group is just a hydrogen atom. In others, such as the three in Figure B, it is one or more carbon atoms with various functional groups attached. The structure of the R group determines the specific properties of each of the 20 amino acids in proteins.

The amino acids in Figure B represent two main types. Leucine (abbreviated Leu) is an example of a type of amino acid whose R group is nonpolar and hydrophobic. Serine (Ser), with a hydroxyl group in its R group, represents a type with polar, hydrophilic R groups. Cysteine (Cys) is one of two amino acids whose R groups include a sulfur atom (S); it is also hydrophilic. Hydrophilic amino acids help proteins dissolve in solutions within cells.

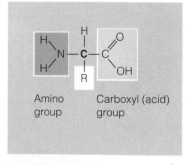

A. General structure of an amino acid

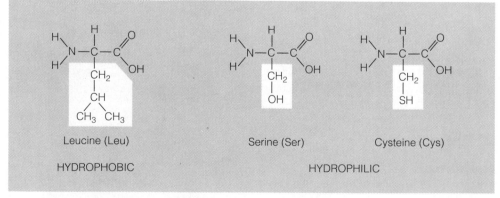

Leucine (Leu)

HYDROPHOBIC

Serine (Ser)

Cysteine (Cys)

HYDROPHILIC

B. Examples of amino acids

Amino acids can be linked by peptide bonds

3.13

Cells link amino acids together by dehydration synthesis. For this linkage to occur, the carboxyl group of one amino acid must be positioned next to the amino group of another, as shown on the left side of the figure below. A water molecule is then removed as the carboxyl-group carbon atom bonds to the amino-group nitrogen of its neighbor. The resulting covalent linkage is called a **peptide bond.** The product of the reaction in the figure is called a *di*peptide, because it was made from *two* amino acids. Additional amino acids can be added by the same process to form a chain of amino acids, a **polypeptide.** To release amino acids from the polypeptide, H_2O must be added back to each peptide bond by hydrolysis.

Polypeptides range in length from a few monomers to a thousand or more. Each polypeptide has a unique sequence of amino acids and assumes a unique three-dimensional shape in a protein, as we will see next.

A protein's specific shape determines its function

3.14

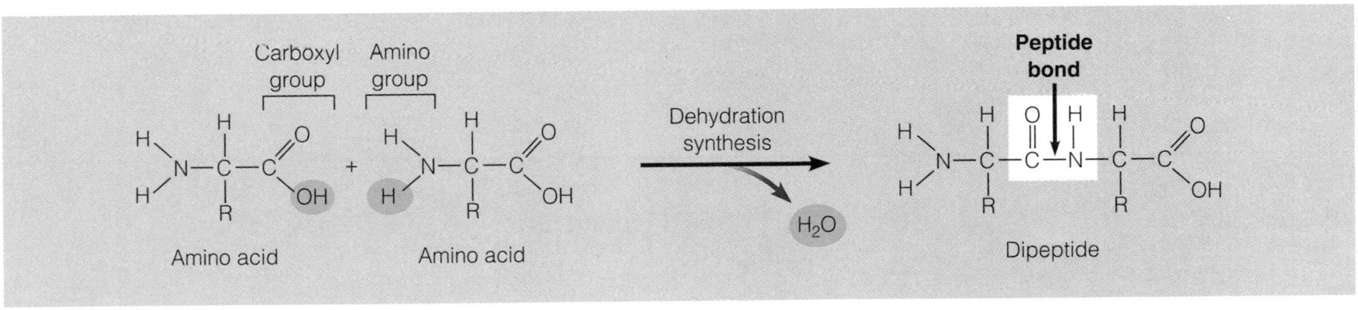

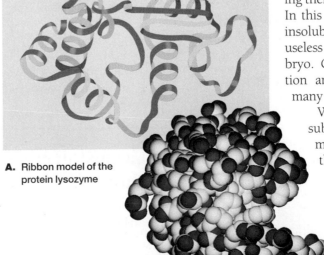

A protein consists of one or more polypeptide chains folded into a unique shape. Figure A is a model of the protein lysozyme, an enzyme found in our tears and white blood cells. Lysozyme consists of one long polypeptide chain, represented by the purple ribbon. Roughly spherical, lysozyme's general shape is called globular. This overall shape is more apparent in Figure B, a space-filling model of lysozyme. Most proteins are globular, although structural proteins are typically long and thin—fibrous.

General shape is one thing; specific shape is another. The coils and twists of lysozyme's polypeptide ribbon appear haphazard, but they represent the molecule's specific, three-dimensional shape, and this shape is what determines its specific function. Nearly all proteins must recognize and bind to some other molecule in order to function. Lysozyme, for example, can destroy bacterial cells, but first it must bind to specific molecules on the bacterial cell surface. Lysozyme's specific shape enables it to recognize and attach to its molecular target.

The dependence of function on specific shape becomes clear when proteins are altered. In a process called **denaturation,** a protein may unravel, losing its specific shape and, as a result, its function. For example, visualize what happens when you fry an egg. Heat quickly denatures the clear proteins surrounding the yolk, making them solid, white, and opaque. In this state, the proteins become insoluble in water and would be useless to a developing bird embryo. Changes in salt concentration and pH can also denature many proteins.

We examine the important subject of protein structure more closely in the modules that follow.

A. Ribbon model of the protein lysozyme

B. Space-filling model of lysozyme

3.15 A protein's primary structure is its amino acid sequence

The specific shape that determines a protein's function comprises four successive levels of structure, each determining the next one. We describe these levels in the four modules on this page. The figures on the facing page accompany these modules, with Figures A–D illustrating the four levels of structure in a single protein called prealbumin. Found in blood, prealbumin is an important transport protein. It is a globular molecule whose specific shape enables it to transport two key chemicals throughout the body, a hormone from the thyroid gland and vitamin A.

Every kind of protein, including prealbumin, has a unique **primary structure,** which is the sequence of amino acids forming its polypeptide chains (or chain, if, like lysozyme, it has only one). Figure A on the facing page shows part of the primary structure of prealbumin. The three-letter abbreviations represent amino acids. A complete molecule of prealbumin has four polypeptide chains, each made up of 127 amino acids. In order for this or any other protein to perform its specific function, it must have the correct collection of amino acids arranged in a precise order. Even a slight change in a protein's primary structure may affect its overall shape and its ability to function. For instance, a single amino acid change in hemoglobin, the oxygen-carrying blood protein, causes sickle-cell anemia, a serious blood disorder.

3.16 Secondary structure is polypeptide coiling or folding produced by hydrogen bonding

In the second level of protein structure, the polypeptide coils or folds into regular patterns called **secondary structure.** As you can see in Figure B, coiling of a polypeptide chain results in a secondary structure called an **alpha helix;** folding leads to a **pleated sheet.** Both of these patterns are maintained by hydrogen bonds between the $-N-H$ groups and $-C=O$ groups along the polypeptide chain.

Notice that each hydrogen bond (represented by three dots) links the $-N-H$ of one amino acid with the $-C=O$ of *another* amino acid further down the chain. (Because the R groups of the amino acids are not important in the secondary structure, they are omitted from the diagrams, as are some of the hydrogen atoms.)

3.17 Tertiary structure is the overall shape of a polypeptide

The term **tertiary structure** refers to the overall, three-dimensional shape of a polypeptide. Most tertiary structures can be roughly described as either globular or fibrous. A prealbumin polypeptide has a generally globular shape, which results from the compact combination of both helical and pleated-sheet regions, as you can see in Figure C. The indentations and bulges arising from its particular arrangement of coils and folds give the polypeptide the specific shape appropriate to its function. Many proteins with globular tertiary structure have both helical and pleated-sheet regions. In contrast, many fibrous proteins, including the capture-strand silk of spiderwebs and our hair proteins, are almost entirely helical. Tertiary structure is maintained by bonding (mostly hydrogen bonding and ionic bonding) between the R groups of various amino acids in the polypeptide chain.

3.18 Quaternary structure is the relationship among multiple polypeptides of a protein

Many proteins consist of two or more polypeptide chains, or subunits. Such proteins have a **quaternary structure,** resulting from bonding interactions among the subunits. Figure D shows a complete prealbumin molecule with its four subunits; all four are identical. Many other proteins have subunits that are different from one another. For example, the oxygen-transporting molecule hemoglobin has four subunits of two distinct types (see Figure 22.10B).

LEVELS OF PROTEIN STRUCTURE

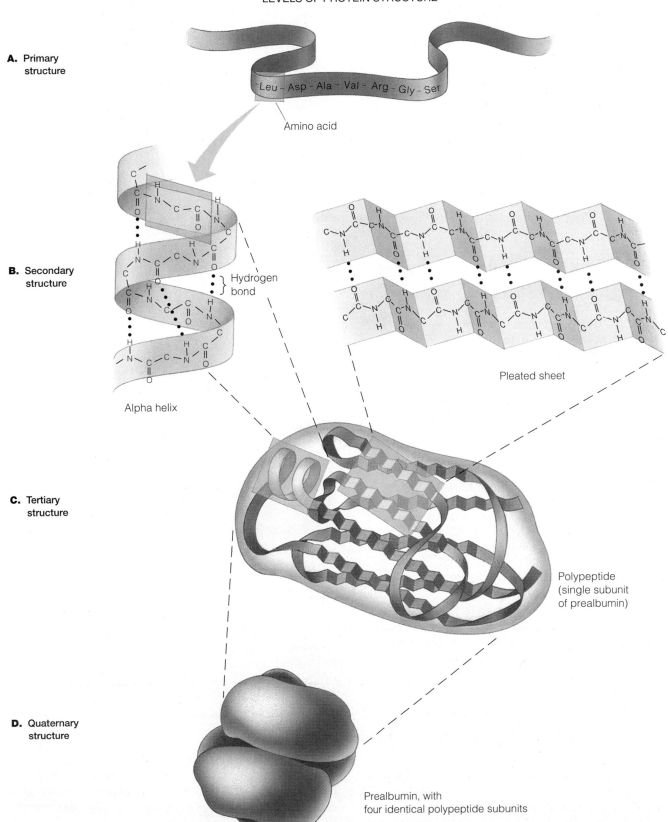

A. Primary structure

- Leu - Asp - Ala - Val - Arg - Gly - Ser

Amino acid

B. Secondary structure

Hydrogen bond

Alpha helix

Pleated sheet

C. Tertiary structure

Polypeptide (single subunit of prealbumin)

D. Quaternary structure

Prealbumin, with four identical polypeptide subunits

Linus Pauling has contributed to our understanding of the chemistry of life

Often associated with his controversial belief that vitamin C can help prevent the common cold, cancer, and other diseases, 92-year-old Linus Pauling has made extraordinary contributions to both science and human affairs. His connection with this chapter is his work on protein structure and function.

Driven by the desire "to understand the world," Pauling started out in chemistry and physics. While a professor at the California Institute of Technology, he published a series of papers on chemical bonding that eventually led to a Nobel Prize in chemistry in 1954. By that time, Pauling was also studying biological molecules. In an interview several years ago, we asked him about the value of studying molecules for understanding life, and he responded by talking about his work on hemoglobin:

A. Linus Pauling in 1993

> Life is too complicated to permit a complete understanding through the study of whole organisms. Only by simplifying the problem—breaking it down into a multitude of individual problems—can you get the answers. In 1935, Dr. Charles Coryell and I made our discovery about how oxygen molecules are attached to the iron atoms of hemoglobin, not by getting a cow and putting it through our magnetic apparatus, but by getting some blood from the cow and studying this blood and the hemoglobin from it. . . .
>
> Of course, the study of different parts of an organism leads to the question, Do these parts interact? Can we learn more about the living organism by putting two parts together to see to what extent the properties of the combination are different from those of the two separated parts? This approach permits further progress in understanding the organism. And yet I, myself, have confidence that all of the properties of living organisms could ultimately be discovered by the process of attempting to reduce the organism . . . to a combination of the different parts: essentially, the molecules that make up the organism.

B. Pauling with a model of the alpha helix in 1948

Besides finding out how hemoglobin carries oxygen, Pauling discovered how an abnormal hemoglobin molecule causes the disease sickle-cell anemia. And it was Pauling who first described the two fundamental secondary structures of proteins, the alpha helix and the pleated sheet. His recounting of that work reveals more about his scientific attitudes and efforts:

> I began trying to find the structure of proteins in 1937, and didn't succeed. So I began working with my collaborators to determine the three-dimensional structures of amino acids and simple peptides. In 1937, no one had yet determined such structures. . . . And then in 1948 I found the alpha-helix and pleated sheet structures in proteins. . . . I'm surprised that nobody else had done this job in the 11 years that intervened—in a sense, surprised that I hadn't done it in '37, when my ideas were all the right ones. I just hadn't worked hard enough.

Pauling's efforts were not limited to science. He also became the scientific community's leading advocate for halting the testing of nuclear weapons, resulting in the accusation that he was a Communist and the revocation of his passport. Nevertheless, in 1963 he received the Nobel Peace Prize for helping produce a ban on nuclear testing. Pauling is the only person who has ever received two unshared Nobel Prizes.

The cancellation of Pauling's passport in 1952 kept him from working directly with scientists in Europe. This may have prevented Pauling from earning a third Nobel Prize, for at that time he was hot on the trail of the structure of DNA. But his triple-helix model was wrong, and it was James Watson and Francis Crick, working in England, who came up with the correct solution—a double helix, which we describe in the next module.

Nucleic acids are polymers of nucleotides

The **nucleic acids** are polymers that serve as the blueprints for proteins. There are two types: **deoxyribonucleic acid (DNA)** and **ribonucleic acid (RNA).** The genetic material that organisms inherit from their parents consists of DNA. Within the DNA are **genes,** specific stretches of the molecule that program the amino acid sequences (primary structure) of proteins. In determining primary structure, genes determine the specific three-dimensional structures and therefore the functions of proteins. Thus, through the actions of proteins, DNA controls the life of the cell and the organism.

DNA does not put its genetic information to work directly. It works through an intermediary—RNA. DNA's information is transcribed into RNA, which is then translated into the primary structure of proteins. We will return to this chain of command and the functions of DNA and RNA later in the book. Here, we just want to introduce the structure of nucleic acids.

The monomers that make up nucleic acids are called **nucleotides.** As indicated in Figure A, each nucleotide has three parts. One part is a five-carbon sugar (blue); DNA has the sugar deoxyribose, whereas RNA has a closely related sugar called ribose. Linked to one end of the sugar in both types of nucleic acid is a functional group called a **phosphate group** (yellow). At the other end of the sugar is one of a number of chemical units called **nitrogenous bases** (like the one in Figure A, they all contain nitrogen). DNA has the nitrogenous bases adenine (A), thymine (T), cytosine (C), and guanine (G). RNA also has A, C, and G, but instead of thymine has uracil (U).

Like polysaccharides and polypeptides, a nucleic acid polymer—a polynucleotide—forms from its monomers by dehydration synthesis. In this process, the phosphate group of one nucleotide bonds to the sugar of the next monomer. The result is a repeating sugar-phosphate backbone in the polymer, as shown in Figure B.

RNA usually consists of a single polynucleotide strand, but DNA is a **double helix,** in which two polynucleotides wrap around each other (Figure C). The nitrogenous bases protrude from the two sugar-phosphate backbones into the center of the helix. There they always pair up as shown: A pairs with T, and C pairs with G.

The two DNA chains are held in a double helix by hydrogen bonds (dotted lines) between their paired bases. Most DNA molecules are very long, with thousands or even millions of base pairs. One long DNA molecule contains many genes, each a specific series of hundreds or thousands of nucleotides along one of the polynucleotide strands.

This introduction to nucleic acids concludes our look at the four major classes of biological molecules. In the next chapter, we set the scene of molecular action—the cell.

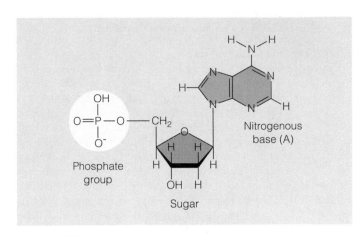

A. Nucleotide

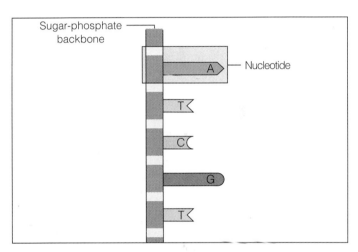

B. Polynucleotide

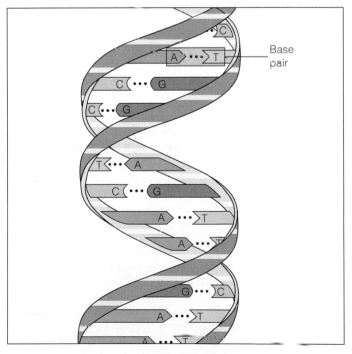

C. DNA double helix

Begin your review by rereading the module headings and scanning the figures before proceeding to the Chapter Summary and questions.

Chapter Summary

Introduction–3.2 Living organisms are made of a variety of organic—that is, carbon-containing—compounds. A carbon atom forms four covalent bonds, so it can join with other carbon atoms to make chains or rings. Carbon chains vary in length, degree of branching, number and location of single and double bonds, and in the presence or absence of other elements. Functional groups are particular groupings of atoms attached to the carbon skeleton; they give organic molecules specific properties. For example, hydroxyl groups are characteristic of alcohols, and carboxyl groups are acidic.

3.3 Three of the four kinds of large molecules (macromolecules) in living things are polymers, long chains of smaller molecular units called monomers. A huge number of polymers can be made from a small number of monomers. Cells link monomers to form polymers by dehydration synthesis. One monomer loses a hydroxyl group, the other loses a hydrogen atom, a covalent bond links the monomers, and a molecule of water is released. Macromolecules are broken down to their component monomers by the reverse process, called hydrolysis.

3.4–3.7 Carbohydrates range from small sugars to large polysaccharides. A monosaccharide, or single sugar, typically has a formula that is a multiple of CH_2O and contains hydroxyl groups and a carbonyl group. The monosaccharides glucose and fructose, both with the formula $C_6H_{12}O_6$, are isomers, that is, they contain the same atoms but in different arrangements, and thus they have different properties. Monosaccharides are the fuels for cellular work. They can also join to form disaccharides, such as sucrose (table sugar), and polysaccharides. Starch and glycogen are polysaccharides that store sugar for later use; cellulose is a polysaccharide that forms plant cell walls.

3.8–3.10 Lipids, compounds composed largely of carbon and hydrogen, are not true polymers. These varied molecules are grouped together because they do not mix with water. Fats, also called triglycerides, are lipids whose main function is energy storage. A triglyceride molecule consists of glycerol linked to three fatty acids. The fatty acids of unsaturated fats such as plant oils contain double bonds, which prevent them from solidifying at room temperature. Saturated fats, such as those in lard, lack double bonds and are solid. Other lipids include phospholipids (found in cell membranes), waxes (which form waterproof coatings), and steroids (including sex hormones).

3.11– 3.18 Proteins are involved in cellular structure, movement, defense, transport, and communication; and, as enzymes, they regulate chemical reactions. A typical protein consists of hundreds of amino acids in one or more polypeptide chains. Each amino acid contains an amino group, a carboxyl group, and an R group. The R groups in proteins distinguish 20 different amino acids, each with specific properties. Cells link amino acids together by dehydration synthesis; the bonds between amino acid monomers are called peptide bonds. Each polypeptide chain has a particular number and sequence of amino acids, its primary structure. The polypeptide chain folds into a specific shape, largely determined by the properties of its amino acids. Secondary structure consists of helical coiling or pleated-sheet folding, stabilized by hydrogen bonds between amino acids. Tertiary structure is the overall three-dimensional shape of a polypeptide, resulting from bonding between R groups. Many proteins consist of more than one polypeptide chain and thus display quaternary structure. The function of a protein molecule depends on its overall shape. Changes in the environment can ruin a protein's shape. This process, called denaturation, also destroys the protein's function.

3.20 Nucleic acids—DNA and RNA—control the life of a cell. The monomers of nucleic acids are nucleotides. Each nucleotide is composed of a sugar, a phosphate, and a nitrogenous base. DNA consists of two polynucleotides twisted around each other in a double helix. The sequence of the four kinds of nitrogenous bases in DNA carries genetic information. Stretches of a DNA molecule called genes program the amino acid sequences of proteins. DNA information is transcribed into RNA, a single-stranded nucleic acid, which is then translated into the primary structure of proteins.

Testing Your Knowledge

Multiple Choice

1. A glucose molecule is to starch as (*Explain your answer.*)
   a. a steroid is to a lipid
   b. a protein is to an amino acid
   c. a nucleic acid is to a polypeptide
   d. a nucleotide is to a nucleic acid
   e. an amino acid is to a nucleic acid

2. What makes a fatty acid an acid?
   a. It does not dissolve in water.
   b. It is capable of bonding with other molecules to form a fat.
   c. It has a carboxyl group that donates a hydrogen ion to a solution.
   d. It contains only two oxygen atoms.
   e. It is a polymer made of many smaller subunits.

3. Where in the three-dimensional structure of a protein would you be most likely to find a hydrophobic amino acid R group?
   a. at both ends of the polypeptide chain
   b. on the outside, in the water
   c. covalently bonded to another R group
   d. on the inside, away from water
   e. covalently bonded to the amino group of the next amino acid

4. The enzyme called pancreatic amylase is a protein whose job is to attach to starch molecules in food and help break them down to disaccharides. Amylase cannot break down cellulose. Why not?
   a. Cellulose is a kind of fat, not a carbohydrate like starch.
   b. Cellulose molecules are much too large.
   c. Starch is made of glucose; cellulose is made of other sugars.
   d. The bonds between sugars in cellulose are much stronger.
   e. The sugars in cellulose bond differently, giving cellulose a different shape.

5. A shortage of phosphorus in the soil would make it especially difficult for a plant to manufacture
   a. DNA
   b. proteins
   c. cellulose
   d. fatty acids
   e. sucrose

6. Lipids differ from other large biological molecules in that they
   a. are much larger
   b. are not truly polymers
   c. do not have specific shapes
   d. do not contain carbon
   e. contain nitrogen atoms

7. Which functional group(s) act(s) as an acid?

  a. carbonyl
  b. amino
  c. hydroxyl
  d. carboxyl
  e. all of the above

True/False (*Change false statements to make them true.*)

1. The sugars circulating in your blood are primarily disaccharides.

2. The fats in corn oil contain less hydrogen and more double bonds than the fats in butter.

3. Genes are composed of DNA.

4. The hormone testosterone is a steroid, a type of protein.

5. When amino acids are linked to form a polypeptide, water is produced as a by-product.

6. All lipids are hydrophilic.

7. The three-dimensional folding of a protein molecule results from attractions and bonds between its amino acid R groups.

8. When a protein is denatured, its primary structure is the level of structure most likely to be disrupted.

9. Users of anabolic steroids can experience reduced sex drive and infertility.

10. Fat molecules store energy in the cell.

Describing, Comparing, and Explaining

1. List four different kinds of lipids and briefly describe their functions.

2. Explain why heating, changes in pH, and other environmental changes can interfere with the function of proteins.

3. How can a cell make so many different kinds of protein out of only 20 amino acids? Of the myriad possibilities, how does the cell "know" which proteins to make?

4. Briefly describe the various functions performed by proteins in a cell.

Thinking Critically

1. Pentane is a hydrocarbon molecule with the molecular formula C_5H_{12}. See if you can sketch two isomers of pentane. Make sure that each of the atoms in the molecules forms the correct number of bonds.

2. Lysozyme is a small protein consisting of a single polypeptide chain of 129 amino acids. How could you calculate the number of possible different proteins 129 amino acids long that could be built using 20 amino acids? (*Hint:* How many different choices are there for the first amino acid? How many choices for the second? Then how many possible proteins could there be with only two amino acids? Can you extend this logic to 129 amino acids?)

3. When you eat a candy bar, the disaccharide sucrose in the candy is broken down in your intestine to produce two monosaccharide molecules (glucose and fructose), which are then absorbed into your blood. Starting with the sucrose molecule illustrated here, show how it would break down to produce glucose and fructose. What is the name of this reaction?

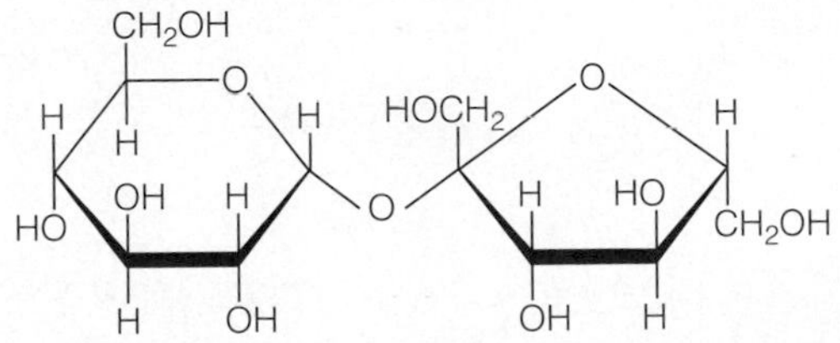

Sucrose

Science, Technology, and Society

1. Each year, industrial chemists develop and test thousands of new organic compounds for use as pesticides, such as insecticides, fungicides, and weed killers. In what ways are these chemicals useful and important to us? In what ways can they be harmful? Is your opinion of pesticides positive or negative, in general? What influences have shaped your feelings about these chemicals?

2. Linus Pauling believes that large doses of vitamin C can help prevent the common cold, cancer, and other diseases. Imagine you have been given a research grant by the National Institutes of Health to evaluate Pauling's claims. How would you go about setting up an experimental study to determine whether vitamin C can prevent colds? How would you evaluate the results of your study?

3. Module 3.6 mentions that the economies of some developing tropical countries were hurt when food and beverage companies in the developed countries switched from using sucrose (derived from sugarcane) to high-fructose corn syrup (from corn grown in the developed countries). Many developing countries are dependent on single cash crops like sugarcane or coffee for most of their export income. Do the developed countries have a responsibility to protect the economies of developing countries from the impact of major product shifts like this one? Why or why not? Do consumers of the products have any say in the matter? Why or why not?

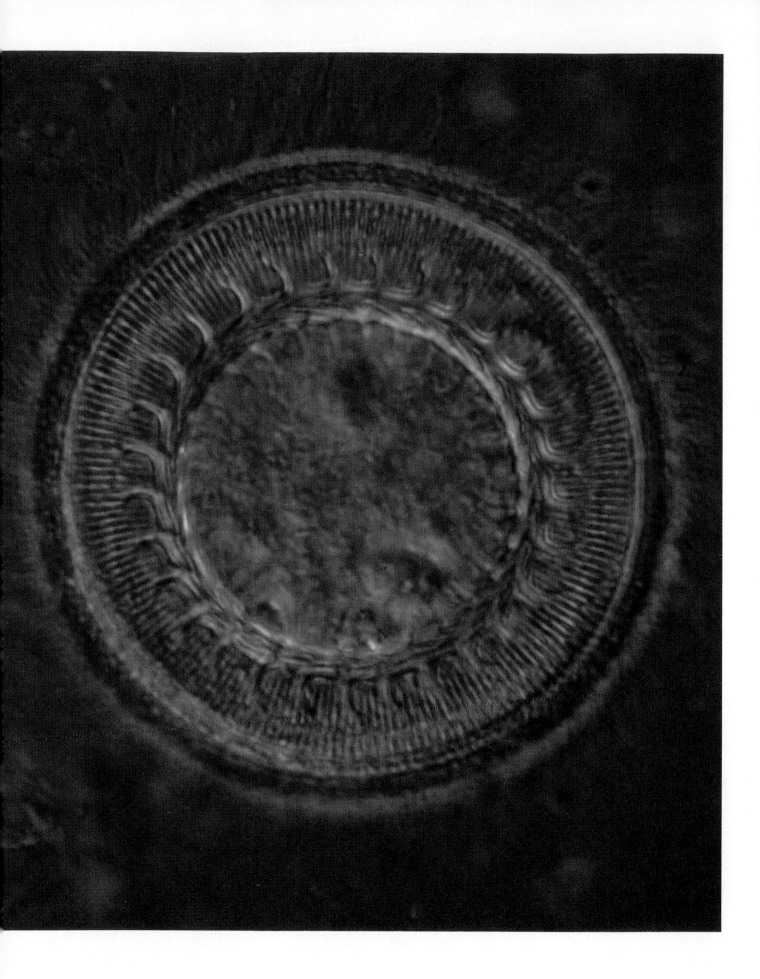

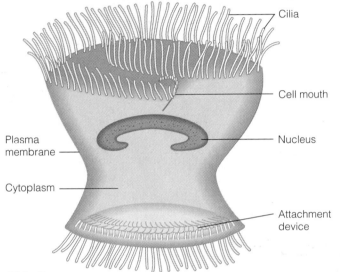

Plasma membrane

Cytoplasm

Cilia

Cell mouth

Nucleus

Attachment device

Trichodina

S pinning like a top in the water, a single-celled organism called *Trichodina* (Greek for whirling hairs) skims over the slippery surface of a fish and touches down. You see the underside of *Trichodina* in the photograph at the left—a complex circular clamp made of toothlike hooks and spines well suited for gripping fish skin. Fortunately for the fish, *Trichodina* eats mostly bacteria. Though the organism looks like a monster in this enlarged photograph, *Trichodina* is microscopic; about 100 of them would fit inside one of the o's on this page.

Trichodina is common in aquatic environments and often hitches a ride on fish. It is usually harmless, but sometimes its numbers build up to the point where it virtually covers a fish. When this happens, the host fish may die from tiny skin wounds inflicted by *Trichodina* or from suffocation. When *Trichodina* coats a fish, it may cover the gills and prevent them from absorbing oxygen from the water.

With its specialized attachment device, *Trichodina* is one of the most complex of all single-celled organisms. But this protist also has a number of features in common with the cells that make up our bodies. In the drawing on this page, notice the rows of hairlike cilia ringing the organism's upper and lower surfaces. Beating back and forth, the cilia propel the cell through the water and also sweep bacteria into its cell mouth. Cells in the human body lack a cell mouth, but many, including those that line the windpipe, have cilia.

Another feature common to both *Trichodina* and our body cells is a nucleus containing DNA. The DNA specifies the proteins made in the cell and thus controls what the organism looks like and what it does. For instance, certain genes in *Trichodina*'s DNA specify special hard proteins that become assembled into the hooks and spines of the organism's attachment device. Other genes specify tubular proteins that function in the beating of the cilia.

Every cell, *Trichodina* included, also has a **plasma membrane** forming its outside border and setting it off from its watery environment. More than just a boundary, the plasma membrane enables the cell to take up needed molecules from the surroundings, provides sites where important chemical reactions occur, and disposes of wastes. In doing so, it helps maintain a life-sustaining collection of molecules inside the cell quite different from that of the outside environment.

Everything inside a cell between the plasma membrane and the nucleus is called the **cytoplasm.** The cytoplasm consists of a semifluid medium and various structures called organelles suspended in it. An **organelle** (Latin for small organ) is a cellular structure whose anatomy gives it a specific role to play in the life of the cell. The nucleus, specialized for controlling cellular activities, and the cilia, specialized for movement, are both organelles. Cilia are enclosed in extensions of the plasma membrane.

In this chapter, we examine the structure of life at the cellular level. We begin by looking at the main kinds of microscopes that biologists use to explore the aspects of life we can't see with the unaided eye. We then show the cellular structures that microscopes have revealed and describe their functions in the cell.

Microscopes provide windows to the world of the cell

Before microscopes were first used about 330 years ago, no one knew for certain that living organisms were composed of cells. The first microscopes, like the ones you may have used in a biology laboratory, were light microscopes. A **light microscope** (LM) works by passing visible light through a specimen, such as a living *Trichodina* or a piece of animal or plant tissue. As Figure A shows, glass lenses in the microscope bend the light to magnify the image of the specimen and project the image into the viewer's eye or onto photographic film. A photograph taken through a microscope is called a **micrograph.** The notation "LM 109×" printed along the right edge of the micrograph of *Trichodina* in Figure A is in a form we use throughout this book. It tells you that the photograph was taken through a light microscope and that this image is about 109 times the actual size of the organism. *Trichodina* is actually about ¹⁄₂₀ of a millimeter in diameter. This image could be magnified many more times than shown here. Beyond a certain point, though, the image would begin to blur, and additional magnification would only cause more blurring. Light microscopes can magnify objects only about 1000 times without causing blurriness.

Magnification, the increase in the apparent size of an object, is only one important factor in microscopy (the use of a microscope). Also important is **resolving power,** a measure of the clarity of an image. Resolving power is the ability of an optical instrument to show two objects as separate. For example, what looks to your unaided eye like a single star in the sky may be resolved as two stars with the help of a telescope. Any optical device is limited by its resolving power. The light microscope cannot resolve detail finer than 0.2 micrometer (abbreviated μm; 1 μm = ¹⁄₁₀₀₀ mm), about the size of the smallest bacterium. Consequently, no matter how many times its image of such a bacterium is magnified, the light microscope cannot show the details of the cell's internal structure. (The μ in the abbreviation for micrometer is the Greek letter mu.)

From the year 1665, when English microscopist Robert Hooke discovered cells, until the middle of this century, biologists had only light microscopes for viewing cells. But they discovered a great deal, including the cells composing animal and plant tissues, microscopic organisms—for example, *Trichodina* was discovered about 200 years ago—and some of the structures within cells. By the mid 1800s, these discoveries led to the **cell theory,** which states that all life is composed of cells and that all cells come from other cells.

Our knowledge of cell structure took a giant leap forward as biologists began using the electron microscope in the 1950s. Instead of light, the **electron microscope (EM)** uses a beam of electrons. The EM has a much higher resolving power than the light microscope. In fact, the most powerful modern EMs can distinguish objects as small as 0.2 nanometer (abbreviated nm; 1 nm= ¹⁄₁,₀₀₀,₀₀₀ mm), a thousandfold

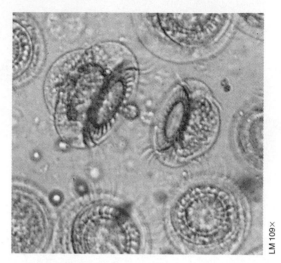

LM 109×

Light micrograph

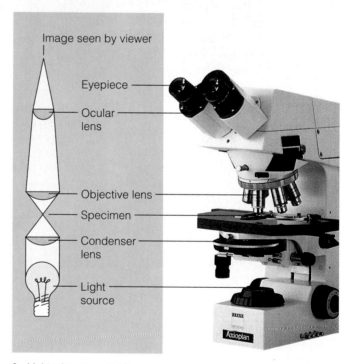

A. Light microscope (LM)

improvement over the light microscope. The period at the end of this sentence is about a million times bigger than an object 0.2 nm in diameter, which is the size of a large atom. Only under special conditions can EMs detect individual atoms. However, cells, cellular organelles, and even molecules like DNA and protein are much larger than single atoms. The highest power electron micrographs you will see in this book have magnifications of about 100,000 times.

Figures B and C show two kinds of electron microscopes, along with images they have produced of cilia. (The cilia in the photographs are from cells of a rabbit, but

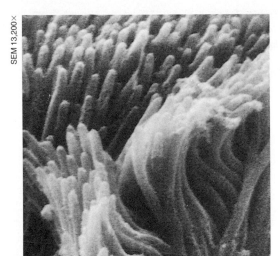

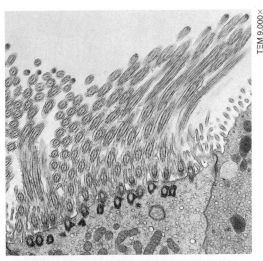

Scanning electron micrograph

Transmission electron micrograph

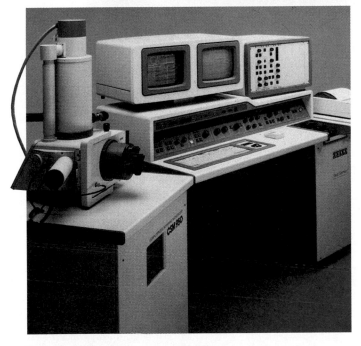

B. Scanning electron microscope (SEM)

C. Transmission electron microscope (TEM)

they are identical in structure to the cilia of *Trichodina* and other single-celled organisms.) Biologists use the **scanning electron microscope (SEM)** to study the detailed architecture of cell *surfaces*. The SEM uses an electron beam to scan the surface of a cell or group of cells that have been coated with metal. The metal stops the beam from going through the cells. When the metal is hit by the beam, it emits electrons. The electrons are focused to form an image of the outside of the cells. The scanning electron micrograph in Figure B shows the shapes and arrangement of cilia covering a cell. Many structural details of cell surfaces have been discovered using the SEM. As you can see, the SEM produces images that look three-dimensional.

The **transmission electron microscope (TEM)** is used to study the details of *internal* cell structure. Specimens are cut into extremely thin sections, and the TEM aims an electron beam through a section, just as a light microscope aims a beam of light through a specimen. However, instead of lenses

made of glass, the TEM uses electromagnets as lenses, as do all electron microscopes. The electromagnets bend the electron beam to magnify and focus an image onto a viewing screen or photographic film. The micrograph in Figure C shows internal details of the cilia of a rabbit cell as seen with the TEM.

Electron microscopes have truly revolutionized the study of cells and cell organelles. Nonetheless, they have not replaced the light microscope. One problem with electron microscopes is that they cannot be used to study living specimens because the specimen must be held in a vacuum chamber; that is, all the air and liquid must be removed. For a biologist studying a living process, such as the whirling movement of *Trichodina*, a light microscope equipped with a video camera might be better than either an SEM or a TEM. Thus, the light microscope remains a useful tool, especially for studying living cells. The size of a cell often determines the type of microscope a biologist uses to study it. The next two modules discuss the subject of cell size.

Chapter 4   A Tour of the Cell   **53**

4.2 Cell sizes vary with their function

The figure at the right shows the size range of cells compared with objects both larger and smaller. The smallest cells are bacteria called mycoplasmas, with diameters between 0.1 μm and 1.0 μm. The bulkiest cells are bird eggs, and the longest cells are certain muscle and nerve cells. Most cells lie well within these extremes, in the range indicated by the white area on the figure. *Trichodina*, for instance, is about 50 μm in diameter. The scale is logarithmic, with the length labels on the left ascending in powers of ten, to accommodate the range of sizes shown. Thus, except for the lengthy nerve and muscle cells, and the egg cells of many animals, the biggest plant and animal cells, with diameters of about 100 μm, are ten times larger than the smallest, at about 10 μm.

Cell size and shape are related to cell function. Bird eggs are bulky because they contain a large amount of nutrient material for the developing young. Long muscle cells are efficient in pulling different body parts together. Lengthy nerve cells can transmit nerve impulses rapidly between distant parts of an animal's body. On the other hand, small size also has many benefits. For example, human red blood cells are only about 8 μm in diameter and therefore can fit through our tiniest blood vessels.

Below is a list of the most common units of length biologists use. As you can see, they are metric, so interconversions are easy.

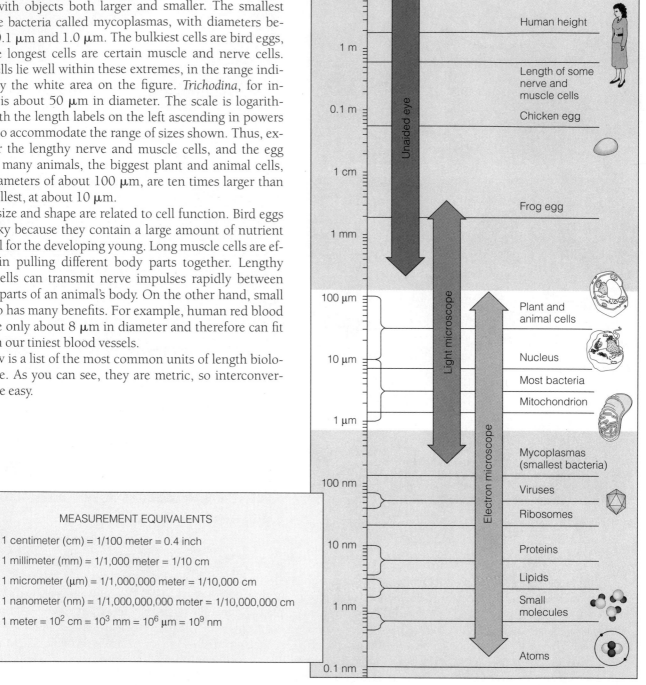

MEASUREMENT EQUIVALENTS

1 centimeter (cm) = 1/100 meter = 0.4 inch

1 millimeter (mm) = 1/1,000 meter = 1/10 cm

1 micrometer (μm) = 1/1,000,000 meter = 1/10,000 cm

1 nanometer (nm) = 1/1,000,000,000 meter = 1/10,000,000 cm

1 meter = 10^2 cm = 10^3 mm = 10^6 μm = 10^9 nm

4.3 Natural laws limit cell size

There are lower and upper limits to cell size. At minimum, a cell must be able to house enough DNA, protein molecules, and internal structures to survive and reproduce. The maximum size of a cell is limited by its requirement for enough surface area to obtain adequate nutrients from the environment and dispose of wastes. Large cells have more surface area than small cells do, but large cells have much less surface area *relative to their volume* than do small cells of the same shape.

The figure at the right illustrates the surface-to-volume relationship. It shows one large cube-shaped cell and 27 small ones. (The purple spheres are cell nuclei.) In both cases, the total volume is the same:

$$\text{Volume} = 30\ \mu m \times 30\ \mu m \times 30\ \mu m = 27{,}000\ \mu m^3$$

In contrast to the total volume, the total surface areas are very different. Because a cube has six sides, its surface area is six times the area of one side. The surface areas of the cubes are as follows:

$$\text{Area of large cube} = 6 \times (30\ \mu m \times 30\ \mu m) = 5400\ \mu m^2$$

$$\text{Area of small cube} = 6 \times (10\ \mu m \times 10\ \mu m) = 600\ \mu m^2$$

For all 27 of the small cubes, the total surface area is $27 \times 600\ \mu m^2$, which equals $16{,}200\ \mu m^2$—three times the surface area of the large cube.

Thus, we see that a large cell has a much smaller surface area relative to its volume than smaller cells have. In fact, *it is the ratio of cell surface to cell volume that imposes upper limits on cell size.* If these were living cells, the plasma membranes of the small cells would service their small volumes of cytoplasm much more easily than would the membrane of the large cell, with its single large volume. As living cells evolved, only those with sufficient surface area to serve their volume survived and reproduced. Of course, no cells are perfect cubes or spheres, and many cell shapes have evolved that affect the size restriction. Muscle and nerve cells can be very long because they are thin and therefore have more surface area per unit of volume than spherical cells.

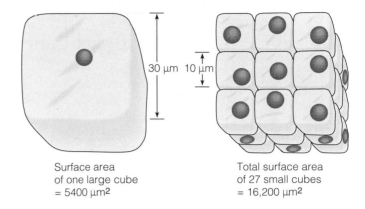

Surface area
of one large cube
= 5400 μm²

Total surface area
of 27 small cubes
= 16,200 μm²

Prokaryotic cells are small and structurally simple 4.4

Two very different kinds of cells have evolved over time. Members of the kingdom Monera, the bacteria, consist of **prokaryotic cells.** All other forms of life are composed of **eukaryotic cells.** This module and the next give overviews of the structures of prokaryotic and eukaryotic cells.

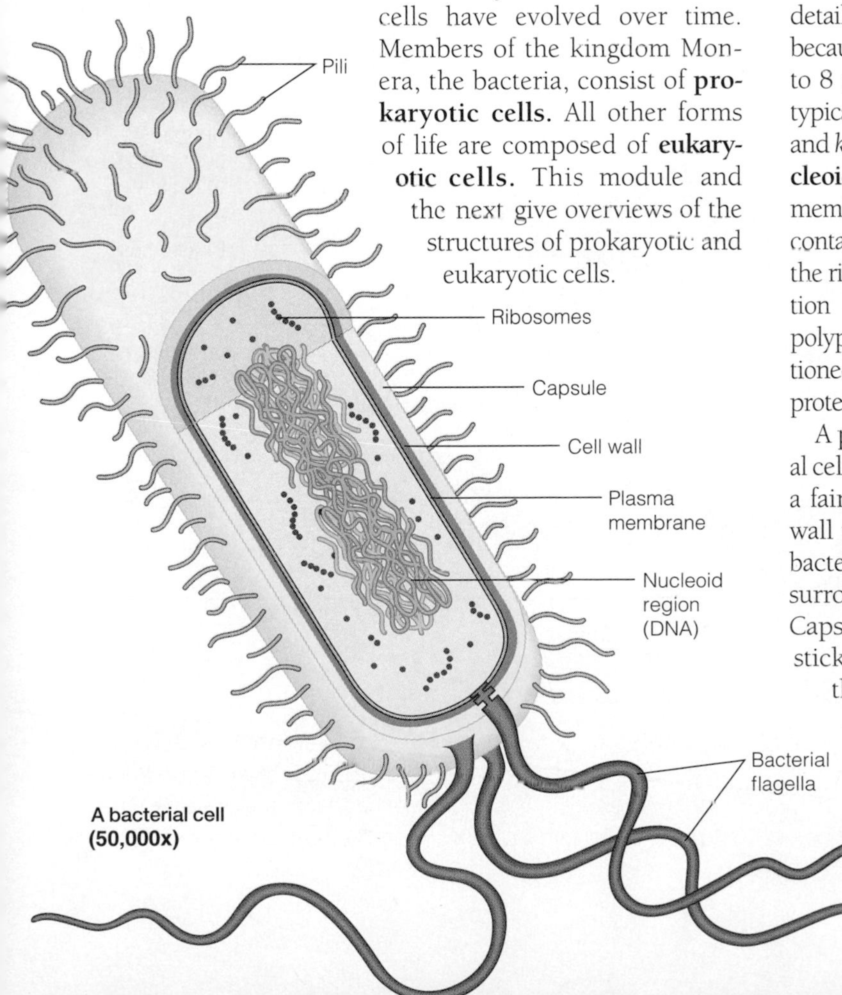

It takes an electron microscope to clearly see the structural details of any cell, and this is especially true of prokaryotic cells because they are so small. Most prokaryotic cells range from 2 to 8 μm in length, averaging about one-tenth of the size of a typical eukaryotic cell. A prokaryotic cell (Greek *pro*, before, and *karyon*, kernel) lacks a nucleus; its DNA is coiled into a **nucleoid** (nucleuslike) **region,** as shown in the figure. Because no membrane surrounds the nucleoid region, the DNA is in direct contact with the rest of the cell contents, the cytoplasm. Notice the ribosomes (brown dots) in the cytoplasm. Under the direction of the DNA, **ribosomes** assemble amino acids into polypeptides, the polymers that make up proteins. As mentioned in Chapter 3, DNA controls all cells by controlling what proteins are made.

A plasma membrane encloses the cytoplasm of the bacterial cell. Surrounding the plasma membrane of most bacteria is a fairly rigid, chemically complex **bacterial cell wall.** The wall protects the cell and helps maintain its shape. In some bacteria, another layer, a sticky outer coat called a **capsule,** surrounds the cell wall and further protects the cell surface. Capsules also help glue some bacteria to surfaces, such as sticks and rocks in fast flowing streams, or tissues within the human body. In addition to outer coats, some bacteria have surface projections. Short projections called **pili** (singular, *pilus*) help attach bacteria to surfaces. Longer projections called **bacterial flagella** (singular, *flagellum*) propel the cell through its liquid environment.

Pili

Ribosomes

Capsule

Cell wall

Plasma membrane

Nucleoid region (DNA)

Bacterial flagella

A bacterial cell (50,000x)

Eukaryotic cells are partitioned into functional compartments

All eukaryotic cells (Greek *eu*, true, and *karyon*, kernel, referring to the nucleus)—whether from animals, plants, protists, or fungi—are fundamentally similar to one another and profoundly different from prokaryotic cells. Let's look at an animal cell and a plant cell as representative of the eukaryotes.

Figure A illustrates an idealized animal cell, showing the details visible with the transmission electron microscope. Just a glance at the figure confirms that eukaryotic cells are much more complex than prokaryotic cells. The most obvious difference is the variety of structures in the cytoplasm. Notice that most of these structures are composed of membranes (recognizable in the drawing as white lines in cross section). In eukaryotes, membranes partition the cytoplasm into a maze of compartments, which biologists call *membranous* organelles (their names are underlined in the figures, as is the term plasma membrane). The membranous organelles shown in Figure A are the nucleus, endoplasmic reticulum, Golgi apparatus, mitochondria, lysosomes, and microbodies. Although most organelles are colorless, we color-code them here and in many places

throughout the book for easier identification. For example, we consistently use purple for the nucleus, the control center of the eukaryotic cell, when it is featured in a figure.

Many of the chemical activities of cells—activities known collectively as **cellular metabolism**—occur in the fluid-filled spaces within membranous organelles. These spaces are important as sites where specific chemical conditions are maintained, conditions that vary from one organelle to another. Metabolic processes that require different conditions can take place simultaneously in a single cell because they occur within separate organelles. For example, while the endoplasmic reticulum is engaged in making a steroid hormone, neighboring microbodies may be making hydrogen peroxide (H_2O_2) as a poisonous by-product of certain metabolic reactions. Only because the H_2O_2 is confined within the microbodies until it can be detoxified are the steroids protected from destruction.

Another benefit of internal membranes is that they greatly increase a eukaryotic cell's total surface area. A typical eukaryotic cell, with a diameter about ten times greater than that of a typical prokaryotic cell, has a thousand times the

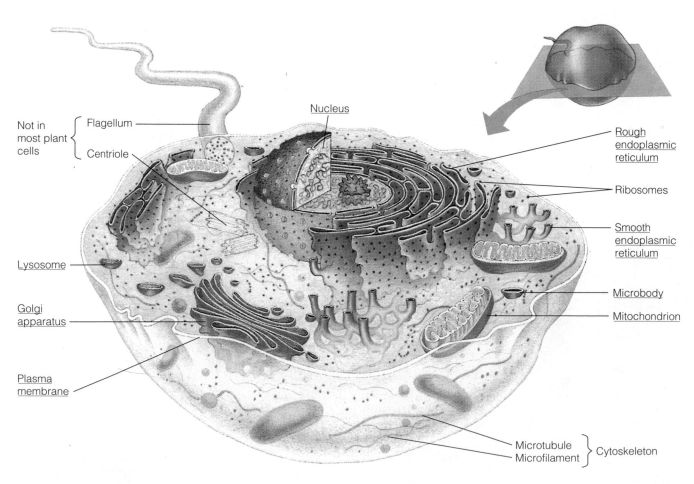

A. An animal cell (8,000x)

cytoplasmic volume but only a hundred times the outer surface area of the typical prokaryotic cell. In eukaryotic cells, internal (cytoplasmic) membranes are where many important metabolic processes occur; in fact, many enzymatic proteins essential for metabolic processes are components of cytoplasmic membranes. Without their internal membranes, eukaryotic cells probably would not have enough membrane surface area to meet their metabolic needs.

All the organelles we have mentioned so far are present in the cells of both animals and plants, as you can see by comparing Figure A and Figure B. But there are also differences between plant and animal cells. One difference is that an animal cell has a pair of centrioles, which plant cells lack. An animal cell may also have a flagellum, and many animal cells have more than one. In contrast, among the plants, only sperm cells in a few species have flagella. (The eukaryotic flagellum is different from the bacterial flagellum in both structure and operation.)

A plant cell has some structures that an animal cell lacks. For one, a plant cell has a rigid, rather thick cell wall (as do the cells of fungi and many protists). Cell walls protect cells and help maintain their shape. Chemically different from bacterial cell walls, plant cell walls contain the polysaccharide cellulose. Unlike typical animal cells, many mature plant cells have the polygonal shape shown in Figure B.

Another organelle found in plant cells but not in animal cells is the chloroplast, where photosynthesis occurs. (Chloroplasts also are found in some protists.) Unique to plant cells is a large central vacuole, a sac that stores water and a variety of other chemicals. Evident in most mature plant cells, the central vacuole contains enzymes that carry out cellular digestion. Furthermore, by taking up additional water and expanding, the central vacuole can help the cell enlarge.

Although we have emphasized membranous organelles, eukaryotic cells contain nonmembranous structures as well (those with labels not underlined). Among these are the centrioles, flagellum, and cytoskeleton, all of which contain protein tubes called microtubules. Also, you can see by the many brown dots in both figures that ribosomes, the sites of protein synthesis, occur throughout the cytoplasm, as they do in prokaryotic cells. Eukaryotic cells also have many ribosomes attached to the membrane of the endoplasmic reticulum (making it "rough") and to the outside of the nucleus.

In the remaining modules of this chapter, we discuss in more detail the organelles of eukaryotic cells, starting with the nucleus.

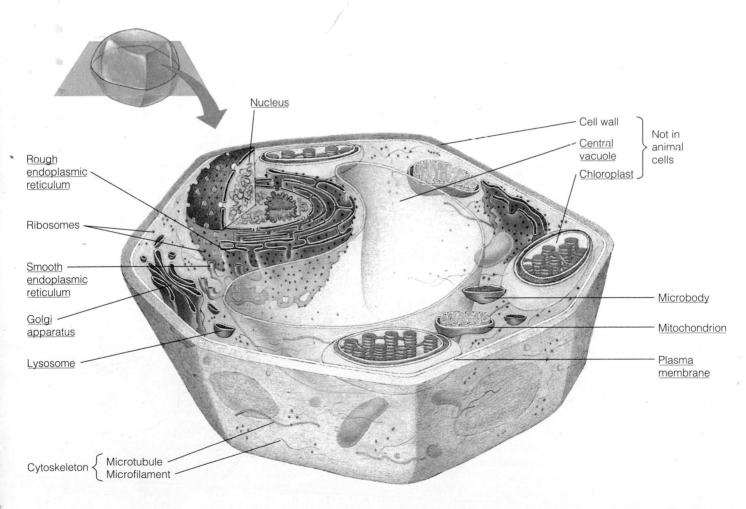

B. A plant cell (8,000x)

4.6 The nucleus is the cell's genetic control center

The **nucleus** is the genetic control center of a eukaryotic cell. Its DNA is the cell's hereditary blueprint, and it directs the cell activities, as does the DNA of a prokaryote's nucleoid region. Most nuclear DNA is attached to proteins, forming very long fibers called **chromatin** (the purple threads in the drawing). During a cell's reproduction, chromatin coils up into structures called **chromosomes**, which are thick

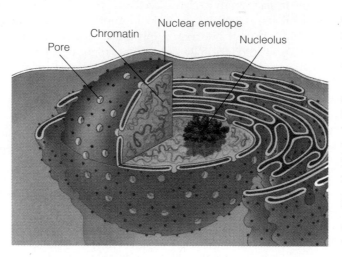

Pore
Chromatin
Nuclear envelope
Nucleolus

enough to be seen with a light microscope. (Chromosomes will be discussed in detail in Chapters 8 and 11.)

Enclosing the nucleus and separating it from the rest of the cell is a **nuclear envelope,** which is a double membrane perforated with pores. Adjoining the chromatin within the nucleus is a mass of fibers and granules called the **nucleolus.** The nucleolus, a combination of DNA, RNA, and proteins, is where ribosomes are made.

4.7 Many cell organelles are related through the endomembrane system

The next eight modules focus on eukaryotic organelles that are formed of interrelated membranes. Some of these membranes are physically connected and some are not, but collectively they constitute a cytoplasmic network that biologists call the **endomembrane system.** Many of the organelles of this system work together in the synthesis, storage, and export of important molecules.

One of the organelles, the endoplasmic reticulum (ER), is a good example of the direct interrelatedness of many parts of the endomembrane system. (The term endoplasmic reticulum comes from Greek words meaning "network

within the cell.") As we will discuss, there are two kinds of ER: rough ER and smooth ER. These organelles differ in structure and function, but the membranes that form them are continuous. Membranes of the rough ER are also continuous with the outer membrane of the nuclear envelope. The space within the ER is separated from the cytoplasmic fluid by the ER membrane (see Figure 4.8A). Thus, the interconnected membranes of the ER and nuclear envelope partition the inside of the cell into two separate compartments. Dividing the cell into compartments is a major function of the endomembrane system.

4.8 Rough endoplasmic reticulum makes membrane and proteins

The "rough" in **rough endoplasmic reticulum,** or rough ER, refers to the appearance of this organelle in electron micrographs. As Figure A shows, the roughness results from ribosomes, which stud the membranes of the organelle. Rough ER is a network of interconnected flattened sacs, with two main functions. One is to make more membrane. Some of the proteins made by ER ribosomes are inserted into the ER membrane, enlarging it, and some of this membrane later ends up in other organelles.

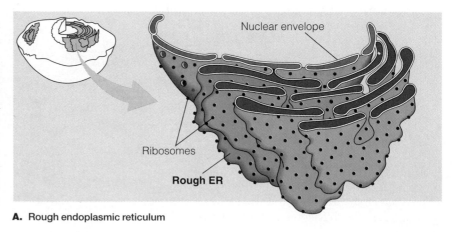

Nuclear envelope

Ribosomes

Rough ER

A. Rough endoplasmic reticulum

The other major function of rough ER is to produce proteins that are secreted by the cell. An example of such a **secretory protein** is an antibody, a defensive molecule made and secreted by white blood cells. Ribosomes of the rough ER synthesize the antibody's polypeptides, which assemble into functional proteins inside the ER. Figure B shows the synthesis and packaging of a typical secretory protein made of a single polypeptide. ① As the polypeptide is synthesized, it passes into the ER. ② Short chains of sugars are then linked to the polypeptide, making the final molecule a **glycoprotein** (*glyc-* means sugar). When the molecule is ready for export, ③ the ER packages it in a tiny sac called a **transport vesicle**. This vesicle ④ buds from the ER membrane and eventually makes its way to the plasma membrane for release from the cell.

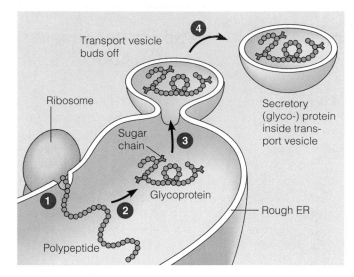

B. Synthesis and packaging of a secretory protein by the rough ER

Smooth endoplasmic reticulum has many functions

4.9

As the figure here indicates, **smooth endoplasmic reticulum**, or smooth ER, is continuous with rough endoplasmic reticulum. Smooth ER is a network of interconnected tubules that lack ribosomes. Much of its activity results from enzymes embedded in its membranes. One of the most important functions of smooth ER is the synthesis of lipids, including fats, phospholipids, and steroids. Each of these products is made by particular kinds of cells. In mammals, for example, smooth ER in cells of the ovaries and testes synthesizes the steroid sex hormones.

Our liver cells also have large amounts of smooth ER, with additional kinds of functions. Certain enzymes in the smooth ER of liver help regulate the amount of sugar released from liver cells into the bloodstream. Other liver enzymes help break down drugs and other potentially harmful substances. The drugs detoxified by these enzymes include, among others, sedatives such as barbiturates, stimulants such as amphetamines, and certain antibiotics.

Undesirable complications may result when liver cells respond to drugs. As the cells are exposed to such chemicals, the amounts of smooth ER and its detoxifying enzymes increase, thereby increasing the body's tolerance to the drugs. This means that higher and higher doses of a drug are required to achieve a particular effect, such as sedation. Another complication is that detoxifying enzymes often cannot distinguish among chemicals and therefore respond to many of them in virtually the same way, breaking down a wide variety of foreign substances in the blood. As a result, the growth of smooth ER in response to one drug can increase tolerance to other drugs, including important medicines. Barbiturate use, for example, may decrease the effectiveness of certain antibiotics.

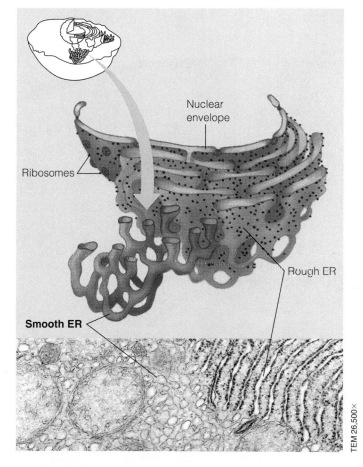

Smooth ER in muscle cells has yet another function: It stores up calcium ions, which are necessary for muscle contraction. When a nerve impulse stimulates a muscle cell, calcium ions leak from the smooth ER into the cytoplasmic fluid, where they trigger contraction of the cell.

The Golgi apparatus finishes, stores, and ships cell products

The **Golgi apparatus** was named after Italian biologist and physician Camillo Golgi, whose career spanned the turn of the twentieth century. Using the light microscope, Golgi and his contemporaries discovered this organelle and a number of others in animal and plant cells. The electron microscope has revealed that a Golgi apparatus is a stack of flattened sacs formed of membranes. As you can see in the illustration below, the sacs are not interconnected like ER sacs. A cell may contain only a few Golgi stacks or hundreds. The number of Golgi stacks correlates with how active the cell is in secreting proteins—a multistep process that, as we have just seen, also involves the endoplasmic reticulum.

The Golgi apparatus performs several functions in close partnership with the ER. Serving as a molecular warehouse

and finishing factory, a Golgi apparatus receives and modifies substances manufactured by the ER. One side of a Golgi serves as a receiving dock for transport vesicles produced by the ER, as you can see in the drawing. When a Golgi receives transport vesicles containing glycoprotein molecules, for instance, it takes in the materials and may then proceed to modify them chemically. One function of this chemical modification seems to be to mark and sort the molecules into different batches for different destinations. The opposite side of the Golgi serves as a shipping depot from which finished products, also packaged in transport vesicles, move to the plasma membrane for export from the cell. Alternatively, finished products may become part of the plasma membrane itself or of another organelle, such as a lysosome.

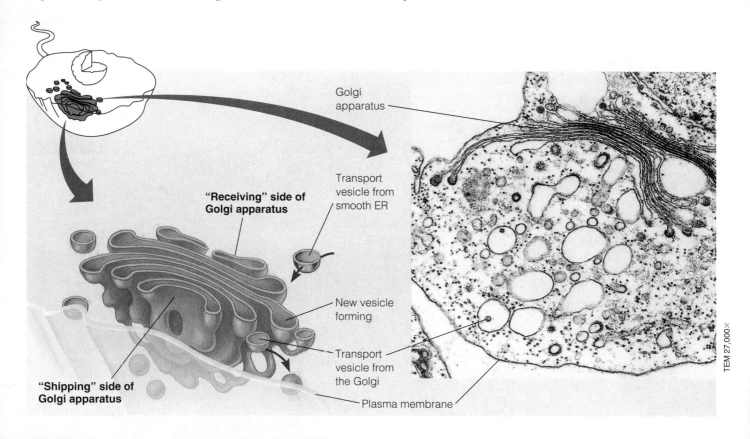

Golgi apparatus

"Receiving" side of Golgi apparatus

Transport vesicle from smooth ER

New vesicle forming

Transport vesicle from the Golgi

"Shipping" side of Golgi apparatus

Plasma membrane

TEM 27,000×

Lysosomes digest the cell's food and wastes

A fourth component of the endomembrane system, the lysosome, is produced by the rough ER and the Golgi apparatus. The name **lysosome** is derived from two Greek words meaning "breakdown body," and lysosomes consist of digestive (hydrolytic) enzymes enclosed in a membranous sac

(Figure A on the next page). The formation and functions of lysosomes are diagrammed in Figure B. First (bottom of figure), the rough ER puts the enzymes and membranes together; then a Golgi apparatus chemically refines the enzymes and releases mature lysosomes. Lysosomes illustrate the main theme of eukaryotic cell structure—compartmentalization. The lysosomal membrane encloses a compartment where digestive enzymes are stored and safely isolated from the rest of the cytoplasm. Without lysosomes, a cell could not contain active hydrolytic enzymes without digesting itself.

Lysosomes have several types of digestive functions, as Figure B shows. Many cells engulf nutrients into tiny cytoplasmic sacs called food vacuoles. Lysosomes fuse with the food vacuoles, exposing the nutrients to hydrolytic enzymes that digest them. In the same way, lysosomes help destroy harmful bacteria. Our white blood cells ingest bacteria into vacuoles, and lysosomal enzymes emptied into these vacuoles rupture the bacterial cell walls. Lysosomes also serve as recycling centers for worn-out or damaged organelles. Without harming the cell, a lysosome can engulf and digest parts of another organelle, making its molecules available for the construction of new organelles. Lysosomes also play vital roles in embryonic development. For example, lysosomal enzymes destroy cells of the webbing that joins the fingers of early human embryos.

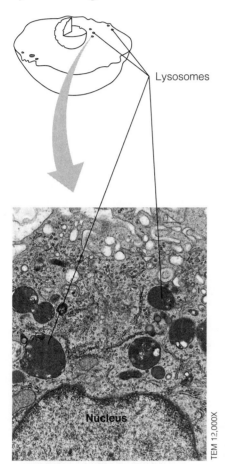

A. Appearance of lysosomes in a white blood cell

B. Lysosome formation and functions

Abnormal lysosomes cause fatal diseases 4.12

The importance of lysosomes to cell function and human health is made strikingly clear by the serious hereditary disorders called **lysosomal storage diseases.** A person afflicted with a lysosomal storage disease is missing one of the hydrolytic enzymes of the lysosome. The abnormal lysosomes become engorged with indigestible substances, which begin to interfere with other cellular functions.

Most of these diseases are fatal in early childhood. In Pompe's disease, harmful amounts of the polysaccharide glycogen accumulate in liver cells because lysosomes lack a glycogen-digesting enzyme. Tay-Sachs disease ravages the nervous system. In this disorder, lysosomes lack a lipid-digesting enzyme, and nerve cells in the brain are damaged as they accumulate lipids. Fortunately, storage diseases are rare in the general population. For Tay-Sachs disease, carriers of the abnormal gene that causes it can be identified before they decide whether to have children.

4.13 Vacuoles function in general cell maintenance

Like lysosomes, **vacuoles** are membranous sacs that belong to the endomembrane system. Vacuoles come in different shapes and sizes and have a variety of functions. In Module 4.11, we saw that the food vacuole functions in collaboration with a lysosome. Here, in Figure A, we see a plant cell's **central vacuole**, which can serve as a large lysosome. The central vacuole

may also help the plant cell grow in size by absorbing water, and it can store vital chemicals or waste products of cell metabolism. Central vacuoles in flower petals may contain pigments that attract pollinating insects. Others contain poisons that protect against plant-eating animals.

Figure B shows a very different kind of vacuole, in the protist *Paramecium*. (Like *Trichodina* in this chapter's introduction, *Paramecium* eats bacteria and moves by means of cilia.)

Notice the two contractile vacuoles, looking somewhat like wheel hubs with radiating spokes. The "spokes" collect excess water from the cell, and the hub expels it to the outside. This function is necessary for freshwater protists because they constantly take up water from their environment. Without a way to get rid of the excess water, the cell fluid would become too dilute to support life, and eventually the cell would swell and burst. Thus, the contractile vacuole is vital in maintaining the cell's internal environment.

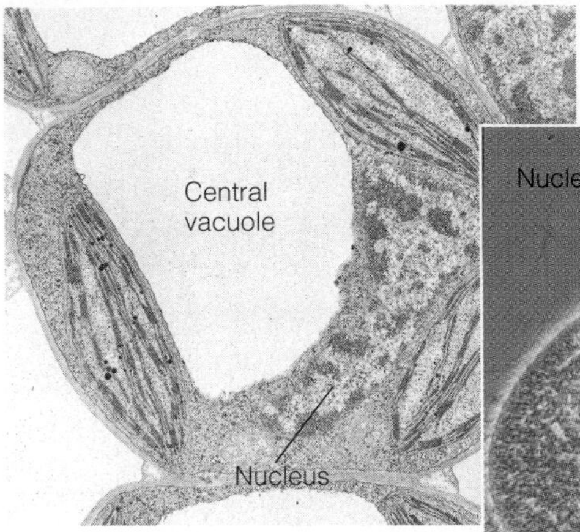

TEM 3,600×

Central vacuole

Nucleus

A. Central vacuole in a plant cell

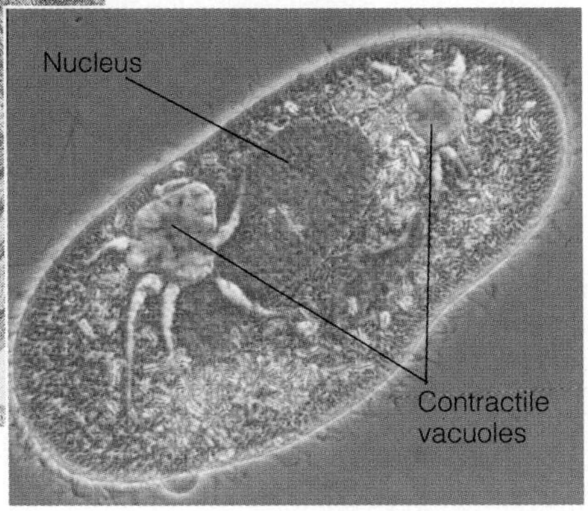

LM 880×

Nucleus

Contractile vacuoles

B. Contractile vacuoles in a protist

4.14 A review of the endomembrane system

The drawing here summarizes the relationships among the major organelles of the endomembrane system (all shown in gray). You can see the direct *structural* connections between the nuclear envelope, rough ER, and smooth ER. The red arrows show the *functional* connections within the endomembrane system, as transport vesicles made by the ER and Golgi develop into the lysosomes and vacuoles that carry out digestion and other processes in the cell.

The drawing also shows that some transport vesicles born in the ER and Golgi fuse with the plasma membrane and contribute their membranes to it. The blue color filling the spaces inside these organelles and outside the cell highlights the fact that an ER product can get out of the cell without ever actually crossing a membrane. A transport vesicle containing the product fuses with the plasma membrane and releases the product to the outside.

Next we look at two membranous organelles that are not part of the endomembrane system—the chloroplast

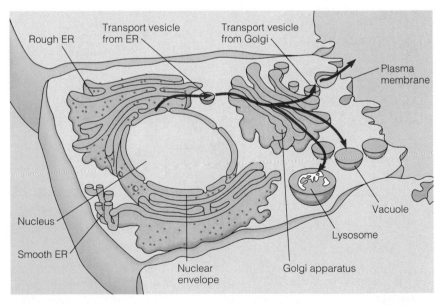

Rough ER

Transport vesicle from ER

Transport vesicle from Golgi

Plasma membrane

Nucleus

Smooth ER

Nuclear envelope

Golgi apparatus

Lysosome

Vacuole

and the mitochondrion. Both of these organelles contain some DNA and ribosomes and make some of their own proteins; the rest are made by free ribosomes in the cytoplasm. Chloroplasts and mitochondria are both fuel processors: They convert energy into forms that living cells can use.

Chloroplasts convert solar energy to chemical energy

It is not possible to overemphasize the importance of chloroplasts. Found in plants and some protists, **chloroplasts** carry out photosynthesis, absorbing solar energy and converting it to chemical energy in sugar molecules. Most of the living world runs on the energy provided by photosynthesis. In viewing a chloroplast's internal structure, made visible by the electron microscope, we see a solar power system much more successful than anything yet produced by human ingenuity.

Befitting an organelle that carries out complex, multistep processes, internal membranes partition the chloroplast into three major compartments. As you can see in the drawing below, the narrow intermembrane space, between the outer and inner membranes of the chloroplast, is one compartment. A second compartment, the space enclosed by the inner membrane, contains a thick fluid called **stroma** and a network of tubules and hollow disks formed of membranes. The space inside the membranous tubules and disks constitutes a third compartment. Notice that the disks occur in stacks, each called a **granum** (plural, *grana*). The grana are the chloroplast's solar power packs—the sites where chlorophyll actually traps solar energy. As we will discuss in Chapter 7, each part of the chloroplast plays a particular role in converting solar energy to chemical energy.

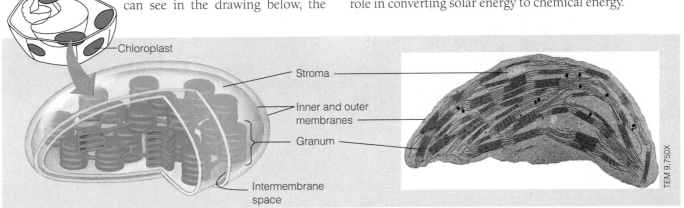

Chloroplast

Stroma

Inner and outer membranes

Granum

Intermembrane space

TEM 9,750X

Mitochondria harvest chemical energy from food

Mitochondria (singular, *mitochondrion*) are somewhat simpler in structure than chloroplasts, perhaps because they convert energy from one chemical form to another, rather than from solar energy to chemical energy. Mitochondria carry out the process of cellular respiration, in which the chemical energy of foods such as sugars is converted into the chemical energy of a cellular fuel molecule called ATP (adenosine triphosphate).

As with other organelles, the structure of the mitochondrion suits its function. You can see in the drawing that the mitochondrion, like the chloroplast, is enclosed by two membranes. However, the mitochondrion has only two compartments. The intermembrane space forms one fluid-filled compartment. The inner membrane encloses the second compartment, containing a fluid called the **mitochondrial matrix.** (The intermembrane fluid and the matrix contain different sets of enzymes.) The inner membrane is highly folded, and the enzyme molecules that actually make ATP are embedded in it. The folds, called **cristae** (singular, *crista*), greatly increase the membrane's surface area, enhancing the mitochondrion's ability to produce ATP. We will discuss the role of mitochondria in cellular respiration in more detail in Chapter 6.

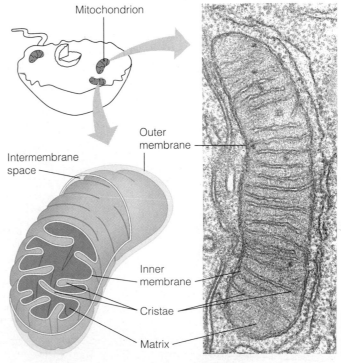

Mitochondrion

Intermembrane space

Outer membrane

Inner membrane

Cristae

Matrix

TEM 44,880×

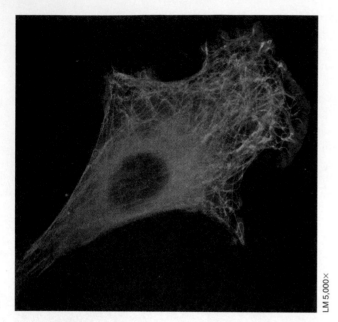

A. Cytoskeleton

LM 5,000×

Many of the organelles we have already described provide some structural support for cells. In addition, eukaryotic cells contain a supportive meshwork of fine fibers, collectively called the **cytoskeleton.** The light micrograph in Figure A shows a cell injected with dyes that highlight the cytoskeleton. The dark, undyed sphere near the center is the nucleus. The fibers of the cytoskeleton extend throughout the cytoplasm.

At the top of Figure B is an electron micrograph of part of a cytoskeleton. Three main kinds of fibers make up the cytoskeleton: microfilaments, the thinnest type of fiber; microtubules, the thickest; and intermediate filaments, in between in thickness.

Microfilaments are solid helical rods composed mainly of a globular protein called actin. Notice at the bottom left of Figure B that each microfilament consists of a twisted double chain of actin molecules. Actin microfilaments can help cells change shape and move by assembling (adding subunits) at one end while disassembling (losing subunits) at the other. The amoeboid (oozing) movement of the protist *Amoeba* and certain of our white blood cells depends mainly on this sort of process. In addition, actin microfilaments often interact with other kinds of protein filaments to make cells contract. This function of microfilaments is best known from studies of muscle cells, as we will see in Chapter 30.

Intermediate filaments are a varied group. As shown in Figure B, they are made of fibrous proteins rather than globular ones and have a ropelike structure. Intermediate filaments serve mainly as reinforcing rods for bearing tension but also help anchor certain organelles. For instance, the nucleus is often held in place by a basket of intermediate filaments.

Microtubules are straight, hollow tubes composed of globular proteins called tubulins, as shown in Figure B. Microtubules elongate by adding doublets of tubulin. They are readily disassembled in a reverse manner, and the tubulin subunits can then be reused in another microtubule. Microtubules that provide rigidity and shape in one area may disassemble and then reassemble elsewhere in the cell.

Other important functions of microtubules are to provide anchorage for organelles and to act as tracks along which organelles can move within the cytoplasm. For example, a lysosome might move along a microtubule to reach a food vacuole. Microtubules also guide the movement of chromosomes when cells divide, and, as we will see next, they are the basis of ciliary and flagellar movement.

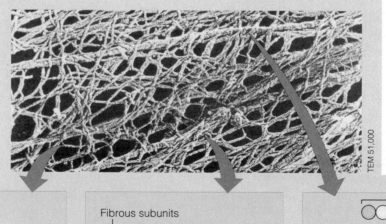

TEM 51,000

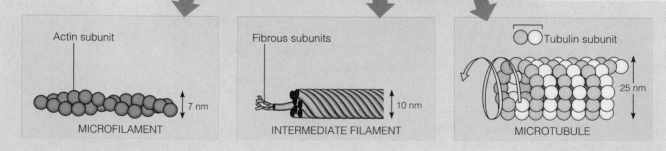

Actin subunit — 7 nm — MICROFILAMENT

Fibrous subunits — 10 nm — INTERMEDIATE FILAMENT

Tubulin subunit — 25 nm — MICROTUBULE

B. Fibers of the cytoskeleton

Cilia and flagella move when microtubules bend

The role of the cytoskeleton in movement is clearly seen in eukaryotic flagella and cilia, the locomotor appendages that protrude from certain cells. Eukaryotic flagella and cilia have a common structure and mechanism of movement. The short, numerous appendages that propel protists such as *Trichodina* through water are called **cilia** (singular, *cilium*). Longer, generally less numerous appendages on other protists are called **flagella.** Some cells of multicellular organisms also have cilia or flagella. For example, cilia on cells lining the human windpipe sweep mucus with trapped debris out of our breathing tubes. Most animals and some plants have flagellated sperm.

As you can see in Figure A, a cilium or flagellum is composed of a core of microtubules wrapped in an extension of the plasma membrane. A ring of nine microtubule doublets surrounds a central pair of microtubules. This arrangement, found in nearly all eukaryotic flagella and cilia, is called the 9 + 2 pattern. Notice that this pattern extends through the length of the organelle but is different at the base. Here, the nine doublets extend into an anchoring structure called a **basal body,** which has a pattern of nine microtubule triplets. The central pair of microtubules terminates above the basal body. When a cilium or flagellum begins to grow, the basal body may act as a foundation for microtubule assembly from tubulin subunits. Basal bodies are identical in structure to **centrioles** (shown in Figure 4.5A). As we will see in Chapter 8, centrioles are involved in animal cell division.

In cilia and flagella, microtubules provide both support and the locomotor mechanism underlying the whipping action of these organelles. Figure B shows the position of two microtubule doublets in a flagellum that is stationary (left) and in the process of bending (right). Bending involves protein knobs attached to each microtubule doublet—the **dynein arms** colored red in the drawings. Using energy from ATP, the dynein arms grab and pull at an adja-

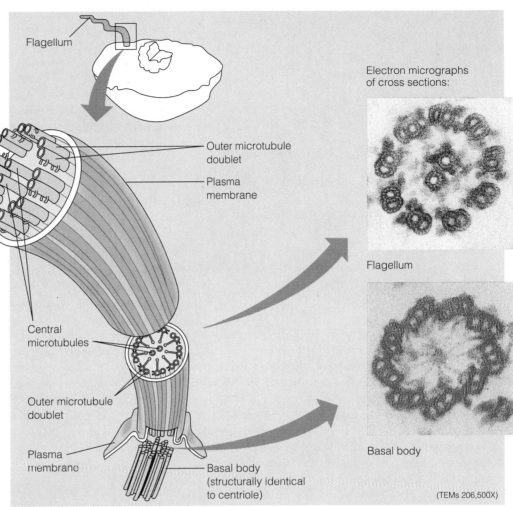

Electron micrographs of cross sections:

Outer microtubule doublet

Plasma membrane

Central microtubules

Outer microtubule doublet

Plasma membrane

Basal body (structurally identical to centriole)

Flagellum

Basal body

(TEMs 206,500X)

A. Structure of eukaryotic flagellum or cilium

cent doublet. The doublets are held together by cross-links (not illustrated); if they were not held in place, the pulling action would make one doublet slide past the other. When the dynein arms grab and pull, the microtubules (and consequently the flagellum or cilium) bend.

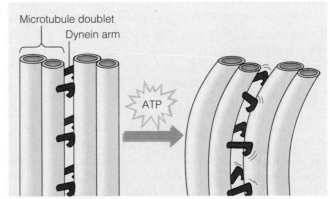

Microtubule doublet

Dynein arm

ATP

B. The mechanism of microtubule bending in cilia and flagella

Cell surfaces protect, support, and join cells

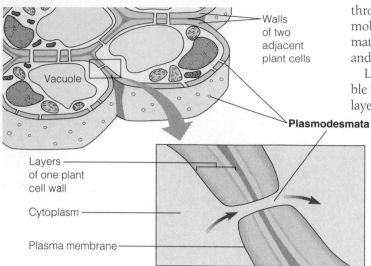

A. Plant cell walls and cell junctions

Walls of two adjacent plant cells

Vacuole

Plasmodesmata

Layers of one plant cell wall

Cytoplasm

Plasma membrane

You might guess that the delicate plasma membrane alone could not handle all the challenges of the environment outside a cell. In fact, most cells have additional surface coverings surrounding the plasma membrane. We introduced the cell walls and capsules of prokaryotes in Module 4.4. Because most prokaryotes exist as single cells or as loose aggregates of cells, their surface coverings interact mainly with noncellular surroundings. In contrast, most eukaryotes are composed of many cells, which are organized into a single, functional organism.

In plants, rigid cell walls not only protect the cells but provide the skeletal support that keeps plants upright on land. Typically 10 to 100 times thicker than the plasma membrane, plant cell walls consist of fibers of the polysaccharide cellulose embedded in a matrix of other polysaccharides and proteins. This tough, fibers-in-a-matrix construction resembles that of fiberglass, also noted for its strength. Figure A shows how the walls of plant cells are arranged. Notice that the cell walls are multilayered. Between the walls of adjacent cells is a layer of sticky polysaccharides (brown) that glues the cells together. The walls of mature plant cells may be very strong; they are the main component of wood, for instance.

Despite their thickness, plant cell walls do not totally isolate the cells from each other. To function in a coordinated way as part of a tissue, the cells must have **cell junctions,** structures that connect them to one another. As Figure A shows, numerous **plasmodesmata** (singular, *plasmodesma*), channels between adjacent plant cells, form a circulatory and communication system connecting the cells in plant tissues. Notice that the plasma membrane (thin black line) and the cytoplasm of the cells in the drawing extend through the plasmodesmata, so that water and other small molecules can pass from cell to cell. Through plasmodesmata, the cells of a plant tissue share water, nourishment, and chemical messages.

Lacking rigid walls, animal cells are generally more flexible than plant cells. Most animal cells are covered by a sticky layer of polysaccharides and proteins that helps hold cells together in tissues. In Figure B, the brown layer at the top designates the surface coating of the cells lining our digestive tract. This coating helps protect the cells from being digested by acids and enzymes in the tract.

Adjacent cells in many animal tissues also connect by cell junctions; there are three general types. **Tight junctions** bind cells together, forming a leakproof sheet. Such a sheet of tissue lines the digestive tract, preventing the contents from leaking into surrounding tissues. **Anchoring junctions** attach adjacent cells to each other or to an extracellular matrix, the substance in which tissue cells are embedded. Anchoring junctions rivet cells together with cytoskeletal fibers but still allow materials to pass along the spaces between cells. **Communicating junctions** are channels similar in function to the plasmodesmata of plants; they allow water and other small molecules to flow between neighboring cells. In Chapter 28, we will explore a type of communicating junction called the synapse, across which adjacent nerve cells pass chemical messages.

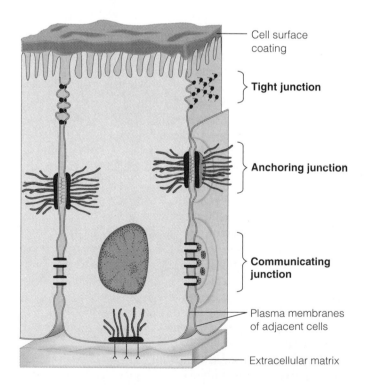

B. Animal cell surface and cell junctions

Cell surface coating

Tight junction

Anchoring junction

Communicating junction

Plasma membranes of adjacent cells

Extracellular matrix

Eukaryotic organelles comprise four functional categories

We have introduced many important cell structures in this chapter. To provide a framework for this information and to reinforce the theme that structure is correlated with function, the table below groups the eukaryotic cell organelles into four categories by general function.

The first category is manufacture. Here we include not only the synthesis of molecules, but also their transport within the cell. The second category includes three organelles that break down and recycle materials that are harmful or no longer needed. (Vacuoles are included here, although, being multifunctional, they do not fit neatly into any one category here.) The third category contains the two energy-processing organelles. Finally, the fourth category is support, movement, and intercellular communication. These three functions are related, because for movement to occur, there must be some sort of rigid support against which force can be applied. And when a supporting structure forms the cell's outer boundary, it is necessarily involved in the cell's communication with its neighbors.

Within each of the four categories a structural similarity underlies the general function of the organelles. In the first category, manufacture depends heavily on a network of metabolically active membranes. In the second category, all the organelles listed are composed of single membranous sacs, inside which materials can be broken down. In the third category, expanses of metabolically active membranes within the organelles allow chloroplasts and mitochondria to perform complex energy conversions that power the cell. Even in the diverse fourth category, there is a common structural theme in the various fibers involved in the functioning of most of the organelles listed.

We can summarize further by emphasizing that these four categories of organelles form an integrated team, and that properties of life at the cellular level emerge from the coordinated functions of the team members. Cell movement is one such emergent property. In that case, mitochondria provide the energy that makes microtubules bend and cilia or flagella beat. As we will see in later chapters, the coordinated actions of cellular organelles underlie most of the emergent properties of life.

Eukaryotic Organelles and Their Functions

General Function: Manufacture

Nucleus	DNA synthesis; RNA synthesis; assembly of ribosomal subunits (in nucleolus)
Ribosomes	Polypeptide (protein) synthesis
Rough ER	Synthesis of membrane proteins and secretory proteins; formation of transport vesicles
Smooth ER	Lipid synthesis; carbohydrate metabolism in liver cells; detoxification in liver cells; calcium ion storage in muscle cells
Golgi apparatus	Modification, temporary storage, and transport of macromolecules; formation of transport vesicles

General Function: Breakdown

Lysosomes	Digestion of nutrients, foreign materials, and damaged organelles
Microbodies	Diverse metabolic processes, such as breakdown of H_2O_2
Vacuoles	Digestion (like lysosome); storage of chemicals; cell enlargement

General Function: Energy Processing

Choroplasts (in plants and some protists)	Conversion of light energy to chemical energy (sugar)
Mitochondria	Conversion of chemical energy of food to chemical energy of ATP

General Function: Support, Movement, and Communication Between Cells

Cytoskeleton (including cilia, flagella, and centrioles in animal cells)	Maintenance of cell shape; anchorage for organelles; movement of organelles within cells; cell movement
Cell walls (in plants, fungi, and some protists)	Maintenance of cell shape and skeletal support; surface protection; binding of cells in tissues
Cell surfaces (in animals)	Surface protection; binding of cells in tissues
Cell junctions	Communication between cells; binding of cells in tissues

4.21 All life forms share fundamental features

Cells are complicated, as this colorized electron micrograph of a white blood cell confirms. Nevertheless, we could argue that cell structure is simpler than we might expect, considering the complex functions that cells perform and the diversity of life forms that exist on Earth. At the cellular level, the living world has only two basic subdivisions—prokaryotic and eukaryotic.

The study of life would be more complicated if there were more than these two fundamental forms of life. In fact, although it is almost certain that Earth is the only life-bearing planet in our solar system, it is conceivable that conditions on planets elsewhere have resulted in the evolution of life. However, organisms on other planets may be very different from the ones we know.

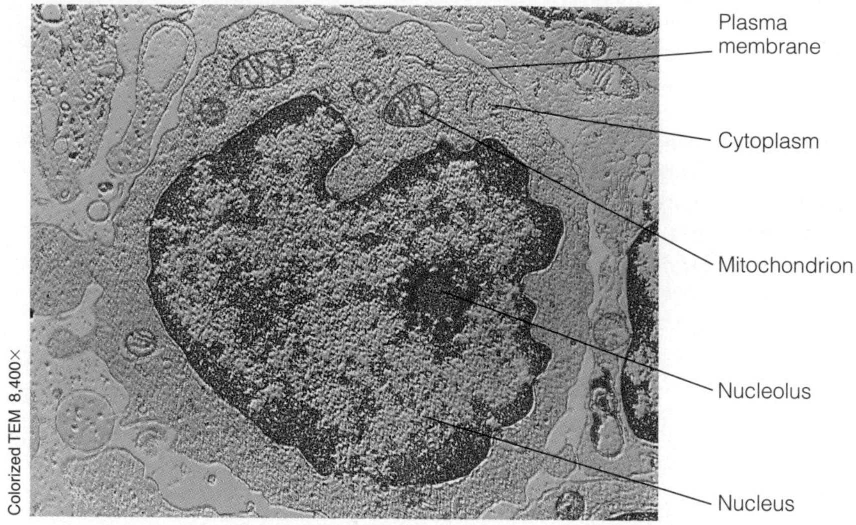

The complex interior structure of a white blood cell

Speculation about the nature of extraterrestrial life is mostly the realm of science fiction, but suppose we learned that an alien organism actually does exist. What could we predict about it from our knowledge of prokaryotes and eukaryotes? We could, at least, predict that it would be highly structured and would exhibit the structure–function theme common to Earth's organisms. We could also expect it would consist of some type of fundamental units, like cells, that are set off from their environment by some sort of membrane. To enable the organism to reproduce and adapt to its environment, it would have to have some type of genetic machinery in each "cell," perhaps housed in a control center like a nucleus. The molecular basis of heredity on another planet might not be DNA; the biological molecules there might even be based on an element other than carbon. But whatever the details of their chemistry, the alien units would have to carry out metabolism.

Speculating about extraterrestrial life helps us summarize what we have learned in this chapter about cellular life as we do know it: All life forms on our planet share the fundamental features of (1) being made of cells enclosed by a membrane that maintains internal conditions very different from the surroundings; (2) having DNA as the genetic material; and (3) carrying out metabolism, which involves the interconversion of different forms of energy and of chemical materials. We expand on the subjects of membranes and metabolism in Chapter 5.

Chapter Review

Begin your review by rereading the module headings and Module 4.20, and scanning the figures before proceeding to the Chapter Summary and questions.

Chapter Summary

Introduction–4.1 All living things are made of units called cells. The light microscope enables us to see the overall shape and structure of a cell. The greater resolving power and magnifying power of electron microscopes reveal cellular details.

4.1 At minimum, a cell must be large enough to house the parts it needs to survive and reproduce. The maximum size of a cell is limited by the amount of surface needed to obtain nutrients from its environment and dispose of wastes. A small cell has a greater ratio of surface area to volume than a large cell of the same shape.

4.1–4.5 There are two kinds of cells: prokaryotic and eukaryotic. Prokaryotic cells—bacteria—are small, relatively simple cells that do not have a nucleus. A bacterium is enclosed by a plasma membrane and usually encased in a rigid bacterial cell wall. The cell wall may be covered with a sticky capsule. Inside the cell, its DNA and other parts are suspended in semifluid cytoplasm. Projections called bacterial flagella may propel the bacterium. All other forms of life are composed of one or more larger and more complex eukaryotic cells, which are distinguished by the presence of a true nucleus. A eukaryotic cell is enclosed in a plasma membrane that controls the cell's interaction with its environment. The cytoplasm of a eukaryotic cell contains complex structures called organelles. Membranes form the boundaries of many of the organelles, compartmentalizing the interior of the cell and thereby enabling the cell to carry out a variety of metabolic activities simultaneously.

4.6 The largest organelle is the nucleus, which is separated from the cytoplasm by a porous nuclear envelope. The nucleus is the cellular control center, containing the DNA that carries the cell's hereditary blueprint and directs its activities.

4.7–4.10, 4.14 The endomembrane system is a group of membranous organelles that manufacture cell products. The endoplasmic reticulum (ER) is a membranous network consisting of rough ER and smooth ER. Rough ER manufactures membranes, and ribosomes on its surface produce proteins that are secreted from the cell. Smooth ER synthesizes lipids. In liver cells, smooth ER also regulates carbohydrate metabolism and breaks down toxins and drugs. In muscle, it plays a role in contraction. The Golgi apparatus, a stack of membranous sacs, receives and modifies ER products and membranes, then sends them on to other organelles or to the cell surface.

4.11–4.13 Lysosomes are sacs of digestive enzymes budded off the Golgi. Lysosomal enzymes digest food, destroy bacteria, recycle damaged organelles, and function in embryonic development. Plant cells contain a large central vacuole, which has both lysosomal and storage functions and also can help the cell enlarge. Protists have contractile vacuoles that pump out excess water.

4.15–4.16 Two organelles are involved in energy conversion. Mitochondria carry out cellular respiration, a process that uses the energy in food to make ATP, which powers most cellular activities. Chloroplasts, found in plants and some protists, absorb solar energy and convert it to chemical energy in sugars.

4.17–4.18 A network of fine protein fibers makes up the cytoskeleton, the cell's structural framework. Thin microfilaments enable cells to change shape and move. They also function in muscle contraction. Intermediate filaments reinforce the cell and anchor certain organelles. Microtubules are thick, hollow tubes that give the cell rigidity, provide anchorage for organelles, and act as tracks for the movement of organelles. Eukaryotic cilia and flagella are elongated locomotor appendages that protrude from certain cells, such as sperm. Clusters of microtubules are responsible for their whipping action.

4.19 Cells interact with their environments and with each other via their surfaces. Plant cells are supported by rigid cell walls made largely of cellulose. Plant cells connect by plasmodesmata, channels that allow them to share water, food, and chemical messages. Animal cells, which are more flexible, are covered by polysaccharides and protein. Tight junctions can bind them together into leakproof sheets. Anchoring junctions link animal cells but allow substances to travel through the spaces between the cells. Communicating junctions allow substances to flow from cell to cell.

Testing Your Knowledge

Multiple Choice

1. Sara would like to film the movement of chromosomes when a cell divides. Which of the following would work best for this purpose, and why?

a. a light microscope, because of its greater resolving power
b. a transmission electron microscope, because of its greater magnification
c. a scanning electron microscope, because the specimen is alive
d. a transmission electron microscope, because of its greater resolving power
e. a light microscope, because the specimen is alive

2. The cells of an ant and a horse are, on average, the same small size; a horse just has more of them. Why are cells so small?

a. If cells were any larger, they would burst.
b. Small cells are easy to replace if damaged or diseased.
c. Small cells are able to absorb what they need.
d. It takes little energy and materials to make small cells.
e. Small cells can change shape easily.

3. Which of the following clues would tell you whether a cell is prokaryotic or eukaryotic?

a. the presence or absence of a rigid cell wall
b. whether or not the cell is partitioned by internal membranes
c. the presence or absence of ribosomes
d. whether or not the cell carries out cellular metabolism
e. whether or not the cell contains DNA

4. Which of the following is *not* directly involved in cell support or movement?

a. microfilament   c. microtubule   e. cell wall
b. flagellum   d. lysosome

5. A type of cell called a lymphocyte makes proteins that are exported from the cell. It is possible to track the path of these proteins as they leave the cell by labeling them with radioactive isotopes. Which of the following might be the path of proteins from where they are made to the lymphocyte's plasma membrane?

a. chloroplast . . . Golgi . . . plasma membrane
b. Golgi . . . rough ER . . . plasma membrane
c. rough ER . . . Golgi . . . plasma membrane
d. smooth ER . . . lysosome . . . plasma membrane
e. nucleus . . . Golgi . . . rough ER . . . plasma membrane

Describing, Comparing, and Explaining

1. Briefly describe the three kinds of junctions between animal cells and compare their functions.

2. What general function do the chloroplast and mitochondrion have in common? How are their functions different?

3. How does internal compartmentalization by membranes help a eukaryotic cell?

Thinking Critically

1. For a class project, Tom made two electron micrographs of mouse cells, two of bean leaf cells, and two of *E. coli* bacteria. He forgot to label the pictures, and on the way to class he got them mixed up. The micrographs are closeups; only the structures listed below are visible. Which pictures can you positively identify? Is it possible to sort out the others? How? Identify as many as you can.

Picture A: Chloroplast, ribosomes, nucleus
Picture B: Cell wall, plasma membrane
Picture C: Mitochondrion, cell wall, plasma membrane
Picture D: Microtubules, Golgi apparatus
Picture E: Plasma membrane, ribosomes
Picture F: Nucleus, rough ER

2. Imagine a cell shaped like a cube, 10 μm on each side. What is the surface area of the cell, in μm^2? What is its volume, in μm^3? What is the ratio of surface area to volume for this cell? (Sketches may help.) Now imagine a second cell, this one 20 μm on each side. What are its surface area, volume, and surface-to-volume ratio? Compare the surface-to-volume ratios of the two cells. How is this comparison significant to the functioning of cells?

Science, Technology, and Society

Doctors at UCLA removed John Moore's spleen, standard treatment for his type of leukemia, and the disease did not recur. Researchers kept the spleen cells alive in a nutrient medium. They found that some of them produced a blood protein called GM-CSF, which they are now testing as a treatment for cancer and AIDS. The researchers patented the cells. Moore sued, claiming a share in profits from any products derived from his cells. The California Supreme Court ruled against Moore (and the U.S. Supreme Court agreed), stating that his suit "threatens to destroy the economic incentive to conduct important medical research." Moore argued that the ruling left patients "vulnerable to exploitation at the hands of the state." Do you think Moore was treated fairly? Is there anything else you would like to know about this case that might help you decide?

The Working Cell 5

A humid summer evening at the edge of a cornfield in Wisconsin—but we could be almost anywhere in the eastern or central United States. The light display comes from insects commonly known as fireflies or lightning bugs. Most of the lights you see are from males on the wing. Females are perched on leaves close to the ground.

When a female sees flashes of light from a male of her species, she reacts with flashes of her own. If the male sees her flashes, he automatically gives another display and flies in the female's direction. Both sexes are responding to a particular pattern of light flashes characteristic of their species. Mating occurs when the female's display leads a male to her, and most females stop flashing after they mate. But in a few species, a mated female will continue to flash, using a pattern that attracts males of *other* firefly species. A veritable *femme fatale,* she waits until an alien male gets close, then grabs and eats him. In the photograph on this page, the firefly hanging on to the leaf is a female dining on a luckless male of another species.

Each of the 2000 or so species of firefly has its own way to signal a mate. Some flash more often than others or during different hours, while other species give fewer but longer flashes. Many species produce light of a characteristic color: yellow, bluish-green, or reddish. In areas where fireflies congregate—often on lawns, golf courses, or open meadows—you can usually see several different species signaling. Because the insects respond instinctively to specific flash patterns, you can even attract males to artificial light if you flash it a certain way.

From a biological standpoint, fireflies are poorly named. They are beetles, not flies, and

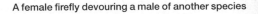
A female firefly devouring a male of another species

their light is almost cold, not fiery. In fact, in emitting light, they give off only about one hundred-thousandth of the amount of heat that would be produced by a candle flame of equal brightness.

What happens in a firefly that makes this light? The light comes from a set of chemical reactions that occur in light-producing organs at the rear of the insect. Light-emitting cells in these organs contain an acidic substance called luciferin and an enzyme called luciferase (both named from the Latin for "light bearer"). In the presence of oxygen (O_2) and chemical energy from ATP molecules, luciferase catalyzes the conversion of luciferin to a molecule that emits light.

The luciferin–luciferase system is one example of how living cells put energy to work by means of enzyme-controlled chemical reactions. Light is a form of energy, and the firefly's ability to make light energy from chemical energy is an example of life's dependence on energy conversions.

Many of the enzymes that control a cell's chemical reactions, including the firefly's luciferase, are located within membranes. Membranes thus serve as sites where chemical reactions can occur in an orderly manner. Also, as we mentioned in Chapter 4, membranes control the passage of crucial substances into and out of cells, and they partition the eukaryotic cell into useful compartments. In fact, the firefly could not produce light, nor could any cell or organism survive, without membranes. Everything that happens in the firefly's light organs has some relation to this chapter's subject: the way working cells use energy, enzymes, and membranes.

5.1 Energy is the capacity to perform work

Energy can only be described and measured by how it affects matter. **Energy** is the capacity to perform work—that is, to move matter in a direction it would not move if left alone. Energy makes change possible, such as the movement of an animal from one place to another. And all organisms require energy to stay alive.

The photographs here compare two forms of energy. **Kinetic energy** is energy that is actually doing work, such as pedaling a bicycle. A mass of matter that is moving performs work by transferring its motion to other matter, whether it is leg muscles pushing bicycle pedals or firefly wings moving air as they beat. **Heat,** the energy associated with the movement of molecules in a body of matter, is one kind of kinetic energy. Light is another kind of kinetic energy.

The second form of energy is stored energy, or **potential energy.** This is the capacity to perform work that matter possesses as a result of its location or arrangement. A cyclist motionless at the top of a hill and water behind a dam both have potential energy due to their altitude. Likewise, the negatively charged electrons of an atom have potential energy due to their positions in electron shells at a distance from the positively charged nucleus. The molecules in a living cell have potential energy due to the arrangement of their atoms; this potential energy is the chemical energy that does the work of the cell. **Chemical energy** is the potential energy of molecules, the most important type of energy for living organisms.

Kinetic energy

Potential energy

5.2 Two laws govern energy conversion

Life depends on the fact that energy can be converted from one form to another. Chemical energy is tapped when a chemical reaction rearranges the atoms of molecules in such a way that potential energy is transformed into kinetic energy. This kind of transformation occurs in an automobile engine when gasoline (containing chemical energy) burns (reacts with oxygen), releasing kinetic energy that pushes the pistons. An organism taps chemical energy when chemical reactions in its cells rearrange atoms in the sugar glucose, converting it to other molecules. In the process, some of the energy stored in the sugar is made available for cellular work. Whether energy transformations occur in a car engine or an organism, they are governed by the same two laws of the universe, known as the first and second laws of thermodynamics.

Thermodynamics is the study of energy transformations that occur in a collection of matter. In discussing energy transformations, we call the collection of matter under study the system and the rest of the universe the surroundings. A system could be an automobile engine, a single cell, or the entire planet. Like those examples, a living organism is an open system; that is, it exchanges both energy and matter with its surroundings. The firefly, for instance, takes in food and oxygen and releases heat, light, and chemical waste products, such as carbon dioxide.

According to the **first law of thermodynamics**, also known as the law of energy conservation, the total amount of energy in the universe is constant. Energy can be transferred and transformed, but it cannot be created or destroyed. An electric company does not make energy; it merely converts it to a form that is convenient to use. The car in Figure A will transform some of the chemical (potential) energy in its gasoline fuel to kinetic energy of movement, and the cells in the woman's body will do the same with her sugar "fuel."

What actually happens to gasoline fuel as a car runs down the road? Some of the energy in the fuel remains in chemical form, in unburned hydrocarbons spewed into the air. The rest of the fuel energy is converted by the car's engine into kinetic energy. The second law of thermodynamics tells us what happens to this kinetic energy during the conversion.

The **second law of thermodynamics** states that energy conversions reduce the order of the universe. Put another way, energy changes, such as the conversion of chemical energy to kinetic energy, are accompanied by an increase in disorder, or randomness. The amount of disorder in a system is called **entropy.** Heat, which is due to random molecular motion, is one form of disorder. The more heat that is generated when one form of energy is converted to another, the more the entropy of the system increases. In our automobile example, heat is generated by the engine and by friction between the tires and the road. Thus, some of the energy of the original system (automobile with a tank of gasoline) becomes disordered (converted to heat). This disordered energy is lost to the surroundings.

In a broad sense, the second law of thermodynamics tells us that the entropy of the universe as a whole is increasing. The second law also implies that if a particular system does become more ordered, its surroundings become more disordered. For instance, if the student in Figure B expends energy to clean up her messy room, she decreases the room's entropy. However, at the same time, disorder is added to the surroundings, as heat generated by her body and as trash.

The second law of thermodynamics has direct application to cellular activities. Because of the second law, energy cannot be transferred or transformed in the cell with 100% efficiency. When a chemical reaction occurs in the cell, chemical energy is transferred between molecules; when light energy is trapped for photosynthesis, it is converted to chemical energy; when a flagellum moves, chemical energy is converted to the kinetic energy of movement. As any such transfer or energy conversion occurs, some energy always escapes from the system as heat. Cells do not have the machinery necessary to put the disordered molecular movement of heat energy to work, and even if they did, some of the heat would still be lost to the surroundings. As we will see in the modules that follow, the work of cells is powered by the potential energy contained in molecules.

B. Second Law: The entropy (disorder) of the universe is increasing

A. First Law: Energy can be transformed but not created or destroyed

5.3 Life's chemical reactions either store or release energy

Chemical reactions, including those that occur in cells, are of two types. One type, called **endergonic reactions,** requires a net input of energy (endergonic means "energy-in"). Endergonic reactions yield products that are rich in potential energy. As you can see in Figure A, an endergonic reaction starts out with reactant molecules that contain relatively little potential energy. Energy is absorbed from the surroundings as the reaction occurs, so that the products of an endergonic reaction store more energy than the reactants. The energy is actually stored in the covalent bonds of the product molecules. And as the graph shows, the amount of additional energy stored in the products equals the difference in potential energy between the reactants and the products.

Photosynthesis, the process whereby plant cells make sugar molecules, is one example of a strongly endergonic process. Photosynthesis starts with energy-poor reactants (carbon dioxide and water molecules) and, using energy absorbed from sunlight, produces energy-rich sugar molecules.

Some other chemical processes are exergonic. An **exergonic reaction** is a chemical reaction that releases energy (exergonic means "energy-out"). As indicated in Figure B, an exergonic reaction begins with reactants whose covalent bonds contain more energy than those in the products. The reaction releases to the surroundings an amount of energy equal to the difference in potential energy between the reactants and products.

As an example of an exergonic reaction, consider what happens when wood burns. One of the major components of wood is cellulose, a large carbohydrate composed of many glucose monomers. Each glucose monomer is rich in potential energy. When wood burns, the potential energy is released as heat and light. Carbon dioxide and water are the products of the reaction.

Burning is one way to release energy from chemicals. Cells release energy by means of a different exergonic process, called cellular respiration. **Cellular respiration** is the energy-releasing chemical breakdown of glucose molecules and the storage of the energy in a form that the cell can use to perform work. Burning and cellular respiration are alike in being exergonic. They differ in that burning is essentially a one-step process that releases all of a substance's energy at once. Cellular respiration, on the other hand, involves many steps, each a separate chemical reaction; you could think of it as a "slow burn." Some of the energy released from glucose by cellular respiration escapes as heat, but a substantial amount of released energy is converted into the chemical energy of molecules of ATP, which we discuss in the next module. Cells use ATP as an immediate source of fuel.

Every working cell in every organism carries out thousands of endergonic and exergonic reactions, the sum of which is known as **cellular metabolism.** In a firefly, for instance, the light display discussed earlier is exergonic. In this case, light energy is released as reactants are converted to product molecules with less energy than the reactants. In addition to generating light, fireflies, like all animals, must find, eat, and digest food, escape predators, repair damage to the body, grow, and reproduce. All these activities require energy, which is obtained from sugar and other food molecules by the exergonic reactions of cellular respiration. Cells then use that energy in endergonic reactions to make molecules that perform specific tasks. To digest food, for instance, an animal's cells use chemical energy to synthesize digestive enzymes. To repair damaged tissues, cells make other proteins that seal up wounds. In the next module, we see that the connection between exergonic and endergonic reactions in cellular metabolism is made by ATP molecules.

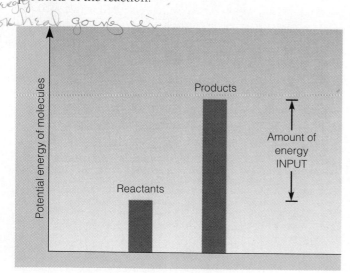

A. Endergonic reaction (energy required)

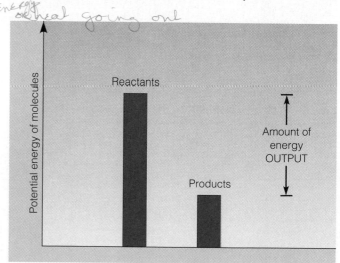

B. Exergonic reaction (energy released)

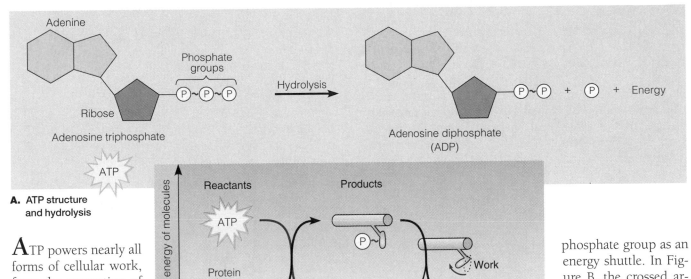

A. ATP structure and hydrolysis

B. How ATP powers cellular work

ATP powers nearly all forms of cellular work, from the generation of light by fireflies to the movements of muscle cells that enable you to pedal a bicycle. Even as you read this page, nerve cells in your brain are using the chemical energy of ATP, energy your cells obtained earlier from food molecules.

When a cell uses chemical energy to perform work, it couples an exergonic reaction with an endergonic one. It first obtains chemical energy from an exergonic reaction and then uses the energy to drive an endergonic reaction. **Energy coupling**—using energy released from exergonic reactions to drive essential endergonic reactions—is a crucial ability of all cells. ATP molecules are the key to energy coupling. The usable energy released by most exergonic reactions, such as the breakdown of glucose molecules, is stored in ATP, and the energy used in most endergonic reactions comes from ATP. Let's see how the structure of ATP fits this function.

As shown in Figure A, ATP (adenosine triphosphate) has three parts, connected by covalent bonds: (1) adenine, a nitrogenous base; (2) ribose, a five-carbon sugar; and (3) a chain of three phosphate groups (symbolized Ⓟ).

The covalent bonds connecting the second and third phosphate groups of ATP are unstable, as symbolized in the figure by a red wavy line: ~. These bonds can readily be broken by hydrolysis (see Module 3.3). Notice in Figure A that three things happen when the third phosphate bond breaks:

1. A phosphate is removed.
2. ATP becomes ADP (adenosine diphosphate).
3. Energy is released.

We call the hydrolysis of ATP an exergonic reaction because it releases energy. The cell can couple this reaction to an endergonic reaction. It does so by using the third phosphate group as an energy shuttle. In Figure B, the crossed arrows indicate two coupled reactions. In the figure, a protein molecule gains energy when it acquires a phosphate from ATP, changing shape in the process. The added energy allows the molecule to perform work, here changing back to its original shape. This kind of shape change is what helps a muscle cell contract. Notice that the energized molecule loses the phosphate in performing the work.

The transfer of a phosphate group to a molecule is called **phosphorylation**, and most cellular work depends on ATP energizing other molecules by phosphorylating them. Work can be sustained because ATP is a renewable resource that cells can regenerate. Figure C shows the ATP cycle. Energy released in exergonic reactions, such as glucose breakdown during cellular respiration, is used to regenerate ATP from ADP. In the process, a phosphate group is bonded to ADP, forming ATP by dehydration synthesis. This is an endergonic (energy-storing) reaction, just the reverse of a hydrolysis reaction. An organism or cell at work uses ATP continuously. In fact, a working cell consumes and regenerates its entire pool of ATP about once each minute.

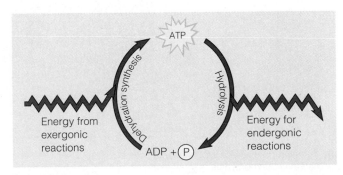

C. The ATP cycle

5.5 Enzymes speed up the cell's chemical reactions by lowering energy barriers

A cell's supply of ATP is like money in a checking account, available for immediate use whenever it is needed. ATP breaks down easily, so easily, in fact, that a cell's ATP supply might decompose spontaneously if not for what is called an energy barrier. An energy barrier is an amount of energy, called the **energy of activation** (E_A), that reactants must absorb to start a chemical reaction. In the case of ATP breakdown, for instance, E_A is the amount of energy needed to break the bond between the second and third phosphate (see Figure 5.4A). Since most reactions require energy to get started, ATP and most other vital molecules in our cells do not break down spontaneously.

Figure A illustrates the E_A concept with an analogy involving Mexican jumping beans. These are seeds of certain desert shrubs, each containing an insect larva whose wriggling makes the bean jump. Jumping beans in the left-hand chamber of container 1 represent reactant molecules in a chemical reaction; those in the right-hand chamber represent product molecules. The partition symbolizes the E_A of the chemical reaction. Notice that the level of the left-hand chamber is higher than that of the right-hand chamber. Because they have a higher position, beans on the left have more potential energy than those on the right. But even though the beans that reach the lower chamber end up with less energy, they first had to have extra energy to make it over the partition.

Jumping beans vary in how much energy they have. Those with the most actively wriggling larvae jump the most often and the highest. At any particular instant, only a few beans may jump high enough to clear the partition between the two chambers and become "products." Thus, it may take a very long time for a significant number of reactant beans to reach the right-hand chamber.

Now we have a dilemma. Most of the essential reactions of metabolism must occur quickly and precisely for a cell to survive. If the chemical reactants of metabolism are like jumping beans faced with an energy barrier that cannot readily be cleared, a cell may die because it cannot make vital products fast enough. The solution lies in the capabilities of enzymes. An **enzyme** is a protein molecule that serves as a biological catalyst, increasing the rate of a reaction without itself being changed into a different molecule. An enzyme does not add energy to a cellular reaction; it speeds up a reaction by lowering the E_A barrier. Without enzymes, many metabolic reactions would occur too slowly to sustain life.

In Figure A, container 2 symbolizes the same chemical reaction as container 1, but catalyzed by an enzyme. The enzyme, in effect, pushes down the partition, making it possible for beans with less energy to clear it. As a result, in a given period of time, more beans will end up in the right-hand, product chamber than would without the enzyme.

In Figure B, a graph shows the effect of an enzyme on the reaction it catalyzes. The black curve represents the course of the reaction without an enzyme; the E_A barrier is higher than in the reaction with an enzyme (red curve). Notice that the net change in energy from start to finish, analogous to the difference in heights of the two chambers in the jumping-bean model, is the same for both curves.

In catalyzing metabolic reactions in cells, enzymes are essential to life. Enzymes lower the energy of activation by changing the shape of the reactant molecules, as we see next.

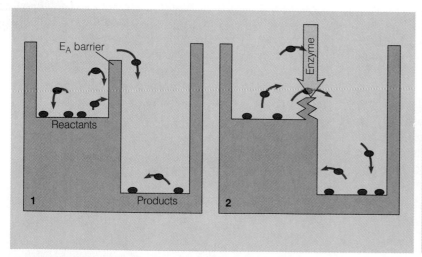

A. Jumping-bean analogy for energy of activation (E_A) and the role of enzymes

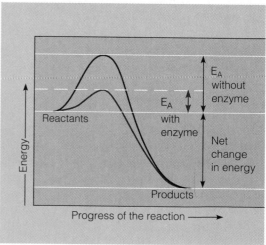

B. The effect of an enzyme on E_A

Enzymes are very selective in the reactions they catalyze, and their selectivity determines which chemical processes occur in a cell at any particular time. As a protein, an enzyme has a unique three-dimensional shape, and that shape determines the enzyme's specificity. A substance that an enzyme acts on—a reactant in a chemical reaction—is called the enzyme's **substrate.** Each enzyme recognizes only the specific substrate or substrates of the reaction it catalyzes. It therefore takes many different kinds of enzymes to catalyze all the reactions that occur in a cell.

In catalyzing a reaction, an enzyme binds to its substrate. While the two are joined, the substrate changes into the product (or products) of the reaction. Only a small part of an enzyme molecule, called the **active site,** actually binds to a substrate. The active site is typically a small pocket or groove on the surface of the enzyme. The enzyme is specific because its active site fits only one kind of substrate molecule.

The figure here illustrates how an enzyme works. We use the enzyme sucrase as an example. Its specific substrate is table sugar (sucrose), and the reaction it catalyzes is the hydrolysis of sucrose to glucose and fructose. ① Sucrase starts with an empty active site. ② Sucrose enters the active site, attaching by weak bonds. The interaction with sucrose induces the enzyme to change shape slightly so that the active site fits even more snugly around the sucrose. This "induced fit" is like a clasping handshake. It holds the substrate in a position that facilitates the reaction, and ③ the substrate is converted into the products, glucose and fructose.

④ The enzyme releases the products and emerges unchanged from the reaction. Its active site is now available for another substrate molecule, and another round of the cycle can begin. A single enzyme molecule may act on thousands or even millions of substrate molecules per second.

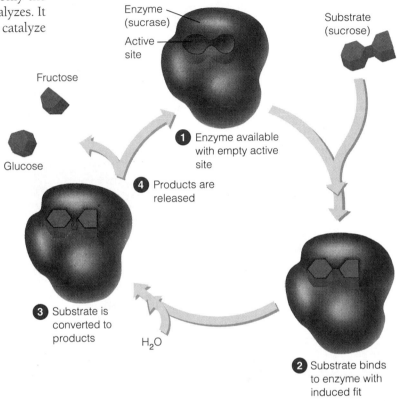

The activity of an enzyme is affected by its environment, and for every enzyme there are conditions under which it is most effective. Temperature, for instance, affects molecular motion, and an enzyme's optimal temperature produces the highest rate of contact between reactant molecules and the enzyme's active site. Higher temperatures denature the enzyme, altering its specific three-dimensional shape and destroying its function. Most human enzymes have temperature optima of 35°–40°C (close to our normal body temperature).

Salt concentration and pH also influence enzyme activity. Few enzymes can tolerate extremely salty solutions because the salt ions interfere with some of the chemical bonds that maintain protein structure. The optimal pH for most en-zymes is near neutrality, in the range of 6 to 8. Outside this range, enzyme action, and thus normal chemical functioning of cells, may be impaired. Entire lakes are threatened by acid precipitation, which can make the water so acidic that aquatic organisms experience enzyme failure (see Module 2.16).

Many enzymes will not work unless they are accompanied by nonprotein helpers called **cofactors.** Cofactors may be inorganic substances such as atoms of zinc, iron, or copper. If the cofactor is an organic molecule, it is called a **coenzyme.** Most coenzymes are compounds made from vitamins or are vitamins themselves.

Next we discuss chemicals that interfere with enzyme activity.

5.8 Enzyme inhibitors block enzyme action

A chemical that interferes with an enzyme's activity is called an inhibitor. There are two types of enzyme inhibitors. A **competitive inhibitor** resembles the enzyme's normal substrate and competes with the substrate for the active site on the enzyme. As shown in the figure below, when a competitive inhibitor sits in the active site, it blocks the substrate from entering and thereby prevents the enzyme from acting.

A **noncompetitive inhibitor** does not enter the active site. As the figure shows, a noncompetitive inhibitor binds to the enzyme somewhere outside the active site, but its binding changes the shape of the enzyme so that the active site no longer fits the substrate.

The action of any inhibitor can be irreversible or reversible, depending on the kind of bonds formed with the enzyme. Inhibition is irreversible when covalent bonds form between inhibitor and enzyme. Inhibition is reversible when only weak bonds, such as hydrogen bonds, form. Weak bonds eventually break as neighboring molecules jostle together. When the concentration of substrate is higher than that of the inhibitor, it is more likely that a substrate mole-cule will be nearby when an active site becomes vacant, and the reaction will proceed. Conversely, when there is more inhibitor than substrate, the reaction will slow down.

Enzyme inhibitors, especially reversible ones, are important regulators of cell metabolism. In some cases, the inhibitor of an enzyme is the very substance the reaction produces. For instance, when a cell's supply of ATP exceeds demand, ATP itself acts as a noncompetitive inhibitor, interfering with the enzymes that drive ATP synthesis. This sort of inhibition, whereby a metabolic reaction is blocked by its products, is called **negative feedback,** and it is one of the most important mechanisms that regulate metabolism. As we will see in Chapter 6, cells actually make most of their molecules by sequences of reactions, called metabolic pathways. Negative feedback controls many such pathways.

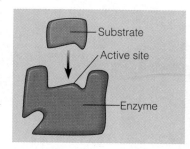

Normal binding of substrate

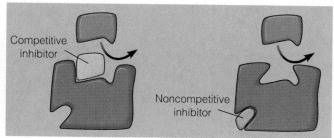

Enzyme inhibitors

5.9 Some pesticides and antibiotics inhibit enzymes

When an inhibitor, especially an irreversible one, prevents an enzyme from catalyzing a crucial metabolic reaction, an organism may be poisoned. This effect can be turned to human advantage. Certain pesticides are toxic to insects because they irreversibly inhibit key enzymes in the nervous system. The insecticide malathion, for instance, inhibits a nervous system enzyme called acetylcholinesterase. The inhibition prevents nerve cells from transmitting signals and kills the insect. For the same reason, malathion can also be toxic to other animals, including humans, though doses that kill insects are not generally harmful to people.

Many antibiotics also work by inhibiting enzymes—in this case, enzymes that are essential to the survival of disease-causing bacteria. Penicillin, for example, inhibits an enzyme that bacteria use in making cell walls. Since humans lack this enzyme, we are not harmed by the drug. The actions of pesticides and antibiotics help us appreciate how important enzymes are in living cells. Let's now turn to a discussion of membranes, which help organize the enzyme-catalyzed reactions that make up the cell's metabolism.

5.10 Membranes organize the chemical activities of cells

So many metabolic reactions occur simultaneously in a cell that utter chaos would result if the cell were not highly organized and able to time its metabolic reactions precisely. Some of the organization comes from teams of enzymes that function like assembly lines. A product from one enzyme-catalyzed reaction becomes the substrate for a neighboring enzyme, and so on, until a final product is made. For a cell's assembly lines to operate, the right enzymes have to be present at the right time and in the right place.

Membranes provide the structural basis for metabolic order. In eukaryotes, membranes form most of the cell's organelles and partition the cell into compartments that contain enzymes in solution. For all cells, the plasma membrane is literally the edge of life, forming a fine boundary between the living cell and its surroundings and controlling the traffic of molecules into and out of the cell. Like all cell membranes, the plasma membrane exhibits **selective permeability;** that is, it allows some substances to cross more easily than others and blocks passage of some substances altogether. The plasma membrane takes up substances the cell needs and disposes of the cell's wastes.

For a structure that separates life from nonlife, the plasma membrane is amazingly thin, twenty times too thin to be resolved by the light microscope. It shows up quite clearly in electron micrographs, however. Figure A displays part of a red blood cell in cross section, magnified 100,000 times. At this magnification, you can see that the plasma membrane has three zones: an out-

ermost dark band (upper arrow), a light zone just inside it, and a second dark band (lower arrow) bordering the outer edge of the cytoplasm.

Much of our knowledge about plasma membranes comes from studies of red blood cells. Cell biologists can prepare large amounts of pure plasma membrane by breaking open these cells and washing away the cytoplasm. The empty sacs remaining are called ghosts (Figure B). Chemical studies on these ghosts have revealed that the three layers seen in electron micrographs come from the way membrane lipid molecules are arranged, as we see next.

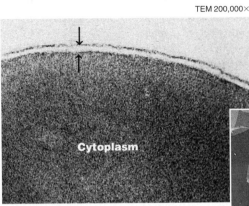

TEM 200,000×

Cytoplasm

A. Plasma membrane in cross section

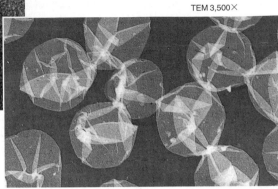

TEM 3,500×

B. "Ghosts" of red blood cells

Membrane phospholipids form a bilayer

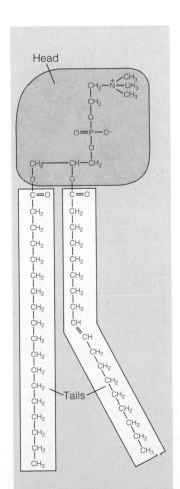

Head

Tails

A. Phospholipid molecule

Lipids, mainly phospholipids, are the main structural components of membranes. Figure A illustrates the structure of a phospholipid. The structural formula on the left highlights the two parts of the molecule, which have opposite interactions with water. The head (gray) is polar and therefore hydrophilic. The double tail (yellow) is nonpolar and hydrophobic. Below is the phospholipid symbol that is commonly used in depicting membranes.

Symbol

The structure of phospholipid molecules is well suited to their role in membranes. In water, phospholipids spontaneously form a stable two-layer framework called a phospholipid bilayer (Figure B). Their hydrophilic heads face outward, exposed to the water, and their hydrophobic tails point inward, shielded from the water. This is the arrangement that membrane phospholipids have in the watery environment within living organisms.

The hydrophobic interior of the bilayer is one reason why membranes are selectively permeable. Nonpolar, hydrophobic molecules are soluble in lipids and can easily pass through membranes. In contrast, polar, hydrophilic molecules are not soluble in lipids. Whether polar molecules pass through the membrane depends on protein molecules in the phospholipid bilayer. In fact, much of a membrane's selective permeability depends on membrane proteins, which we examine in the next module.

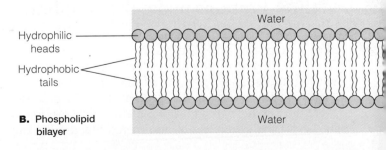

Water

Hydrophilic heads

Hydrophobic tails

B. Phospholipid bilayer

Water

5.12 The membrane is a fluid mosaic of phospholipids and proteins

The drawing below shows the structure of the plasma membrane. Like other cellular membranes, it is commonly described as a **fluid mosaic.** The word mosaic denotes a surface made of small fragments, like pieces of colored tile cemented together in a mosaic floor or picture. A membrane is a *mosaic* in having diverse protein molecules embedded in a framework of phospholipids. Yet, the membrane mosaic is *fluid* in that most of the individual proteins and phospholipid molecules can drift laterally in the membrane even though the phospholipid bilayer remains intact.

Notice the kinked tails of many of the phospholipids. As we saw in Module 3.8, kinks result from double bonds in lipid tails. Kinks make the membrane more fluid by keeping adjacent phospholipids from packing tightly together. The steroid cholesterol (see Module 3.9), wedged into the bilayer, helps stabilize the phospholipids at body temperature (about 37°C for humans) but helps keep the membrane fluid at lower temperatures. In a living cell, the phospholipid bilayer remains about as fluid as salad oil at room temperature.

The two surfaces of any membrane are different. For the plasma membrane, the most striking difference is that the outside surface has carbohydrates (chains of sugars, shown here in green) covalently bonded to proteins and lipids embedded in the membrane. A protein with sugars attached to it is called a glycoprotein, whereas a lipid with sugars is called a glycolipid. The carbohydrate portions of glycoproteins and glycolipids vary from species to species, from one individual to another in the same species, and even from one cell type to another in a single individual. They seem to function as identification tags on cells, enabling a cell to determine if other cells it encounters are like itself or not. The ability to distinguish cell types is crucial to life. It allows cells in an embryo to sort themselves into tissues and organs. It also enables cells of the immune system to recognize and reject foreign cells, such as infectious bacteria. The protein portions as well as the carbohydrates of glycoproteins play a role in making cell surfaces recognizable. This is one of several membrane-protein functions that we look at next.

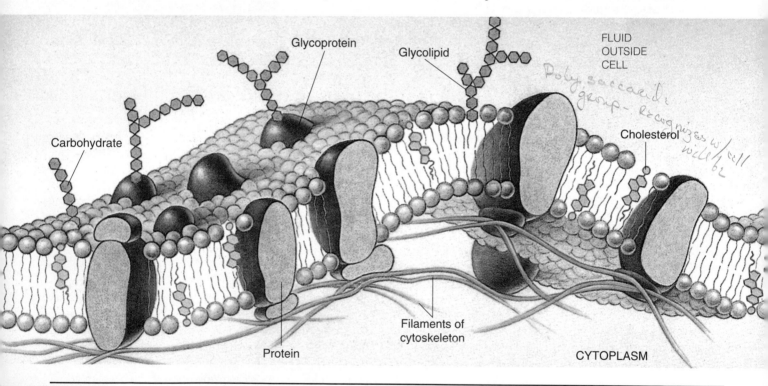

5.13 Proteins make the membrane a mosaic of function

The word mosaic as applied to a membrane refers not only to the positioning of proteins in the phospholipid bilayer but also to the varied activities of these proteins. In fact, proteins perform most of the functions of the membrane. More than 50 different kinds of proteins have been found in the plasma membrane of human red blood cells, and

there are probably many more that are just too scarce to have been detected so far.

What do membrane proteins do besides act as identification tags on cells? Many membrane proteins are enzymes, which may function in catalytic teams for molecular assembly lines (Figure A, facing page). Some membrane proteins, as

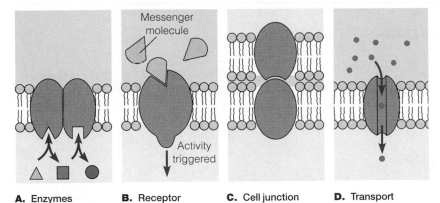

A. Enzymes   **B.** Receptor   **C.** Cell junction   **D.** Transport

just as an enzyme fits its substrate. The attachment of the messenger to the receptor triggers a specific cell activity (Figure B).

Other membrane proteins act like glue, forming cell junctions where adjacent cells stick together (Figure C; also see Module 4.19). Some proteins that project into the cytoplasm attach to the cytoskeleton and help maintain cell shape.

Finally, some membrane proteins help move substances across the membrane (Figure D). Although small molecules such as O_2 pass freely through the phospholipid bilayer, many essential molecules need assistance from proteins to enter or leave the cell. In the next six modules, we look at molecular transport of all sorts.

we will discuss later, function as **receptors** for molecular messengers from other cells. A receptor protein has a shape that fits the shape of a specific messenger, such as a hormone,

---

Passive transport is diffusion across a membrane 5.14

The nature of phospholipids and the kinds of proteins in a membrane determine whether a particular substance can cross the membrane. Some molecules can cross without the cell doing any work, but even in these cases, physical principles dictate when they can cross and in which direction they will go.

Diffusion is the tendency for particles of any kind to spread out spontaneously from where they are more concentrated to where they are less concentrated. Diffusion requires no work; it results from the random motion (kinetic energy) of atoms and molecules and is driven by the universal tendency of order to deteriorate into disorder (entropy). Because a cell does not perform work when molecules diffuse across its membrane, the diffusion of a substance across a biological membrane is called **passive transport.**

The figures here illustrate passive transport, showing how concentration affects the direction in which a substance crosses a membrane. In Figure A, molecules of a colored dye in solution tend to diffuse from the side of the membrane where they are more concentrated to the side where they are less concentrated, until the solutions on both sides have equal concentrations of dye. Put another way, the dye diffuses down its **concentration gradient** until equilibrium is reached. At equilibrium, molecules ·continue to move back and forth, but there is no *net* change in concentration on either side of the membrane. Figure B illus-

trates the important point that two or more substances diffuse independently of each other; that is, each diffuses down its own concentration gradient.

Passive transport is extremely important to all cells. In our lungs, for example, passive transport driven by concentration gradients is the sole means by which oxygen (O_2, essential for metabolism) enters red blood cells and carbon dioxide (CO_2, a metabolic waste) passes out of them. Water is another substance that crosses membranes by passive transport, and we focus on this topic next.

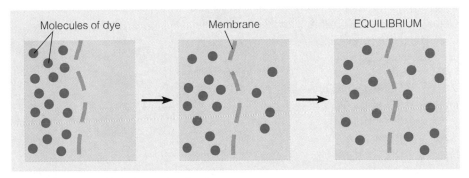

A. Passive transport of one type of molecule

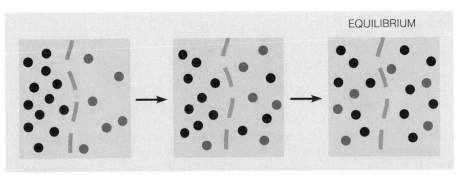

B. Passive transport of two types of molecules

5.15 Osmosis is the passive transport of water

Because a cell contains an aqueous solution and is surrounded by one, and because the plasma membrane is permeable to water, water molecules can readily pass into and out of cells. Diffusion of water molecules across a selectively permeable membrane is a special case of passive transport called **osmosis.**

Figure A shows what happens if a membrane permeable to water but not to solutes separates two solutions with different concentrations of solute, here a molecule such as glucose. The solution with a higher concentration of solutes is said to be **hypertonic** (*hyper*, above; *tonos*, tension). The solution with the lower solute concentration is **hypotonic** (*hypo*, below). Water crosses the membrane until the solute concentrations (molecules per milliliter solution) are equal on both sides.

In the close-up view in Figure B, you can see what happens at the molecular level. The polar solute molecules attract clusters of water molecules, so that fewer water molecules are free to diffuse across the membrane. With fewer solute molecules, the hypotonic solution has more free water molecules, and there is a net movement of water from the hypotonic solution to the hypertonic one. The result is the difference in water levels you see in Figure A.

Here we show only one type of solute molecule, but the same net movement of water would occur no matter how many kinds of so-

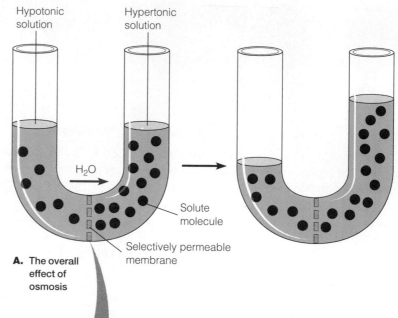

Hypotonic solution   Hypertonic solution

H_2O

Solute molecule

Selectively permeable membrane

A. The overall effect of osmosis

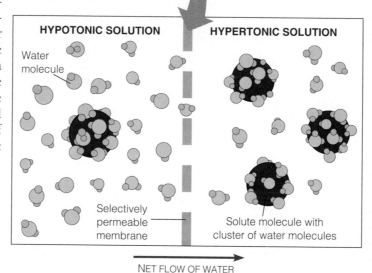

HYPOTONIC SOLUTION   **HYPERTONIC SOLUTION**

Water molecule

Selectively permeable membrane

Solute molecule with cluster of water molecules

NET FLOW OF WATER

B. Osmosis at the molecular level

lutes were present. The direction of osmosis is determined only by the difference in *total* solute concentration, not by the nature of the solutes. For example, seawater has a great variety of solutes, but it will lose water to a solution containing a high enough concentration of a single solute. Only if the total solute concentrations are the same on both sides of the membrane will water molecules move at the same rate in both directions. Solutions of equal solute concentration are said to be **isotonic** (*isos*, equal).

5.16 Water balance between cells and their surroundings is crucial to organisms

The figure at the top of the next page illustrates how the principle of osmosis applies to living cells. When an animal cell, such as the red blood cell shown in the figure, is immersed in an isotonic solution, the cell's volume remains constant because the cell gains water at the same rate that it loses it (part 1). In this situation, the cell and its surroundings are in equilibrium because the two solutions have the

same total concentration of solutes. We describe an organism or a cell in this situation as being isotonic to the surrounding solution. Many marine animals, such as sea stars and crabs, are isotonic to seawater. Part 2 shows what can happen to an animal cell in a hypotonic solution, which has a lower solute concentration than the cell. The cell gains water, swells, and may pop (lyse) like an overfilled water

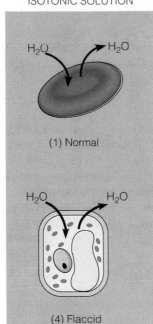
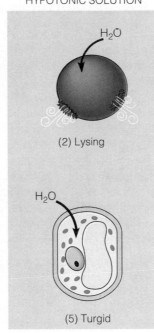
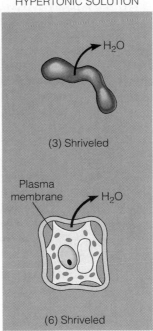

ISOTONIC SOLUTION | HYPOTONIC SOLUTION | HYPERTONIC SOLUTION

ANIMAL CELL

H_2O → H_2O

(1) Normal

H_2O

(2) Lysing

H_2O

(3) Shriveled

PLANT CELL

H_2O → H_2O

(4) Flaccid

H_2O

(5) Turgid

Plasma membrane → H_2O

(6) Shriveled

How cells behave in different solutions

balloon. Part 3 shows the opposite situation: An animal cell in a hypertonic solution shrivels and can die from water loss.

For an animal to survive if its cells are exposed to a hypertonic or hypotonic environment, the animal must have a way to prevent excessive uptake or excessive loss of water. The control of water balance is called **osmoregulation.** For example, a freshwater fish, which lives in a hypotonic environment, has kidneys and gills that work constantly to prevent an excessive buildup of water in the body. (We discuss osmoregulation further in Chapter 25.)

Water balance problems are somewhat different for plant cells because of their rigid cell walls. A plant cell immersed

in an isotonic solution (part 4) is flaccid, and a plant wilts in this situation. In contrast, a plant cell is turgid, and plants are healthiest, in a hypotonic environment, as in part 5. To become turgid, a plant cell needs a net inflow of water. Although the somewhat elastic cell wall expands a bit, the pressure it exerts prevents the cell from taking in too much water and bursting, as an animal cell would in this environment. Part 6 shows that in a hypertonic environment a plant cell is no better off than an animal cell. As a plant cell loses water, it shrivels, and its plasma membrane pulls away from the cell wall. This usually kills the cell.

Specific proteins facilitate diffusion across membranes 5.17

In discussing traffic across membranes, we have focused so far on substances, including water, that diffuse freely across biological membranes. Numerous substances that do not diffuse freely across membranes can cross by special mechanisms that promote and regulate their movement.

Many molecules move across a membrane with the help of transport proteins in the membrane. When one of these proteins makes it possible for a substance to move down its concentration gradient, the process is called **facilitated diffusion**. Without the protein, the substance does not cross the membrane, or it diffuses across it too slowly to be useful to the cell. Like osmosis, facilitated diffusion is a type of passive transport because it does not expend energy.

The figure here shows the most common way membrane proteins facilitate diffusion. The transport protein (violet) spans the membrane and provides a pore for the passage of a particular solute (orange). The rate of facilitated diffusion depends on how many transport-protein molecules are available in the membrane and how fast they can move their specific solute.

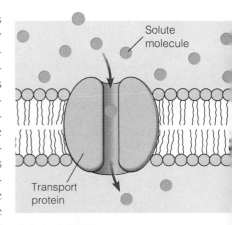

Transport protein providing a pore for solute passage

5.18 Cells expend energy for active transport

In contrast to passive transport, **active transport** requires that a cell expend energy to move molecules across a membrane. In this situation, a transport protein actively pumps a specific solute across a membrane *against* the solute's concentration gradient—that is, away from the side where it is less concentrated. Membrane proteins usually use ATP as their energy source for active transport.

The figure below shows an active transport system involving the passage of two different solutes across a membrane, in opposite directions. The transport protein has a separate binding site for each of the solutes. ① Active transport begins when one of the solutes binds to the transport protein. ② After the binding, a phosphate group is transferred from ATP to the pro-

tein. ③ An energy surge from the transfer (phosphorylation) makes the protein change shape and release the solute molecule on the other side of the membrane. ④ As the protein releases the first solute, its shape and position allow the second solute to bind to it. ⑤ The phosphate group is released. This causes the protein to return to its original shape, ⑥ releasing the second solute on the opposite side of the membrane.

Active transport systems like this, coupling the passage of two solutes, are common in cell membranes, and they transport ions as well as uncharged molecules. A very important one is the sodium–potassium (Na^+–K^+) pump (discussed in Chapter 29), which helps nerve cells generate nerve signals.

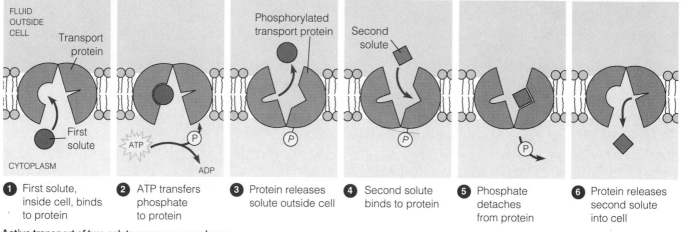

1 First solute, inside cell, binds to protein

2 ATP transfers phosphate to protein

3 Protein releases solute outside cell

4 Second solute binds to protein

5 Phosphate detaches from protein

6 Protein releases second solute into cell

Active transport of two solutes across a membrane

5.19 Exocytosis and endocytosis transport large molecules

So far we have focused on how water and small solutes enter and leave cells by moving through the plasma membrane. The story is quite different for very large molecules such as proteins.

As shown in Figure A, a cell uses the process of **exocyto-**

sis (from the Greek *exo*, outside, and *kytos*, cell) to export bulky materials from its cytoplasm. In the first step of this process, a membrane-bounded vesicle filled with macromolecules (green) moves to the plasma membrane. Once there, the vesicle fuses with the plasma membrane, and the

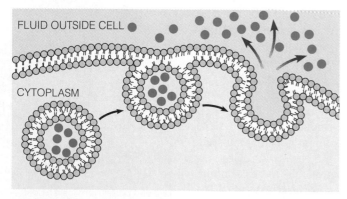

A. Exocytosis

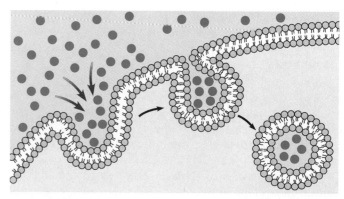

B. Endocytosis

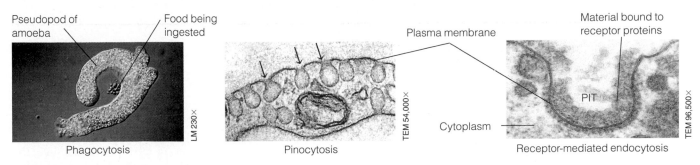

Pseudopod of amoeba | Food being ingested

LM 230×

Phagocytosis

Plasma membrane

TEM 54,000×

Pinocytosis

Material bound to receptor proteins

PIT

Cytoplasm

TEM 96,500×

Receptor-mediated endocytosis

C. Three kinds of endocytosis

vesicle's contents spill out of the cell. When we weep, for instance, cells in our tear glands use exocytosis to export a salty solution containing proteins. In another example, certain cells in the pancreas manufacture the hormone insulin and secrete it into the blood stream by exocytosis.

Figure B illustrates a transport process that is basically the opposite of exocytosis. In **endocytosis** (*endo,* inside), a cell takes in macromolecules or other particles by forming vesicles or vacuoles from its plasma membrane. Figure C shows three kinds of endocytosis. The micrograph on the left shows an amoeba taking in a food particle, an example of **phagocytosis,** or "cellular eating." The amoeba engulfs its prey by wrapping extensions, called pseudopodia, around it and packaging it within a vacuole. As Module 4.11 described, the vacuole then fuses with a lysosome, and the lysosome's hydrolytic enzymes digest the contents.

The center micrograph shows **pinocytosis,** or "cellular drinking." The cell is in the process of taking droplets of fluid from its surroundings into tiny vesicles (arrows). Pinocytosis is not specific, in that it takes in any and all solutes dissolved in the droplets.

In contrast to pinocytosis, a third type of endocytosis, called **receptor-mediated endocytosis,** is highly specific. The micrograph on the right shows one stage of the process. The plasma membrane has indented to form a pit. The pit is lined with receptor proteins (not visible) that have picked up particular molecules from the surroundings. The pit will pinch closed to form a vesicle that will carry the molecules into the cytoplasm. In the next module, we examine a human hereditary condition in which receptor-mediated endocytosis malfunctions.

Faulty membranes can overload the blood with cholesterol | 5.20

Few molecules make the news as often as cholesterol does. Cholesterol is notorious as a possible factor in heart disease, yet it is essential for the normal functioning of all our cells. As mentioned earlier, cholesterol is a component of membranes as well as a starting material for making other steroids, such as sex hormones. Problems tend to arise only when we have too much cholesterol in our blood.

In most of us, cells in the liver remove excess cholesterol from the blood by receptor-mediated endocytosis. Consequently, we have only small amounts of the substance in our blood. Cholesterol circulates in the blood mainly in particles called low-density lipoproteins, or LDLs. As the figure here shows, an LDL is a globule of cholesterol (and other lipids) surrounded by a single layer of phospholipids in which proteins are embedded. One of the LDL proteins (pink) fits a specific type of receptor protein (purple) on cell membranes. Normally, liver cells take up LDLs from the blood by receptor-mediated endocytosis in which the receptors are these LDL receptor proteins.

Worldwide, about one in 500 human babies inherits a disease called **hypercholesterolemia,** which is characterized by an excessively high level of cholesterol in the blood. In severe cases (about one in a million people), there are no functional LDL receptors. People with a milder form of the disease have functional receptors but in low numbers. In either case, LDLs tend to accumulate in the blood. High levels of LDLs in the

blood are life-threatening because the LDLs can deposit cholesterol on the lining of blood vessels, eventually obstructing them. The blockage of blood vessels that nourish the heart leads to heart disease. Individuals with severe hypercholesterolemia may die from heart disease in early childhood. Milder cases are also dangerous, but they can be treated with medications and a low-cholesterol diet. We will discuss the genetics of hypercholesterolemia in Module 9.10.

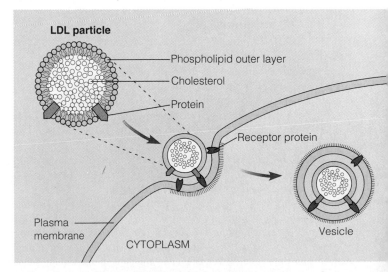

LDL particle

Phospholipid outer layer

Cholesterol

Protein

Receptor protein

Vesicle

Plasma membrane

CYTOPLASM

A cell using receptor-mediated endocytosis to take up an LDL

5.21 Chloroplasts and mitochondria make energy available for cellular work

This chapter's main topics—energy, enzymes, and membranes—lead us to see connections that are important to all organisms. Two organelles, the chloroplast and the mitochondrion, are both composed of membranes with enzyme assembly lines. Both play central roles in harvesting energy and making it available for cellular work. As shown in the diagram here, chloroplasts in photosynthetic organisms use solar energy to make glucose from carbon dioxide and water. In the process of photosynthesis, light energy is converted to chemical energy, and oxygen gas escapes as a by-product. Mitochondria, present in all eukaryotes, consume oxygen in cellular respiration, using the chemical energy stored in glucose to make ATP.

Nearly all the chemical energy that organisms use comes ultimately from sunlight. Chloroplasts in plants and photosynthetic protists convert light energy to chemical energy. No animal can do this. Some animals obtain chemical energy by eating plants. Others, like the firefly, obtain it by eating other animals. In either case, the chemical energy was originally derived from sunlight through the process of photosynthesis.

The figure also makes the important point that chemicals recycle among living organisms and their environment. Even the oxygen that plants give off is not lost, but remains available in the air for mitochondrial use. Energy, however, does not recycle. In carrying out the activities of life, organisms dissipate some energy to the environment in the form of heat. This constitutes an energy loss, because heat cannot be used to make ATP. For this reason, organisms must be constantly supplied with energy. In the next chapter, we discuss how cells maintain their supplies of ATP.

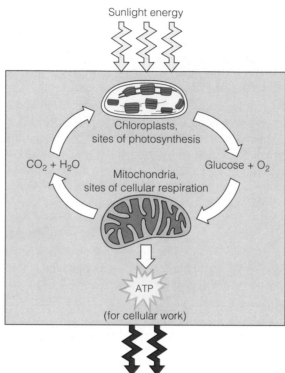

Sunlight energy

Chloroplasts, sites of photosynthesis

$CO_2 + H_2O$

Mitochondria, sites of cellular respiration

Glucose + O_2

ATP
(for cellular work)

Heat energy

Chapter Review

Begin your review by rereading the module headings and scanning the figures before proceeding to the Chapter Summary and questions.

Chapter Summary

5.1–5.2 In order to stay alive, all organisms require energy, which is the capacity to perform work. Kinetic energy is energy that is actually doing work. Potential energy is stored energy, energy that matter possesses because of its location or arrangement. According to the laws of thermodynamics, energy can be changed from one form to another, but it cannot be created or destroyed. Energy changes are not 100% efficient; they increase disorder, or entropy, and some energy is always lost as heat.

5.3–5.4 Cells carry out thousands of chemical reactions, the sum of which constitutes cellular metabolism. There are two types of chemical reactions. An endergonic reaction absorbs energy and yields products rich in potential energy. An exergonic reaction releases energy; its products contain less potential energy than its reactants. Burning and cellular respiration are both exergonic. In cellular respiration, some energy is stored in ATP molecules. When the bond joining a phosphate group to the rest of an ATP molecule is subsequently broken by hydrolysis, the reaction supplies energy for cellular work.

5.5–5.6 Reactants must absorb some energy, called the energy of activation (E_A), in order for a reaction to begin. This represents an energy barrier that helps prevent molecules from breaking down spontaneously. A protein catalyst called an enzyme can decrease the energy barrier. Enzymes are selective, and this selectivity determines which chemical reactions occur in a cell. The reactant an enzyme acts on—its substrate—attaches to a specifically shaped active site on the enzyme. The substrate is held in a position that facilitates the reaction and is converted into a product, which is then released. The enzyme is unchanged and can rapidly repeat the process.

5.7–5.9 Temperature, salt concentration, and pH can influence enzyme activity. Some enzymes require nonprotein cofactors. Inhibitors interfere with enzymes. A competitive inhibitor takes the place of the substrate in the active site. A noncompetitive inhibitor alters an enzyme's function by changing its shape. Some pesticides and antibiotics are enzyme inhibitors.

5.10–5.13 Membranes order metabolism. They are selectively permeable, controlling the flow of substances into and out of a cell and among compartments within a cell. They can also hold teams of enzymes that function in metabolism. The membrane is a fluid mosaic: Phospholipid molecules form a flexible bilayer, with cholesterol and protein molecules embedded in it. Attached carbohydrates act as cell identification tags. Some of the proteins are enzymes, and others are receptors for chemical messages. Still other membrane proteins form cell junctions, and some transport substances across the membrane.

5.14–5.16 In passive transport, substances simply diffuse through membranes, spreading from areas of higher concentration to areas of lower concentration. Osmosis is the passive transport of water, from a solution of lower solute concentration to one of higher solute concentration. Osmosis causes cells to shrink in a hypertonic solution and swell in a hypotonic solution. The control of water balance is important for organisms.

5.17–5.19 Small and nonpolar molecules diffuse freely though the phospholipid bilayer. Other molecules may pass through selective protein pores in the process of facilitated diffusion. Transport proteins can move solutes against a concentration gradient; this active transport process requires ATP energy. To move large molecules or particles through the membrane, a vesicle may fuse with the membrane and expel its contents (exocytosis), or the membrane may fold inward, trapping material from the outside (endocytosis). Receptor-mediated endocytosis traps specific substances.

5.21 Enzymes and membranes are central to the processes that make energy available to a cell. Chloroplasts carry out photosynthesis, using solar energy to make glucose and oxygen from carbon dioxide and water. Mitochondria consume oxygen in cellular respiration, using the energy stored in glucose to make ATP.

Testing Your Knowledge

Multiple Choice

1. Consider the following situations: chemical bonds in the gasoline in a car's gas tank, and the movement of the car along the highway; the tension of a stretched rubber band, and the pull that stretched it; a climber poised at the top of a hill, and the hike by which he got there. In each of these situations, the first part illustrates _____, and the second part illustrates _____.

    a. the first law of thermodynamics . . . the second law
    b. kinetic energy . . . potential energy
    c. an exergonic reaction . . . an endergonic reaction
    d. potential energy . . . kinetic energy
    e. the second law of thermodynamics . . . the first law

2. Which of the following best describes the general structure of a cell membrane?

    a. proteins sandwiched between two layers of phospholipid
    b. proteins embedded in two layers of phospholipid
    c. a layer of protein coating a layer of phospholipid
    d. phospholipids sandwiched between two layers of protein
    e. phospholipids embedded in two layers of protein

3. The concentration of solutes in a red blood cell is about 2%. Sucrose cannot pass through the membrane, but water and urea can. Osmosis would cause red blood cells to shrink the most when immersed in which of the following solutions?

    a. a hypertonic sucrose solution
    b. a hypotonic sucrose solution
    c. a hypertonic urea solution
    d. a hypotonic urea solution
    e. pure water

4. The concentration of calcium in a cell is 0.3%. The concentration of calcium in the surrounding fluid is 0.1%. How could the cell obtain more calcium? (*Explain your answer.*)

    a. passive transport
    b. diffusion
    c. active transport
    d. osmosis
    e. any of the above

Describing, Comparing, and Explaining

1. Only a few kinds of plants have adapted to living in the salty soils around desert lakes and salt flats. Explain why having their roots exposed to salt water would be a problem for most plants.

2. Food rots when microorganisms break down food molecules. Various food preservation methods interfere with the enzyme activity of microorganisms and prevent them from surviving. Explain how each of the following would interfere with enzyme activity: canning (heating), freezing, pickling (soaking in acetic acid), salting.

Thinking Critically

A biologist isolated a sample of an enzyme called lactase from the intestinal lining of a calf. The substrate of lactase is the disaccharide lactose. The enzyme breaks a lactose molecule in two, producing a molecule of glucose and a molecule of galactose. The biologist performed two series of experiments. First she made up 10% lactose solutions with different concentrations of enzyme and measured the rate at which galactose was produced (in terms of grams of galactose produced per minute; a bigger number means a faster breakdown of lactose). Results of these experiments are shown in the top part of the table below. In the second series of experiments, she prepared 2% enzyme solutions with different concentrations of lactose and again measured the rate of galactose production. Results are shown in the bottom part of the table below.

Rate and Enzyme Concentration

Lactose concentration	10%	10%	10%	10%	10%
Enzyme concentration	0	1%	2%	4%	8%
Reaction rate	0	25	50	100	200

Rate and Substrate Concentration

Lactose concentration	0	5%	10%	20%	30%
Enzyme concentration	2%	2%	2%	2%	2%
Reaction rate	0	25	50	65	65

Explain the relationship between the rate of the reaction and the enzyme concentration. Explain the relationship between the rate of the reaction and the substrate concentration. (*Hint:* It is helpful to sketch graphs of rate versus concentration.) State your explanations in terms of what might be happening between individual enzyme and substrate molecules. How did the results of the two experiments differ and why?

Science, Technology, and Society

Lead in the environment acts as an enzyme inhibitor, and it can interfere with the development of the nervous system. The Johnson Controls Company, a maker of lead-acid automobile batteries, had a "fetal protection policy" that banned female employees of child-bearing age from working in areas where they might be exposed to high levels of lead. Under this policy, women were involuntarily transferred to lower-paying jobs in lower-risk areas. A group of Johnson Controls employees challenged the policy in court, claiming that it deprived women of job opportunities available to men. In 1991, the U.S. Supreme Court unanimously decided that this policy was sex discrimination and ruled it illegal. While the decision is clearly a victory for equal employment rights, many people are uncomfortable about a woman's, or anyone's, "right" to work in an unsafe environment. What rights and responsibilities of employers, employees (male and female), employees' children, and government agencies are in conflict in this situation? Whose responsibility is it to determine what constitutes a safe environment and who should or should not work there? What criteria should be used to decide?

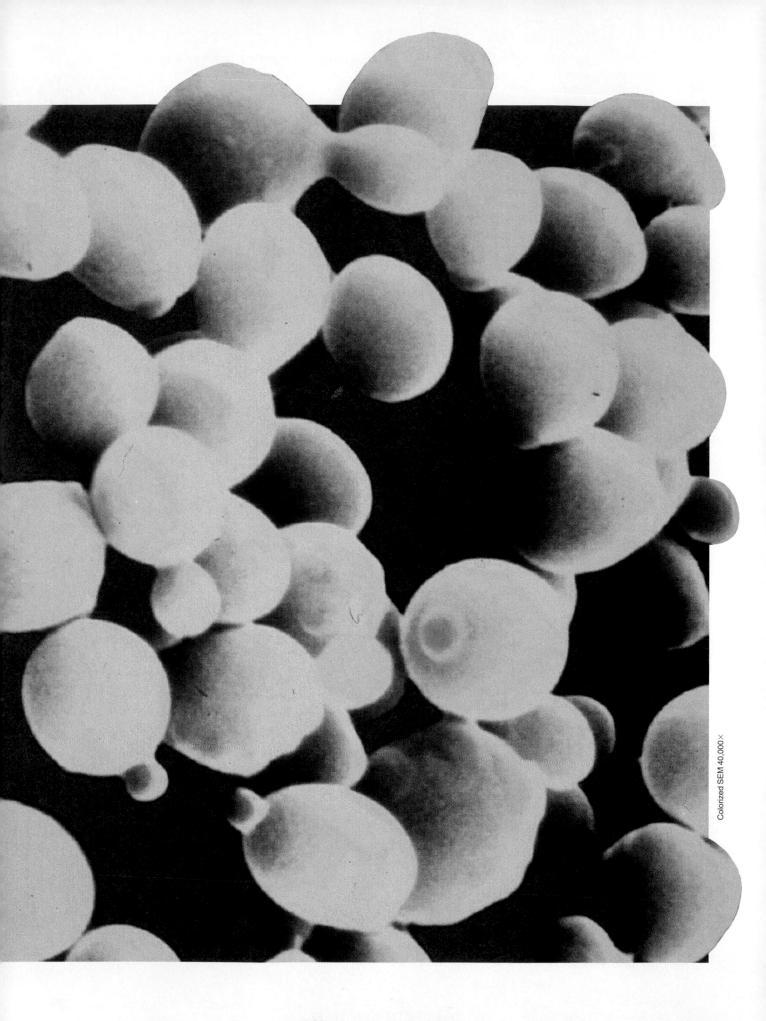

How Cells Harvest Chemical Energy

We debated. The meeting broke up and we thought about it overnight. We met again and debated some more. Should we really begin this chapter with a photograph of yeast cells?

"They're just not exciting to look at; at best, they look like a bunch of beige blobs."

"But maybe we can use a colorized electron micrograph, and seeing them bud is kind of neat."

"Well, even if the photo's not very interesting, what better organism to introduce this chapter?"

We finally decided—yeasts had to be our focus organism for this chapter. Even though yeasts are among the simplest eukaryotes, they perform all the activities this chapter covers. At the molecular level, yeasts are quite similar to other eukaryotes. This similarity has made them important research organisms, even for uncovering clues to mysteries of the human cell.

Members of the kingdom Fungi, yeasts grow on moist surfaces and in liquids that are rich in nutrients. The cells in the photograph at the left were growing in a sugar solution. Yeasts exist as single cells, as cell clusters, and as short chains of cells. They usually reproduce by budding, a cell division process that produces a small new cell (the bud) from the larger parent cell. The yeast cells here belong to the species *Saccharomyces cerevisiae*, which has been used to make bread, beer, and wine throughout recorded history. You may recognize part of the name from our earlier discussion of carbohydrates; *Saccharomyces* means "sugar fungus" in Greek.

Saccharomyces cerevisiae and other yeasts normally live in **aerobic** (oxygen-containing) environments, but they can also live in **anaerobic** (oxygen-lacking) environments. When yeast is first added to bread dough or a wine vat, the cells are in an aerobic environment. As they grow, the cells consume oxygen (O_2) while obtaining energy from sugars in the dough or grape juice. They give off carbon dioxide (CO_2) gas as a waste product. For the baker, the CO_2 bubbles raise the dough; the yeast cells are killed by the heat when the bread is baked.

Yeast functions somewhat differently in winemaking, which requires anaerobic conditions. The yeast cells still give off CO_2 but also produce ethyl alcohol as a waste product. When the alcohol content reaches about 12–16%, the cells die, poisoned by their own wastes.

In being able to thrive in both aerobic and anaerobic environments, yeast cells are metabolically versatile. However, they can harvest much more energy from food molecules by using oxygen than they can without it.

Like yeast cells, our own muscle cells thrive when they have plenty of oxygen. But unlike yeasts, our muscle cells function efficiently for only a brief time without oxygen, even if sugars are available. When we exercise moderately—by jogging, for instance— our muscles use O_2 in harvesting energy from sugar molecules the same way yeasts do in an aerobic environment. The jogging is "aerobic exercise," meaning that the workout is slow enough that the blood can deliver O_2 to the muscle cells at least as fast as they use it. In contrast, when we exercise very strenuously, our muscle cells soon use up their O_2 supply and become anaerobic. They continue to function for a while, but, like yeasts, they produce a toxic waste. In muscles, the waste is lactic acid, which makes the muscles ache and fatigue quickly.

We begin this chapter on how cells harvest energy from food with a look at how our own cells obtain O_2 and dispose of CO_2.

Aerobic exercise

6.1 Breathing supplies oxygen to our cells and removes carbon dioxide

We often use the word "respiration" as a synonym for breathing, the meaning of its Latin root. In this sense, respiration refers to an exchange of gases: An organism obtains O_2 from its environment and gives off CO_2. Biologists also define respiration as the aerobic harvesting of energy from food molecules by cells. To distinguish it from breathing, this process is called **cellular respiration.**

Breathing and cellular respiration are closely related. As the gymnast below goes through her routine, her lungs take up O_2 from the air and pass it to her bloodstream. The bloodstream carries the O_2 to her muscle cells. Mitochondria in the muscle cells use the O_2 in cellular respiration, harvesting energy from sugar and other organic molecules the gymnast obtained from food several hours before. The muscle cells then use the energy to contract. Simultaneous contraction of many thousands of muscle cells, precisely controlled by the nervous system, make the gymnast's body move. Her bloodstream and lungs also perform the vital function of disposing of the CO_2 waste produced by cellular respiration in the muscle cells.

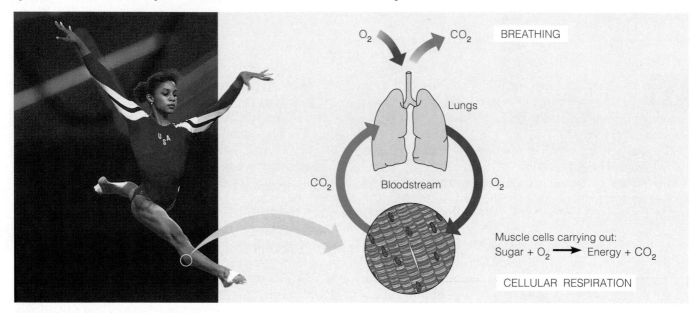

The connection between breathing and cellular respiration

6.2 Cellular respiration banks energy in ATP molecules

As the gymnast example indicates, oxygen usage is only a means to an end. Harvesting energy is the fundamental function of cellular respiration. The chemical equation below summarizes cellular respiration as carried out by muscle cells, yeasts, and all other cells that use O_2 to harvest energy from the sugar glucose. Throughout this chapter, we use glucose as a convenient representative food molecule, although cells also "burn" many other organic molecules in cellular respiration. In brief, when a molecule of glucose interacts chemically with O_2, the glucose splits apart, and its atoms are regrouped into CO_2 and H_2O. In the process, the cell extracts some of glucose's chemical-bond energy and stores (or "banks") it in the chemical bonds of ATP.

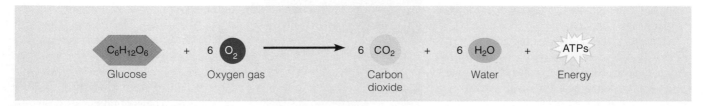

A. Summary equation for cellular respiration

Glucose contains a lot of chemical energy. For instance, 10 grams (roughly a tablespoon) contain about 40 kilocalories (kcal) of energy that can be made available for work. An adult human might use about this much energy by exercising vigorously for 15 minutes. A cell deals in tiny parcels of energy. Each ATP molecule it makes contains only about 1% of the amount of chemical energy in a single glucose molecule.

Does this 1% figure mean that the cell makes 100 molecules of ATP from breaking down one molecule of glucose? Unfortunately not. Cellular respiration is not able to harvest all of the energy of glucose in usable form. When glu-cose is burned in a chemistry lab, 100% of its energy is released (Figure B, left). By contrast, a typical cell banks only about 40% of glucose's energy in ATP molecules (Figure B, center). Most of the rest is converted to heat. This may seem inefficient, but it compares very well with the efficiency of most energy-conversion systems. The average automobile engine, for instance, is able to convert only about 25% of the energy in gasoline into kinetic energy of movement. Cellular respiration is also much more efficient than any process a cell can perform without O_2. For example, a yeast cell in an anaerobic environment harvests only about 2% of the energy in glucose.

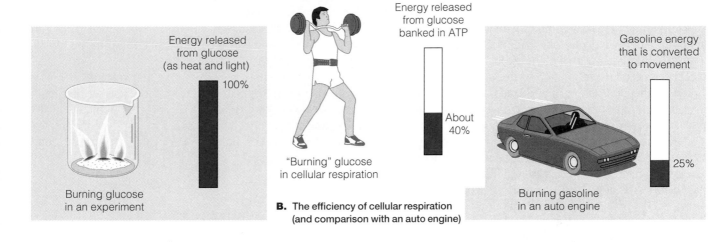

B. The efficiency of cellular respiration (and comparison with an auto engine)

The human body uses energy from ATP for all its activities | 6.3

Our body cells require a continuous supply of energy just to stay alive—to keep the heart pumping blood, to breathe, to maintain body temperature, and to digest food. These and other life-sustaining activities typically use as much as 75% of the energy a person takes in during a typical day. At any particular time, whether you are sleeping or active, most of your cells are busy with cellular respiration, producing ATP just to maintain the body.

Above and beyond the energy we need for body maintenance, the ATP made during cellular respiration provides energy for voluntary activities. The table shows the amounts of energy it takes to perform some of these activities. The energy units are kilocalories (kcal), commonly referred to simply as calories. The values shown do not include the energy the body consumes for its life-sustaining activities. Thus, sleeping or lying still does not consume any energy except the energy used in maintenance.

The U.S. National Academy of Sciences estimates that the average adult human needs to take in food that provides about 2200 kcal of energy per day. This is a very general estimate of the total amount of energy a person of average weight expends, or "burns," in both maintenance and voluntary activity.

Energy Consumed by Various Activities (in kcal)	
Activity	Kcal consumed per hour by a 67.5-kg (150-lb) person*
Bicycling (racing)	514
Bicycling (slowly)	170
Dancing (slow)	202
Dancing (fast)	599
Driving a car	61
Eating	28
Gymnastics	186
Laboratory work	73
Piano playing	73
Running (7 min/mi)	865
Sitting (writing)	28
Sitting (playing chess)	30
Sleeping or lying still	0
Standing (relaxed)	32
Swimming (2 mph)	535
Walking (3 mph)	158
Walking (4 mph)	231

* Not including kcal necessary for life maintenance

6.4 Cellular respiration harvests energy by rearranging electrons in chemical bonds

Let's now begin examining the process that enables our cells to harvest energy from a glucose molecule. In the next nine modules, we study the main events of cellular respiration, the chemical process that changes glucose and oxygen into carbon dioxide and water and allows the cell to harvest some of glucose's energy and store it in ATP.

As we discussed in Chapter 5, there is a common theme to the way cells carry out their energy maneuvers. A cell transfers energy from one molecule to another by coupling an exergonic (energy-releasing) chemical reaction to an endergonic (energy-storing) one. The energy released from a molecule like glucose is the energy that was stored in the specific arrangement of the molecule's covalent bonds.

In a nutshell, cellular respiration taps the energy of glucose by rearranging electrons in chemical bonds. It dismantles glucose in a series of steps and shuttles electrons removed from glucose through a sequence of exergonic reactions. At each step, electrons move from a chemical bond in a molecule where they have more energy to a bond where they have less energy. As the exergonic reactions occur, energy is released in small amounts, and the cell stores some of the energy in ATP.

In the cellular respiration equation here, you cannot see any electron transfers. What you can see are changes in hydrogen atom distribution. Glucose loses hydrogen atoms as it is converted to carbon dioxide; simultaneously, molecular oxygen (O_2) gains hydrogen atoms in being converted to water. These hydrogen movements represent electron transfers because, as we saw in Chapter 2, each hydrogen atom consists of an electron and a proton. As we will see later, the O_2, which we talked about so much in the previous modules, serves as the ultimate recipient of electrons in cellular respiration. O_2 can play this role because its atoms are very electronegative and tend to pull electrons away from organic molecules. The electrons stripped from glucose during the harvesting of energy end up in the hydrogen atoms in H_2O.

It is important to recognize that the reaction below is only a summary of cellular respiration. Respiration breaks glucose down gradually, and the process depicted in the equation actually occurs in a series of steps, which involve other chemical compounds. We will get to those compounds in a few pages.

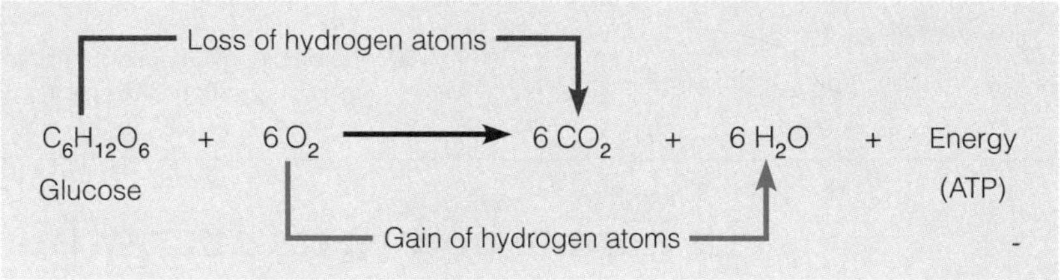

Loss of hydrogen atoms

$$C_6H_{12}O_6 \ + \ 6\,O_2 \longrightarrow 6\,CO_2 \ + \ 6\,H_2O \ + \ \text{Energy (ATP)}$$
Glucose

Gain of hydrogen atoms

6.5 Hydrogen carriers shuttle electrons in redox reactions

What actually happens in cellular respiration when electrons are shuttled from molecule to molecule? Electron shuttling involves a type of chemical process called oxidation-reduction, or **redox** for short. During redox, electrons are lost from one substance (a process called **oxidation**) and added to another substance (**reduction**). A molecule is said to be oxidized when it loses one or more electrons and reduced when it gains one or more electrons. Because an electron transfer requires both a donor and an acceptor, oxidation and reduction always go together. An electron leaves one molecule only when it contacts another molecule that attracts it more strongly. You can see the overall results of redox in Figure 6.4 above: Glucose loses electrons (in H atoms), while O_2 gains electrons (in H atoms).

The top equation in the figure here depicts one of the organic molecules involved in cellular respiration being oxidized. We show only its four carbon atoms (gray balls) and a few of its other atoms. Two key players in the process are an enzyme called **dehydrogenase** and a coenzyme called NAD$^+$. **NAD$^+$** (nicotinamide adenine dinucleotide) is an organic molecule that the cell itself makes and uses to shuttle electrons in redox reactions. With the help of a de-

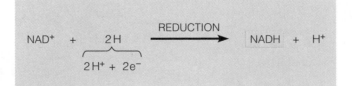

A pair of redox reactions

hydrogenase, NAD$^+$ is able to remove electrons from food molecules. The enzyme and coenzyme oxidize our example compound by stripping electrons (actually two H atoms) from it.

Meanwhile, as shown in the lower equation, NAD$^+$ is reduced (picks up the electrons). When the dehydrogenase strips two hydrogen atoms from an organic molecule, it is actually removing two protons (written here as 2H$^+$) and

two electrons (2e^-). NAD$^+$ picks up the two electrons and one of the protons and becomes NADH. The other proton (H$^+$) is released into the surrounding solution in the cell.

The electrons added to NAD$^+$ in making NADH are the important ones. They carry energy that the cell has harvested and can eventually use. In the next module, we see how the cell processes the energy in NADH in preparation for making ATP.

The cell banks energy from NADH during an electron cascade

Glucose is like an electron bank; it is a rich source of electrons for the energy-yielding redox reactions in cells. NAD$^+$ and dehydrogenase enzymes form an efficient system for extracting these electrons. The NADH molecules that result are loaded with energy, though each contains much less energy than glucose. NADH conveys H atoms containing electrons from glucose to other molecules in the cell.

Figure A shows NADH delivering its electron load to an **electron carrier** molecule, represented by the blue sphere at the upper left. In doing so, NADH participates in a redox reaction; it gives up electrons and becomes NAD$^+$ (and is recycled), while an electron carrier gains the electrons. This redox reaction starts an electron cascade, a slide of electrons down an energy "hill" through a series of electron carriers. Each of the carriers is a different molecule, most of them proteins. The first one is oxidized as the next one is reduced, and so on, down to the last molecule, which is O$_2$,

at the bottom of the hill. What keeps the electrons moving is that each carrier molecule has a greater affinity for electrons than its "uphill" neighbor. In a sense, O$_2$, with the greatest affinity for electrons of all the carriers, pulls electrons down the energy hill. All along the way, the redox steps in the cascade allow energy to be released in small enough parcels to be used by the cell. Figure B shows what happens if oxygen is reduced all in one step, by reacting directly with hydrogen. An explosion releases all the energy as heat and light, which a cell could not use.

Figure A is only a simple model, but series of electron carrier molecules do exist in the cell. Called **electron transport chains,** these ordered groups of molecules are embedded in membranes in the eukaryotic cell's mitochondria (in bacteria, they are in the plasma membrane). As electrons pass along the chains, they gradually lose energy, and the cell banks some of the energy as ATP.

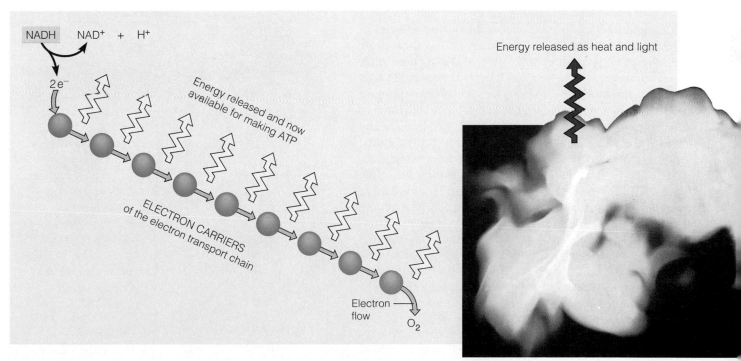

A. Cascading electrons release energy in small increments and finally reduce O$_2$

B. O$_2$ reduced in one explosive step

Two mechanisms generate ATP

Virtually every cell in every organism relies on energy from ATP molecules. Every movement we make, every thought or memory we have, and every molecule our cells manufacture depend directly on ATP energy. Recall from Module 5.4 that cells generate ATP by phosphorylation, that is, by adding a phosphate group to ADP. A cell has two ways to do this: chemiosmotic phosphorylation, also called chemiosmosis, and substrate-level phosphorylation.

In 1978, British biochemist Peter Mitchell was awarded the Nobel Prize for developing the theory of **chemiosmosis**. Mitchell's theory describes how cells use the potential energy in concentration gradients to make ATP. A concentration gradient of a solute stores energy due to the tendency of the solute molecules to diffuse from where they are more concentrated to where they are less concentrated. The theory of chemiosmosis centers on membranes, and in particular on the activity of **ATP synthases**, protein complexes (clusters) that reside in membranes. ATP synthases synthesize ATP using the energy stored in concentration gradients of H^+ ions (that is, protons) across membranes. Cells generate most of their ATP in this way.

Let's look at an overview of the relationship between membrane structure and chemiosmotic ATP synthesis. As shown in Figure A, ATP synthase is built into the same membrane as the molecules of an electron transport chain. This structural connection allows the energy that NADH delivers to the electron transport chain to drive the production of ATP by the ATP synthase. Not shown in the figure are the details of the electron transport chain that make this energy transfer possible. Within the chain, redox reactions release energy from electrons cascading down the series of electron carriers mentioned in Module 6.6. As these exergonic reactions release energy, some of the proteins built into the chain use the energy to actively transport H^+ ions across the membrane. This flow results in a concentration gradient of H^+ ions across the membrane (notice the higher concentration of H^+ on the top side of membrane). The ATP synthase then uses the potential energy in the concentration gradient to drive the endergonic (energy-storing) reaction that generates ATP from ADP and phosphate ($\textcircled{P}$). We will discuss the details of this process in Module 6.12.

Substrate-level phosphorylation (Figure B) is much simpler than chemiosmosis and does not involve a membrane. In substrate-level phosphorylation, an enzyme transfers a phosphate group from an organic substrate molecule to ADP. The substrate is one of several substances produced as cellular respiration converts glucose to CO_2. The reaction occurs because the bond holding the phosphate group in the substrate molecule is less stable than the new bond holding it in ATP. The reaction products are a new organic molecule and a molecule of ATP. Substrate-level phosphorylation accounts for only a small percentage of the ATP that a cell generates.

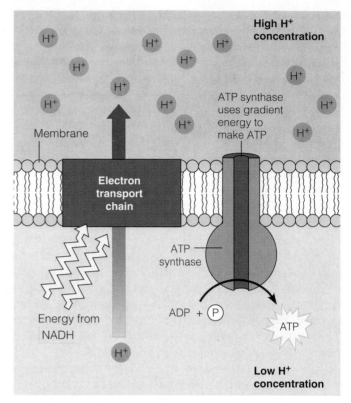

A. Chemiosmosis

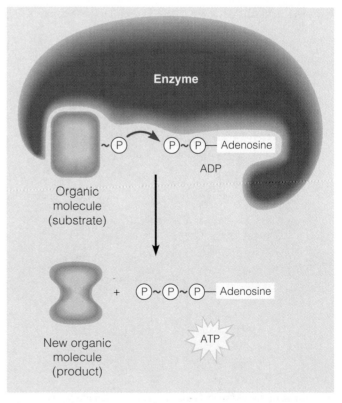

B. Substrate-level phosphorylation

Respiration occurs in three main stages

We have now completed an overview of cellular respiration, starting with energy stored in glucose and ending with ATP. The harvesting of energy is a continuous process, but for study purposes we can look at it in three main stages. The figure below summarizes the stages of cellular respiration and shows where they occur in the cell.

The first two stages, glycolysis and the Krebs cycle, are exergonic (energy-releasing) processes that break down glucose and other organic fuels. **Glycolysis** (light blue throughout the chapter) occurs in the cytoplasmic fluid of the cell, that is, outside the organelles. Glycolysis begins respiration by breaking glucose into two molecules of a compound called pyruvic acid.

The **Krebs cycle** (salmon color), which takes place within the mitochondria, completes the breakdown of glucose by decomposing a derivative of pyruvic acid to carbon dioxide. As suggested by the smaller yellow symbols in the diagram, the cell makes a small amount of ATP (by substrate-level phosphorylation) during glycolysis and the Krebs cycle. The main function of glycolysis and the Krebs cycle, however, is to supply the third stage of respiration with electrons (gold arrows). The third stage is the electron transport chain (purple). As we have seen, the electron transport chain obtains electrons from the hydrogen carrier NADH, the reduced form of NAD$^+$. A related hydrogen carrier called FAD (flavin adenine dinucleotide) also shuttles some electrons from the Krebs cycle to the electron transport chain. The reduced form of FAD is FADH$_2$.

Thus, glycolysis and the Krebs cycle are energy-releasing stages in cellular respiration. Both extract electrons from food molecules while breaking these molecules down to CO$_2$. NAD$^+$ and FAD temporarily capture the electrons and relay them to the top of the electron transport chain. The chain then uses the downhill flow of electrons (from NADH and FADH$_2$ to O$_2$) to pump H$^+$ ions across a membrane. This process stores energy that ATP synthase uses to make most of the cell's ATP by chemiosmosis. We examine the three stages of metabolism in detail in Modules 6.9–6.12.

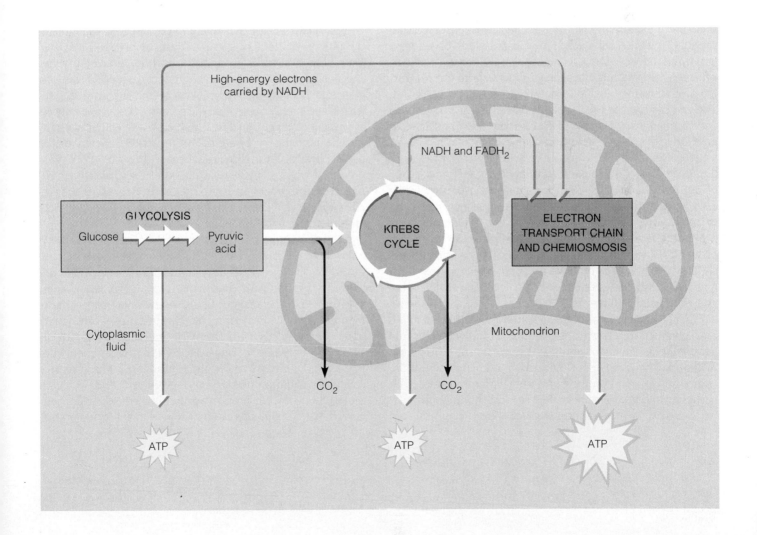

6.9 Glycolysis oxidizes glucose to pyruvic acid

Now that we have introduced the major players and processes in cellular respiration, it is time to focus on the individual stages in the breakdown of a fuel molecule. The term for the first stage, glycolysis, means "splitting of sugar," and that's exactly what happens during this phase of respiration.

Figure A gives an overview of glycolysis in terms of input and output. Glycolysis begins with a single molecule of glucose and concludes with two molecules of another organic compound, pyruvic acid. The gray balls represent the carbon atoms in each molecule; glucose has six, and these same six end up in the two pyruvic acid molecules (three in each). The straight arrow from glucose to pyruvic acid represents nine chemical steps. During these steps, a number of organic compounds form, as enzymes catalyze the rearrangement of chemical bonds and the splitting of the carbon skeleton of glucose in half. As these reactions occur, the cell produces two molecules of ATP by substrate-level phosphorylation and reduces two molecules of NAD^+, forming two molecules of NADH. Thus, the energy extracted from glucose during glycolysis is banked in a combination of ATP and NADH. The cell can use the energy in ATP immediately, but for it to use the energy banked in NADH, electrons from NADH must pass down the electron transport chain.

Glycolysis is the universal energy-harvesting process of life. If we looked inside a yeast or bacterial cell, inside one of our own body cells, or inside virtually any other living cell, we would find the metabolic machinery of glycolysis in full swing. Because glycolysis occurs universally, it is thought to be an ancient metabolic system. In fact, what we call glycolysis today may be very similar to the process some of the first cells on Earth used to extract energy from their environment. Let's take a closer look at this venerable and vital metabolic system.

Figure B shows all the organic compounds that form in the nine chemical reactions of glycolysis. Commentary on the left highlights the main features of the reactions. As in Figure A, the gray balls represent the carbon atoms in each of the compounds named on the right. The compounds that form between the initial reactant, glucose, and the final product, pyruvic acid, are called **intermediates**. Each chemical step leads to the next one. For instance, the intermediate glucose 6-phosphate is the product of step 1 and the reactant for step 2. Similarly, fructose 6-phosphate is the product of step 2 and the reactant for step 3, and so on down to pyruvic acid.

In Figure B, you can see exactly what materials are needed for glycolysis and where they enter the pathway. As Figure A indicated, these starting materials include (1) glucose (the fuel), (2) ADP (and inorganic phosphate, or Ⓟ), and (3) the hydrogen-shuttle molecule NAD^+. Notice that ATP is also needed as a starting material; this tells us that the cell must expend some energy to get glycolysis started. Also essential are specific enzymes that catalyze each of the chemical steps; however, to make the figures as simple as possible, we have not included enzymes in the diagrams.

As indicated by the two blue background areas in Figure B, the individual steps of glycolysis can be grouped into two main phases. Steps 1–4, the first phase, are preparatory and *consume* energy. In this phase, ATP energy is used to split one glucose molecule into two small sugars that are primed to release some energy. Steps 5–9, the second phase, *yield an energy payoff* for the cell. Because the glucose was split in two during the preparatory phase, all the reactions shown in the payoff phase occur in duplicate. During these steps, NADH is produced when a sugar molecule is oxidized, and four ATP molecules are generated. Since the preparatory steps use two ATPs, *the net gain to the cell is actually two ATPs for each glucose molecule that enters glycolysis.*

The net gain of two ATPs from glycolysis accounts for only 5% of the energy that a cell can harvest from a glucose molecule. The two NADHs generated during steps 5–9 account for another 16%, but their stored energy is not available for use in the absence of O_2. A few organisms—yeasts in an anaerobic environment, for instance—can satisfy their energy needs with the ATP produced by glycolysis alone. Most organisms, however, have far higher energy demands and cannot live on glycolysis alone. The stages of cellular respiration that follow glycolysis release much more energy. In the next two modules, we see what happens in most organisms after glycolysis forms pyruvic acid.

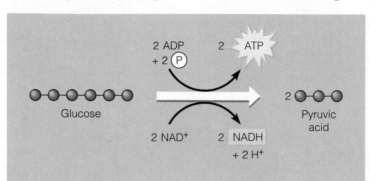

A. An overview of glycolysis

Step 1 – 3 A fuel molecule is energized, using ATP. A sequence of three chemical reactions converts glucose to a molecule of the intermediate fructose 1,6-diphosphate. The coupled arrows indicate the transfer of a phosphate group with high energy content from ATP to another molecule. In these preparatory steps, the cell invests two ATP molecules, one at step 1 and one at step 3, to energize a fuel molecule. In becoming energized, the molecule becomes less stable and thus more reactive.

Step 4 A six-carbon intermediate splits into two three-carbon intermediates. Fructose 1,6-diphosphate is highly reactive and breaks into two three-carbon intermediates. Two molecules of glyceraldehyde 3-phosphate (G3P) emerge from each glucose molecule that enters glycolysis. The two G3P molecules enter step 5, so steps 5–9 occur twice per glucose molecule.

Step 5 A redox reaction generates NADH. The first step in the payoff phase of glycolysis is a redox reaction. The cell harvests its first parcel of energy. The coupled arrows indicate the transfer of hydrogen atoms (containing high-energy electrons) as G3P is oxidized and NAD^+ is reduced to NADH.

Steps 6 – 9 ATP and pyruvic acid are produced. This series of four chemical reactions completes glycolysis, producing two molecules of pyruvic acid for each initial molecule of glucose. During steps 6–9, specific enzymes make four molecules of ATP (per glucose molecule) by substrate-level phosphorylation, and water is produced (at step 8 as a by-product).

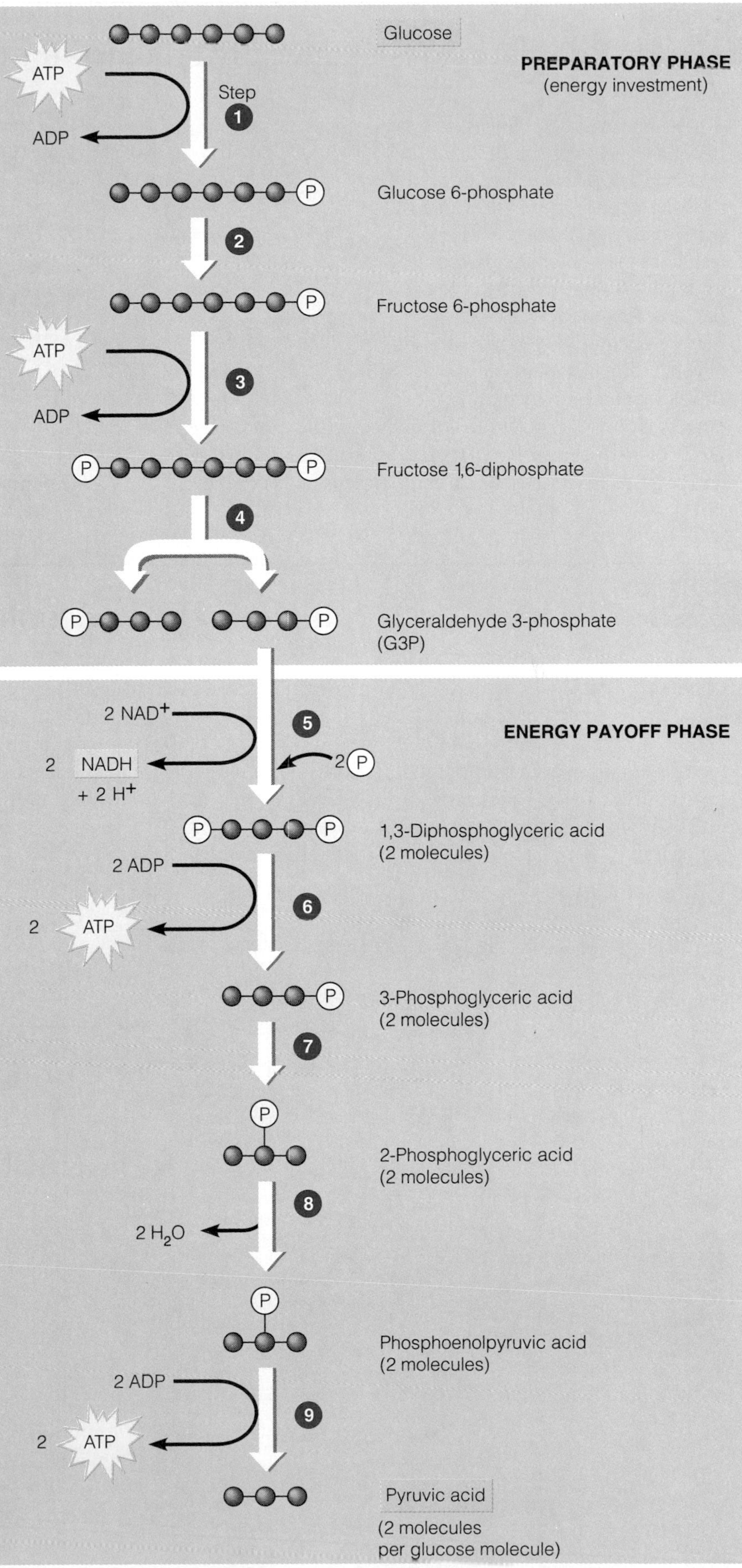

B. Details of glycolysis

Glucose

PREPARATORY PHASE
(energy investment)

ATP

ADP

Step 1

Glucose 6-phosphate

2

Fructose 6-phosphate

ATP

ADP

3

Fructose 1,6-diphosphate

4

Glyceraldehyde 3-phosphate (G3P)

ENERGY PAYOFF PHASE

2 NAD$^+$

5

2 NADH + 2 H$^+$

2 P

1,3-Diphosphoglyceric acid (2 molecules)

2 ADP

2 ATP

6

3-Phosphoglyceric acid (2 molecules)

7

2-Phosphoglyceric acid (2 molecules)

8

2 H$_2$O

Phosphoenolpyruvic acid (2 molecules)

2 ADP

2 ATP

9

Pyruvic acid
(2 molecules per glucose molecule)

6.10 Pyruvic acid is chemically groomed for the Krebs cycle

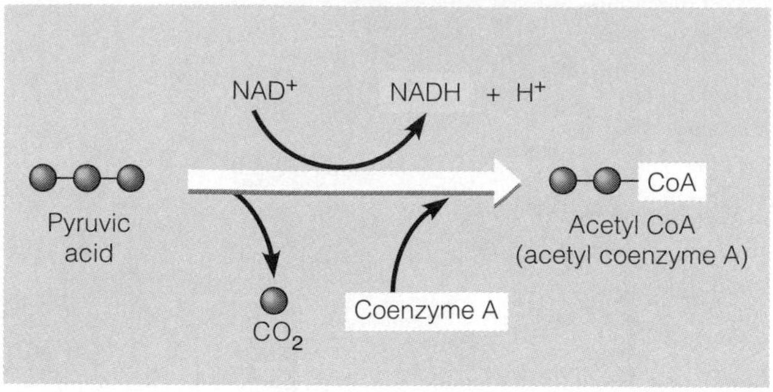

As pyruvic acid forms at the end of glycolysis, it diffuses from the cytoplasmic fluid into the mitochondria, the sites of the Krebs cycle. Pyruvic acid itself does not enter the Krebs cycle. As shown in the figure here, it first undergoes some major chemical "grooming." Several things happen almost simultaneously to each molecule of pyruvic acid: (1) It is oxidized while a molecule of NAD^+ is reduced to NADH; (2) a carbon atom is removed and released in CO_2; and (3) a compound called coenzyme A,

derived from a B vitamin, joins with the two-carbon fragment remaining from pyruvic acid to form a molecule called acetyl coenzyme A.

These grooming steps, amounting to a chemical "haircut and conditioning" of pyruvic acid, set up the second major stage of cellular respiration. Acetyl coenzyme A, abbreviated **acetyl CoA**, is a high-energy fuel molecule for the Krebs cycle. For each molecule of glucose that entered glycolysis, two molecules of acetyl CoA enter the Krebs cycle.

6.11 The Krebs cycle completes the oxidation of organic fuel

The Krebs cycle was named for German-British researcher Hans Krebs, who worked out much of this cyclical phase of cellular respiration in the 1930s. As we did with glycolysis, we present an overview figure first, followed by a more detailed outline of the Krebs cycle.

As shown in Figure A, only the two-carbon acetyl part of the acetyl CoA molecule actually participates in the Krebs

cycle. Coenzyme A helps the acetyl fragment enter the cycle and then splits off and is recycled. Not shown in this figure are the multiple steps that follow, each catalyzed by a specific enzyme in the mitochondrial matrix. The cycle completely disassembles acetyl CoA, stripping away its electrons and casting off two carbon atoms as CO_2 for every acetyl fragment that enters the Krebs cycle.

Compared with glycolysis, the Krebs cycle pays big energy dividends to the cell. Each turn of the cycle makes one ATP molecule by substrate-level phosphorylation (shown at the bottom in Figure A). It also produces four other energy-rich molecules: three NADH molecules and one molecule of the cell's other hydrogen carrier, $FADH_2$. Since the Krebs cycle processes two molecules of acetyl CoA for each initial molecule of glucose, the overall yield per molecule of glucose is: 2 ATP, 6 NADH, and 2 $FADH_2$. This yield is considerably more than the 2 ATP plus 2 NADH produced by glycolysis alone. In fact, the difference between the energy the Krebs cycle releases from organic fuel and the energy output of glycolysis is like the difference between the energy released by a blast furnace and a tiny campfire.

Overall, how many energy-rich molecules has the cell gained by processing one molecule of glucose through glycolysis and the Krebs cycle? The cell has gained a total of 4 ATP, 10 NADH, and 2 $FADH_2$. Still, to be able to use all the energy banked in these molecules, the cell has to have all the energy stored in ATP. But before we look at how that comes about, you may want to examine the inner workings of the Krebs "furnace" in Figure B, at the right.

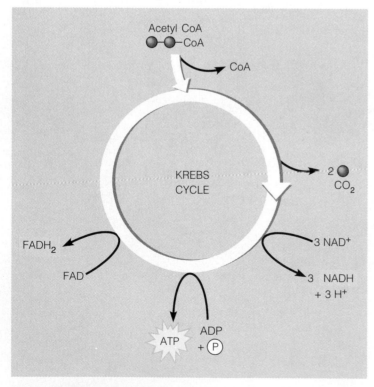

A. An overview of the Krebs cycle

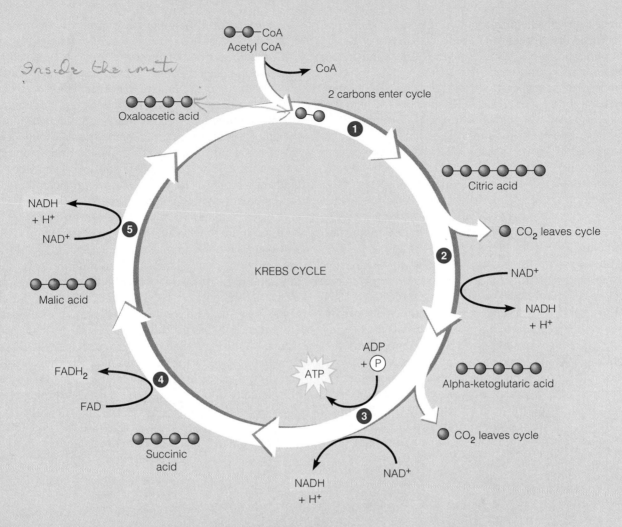

Inside the mito

CoA—Acetyl CoA

CoA

2 carbons enter cycle

Oxaloacetic acid

KREBS CYCLE

Citric acid

CO$_2$ leaves cycle

NADH + H$^+$

NAD$^+$

NAD$^+$

NADH + H$^+$

Malic acid

Alpha-ketoglutaric acid

FADH$_2$

FAD

ADP + (P)

ATP

CO$_2$ leaves cycle

Succinic acid

NADH + H$^+$

NAD$^+$

Step ❶

Acetyl CoA stokes the furnace.

A turn of the Krebs cycle begins (top center) as enzymes strip the CoA portion from acetyl CoA and combine the remaining two-carbon acetyl fragment with oxaloacetic acid (top left) already present in the mitochondrion. The product of this reaction is the six-carbon molecule citric acid, which contains most of the acetyl's fuel energy. Adding acetyl to the cycle is like shoveling coal into a furnace: The Krebs cycle "furnace" runs on the chemical energy that was stored in the acetyl group.

Steps ❷ and ❸

NADH, ATP, and CO$_2$ are generated during redox reactions.

Successive redox reactions harvest some of the energy of the acetyl group by stripping hydrogen atoms from organic acid intermediates (such as alpha-ketoglutaric acid) and producing energy-laden NADH molecules. The redox reactions dispose of two carbon atoms that came from oxaloacetic acid. The carbons are completely oxidized and released as two molecules of CO$_2$. Energy is also harvested by substrate-level phosphorylation of ADP to produce ATP. A four-carbon compound called succinic acid emerges at the end of step 3.

Steps ❹ and ❺

Redox reactions generate FADH$_2$ and NADH.

Enzymes rearrange chemical bonds, eventually completing the cycle by regenerating oxaloacetic acid. Concurrently, the hydrogen (electron) carriers FAD and NAD$^+$ are reduced to FADH$_2$ and NADH, respectively. One turn of the Krebs cycle is completed with the conversion of a molecule of malic acid to one of oxaloacetic acid. This compound is then ready to start the next turn of the cycle by accepting another acetyl group from acetyl CoA.

B. Details of the Krebs cycle

6.12 Chemiosmosis powers most ATP production

The final stage of cellular respiration is the electron transport chain and the synthesis of ATP by chemiosmosis. As we discussed in Module 6.7, chemiosmosis is a clear illustration of structure fitting function: The spatial arrangement of membrane proteins makes it possible for the mitochondrion to use chemical energy to create an H^+ gradient and then use the energy stored in that gradient to drive ATP synthesis.

The figure below expands on our earlier discussion of chemiosmosis. It shows that the electron transport chain is built into the inner membrane of the mitochondrion. The folds (cristae) of this membrane enlarge its surface area, providing space for many electron transport chains and ATP synthase complexes—and thus making it possible for a mitochondrion to produce many ATP molecules simultaneously.

The blue spheres in the diagram represent the electron carrier molecules. Starting on the left, the gold arrow traces the path of electron flow from the shuttle molecule NADH down the electron transport chain to O_2, the final electron acceptor in the chain. Each of the oxygen atoms in O_2 combines with two electrons and with two H^+ ions (from the surrounding solution) to form H_2O, one of the final products of cellular respiration. Most of the carrier molecules reside in three protein complexes (purple spheres), which span the inner membrane of the mitochondrion. Two mobile carriers in the chain (the ones outside the protein complexes)

transport electrons between the complexes. All of the carriers bind and release electrons in redox reactions.

As redox occurs, the protein complexes use energy released from the electrons to actively transport H^+ ions from one side of the membrane to the other. The red vertical arrows indicate that the H^+ ions are transported from the matrix of the mitochondrion (its innermost compartment) into the mitochondrion's intermembrane space. Though the diagram shows only four H^+ ions in the intermembrane space, in fact many of them are stockpiled there. The resulting H^+ gradient—more H^+ ions on one side of the membrane than on the other—stores (potential) energy. With this potential energy, the mitochondrion is poised for the final act of cellular respiration: the production of a large amount of ATP.

If you follow the dashed arrows across the top of the figure, you will arrive at the site of ATP production. The H^+ ions tend to be pushed back across the membrane into the matrix by the energy of the gradient. However, the membrane is not very permeable to H^+, and H^+ ions can only cross back by passing through a special protein port. As shown on the right side of the diagram, ATP synthase, the protein complex that synthesizes ATP, provides that port—an actual channel for the passage of H^+ ions. ATP synthase also contains the enzyme that catalyzes the phosphorylation of ADP to form ATP. As the H^+ ions move through the port, their flow drives the synthesis of ATP. Thus, by means of chemiosmosis, a cell couples the exergonic reactions of electron transport to the endergonic synthesis of ATP.

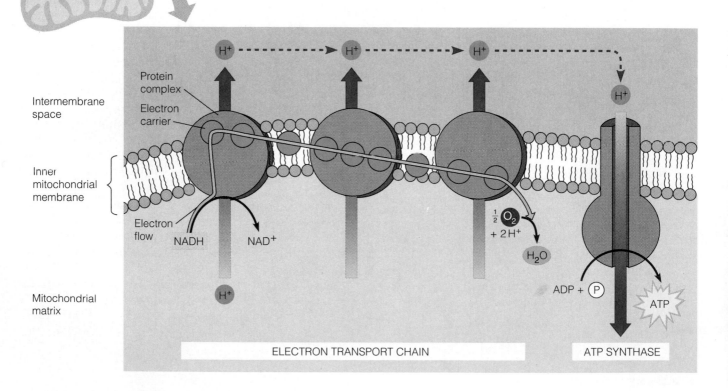

Certain poisons interrupt critical events in cellular respiration

A number of poisons produce their deadly effects by interfering with some of the events we have just discussed. The figure here shows the places where three different categories of poisons obstruct cellular respiration.

Poisons in one category block the electron transport chain. A substance called rotenone, for instance, binds tightly with one of the electron carrier molecules in the first protein complex, preventing electrons from passing to the next carrier molecule. Rotenone is commonly used to kill pest insects and fish. By blocking the electron transport chain near its start and thus preventing ATP synthesis, rotenone literally starves an organism's cells of energy. Two other electron transport blockers, cyanide and carbon monoxide, bind

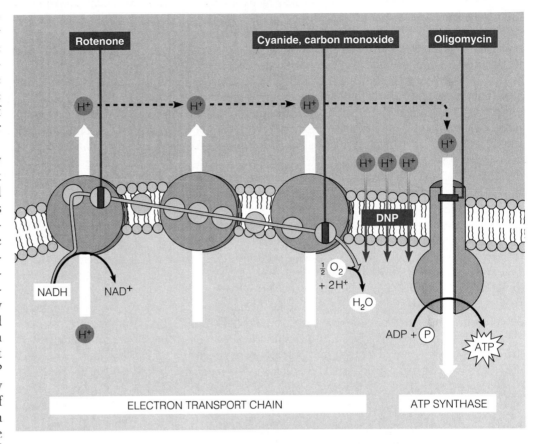

with an electron carrier in the third protein complex. Here they block the passage of electrons to oxygen. This blockage is like turning off a faucet; electrons cease to flow throughout the "pipe." The result is similar to that of rotenone: No H⁺ gradient is generated, and no ATP is made.

A second kind of respiratory poison inhibits ATP synthase. On the right side of the figure, the antibiotic oligomycin blocks the passage of H⁺ ions through the channel in ATP synthase. Oligomycin is used on the skin to combat fungal infections. It kills fungal cells by preventing them from using the potential energy of the H⁺ gradient to make ATP. (Because the drug cannot get into the living skin cells, they are protected from its effects.)

A third kind of poison, collectively called uncouplers, makes the membrane of the mitochondrion leaky to H⁺ ions. Electron transport continues, but ATP cannot be made because leakage of H⁺ ions through the membrane abolishes the H⁺ gradient. Cells continue to consume oxygen, often at a higher than normal rate—but to no avail, for they cannot make any ATP.

One uncoupler, dinitrophenol (DNP), is highly toxic to humans. DNP poisoning produces an enormous increase

in metabolic rate, profuse sweating as the body attempts to dissipate excess heat, collapse, then death. For a short time in the 1940s, some physicians prescribed DNP in low doses as weight-control pills, but fatalities soon made it clear that there were far safer ways to lose weight. The pills caused weight loss by making the body's cells break down *all* fuel molecules in the diet, as well as stored molecules, including fat. When DNP is present, all steps of cellular respiration except chemiosmosis continue to run, consuming fuel molecules, even though almost all the energy is lost as heat.

Poisons do have some good points. Substances that are toxic to certain organisms may be medically beneficial as antibiotics, as is true for oligomycin. In addition, discovering exactly what these toxic substances do to the cell's respiratory machinery has, in many cases, helped biochemists understand how the machinery works. The effects of uncouplers, for example, made it clear that ATP synthesis is a complicated activity involving the distinct, but related, processes of electron transport and generation of a membrane H⁺ gradient.

6.14 Review: Each molecule of glucose yields many molecules of ATP

Now that we have looked at all stages of aerobic respiration, let's review what the cell accomplishes by oxidizing a molecule of glucose. The figure below puts all the stages together and indicates where they occur in the cell. We also include a tally of ATP molecules (yellow), showing the potential energy payoff for a typical working cell. If you wish to refer back to earlier modules, this diagram summarizes the energy payoff stage of glycolysis (steps 4 through 9 in Figure 6.9B), the chemical grooming of pyruvic acid (Figure 6.10), and the Krebs cycle (Figure 6.11B), all of which occur *twice per glucose molecule.*

Starting on the left, glycolysis, occurring in the cytoplasmic fluid, and the Krebs cycle, occurring in the mitochondrial matrix, contribute a total of 4 ATP molecules per glucose molecule by substrate-level phosphorylation. The cell harvests much more energy than this from the carrier molecules NADH and $FADH_2$ (gold), which are produced by glycolysis, the grooming of pyruvic acid, and the Krebs cycle. This energy is used to make numerous molecules of ATP (an estimated 34 in the diagram) when acted on by the electron transport chain and chemiosmosis. The actual number of ATPs made by chemiosmotic phosphorylation varies somwhat from cell to cell and with a cell's functional state.

Let's see where the numbers in the diagram come from. Our model assumes that each NADH that transfers a pair of high-energy electrons from a food molecule to the electron transport chain contributes enough to the mitochondrion's H^+ gradient to generate 3 ATP. (Actually, the ATP yield per NADH varies between 2 and 3.) As indicated in the diagram, the cell also loses about 2 ATP per glucose molecule in transporting the NADH made by glycolysis from the cytoplasmic fluid into the mitochondrion. The model also assumes that each $FADH_2$ molecule yields 2 ATP (the yield varies, with 2 being maximum).

More important than the actual numbers of ATP molecules is the point that a cell can harvest a great deal of energy from glucose—up to about 40% of the molecule's potential energy—when oxygen is available for the electron transport chain and chemiosmosis can proceed.

Because most of the ATP generated by cellular respiration results from chemiosmosis, the figure's estimate of 36 ATP per glucose molecule depends on an adequate supply of oxygen to the cell. Without oxygen to function as the final electron acceptor in the electron transport chain, chemiosmosis ceases, and death may ensue from energy starvation. However, as we discussed in this chapter's introduction, yeasts and certain other cells have a way to survive when they run out of oxygen. We look at that situation next.

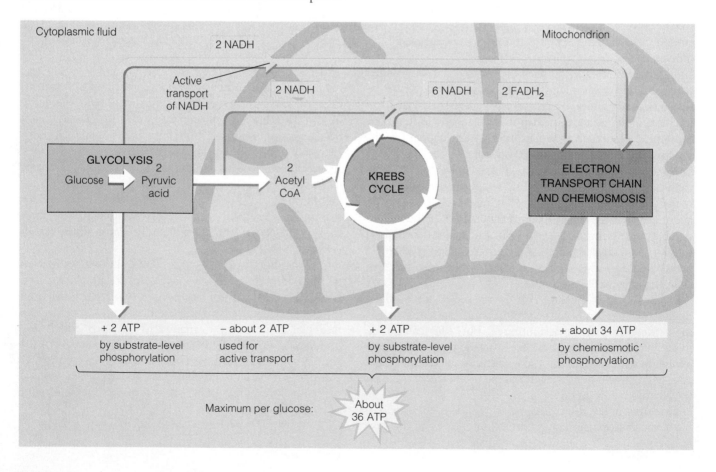

Fermentation is an anaerobic alternative to aerobic respiration

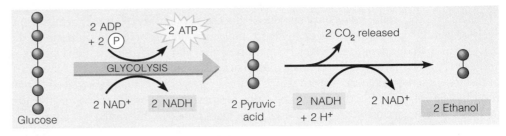

A. Alcoholic fermentation

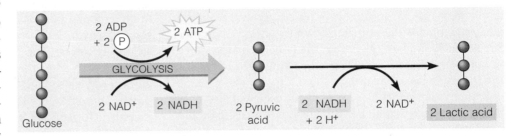

B. Lactic acid fermentation

Yeasts normally use aerobic respiration to process food, harvesting about 36 ATP per glucose molecule the way most organisms do. They are also able to survive without oxygen, on the two molecules of ATP per glucose molecule that come from glycolysis. Remember that glycolysis uses no oxygen; it simply generates a net gain of 2 ATP while converting glucose to two molecules of pyruvic acid and reducing NAD$^+$ to NADH. This is an inefficient way to use fuel—something like running an automobile on one poorly tuned cylinder—but yeasts can thrive on it in anaerobic environments where there is plenty of glucose to keep glycolysis operating.

There is one catch in using glycolysis as the sole means of producing ATP: A cell must have a way of replenishing its supply of NAD$^+$ as this molecule is reduced. Yeasts and certain bacteria do this by fermenting the pyruvic acid produced by glycolysis to ethanol (ethyl alcohol). The process is called **alcoholic fermentation,** and it is catalyzed by specific microbial enzymes.

As indicated in Figure A, during alcoholic fermentation, CO_2 is removed from pyruvic acid and NADH is oxidized, thus recharging the cell with a supply of NAD$^+$ to keep glycolysis working. Notice that ethanol, formed when NADH is oxidized, is a reduced molecule. Thus, unlike the energy-poor H_2O and CO_2 molecules remaining after aerobic respiration, ethanol is energy-rich. However, it is toxic to the organisms that produce it. Yeasts release their alcohol wastes to their surroundings but die if the alcohol becomes too concentrated.

Figure B shows a different type of fermentation, used by many kinds of cells. It is called **lactic acid fermentation** because lactic acid, rather than alcohol, is produced when NADH from glycolysis is oxidized. The ATP yield is the same as in alcoholic fermentation, since glycolysis generates it, but no CO_2 is given off, and lactic acid retains all three carbons from pyruvic acid. Lactic acid fermentation is used in the dairy industry to make cheese and yogurt.

As we mentioned in the chapter's introduction, human muscle cells can also make ATP by lactic acid fermentation when oxygen is scarce, as it often is during strenuous exercise. The accumulation of lactic acid is the cause of muscle fatigue and pain, but eventually the lactic acid is carried in the blood to the liver, where it is converted back to pyruvic acid.

Unlike muscle cells and yeasts, many bacteria that live in stagnant ponds and deep in the soil are **strict anaerobes,**

C. Fermentation vats for wine

meaning they require anaerobic conditions and are poisoned by oxygen. Yeasts and many other bacteria, including one called *E. coli* that thrives in the human intestine, are facultative anaerobes. A **facultative anaerobe** can make ATP either by fermentation or by chemiosmosis, depending on whether O_2 is available.

For a facultative anaerobe, pyruvic acid is a fork in the metabolic road. If oxygen is available, the organism will always use the more efficient process—aerobic respiration. Thus, to make wine and beer, yeasts must be grown anaerobically so that they will ferment sugars and produce ethanol. For this reason, the large fermentation vats in Figure C are equipped with one-way gas valves that vent off CO_2 but keep air out.

6.16 Cells use many organic molecules as fuel for cellular respiration

Throughout this chapter, we have spoken of glucose as the fuel for cellular respiration. But free glucose molecules are not common in our diet. In fact, we obtain most of our calories as fats, proteins, sucrose and other disaccharide sugars, and starch, a polysaccharide. You consume all these types of food molecules when you eat a bag of peanuts, for instance.

This figure illustrates how the cell uses three main kinds of food molecules to make ATP. A cell can funnel a wide range of polysaccharides and sugars into glycolysis, as shown by the pathway on the far left in the diagram. For example, enzymes in our digestive tract hydrolyze starch to glucose, which is then broken down by glycolysis and the Krebs cycle. Similarly, glycogen, the polysaccharide stored in our liver and muscle cells, can be hydrolyzed to glucose to serve as fuel between meals.

Proteins (far right) can be used for fuel, but first they must be digested to their constituent amino acids. Typically, a cell will use most of the amino acids to make its own pro-teins, but enzymes will convert excess amino acids to other organic compounds. During the conversion, the amino groups, which can be toxic, are stripped off and disposed of in urine. The other parts of the amino acid molecules are usually converted to pyruvic acid, acetyl CoA, or one of the organic acids in the Krebs cycle, and their energy is then harvested by cellular respiration.

Fats make excellent cellular fuel because they contain many hydrogen atoms and thus many energy-rich electrons. As the diagram shows (center), the cell first hydrolyzes fats to glycerol and fatty acids. It then converts the glycerol to glyceraldehyde 3-phosphate (G3P), one of the intermediates in glycolysis. The fatty acids are changed into acetyl CoA and then enter the Krebs cycle. Processed this way, a gram of fat yields more than twice as much ATP as a gram of starch. This explains why it is so difficult for a dieter to lose excess fat. To get rid of fat, we have to expend the same large amount of energy we stored in the fat in the first place.

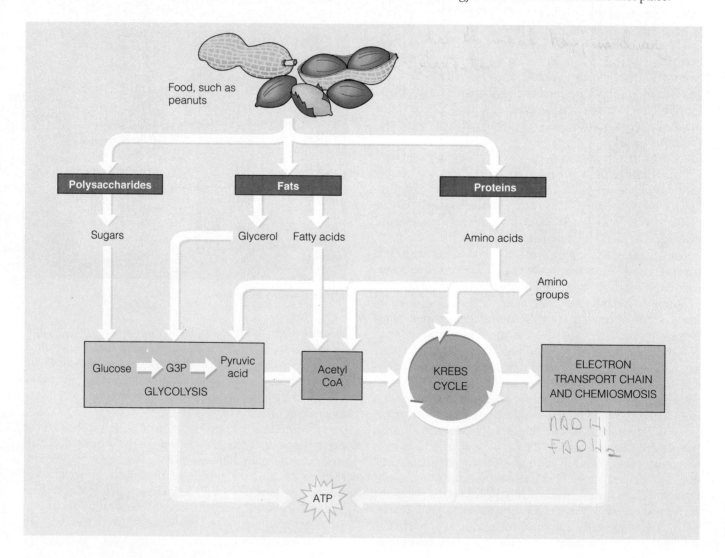

Food molecules provide raw materials for biosynthesis

Because cells need matter as well as energy, not all food molecules are destined to be oxidized as fuel for making ATP. Food also provides the raw materials a cell uses for biosynthesis, to make its own molecules for repair and growth. Cells obtain some raw materials directly from food. Amino acids, for instance, can be incorporated without further change into an organism's own proteins. However, the body also needs certain molecules that are not present in the same form in food. Our cells make these molecules using some of the compounds of glycolysis and the Krebs cycle.

The diagram below outlines biosynthetic pathways by which cells make three classes of macromolecules, using some of the small organic molecules produced in glycolysis

and the Krebs cycle. Notice that these pathways consume ATP energy, rather than generate it. The pathways are almost the reverse of those in Figure 6.16. But there is one important difference. Notice that the label "Glucose synthesis" here replaces "Glycolysis" on Figure 6.16. Although some of the key components are the same in the two processes, the pathway of glucose synthesis from pyruvic acid is *not* the exact opposite of glycolysis.

Thus, we see an important distinction as well as some clear connections between two aspects of metabolism that are central to life: the energy-harvesting process of cellular respiration and the biosynthetic pathways used to construct all parts of the cell.

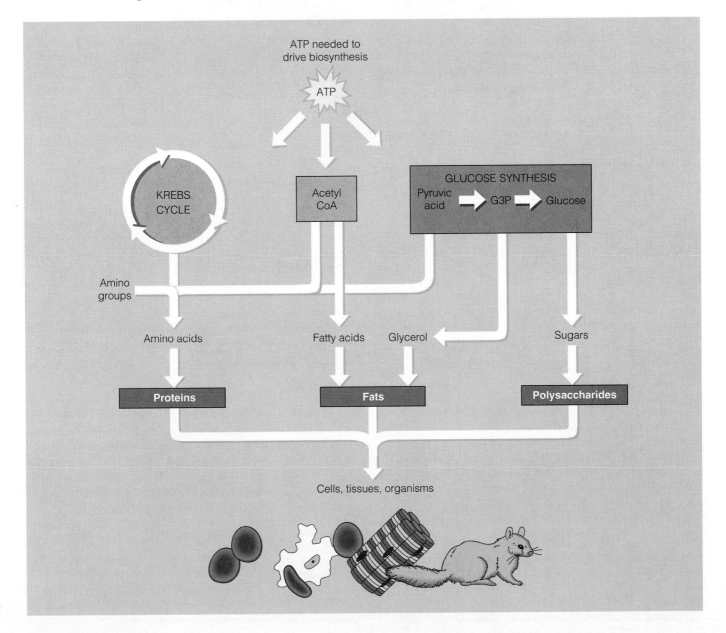

6.18 The fuel for respiration comes from photosynthesis

A giant panda eating its favorite food, a bamboo shoot, is an appropriate ending to this chapter and a good starting point for the next one. Almost entirely a vegetarian, a panda can get all the energy and nutrients it needs by eating a few armloads of bamboo leaves and shoots each day. The panda's digestive tract breaks the plant material down to fuel molecules, and its cells harvest energy from the molecules using cellular respiration.

The cells of all living organisms—those of pandas and bamboo plants included—have the ability to *harvest* energy from the breakdown of organic fuel molecules. When the breakdown process is cellular respiration, the atoms of the starting materials end up in CO_2 and H_2O. In contrast, the ability to *make* organic molecules from CO_2 and H_2O is not universal. Giant pandas—in fact, all animals—lack this ability, but plants have it. Animal cells can only convert the energy in the chemical bonds of organic molecules to other chemical forms, but plant cells can actually produce organic molecules from inorganic ones using the energy of sunlight. This process, photosynthesis, is the subject of Chapter 7.

Chapter Review

Begin your review by rereading the module headings and scanning the figures before proceeding to the Chapter Summary and questions.

Chapter Summary

Introduction–6.7 Every organism requires a continuous supply of energy. In each cell, the process of cellular respiration breaks down food molecules and banks their energy in ATP, which powers all cell and body activities. Cellular respiration harvests energy by oxidizing glucose molecules: Enzymes remove electrons (as part of hydrogen atoms) from glucose (oxidation) and transfer them to a coenzyme, such as NAD^+ (reduction). The NADH formed then delivers its electrons to a series of electron carriers in an electron transport chain. As electrons move from carrier to carrier, their energy is used to pump H^+ ions across a membrane, creating an H^+ gradient. The energy of the gradient is harnessed to make ATP from ADP and phosphate by a process called chemiosmosis. ATP can also be made by transferring phosphate groups from organic molecules to ADP. This process is substrate-level phosphorylation.

6.8–6.10 Cellular respiration oxidizes sugar and produces ATP in three main stages: glycolysis in the cytoplasm, and the Krebs cycle and the electron transport chain in mitochondria. In the first phase of glycolysis, ATP energy is used to split a glucose molecule in two. In the second phase, these parts give up some of their energy, yielding some ATP and giving up electrons and H^+ to NAD^+. Glycolysis converts each original glucose into two molecules of pyruvic acid. Each pyruvic acid molecule then breaks down to form CO_2 and a two-carbon acetyl group, which enters the Krebs cycle.

6.11 The Krebs cycle is a series of reactions in which enzymes strip away the remaining electrons and H^+ from each acetyl group. The carbons are completely oxidized and released as CO_2. Two ATP molecules are produced directly, but most of the glucose energy is captured in NADH and $FADH_2$, which carry electrons and H^+ to the electron transport chain.

6.12–6.13 The electrons from NADH and $FADH_2$ travel down the electron transport chain to oxygen, which combines with the electrons and H^+ to form water. Energy released by the electrons' journey is used to pump H^+ into the space between the membranes of the mitochondrion. In chemiosmosis, the H^+ ions diffuse back across the inner membrane (down their concentration gradient) by passing through ATP synthase complexes, which capture their energy to make ATP. Chemiosmosis produces up to about 36 ATP molecules for every glucose molecule that enters cellular respiration. Various poisons block the movement of electrons, block the flow of H^+ through ATP synthase, or allow H^+ to leak through the membrane. These effects provide clues about the function of the respiratory machinery.

6.14–6.15 Under anaerobic (oxygen-lacking) conditions, many organisms, such as yeasts and bacteria, can use glycolysis alone to produce small amounts of ATP from each glucose molecule. The pyruvic acid that is produced may be converted to other substances, such as alcohol and CO_2 (alcoholic fermentation) or lactic acid (lactic acid fermentation), recycling the NAD^+ needed to keep glycolysis working. Human muscle cells can use lactic acid fermentation to make ATP for short periods, when oxygen is in short supply.

6.16 Molecules other than glucose can fuel cellular respiration. Polysaccharides can be hydrolyzed to monosaccharides and then converted to glucose for glycolysis. Proteins can be digested to amino acids, their amino groups disposed of in urine, and their remains oxidized in the Krebs cycle. Fats, rich in hydrogen, electrons, and energy, are broken up and fed into glycolysis and the Krebs cycle.

6.17–6.18 In addition to energy, cells need raw materials for growth and repair. Some are obtained directly from food. Others are made from intermediates in glycolysis and the Krebs cycle. Biosynthesis consumes ATP. All organisms have the ability to harvest energy from organic molecules. Plants (but not animals) can also make these molecules from inorganic sources, by the process of photosynthesis.

Testing Your Knowledge

Multiple Choice

1. Which of the following processes produces the most ATP molecules per glucose molecule consumed?
   - a. lactic acid fermentation
   - b. Krebs cycle
   - c. electron transport and chemiosmosis
   - d. alcoholic fermentation
   - e. glycolysis

2. When a poison such as cyanide blocks the electron transport chain, glycolysis and the Krebs cycle soon grind to a halt as well. Why do you think they stop? *(Explain your answer.)*
   - a. They run out of ATP.
   - b. The buildup of unused oxygen interferes with glycolysis and the Krebs cycle.
   - c. They run out of NAD^+ and FAD.
   - d. Electrons are no longer available from the electron transport chain.
   - e. They run out of ADP.

3. A biochemist wanted to study how various substances were used and changed in cellular respiration. In one experiment, he allowed a mouse to breathe air containing a particular isotope of oxygen, O_2* (a procedure harmless to the mouse). In the mouse, the "labeled" oxygen atoms, identified by *, first showed up in
   - a. ATP*
   - b. glucose, $C_6H_{12}O_6$*
   - c. NADH*
   - d. carbon dioxide, CO_2*
   - e. water, H_2O*

4. In glycolysis, _____ is oxidized and _____ is reduced.
   - a. NAD^+ . . . glucose
   - b. glucose . . . oxygen
   - c. ATP . . . ADP
   - d. glucose . . . NAD^+
   - e. ADP . . . ATP

5. Which of the following is the immediate source of the energy used to make most of the ATP in your cells?
   - a. the breakdown of ADP
   - b. the transfer of phosphate groups from glucose breakdown products to ADP
   - c. the movement of hydrogen ions through a membrane
   - d. the splitting of glucose into two molecules of pyruvic acid
   - e. the movement of electrons along the electron transport chain

6. Sports physiologists at an Olympic training center wanted to monitor athletes to determine at what point their muscles were functioning anaerobically. They could do this by checking for a buildup of
   - a. ATP
   - b. lactic acid
   - c. carbon dioxide
   - d. ADP
   - e. oxygen

7. Which of the following might be a strict anaerobe?
   - a a bacterial cell
   - b. a yeast cell
   - c. a human muscle cell
   - d. a plant cell
   - e. a human skin cell

True/False *(Change false statements to make them true.)*

1. A molecule is oxidized when it loses an electron.

2. Glycolysis takes place in the mitochondria of eukaryotic cells.

3. FAD and NAD^+ carry electrons from the electron transport chain to the Krebs cycle.

4. If oxygen is in short supply, some human cells are still able to make some ATP.

5. Some ATP is made in the Krebs cycle, without the help of the electron transport chain.

6. The Krebs cycle converts glucose to pyruvic acid.

7. ATP is used to get cellular respiration started.

8. Oxygen's role in cellular respiration is to combine with carbon from glucose to make carbon dioxide.

Describing, Comparing, and Explaining

1. What is the biggest disadvantage of making ATP by means of fermentation? What is the biggest advantage of making ATP this way?

2. Which of the three stages of cellular respiration is considered to be the most ancient? What is the reasoning behind this conclusion?

3. Explain in terms of cellular respiration why we need oxygen and why we exhale carbon dioxide.

Thinking Critically

1. An average adult human requires 2200 kcal of energy per day for maintenance and voluntary activities. Suppose your diet provides an average of 2300 kcal per day. To avoid storing the extra calories as fat and gaining weight, you need additional exercise. How many hours per week would you have to walk to burn off the extra calories? Swim? Run? (See the table in Module 6.3.)

2. As you may know from experience, it is possible for the body to convert excess carbohydrates in the diet (from too many sweets, for example) into fats, resulting in weight gain. It is not possible to convert a carbohydrate or fat alone into protein; some input of protein from the diet is needed to assist in this conversion. What does the dietary protein contribute?

3. Your body makes NAD^+ and FAD from two B vitamins, niacin and riboflavin. You need only tiny amounts of vitamins; the recommended daily allowance for niacin is 20 mg and for riboflavin, 1.7 mg. These amounts are thousands of times less than the amount of glucose your body needs each day to fuel its energy needs. How many NAD^+ and FAD molecules are needed for the breakdown of each glucose molecule? Why do you think your daily requirement for these substances is so small?

Science, Technology, and Society

Research shows that consumption of alcohol by a pregnant woman can cause a complex of birth defects called fetal alcohol syndrome (FAS). Symptoms of FAS include head and facial irregularities, heart defects, mental retardation, and behavioral problems such as hyperactivity. The U.S. Surgeon General's Office recommends that pregnant women abstain from drinking alcohol, and the government has mandated that a warning label be placed on liquor bottles: "Women should not drink alcoholic beverages during pregnancy because of the risk of birth defects." Imagine the dilemma of two young servers in a Seattle restaurant in 1991. An obviously pregnant woman ordered a strawberry daiquiri. The server, unsure about what to do, approached a second server, who showed a warning label to the customer and asked her, "Ma'am, are you sure you want this drink?" The customer became angry, demanded her drink, and went to the restaurant manager, who fired the two servers. What would you have done, if you were in the servers' place? Serve the customer, tell her of the possible harm her drinking alcohol might cause, or refuse to serve her, and risk losing your job? Is a restaurant responsible for monitoring the health habits of its customers?

Photosynthesis: Using Light to Make Food 7

Life on Earth is almost entirely solar-powered, with nearly all organisms depending ultimately on food made by photosynthesis, which uses energy from sunlight. Knowledge about photosynthesis has accumulated rapidly in recent years, although the process was not well defined before this century. The ancient Greeks believed that the soil satisfies all of a plant's needs, and this idea was generally accepted as fact for nearly 2000 years.

The two types of plants in the photographs here played key roles in early research on photosynthesis. The tree at the left is a weeping willow, a native of Northern China that is now widely grown in Europe and the Western Hemisphere. The small plant above is a member of the mint family, a group of fragrant herbs that includes peppermint, spearmint, thyme, basil, oregano, and sage.

The origins of our understanding of photosynthesis extend back to a mid-seventeenth century Belgian physician and part-time chemist named Jan Baptista van Helmont. Curious about whether plants really do get all their nourishment from the soil, van Helmont grew a small willow tree in a pot, adding nothing to the soil except water as the tree grew. Five years later, he found that the tree had gained nearly 75 kilograms while the soil had lost only about 60 grams. Van Helmont concluded correctly that a plant does not gain most of its substance from the soil. However, he concluded incorrectly that his willow tree gained most of its substance from the water he gave it.

A mint plant entered the picture in the early 1770s. English clergyman-chemist Joseph Priestley discovered that, although a candle burned out in a closed container, when he added a living sprig of mint to the container, the candle would continue to burn. At the time, Priestley did not know of O_2, but he correctly concluded that the mint sprig "restored" the air that the burning candle had depleted. Priestley also noted that his result was not always repeatable; in some of his experiments, the plant did not restore the air.

It was left to Dutch physician Jan Ingenhousz later in the decade to discover that plants require light and to explain Priestley's conflicting results. Ingenhousz demonstrated that plants can restore air only when their green parts are exposed to light. Although Ingenhousz did not point it out directly, he understood that Priestley had not provided adequate light for the plants he used in the failed experiments.

Van Helmont, Priestley, and Ingenhousz set the stage for understanding photosynthesis, but it was not until well into the twentieth century that the details of the process were sorted out. The equation below summarizes photosynthesis. It shows that green plants—willows, mints, and all others—need only energy from sunlight, carbon dioxide from the air, and water from the soil to make food molecules. The carbohydrate glucose ($C_6H_{12}O_6$) and O_2 are the products of photosynthesis.

Photosynthesis is thus the process by which plants use light energy to make food molecules from carbon dioxide and water. It is the most important chemical process on Earth, for it provides the food supply for virtually all organisms—plants, animals, protists, fungi, and bacteria alike. We examine this remarkable food-making process in this chapter.

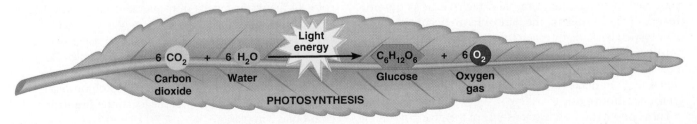

$$6\ CO_2 + 6\ H_2O \xrightarrow{\text{Light energy}} C_6H_{12}O_6 + 6\ O_2$$

Carbon dioxide   Water   Glucose   Oxygen gas

PHOTOSYNTHESIS

Autotrophs are the producers of the biosphere

A. Oak tree

B. Cactus

C. Kelp, a large alga (a photosynthetic protist)

D. Photosynthetic bacteria in a tidal mud flat

Plants are **autotrophs** (meaning "self-feeders" in Greek) in that they sustain themselves without eating other organisms or even organic molecules. The chloroplasts of plant cells capture light energy that has traveled 150 million kilometers from the sun and convert it to chemical energy that is stored in glucose and other organic molecules made from carbon dioxide and water.

Because plants make organic food molecules from inorganic raw materials, they are often referred to as the **producers** of the biosphere, the part of Earth occupied by living organisms. Actually, plants are not the only producers in this sense; certain bacteria and protists also make food molecules from inorganic materials. All organisms that use light energy to make food molecules are said to be photosynthetic autotrophs.

These photographs illustrate some of the diversity among photosynthetic autotrophs. On land, plants, such as the oak tree in Figure A and the cactus in Figure B, are the predominant producers. However, in aquatic environments, algae (photosynthetic protists) and photosynthetic bacteria are the main food producers. Figure C shows part of a large alga (a kelp) that forms extensive underwater and floating "forests" off the coast of California. Figure D is an aerial photograph showing a heavy growth of purple photosynthetic bacteria in a tidal mud flat in Massachusetts. The purple color of the water results from pigment molecules that the bacteria use in photosynthesis.

Plants, algae, and photosynthetic bacteria all use light energy to drive the synthesis of organic molecules from carbon dioxide and water. In plants and algae, this process goes on in the cellular organelles called chloroplasts. The next module is an overview of the locations and structure of chloroplasts in plants.

All green parts of a plant have chloroplasts and can carry out photosynthesis although in most plants, the leaves have the most chloroplasts and are the major sites of the process. The green color in plants is from chlorophyll pigment in the chloroplasts. Chlorophyll absorbs the light energy that the chloroplast puts to work in making food molecules.

The figure below zooms in on a willow leaf to show the actual sites of photosynthesis. The top center drawing is a cross section (slice) of a leaf as it would look with a light microscope. Chloroplasts are concentrated in the cells of the **mesophyll**, the green tissue in the interior of the leaf. Carbon dioxide enters the leaf, and oxygen exits, by way of tiny pores called **stomata** (singular, *stoma*). As shown in the drawing and in the micrograph of a single mesophyll cell, each mesophyll cell has numerous chloroplasts.

The bottom micrograph and drawing show the structures in a chloroplast that constitute the machinery of photosynthesis. Membranes in the chloroplast form the structural framework where many of the reactions of photosynthesis occur, just as mitochondrial membranes do for the energy-harvesting machinery we discussed in Chapter 6. Like the mitochondrion, the chloroplast has an outer membrane and an inner membrane, with an intermembrane space between them. The chloroplast's inner membrane encloses a second compartment, which is filled with **stroma**, a thick fluid. The stroma is where sugars are made from carbon dioxide. In the stroma is suspended an elaborate system of disklike membranous sacs, called **thylakoids**, which contain the third chloroplast compartment. The thylakoids are concentrated in stacks called **grana** (singular, *granum*). Built into the thylakoid membranes are the chlorophyll molecules that capture light energy. The thylakoid membranes also house much of the molecular machinery that converts light energy to chemical energy.

Later in the chapter, we examine the function of these structures in more detail. But first, let's look more closely at the general equation for photosynthesis.

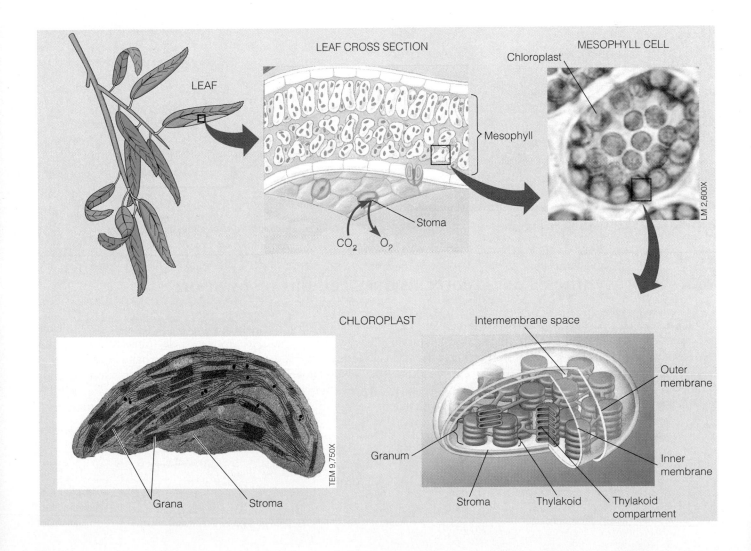

LEAF CROSS SECTION

MESOPHYLL CELL

Chloroplast

LEAF

Mesophyll

Stoma

CO_2 O_2

LM 2,600X

CHLOROPLAST

Intermembrane space

Outer membrane

Granum

Inner membrane

Grana   Stroma

TEM 9,750X

Stroma   Thylakoid   Thylakoid compartment

7.3 Plants produce O_2 gas by splitting water

The leaves of plants that live in lakes and ponds are often covered with bubbles like the ones in Figure A. The bubbles are oxygen (O_2) gas produced during photosynthesis.

In the 1700s, Ingenhousz suggested that plants produce O_2 by extracting it from CO_2. In the 1950s, scientists tested this hypothesis by using an isotope of oxygen, ^{18}O, to follow the fate of oxygen atoms during photosynthesis. (To review isotopes and their use as tracers, see Module 2.5.) In the photosynthesis equations in Figure B, the red type denotes ^{18}O. The photosynthesis equations here (and the photosynthesis and respiration equations in the next module) are written in a slightly more detailed form than you saw earlier. Notice that water is shown as both a reactant and a product.

In Experiment 1, a plant given carbon dioxide containing ^{18}O gave off no labeled (^{18}O-containing) oxygen gas. But in Experiment 2, a plant given water containing ^{18}O did produce labeled O_2 gas. These experiments showed that the O_2 produced during photosynthesis comes from water—not,

as Ingenhousz predicted, from CO_2. It takes two water (H_2O) molecules to make each molecule of O_2.

Knowing where the O_2 comes from gives us a hint of what else happens during photosynthesis. Additional experiments have revealed that the oxygen atoms in CO_2 and the hydrogens in the reactant H_2O molecules end up in the sugar molecule and in water that is formed anew. The carbon in CO_2 ends up in the sugar molecule. Figure C summarizes the fates of all the atoms that start out in the reactant molecules of photosynthesis.

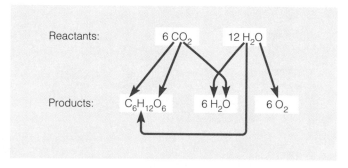

B. Experiments to follow the oxygen atoms in photosynthesis

A. Oxygen bubbles on the leaves of an aquatic plant

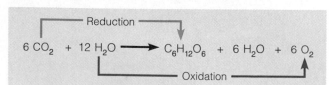

C. Fates of all the atoms in photosynthesis

7.4 Photosynthesis uses redox, as does cellular respiration

What actually happens when photosynthesis converts CO_2 and water into sugars and new water molecules? Photosynthesis is a redox (oxidation–reduction) process, just as cellular respiration is (see Module 6.5). As indicated in the photosynthesis equation below (Figure A), when water molecules are split apart, yielding O_2, they are actually oxidized; that is,

they lose electrons, along with hydrogen ions (H^+). Meanwhile, CO_2 is reduced to sugar as electrons and H^+ ions are added to it. Oxidation and reduction go hand in hand.

Now move from the food-producing equation for photosynthesis to the energy-releasing equation for cellular respiration (Figure B). Overall, cellular respiration har-

A. Photosynthesis (uses light energy)

B. Cellular respiration (releases chemical energy)

vests energy stored in a glucose molecule by oxidizing the sugar and reducing O_2 to H_2O. This process involves a number of energy-releasing redox reactions, with electrons losing potential energy as they travel down an energy hill from sugar to O_2. Along the way, the mitochondrion uses some of the energy to synthesize ATP, as we saw in Chapter 6.

In contrast, the food-producing redox reactions of photosynthesis involve an uphill climb. As water is oxidized and CO_2 is reduced during photosynthesis, electrons gain energy by being boosted up an energy hill. The light energy captured by chlorophyll molecules in the chloroplast provides the boost for the electrons. Photosynthesis converts the light energy to chemical energy and stores it in sugar molecules.

7.5 Photosynthesis occurs in two stages linked by ATP and NADPH

The equation for photosynthesis is a simple summary of a very complex process. Actually, photosynthesis is not a single process, but has two stages, each with multiple steps. The steps of the first stage are known as the **light reactions;** these are the reactions that convert light energy to chemical energy and produce O_2 gas as a waste product. The steps of the second stage are known as the **Calvin cycle;** this is a cyclic series of reactions that assemble sugar molecules using CO_2 and the energy-containing products of the light reactions. The second stage of photosynthesis is named for American biochemist and Nobel laureate Melvin Calvin. In the 1940s, Calvin and his colleagues traced the path of carbon in the cycle, using the radioactive isotope ^{14}C to label the carbon from CO_2. The word "photosynthesis" capsulizes the two stages. *Photo-,* from the Greek word for light, refers to the light reactions; *synthesis,* meaning "putting together," refers to sugar construction by the Calvin cycle.

As indicated in the diagram here, the light reactions of photosynthesis occur in the thylakoid membranes of the chloroplast's grana. Light absorbed by chlorophyll in the thylakoid membranes furnishes the energy that eventually powers the food-making machinery of photosynthesis. Light energy is used to make ATP from ADP and phosphate. It is also used to drive a transfer of electrons from water to $NADP^+$, a hydrogen carrier similar to the NAD^+ that carries hydrogens in cellular respiration. Enzymes reduce $NADP^+$ to NADPH by adding a pair of light-excited electrons along with an H^+. This reaction temporarily stores the energized electrons. As $NADP^+$ is reduced to NADPH, water is split (oxidized), giving off O_2.

In summary, the light reactions of photosynthesis are the steps that absorb solar energy and convert it

into chemical energy stored in ATP and NADPH. Notice that these reactions produce no sugar; sugar is not made until the Calvin cycle, the second stage of photosynthesis.

The Calvin cycle occurs in the stroma of the chloroplast. The incorporation of carbon from CO_2 into organic compounds, shown in the figure as CO_2 entering the Calvin cycle, is called **carbon fixation.** After carbon fixation, enzymes of the cycle make sugars by further reducing the fixed carbon—by adding high-energy electrons to it, along with H^+.

As the figure suggests, it is NADPH produced by the light reactions that provides the high-energy electrons for reduction in the Calvin cycle. And ATP from the light reactions provides chemical energy that powers several of the steps of the Calvin cycle. The Calvin cycle does not require light directly. However, in most plants, the Calvin cycle runs during daytime, when the light reactions power the cycle's sugar assembly line by supplying it with NADPH and ATP.

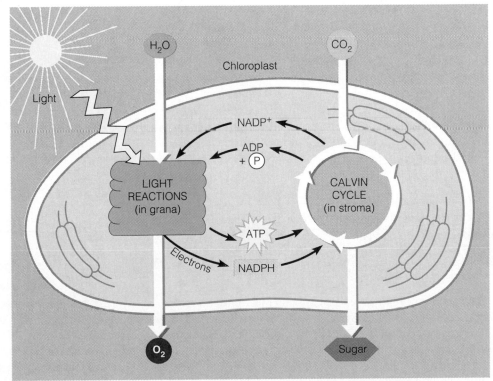

An overview of photosynthesis

7.6 Visible radiation drives the light reactions

What exactly do we mean when we say that photosynthesis is powered by light energy from the sun? Sunlight is a type of energy called radiation, or **electromagnetic energy.** Electromagnetic energy travels in space as rhythmic waves analogous to those made by a pebble dropped in a puddle of water. The distance between the crests of two adjacent waves is called a **wavelength.**

Figure A shows the full range of electromagnetic wavelengths, known as the electromagnetic spectrum. Visible light forms only a small fraction of the spectrum. It consists of different wavelengths (measured in nanometers, or nm) that our eyes see as different colors. As sunlight shines on a plant leaf, the light of some wavelengths is absorbed and put to use in photosynthesis, while the light of other wavelengths is reflected back from the leaf or transmitted through it. The light reactions of photosynthesis use only certain components (wavelengths, or colors) of visible light.

Figure B shows what happens to visible light in the chloroplast. Light-absorbing molecules called pigments in the membranes of a granum absorb mainly blue-violet and red-orange wavelengths. We do not see these absorbed wavelengths. What we see when we look at a leaf are the green wavelengths that the pigments transmit and reflect.

Different pigments absorb light of different wavelengths, and chloroplasts contain several kinds of pigments. One, chlorophyll *a,* absorbs mainly blue-violet and red light. Chlorophyll *a* participates directly in the light reactions. It looks grass-green because it reflects mainly green light. A very similar molecule, chlorophyll *b,* absorbs mainly blue and orange light and reflects (appears) yellow-green. Chloroplasts also contain a family of yellow-orange pigments called carotenoids, which absorb mainly blue-green light. Chlorophyll *b* and the carotenoids do not participate directly in the light reactions, but they do broaden the range of light that a plant can use. These pigments convey the light energy they absorb to chlorophyll *a,* which then puts the energy to work in the light reactions.

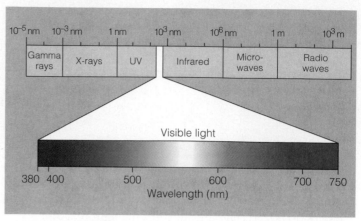

A. The electromagnetic spectrum

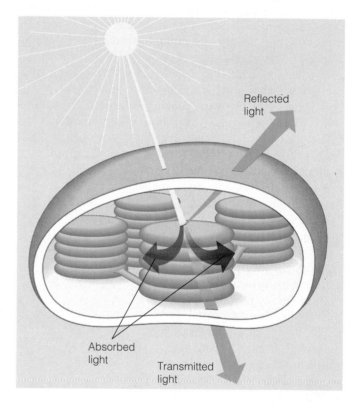

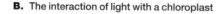

B. The interaction of light with a chloroplast

7.7 Photosystems capture solar power

The theory of light as waves explains most of light's properties relative to photosynthesis. However, light also behaves as discrete packets of energy called photons. A **photon** is a fixed quantity of light energy, and the shorter the wavelength, the greater the energy. For example, a photon of violet light packs nearly twice as much energy as a photon of red light.

What happens when chlorophyll and the other pigments in a chloroplast absorb photons? When a pigment molecule absorbs a photon, one of the pigment's electrons gains potential energy and is said to have been raised from a ground state to an excited state. The excited state is very unstable, and generally the electron falls back to its ground state almost immediately, releasing its excess energy as

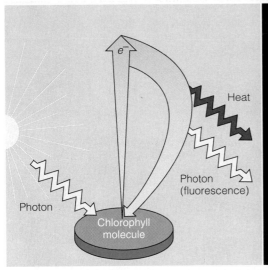
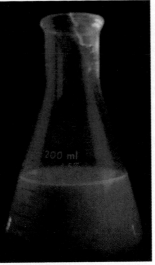

A. Fluorescence of isolated chlorophyll in solution

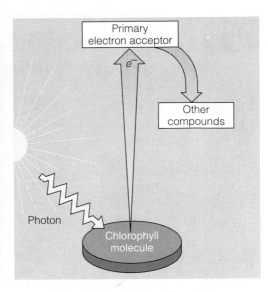

B. Excitation of chlorophyll in a chloroplast

heat. This conversion of light energy to heat is the same process that makes the top of an automobile hot on a sunny day.

Some pigments emit light as well as heat after absorbing photons. In this case, the excited electron gives off a photon, in addition to heat, as it reverts to the ground state. We can demonstrate this phenomenon in the laboratory with a solution of chlorophyll isolated from plant cells, as shown in Figure A (above). When illuminated, the chlorophyll will emit heat and photons of light that produce a reddish afterglow, as electrons fall from an excited state to the ground state. This afterglow is called **fluorescence.**

In contrast to pure chlorophyll in solution, illuminated chlorophyll in an intact chloroplast passes its excited electron to a neighboring molecule (Figure B). The neighboring molecule, called the primary electron acceptor, is reduced as chlorophyll is oxidized.

The solar-powered electron transfer from chlorophyll to the primary electron acceptor is the first step in the light reactions and the first of many redox reactions in photosynthesis. The box labeled "other compounds" in Figure B represents the molecular machinery that uses the later redox reactions to make ATP and NADPH. (We describe these reactions in Module 7.8.)

Chlorophyll *a*, chlorophyll *b*, and the carotenoid pigments are clustered in the thylakoid membrane of each chloroplast in assemblies of 200–300 pigment molecules. Evidence suggests that only a single pair of the many chlorophyll *a* molecules in each assembly actually donates excited electrons to the primary electron acceptor, thus triggering the light reactions. This pair of chlorophyll *a* molecules is called the **reaction center** of the pigment assembly (Figure C). The other pigment molecules function collectively as a light-gathering antenna that absorbs photons and passes the energy from molecule to molecule until it reaches the reaction center. The combination of the antenna molecules, the reaction center, and the primary electron acceptor is called a **photosystem.** This is the light-harvesting unit of the chloroplast's thylakoid membrane.

Two types of photosystems have been identified. They are referred to as photosystem I and photosystem II, in order of their discovery. In photosystem I, the chlorophyll *a* molecule of the reaction center is called P700 because the light it absorbs best is red light with a wavelength of 700 nm. The reaction-center chlorophyll of photosystem II is called P680 because the wavelength of light it absorbs best is 680 nm (a slightly more orange shade of red). These two reaction-center pigments are actually identical chlorophyll *a* molecules, but their association with different proteins in the thylakoid membrane accounts for the slight difference in their light absorption.

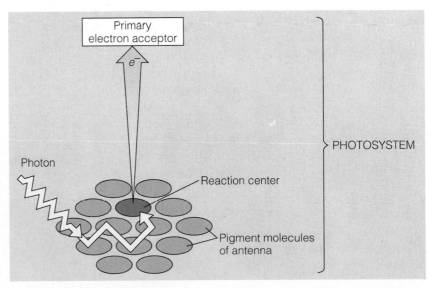

C. The parts of a photosystem

Cyclic electron flow generates ATP using only one photosystem

There are two possible routes for electron flow during the light reactions: cyclic flow and noncyclic flow. **Cyclic electron flow** is the simpler of the two. As shown in the diagram below, it involves only one photosystem and generates only ATP—no NADPH or O_2. It is called cyclic because the high-energy electrons that leave the reaction-center chlorophyll return to the reaction center after passing through an electron transport chain.

This electron transport chain is similar to the one that functions in cellular respiration. It consists of a series of electron-carrier molecules arranged in a membrane. In cellular respiration, the chain is situated in the inner membrane of the mitochondrion. In photosynthesis, it is in the thylakoid membrane of the chloroplast. After the antenna assembly is energized by a photon of light, the first step of the cycle is the transfer of a high-energy electron from the reaction-center chlorophyll to the primary electron acceptor. Next, the primary electron acceptor is oxidized as it donates the excited electron to the first electron carrier of the electron transport chain. The electron is then shuttled from one electron carrier molecule to the next by additional redox reactions. At each link in the chain, the electron loses energy, finally returning to the reaction center in its low-energy, ground state. Some of the energy given up by the electron during the redox reactions is used to generate ATP by chemiosmosis, a process we will review in Module 7.10.

It is likely that cyclic electron flow was important in the evolution of autotrophic organisms. As a relatively simple way to convert sunlight energy to chemical energy, it may have been the earliest type of light-harvesting system to evolve. The first photosynthetic organisms (primitive bacteria) may have used cyclic electron flow to generate ATP.

Today, cyclic electron flow is still important in certain photosynthetic bacteria. It also operates along with noncyclic electron flow in plants. Noncyclic flow, which we examine in Module 7.9, produces roughly equal quantities of ATP and NADPH. However, the Calvin cycle uses more ATP than NADPH. The ATP made by cyclic electron flow may make up the difference.

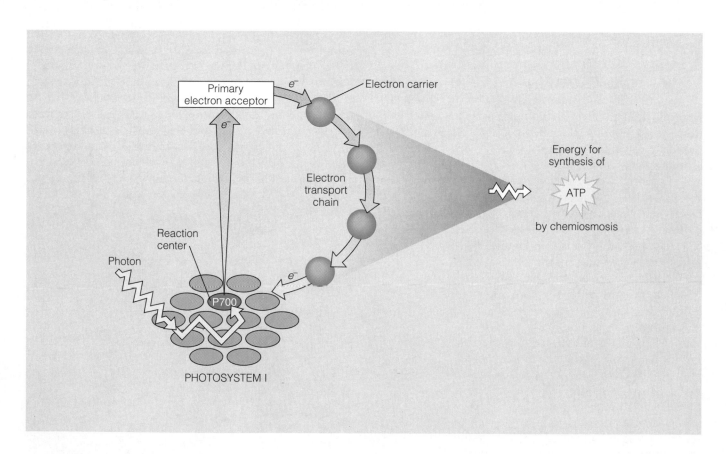

Noncyclic electron flow uses two photosystems and generates ATP, NADPH, and O₂

Noncyclic electron flow occurs in many photosynthetic bacteria and in all plants and algae. In contrast to cyclic electron flow, noncyclic flow uses both photosystem I and photosystem II. In the figure below, you can see electrons (gold arrows) passing from water into photosystem II (lower left), from there by way of an electron transport chain to photosystem I, and then from photosystem I via another electron transport chain to NADP⁺ (upper right). The process is called noncyclic because electrons do not cycle back to the starting point. Since it takes two electrons to reduce NADP⁺ to NADPH, we track two electrons through the diagram.

It is easiest to see what happens in noncyclic electron flow by starting with photosystem I, on the right in the figure. Photons (entering at the lower right) energize the antenna assembly, and high-energy electrons pass from the reaction-center chlorophyll to the primary electron acceptor. From here, the two electrons pass through a short electron transport chain and are temporarily stored as high-energy electrons in NADPH. This leaves P700 with two missing electrons that must be replaced.

It is photosystem II that replaces the electrons lost by P700. Absorption of light drives the transfer of two electrons from P680 to the primary electron acceptor of photosystem II. The electrons then pass along an electron transport chain (the downhill cascade shown in the center of the diagram). During the cascade, the electrons lose energy, some of which is used to make ATP. At the end of the cascade, the electrons reach P700, filling its electron vacancies.

So far, we have seen when noncyclic electron flow generates NADPH and ATP and how it restores electrons to photosystem I. We have two major events yet to go: the splitting of water, which results in the production of O₂, and the restoration of electrons to photosystem II. These events are related. When photosystem II loses the two electrons that go to P700, P680 develops a strong attraction for electrons. As shown on the far left in the diagram, P680 replaces its lost electrons by splitting (oxidizing) a water molecule. When two electrons are removed from H_2O, two hydrogen ions (H^+) and an oxygen atom ($\frac{1}{2}\ O_2$) remain. The H^+ ions remain in the chloroplast. The oxygen atom immediately combines with a second oxygen atom from another water molecule to form a molecule of O_2, which diffuses out of the plant cell and leaves the leaf through a stoma.

The formation of NADPH, ATP, and O₂ by noncyclic electron flow marks the end of the light reactions. We have noted that the high-energy molecule NADPH and the waste product O₂ both result directly from redox reactions. The synthesis of ATP is different. In both noncyclic and cyclic electron flow, ATP synthesis is driven by chemiosmosis—the same mechanism that generates ATP in cellular respiration.

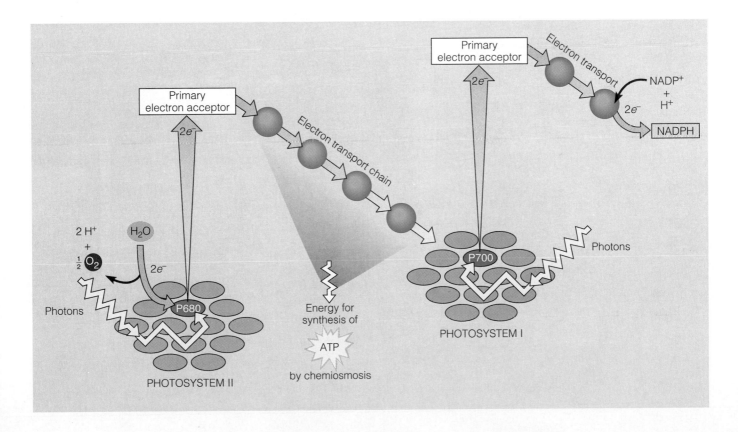

Chemiosmosis powers ATP synthesis in the light reactions

The figure below illustrates, in a highly diagrammatic way, the relationship between chloroplast structure and function in photosynthesis. It shows the two photosystems and electron transport chains of noncyclic electron flow, all located within the thylakoid membrane of a chloroplast. Though the illustration is very schematic, it makes the point that the photosystems are arranged in such a way that energy released during electron flow drives the transport of hydrogen ions (H^+) across the thylakoid membrane. The arrangement is very much like the one in our model for the electron transport chain of cellular respiration in the mitochondrion (see Figure 6.12). In both cases, excited electrons (gold line) pass along a series of electron carriers (blue circles) within a membrane, as redox reactions occur. The electrons give up energy on the way, and some of the energy is used to make ATP by chemiosmosis.

Although photosynthesis is a food-making process and cellular respiration is an energy-harvesting one, electron transport in the chloroplast drives chemiosmosis the same way it does in the mitochondrion. Specifically, some of the electron carriers use energy released from the electrons to actively transport H^+ ions from one side of a membrane to the other. In the chloroplast, the carriers move H^+ across the thylakoid membrane from the stroma into the thylakoid compartment. This generates a concentration gradient of H^+ across the membrane. (The higher H^+ concentration is indicated by the darker shade of gray in the diagram.) As in the mitochondrion, energy stored in this concentration gradient is used to drive ATP synthesis.

The flask-shaped structure on the right in the figure represents the protein complex ATP synthase, which is like the one we saw in the mitochondrion. ATP synthase provides a port through which H^+ can diffuse back into the stroma from the thylakoid compartment. The energy of the H^+ gradient drives H^+ back across, and energy is released in the process. ATP synthase uses some of this energy to phosphorylate ADP, making ATP. In photosynthesis, the chemiosmotic production of ATP is called **photophosphorylation** because the initial energy input is light energy.

Notice that the final electron acceptor is $NADP^+$, not O_2 as in cellular respiration. Rather than being consumed, O_2 is produced, as explained in Module 7.9, when water is split to provide replacement electrons for photosystem II.

We have now examined the major events of the light reactions. The ATP and NADPH that are produced during these reactions are used in the next stage of photosynthesis, the Calvin cycle. Module 7.11 describes how the Calvin cycle makes sugar.

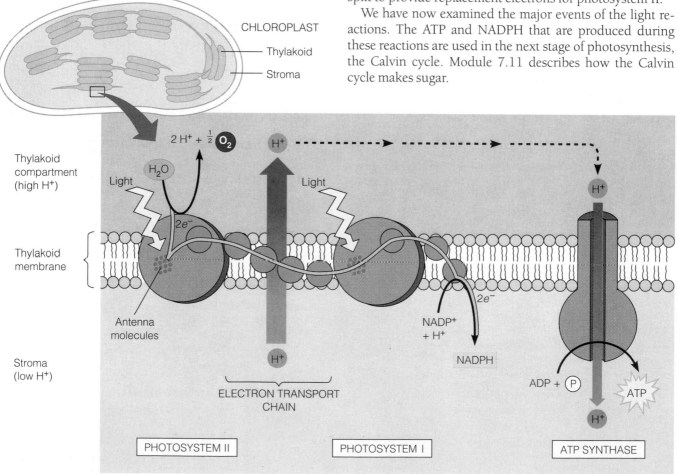

ATP and NADPH power sugar synthesis in the Calvin cycle

The Calvin cycle functions like a sugar factory within a chloroplast. As Figure A shows, inputs to this all-important food-making process are CO_2 (from the air) and ATP and NADPH (both generated by the light reactions). Using carbon from CO_2, energy from ATP, and high-energy electrons from NADPH, the Calvin cycle constructs an energy-rich sugar molecule, glyceraldehyde 3-phosphate (G3P). The

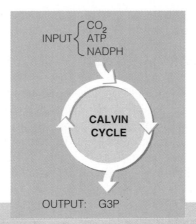

A. Overview of the Calvin cycle

plant cell can use G3P to make glucose or other organic molecules as needed.

Figure B presents the details of the Calvin cycle. To make a molecule of G3P, the cycle must incorporate the carbon atoms (gray balls) from three molecules of CO_2. The cycle actually incorporates one carbon at a time, but we show it starting with three CO_2 molecules so that we end up with a complete G3P molecule.

Step 1 Carbon fixation. An enzyme called rubisco combines three molecules of CO_2 with three molecules of a five-carbon sugar called ribulose bisphosphate (abbreviated RuBP). Six molecules of the three-carbon organic acid 3-phosphoglyceric acid (3-PGA) result.

Step 2 Energy consumption and redox. Two chemical reactions (indicated by the two arrows) consume energy from six molecules of ATP and oxidize six molecules of NADPH. Six molecules of 3-PGA are reduced, producing six molecules of the energy-rich three-carbon G3P.

Step 3 Release of one molecule of G3P. Five of the G3Ps from step 2 remain in the cycle. The single molecule of G3P you see leaving the cycle is the net product of photosythesis. A plant cell uses two G3P molecules to make one molecule of glucose, which has six carbons. Since the Calvin cycle incorporates only one molecule of CO_2—and thus only one carbon—at a time, it takes six complete turns of the cycle to make the two molecules of G3P that go into one glucose molecule.

Step 4 Regeneration of RuBP. A series of chemical reactions uses energy from ATP to rearrange the atoms in the five G3P molecules, forming three RuBP molecules. These can start another turn of the cycle.

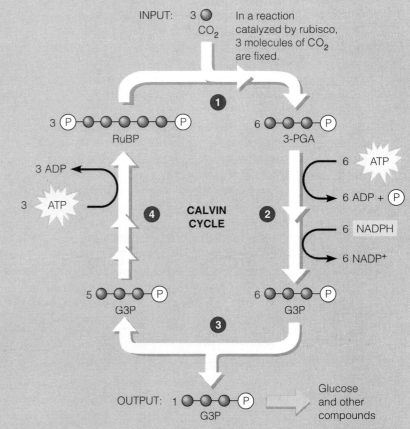

B. Details of the Calvin cycle

The completion of the Calvin cycle, the final phase of photosynthesis, is an appropriate place to reflect on the metabolic ground we have covered in this chapter and in the previous one. In Chapter 6, we saw that virtually all organisms, plants included, use cellular respiration to obtain the energy they need from fuel molecules such as glucose. We followed the chemical pathways of glycolysis and the Krebs cycle, which break glucose down and release energy

from it. We have now come full circle, seeing how plants trap sunlight energy and use it to make glucose from the raw materials carbon dioxide and water.

In tracing glucose synthesis and its breakdown, we have also seen that cells use several of the same mechanisms—electron transport, redox reactions, and chemiosmosis—in energy storage (photosynthesis) and energy harvest (cellular respiration).

7.12 C₄ and CAM plants have special adaptations that save water

Plants are the main food-producing organisms on land, and they thrive in diverse environments, on all continents except Antarctica. One of the reasons for their great success has been the evolution of different ways of fixing CO_2 and saving water during photosynthesis.

Plants in which the Calvin cycle uses CO_2 directly from the air are called **C₃ plants** because the first organic compound produced is the three-carbon compound 3-PGA (step 1 in Figure 7.11B). C₃ plants are common and widely distributed, and some of them, such as soybeans, oats, wheat, and rice, are important in agriculture. One of the problems that farmers face in growing C₃ plants, however, is that dry weather can reduce the rate of photosynthesis and decrease crop productivity. On a hot, dry day, a C₃ plant closes its stomata, the pores in the leaf surface.

Closing stomata is an adaptation that reduces water loss, but it also prevents CO_2 from entering the leaf and O_2 from leaving. As a result, CO_2 levels can get very low in the leaf, while O_2 from the light reactions builds up. When this happens, the first enzyme of the Calvin cycle (called rubisco) incorporates O_2 instead of CO_2 (Figure A), and the Calvin cycle produces a two-carbon (2-C) compound instead of its usual three-carbon product. The plant cell then breaks the two-carbon compound down to CO_2 and H_2O. The entire process, starting with the fixation of O_2, is called **photorespiration.** Unlike photosynthesis, photorespiration yields no sugar molecules. Unlike cellular respiration, it produces no ATP.

In contrast to C₃ plants, so-called **C₄ plants** have special adaptations that save water and also prevent photorespiration. When the weather is hot and dry, a C₄ plant keeps its stomata closed most of the time, thus conserving water. At the same time, it continues making sugars by photosynthesis, using the route shown in Figure B. A C₄ plant has an enzyme that fixes carbon into a four-carbon (4-C) compound instead of into 3-PGA. This enzyme, unlike the carbon-fixing enzyme in C₃ plants, cannot fix O_2. As a result, the C₄ enzyme can continue to fix carbon even when the CO_2 concentration in the leaf is much lower than the O_2 concentration. The four-carbon compound acts as a carbon shuttle; it donates the CO_2 to the Calvin cycle in a nearby cell, which therefore keeps on making sugars even though the plant's stomata are closed most of the time. Corn, sorghum, and sugarcane are examples of agriculturally important C₄ plants. All three evolved in the tropics, and their method of carbon fixation is advantageous in hot, dry climates.

A third mode of carbon fixation and water conservation has evolved in pineapples, many cacti, and most of the so-called succulent plants (those with very juicy tissues), such as ice plants and jade plants. Collectively called **CAM plants,** most such species are adapted to very dry climates. A CAM plant (Figure C) conserves water by opening its stomata and admitting CO_2 only at night. When CO_2 enters the leaves it is fixed into a four-carbon compound, as in C₄ plants. The four-carbon compound in a CAM plant banks CO_2 at night and releases it to the Calvin cycle during the day. This keeps photosynthesis operating during the day, even though the leaf admits no more CO_2. CAM stands for crassulacean acid metabolism, after the plant family Crassulaceae (jade plants and others), in which this important water-saving adaptation was first discovered.

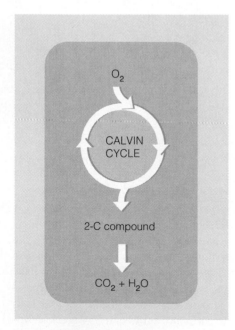

A. Photorespiration in a C₃ plant

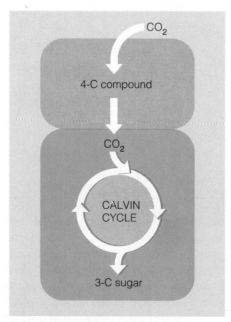

B. Carbon fixation in a C₄ plant

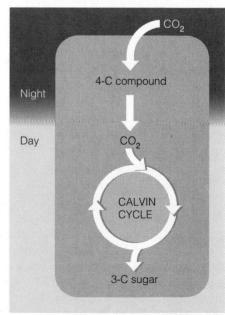

C. Carbon fixation in a CAM plant

Photosynthesis moderates the greenhouse effect; deforestation can intensify it

The trees in the photograph (Figure A) form part of an **old-growth forest** in the Pacific Northwest. Such a forest is an ancient one that has never been seriously disturbed by humans. Many of the trees—mainly Douglas fir here—are thousands of years old. The forest they dominate is one of a few remaining undisturbed areas that contain harvestable timber in the United States.

Old-growth forests are the focus of a long-standing controversy between the timber industry and conservationists. Trees like these contain a lot of marketable lumber, and economic opinion favors harvesting them. On the other side of the controversy, conservationists argue that old-growth forests are home to many species that can survive nowhere else, and that we should save these remnants of our ancient forests for future generations.

The photosynthesis carried out by old-growth forests has direct bearing on the controversy. At center stage is CO_2, the gas that plants use to make sugars in photosynthesis and that all organisms give off as waste from cellular respiration.

Carbon dioxide normally makes up about 0.03% of the air we breathe. This amount of CO_2 in the atmosphere provides plants with plenty of carbon. It also helps moderate world climates, because CO_2 retains heat from the sun that would otherwise radiate from Earth back into space. The CO_2-induced warming is called the **greenhouse effect** because atmospheric CO_2 traps heat and warms the air just as

A. Old-growth forest in the Pacific Northwest

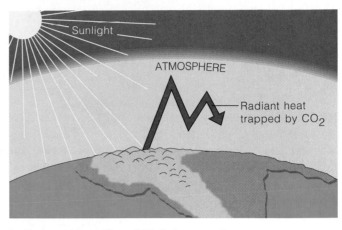

B. The greenhouse effect of CO_2 in the atmosphere

clear glass does in a greenhouse (Figure B). The greenhouse effect keeps the average temperature on Earth some 10°C warmer than it would be otherwise.

Ironically, planet Earth may now be in danger of overheating from the greenhouse effect. The amount of atmospheric CO_2 has been on the rise in the past century, mainly because of worldwide industrialization and increased use of oil, gas, coal, and wood as fuels. When these substances are burned, the carbon in them is released as CO_2. As a result of the steady increase in atmospheric CO_2, global temperatures may be increasing. Photosynthesis consumes CO_2, tending to counteract the greenhouse effect, but the global rate of photosynthesis may decline as we continue to clear huge tracts of forest for farming and urban expansion. Deforestation continues in the tropics, the northwestern United States, Canada, and Siberia.

The greenhouse effect has come up on both sides of the old-growth forest controversy. Those who favor saving the old-growth trees argue that these large photosynthesizers remove large amounts of potentially harmful CO_2 from the atmosphere. Those favoring harvesting argue that replacing the old trees with seedlings would actually increase photosynthesis and reduce atmospheric CO_2.

Relative to their size and weight, young, rapidly growing trees do, in fact, take up CO_2 at a faster rate than old ones. However, when old trees are harvested, much less than half of their bulk becomes lumber. All the roots and many branches are left behind to decompose. Most of the wood itself is turned into paper, sawdust, or fuel—products that usually decompose or are burned within a few years. Decomposition and burning turn the carbon compounds that were in the tree into CO_2, and this puts much more CO_2 into the atmosphere than young trees can take up.

We will return to timber controversies in Module 18.10 and to the greenhouse effect and some of its potentially dire consequences in Module 38.13.

Review: Photosynthesis uses light energy to make food molecules

As we have discussed, most of the living world depends on the food-making machinery of photosynthesis. The figure here summarizes the two stages of this vital process and reviews where they occur in the chloroplast.

Starting on the left in the diagram, you see a summary of the light reactions, which occur in the thylakoid membranes (green). Two photosystems in the membranes capture solar energy, using it to energize electrons. Simultaneously, water is split, and O_2 is released. The photosystems transfer energized electrons to electron transport chains where energy is harvested and used to make the high-energy molecules NADPH and ATP.

The chloroplast's sugar factory is the Calvin cycle, the second stage of photosynthesis. In the stroma, enzymes of the cycle combine CO_2 with RuBP and produce G3P. Sugar molecules made from G3P are a plant's own food supply. Plants use sugars as fuel for cellular respiration and as starting material for making other organic molecules, such as the

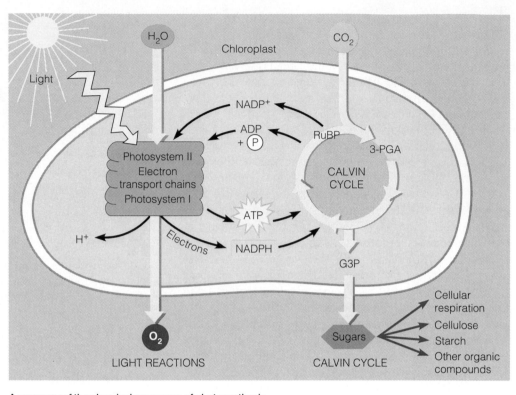

A summary of the chemical processes of photosynthesis

structural molecule cellulose. Most plants make considerably more sugar than they need. They stockpile the excess sugar as starch, storing some in chloroplasts and some in roots, tubers, and fruits. These stored sugars are a major source of food for many animals.

Photosynthesis feeds the world

On a global scale, photosynthesis by plant chloroplasts creates billions of tons of organic matter each year, an output that no other chemical process on Earth can match. And no other process is more important to life than photosynthesis.

Plants (and other photosynthesizers) not only feed themselves, they are also the ultimate source of food for virtually all other organisms. Humans and other animals, for example, make none of their own food and are totally dependent on the organic matter made by photosynthesizers. Even the energy we acquire when we eat meat was originally captured by photosynthesis. The energy in a hamburger, for instance, came from sunlight that was originally converted to a chemical form in chloroplasts in the cells of the grasses eaten by cattle.

*   *   *

Photosynthesis, the process that sustains life at virtually every level, is an appropriate ending to this unit on basic structures and functions common to all life. From atoms and molecules, the elemental ingredients of life, to the various kinds of cells in which cellular respiration and photosynthesis occur, we have seen the importance of structure in determining function. As we explore other topics, we will see the structure–function theme repeated over and over. In the next unit, we study cellular reproduction and inheritance, examining the cellular structures that provide for the continuity of life.

Begin your review by rereading the module headings and Module 7.14 and scanning the figures before proceeding to the Chapter Summary and questions.

Chapter Summary

Introduction Photosynthesis is the process by which autotrophic organisms use light energy to make sugar and oxygen gas from carbon dioxide and water. It is the most important chemical process on Earth. The chemical equation for photosynthesis is:

$$6\,CO_2 + 6\,H_2O + \text{light energy} \longrightarrow C_6H_{12}O_6 + 6\,O_2$$

7.1 Plants, algae, and some bacteria are photosynthetic autotrophs, the ultimate producers of food consumed by virtually all organisms.

7.2 In most plants, photosynthesis occurs primarily in the leaves, in green cellular organelles called chloroplasts. A chloroplast contains a fluid called stroma and structures called grana. Grana are made up of thylakoid membranes containing chlorophyll, the green pigment that captures sunlight.

7.3–7.5 Photosynthesis is a two-stage process, consisting of two linked sets of reactions: the light reactions and the Calvin cycle. In the light reactions, water molecules are split, and electrons and H^+ ions are removed (oxidation), leaving O_2 gas. In the Calvin cycle, these electrons and H^+ are transferred to CO_2 (reduction), producing sugar. Thus, like cellular respiration, photosynthesis is a redox process. But the food-producing reactions of the Calvin cycle consume energy, whereas cellular respiration releases energy.

7.6–7.9 The light reactions of photosynthesis are driven by certain wavelengths of visible light, captured by chlorophyll and other pigments in photosystems in the thylakoid membranes. The process of cyclic electron flow, which may have evolved early in the history of life, generates ATP using a single photosystem. In most photosynthetic organisms, the food-making machinery is powered primarily by non-cyclic electron flow, which involves two photosystems. Light excites electrons in photosystem I. Chlorophyll molecules at the reaction center donate excited electrons to a neighboring molecule. The electrons pass along an electron transport chain and end up as high-energy electrons in NADPH. At the same time, light excites electrons in photosystem II. These electrons travel down an electron transport chain and end up replacing the electrons lost by photosystem I. Photosystem II finally regains its electrons by splitting water, leaving O_2 gas as a by-product.

7.10 The electron transport chains associated with both photosystems are arranged in the thylakoid membranes, much like the electron carriers in the mitochondrion. Electron transport in the chloroplast drives the transport of H^+ through the thylakoid membrane. The flow of H^+ back through the membrane is harnessed by ATP synthase to make ATP. In photosynthesis, this chemiosmotic process is called photophosphorylation. The H^+ ions, along with electrons from the electron transport chain, join with $NADP^+$ to form NADPH.

7.11 The NADPH and ATP produced in the light reactions are used in the next stage of photosynthesis, the Calvin cycle, which occurs in the stroma. This is where carbon fixation takes place and sugar is manufactured. Using carbon from atmospheric CO_2, high-energy electrons and H^+ from NADPH, and energy from ATP, enzymes of the Calvin cycle construct G3P, an energy-rich sugar. The G3P, in turn, is used to build glucose and other organic molecules, such as the structural molecule cellulose. Many plants make more sugar than they need; the excess is stored in roots, tubers, and fruits, and is a major food source for animals.

7.12 Most plants are C_3 plants, which take carbon directly from CO_2 in the air and use it to build a three-carbon organic molecule. In such plants, stomata in the leaf surface usually close when it is hot and dry. Although this mechanism saves water, it can cause a drop in CO_2 and a buildup of O_2 in the leaf, diverting the Calvin cycle to an inefficient process called photorespiration. Some plants have special adaptations that enable them to save water and avoid photorespiration. Special cells in C_4 plants, such as corn and sugarcane, incorporate CO_2 into a four-carbon compound that can donate CO_2 to the Calvin cycle in another kind of cell, compensating for the shortage of CO_2 when the stomata are closed. Pineapples, many cacti, and most succulents—the so-called CAM plants—employ a different mechanism. They open their stomata at night and make a four-carbon compound that is used as a CO_2 source by the same cell during the day.

7.13–7.15 Because of the increased burning of oil, gas, coal, and wood as fuels, atmospheric CO_2 is increasing. CO_2 warms the Earth by trapping heat in the atmosphere; this is called the greenhouse effect. An excess of CO_2 may cause the planet to overheat. Because photosynthesis removes CO_2 from the atmosphere, it could moderate the greenhouse effect; unfortunately, deforestation may cause a decline in the global rate of photosynthesis. Photosynthesis produces billions of tons of organic matter each year—including all our food—and sustains almost all life on the planet.

Testing Your Knowledge

Multiple Choice

1. The process of photosynthesis consumes ———— and produces ————.
   - **a.** chlorophyll . . . H_2O
   - **b.** H_2O . . . CO_2
   - **c.** CO_2 . . . chlorophyll
   - **d.** H_2O . . . O_2
   - **e.** glucose . . . O_2

2. Which of the following are produced by reactions that take place in the thylakoids and consumed by reactions in the stroma?
   - **a.** CO_2 and H_2O
   - **b.** $NADP^+$ and ADP
   - **c.** ATP and NADPH
   - **d.** glucose and O_2
   - **e.** CO_2 and ATP

3. In photosynthesis, ———— is oxidized and ———— is reduced.
   - **a.** glucose . . . oxygen
   - **b.** carbon dioxide . . . water
   - **c.** water . . . carbon dioxide
   - **d.** glucose . . . carbon dioxide
   - **e.** water . . . oxygen

4. Why is it difficult for most plants to carry out photosynthesis in very hot, dry environments such as deserts?
   - **a.** The light is too intense and overpowers pigment molecules.
   - **b.** The closing of stomata keeps CO_2 from entering and O_2 from leaving the plant.
   - **c.** They are forced to rely on cyclic electron flow to make ATP.
   - **d.** The greenhouse effect is intensified in a desert environment.
   - **e.** CO_2 builds up in the leaves, blocking carbon fixation.

5. When light strikes chlorophyll molecules, they lose electrons, which are ultimately replaced by
   - **a.** splitting water
   - **b.** breaking down ATP
   - **c.** removing them from NADPH
   - **d.** fixing carbon
   - **e.** oxidizing glucose

6. Which of the following is a relatively simple way to capture the energy of sunlight and may have been important to early photosynthetic organisms?

a. the Calvin cycle
b. cyclic electron flow
c. crassulacean acid metabolism (CAM)
d. energy transfer to NADPH
e. noncyclic electron flow

7. What is the role of $NADP^+$ in photosynthesis?

a. It assists chlorophyll in capturing light.
b. It acts as the primary electron acceptor for the photosystems.
c. As part of the electron transport chain, it manufactures ATP.
d. It assists photosystem II in the splitting of water.
e. It carries electrons to the Calvin cycle.

8. The phase of photosynthesis in which sugar is manufactured is called

a. the Calvin cycle
b. photophosphorylation
c. the light reactions
d. cyclic electron flow
e. the electron transport chain

9. The reactions of the Calvin cycle are not directly dependent on light, but they usually do not occur at night. Why? *(Explain your answer.)*

a. It is often too cold at night for these reactions to take place.
b. Carbon dioxide concentrations decrease at night.
c. The Calvin cycle depends on products of the light reactions
d. Plants usually open their stomata at night.
e. At night, plants cannot produce the water needed for the Calvin cycle.

10. Which of the following correctly ranks the following structures in terms of size, largest to smallest? Chloroplast (C), mesophyll cell (MC), photosystem (P), chlorophyll molecule (M), thylakoid (T)

a. P-MC-T-C-M
b. MC-C-T-P-M
c. P-MC-C-T-M
d. MC-T-P-C-M
e. C-MC-T-P-M

True/False *(Change false statements to make them true.)*

1. Only plants are able to perform photosynthesis.

2. If photosynthesis ceased, herbivores would die but carnivores might survive.

3. The greenhouse effect results from the buildup of O_2 in the atmosphere.

4. Stomata let CO_2 enter a leaf and allow O_2 to exit.

5. The oxygen in glucose ($C_6H_{12}O_6$) comes from H_2O.

6. Planting trees on barren land could slow down greenhouse warming.

7. Photorespiration is a backup method for making ATP.

8. Some plants store CO_2 during the day and use it to make sugar at night.

9. Most raw materials for photosynthesis come from the soil.

10. A leaf looks green because it absorbs green light.

Describing, Comparing, and Explaining

1. Create a diagram identifying the two major stages of photosynthesis. Show with arrows the major inputs and outputs of each process, including the products of one process that are consumed in the other.

2. Which take up CO_2 faster (per unit of weight), young, rapidly growing trees or old ones? What effect does harvesting old trees and replacing them with seedlings have on the greenhouse effect? Why?

3. Compare the electron transport chain in the thylakoid of a chloroplast with the electron transport chain in a mitochondrion: Where do the electrons come from? Where do the electrons get their energy? What picks up the electrons at the end of the chain? How is the energy given up by the flow of electrons used?

4. Briefly describe what plants do with the sugar they produce in photosynthesis.

Thinking Critically

1. Tropical rain forests cover only about 3% of the Earth's surface, but they are estimated to be responsible for more than 20% of global photosynthesis. For this reason, rain forests are often referred to as the "lungs" of the planet, providing oxygen for life all over the Earth. However, most experts believe that rain forests make little or no *net* contribution to global oxygen production. From your knowledge of photosynthesis and cellular respiration, can you explain why they might think this? (What happens to the food produced by a rain forest tree when it is eaten by animals or when the tree dies?)

2. Red algae can live in deeper water than other kinds of "seaweeds." This ability may be due to their red pigments, which allow red algae to carry out photosynthesis using the wavelengths of light that best penetrate seawater. Perhaps you have snorkeled in the ocean or seen an underwater film. What color is the light under the sea? Why might red algae be more efficient at absorbing these wavelengths than green algae?

3. Do the oxygen atoms in the glucose produced by photosynthesis come from water or carbon dioxide? If you did not already know the answer, how could you use a radioactive isotope to find out? If the oxygen comes from water, how would your experiment turn out? How would it turn out if the oxygen comes from carbon dioxide?

Science, Technology, and Society

Many scientists believe the Earth is getting warmer, and the cause could be an intensified greenhouse effect resulting from increased CO_2 from industry, vehicles, and the burning of forests. The year 1990 was the warmest in nearly two centuries. Seven of the 10 warmest years in the last century have occurred since 1980. Global warming could influence agriculture and perhaps even melt polar ice and flood coastal regions. In response to the possible greenhouse threat, several European countries, the United States, Australia, and New Zealand have made a commitment to reduce carbon dioxide emissions significantly by the year 2000. However, some policy experts oppose taking strong actions at this time. Several reasons are cited: First of all, though the temperature during the 1980s was one-third of a degree higher than the average temperature for the previous 30 years, there is some uncertainty whether this really represents a warming trend or is just a random fluctuation in temperature. Second, if the temperature increase is real, it has yet to be proven that it is caused by increased CO_2. Some people also believe that it would be difficult to cut CO_2 emissions without sacrificing comfort, convenience, and economic growth. Do you think we should have more evidence that greenhouse warming is real before taking action? Or is it better to play it safe and act now to cut back CO_2? What are the possible costs and benefits of following each of these two strategies? Which do you favor, and why?

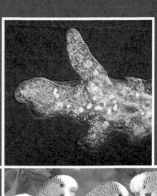

Unit 2
Genetics

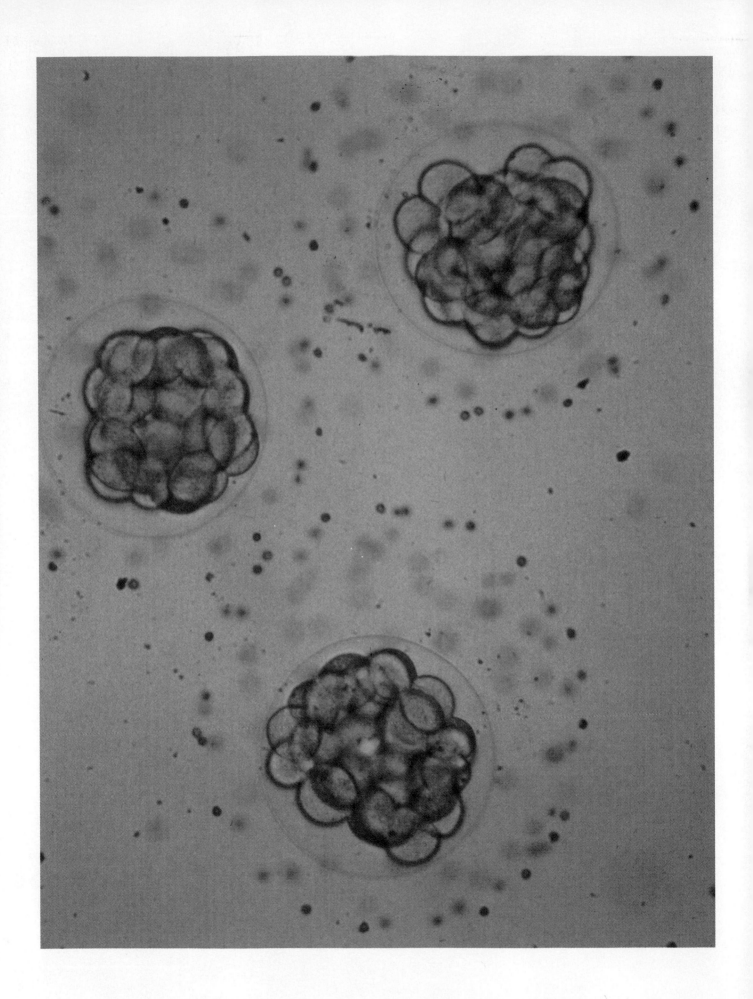

The Cellular Basis of Reproduction and Inheritance

8

What would you say if you were asked to describe the objects in the photograph on the left page? Early microscopists called these morulae (singular, morula), from the Latin word for mulberries, and the name stuck. In fact, the photo has nothing to do with berries or even plants. What you are seeing are three sand-dollar embryos, each consisting of 32 cells. Like their close relatives the sea urchins and sea stars, sand dollars belong to a group of marine animals called echinoderms. Most animals, echinoderms and humans included, have a morula stage early in their embryonic development.

A morula looks static in a photograph, but in real life it is a dynamic stage in animal development, part of a process that ensures the continuation of life from generation to generation. For a sand dollar or a sea star, the process begins when an adult female casts thousands or millions of eggs into the ocean and they are joined by sperm discharged by a nearby male. The chance meeting of one sperm cell and one egg cell results in a fertilized egg. Within about an hour, the fertilized egg divides in two. The resulting cells stay together and divide again, their four descendants divide in turn, and so on. A morula, like the ones in the photograph, is a cluster of 16 to 64 cells. Every one of its cells is about to divide again. As development continues, the ball of cells will transform into an echinoderm larva and eventually into an adult.

Development, from a fertilized egg to a new adult organism, is one phase of a multicellular organism's **life cycle**, the sequence of life stages leading from the adults of one generation to the adults of the next. The other phase of the life cycle

A sea star regenerating an arm

is reproduction, the formation of new individuals from pre-existing ones; for animals such as sea stars, a new individual first appears as a fertilized egg. The reproductive phase of the life cycle entails the transfer of genetic information in the form of DNA from parents to offspring. The reproductive process that includes the union of a sperm and an egg is called **sexual reproduction.** Because both sperm and egg carry DNA, the offspring of sexual reproduction inherit traits from two parents.

The life cycles of many organisms involve asexual reproduction, either instead of sexual reproduction or in addition to it. **Asexual reproduction** is the production of offspring by a single parent, without the participation of sperm and egg. Sea stars, for example, can reproduce asexually by regeneration, the remarkable ability to regrow body parts. The sea star in the photograph here is regenerating an arm it lost, probably in an encounter with a predator. If a sea star is broken apart, whole new individuals may develop from some of the larger pieces. Offspring produced by asexual reproduction inherit DNA from only one parent.

A sea star's life cycle illustrates the main theme of this chapter: Cell division is at the heart of reproduction. Regeneration of new sea stars from a fragmented parent results from repeated cell divisions, as does the development of a multicellular larva from a fertilized egg. In this chapter, we discuss various aspects of cell division, concentrating particularly on the mechanisms for the transfer of information, which ensure that new cells receive identical copies of the original cell's DNA.

Like begets like, more or less

Only sea stars produce more sea stars, only people make more people, and only maple trees produce more maple trees. These simple facts of life have been recognized for thousands of years and are summarized by the age-old saying, "Like begets like."

In a strict sense, "Like begets like" applies only to asexual reproduction, such as sea star regeneration. In this case, because offspring inherit all their DNA from a single parent, they are exact genetic replicas of that one parent and of each other.

Single-celled organisms, such as *Amoeba,* also reproduce asexually. The amoeba in Figure A is reproducing by dividing in half. The amoeba's **chromosomes,** the structures that contain most of the organism's DNA, have been duplicated, and identical chromosomes have been allocated to opposite sides of the parent cell. When the parent cell divides, the two daughter amoebas that result will be genetically identical to each other and to the original parent. (Biologists traditionally use the word "daughter" in this context only to indicate offspring, not to imply gender.)

The eight photographs in Figure B make the point that, in a *sexually* reproducing species, like does not precisely beget like. Offspring produced by sexual reproduction generally resemble their parents more closely than they resemble unrelated individuals of the same species, but they are not identical to their parents or to each other. Each offspring inherits a unique combination of genes from its two parents, and this one-and-only set of genes programs a unique combination of traits. As a result, sexual reproduction can produce great variation among offspring. The top four photographs in Figure B show two sets of parents. Their four children are shown in random order underneath. (All the subjects were the same age when these pictures were taken.) Two children belong to each set of parents. Can you match the offspring with their parents?*

Long before anyone knew about genes, chromosomes, or the underlying principles of inheritance, people recognized that individuals of sexually reproducing species are highly varied. By manipulating sexual reproduction, humans have developed domestic breeds of animals and plants that have specific traits. Domestic breeds exhibit much less variability than the species as a whole. In producing the many kinds of domestic dogs, for example, breeders selected from a varied population of dogs certain individuals that exhibited specific traits and allowed these individuals to mate and produce offspring. The an-

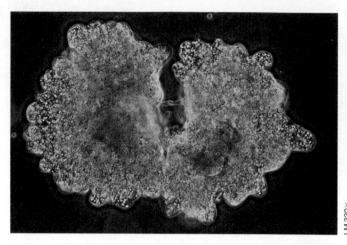

LM 330×

A. An amoeba producing genetically identical offspring

cestry of a dog breed such as the German shepherd can be traced back for many generations, in which breeders reduced variability in the breed by rigidly selecting only those dogs with specific (German shepherd) traits for mating. In a sense, selective breeding is an attempt to make like beget like more than it does in nature.

B. Two sets of human parents and their nonidentical offspring, produced by sexual reproduction. (Can you match parents with children?)

*Answer: The boy and girl on either end of the bottom row are children of the parents on the right. The boy and girl in the center of the bottom row are children of the other parents.

Cells arise only from preexisting cells

In 1858, German physician Rudolf Virchow formulated an important biological principle, to which we alluded in the chapter's introduction: All cells come from cells. Virchow's principle is illustrated graphically by a dividing amoeba, a regenerating sea star, and any fertilized egg that has begun to divide.

Like many important ideas that we now take for granted, Virchow's principle is both simple and profound. It tells us that the perpetuation of life, including all aspects of reproduction and inheritance, is based on the reproduction of cells. We commonly refer to cellular reproduction as **cell division**.

Cell division plays two main roles in the life cycle of animals and other multicellular organisms. First, as we mentioned earlier, cell division makes it possible for a fertilized egg to develop through various embryonic stages, and for an embryo to develop into an adult organism. Second, cell division ensures the continuity of life from generation to generation; it is the basis of both asexual reproduction and the formation of sperm and eggs in sexual reproduction.

So far, we have discussed only the division of eukaryotic cells, and we will emphasize them in this chapter. Prokaryotes, however, also illustrate Virchow's principle, and in the next module, we take a brief look at prokaryotic cell division.

Bacteria reproduce by binary fission 8.3

Bacteria (prokaryotes) reproduce by a type of cell division called **binary fission**, meaning "dividing in half." In bacteria, most genes are carried on a circular DNA molecule that, with associated proteins, constitutes the bacterium's single chromosome. (Bacterial chromosomes are much less complicated than those of eukaryotes, though both types consist of DNA and proteins.) Bacteria are generally smaller and simpler than eukaryotic cells. Nonetheless, replicating their DNA in an orderly fashion and distributing the copies equally to two daughter bacteria is a formidable task. Consider, for example, that, when stretched out, the chromosome of the bacterium *Escherichia coli* is some 500 times longer than the cell itself. Accurately duplicating this molecule when it is coiled and packed inside the cell is no small achievement.

Figure A illustrates how a bacterial cell allocates genes to its daughter cells. When the DNA in a bacterial cell is replicating (top), it is attached to the plasma membrane. After replication, the duplicate chromosomes are attached to the membrane at two separate points. Continued growth of the cell gradually separates the chromosomes, which are still attached to the membrane. Eventually, the plasma membrane and bacterial cell wall grow inward, dividing the cell in two. Figure B shows an electron micrograph of a dividing bacterium, at a stage similar to the third step in Figure A.

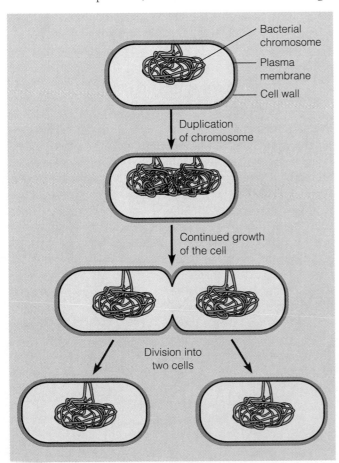

A. Binary fission in bacteria

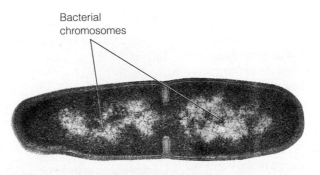

B. Electron micrograph of a dividing bacterium

Chapter 8   The Cellular Basis of Reproduction and Inheritance   **129**

8.4 Eukaryotes have multiple chromosomes that are large and complex

Eukaryotic cells are more complex and generally much larger than bacteria, and they have many more genes. Human cells, for example, carry about 100,000 genes, versus about 3000 for a typical bacterium. The genes in human cells, and in all other eukaryotes, are grouped into multiple chromosomes, which are found in the cell nucleus. Chromosomes get their name (Greek *chroma*, colored, and *soma*, body) from their affinity for certain stains used in microscopy. In the micrograph at the right, which shows a plant cell (of the African blood lily) that is about to divide, the chromosomes are stained dark purple. Each dark thread is an individual chromosome. Chromosomes are clearly visible under the light microscope as individual structures like these only when a cell is in the process of dividing. The rest of the time, the chromosomes form a diffuse mass of very long, very thin

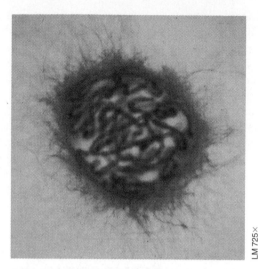

LM 725×

A plant cell just before division

fibers of **chromatin**, which is a complex of DNA and protein molecules. As a cell prepares to divide, its chromatin coils up, forming compact, distinct chromosomes.

Like the bacterial chromosome, each eukaryotic chromosome contains one very long DNA molecule bearing thousands of genes. However, the eukaryotic chromosome has a much more complex structure than the bacterial chromosome. This structure involves many more protein molecules, which help maintain the chromosome's shape and control the activity of its genes.

In multicellular species, the **somatic cells** (all body cells except sperm or eggs) contain a characteristic number of chromosomes. This number is usually twice the number in sperm or eggs of the same organism. For example, each human somatic cell has 46 chromosomes; our sperm cells or egg cells each have 23.

8.5 Chromosomes duplicate and then split in two as a cell divides

Well before a eukaryotic cell begins to divide, it duplicates all of its chromosomes. Resulting from this duplication are two identical structures called **sister chromatids.** Figure A is an electron micrograph of a human chromosome that has duplicated. The chromosome now consists of two sister chromatids, containing identical genes. The chromatids are joined together at a specialized region called the **centromere.** The fuzzy appearance of the chromosome comes from the intricate coils and twists of its chromatin fibers.

Figure B is a highly simplified diagram showing duplication of a chromosome.

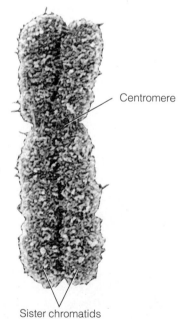

Centromere

Sister chromatids

A. Electron micrograph of a duplicated chromosome (TEM 36,540×)

When the cell divides, the sister chromatids of a duplicated chromosome separate from each other. As indicated at the bottom in the figure, one chromatid goes to one daughter cell, and the other goes to the other daughter cell. Once separated from its sister, each chromatid is actually called a chromosome; it is identical to the chromosome we started with. In this way, each daughter cell receives a complete set of single chromosomes. In humans, for example, a dividing cell has 46 duplicated chromosomes, and each of the two daughter cells that result from it has 46 single chromosomes.

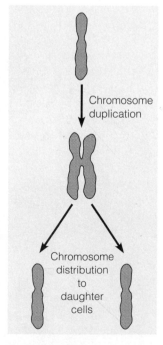

Chromosome duplication

Chromosome distribution to daughter cells

B. Chromosome duplication and distribution

Cell division is essential to life. It enables a multicellular organism to grow to adult size. It also replaces worn-out or damaged cells, keeping the total cell number in a mature individual relatively constant. In your own body, for example, millions of cells must divide every second to maintain the total number of about 60 trillion cells. Some cells divide once a day, others less often, and highly specialized cells, such as our nerve and muscle cells, not at all. As we have already mentioned, cell division is also the basis of reproduction in every organism, including single-celled organisms such as *Amoeba*.

Eukaryotic cells that divide undergo a **cell cycle,** an orderly sequence of events that extends from the time a cell divides to form two daughter cells to the time those daughter cells divide again. Before cell division occurs, the cell roughly doubles everything in its cytoplasm and precisely duplicates its DNA in preparation for division.

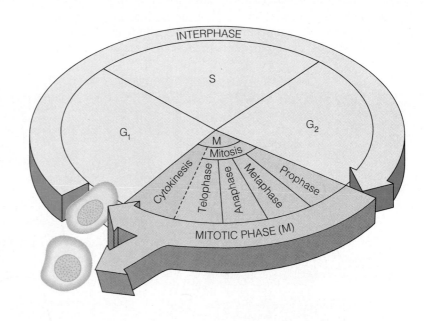

The cell cycle

As the figure shows, most of the cell cycle is spent in **interphase** (blue area). Under the microscope, a cell in interphase may appear to be resting, but this is not at all the case. Interphase is a time when a cell's metabolic activity is very high. Chromosomes duplicate during this period, many cell parts are made, and the cell does most of its growing. Typically, interphase lasts for at least 90% of the total time required for the cell cycle.

DNA synthesis, the main event in chromosome duplication, occurs in the middle of interphase and serves as the basis for dividing interphase into three subphases. These subphases were named by early microscopists before much was understood about the cell cycle, so their names are not especially revealing. The first subphase, designated G_1, is the period before DNA synthesis begins. G stands for *gap,* and G_1 refers to the gap between cell division and DNA synthesis. We now know that this subphase is anything but a gap in cell activity. G_1 is the time when the cell increases its supply of proteins, increases the numbers of many of its organelles (such as mitochondria and ribosomes), and grows in size.

Following G_1 is the subphase called the S phase, when DNA *synthesis* (replication) actually occurs. At the beginning of the S phase, each chromosome is single. At the end of this phase, after DNA replication, the chromosomes are double, each consisting of two sister chromatids.

The third subphase, called G_2, spans the time from the completion of DNA synthesis to the onset of cell division.

Like G_1, G_2 is a time of metabolic activity. Among the proteins synthesized during G_2 are some that are essential to cell division.

Cell division usually involves two discrete processes. First, by a process called **mitosis** (pink area in the figure), the nucleus and its contents, including the duplicated chromosomes, divide and are evenly distributed into two daughter nuclei. Then, in most cells, the cytoplasm divides in two by a process called **cytokinesis** (purple in the figure). The combination of mitosis and cytokinesis produces two separate daughter cells, each with a single nucleus and surrounding cytoplasm. Taken together, mitosis and cytokinesis make up the mitotic phase (M) of the cell cycle.

Mitosis is unique to eukaryotes and may be an evolutionary solution to the problem of allocating identical copies of a large amount of genetic material, in a number of separate chromosomes, to two daughter cells. Mitosis is a remarkably accurate mechanism. Experiments with yeast, for example, indicate that an error in chromosome distribution occurs only once in about 100,000 cell divisions.

A living cell viewed through a light microscope undergoes dramatic changes in appearance during the mitotic phase. During interphase, the cell's individual chromosomes are not distinguishable because they remain in the form of loosely packed chromatin fibers. With the onset of mitosis, however, striking changes are visible in the chromosomes and other structures, as we see in the next module.

Cell division is a continuum of dynamic changes

The photographs here, taken with a light microscope (600×), illustrate the cell cycle for an animal cell, in this case, from a fish. Interphase is shown, but the emphasis is on the dramatic changes that occur during cell division, the mitotic phase. Mitosis is a continuous process, but biolo- gists distinguish four main stages: **prophase, metaphase, anaphase,** and **telophase.** The drawings show details not visible in the micrographs. For simplicity, only four chromosomes are drawn. Descriptions below the drawings explain what is happening at each stage.

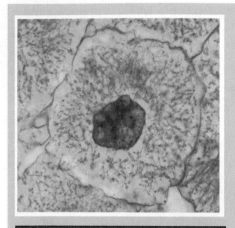

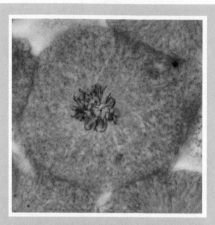

INTERPHASE

PROPHASE

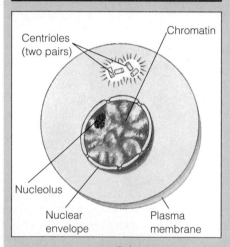

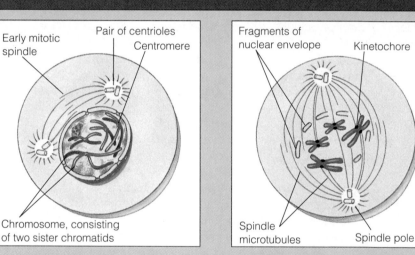

Interphase

Interphase is the period of cell growth, when the cell synthesizes new molecules and organelles. At the point shown here, late interphase (G_2), the cell looks much the same as it does throughout interphase. Nonetheless, by the G_2 stage the cell has doubled much of its earlier contents. The cytoplasm contains two pairs of centrioles (plant cells lack centrioles). Within the nucleus, the chromosomes are duplicated, but they cannot be distinguished individually because they are still in the form of loosely packed chromatin. The nucleus also contains one or more nucleoli, an indication that the cell is actively making proteins. Nucleoli make ribosomes, which are the cell's factories for protein synthesis.

Prophase

During prophase, changes occur in both the nucleus and the cytoplasm. Within the nucleus, the chromatin fibers become more tightly coiled and folded, forming discrete chromosomes that are observable with the light microscope. Each duplicated chromosome appears as two identical sister chromatids joined at the centromere. In the cytoplasm, the mitotic spindle forms out of microtubules and associated proteins. In animal cells, the spindle forms between the two pairs of centrioles, as they move away from each other during prophase.

Late in prophase, the nuclear envelope fragments, and the nucleoli disappear.

With the nuclear envelope gone, the mitotic spindle invades the nucleus. Microtubules extend from the poles of the spindle toward the chromosomes, which, by this time, are highly condensed. At the centromere region, each sister chromatid has a protein structure called a kinetochore (shown as a black dot). Some of the spindle microtubules attach to the kinetochores, throwing the chromosomes into agitated motion, which can be seen with a microscope. As we will discuss in Module 8.8, forces exerted by the microtubules tend to move the chromosomes toward the center of the cell.

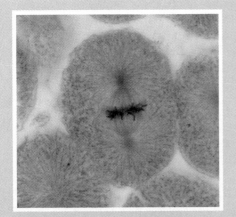

| METAPHASE | ANAPHASE | TELOPHASE AND CYTOKINESIS |

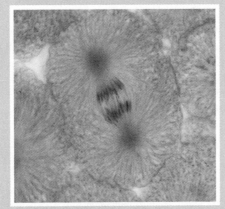

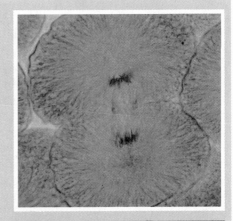

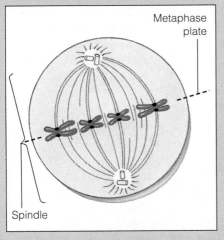

Metaphase plate

Spindle

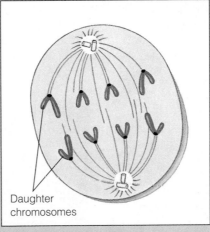

Daughter chromosomes

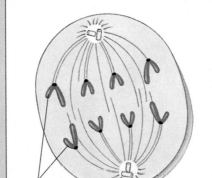

Nucleolus forming

Cleavage furrow

Nuclear envelope forming

Metaphase

At metaphase, the mitotic spindle is fully formed, with spindle poles at opposite ends of the cell. The chromosomes, each consisting of two sister chromatids, convene on the metaphase plate, an imaginary plane equidistant between the two poles of the mitotic spindle. (The metaphase plate is represented by the dashed black line.) The centromeres of all the chromosomes are lined up on the metaphase plate. At this point, the chromosomes are roughly perpendicular to the spindle microtubules. For each chromosome, the kinetochores of the two sister chromatids face opposite poles of the spindle.

Anaphase

Anaphase begins when the paired centromeres of each chromosome divide, separating the sister chromatids from each other. Once separated from its sister, each sister chromatid is considered a full-fledged chromosome (called a daughter chromosome). Now the spindle microtubules attached to the kinetochores shorten. As this happens, the daughter chromosomes move centromere-first toward opposite poles of the cell. Meanwhile, the spindle microtubules not attached to chromosomes lengthen. Their lengthening seems to move the poles farther apart and elongate the cell. Anaphase is finished when equivalent—and complete—collections of chromosomes are located at the two poles of the cell.

Telophase and Cytokinesis

Telophase is roughly the reverse of prophase. The cell elongation that started in anaphase continues, and daughter nuclei begin to form at the two poles of the cell. Nuclear envelopes form around the chromosomes, nucleoli reappear, and the chromatin fiber of each chromosome uncoils. At the end of telophase, the mitotic spindle disappears. Mitosis, the equal division of one nucleus into two genetically identical daughter nuclei, is now completed.

Cytokinesis, the division of the cytoplasm, usually occurs along with telophase, with two daughter cells completely separating soon after the end of mitosis. In animal cells, cytokinesis involves formation of a cleavage furrow, which pinches the cell in two.

8.8 The mitotic spindle orchestrates mitosis

Many of the events of mitosis depend on the actions of the mitotic spindle. In animal cells, the first sign of the spindle is the appearance of two clusters of microtubules, called asters, surrounding the centrioles. You can see asters in the diagram below, which shows the spindle of a cell at metaphase. In the micrograph, the asters are not clearly visible, but you can see the longer spindle microtubules that seem to radiate from the pair of centrioles at each pole of the cell. These microtubules actually emerge from the area around the centrioles (gray in the diagram). This region, called the microtubule organizing center, is where the assembly of microtubules of the mitotic spindle begins. After assembly begins, the microtubules lengthen at the ends *away from* the organizing center. The roles of centrioles and asters in the formation of the mitotic spindle are still a mystery. Plant cells lack both centrioles and asters, and animal cells whose centrioles have been destroyed in laboratory experiments can still form a mitotic spindle and carry out mitosis.

The diagram illustrates the relationship between the chromosomes and the mitotic spindle at metaphase. In both plant cells and animal cells, many of the spindle microtubules (colored gold) overlap at the metaphase plate of the cell. Other spindle microtubules (brown here) are attached to the kinetochores (blue) on chromatids. Notice that the microtubules attached to a particular chromatid all come from one pole of the spindle and that the microtubules attached to its sister chromatid come from the opposite pole.

The structure of the mitotic spindle fits its function of moving chromosomes in mitosis. During prophase, microtubules attached to opposite sides of a chromosome are positioned to exert opposing forces on the chromosome, analogous to a tug-of-war. The tug-of-war seems to end in a draw when the chromosomes come to rest at the metaphase plate. Anaphase begins when each chromosome's centromeres divide, and the sister chromatids separate and head toward opposite poles of the cell. Their movement is accompanied by the disassembly of microtubules near the kinetochores. (Microtubule assembly and disassembly were discussed in Module 4.17). The kinetochores seem to cling to the tips of the shortening microtubules, and the sister chromatids move poleward as a result. Perhaps the kinetochores "walk" their way poleward along their microtubules just ahead of the microtubule disassembly. The mechanism of kinetochore walking may resemble the one used by the dynein arms of microtubules in the bending of flagella and cilia (see Module 4.18).

What about the spindle microtubules that are not attached to kinetochores (the gold ones in the diagram)? Biologists think these microtubules play a role in the cell elongation that occurs during anaphase and telophase. What happens is not fully understood, but one hypothesis is that the poles of the cell are moved apart as the microtubules lengthen by adding new subunits, and as overlapping microtubules slide against each other. The dynein walking mechanism may be involved here also.

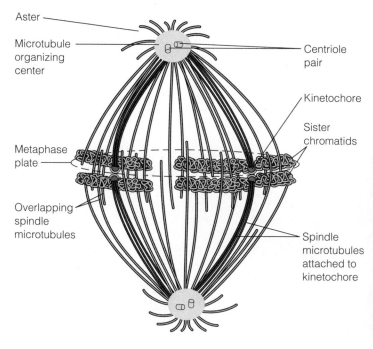

Aster
Microtubule organizing center
Centriole pair
Kinetochore
Sister chromatids
Metaphase plate
Overlapping spindle microtubules
Spindle microtubules attached to kinetochore

The mitotic spindle at metaphase

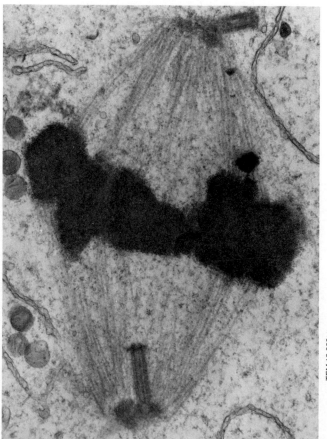

TEM 12,000×

Cytokinesis, or division of the cell into two, typically occurs with telophase, although it may actually begin in late anaphase. In animal cells, cytokinesis occurs by a process known as **cleavage.** As shown in Figure A, the first sign of cleavage is the appearance of a cleavage furrow, which begins as a shallow groove in the cell surface. At the site of the furrow, the cytoplasm has a ring of microfilaments made of actin, a protein that functions in contraction. The ring contracts much like the pulling of drawstrings, deepening the cleavage furrow and eventually pinching the parent cell in two.

Cytokinesis in a plant cell occurs differently, as Figure B shows. First, membrane-bounded vesicles containing cell wall material collect at the midline of the parent cell. The vesicles then fuse, forming a membrane-bounded disc called the cell plate. The cell plate grows outward, accumulating more cell wall materials as more vesicles fuse with it. Eventually, the cell plate membrane fuses with the plasma membrane. The result is two daughter cells, each bounded by its own continuous plasma membrane and cell wall.

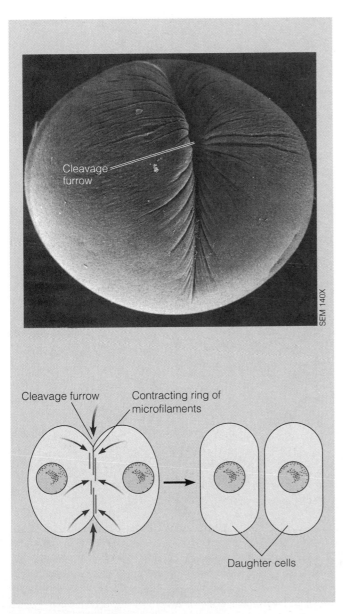

A. Cleavage of an animal cell

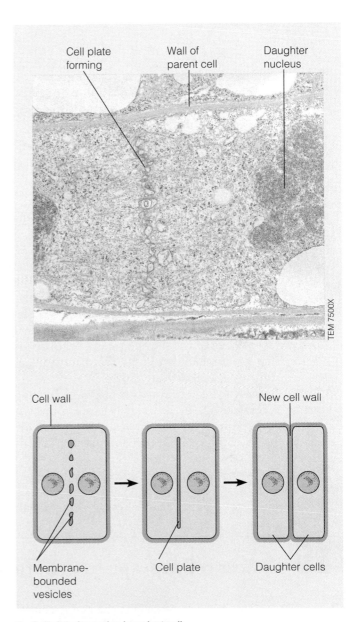

B. Cell plate formation in a plant cell

8.10 Cell density, chemical growth factors, and anchorage affect cell division

For a plant or animal to grow and develop normally, and to maintain its tissues once full-grown, it must be able to control the timing and rate of cell division in different parts of its body. In the adult human, for example, skin cells and the cells lining the digestive tract divide frequently throughout life, replacing cells that are constantly being abraded and sloughed off. In contrast, cells in the human liver may not divide unless the liver is damaged. Cell division in this case repairs wounds.

Many questions about how cell division is controlled are still unanswered, but biologists have learned a great deal by studying cells grown in laboratory cultures. They have discovered, for instance, that animal cells growing on the surface of a laboratory vessel will grow in a single layer, dividing continuously until they touch one another; then cell division stops. This phenomenon is called **density-dependent inhibition**, implying that the rate of cell division generally decreases as the cell population becomes more dense. As shown in the figure, when some of the cells are scraped off a cell-covered surface in a culture vessel, cells bordering the open space begin dividing again. They continue to do so until the space is filled, when the cell density inhibits further division.

Clearing a space in a cell culture is analogous to cutting your skin. When you cut yourself, skin cells all around the cut immediately begin dividing, and healing occurs as the cells fill in the space. The cells stop dividing when they encounter other cells.

What actually causes density-dependent inhibition? Studies of cultured cells indicate that specific proteins called growth factors, rather than physical contact with other cells, are the main factor. A cell requires growth factors for division, and it stops dividing when it runs out of these substances. Apparently, density-dependent inhibition occurs when cultured cells become so crowded that they use up the supply of growth factors in their environment. Density-dependent inhibition is probably an important regulatory mechanism in the body's tissues as well as in cell culture. It may help keep cell numbers at optimal levels in the tissues.

But growth factors do not entirely explain why cells start or stop dividing. Most animal and plant cells will not divide at all unless they are in contact with a solid surface. If cells are suspended in a liquid medium, they will rarely divide no matter how many growth factors are available. However, when the same cells are poured onto a solid medium and allowed to attach, or "anchor" themselves, they will start dividing immediately. In the body, this mechanism may keep cells that become separated from their normal surroundings from dividing uncontrollably.

For most normal cells, the crucial events in controlling cell division seem to occur during the G_1 phase of the cell cycle—before DNA replication occurs. Exactly what these events are, however, remains to be discovered, although they undoubtedly involve growth factors. Apparently, the control mechanism somehow determines whether a cell will go through a critical stage, called the restriction point, late in the G_1 phase. A cell will go on through its cycle and divide if it passes the restriction point, but if it does not pass this point in the cell cycle, it will switch into a nondividing state. Our nerve cells and muscle cells are in this nondividing state permanently.

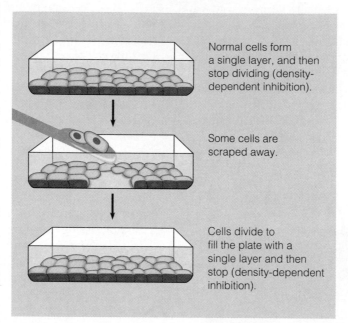

An experiment demonstrating density-dependent inhibition

- Normal cells form a single layer, and then stop dividing (density-dependent inhibition).
- Some cells are scraped away.
- Cells divide to fill the plate with a single layer and then stop (density-dependent inhibition).

8.11 Cancer cells escape from normal control mechanisms

Some cells do not respond normally to the body's control mechanisms. These **cancer cells** divide excessively, invading other tissues; if unchecked, they can kill the whole organism. The micrographs in Figure A at the top of the facing page compare a slice of normal liver tissue with a slice of cancerous liver tissue. Notice the disorderly arrangement of the cancer cells, resulting from uncontrolled division, contrasted with the more orderly array of the normal cells.

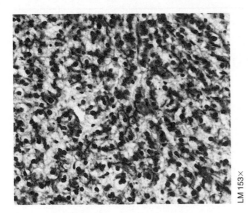

Normal liver tissue

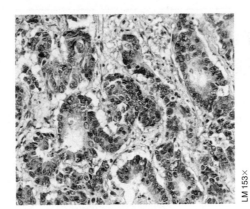

Cancerous liver tissue

A. A comparison of cell growth patterns in normal and cancerous tissue

By studying cancer cells in culture, researchers have learned that these cells are unaffected, or affected to a much lesser degree, by the factors that control division in normal cells. Cancer cells grown in culture vessels often seem unaffected by density-dependent inhibition. Indeed, they seem to have less need for growth factors than normal cells and will continue to divide at high densities, eventually piling up on one another (Figure B). As we discuss in Module 11.16, some cancer cells synthesize growth factors that make them divide continuously. Cancer cells also have a reduced need for anchorage and often grow without being attached to a solid surface.

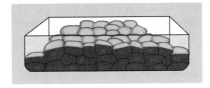

B. Absence of density-dependent inhibition

Whereas normal mammalian cells will grow and divide in culture for only a limited number of generations (usually 20–50), cancer cells can go on dividing indefinitely, as long as they have a supply of nutrients. If cancer cells do stop dividing, they seem to do so at random points in the cell cycle, rather than just at the restriction point.

Two of the main treatments for cancer, chemotherapy and radiation therapy, attempt to halt the spread of cancer cells by stopping them from dividing. In radiation therapy (Figure C), parts of the body that have cancerous growths are exposed to high-energy radiation, which disrupts cell division. Because cancer cells divide more often than most normal cells, they are more likely to be undergoing division at any given time. So radiation can sometimes destroy cancer cells without seriously injuring the normal cells of the body. However, there is often enough damage to normal body cells to produce harmful side effects. For example, damage to cells of the ovaries or testes can lead to sterility.

Chemotherapy employs the same strategy as radiation; here drugs that disrupt cell division are administered to the patient. These drugs work in a variety of ways. Some, called antimitotic drugs, prevent cell division by interfering with the mitotic spindle. One of these, vinblastin, disrupts microtubules, thus preventing formation of the spindle. Now manufactured by drug companies, vinblastin was first obtained from the periwinkle, a flowering plant native to tropical rain forests in Madagascar.

Another antimitotic drug, taxol, immobilizes microtubules, thereby preventing the mitotic spindle from undergoing the dynamic changes necessary for mitosis. Taxol is especially promising because it causes fewer side effects than many anticancer drugs and seems to be effective against some hard-to-treat cancers of the ovary and breast. Taxol is extracted from the bark of the Pacific yew, a tree found mainly in the northwestern United States. Pacific yews are scarce, and harvesting the bark kills the tree. In fact, treating a single cancer patient may require killing at least three trees over 100 years old. As a result, taxol has been a subject of debate. Fortunately, researchers are making progress toward using other species of yew, synthesizing taxol in the laboratory, and growing yew cells that will produce it in culture. And recently they have found a fungus growing on yew trees that itself makes small amounts of taxol. But it may be years before wild Pacific yews are no longer used to make the drug.

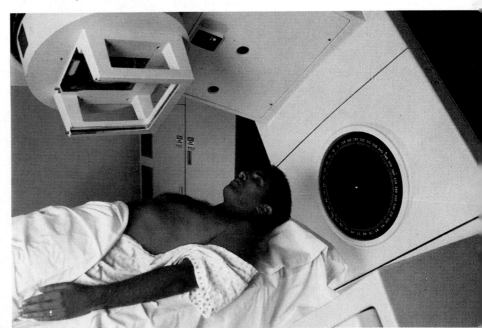

C. Radiation therapy

Review of the functions of mitosis: Growth, cell replacement, and asexual reproduction

We have seen that the perpetuation of life depends on cell division. The pictures here summarize the roles that mitotic cell division plays in the lives of multicellular organisms. Figure A shows some of the cells in the tip of an onion plant root. Notice the large number of cells whose nuclei are in various stages of mitosis. (You can distinguish several of the stages of mitosis in these cells.) This root was growing rapidly when it was harvested and prepared for microscopy. Cell division in the root tip produces new cells, which elongate to bring about growth of the root.

Figure B shows a cross section of human skin. The layers of dead cells on the surface protect the body from injuries and infections and help prevent it from drying out. These surface layers are constantly abraded and sloughed off, but they are replaced by cells from the living layers underneath. This regeneration of the skin goes on throughout a person's life. New cells generated near the base of the epidermis are constantly moving outward toward the skin surface. As they do, they gradually flatten, become hardened, and die, regenerating the protective surface layers.

Figure C is a photograph of *Hydra*, a common inhabitant of freshwater lakes. *Hydra* is a tiny multicellular animal that reproduces by either sexual or asexual means. This individual is reproducing asexually by budding. A bud starts out as a mass of mitotically dividing cells growing on the side of the parent. The bud develops into a small hydra like the one in the photograph here. Eventually, the offspring detaches from the parent and takes up life on its own. The offspring is literally a "chip off the old block," being genetically identical to its parent.

In all three of these cases—budding *Hydra*, regenerating human skin, and growing onion—the new cells have exactly the same number and types of chromosomes as the parent cells because of the way duplicated chromosomes divide during mitosis. Mitosis makes it possible for organisms to grow, regenerate and repair tissues, and reproduce asexually by producing cells that carry the same genes as the parent cells.

If we examine the somatic cells of any individual organism, we see that they all contain the same number and types of chromosomes. Likewise, if we examine cells from different individuals of any one species, we see that they have the same number and types of chromosomes. We take a closer look at the numbers and types of chromosomes in somatic cells in the next module.

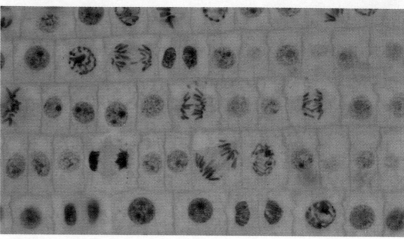

A. Growth (in an onion root)

LM 600×

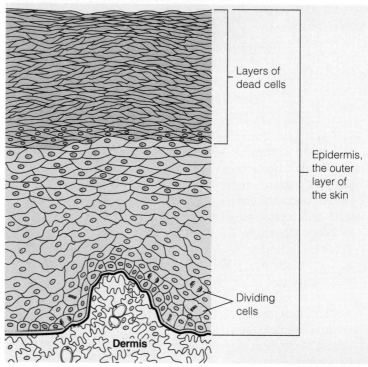

Layers of dead cells

Epidermis, the outer layer of the skin

Dividing cells

Dermis

B. Cell replacement (in skin)

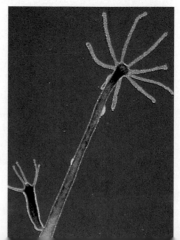

10×

C. Asexual reproduction (of *Hydra*)

Chromosomes are matched in homologous pairs

A human somatic cell has a total of 46 chromosomes. If we examine our chromosomes under a microscope, we see that they can be matched up as 23 pairs. Other species have different numbers of chromosomes, but their chromosomes also match in pairs. In almost every case, the chromosomes of each pair are similar in length, centromere position, and staining pattern (represented by the colored stripes here). The drawing here shows a pair of metaphase chromosomes; each consists of two sister chromatids joined at the centromere. The two chromosomes making up a matched pair are called **homologous chromosomes** because they both carry genes controlling the same inherited traits. For example, if a gene for eye color is located at a particular place, or **locus** (plural, *loci*), on one chromosome, then the homologous chromo-

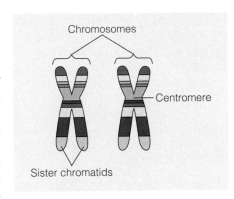

A homologous pair of chromosomes

some will also have a gene for eye color (though perhaps for a different color) at that locus. We inherit one chromosome of each homologous pair from our mother and the other chromosome of the pair from our father.

Our 23 pairs of homologous chromosomes are of two general types. Twenty-two pairs consist of chromosomes called **autosomes**, found in both males and females. The other pair of chromosomes, the **sex chromosomes**, determine a person's sex. Human females have a pair of sex chromosomes called *X* chromosomes. By contrast, human males have one *X* and one *Y*. The *X* and *Y* chromosomes are an important exception to the rule of homologous chromosomes. They differ in size and shape, and most of the genes carried on the *X* chromosome do not have counterparts on the *Y* chromosome.

Gametes have a single set of chromosomes

Having two sets of chromosomes, one inherited from each parent, is a key factor in the human life cycle, outlined in the figure here, and in the life cycles of all other species that reproduce sexually. (We examine other life cycles in Unit 3.)

Cells whose nuclei contain two homologous sets of chromosomes are called **diploid cells,** and the total number of chromosomes is called the diploid number (abbreviated 2*n*). In humans, the diploid number is 46 (that is, 2*n* = 46). Humans are said to be diploid organisms because almost all our cells are diploid. The exceptions are the egg and sperm cells, collectively known as **gametes**. Each gamete has a single set of chromosomes: 22 autosomes plus a single sex chromosome, either *X* or *Y*. A cell with a single chromosome set is called a **haploid cell.** For humans, the haploid number (abbreviated *n*) is 23 (*n* = 23).

In the human life cycle, sexual intercourse allows a haploid sperm cell from the father to

reach and fuse with a haploid egg cell of the mother in the process of **fertilization.** The resulting fertilized egg, called a **zygote,** is diploid. It has two haploid sets of chromosomes: one set from the mother and a homologous set from the father. The life cycle is completed as a sexually mature adult develops from the zygote. Mitotic cell division ensures that all somatic cells of the body receive copies of all of the zygote's 46 chromosomes.

All sexual life cycles, including our own, involve an alternation of diploid and haploid stages. Having haploid gametes keeps the chromosome number from doubling in each succeeding generation. Haploid gametes are produced by a special sort of cell division called **meiosis,** which occurs only in reproductive organs (ovaries and testes in animals). Whereas mitosis produces daughter cells with the same numbers of chromosomes as the parent cell, meiosis reduces the chromosome number by half. We turn to meiosis next.

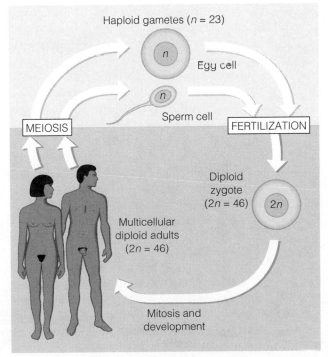

The human life cycle

Meiosis reduces the chromosome number from diploid to haploid

Meiosis, the process that produces haploid gametes, resembles mitosis, following the same phases. But there is a significant difference. In meiosis, a cell undergoes two consecutive divisions, called meiosis I and meiosis II. Four daughter cells result from these divisions rather than just two, as in mitosis. The two divisions of meiosis are preced-ed by only a single duplication of the chromosomes. As a result, each of the four daughter cells has only half as many chromosomes as the starting cell. The actual halving occurs during meiosis I. The drawings here show the two meiotic divisions for an animal cell whose diploid number is 4.

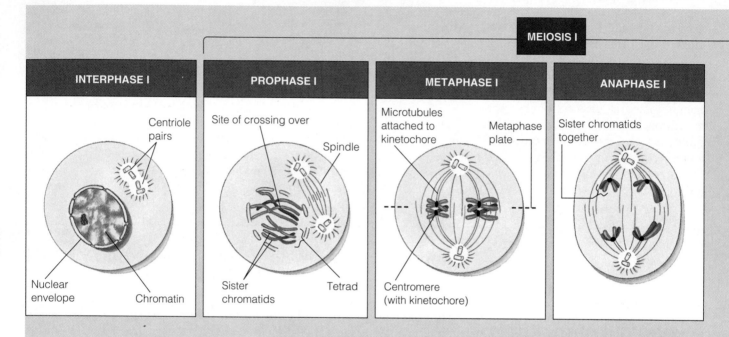

MEIOSIS I

INTERPHASE I — Centriole pairs, Nuclear envelope, Chromatin

PROPHASE I — Site of crossing over, Spindle, Sister chromatids, Tetrad

METAPHASE I — Microtubules attached to kinetochore, Metaphase plate, Centromere (with kinetochore)

ANAPHASE I — Sister chromatids together

Interphase I

Like mitosis, meiosis is preceded by an interphase, during which the chromosomes duplicate. At the end of this interphase, each chromosome consists of two, genetically identical sister chromatids, attached at their centromeres. But at this stage, the chromosomes are not yet visible under the microscope except as a mass of chromatin. The centrioles, which take part in meiosis in animal cells, are also duplicated by the end of this interphase.

Prophase I

Prophase I is the most complex phase of meiosis and typically occupies over 90% of the time required for meiotic cell division. Early in this phase, the chromatin forms into long thin chromosomes, the ends of which attach to the inside of the nuclear envelope. Then, in a process called synapsis, homologous chromosomes, each composed of two sister chromatids, pair up, forming a structure called a tetrad. Notice that each tetrad has four chromatids. During synapsis, chromatids of homologous chromo-somes exchange segments, in a process called **crossing over.** As we discuss in Module 8.18, the genetic information on a chromosome (or one of its chromatids) is somewhat different from the information on its homologue. By rearranging genetic information, crossing over can make an important contribution to the genetic variability resulting from sexual reproduction.

As prophase I continues, the chromosomes condense and detach from the nuclear envelope. Now the cell prepares further for nuclear division in a manner similar to that in mitosis. The centriole pairs move away from each other, and spindle microtubules form between them. The nuclear envelope and nucleoli disappear, and the chromosome tetrads, attached to spindle microtubules, move toward the center of the cell.

Metaphase I

By metaphase I, the chromosome tetrads are aligned on the metaphase plate, midway between the two poles of the spindle. Each chromosome is condensed and thick, with its sister chromatids attached at prominent centromeres. Spindle microtubules are attached to kinetochores at the centromeres. Notice that, for each tetrad, the spindle microtubules attached to one of the homologous chromosomes extend to one pole of the cell, and the microtubules attached to the other homologous chromosome extend to the opposite pole. With this arrangement, the homologous chromosomes of each tetrad are poised to move toward opposite poles of the cell.

Anaphase I

Like anaphase of mitosis, anaphase I of meiosis is marked by the migration of chromosomes toward the two poles of the cell. In contrast to mitosis, however, the sister chromatids making up each doubled chromosome remain attached at their centromeres. Only the tetrads (pairs of homologous chromosomes) split up. Thus, in the drawing you see two still-doubled chromosomes (the haploid number) moving toward each spindle pole. If this were anaphase of mitosis, you would see four daughter chromosomes moving toward each pole.

| TELOPHASE I AND CYTOKINESIS | PROPHASE II | METAPHASE II | ANAPHASE II | TELOPHASE II AND CYTOKINESIS |

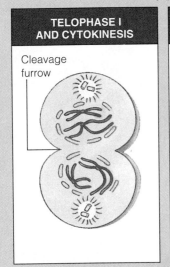

Cleavage furrow

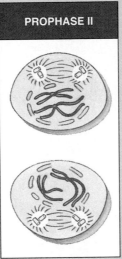

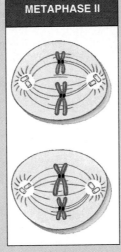

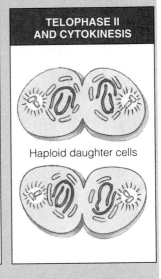

Haploid daughter cells

Telophase I and Cytokinesis

Early in telophase I, the chromosomes continue to move toward opposite poles of the cell. When the chromosomes finish their journey, each pole of the cell has a haploid chromosome set. Notice that each chromosome still consists of two sister chromatids. Usually, cytokinesis occurs along with telophase I, and two haploid daughter cells are formed.

Following telophase I in some organisms, nuclear membranes and nucleoli re-form, and there is a period of time called interphase II that precedes meiosis II. In other species, daughter cells of telophase I immediately begin preparation for the second meiotic division. In either case, no chromosome duplication occurs between telophase I and the onset of meiosis II.

Meiosis II

In species having interphase II, the nuclear membrane and nucleoli disperse during prophase II; if there is no interphase II, this is unnecessary. In both cases, meiosis II is essentially the same as mitosis. The important difference is that meiosis II starts with a haploid cell.

During prophase II, a spindle appears, and the chromosomes move toward the middle of the cell. During metaphase II, the chromosomes are aligned on the metaphase plate as they are in mitosis, with the kinetochores of the sister chromatids of each chromosome pointing toward opposite poles. In anaphase II, the centromeres of sister chromatids separate, and the sister chromatids of each pair, now individual daughter chromosomes, move toward opposite poles of the cell. In telophase II, nuclei form at opposite poles of the cell, and cytokinesis occurs. There are now four daughter cells, each with the haploid number of (single) chromosomes.

Review: A comparison of mitosis and meiosis

We have now described the two ways that cells in eukaryotic organisms divide. Mitosis, which provides for growth, tissue repair, and asexual reproduction, produces daughter cells that are genetically identical to the parent cell. Meiosis, required for sexual reproduction, produces daughter cells that are haploid—that have only one member of each homologous chromosome pair. For both mitosis and meiosis, the chromosomes replicate only once, during interphase (for meiosis, interphase I). Mitosis involves only one division of the nucleus, and it is usually accompanied by cytokinesis, producing two diploid cells. Meiosis entails two nuclear and cellular divisions, and yields four haploid cells.

This figure compares the main events in mitosis and meiosis, tracing these two types of cell division for a diploid parent cell with four chromosomes. Homologous chromosomes are those shown as matching in size.

All the events unique to meiosis occur during meiosis I. In prophase I, duplicated homologous chromosomes pair to form tetrads, and crossing over occurs between homologous (nonsister) chromatids. In metaphase I, tetrads (rather than individual chromosomes) are aligned at the metaphase plate. During anaphase I, sister chromatids of each duplicated chromosome stay together and go to the same pole of the cell. Consequently, meiosis I separates homologous pairs of chromosomes, and in doing so produces two haploid cells. In the case shown here, each haploid cell has two duplicated chromosomes.

The second meiotic division, meiosis II, is virtually identical to mitosis, separating sister chromatids of the haploid number of chromosomes in the two starting cells. The resulting four daughter cells have the haploid number of single daughter chromosomes.

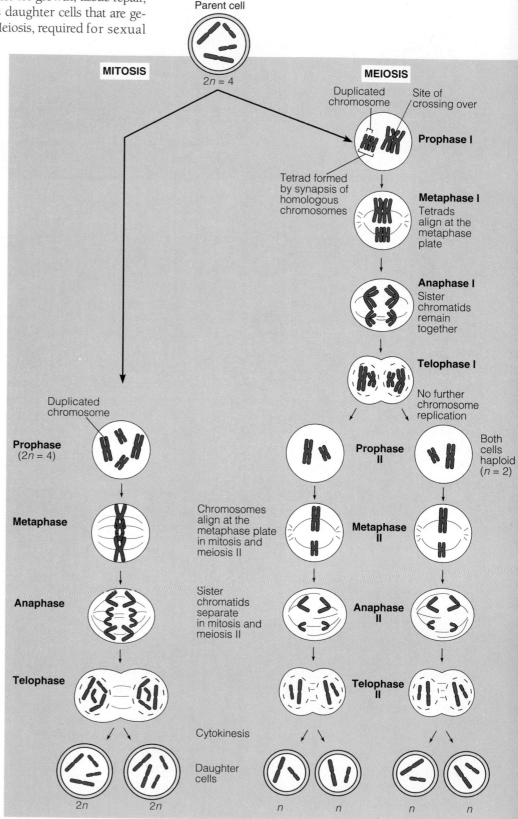

Independent orientation of chromosomes in meiosis and random fertilization lead to varied offspring

As we discussed in Module 8.1, off-spring that result from sexual reproduction are highly varied; they are genetically different from their parents and from one another. When we discuss natural selection and evolution in Unit 3, we will see that this genetic variety in offspring is the raw material for natural selection. For now, let's take a closer look at meiosis and fertilization to see how genetic variety arises.

The figure here illustrates one way in which the process of meiosis contributes to genetic differences in gametes. The figure shows how the arrangement of homologous chromosome pairs at metaphase of meiosis I affects the resulting gametes. Once again, our example is an organism with a diploid chromosome number of four ($2n = 4$). Here we use red and blue to distinguish chromosomes that come from two parents. The colors highlight the impor-

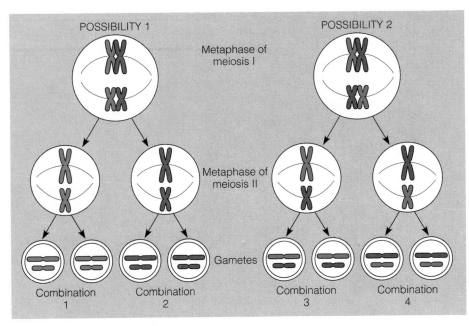

POSSIBILITY 1   POSSIBILITY 2

Metaphase of meiosis I

Metaphase of meiosis II

Gametes

Combination 1   Combination 2   Combination 3   Combination 4

Results of the independent orientation of chromosomes at metaphase

tant fact that each maternal chromosome differs genetically from its paternal homologue, although the two look alike under a microscope. (In fact, each chromosome you received from your mother carries many genes that are different versions from those on the homologous chromosome you received from your father.)

The orientation of the homologous pairs of chromosomes (tetrads) at metaphase I is a matter of chance, like the flip of a coin. In this example, there are two possible ways that the two tetrads can align during metaphase I. In possibility 1, the tetrads are oriented with both red chromosomes on the same side of the metaphase plate. In this case, one of the two daughter cells of meiosis I receives only red chromosomes, and the other cell receives only blue chromosomes (you can see these cells, at the metaphase II stage, in the second row of the figure). Therefore, the gametes produced in possibility 1 can each have only red *or* blue chromosomes (bottom row, combinations 1 and 2).

In possibility 2, the tetrads are oriented differently, with one red and one blue chromosome on each side of the metaphase I plate. This arrangement produces gametes that each have one red and one blue chromosome. Furthermore, half the gametes have a big blue chromosome and a small red one (combination 3), and half have a big red and a small blue (combination 4).

So we see that for this example a total of four chromosome combinations is possible in the gametes, and in fact the organism will produce gametes of all four types. This variety in gametes arises because each homologous pair of

chromosomes orients itself on the metaphase I plate independently of the other pair. For a species with more than two pairs of chromosomes, such as the human, *all* the chromosome pairs orient independently at metaphase I. (Chromosomes *X* and *Y* behave as a homologous pair in meiosis.)

For any species, the total number of combinations of chromosomes that meiosis can package into gametes is 2^n, where *n* is the haploid number. For the organism in this figure, $n = 2$, so the number of chromosome combinations is 2^2, or 4. For a human ($n = 23$), there are 2^{23}, or about 8 million, possible chromosome combinations. This means that every gamete a human produces contains one of about 8 million possible combinations of maternal and paternal chromosomes.

What are the possibilities when a gamete from one individual unites with a gamete from another individual in fertilization? A human egg cell, representing one of about 8 million possibilities, will be fertilized at random by one sperm cell, representing one of about 8 million other possibilities. By multiplying 8 million times 8 million, we find that a man and a woman can produce a diploid zygote with any of 64 trillion combinations of chromosomes! Thus, the random nature of fertilization adds a huge amount of potential variability to the offspring of sexual reproduction.

These large numbers suggest that independent orientation of chromosomes at metaphase I and random fertilization could account for all the variety we see among people. Actually, these two events are only part of the picture, as we see in the next module.

The two chromosomes in a homologous pair carry different genetic information

So far, we have examined genetic variability in gametes and zygotes at the whole-chromosome level. We have yet to focus on the actual genetic information—the genes—contained in the chromosomes of gametes and zygotes. The question we need to answer now is this: What is the significance of the independent orientation of metaphase chromosomes at the level of genes?

Let's take a simple example, the single tetrad in Figure A. The letters on the homologous chromosomes represent genes. Recall that the position of a particular gene along the length of a chromosome is called its locus. As indicated here, homologous chromosomes have genes for the same traits at corresponding loci. In this example, C and c are different versions of a gene for one trait in mice, coat color. And genes E and e are different versions of a gene for another trait, eye color. Letter C represents the gene for agouti, the brownish color of the mice in the left photograph in Figure B. Letter c represents the gene for white (albino) coat color (Figure B, right). Notice that gene C occurs at the same locus on the red homologue as c does on the blue one. Likewise, gene E (for black eyes, left photo) is at the same locus as e (pink eyes, right photo).

The fact that homologous chromosomes can bear different genetic information for the same traits (for example,

agouti versus white coat color) at corresponding loci is what really makes gametes—and therefore offspring—different from one another. After meiosis, in our example, a gamete carrying a red chromosome would have genes specifying agouti coat color and black eye color, while a gamete with the homologous blue chromosome would have genes for albino coat color and pink eyes.

Thus, we see how a hypothetical tetrad with genes for only two traits can yield two genetically different kinds of gametes. In the next module, we go a step further and see how this same tetrad can actually yield *four* different kinds of gametes.

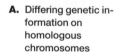

Coat-color genes   Eye-color genes

Agouti   Black
C   E

Meiosis →

c   e
Albino   Pink

Tetrad
(homologous pair of
duplicated chromosomes)

C   E
C   E
c   e
c   e

Chromosomes of
the four gametes

A. Differing genetic information on homologous chromosomes

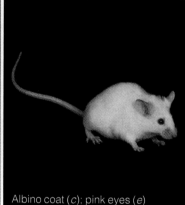

Agouti coat (*C*); black eyes (*E*)   Albino coat (*c*); pink eyes (*e*)

B. Coat-color and eye-color traits in mice

Crossing over further increases genetic variability

Crossing over is the exchange of corresponding segments between two homologous chromosomes. The micrograph and drawing in Figure A show the results of crossing over between two homologous chromosomes during prophase I of meiosis. The chromosomes are a tetrad—four separate chromatids, with each pair of sister chromatids linked at their centromeres. The sites of crossing over appear as X-shaped regions; each is

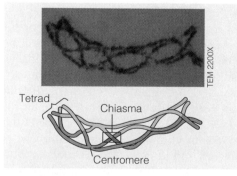

Tetrad
Chiasma
Centromere
TEM 2200X

A. Chiasmata

called a chiasma (the Greek word for cross). A **chiasma** (plural, *chiasmata*) is a place where two homologous (nonsister) chromatids are attached to each other.

Crossing over can take place because during synapsis homologous chromosomes pair up very closely, with a precise, gene-by-gene alignment. When homologous chromosomes exchange segments, the genetic variability provided by the independent orientation of chromosomes at metaphase I is compounded.

Figure B illustrates how crossing over can produce new combinations of genes, using as examples the mouse genes mentioned in the previous module. The process begins during prophase I of meiosis. At the top of the figure is a tetrad with coat-color (*C, c*) and eye-color (*E, e*) genes labeled. In step ①, a chromatid from each homologous chromosome breaks in two; notice that the two chromatids break at corresponding points. ② Next, the two broken chromatids join together; the result is a chiasma. ③ When the homologous chromosomes separate in anaphase I, the joined homologous chromatids come completely apart. However, as the colors indicate, crossing over has changed the content of these two chromatids. A segment of one chromatid has changed place with the equivalent segment of its homologue. ④ Finally, in meiosis II, the sister chromatids separate, each going to a different gamete.

In this example, if there were no crossing over, meiosis could produce only two genetic types of gametes. These would be the ones with the "parental" types of chromosomes shown at the bottom in the figure, carrying either genes *C* and *E* (agouti coat, black eyes) or genes *c* and *e* (white coat, pink eyes). These are the same two kinds of gametes we saw in Figure 8.18A. With crossing over, two other types of gametes can result. One of these carries genes *C* and *e* (agouti coat, pink eyes), and the other carries genes *c* and *E* (white coat, black eyes). The chromosomes carried by these gametes are called "recombinant" because they result from **genetic recombination,** the production of gene combinations different from those carried by the original chromosomes.

But our two-gene model oversimplifies the situation. The recombined segments of these chromatids carry many genes rather than just two; consequently, a single crossover event would affect many genes. When we consider that most chromosomes contain thousands of genes, it's no wonder that gametes and the offspring that result from them can be so different. In fact, it's surprising that even siblings resemble one another as much as they do.

We have now examined three mechanisms that produce genetic variability in sexually reproducing organisms: independent orientation of chromosomes at metaphase I, random fertilization, and crossing over. When we take up molecular genetics in Chapter 10, we will see yet another source of variability—mutations, or rare changes in the DNA of genes. The different versions of genes that homologous chromosomes may have at each locus originally arise from mutations, so mutations are ultimately responsible for genetic diversity in living organisms. In Chapter 9, we continue our study of inheritance, looking first at the historical development of the science of genetics and then at the principles governing the way traits are passed from parents to offspring.

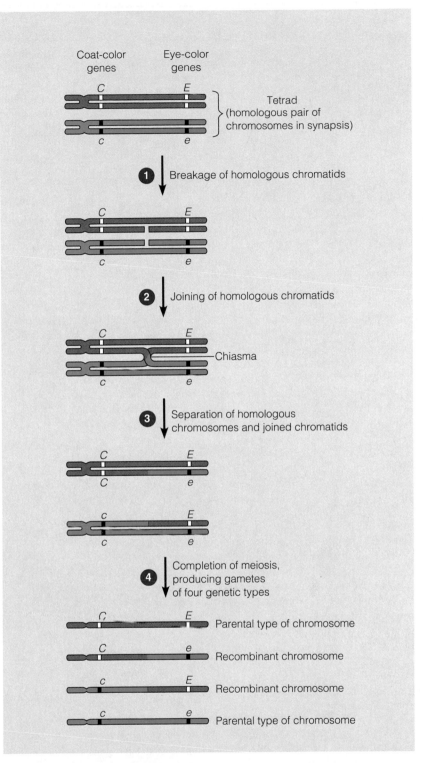

B. How crossing over leads to genetic recombination

Begin your review by rereading the module headings and Module 8.16 and scanning the figures before proceeding to the Chapter Summary and questions.

Chapter Summary

Introduction–8.2 Cell division is at the heart of the reproduction of cells and organisms, because cells come only from cells. Some organisms reproduce asexually, and their offspring are all exact genetic copies of the parent and of each other. Others reproduce sexually, producing a variety of offspring, each with a unique combination of characteristics.

8.3 Bacterial cells divide asexually by binary fission. A bacterium possesses a single chromosome—the structure containing its genes, consisting of DNA. The cell replicates the DNA, leaving both copies attached to the cell membrane. Growth of the cell separates the chromosomes, and the membrane and cell wall divide the cell in two.

8.4–8.6 In eukaryotes, cell division is one phase of a more complex cell cycle. A eukaryotic cell has many more genes than a bacterium, grouped into multiple chromosomes in the nucleus. Each chromosome contains a long DNA molecule with thousands of genes. During the interval between divisions, called interphase, chromosomes are not visible, because they remain in the form of thin, loosely packed chromatin fibers. During interphase, the chromosomes replicate, producing sister chromatids (containing identical genes) joined together at the centromere. Before and after the period of DNA synthesis, the cell goes through phases of metabolic activity and growth.

8.6–8.9, 8.12 Eukaryotic cell division consists of two processes: mitosis and cytokinesis. In mitosis, the duplicated chromosomes of one nucleus are distributed into two daughter nuclei. After the chromosomes thicken (they become visible with a light microscope), a mitotic spindle made of microtubules moves them to the middle of the cell. The microtubules then separate the sister chromatids and move them to opposite poles of the cell, where two new nuclei form. The daughter cells receive identical sets of chromosomes. The process of cytokinesis, in which the cell divides in two, overlaps the end of mitosis. In animals, cleavage pinches the cell apart. In plants, a membranous cell plate splits the cell in two. Mitosis produces identical cells for growth, cell replacement, and asexual reproduction.

8.10–8.11 Some cells divide regularly, other cells only when stimulated, and others not at all. In laboratory cultures, most normal cells divide only on a solid surface, and only for 20–50 generations. They continue dividing only until they touch one another, a phenomenon called density-dependent inhibition. A cell will divide only if it passes a crucial stage called the restriction point, which occurs just before DNA replication. Cancer cells fail to respond to these controls. They divide excessively, and can kill the organism. Radiation and chemotherapy are effective as cancer treatments because they interfere with cell division.

8.13–8.14 The somatic, or body, cells of each species contain a specific number of chromosomes; human cells have 46. These cells are said to be diploid, because they contain two homologous sets (23 pairs) of chromosomes. The chromosomes of a homologous pair carry genes for the same traits at the same place, or locus. Gametes—eggs and sperm—are haploid cells. Each gamete contains a single set of chromosomes; human gametes have 23. At fertilization, a sperm fuses with an egg, forming a diploid zygote. Repeated mitotic cell divisions lead to a multicellular adult made of diploid cells. The diploid adult produces haploid gametes by meiosis, a kind of cell division that reduces the chromosome number by half. Sexual life cycles involve the alternation of haploid and diploid stages.

8.15–8.16 Meiosis, like mitosis, is preceded by chromosome duplication, but in meiosis the cell divides twice to form four daughter cells. The first division, meiosis I, starts with synapsis, the pairing of homologous chromosomes. Crossing over causes homologous chromosomes to exchange corresponding segments. Meiosis I separates each homologous pair, producing two daughter cells, each with one set of chromosomes. Meiosis II is essentially the same as mitosis. In each of the cells, the sister chromatids of each chromosome separate; a total of four haploid cells is produced.

8.17–8.19 Homologous chromosomes have genes for the same traits at corresponding loci. But the two chromosomes can bear different genetic information for the same traits at corresponding loci. The great number of possible arrangements of chromosome pairs at metaphase of meiosis I can produce many different combinations of chromosomes in eggs and sperm. Genetic recombination further increases variability. Finally, random fertilization of eggs by sperm greatly multiplies the number of possible combinations of chromosomes and genes in the offspring.

Testing Your Knowledge

Multiple Choice

1. If a muscle cell in a grasshopper contains 24 chromosomes, a grasshopper sperm cell would contain ———— chromosomes.
   a. 3   d. 24
   b. 6   e. 48
   c. 12

2. Which of the following phases of mitosis is essentially the opposite of prophase in terms of nuclear changes?
   a. telophase   d. interphase
   b. metaphase   e. anaphase
   c. S phase

3. A biochemist measured the amount of DNA in cells growing in the laboratory and found that the quantity of DNA in a cell doubled
   a. between prophase and anaphase of mitosis
   b. between the G_1 and G_2 phases of the cell cycle
   c. during the M phase of the cell cycle
   d. between prophase I and prophase II of meiosis
   e. between anaphase and telophase of mitosis

4. Which of the following is *not* a function of mitosis in humans?
   a. repair of wounds
   b. growth
   c. production of gametes
   d. replacement of lost or damaged cells
   e. multiplication of somatic cells

5. A micrograph of a dividing cell from a mouse showed 19 chromosomes, each consisting of two sister chromatids. During which of the following stages of cell division could this picture have been taken? (*Explain your answer.*)
   a. prophase of mitosis
   b. telophase II of meiosis
   c. prophase I of meiosis
   d. anaphase of mitosis
   e. prophase II of meiosis

6. Cytochalasin B is a chemical that disrupts microfilament formation. This chemical would interfere with

    a. DNA replication

    b. formation of the mitotic spindle

    c. cleavage

    d. formation of the cell plate

    e. crossing over

7. It is difficult to observe individual chromosomes during interphase because

    a. the DNA has not been replicated yet

    b. they uncoil to form long, thin strands

    c. they leave the nucleus and are dispersed to other parts of the cell

    d. homologous chromosomes do not pair up until division starts

    e. the spindle must move them to the metaphase plate before they become visible

8. A fruit fly somatic cell contains 8 chromosomes. This means that _____ different combinations of chromosomes are possible in its gametes.

    a. 4   **d.** 32

    b. 8   **e.** 64

    c. 16

True/False *(Change false statements to make them true.)*

1. Homologous chromosomes can carry different information for the same traits.

2. Microtubules move the chromosomes in mitosis and meiosis.

3. Sister chromatids separate from each other in prophase of mitosis.

4. The *X* and *Y* chromosomes are called autosomes.

5. Homologous chromosome pairs split up in the second division of meiosis.

6. The haploid phase of the human life cycle begins with fertilization.

7. Chromosomes may swap genes in the process of crossing over.

8. Nerve cells, which do not divide, stay in the G_2 phase of the cell cycle.

9. In humans, $n = 46$.

10. Crossing over occurs during prophase I of meiosis.

Describing, Comparing, and Explaining

1. Briefly describe how three different processes that occur during the sexual life cycle increase the genetic diversity of offspring.

2. What are the major differences between mitosis and meiosis?

3. In the light micrograph below of dividing cells near the tip of an onion root, identify a cell in interphase, prophase, metaphase, anaphase, and telophase. Describe the major events occurring at each stage.

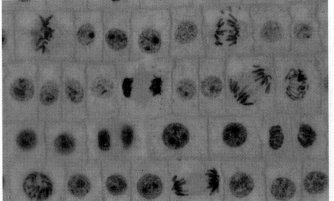

LM 600×

4. Discuss the factors that control the division of eukaryotic cells grown in the laboratory. Cancer cells are easier to grow in the lab than other cells. Why do you suppose this is the case?

5. Compare cytokinesis in plant and animal cells.

Thinking Critically

1. If a human gamete has 100,000 genes, how many genes are carried by the average human chromosome?

2. Bacteria are able to divide on a much faster schedule than eukaryotic cells. Some bacteria can divide every 20 minutes, while the minimum time required by eukaryotic cells in a rapidly developing embryo is about once per hour, and most cells divide much less often than that. What are some possible reasons why bacteria can divide at a faster rate than eukaryotic cells?

3. Red blood cells, which carry oxygen to body tissues, live for only about 120 days. Replacement cells are produced by cell division in bone marrow. How many cell divisions must occur each second in your bone marrow just to replace red blood cells? Here is some information to use in calculating your answer. There are about 5 million red blood cells per mm^3 of blood. An average adult has about 5 L (5000 cm^3) of blood. (*Hints:* What is the total number of red blood cells in the body? What fraction of them must be replaced each day if all are replaced in 120 days?)

4. A mule is the offspring of a horse and a donkey. A donkey sperm contains 31 chromosomes and a horse egg 32 chromosomes, so the zygote contains a total of 63 chromosomes. The zygote develops normally. The combined set of chromosomes is not a problem in mitosis, and the mule combines some of the best characteristics of horses and donkeys. However, a mule is sterile; meiosis cannot occur normally in its testes (or ovaries). Explain why mitosis is normal in cells containing both horse and donkey chromosomes, but the mixed set of chromosomes interferes with meiosis.

Science, Technology, and Society

Every year about a million Americans are diagnosed as having cancer. This means that about 75 million Americans now living will eventually have cancer, and one in five will die of the disease. There are many kinds of cancers and many causes of the disease. For example, smoking causes most lung cancers. Overexposure to ultraviolet rays in sunlight causes most skin cancers. There is evidence that a high-fat, low-fiber diet is a factor in breast, colon, and prostate cancers. And agents in the workplace such as asbestos and vinyl chloride are also implicated as causes of cancer. Hundreds of millions of dollars are spent each year in the search for effective treatments for cancer; far less money is spent on preventing cancer. Why might this be the case? What kinds of lifestyle changes could we make to help prevent cancer? What kinds of prevention programs could be initiated or strengthened to encourage these changes? What factors might impede such changes and programs? Should we devote more of our resources to treating cancer or preventing it? Why?

Patterns of Inheritance 9

udgies, also called parakeets, are small, long-tailed parrots native to Australia and kept as pets throughout the world. Female budgies (short for budgerigars) have pinkish legs and a brownish cere, the featherless area above the beak. Male budgies have gray-blue legs and a bluish cere. Wild budgies typically have green underparts and yellow upperparts barred with black, like the bird in the top row (second from right) in the photograph at the left. The white, blue, and yellow birds are uncommon in nature, but large numbers of these and other color variants are raised in captivity and are available in most pet stores. Blue budgies called sky-blues are especially popular.

Feather color in budgies is an inherited trait, and pet breeders can predict which color variants will result from which kinds of parents. For instance, if two wild-type (green and yellow) budgies are mated, as shown in Figure A below, it is likely that all of their offspring will be green and yellow. The same is true when a wild-type bird and a sky-blue are mated (Figure B). However, if two *offspring* of a wild budgie and a sky-blue bird are mated, their offspring, the "second generation" in Figure B, will most likely include both green and yellow birds and sky-blues. Knowing how these feather-color traits are inherited—that is, their inheritance pattern—a breeder would predict that about one out of four second-generation offspring would be sky-blue birds.

The heritable variation we see in budgies is minor compared to some of the dramatic differences that breeders have developed in certain other species. Of the hundreds of domestic dog breeds, for instance, some, such as Saint Bernards and Chihuahuas, are so different that they can hardly be recognized as members of the same species. In the plant kingdom, our domestic corn plant is so different from its original form that biologists have only recently identified its ancestor as a species of wild Mexican grass called teosinte.

Every species has a spectrum of heritable variation, and people have exploited this variation for centuries in developing domestic strains of plants and animals. The historical roots of **genetics,** the science of heredity, date back to the earliest attempts at selective breeding. In this chapter, we examine the rules that govern how inherited traits, such as feather color in budgies, are passed from parents to offspring. We will look at several inheritance patterns, including the ones shown here, and see how to predict the ratios of offspring with particular traits. Most important of all, we will uncover a basic biological concept—that the behavior of chromosomes during gamete formation and fertilization, which we discussed in Chapter 8, accounts for patterns of inheritance.

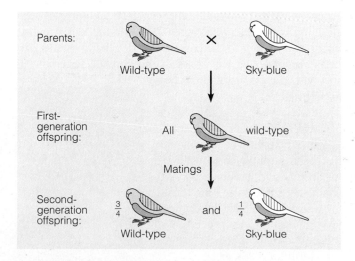

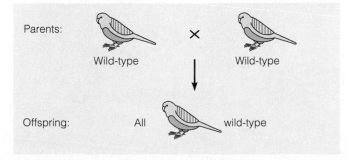

A. Offspring from the mating of two wild-type birds

B. Two generations of offspring from the mating of wild-type with sky-blue birds

9.1 The science of genetics has ancient roots

Attempts to explain inheritance date back at least to ancient Greece. Hippocrates, called the father of medicine, suggested an explanation called pangenesis. According to this idea, particles, called pangenes, from each part of an organism's body collect in the eggs or sperm and are then passed to the next generation. If pangenesis were true, changes that occur in various parts of the body during an organism's life could be passed on to the next generation. The Greek philosopher Aristotle rejected this idea as too simplistic, saying that what is inherited is the potential to produce body features, rather than particles of the features themselves.

Actually, pangenesis proves incorrect on several counts. The reproductive cells are not composed of particles from somatic (body) cells, and changes in somatic cells do not influence eggs and sperm. For instance, no matter how much you might enlarge your biceps by lifting weights, muscle cells in your arms do not transmit genetic information to your gametes. Accordingly, neither your gametes nor your offspring will be changed by your weight-lifting efforts. None of the changes that occur in parts of the body during an organism's lifetime are inherited by the next generation. This may seem like common sense today, but the pangenesis hypothesis and the idea that traits acquired during an individual's lifetime are passed on to offspring prevailed well into the nineteenth century.

By carefully tracking inheritance patterns in ornamental plants, biologists of the early nineteenth century established that offspring inherit traits from both parents. The favored explanation of inheritance then became the "blending" hypothesis, the idea that the hereditary materials contributed by the male and female parents mix in forming the offspring, much the way blue and yellow paints blend to make green. According to this hypothesis, if a blue budgie mated with a yellow budgie, all the offspring would be green, and once the colors blended in the hereditary material, they would be as inseparable as paint pigments. Consequently, all offspring of the green budgies would be green, and eventually all budgies everywhere would be green. As we saw in the introduction, however, green parakeets can produce blue offspring. The blending hypothesis was generally accepted for much of the nineteenth century, but it was finally found to be untenable because it does not explain how traits that disappear in one generation can reappear in later ones.

9.2 Experimental genetics began in an abbey garden

The modern science of genetics began in the 1860s, when an Augustinian monk named Gregor Mendel discovered the fundamental principles of genetics by breeding garden peas (Figure A). Mendel lived and worked in an abbey in Brunn, Austria (now Brno, in the Czech Republic). In a paper published in 1866, Mendel correctly argued that parents pass on to their offspring discrete heritable factors. He stressed that heritable factors (we call them genes today) retain their individuality generation after generation. Mendel's explanation contrasted both with the pangenesis idea that heritable particles are affected by changes in the parents' somatic cells and with the erroneous blending hypothesis.

Mendel's work is a classic in the history of biology. While studying at the University of Vienna, he had been strongly influenced by his physics, mathematics, and chemistry professors. As a result, his research was both experimental and numerically precise, and these qualities were largely responsible for his success.

Mendel probably chose to study garden peas because they were easy to grow and available in many readily distinguishable varieties. Also, with pea plants, Mendel was able to exercise strict control over which plants mated with which. As Figure B shows, the petals of the pea flower almost completely enclose the female and male parts—carpel and stamens, respectively. (In the drawing, we have made one of the petals transparent.) Consequently, in nature, the plants usually **self-fertilize**, when sperm-carrying pollen grains released from the stamens land on the egg-containing carpel of the same flower. Mendel could ensure self-

A. Gregor Mendel in his garden

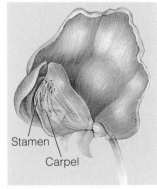

B. Anatomy of a pea flower

Stamen

Carpel

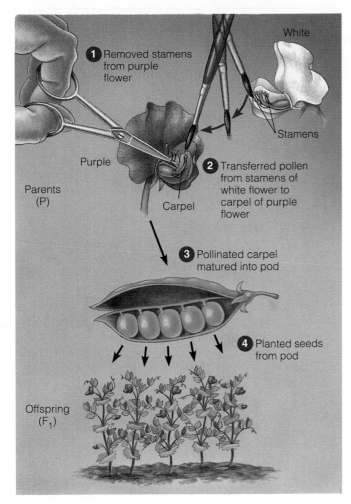

C. Mendel's technique for cross-fertilization of pea plants

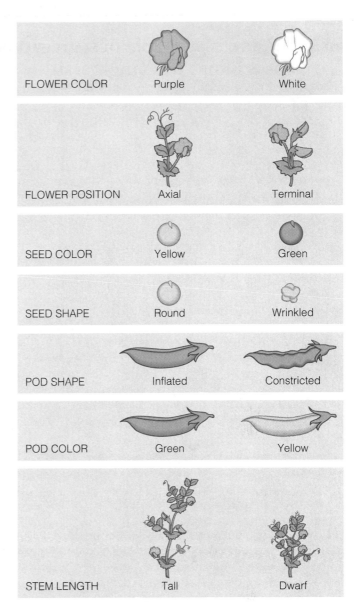

D. The seven pea characteristics studied by Mendel

fertilization by covering a flower with a small bag so that no pollen from another plant could reach the carpel. When he wanted **cross-fertilization** (fertilization of one plant by pollen from a different plant), he used the method shown in Figure C. ① He first prevented self-fertilization by cutting off the stamens from an immature flower before it produced pollen. This stamenless plant would be the female parent in the experiment. ② To cross-fertilize this female, he dusted its carpel with pollen from another plant. After pollination, ③ the carpel developed into a pod, containing seeds (peas) that ④ he planted. The seeds grew into offspring plants.

Thus, Mendel could either let a pea plant self-fertilize or cross-fertilize it with a known source of pollen. In either case, he could always be sure of the parentage of new plants.

Mendel's success was due not only to his experimental approach and choice of organism, but also to his selection of traits to study. He chose to follow seven traits, each of which occurs in two distinct forms (Figure D). Mendel worked with his plants until he was sure he had **true-breeding** varieties—that is, varieties for which self-fertilization produced offspring all identical to the parent. For instance, he identified a round-seed variety that, when self-fertilized, produced plants with round seeds only.

Now Mendel was ready to ask what would happen when he crossed his different varieties. For example, what offspring would result if plants with purple flowers and plants with white flowers were cross-fertilized, as shown in Figure C? In the language of plant and animal breeders and geneticists, the offspring of two different varieties, such as these, are called **hybrids,** and the cross-fertilization itself is referred to as a **hybridization,** or simply **a cross.** The parental plants are called the **P generation** (P for parental), and their hybrid offspring are the **F₁ generation** (F for filial, from the Latin word for son). In the next module, we discuss the F_1 generation results of Mendel's flower-color crosses. We also see what happened when he allowed the F_1 offspring to self-fertilize to produce the next generation, the **F₂ generation.**

9.3 Mendel's principle of segregation describes inheritance of a single trait

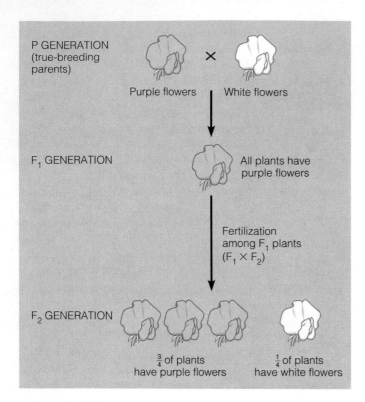

A. A monohybrid cross

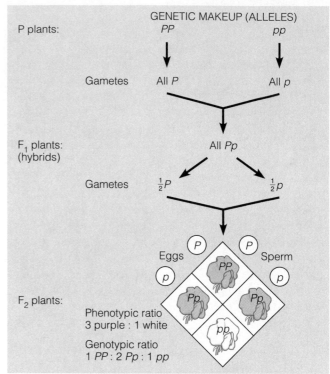

B. Explanation of the monohybrid cross in Figure A

A breeding experiment that tracks the inheritance of a single trait, such as flower color, is called a **monohybrid cross**. After performing many such crosses, Mendel formulated several hypotheses about inheritance. Let's look at some of his experimental results and follow the reasoning that led to his conclusions.

Figure A illustrates the results of a monohybrid cross between a pea plant with purple flowers and one with white flowers. Mendel discovered that F_1 plants produced by these two true-breeding parents all had truly purple flowers. They were not a lighter purple, as predicted by the blending hypothesis. Was the heritable factor for white flowers now lost as a result of the hybridization? By crossing two F_1 plants, Mendel found the answer to be no. Out of 929 F_2 plants, Mendel found that 705 (about $\frac{3}{4}$) had purple flowers and 224 (about $\frac{1}{4}$) had white flowers—in other words, a ratio of about three plants with purple flowers to every one with white flowers in the F_2 generation. Mendel concluded that the heritable factor for white flowers did not disappear in the F_1 plants, but only the purple-flower factor was affecting flower color in the F_1. He also deduced that, since all the F_1 plants had purple flowers but passed the white trait on to some of their offspring, the F_1 plants must have carried two factors for the flower-color trait, one for purple flowers and one for white. From these results and many others, Mendel developed four hypotheses:

1. *There are alternative forms of genes, the units that determine heritable traits.* For example, the gene for flower color exists in one form for purple and another for white. We now call alternative forms of genes **alleles** (from the Greek *allelon,* "of each other").

2. *For each inherited trait, an organism has two genes, one from each parent. These genes may both be the same allele, or they may be different alleles.*

3. *A sperm or egg carries only one allele for each inherited trait, because allele pairs separate (segregate) from each other during the production of gametes.* Mendel also theorized that when sperm and egg unite at fertilization, each contributes its allele, thus restoring the paired condition in the offspring.

4. *When the two genes of a pair are different alleles, one is fully expressed and the other has no noticeable effect on the organism's appearance. These are called the **dominant allele** and the **recessive allele,** respectively.*

Figure B illustrates how Mendel's hypotheses explain the results shown in Figure A. Geneticists often use uppercase and lowercase letters to distinguish between dominant and recessive alleles. These abbreviations help us see the connection between Mendel's results and his four hypotheses. Here, *P* represents the dominant allele, for purple flowers, and *p* stands for the recessive allele, for white flowers. At the top in Figure B, you see the alleles carried by the parental plants. Both plants were true-breeding, and Mendel's first two hypotheses propose that one parental variety had two alleles for purple flowers (*PP*), while the other variety had two alleles for white flowers (*pp*). A true-breeding organism, which has a pair of identical alleles, such as *PP* or *pp*, for a trait, is said to be **homozygous** (Greek *homos*, alike, and *-zygos*, joined) for that trait.

Consistent with hypothesis 3, the gametes of Mendel's parental plants each carried one allele; thus, the parental gametes in Figure B are either *P* or *p*. As a result of fertilization, the F_1 hybrids each inherited one allele for purple flowers and one for white flowers. Hypothesis 4 explains why all of the F_1 hybrids (designated *Pp*) had purple flowers: The dominant *P* allele is fully expressed, while the recessive *p* allele has no effect on flower color. An organism with two different alleles for a trait, such as a pea plant with alleles *P* and *p*, is said to be **heterozygous** (Greek *heteros*, different) for that trait.

Mendel's hypotheses also explain the 3:1 ($\frac{3}{4}$ purple flowers to $\frac{1}{4}$ white flowers) ratio in the F_2 generation. Because the F_1 hybrids are *Pp*, they will produce gametes *P* and *p* in equal numbers. The diamond at the bottom in Figure B, called a **Punnett square**, shows the four possible combinations of these gametes.

The Punnett square indicates the proportions of F_2 plants predicted by Mendel's hypotheses. If a sperm carrying allele *P* fertilizes an egg carrying allele *P*, the offspring (*PP*) will produce purple flowers. Mendel's hypotheses predict that this combination will occur in $\frac{1}{4}$ of the offspring. As shown

in the Punnett square, the hypotheses also predict that $\frac{2}{4}$ of the offspring will inherit one *P* allele and one *p* allele. These offspring (*Pp*) will all have purple flowers because *P* is dominant. The remaining $\frac{1}{4}$ of the F_2 plants will inherit two *p* alleles and will produce white flowers.

Because an organism's appearance does not always reveal its genetic composition, geneticists distinguish between an organism's expressed, or physical, traits, called its **phenotype** (such as purple or white flowers), and its genetic makeup, its **genotype** (in this example, *PP*, *Pp*, or *pp*). So now we see that Figure A shows the phenotypes and Figure B the genotypes involved in our sample cross. For the offspring of our cross, the ratio of plants with purple flowers to those with white flowers (3:1) is called the phenotypic ratio. By contrast, the genotypic ratio, as shown by the Punnett square, is 1(*PP*) : 2(*Pp*) : 1(*pp*).

Mendel found that each of the seven traits he studied exhibited the same inheritance pattern. (In Figure 9.2D, the traits in the left column are all determined by dominant alleles, in the right column by recessive alleles.) One parental trait disappears in the F_1 generation of heterozygotes, only to reappear in one-fourth of the F_2 offspring. The mechanism underlying this pattern is stated by Mendel's **principle of segregation:** *Pairs of genes segregate (separate) during gamete formation; the fusion of gametes at fertilization pairs genes once again.* Research since Mendel's time has established that the principle of segregation applies to all sexually reproducing organisms, including humans. We return to Mendel and his experiments with pea plants in Module 9.5. Before that, let's see how some of the concepts we discussed in Chapter 8 fit with what we have said about genetics so far.

Homologous chromosomes bear the two alleles for each trait 9.4

The figure at the right shows a matched pair of unduplicated chromosomes. Recall from Chapter 8 that the two chromosomes making up a matched pair are called homologous chromosomes, and that every individual, whether pea plant or human, has two sets of homologous chromosomes. One set comes from the female parent, the other from the male parent.

The chromosomes shown here represent one homologous pair in a pea plant. The labeled bands on the chromosomes represent a few gene loci, specific locations of genes along the length of the chromosome. The matching colors of corresponding loci on the two homologues highlight the fact that homologous chromosomes have genes for the same traits located at the same positions along their lengths. However, as the uppercase and lowercase letters next to the loci indicate, the two chromosomes may bear either the same alleles or different ones. Thus, we see the connection between Mendel's principles and homologous chromosomes: *Alleles (alter-*

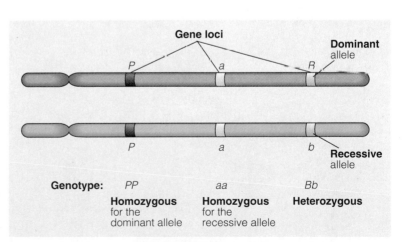

native forms) of a gene reside at the same locus on homologous chromosomes.

The diagram here also serves as a review of some of the genetic terms we have encountered to this point. We will return to the chromosomal basis of inheritance in more detail beginning with Module 9.14.

9.5 The principle of independent assortment is revealed by tracking two traits at once

Two of the seven traits Mendel studied were seed shape and seed color. Mendel's seeds were either round or wrinkled in shape, and they were either yellow or green in color. From monohybrid crosses, Mendel knew that the allele for round shape (designated *R*) was dominant to the allele for wrinkled shape (*r*) and that the allele for yellow seed color (*Y*) was dominant to the allele for green seed color (*y*). What would result from a mating of parental varieties differing in two traits—a *dihybrid cross*? Mendel performed a dihybrid cross between homozygous plants having round yellow seeds (genotype *RRYY*) and plants with wrinkled green seeds (*rryy*). As shown in Figure A below, the union of *RY* and *ry* gametes produced hybrids heterozygous for both traits (*RrYy*). As we would expect, all of these offspring, the F_1 generation, had round yellow seeds. But were the two traits transmitted from parents to offspring as a package, or was each trait inherited independently of the other?

The question was answered when Mendel allowed the F_1 plants to self-fertilize. If the genes for the two traits were inherited together, as shown on the left in the figure, then the F_1 hybrids could produce only the same two kinds of gametes that they received from their parents. In that case, the F_2 generation would show a 3:1 phenotypic ratio (three plants with round yellow seeds for every one with wrinkled green seeds), as in the left Punnett square. If, however, the two seed traits segregated independently, then the F_1 generation could produce four types of gametes—*RY, rY, Ry,* and *ry*—in equal quantities. The Punnett square on the right shows all possible combinations of alleles that can result in the F_2 generation from the union of four kinds of sperm with four kinds of eggs. You can see that there are nine different genotypes in the F_2. However, there are only four kinds of phenotypes, with a ratio of 9:3:3:1 (9 round yellow to 3 wrinkled yellow to 3 round green to 1 wrinkled green).

The right-hand Punnett square also reveals that a dihybrid cross is equivalent to two monohybrid crosses occurring simultaneously. From the 9:3:3:1 ratio, we can see that there are 12 plants with round seeds to 4 with wrinkled seeds, and 12 yellow-seeded plants to 4 green-seeded ones. These 12:4 ratios each reduce to 3:1, which is the F_2 ratio for a monohybrid cross. Mendel tried his seven pea traits in various dihybrid combinations and always observed a 9:3:3:1 ratio (or two simultaneous 3:1 ratios) of phenotypes in the F_2 generation. These results supported the hypothesis that *each pair of alleles segregates independently during gamete formation.* This behavior of genes is called Mendel's **principle of independent assortment.**

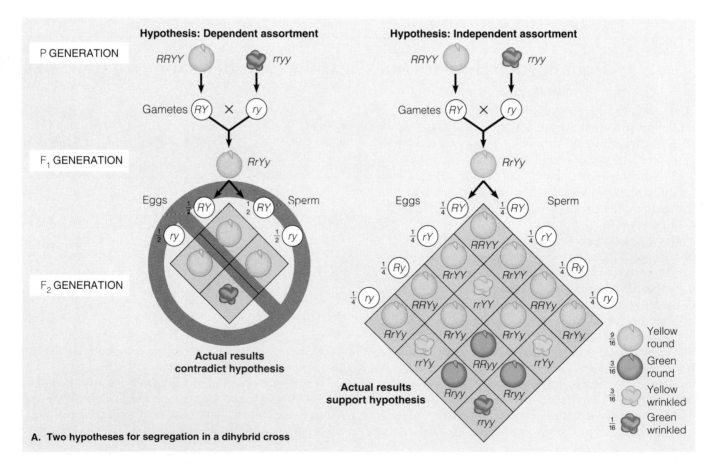

A. Two hypotheses for segregation in a dihybrid cross

Phenotypes				
	Green	Blue ("sky-blue")	Yellow ("black-eyed yellow")	White
Genotypes	B_C_	bbC_	B_cc	bbcc

Mating of heterozygotes	(Green) BbCc × BbCc (Green)

Phenotypic ratio of progeny	9 Green	3 Blue	3 Yellow	1 White

B. Independent assortment in the budgie

Figure B shows how this principle applies to the inheritance of feather color in budgies. Four alleles (B, b, C, and c) are involved. Birds with the B allele have yellow pigment in their feathers. Those with the C allele have a pigment called melanin, which gives feathers a blue or blackish color. The recessive alleles c and b produce no pigments in the feathers. Thus, the white budgie's genotype is bbcc. The other three budgies have color in their feathers because they have at least one B or one C allele. (The blanks indicate that the other alleles are unknown.) The all-yellow bird has the B allele but no C, thus no melanin pigment in its feathers. A sky-blue budgie has C but no B, thus no yellow. The green and yellow bird has both B and C; its green feathers result from the combination of yellow pigment and melanin. Its upper parts are colored by the separate expression of the B and C alleles—yellow color from B and black stripes from C.

The lower part of Figure B shows what happens when we mate two BbCc heterozygous, green and yellow budgies. The F_2 phenotypic ratio will be 9 green-and-yellows to 3 sky-blues to 3 pure yellows to 1 white. These results are analogous to the results in Figure A and show that the B and C genes are inherited independently.

Geneticists use the testcross to determine unknown genotypes

Suppose you have a sky-blue budgie and you want to determine its genotype. You know that it carries two copies of the recessive allele (b) because it shows no evidence of yellow pigment. But is it homozygous (CC) or heterozygous (Cc) for the melanin allele? To answer this question, you would need to perform what geneticists call a **testcross**, a mating between an individual of unknown genotype (your bird) and a homozygous recessive individual—in this case, a white budgie.

The figure here shows the offspring that could result from a mating between a white, bbcc budgie and a sky-blue one (bbCC or bbCc). If, as shown on the left, the sky-blue parent's genotype is CC, we would expect all the offspring to be sky-blue, because a cross between genotypes CC and cc can produce only Cc offspring. On the other hand, if the sky-blue parent is Cc, we would expect half the offspring to be sky-blue (Cc) and half white (cc). Thus, the appearance of the offspring reveals the sky-blue bird's genotype. The figure also shows that you could expect the white and sky-blue offspring of a Cc × cc cross to exhibit a 1:1 (1 sky-blue to 1 white) phenotypic ratio.

Mendel used testcrosses to determine whether he had true-breeding varieties of plants. The testcross continues to be an important tool of geneticists.

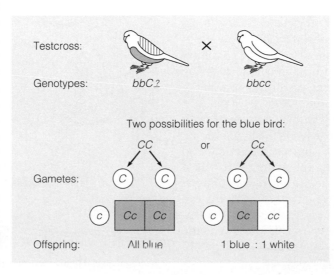

Testcross:

Genotypes: bbC? × bbcc

Two possibilities for the blue bird:

CC or Cc

Gametes: C C | C c

	C	C
c	Cc	Cc

	C	c
c	Cc	cc

Offspring: All blue | 1 blue : 1 white

Mendel's principles reflect the rules of probability

Mendel had a firm background in mathematics, which served him well in his studies of inheritance. He understood, for instance, that the segregation of allele pairs during gamete formation and the re-forming of pairs at fertilization obey the rules of probability—the same rules that apply to the tossing of coins, the rolling of dice, and the drawing of cards. Mendel also appreciated the statistical nature of inheritance. He knew that he needed to obtain large samples—count many offspring from his crosses—before he could begin to interpret inheritance patterns.

Let's see how the rules of probability apply to inheritance. The probability scale ranges from 0 to 1. An event that is certain to occur has a probability of 1, while an event that is certain not to occur has a probability of 0. The probabilities of all possible outcomes for an event must add up to 1. With a coin, the chance of tossing heads is $\frac{1}{2}$, and the chance of tossing tails is $\frac{1}{2}$. In a standard deck of 52 playing cards, the chance of drawing a jack of diamonds is $\frac{1}{52}$, and the chance of drawing any card other than the jack of diamonds is $\frac{51}{52}$.

An important lesson we can learn from coin tossing is that for each and every toss of the coin, the probability of heads is $\frac{1}{2}$. In other words, the outcome of any particular toss is unaffected by what has happened on previous attempts. Each toss is an *independent event*.

If two coins are tossed simultaneously, the outcome for each coin is an independent event, unaffected by the other coin. What is the chance that both coins will land heads up? The probability of such a compound event is the *product* of the separate probabilities of the independent events—for the coins, $\frac{1}{2} \times \frac{1}{2} = \frac{1}{4}$. This is called the **rule of multiplication,** and it holds true for genetics as well as coin tosses, as illustrated in the figure here. In our dihybrid cross of budgies, the genotype of the F_1 birds heterozygous for the melanin pigment was *Cc*. What is the probability that a particular F_2 bird will have the *cc* genotype? To produce a *cc* offspring, both egg and sperm must carry the *c* allele. The probability that an egg will have the *c* allele is $\frac{1}{2}$, and the probability that a sperm will have the *c* allele is also $\frac{1}{2}$. By the rule of multiplication, the probability that two *c* alleles will come together at fertilization is $\frac{1}{2} \times \frac{1}{2} = \frac{1}{4}$. This is exactly the answer given by the Punnett square. If we know the genotypes of the parents, we can predict the probability for any genotype among the offspring.

Now let's consider the probability that an F_2 budgie will be heterozygous for the melanin gene. As the figure shows, there are two ways in which F_1 gametes can combine to produce a heterozygous offspring. The dominant (*C*) allele can come from the egg and the recessive (*c*) allele from the sperm, or vice versa. The probability that an event can occur in two or more alternative ways is the *sum* of the separate probabilities of the different ways; this is known as the **rule of addition.** Using this rule, we can calculate the probability of an F_2 heterozygote as $\frac{1}{4} + \frac{1}{4} = \frac{1}{2}$.

By applying the rules of probability to segregation and independent assortment, we can solve some rather complex genetics problems. For instance, we can predict the results of trihybrid crosses, in which three different traits are involved. Consider a cross between two organisms that both have the genotype *AaBbCc*. What is the probability that an offspring from this cross will be a recessive homozygote for all three traits (*aabbcc*)? Since each allele pair assorts independently, we can treat this trihybrid cross as three separate monohybrid crosses:

Aa × *Aa*: Probability of *aa* offspring = $\frac{1}{4}$

Bb × *Bb*: Probability of *bb* offspring = $\frac{1}{4}$

Cc × *Cc*: Probability of *cc* offspring = $\frac{1}{4}$

Because the segregation of each allele pair is an independent event, we use the rule of multiplication to calculate the probability that the offspring will be *aabbcc:*

$$\tfrac{1}{4}aa \times \tfrac{1}{4}bb \times \tfrac{1}{4}cc = \tfrac{1}{64}$$

We could reach the same conclusion by constructing a 64-section Punnett square, but that would be a very laborious process.

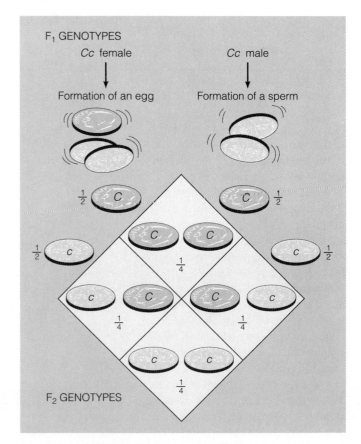

F_1 GENOTYPES

Cc female   *Cc* male

Formation of an egg   Formation of a sperm

F_2 GENOTYPES

Segregation and fertilization as chance events

Mendel's principles apply to the inheritance of many human traits

Garden peas, budgies, and many other species are more convenient subjects for genetic research than humans. The human generation time is about 20 years, and parents produce relatively few offspring compared to most other species. Furthermore, well-planned breeding experiments like the ones Mendel performed are impossible (or at least ethically unacceptable) with humans. Despite these difficulties, the study of human genetics continues to advance, driven by our desire to understand our own inheritance. Recently, geneticists have made great strides using the techniques of cell culture and molecular biology. We discuss this work in some detail in Chapter 13. Here, we look at a few examples of human genetics in light of Mendel's principles. Since Chapter 13 discusses genetic diseases, this module focuses on nondisease traits.

The figure here illustrates five human traits that are thought to be determined by simple dominant-recessive inheritance. If we call the dominant allele of any such gene A, the dominant phenotypes result from either the homozygous genotype AA or the heterozygous genotype Aa. Recessive phenotypes result from the homozygous genotype aa. In genetics, the word "dominant" does not imply that a phenotype is either normal or more common than a recessive phenotype. In genetics, dominance means that an allele is expressed in a heterozygote, as well as in an individual homozygous for that allele. By contrast, the corresponding recessive allele is expressed only in a homozygote. Recessive traits are often more common in the population than dominant ones. For example, tight thumb ligaments and the absence of freckles are more common than double-jointed thumbs and freckles. As we will see in Chapter 13, some human diseases are caused by dominant alleles that are much less common than the recessive ones.

Mendel's principles can be applied to predict inheritance patterns for many human traits that involve simple dominance and recessiveness. For example, suppose a man with hair on the backs of his fingers marries a woman with hairless fingers. We know that the woman is homozygous (aa) and that her eggs all carry the a allele. But the man could be either homozygous (AA) or heterozygous (Aa). If the man is homozygous, all his sperm must carry the A allele. As a result, we would expect all of this couple's children to have the Aa genotype and finger hair.

What is the expectation if the man is heterozygous? In this case, the man produces sperm with alleles A and a in equal numbers. We would predict that this man and his aa wife would have equal numbers of children with and without finger hair. If the couple decides to have only two children, what is the probability that both children will have finger hair? Because each child represents an independent event, we can apply the rule of multiplication and determine that this probability is $\frac{1}{2} \times \frac{1}{2} = \frac{1}{4}$.

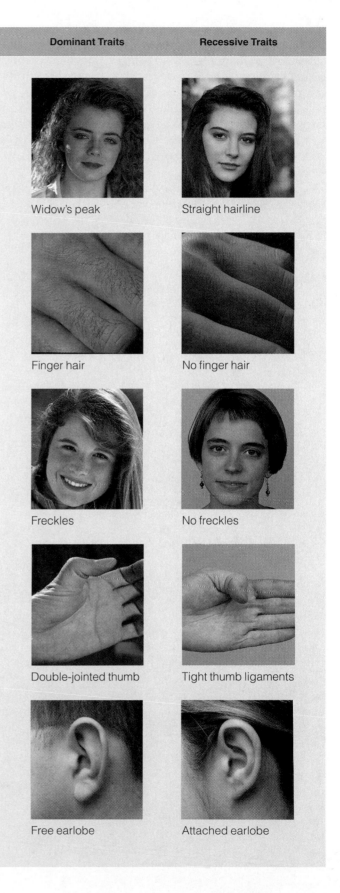

Dominant Traits **Recessive Traits**

Widow's peak | Straight hairline

Finger hair | No finger hair

Freckles | No freckles

Double-jointed thumb | Tight thumb ligaments

Free earlobe | Attached earlobe

The relationship of genotype to phenotype is rarely simple

Mendel's two principles explain inheritance in terms of discrete factors—genes—that are passed along from generation to generation, according to simple rules of probability. Mendel's principles are valid for all sexually reproducing organisms, including garden peas, birds, and human beings. But the patterns of inheritance we have described so far are simpler than most. Other inheritance patterns include cases in which one allele is not completely dominant to the other allele; cases in which there are more than two alternative alleles for a trait; and still other cases in which the genotype does not always dictate the phenotype in the way that Mendel's principles prescribe. Modules 9.10–9.13 consider some of these more complex inheritance patterns.

Incomplete dominance results in intermediate phenotypes

The F₁ offspring of Mendel's pea crosses always looked like one of the two parental varieties because only the dominant alleles were expressed. But for some traits, the F₁ hybrids have an appearance somewhat in between the phenotypes of the two parental varieties, an effect called **incomplete dominance.** For instance, as Figure A illustrates, when red snapdragons are crossed with white snapdragons, all the F₁ hybrids have pink flowers. In this case, both the red allele and the white allele are expressed in the heterozygote flowers.

Incomplete dominance does not support the blending hypothesis, which would predict that the red or white traits could never be retrieved from the pink hybrids. As the Punnett square in Figure A shows, the F₂ offspring appear in a phenotypic ratio of 1 red to 2 pink to 1 white. This F₂ generation results from segregation of the red and white alleles in the gametes of the pink F₁ hybrids. In all cases of incomplete dominance, heterozygotes are phenotypically distinct from the two homozygous varieties, and as a result, the genotypic ratio and the phenotypic ratio are both 1:2:1.

We also see examples of incomplete dominance in humans. One case involves a recessive allele (*h*) responsible for hypercholesterolemia (dangerously high levels of cholesterol in the blood), a disease we introduced in Module 5.20. Normal individuals are genotype *HH*. Heterozygotes (*Hh*; about one in 500 people) have blood cholesterol levels about twice as high as normal. They are unusually prone to atherosclerosis, the blockage of arteries by a buildup of cholesterol in artery walls, and they may have heart attacks from blocked heart arteries by their mid-thirties. Hypercholesterolemia is even more serious in homozygous individuals (*hh*; about one in a million people). Homozygotes have about five times the normal amount of blood cholesterol and may have heart attacks as early as age 2.

Figure B illustrates the molecular basis for hypercholesterolemia. The dominant allele, which normal individuals carry in duplicate (*HH*), specifies a cell-surface protein called an LDL receptor (purple in the diagram). LDLs, or low-density lipoproteins, are cholesterol-containing particles in the blood. The LDL receptors pick up LDL particles

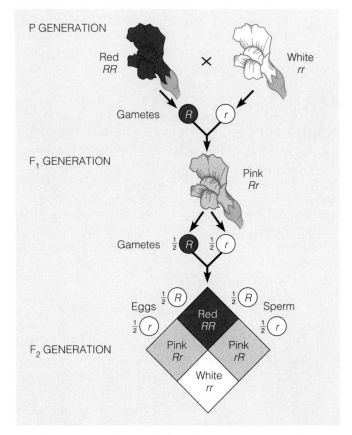

A. Incomplete dominance in snapdragon color

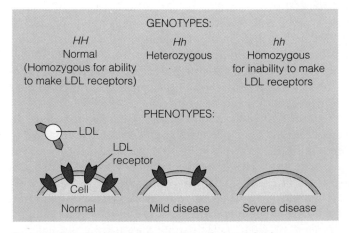

B. Incomplete dominance in human hypercholesterolemia

from the blood and promote their uptake by cells that break down the cholesterol. This process helps prevent the accumulation of cholesterol in arteries. Without the receptors, lethal levels of LDL build up in the blood. Heterozygotes (*Hh*) have only half the normal number of LDL receptors, and homozygotes (*hh*) have none.

Many genes are represented by more than two alleles

So far, we have discussed inheritance patterns involving only two alleles per gene. But many genes have *multiple alleles*. Although each individual carries, at most, two different alleles for a particular trait, in cases of multiple alleles, more than two possible alleles exist. The ABO blood groups in humans are one example of multiple alleles. There are four phenotypes for this trait: A person's blood type may be O, A, B, or AB. These letters refer to two carbohydrates, designated A and B, which are found on the surface of red blood cells. A person's red blood cells may be coated with one substance or the other (type A or B), with both (type AB), or with neither (type O). Matching compatible blood groups is critical for blood transfusions. If a donor's blood cells have a carbohydrate (A or B) that is foreign to the recipient, then the recipient produces blood proteins called antibodies that bind specifically to the foreign carbohydrates and cause the donor blood cells to clump together. This clumping can kill the recipient. The table shows which combinations of blood groups result in clumping. (You'll see blood-cell clumping if you type your blood in the laboratory.)

The four blood types result from various combinations of three different alleles, symbolized as I^A (for the ability to make substance A), I^B (for B), and i (for neither A nor B). Each person inherits one of these alleles from each parent. Because there are three alleles, there are six possible genotypes, as listed in the table. Both the I^A and I^B alleles are dominant to the i allele. Thus, $I^A I^A$ and $I^A i$ people have type A blood, and $I^B I^B$ and $I^B i$ people have type B. Recessive homozygotes, ii, have type O blood because neither the A nor the B substance is produced. The I^A and I^B alleles are said to be **codominant,** meaning that both alleles are expressed in heterozygous individuals ($I^A I^B$), who have type AB blood.

Because a person's ABO blood group can be determined by a simple test, it is sometimes used as evidence in paternity suits. Such evidence can prove that a man is not the father of a certain baby, and it can suggest that he could be. For example, if a baby has type AB blood (genotype $I^A I^B$) and a man has type O (genotype ii), he cannot possibly be the father. If the man has type B (genotype $I^B I^B$ or $I^B i$), he *could* be the father—but his blood type does not prove that he is.

Multiple Alleles for the ABO Blood Groups						
Blood Group Phenotypes	Genotypes	Antibodies Present in Blood	Reaction When Blood from Groups Below Is Mixed with Antibodies from Groups at Left			
			O	A	B	AB
O	*ii*	Anti-A Anti-B	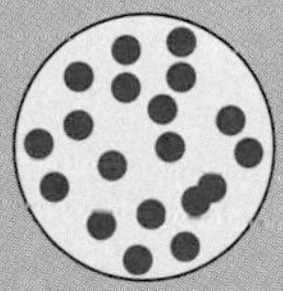	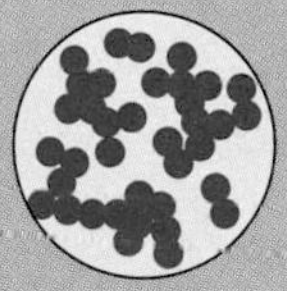	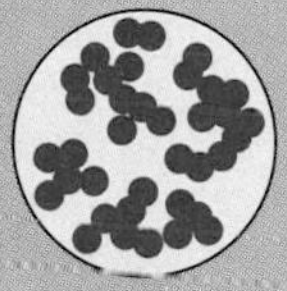	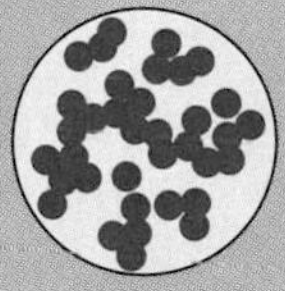
A	$I^A I^A$ or $I^A i$	Anti-B	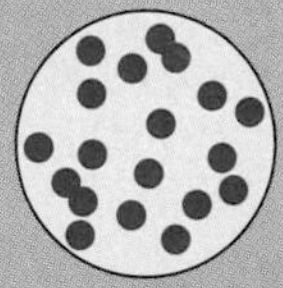	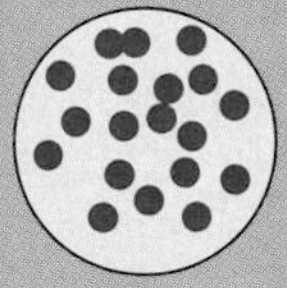	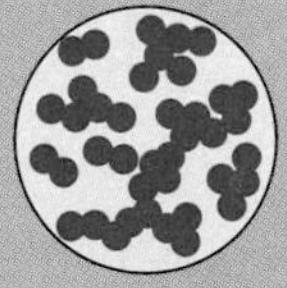	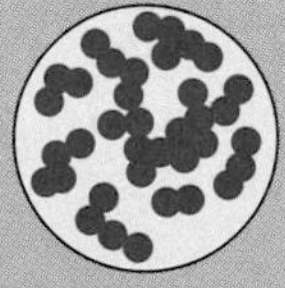
B	$I^B I^B$ or $I^B i$	Anti-A	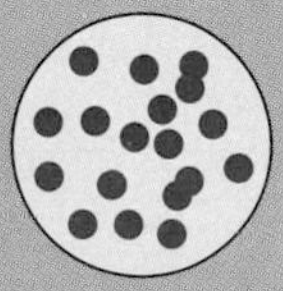	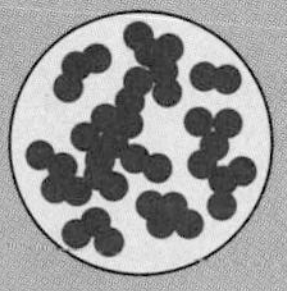	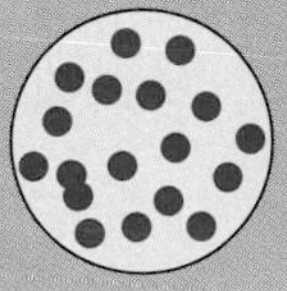	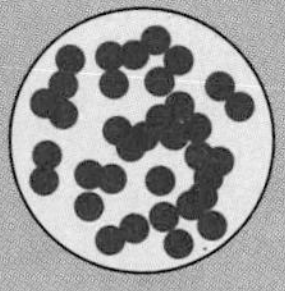
AB	$I^A I^B$	—	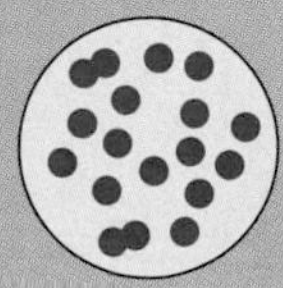	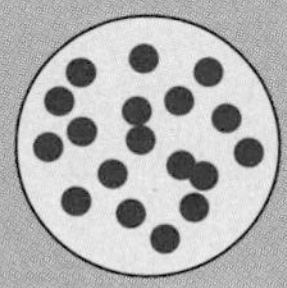	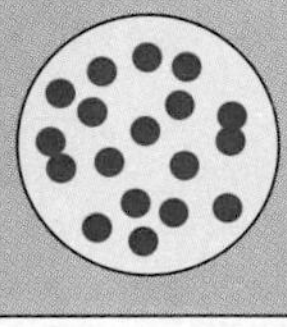	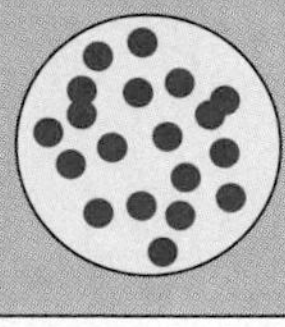

A single gene may affect many phenotypic characteristics

All of our genetic examples to this point have been cases in which each gene specified a single trait. But in many cases, one gene influences several traits. The impact of a single gene on more than one trait is called **pleiotropy** (from the Greek *pleion*, more).

An example of pleiotropy in humans is sickle-cell anemia, a disease characterized by the diverse symptoms shown below. All of these possible phenotypic effects result from the action of a single allele when it is present on both homologous chromosomes. The direct effect of the sickle-cell allele is to make red blood cells produce abnormal hemoglobin molecules. These molecules tend to link together and crystallize, especially when the oxygen content of the blood is lower than usual because of high altitude, overexertion, or respiratory ailments. As the hemoglobin crystallizes, the normally disk-shaped red blood cells deform to a sickle shape with jagged edges, as shown in the micrograph. Sickled cells are destroyed rapidly by the body, and the destruction of these cells may seriously lower the individual's red cell count, causing anemia and general weakening of the body. Also, because of their an-gular shape, sickled cells do not flow smoothly in the blood and tend to accumulate and clog tiny blood vessels. Blood flow to body parts is reduced, resulting in periodic fever, severe pain, and damage to various organs, including the heart, brain, and kidneys. Sickle cells accumulate in the spleen, damaging it. Blood transfusions and certain drugs may relieve some of the symptoms, but there is no cure, and sickle-cell anemia kills about 100,000 people in the world annually.

In most cases, only people who are homozygous for the sickle-cell allele suffer from the disease. Heterozygotes, who have one sickle-cell allele and one nonsickle allele, are usually healthy, although in rare cases they may experience some pleiotropic effects when oxygen in the blood is severely reduced, such as at very high altitudes. These effects may occur because the nonsickle and sickle-cell alleles are codominant: Both alleles are expressed in heterozygous individuals, and their red blood cells contain both normal and abnormal hemoglobin. Heterozygotes are said to have "sickle-cell trait."

Sickle-cell anemia is by far the most common inherited disease among black people, striking one in 500 African-American children born in the United States. About one in ten African-Americans is a heterozygote, whereas the sickle-cell allele is extremely rare in Caucasian Americans. Individuals who are heterozygous for a genetic disorder but do not themselves show symptoms are called **carriers** because they "carry" the disease-causing allele and may transmit it to their offspring.

One in ten is an unusually high frequency of carriers for an allele with such harmful effects in homozygotes. We might expect that the frequency of the sickle-cell allele in the population would be much lower because many homozygotes die before passing their genes to the next generation. The high frequency appears to be a vestige of the roots of African-Americans. Sickle-cell anemia is most common in tropical Africa, where the deadly disease malaria is also prevalent. The protistan parasite that causes malaria spends part of its life inside red blood cells. When it enters those of a person with the sickle-cell allele, it triggers sickling. The body destroys most of the sickled cells, and the parasite does not grow well in those that remain. Consequently, sickle-cell carriers are resistant to malaria, and in many parts of Africa they live longer and have more offspring than noncarriers who are exposed to malaria. In this way, malaria has kept the frequency of the sickle-cell allele relatively high in much of the African continent.

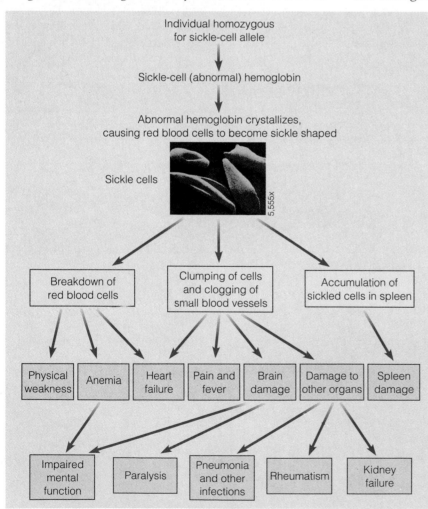

Sickle-cell anemia, multiple effects of a single human gene

A single trait may be influenced by many genes

Mendel studied genetic traits that could be classified on an either-or basis, such as purple or white flower color in peas. However, many traits, such as human skin color and height, vary in a population along a continuum. Many such features result from **polygenic inheritance,** the additive effect of two or more genes on a single phenotypic characteristic. (This is the converse of pleiotropy, in which a single gene affects several phenotypic traits.)

The figure at the right is a model that illustrates how three hypothetical genes could produce some of the continuous variation in human skin pigmentation. The three genes are inherited separately, like Mendel's pea genes. The "dark-skin" allele for each gene (*A, B, C*) contributes one "unit" of darkness to the phenotype and is incompletely dominant to the other alleles (*a, b, c*). An *AABBCC* person would be very dark, while an *aabbcc* individual would be very light. An *AaBbCc* person (resulting, for example, from a mating between an *AABBCC* person and an *aabbcc* person) would have skin of an intermediate shade. Because the alleles have an additive effect, the genotype *AaBbCc* would produce the same skin color as any other genotype with just three dark-skin alleles, such as *AABbcc*.

The Punnett square shows all possible offspring from a mating of two triple heterozygotes (the F₁ generation here). The row of diamonds below the Punnett square shows the seven skin-pigmentation phenotypes that would theoretically result from this mating. The seven bars in the graph at the bottom of the figure depict the relative numbers of each of the phenotypes in the population.

If this were a real human population, the range of skin color would probably be even more of a continuum, perhaps similar to the entire spectrum of color under the bell-shaped curve in the graph. Most likely, there would be intermediate types between each of our seven color variants. Such intermediate colors would result from the effects of environmental factors, such as sun-tanning, on the seven variants. Of course, any variation resulting from environmental factors rather than genes would not be passed on to the next generation.

* * *

To this point in the chapter, we have presented four types of inheritance patterns that are extensions of Mendel's principles of inheritance: incomplete dominance, codominance, pleiotropy, and polygenic inheritance. It is important to realize that these patterns are extensions of Mendel's model, rather than exceptions to it. From Mendel's garden pea experiments came data supporting a particulate theory of inheritance, with the particles (genes) being transmitted according to the same rules of chance that govern the tossing of coins. The particulate theory holds true for all types of inheritance patterns.

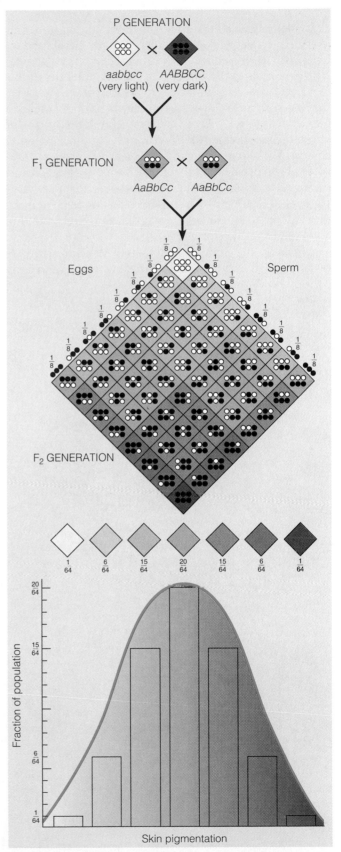

A simplified model for polygenic inheritance of skin color

Chromosome behavior accounts for Mendel's principles

Mendel published his results in 1866, but not until long after he died did biologists understand the significance of his work. Cell biologists worked out the processes of mitosis and meiosis in the late 1800s. Then, around the turn of the century, researchers began to notice parallels between the behavior of chromosomes and the behavior of Mendel's heritable factors. One of biology's central concepts—the **chromosome theory of inheritance**—was beginning to emerge. The chromosome theory states that genes are located on chromosomes and that the behavior of chromosomes during meiosis and fertilization accounts for inheritance patterns. Indeed, it is chromosomes that undergo segregation and independent assortment during meiosis and thus account for Mendel's principles.

We can see the chromosomal basis of Mendel's principles by following the fate of two genes during meiosis and fertil-

ization in pea plants. In the figure below, we picture the genes for seed shape (alleles *R* and *r*) and seed color (alleles *Y* and *y*) as black bars on different chromosomes. We start with the F$_1$ generation, in which all individuals have the *RrYy* genotype (also shown in Figure 9.5A). To simplify the diagram, we show only two of the seven pairs of pea chromosomes and three of the stages of meiosis: metaphase I, anaphase I, and metaphase II.

To see the chromosomal basis of the principle of segregation, let's follow just the pair of long chromosomes, the ones carrying *R* and *r*, taking either the left or the right branch from the F$_1$ cell. Whichever arrangement the chromosomes assume at metaphase I, the two alleles *segregate* as the homologous chromosomes separate in anaphase I. And at the end of meiosis II, a single long chromosome ends up in each of the gametes. Random fertilization then leads to

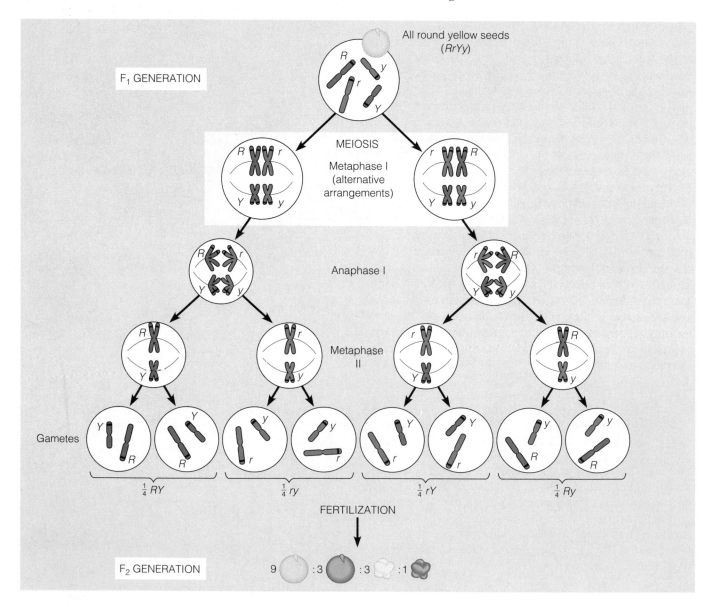

F$_2$ offspring with the 3:1 (12 round to 4 wrinkled) ratio of phenotypes that Mendel observed.

To see the chromosomal basis of the principle of independent assortment, follow both the long and the short chromosomes from metaphase I in the F$_1$ generation. Two alternative, equally likely arrangements of tetrads can occur at this stage of meiosis. The nonhomologous chromosomes and the genes they carry assort independently, forming gametes of four genotypes. Random fertilization leads to the 9:3:3:1 phenotypic ratio in the F$_2$ generation.

Genes on the same chromosome tend to be inherited together 9.15

In 1908, British biologists William Bateson and Reginald Punnett (originator of the Punnett square) discovered an inheritance pattern that seemed totally inconsistent with Mendelian principles. Bateson and Punnett were working with two traits in sweet peas, flower color and pollen shape. They crossed heterozygous plants (*PpLl*) that exhibited the dominant traits: purple flowers (expression of the *P* allele) and long pollen grains (expression of the *L* allele). The corresponding recessive traits are red flowers (*p*) and round pollen (*l*).

The top part of the figure here illustrates Bateson and Punnett's experiment. When they looked at just one of the two traits (that is, either cross *Pp* × *Pp* or cross *Ll* × *Ll*), they found that the dominant and recessive alleles segregated, producing a phenotypic ratio of approximately 3:1 for the offspring, in agreement with Mendel's segregation principle. However, when the biologists combined their data for the two traits, they did not see the 9:3:3:1 ratio they predicted for a dihybrid cross. Instead, as shown in the table, they found a disproportionately large number of plants (339 of 381 observed offspring) with either purple flowers and long pollen (284 of 381, almost 75% of the total) or red flowers and round pollen (55 of 381, about 14% of the total). These results were not explained until several years later, when other studies revealed that the genes for flower color and pollen shape are on the same chromosome.

The number of genes in a cell is far greater than the number of chromosomes; in fact, each chromosome has thousands of genes. Genes that are located close together on the same chromosome, called **linked genes,** tend to be inherited together. As a result, they generally do not follow Mendel's principle of independent assortment. As shown in the "Explanation" part of the figure, meiosis in the heterozygous (*PpLl*) sweet-pea plant yields two predominant types of gametes (*PL* and *pl*) rather than equal numbers of the four types of gametes that would result if the flower-color and pollen-shape genes were not linked. The large numbers of plants with purple long and red round traits in the Bateson-Punnett experiment resulted from fertilization among the *PL* and *pl* gametes. But what about the smaller numbers of plants with purple round and red long traits? As we see in the next module, the phenomenon of crossing over accounts for these offspring.

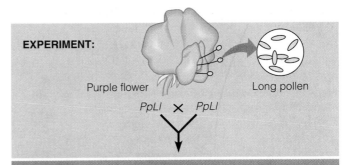

EXPERIMENT:

Purple flower   Long pollen

PpLl × *PpLl*

PHENOTYPES	OBSERVED OFFSPRING	PREDICTION (9 : 3 : 3 : 1)
Purple long	284	215
Purple round	21	71
Red long	21	71
Red round	55	24

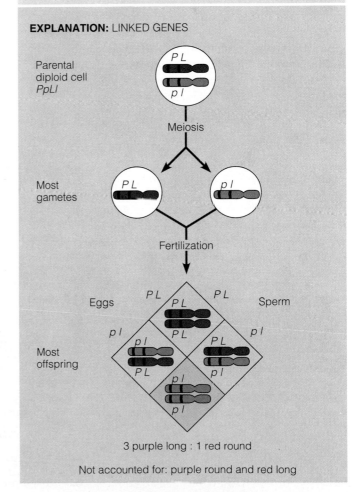

EXPLANATION: LINKED GENES

Parental diploid cell *PpLl*

P L
p l

Meiosis

Most gametes   *P L*   *p l*

Fertilization

Eggs   *P L*   *P L*   Sperm

p l   *p l*

Most offspring

3 purple long : 1 red round

Not accounted for: purple round and red long

Crossing over produces new combinations of alleles

In Module 8.19, we saw that during meiosis, crossing over between homologous chromosomes produces new combinations of alleles in gametes. Here, Figure A reviews this process, showing that four different gamete genotypes can result from two linked genes. Gametes with genotypes *AB* and *ab* carry parental-type chromosomes that have not been altered by crossing over. In contrast, gametes with genotypes *Ab* and *aB* are recombinant gametes. They carry new combinations of alleles that result from the exchange of chromosome segments during crossing over. Crossing over *recombines* linked genes into assortments of alleles not found in parents.

The discovery of the way crossing over creates gamete diversity confirmed the relationship between chromosome behavior and inheritance. Some of the key experiments that first demonstrated the effects of crossing over were performed in the laboratory of American embryologist T. H. Morgan in the early 1900s. Morgan and his colleagues used the fruit fly *Drosophila melanogaster* in many of their experiments. Often seen buzzing around overripe fruit, *Drosophila* is an ideal research animal for studies of inheritance. It can be grown in small containers on a mixture of cornmeal and molasses and will produce hundreds of offspring in a few weeks. Using fruit flies, geneticists can trace the inheritance of a trait through several generations in a matter of months.

Figure B shows one of Morgan's experiments, a cross between a normal fruit fly (gray body and long wings) and an abnormal fly (black body and undeveloped, or vestigial, wings). Morgan knew the genotypes of these flies from previous studies. In the figure here, we use the following gene symbols:

G = gray body (dominant)

g = black body (recessive)

L = long wings (dominant)

l = vestigial wings (recessive)

In mating a gray fly with long wings (genotype *GgLl*) with a black fly with vestigial wings (genotype *ggll*), Morgan performed a testcross (see Module 9.6). If the genes had not been linked, then independent assortment would have produced offspring in a

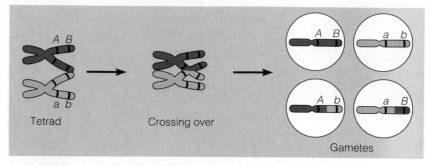

A. Review: Production of recombinant gametes

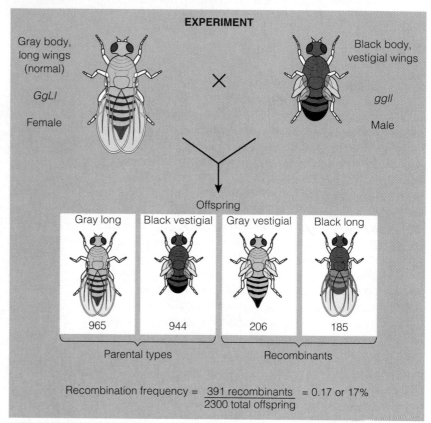

EXPERIMENT

Gray body, long wings (normal)

GgLl

Female

Black body, vestigial wings

ggll

Male

Offspring

Gray long	Black vestigial	Gray vestigial	Black long
965	944	206	185

Parental types · Recombinants

$$\text{Recombination frequency} = \frac{391 \text{ recombinants}}{2300 \text{ total offspring}} = 0.17 \text{ or } 17\%$$

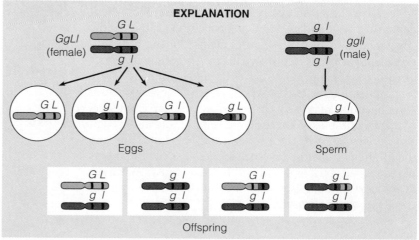

EXPLANATION

GgLl (female)   *ggll* (male)

Eggs

Sperm

Offspring

B. Fruit-fly experiment demonstrating the role of crossing over in inheritance

phenotypic ratio of 1:1:1:1 ($\frac{1}{4}$ gray body, long wings; $\frac{1}{4}$ black body, vestigial wings; $\frac{1}{4}$ gray body, vestigial wings; and $\frac{1}{4}$ black body, long wings). But because these genes were linked, Morgan obtained the results shown here: Most of the offspring had parental phenotypes, but 17% of the offspring flies were recombinants. The percentage of recombinants is called the recombination frequency.

When Morgan first obtained these results, he did not know about crossing over. To explain the ratio of offspring, he hypothesized that the genes were linked and that some mechanism occasionally broke the linkage. Tests of the hypothesis proved him correct, establishing that crossing over was the mechanism that "breaks linkages" between genes.

The lower part of Figure B explains Morgan's results in terms of crossing over. A crossover between chromatids of homologous chromosomes in parent *GgLl* broke linkages between the *G* and *L* alleles and between the *g* and *l* alleles, forming the recombinant chromosomes *Gl* and *gL*. Later steps in meiosis distributed the recombinant chromosomes to gametes, and random fertilization produced the four kinds of offspring Morgan observed.

Geneticists use crossover data to map genes   9.17

Working mostly with *Drosophila,* Morgan and his students produced a virtual explosion in our understanding of genetics. In the photograph, Figure A, Morgan (back row, far right) and several students are celebrating the return of Alfred H. Sturtevant (left foreground) from World War I military service. One of Sturtevant's major contributions to genetics was an approach for using crossover data to map gene loci. Assuming that the chance of crossing over is approximately equal at all points on a chromosome, Sturtevant hypothesized that the farther apart two genes are on a chromosome, the higher the probability that a crossover would occur between them. His reasoning was simple: The greater the distance between two genes, the more points there are between them where crossing over can occur. With this principle in mind, Sturtevant began using recombination data from fruit-fly crosses to assign to genes relative positions on chromosomes—that is, to *map* genes.

Figure B represents a part of the chromosome that carries the linked genes for black body (*g*) and vestigial wings (*l*) that we described in Module 9.16. This same chromosome also carries a gene that has a recessive allele (*c*) determining cinnabar eye color, a much brighter red than normal. The diagram shows the actual cross-

over (recombination) frequencies between these alleles, taken two at a time: 17% between the *g* and *l* alleles, 9% between *g* and *c*, and 9.5% between *l* and *c*. Sturtevant reasoned that these values represent the relative distances between the genes. Because the crossover frequencies between *g* and *c* and between *l* and *c* are approximately half that between *g* and *l*, gene *c* must lie roughly midway between *g* and *l*. Thus, the sequence of these genes on one of the fruit-fly chromosomes must be *g–c–l*.

Years later it was learned that Sturtevant's assumption that crossovers are equally likely at all points on a chromosome was not exactly correct. Nevertheless, his method of mapping genes on chromosomes worked, and it proved extremely valuable in establishing the relative positions of many other fruit-fly genes. Eventually, enough data were accumulated to show that *Drosophila* has four groups of genes, corresponding to its four pairs of chromosomes. More recently, geneticists have actually been able to measure the distances (in nanometers) between genes on chromosomes. Their results, establishing absolute locations of genes on chromosomes, confirm the relative positions established by Sturtevant's method.

A. A party in Morgan's fly room

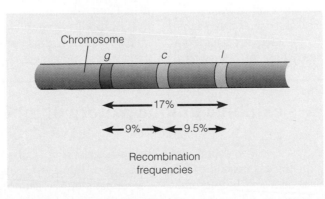

B. Mapping genes from crossover data

Chromosomes determine sex in many species

Many animals, including fruit flies and humans, have a pair of **sex chromosomes**, designated *X* and *Y,* that determine an individual's sex. Figure A shows the different-sized *X* and *Y* chromosomes of humans. Individuals with one *X* and one *Y* are males; *XX* individuals are females. Figure A indicates that human males and females both have 44 autosomes (nonsex chromosomes). As a result of chromosome segregation during meiosis, each gamete contains one sex chromosome and a haploid set of autosomes (22 in humans). All eggs contain a single *X* chromosome. Of the sperm cells, half contain an *X* chromosome and half contain a *Y* chromosome. An offspring's sex depends on whether the sperm cell that fertilizes the egg bears an *X* or a *Y*. The genetic basis of sex determination in humans is not yet well understood, but one or more genes on the *Y* chromosome are likely to be crucial. In 1990, a British research team discovered a gene on the Y chromosome that seems to trigger testis development. In the absence of this gene, an individual develops ovaries rather than testes.

The *X-Y* system is only one of several sex-determining systems. Grasshoppers, crickets, and roaches, for example, have an *X-O* system, in which *O* stands for the absence of a sex chromosome (Figure B). Females have two *X* chromosomes (*XX*); males have only one sex chromosome, giving them genotype *XO.* Males produce two classes of sperm (*X*-bearing and lacking any sex chromosome), and sperm cells determine the sex of the offspring at fertilization.

In contrast to the *X-Y* and *X-O* systems, *eggs* determine sex in certain fishes, butterflies, and birds (Figure C). The sex chromosomes in these animals are designated *Z* and *W.* Males have the genotype *ZZ;* females are *ZW.* Sex is determined by whether the egg carries a *Z* or a *W.*

Some organisms lack sex chromosomes altogether. In most ants and bees, sex is determined by chromosome *number,* rather than by sex chromosomes (Figure D). Females develop from fertilized eggs and thus are diploid. Males develop from unfertilized eggs—they are fatherless—and are haploid.

Most animals have two separate sexes; that is, individuals are either male or female. Many plants also have separate sexes, with male and female flowers borne on different individuals. Some plants with separate sexes, such as date palms, spinach, and marijuana, have the *X-Y* system of sex determination; others, such as the wild strawberry, have the *Z-W* system.

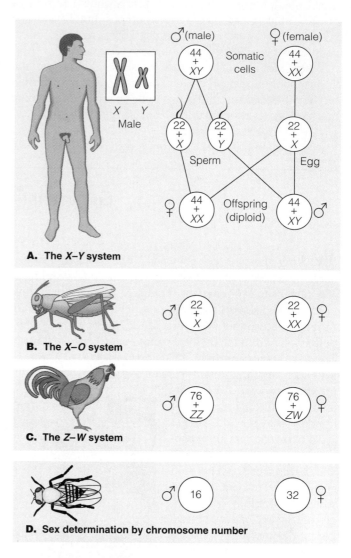

A. The *X–Y* system

B. The *X–O* system

C. The *Z–W* system

D. Sex determination by chromosome number

But not all organisms have separate sexes. Most plant species and some animal species have individuals that produce both sperm and eggs. Plants of this type—corn, for example—are said to be **monoecious** (from the Greek *monos,* one, and *oikos,* house). Animals of this type, such as earthworms and garden snails, are said to be **hermaphroditic** (from the names of the Greek god Hermes and goddess Aphrodite). In monoecious plants and hermaphroditic animals, all individuals of a species have the same complement of chromosomes.

Sex-linked genes exhibit a unique pattern of inheritance

Besides bearing genes that specify sex, the so-called sex chromosomes also contain genes for traits unrelated to femaleness or maleness. Genes located on a sex chromosome are called **sex-linked genes**. In humans, fruit flies, and certain fishes, some sex-linked genes are carried on the *Y* chromosome. In humans, for instance, a gene that produces long, coarse hairs on the earlobes seems to reside only on the *Y* chromosome. Such a *Y*-linked trait appears only in men and is passed from father to son. Actually, little is known about the *Y* chromosome in humans or in other species. Most known sex-linked genes reside on the *X* chromosome.

The figures below illustrate inheritance patterns for a sex-linked (actually *X*-linked) recessive trait, white eye color in the fruit fly. Normal fruit flies have red eyes; white eyes are abnormal (Figure A). Uppercase letter *R* represents the dominant allele for red eye; lowercase *r* represents the recessive white-eye allele. Because these alleles are carried on the *X* chromosome, we show them as superscripts to the letter *X*. Thus, normal male fruit flies have the genotype X^RY; white-eyed males are X^rY. The *Y* chromosome does not have a gene locus for eye color; therefore, the male's phenotype results entirely from his single *X*-linked gene. In the female, X^RX^R and X^RX^r flies have normal red eyes, and X^rX^r flies have white eyes.

A white-eyed male (X^rY) will transmit his X^r to all of his female offspring, but to none of his male offspring. This is because his daughters can inherit only his *X* chromosome and his sons can inherit only his *Y* chromosome. As shown in Figure B, when the female parent is a dominant homozygote (X^RX^R) and the male parent is X^r, all the offspring have normal red eyes, but the female offspring are all carriers of the abnormal allele for white eyes.

In Figure C, we see that when a heterozygous, carrier female (X^RX^r) mates with a normal red-eyed male (X^RY), half the male offspring are white-eyed and half are red-eyed. All the female offspring of this cross have normal red eyes because they inherit at least one dominant allele (from their father). Half of the female offspring are homozygous dominant because they inherit their mother's *R* allele. The other half of the females are carriers (X^rX^R) like their mother.

As Figure D indicates, if a carrier female mates with a white-eyed male, there is a 50% chance that each offspring will have white eyes (resulting from genotype X^rX^r or X^rY), regardless of sex. Daughters with normal phenotypes are carriers, whereas normal male offspring are completely free of the recessive allele.

Fruit-fly genetics has taught us much about human inheritance. A number of human conditions, including red-green color blindness, a type of muscular dystrophy, and hemophilia (all discussed in Chapter 13), result from sex-linked recessive alleles that are inherited in the same way as the white-eye trait in fruit flies. The fruit-fly model also shows us why recessive sex-linked traits are expressed much more frequently in men than in women. Just like male fruit flies, if a man inherits only one sex-linked recessive allele—from his mother—the allele will be expressed. In contrast, a woman has to inherit two such alleles—one from each parent—to exhibit the trait.

The discovery of the white-eye allele and its inheritance pattern in fruit flies was one of many breakthroughs in understanding how genes are passed from one generation to the next. During the first half of the twentieth century, Mendel's work was rediscovered, his principles were reinterpreted in light of chromosomal behavior during meiosis, and the chromosome theory of inheritance was firmly established. The chromosome theory set the stage for another explosion of experimental work in the second half of the twentieth century. This work was mostly in molecular genetics, an area we explore in the next four chapters.

A. Fruit-fly eye color, a sex-linked trait

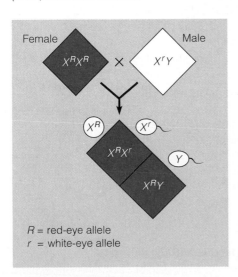

B. Homozygous, red-eyed female × white-eyed male

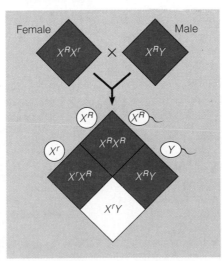

C. Heterozygous female × red-eyed male

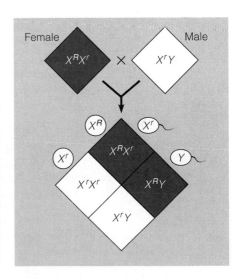

D. Heterozygous female × white-eyed male

Begin your review by rereading the module headings and scanning the figures before proceeding to the Chapter Summary and questions.

Chapter Summary

Introduction–9.1 Ideas about inheritance date back to ancient times. Hippocrates believed that hereditary particles could transmit bodily changes to offspring. Many nineteenth-century biologists believed that characteristics acquired during an organism's lifetime could be passed on, and that characteristics of both parents blended in their offspring.

9.2–9.6, 9.14 Modern genetics—the science of heredity—began with Gregor Mendel's careful and quantitative experiments. Mendel crossed pea plants that differed in traits such as flower color and seed shape, and he traced these traits from generation to generation. He hypothesized that there are alternative forms of genes (although he did not use that term), the units that determine heritable traits. An organism has two genes for each trait, one from each parent. A sperm or egg carries only one gene for each trait, because gene pairs separate when gametes form—the principle of segregation. Mendel also noted that when two genes of a pair are different forms, or alleles, one (the dominant allele) is expressed and one (the recessive allele) is hidden. By looking at two traits at once, Mendel found that each pair of alleles segregates and is inherited independently—the principle of independent assortment. We now know that the alleles of a gene reside at the same locus, or position, on homologous chromosomes. It is the behavior of chromosomes that explains the segregation and independent assortment of genes; this concept is the chromosome theory of inheritance.

9.7–9.8 Inheritance follows the rules of probability. The chance of inheriting a recessive allele (b) from a heterozygous (Bb) parent is $\frac{1}{2}$. The chance of inheriting it from both of two heterozygous parents is $\frac{1}{2} \times \frac{1}{2} = \frac{1}{4}$, illustrating the rule of multiplication for calculating the probability of two independent events. There are two ways a dominant and recessive allele from heterozygous parents can combine in offspring: B from the father and b from the mother, or b from the father and B from the mother. The probability of this occurring is $\frac{1}{4} + \frac{1}{4} = \frac{1}{2}$, illustrating the rule of addition for calculating the probability of an event that can occur in alternative ways. The inheritance of many human traits, from freckles to genetic diseases, follows Mendel's principles and these rules of probability.

9.9–9.13 There are a number of variations on Mendel's simple principles. When an offspring's phenotype—flower color, for example—is halfway between the phenotypes of its parents, it exhibits incomplete dominance. There are often multiple alleles for a trait, such as the three alleles for the ABO blood groups. In addition, the alleles determining the A and B blood factors are codominant; that is, both are expressed in a heterozygous individual. A single gene, such as the allele for sickle-cell anemia, may affect phenotype in several ways, a phenomenon called pleiotropy. Or a single trait, such as skin color, may be affected by several genes, creating a continuum of phenotypes.

9.15–9.17 Biologists working with flowers and fruit flies found that certain genes are linked; they tend to be inherited together because they reside close together on the same chromosome. In fact, there are many genes on each chromosome. Crossing over can separate linked alleles, producing gametes with recombinant chromosomes. Because crossing over is more likely to occur between genes that are farther apart, recombination frequencies can be used to map the relative positions of genes on chromosomes.

9.18–9.19 A pair of sex chromosomes determines sex in many species. In humans (and many other organisms, including fruit flies),

a male has one *X* and one *Y* sex chromosome, and a female has two *X* chromosomes. Whether a sperm cell contains an *X* or a *Y* determines the sex of the offspring. The *Y* chromosome has genes for the development of testes, whereas an absence of the *Y* allows ovaries to develop. (Other systems of sex determination exist in other animals and plants.) The sex chromosomes also carry genes unrelated to maleness or femaleness. Most of these sex-linked genes are on the *X* chromosome. Sex-linked traits such as color blindness are more common in males because a male need inherit only a sex-linked recessive gene from his mother to exhibit the trait, while a female must inherit one such allele from each parent, an event that is much less likely.

Testing Your Knowledge

Multiple Choice

1. Edward was found to be heterozygous (*Ss*) for the sickle-cell anemia trait. The alleles represented by the letters *S* and *s* are

  a. on the *X* and *Y* chromosomes
  b. linked
  c. on homologous chromosomes
  d. both present in each of Edward's sperm cells
  e. on the same chromosome

2. Whether an allele is dominant or recessive depends on

  a. how common the allele is, relative to other alleles
  b. whether it is inherited from the mother or the father
  c. which chromosome it is on
  d. whether it or another allele is expressed when they are both present
  e. whether or not it is linked to other genes

3. Two fruit flies with red eyes are crossed and their offspring are as follows: 77 red-eyed males, 71 ruby-eyed males, 152 red-eyed females. The allele for ruby eyes is (*Explain your answer.*)

  a. autosomal (carried on an autosome) and dominant
  b. autosomal and recessive
  c. sex-linked and dominant
  d. sex-linked and recessive
  e. impossible to determine without more information

4. All the offspring of a white hen and a black rooster are gray. The simplest explanation for this pattern of inheritance is

  a. pleiotropy   **d.** independent assortment
  b. sex linkage   **e.** incomplete dominance
  c. linkage

5. In some of his experiments, Mendel studied the inheritance patterns of two traits at once—flower color and pod color, for example. He did this to find out

  a. whether factors (what we call genes) for different traits are inherited together or separately
  b. how many genes are responsible for determining a particular trait
  c. whether genes are on chromosomes
  d. the distance between genes on a chromosome
  e. how many different genes a pea plant has

6. A man who has type B blood and a woman who has type A blood could have children of which of the following phenotypes?

  a. A or B only   **d.** A, B, or O
  b. AB only   **e.** A, B, AB, or O
  c. AB or O

Describing, Comparing, and Explaining

1. Describe a genetic cross that illustrates the principle of independent assortment. How do the offspring of your cross show that genes assort independently?

2. Explain why fruit flies (*Drosophila melanogaster*) make good subjects for genetics experiments.

3. Red-green color blindness is a sex-linked trait, carried on the *X* chromosome. Why are there more color-blind men than color-blind women?

4. In fruit flies, the genes for wing shape and body stripes are linked. In a fly whose genotype is *WwSs*, *W* is linked to *S*, and *w* is linked to *s*, as in this diagram:

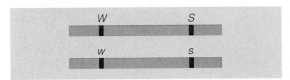

Show how this fly can produce gametes containing four different combinations of genes. Which are parental-type gametes? Which are recombinant gametes? What process produces recombinant gametes?

Thinking Critically

1. Adult height in humans is at least partially hereditary; tall parents tend to have tall children. But humans come in a range of sizes, not just tall and short. Explain an extension of Mendel's model that could produce the hereditary variation in human height.

2. In 1981, a stray black cat with unusual rounded, curled-back ears was adopted by a family in Lakewood, California. Hundreds of descendants of this cat have since been born, and cat fanciers hope to develop the "curl" cat into a show breed. The curl gene is apparently dominant and autosomal (carried on an autosome). Suppose you owned the first curl cat and wanted to breed it in order to develop a true-breeding variety. How would you figure out whether the curl gene is dominant or recessive? Autosomal or sex-linked? How would you go about breeding true-breeding cats? How would you know they are true-breeding? What is the genotype of a true-breeding curl cat?

Genetics Problems

1. A brown mouse is repeatedly mated with a white mouse, and all their offspring are brown. If two of these brown offspring are mated, what fraction of the F_2 mice will be brown?

2. How could you determine the genotype of one of the brown F_2 mice in problem 1? How would you know whether a brown mouse is homozygous? Heterozygous?

3. Tim and Carolyn both have freckles (see Module 9.8), but their son Michael does not. Show with a Punnett square how this is possible. If Tim and Carolyn have two more children, what is the probability that both of them will have freckles?

4. Both Tim and Carolyn (problem 3) have a widow's peak (Module 9.8), but Michael has a straight hairline. What are their genotypes? What is the probability that Tim and Carolyn's next child will have freckles and a straight hairline?

5. In rabbits, black hair is dependent on a dominant allele, *B*, and brown upon a recessive allele, *b*. Short hair is due to a dominant allele, *S*, and long hair to a recessive allele, *s*. If a true-breeding black, short-haired male is mated with a brown, long-haired female, what will their offspring look like? What will be the genotypes of the offspring? If two of these F_1 rabbits are mated, what phenotypes would you expect among their offspring, in what proportions?

6. Incomplete dominance is seen in the inheritance of hypercholesterolemia. Mack and Toni are both heterozygous for this trait, and both have elevated levels of cholesterol. Their daughter Zoe has a cholesterol level six times normal; she is apparently homozygous, *hh*. What fraction of Mack and Toni's children are likely to have elevated but not extreme levels of cholesterol, like their parents? If Mack and Toni have one more child, what is the probability that the child will suffer from the more serious form of hypercholesterolemia seen in Zoe?

7. A man with type O blood has a sister with type AB blood. What are the genotypes and phenotypes of their parents?

8. A fruit fly with a gray body and red eyes (genotype *BbPp*) is mated with a fly having a black body and purple eyes (genotype *bbpp*). What offspring, in what proportions, would you expect if the body color and eye color genes are on different chromosomes (unlinked)? When this mating is actually carried out, most of the offspring look like the parents, but 3% have gray body and purple eyes, and 3% have black body and red eyes. Are these genes linked or unlinked? What is the recombination frequency?

9. A series of matings shows that the recombination frequency between the black-body gene (problem 8) and the gene for dumpy (shortened) wings is 36%. The recombination frequency between purple eyes and dumpy wings is 41%. What is the sequence of these three genes on the chromosome?

10. A female fruit fly whose body is covered with forked bristles is mated with a male fly with normal bristles. Their offspring are 121 females with normal bristles and 138 males with forked bristles. Explain the inheritance pattern for this trait.

Science, Technology, and Society

1. Understanding the principles of genetics has increased our ability to breed animals to suit our needs and desires. Many animals quite unlike any that occur in nature are bred to be kept as pets. In addition to the curl cat shown at the left, there are goggle-eyed exotic goldfish, pigeons with greatly elongated feathers, miniature horses the size of dogs, and a breed of "fainting" goats that lose consciousness when startled by loud noises. Can you think of other examples? Why do people breed these unusual pets? Are there any limits to the sizes and shapes of animals that might be produced? What principles do you think should guide animal breeders as they consider developing exotic pets?

2. Gregor Mendel never saw a gene, yet he concluded that "heritable factors" were responsible for the patterns of inheritance that he observed in peas. Similarly, T. H. Morgan and his colleague Alfred Sturtevant never actually observed the linkage of genes on chromosomes. Their maps of *Drosophila* chromosomes (and the very idea that genes are carried on chromosomes) were conceived by observing the patterns of inheritance of linked genes, not by observing the genes directly. The actual positions of linked genes were not measured until relatively recently. Is it legitimate for biologists to claim the existence and map the locations of objects and processes they cannot actually see? Does this happen in other fields of science? Do scientists usually see something happen and then try to explain it, or do they devise an explanation and then look to see whether things seem to happen that way? How do they know whether an explanation is correct?

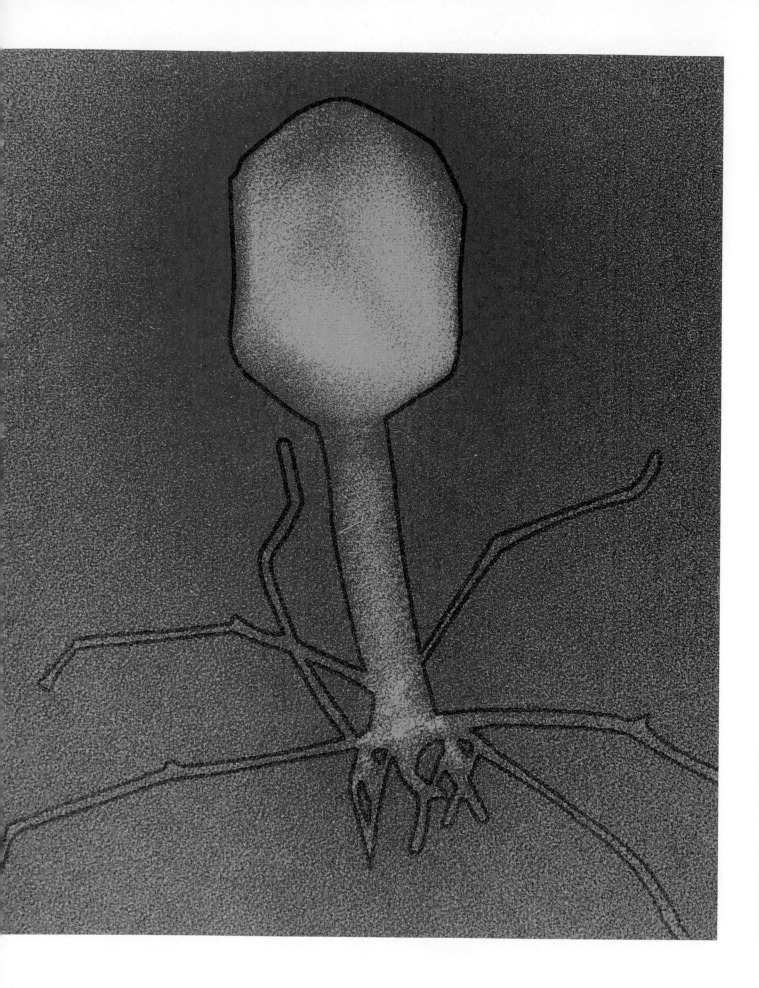

Molecular Biology of the Gene $\quad$ 10

The landing craft drifts slowly down to the surface. The life forms—whatever they are—must be in the 20-sided capsule at the top. Maybe the tubelike underpart is an entryway or an exit. The six legs look as if they would be stable on any kind of surface.

Microscopists often say that viewing tiny objects and organisms magnified thousands of times gives them a sense of peering in on other worlds. In the large picture at the left, we're actually looking at a virus about to land on the surface of a bacterial cell. The virus, known as T4, is magnified 715,000 times with an electron microscope. The drawing on this page shows how the T4 virus infects a bacterium. The "legs" (called tail fibers) bend when they touch the cell surface. The tail is a hollow rod enclosed in a springlike sheath. As the legs bend, the spring compresses, and the bottom of the rod touches the cell surface, attaching to a protein landing pad (the purple "donut"). The rod punctures the cell membrane, and the viral DNA passes from inside the head of the virus through the hollow rod and into the cell.

The scale here is very small indeed. The virus is only about 200 nanometers (2 ten-thousandths of a millimeter) long. The bacterium, called *Escherichia coli* (*E. coli*) is only 1 or 2 μm (1 or 2 thousandths of a mm) across. Over a million *E. coli* would fit on the head of a pin.

On a smaller scale yet—at the molecular level—the T4 virus performs a deadly act. It takes control of the bacterial cell and then kills it. The DNA that it injects is the virus's genetic material. Once inside the bacterial cell, the DNA uses the cell's resources to produce hundreds of new T4s. The new viruses eventually break open (lyse) their host cell and infect any other *E. coli* cells they encounter. Viruses that attack bacteria are called **bacteriophages** (meaning "bacteria-eaters"), or **phages** for short.

T4 and other viruses sit on the fence, so to speak, between life and nonlife. A virus is lifelike in having DNA and a highly organized structure, but it differs from a living organism in not being cellular or able to reproduce on its own. A virus can survive only by infecting a living cell and using the cell's molecular machinery to make more viruses. T4 and other viruses are essentially genetic material enclosed in a protein coat. Because viruses are so simple, they are relatively easy to study on the molecular level—far easier than Mendel's peas or Morgan's fruit flies. For this reason, we owe our first glimpses of the functions of DNA, the molecule that carries hereditary traits, to bacteriophages and bacteria.

This chapter is about the DNA molecule and how it serves as the molecular basis of heredity. Here we explore the structure of DNA, how it replicates (the molecular basis of why offspring resemble their parents), how it controls the cell by directing protein synthesis, and how it can change. We also look at some viruses that infect animals and plants. We begin with the story of how we know that DNA is the genetic material, a story in which a close relative of T4 played a major role.

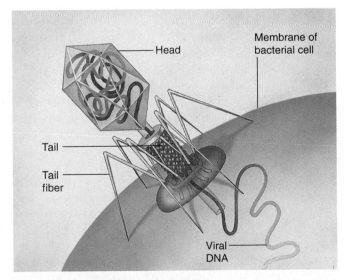

Bacteriophage T4 infecting a bacterial cell

Experiments showed that DNA is the genetic material

Our knowledge of T4 and DNA is relatively new. Viruses were discovered as agents of disease a century ago but were not actually seen with a microscope until 1942. Similarly, DNA was known as a substance in cells 100 years ago, but Mendel, Morgan, and other early geneticists did all their work without any knowledge of DNA's role in heredity.

We can trace the discovery of the genetic role of DNA back to 1928. That year, English bacteriologist Frederick Griffith reported results of experiments on a species of bacterium that causes pneumonia. Griffith studied two varieties of the bacterium, a pathogenic (disease-causing) one and a variant that was harmless. He found that when he killed the pathogenic bacteria with heat and then mixed the cell remains with living bacteria of the harmless variety, some of the living cells were converted to the pathogenic form. Furthermore, this new trait of pathogenicity was inherited by all the descendants of the transformed bacteria. Clearly, some chemical component of the dead pathogenic cells—some "transforming factor"—caused this heritable change, but the identity of the factor was not known.

By the late 1930s, other experimental studies had convinced most biologists that a specific kind of molecule, rather than some complex chemical mixture, was at the root of inheritance. Attention focused on the chemical nature of chromosomes, which had been known since the turn of the century to carry hereditary information. By the 1940s, scientists knew that eukaryotic chromosomes consisted of two substances, DNA and protein. Most researchers thought the protein was the material of genes. Proteins were known to be made of 20 kinds of building blocks (amino acids) and to have elaborate and varied structures and functions. DNA, on the other hand, has only four types of building blocks (nucleotides), and its chemical properties seemed far too monotonous to account for the multitude of traits inherited by every organism. These arguments were so persuasive that even after researchers established in 1944 that Griffith's transforming factor was DNA, the scientific community remained skeptical.

Gradually, the evidence in support of DNA built up. In 1952, American biologists Alfred Hershey and Martha Chase performed one of the most convincing experiments. They demonstrated that DNA is the genetic material of a bacteriophage called T2 (a close relative of T4). At that time, biologists already knew that T2 was composed solely of DNA and protein, and that it could somehow reprogram its host cell to produce new phages. But they did not know which phage component—DNA or protein—was responsible.

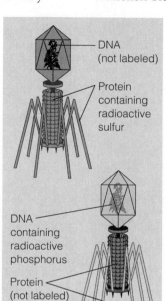

DNA (not labeled)

Protein containing radioactive sulfur

DNA containing radioactive phosphorus

Protein (not labeled)

A. Labeled phage

Hershey and Chase found the answer by devising an experiment to determine what the phage transferred to *E. coli* during infection. Their experiment used only a few, relatively simple tools: chemicals containing radioactive isotopes (see Module 2.5); a device for detecting radioactivity; a kitchen blender; and a centrifuge, which is an extremely fast merry-go-round for test tubes that is used to separate particles of different weights. (These are still basic tools of molecular biology.)

Hershey and Chase used different radioactive isotopes to label the DNA and protein in T2. First, they grew T2 with *E. coli* in a growth medium containing radioactive sulfur (pink in Figure A). Protein contains sulfur but DNA does not, so as the phages multiplied, the radioactive sulfur atoms were incorporated only into their proteins (the head exterior, tail, and tail fibers). Next, the researchers grew a separate batch of phages in medium containing radioactive phosphorus (blue in Figure A). Because nearly all the phages' phosphorus is in their DNA, this procedure labeled the phage DNA.

Armed with radioactively labeled T2, Hershey and Chase were ready to perform the experiment outlined in Figure B on the facing page. ① They allowed the protein-labeled and DNA-labeled batches of T2 to infect separate samples of nonradioactive bacteria. ② Shortly after the onset of infection, they agitated the cultures in a blender to shake loose any parts of the phages that remained outside the bacterial cells. ③ They then spun the mixtures in a centrifuge. The bacterial cells, being heavier, were deposited as a pellet at the bottom of the centrifuge tubes, but the phages and phage parts remained suspended in the liquid. ④ The researchers then measured and compared the radioactivity in the pellet and the liquid.

Hershey and Chase found that when the bacteria had been infected with the T2 phages containing labeled proteins, most of the radioactivity was in the liquid, which contained phages but not bacteria. This result suggested that the phage protein did not enter the host cells. But when the bacteria had been infected with phages whose DNA was tagged with radioactive phosphorus, then most of the radioactivity was in the pellet, made up of bacteria. When these bacteria were returned to liquid growth medium, the bacterial cells lysed and released new phages containing radioactive phosphorus in their DNA but no radioactive sulfur in their proteins.

Hershey and Chase concluded that the phage DNA was injected into the host cell but that most of the phage proteins remained outside. More important, they showed that it is the injected DNA molecules that cause the cells to produce additional phage DNA and proteins—indeed, new, complete phages. Figure C outlines the reproductive cycle for phages T2 and T4.

The Hershey-Chase results, added to earlier evidence, convinced the scientific world that DNA was the hereditary material. What happened next was one of the most celebrated quests in the history of science—the effort to figure out the structure of DNA and how this structure enables the molecule to store genetic information and transmit it from parents to offspring.

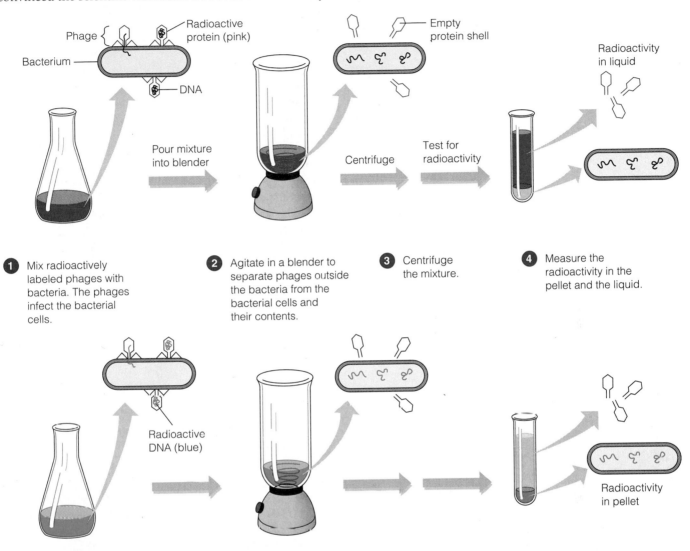

① Mix radioactively labeled phages with bacteria. The phages infect the bacterial cells.

② Agitate in a blender to separate phages outside the bacteria from the bacterial cells and their contents.

③ Centrifuge the mixture.

④ Measure the radioactivity in the pellet and the liquid.

B. The Hershey-Chase experiment

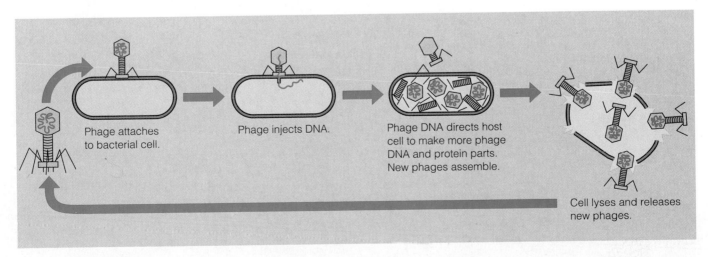

C. Phage reproductive cycle

Nucleotides are the chemical units of DNA and RNA polymers

By the time Hershey and Chase performed their experiments, a good deal was already known about DNA. Scientists had identified all its atoms and knew how they were covalently bonded to one another. What was not understood was the specific arrangement of parts that gave DNA its unique properties—the capacity to store genetic information, copy it, and pass it from generation to generation. However, it was only one year after Hershey and Chase published their results that scientists figured out DNA's three-dimensional structure and the basic strategy of how it works. We examine that momentous discovery in Module 10.3. First, let's look at the underlying chemical structure of DNA and its chemical cousin RNA.

Recall from Module 3.20 that DNA and RNA are nucleic acids, long chains (polymers) of chemical units (monomers) called **nucleotides.** A very simple diagram of such a polymer, or **polynucleotide,** is shown on the left in Figure A. This sample polynucleotide chain shows only one possible arrangement of the four different types of nucleotides (abbreviated A, C, T, and G) that make up DNA. Since nucleotides can occur in a polynucleotide in any sequence and polynucleotides vary in length from long to very long, the number of possible polynucleotides is very great.

Zooming in on our polynucleotide, we see in the center of Figure A that each nucleotide consists of three components: a nitrogenous base (gray), a sugar (blue), and a phosphate group (yellow). The nucleotides are joined by covalent bonds between the sugar of one nucleotide and the phosphate of the next. This results in a **sugar-phosphate backbone,** a repeating pattern of sugar-phosphate-sugar-phosphate. The nitrogenous bases are arranged as appendages all along this backbone.

Examining a single nucleotide even more closely (on the right in Figure A), we note the chemical structure of its three components. The phosphate group has a phosphorus atom (P) at its center and is acidic; it is the *acid* in nucleic acid. The sugar has five carbon atoms (shown in red)—four in its ring and one extending above the ring. The ring also includes a single oxygen atom. The sugar is called deoxyribose because, compared to the sugar ribose, it is missing an oxygen atom. (Notice that the C atom in the lower right corner of the ring is bonded to an H atom instead of to an —OH group, as it is in ribose.) The full name for DNA is deoxyribonucleic acid, with the "nucleic" part coming from DNA's location in the nuclei of eukaryotic cells. The nitrogenous base (thymine, in our example) has a

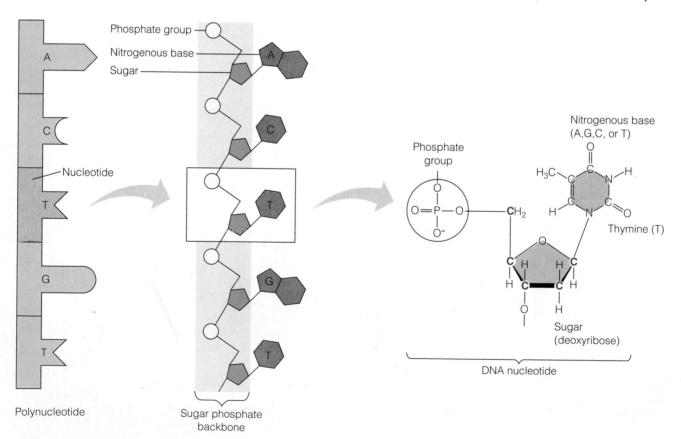

A. DNA polynucleotide

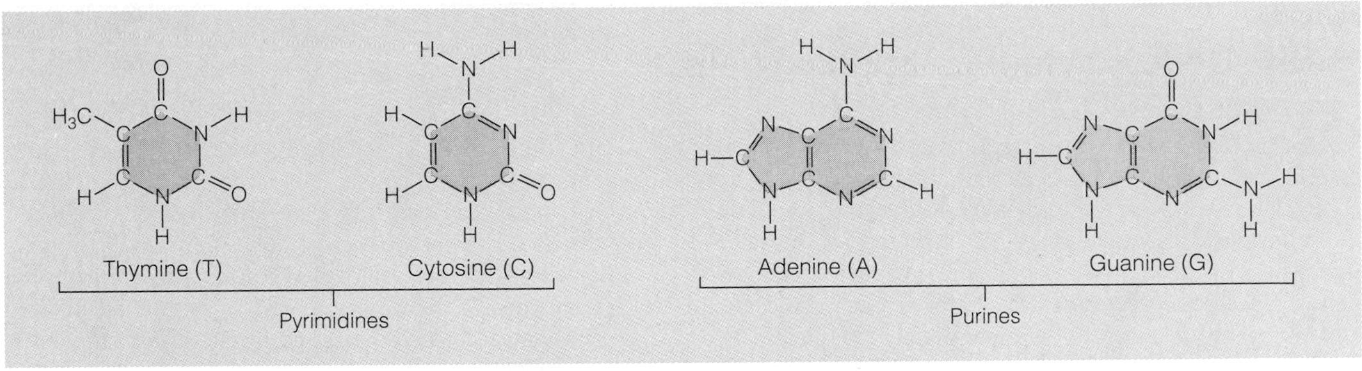

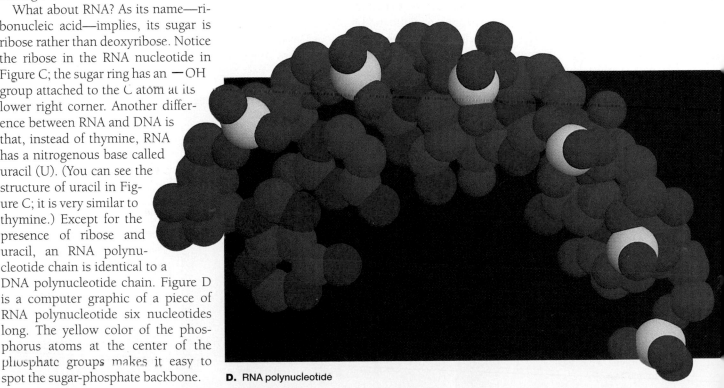

B. Nitrogenous bases of DNA

Thymine (T) Cytosine (C)

Pyrimidines

Adenine (A) Guanine (G)

Purines

ring of nitrogen and carbon atoms with various functional groups attached. In contrast to the acidic phosphate group, nitrogenous bases are basic, hence their name.

The four nucleotides found in DNA differ only in their nitrogenous bases. Figure B shows the structures of DNA's four nitrogenous bases. At this point, the structural details are not as important as the fact that the bases are of two types. Thymine (T) and cytosine (C) are single-ring structures called **pyrimidines.** Adenine (A) and guanine (G) are larger, double-ring structures called **purines.** Note that the one-letter abbreviations can be used for either the bases alone or for the nucleotides containing them.

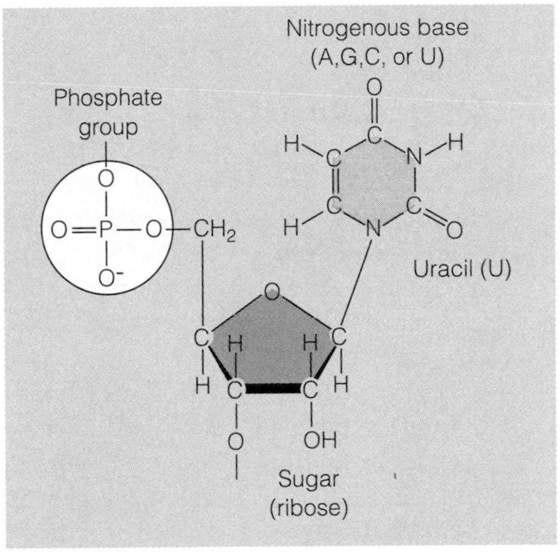

What about RNA? As its name—ribonucleic acid—implies, its sugar is ribose rather than deoxyribose. Notice the ribose in the RNA nucleotide in Figure C; the sugar ring has an —OH group attached to the C atom at its lower right corner. Another difference between RNA and DNA is that, instead of thymine, RNA has a nitrogenous base called uracil (U). (You can see the structure of uracil in Figure C; it is very similar to thymine.) Except for the presence of ribose and uracil, an RNA polynucleotide chain is identical to a DNA polynucleotide chain. Figure D is a computer graphic of a piece of RNA polynucleotide six nucleotides long. The yellow color of the phosphorus atoms at the center of the phosphate groups makes it easy to spot the sugar-phosphate backbone.

Phosphate group

Nitrogenous base (A,G,C, or U)

Uracil (U)

Sugar (ribose)

C. RNA nucleotide

D. RNA polynucleotide

DNA is a double-stranded helix

Once biologists were convinced that DNA was the genetic material, knowing the molecule's building blocks was not enough. A race was on to determine how the structure of this molecule could account for its role in heredity. The three-dimensional structures of proteins were starting to yield fascinating clues about the functions of those macromolecules, and biologists hoped the three-dimensional structure of DNA might do the same. Among the scientists working on the prob-

A. Rosalind Franklin

lem were several who had already made discoveries in deciphering protein structure: Linus Pauling in California, and Maurice Wilkins and Rosalind Franklin (Figure A) in London. First to the finish line with DNA, however, were two scientists who were relatively unknown at the time—American James D. Watson (left in Figure B) and Englishman Francis Crick. In 1962, Watson, Crick, and Wilkins received the Nobel Prize for their work. (Franklin probably would have received the prize as well, but for her untimely death from cancer in 1958.)

The celebrated partnership that determined the structure of DNA began soon after the 23-year-old Watson journeyed to Cambridge University, where Crick was studying protein structure with a technique called X-ray crystallography. While visiting the laboratory of Maurice Wilkins at King's College in London, Watson saw an X-ray crystallographic photograph of DNA, produced by Wilkins's colleague Rosalind Franklin. The photograph clearly revealed the basic shape of DNA to be a helix. On the basis of Watson's later recollection of the photograph, he and Crick deduced that the helix had a uniform diameter of 2 nm, with its nitrogenous bases stacked about a third of a nanometer apart. (For comparison, recall that the plasma membrane of a cell is about 8 nm thick.) The diameter of the helix suggested that it was made up of two polynucleotide strands. The presence of two strands accounts for the now-familiar term **double helix**.

B. Watson and Crick in 1953 with their model of the DNA double helix

Using wire models of the nucleotides, Watson and Crick began trying to construct a double helix that would conform both to Franklin's data and to what was then known about the chemistry of DNA. After failing to make a satisfactory model that placed the sugar-phosphate backbones inside the double helix, Watson tried putting the backbones on the outside and forcing the nitrogenous bases to swivel to the interior of the molecule. It occurred to him that the four kinds of bases might pair in a specific way. This idea of specific base pairing was a flash of inspiration that enabled Watson and Crick to solve the DNA puzzle.

At first, Watson imagined that the bases paired like with like—for example, A with A, and C with C. But that kind of pairing did not fit with the fact that the DNA molecule has a *uniform* diameter. An A-A pair (made of double-ringed bases) would be almost twice as wide as a C-C pair, causing bulges in the molecule. It soon became apparent that a double-ringed base (purine) must always be paired with a single-ringed base (pyrimidine) on the opposite strand. Moreover, Watson and Crick realized that the individual structures of the bases dictated the pairings even more specifically. Each base has chemical side groups that can best form hydrogen bonds with one appropriate partner (to review the hydrogen bond, see Module 2.10). Adenine can best form hydrogen bonds with thymine, and guanine with cytosine. In the biologist's shorthand, A pairs with T, and G pairs with C. A is also said to be "complementary" to T, and G to C.

Watson and Crick's pairing scheme not only fit what was known about the physical attributes and chemical bonding of DNA, it also explained some data obtained several years earlier by American biochemist Erwin Chargaff. Chargaff and his co-workers had discovered that the amount of adenine in the DNA of any one species was equal to the amount of thymine, and that the amount of guanine was equal to that of cytosine. Chargaff's rule, as it is called, is explained by the fact that A on one of DNA's polynucleotide chains always pairs with T on the other polynucleotide chain, and G on one chain pairs only with C on the other chain.

You can picture the model of the DNA double helix proposed by Watson and Crick as a rope ladder having rigid, wooden rungs, with the ladder twisted into a spiral (Figure C). The side ropes are the equiva-

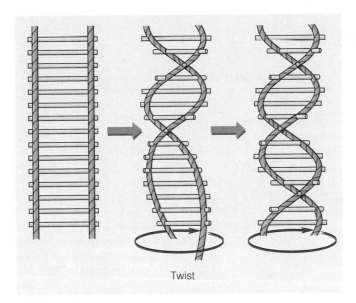

Twist

C. A rope-ladder model for the double helix

that the sugars on the two strands are upside-down with respect to each other.) On the right is a computer graphic showing every atom of part of a double helix. Atoms that compose the deoxyribose sugars are blue, phosphate groups are yellow, and nitrogenous bases are shades of green and orange.

Although the Watson-Crick base-pairing rules dictate the side-by-side combinations of nitrogenous bases that form the rungs of the double helix, they place no restrictions on the *sequence* of nucleotides along the length of a DNA strand. In fact, the sequence of bases can vary in countless ways. Consequently, it is not surprising that the DNA of different species, which have different genes, have different proportions of the bases in their DNA.

In April 1953, Watson and Crick shook the scientific world with a succinct, two-page announcement of their molecular model for DNA in the journal *Nature*. Few milestones in the history of biology have had as broad an impact as their double helix, with its A-T and C-G base pairing.

The Watson-Crick model gave new meaning to the words genes and chromosomes—and to the chromosomal theory of inheritance. With a complete picture of DNA, we can see that the genetic information in a chromosome must be encoded in the nucleotide sequence of the molecule. The structure of DNA also suggests a molecular explanation for life's unique properties of reproduction and inheritance, as we see next.

lent of the sugar-phosphate backbones, and the rungs represent pairs of nitrogenous bases joined by hydrogen bonds.

Figure D shows three representations of the double helix. The ribbonlike diagram on the left symbolizes the bases with shapes that emphasize their complementarity. In the center is a more chemical version, with the helix untwisted and the hydrogen bonds specified by dotted lines. Here you can see that the two sugar-phosphate backbones of the double helix are oriented in opposite directions. (Notice

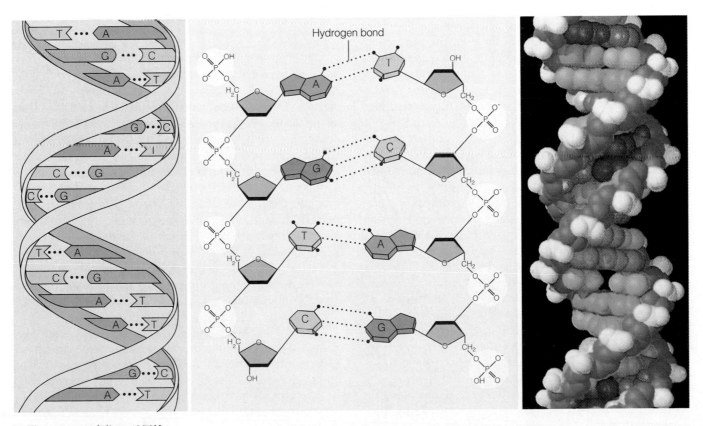

Hydrogen bond

D. Three representations of DNA

DNA replication depends on base pairing

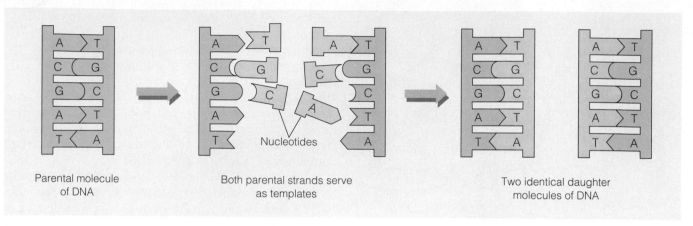

Parental molecule of DNA

Both parental strands serve as templates

Nucleotides

Two identical daughter molecules of DNA

A. A template model for DNA replication

An essential aspect of reproduction and inheritance is that a complete set of genetic instructions passes from one generation to the next. In order for this to occur, there must be a means of copying the instructions. Long before DNA was identified as the genetic material, some people argued that gene replication must be based on a concept called complementary surfaces. According to this idea, when a gene replicates, a "negative image" is created along the original ("positive") surface, just as clay or plaster forms a "negative" shape when it is packed around an object. The gene's negative image, like the plaster, could serve as a template (mold) for making copies of the original positive image. Photography provides another example of the template principle: A print can be used to make a negative, which can then be used to make copies of the original print. Until 1953, the template idea was discounted by many geneticists. However, Watson and Crick's model for DNA structure suggested a template mechanism for DNA replication. As they said in the conclusion of their first paper, "It has not escaped our notice that the specific pairing we have postulated immediately suggests a possible copying mechanism for the genetic material."

The logic behind the Watson-Crick proposal for how DNA is copied—by specific pairing of complementary bases—is quite simple. You can see this by covering one of the strands in the parental DNA molecule in Figure A with a piece of paper. You can determine the sequence of bases in the covered strand by applying the base-pairing rules to the unmasked strand: A pairs with T, G with C. Watson and Crick predicted that a cell applies the same rules when copying its genes. Figure A illustrates the template hypothesis for DNA replication. First, the two strands of parental DNA separate, and each becomes a template for the assembly of a complementary strand from a supply of free nucleotides. The nucleotides line up one at a time along the template strand in accordance with the base-pairing rules. Enzymes then link the nucleotides to form the new DNA

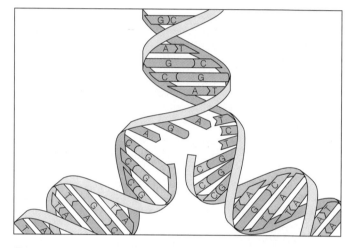

B. Untwisting and replication of DNA

strands. The completed new molecules, identical to the parent molecule, are known as daughter DNA. This hypothesis was confirmed by experiments performed in the 1950s.

Although the general mechanism of DNA replication is conceptually simple, the actual process involves complex biochemical gymnastics. Some of the complexity arises from the fact that the helical DNA molecule must untwist as it replicates and must copy its two strands roughly simultaneously (Figure B, above). Another challenge is the speed of the process. Nucleotides are added at a rate of about 50 per second in mammals and 500 per second in bacteria. Yet despite its speed, replication is amazingly accurate; typically, only about one in a billion nucleotides in DNA is incorrectly paired.

In achieving such speed and accuracy, DNA replication requires the cooperation of more than a dozen enzymes and other proteins. The chief enzymes responsible for assembling DNA polynucleotides are called **DNA polymerases.** These enzymes carry out a number of functions, including a proofreading step that quickly removes incorrectly paired nucleotides.

The DNA genotype is expressed as proteins, which are the molecular basis of phenotypic traits

With our knowledge of DNA, we can now define genotype and phenotype more precisely than we did in Chapter 9. An organism's genotype, its genetic makeup, is the heritable information contained in its DNA. The phenotype is the organism's specific traits. The molecular basis of the phenotype lies in proteins with a variety of functions. For example, structural proteins help make up the body of an organism, and enzymes catalyze its metabolic activities.

What is the connection between the genotype and the protein molecules that directly determine the phenotype? The answer is that DNA specifies the synthesis of proteins. A gene does not build a protein directly, but rather dispatches instructions in the form of RNA, which in turn programs protein synthesis. This central concept in biology is summarized in Figure A. The chain of command is from DNA in the nucleus of the cell (purple area) to RNA to protein synthesis in the cytoplasm (tan area). The two main stages are **transcription**, the transfer of genetic information from DNA into an RNA molecule, and **translation**, the transfer of the information in the RNA into a

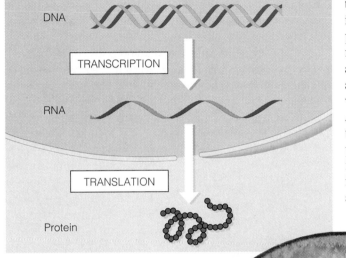

A. The flow of genetic information in a eukaryotic cell

protein. The next eight modules describe the steps in this flow of molecular information from gene to protein.

The relationship between genes and proteins was first proposed in 1909, when English physician Archibald Garrod suggested that genes dictate phenotypes through enzymes, the proteins that catalyze chemical processes in the cell. Garrod's idea came from his observations of inherited diseases. He hypothesized that an inherited disease reflects a person's inability to make a particular enzyme, and he referred to such diseases as "inborn errors of metabolism." He gave as one example the hereditary condition called alkaptonuria, in which the urine appears dark red because it contains a chemical called alkapton. Garrod reasoned that normal individuals have an enzyme that breaks down alkapton, whereas alkaptonuric individuals lack the enzyme. Garrod's hypothesis was ahead of its time, but research conducted several decades later proved him right. In the intervening decades, biochemists accumulated evidence that cells make and break down biologically important molecules via metabolic pathways, as in the synthesis of an amino acid or the breakdown of a sugar. As we described in Unit 1, each step in a metabolic

pathway is catalyzed by a specific enzyme. So individuals lacking one of the enzymes for a pathway are unable to complete the pathway.

The major breakthrough in demonstrating the relationship between genes and enzymes came in the 1940s, from the work of American geneticists George Beadle and Edward Tatum with the orange bread mold *Neurospora crassa* (Figure B). Beadle and Tatum studied strains of the mold that were unable to grow on the usual simple growth medium. Each of these so-called nutritional mutants turned out to lack an enzyme in a metabolic pathway that produced some molecule the mold needed, such as an amino acid. Beadle and Tatum also showed that each mutant was defective in a single gene. Accordingly, they formulated the one gene–one enzyme hypothesis, which stated that the function of an individual gene is to dictate the production of a specific enzyme.

B. *Neurospora crassa* growing in a culture dish

The one gene–one enzyme hypothesis has been amply confirmed, but with some important modifications. First it was extended beyond enzymes to include *all* types of proteins. For example, alpha-keratin, the structural protein of your hair, and the structural proteins that make up the outside of phage T4 are as much the products of genes as enzymes are. So biologists soon began to think in terms of one gene–one *protein*. Then it was discovered that many proteins consist of two or more different polypeptide chains (see Module 3.18), and each polypeptide may be specified by its own gene. Thus, Beadle and Tatum's hypothesis has come to be restated as one gene–one *polypeptide*.

10.6 Genetic information is written as codons and translated into amino acid sequences

Stating that genetic information in DNA is transcribed into RNA and then translated into the polypeptides that make up proteins does not tell us *how* these processes occur. Transcription and translation are linguistic terms, and it is useful to think of nucleic acids and polypeptides as having languages. But to understand how genetic information passes from genotype to phenotype, we need to see how the chemical language of DNA is translated into the different chemical language of polypeptides.

What, exactly, is the language of nucleic acids? Both DNA and RNA are polymers made of monomers in specific sequences that carry information, much as specific sequences of letters carry information in English or Spanish, for example. In DNA, the monomers are the four types of nucleotides, which differ in their nitrogenous bases (A, T, C, and G). The same is true for RNA, although it has the base U instead of the T in DNA.

The figure here focuses on a small region of one of the genes ("Gene 3," shown in dark blue) carried by a DNA molecule. DNA's language is written as a linear sequence of nucleotide bases on a polynucleotide, a sequence such as the one you see on the enlarged DNA strand in the figure. Specific sequences of bases, each with a beginning and an end, make up the genes on a DNA strand. A typical gene consists of hundreds or thousands of nucleotides, and a molecule of DNA may contain thousands of genes.

The pink strand underneath the enlarged DNA region rep-

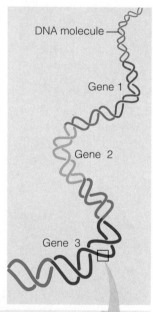

DNA molecule

Gene 1

Gene 2

Gene 3

resents the results of transcription: an RNA molecule. The process is called transcription because the nucleic acid language of DNA has simply been rewritten as a sequence of bases on RNA; the language is still that of nucleic acids. Notice that the nucleotide bases on the RNA molecule are complementary to those on the DNA strand. As we will see in Module 10.7, this is because the RNA was synthesized using the DNA as a template.

The purple chain represents the results of translation, the conversion of the nucleic acid language into the polypeptide language made up of amino acids. Like nucleic acids, polypeptides are polymers, but the monomers that make them up—the letters of the polypeptide alphabet—are the 20 amino acids common to all organisms. Again, the language is written in a linear sequence, and the sequence of nucleotides of the RNA molecule dictates the sequence of amino acids of the polypeptide. RNA is only a messenger; the genetic information that dictates the amino acid sequence is based in DNA.

The brackets below the RNA indicate how genetic information is coded in nucleic acids. Notice that each bracket encloses *three* nucleotides on RNA. Recall that there are only four different kinds of nucleotides in DNA and RNA. In translation, these four must somehow specify 20 amino acids. If each nucleotide base specified one amino acid, only four of the 20 amino acids could be accounted for. What if the language consisted of two-letter code words? If we read the bases of a gene two at a time, AG, for example, could specify one amino acid, while AT could designate a different amino acid. However, when the 4 bases are taken in doublets, there are only 16 (that is, 4^2) possible arrangements—still not enough to specify all 20 amino acids.

Triplets of bases are the smallest "words" of uniform length that can specify all the amino acids. Suppose each code word in DNA consists of a triplet, with each arrangement of three consecutive bases specifying an amino acid. Then there can be 64 (that is, 4^3) possible code words—more than enough to specify the 20 amino acids. Indeed, there are enough triplets to allow more than one coding for each amino acid. For example, the base triplets AAT and AAC could both code for the same amino acid—and, in fact, they do.

Experiments have verified that the flow of information from gene to protein is based on a **triplet code:** The genetic instructions for the amino acid sequence of a polypeptide chain are written in DNA and RNA as a series of three-base words, called **codons.** Notice in the figure that three-base codons in the DNA are transcribed into complementary three-base codons in the RNA, and then the RNA codons are translated into amino acids that form a polypeptide.

We return to the codons themselves in Module 10.14. In the next several modules, we examine transcription and translation in more detail.

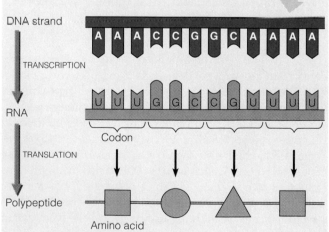

DNA strand

A A A C C G G C A A A A

TRANSCRIPTION

RNA

U U U G G C C G U U U U

Codon

TRANSLATION

Polypeptide

Amino acid

Transcription, the transfer of genetic information from DNA to RNA, occurs in the cell nucleus, as we indicated in Figure 10.5A. An RNA molecule is transcribed from a DNA template by a process that resembles the synthesis of a DNA strand during DNA replication. Figure A is a closeup view of this process. As with replication, the two DNA strands must first separate at the place where the process will start. In transcription, however, only one of the DNA strands serves as a template for the newly forming molecule. The nucleotides that make up the new RNA molecule take their places one at a time along the DNA template strand by forming hydrogen bonds with the nucleotide bases there. Notice that the RNA nucleotides follow the same base-pairing rules that govern DNA replication, except that U, rather than T, pairs with A. The RNA nucleotides are linked by the transcription enzyme **RNA polymerase,** symbolized in the figure by the large gray shape in the background.

Figure B is an overview of the transcription of an entire gene. RNA polymerase must be instructed where to start and where to stop the transcribing process. The "start transcribing" signals are specific nucleotide sequences called **promoters,** which are located in the DNA next to the beginning of the gene. The first phase of transcription, called initiation, occurs when RNA polymerase attaches to the promoter DNA. For any gene, the promoter region signals only one of the two DNA strands to be transcribed (the particular strand varying from gene to gene).

During a second phase of transcription, the RNA elongates. As elongation continues, the RNA peels away from its DNA template, allowing the two separated DNA strands to come back together in the region already transcribed.

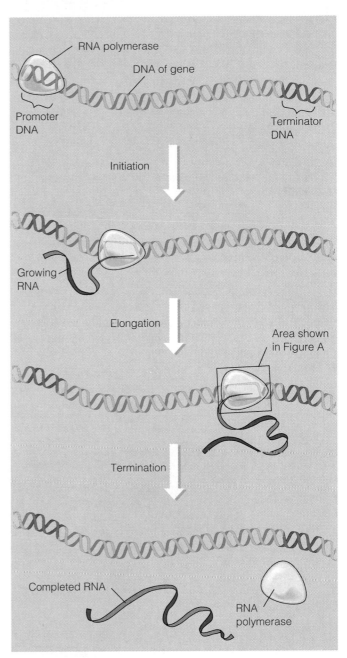

B. Transcription of a gene

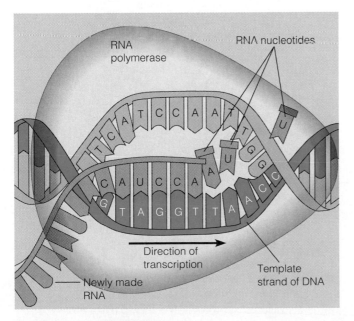

A. A closeup view of transcription

Finally, in the third phase, termination, the RNA polymerase reaches a special sequence of bases in the DNA template called a **terminator.** This sequence signals the end of the gene; at that point, the polymerase molecule detaches from the RNA molecule and the gene.

In addition to producing RNA that encodes amino acid sequences, transcription makes two other kinds of RNA that are involved in building polypeptides. We discuss these three kinds of RNA in the next three modules.

10.8 Genetic messages are translated in the cytoplasm

The kind of RNA that encodes amino acid sequences is called **messenger RNA (mRNA)** because it conveys genetic information from DNA to the translation machinery of the cell. Messenger RNA is transcribed from DNA, and the genetic message the RNA carries is then translated into polypeptides. In prokaryotes, which lack a nucleus, transcription and translation both take place in the cytoplasm. As we indicated in Figure 10.5A, the situation is different in eukaryotic cells. Because DNA stays in the nucleus, eukaryotic mRNA molecules and the other RNA molecules required for translation must cross the nuclear envelope into the cytoplasm, where the equipment for polypeptide synthesis is located.

The figure below shows the results of an experiment that demonstrates the migration of RNA from the nucleus to the cytoplasm. Researchers first made radioactive RNA nucleotides and then grew eukaryotic cells in culture with the nucleotides. The cells used the radioactive nucleotides in synthesizing RNA. Radioactive emissions from the resulting RNA produced the black spots on the light micrographs shown here. Thus, the dots pinpoint the location of radioactive RNA in the cells.

The cell on the left was killed after 15 minutes of growing with the radioactive nucleotides, and the excess nucleotides were washed away. As you can see, the newly made RNA is heavily concentrated in the nucleus.

The cell on the right was allowed to make radioactive RNA for 15 minutes. It was then washed free of excess radioactive nucleotides but not killed. Instead, it was transferred for an additional 90 minutes to a growth liquid containing no radioactive nucleotides. As you can see, the radioactive RNA moved into the cytoplasm.

We are now ready to begin examining how the translation process works. Translation is a conversion between different languages—from nucleic acid language to protein language—and it involves more complex machinery than transcription. The machinery required for translation includes the following:

- Transfer RNA, another kind of RNA molecule
- Ribosomes, the organelles where polypeptide synthesis occurs
- Enzymes and a number of protein "factors"
- Sources of chemical energy, such as ATP

In the next two modules, we take a closer look at transfer RNA and ribosomes.

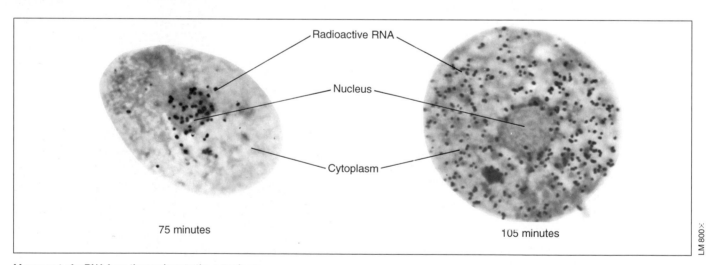

Movement of mRNA from the nucleus to the cytoplasm

10.9 Transfer RNA molecules serve as interpreters during translation

Translation of any language into another language requires an interpreter, someone who can recognize the words of one language and convert them into the other. Translation of the genetic message carried in mRNA into the amino acid language of proteins also requires an interpreter. To convert the three-letter words (codons) of nucleic acids to the one-letter, amino acid words of proteins, a cell employs a molecular interpreter, a special type of RNA called **transfer RNA (tRNA)**.

A cell that is ready to have some of its genetic information translated into polypeptides has a supply of amino acids in its cytoplasm. It has either synthesized them from other chemicals or obtained them from food. The amino acids themselves cannot recognize the codons arranged in sequence along messenger RNA. The amino acid tryptophan, for example, is no more attracted by codons for tryptophan than by any other codons. It is up to the cell's molecular interpreters, tRNA molecules, to match amino acids to the appropriate codons to form the new polypeptide. To perform this task, tRNA molecules must carry out two distinct functions: (1) picking up the appropriate amino acids, and (2) recognizing the appropriate codons in the mRNA. The unique structure of tRNA molecules allows them to perform both tasks.

As shown in Figure A, a tRNA molecule is made of a single strand of RNA—one polynucleotide chain—consisting of only about 80 nucleotides. By twisting and folding upon itself, tRNA forms several double-stranded regions in which short stretches of RNA base pair with other stretches. A single-stranded loop at one end of the folded molecule contains a special triplet of bases called an **anticodon.** The anticodon triplet is complementary to a codon triplet on mRNA. During translation, the anticodon on tRNA recognizes a particular codon on mRNA by using base-pairing rules. At the other end of the tRNA molecule is a site where an amino acid can attach.

In the modules that follow, in which we trace the process of translation, we represent tRNA with the simplified shape shown in Figure B (above right). This symbol emphasizes the two parts of the molecule—the anticodon and the amino acid attachment site—that give tRNA its ability to match up a particular nucleic acid word (codon) with its corresponding protein word (amino acid). Although all tRNA molecules are similar, there is a slightly different variety of tRNA for each amino acid.

Translating any language is a complex task. Likewise, the conversion of genetic information into the exact polypeptide it specifies is a complex process—too complex, in fact, for the cell's tRNA interpreters to carry out by themselves. By itself, a tRNA molecule cannot directly recognize an amino acid. What ensures that the appropriate amino acid attaches to a tRNA is an enzyme. There is a whole family of these enzymes, with at least one enzyme for each amino acid. Each enzyme specifically binds one type of amino acid to the appropriate tRNA molecule, using a molecule of ATP as energy to drive the reaction. The resulting amino acid–tRNA complex can then furnish

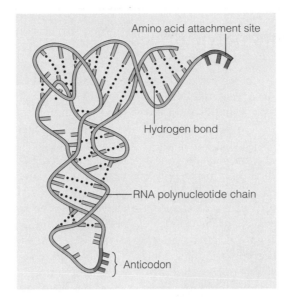

Amino acid attachment site

Hydrogen bond

RNA polynucleotide chain

} Anticodon

A. The structure of tRNA

Amino acid attachment site

Anticodon

B. The symbol for tRNA used in this book

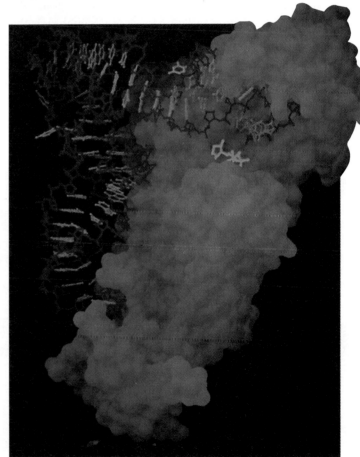

C. A molecule of tRNA binding to an enzyme molecule

its amino acid to a growing polypeptide chain, a process we describe in Module 10.10.

The computer graphic in Figure C shows a tRNA molecule (red and yellow) and an ATP molecule (green) bound to the enzyme molecule (blue). In this picture, you can see the proportional sizes of these three molecules. The amino acid that would attach to the tRNA is not shown; it would be less than half the size of the ATP.

10.10 Ribosomes build polypeptides

We have now looked at many of the things a cell needs to carry out translation: instructions in the form of mRNA molecules, tRNA to interpret the instructions, a supply of amino acids, enzymes for attaching amino acids to tRNA, and ATP for energy. Still needed are the actual polypeptide "factories"—organelles in the cytoplasm that coordinate the functioning of the mRNA and tRNA and actually make polypeptides. Ribosomes are these factories.

A ribosome consists of two subunits, each made up of proteins and a considerable amount of yet another kind of RNA, **ribosomal RNA (rRNA).** In Figure A, you can see the shapes and relative sizes of the ribosomal subunits. You can also see where mRNA and the growing polypeptide are located while translation proceeds.

The simplified drawings in Figures B and C indicate how tRNA anticodons and mRNA codons are brought together

on ribosomes. As Figure B shows, each ribosome has a binding site for mRNA on its small subunit. In addition, it has two binding sites for tRNA. The P site holds the tRNA carrying the growing polypeptide chain, while the A site holds a tRNA carrying the next amino acid to be added to the chain. Figure C shows tRNA molecules occupying the two sites. The anticodon on each tRNA base-pairs with a codon on mRNA. The subunits of the ribosome act like a vise, holding the tRNA and mRNA molecules close together. The ribosome can then connect the amino acid from the A-site tRNA to the growing polypeptide.

We have now sketched the main events in the conversion of genetic information to polypeptides. Before moving ahead, we need to examine the first stage of translation in a bit more detail.

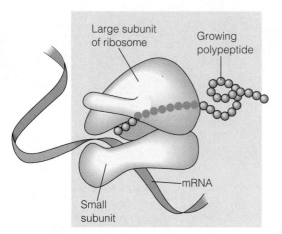

A. The true shape of a functioning ribosome

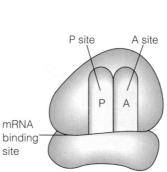

B. The binding sites of a ribosome

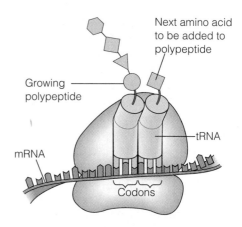

C. A ribosome with occupied binding sites

10.11 An initiation codon marks the start of an mRNA message

Translation can be divided into the same three phases as transcription: initiation, elongation, and termination. The process of polypeptide initiation brings together the mRNA, the first amino acid with its attached tRNA, and the two subunits of a ribosome.

As indicated in Figure A, an mRNA molecule transcribed from DNA is longer than the genetic message it carries. A sequence of nucleotides (light pink) at either end of the molecule is not part of the message but helps the mRNA bind to the ribosome. The role of the initiation process is to determine exactly where translation will begin, so that the mRNA codons

will be translated into the correct sequence of amino acids.

Initiation occurs in two steps, as shown in Figure B at the top of the next page:

Step ① An mRNA molecule binds to a small ribosomal subunit. A special initiator tRNA locates and binds to the specific codon, called the **start codon,** where translation is to begin on the mRNA molecule. The initiator tRNA usually carries the amino acid methionine (Met); its anticodon, UAC, binds to the start codon, AUG.

Step ② A large ribosomal subunit binds to the small one, creating a functional ribosome. The initiator tRNA fits into the P site on the ribosome.

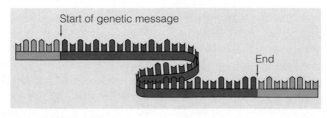

Start of genetic message

End

A. A molecule of mRNA

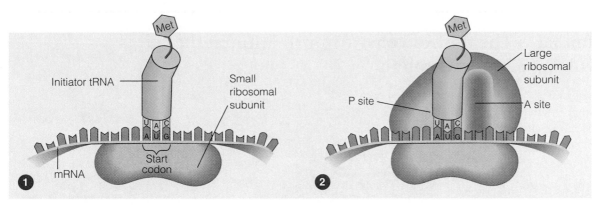

B. The initiation of translation

Elongation adds amino acids to the polypeptide chain until a stop codon terminates translation 10.12

Once initiation is complete, amino acids are added one by one to the initial amino acid. Each addition occurs in a three-step elongation process, shown in the figure here. (Red arrows indicate movement.)

Step ① Codon recognition. The anticodon of an incoming tRNA molecule, carrying its amino acid, pairs with the mRNA codon in the A site of the ribosome.

Step ② Peptide bond formation. The polypeptide separates from the tRNA to which it was bound (the one in the P site) and attaches by a peptide bond to the amino acid carried by the tRNA in the A site. An enzyme in the ribosome catalyzes formation of the bond. Thus, one more amino acid is added to the chain.

Step ③ Translocation. The P-site tRNA now leaves the ribosome, and the A-site tRNA, carrying the growing polypeptide, is translocated (moved) to the P site. The codon and anticodon remain bonded, and the mRNA and tRNA move as a unit. This movement brings into the A site the next mRNA codon to be translated, and the process can start again with step 1.

Elongation continues until a **stop codon** reaches the ribosome's A site. Stop codons—UAA, UAG, and UGA—do not code for amino acids but instead signal translation to stop. This is the termination stage of translation. The completed polypeptide, typically about 100 amino acids long, is freed from the last tRNA and from the ribosome, which then splits into its subunits.

During and after translation, the polypeptide coils and folds, assuming a three-dimensional shape, its tertiary structure. Lastly, several polypeptides may come together, forming a protein with quaternary structure (see Module 3.18).

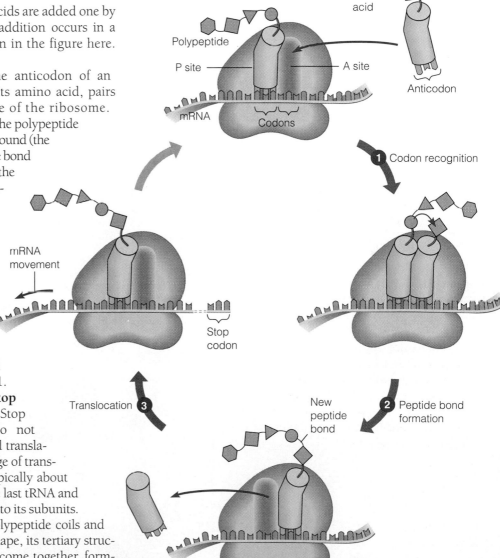

Summary: The flow of genetic information in the cell is DNA $\longrightarrow$ RNA $\longrightarrow$ protein

This figure summarizes the key steps in the flow of genetic information from DNA to protein. These steps are common to all cells.

In transcription (DNA $\longrightarrow$ RNA), the genetic messenger mRNA is synthesized on a DNA template (step ①). In eukaryotic cells, transcription occurs in the nucleus, and the mRNA must travel from the nucleus to the cytoplasm.

Translation (RNA $\longrightarrow$ protein) can be divided into four steps (numbered ② through ⑤), all of which occur in the cytoplasm. When the polypeptide is complete, the two ribosomal subunits come apart, and the tRNA and mRNA are released (not shown in figure).

The translation process is very rapid; a single ribosome can make an average-sized polypeptide in less than a minute. Typically, an mRNA molecule is translated simultaneously by a number of ribosomes. Once the start codon emerges from the first ribosome, a second ribosome can attach to it, and thus several ribosomes (collectively known as a polyribosome) may trail along on the same mRNA molecule.

What is the overall significance of transcription and translation? These are the processes whereby genes control the structures and activities of cells, or, more broadly, the way the genotype produces the phenotype. The chain of command originates with the information in a gene, a specific linear sequence of nucleotides in DNA. The gene serves as a template, dictating transcription of a complementary sequence of nucleotides in mRNA. In turn, mRNA dictates the linear sequence in which amino acids appear in a specific polypeptide. Finally, the proteins that form from the polypeptides determine the appearance and the capabilities of the cell and organism.

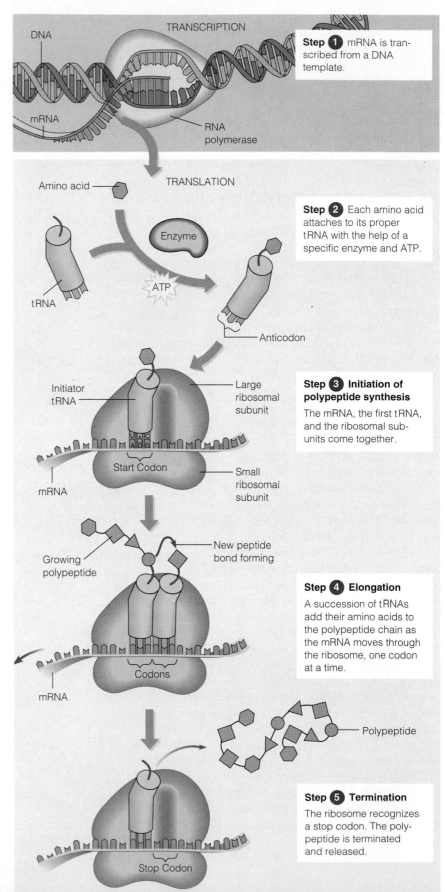

TRANSCRIPTION

DNA

mRNA

RNA polymerase

Step ① mRNA is transcribed from a DNA template.

TRANSLATION

Amino acid

Enzyme

ATP

tRNA

Anticodon

Step ② Each amino acid attaches to its proper tRNA with the help of a specific enzyme and ATP.

Initiator tRNA

Large ribosomal subunit

Start Codon

mRNA

Small ribosomal subunit

Step ③ Initiation of polypeptide synthesis

The mRNA, the first tRNA, and the ribosomal subunits come together.

Growing polypeptide

New peptide bond forming

Codons

mRNA

Step ④ Elongation

A succession of tRNAs add their amino acids to the polypeptide chain as the mRNA moves through the ribosome, one codon at a time.

Polypeptide

Stop Codon

Step ⑤ Termination

The ribosome recognizes a stop codon. The polypeptide is terminated and released.

In 1799, a large stone tablet was found in Rosetta, Egypt, carrying the same lengthy inscription in three ancient languages: Egyptian written in hieroglyphics and two forms of Greek. This stone provided the key that enabled scholars to crack the previously indecipherable hieroglyphic code.

In cracking the genetic code, scientists wrote their own Rosetta stone. It was based on information gathered from a series of elegant experiments that disclosed the amino acid translations of each of the nucleotide-triplet code words. The first codon was deciphered in 1961 by American biochemist Marshall Nirenberg. He had synthesized an artificial mRNA by linking together identical RNA nucleotides having uracil as their base. No matter where this message started or stopped, it could contain only one type of triplet codon: UUU. Nirenberg added this "poly U" to a test-tube mixture containing ribosomes and the other ingredients required for polypeptide synthesis. This mixture translated the poly U into a polypeptide containing a single kind of amino acid, phenylalanine. Thus, Nirenberg learned that the mRNA codon UUU specifies the amino acid phenylalanine (Phe). Soon the amino acids specified by all the codons were determined by similar methods.

As Figure A shows, 61 of the 64 triplets code for amino acids. The triplet AUG has a dual function: It not only codes for the amino acid methionine (Met) but also can provide a signal for the start of a polypeptide chain. Three of the other codons (in white boxes in the figure) do not designate amino acids. They are the stop codons that instruct the ribosomes to end the polypeptide.

Notice in Figure A that there is redundancy in the code but no ambiguity. For example, although codons UUU and UUC both specify phenylalanine (redundancy), neither of them ever represents any other amino acid (no ambiguity). The codons in Figure A are the triplets found in mRNA. They have a straightforward, complementary relationship to the codons found in DNA. The nucleotides making up

the codons occur in a linear order along the DNA and RNA, with no gaps or "punctuation" separating the codons.

As an exercise in translating the genetic code, consider the 12-nucleotide segment of DNA in Figure B. Let's read this as a series of triplets. Using the base-pairing rules (C-G and A-T in DNA, with U replacing T in RNA), we see that the RNA codon corresponding to the first transcribed DNA triplet, TAC, is AUG. AUG says, "Place Met as the first amino acid in the polypeptide." The second DNA triplet, TTC, dictates RNA codon AAG, which designates lysine (Lys) as the second amino acid. We continue reading triplets until we reach the stop codon.

Almost all of the genetic code is shared by all organisms, from the simplest bacteria to humans. For instance, the RNA codon CCG is translated as proline in all organisms. In laboratory experiments, bacterial cells can translate genetic messages obtained from human cells, and human cells can translate bacterial genes (see Chapter 12). The universal consistency of the genetic vocabulary implies that the genetic code was established very early in evolution.

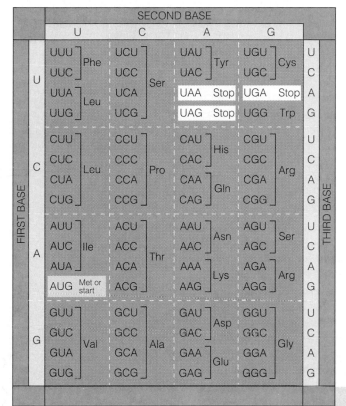

A. Dictionary of the genetic code

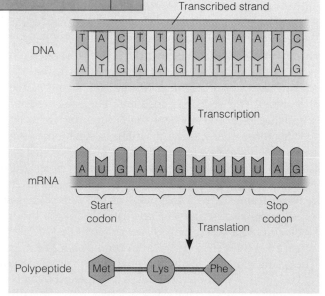

B. Deciphering the genetic information in DNA

Mutations can change the meaning of genes

Since discovering how genes are translated into proteins, scientists have been able to describe many heritable differences in molecular terms. For instance, when a child is born with sickle-cell anemia (see Module 9.12), the condition can be traced back through the difference in a protein to one tiny change in a gene. In one of the two kinds of polypeptides in the hemoglobin molecule, the sickle-cell child has a single different amino acid, a Val instead of a Glu. This difference is caused by the change of a single nucleotide in the coding strand of DNA (Figure A). In the double helix, a base *pair* is changed.

The sickle-cell allele is not a unique case. We now know that the alternative alleles of many genes result from changes in single base pairs in DNA. Any change in the nucleotide sequence of DNA is called a **mutation.** Mutations can involve large regions of a chromosome or just a single nucleotide pair, as in sickle-cell anemia. Here we consider how mutations involving only one or a few nucleotide pairs can affect gene translation.

Mutations within a gene can be divided into two general categories: base substitutions and base insertions or deletions (Figure B). A base substitution is the replacement of one nucleotide with another. In the second row in Figure B, A replaces G in the fourth codon of the mRNA molecule. Depending on how a base substitution is translated, it can result in no change in the protein, in an insignificant change, or in a change that might be crucial to the life of the organism. It is because of the redundancy of the genetic code that some substitution mutations have no effect. For example, if a mutation causes an mRNA codon to change from GAA to GAG, no change in the protein product would result, because GAA and GAG both code for the same amino acid (Glu). Other changes of a single nucleotide may alter an amino acid but have little effect on the function of the protein.

Some base substitutions, as we saw in the sickle-cell case, cause an important change in the protein, preventing it from performing normally. Occasionally, a base substitution will lead to an improved protein or one with new capabilities that enhance the success of the mutant organism and its descendants. Much more often, such mutations are harmful.

Mutations involving the insertion or deletion of one or more nucleotides in a gene often have disastrous effects. Because mRNA is read as a series of nucleotide triplets during translation, adding or subtracting nucleotides may alter the **reading frame** (triplet grouping) of the genetic message. All the nucleotides that are "downstream" of the insertion or deletion will be regrouped into different codons, as you can see for the deletion shown at the bottom in Figure B. The result will most likely be a nonfunctional polypeptide.

The creation of mutations, called **mutagenesis,** can occur in a number of ways. Mutations resulting from errors during DNA replication or recombination are called spontaneous

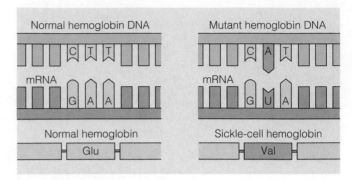

A. The molecular basis of sickle-cell anemia

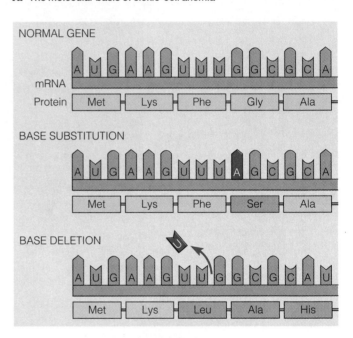

B. Types of mutations and their effects

mutations, as are other mutations of unknown cause. Another source of mutation is a physical or chemical agent, called a **mutagen.** The most common physical mutagen in nature is high-energy radiation, such as X-rays and ultraviolet light. Chemical mutagens are of various types. One type, for example, consists of chemicals that are similar to normal DNA bases but that pair incorrectly. Mutations induced by chemicals may lead to cancer (see Module 11.17).

Although mutations are often harmful, they are also extremely useful, both in nature and in the laboratory. It is because of mutations that there is such a rich diversity of genes in the living world, a diversity that makes evolution by natural selection possible. In addition, mutations are essential tools for geneticists. Whether naturally occurring (in Mendel's peas, for example) or created in the laboratory (Morgan used X-rays to make most of his fruit-fly mutants), mutations create the different alleles that are necessary for gene study.

As we discussed at the beginning of the chapter, viruses provided some of the first glimpses into the molecular mechanisms of heredity. Now that we have examined those mechanisms, let's take a closer look at viruses, focusing on the relationship between viral structure and the processes of nucleic acid replication, transcription, and translation.

In a sense, viruses are nothing more than packaged genes. For a researcher in molecular genetics, they may seem almost ideal tools—some nucleic acid wrapped in a protein shell—all you need to study the flow of genetic information and nothing more. It's really not that simple, however, because viruses can reproduce only by infecting cells. In fact, the host cell provides most of the components necessary for transcription and translation (ribosomes, enzymes, tRNA, etc.) and actually replicates, transcribes, and translates the viral nucleic acid.

Earlier in the chapter, we described what happens after the nucleic acid of phage T2 or T4 enters a host cell. The way T phages reproduce is called a **lytic cycle** because it always leads to the lysis (breaking open) of the host cell. The study of another phage of E. coli, called lambda, led to the discovery that some viruses can also reproduce by an alternative route called a lysogenic cycle. During a **lysogenic cycle**, viral DNA replication occurs without phage production or the death of the host cell.

Below you see the two kinds of cycles for phage lambda. Like T4, lambda has a head containing DNA and a tail, but no long tail fibers. Both cycles begin when the phage DNA enters the bacterium (top of the figure) and forms a circle (center). The DNA then embarks on one of the two pathways. In the lytic cycle (left), lambda's DNA immediately turns the cell into a virus-producing factory. In the lysogenic cycle, the DNA inserts by genetic recombination into the bacterial chromosome (bottom right). Once inserted, the phage DNA is referred to as a **prophage,** and most of its genes are inactive. Survival of the prophage depends on reproduction of the host cell. The host cell replicates the prophage DNA along with its cellular DNA (right side of figure) and then, upon dividing, passes on both the prophage and the cellular DNA to its two daughter cells. A single infected bacterial cell can quickly give rise to a large population of bacteria carrying prophages. The prophages may remain in the bacterial cells indefinitely. Occasionally, however, one leaves its host chromosome (top right); this event may be triggered by environmental conditions. Once separate, the lambda DNA usually switches to the lytic cycle.

Sometimes, the few prophage genes active in a bacterial cell can cause medical problems. For example, the bacteria that cause diphtheria, botulism, and scarlet fever would be harmless to humans if it were not for prophage genes they carry. Certain of these genes direct the bacteria to produce the toxins that are directly responsible for making people ill.

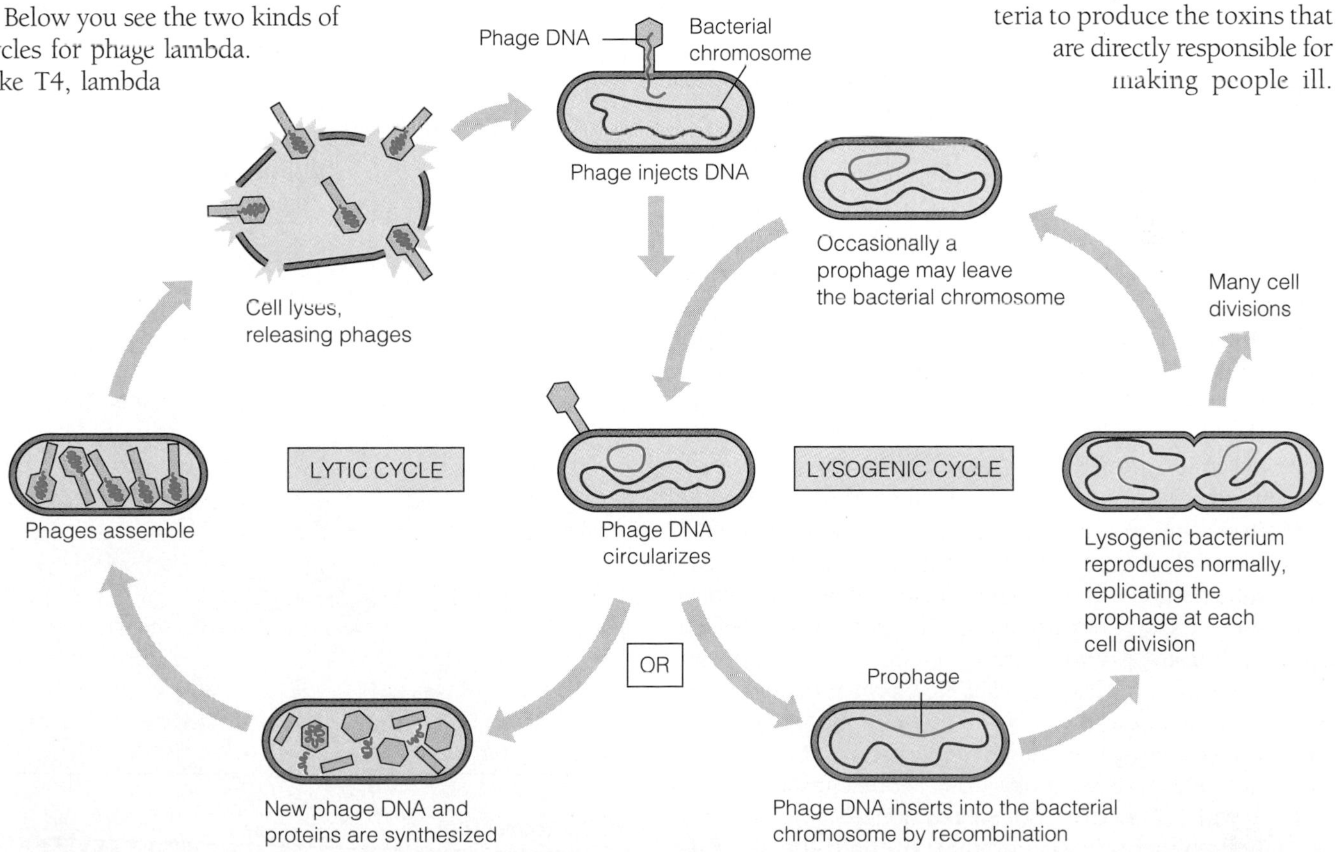

Two types of viral reproductive cycles

A. Tobacco mosaic disease and the tobacco mosaic virus

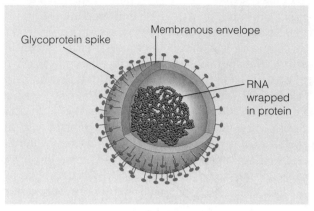

B. An influenza virus

Viruses that infect plant or animal cells are common causes of disease. Plant viruses can be serious agricultural pests that stunt plant growth and diminish crop yields. Many of them, like the tobacco mosaic virus in Figure A, are rod-shaped, with a spiral arrangement of proteins surrounding the nucleic acid. Unlike cells, viruses may have RNA rather than DNA as their genetic material. The nucleic acid in most plant viruses, including tobacco mosaic virus, is RNA.

We have all suffered from infections by RNA viruses, because these include the ones that cause the common cold and influenza ("flu"). RNA viruses are also responsible for some more serious human diseases, such as AIDS and polio. Polio is no longer epidemic in the United States, but it remains a major disease in many areas of the world.

Figure B shows the structure of an influenza virus. Like many animal viruses, this one has a membranous envelope surrounding its protein coat and projecting spikes made of glycoprotein. The envelope helps the virus enter and leave the host cell.

Figure C shows the reproductive cycle of an enveloped RNA virus (the one that causes mumps). When the virus contacts a host cell, the glycoprotein spikes protruding from the viral envelope attach to receptor proteins on the cell's plasma membrane. The envelope fuses with the cell's membrane, allowing the protein-coated RNA to ① enter the cytoplasm. ② Enzymes then remove the protein coat. ③ The viral RNA then serves as a template for making complementary strands of RNA (purple). The new strands have two functions: ④ They serve as mRNA for the synthesis of new viral proteins, and ⑤ they serve as templates for synthesizing new viral RNA. ⑥ The new coat proteins assemble around the new viral RNA. ⑦ Finally, the viruses leave the cell by cloaking themselves in plasma membrane. Thus, the virus obtains its envelope from the host cell and leaves the cell without necessarily lysing it.

Not all animal viruses reproduce in the cytoplasm. The herpesviruses that cause chickenpox, shingles, mononucle-

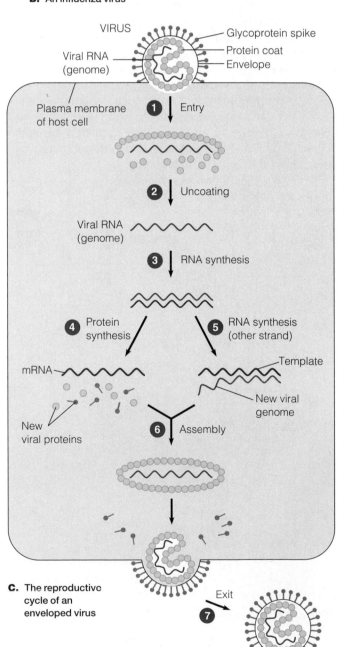

C. The reproductive cycle of an enveloped virus

osis, cold sores, and genital herpes, for example, are DNA viruses that reproduce in the host cell's nucleus, and they get their envelopes from the cell's nuclear membranes. While inside the nucleus, herpesvirus DNA may be able to insert itself into the cell's DNA as a provirus, similar to a bacterial prophage, so that the virus remains latent within the body. From time to time, physical stress, such as a cold or sunburn, or emotional stress, may cause the herpes proviruses to begin manufacturing the complete virus, resulting in unpleasant symptoms. Once acquired, many herpes infections tend to flare up repeatedly throughout a person's life.

The amount of damage a virus causes our body depends partly on how quickly our immune system responds to fight the infection and partly on the ability of the infected tissue to repair itself. We usually recover completely from colds because our respiratory tract tissue can efficiently replace damaged cells by mitosis. In contrast, the poliovirus attacks nerve cells, which do not divide. The damage to such cells by polio, unfortunately, is permanent. In such cases, we try to prevent the disease with vaccines (see Chapter 24). The antibiotic drugs that help us recover from bacterial infections are powerless against viruses. The development of antiviral drugs has been slow, because it is difficult to find ways to kill a virus without killing its host cell.

The AIDS virus makes DNA on an RNA template 10.18

The devastating disease known as AIDS is caused by a type of RNA virus with some special twists we have not yet discussed. In outward appearance, the AIDS virus, called HIV, resembles the flu or mumps virus. As illustrated in Figure A, HIV has a membranous envelope and glycoprotein spikes. These components allow HIV to enter and leave a host cell much the way the mumps virus does. Notice, however, that HIV contains two copies of its RNA instead of one. HIV also has a different mode of reproduction. It is a **retrovirus**, an RNA virus that reproduces by means of a DNA molecule. Retroviruses are so named because they reverse the usual DNA—>RNA flow of genetic information. They carry molecules of an enzyme called **reverse transcriptase**, which catalyzes reverse transcription, the synthesis of DNA on an RNA template.

Figure B illustrates what happens after HIV RNA is uncoated in the cytoplasm of a host cell. The reverse transcriptase (yellow) ① uses the RNA as a template to make a DNA strand and then ② adds a second, complementary DNA strand. ③ The resulting double-stranded DNA then enters the host cell nucleus and inserts itself into the chromosomal DNA, becoming a provirus. The provirus usually remains part of the cell's DNA. But occasionally the provirus is ④ transcribed into RNA and ⑤ translated into viral proteins. ⑥ New viruses assembled from these components eventually leave the cell. They can then infect other cells. This is the standard reproductive cycle for retroviruses.

AIDS stands for "acquired immune deficiency syndrome," and HIV for "human immunodeficiency virus," and these terms describe the main mechanism of the disease. HIV infects and eventually kills several kinds of white blood cells that are important in our immune system. We discuss AIDS and its deadly effects in more detail when we take up the immune system in Chapter 24.

A. A model of HIV structure

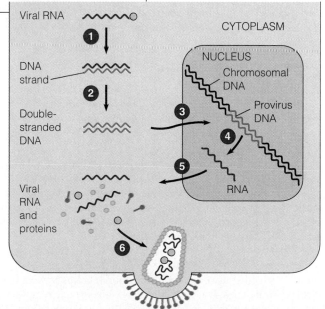

B. The behavior of HIV nucleic acid in a host cell

Molecular geneticists have a love-hate relationship with viruses. On one hand, studies of viruses essentially launched the science of molecular genetics some 40 years ago. The Hershey-Chase experiments, confirming that the DNA of phage T2 infects and usurps control of bacterial cells, focused attention on the central role of nucleic acids in inheritance and set the stage for Watson and Crick. After the structure of DNA was defined, studies of viral reproduction helped show how genes function and how genetic information flows from DNA to proteins.

The micrograph here represents the other side of the virus–molecular genetics relationship: a battle instead of a love affair. The blue dots are AIDS viruses (HIV) attacking a human white blood cell. HIV invades and kills cells like this one, weakening the immune system of the infected person. AIDS is one of the deadliest diseases humankind has ever faced and also one of the most difficult to combat. Fighting viruses that cause disease is an important practical goal of molecular genetics. Effective vaccines have already been developed against some viral diseases, including mumps, measles, and polio. It is likely that when we find the means to control HIV and other deadly viruses (including some that cause certain types of cancer), research in molecular genetics will be responsible for their discovery.

We continue our study of molecular genetics in Chapter 11, where we explore what is known about how genes themselves are controlled.

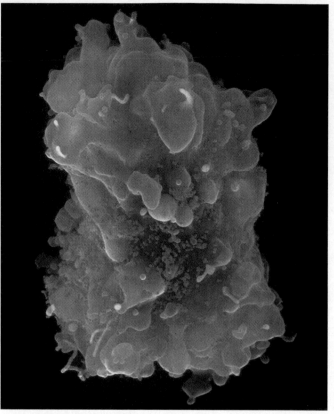

SEM 14,000×

HIV (blue dots) attacking a white blood cell

Begin your review by rereading the module headings and Module 10.13, and scanning the figures before proceeding to the Chapter Summary and questions.

Chapter Summary

Introduction–10.1 Genes are made of DNA, as was revealed when DNA from pathogenic bacteria was found to transform harmless bacteria into pathogenic ones. Early experiments with bacteriophages (viruses that infect bacteria) showed that certain phages reprogram their host cells by injecting their DNA.

10.2 DNA is a nucleic acid, made of long chains of nucleotide monomers. Each nucleotide consists of a sugar connected to a phosphate group and a nitrogenous base. Alternating sugars and phosphates form a backbone, with the bases as appendages. There are four kinds of bases, abbreviated A, T, C, and G. RNA is also a nucleic acid, with a different sugar and U instead of T.

10.3–10.4 James Watson and Francis Crick worked out the structure of DNA, which is two polynucleotide strands wrapped around each other in a double helix. Hydrogen bonds between bases hold the strands together. Each base pairs with only one complementary partner—A with T, and G with C. During DNA replication, the strands separate, and enzymes use each strand as a template to as-

semble new nucleotides into a complementary strand. The resulting two daughter DNA molecules are identical to the parent molecule. This is how genetic instructions are copied for the next generation.

10.5–10.7 The information constituting an organism's genotype is carried in the sequence of its DNA bases. Studies of inherited metabolic defects first suggested that phenotype is expressed through proteins. A particular gene—a linear sequence of many nucleotides—specifies a polypeptide. The "words" of the DNA "language" are triplets of bases called codons. The codons in a gene specify the amino acid sequence of a polypeptide.

10.8–10.14 Transcription of the DNA message into RNA is the first stage in making a protein. The DNA helix unzips, and RNA nucleotides line up and hydrogen-bond along one strand of the DNA, following the base-pairing rules. As the single-stranded messenger RNA (mRNA) peels away from the gene, the DNA strands rejoin. A ribosome attaches to the mRNA and assembles a specific polypeptide, aided by transfer RNAs (tRNAs) that act as interpreters. Each tRNA is a folded molecule bearing a base triplet called an anticodon on one end; a specific amino acid is added to the other end. The mRNA moves relative to the ribosome a codon at a time, and a tRNA with a complementary anticodon pairs with each codon, adding its amino acid to the peptide chain. Thus, the sequence of codons in DNA, via the sequence of codons in mRNA, spells out the structure of a polypeptide. Polypeptides form proteins that determine the appearance and functions of the cell and organism. Virtually all organisms share the same genetic code.

10.15 Mutations are changes in the DNA base sequence, caused by errors in DNA replication or by mutagens. Substituting, inserting, or deleting a nucleotide alters a gene, with varying effects on the organism.

10.16 Viruses can be regarded as simply packaged genes. When phage DNA enters a bacterium, the host cell replicates, transcribes, and translates the viral nucleic acid. In the lytic cycle, new phages quickly burst from the host cell. In the lysogenic cycle, phage DNA joins the host chromosome and is passed on to daughter cells. Much later, it may initiate phage production.

10.17–10.19 Many viruses cause disease when they invade plant or animal cells. Some, such as flu viruses, contain only RNA, rather than DNA. Some viruses steal a bit of host cell membrane as a protective envelope. Some insert their DNA into a host chromosome and remain latent for long periods of time. HIV, the virus that causes AIDS, is a retrovirus; inside a cell it uses its RNA as a template for making DNA, which then inserts into a host chromosome. Virus studies helped establish molecular genetics. Now molecular genetics helps us fight AIDS and other viral diseases.

Testing Your Knowledge

Multiple Choice

1. Scientists have discovered how to put together a bacteriophage with the protein coat of phage T2 and the DNA of phage T4. If this composite phage were allowed to infect a bacterium, the phages produced in the host cell would have (*Explain your answer.*)

    a. the protein of T2 and the DNA of T4
    b. the protein of T4 and the DNA of T2
    c. a mixture of the DNA and proteins of both phages
    d. the protein and DNA of T2
    e. the protein and DNA of T4

2. A geneticist found that a particular mutation had no effect on the polypeptide coded by a gene. This mutation probably involved

    a. deletion of one nucleotide
    b. alteration of the start codon
    c. insertion of one nucleotide
    d. deletion of the entire gene
    e. substitution of one nucleotide

3. Which of the following correctly ranks the structures in order of size, from largest to smallest?

    a. gene-chromosome-nucleotide-codon
    b. chromosome-gene-codon-nucleotide
    c. nucleotide-chromosome-gene-codon
    d. chromosome-nucleotide-gene-codon
    e. gene-chromosome-codon-nucleotide

4. The nucleotide sequence of a DNA codon is GTA. A messenger RNA molecule with a complementary codon is transcribed from the DNA. In the process of protein synthesis, a transfer RNA pairs with the mRNA codon. What is the nucleotide sequence of the tRNA anticodon?

    a. CAT   **c.** GUA   **e.** GT
    b. CUT   **d.** CAU

Matching

The following story is an analogy that will test your knowledge of RNA transcription and translation of the RNA message into protein. Read the story, then match each element in the story with a component of the protein-making process in a cell.

Joe works at the Cakes Alive Bakery. Each day when he comes to work, he finds a photocopy of a recipe for a particular kind of cake on the counter. Joe combines flour, milk, eggs, and other ingredients as directed by the recipe. A cookbook with master copies of all the cake recipes is stored in the main office of the bakery, and each day Joe's boss copies one of the recipes for him to follow.

    1. protein   **a.** Joe
    2. chromosome   **b.** flour
    3. messenger RNA   **c.** bakery
    4. RNA polymerase enzyme **d.** cookbook
    5. ribosome   **e.** photocopy machine
    6. amino acid   **f.** photocopy of recipe
    7. cell   **g.** recipe in cookbook
    8. gene   **h.** cake

Describing, Comparing, and Explaining

1. Sketch a simple picture of part of a DNA molecule consisting of five nucleotide pairs. Make the sugars triangles, the phosphate groups circles, and the bases rectangles. Label these parts, and show hydrogen bonds and complementary base pairing.

2. Describe the process of DNA replication: the "ingredients" needed, steps in the process, and the final product.

3. Describe the process by which the information in a gene is transcribed and translated into a protein. Correctly use these words in your description: tRNA, amino acid, start codon, transcription, mRNA, gene, codon, RNA polymerase, ribosome, translation, anticodon, peptide bond, stop codon.

4. Outline the life cycle of an enveloped RNA virus such as the mumps virus. How does the virus enter the cell? What two functions are served by its RNA? What components assemble to form new viruses? How do the viruses leave the cell? Why might this process cause illness?

Thinking Critically

1. A cell containing a single chromosome is placed in a medium containing radioactive phosphate, so that any new DNA strands formed by DNA replication will be radioactive. The cell replicates its DNA and divides. Then the daughter cells (still in the radioactive medium) replicate their DNA and divide, so that a total of four cells are present. Sketch the DNA molecules in all four cells, showing a normal (nonradioactive) DNA strand as a solid line and a radioactive DNA strand as a dashed line.

2. When bacteria infect an animal, the number of bacteria in the body increases gradually. A graph of the growth of the bacterial population is a smoothly increasing curve. A viral infection shows a different pattern. For a while, there is no evidence of infection, and then there is a sudden rise in the number of viruses. The number stays constant for a time, then there is another sudden increase. The graph of virus population growth looks like a series of steps. Explain the difference between these two types of growth curves.

3. The base sequence of the gene coding for a short polypeptide is C T A C G C T A G G C G A T T G A C T. What would be the base sequence of the mRNA transcribed from this gene? Using the genetic code chart in Figure 10.14A, give the amino acid sequence of the polypeptide translated from this mRNA. (*Hint:* What is the start codon?)

Science, Technology, and Society

Researchers on the Human Genome Project are determining the nucleotide sequences of human genes. In some cases, they are identifying the proteins encoded by the genes. Scientists at the National Institutes of Health (NIH) have worked out thousands of sequences, and similar analysis is being carried out at private companies. Knowledge of the nucleotide sequence of genes might be used to treat genetic defects or produce life-saving medicines. The NIH and U.S. biotechnology companies have applied for patents on their discoveries. In Britain, the courts have ruled that a naturally occurring gene cannot be patented. Do you think individuals and companies should be able to patent genes and gene products? Before answering, consider the following: What are the purposes of a patent? How might the discoverer of a gene benefit from a patent? How might the public benefit? What negative impacts might result from patenting genes?

SEM 55×

The Control of Gene Expression $\quad$ 11

Fruit flies of the species *Drosophila melanogaster* are only about 3 millimeters long, and they often go unnoticed as they hover around fruit in the kitchen or grocery store. Fruit growers consider them a nuisance but not a threat, because they eat the yeasts that grow on ripe fruit, not the fruit itself. (By contrast, the Mediterranean fruit fly, or medfly, a distant relative of *Drosophila*, is a serious threat to fruit crops.) *Drosophila* has proved to be of immense value to our understanding of genetics.

Something's wrong with the fruit fly in the photograph at the left. It is similar to the ones we saw in Chapter 9, but there's something different about it. One way you can tell that an insect is a fly—a fruit fly, house fly, gnat, mosquito, or any other kind of fly—is that it has two wings. Other insects, such as bees, beetles, butterflies, and moths have four wings. The fruit fly in the drawing on this page is normal. The four-winged fly in the photo is a mutant, and it has become a valuable model for biologists attempting to uncover the secrets of how gene activity controls embryonic development, as we'll see in this chapter.

What determines that a zygote will develop into a particular kind of organism? We know that DNA in the zygote nucleus contains all the genes that dictate what an organism will look like. But this is not the whole story. Every cell that develops from a zygote by mitotic division—that is, every cell except the organism's gametes—has DNA that is virtually identical to the DNA in the zygote. If all that DNA in every cell in a developing embryo were constantly active, every cell would look like every other cell. The only way that cells with the same genetic information can develop into cells with different structures and functions is if gene activity is regulated. Some sort of control mechanisms must determine that certain genes will be active while others remain inactive in a particular cell. For instance, for cells in a fruit fly's eyes to develop into a lens, genes that produce lens proteins must be turned on, and other genes must be turned off.

Learning how genes function and are regulated during embryonic development is one of the major challenges of biology today. A combination of two powerful approaches is making rapid progress possible. One approach is biochemical: the use of recently developed recombinant DNA methods, which we discuss in Chapter 12. The other, older, approach is genetic: the use of mutants that look or behave abnormally as clues to the normal. The four-winged fruit fly is one such mutant. At some point in this fly's development from a zygote, its DNA directed the formation of twice as many wings as normal. Although wing formation involves many genes, the genetic basis for this abnormality turns out to be surprisingly simple—mutation of a single gene. Because that one gene regulates the activity of many other genes in fruit-fly development, its mutation, and consequent malfunctioning, can bring about a large-scale abnormality.

Gene regulation—how cells control the expression of their genes—is the subject of this chapter. This topic is closely related to some of our most pressing medical problems, notably cancer and birth defects, and we look at cancer in some detail at the end of the chapter.

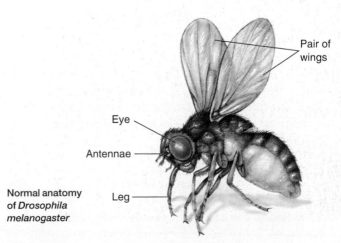

Normal anatomy of *Drosophila melanogaster*

Pair of wings

Eye

Antennae

Leg

11.1 In both eukaryotes and prokaryotes, cell specialization depends on the selective expression of genes

What do we actually mean when we say that genes are turned on or off? As we discussed in Chapter 10, genes direct the synthesis of specific RNA and protein molecules. Thus, a gene that is turned on is directing the transcription of its genetic information into RNA, and that message is being translated into specific protein molecules. The overall process by which genetic information flows from genes to proteins—that is, from genotype to phenotype—is called **gene expression.**

The turning on and off of genes is the main way that gene expression is regulated, and it is the basis of the cell specialization that occurs as a multicellular organism develops from a zygote. Turned-on genes in the cells of a developing fly wing, for example, are expressed as proteins that make the cells flat and smooth, forming a strong, thin flight surface resembling clear plastic. In contrast, turned-on genes in the cells of a developing fly eye are expressed as proteins that form light-focusing lenses. This specialization in the structure and function of cells is called **cellular differentiation.** Cellular differentiation helps transform the cells arising from a zygote into an adult organism. It also occurs in an adult whenever new cells replace old or damaged ones. Cells become specialized because, of all the genes present, only certain ones are expressed. The control of gene expression makes it possible for cells to produce specific kinds of proteins when and where they are needed.

Our earliest understanding of gene control came not from multicellular organisms but from the bacterium *Escherichia coli*. Although *E. coli* exists as only a single cell, it must change its activities from time to time to suit changes in its environment. Let's look at one way *E. coli* accomplishes these changes.

11.2 Proteins interacting with DNA can turn genes on or off in response to environmental changes

Picture an *E. coli* cell living in your intestine. If you eat only a sweet roll with butter for breakfast, the bacterium will be bathed in the sugars glucose and fructose and the digestion products of fats. Later on, if you have a fat-free milkshake and a salad for lunch, *E. coli*'s environment will change drastically.

Let's focus on your milkshake for a moment. One of the main nutrients in milk is the sugar lactose. When lactose is plentiful in the intestine, *E. coli* can make the enzymes necessary to absorb the sugar and use it as an energy source.

Remember that enzymes are proteins; their production is an example of gene expression. *E. coli* can make lactose-utilization enzymes because it has genes that code for these enzymes. In 1961, French biologists François Jacob and Jacques Monod proposed a model for how an *E. coli* cell could adjust its enzyme production in response to frequent changes in its environment. The Jacob and Monod model, shown in Figures A and B on the next page, explains how genes coding for lactose-utilization enzymes are turned off or on, depending on whether lactose is available.

E. coli uses three enzymes to take up and start metabolizing lactose, and the genes that code for these enzymes are regulated as a unit. Notice in Figure A, which represents a piece of *E. coli* DNA, that the three genes that code for the lactose-utilization enzymes (light blue) are next to each other.

Adjacent to the group of lactose enzyme genes are short sections of DNA that help control them. One specific sequence of nucleotides, a promoter (orange), marks the transcription starting point for all three genes. Between the promoter and the enzyme genes, a sequence of nucleotides called an **operator** (yellow) acts as a switch. The operator determines whether the transcription enzyme, RNA polymerase, can attach to the promoter and move along the genes.

Such a cluster of genes with related functions, along with a promoter and an operator, is called an **operon**; operons exist only in prokaryotes. The key advantage to the grouping of genes into operons is that the expression of these genes can be easily coordinated. The operon discussed here is called the *lac* operon, short for lactose operon. When an *E. coli* encounters lactose, all the enzymes needed for its use are made at once, because the operon's genes are all controlled by a single switch, the operator. But what determines whether the operator says on or off?

Figure A shows the *lac* operon in "off" mode, its status when there is no lactose in the cell's environment. Transcription is turned off by a molecule called a **repressor** (red), a protein that functions by binding to the operator and blocking attachment of RNA polymerase to the promoter. The left side of Figure A indicates where the repressor comes from. A gene called a **regulatory gene** (dark blue), located outside the operon, codes for the repressor. The regulatory gene is expressed continually, so the cell always has a supply of repressor molecules.

How can an operon be turned on if its repressor is always present? As indicated in Figure B, lactose interferes with the

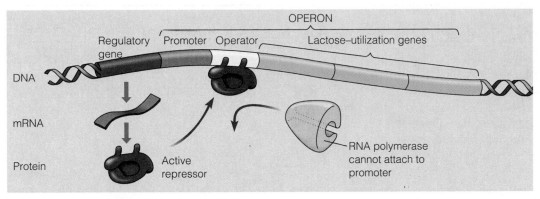

A. *Lac* operon, turned off

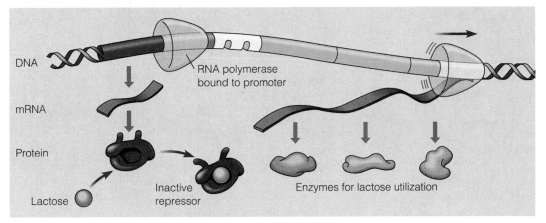

B. *Lac* operon, turned on (repressor inactivated by lactose)

functioning of the *lac* repressor by binding to the repressor and changing its shape. With its new shape, the repressor cannot bind to the operator, and the operator switch remains on. RNA polymerase can now bind to the promoter and from there slide along the genes of the operon. The mRNA produced is a single, long molecule that carries coding sequences for all three enzymes (purple) needed to absorb and use lactose. The cell can translate this message into separate polypeptides because the mRNA has codons signaling the start and stop of translation.

Operons come in several different varieties

The *lac* operon is only one type of operon in bacteria. Other types also have a promoter, an operator, and several adjacent genes, but they differ in the way the operator switch is controlled. The figure below shows two types of repressor-controlled operons. As we discussed, the *lac* operon's repressor is active alone and inactive when bound to lactose. A second type of operon, represented here by the *trp* operon, is controlled by a repressor that is *inactive* alone. To be active, this type of repressor must combine with a specific small molecule. In our example, the small molecule is tryptophan, an amino acid essential for protein synthesis. *E. coli* can make tryptophan from scratch if it has to, using enzymes encoded in the *trp* operon. But it will stop making tryptophan and simply absorb it from its surroundings whenever possible. When *E. coli* is swimming in tryptophan (it occurs in large amounts in such foods as milk and poultry), tryptophan binds to the repressor of the *trp* operon. This activates the *trp* repressor, enabling it to switch off the operon. Thus, this type of operon allows bacteria to stop making certain essential molecules when the molecules are already present in the environment, saving materials and energy for the cells.

A third type of operon uses **activators,** proteins that turn operons *on* by binding to DNA. These proteins act by somehow making it easier for RNA polymerase to bind to the promoter, rather than by blocking RNA polymerase, as repressors do. Armed with a variety of operons, regulated by repressors and activators, *E. coli* and other prokaryotes can thrive in frequently changing environments.

Two types of repressor-controlled operons

11.4 Differentiation produces a variety of specialized cells

As we saw in Chapter 4, eukaryotic cells are more complex structurally than prokaryotes. Even a single-celled eukaryote like an amoeba is more complex than any bacterium. The cell of an amoeba is packed with organelles, such as mitochondria and endoplasmic reticulum, that are specialized to perform particular functions. Producing these organelles and regulating their functions require a complex network of gene control.

The situation is even more complicated in multicellular eukaryotes. During the repeated cell divisions that lead from a unicellular zygote to a multicellular organism, individual cells differentiate; that is, they become the specialized cells that belong to different tissues. This cellular differentiation requires more complex gene control systems than single-celled organisms need.

The light micrographs here show four examples of the many different types of human cells. Each sample was stained with dyes to bring out structural details of the cells, and all four micrographs are shown at the same magnification (750×). Figure A shows short segments of two muscle cells, which are very long fibers with multiple nuclei (which appear as dark blue rods here). The stripes are structures that help the cells contract. The large orange shape in Figure B is

part of a nerve cell, with its nucleus appearing as a lighter orange disk. The dark spot in the nucleus is a nucleolus. The cell's three long extensions carry nerve signals from one part of the body to another. Figure C shows sperm cells; their long flagella can propel them through a woman's reproductive tract to meet an egg cell. Figure D shows a single white blood cell (with a three-part purple nucleus) surrounded by red blood cells. The small size and disklike shape of red blood cells help them pass through blood vessels and transport oxygen molecules. Mammalian red blood cells lack nuclei and other organelles, having lost them during differentiation. Without organelles taking up space, a mature red blood cell can be literally packed with oxygen-carrying hemoglobin molecules.

Thus, each type of differentiated cell has a unique structure appropriate for carrying out its function, and together the various types of cells enable a multicellular organism to function as a whole. How do cells become differentiated and form the specialized tissues and organs making up a multicellular organism? An important part of the answer lies in the control of gene expression, the turning on and off of particular genes. In the next module, we look at some of the evidence.

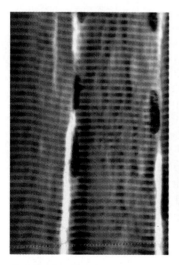
A. Muscle cells

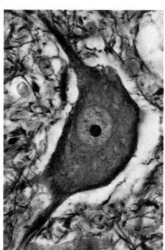
B. Nerve cell

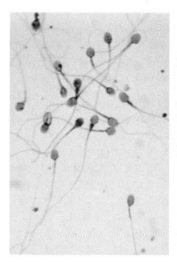
C. Sperm cells

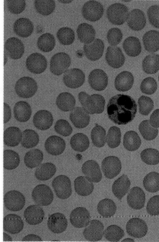
D. Blood cells

11.5 Specialized cells may retain all of their genetic potential

The zygote has a complete set of genes for making all types of specialized cells as an organism develops. What happens to these genes as a cell differentiates? A red blood cell loses its nucleus and with it virtually all its DNA. But most differentiated cells retain a nucleus and most, if not all, of their

DNA. Are all the genes still present, and, if they are, do the differentiated cells retain the potential to express them?

One way to approach these questions is to replace the nucleus of an egg or zygote with the nucleus from a differentiated cell. If genes are lost or irreversibly turned off during differentiation, the transplanted nucleus will not be able to guide the development of a normal embryo. The pioneering experiments in nuclear transplantation were performed by American embryologists Robert Briggs and Thomas King in the 1950s. You can follow one experiment in Figure A. The researchers destroyed the nuclei of frog egg cells with ultraviolet (UV) light and then transplanted nuclei from tadpole intestinal cells into the eggs. (The tadpole is the larval stage of the frog; its intestinal cells are already differentiated.) Many of the eggs containing transplanted nuclei started to develop; a few even developed into normal tadpoles. From such transplantation studies, researchers concluded that nuclei from differentiated cells retain most, if not all, of their genetic potential.

Another indication that differentiation does not impair a cell's genetic potential is that differentiation is sometimes reversible. In other words, after a cell of a multicellular organism has become structurally and functionally specialized and has stopped dividing, the cell may be able to dedifferentiate and undergo mitosis. Its daughter cells may then differentiate into the specialized form of the original cell or even other cell types. We see a type of reversible differentiation in regeneration, the replacement of lost body parts. When a salamander loses a leg, for example, certain cells in the leg stump dedifferentiate, divide, and then redifferentiate to give rise to a new leg. Many animals can regenerate lost parts, and in a few, isolated differentiated cells can be made to dedifferentiate and then re-form an entire organism.

The story is different in plants, however. In many species, whole plants can regenerate from single differentiated cells. Figure B illustrates this process using a carrot. A single cell removed from the root and placed in culture medium will begin dividing and eventually grow into an adult plant. This technique can be used to clone plants, reproducing hundreds or thousands of genetically identical organisms

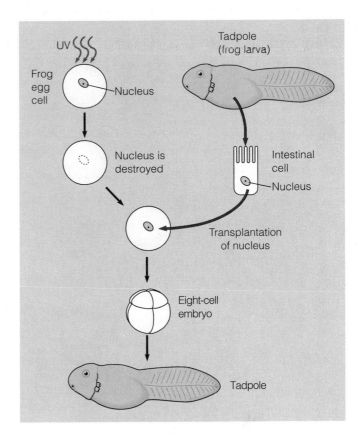

A. Nuclear transplantation experiment

from the somatic cells of a single plant. In this way, it is possible to propagate large numbers of plants that have especially desirable traits, such as high fruit yield or resistance to disease. The fact that a mature plant cell can dedifferentiate and then give rise to all the specialized cells of a new plant supports the conclusion that differentiation does not necessarily involve irreversible changes in the DNA.

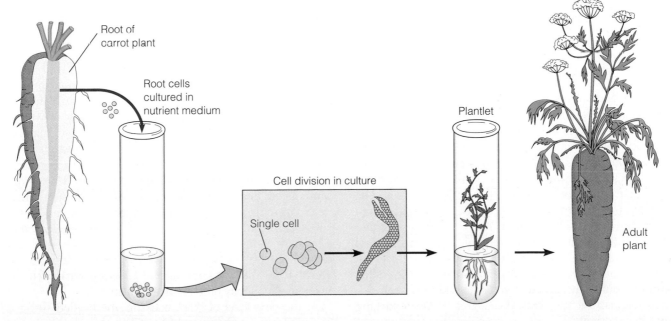

B. Growth of a carrot plant from a differentiated cell

11.6 Each type of differentiated cell has a particular pattern of expressed genes

If all the differentiated cells in an individual organism—say, a carrot plant or a fruit fly—contain the same genes, and all the genes have the potential of being expressed, how do cells become specialized? The great differences among cells in an organism result from the selective expression of genes. As a developing embryo undergoes successive cell divisions, different genes must be activated in different cells at different times. Groups of cells follow diverging developmental pathways, and each group develops into a particular kind of tissue. Finally, in the mature organism, each cell type—nerve or muscle, for instance—has a different pattern of turned-on genes.

The figure here illustrates patterns of gene expression for a few genes in three different specialized cells. The genes for the enzymes of the metabolic pathway of glycolysis are active in all metabolizing cells, including the pancreas, embryonic eye lens, and nerve cells shown here. On the other hand, the genes for specialized proteins are expressed only by particular cells.

The specialized proteins we use as examples are the trans-

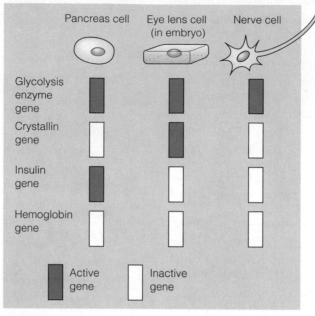

parent protein crystallin, which forms the lenses of our eyes; the hormone insulin; and the oxygen-transport protein hemoglobin. Notice that the hemoglobin genes are not active in any of the cell types shown here. They are turned on only in cells that are developing into red blood cells. Insulin genes are activated only in the cells of the pancreas that produce that hormone. Nerve cells express genes for other specialized proteins not shown here. Mature lens cells, like mature red blood cells, achieve the ultimate in differentiation. After a frenzy of expressing crystallin genes and accumulating the protein products, these cells lose their nuclei, and thus all their genes.

So we see that eukaryotic cells become specialized because they express only certain genes. Thus, cellular differentiation in multicellular organisms results from selective gene expression, as does the ability of bacteria to produce different enzymes as needed. Now it's time to examine some of the ways eukaryotic cells control the expression of their genes. We begin with a look at how DNA is arranged in a eukaryotic cell.

11.7 DNA packing in eukaryotic chromosomes affects gene expression

The nucleus in any one of your cells would be the envy of a computer manufacturer. Computers can store large amounts of information on a tiny microchip, but they are still a long way from the extraordinary storage capacity of a eukaryotic cell. Eukaryotic chromosomes contain an enormous amount of DNA relative to their size. Each one contains a single DNA molecule that is thousands of times longer than the 5-μm diameter of a typical nucleus. The total DNA in a human cell's 46 chromosomes would stretch for 3 meters! All this DNA can fit into the nucleus because of an elaborate, multilevel system of folding, or packing, of the DNA in each chromosome. A crucial aspect of packing is the association of DNA with special small proteins called **histones,** found only in eukaryotes. (Bacteria have analogous proteins, but they lack the degree of DNA packing found in eukaryotes.)

Figure A outlines the main levels of DNA packing. As shown at the top, the double helix has a 2-nm diameter, and this is not altered by packing. As we move down the figure, at the first level of packing we see histones attach to the DNA. In electron micrographs, the DNA-histone complex has the appearance of beads on a string. Each "bead," called a **nucleosome,** consists of DNA wound around a protein core of eight histone molecules. Nucleosomes may help control gene expression by limiting the access of the cell's transcription enzymes to the DNA. Most nucleosomes also have nonhistone proteins attached, although our drawing doesn't show them. A typical eukaryotic cell has about 1000 different nonhistone chromosomal proteins, many of which may help regulate genes.

At the next level of DNA packing, the beaded string is wrapped into a tight helical fiber. Then this fiber coils further into a thick supercoil with a diameter of about 200 nm. More looping and folding can further compact the DNA, as you can see in the metaphase chromosome at the bottom of the figure. Viewed as a whole, Figure A gives a sense of how the successive levels of folding enable one chromosome to contain a huge amount of DNA.

The roles of DNA packing in the control of gene expression are mostly unknown, except for a few intriguing cases. One such case is the **Barr body** found in the cells of female mammals. A Barr body is a highly compacted X chromosome in which almost all the genes are inactive. When a female is born, one of the two X chromosomes in every one of her somatic cells is a Barr body and therefore inactive. Which X chromosome is inactive in any cell is random. As a consequence, in a female heterozygous for genes on the X chromosome, different cells express different X-chromosome alleles.

A striking example of the phenotypic effect of the Barr body is the calico cat (Figure B). The genes that determine calico (orange and black patches on a white background) are on the X chromosome, and it takes two different alleles (one for orange fur and one for black) to make calico. If a female carries both alleles, she is calico. An orange patch is formed by a population of cells in which the X chromosome with the orange allele is active; black patches have cells in which the X chromosome with the black allele is active. The color of each chromosome in the diagram indicates the allele it carries.

Because a cat must have some cells with an active orange allele and others with an active black allele to be calico, we would expect that only female cats, which are XX, could be calico. Male cats have only one X chromosome (they are XY), so they cannot have both orange and black alleles. In fact, however, calico male cats are occasionally seen. Those that have been examined are sterile, and most likely they have the abnormal genotype XXY.

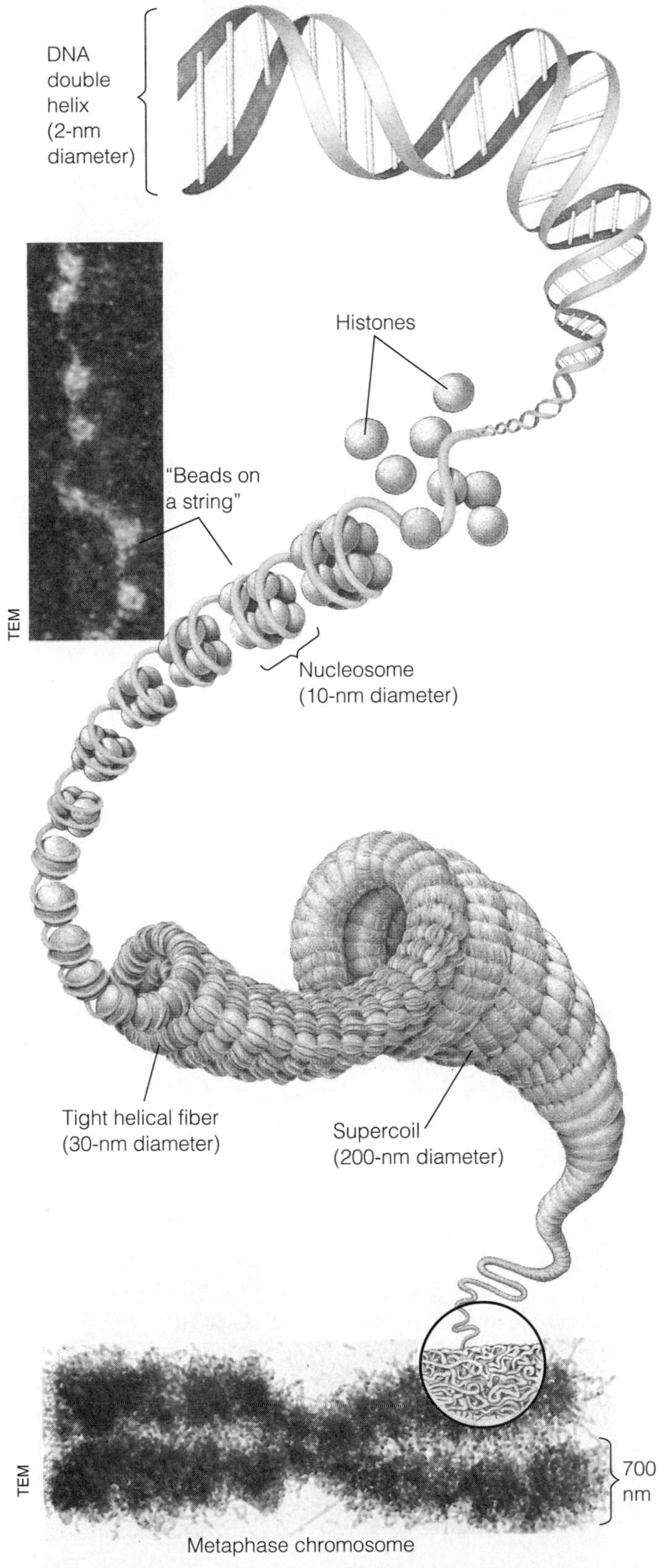

DNA double helix (2-nm diameter)

Histones

"Beads on a string"

TEM

Nucleosome (10-nm diameter)

Tight helical fiber (30-nm diameter)

Supercoil (200-nm diameter)

TEM

700 nm

Metaphase chromosome

A. DNA packing in a eukaryotic chromosome

Active X chromosome

Barr body (inactive X chromosome)

Cells

B. A calico cat, a demonstration of X-chromosome inactivation

Eukaryotic gene expression can be regulated at transcription

Little is known about how DNA packing affects gene expression, but it is likely that control often occurs at the transcription stage, as in prokaryotes. Our old friend the fruit fly has provided one type of evidence of transcriptional control and its role in the development of eukaryotes. One of the reasons fruit flies have proved so useful is that some of their larval cells, for example in the salivary glands, contain giant chromosomes. Such salivary-gland cells do not divide, but they do replicate their DNA repeatedly. Their chromosomes become very thick—much thicker and therefore easier to study than the chromosomes in most other kinds of cells.

Figure A shows giant chromosomes from a salivary-gland cell of a larval fruit fly, viewed with a light microscope. The dark and light bands are regions of tighter and looser DNA packing, respectively. At certain stages in the larva's development, parts of the DNA loop out from the chromosome at specific places. The loops form structures known as puffs, which you can see clearly in Figure B.

Puffing may make the DNA in that region more accessible to the transcription enzyme RNA polymerase, and, in fact, puffs do correspond to regions of intense RNA synthesis. The locations of puffing along a chromosome change as a fruit-fly larva develops. When the larva gets ready to molt (shed its body covering) in preparation for its next developmental stage, some puffs disappear, and others form in new places. The shifting puffs are visual indicators of the switching on and off of specific genes during development.

Experiments have shown that this regulation of gene activity is responsive to chemical signals. Changes in puffing patterns can be brought about by the insect hormone that initiates molting. This hormone, a steroid, regulates the transcription of certain genes by binding to and activating a regulatory protein. In this case, the regulatory protein is an activator, turning genes on by attaching to specific DNA nucleotide sequences.

A number of eukaryotic activator proteins have been identified. Like prokaryotic regulatory proteins, many eukaryotic activators work by attaching to DNA and somehow influencing the ability of RNA polymerase to start transcription. Figure C shows a highly simplified model illustrating the basic concept of proteins interacting with each other and with DNA. Here, the RNA polymerase recognizes an activator protein bound to DNA near a gene's promoter. Notice how the shapes of the activator and polymerase fit together. The activator helps position the RNA polymerase on the promoter DNA and stimulates the polymerase to start transcribing the gene. (In more complete models, the activator protein interacts with the polymerase via still other proteins.)

In summary, eukaryotes and prokaryotes regulate transcription in similar ways: Both employ regulatory proteins that bind to DNA and turn a gene (or genes) on or off. There also are

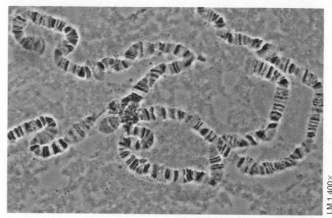

A. Giant chromosomes from a fruit-fly salivary gland

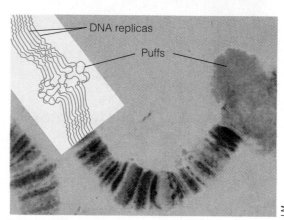

B. Chromosome puffs in a giant chromosome

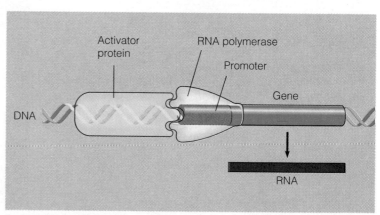

C. Simplified model for the effect of an activator on transcription

some differences. Whereas prokaryotic genes under the control of a particular regulatory protein are often grouped into an operon, eukaryotic genes controlled by a single regulator tend to be separated. Also, the number of different proteins that work together to control transcription is larger in eukaryotes.

As we see next, transcriptional control is only part of the story of eukaryotic gene regulation.

Noncoding segments interrupt many eukaryotic genes

Both prokaryotes and eukaryotes regulate gene expression at the transcriptional level, but the eukaryotic cell has additional opportunities for control *after* transcription. Structural compartmentalization—the separation of transcription from translation by the nuclear envelope—provides these post-transcriptional opportunities. Because translation occurs in the cytoplasm, the cell can make RNA molecules and modify them in the nucleus before they become available to ribosomes in the cytoplasm.

One opportunity for post-transcriptional gene regulation arises from the fact that eukaryotic genes usually include noncoding segments. Surprisingly, much of the DNA of most organisms does not code for polypeptides. In eukaryotes, noncoding stretches of DNA that fall *within* genes provide a way for cells to regulate gene expression after transcription. Noncoding stretches within genes interrupt the DNA that codes for the amino acid sequence of a polypeptide. At first, researchers found such unexpected interruptions very puzzling. It was as if unintelligible sequences of letters were randomly interspersed in an otherwise intelligible written document.

Most genes of plants and animals, it turns out, are interrupted by long noncoding regions, which are called intervening sequences, or **introns**. The coding regions—the parts of a gene that are expressed—are called **exons**. As Figure A illustrates, both exons and introns are transcribed from DNA into RNA. However, before the RNA leaves the nucleus, the introns are removed, and the exons that flanked them are joined to produce an mRNA molecule with a continuous coding sequence. This process is called **RNA splicing.** It is used by eukaryotic cells in producing transfer RNA and ribosomal RNA, as well as messenger RNA.

Biologists have learned that RNA splicing can occur by two mechanisms. In mRNA splicing, the process is usually catalyzed by a complex of proteins and small RNA molecules. In certain other cases, splicing occurs completely without proteins or extra RNA molecules; the RNA transcript itself catalyzes the process. In other words, RNA acts as an enzyme.

What are the biological functions of introns? We don't have the complete answer yet. Some intron DNA includes nucleotide sequences that somehow regulate gene activity. And the splicing process itself may help control the flow of mRNA from nucleus to cytoplasm. Moreover, in some cases, the cell can carry out splicing in more than one way and thus generate different mRNA molecules from the same RNA transcript (Figure B). In this fashion, an organism can get more than one type of polypeptide from a single gene. The fruit fly provides an interesting example of two-way mRNA splicing. In this animal, the differences between males and females are controlled by two sets of regulatory proteins, one set specific to the male and one to the female. The two sets of proteins are coded by the same set of genes; the sex differences are largely due to different patterns of RNA splicing.

Introns may also contribute to genetic diversity. As we discussed in Module 9.16, crossing over is one of the main sources of genetic variety in eukaryotes. The presence of introns makes DNA longer, thereby increasing the number of points where crossing over can occur between two alleles of a gene. An increase in genetic diversity resulting from this added crossing over may have been important in the evolution of eukaryotes.

A. The splicing together of RNA exons

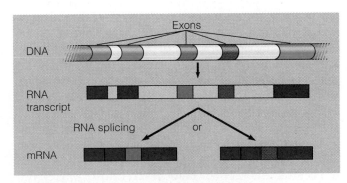

B. Production of two different mRNAs from the same gene

11.10 Eukaryotic mRNA is capped, tailed, and sometimes edited

Before they leave the nucleus, eukaryotic mRNA molecules undergo other changes in addition to splicing. To the two ends are added extra nucleotides, called the cap and the tail. As you can see in the figure below, the cap is added in front of the start codon, and the tail is added after the stop codon. These additions seem to help protect the mRNA from attack by cellular enzymes and may enhance translation. Neither the cap, the tail, nor the sequences that lie between them and the start and stop codons are translated.

In addition to getting a cap and a tail, eukaryotic mRNA molecules may be changed in ways that *are* translated. These changes, called **RNA editing,** are alterations in the sequence of coding nucleotides. They include nucleotide substitutions, insertions, and deletions. Collectively, these editorial changes produce what we might call a "final draft" of mRNA.

RNA editing can lead to massive discrepancies between the gene and the protein. In one case (in a protozoan parasite), so many nucleotides are added that the final draft of mRNA is twice the length of the original gene. Instances of RNA editing have been discovered in plants and animals, but biologists do not yet know how widespread the phenomenon is or understand much about how it works. Also unclear is its evolutionary significance. What is clear is that, where RNA editing occurs, gene expression in eukaryotes is much more complicated than in prokaryotes. In prokaryotes, the nucleotide sequence of a gene is translated directly into the amino acid sequence of a polypeptide, with virtually no change.

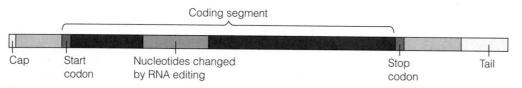

The final version of a eukaryotic mRNA molecule

Cap | Start codon | Nucleotides changed by RNA editing | Coding segment | Stop codon | Tail

11.11 Gene expression is also controlled during and after translation

After the "final draft" of eukaryotic mRNA is produced and transported to the cytoplasm, there are additional opportunities for gene regulation. The lifetime of mRNA molecules is an important factor regulating the amounts of various proteins that are produced in the cell. Long-lived mRNAs are generally translated into much more protein than short-lived ones. Prokaryotic mRNAs have very short lifetimes; they are degraded by enzymes after only a few minutes. This is one reason bacteria can change their proteins relatively quickly in response to environmental changes. In contrast, the mRNA of eukaryotes can have lifetimes of hours or even weeks.

A striking example of long-lived mRNA is found in vertebrate red blood cells, which act like factories for manufacturing the protein hemoglobin. In most species of vertebrates, the mRNAs for hemoglobin are unusually stable. They probably last as long as the red blood cells that contain them—about a month in birds and perhaps longer in reptiles, amphibians, and fishes—and are translated repeatedly. Mammals are an exception. When their red blood cells are mature, they lose their ribosomes (along with their other organelles), and thus cease to make new hemoglobin. However, mammalian hemoglobin itself lasts about as long as the red blood cells last, about four months.

There are ample opportunities for control of gene expression during translation. Among the molecules required for translation are a great many proteins that have regulatory functions. Red blood cells, for instance, have an inhibitory protein that blocks translation of hemoglobin mRNA unless the cell has a supply of heme (an iron-containing chemical group essential for hemoglobin function). The heme blocks the inhibitor, allowing hemoglobin to be synthesized.

The final opportunities for the control of gene expression are after translation. Post-translational control mechanisms in eukaryotes often involve cutting eukaryotic polypeptides into smaller, active final products. The hormone insulin, for example, is synthesized as one long polypeptide, which is not active as a hormone. Then, as illustrated here, a large

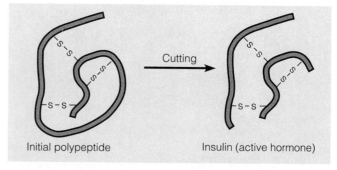

Initial polypeptide   Cutting   Insulin (active hormone)

Polypeptide cutting, the final step in forming the active insulin protein center portion is cut away, leaving two shorter chains, held together by chemical bonds that link up sulfur atoms ("S" in the figure stands for sulfur). This combination of two shorter polypeptides is the active form of insulin.

Another control mechanism operating after translation is the selective degradation of proteins. Though mammalian hemoglobin may last as long as the red blood cell housing it, the lifetimes of many other proteins are closely regulated.

This regulation allows a cell to adjust the kinds and amounts of its proteins in response to changes in its environment. It also enables the cell to maintain its proteins in prime working order. Indeed, when some proteins are altered chemically, by a change in the cell's environment, for example, they are degraded immediately and replaced by new ones that function properly.

Review: Multiple mechanisms regulate gene expression in eukaryotes

The figure at the right provides an overview of eukaryotic gene expression and highlights the multiple control points where the process can be turned on or off, speeded up, or slowed down. Picture the series of pipes that carry water from your local water supply, perhaps a reservoir, to a faucet in your home. At various points, valves control the flow of water. We use this model in the figure to illustrate the flow of genetic information from a chromosome—a reservoir of genetic information—to an active protein that is synthesized in the cell's cytoplasm. The multiple mechanisms that control gene expression are analogous to the control valves in water pipes. In the figure, each gene-expression "valve" is indicated by a "control knob."

Of course, this diagram oversimplifies the control of gene expression. What it does not show is the complicated web of control that connects genes and their products. We have seen examples in both prokaryotes and eukaryotes of the actions of gene products (usually proteins) on other genes or on other gene products within the same cell. The genes of operons in E. coli, for instance, are controlled by repressor or activator proteins encoded by regulatory genes on the same DNA molecule. In eukaryotes, many genes are controlled by proteins encoded by regulatory genes on different chromosomes.

In a multicellular eukaryote, particularly a developing one, the web of regulation crosses cell boundaries. The genes of one cell can produce chemicals such as hormones that induce a neighboring or distant cell to develop along a certain pathway. In this web of interactions, some types of genes stand out as having a major impact on cell differentiation; they have been identified in situations where normal development has gone awry. In the next several modules, we look at two such classes of genes: genes whose malfunctioning causes large-scale developmental abnormalities, and genes that cause cancer.

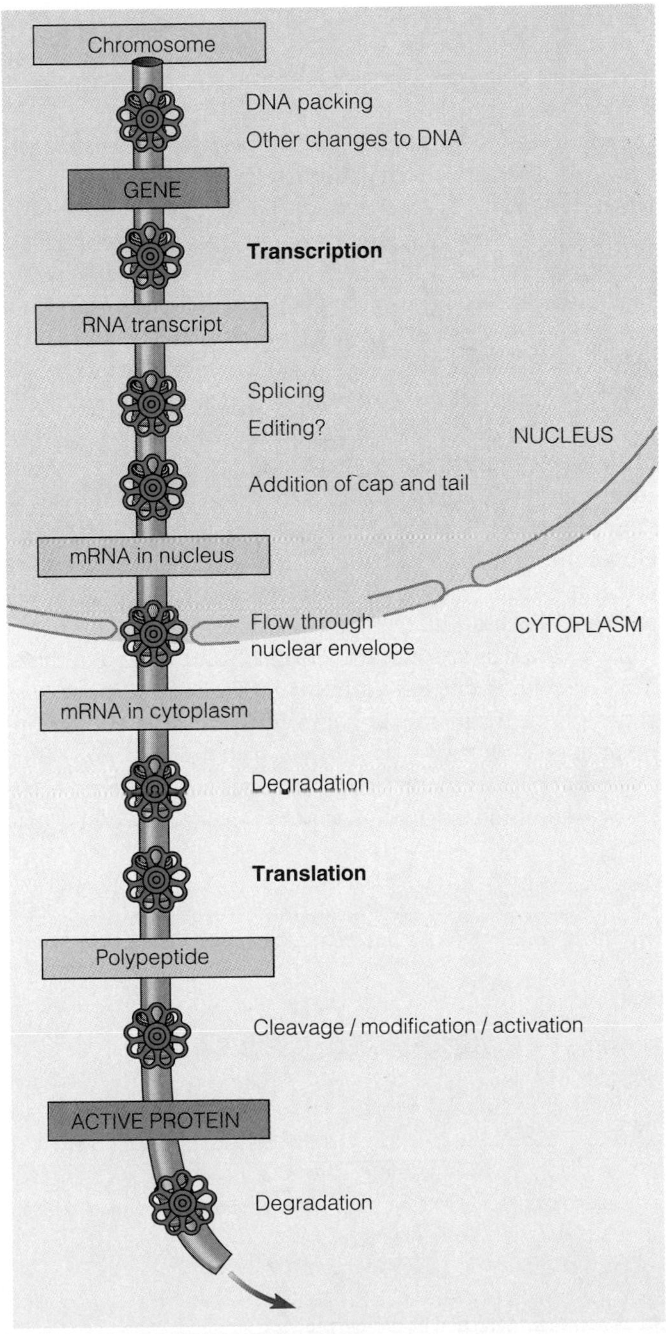

The gene-expression "pipeline" in a eukaryotic cell

Chapter 11   The Control of Gene Expression   **205**

We began this chapter with a picture of a fruit fly with an extra set of wings. The micrographs in Figure A illustrate another fruit-fly oddity. The fly head on the left is normal; notice the two short sensory structures called antennae on the front of its head. The fly head on the right is from a mutant; it exhibits a condition called antennapedia (antenna-feet), meaning it has two legs on the front of its head where antennae should be. If these mutant fruit flies do not sound strange to you, imagine a human with analogous body features—a person with four arms or with legs dangling from the head! Clearly, in a fruit fly, human, or any other organism, these are major abnormalities, significant changes in the organism's overall body plan.

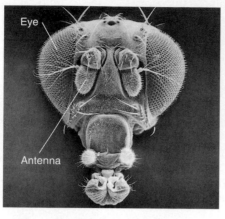
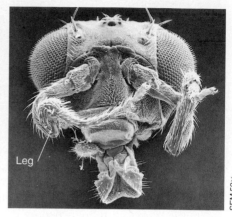

Eye

Antenna

Leg

SEM 52×

Head of a normal fruit fly

Head of a homeotic mutant

A. Antennapedia in the fruit fly, which results from a homeotic mutation

Fruit flies like the antennapedia mutant gave researchers their first glimpse of the underlying cause of major body-plan changes. We now know that such changes result from mutations in what are called homeotic genes. A **homeotic gene** is a master control gene that functions during embryonic development. It determines a major feature of the body plan by controlling the developmental fate of groups of cells. Antennapedia in fruit flies, for instance, results from a mutation in a homeotic gene that controls the differentiation of cells to form the insect's head and forelegs. Homeotic genes are currently receiving a great deal of attention from biologists who study development. Researchers have found these master control genes in organisms as varied as yeasts, earthworms, fruit flies, sea urchins, frogs, chickens, mice, and humans. Biologists think these genes play a major role in controlling patterns of development in eukaryotes.

Researchers studying homeotic genes in fruit flies discovered a common structural feature: Every homeotic gene they examined contained a common sequence of 180 nucleotides. Very similar sequences have since been found in homeotic genes in virtually every eukaryotic organism that has been examined. These nucleotide sequences are called **homeoboxes.** What do homeoboxes do? As illustrated in Figure B, a homeobox is translated into a small segment (60 amino acids long) of the homeotic gene's protein product. (The 180 nucleotides of the homeobox make up 60 codons, which translate into 60 amino acids.) The homeobox segment apparently binds to DNA at sites where genes are regulated (Figure C). There it can help the homeotic protein activate or repress gene transcription. Presumably, homeotic proteins regulate development by coordinating the transcription of batteries of developmental genes, switching them on or off together.

While homeoboxes and the amino acid sequences they encode vary somewhat from gene to gene and from organism to organism, they are all remarkably similar. For example, a homeobox sequence in the protein that produces antennapedia in fruit-fly mutants differs from a homeobox sequence in a frog by only a single amino acid. Considering that flies and frogs have evolved separately for over 600 million years, this similarity suggests that homeoboxes arose very early in the history of life, and that their nucleotide sequence has remained remarkably unchanged for eons of eukaryotic evolution. The way homeobox proteins function echoes an even older theme, one common to eukaryotes and prokaryotes alike: the coordinated control of multiple genes by interaction between regulatory proteins and specific sites on DNA.

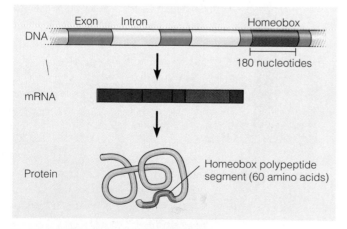

DNA

Exon   Intron   Homeobox

180 nucleotides

mRNA

Protein

Homeobox polypeptide segment (60 amino acids)

B. A homeotic gene and its protein product

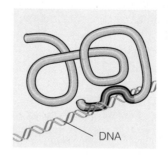

DNA

C. A homeotic protein bound to DNA

Growing out of control, cancer cells produce malignant tumors

How is the subject of cancer related to the regulation of genes? Cancer can develop when cells have defective gene-regulating mechanisms, and there is now ample evidence that cancer in all its forms is actually a disease of gene expression. Before we focus on the genetic basis of cancer, let's examine the general features of this disease.

In normal differentiated cells, cellular control mechanisms limit growth and division. But cells sometimes escape from these controls and multiply excessively. Such excessive growth can result in a **tumor,** an abnormal mass of cells within otherwise normal tissue. There are two general types of tumors. A **benign tumor** is an abnormal mass that remains at its original site in the body. Benign tumors can cause problems if they grow in certain organs such as the brain, but usually they can be completely removed by local surgery.

In contrast to benign tumors, a **malignant tumor** is an abnormal tissue mass capable of spreading into neighboring tissue and often to other parts of the body. Malignant tumors are composed of cancer cells—cells that are free of normal control mechanisms (see Module 8.11).

As illustrated here, a malignant tumor arises from a single cancer cell and displaces normal tissue as it grows. If not killed or removed, some of the cancer cells will spread into surrounding tissues, and the original tumor will expand. Cells may also split off from the tumor, invade the circulatory system (lymph vessels and blood vessels), and be carried to new locations, where they form new tumors. The spread of cancer cells beyond their original site is called **metastasis.** If a tumor metastasizes, physicians usually treat the patient with high-energy radiation and/or chemicals that are toxic to the actively dividing cancer cells.

Cancers are named according to the type of organ or tissue in which they appear, and over 200 different types are recognized in humans. For simplicity, they are grouped into four categories. **Carcinomas** are cancers that originate in the coverings of the body, such as the skin or the linings of the intestinal tract or breathing tubes. **Sarcomas** arise in tissues that support the body, such as bone and muscle. Cancers of blood-forming tissues, such as bone marrow, spleen, and lymph nodes, are called **leukemias** and **lymphomas.**

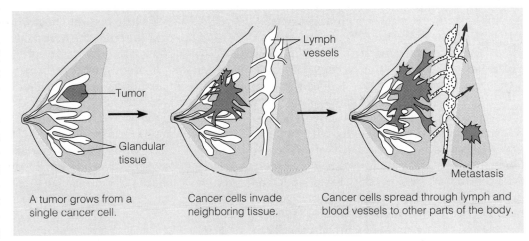

A tumor grows from a single cancer cell.

Cancer cells invade neighboring tissue.

Cancer cells spread through lymph and blood vessels to other parts of the body.

Growth and metastasis of a malignant tumor of the breast

Virus-induced genetic changes cause some forms of cancer

The odd behavior of cancer cells was observed years before much was known about what makes cells cancerous. One of the earliest clues to the cancer puzzle was the discovery, in 1911, of a virus that causes cancer in chickens. Since then, researchers have discovered that a number of viruses can cause cancer in animals, including humans. Viruses linked to human cancer include the hepatitis B virus and the human papillomavirus. Long-term infections by hepatitis B virus can result in liver cancer; and the human papillomavirus, which causes genital warts, is associated with cancer of the cervix. These and other viruses are thought to account for about 15% of human cancer cases worldwide.

How do viruses cause cancer? Remember, viruses are simply some DNA or RNA coated with protein and perhaps a membranous envelope (we referred to them as "packaged genes" in Chapter 10). Viruses that cause cancer can become permanent residents in host cells, often by inserting their nucleic acid into the DNA of host chromosomes. It is now known that a number of cancer viruses carry specific cancer-causing genes in their nucleic acid. When inserted into a host cell, these genes can make the cell cancerous.

A gene that causes cancer is called an **oncogene** (Greek *onkos,* tumor). Some oncogenes come from cancer-causing viruses. Surprisingly, many others arise within host cells themselves, as we see next.

Cancer results from mutations in somatic cells

In 1976, American molecular biologists J. Michael Bishop, Harold Varmus, and their colleagues made a startling discovery. They found that a cancer-causing virus of chickens contains an oncogene that is an altered form of a gene found in normal chicken cells. Apparently, the virus picked up the gene from a former host cell. Does this mean that organisms have genes that can make their own cells cancerous? In

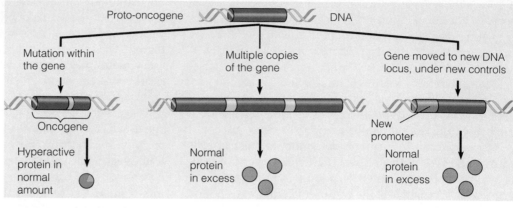

A. Alternative ways to make oncogenes from a proto-oncogene

fact, subsequent research has shown that the chromosomes of many animals, including humans, contain genes that can be converted to oncogenes. A normal gene with the potential to become an oncogene is called a **proto-oncogene.** About sixty proto-oncogenes have been identified in mammalian cells. Thus, organisms have genes with the potential to cause cancer. In 1989, Bishop and Varmus received the Nobel Prize for the pioneering work that led to this understanding.

What is the normal role of proto-oncogenes in the cell? Many proto-oncogenes code for growth factors—proteins that stimulate cell division—or for other proteins that affect growth-factor synthesis or function. When functioning normally, in the right amounts at the right times, these proteins control cell division and differentiation appropriately.

For a proto-oncogene to become an oncogene, a change, or mutation, must occur in the cell's DNA. Mutations that produce most types of cancer occur in somatic cells, those not involved in gamete formation. (Thus, cancer-causing mutations are generally not inherited.) Figure A illustrates three kinds of changes in somatic cell DNA that can produce active oncogenes. Let's assume that the starting proto-oncogene codes for a protein that stimulates cell division. On the left in the figure, a mutation in the proto-oncogene itself creates an oncogene that codes for a hyperactive protein, one whose stimulating effect is stronger than normal. In the center, an error in DNA replication or recombination generates multiple copies of the gene, which are all transcribed and translated; the result is an excess of the normal stimulatory protein. On the right, the proto-oncogene has been moved from its normal location in the cell's DNA to another location. At its new site, the gene is under the control of a different promoter and other genetic control elements, which cause it to be transcribed more often than normal; and the normal protein is again made in excess. So in all three cases, normal gene expression is changed, and the cell is stimulated to divide excessively.

Changes in genes whose products *inhibit* cell division can also be involved in cancer. These genes are called **tumor-suppressor genes,** because the proteins they encode normally help prevent uncontrolled cell growth. Any mutation that prevents the normal tumor-repressor protein from being made may contribute to uncontrolled cell division.

Mounting evidence suggests that more than one somatic mutation is needed to produce a full-fledged cancer cell. One of the best understood types of human cancer, cancer of the large intestine (colon), illustrates this principle (Figure B). As in many cancers, the development of a metastasizing colon cancer is gradual. ① The first sign is unusually frequent division of apparently normal cells in the colon lining; ② later a benign tumor (polyp) appears in the colon wall, and ③ eventually the malignant carcinoma. In most cases, these cellular changes are paralleled by three changes at the DNA level: the activation of a cellular oncogene and the inactivation of two tumor-suppressor genes.

What about virus-induced cancers? A virus may add an oncogene to the cell or disrupt the cell's DNA in a way that affects a proto-oncogene or a tumor-suppressor gene. However, the change in the cell's DNA induced by the virus is probably only one in a series of changes required for cancer. The requirement for several genetic changes could explain why virus-associated cancers can take a long time to develop.

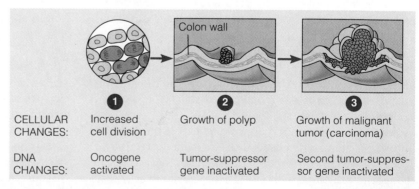

	1	2	3
CELLULAR CHANGES:	Increased cell division	Growth of polyp	Growth of malignant tumor (carcinoma)
DNA CHANGES:	Oncogene activated	Tumor-suppressor gene inactivated	Second tumor-suppressor gene inactivated

B. Three-step development of a typical colon cancer

By the year 2000, it is likely that cancer will be the number-one cause of death in most developed countries, including the United States. Death rates due to certain forms of cancer (for example, stomach, cervical, and uterine cancers) have decreased, but the overall cancer death rate is still on the rise, currently increasing at about 1% per decade.

Cancer-causing agents, factors that alter DNA and make cells cancerous, are called **carcinogens.** In general, mutagens are carcinogens. Two of the most potent carcinogens were mentioned as mutagens in Module 10.15: X-rays and ultraviolet radiation in sunlight. X-rays are a major cause of leukemia and brain cancer. And exposure to UV radiation from the sun is known to cause skin cancer, including a deadly type called melanoma.

The largest group of carcinogens are mutagenic chemicals and substances containing them. Among the most important carcinogens, the one substance known to cause more cases and types of cancer than any other single agent is tobacco. In 1900, lung cancer was a rare disease. Since then, largely because of an increase in cigarette smoking, lung cancer has steadily increased, and today more people die of lung cancer than any other form of cancer. Most tobacco-related cancers come from actually smoking, but passive inhalation of smoke is also a risk. As the table here indicates, tobacco use, sometimes in combination with alcohol consumption, causes a number of other types of cancer in addition to lung cancer. In nearly all cases, cigarettes are the main culprit, but smokeless tobacco products (snuff and chewing tobacco) are linked to cancer of the mouth and throat.

How do carcinogens cause cancer? As we mentioned in Module 11.16, most cancers result from multiple genetic changes, including activation of oncogenes and inactivation of tumor-suppressor genes. In many cases, these changes result from decades of exposure to the mutagenic effects of carcinogens. Carcinogens can also produce their effect by promoting cell division. Generally, the higher the rate of cell division, the greater the chance for mutations resulting from errors in DNA replication or recombination. Some carcinogens seem to have both effects. For instance, the hormones that cause breast and uterine cancers promote cell division and may also cause genetic changes that lead to cancer. In other cases, several different agents, such as viruses and one or more carcinogens, may together produce cancer.

We are still a long way from knowing all the factors that contribute to cancer. The effects of many environmental pollutants have yet to be evaluated, although some substances, such as asbestos, are definitely known to be carcinogens. Exposure to some of the most important known carcinogens are often matters of individual choice. Tobacco use, consumption of animal fat and alcohol, and time spent in the sun, for example, are all behavioral factors that affect our cancer risk.

Avoiding carcinogens is not the whole story, for there is growing evidence that some food choices significantly reduce cancer risk. For instance, eating 20–30 grams of plant fiber daily (about twice the amount the average American consumes) and at the same time reducing animal-fat intake may help prevent colon cancer. Evidence is also strong that the plant pigment beta-carotene can protect against lung cancer and probably other malignancies, and vitamins C and E may act similarly. Carrots, parsley, and celery are especially rich in beta-carotene. Other foods that seem to offer protection against cancer include fresh garlic, onion, and members of the cabbage family such as broccoli and cauliflower. Determining dietary recommendations for cancer prevention has become a major research goal.

The battle against cancer is being waged on many fronts, and there is reason for optimism in the progress being made. Research on the genetic basis of cancer is advancing rapidly and may soon lead to new treatments that target tumor cells very specifically. It is especially encouraging that we can help reduce our risk of acquiring some of the most common forms of cancer by the choices we make in daily life.

Cancer in the United States		
Cancer	**Examples of Known or Likely Carcinogens**	**Estimated No. of New Cases in 1990**
Lung	Tobacco	157,000
Colon	Animal fat; low dietary fiber	155,000
Breast	Estrogen; other ovarian hormones	150,900
Prostate	Testosterone	106,000
Bladder	Tobacco	49,000
Lymphomas	Virus (for one type)	43,000
Uterus	Estrogen	33,000
Mouth and throat	Tobacco	30,500
Pancreas	Tobacco	28,100
Leukemia	X-rays	27,800
Melanoma	Ultraviolet light	27,600
Kidney	Tobacco	24,000
Stomach	Table salt; tobacco	23,200
Ovary	Ovulation	20,500
Brain and nerve	Trauma; X-rays	15,600
Cervix	(Possibly viruses)	13,500
Liver	Alcohol; hepatitis viruses	13,100
All others	Multiple factors	122,200

Cancer is a reminder of our mortality. In a less threatening way, so are our family photo albums. Though we often hear that our population is living longer, what has actually increased is the number of people who survive long enough to die of old age, rather than our maximum life span.

The series of photographs here, showing former President Nixon at three different ages, illustrates a principle we take for granted: Organisms continue to change after completing embryonic development. The most obvious example is our own growth and maturation. We can also regard aging and even death as forms of developmental change in an organism. As we have seen repeatedly in this chapter, genes program cell growth, division, and differentiation—key processes in development. The process of aging also seems to be genetically programmed and may result from changes in genetic control mechanisms.

Some of the evidence for a connection between aging and changes in

genetic control mechanisms comes from hormone studies. A recent study looked at the effects of injecting elderly men with human growth hormone (GH). The GH in our blood declines naturally with age, but injections of GH dramatically reversed muscle loss and other signs of aging in the experimental subjects. Researchers think that GH may somehow regulate the activity of genes involved in aging.

At least two general hypotheses have been proposed to explain aging. In one view, the cumulative effects of nonlethal mutations and other damage to cells cause cells to lose function gradually. The other hypothesis suggests that aging is an innate property of cells, brought about either by the expression of specific genes or by scheduled changes that affect all genes. These changes might include a decline in the ability to replicate or repair DNA.

Overall, our understanding of gene regulation and its role in growth, development, and aging is still scant. Nonetheless, we know much more now than even five short years ago. The recent explosion of knowledge has been fueled by the invention of powerful new molecular methods—the subject of the next chapter.

Chapter Review

Begin your review by rereading the module headings and Module 11.12 and scanning the figures before proceeding to the Chapter Summary and questions.

Chapter Summary

11.1–11.2 Gene expression—the flow of information from genes to proteins—is controlled mainly by the turning of genes on and off. In prokaryotes such as *E. coli*, genes for related enzymes are often regulated together by being grouped into operons. The *lac* operon, for example, produces enzymes that break down lactose. A repressor protein locks onto the DNA and prevents transcription of these genes unless lactose is present. The *trp* operon produces enzymes for a metabolic pathway whose product, tryptophan, helps turn that operon off. Other operons are turned on by protein activators.

11.4–11.6 Eukaryotes, and especially multicellular ones, require more complex genetic control than prokaryotes. As a zygote develops into a multicellular organism, cells become specialized, or differentiated. Most differentiated cells retain a complete set of genes, so a carrot plant, for example, can be made to grow from a

single carrot cell. In general, the somatic cells of a multicellular organism all have the same genes; the cells are different because different genes are active.

11.7–11.8 Packing of DNA in chromosomes affects gene expression. A chromosome contains a DNA double helix wound around clusters of histone proteins, forming a string of beadlike nucleosomes. This beaded fiber is further wound and folded, packing 3 m of DNA into the nucleus of each human cell. Packing may restrict access of transcription enzymes to DNA, as happens with the inactive *X* chromosome in the cells of female mammals. In fruit-fly larvae, chromosome puffs make DNA more accessible for transcription; the pattern of puffs changes as the larva develops, showing that genes are switching on and off. A hormone activates RNA transcription at the puffs by binding to an activator protein.

11.9–11.12 Gene expression is also controlled after transcription. Most plant and animal genes are interrupted by noncoding introns, which may help regulate information flow. After transcription, the introns are cut out of the RNA, leaving exons that code for protein. Nucleotides are added to both ends of the RNA, and RNA editing may change the coding sequence, producing a "final draft" mRNA that is translated into a polypeptide. Protein factors control the translation process, and the lifetime of the RNA molecule helps determine how much protein is made. Finally, the cell may activate and later degrade the finished protein.

11.13 Bizarre mutations in fruit flies first suggested that master genes determine overall body plan. These so-called homeotic genes contain nucleotide segments called homeoboxes, which code for amino acid sequences that bind to DNA. Homeotic proteins apparently activate or repress batteries of genes, shaping development in major ways. Homeoboxes are similar in diverse organisms, suggesting they appeared early in the history of life.

11.14–11.17 Control mechanisms normally regulate the growth and division of differentiated cells, but sometimes cells escape these restraints and multiply into masses called tumors. Malignant tumors that grow from cancer cells can invade other tissues and metastasize to new locations. Cancer results from mutations affecting proto-oncogenes, which normally regulate cell division. A mutation can change a proto-oncogene into an oncogene, which causes cells to divide excessively. Mutations can also inactivate tumor-suppressor genes. Viruses can cause cancer by altering host genes or inserting oncogenes. Carcinogens include X-rays, ultraviolet radiation, and chemicals, such as the substances in tobacco that cause lung cancer. Reduction of exposure to carcinogens and certain other lifestyle choices can help reduce cancer risk.

11.18 Aging seems to be a developmental process and may be governed by genes. Perhaps mutations cause cells to gradually lose proper functioning, or perhaps aging and death are caused by programmed changes in gene expression.

Testing Your Knowledge

Multiple Choice

1. The control of gene expression is much more complex in multicellular eukaryotes than in prokaryotes because (*Explain your answer.*)
   a. eukaryotic cells are much smaller
   b. in a multicellular eukaryote, different cells are specialized for different functions
   c. the environment around a multicellular eukaryote is always changing
   d. eukaryotes have fewer genes, so each gene must do several jobs
   e. the genes of eukaryotes are information for making proteins

2. The proteins in the white of an egg are produced in a hen's liver. A rooster does not normally produce these proteins, but his liver will start making them if the rooster is injected with estrogen, a female hormone. Estrogen is known to enter the nuclei of liver cells. There it probably
   a. attaches to the operator site of an operon
   b. behaves as a homeotic protein
   c. is a carcinogen, causing a mutation in a proto-oncogene
   d. interacts with an activator protein that triggers transcription
   e. triggers RNA splicing, altering the rooster's genetic program

3. Your bone cells, muscle cells, and skin cells look different because
   a. different kinds of genes are present in each kind of cell
   b. they are present in different organs
   c. different genes are active in each kind of cell
   d. they contain different numbers of genes
   e. different mutations have occurred in each kind of cell

4. Which of the following methods of gene regulation do eukaryotes and prokaryotes appear to have in common?
   a. elaborate packing of DNA in chromosomes
   b. activator and repressor proteins, which attach to DNA
   c. addition of a cap and tail to mRNA after transcription
   d. editing of mRNA
   e. removal of noncoding portions of RNA

5. A eukaryotic gene was inserted into the DNA of a bacterium. The bacterium then transcribed this gene into mRNA and translated the mRNA into protein. The protein produced was useless; it contained many more amino acids than the protein made by the eukaryotic cell, and even the sequence of amino acids was somewhat different. Why?
   a. The mRNA was not spliced and edited as it is in eukaryotes.
   b. Eukaryotes and prokaryotes use different genetic codes.
   c. Repressor proteins interfered with transcription and translation.
   d. A proto-oncogene was changed into an oncogene.
   e. Ribosomes were not able to find the start codon on the mRNA.

6. All your cells contain proto-oncogenes, which can change into cancer-causing genes. Why do cells possess such potential time bombs?
   a. Viruses infect cells with proto-oncogenes.
   b. Proto-oncogenes are necessary for normal cell division.
   c. Proto-oncogenes are genetic "junk" with no known function.
   d. Proto-oncogenes are unavoidable environmental carcinogens.
   e. Cells produce proto-oncogenes as a by-product of the aging process.

Describing, Comparing, and Explaining

1. What kinds of evidence demonstrate that differentiated cells in a plant or animal retain the same full genetic potential?

2. A mutation in a homeotic gene causes a major change in the body of a fruit fly, such as an extra pair of legs or wings. Yet it probably takes the combined action of hundreds or thousands of genes to produce a wing or leg. How can a change in just one gene cause such a big change in the body?

3. Explain how operons enable a bacterium to adjust efficiently to a changing environment.

4. A single type of cancer such as leukemia can have alternative causes: X radiation, chemicals, or even viruses. Explain how such different factors can all trigger the same disease, with the same characteristics and symptoms.

Thinking Critically

Study the illustrations of the *lac* operon in Module 11.2. Normally, the genes are turned off when lactose is not present. Lactose activates the genes, which code for enzymes that then break down lactose. Mutations can alter the function of this operon; in fact, it was the effects of various mutations that enabled Jacob and Monod to figure out how the operon works. Test your understanding of the *lac* operon by predicting how the following mutations would affect its function in the presence and absence of lactose:
   a. Mutation of regulatory gene; repressor will not bind to lactose.
   b. Mutation of operator; repressor will not bind to operator.
   c. Mutation of regulatory gene; repressor will not bind to operator.
   d. Mutation of promoter; RNA polymerase will not attach to promoter.

Science, Technology, and Society

In 1993, the U.S. Environmental Protection Agency (EPA) issued a report summarizing scientific studies on environmental ("secondhand") tobacco smoke. The EPA concluded that secondhand smoke is responsible for the lung cancer deaths of an estimated 3000 nonsmokers each year. The report also determined that secondhand smoke increases the risk of pneumonia, bronchitis, and middle-ear disorders, and increases the severity of asthma. Can nonsmokers be protected from tobacco smoke in the environment? How? Do you think a nonsmoker has the right to breathe clean air? Why or why not? Do you think a smoker has the right to light up in a public place? Why or why not? How is the EPA report likely to tip the balance in this clash? Does a smoker have the right to smoke at home if it might affect the health of his or her children?

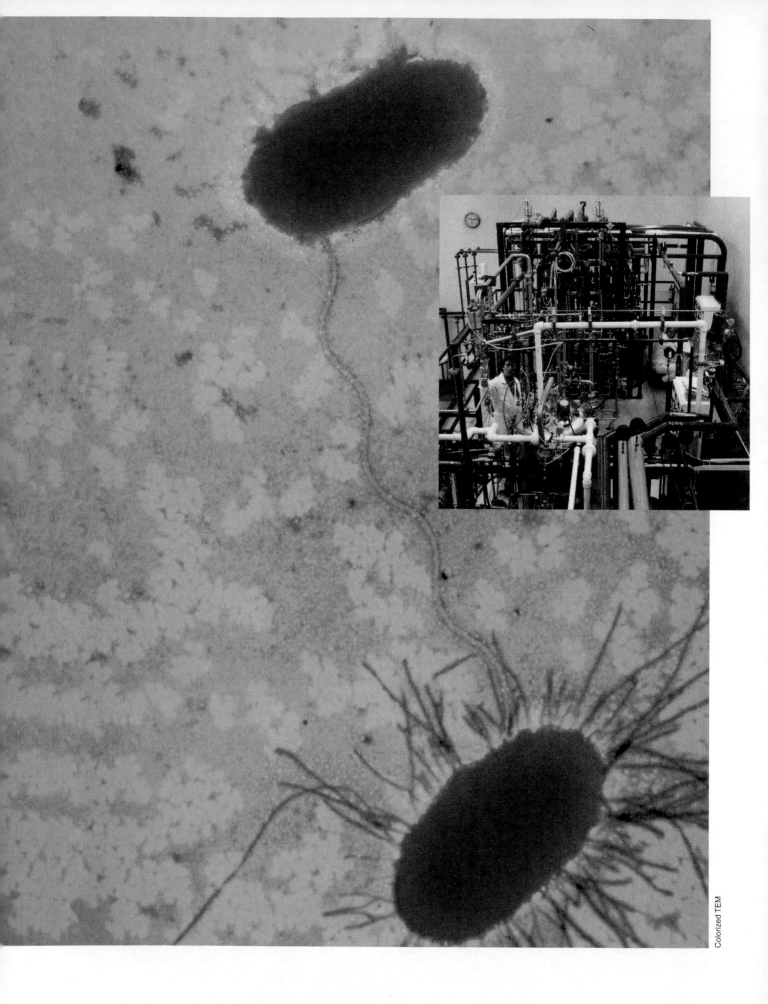

Recombinant DNA Technology **12**

The scene in the large photograph at the left could be taking place right now in your large intestine. It is a pair of cells of the bacterium *Escherichia coli* in the process of mating. Bacteria have a sex life? Yes, in fact *E. coli* bacteria have a very active one. The cells in this picture are joined by a long, threadlike sex pilus, an appendage of the "male" bacterium (the lower one in the photo). A bridge will soon form between the cells, and the "male" will pass part of its DNA to the "female."

Literally billions of *E. coli* inhabit the human large intestine (colon), where, along with countless other bacteria, they thrive on nutrients left over from our food. There are a number of different strains, or varieties, of *E. coli* that are differentiated by their dietary requirements. Studies of such strains led to the discovery of bacterial sex and what it accomplishes. In 1946, American geneticists Joshua Lederberg and Edward Tatum concluded from experiments with *E. coli* that these bacteria have a sexual mechanism that results in the combining of genes from two different cells.

In their ground-breaking experiment, Lederberg and Tatum used two strains of *E. coli*. Each strain carried several mutations that prevented it from synthesizing some of the amino acids it needed. Because the mutated genes differed in the two strains, the strains required the addition of different amino acids to their growth media. However, Lederberg and Tatum found that after mixing the two strains, cells appeared that did not require *any* amino acids for growth. Apparently, cells of the two strains had somehow combined genes, creating recombinant bacteria with functional versions of all the amino acid synthesis genes.

Lederberg, Tatum, and their colleagues were pioneers of bacterial genetics, a field that within twenty years made *E. coli* the most thoroughly understood of all organisms at the molecular level. Eventually, in the mid-1970s, research on this bacterium led to the development of **recombinant DNA technology**. This is a set of techniques for combining in a test tube genes from different sources—even different species—and transferring the recombinant DNA into cells, where it can be replicated and may be expressed.

In the past decade, recombinant DNA technology has paid high dividends in basic research. For example, introns, the noncoding sequences within eukaryotic genes, were discovered through recombinant DNA technology. This technology also helped reveal the role of oncogenes in cancer.

Furthermore, recombinant DNA technology is being used to engineer the genes of cells for practical purposes. Almost daily we read in the news about the potential of genetic engineering for improving human welfare. Scientists have already created genetically engineered bacteria that can mass-produce useful chemicals, including the hormone insulin and certain cancer drugs; the inset on the facing page shows an industrial apparatus for growing engineered bacteria. Ongoing work may lead to cures for many diseases, ways to clean up toxic wastes, and improved crop plants. In the photograph below, a researcher examines a cotton plant that has been genetically engineered to resist insect attack.

This chapter examines recombinant DNA methods, current and future applications of these methods, and some of the social and ethical issues that this new technology has raised. But before entering the recombinant DNA laboratory, let's take a closer look at some crucial aspects of bacterial genetics, the foundation of this modern revolution.

Researcher with a genetically engineered cotton plant

Bacteria can transfer DNA in three ways

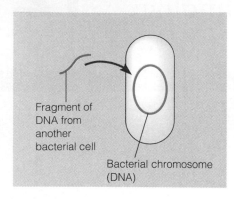

A. Transformation

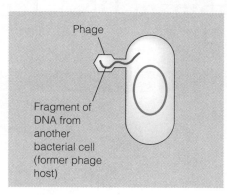

B. Transduction

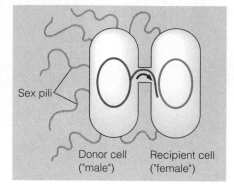

C. Conjugation

The sexual process described on the preceding page involves the fusion of two bacterial cells and the transfer of DNA from one cell to another. After the transfer, the two cells separate. The process is not reproductive, for the number of individuals does not increase. In contrast, the sexual process in plants and animals is a reproductive one. As we saw in Chapter 8, it involves the formation of gametes by meiosis and the fusion of gametes to form new individuals. Meiosis and fertilization both provide opportunities for genetic recombination, the creation of new combinations of genes originating from two individuals. Bacteria do not have meiosis, gametes, or fertilization, but they do not lack for ways to produce new combinations of DNA. In fact, in the bacterial world there are three mechanisms for transferring genes from one cell to another. These mechanisms of gene transfer are called transformation, transduction, and conjugation.

Most of a bacterium's DNA is found in a single bacterial chromosome, which is a closed loop of DNA with associated proteins. For convenience, in the figures here, we show the chromosome much smaller than it actually is relative to the cell. A bacterial chromosome is hundreds of times longer than its cell; it fits inside the cell because it is highly folded.

Figure A illustrates **transformation**, which is the taking up of DNA from the fluid surrounding a cell. This is what happened in the famous experiment with pneumonia-causing bacteria performed by Frederick Griffith in the 1920s (see Module 10.1). A harmless strain of bacteria took up pieces of DNA left from the dead cells of another, disease-causing, strain. The DNA from the pathogenic bacteria carried a gene for the ability to cause pneumonia, and the previously harmless bacteria became pathogenic when they acquired this gene.

Bacteriophages provide the second means of bringing together genes of different bacteria. The transfer of bacterial genes by a phage is called **transduction.** In this process, a fragment of DNA belonging to a phage's host cell is accidentally packaged within the phage's coat. When the phage infects a new bacterial cell, the DNA stowaway from the former host cell is injected into the new host (Figure B).

Figure C shows what happens at the DNA level when two bacterial cells mate, as we described in the chapter's introduction. This union and the DNA transfer between the cells is called **conjugation** (Latin *conjugatus,* united). Notice that the "male," donor cell has sex pili, some of which are attached to the "female," recipient cell. The outside layers of the cells have fused, and a cytoplasmic bridge has formed between them. Through this bridge, donor-cell DNA (gray in the figure) passes to the recipient cell. The donor cell replicates its DNA as it transfers it, so the cell doesn't end up lacking any genes. The DNA replication is a special type that allows one copy to peel off and transfer into the recipient cell.

By whatever mechanism new DNA gets into a bacterial cell, part of it can then integrate into the recipient's chromosome. As Figure D indicates, integration occurs by crossing over between the two DNA molecules, a process similar to crossing over between eukaryotic chromosomes (see Module 8.19). Here we see that two crossovers result in some of the donated DNA's replacing part of the recipient cell's original DNA. The leftover pieces of DNA are degraded by enzymes, leaving the recipient bacterium with a changed chromosome. The resulting cell is a recombinant; it carries a new combination of genes, some of which came from another cell. We examine gene transfer by conjugation further in the next module.

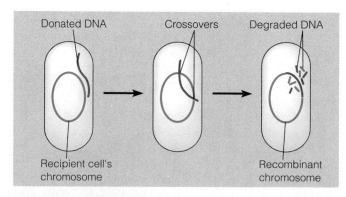

D. Integration of donated DNA into the recipient cell's chromosome

The ability of a male *E. coli* cell to carry out conjugation is usually due to a specific segment of DNA called the **F factor** (F for fertility). The F factor carries genes for making sex pili and other things needed for conjugation; it also contains a site where DNA replication can start.

Let's see what the F factor actually does during conjugation. In Figure A, the F factor (dark gray line segment) is an integrated part of a male bacterium's chromosome. When this cell joins to a recipient cell, as shown in the second part of Figure A, the F factor starts replication. The site where replication starts, indicated by the arrowhead in the donor cell, becomes the leading end of the transferred DNA. This end of the F factor peels off the chromosome and heads into the donor cell, dragging some of the genes from the donor's original chromosome along with it. The rest of the F factor stays in the donor cell. Once inside the recipient cell, the transferred donor genes can recombine with the homologous part of the recipient chromosome by crossing over. In this case, the recipient cell is genetically changed, but it usually remains female because the two cells break apart before the rest of the F factor transfers.

Alternatively, as Figure B illustrates, an F factor can exist as a **plasmid**, a small, circular DNA molecule separate from the much larger bacterial chromosome in a bacterial cell. A plasmid can replicate itself within the cell, and some, including the F-factor plasmid, can serve as carriers (also called **vectors**) that move genes from one cell to another. When the male cell in Figure B is joined to a recipient cell, the F factor starts to replicate and at the same time transfers one whole copy of itself, in linear rather than circular form, to the recipient cell. The transferred plasmid forms a circle in the recipient cell, and the cell is converted to male.

E. coli and other bacteria have many different kinds of plasmids. You can see several from one bacterial cell in Figure C, along with part of the bacterial chromosome. Some plasmids carry particular genes that can affect the survival of the cell. Plasmids of one class, called **R plasmids**, pose serious problems for human medicine. Transferable R plasmids carry genes for conjugation and also genes for enzymes that destroy antibiotics such as penicillin and tetracycline. Bacteria carrying R plasmids are resistant (hence the designation R) to antibiotics that would otherwise kill them. The widespread use of antibiotics in medicine and agriculture has tended to kill off bacteria that lack R plasmids, while those with R plasmids have multiplied. As a result, an increasing number of bacteria that cause human diseases, including food poisoning and gonorrhea, are becoming resistant to antibiotics. The transfer of R plasmids from resistant to nonresistant bacteria has contributed to the problem.

R plasmids can cause difficulties, but they are useful for genetic engineering, as we will see in Module 12.4. First, let's take a brief look at another kind of mobile genetic element, one that was first detected in eukaryotes.

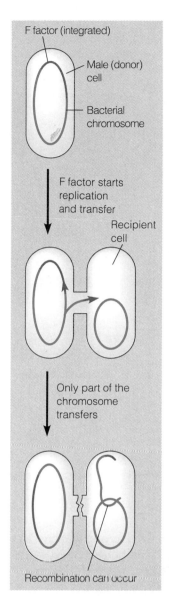

F factor (integrated)
— Male (donor) cell
— Bacterial chromosome

F factor starts replication and transfer

Recipient cell

Only part of the chromosome transfers

Recombination can occur

A. Transfer of chromosomal DNA by an integrated F factor

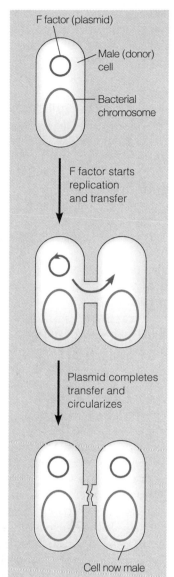

F factor (plasmid)
— Male (donor) cell
— Bacterial chromosome

F factor starts replication and transfer

Plasmid completes transfer and circularizes

Cell now male

B. Transfer of an F-factor plasmid

TEM 2,000×

C. Plasmids and part of a bacterial chromosome released from a ruptured *E. coli* cell

TALKING ABOUT SCIENCE

Barbara McClintock discovered jumping genes

In the 1940s, studying inheritance in corn plants, American geneticist Barbara McClintock (Figure A) made a startling discovery: Certain genetic elements can move from one location to another in a chromosome or even from one chromosome to another. McClintock found that these mobile elements, now sometimes called "jumping genes," may land in the middle of other genes and disrupt them. In Indian corn, for instance, genetic changes caused by mobile elements produce spotted kernels like the ones in Figure B. Certain pigment genes are expressed in most of the cells in these kernels. In other cells, however, some pigment genes are not expressed because they have been disrupted by mobile elements.

McClintock discovered jumping genes two decades before other geneticists began to grasp their importance. Indeed, few scientists understood McClintock's findings until the 1970s, after mobile genes were discovered in *E. coli* and our knowledge of molecular genetics was well advanced. At last, in 1983, at age 81, Barbara McClintock received the Nobel Prize for her pioneering work.

A. Barbara McClintock in 1947

Strong evidence now suggests that all cells have segments of DNA, officially called transposable elements or **transposons**, that are able to move from one site to another. Part ① in Figure C shows the simplest kind of transposon, which includes one gene (purple), for an enzyme that catalyzes movement of the transposon. Identical noncoding nucleotide sequences (yellow) mark the ends of the transposon. In part ②, the enzyme cuts out the transposon. The transposon then moves to another site on the chromosome, where the enzyme cuts open the DNA and catalyzes insertion of the transposon. ③ Insertion of the transposon in its new location disrupts the nucleotide sequence of the gene shown in green.

McClintock found that a transposon inserted into the middle of a gene can make the gene nonfunctional. She showed that these mobile genetic units can act as natural mutagens that generate genetic diversity. Thus, she saw that, in a broad sense, transposons could be a powerful mechanism in evolution.

After a lifetime of important contributions to genetics, Barbara McClintock died at age 90 in 1992. She is now recognized as one of the twentieth century's truly great scien-

tists. For years, McClintock worked alone and virtually shunned social contact, even with her colleagues. How was she able to make significant discoveries, often years before other scientists? Her biographer, Evelyn Fox Keller, wrote in *A Feeling for the Organism: The Life and Work of Barbara McClintock*, "Over and over again, [McClintock] tells us one must have the time to look, the patience to 'hear what the material has to say to you,' the openness to 'let it come to you.' Above all, one must have 'a feeling for the organism.'"

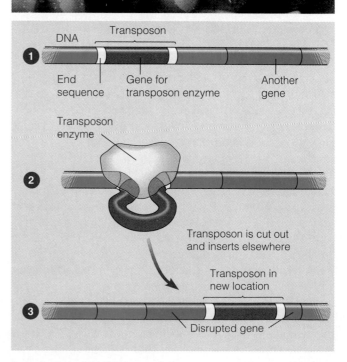

B. Corn kernels with spots caused by mobile genetic elements

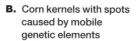

C. Transposon movement

Let's return now to bacterial plasmids. Because they can carry virtually any gene and can self-replicate inside bacteria, they are key tools in recombinant DNA technology. Below is an overview of how plasmids can be used to custom-build a bacterial cell. Starting at the top left, we see that ① a plasmid is first isolated from a bacterium such as *E. coli*. At the same time (top right), ② DNA carrying a particular gene of interest is removed from another cell—an animal cell (as here), a plant cell, or another bacterium. The gene we are interested in might be a human gene that encodes a hormone such as insulin or a plant gene that confers resistance to pest insects. ③ A piece of DNA containing the gene is inserted into the plasmid, producing recombinant DNA, and ④ the bacterial cell takes up the plasmid by transformation. ⑤ This geneti-

cally engineered, recombinant bacterium is then cloned in large numbers to make multiple copies of the gene.

The bottom of the figure illustrates a few of the current applications of genetically engineered bacteria. In some cases, such as those on the left, copies of the original gene are the desired product. In other cases, useful protein products encoded by the gene are harvested in large quantities from the bacterial cultures. Many pharmaceutical companies employ molecular biologists to design bacteria that will produce new marketable products. In fact, genetic engineering has launched a revolution in **biotechnology,** the use of living organisms to perform practical tasks. In the next several modules, we examine in more detail the tools and procedures of this exciting new technology.

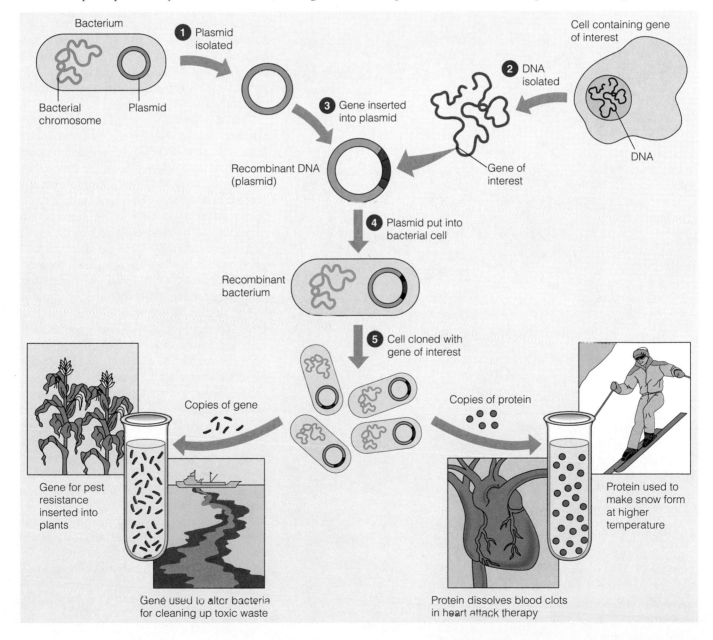

Enzymes are used to "cut and paste" DNA

Extracting a gene from one cell and inserting it into another require precise "cutting and pasting." To carry out the procedure outlined in Module 12.4, a specific piece of DNA containing the gene of interest must be cut out of a chromosome and "pasted" into a bacterial plasmid.

The cutting tools of recombinant DNA technology are bacterial enzymes called **restriction enzymes,** which were first discovered in the late 1960s. In nature, these enzymes protect bacteria against intruding DNA from phages and from other organisms. They work by cutting up the foreign DNA, a process called restriction because it *restricts* DNA from other organisms from surviving in the cell. Other enzymes chemically modify the cell's own DNA in a way that protects it from the restriction enzymes.

Most restriction enzymes recognize short nucleotide sequences in DNA molecules and cut at specific points within these so-called recognition sequences. There are several hundred restriction enzymes and about a hundred different recognition sequences known. In the figure here, part ① shows two recognition sequences in a piece of DNA. In this case, the restriction enzyme will cut the DNA strands in the four places where the bases A and G lie next to each other. (The places where DNA is cut are called restriction sites.) ② The result is a set of double-stranded DNA fragments with single-stranded ends, called "sticky ends." Sticky ends are the key to joining DNA restriction fragments originating from different sources, even from different species. These short extensions will form hydrogen-bonded base pairs with complementary single-stranded stretches of DNA.

Part ③ shows a piece of DNA (gray) from another source. Notice that the gray DNA has single-stranded ends identical in base sequence to the sticky ends on the blue DNA. The gray, "foreign" DNA has ends with this particular base sequence because it was cut from a larger molecule with the same restriction enzyme used to cut the blue DNA. ④ The complementary ends on the blue and gray fragments allow them to stick together by base pairing. (The hydrogen bonds that hold the base pairs together are not shown.)

The union between the blue and gray DNA fragments shown in part 4 is only temporary, because only a few hydrogen bonds hold the fragments together. The union can be made permanent, however, by the "pasting" enzyme, called **DNA ligase.** This enzyme, which the cell normally uses in DNA replication, catalyzes the formation of covalent bonds between adjacent nucleotides, sealing the breaks in the DNA strands. ⑤ The final outcome is **recombinant DNA,** a DNA molecule carrying a new combination of genes.

Now, with an enzymatic method for cutting and pasting DNA, plasmids for carrying that DNA into cells, and bacterial cells for replicating it, we are ready to make multiple copies of a gene.

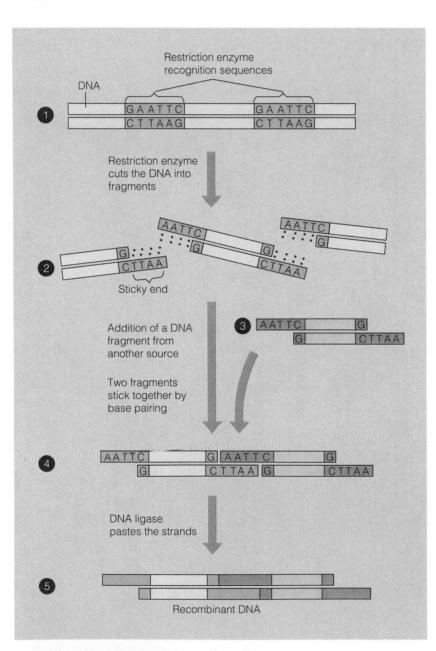

Creating recombinant DNA using a restriction enzyme and DNA ligase

Genes can be cloned in recombinant plasmids

Making recombinant DNA in large enough amounts to be useful requires several steps. Consider a typical genetic engineering challenge: A molecular biologist at a pharmaceutical company has identified a human gene that codes for a valuable product—a hypothetical substance called protein V that kills certain human viruses. The biologist wants to set up a system for making large amounts of the gene, so that the protein can be manufactured on a large scale. This figure illustrates a way to make many copies of the gene using the techniques we have been discussing.

In step ①, the biologist isolates two kinds of DNA: the bacterial plasmid that will serve as the vector, and eukaryotic DNA containing the protein-V gene. In this hypothetical example, the DNA containing the gene of interest comes from human tissue cells that have been growing in laboratory culture. The plasmid comes from the bacterium *E. coli*.

In step ②, the researcher treats both the plasmid and the human DNA with the same restriction enzyme. The enzyme cuts the plasmid DNA at one specific restriction site. It also cuts the human DNA, generating many thousands of fragments; one of these fragments carries the protein-V gene. In making the cuts, the restriction enzyme creates sticky ends on both the human DNA fragments and the plasmid. For simplicity, the figure here shows the step-by-step processing of one human DNA fragment and one plasmid, but actually millions of plasmids and human DNA fragments are treated simultaneously.

In step ③, the human DNA is mixed with the cut plasmid. The sticky ends of the plasmid base pair with the complementary sticky ends of the human DNA fragment. In step ④, the enzyme DNA ligase joins the two DNA molecules by covalent bonds, and the result is a recombinant DNA plasmid containing gene V.

In step ⑤, the recombinant plasmid is added to a bacterium. Under the right conditions, the bacterium will take up the plasmid DNA from solution by the process of transformation (see Module 12.1).

The last step shown here, step ⑥, is the actual **gene cloning**, the production of multiple copies of the gene. The bacterium, with its recombinant plasmid, is allowed to reproduce. As the bacterium forms a cell clone (a group of identical cells descended from a single ancestral cell), any genes carried by the recombinant plasmid are also cloned. In our example, the biologist would develop a cell clone to produce protein V in marketable quantities.

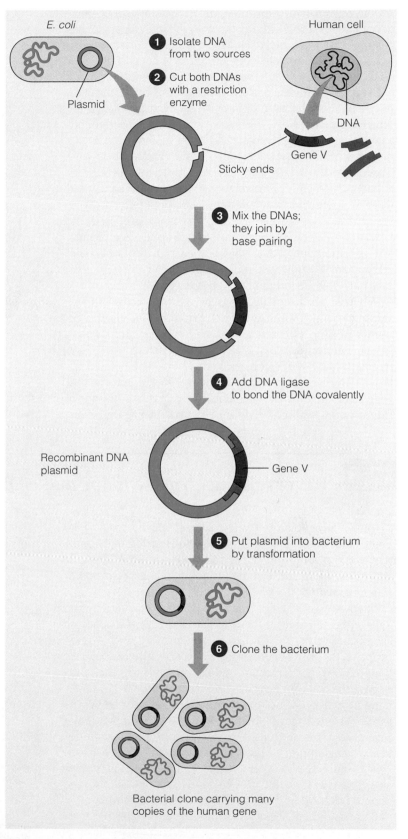

Cloning a gene in a bacterial plasmid

12.7 Cloned genes can be stored in genomic libraries

As in our protein-V example, the process of genetic engineering often starts with the discovery that an organism has a gene that can be put to practical use. Scientists then isolate the DNA from cells containing the gene, and use a restriction enzyme to cut the DNA into thousands of pieces, each typically the size of a few genes. At this point, the DNA fragment with the gene of interest is mixed in with all the other fragments. Next, every DNA fragment is inserted into a separate molecular vector, and each vector is put into a bacterial cell. This is called the shotgun approach to gene cloning, because it does not target any particular gene.

Bacterial plasmids are one type of vector that can be used in the shotgun cloning of genes, but not the only type. Phages are also useful vectors. When a phage is used, the DNA to be cloned is inserted into the phage DNA molecule. The recombinant phage DNA is then put into a bacterial cell, where it replicates and produces many new phages, each carrying the foreign DNA "passenger."

Whether phages or plasmids are used as vectors, the result of cloning DNA taken directly from living cells is a **genomic library**. This is a set of DNA fragments from an organism's **genome**, which is a complete set of its genes. Each fragment is carried by a plasmid or phage. The figure illustrates two types of genomic libraries. On the left, the red, yellow, and green DNA segments (three of the thousands of "library books") are shelved in plasmids in bacterial cells.

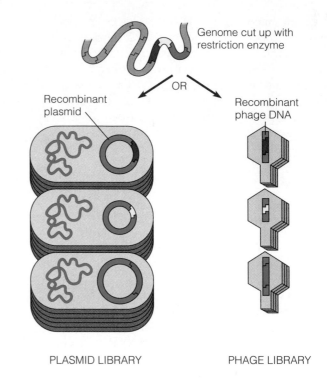

Genome cut up with restriction enzyme

OR

Recombinant plasmid

Recombinant phage DNA

PLASMID LIBRARY   PHAGE LIBRARY

On the right, the same books are shelved in phage DNA. In Module 12.9, we learn how to find a book in this library.

12.8 Reverse transcriptase helps make DNA for cloning

Not all DNA that is cloned comes directly from cells. Some eukaryotic genes are too large to clone easily because they contain long noncoding regions (introns). When this is the case, an intron-lacking version of the gene can sometimes be made in the laboratory.

The first two steps of this process are performed by the eukaryotic cell that is the original source of the gene. As shown in the figure, ① the cell first makes an RNA transcript of the intron-containing gene. ② Cellular enzymes then remove the introns and splice the exons together, producing mRNA. ③ The researcher isolates the mRNA molecules from the cell and uses them as templates for DNA synthesis. This synthesis of DNA on an RNA template is the reverse of transcription. It is catalyzed by the enzyme reverse transcriptase, which is obtained from retroviruses (see Module 10.18). After a single strand of DNA is synthesized, ④ the RNA and DNA strands separate, and ⑤ the second DNA strand is made using the first as a template.

Because the mRNA lacks introns, the artificial gene created by this method is more manageable in size than the original gene. It also has the potential for being transcribed and

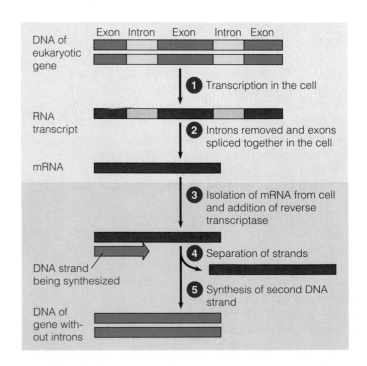

DNA of eukaryotic gene

Exon   Intron   Exon   Intron   Exon

① Transcription in the cell

RNA transcript

② Introns removed and exons spliced together in the cell

mRNA

③ Isolation of mRNA from cell and addition of reverse transcriptase

DNA strand being synthesized

④ Separation of strands

⑤ Synthesis of second DNA strand

DNA of gene without introns

translated by bacterial cells, which do not have RNA-splicing machinery. So bacteria may be able to manufacture the eukaryotic protein product.

Because the mRNA molecules for one particular gene usually cannot be isolated from the cell's other mRNA molecules, the reverse transcription method also produces libraries. Such libraries, however, represent only part of the cell's DNA content. In fact, they represent only those genes that the cell expresses (transcribes). This is an advantage if the desired gene is one of only a few expressed in the cell.

Molecular probes identify clones carrying specific genes 12.9

Often the most difficult task in gene cloning is finding the right shelf in a genomic library—that is, identifying the bacterial or phage clone containing a desired gene. If bacterial clones containing a specific gene actually translate the gene into protein, they can be identified by screening for the protein. This is not always the case, however. Fortunately, researchers can also test directly for the gene itself.

Methods for detecting genes all depend on base pairing between the gene and a complementary sequence on another nucleic acid molecule, either DNA or RNA. When at least part of the nucleotide sequence of a gene can be guessed, this information can be used to make a suitable probe. Recall that there is a precise relationship between the nucleotide sequence of a gene and the amino acid sequence of the protein it encodes. For example, suppose our hypothetical protein V contains an amino acid sequence encoded by the nucleotide sequence TAGGCT. Knowing that the gene for protein V probably contains this sequence, a biochemist uses nucleotides labeled with a radioactive isotope to synthesize a short RNA molecule with a complementary sequence (AUCCGA). This labeled molecule is called a **probe** because it is used to find a specific gene. (In practice, a probe molecule would be considerably longer.)

The figure here shows how the probe works. Once the

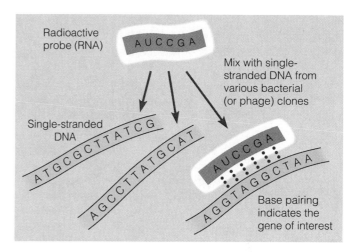

Using an RNA probe to find a gene

probe is ready, single-stranded DNA is prepared from each bacterial or phage clone to be tested. When the probe is mixed with the DNA samples, the radioactive RNA tags the correct clone—finds the needle in the haystack—by hydrogen-bonding to the complementary sequence in gene V. Once the clone carrying the gene is identified, gene V can be isolated in large amounts.

Automation makes rapid DNA synthesis and DNA sequencing possible 12.10

In some cases, genes for cloning are chemically synthesized in the laboratory, without using mRNA or any other nucleic acid template. If the order of amino acids in a protein is known, it is easy to use the genetic code (see Figure 10.14) to determine a nucleotide sequence that would code for it. More difficult is actually synthesizing the DNA. Among the first genes synthesized artificially were two that encoded the two polypeptide chains making up human insulin. (In the cell, the insulin polypeptides are encoded in one long gene; see Module 11.11.) The task of making the first artificial genes was long and laborious, feasible only because the two genes together had only about 150 nucleotides. DNA-synthesizing machines are now available that can rapidly produce genes hundreds of nucleotides long.

Automation has also revolutionized DNA sequencing, the process of determining the exact sequence of nucleotides in

a stretch of DNA. Sequencing even a short gene used to be very difficult, but thanks to methods developed during the late 1970s in the U.S. and England, the sequences of even large genes can now often be determined in a few days. DNA sequencing is made possible by restriction enzymes, which are used to cut the very long DNA found in cells into discrete fragments that can be analyzed by machine.

Thousands of DNA sequences are now being collected in computer data banks, and they are proving of great value for understanding genes and gene control, as well as for biotechnology. With the help of computers, long sequences can be scanned to find particular genes or control sequences, such as promoters. Automated sequencing and the use of computers also help researchers compare the DNA of two or more species, to determine how similar or different they are at the molecular level.

Bacteria, yeasts, and mammalian cells are used to mass-produce gene products

Making a gene or transferring a gene from one kind of cell into another is one thing; getting a gene to function in a new setting is quite another. Scientists expected it would be very difficult to get bacteria to express eukaryotic genes. After all, many details of transcription and translation are different in prokaryotes and eukaryotes. However, researchers were pleasantly surprised to find that, with a little genetic trickery, they could get bacteria to produce eukaryotic proteins.

Bacteria are often the best organisms for making large amounts of a gene product. A major advantage is that bacteria can be grown rapidly and cheaply in large tanks. Bacteria can be engineered so that they will produce large amounts of desired proteins and secrete their protein products into the medium in which they are grown. Secretion into the growth medium simplifies the task of collecting and purifying the products. As the table below shows, a number of proteins of importance in human medicine and agriculture are being produced in bacteria, mostly *E. coli.*

Despite the advantages of bacteria, for some commercial applications, it's necessary to use eukaryotic cells to produce the protein. Often the first-choice eukaryotic organism for protein production is the yeast *Saccharomyces cerevisiae,* the single-celled fungus featured in Chapter 6. As bakers and brewers have recognized for centuries, yeast cells are easy to grow. And like *E. coli,* yeast cells can take up foreign DNA and integrate it into their genomes. Yeasts also have plasmids that can be used as gene vectors, and sometimes yeasts are better than bacteria at synthesizing and secreting eukaryotic proteins. *S. cerevisiae* is currently used to produce a number of proteins. In some cases, the same product (for example, interferons used in cancer research) can be made in either yeast or bacteria. In other cases, such as the hepatitis B vaccine, yeast alone is used.

The cells of choice for making some gene products come from mammals (see the table). Genes for these products are often cloned in bacteria as a preliminary step. For example, the genes for colony-stimulating factor (CSF) and for two proteins that affect blood clotting, Factor VIII and t-PA, are cloned in a bacterial plasmid before transfer to mammalian cells for large-scale production. Many proteins that mammalian cells normally secrete are glycoproteins, meaning that they have chains of sugars attached to the protein. Because only mammalian cells can attach the sugars correctly, mammalian cells must be used for making these products.

Some Protein Products of Recombinant DNA Technology

Product	Made in	Use
Human insulin	*E. coli*	Treatment for diabetes
Human growth hormone (GH)	*E. coli*	Treatment for growth defects
Epidermal growth factor (EGF)	*E. coli*	Treatment for burns, ulcers
Tumor necrosis factor	*E. coli*	Killing of certain tumor cells
Interleukin-2 (IL-2)	*E. coli*	Possible treatment for cancer
Prourokinase	*E. coli*	Treatment for heart attacks
Porcine growth hormone (PGH)	*E. coli*	Improving weight gain in hogs
Bovine growth hormone (BGH)	*E. coli*	Improving weight gain in cattle
Cellulase	*E. coli*	Breaking down cellulose for animal feeds
Snomax ®	*Pseudomonas* bacterium	Making snow for ski resorts
Interferons (alpha and gamma)	*S. cerevisiae; E. coli*	Possible treatment for cancer and virus infections
Hepatitis B vaccine	*S. cerevisiae*	Prevention of hepatitis-virus infection
Colony-stimulating factor (CSF)	Mammalian cells	Treatment for leukemia; boosts resistance to AIDS and other infectious diseases
Erythropoietin (EPO)	Mammalian cells	Treatment for anemia
Factor VIII	Mammalian cells	Treatment for hemophilia
Tissue plasminogen activator (t-PA)	Mammalian cells	Treatment for heart attacks

DNA technology is revolutionizing the pharmaceutical industry and human medicine

Commercial applications of recombinant DNA technology have not proceeded as far or fast as some biologists predicted a decade ago (to the dismay of certain Wall Street investors). Nevertheless, the new technology has already had a major impact on the pharmaceutical industry and on human medicine.

Consider the first two products in the table on the previous page, human insulin and human growth hormone (GH). These were the first pharmaceutical products made using recombinant DNA technology. In the U.S. alone, about two million people with diabetes depend on insulin treatment. Before 1982, the main sources of this hormone were pig and cattle tissues obtained from slaughterhouses. Insulin extracted from these animals is chemically similar, but not identical, to human insulin, and it causes harmful side effects in some people. Genetic engineering has largely solved this problem by developing bacteria that actually synthesize and secrete human insulin.

GH was harder to produce than insulin because the GH molecule is about twice as big. Since growth hormones from other animals are not effective growth stimulators in humans, however, GH was urgently needed. The gene that codes for GH consists of nearly 600 nucleotides. In 1985,

molecular biologists produced GH in *E. coli* after combining a DNA fragment taken from human cells with a chemically synthesized piece of DNA. The photograph on this page shows vials of GH coming off the assembly line at a pharmaceutical company. Before this genetically engineered version became available in 1985, children with a GH deficiency had to rely on scarce supplies from human cadavers, or else face dwarfism. Increased availability of GH has led some parents of short but hormonally normal children to seek it. The long-term effects of giving GH to such children are unclear, and GH use in these cases raises ethical questions.

Recombinant DNA technology has also given a boost to the development of vaccines. A **vaccine** is a harmless variant or derivative of a pathogen (usually a bacterium or virus) that is used to prevent an infectious disease. When a person—a potential host of the pathogen—is inoculated, the vaccine stimulates the immune system to develop lasting defenses against the pathogen. Especially for the many viral diseases for which there is no effective drug treatment, prevention by vaccination is virtually the only medical way to fight the disease.

Genetic engineering can be used in several ways to make vaccines. One approach is to use genetically engineered cells to produce large amounts of a protein molecule that is found on the pathogen's outside surface. This method has been used to make the vaccine against hepatitis B virus. Hepatitis is a disabling and sometimes fatal liver disease, and the hepatitis B virus can cause liver cancer.

Another way to genetically engineer a vaccine is to make a harmless artificial mutant of the pathogen by altering one or more of its genes. When a harmless mutant is used as a vaccine, it may trigger a stronger immune response than the protein-molecule type of vaccine. Artificial-mutant vaccines may cause fewer side effects than those that have traditionally been made from natural mutants.

Yet another scheme for genetically engineering a vaccine employs a virus related to the one that causes smallpox. Smallpox was once a dreaded human disease, but it was eradicated worldwide in the 1970s by widespread vaccination with a harmless variant (natural mutant) of the smallpox virus. Using this harmless virus, genetic engineers can replace some of the genes encoding proteins that induce immunity to smallpox with genes that induce immunity to other diseases. In fact, the virus can be engineered to carry the genes needed to vaccinate against several diseases simultaneously. In the future, a single inoculation may prevent as many as a dozen diseases.

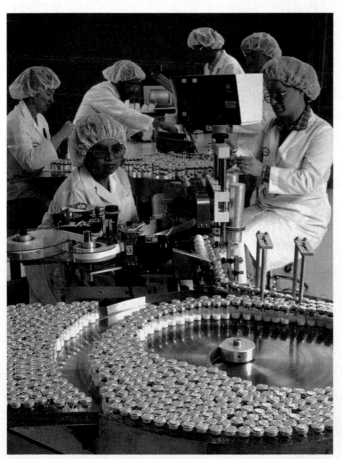

The manufacture of human growth hormone

12.13 Genetic engineering involves some risks

The capability of changing organisms genetically has enormous potential benefits. But as soon as scientists realized the power of this new technology, they also began to worry about potential dangers. Early concerns focused on the possibility that genetic engineering might create hazardous new pathogens. What might happen, for instance, if cancer cell genes were transferred into bacteria or viruses? American scientists developed a set of guidelines that became a formal program administered by the U.S. government.

Two types of safety measures have been put into practice. The first is a set of strict laboratory procedures designed to protect researchers from infection by engineered microbes, and to prevent the microbes from accidentally leaving the laboratory. These procedures are based on the standard microbiological methods that protect scientists and medical personnel who work with naturally occurring pathogens. The second type of safety measure is biological. Strains of microorganisms to be used in recombinant DNA experiments are genetically crippled, to ensure that they can't possibly survive outside the laboratory. In addition, certain obviously dangerous experiments have been banned.

With safeguards in place against the escape of potentially dangerous organisms from the laboratory, the focus of

Testing "ice-minus" bacteria on plants

concern has switched to organisms *designed* to go into the environment. Introducing genetically engineered organisms into the environment could conceivably cause harm in several ways. The new organisms might hurt other species directly, or they might multiply rampantly and displace native species. Another possibility is that engineered organisms might transfer genes to other organisms, genes that might make the recipients harmful.

An early case of a microbe designed for environmental release was the "ice-minus" bacterium, which was engineered to protect crop plants against damage from freezing. You may have heard of the controversy over Frostban®, the trade name for the ice-minus bacterium. Tests of Frostban® outside the laboratory were delayed for months by arguments about possible dangers. Safety tests were eventually completed in test plots (see the photograph), and Frostban® is now commercially available.

In the next module, we examine a number of potential applications of genetically engineered organisms in agriculture. All such projects in the U.S. are now closely scrutinized and evaluated for potential risks by the U.S. Department of Agriculture, the Food and Drug Administration, the Environmental Protection Agency, and the National Institutes of Health.

12.14 Agriculture will make increasing use of genetic engineering

Scientists concerned with feeding the Earth's human population hope to use recombinant DNA technology to improve the productivity of plants and animals important to agriculture.

One of the most exciting potential uses of genetic engineering in agriculture involves nitrogen fixation. Nitrogen fixation is the conversion of atmospheric nitrogen gas, which plants cannot use, into nitrogen compounds in the soil that plants can take up and use in making essential molecules such as amino acids. In nature, nitrogen fixation is performed by certain bacteria that live in the soil or in plant roots. Even so, the level of nitrogen compounds in the soil is often so low that fertilizers must be applied in order for crops to grow. Nitrogen-providing fertilizers are costly and contribute to water pollution. Recombinant DNA technology offers ways to increase the nitrogen fixation carried out by bacteria.

Although the introduction of functional nitrogen-fixing genes directly into plants does not now seem feasible, agricultural scientists are already engineering crop plants to carry genes for other desirable traits, such as delayed ripening and resistance to disease. How are genetically engineered plants produced? It helps that many plants can be regenerated from a single cell grown in culture, as we saw in Module 11.5. This capability is important because it is far easier to manipulate genes in a single cell than in a whole multicellular organism. Crop plants readily grown from single cells include tomatoes, potatoes, citrus fruits, and carrots.

Genetic engineers often use a plasmid vector to introduce new genes into plant cells. The plasmid they use is from the soil bacterium *Agrobacterium tumefaciens*. When

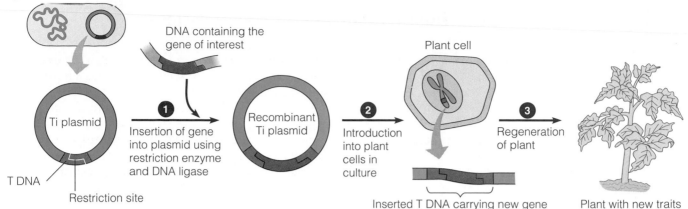

Agrobacterium tumefaciens

Using the Ti plasmid as a vector for genetically engineering plants

susceptible plants are infected by this bacterium, they develop tumors. The tumors are induced by the bacterium's **Ti plasmid** (Ti for tumor-inducing), which can enter the plant's cells. Researchers have developed ways to eliminate the plasmid's tumor-causing properties while keeping its ability to transfer DNA into plant cells.

The figure above shows how a new plant can be created using the Ti plasmid. ① With the help of a restriction enzyme and DNA ligase, a gene of interest (red) is inserted into a segment of the plasmid called T DNA (black). ② Then the recombinant plasmid is put into a plant cell, where the T DNA carrying the new gene integrates into a plant chromosome. ③ Finally, the recombinant cell is cultured and grows into a whole new plant. Several kinds of plants have been engineered to carry and express genes from other species.

One such gene makes tobacco plants resistant to chemical herbicides that might otherwise kill them along with nearby weeds. An organism that contains genes from another species is called a **transgenic organism.**

Progress in using plasmids to create transgenic crop plants has been slow, mainly because the Ti vector does not work with many grain-producing species. A newer technique that uses a "gene gun" to fire pieces of foreign DNA directly into plant cells is being increasingly used for plant engineering.

Animal husbandry is also starting to use recombinant DNA technology. Engineered vaccines are now available, and useful growth hormones can be made cheaply in large amounts. Yet another spinoff of the new technology is the creation of transgenic animals.

Transgenic animals are especially useful in research 12.15

Molecular geneticists can now introduce particular genetic traits into animals. They first remove egg cells from a female and fertilize them in a test tube. Meanwhile, they clone the desired gene using recombinant DNA technology. Then, using a needle of microscopic fineness, they inject the DNA directly into the nuclei of the eggs. Some of the cells integrate the foreign DNA into their genomes and are able to express the foreign gene. The engineered eggs are then surgically implanted into a female that will serve as a foster mother. If an embryo develops successfully, the result is a transgenic animal, containing a gene from a third "parent," which may even be of another species.

In a number of experiments, animals have been engineered to produce high levels of a foreign growth hormone that caused accelerated growth. This photograph shows the results of the first such experiment. The transgenic mouse on the left, carrying a gene for rat growth hormone, grew much faster than its nontransgenic sibling. Transferring genes to farm animals has been problematic, however. In one experiment, hogs were engineered to produce high levels of bovine (cattle) growth hormone. The transgenic hogs gained weight well and were less fatty than normal, but these good effects were offset by low fertility and an increased susceptibility to disease.

Meanwhile, in basic research and in medical research, transgenic laboratory animals are proving extremely valuable. For example, researchers have created a transgenic mouse carrying human genes that make it susceptible to HIV, the AIDS virus. This creature should accelerate the discovery of weapons to fight AIDS.

Recombinant DNA technology has opened the door to a new world, one of novel organisms and gene products created by scientists instead of by the mechanisms of evolution. The likely benefits to us are becoming more and more obvious. And it's reassuring that the potential risks are being carefully weighed. But there are some important questions that can give us pause.

The child in this photograph is growing at a normal rate, thanks to regular injections of human growth hormone made by genetically engineered *E. coli*. Like any new drug, this GH was subjected to exhaustive laboratory tests before it was released for human use. But it is impossible to know what side effects it may produce over the long term. A similar issue arises with the growth hormones administered to the animals that give us beef, pork, and dairy products. It is unlikely that eating these products is dangerous because growth hormones are broken down to amino acids in our digestive system. But the argument remains that we do not really know if there will be any long-term effects.

In a broader sense, how do we really feel about wielding one of nature's singular powers—the ability to make new microorganisms, plants, and even animals? Some

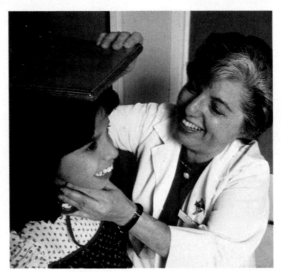

A child treated with growth hormone made by bacteria

might ask: Do we have any right to alter an organism's genes? Or to add our new creations to an already beleaguered environment?

Some situations may almost force us to use genetically engineered organisms once they are available. Bacteria are being genetically engineered to clean up oil spills and detoxify a number of industrial and domestic wastes that threaten our soil, water, and air. These organisms may be the only feasible solutions to some of our most pressing environmental problems.

Another door that recombinant DNA technology is opening, and that seems destined to stay open, is human gene therapy, the alteration of genes in a person's cells in order to remedy a genetic disease. As this book goes to press, researchers have already had some success using genetic engineering to treat the genetic disease cystic fibrosis. As more diseases are treated this way, what new safety and ethical issues will arise? Advances in DNA-sequencing methods, mentioned in Module 12.10, will make it possible for the entire human genome to be sequenced in our lifetime. What use should be made of this information? We will discuss these and related matters further in the next chapter.

Chapter Review

Begin your review by rereading the module headings and scanning the figures before proceeding to the Chapter Summary and questions.

Chapter Summary

Introduction Recombinant DNA technology is a set of techniques for combining genes from different sources and transferring this new DNA into cells where the genes may be expressed. These techniques have enabled biologists to learn much about genes, and they hold promise for the manufacture of useful products, improved agriculture, and the treatment of disease.

12.1–12.5 In nature, bacterial genes can transfer from cell to cell by three processes: transformation, transduction, and conjugation. (In addition, pieces of DNA called transposons can move about within a cell's genome.) Recombinant DNA technology exploits transformation, in which a bacterium takes up DNA from its surroundings. Genetic engineers often use plasmids, small circular DNA molecules separate from the bacterial chromosome, as vectors for the transfer of genes. Restriction enzymes (which cut DNA at specific points) and DNA ligase (which "pastes" DNA fragments together) are used to insert genes into plasmids,

creating recombinant DNA. Recombinant plasmids are taken up by bacteria. As the bacteria reproduce, the plasmids are cloned— that is, replicated—and passed on to the descendants of the original bacteria. Copies of a gene or quantities of a gene's protein product may then be harvested.

12.6–12.10 There are two main approaches to gene cloning. In one approach, all the DNA from a cell is cut into fragments and cloned. Alternatively, reverse transcriptase is used to make DNA genes from eukaryotic mRNA. In either procedure, each piece of DNA is inserted into a vector, either a plasmid or a phage. The vectors replicate in recipient cells, producing a set of cloned DNA pieces called a genomic library. A molecular probe can be used to find a desired gene in the library. The probe is a short segment of radioactively labeled DNA or RNA whose nucleotide sequence is complementary to part of the gene. The probe hydrogen bonds with the gene, thus tagging the desired clone. Automation speeds up the sequencing and synthesis of genes for recombinant DNA work.

12.11–12.12 Recombinant bacteria (usually *E. coli*) are used to mass-produce gene products such as human insulin and human growth hormone. Yeasts are often better than bacteria at producing eukaryotic proteins, such as those used to make vaccines. Mammalian cells manufacture such medical products as Factor VIII (used to treat hemophilia). Harmless viruses can be engineered to carry genes from disease-causing organisms and used to make vaccines.

12.13 Genetic engineering involves some risks, such as a harmful recombinant organism escaping into the environment. In the U.S., this risk is reduced by laboratory containment procedures, use of microorganisms that cannot survive outside the lab, and a government ban on dangerous experiments. Recombinant organisms designed to be released into the environment are strictly regulated.

12.14–12.15 Recombinant DNA technology is being used to produce new genetic varieties of plants and animals. A plasmid from a tumor-inducing bacterium or direct injection is used to introduce new genes into plant cells. The new gene might, for example, equip a transgenic plant to resist chemical herbicides. Bits of DNA can also be injected into animal eggs. A transgenic animal can combine characteristics of different species.

12.16 How do we feel about using genetic engineering to alter plants and animals? Do we have the right to change the genes of species or to release altered organisms into the environment? Should these techniques be used to alter human genes? These are some ethical questions raised by recombinant DNA technology.

Testing Your Knowledge

Multiple Choice

1. Which of the following would be considered a transgenic organism?
   a. a bacterium that has received genes via conjugation
   b. a human given a corrected human blood-clotting gene via gene therapy
   c. a fern grown in cell culture from a single fern root cell
   d. a rat with rabbit hemoglobin genes
   e. a human treated with insulin produced by E. coli bacteria

2. A microbiologist found that some bacteria infected by phages had developed the ability to make a particular amino acid that they could not make before. This new ability was probably a result of
   a. transformation
   b. natural selection
   c. conjugation
   d. mutation
   e. transduction

3. When a typical restriction enzyme cuts a DNA molecule, the cuts are uneven, so that the DNA fragments have single-stranded ends. This is important in recombinant DNA work because
   a. it allows a cell to recognize fragments produced by the enzyme
   b. the single-stranded ends serve as starting points for DNA replication
   c. the fragments will bond to other fragments with complementary single-stranded ends
   d. it enables researchers to use the fragments as molecular probes
   e. only single-stranded DNA segments can code for proteins

4. Researchers have built viruses that combine genes from several disease-causing microbes. These viruses
   a. are used to test laboratory safety precautions
   b. can be used to make vaccines
   c. are used as vectors in gene cloning
   d. are being used for human gene therapy
   e. can be used as probes to test for the presence of disease

5. A biologist isolated a gene from a human cell, attached it to a plasmid, and inserted the plasmid into a bacterium. The bacterium made a new protein, but it was nothing like the protein normally produced in a human cell. Why? (*Explain your answer.*)
   a. The bacterium had undergone transformation.
   b. The gene did not have sticky ends.
   c. The gene contained introns.
   d. The gene did not come from a genomic library.
   e. The biologist should have cloned the gene first.

6. Which of the following most accurately describes the "sex lives" of bacteria?
   a. Bacteria do not carry out any sexual processes.
   b. Bacteria can exchange genes but do not reproduce sexually.
   c. Bacteria can exchange genes but only with the help of virus vectors.
   d. Bacteria can reproduce sexually and asexually.
   e. Bacteria can only reproduce sexually.

Describing, Comparing, and Explaining

1. Describe three different ways in which genes may be transferred between bacteria.

2. Explain how to engineer E. coli to produce human growth hormone, using the following: E. coli containing plasmids, DNA carrying a gene for GH, DNA ligase, a restriction enzyme, equipment for manipulating and growing bacteria, a method for extracting and purifying the hormone.

3. Recombinant DNA techniques are used to custom-build bacteria for two main purposes: to obtain multiple copies of certain genes, and to obtain useful proteins produced by certain genes. Give an example of the use of each of these in medicine. Give an example of the use of each in agriculture.

4. Explain some of the possible dangers of introducing genetically engineered organisms into the environment.

Thinking Critically

1. A biochemist hopes to find a gene in human liver cells that codes for an important blood-clotting protein. She knows that the nucleotide sequence of a small part of the gene is CTGGACTGACA. Briefly explain how to obtain the desired gene.

2. There has been some concern about the safety of "shotgun" recombinant DNA experiments, in which entire genomes of DNA are cut up and cloned in bacteria. What is the possible danger in these kinds of experiments? How might this danger be reduced?

3. After recombinant plasmids have been produced, they are mixed with bacteria to allow the bacteria to take up the plasmids from solution. Normally, only a small fraction of the bacteria present actually take up the plasmids. Given this information, why do you think that R plasmids, which confer antibiotic resistance, are often used for recombinant DNA work?

Science, Technology, and Society

In the early days of genetic engineering, there was much public concern over control of the new technology. Most of the debate concerned safety and safety regulations—whether scientists should develop their own research guidelines or government rules should be imposed. Some local governments actually banned recombinant DNA experiments. Since that time, public interest in the recombinant DNA decision-making process has waned. In the U.S., several government agencies enforce rules covering federally funded research, and private industry voluntarily adheres to government guidelines. Today, transgenic plants and animals are being made, recombinant organisms are being released into the environment, and researchers are experimenting with human gene therapy. What are some important safety and ethical issues raised by the use of recombinant DNA technology? What are some reasons for and against leaving decisions in this area to scientists and business executives? What are some reasons for and against more public involvement? How might these decisions affect you? How do you think these decisions should be made?

The Human Genome 13

The photograph at the left, taken around 1900, shows a scene in the village of Chilmark, on the island of Martha's Vineyard, near the Massachusetts coast (see the inset map). Of the people in the picture, one or more may have been totally deaf from birth. Deafness was so common on the island, however, that it was not regarded as a serious affliction, and the deaf were fully integrated into the community. This was possible because almost everyone on Martha's Vineyard, hearing and deaf alike, used sign language. At home, at work, at social events, and in church, English and sign language were used interchangeably. Children grew up bilingual.

For over 200 years, from about 1700 to 1900, Martha's Vineyard had an unusually high incidence of congenital deafness (deafness existing from birth). Whereas the number of deaf people in the United States as a whole was about one in 5700 during the nineteenth century, on Martha's Vineyard the number was one in 155. In Chilmark, the most isolated town on the island, the number was one in 25.

You have probably guessed that the deafness was hereditary, but the cause was not clear back in the nineteenth century, before Mendel's genetic principles became known. The deafness frequently skipped generations, and it usually affected only *some* of the children from a marriage, not all. Consequently, no one—not even Alexander Graham Bell, who directed a study of deafness on the island—could come up with a coherent explanation.

When Mendel's principles became known in the early years of the twentieth century, the genetic explanation for the Vineyard deafness was immediately obvious: The deafness was caused by a recessive allele, and inbreeding by carriers of that allele had increased the frequency of deafness in the population. In the late 1970s, American sociologist Nora Groce began investigating the origins of the deafness allele. (For the full story, see her book, *Everyone Here Spoke Sign Language,* Harvard University Press, 1985.) Although the last islander with hereditary deafness had died in 1952,

Groce was able to piece together the story from historical records and interviews with elderly residents.

The allele apparently originated with a mutation that occurred hundreds of years ago in the English county of Kent, and more specifically within a small area in Kent called the Weald. In the era before the industrial revolution and the growth of big cities, most people lived in communities of ten or twelve families that remained in the same locale for many generations. Marriage partners almost invariably came from the same town, and cousins carrying the recessive deafness allele commonly intermarried. In this inbred society, the deafness resulting when two copies of the recessive allele were present in the same individual became quite common.

In the seventeenth century, a number of families from the Weald migrated as a group to America, settling in several towns in Massachusetts. Seeking more farmland, several young families from the original group moved to Martha's Vineyard between 1642 and 1710. Once settled on the island, the immigrants were largely isolated from the mainland; travel was difficult, and unnecessary because the island was almost entirely self-sufficient. For the next two centuries, few new immigrants settled on the island, but the residents intermarried, and the population grew to about 3000—nearly all related by direct descent or marriage to the Kentish settlers. In this isolated environment, as in the isolated communities of the Weald, the incidence of hereditary deafness grew. On Martha's Vineyard, it did not decline until a new wave of residents arrived in the twentieth century.

The gene whose mutation caused the Martha's Vineyard deafness is but one of some hundred thousand in the human genome. In this chapter, we explore what is now known about the human genome. The powerful combination of the classical genetic approach (used by Nora Groce) and recombinant DNA technology has put within our grasp the exciting possibility of completely deciphering the "book of life" encoded in our DNA. We start by focusing on the DNA itself.

13.1 Three billion nucleotide pairs are packed into the human genome

Within the haploid set of 23 chromosomes making up the human genome, there are approximately 3 billion nucleotide pairs of DNA. The diploid set of chromosomes found in each somatic cell has twice this number. Let's try to get a sense of this quantity of DNA compared to the size of a typical cell. The cell in the light micrograph below—a human white blood cell—is enlarged about 1000 times. At this scale, the total length of all the DNA in the cell's nucleus would be 3 km (almost 2 miles)—about the distance from the Lincoln Memorial to the Capitol in Washington, DC! At actual size, a human cell's DNA totals about 3 m in length, about 300,000 times longer than the cell's diameter.

As we saw in Module 11.7, an immense length of DNA can be packed into a cell nucleus because DNA is extremely thin and can be coiled very tightly. Even at our enlarged scale, the thickness of the DNA is still microscopic, about the thickness of a bacterium (2 μm).

The amount of DNA in a human cell is about 1000 times greater than the DNA in *E. coli*. Does this mean humans have 1000 times as many genes as the 2000 in *E. coli*? The answer is probably no; the human genome is thought to carry between 50,000 and 100,000 genes, which code for various proteins (as well as for tRNA and rRNA). In addition to these genes, humans have a huge amount of noncoding DNA, probably more than 80% of the total DNA. As we discussed in Module 11.9, noncoding DNA includes introns (whose total length may be ten times greater than the coding portion of the gene) and regulatory sequences, such as promoters. Also, there are noncoding stretches of DNA located between genes, including some short sequences of nucleotides that are repeated many hundreds of times. A major portion of this highly repetitive DNA is in the centromere region of the chromosomes. Here it is thought to play a role in chromosome structure, and it may help keep the DNA properly organized during DNA replication and mitosis.

The sheer quantity of DNA in our cells is truly astounding, but consider what is contained in the approximately 3-billion-nucleotide human genome. Every nucleus in every one of our somatic cells contains a full inventory of instructions for building and maintaining a complete human being.

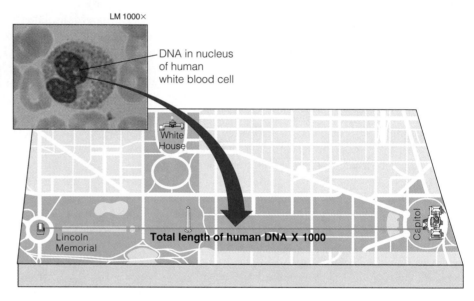

LM 1000×

DNA in nucleus of human white blood cell

White House

Lincoln Memorial

Total length of human DNA X 1000

Capitol

13.2 A karyotype is a photographic inventory of an individual's chromosomes

If we look through a microscope at a human cell in metaphase of mitosis, we see that the chromosomes vary in size and shape. By photographing them and arranging the images in an orderly display, we get an overview of a person's genome called a **karyotype.**

To prepare a karyotype, medical scientists often use lymphocytes, a type of white blood cell. A blood sample is treated with a chemical that stimulates mitosis of the white blood cells, which are grown in culture for several days. The cells are then treated with another chemical to arrest the cell cycle at metaphase, when the chromosomes, each consisting of two joined sister chromatids, are most highly condensed. The figure on the facing page outlines the steps in one method for the preparation of a karyotype from a blood sample.

The photograph in step 7 shows the karyotype of a normal human male. The 46 chromosomes of a single, diploid cell are arranged in 23 homologous pairs: autosomes numbered from 1 to 22, and one pair of sex chromosomes. Because this is a male karyotype, it has one *X* and one *Y* chromosome. The chromosomes have been stained to reveal band patterns, which are helpful in differentiating the chromosomes. Karyotyping is used to screen for abnormal numbers of chromosomes and for defective ones.

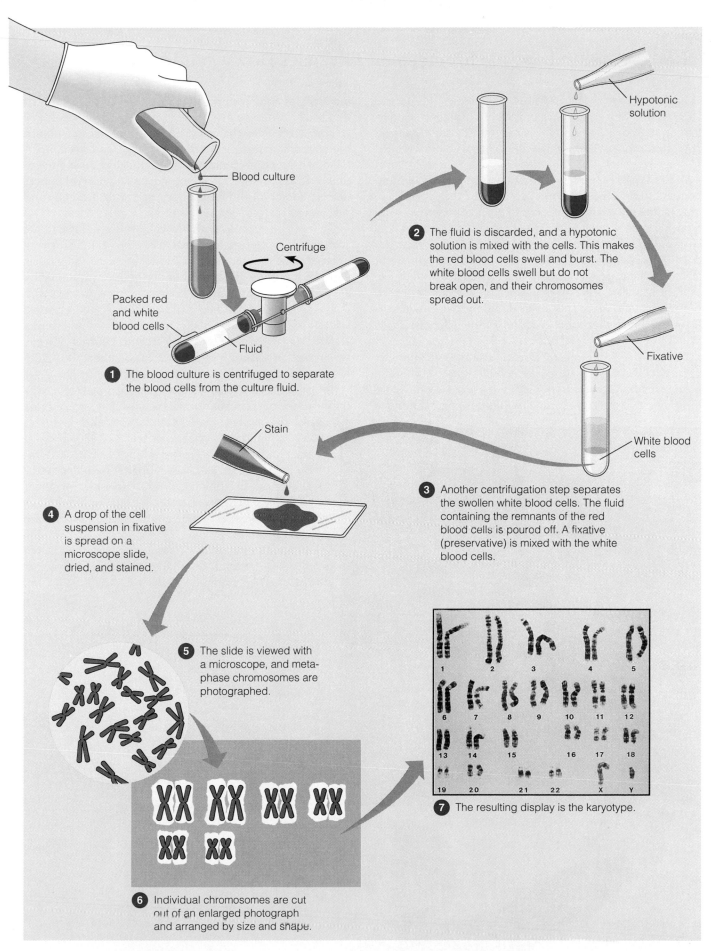

Blood culture

Centrifuge

Packed red and white blood cells

Fluid

1 The blood culture is centrifuged to separate the blood cells from the culture fluid.

Hypotonic solution

2 The fluid is discarded, and a hypotonic solution is mixed with the cells. This makes the red blood cells swell and burst. The white blood cells swell but do not break open, and their chromosomes spread out.

Fixative

White blood cells

3 Another centrifugation step separates the swollen white blood cells. The fluid containing the remnants of the red blood cells is poured off. A fixative (preservative) is mixed with the white blood cells.

Stain

4 A drop of the cell suspension in fixative is spread on a microscope slide, dried, and stained.

5 The slide is viewed with a microscope, and metaphase chromosomes are photographed.

7 The resulting display is the karyotype.

6 Individual chromosomes are cut out of an enlarged photograph and arranged by size and shape.

Preparation of a karyotype from a blood sample

An extra copy of chromosome 21 causes Down syndrome

The karyotype on the previous page shows the normal complement of 23 pairs of homologous chromosomes. The karyotype shown in Figure A is different; notice that there are three number-21 chromosomes. This condition is called trisomy 21.

In most cases, a human offspring with an abnormal number of chromosomes is spontaneously aborted (miscarried) long before birth. However, certain types of abnormal chromosome number, including the type with an extra copy of chromosome 21, appear to upset the genetic balance less drastically, so that individuals carrying them are born occasionally. These people usually have a characteristic set of symptoms, called a syndrome. A person with an extra copy of chromosome 21, for instance, is said to have **Down syndrome** (named after John Langdon Down, who characterized it in 1866).

Trisomy 21 is the most common chromosome-number abnormality. Affecting about one out of every 700 children born, it makes Down syndrome the most common serious birth defect in the United States. Chromosome 21 is one of our smallest chromosomes, but an extra copy affects the individual's phenotype in a number of ways. Down syndrome (Figure B) includes characteristic facial features, including a broad, rounded face, flattened nose bridge, and small, irregular teeth, as well as short stature, heart defects, and susceptibility to respiratory infection, leukemia, and Alzheimer's disease. (Some of the genes associated with leukemia and Alzheimer's disease are also on chromosome 21.)

On average, people with Down syndrome have a life span much shorter than normal. They also exhibit varying degrees of mental retardation. However, some individuals with the syndrome live to middle age or beyond, and many are socially adept and able to hold jobs. A few women with Down syndrome have had children, though most people with the syndrome are sexually underdeveloped and sterile. Half the eggs produced by a woman with Down syndrome will have the extra chromosome 21, so there is a 50% chance that she will transmit the syndrome to her child.

As indicated in Figure C, the incidence of Down syndrome in the offspring of normal parents increases with the age of the mother. Down syndrome strikes less than 0.05% of children (fewer than one in 2000) born to women under age 30. The risk climbs to 1% for mothers in their late 30s and is even higher for older mothers. Because of this relatively high risk, pregnant women over 35 are candidates for fetal testing for trisomy 21 and other major chromosomal defects (see Module 13.11).

What causes trisomy 21 and, consequently, Down syndrome? In search of an answer, we need to revisit the process of meiosis.

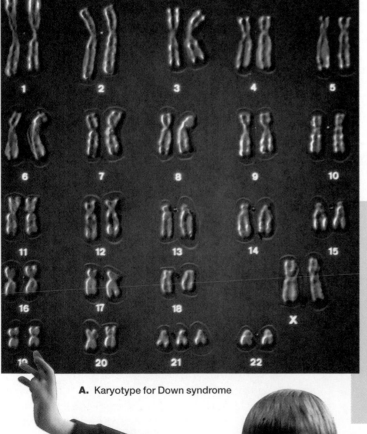

A. Karyotype for Down syndrome

B. A child with Down syndrome

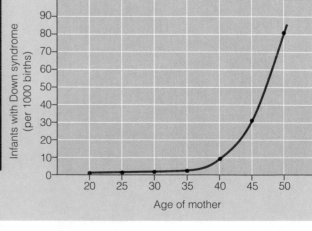

C. Maternal age and Down syndrome

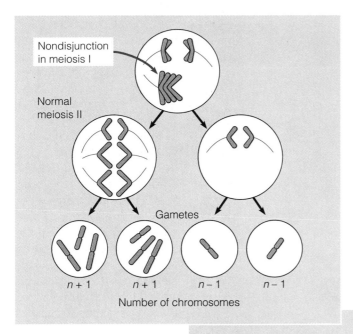

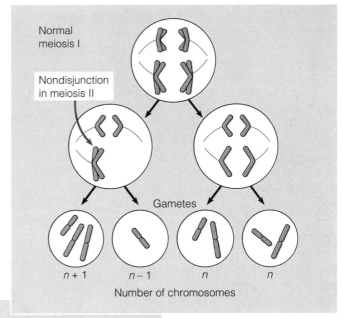

A. Nondisjunction in meiosis I

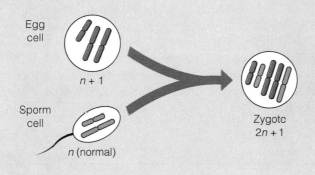

C. Fertilization after nondisjunction in the mother

B. Nondisjunction in meiosis II

Meiosis occurs repeatedly in our lifetime, as our testes or ovaries produce gametes. Almost always, the meiotic spindle distributes chromosomes to daughter cells without error. But occasionally there is an accident, called a **nondisjunction,** in which the members of a chromosome pair fail to separate. Figures A and B illustrate two ways that nondisjunction can occur. For simplicity, we use a hypothetical organism whose diploid chromosome number is 4. In both figures, the cell at the top is diploid ($2n$), with two pairs of homologous chromosomes undergoing meiosis I.

Sometimes, as in Figure A, a pair of homologous chromosomes does not separate during meiosis I. In this case, even though the rest of meiosis occurs normally, all the resulting gametes end up with abnormal numbers of chromosomes. Two of the gametes have three chromosomes, two of which are the same; the other two gametes have only one chromosome each. In Figure B, meiosis I is normal, but a pair of sister chromatids fails to move apart in one of the cells during meiosis II. In this case, two gametes have the normal complement of two chromosomes each, but the other two gametes are abnormal.

Figure C shows what happens when an abnormal gamete produced by nondisjunction unites with a normal gamete in fertilization. Here, an egg cell with two copies of one of its chromosomes (a total of $n + 1$ chromosomes) is fertilized by a nor-

mal sperm cell (n). The resulting zygote has an extra chromosome (a total of $2n + 1$ chromosomes). Mitosis will then transmit the abnormality to all embryonic cells. If this were a real organism and it survived, it would have an abnormal karyotype and probably a syndrome of disorders caused by the abnormal number of genes.

Nondisjunction can lead to an abnormal chromosome number in either sex of any sexually reproducing, diploid organism, including humans. If, for example, nondisjunction affects human chromosome 21 during meiosis I, half of the resulting gametes will carry an extra copy of chromosome 21. Then, if one of these gametes unites with a normal gamete, trisomy 21 will result.

Nondisjunction explains how abnormal chromosome numbers come about, but what causes nondisjunction in the first place? We do not yet know the answer, nor do we fully understand why offspring with trisomy 21 are more likely to be born as a woman ages. We do know, however, that meiosis begins in a woman's ovaries before she is born but is not completed until years later, at the time of an ovulation. Because only one egg cell usually matures each month, a cell might remain arrested in the mid-meiosis state for decades. Perhaps damage to the cell during this time leads to meiotic errors. It seems that the longer the time lag, the greater the chance that there will be errors such as nondisjunction when meiosis is completed.

Nondisjunction can alter the number of sex chromosomes

Nondisjunction in meiosis does not affect just autosomes, such as chromosome 21. It can also lead to abnormal numbers of sex chromosomes, X and Y. Unusual numbers of sex chromosomes seem to upset the genetic balance less than unusual numbers of autosomes, however. This may be because the Y chromosome carries fewer genes than other chromosomes, and because extra copies of the X chromosome become inactivated as Barr bodies in the cells (see Module 11.7). The table here lists the most common sex-chromosome abnormalities.

Abnormalities of Sex-Chromosome Number in Humans			
Genotype	Phenotype	Origin of Nondisjunction	Frequency in Population
XO	Turner syndrome (female)	Meiosis in egg or sperm formation	$\frac{1}{5000}$
XXX	Metafemale	Meiosis in egg formation	$\frac{1}{1000}$
XXY	Klinefelter syndrome (male)	Meiosis in egg or sperm formation	$\frac{1}{2000}$
XYY	Normal male	Meiosis in sperm formation	$\frac{1}{2000}$

An extra X chromosome in a male, producing an XXY genotype, occurs approximately once in every 2000 live births (once in every 1000 male births). Men with this disorder, called Klinefelter syndrome, have male sex organs, but the testes are abnormally small and the individual is sterile. The syndrome often includes breast enlargement and other feminine body contours. The person is usually of normal intelligence. Klinefelter syndrome is also found in individuals with more than one additional sex chromosome, such as XXYY, XXXY, or XXXXY. Such individuals, whose genotypes result from multiple nondisjunctions of sex chromosomes, are more likely to be mentally retarded than XY or XXY individuals.

Human males with a single extra Y chromosome (XYY) do not have any well-defined syndrome, although they tend to be taller than average. Females with an extra X chromosome (XXX) are called metafemales; they have limited fertility but are otherwise apparently normal. Females who are lacking an X chromosome (genotype XO) are sterile, because their sex organs do not mature at adolescence. These women, said to have Turner syndrome, also fail to develop secondary sex characteristics, and they are short in stature.

The sex-chromosome abnormalities described here illustrate the crucial role of the Y chromosome in determining a person's sex. In general, a single Y chromosome is enough to produce "maleness," even in combination with several X chromosomes. The absence of a Y chromosome results in "femaleness."

Alterations of chromosome structure may cause serious disorders

Even if all chromosomes are present in normal numbers, abnormalities in chromosome structure may cause disorders. Breakage of a chromosome can lead to a variety of rearrangements affecting the genes of that chromosome. Figure A shows three types of rearrangement. (The small red arrows indicate chromosome breaks.) If a fragment of a chromosome is lost, the remaining chromosome will then have a **deletion**. If a fragment from one chromosome joins to a homologous chromosome, it will produce a **duplication** there. If a fragment reattaches to the original chromosome but in the reverse direction, an **inversion** results.

Inversions are less likely than deletions or duplications to produce harmful effects, because in inversions all genes are still present in their normal number. Deletions tend to have the most serious effects. One example is a specific deletion in chromosome 5 that causes the cri du chat ("cat-cry") syndrome. A child born with this syndrome is mentally retarded and has a small head and a cry like the mewing of a cat. Death usually occurs in infancy or early childhood.

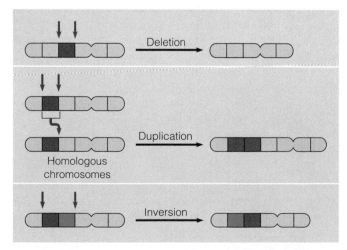

A. Alterations of chromosome structure involving one chromosome or a homologous pair

Another type of chromosomal change is chromosomal **translocation**, the attachment of a chromosomal fragment to a nonhomologous chromosome. Figure B shows a translocation that is reciprocal; that is, two nonhomologous chromosomes exchange segments. Like inversions, translocations may or may not be harmful. Some people with Down syndrome have only part of a third chromosome 21; as the result of a translocation, it is attached to another (nonhomologous) chromosome.

Whereas chromosomal changes present in sperm or egg can cause congenital disorders, such changes in an individual's somatic cells may contribute to the development of cancer. For example, as we discussed in Module 11.16, duplication of a proto-oncogene whose protein product promotes cell division can lead to the uncontrolled cell division characteristic of cancer.

Translocations of somatic cell chromosomes are also associated with cancer. One example is seen in chronic myelogenous leukemia (CML), a cancer affecting the cells that give rise to white blood cells. In the cancerous cells of CML

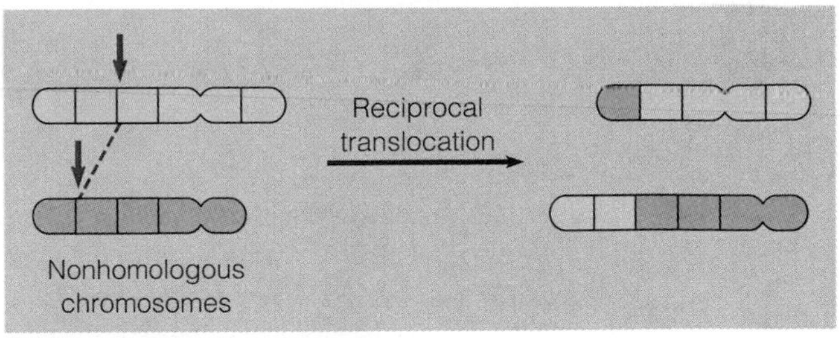

B. Chromosomal translocation between nonhomologous chromosomes

patients, a part of chromosome 22 has switched places with a small fragment from chromosome 9. This reciprocal translocation activates an oncogene, leading to leukemia.

Because the chromosomal changes in cancer are usually confined to somatic cells, cancer is not usually inherited. The next module returns to inherited conditions, showing how we study genetic diseases in families.

Pedigrees track genetic traits through a family history   13.7

Unlike fruit-fly researchers, geneticists who study humans cannot control the mating of their subjects. Instead, they must analyze the results of matings that have already occurred. Suppose you wanted to study the inheritance of the deafness trait on Martha's Vineyard. First you would collect as much information as possible about a family's history for this trait. Then you would assemble this information into a family tree representing the interrelationships among parents and children across the generations—the family **pedigree**. Finally, you would use Mendel's principles to analyze the pedigree.

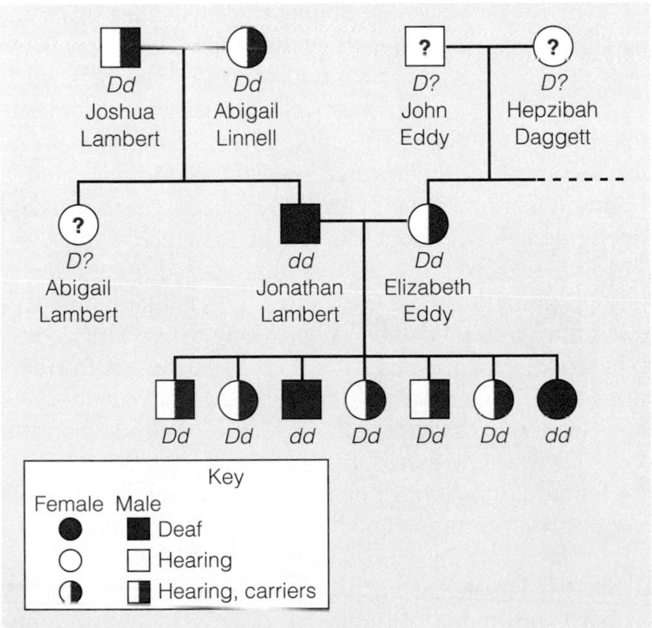

The figure here shows part of an actual pedigree from one Vineyard family described by Nora Groce in *Everyone Here Spoke Sign Language.* Squares represent males, and circles represent females; deafness is indicated by a fully colored symbol. The earliest generation studied is at the top of the pedigree. Notice that deafness does not appear in this generation and that it shows up in only two of the seven children in the generation at the bottom. By applying Mendel's principles, we can deduce that the deafness allele is recessive.

Mendel's principles also allow us to deduce the genotypes that are shown for each of the people in the pedigree. The letter *D* stands for the hearing allele, and *d* symbolizes the recessive allele for deafness. Half-colored squares and circles indicate heterozygotes. The first deaf individual to appear in this pedigree is Jonathan Lambert. Because only people who are homozygous for the recessive allele are deaf, his genotype must have been *dd*. Therefore, both his parents must have carried a *d* allele along with the *D* allele that gave them normal hearing. Since two of Jonathan's children were deaf (*dd*), his wife, who had normal hearing, must also have carried a *d* allele. Likewise, all of the couple's normal children must have been heterozygous (*Dd*).

The only individuals whose genotypes are uncertain are Elizabeth Eddy's parents and Jonathan's sister Abigail. These three people had normal hearing, so they must have carried at least one *D* allele. And at least one of Elizabeth's parents must have had a *d* allele. But more than that we cannot say without additional information.

A pedigree not only helps us understand the past; it can help us predict the future, as we see next.

Many human disorders are inherited as Mendelian traits

Over 1000 human genetic disorders are known to be inherited as Mendelian traits; ten examples are listed in the table on the next page. By Mendelian trait, we mean one that is controlled by a single gene locus and shows a simple inheritance pattern like the traits Mendel studied in pea plants.

RECESSIVE DISORDERS Most of the genetic disorders known in humans are recessive. These disorders range in severity from relatively harmless conditions, such as albinism (lack of pigmentation), to deadly diseases. The hereditary deafness of Martha's Vineyard falls somewhere in between.

The vast majority of people afflicted with recessive disorders are born to normal parents who are both heterozygotes, that is, who are carriers of a recessive allele for a disorder but are phenotypically normal. For example, the hearing parents of Jonathan Lambert, in the last module, were both carriers (see Figure 13.7).

Using Mendel's principles, we can predict the fraction of affected offspring likely to result from a marriage between two carriers. Suppose one of Jonathan Lambert's hearing sons (*Dd*) married a hearing cousin whose pedigree indicated that her genotype was also *Dd*. What is the probability that they would have deaf children? As the Punnett square in Figure A shows, each child of two carriers has a $\frac{1}{4}$ chance of inheriting two recessive alleles. Thus, we can say that about one-fourth of the children of this marriage are likely to be deaf. We can also say that a hearing ("normal") child from such a marriage has a $\frac{2}{3}$ chance of being a carrier (that is, 2 out of 3 of the offspring with hearing phenotype are likely to be carriers). We can apply this same method of pedigree analysis and prediction to any genetic trait controlled by a single locus.

The most common lethal genetic disease in the United States is **cystic fibrosis.** Though the disease affects only one in 17,000 African-Americans and one in 90,000 Asian-Americans, it occurs in one out of every 1800 Caucasian births. The cystic fibrosis allele is recessive and is carried by one out of every 25 Caucasians. A person with two copies of this allele has cystic fibrosis, which is characterized by excessive secretion of very thick mucus from the lungs, pancreas, and other organs. This mucus can interfere with breathing, digestion, and liver function, and make the person vulnerable to pneumonia and other infections. Untreated, most children with cystic fibrosis die by the time they are 5 years old. However, a special diet, antibiotics to prevent infection, frequent pounding of the chest and back to clear the lungs, and other treatments can prolong life to adulthood.

Like cystic fibrosis, most genetic disorders are not evenly distributed across all racial and cultural groups. Such uneven distribution is the result of prolonged geographic isolation of various populations. The isolated lives of the Martha's Vineyard inhabitants and their Kentish ancestors led to frequent marriage between close relatives; thus, the frequency of deafness remained high, and the deafness allele was not transmitted to outsiders.

With the increased mobility in most societies today, it is relatively unlikely that two carriers of the same rare harmful allele will meet and mate. However, the probability increases greatly if close relatives, such as first cousins, marry and have children. People with recent common ancestors are more likely to carry the same recessive alleles than are unrelated people. Therefore, a mating of close relatives is more likely to produce offspring homozygous for a harmful recessive trait.

Most societies have taboos and laws forbidding marriages between close relatives. These rules may have arisen out of the observation that stillbirths and birth defects are more common when parents are closely related. Such effects can also be observed in many types of inbred domesticated and zoo animals. For example, dogs that have been inbred for appearance may have serious genetic disorders, such as weak hip joints and undesirable behaviors. The detrimental effects of inbreeding are also seen in some endangered species (see Module 14.16).

Although inbreeding is clearly dangerous, there is debate among geneticists about the extent to which human inbreeding increases the risk of inherited diseases. Many harmful mutations have such severe effects that a homozygous embryo spontaneously aborts long before birth—perhaps so early that the miscarriage goes undetected. Furthermore, as some geneticists argue, marriage between relatives is just as likely to concentrate favorable alleles as harmful ones. There are, in fact, some human populations, such as the Tamils of India, in which marriage between first cousins has long been common and has *not* produced ill effects.

DOMINANT DISORDERS Although most harmful alleles are recessive, a number of human disorders are due to domi-

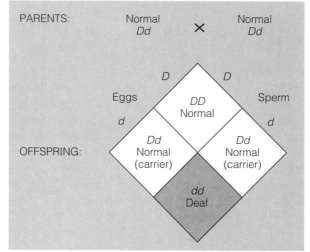

A. Offspring produced by parents who are both carriers for a recessive disorder

PARENTS: Normal *Dd*   ✕   Normal *Dd*

D   *D*

Eggs   *DD* Normal   Sperm

d   *d*

OFFSPRING: *Dd* Normal (carrier)   *Dd* Normal (carrier)

dd Deaf

nant alleles. Several are nonlethal handicaps, such as extra fingers and toes, or fingers and toes that are webbed.

One serious disorder caused by a dominant allele is **achondroplasia,** a form of dwarfism. The head and torso of the body develop normally, but the arms and legs are short. (Figure B shows David Rappaport, an actor.) Approximately one out of every 25,000 people has achondroplasia. Only heterozygotes, individuals with a single copy of the defective allele, have this disorder. The homozygous dominant genotype causes death of the embryo. Therefore, all those who do *not* have achondroplasia, more than 99.99% of the population, are homozygous for the normal, recessive allele. This example makes it clear that the term "dominant" does not imply that a dominant allele is somehow better than the corresponding recessive allele, or that a dominant allele will be more plentiful in a population.

Dominant alleles that are lethal are, in fact, much less common than lethal recessives. One reason for this difference is that the dominant lethal allele cannot be carried in a masked form by heterozygotes. Many lethal dominant alleles result from mutations in a sperm or egg that subsequently kill the embryo. And if the afflicted individual is born but does not

B. Achondroplasia, a dominant trait

survive long enough to reproduce, he or she will not pass on the lethal allele. This is in contrast to lethal recessive mutations, which are perpetuated from generation to generation by the reproduction of heterozygous carriers.

A lethal dominant allele can escape elimination, however, if it does not cause death until a relatively advanced age. By the time the symptoms become evident, the afflicted individual may have already transmitted the lethal gene to his or her children. **Huntington's disease,** which causes degeneration of the nervous system, is one example. Usually there are no obvious symptoms until the affected individual is about 40 years old. As the disease progresses, patients experience uncontrollable movements in all parts of the body. Loss of brain cells leads to loss of memory and judgment, and contributes to depression. Loss of motor skills eventually prevents swallowing and speaking. Death usually ensues 10 to 20 years after the onset of symptoms. (We discuss Huntington's disease in more detail in Module 13.13.)

The disorders discussed in this module all arise from genes located on autosomes (chromosomes other than X and Y). Next we discuss sex-linked disorders in humans.

Some Autosomal Disorders in Humans

Disorder	Major Symptoms	Incidence	Comments
Recessive disorders			
Albinism	Lack of pigment in skin, hair, and eyes	$\frac{1}{22,000}$	Very easily sunburned
Cystic fibrosis	Excess mucus in lungs, digestive tract, liver; increased susceptibility to infections; death in infancy unless treated	$\frac{1}{1800}$ Caucasians	See Modules 13.8, 13.12, and 13.17
Galactosemia	Accumulation of galactose in tissues; mental retardation; eye and liver damage	$\frac{1}{100,000}$	Treated by eliminating galactose from diet
Phenylketonuria (PKU)	Accumulation of phenylalanine in blood; lack of normal skin pigment; mental retardation	$\frac{1}{10,000}$ in U.S. and Europe	See Module 13.10
Sickle-cell anemia (homozygous form)	Sickled red blood cells; damage to many tissues	$\frac{1}{500}$ African-Americans	Alleles are codominant; see Module 9.12
Tay-Sachs disease	Lipid accumulation in brain cells; mental deficiency; blindness; death in childhood	$\frac{1}{3500}$ Jews from central Europe	See Module 4.12
Dominant disorders			
Achondroplasia	Dwarfism	$\frac{1}{25,000}$	See Module 13.8
Alzheimer's disease (one type)	Mental deterioration; usually strikes late in life	Not known	
Huntington's disease	Mental deterioration and uncontrollable movements; strikes in middle age	$\frac{1}{25,000}$	See Module 13.13
Hypercholesterolemia	Excess cholesterol in blood; heart disease	$\frac{1}{500}$ is heterozygous	Incomplete dominance; see Module 9.10

13.9 Sex-linked disorders affect mostly males

Sex-linked disorders—those resulting from alleles of genes located on sex chromosomes—exhibit distinctive patterns of inheritance. We use the terms "sex-linked" and "X-linked" interchangeably here, because most of the known sex-linked disorders involve genes on the X chromosome. As we discussed in Module 9.19, fathers pass X-linked genes to all their daughters but to none of their sons, because males receive their one X chromosome from their mothers. In contrast, mothers can pass X-linked genes to both sons and daughters.

If a sex-linked disorder is due to a recessive allele, as most are, any male receiving the harmful X-linked allele from his mother will have the disorder; the allele on his one X chromosome determines his phenotype. However, a female will exhibit the disorder only if she is a homozygote. Therefore, more males than females have disorders that are inherited as sex-linked recessives.

Red-green color blindness is a common sex-linked disorder characterized by a malfunction of light-sensitive cells in the eyes. It is actually a complex of disorders, involving several X-linked genes. A person with normal color vision can see more than 150 colors. In contrast, someone with red-green color blindness can see fewer than 25. For some affected people, red hues appear gray; others see gray instead of green; still others are green-weak or red-weak, tending to confuse shades of these colors. Mostly males are affected, but heterozygous females have some defects. (If you have red-green color blindness, you probably cannot see the numeral 7 in Figure A.)

Hemophilia is a sex-linked recessive trait with a long, well-documented history. Hemophiliacs bleed excessively when injured because they have inherited an abnormal allele for a factor involved in blood clotting. The most seriously affected individuals may bleed to death after relatively minor bruises or cuts.

A high incidence of hemophilia has plagued the royal families of Europe. The first royal hemophiliac seems to have been a son of Queen Victoria (1819–1901) of England. It is likely that the hemophilia allele arose through a mutation in one of the gametes of Victoria's mother or father, making Victoria a carrier of the deadly allele. Hemophilia was eventually introduced into the royal families of Prussia, Russia, and Spain through the marriages of two of Victoria's daughters who were carriers. Thus, the age-old practice of strengthening international alliances by marriage effectively spread hemophilia through the royal families of several nations. The photograph in Figure B shows Queen Victoria's granddaughter Alexandra, her husband Nicholas, who was the last czar of Russia, and their son Alexis.

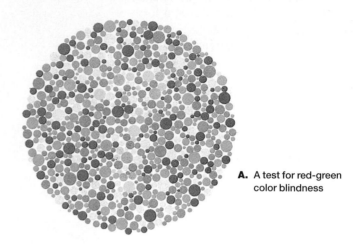

A. A test for red-green color blindness

As you can see in the pedigree, Alexandra, like her mother and grandmother, was a carrier, and Alexis had the disease.

Another sex-linked recessive disorder is muscular dystrophy, a condition characterized by progressive weakening and loss of muscle tissue. **Duchenne muscular dystrophy** is the most devastating type of the disease. Almost all cases are males, and the first symptoms appear in early childhood, when the child begins to have difficulty standing up and rises to a standing position in a characteristic way (Figure C). He is inevitably wheelchair-bound by age 12. Eventually, he becomes severely wasted, and normal breathing becomes difficult. Death usually occurs by age 20.

For such a severe disease, Duchenne muscular dystrophy is relatively common. In the general U.S. population, about one in 3500 male babies is affected, and the disease is even more common in some inbred populations. In one Amish community in Indiana, for instance, one out of every 100 males is born with the disease.

B. Hemophilia in the royal family of Russia

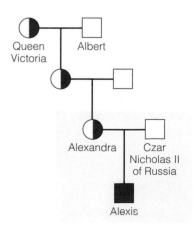

With the help of recombinant DNA techniques, the gene whose mutation causes Duchenne muscular dystrophy has been mapped at a particular point on the X chromosome. Consisting of about 3,000,000 nucleotides, it is the longest human gene found to date. About 99% of the gene consists of introns. The other 1% codes for a protein called dystrophin, which is present in normal muscle but missing in Duchenne patients.

C. Duchenne muscular dystrophy (characteristic rising position)

Genetic counselors can help prospective parents

Couples who think they may be at risk for having a child with a serious genetic disorder such as muscular dystrophy can receive guidance from **genetic counselors,** who are on the staffs of many major hospitals. Genetic counselors gather information about the family histories of the couple, analyze this information, and then educate the couple about their risk of having an affected child. They also provide information about possible medical tests for defining the risk more precisely, the disorder itself, and resources available for dealing with an affected child. Thus, genetic counselors give couples both a framework for making reproductive decisions and help coping with an affected child, if one is born.

Let's consider a hypothetical couple, John and Carol, who are planning to have their first child. They are seeking genetic counseling because they both have family histories of the same serious disease, **phenylketonuria (PKU).** Children with phenylketonuria cannot properly break down the amino acid phenylalanine, and this chemical and one of its by-products accumulate to toxic levels in the blood,

A genetic counselor and physician with clients

causing mental retardation. Although neither John, Carol, nor any of their parents has PKU, John and Carol each have a brother with the disease. They want to determine their risk of having a child with PKU.

To analyze their situation, the genetic counselor uses the same principles we applied to the issue of deafness on Martha's Vineyard. The counselor knows that PKU is recessively inherited, so she uses p to represent the PKU allele and P for the normal, dominant allele. Since both John and Carol have a brother with PKU, both parents of John and both parents of Carol must be carriers (Pp). What is the probability that John and Carol are also carriers? (A test for PKU carriers is now available, but let's see how a counselor would proceed without it.) We know that John and Carol are not pp, so they must be either PP or Pp. Because John's parents are both carriers, there is a $\frac{2}{3}$ chance that John is a carrier; (the Punnett square in Figure 13.8A, predicting the results of a mating between two carriers of the recessive deafness allele, shows how the probability of $\frac{2}{3}$ is derived). The same is true for Carol. We can use the rule of multiplication (see Module 9.7) to find out that the probability that both John and Carol are carriers is $\frac{2}{3} \times \frac{2}{3} = \frac{4}{9}$. The counselor would calculate the overall probability of their child's having PKU by multiplying $\frac{4}{9}$ (the probability that they're both carriers) by $\frac{1}{4}$ (the probability of two carriers having a child with the disease). The result, she would tell the couple, is $\frac{4}{36}$, or $\frac{1}{9}$.

On the basis of this information, John and Carol may decide to go ahead and have a baby; after all, there is an $\frac{8}{9}$ chance that the baby will be normal. The genetic counselor has told them that PKU, like a number of other genetic disorders, can be detected in a newborn infant by simple biochemical tests that are now routinely performed in most hospitals in the United States. PKU is, in fact, the genetic disorder most widely tested for; the test is mandated by law in 40 states and available in all others. If PKU is detected in the newborn, retardation can be prevented with a special diet that is low in phenylalanine. Thus, screening newborns for PKU and other treatable disorders is vitally important. Unfortunately, very few genetic disorders are treatable at the present time, and prospective parents must often make difficult choices, as we discuss further in the next module.

Fetal testing can spot many disorders early in pregnancy

Many genetic disorders, including PKU, can be detected before birth. Tests done in conjunction with a technique called **amniocentesis** (Figure A) can determine, between the fourteenth and sixteenth weeks of pregnancy, whether the developing fetus has the condition. By this time, the fetus is about 15 cm (6 inches) long and is surrounded by a pool of liquid called amniotic fluid. To perform amniocentesis, a physician first determines the position of the fetus and then carefully inserts a needle through the mother's abdomen into her uterus, avoiding the fetus. The physician extracts a small sample—about 10 milliliters (2 teaspoonsful)—of the amniotic fluid. The sample is then centrifuged to separate the fluid from cells suspended in it. The cells come from the fetus, sloughed off from the skin and mouth cavity.

Once the sample is taken, some genetic disorders, such as PKU, can be detected by immediate biochemical tests be-cause of the presence of certain telltale chemicals in the amniotic fluid. Tests for other disorders are performed on the fetal cells. Before testing, these cells are cultured in the laboratory for several weeks. By then, enough dividing cells can be harvested so that karyotyping can be done and chromosomal abnormalities such as Down syndrome can be detected. Biochemical tests can also be performed on the cultured cells, revealing conditions such as Tay-Sachs disease (see the table in Module 13.8).

In a newer technique called **chorionic villi sampling** (Figure B), the physician inserts a narrow, flexible tube through the mother's vagina and cervix into the uterus, and suctions off a small amount of fetal tissue (chorionic villi) from the placenta, the organ that transmits nourishment and wastes between the fetus and the mother. Because the cells of chorionic villi are proliferating rapidly, enough cells

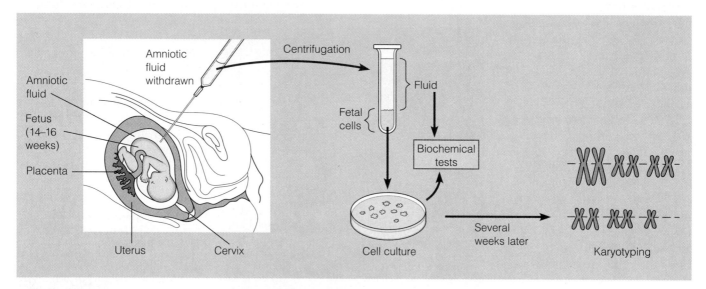

A. Amniocentesis

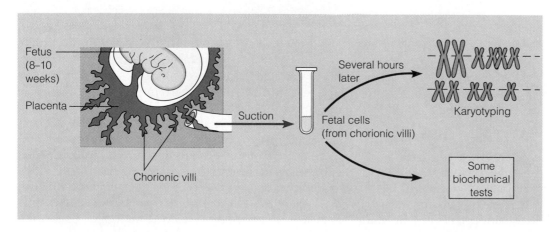

B. Chorionic villi sampling

are undergoing mitosis to allow karyotyping to be carried out immediately. Results of the karyotyping, along with some biochemical tests, are available within a few hours. The speed of this method is an advantage over amniocentesis. Another advantage of chorionic villi sampling is that it can be performed early, between the eighth and tenth weeks of pregnancy. It is appropriate for some biochemical tests, but not for those requiring amniotic fluid. Chorionic villi sampling is less widely available than amniocentesis.

Other techniques enable a physician to examine a fetus directly for anatomical deformities. One such technique is **ultrasound imaging**, which uses sound waves to produce an image of the fetus. Figure C shows an ultrasound scanner, which produces high-frequency sounds, beyond the range of hearing. When the sound waves bounce off the fetus, the echoes produce an image on the monitor. The color-enhanced image in Figure D shows a fetus at about eighteen weeks. Ultrasound imaging is noninvasive (no foreign objects are inserted into the mother's body) and has no known risk. This procedure is employed during amniocentesis and chorionic villi sampling to determine the position of the fetus and of the needle or tube. With another technique, **fetoscopy,** a needle-thin tube containing a viewing scope is inserted into the uterus, giving the physician a direct view of the fetus.

Fetoscopy, amniocentesis, and chorionic villi sampling all involve some risk of complications, such as maternal bleeding, miscarriage, or premature birth. The complication rate for fetoscopy is the highest (up to about 10%). Complication rates for chorionic villi sampling and amniocentesis are about 2% and 1%, respectively. Because of the risks, all these techniques are usually reserved for situations in which the possibility of a genetic disorder or other type of birth defect is significant. For example, amniocentesis or chorionic villi sampling to test for Down syndrome is usually reserved for pregnant women age 35 or older.

Family histories, blood tests, genetic counseling, and fetal testing offer modern couples a great deal of information about the condition of their unborn children. If fetal tests reveal a serious genetic disorder, the parents must choose between terminating the pregnancy and preparing themselves for a baby with a birth defect. Chorionic villi sampling provides a chance to make a decision while the fetus is still quite young.

For some, no matter how undeveloped the fetus, an abortion for any reason is an unthinkable waste of human life. For others, it is a regrettable but necessary way to prevent suffering and to avoid a situation that could adversely affect their own lives, the life of the child, and the lives of other family members and close friends. Indeed, even a treatable genetic disorder usually requires constant parental vigilance. PKU, for instance, demands maintaining a strict diet during early childhood and reduced phenylalanine intake throughout life. A disorder such as Down syndrome can lead to the agony of childhood leukemia and heart attack, in addition to the symptoms of the disease itself.

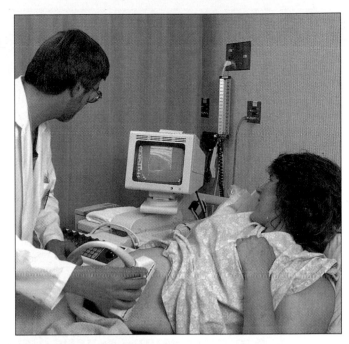

C. Ultrasound scanning of a fetus

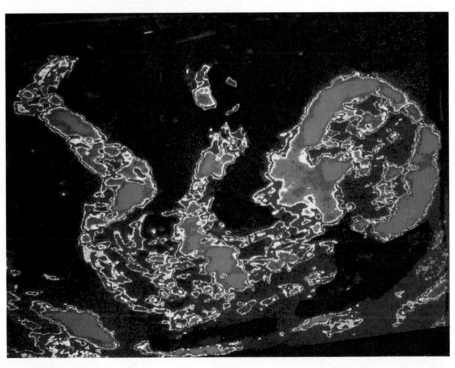

D. Ultrasound image

Carrier recognition is a key strategy in human genetics

Most children with recessive genetic disorders, which are the most common type, are born to parents with normal phenotypes. So the key to assessing genetic risk for a recessive disease is determining if the prospective parents are heterozygous carriers of the recessive trait. Recall John and Carol from Module 13.10. If they could determine that neither or only one of them was a carrier, they could go ahead with a pregnancy knowing that their baby would not have PKU. However, as in most such cases, their family pedigrees do not make their genotypes clear.

Fortunately, for some heritable disorders we now have tests that can distinguish between heterozygotes and normal homozygotes, and the number of such tests is increasing. For instance, relatively simple and inexpensive biochemical tests can identify carriers of the alleles for Tay-Sachs disease, sickle-cell anemia, and PKU. For cystic fibrosis, a new method allows researchers to directly probe the DNA of a suspected carrier for harmful alleles.

A similar DNA-probing method can be used, in many cases, to detect the presence of the dominant Huntington's disease allele before the onset of the disease. We describe this method in Module 13.14. But first we'll tell you more of the Huntington's disease story, from the viewpoint of a researcher who is at risk for the disease herself.

TALKING ABOUT SCIENCE

Nancy Wexler: A top researcher discusses a mind-ravaging genetic disease

Nancy Wexler is a world leader in human genetic research. An associate professor at Columbia University in New York City, she is also president of the Hereditary Disease Foundation in Los Angeles. Dr. Wexler earned her Ph.D. in clinical psychology but turned to human genetics for profound personal reasons: Her mother had Huntington's disease, and she herself has a 50 percent chance of developing it.

Because Huntington's disease (HD) results from a dominant allele, any child born to a parent with the disease has a 50–50 chance of being a heterozygote and developing HD. Finding people with the disease, tracking it through families, and encouraging other scientists to learn more about how the allele causes HD have become Nancy Wexler's life's work. She works closely with scientists and patients all over the world. Since the early 1980s, her research team has focused on the inheritance of Huntington's disease in several villages along Lake Maracaibo in northwestern Venezuela. From studies that had been done in the 1950s, Wexler learned that HD was especially common there and that most of the people in the villages were related. One huge family caught her attention early on. In an interview a few years ago, she explained that "the parents had 14 children, and they have 68 grandchildren and great grandchildren. Finding that family really persuaded us to do the study."

Since 1981, the Wexler team has pieced together information that links nearly 10,000 Venezuelan people in one

giant pedigree. In the photograph on the facing page, Dr. Wexler stands in front of a wall covered with part of that pedigree. She explains the pedigree's significance:

The Venezuelan Pedigree has proved spectacular, not only for research on Huntington's disease, but for understanding inheritance in general. The family trees we've collected are perfectly structured for doing genetic linkage studies. Genetic maps of the locations of genes on chromosomes can be made by studying the Venezuelan families.

By comparing genetic maps (see Module 9.17) from people with HD to those of related individuals who are unaffected, Wexler and her associates determined in 1983 that the HD gene lies on chromosome 4. Using recombinant DNA techniques, molecular geneticists were able to establish the gene's approximate position on the chromosome.

To localize a gene, a molecular geneticist first finds some detectable segment of DNA, called a marker, that is inherited by people with the disease but not by others. Recall from Chapter 9 that if two genes are very close together on the same chromosome, they are almost invariably inherited together in successive generations. A gene that is already known may therefore serve as a genetic marker for another gene under investigation, if the two genes happen to lie close enough together. Alternatively, the marker may be

Nancy Wexler with the pedigree of a Venezuelan family

simply a particular sequence of nucleotides, which may or may not be part of a gene. In Wexler's words:

> A marker is really just what it sounds like; it just sits like a landmark and marks a spot. A DNA marker is a tiny variant in the DNA itself that has a specific home on a chromosome and is inherited from generation to generation, exactly like a gene. Genes near the marker on the chromosome are usually inherited along with the marker.

The 1983 discovery of genetic markers close to the HD gene enabled scientists to develop an indirect, but fairly accurate, test for the HD allele. Researchers were confident that the markers would also lead them quickly to the HD gene itself. To their surprise, it was 10 long years before molecular biologists isolated the gene. The HD gene is a so-called expanding gene. It contains a short nucleotide sequence (CAG) that is repeated from 11 to 100 or more times. The normal HD allele has 11–34 repeats; the disease allele has more. Furthermore, the number of repeats seems to correlate with the severity of the disease and the age of onset.

The first practical use of the gene will be the development of a direct test for the HD allele. But having a simpler and more reliable test will not mean that everyone at risk will want to be tested. As Wexler pointed out in regard to the earlier test:

> The availability of the test has provoked people to consider in earnest what it would be like to know that they are inevitably going to die, at some unspecified time, of a degenerative disease of the brain. You want to know that you *don't* have the disease, but you don't necessarily want to know that you *do* have the disease. You can't have the opportunity of hearing one answer without the risk of hearing the other. I think that most people who have lived their lives at risk have built certain defenses with respect to having the disease. Knowing, in fact, that you *are* carrying the gene and *will* develop the disease is a very different psychological experience from knowing that there is a *possibility* of developing the disease. Many people have decided not to take the test until there is a treatment. However, there are understandable, pressing reasons for taking the test in some cases. People may want to have children free of disease or to clarify what the risk is for children already born by testing themselves. (The child of a person with the 50% risk has a 25% risk of inheriting HD.) And some people may just want to resolve that ambiguity. Counseling prior to the test is absolutely essential. People need to know what they're doing when they take this test because you can't erase the results once you've heard them.

Wexler herself has declined to say whether she has been tested.

Isolating the HD gene is a major milestone in the study of Huntington's disease, but there is still much to do. Scientists have no idea what the function of the normal gene is in the body, how the extra CAG repeats are produced, or what goes wrong as a result. The ultimate goals, of course, are treatment and a cure for Huntington's disease. As Nancy Wexler said when the gene was discovered (*Science,* 4/2/93, p. 30):

> We're going to keep going, working and raising money to find a cure. If it turns out that our grandkids are saved because of the work we're doing, I'll be in seventh heaven!

In the next module, we examine a method that has played an important role in the HD story, the method used to identify specific genetic markers and alleles.

Restriction fragment analysis is a powerful method for carrier detection

Unless you have an identical twin, your DNA is different from everyone else's; its total nucleotide sequence is unique. Some of your DNA consists of your particular set of alleles for all the human genes, and even more of it is composed of noncoding regions. Whether a sequence of DNA nucleotides is within a gene or not, it is inherited in Mendelian fashion just like any other part of a chromosome. Thus, the nucleotide sequences in your DNA may be very similar to those of other members of your family but very different from those of unrelated individuals.

With the help of restriction enzymes, a molecular geneticist can demonstrate the uniqueness of your DNA and the genotypic features that earmark your family. To do this, DNA is extracted from some of your cells and cut up with a restriction enzyme. The sizes and numbers of the pieces—called **restriction fragments**—reflect the specific sequence of nucleotides in your DNA. The differences in homologous DNA sequences that result in restriction fragments of different lengths have been dubbed restriction fragment length polymorphisms, or **RFLPs** (pronounced "riflips").

The figures below illustrate the principles underlying restriction fragment (RFLP) analysis. In Figure A, we see corresponding segments of DNA, *1* and *2,* from two homologous chromosomes. Segment *1* might be a piece of DNA from one of your chromosomes; *2* might be a matching piece from the same chromosome from one of your cousins. Notice that the segments differ by a single base pair (highlighted in yellow). In this case, the restriction enzyme cuts the DNA between two cytosine (C) bases in the sequence CCGG or in its complement, GGCC. Thus, the enzyme cleaves segment *1* in two places (called restriction sites), producing three restriction fragments (labeled *w*, *x*,

and *y*). Segment *2*, however, it cuts at only one restriction site, producing two restriction fragments (*z* and *y*). Notice that fragments *w* and *x* from segment *1* differ in length from both of the segment *2* fragments. In this way, the lengths of restriction fragments, as well as their numbers, differ depending on the exact sequence of bases in the DNA.

To make practical use of RFLP analysis, we need to have an easy method for separating the restriction fragments in a mixture and determining their lengths. In our simplified example, we need to tell the difference between the mixture of fragments from segment *1* and the mixture from segment *2*. We can do this using a separation technique called **electrophoresis** (Figure B). With electrophoresis we can separate macromolecules, such as nucleic acids or proteins, in a mixture on the basis of their size and electric charge. In our case, the macromolecules are the restriction fragments of DNA from Figure A. The two mixtures of fragments are placed at one end of a flat, rectangular gel (a thin slab of jellylike material). Then, a negatively charged electrode is attached to that end of the gel and a positive electrode to the other end. The phosphate groups on DNA give the fragments a negative charge, so they move through the gel toward the positive pole. Less impeded by the gel, the smaller fragments move faster, and therefore farther, than the larger ones. In our example, the restriction fragments from segment *1* separate into three bands in the gel, while those from segment *2* separate into only two bands. Notice that the smallest fragment from segment *1* (*y*) produces a band at the same location as the identical small fragment from segment *2*. Thus, electrophoresis allows us to see similarities as well as differences between mixtures of restriction fragments—and thus similarities as well as differences between the base sequences in DNA from two individuals.

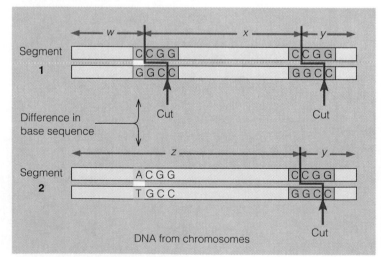

A. Restriction-site differences between two homologous DNA molecules

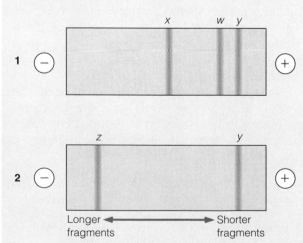

B. Electrophoresis of restriction fragments

Restriction fragment analysis is especially important because it can be used to detect potentially harmful alleles in heterozygotes who are free of symptoms, either because the harmful allele is recessive (as in cystic fibrosis) or because it is not yet expressed (as in Huntington's disease). The key is that, within a particular family, individuals with a disease allele usually have a specific set of restriction sites that are close to, or within, the allele, and are therefore inherited with it. Relatives without the disease allele have a different set of associated restriction sites.

Figure C shows two DNA segments from the same family, one carrying a disease-causing allele and one its normal counterpart. The disease allele is earmarked by two nearby restriction sites, whereas the normal allele has only one restriction site in the same region. The sections of DNA containing the differing restriction sites thus serve as **genetic markers** (specifically, RFLP markers) for the two alleles.

It was RFLP analysis that enabled researchers studying Huntington's disease in a large family in Venezuela to find a genetic marker for the defective HD allele in 1983. They discovered that, in that family, individuals with the disease have one particular marker close to the HD gene on chromosome 4. Family members without the defective gene do not have the marker.

Once a genetic marker is known for a particular disease allele, RFLP analysis can be used to test family members who are suspected carriers of that allele. In Figure D you can follow how this is done using blood samples from three relatives: a known carrier and two suspected carriers.

RFLP markers for genetic diseases are powerful tools. They not only enable diagnosticians to test for carriers of a defective gene; they also help researchers pinpoint the location of the defective gene on a chromosome. Once a gene is located, it can be cloned, and researchers can study the cloned gene to learn its normal role in the body, how the defective version causes disease, and, eventually, perhaps, how to prevent or treat the disease itself.

Restriction fragment analysis does not require very much DNA; about 1 μg—the amount in a small drop of blood—is enough. Still, even this amount of DNA may not be available. Fortunately, another powerful method, the topic of the next module, can help out in those situations.

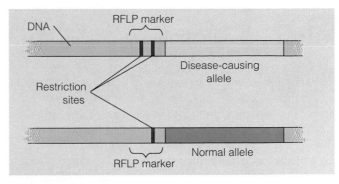

C. Restriction-site markers close to a gene

1 **Restriction fragment preparation**. DNA is extracted from white blood cells taken from individuals I, II, and III. Individual I is known to carry a disease allele. A restriction enzyme is added to the three samples of DNA to produce restriction fragments.

2 **Electrophoresis**. The mixture of restriction fragments from each sample is submitted to electrophoresis. Each forms a characteristic pattern of bands. (Since all the DNA in the blood cells was used, there would be many more bands than shown here.)

3 **Radioactive probe**. A radioactive probe (see Module 12.9) is added to the DNA bands. The probe is a single-stranded DNA molecule that is complementary to the DNA making up the restriction fragments of the genetic marker. The probe attaches to the fragments by base pairing.

4 **Detection of radioactivity (autoradiography)**. After the probe is added, a sheet of photographic film is laid over the entire pattern of DNA bands. The radioactivity in the probe exposes the film to form an image corresponding to specific DNA bands—the bands containing DNA that base pairs with the probe. In this example, the band pattern is the same for individuals I and II but different for individual III. Since we know that individual I carries the disease allele, the test shows that individual II is also a carrier but that individual III is not.

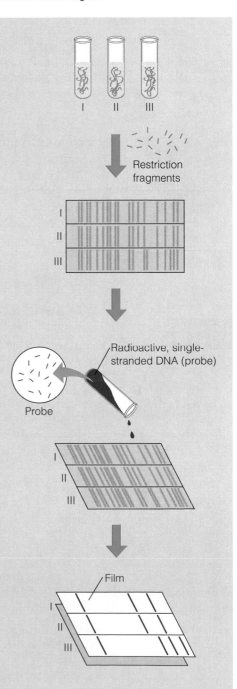

D. The procedure for restriction-fragment analysis

13.15 The PCR method is used to amplify DNA samples

Polymerase chain reaction (PCR) is a technique by which any piece of DNA can be amplified (copied many times) in a test tube. In principle, PCR is surprisingly simple. The DNA is mixed with the DNA-replication enzyme DNA polymerase, nucleotide monomers, and a few other ingredients. In this mixture, the DNA replicates, producing two daughter DNAs; the daughter molecules, in turn, replicate; and the process continues as long as nucleotides remain. Each time replication occurs, the amount of DNA doubles, as indicated in the figure. Starting with a single DNA molecule, PCR can generate 100 billion similar molecules in a few hours—enough for RFLP analysis or other uses. This is a much shorter time than the weeks it usually takes to clone a piece of DNA by attaching it to a plasmid or viral genome.

PCR is already being applied in a number of exciting ways. As an aid in evolution research, the technique has been used to amplify DNA pieces recovered from an ancient mummified human brain, from a 40,000-year-old woolly mammoth frozen in a glacier, and from a 30-million-year-old plant fossil. In the area of medical diagnosis, PCR has been used to analyze DNA from single embryonic cells for rapid prenatal diagnosis, and it has made possible the detection of viral genes in cells infected with the AIDS virus. PCR also has an important potential role in the legal arena, as we discuss next.

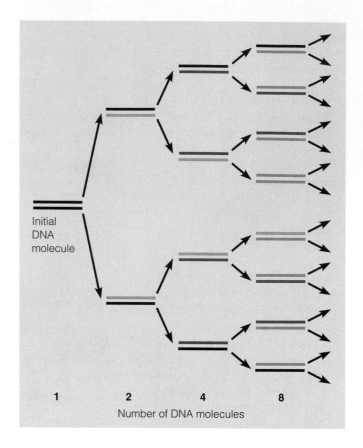

Initial
DNA
molecule

| 1 | 2 | 4 | 8 |

Number of DNA molecules

13.16 DNA technology and the law

"**D**NA FINGERPRINTING SOUGHT IN MURDER PROSECUTION" blare the headlines. In the last few years, the news media have heralded the use of DNA technology as a new tool for forensic (legal) science. From small-town attorneys' offices to the FBI, lawyers are having to learn about the new molecular techniques.

In the words of one district attorney:

> For years, our office has used blood and tissue typing to help solve violent crimes and settle paternity cases. We're still using these older techniques most of the time, but now we're also using the new DNA testing methods when we need more evidence. So we're all having to do some studying.

In violent crimes, blood or small fragments of other tissue may be left at the scene of the crime or on the clothes or other possessions of the victim or assailant. If rape is involved, small amounts of semen may be recovered from the victim's body. If enough tissue or semen is available, forensic laboratories can perform tests to determine the blood type or tissue type. However, such tests have limitations.

First, they require fairly fresh tissue in a sufficient amount for testing. Second, because there are many people in the population with the same blood type or tissue type, this approach can only *exclude* a suspect; it cannot prove guilt.

DNA testing, on the other hand, can theoretically identify the guilty individual with certainty because the DNA base sequence of every individual is unique (except for identical twins). The most frequently used DNA-analysis method is RFLP analysis. In the courtroom scene in Figure A, DNA data are being introduced as evidence in a murder trial; the paper on display is a restriction-fragment autoradiograph. The autoradiograph shows electrophoresis bands of DNA from a murder suspect, from the victim, and from a small amount of blood found at the scene of the crime.

An individual's band pattern on such an autoradiograph is called a **DNA fingerprint**. Not only is a DNA fingerprint unique (like a traditional fingerprint), but the process for producing it requires far less tissue than for blood or tissue typing. And DNA lasts much longer in the environment than the proteins on which the older tests are based. Dried blood stains even three or four years old can be used as sources of DNA.

A. Testifying on DNA data at a murder trial

In addition to its uses in solving crimes of violence, DNA fingerprinting is valuable in paternity cases. Figure B shows DNA fingerprints of three individuals involved in a paternity dispute. Is the alleged father really the child's parent? If so, the bands in the child's DNA fingerprint that do not match any of those in the known mother's column should match the father's. You can see that the child's DNA shows bands that are present in the mother's DNA. The child's DNA also shares bands with the DNA of the alleged father. In fact, the three bands in the child's column that are clearly not shared by the mother are all found in the alleged father's column (yellow highlights) Hence, the DNA evidence indicates that the man *is* the father.

Just how reliable is DNA fingerprinting? When we say that the DNA fingerprint for each individual is absolutely unique, we are speaking of the theoretical situation in which RFLP analysis is done on the entire genome. In practice, forensic DNA tests focus on only five or ten tiny regions of the genome. However, the DNA regions chosen are ones known to be highly variable from one person to another. Therefore, the probability is very small that two people will have exactly the same sequences—and DNA fingerprints—for the regions tested.

However small the elements of doubt in the reliability of DNA evidence, a heated debate now rages over whether it should be used in court. Some legal experts contend that juries are so heavily influenced by scientific evidence that

B. DNA fingerprints from a paternity dispute

such evidence must be virtually flawless before it is admissible. They argue that jurors without specialized scientific training cannot independently evaluate such evidence.

Questions have also been raised about the reliability of electrophoresis and the possibility of misinterpretation of the results it provides. Electrophoresis is a very precise way to match samples of DNA, but it is not perfect. Electrophoretic band patterns from two sources of DNA often match very closely but not exactly. Is a close match—even one that has only one chance in 10,000 or more of being incorrect—good enough evidence to convict a suspect? What about the potential for human error in carrying out the electrophoresis or in interpreting the electrophoretic band patterns? Occasionally, restriction fragment bands do not separate completely, or they end up at slightly different positions in different gels. Such anomalies may become important when the amount of DNA found at the scene of a crime is so small that an electrophoretic analysis cannot be adequately repeated.

Knowing that the amount of tissue available for analysis in criminal cases may be very tiny, perhaps too tiny even for restriction fragment analysis, you may wonder if PCR is being used to amplify the available DNA. The PCR technique could allow criminologists to make plenty of DNA for restriction fragment analysis from the minute amounts of DNA in cells attached to human hairs or in tiny amounts of blood or other tissues. At present, PCR is not used because the technique is so sensitive it could amplify contaminating DNA (say, a few stray molecules from a lab technician) along with the DNA of interest, yielding misleading results. Scientists are working on techniques for getting around this problem.

Aside from questions of reliability, there are other, ethical questions about the use of DNA fingerprints. Some states are now saving DNA fingerprints from convicted sex offenders. The existence of such data raises fears about potential abuse. The possibility of uncontrolled access to DNA data on individuals conjures up specters of new kinds of privacy invasion and discrimination. We discuss some of the ethical issues associated with the new DNA technology in more detail in the next three modules.

Gene therapy is already a reality, and it raises serious ethical questions

We have mentioned the use of new DNA techniques in medical diagnosis. Genetic engineering has the potential for actually correcting many genetic disorders. In people suffering from disorders traceable to a single defective gene, it is theoretically possible to replace or supplement the defective gene with a normal gene. Ideally, the normal gene would be put into cells that multiply throughout a person's life. Bone marrow cells, which include the "stem cells" that give rise to all the cells of the blood and immune system, are prime candidates.

The figure here outlines one procedure for a situation in which bone marrow cells are failing to produce a vital protein product because of a defective gene. ① The normal gene is cloned by recombinant DNA techniques. It is then spliced into the nucleic acid of a retrovirus that has been engineered to be harmless. ② Bone marrow cells are taken from the patient and infected with the virus. ③ The virus inserts its nucleic acid, including the human gene, into the cells' DNA (Module 10.18 describes this process). ④ The engineered cells are then injected back into the patient. If the procedure succeeds, the cells will multiply throughout the patient's life and express the normal gene. The engineered cells will supply the missing protein, and the patient will be cured.

Gene therapy tests are already under way in humans with well-defined, life-threatening genetic diseases. In one case, researchers have injected bone marrow cells engineered to carry a normal allele into several children with severe combined immunodeficiency syndrome. This is an autosomal recessive disease caused by a single defective gene and is potentially fatal because the patient's immune system cannot fight infections. If engineered stem cells suc-cessfully colonize the bone marrow, the children should be cured. Research is also under way to use gene therapy to treat other single-gene disorders, including cystic fibrosis, Duchenne muscular dystrophy, and several rare disorders affecting the blood and immune system. Eventually, gene therapy may also be used to treat diseases affected by multiple genes, such as certain types of heart disease and cancer.

Human gene therapy raises both technical and ethical issues. One important technical question is whether scientists can make the transferred gene turn on appropriately and remain turned on throughout life. Another technical problem is that, at present, gene therapy cannot be applied to diseases affecting cells that do not multiply throughout life, such as nerve cells.

The most difficult ethical question is whether we should try to treat human germ cells (the cells in our testes or ovaries that develop into gametes) in the hope of eliminating genetic defects in future generations. If we tamper with our genes in this way, we may change the genetic makeup of future generations; in effect, we would have a new way to alter the evolution of our species. Do we have the right to do this? Whatever our answer, from a biological perspective, the elimination of unwanted alleles from the gene pool could backfire. Genetic variety is a necessary ingredient for the survival of a species as environmental conditions change with time. Genes that are damaging under some conditions may be advantageous under others. Are we willing to risk making genetic changes that could be detrimental to our species in the future? We may have to face this question soon: Scientists have already succeeded in transferring foreign genes into the germ line of laboratory mice.

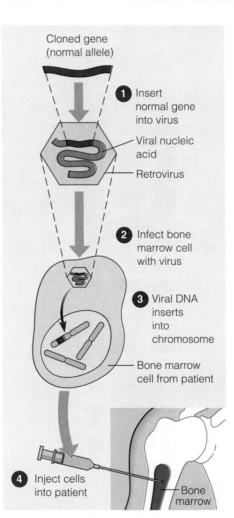

Cloned gene (normal allele)

1 Insert normal gene into virus

Viral nucleic acid

Retrovirus

2 Infect bone marrow cell with virus

3 Viral DNA inserts into chromosome

Bone marrow cell from patient

4 Inject cells into patient

Bone marrow

One type of gene therapy procedure

The Human Genome Project is a scientific adventure

Imagine an immense encyclopedia containing the secrets of human life, written in code. We know how to decipher the basic elements of this code, and we have already read a few, small parts of the encyclopedia successfully. But we suspect that deep within its volumes may be provocative new concepts, some beyond our current ability to comprehend. The characters in which the encyclopedia is written are too small to be seen by even the most powerful microscope, but if printed at the same size as the letters in this book, they would fill 500 volumes. Our goal is to decipher the entire encyclopedia in order to better understand our species.

Such an encyclopedia actually exists in the form of the human genome, the DNA that is the blueprint for human life. Only a few years ago, the task of figuring out the complete nucleotide sequence of human DNA seemed unimaginably difficult. But with advances in DNA technology and automation, scientists expect the complete sequence to be determined within the next decade or two, and they are looking forward with excitement to deciphering its meaning. An international effort to map the human genome is now under way.

The Human Genome Project involves four major lines of work:

1. *Genetic (linkage) mapping of the human genome.* In Module 9.17, we described how geneticists use data from genetic crosses to determine the order of linked genes on a chromosome and the relative distances between the genes. Going further, to map the human genome, scientists combine pedigree analysis of large families with newer techniques such as restriction fragment analysis—the same method used to find genetic markers for Huntington's disease. The initial goal is to locate several thousand genetic markers spaced evenly throughout the chromosomes. This map will enable researchers to map other genes easily, by testing for genetic linkage to the known markers.

2. *Physical mapping of the human genome.* This is done by breaking each chromosome into a number of identifiable fragments (with the help of restriction enzymes) and then determining the order of the fragments in the chromosome.

3. *Sequencing the human genome.* This is the process of determining the exact order of the nucleotide pairs of each chromosome. Since a haploid set of human chromosomes contains about 3 billion nucleotide pairs, this will probably be the most time-consuming part of the project.

4. *Analyzing the genomes of other species.* These organisms include *E. coli,* yeast, a plant named *Arabidopsis, Drosophila,* and mice. Comparative analysis of the genes of other species will help us interpret the human data. For example, if scientists find a nucleotide sequence in the human genome similar to a yeast gene whose function is known, they will have a valuable clue about the function of the human sequence.

The data from these four approaches will come together to give a complete map of the human genome. The potential benefits of such information are huge. For basic science, the information will give insight into such fundamental mysteries as embryonic development and evolution. For human health, the identification of genes will aid in the diagnosis, treatment, and prevention of a large number of disastrous disorders. Mapping studies hold great promise for individuals affected by the rare genetic diseases. They will also provide insight into many of our more common ailments, including high blood pressure, heart disease, allergies, diabetes, mental illness such as schizophrenia, alcoholism, Alzheimer's disease, and some (if not all) cancers. Most of these diseases involve multiple genes, as well as environmental factors. Mapping the genes is only one of many steps needed to fully understand a disease and find a cure. Still, in the end, the Human Genome Project will probably have its greatest practical impact on these complex and widespread health problems.

The Human Genome Project involves individual researchers and research centers in many countries around the world. In the United States, the National Institutes of Health (NIH) and the Department of Energy (DOE) will provide much of the support for the project. A significant part of the technological power for achieving the project's goals will come from advances in automation. The photograph at the left shows a computer screen from a state-of-the-art DNA sequencer. Like restriction fragment analysis, sequencing techniques use electrophoresis; notice the band patterns on the screen. Sophisticated computer skills are required to analyze the mountains of data gathered in the Human Genome Project. The ultimate success of the project may depend equally on computer scientists and DNA specialists.

Mapping the human genome will have implications for many aspects of our lives, and many citizens, scientist and nonscientist alike, are concerned about the ethical issues that will arise. The Human Genome Project already has an ethics committee, which Nancy Wexler chairs. And James Watson, of double-helix fame, has urged that 3% of the NIH project's budget—about $90 million over 15 years—be devoted to research on ethical issues. As we now discuss in the last module of the chapter, these issues concern humankind as a whole.

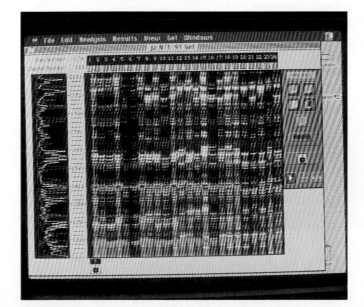

Computer screen from a DNA sequencer

The ethical questions raised by human genetics demand our attention

As our knowledge of human genetics increases, we must face a number of difficult ethical questions. On one hand, as many people argue, the potential benefits for fighting disease provide a strong ethical reason for obtaining this knowledge as fast as possible. On the other hand, the potential for abusing the knowledge makes some people wary of proceeding. As the Human Genome Project presses forward and scientists develop new ways to make practical use of the information it yields, DNA technology will influence our lives more and more. In the words of James Watson (*Science,* April 6, 1990):

We must work to ensure that society learns to use [genetic] information only in beneficial ways and . . . pass laws . . . to prevent invasion of privacy of an individual's genetic background by either employers, insurers, or government agencies and to prevent discrimination on genetic grounds. . . . We have only to look at how the Nazis used leading members of the German human genetics and psychiatry communities to justify their genocide programs, first against the mentally ill and then the Jews and the Gypsies. We need no more vivid reminders that science in the wrong hands can do incalculable harm.

Largely because of the events in Nazi Germany, our society today rejects the notion of **eugenics**—the idea that society should systematically purge all "undesirable" genetic traits from the human population. The possibility of gene therapy on germ cells raises the greatest fears in this regard.

With growing knowledge about our genome, the potential for genetic discrimination is also becoming a thorny issue. Insurance companies and employers are becoming increasingly interested in gene testing. A quote from our interview with Nancy Wexler gets to the heart of the matter:

James D. Watson

Nancy Wexler

Leroy Hood

A very big question is the problem of possible discrimination and stigmatization. An insurance company could refuse coverage or an employer could refuse a job if a person was found to carry the gene for a particular disease. People might also be coerced into taking a test that they wouldn't want in order to be considered for a job or an insurance policy.

To what extent should we allow genetic information to be used this way? If we allow its use at all, how do we prevent the information from being used in a discriminatory manner? These are complex issues. Again, Nancy Wexler says it poignantly:

The question could be asked, Do you want people who have genetic susceptibility in positions in which they could have a major impact on other individuals? For example, do you want an airline pilot with a genetic predisposition toward heart attack? My own feeling is that all of us have genes for something [harmful] that may be quiescent for a major part of our lives. If you start kicking out everybody who has some genetic susceptibility, then you're going to be in tough shape because there won't be anyone left. It's better to provide excellent medical care and preventive measures or early treatment for problems as they arise.

Some people have suggested that the dangers of abusing genetic information are so great that we should cease research in certain areas. But most scientists would probably tend to agree with the following statement by molecular biologist Leroy Hood (as quoted in *Science News,* January 21, 1989): "What science does is give society opportunities. What we have to do is look at these opportunities and then set up the constraints and the rules that will allow society to benefit in appropriate ways."

As citizens in the 1990s, we must all participate in making the decisions that Watson, Wexler, and Hood call for.

Begin your review by rereading the module headings and scanning the figures before proceeding to the Chapter Summary and questions.

Chapter Summary

13.1 The 46 chromosomes in the nucleus of a human cell contain about 3 m of DNA, in 6 billion nucleotide pairs. Each cell is thought to contain between 50,000 and 100,000 genes, which comprise the instructions for a complete human being. Recombinant DNA technology combined with classical genetic techniques present us with the possibility of deciphering the human genome.

13.2–13.6 To study human chromosomes microscopically, researchers stain and display them as a karyotype, which usually shows 22 pairs of autosomes and one pair of sex chromosomes. Sometimes there is an abnormal number of chromosomes, which causes problems. Down syndrome is caused by an extra copy of chromosome 21. The abnormal chromosome count is a product of nondisjunction, the failure of a homologous pair to separate during meiosis. Nondisjunction can also produce gametes with extra or missing sex chromosomes, which cause varying degrees of malfunction. Breakage of chromosomes can lead to rearrangements—deletions, duplications, inversions, and translocations—that can also produce disorders.

13.7–13.11 Pedigrees, family trees that trace traits through the generations, enable researchers to analyze human inheritance. Most human genetic disorders, among them cystic fibrosis, PKU, and sickle-cell anemia, are caused by autosomal recessive alleles. A few disorders, such as achondroplasia and Huntington's disease, are caused by dominant alleles. Most sex-linked disorders, like red-green color blindness and hemophilia, are due to recessive alleles and are seen mostly in males. A male receiving a single X-linked recessive allele from his mother will have the disorder; a female has to receive the allele from both parents to be affected. Genetic counselors help parents assess the risks of genetic disorders. Information about family history, along with genetic and biochemical data and examination of the fetus, can help people make reproductive decisions.

13.12–13.16 The key to genetic risk assessment is determining whether prospective parents are carriers of recessive traits. This task is similar to finding the dominant but latent Huntington's disease allele. Nancy Wexler and her colleagues traced the inheritance of this disease in large families, found a genetic marker (a nearby nucleotide sequence) inherited with the gene, and finally, in 1993, found the gene itself. Markers and genes are located by means of restriction fragment analysis, also called RFLP analysis: DNA is cut up with restriction enzymes, and the fragments are sorted by size using electrophoresis. The patterns of fragments reveal the presence of genes and markers, even those that are not expressed. Once a gene is located, it can be cloned for study. DNA for testing may be amplified using the polymerase chain reaction (PCR), which doubles and redoubles a DNA sample via replication in a test tube. Restriction fragment analysis is being used to identify DNA fingerprints from blood and other tissues in crime investigations and paternity suits, but there are questions about the reliability and interpretation of DNA evidence.

13.17 Genetic engineering has the potential for correcting genetic disorders and has already been used experimentally to treat some patients, but technical problems remain. In addition, there are ethical questions, the most significant being whether we should treat reproductive cells, which would eliminate defects in future generations but could alter the course of evolution.

13.18–13.19 A worldwide effort is under way to map the human genome. The Human Genome Project involves genetic and physical mapping of the chromosomes, DNA sequencing, and comparison of human genes to those of other species. The data will give insight into development and evolution, and will aid in the diagnosis and treatment of disease. Our new genetic knowledge will affect our lives in many ways. There are reasons to press forward, but we must use the information wisely.

Testing Your Knowledge

Multiple Choice

1. DNA fingerprints are being used as evidence to link suspects with blood and tissues found at crime scenes. These fingerprints look something like supermarket bar codes. The pattern of bars in a DNA fingerprint shows
   a. the order of bases in a particular gene
   b. the individual's karyotype
   c. the order of genes along particular chromosomes
   d. the presence of dominant or recessive alleles for particular traits
   e. the presence of various-sized fragments from chopped-up DNA

2. Why do researchers often study genetic disorders in relatively isolated communities, such as nineteenth-century Martha's Vineyard, Massachusetts, and present-day Lake Maracaibo, Venezuela? In these communities,
   a. defective genes are usually dominant
   b. mutations occur more frequently
   c. relatives with similar alleles are likely to mate and remain within the community
   d. poor health care results in a higher incidence of genetic disease
   e. there is less discrimination against victims of genetic disorders

3. A paleontologist has recovered a bit of organic material from the 400-year-old preserved skin of an extinct dodo. She would like to compare DNA from the sample with DNA from living birds. Which of the following would be most useful for increasing the amount of DNA available for testing?
   a. nondisjunction
   b. polymerase chain reaction
   c. amniocentesis
   d. electrophoresis
   e. restriction fragment analysis

4. Why are individuals with an extra chromosome 21, which causes Down syndrome, more numerous than individuals with an extra chromosome 3 or chromosome 16?
   a. There are probably more genes on chromosome 21 than on the others.
   b. Chromosome 21 is a sex chromosome and 3 and 16 are not.
   c. Down syndrome is not more common, just more serious.
   d. Disorders involving the other chromosomes are probably fatal before birth.
   e. Nondisjunction of chromosome 21 probably occurs more frequently.

5. How many genes are there in a human cell?
   a. 23
   b. 46
   c. about 10,000
   d. about 100,000
   e. about 3 billion

Describing, Comparing, and Explaining

1. Why are the most common serious human genetic disorders caused by recessive alleles? Why is Huntington's disease an exception?

2. Sketch a cell with three pairs of chromosomes undergoing meiosis, and show how nondisjunction can result in the production of gametes with extra or missing chromosomes.

3. Explain why inbreeding can increase the incidence of genetic disease.

4. Explain how restriction fragment analysis could be used to produce a DNA fingerprint. How are restriction enzymes used? Why does the DNA from different people yield different restriction fragments? How are the fragments sorted and displayed? How would your DNA fingerprint differ from that of another person?

5. Describe a procedure for carrying out gene therapy— treating a genetic disease by replacing a defective gene. What are some technical obstacles to effective gene therapy?

Thinking Critically

1. The incidence of Down syndrome appears to increase with the increasing age of the mother. Could the increasing age of the father also be a factor? What kinds of data could you look at to determine the role of the father's age in Down syndrome?

2. When crossing over occurs between two homologous chromosomes, the chromosomes usually break at corresponding points, and the pieces exchanged are identical in size. However, crossing over is sometimes uneven; the chromosomes break at different points, and the pieces exchanged are different sizes, bearing different numbers of genes. Which of the following terms would describe the abnormalities of the two resulting chromosomes? Inversion, deletion, translocation, duplication, nondisjunction. (*Hint:* It helps to make a sketch, with letters representing corresponding genes on the two chromosomes.)

3. Will there be anything left for genetic researchers to do once the Human Genome Project has determined the nucleotide sequences of all of the human chromosomes? Explain.

Genetics Problems

1. Evan and Janet have normal skin pigmentation, but they have a daughter with albinism, a lack of pigmentation. If this is like most genetic disorders, what is the probability that a second child will also suffer from the disorder?

2. Evan and Janet from problem 1 decide to have another child. It is a boy whose skin pigmentation is normal, but he suffers from galactosemia, an autosomal recessive disorder. If the couple decide to have a third child, what is the probability that the child will be free of both albinism and galactosemia?

3. Juan has six fingers on each hand and six toes on each foot. His wife Alicia and their daughter Maria have the normal number of digits. Extra digits is a dominant trait. What fraction of their children would be expected to have extra digits?

4. The pedigree below traces the inheritance of alkaptonuria, a biochemical disorder. Affected individuals (colored symbols) are unable to break down a substance called alkapton, which colors the urine and stains body tissues. Does alkaptonuria appear to be caused by a dominant or recessive allele? Is the disorder sex-linked or autosomal? What genotypes can you determine with certainty? What genotypes are possible for each of the other individuals?

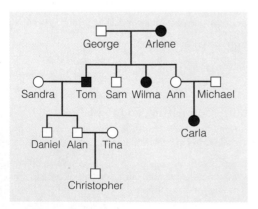

5. Imagine you are a genetic counselor, and a couple who are planning to start a family come to you for information. Charles was married once before, and he and his first wife had a child who suffers from cystic fibrosis. The brother of his current wife Elaine died of cystic fibrosis. What is the probability that Charles and Elaine will have a baby with cystic fibrosis? (This situation is similar to that in Module 13.10, but not exactly the same.)

6. Jon suffers from hereditary deafness of the type that afflicted the people on Martha's Vineyard. He and his wife Christine, who is phenotypically normal, have a child who is deaf. What fraction of their children would be expected to be deaf?

7. A couple are both phenotypically normal, but their son suffers from hemophilia, a sex-linked recessive disorder. What fraction of their children are likely to suffer from hemophilia? What fraction are likely to be carriers?

8. Pierre and Jeanne are both of normal phenotype but have a son with Duchenne muscular dystrophy, a sex-linked recessive disorder. Jeanne is pregnant, and amniocentesis shows that the fetus is a male. What is the probability that the child will suffer from Duchenne muscular dystrophy?

9. Heather was not able to see the numeral 7 in Figure 13.9A and was surprised to discover she suffered from red-green color blindness. She told her biology professor, who said, "Your father is color blind too, right?" How did her professor know this? Why did her professor not say the same thing to the color-blind males in the class?

10. Pseudohypertrophic muscular dystrophy is a gradual deterioration of the muscles. It is seen only in boys born to apparently normal parents and usually results in death in the early teens. Is pseudohypertrophic muscular dystrophy caused by a dominant or recessive allele? How do you know? Does its inheritance appear to be sex-linked or autosomal? How do you know? Explain why this disorder is always seen in boys and never in girls.

Science, Technology, and Society

1. Imagine that one of your parents suffered from Huntington's disease. What would be the probability that you, too, would someday fall victim to the disease? Would you want to be tested for the Huntington's disease allele? Why or why not? Suppose you tested positive for the presence of the Huntington's allele. In what ways would this information change your life?

2. In the not-too-distant future, gene therapy may be an option for the treatment and cure of many inherited disorders. What do you think are the most serious ethical issues that must be dealt with before human gene therapy is used on a large scale? Why do you think these issues are important?

3. One of the benefits of the Human Genome Project will be increased understanding of the connection between genes and disease—not just Mendelian disorders, but also other, more complex conditions affected by both genes and lifestyle, such as heart disease and cancer. For example, genetic markers for susceptibility to lung and liver cancers have been found on chromosomes 10 and 17, and a gene related to heart disease has been traced to chromosome 6. Testing for these genetic markers might warn a person to avoid smoking because he or she is particularly susceptible to lung cancer. Testing could also influence diet or occupational choices. The possibility of extensive genetic testing raises questions about how personal genetic information should be used. For example, should employers or potential employers have access to such information? Why or why not? Should the information be available to insurance companies? Why or why not? Is there any reason for the government to keep genetic files? Is there any obligation to warn relatives who might share a bad gene? Might some people avoid being tested, for fear of being labeled genetic outcasts? Or might they be compelled to be tested against their wishes? Can you think of other reasons to proceed with caution?

Unit 3
Evolution and the Diversity of Life

How Populations Evolve **14**

These are marine iguanas, members of the species *Amblyrhynchus cristatus,* the only oceanic lizard in the world. Marine iguanas thrive on algae that grow in shallow waters off the Galapagos Islands. When not feeding in the ocean, the lizards spend most of their time like this, sunbathing on dark lava rocks. *Amblyrhynchus* (Greek for blunt nose) exhibits some extraordinary adaptations not seen in other reptiles, adaptations that evolved through centuries of interaction with the unusual environment of the Galapagos.

The Galapagos Islands lie on the equator, about 900 km west of the South American coast. In the 1830s, Herman Melville, the author of *Moby Dick*, visited the islands and wrote of them, "Little but reptile life is here found, the chief sound of life is a hiss." Melville was not favorably impressed by either the marine iguana or the islands themselves. British biologist Charles Darwin, who spent some time on the Galapagos just before Melville did, had a different reaction. Though Darwin found the 3-m-long marine iguana "hideous-looking," as a naturalist, he was interested in the animal's unusual appearance and habits. He contrasted its sluggishness on land with its "very graceful and rapid movement" in the water. Darwin noted that the marine iguana's flattened tail, very different from the rounded tail of typical terrestrial iguanas, helps power the animal through the water with side-to-side motions. The marine iguana does not use its limbs much in swimming, although it has partially webbed feet that help steer and push it along the seafloor.

Always interested in animal behavior, Darwin was intrigued by the fact that he could not frighten the iguanas into the sea. When he chased them to the water's edge, they would stop, and he could easily pick them up by the tail. He wrote, "They do not seem to have any notion of biting; but when much frightened they squirt a drop of fluid from each nostril." Darwin tossed one of the lizards into the water repeatedly; each time, the animal quickly returned to its place on the rocky shore. Darwin concluded that the

lizard was safe from predators on land and, except when searching for food, remained on the lava rocks to avoid being eaten by sharks.

Many biologists since Darwin have been fascinated by the marine iguana, and we now know a good deal about this animal's special adaptations. For instance, we know that the iguana's tendency to remain on dark rocks rather than diving into the sea when approached is related more to body temperature regulation than to the danger of shark attack. The ocean around the Galapagos is very cold for the marine iguana. Although it can withstand the water temperature long enough to feed, it must sunbathe to maintain a suitable metabolic rate. When the lizard is out on the warm rocks, blood vessels near its body surface expand, speeding up heat absorption by the blood. Its heart rate also increases, distributing the heat to the rest of its body. When the lizard dives into the ocean, its circulatory system adjusts to cooling. Its surface vessels are constricted and its heart rate slows, keeping much of its blood and heat in deeper tissues.

Biologists have also learned that the fluid Darwin saw squirting from the iguana's nostrils is a highly concentrated salt solution. The marine iguana has salt-excreting glands inside its head just above the eyes. The salt solution drips into its nostrils and passes out in a cloudlike mist as the lizard exhales. This adaptation enables the marine iguana to dispose of excess salt it obtains from eating marine algae.

Darwin saw that there was much to be learned not only from the marine iguana but from all the forms of life on the Galapagos. He recognized that the islands had been formed by geologically recent volcanoes and that the islands' native species, such as the marine iguana, had probably evolved from a few "colonists" that came from the mainland of South America. In this chapter, we examine some of the mechanisms by which such species could have evolved adaptations to life in their new environment.

Evolution accounts for the unity and diversity of life

Today, when we think of an animal or a plant, we tend to envision it in its natural surroundings. Nature films and guides don't just show us interesting organisms; they also tell us how special structures or behaviors help the organisms survive and reproduce in a particular environment. If you visited the Galapagos Islands today, you would see many of the same sights that fascinated Darwin over a century ago. Marine iguanas still bask on the shoreline. Inland, you would see the land iguana (Figure A) and long-necked giant tortoises, which feed on thorny cactuses that grow in the volcanic soil. Awareness of each organism's adaptations and how they fit the particular conditions of the environment would help you appreciate everything you saw and heard. A heightened awareness of adaptations and a greater appreciation for organisms and their close ties with the environment are Charles Darwin's legacy.

Like many important concepts in science, the main ideas Darwin advanced—that species change over time, and that living species have arisen from earlier life forms—can be traced back to the ancient Greeks. About 2500 years ago, the Greek philosopher Anaximander promoted the idea that life arose in water and that simpler forms of life preceded more complex ones. The road from Anaximander to Darwin was long and tortuous. The Greek philosopher Aristotle, whose views had an enormous impact on Western culture, held that species are fixed, or permanent, and do not evolve. Judeo-Christian culture fortified this idea with its belief, as presented in the biblical book of Genesis, that all species were individually designed by the Creator. The ideas that the Earth and all living species are static and perfect and that the Earth is, at most, about 6000 years old dominated the intellectual climate of the Western world for centuries.

In the century prior to Darwin, only a few scientists questioned the biblical story of creation or the belief that species are fixed and perfect. In the mid-1700s, the study of **fossils** (the preserved imprints or remnants of extinct organisms) led French naturalist Georges Buffon to suggest that Earth might be much older than 6000 years. In addition, scientists observed some similarities between fossils and living animals. In 1766, Buffon raised the possibility that different species had arisen by variation from common ancestors, although he went on to argue against the idea. In the early 1800s, French naturalist Jean Baptiste Lamarck suggested that the best explanation for fossils and for the current di-

versity of life is that organisms evolve. Today, we remember Lamarck mainly for his erroneous view of *how* species evolve. He proposed that by using or not using its body parts, an individual tends to develop certain characteristics, which it passes on to its offspring. As an example, he suggested that the giraffe acquired its long neck because its ancestors stretched higher and higher into the trees to reach leaves, and that the animal's increasingly lengthened neck was passed on to its offspring. This mistaken idea, known as the inheritance of acquired characteristics, obscures the important fact that Lamarck helped set the stage for Darwin by strongly advocating evolution and by proposing that species evolve as a result of interactions with their environment.

Charles Darwin (Figure B) was born in 1809, the same year Lamarck published some of his ideas on evolution. In December 1831, at the age of 22, Darwin began a round-the-world sea voyage (Figure C) that profoundly influenced his thinking and eventually the thinking of the entire world. He joined the crew of the surveying ship *Beagle* on a mission to chart poorly known stretches of the South American coastline. Darwin actually spent most of his time on shore, collecting thousands of specimens of fossils and living plants and animals and noting the unique adaptations of the South American organisms. Despite growing up in the antievolutionist climate of Victorian England, Darwin had a questioning mind. He pondered the question of why fossils of the South American continent resembled modern South American species more than fossils of other continents. Could this mean that species in the Americas were more closely related to each other than to species elsewhere? Could contemporary South American species have arisen from South American ancestors? Other questions arose from Darwin's visit to the Galapagos Islands, where the *Beagle* also stopped. Darwin saw that most of the animal species on the Galapagos were ones found nowhere else in the world. Referring to the islands and their inhabitants, he later wrote, "Both in space and time, we seem to be brought somewhat near to that great fact— that mystery of mysteries—the first appearance of new beings on the earth."

While on the voyage, Darwin read and was strongly influenced by the recently published *Principles of Geology*, by Scottish geologist Charles Lyell. Lyell's work led Darwin to realize that natural forces gradually

A. Land iguana and cactus, examples of the diversity of life on the Galapagos Islands

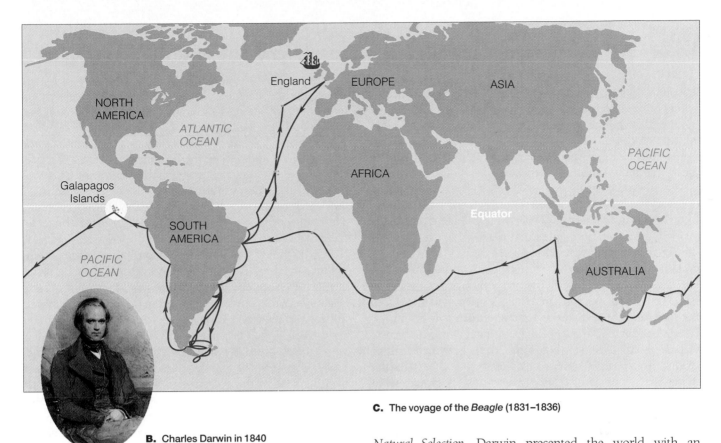

B. Charles Darwin in 1840

C. The voyage of the *Beagle* (1831–1836)

change the Earth's surface and that the forces of the past are still operating in modern times. These ideas ran counter to the prevailing nineteenth-century view that the most important geological events in Earth's history were certain rare catastrophes and sudden changes. Darwin had collected fossils of marine snails in the Andes Mountains. Having read Lyell, he came to believe that slow, natural processes such as the growth of mountains as a result of earthquakes could account for the snails' being there.

By the time Darwin returned to England five years after the *Beagle* first set sail, his experiences and reading had led him to doubt seriously that the Earth and living organisms were unchangeable and had been specially created only a few thousand years earlier. By then he believed that the Earth was very old and constantly changing. These were heretical views, and in accepting them, Darwin had taken an important step toward recognizing that life on Earth had evolved.

By the early 1840s, Darwin had worked out the major features of his theory of evolution and had composed a long essay about it. He realized that his theory would cause a social furor, however, and he delayed publishing it. Then, in the mid-1850s, Alfred Wallace, another British naturalist doing fieldwork in Indonesia, conceived a theory identical to Darwin's. When Wallace sent Darwin a manuscript he had written about his theory, Darwin thought "all my originality will be smashed." However, in 1858, two of Darwin's colleagues presented Wallace's paper and excerpts of Darwin's earlier essay together to the scientific community. With the publication in 1859 of his complete text, *On the Origin of Species by Means of Natural Selection,* Darwin presented the world with an avalanche of evidence and a strong, logical argument for evolution. He also developed a second powerful idea, which he called the theory of natural selection, an explanation of how evolution occurs. Darwin's book was a monumental contribution to science and to human thought in general. Evolution is the great unifying theme of biology, and *The Origin of Species* fueled an explosion in biological research and knowledge that continues today.

In the first edition of his book, Darwin did not actually use the word evolution, referring instead to "descent with modification." This phrase summarized Darwin's view of life. He perceived a unity among species, with all organisms related through descent from some unknown species that lived in the remote past. He postulated that as the descendants of the earliest organism spread into various habitats over millions of years, they accumulated diverse modifications, or adaptations, that accommodated them to diverse ways of life. In the Darwinian view, the history of life is like a tree, with multiple branching and rebranching from a common trunk to the tips of the twigs. At each fork of the evolutionary tree is an ancestor common to all lines of evolution branching from that fork. Species that are closely related, such as the domestic dog and the wolf, share many characteristics because their lineage of common descent extends to the smallest branches of the tree of life.

As we will see in this chapter and the next two, evolutionary theory has been greatly expanded beyond Darwin's basic ideas. Nonetheless, few contributions in all of science have explained as much, withstood so much repeated testing over the years, and stimulated as much other research as those of Darwin.

A mass of evidence validates a Darwinian view of life

Darwin compiled enough support for his theory of descent with modification to convince most of the scientists of his day that organisms evolve without supernatural intervention. Subsequent discoveries, including recent ones from molecular biology, further support this great principle—one that connects an otherwise bewildering chaos of facts about organisms.

It was the geographical distribution of species, known as **biogeography,** that first suggested to Darwin that organisms evolve from common ancestors. Darwin noted that, although the environment of the Galapagos was more like that of certain tropical islands in distant parts of the world than like the environment of the nearby South American mainland, Galapagos animals nonetheless resembled species of the mainland more than they resembled animals on similar but distant islands. The only logical explanation was that the Galapagos species evolved from South American immigrants. Today, it is clearer than ever that biogeography makes sense only in the historical context of evolution. In this module, we look at some of the other scientific evidence for evolution.

The **fossil record**—fossils and the order in which they appear in layers of rock—provides some of the strongest evidence that evolution has occurred. The fossil record shows that animals and plants have appeared in a historical sequence. Moreover, the sequence of fossils is consistent with what is known from other lines of evidence. For example, evidence from biochemistry, molecular biology, and cell biology places prokaryotes (bacteria) as the ancestors of all life. And, indeed, the oldest known fossils, dating from about 3.5 billion years ago, are prokaryotes. Another example is the chronological appearance of the different classes of vertebrates (animals with backbones) in the fossil record. Fishlike fossils are the oldest vertebrates in the fossil record; amphibians are next, followed by reptiles, then mammals and birds. This sequence is consistent with the history of vertebrate descent as revealed by many other types of evidence.

Also providing support for evolution is the science of comparative anatomy, the comparison of body structures in different species. Common descent is evident in anatomical similarities between species grouped in the same taxonomic category. As Figure A shows, the same skeletal elements make up the forelimbs of humans, cats, whales, and bats, all of which are classified as mammals. The functions of these forelimbs differ. A whale's flipper does not do the same job as a bat's wing, so if these structures had been uniquely engineered, it's likely that their basic designs would have been very different. The logical explanation for their similarity is that all mammals descended from a common ancestor having the same basic limb elements. In fact, the arms, forelegs, flippers, and wings of different mammals are variations on a common anatomical theme that has become adapted to different functions. They are **homologous structures,** that is, structures that are similar because of common ancestry. Comparative anatomy testifies that evolution is a remodeling process in which ancestral structures that originally functioned in one capacity become modified as they take on new functions—the kind of process that Darwin called descent with modification.

The study of structures that appear during embryonic development in different organisms, called comparative embryology, is another major source of evidence for the common descent of organisms. Closely related organisms often have similar stages in their embryonic development. For example, all vertebrate embryos pass through a stage in which they have openings called gill pouches on the sides of their throats. At this stage, the embryos of fishes, frogs, snakes, birds, humans—indeed, all vertebrates—look more alike than different. They take on more and more distinctive features as development progresses. In fishes, for example, the gill pouches develop into gills. In terrestrial vertebrates, however, these embryonic structures develop into structures with other functions, such as the Eustachian tubes connecting the middle ear with the throat.

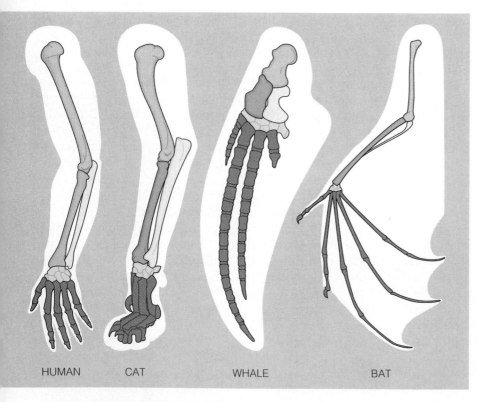

HUMAN   CAT   WHALE   BAT

A. Homologous structures

A great deal of support for evolution has come in recent years from the field of molecular biology. The universality of the genetic code is strong evidence that all life is related. As we saw in Chapter 10, the hereditary background of an organism is documented in the DNA that constitutes its genes, and in its proteins, which are encoded by the genes. Molecular biologists have shown that, consistent with the principle of common descent, related individuals have greater similarity in their DNA and proteins than do unrelated individuals of the same species. And two species judged to be closely related by other criteria have a greater proportion of their DNA and proteins in common than more distantly related species.

Research in molecular biology, especially studies of the amino acid sequences of similar proteins in different species, has been a rich source of data about evolutionary relationships. Figure B illustrates the results of studies on one of the polypeptide chains of hemoglobin, the protein that carries oxygen in the blood of humans and other vertebrates. Researchers have isolated this polypeptide from a number of different animal species and have found that its amino acid sequence differs somewhat from species to species. The graph shows how the polypeptide chain of humans compares with that of five other vertebrate animals. By matching the length of the vertical bar for each animal with the vertical scale on the left, you can see how many amino acids are different from those in the human polypeptide. For example, the rhesus monkey polypeptide differs from the human polypeptide by only 8 amino acids, whereas the lamprey version differs from the human version by 125 amino acids. (Since the human polypeptide is 146 amino acids long, these numbers represent differences of 5% and 86%, respectively.)

These amino acid comparisons lead to a hypothesis about evolutionary relationships: Rhesus monkeys are much more closely related to humans than are lampreys; mice, chickens, and frogs fall in between. It turns out that this hypothesis agrees with earlier conclusions from comparative anatomy and embryology. Furthermore, the hypothesis agrees with fossil evidence. As indicated in Figure B, the lineage that led to humans diverged from the one leading to monkeys only about 26 million years ago, whereas lampreys (some of the most primitive living vertebrates) branched off the trunk of vertebrate evolution some 450 million years ago.

Molecular biology is adding an extensive chapter to the evidence confirming evolution as the basis of life's unity and diversity. But how does evolution operate? We begin to answer that question in the next module.

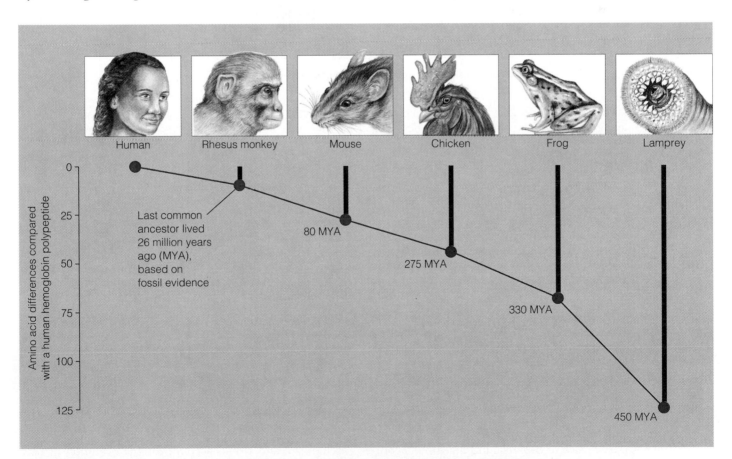

B. Evolutionary relationships between humans and five other vertebrates, based on hemoglobin comparisons

Darwin proposed natural selection as the mechanism of evolution

Darwin devoted much of *The Origin of Species* to the ways that organisms become adapted to their environment. His theory of how this happens arose from several key observations. First of all, Darwin recognized that all species tend to produce excessive numbers of offspring. He had read an influential essay on human population written in 1798 by English economist Thomas Malthus. Malthus contended that much of human suffering—disease, famine, homelessness, and war—was the inescapable consequence of the human population's potential to grow much faster than the rate at which supplies of food and other resources could be produced. It was apparent to Darwin that Malthus's concepts applied to all species. Darwin deduced that, because natural resources are limited, production of more individuals than the environment can support leads to a struggle for existence among the individuals of a population, with only a percentage of offspring surviving in each generation. Many eggs are laid, young born, and seeds spread, but only a tiny fraction complete their development and leave offspring of their own. The rest are starved, eaten, frozen, diseased, unmated, or unable to reproduce for other reasons.

A. Artificial selection in plants produced these vegetables

In addition to overproduction of offspring, two other facts of nature entered into Darwin's thinking: first, that individuals of a population vary extensively in their characteristics, and second, that many of the varying traits are passed from one generation to the next.

What do overproduction of offspring, limited natural resources, and heritable variations have to do with organisms becoming adapted to their environment? Darwin saw that every environment has only a limited supply of resources, and that survival in a limited environment depends in part on the features the organisms inherit from their parents. He concluded that, within a varied population, individuals whose inherited characteristics adapt them best to their environment are most likely to survive and reproduce; these individuals thus tend to leave more offspring than do less fit individuals. Reproduction is central to what Darwin saw as the basic mechanism of evolution, the process he called **natural selection.** Darwin perceived that the essence of natural selection *is* differential, or unequal, reproduction.

Darwin's insight was both simple and profound: Reproduction is unequal, with the individuals that best meet specific environmental demands having the greatest reproductive success. Put another way, differential reproduction (natural selection) is the means by which the environment screens variations, favoring some over others. Individuals better adapted to a particular environment are more likely to survive and leave offspring that in turn will reproduce than are those less well adapted. As Darwin reasoned, natural selection results in favored traits being represented more and more and unfavored ones less and less in ensuing generations. Thus, the unequal ability of individuals to survive and reproduce leads to a gradual change in a population of organisms, with favored characteristics accumulating over the generations.

Darwin found convincing evidence for his ideas in the results of **artificial selection,** the selective breeding of domesticated plants and animals. He saw that by selecting individuals with the desired traits as breeding stock, humans were playing the role of the environment and bringing about differential reproduction. They were, in fact, modifying species. We see evidence of what Darwin was talking about in the vegetables we eat. For example, broccoli, cauliflower, cabbages, brussels sprouts, kale, and kohlrabi (Figure A) are all varieties of a single species of wild mustard, and all were produced by artificial selection. These and many other domesticated plants and animals bear little resemblance to the wild species from which they were derived. Figure B shows some of the enormous diversity that breeders have produced in a few thousand years within a single species, *Canis familiaris,* the domestic dog.

Darwin reasoned that, if so much change could be achieved in a relatively short period of time by artificial selection, then over hundreds or thousands of generations, natural selection should be able to modify species considerably. He reasoned that, with natural selection operating over vast spans of time, heritable changes would gradually accumulate and would account for the evolution of new species, such as the five species of canines shown in Figure C, from preexisting ancestors. In the next module, we examine some documented cases of natural selection.

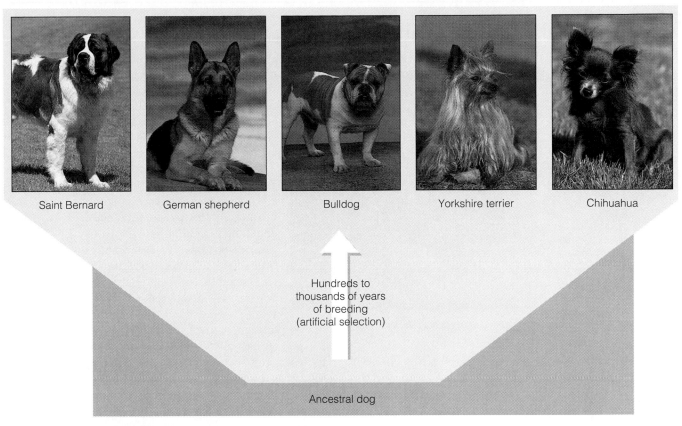

B. Artificial selection in animals: five breeds of dogs

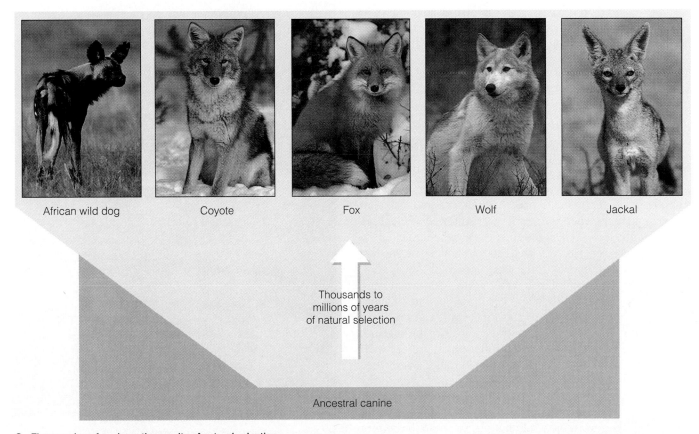

C. Five species of canines, the results of natural selection

Natural selection is a prominent force in nature

A. Different shell colors in snails, resulting from natural selection

There are many well-documented cases of the effects of natural selection. The land snails in Figure A belong to a single species, *Cepaea nemoralis*, found in woodlands and open fields in Great Britain. Their shells differ in color and in the presence or absence of stripes. The snails in the photograph are not usually found together. Snails with light-colored, striped shells are found mainly in well-lighted areas, whereas dark-colored individuals lacking stripes are found in shady places. The shell color pattern camouflages the snails in their respective habitats, hiding them from birds that prey on snails. If a snail appears in an area where its color pattern does not conceal it, birds are more likely to find that snail than a camouflaged snail. Thus, snails whose shells match their environment are likely to survive and reproduce, while those with uncamouflaged shells will likely be eaten.

One of the most extensively documented examples of natural selection in action involves the English peppered moth, *Biston betularia*. Found throughout the English midlands, this insect occurs in two color varieties. The form for which the peppered moth was named is light, with splotches of darker pigment. The other variety is uniformly dark. Peppered moths feed at night and rest during the day, often on trees and rocks encrusted with light-colored lichens. Against this background, shown in the top photograph of Figure B, light moths are camouflaged, whereas the dark moths are very conspicuous and therefore easy prey for birds. Before the Industrial Revolution in the nineteenth century, dark peppered moths were very rare in certain parts of England. No single factor was responsible for the scarcity, but it is likely that many of the dark moths became bird food before they could reproduce and pass the genes for darkness on to the next generation. As industrial pollution in the late 1800s killed large numbers of lichens, uncovering dark tree bark, dark moths became more abundant. As shown in the bottom photograph of Figure B, the dark moth is camouflaged on a

tree trunk that lacks lichens, but the light moth stands out. By the turn of the century, the population of *Biston* near Manchester, England, consisted almost entirely of dark moths. Experiments have supported the hypothesis that natural selection resulting from increased predation by birds on the light-colored moths contributed to this population's shift to a higher proportion of dark individuals.

These and other cases of population change make the point that natural selection is regional and timely, tending to adapt organisms to their local environments. Environmental factors vary from place to place and from time to time. An adaptation in one situation may be useless or even detrimental in different circumstances. In recent years, the case of the peppered moth has taken a satisfying turn: Industrial pollution has been curbed, allowing parts of the countryside to return to their natural coloration. In some places, the light form of *Biston* has made a strong comeback.

B. Natural selection over a short time span in the English peppered moth

A **population** is a group of individuals of the same species living in the same place at the same time. It is the smallest unit that can evolve. We can, in fact, measure evolution as a change in the prevalence of certain heritable traits in a population over a succession of generations. The increasing proportion of the dark-pigment trait in peppered moths in industrial England is one example. When natural selection favored the dark trait, dark moths left more offspring, and the population changed, or evolved, from being mostly light to mostly dark. Note that the individual moths did not evolve. The role of the individual in evolution is its interaction with the environment and its being screened by natural selection. An individual may contribute to population change by its success or failure at reproduction.

Human population centers in the United States

Darwin understood that it is populations that evolve. He saw that natural selection, in favoring some heritable traits over others, changes populations over successive generations. What eluded him was the genetic basis of population change. Without this basis, he could not explain the cause of variation among the individuals making up a population, nor could he account for the perpetuation of parents' traits in their offspring. Today, we know that heritable traits are carried by genes on chromosomes and that mutations may produce new traits. Also, we understand how Mendel's principles of segregation and independent assortment of alleles operate during meiosis (see Module 9.14). In Darwin's time, however, little was known about chromosomes, much less gene mutations or the details of meiosis. The significance of Mendel's work did not come to light until about 20 years after Darwin's death, about 40 years after publication of *The Origin of Species.*

An important turning point for evolutionary theory was the birth in the 1920s of **population genetics,** the science of genetic change in populations. A comprehensive theory of evolution that incorporates genetics was developed in the early 1940s. Known as the **modern synthesis,** this theory focuses on populations as the units of evolution and includes most of Darwin's ideas. Most importantly, it melds population genetics with the theory of natural selection.

Central to the modern synthesis is the relationship between populations and species. A **species** can be defined as a group of populations whose individuals have the potential to interbreed. Each species has a geographical range, within which individuals are usually concentrated in localized populations. One population may be isolated from others of the same species. If individuals in the isolated population interbreed only rarely with those in other populations, there will be little exchange of genes. Such isolation is common for populations confined to widely separated islands, unconnected lakes, or mountain ranges separated by lowlands. However, populations are not always isolated, nor do they necessarily have sharp boundaries. One population center may blur into another in an intermediate region where members of both populations occur but are less numerous.

The photograph above illustrates the tendency for humans to concentrate in localized populations. This is a nighttime satellite photograph showing the lights of population centers—major metropolitan areas—in the United States. We know that these populations are not really isolated; people move around, and there are low-density suburban and rural communities between cities. Nevertheless, people are most likely to choose mates who live in the same city or town, often in the same neighborhood. As a result, for humans and other species, individuals in one population center are, on average, more closely related to one another than to members of other populations.

Understanding what populations are and how they behave sets the stage for understanding how populations and species evolve. To analyze any aspect of a population in an evolutionary context, we must consider the genes in the population. Our next step, then, is to focus on population genetics. But before starting the next module, you might wish to review the basic concepts and terms of Mendelian genetics in Chapter 9 (especially Modules 9.2 and 9.4).

14.6 Microevolution is change in a population's gene pool

In studying evolution at the population level, geneticists focus on what is called the **gene pool**, the total collection of genes in a population at any one time. The gene pool is the reservoir from which members of the next generation of that population derive their genes; it consists of all alleles (alternative forms of genes) in all the individuals making up a population.

For most gene loci, there are two or more alleles in the gene pool, and individuals may be either homozygous (having two identical alleles) or heterozygous (having two different alleles) for each locus. For example, in peppered moths there are two alleles for body color: One allele determines light body color; the other determines dark body color. The relative frequencies of these and other alleles in a gene pool may change over time. In a peppered moth population in an unpolluted region of England, the allele for light color at one time had a much higher frequency in the population than the allele for dark color. But during industrialization, the allele for dark color increased in frequency at the expense of the allele for light color.

When the relative frequencies of alleles in a population change like this over a number of generations, evolution is occurring on its smallest scale. Such a change in a gene pool is called **microevolution.** To understand how microevolution works, it's helpful to begin by examining the genetics of a simple hypothetical population whose gene pool is *not* changing.

14.7 The gene pool of an idealized, nonevolving population remains constant over the generations

Let's return to iguanas and imagine a nonevolving population with two varieties that differ in foot webbing. Moreover, let's imagine that foot webbing is a trait controlled by a single gene. We'll assume that the allele for nonwebbed feet (*W*) is completely dominant to the allele for webbed feet (*w*). (See Figure A.) The term "dominant" may seem to suggest that, over many generations of sexual reproduction, the *W* allele will somehow come to "dominate" the population, becoming more and more common at the expense of the recessive *w* allele. In fact, this is not what happens. In an idealized case, the shuffling of genes that accompanies sexual reproduction does not alter the genetic makeup of the population. In other words, sexual reproduction alone does not lead to microevolution. No matter how many times alleles are segregated into different gametes by meiosis and united in different combinations by fertilization, the frequency of each allele in the gene pool will remain constant unless acted on by other agents. This rule is the Hardy-Weinberg principle, named for the two scientists who derived it in 1908.

To test the Hardy-Weinberg principle, we will look at two generations of our imaginary iguana population. Figure B illustrates the genetic situation in the first generation. Suppose we have a total population of 500 iguanas in which 320 animals have the genotype *WW* (and the phenotype of nonwebbed feet), 20 animals have the genotype *ww* (webbed feet), and 160 have the heterozygous genotype, *Ww* (nonwebbed feet, since *W* is dominant). These numbers are entirely arbitrary; if you wish, you can use different numbers. With the numbers we're using, the genotype frequencies come out 0.64 for *WW* ($\frac{320}{500} = 0.64$), 0.04 for *ww* ($\frac{20}{500} = 0.04$), and 0.32 for *Ww* ($\frac{160}{500} = 0.32$).

From the genotype frequencies, we can now calculate the frequency of each allele in the first-generation population.

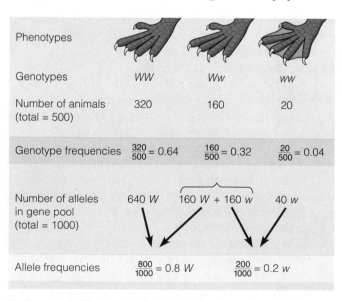

Phenotypes			
Genotypes	*WW*	*Ww*	*ww*
Number of animals (total = 500)	320	160	20
Genotype frequencies	$\frac{320}{500} = 0.64$	$\frac{160}{500} = 0.32$	$\frac{20}{500} = 0.04$
Number of alleles in gene pool (total = 1000)	640 *W*	160 *W* + 160 *w*	40 *w*
Allele frequencies	$\frac{800}{1000} = 0.8\ W$		$\frac{200}{1000} = 0.2\ w$

A. An iguana with nonwebbed feet (hypothetical genotype *WW* or *Ww*)

B. Gene pool of first-generation iguanas

Because every iguana carries two alleles for the foot-webbing trait, there are a total of 1000 genes for that trait in the population. To find the number of W alleles, we add the number carried by the WW iguanas, $2 \times 320 = 640$, to the number carried by the Ww iguanas, 160. The total number of W alleles is thus 800. The frequency of the W allele, which we will call p, is $\frac{800}{1000}$, or 0.8. The frequency of the w allele can be calculated in a similar way; this frequency, called q, is 0.2. (Geneticists routinely use the letters p and q to represent allele frequencies.)

What happens when the first-generation iguanas form gametes? At the end of meiosis, each gamete has one foot-type allele, either w or W. The frequency of the two alleles in the gametes is the same as it is in the parental population, 0.8 for W and 0.2 for w.

Figure C uses these gamete allele frequencies and the rule of multiplication (see Module 9.7) to calculate the frequencies of the three possible genotypes in the second generation. The probability of producing a WW individual (in other words, of combining two W alleles from the pool of gametes) is $p \times p = p^2$, or $0.8 \times 0.8 = 0.64$. Thus, the frequency of WW iguanas in the next generation would be 64%. Likewise, the frequency of ww individuals would be $q^2 = 0.04$, or 4%. For heterozygous individuals, Ww, there are two ways that the genotype can form, depending on whether the sperm or egg supplies the dominant allele. In other words, the frequency of Ww would be $2pq = 2 \times 0.8 \times 0.2 = 0.32$, or 32%. Thus, we see that the frequencies of the three possible genotypes in the second-generation population are the same as they were in the first generation.

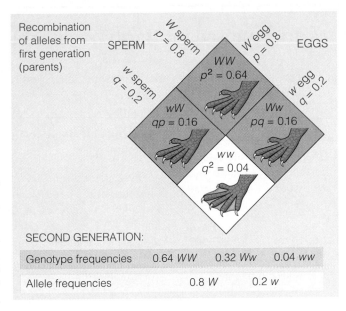

C. Gene pool of second-generation iguanas

Recombination of alleles from first generation (parents)

SPERM

W sperm $p = 0.8$

w sperm $q = 0.2$

W egg $p = 0.8$

w egg $q = 0.2$

EGGS

WW $p^2 = 0.64$

wW $qp = 0.16$

Ww $pq = 0.16$

ww $q^2 = 0.04$

SECOND GENERATION:

Genotype frequencies	0.64 WW	0.32 Ww	0.04 ww
Allele frequencies		0.8 W	0.2 w

Finally, what about the frequency of the alleles in the second generation? Since the genotype frequencies are the same as in the first generation, the allele frequencies, p and q, are the same, too. In fact, we could follow the frequencies through many generations, and the results would continue to be the same. Thus, the gene pool of this population is in a state of equilibrium, referred to as the Hardy-Weinberg equilibrium.

We can now write out a general formula for calculating the frequencies of alleles in a gene pool from the frequencies of genotypes, and vice versa. In our imaginary iguana population, the frequency of the W allele (p) is 0.8, and the frequency of the w allele (q) is 0.2. Note that $p + q = 1$; this is always true, because the combined frequencies of all alleles must account for 100% of the alleles for that gene in the population. If there are only two alleles and we know the frequency of one, we can calculate the frequency of the other one:

$$p = 1 - q \quad \text{or} \quad q = 1 - p$$

Notice in Figures B and C that the frequencies of all possible genotypes in the populations also add up to 1 (that is, $0.64 + 0.32 + 0.04 = 1$). We can represent this symbolically with the **Hardy-Weinberg equation**:

$$\underset{\substack{\text{Frequency} \\ \text{of } WW}}{p^2} + \underset{\substack{\text{Frequency} \\ \text{of } Ww}}{2pq} + \underset{\substack{\text{Frequency} \\ \text{of } ww}}{q^2} = 1$$

The Hardy-Weinberg equation is used in public health $\boxed{14.8}$

The Hardy-Weinberg equation is not only important in population genetics; public health scientists also use it to estimate the percentages of people carrying alleles for certain inherited diseases. Consider the case of the metabolic disorder phenylketonuria (PKU), which is an inherited inability to break down a certain amino acid. PKU occurs in about one out of 10,000 babies born in the United States and, if untreated, results in severe mental retardation.

PKU is due to a recessive allele, and thus the frequency of individuals in the U.S. population born with PKU corresponds to the q^2 term in the Hardy-Weinberg equation.

Given one PKU occurrence per 10,000 births, $q^2 = 0.0001$. Therefore, the frequency of the recessive allele for PKU in the population, q, equals the square root of 0.0001, or 0.01. And the frequency of the dominant allele, p, equals $1 - q$, or 0.99. The frequency of carriers, heterozygous people who are normal but may pass the PKU allele on to offspring, is $2pq$, which equals $2 \times 0.99 \times 0.01$, or 0.0198. Thus, the equation tells us that about 2% (actually 1.98%) of the U.S. population carries the PKU allele. Estimating the frequency of a harmful allele is useful for any public-health program dealing with genetic diseases.

14.9 Five conditions are required for Hardy-Weinberg equilibrium

Hardy-Weinberg equilibrium tells us that something other than the reshuffling processes of sexual reproduction is required to alter a gene pool—that is, to change allele frequencies in a population from one generation to the next. One way to find out what factors *can* change a gene pool is to identify the conditions that must be met if genetic equilibrium is to be maintained. Hardy-Weinberg equilibrium will be maintained in nature only if all five of the following conditions are met:

1. The population is very large.
2. The population is isolated; that is, there is no migration of individuals, and the alleles they carry, into or out of the population.

3. Mutations (changes in genes) do not alter the gene pool.
4. Mating is random.
5. All individuals are equal in reproductive success; that is, natural selection does not occur.

These five conditions are rarely, if ever, met in nature. Consequently, Hardy-Weinberg equilibrium breaks down, and allele frequencies in natural populations change constantly. But Hardy-Weinberg equilibrium give us a basis for comparing idealized, nonevolving populations with actual ones in which gene pools are changing. Let's look now at how real populations evolve.

14.10 There are five potential agents of microevolution

Five agents, each a deviation from one of the five conditions for Hardy-Weinberg equilibrium, can cause changes in gene pools, or microevolution. These agents are genetic drift, gene flow, mutation, nonrandom mating, and natural selection.

Genetic drift is a change in the gene pool of a small population due to chance. As we have seen, in an idealized population that is not evolving, the gene pool will remain the same from one generation to the next. However, if a population of organisms is small, a chance event can have a disproportionately large effect, making the gene pool different in the next generation. For example, consider what could happen in an imaginary iguana population consisting of only ten individuals. Assume that three of the lizards have the genotype *WW* (nonwebbed feet), two are *Ww* (also nonwebbed) and five are *ww* (webbed). Now imagine that three of the iguanas are killed in an earthquake. If, by chance, these three lizards are *WW* individuals, the result will be a marked decrease in the frequency of the *W* allele, with a corresponding increase in the *w* allele frequency. This change in relative allele frequencies would be a case of microevolution caused solely by chance. Because natural selection is not involved, only luck could lead to improved adaptation as a result of such a random shift in a gene pool.

A population must be infinitely large for genetic drift to be ruled out completely as an agent of microevolution; however, many populations are so large that drift has little or no effect on the overall gene pool. The populations in which genetic drift is most likely to play a major role typically have 100 or fewer individuals.

One situation that involves a population small enough for genetic drift is known as a bottleneck. The **bottleneck effect** is genetic drift resulting from a disaster that drastically

reduces population size. Events such as earthquakes, floods, or fires may kill large numbers of individuals unselectively, producing a small surviving population that is unlikely to have the same genetic makeup as the original population. Figure A shows how a bottlenecking event works, representing alleles by balls of different colors. Certain alleles (red balls) will be present at higher frequency in the surviving population than in the original population, others (white balls) will be present at lower frequency, and some (green balls) may not be present at all. In a real example, human hunters in the 1890s reduced the population of northern elephant seals in California to about 20 individuals. Since then, this mammal (pictured in Figure B) has become a protected species, and the population has grown back to

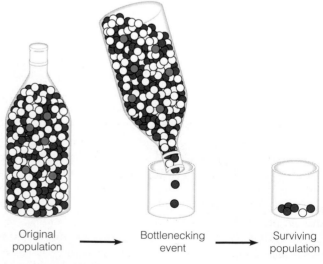

Original population → Bottlenecking event → Surviving population

A. The bottleneck effect

over 30,000 members. However, in examining 24 gene loci in a representative sample of the seals, researchers found *no* variation. For each of the 24 genes they found only one allele, probably because of bottlenecking. In contrast, genetic variation abounds in populations of a closely related species, the southern elephant seal, which was not bottlenecked.

Another situation that can produce a population small enough for genetic drift is the colonization of a new location by a small number of individuals. Random change in the gene pool that occurs in a small colony is called the **founder effect.** In the most extreme case, a single pregnant animal or one plant seed might found a new population. The founder effect was important in the evolution of many species in the Galapagos Islands. For instance, the ancestors of the marine iguana were probably a few strays that arrived there from the South American mainland. The gene pool of the marine iguana would have been derived from those few individuals.

The second agent of microevolution, **gene flow,** is the gain or loss of alleles from a population by the movement of individuals or gametes. Gene flow occurs when fertile individuals move into or out of a population, or when gametes (such as the sperm of plant pollen) are transferred from one population to another. Gene flow tends to reduce genetic differences between populations. Over the history of our own species, for example, local groups have sometimes become reproductively isolated. This isolation has reduced gene flow and has resulted in genetic distinctions among groups of people living in different parts of the world. Reflecting the genetic differences are phenotypic variation in traits such as skin color and other racial and ethnic differences that characterize our species. Working against reproductive isolation has been the influence of migrations and conflicts, which tend to increase interbreeding among groups. Today, air travel makes it possible for people all over the globe to interact, and there is more gene flow among geographically isolated populations than ever before.

A third agent of microevolution is **mutation.** A mutation is a change in an organism's DNA that creates a new allele (see Module 10.5). Mutations of a given gene are rare events, typically occurring only about once per gene locus per 10^5 or 10^6 gametes. As a result, in a large population, mutation alone does not have much effect in a single generation. Generally, if some new allele rapidly increases its frequency in

B. Elephant seals descended from bottleneck survivors

a population, it is not because mutation is generating the allele in abundance, but because individuals carrying the mutant allele are producing a disproportionate number of offspring as a result of genetic drift or natural selection. Over the long run, however, mutation is vital to evolution because it is the only force that actually generates new alleles. Thus, mutation is the ultimate source of the genetic variation that serves as raw material for microevolution.

A fourth agent that can change gene pools, **nonrandom mating,** is the selection of mates other than by chance. In order for Hardy-Weinberg equilibrium to be maintained, every male in the population must have an equal chance of mating with every female in the population. But the fact is, preferential mating is the rule in most populations. In the U.S., for example, short women generally marry short men and tall women, tall men. The photograph above illustrates another example of nonrandom mating. The large elephant seal with the big snout is a male. He has dominated other males in ritual displays and combat, and he alone will mate with the females around him. In this species, a single dominant male prevents any other males in the group from mating.

Darwin paid a lot of attention to species in which body features affect nonrandom mating. Among the features he considered were body size, adornments such as antlers in deer, and the bright colorations of male birds. He saw that if such features gave an individual an edge in gaining a mate, the features would be favored by natural selection.

Natural selection, or differential success in reproduction, is the fifth agent of microevolution. It is *the* factor that is likely to result in adaptive changes in a gene pool.

14.11 Adaptive change results when natural selection upsets genetic equilibrium

One condition for Hardy-Weinberg equilibrium—that all individuals in a population be equal in their ability to reproduce—is probably never met in nature. Populations of sexually reproducing organisms consist of varied individuals, and some variants leave more offspring than others. In our imaginary iguana population, lizards with webbed feet (*ww* genotype) might produce more offspring because they are more efficient at finding food or mates than lizards without webbed feet (genotypes *Ww* or *WW*). This would disturb genetic equilibrium, causing the frequency of the *W* allele to decline in the gene pool. In this way, natural selection results in the accumulation and maintenance of favorable traits in a population. If the environment should change, natural selection would favor traits adapted to the new conditions. The degree of adaptation that can occur is limited by the amount and kind of genetic variation in the population.

14.12 Variation is the rule in populations

We are very conscious of human diversity, but individuality in populations of other animals and plants may escape our notice. Nonetheless, some degree of individual variation occurs in all populations. In addition to anatomical differences, most populations have a great deal of variation that can be detected only by biochemical means. Researchers often use the technique of electrophoresis (see Module 13.14) to study molecular variation in populations.

Not all variation in a population is heritable. The phenotype is the cumulative product of an inherited genotype and a multitude of environmental influences. For instance, a weight-training program can build up your muscle mass above and beyond what would naturally occur from your genetic makeup, but you would not pass this environmentally induced phenotypic change on to your offspring. It is important to remember that only the genetic component of variation can lead to adaptations as a result of natural selection.

Many of the variable traits in a population result from the combined effect of several genes. As we saw in Module 9.13, polygenic inheritance produces traits that vary more or less continuously—in human height, for instance, from very short individuals to very tall ones. By contrast, other traits, such as red versus white flowers in some plants, are determined by a single gene locus with different alleles that produce only distinct phenotypes (red or white flower color, for example); there are no in-between types. In such cases, when a population includes two or more forms of a phenotypic trait, the contrasting forms are called **morphs.** A population is said to be **polymorphic** for a trait if two or more morphs are present in readily noticeable numbers (that is, if neither morph is extremely rare). The photograph here illustrates a striking example of polymorphism in a population of California king snakes. The two morphs differ markedly in their color patterns. Polymorphism is also extensive in human populations, both in physical traits, such as the presence or absence of freckles, and in biochemical features, such as the ABO blood groups (see Module 9.11).

Most species exhibit geographical variation among populations. For example, a wild plant population with two morphs—red flower color (genotypes *RR* and *Rr*) and white color (genotype *rr*)—could have a higher frequency of recessive alleles than another population of the same plant because of a local prevalence of pollinating insects that favor white flowers.

Sometimes geographical variation occurs in what is called a **cline,** a graded change in an inherited trait along a geographical continuum. Clines often occur where there is a gradation in some environmental variable. For example, the body size of many birds and mammals, such as white-tailed deer, increases gradually with increasing latitude and colder climate in North America. Large size is adaptive in northern latitudes because it reduces the ratio of body surface area to volume and helps conserve body heat. Smaller body size helps dissipate heat in warmer regions. How do clines and other inherited variations arise? We address this question in the next module.

Polymorphism in king snakes

Mutations and sexual recombination, which are both random processes, produce genetic variation. As we saw in Module 10.15, mutations can actually create new alleles. For example, in Figure A, a point mutation in a gene substitutes one nucleotide (G) for another (T). This type of change will be harmless if it does not affect the function of the protein the DNA encodes. However, if it does affect the protein's function, the mutation will probably be harmful. An organism is a refined product of thousands of generations of past selection, and a random change in its DNA is not likely to improve its genome any more than shooting a bullet through the hood of a car is likely to improve engine performance.

On rare occasions, however, a mutant allele may actually improve the adaptation of its bearer to the environment and enhance reproductive success. This kind of effect is more likely when the environment is changing in such a way that mutations that were once disadvantageous are favorable under the new conditions. For instance, mutations that endow house flies with resistance to the pesticide DDT also reduce their growth rate. Before DDT was introduced, such mutations were harmful. But once DDT was part of the environment, the mutant alleles were favored, and natural selection increased their frequency in fly populations.

On a larger scale are so-called chromosomal mutations, which are changes involving stretches of DNA long enough to be detected microscopically. A single chromosomal mutation affects many genes and is almost certain to be harmful. Very rarely a translocation (such as the one shown in Figure 13.6B) might bring benefits. For instance, the movement of a piece of one chromosome to another chromosome might link genes that affect the organism in some positive way when they are inherited together as a package.

Organisms with very short generation spans, such as bacteria, can quickly adapt to changing environments—in other words, evolve—by means of mutations alone. Bacteria multiply so rapidly that a beneficial mutation can significantly increase its frequency in descendant populations in a matter of hours or days. Furthermore, bacteria are haploid, meaning that a newly created allele can have an effect immediately, without the possibility of being obscured by the expression of another allele on a homologous chromosome.

For most animals and plants, however, their long generation times and generally diploid condition prevent most mutations from significantly affecting genetic variation from one generation to the next, at least in large populations. Consequently, animals and plants depend mainly on sexual recombination for the genetic variation that makes adaptation possible. As we saw in Modules 8.15 and 8.16, fresh assortments of existing alleles arise every generation from three random components of sexual recombination: independent assortment of homologous chromosomes during meiosis, crossing over during meiosis, and random fertilization of an egg by a sperm. In Figure B, we see the results of a mating of parents with the genotypes A^1A^1 and A^2A^3, where A^1, A^2, and A^3 are three different alleles for a gene. Even without considering independent assortment or crossing over, the offspring have different combinations of alleles than their parents do.

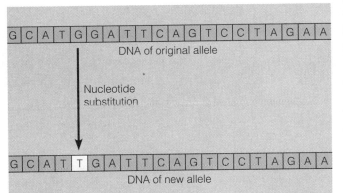

A. Creation of a new allele by mutation

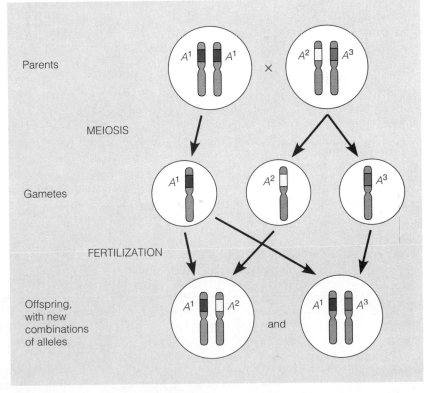

B. Shuffling alleles by sexual recombination

14.14 How natural selection affects variation: an overview

The subject of variation leads us back to Darwin, who devoted two whole chapters of *The Origin of Species* to heritable variations before he discussed natural selection. Following Darwin's lead, we now start to look at the connection between variation and natural selection. The figure here is an idealized model summarizing the effect of natural selection on inherited variation in successive populations. The top row represents an imaginary ancestral iguana population living on a small island. The population is varied, with most individuals having nonwebbed feet and a rounded tail, and two (color-coded dark green) having webbed feet and a flattened tail. Food is scarce on the island itself but abundant in the surrounding ocean. Natural selection favors webbed feet and flattened tails, traits that help the animals swim and exploit the rich food supply at sea. Over the generations, individuals with webbed feet and flattened tails leave proportionately more offspring, and we see the results in the population at the bottom. Natural selection has eliminated most of the individuals with nonwebbed feet and rounded tails. The bottom population is clearly better adapted to life on the island than was the ancestral population.

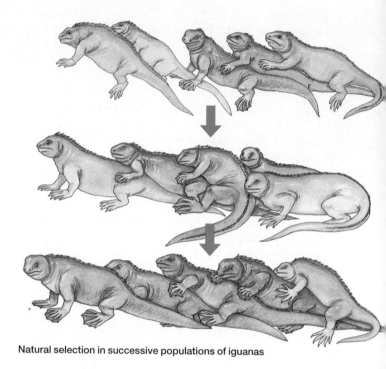

Natural selection in successive populations of iguanas

14.15 Natural selection tends to reduce variation, while the diploid condition helps preserve it

The imaginary model in Figure 14.14 indicates how natural selection adapts populations to the environment. It also makes the point that natural selection tends to reduce variability in populations. Notice that the population at the bottom, adapted to the specific environment of a small oceanic island, is less varied than the ancestral population. In fact, most populations tend to lose genetic variability as a result of the culling effect of natural selection.

Populations are unlikely to become genetically uniform, however. One factor helping to maintain genetic variation is the fact that the effects of recessive alleles are not often displayed in diploid organisms. A recessive allele is subject to natural selection only when it influences the phenotype, and this occurs only when two copies of it appear in a homozygous individual. In a heterozygote, a recessive allele is, in effect, hidden, or protected, from natural selection because only the dominant allele influences the phenotype. In many species, this type of protection helps maintain a large pool of recessive alleles that may lead to variation in later generations. Many recessive alleles are unsuitable for present conditions but may be beneficial if the environment changes.

Genetic variability in diploid organisms can also be preserved by natural selection, the very force that generally reduces it. For some genes, natural selection tends to maintain the two or more alleles that appear in the heterozygote and therefore preserve genetic variability. In such cases, there is a **heterozygote advantage**, that is, heterozygous individuals have greater reproductive success than homozygotes. One example is the resistance to malaria conferred by the recessive sickle-cell allele (see Module 9.12). In areas where malaria is a major cause of death, natural selection favors heterozygotes (carriers of the sickle-cell allele) because they are resistant to malaria.

14.16 Endangered species often have reduced variation

Today, the effects of human activities are extinguishing species at an unprecedented rate, and at least 25,000 species are endangered. Without significant efforts to curb habitat destruction and reduce environmental pollution, the rate will climb even more. Hunting can also push species to extinction. Populations of some whale species have been so severely reduced by commercial hunting that they may not be able to recover, despite the current ban on whaling by most nations. Poaching threatens many endangered species, especially in Africa, where elephants, gorillas, rhinoceroses, cheetahs, and many others are at risk, although most African nations are trying to preserve them.

Endangered species typically have low genetic variability. As their numbers are severely reduced, the diversity of their gene pools also declines. One such species is the cheetah (*Acinonyx jubatus*), pictured here. The fastest of all running animals, the cheetah can run down the swiftest antelope. These magnificent cats were formerly widespread in Africa and Asia. Like many African mammals, their numbers fell drastically during the last ice age some 10,000 years ago, apparently as a result of disease, human hunting, and periodic droughts. In effect, the species suffered a severe bottleneck, which left only two isolated populations, one in East Africa, the other in South Africa, both with reduced genetic variation. Today, both populations are endangered. With an inherently low reproductive rate, they could die out even if they were not plagued by poachers and habitat destruction.

The South African cheetah population is especially precarious. Apparently, this population suffered a second bottleneck during the nineteenth century when South African farmers hunted the animals to near extinction. South African cheetahs exhibit extreme genetic uniformity—even more than some highly inbred strains of laboratory mice. Today, only about 0.07% of this population's gene loci are heterozygous. In contrast, nearly 60 times as many loci (about 4%) are heterozygous in the East African animals, which did not suffer a second bottleneck. With so little variability, South African cheetahs may lack the capacity to adapt to environmental changes. The only way to save them may be through breeding programs that mix the two population's gene pools.

The cheetah, a species with low genetic variability

Not all genetic variation may be subject to natural selection

Some genetic variations in populations seem to have a trivial impact on reproductive success and therefore may not be subject to natural selection. The diversity of human fingerprints, for example, seems to be an example of **neutral variation**—variation in a heritable trait that provides no apparent selective advantage for some individuals over others. The neutral variation hypothesis proposes that species typically have numerous alleles that confer no selective advantage or disadvantage. Some of these supposedly neutral alleles will increase their frequency in the gene pool and others will decrease by the chance effects of genetic drift, but natural selection will not affect them.

There is little consensus among evolutionary biologists about how much genetic variation is neutral. Some researchers do not believe that any variation is truly neutral. They point out that variations appearing to be neutral may influence reproductive success in ways that are difficult to measure. It is possible to show that a particular allele is harmful, but it is impossible to demonstrate that an allele brings no benefits at all to an organism. Also, a variation may be neutral in one environment but not in another. We can never know the degree to which genetic variation is neutral. But we can be certain that even if only a fraction of the extensive variation in a gene pool significantly affects the organism, that is still an enormous resource of raw material for natural selection to act on.

Human fingerprints, probably an example of neutral variation

14.18 What does "survival of the fittest" really mean?

In discussing natural selection in *The Origin of Species*, Darwin wrote: "This preservation of favourable individual differences and variations, and the destruction of those which are injurious, I have called Natural Selection, or the Survival of the Fittest." The phrase "survival of the fittest" caught on and is often quoted when evolution is discussed. As long as we keep in mind what the words "survival" and "fittest" mean in an evolutionary sense, the phrase is still a useful summary of the process of natural selection. The emphasis must be on the survival of genes, not individual organisms. Today, we define **fitness** as the contribution that an individual makes to the gene pool of the next generation relative to the contribution of other individuals. Thus, the fittest individuals in the context of evolution are those that pass the greatest number of genes to the next generation.

We examine the connections among fitness, genes, and individuals in more depth in the next module.

14.19 Natural selection acts on whole organisms and affects genotypes as a result

Evolutionary fitness has to do with genes, but it is the phenotype of an organism—its physical traits, metabolism, and behavior—not the genotype, that is directly exposed to the environment. Natural selection therefore targets the whole organism, which is an integrated composite of many phenotypic traits. Individuals with a high degree of fitness are those whose phenotypic traits enable them to reproduce and contribute genes to more offspring than other individuals. For example, marine iguanas with the flattest tails and most foot webbing may average more offspring because they swim faster and therefore can exploit marine food resources more efficiently.

The figure here illustrates the connections between natural selection, genotypes, and changes in gene pool, using our iguana populations from Figure 14.14 and focusing on the hypothetical foot-webbing gene. Natural selection does not target genes or genotypes directly, but, in favoring certain phenotypes over others, it has the effect of favoring certain genotypes. The favored genotypes are those whose positive phenotypic effects outweigh any harmful effects they may have on the reproductive success of the organism. Put another way, in culling out less fit individuals, natural selection also culls unfavorable genotypes. In this example, you see that the eliminated genotypes are those that produce nonwebbed feet. The result in the population at the bottom is a gene pool in which 90% of the gene loci for foot webbing have the *w* (webbing) allele. In this idealized model, the change in the gene pool is reflected in the relative abundance of the webbed-foot phenotype.

There is one other very important point about genotypes and natural selection: The fitness of a genotype at any one gene locus depends on many other gene loci. For instance, in the marine iguana, alleles that enhance the ability of an individual to swim and search for food at sea may be useless or even harmful in the absence of alleles at other loci that enhance the animal's ability to get rid of excess salt and survive cold temperatures. Overall, the entire genome of an organism is a unit of functionally interrelated genes, in the same sense that the whole organism is an integrated composite of many phenotypic traits. Thus, natural selection culls or favors whole genomes as it targets and acts on whole phenotypes.

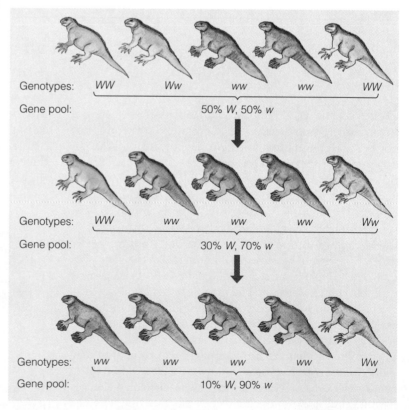

Genotypes:	WW	Ww	ww	ww	WW
Gene pool:			50% *W*, 50% *w*		

Genotypes:	WW	ww	ww	ww	Ww
Gene pool:			30% *W*, 70% *w*		

Genotypes:	ww	ww	ww	ww	Ww
Gene pool:			10% *W*, 90% *w*		

Changes in gene pool as natural selection culls iguanas with nonwebbed feet

Keeping in mind that natural selection targets whole organisms, let's see what its culling effects can actually do to the frequencies of phenotypic variants in a population. Here we use an imaginary population of snails, with individuals varying in the intensity of their shell color. The bell-shaped curve in the top graph of the figure below depicts an idealized population that could result from a polygenic inheritance pattern for variation in shell color. In this case, which we use as our starting population, shell color varies along a continuum from very light (only a few individuals) through various intermediate shades (many individuals) to very dark (few individuals). The other three graphs show three different ways in which natural selection could alter the phenotypic variation in the idealized population.

Stabilizing selection favors intermediate variants. It typically occurs in relatively stable environments where conditions tend to reduce phenotypic variation. In the snail population depicted in the graph on the bottom left, stabilizing selection has eliminated the extremely light and dark individuals, and the population has a greater number of intermediate phenotypes, which are best suited to a stable environment. Stabilizing selection probably prevails most of the time in most populations. For example, this type of selection keeps the majority of human birth weights in the

3- to 4-kg range. For babies much smaller or larger than this size, infant mortality is greater.

Directional selection shifts the overall makeup of the population by acting against individuals at *one* of the phenotypic extremes. For the snail population in the bottom center graph, the trend is toward darker shell color, as might occur if the landscape were blackened by volcanic ash. Directional selection is most common during periods of environmental change, or when members of a species migrate to some new habitat with different environmental conditions. Changes we described in populations of the peppered moth are an example of directional selection.

Diversifying selection typically occurs when environmental conditions are varied in a way that favors individuals at *both* extremes of a phenotypic range over intermediate individuals. For the snails in the graph on the bottom right, individuals with light and dark shells have increased their numbers relative to intermediate variants. We would expect this to occur if the snails had recently colonized a patchy habitat where a background of white sand was studded with dark lava rocks.

The effects of natural selection on populations have direct bearing on some important environmental problems, as we see in the next module.

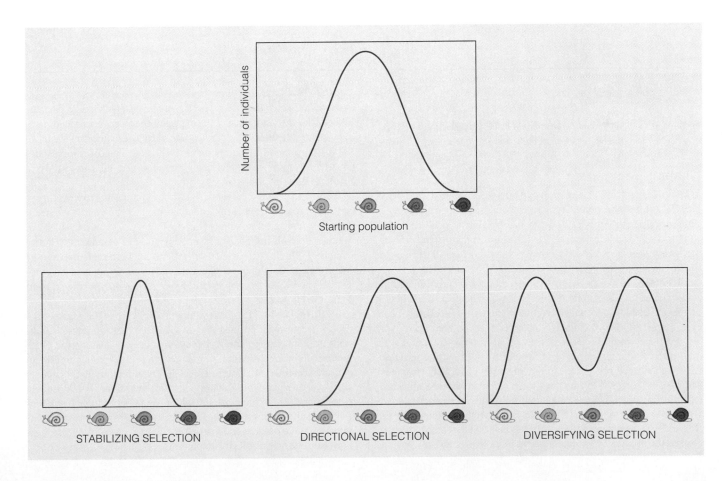

STABILIZING SELECTION · DIRECTIONAL SELECTION · DIVERSIFYING SELECTION

Directional selection has produced resistant populations of pests and parasites

In the past several decades, the world has witnessed some very unsettling examples of natural selection: the development of pesticide resistance in insects, and antibiotic resistance in bacteria that cause human disease. Some of the protists that cause diseases such as malaria have also evolved resistance to drugs used to control them. Pesticide-resistant mosquitoes carrying malarial or bacterial pathogens that may themselves be drug-resistant pose a serious public-health problem with no immediate solutions.

The figure here illustrates how directional selection has produced resistant populations. When a new insecticide, antibiotic, or drug is first used, it is usually effective in small amounts, killing all but a few individuals in a population. The survivors live and reproduce because they have genes that protect them from the poison. Eventually, most of the population consists of resistant individuals, and the poison is ineffective.

When challenged with new problems, any population adapts to the changes, moves, or becomes extinct. Insects and bacteria can usually adapt because they are numerous and varied and can multiply rapidly. As the fossil record shows, however, many more species have become extinct than currently exist. Those populations that do survive crises often change enough to become new species, as we will see in the next chapter.

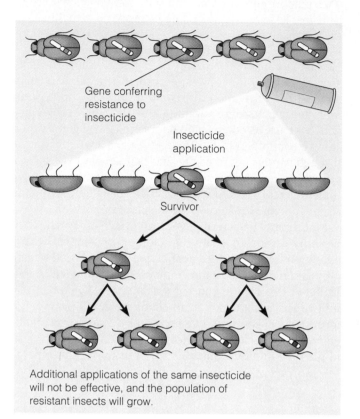

Gene conferring resistance to insecticide

Insecticide application

Survivor

Additional applications of the same insecticide will not be effective, and the population of resistant insects will grow.

Chapter Review

Begin your review by rereading the module headings and scanning the figures before proceeding to the Chapter Summary and questions.

Chapter Summary

14.1 The concept of evolution explains the unity and diversity of life. Charles Darwin noted similarities between living and fossil organisms of South America and the nearby Galapagos Islands. His studies convinced him that the Earth was old and constantly changing. Darwin also realized that living things too could change, or evolve, and that living species are descended from earlier life forms.

14.1, 14.3, 14.4 The idea of evolving species first arose among the ancient Greeks, but Aristotle and the Judeo-Christian culture held that species are fixed. Fossils suggested that living things do change, however. In the early 1800s, Lamarck theorized that acquired phenotypic characteristics could be passed on to offspring. Darwin conceived a different evolutionary mechanism, which he called natural selection. He saw that organisms produce more offspring than the environment can support, organisms vary, and their variations can be inherited. Darwin concluded that the individuals best suited for a particular environment are more likely to survive and reproduce than those less well adapted. As the proportion of individuals with favorable traits increases, the population gradually adapts to its environment.

14.2–14.4 Darwin found evidence for evolution in artificial selection, in which breeders choose organisms with desired traits. Other evidence comes from biogeography, the fossil record, comparative anatomy, and embryology. Conclusions based on comparisons of DNA and proteins generally agree with those based on fossils and anatomy. Natural selection has been observed in populations of snails, insects, microorganisms, and other organisms.

14.5, 14.7–14.9 The modern synthesis connects Darwin's theory with concepts of population genetics, such as the Hardy-Weinberg principle. This principle states that the frequency of various alleles in a gene pool will not change from generation to generation (Hardy-Weinberg equilibrium will exist), as long as the population is large and isolated from other populations, mutations do not alter the population's gene pool, mating is random, and all genotypes are equal in reproductive success.

14.6, 14.10–14.11 Microevolution—a change in the gene pool—occurs because real populations deviate from these idealized conditions. In a small population, chance events cause genetic drift. Gene flow occurs when individuals or gametes move from one population to another. Mutation creates new alleles. Nonrandom mating occurs whenever animals choose mates with particular traits. Finally, in real populations, individuals are not all equal in reproductive success. This differential reproduction, or natural selection, results in an accumulation of favorable genotypes. It is the only agent of microevolution that routinely results in adaptation.

14.12–14.14 Variation may be environmental or genetic, but only genetic changes result in evolutionary adaptation. Many populations exhibit polymorphism—different forms of phenotypic traits—and geographical variation. Variation is generated by mutations and sexual recombination. Most mutations are harmful, but occasionally a mutation enhances reproductive success. Rapidly reproducing bacteria can rely on mutation alone, but plants and animals depend mainly on sexual recombination for the variation that makes adaptation possible.

14.15–14.16 Natural selection tends to reduce variability in populations, while the diploid condition preserves it by "hiding" recessive alleles. Many heritable traits are clearly helpful or harmful, but biologists disagree as to the extent of neutral variation. The shrinking populations of many endangered species have low genetic variability, which may reduce their capacity to adapt to environmental changes.

14.17–14.18 Natural selection acts on the whole phenotype, not directly on genotype. In a stable environment, stabilizing selection tends to favor intermediate forms and weed out extreme forms. During periods of environmental change, directional selection acts against one extreme, while diversifying selection favors extreme types over intermediate forms. In the end, an individual's fitness is the contribution it makes to the gene pool of the next generation.

Testing Your Knowledge

Multiple Choice

1. The processes of _____ and _____ generate variation, and _____ produces adaptation to the environment.
   a. recombination . . . natural selection . . . mutation
   b. mutation . . . recombination . . . genetic drift
   c. genetic drift . . . mutation . . . recombination
   d. mutation . . . natural selection . . . recombination
   e. mutation . . . recombination . . . natural selection

2. Natural selection is sometimes described as "survival of the fittest." Which of the following most accurately measures an organism's fitness?
   a. how strong it is when pitted against others of its species
   b. its mutation rate
   c. how many fertile offspring it has
   d. its ability to withstand environmental extremes
   e. how much food it is able to make or obtain

3. A geneticist studied a grass population growing in an area of erratic rainfall. She found that plants with alleles for curled leaves reproduced better in dry years and plants with alleles for flat leaves reproduced better in wet years. This situation would tend to (*Explain your answer.*)
   a. cause genetic drift in the grass population
   b. preserve the variability in the grass population
   c. lead to directional selection in the grass population
   d. lead to uniformity in the grass population
   e. cause gene flow in the grass population

4. Birds with average-sized wings survived a severe storm more successfully than other birds in the same population with longer or shorter wings. This illustrates
   a. founder effect
   b. stabilizing selection
   c. artificial selection
   d. gene flow
   e. diversifying selection

5. Which of the following is a true statement about Charles Darwin?
   a. He was the first to discover that living things can change, or evolve.
   b. He based his theory on the inheritance of acquired characteristics.
   c. He worked out the principles of population genetics.
   d. He proposed natural selection as the mechanism of evolution.
   e. He was the first to realize that the Earth is billions of years old.

Describing, Comparing, and Explaining

1. Write a paragraph briefly describing the kinds of evidence for evolution.

2. Explain how the Hardy-Weinberg equilibrium is altered in each of the situations below, and how each might lead to microevolution.
   a. In a park, white domestic ducks and mallards interbreed.
   b. Only four pine trees in a secluded valley escape an avalanche.
   c. A mutation causes a gray squirrel to be born with black fur.
   d. Hawks with poor vision catch fewer mice.
   e. Female fruit flies prefer mates with bright red eyes.

3. Sickle-cell anemia, like PKU, is caused by a recessive allele. Roughly one out of every 500 African-Americans (0.2%) is afflicted with sickle-cell anemia. Use the Hardy-Weinberg equation to calculate the percentage of African-Americans who are carriers of the sickle-cell allele. (*Hint:* $0.002 = q^2$).

Thinking Critically

1. An orange grower discovered that most of his trees were infested with destructive mites. He sprayed the trees with insecticide, which killed 99% of the mites. Five weeks later, most of the trees were infested again, so he sprayed again, using the same quantity of insecticide. This time, only about half the mites were killed. Explain why the spray did not work as well the second time.

2. The land snails (*Cepaea nemoralis*) described in Module 14.4 are preyed upon by birds called thrushes. The birds break the snails open on rocks, eat the soft bodies, and leave the shells. The snails occur in both striped and unstriped forms. In one area, researchers counted both live snails and broken shells. Their data are summarized below:

	Striped	Unstriped	Total	Percent Striped
Living	264	296	560	47.1
Broken	486	377	863	56.3

Which of the snail forms seems to be best adapted to this environment? Why? What would you predict for this population?

3. There are two main groups of bats: Smaller "microbats" navigate by using sonar, and larger "megabats" rely on vision. It was once thought that both kinds evolved from insectivorous mammals. But similarities between the visual systems of megabats and primates have led some researchers to think that megabats may have evolved from primates, perhaps lemurs. How could molecular biology help resolve the question of bat ancestry? What would show that the two groups of bats have a common origin? Separate origins?

Science, Technology, and Society

School districts in several states have been criticized by groups demanding that science classes give "equal time" to alternative (usually fundamentalist Christian) interpretations of the origin and history of life. They argue that it is more fair to let students evaluate both evolution and special creation. Do you think religious views about the origin of species should receive the same emphasis as evolution in science courses? Why or why not? Might you feel differently if you were a parent? A school board member? A science teacher?

North America has ten species of skunk. The one most people have seen—or at least smelled—is the abundant and widespread striped skunk *Mephitis mephitis* (from the Latin for "bad odor"). The skunk in the photograph at the left is a spotted skunk, rarely seen but especially interesting because it illustrates some important concepts about biological species. This particular skunk belongs to a species called the western spotted skunk (*Spilogale gracilis,* from the Latin for "slender spotted cat"). The adult is only about the size of a house cat, but it has a potent chemical arsenal that makes up for its small size. Before spraying her potent musk, a female guarding her young usually warns an intruder by raising her tail, stamping her forefeet, raking the ground with her claws, or even doing a handstand, as shown below. When all else fails, she can spray her penetrating odor for 3 meters with considerable accuracy.

The western spotted skunk inhabits a variety of environments in the United States, from the Pacific coast to the western Great Plains. It is closely related to the eastern spotted skunk, *Spilogale putorius* ("foul-smelling cat"), which occurs throughout the southeastern and midwestern United States. The ranges of these two species overlap, and the two species look so much alike that even experts have a difficult time telling them apart. Both are black with broken white stripes and spots. Individuals of the western species are, on average, slightly smaller, and some have a white tip on the tail, but these and other minor differences in body form are not always present.

For many years, biologists debated whether all spotted skunks belong to one species. But in the 1960s, studies of sexual reproduction in these animals showed that they are indeed two species. Reproduction in the eastern spotted skunk is a straightforward affair. Mating occurs in late winter, and young are born between April and July. In marked contrast, the western spotted skunk includes what is called delayed development in its reproductive cycle. Mating takes place in the late summer and early fall, and zygotes begin to develop in the uterus of the female. Further development, however, is temporarily arrested at an early point called the blastocyst stage. The blastocysts remain dormant in the female's uterus throughout the winter months and resume growth in the spring, with the young (usually 5–7) being born in May or June. Because mating occurs at different times of the year for the two species, there is no opportunity for gene flow between populations of eastern and western spotted skunks. Thus, they are separate species, despite the pronounced similarities in their body form and coloration.

Spotted skunks show us that looks can be deceiving. Without knowledge of the mating cycles, we could interpret the minor differences between the two species as insignificant phenotypic variation and conclude that there is only one species of spotted skunk in North America. Many other species are also difficult to distinguish and define. So before we pursue this chapter's main subject—how different species evolve—we need to examine more closely what we mean by the term "species."

A spotted skunk warning an intruder

What is a species?

The word species is from the Latin for "kind" or "appearance," and indeed we learn to distinguish between kinds of plants or animals—between dogs and cats, for instance—from differences in their appearance. **Taxonomy** is the branch of biology concerned with naming and classifying the diverse forms of life. In the eighteenth century, Swedish physician and botanist Carolus Linnaeus developed the two-part, or binomial, system of naming organisms, which we still use. For our own species, the binomial is *Homo sapiens* (Latin for "wise human being"). Linnaeus defined each of the species he named by its physical appearance.

As we saw with spotted skunks, there are potential pitfalls to defining species by their appearance alone. Like the spotted skunks, the two birds in Figure A look much the same, even though they represent two species; the one on the left is an eastern meadowlark (*Sturnella magna*), and the one on the right is a western meadowlark (*Sturnella neglecta*). Though their body shapes and colorations are very similar, the songs of the two species are quite distinct, making it easy to tell them apart.

Whereas the individuals of a species of spotted skunk or meadowlark seem to us to exhibit fairly limited variation in physical appearance, certain other species—our own, for example—seem extremely diverse. If someone did not know that humans all belong to one species, *Homo sapiens*, the physical diversity within our species (partly illustrated in Figure B) might be confusing. Taxonomists usually have to rely on a limited number of phenotypic features (such as physical appearance and songs) to tell one species from another. Whichever features are used, it is important to realize that phenotype reflects both genotype and environmental influences. When all we know about a population are some of the phenotypic traits of its members, we may not be able to tell whether the population should be considered a species.

What actually determines the gap between two species? How can we be sure that a group of organisms constitutes a species? Sexual reproduction is key. A **biological species** is defined as a population or group of populations whose members have the potential to interbreed and produce fertile offspring (offspring who themselves can reproduce). Members of one biological species may mate with members of other species, but any offspring will not be fertile. In effect, a failure of sexual reproduction (an inability to produce fertile offspring) pre-

A. Similarity between species

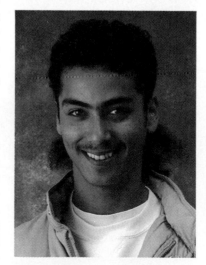

B. Diversity within a species

vents gene flow and determines the gap between species. Western and eastern spotted skunks, for example, remain distinct species even where they share territory, because individuals of these species cannot interbreed and produce fertile hybrids.

When it can be applied, the biological species concept provides a clear-cut definition of a species. Unfortunately, it is not applicable in all cases. For example, classification of extinct forms of life must rely on the appearance and chemical analysis of their fossil remains. Also, two populations that are geographically separated may be placed in the same species because they look identical, but it may not be possible to determine whether they have the potential to interbreed in nature. The criterion of interbreeding is also useless for organisms that are completely asexual in their reproduction, as are bacteria and many single-celled protists. As one of these organisms (a single cell) divides, it produces a lineage of genetically identical cells. Some asexual organisms can exchange genes in processes resembling sex (for instance, bacterial conjugation, described in Chapter 12), but otherwise there is no gene flow among the various lineages resulting from reproduction. Consequently, asexual organisms can be assigned to species only by the grouping of cell lineages that have the same appearance and biochemical features.

The distinction between populations and biological species often blurs

Even if populations reproduce sexually, are not extinct, and are not geographically isolated, there are many cases where the biological species concept cannot be applied unambiguously. For example, consider the deer mouse (*Peromyscus maniculatus*), which is common throughout most of North America. The four names on the map below indicate that there are four populations of this species in the Rocky Mountains. Each population is phenotypically distinct; accordingly, taxonomists call each one a subspecies and give each a third name, as you can see on the map. Notice that the deer mouse populations overlap in certain places. Some interbreeding occurs in these zones; therefore, according to the biological species concept, we would consider the interbreeding populations to belong to a single species.

There is a complication, however. Two of the subspecies—*Peromyscus maniculatus nebrascensis* and *Peromyscus maniculatus artemisiae*—overlap but do not interbreed. Yet their gene pools are not completely isolated, because each of these populations interbreeds with its other neighboring populations (*sonoriensis* and *borealis*), and some genes could flow indirectly from the *artemisiae* population to the *nebrascensis* population. Actually, the extent of gene flow by this roundabout route is probably slight, and if the two populations (*sonoriensis* and *borealis*) connecting them became extinct, *artemisiae* and *nebrascensis* could be named separate species without reservation. This is a situation where two phenotypically distinct populations seem to be on the brink of becoming unable to interbreed. Cases like this may represent **speciation**, the evolution of new species, in progress.

Population biologists are discovering more and more cases where the distinction between populations with limited gene flow and biological species with fully separated gene pools blurs. When there is no genetic exchange at all among populations that are in contact, as with the eastern and western spotted skunks, then they are clearly different species. But when there is a trickle of genes between two quite different populations that are in contact, the biological species concept is difficult to apply. As with the deer mouse, it is as though we are seeing populations at different stages in their evolutionary descent from common ancestors. This is to be expected if new species arise by the gradual divergence of populations.

No single definition of a species can be stretched to cover all cases. Usually, classification based on observable and measurable phenotypic traits is the most practical. However, neither classification by phenotype alone nor the biological species concept is completely applicable to organisms, like the deer mouse, that seem to be in the process of speciating.

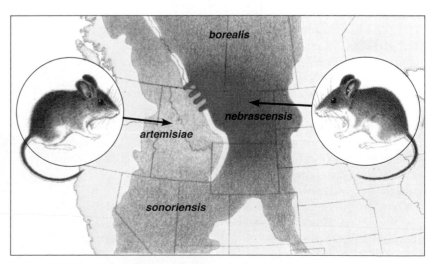

Overlapping subspecies of deer mouse

Reproductive barriers keep species separate

Two different kinds of barriers can prevent closely related species from interbreeding. A geographical barrier isolates two species physically, so they cannot interbreed even though they may have the potential to do so. In contrast, a **reproductive barrier** is a biological feature of the organisms themselves. Reproductive barriers prevent populations belonging to closely related species from interbreeding even when their ranges overlap. As shown in the table at the right, the various types of reproductive barriers that isolate the gene pools of species can be categorized as either prezygotic or postzygotic, depending on whether they function before or after zygotes (fertilized eggs) form.

PREZYGOTIC BARRIERS Prezygotic barriers actually prevent mating or fertilization between species. There are five main types of prezygotic barriers. One type, called **temporal isolation,** occurs when two species breed at different times—during different seasons, at different times of the day, or even in different years. Spotted skunks illustrate seasonal isolation. As mentioned in the chapter's introduction, the eastern and western species overlap in the Great Plains, but the western spotted skunk breeds in the fall and the eastern species in the spring. Many plants also exhibit seasonal differences in breeding time. For example, two species of pine trees, the Monterey pine (*Pinus radiata*) and Bishop's pine (*P. muricata*) inhabit some of the same areas of central California. The two species are reproductively isolated, however, because the Monterey pine releases pollen in February, while the Bishop's pine does so in April. Some plants are temporally isolated because their flowers open at different times of the day, so pollen cannot be transferred from one to another.

In a second type of prezygotic barrier, called **habitat isolation,** two species breed during the same season and in the same general area but not in the same kinds of places. For example, two closely related species of North American toad, *Bufo americanus* and *B. woodhousei,* are isolated by habitat because *B. americanus* breeds in quiet streams, while *B. woodhousei* breeds in temporary rain pools.

In **behavioral isolation,** a third type of prezygotic barrier, there is little or no sexual attraction between females and males of different species. As we discussed in the introduction to Chapter 5, male fireflies of various species signal to females of their kind by flashing their lights in characteristic patterns. Most females respond only to flashes of their own species, by flashing back and attracting the males.

Many animals recognize mates of their species by odor. One individual, usually a female, will emit a perfumelike chemical, called a **pheromone,** into the air. The pheromone signals another individual of the same species to alter its behavior. For example, females of the insect species *Porthetria dispar* (the gypsy moth) attract males by emitting a pheromone. Male gypsy moths are tuned in to the specific

Reproductive Barriers Between Species

PREZYGOTIC BARRIERS: Prevent mating or fertilization	
Temporal isolation:	Mating or flowering occurs at different times for different species.
Habitat isolation:	Species breed in different habitats.
Behavioral isolation:	There is little or no sexual attraction between individuals of different species.
Mechanical isolation:	Structural differences between species (in animal sex organs or plant flowers, for example) prevent copulation or pollen transfer.
Gametic isolation:	Male and/or female gametes die before uniting with gametes of other species, or the gametes fail to unite.

POSTZYGOTIC BARRIERS: Prevent the development of fertile adults	
Hybrid inviability:	Hybrid zygotes fail to develop, or the hybrids fail to reach sexual maturity.
Hybrid sterility:	Hybrids fail to produce functional gametes.
Hybrid breakdown:	Offspring of hybrids are weak or infertile.

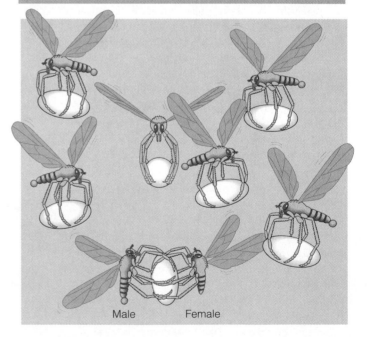

Male   Female

A. Behavioral isolation by courtship ritual

odor of the pheromone produced by females of their species. They do not confuse it with sex attractants given off by females of closely related species.

Figure A (left) shows another form of behavioral isolation, called a courtship ritual. Many species will not copulate until the male and female have performed an elaborate ritual that is unlike that of any other species. The insects in this figure are called dance flies because the males fly up and down in the air in swarms that attract females. The males of this particular species also make balloons from silk secreted by glands in their feet and carry the balloons as they swarm. When a female selects a male from the swarm, the male offers his balloon, and the female accepts it. Only then will the couple leave the "dance floor" and copulate. In other species of dance flies, males offer the females bits of prey or pieces of flower petals, sometimes wrapped in silk. Dance flies are common around streams and lakeshores throughout North America.

A fourth type of prezygotic barrier, called **mechanical isolation,** occurs when female and male sex organs are not compatible, or when flower structure prevents certain animals from pollinating a plant. Mechanical isolation is often important for plants that depend on animal pollinators. Figure B shows a hummingbird obtaining nectar from a flowering plant. As it does so, its head is dusted with pollen, which the bird then transfers to the next flower it visits. In some cases, the beak of a particular species of hummingbird is just the right length for the flower tube of the one plant species it pollinates. Consequently, the bird transfers pollen only among plants of that species.

Gametic isolation is a fifth type of prezygotic barrier. A male and a female from two different species may copulate, but the gametes do not unite to form a zygote. In many mammals, for example, the sperm cannot survive in the female of a different species. Gametic isolation can operate even in species that do not copulate. Male and female sea urchins, for example, release eggs and sperm into the sea, but fertilization occurs only if species-specific molecules on the surfaces of egg and sperm attach to each other.

POSTZYGOTIC BARRIERS   In contrast to prezygotic barriers, postzygotic barriers operate after hybrid zygotes are formed. (Hybrid zygotes are fertilized eggs resulting from the union between gametes of two different species.) In some cases, there is **hybrid inviability;** that is, genes of the two parent species are not compatible, and the hybrids do not survive. Hybrid inviability may occur, for example, in certain frogs of the genus *Rana* that live in the same regions and habitats. Occasional hybrids are produced, but they do not complete development or are extremely frail. Another type of postzygotic barrier is **hybrid sterility,** in which hybrids of two species reach maturity and are vigorous but sterile and therefore unable to bring about gene flow between the parent species. A mule, for example, is the robust offspring of a female horse and a male donkey (Figure C). Smaller than a mule is the hinny, a hybrid of a male horse

B. Mechanical isolation of a plant by the shape of its flowers

and a female donkey. The horse and donkey remain separate species because a mule or a hinny cannot interbreed with either a horse or a donkey. Therefore, the gene pools of the horse and donkey remain isolated. In a third type of postzygotic barrier, called **hybrid breakdown,** the first-generation hybrids are viable and fertile, but when these hybrids mate with one another or with either parent species, the offspring are feeble or sterile. For example, different species of the cotton plant can produce fertile hybrids, but breakdown occurs in the next generation when offspring of the hybrids die in their seeds or grow into weak and defective plants.

In summary, reproductive barriers form the boundaries around closely related species. The process of speciation depends on the formation of these barriers to the spreading of genes. Next, we examine situations that make reproductive isolation, and hence speciation, possible.

Horse   Donkey   Mule

C. Hybrid sterility in the offspring of a horse and a donkey

Geographical isolation can lead to speciation

A crucial episode in the origin of species occurs when a population—with its gene pool—is cut off from other populations of the parent species. With its gene pool isolated, the splinter population can follow its own evolutionary course. Changes in its allele frequencies caused by selection, genetic drift, and mutations occur unaffected by gene flow from other populations. In the formation of many species, the initial block to gene flow seems to have been a geographical barrier that physically isolated a population. This mode of species evolution is called **allopatric speciation** (from the Greek *allos,* other, and *patra,* fatherland). Populations separated by a geographical barrier are known as allopatric populations.

Many factors can isolate a population geographically. For example, a mountain range may emerge and gradually split a population of organisms that can inhabit only lowland lakes; certain fish populations might be isolated this way. Similarly, a creeping glacier may gradually divide a population, or a land bridge such as the Isthmus of Panama may form and separate the marine life on either side.

How formidable must a geographical barrier be to keep allopatric populations apart? The answer depends on the ability of the organisms to disperse. Birds and certain large mammals, such as mountain lions and coyotes, can easily cross mountain ranges, rivers, and canyons. The wind-blown pollen of pine trees is also not hindered by such barriers, and the seeds of many plants may be carried back and forth on animals. In contrast, small rodents may find a deep canyon or a wide river a formidable barrier. For example, the Grand Canyon (pictured below) separates the range of the white-tailed antelope squirrel (*Ammospermophilus leucurus*) from that of the closely related Harris's antelope squirrel (*Ammospermophilus harrisi*). Slightly smaller, with a shorter tail that is white underneath, *A. leucurus* inhabits deserts north of the canyon and west of the Colorado River in southern California. Harris's antelope squirrel has a more limited range, south of the Grand Canyon.

Several factors favor allopatric speciation. An isolated population that is small is more likely than a large population to have its gene pool changed substantially by factors such as genetic drift or natural selection. For example, in less than 2 million years, the small populations of stray animals and plants from the South American mainland that managed to colonize the Galapagos Islands gave rise to all the species that now inhabit the islands.

Geographical isolation creates the opportunity for speciation, but it leads to new species infrequently. For instance, two very large populations of sycamore trees, one in North America, the other in Eurasia, have been separated on their respective continents for about 50 million years. Nevertheless, when specimens from the early British colonies in America were shipped to England, they hybridized with the Eurasian species. Today, the sycamore trees that we see along city streets in America and northern Europe are hybrids.

A. harrisi

A. leucurus

Geographically isolated species of antelope squirrels

When oceanic islands are far enough apart to permit populations to evolve in isolation, but close enough to allow occasional dispersions to occur, they are effectively outdoor laboratories of evolution. Located about 900 km west of Ecuador, the Galapagos Archipelago is one of the world's great showcases of evolution. Each island was born naked from underwater volcanoes and was gradually clothed by plants, animals, and microorganisms derived from strays that rode the ocean currents and winds from other islands and continents. Organisms can also be carried to islands by other organisms, such as sea birds that travel long distances with seeds clinging to their feathers (Figure A).

The species on the Galapagos Islands today, most of which occur nowhere else, descended from stragglers that floated, flew, or were blown over the sea from the South American mainland. For instance, the Galapagos island chain has a total of 13 species of closely related birds called Darwin's finches (because Darwin wrote about them). These birds have many similarities but differ in their feeding habits and their beak type, which is correlated with what they eat. The three examples shown in Figure B illustrate the range of beak shape and size.

Evidence accumulated since Darwin's time indicates that all 13 finch species evolved from a single small population of ancestral birds that colonized one of the islands. Figure C illustrates how this might have happened. Completely isolated on the island, the founder population (species 1 in the figure) may have undergone significant changes in its gene pool and become a new species, which we'll call species 2. Later, a few individuals of species 2 may have been blown by storms to a neighboring island. Isolated on this second island, the second founder population could have evolved into a second new species, species 3. Species 3 could later recolonize the island from which its founding population emigrated, and might coexist there with its ancestral species if reproductive barriers kept the species distinct. In the case of the Galapagos finches, the two species came to rely on different foods, besides being unable to interbreed. In this simplified model, after species 3 evolved on the second island, it colonized a third island, where it adapted and formed species 4. Species 4 then dispersed to the two islands of its ancestors. In actuality, Darwin's finches colonized and speciated repeatedly on the many separate islands of the Galapagos.

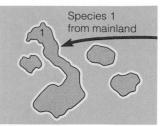
Species 1 from mainland

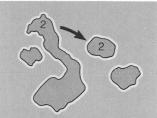

C. Adaptive radiation on an island chain

Today, each of the Galapagos Islands has multiple species of finches, with as many as ten on some islands. In contrast, the island of Cocos, about 700 km north of the Galapagos, has only one finch species, found nowhere else. This single species was apparently derived from an ancestral founder population in the same way as the Galapagos birds, but Cocos Island is so isolated that there apparently has been no opportunity for its finch species to colonize other islands, or for other finches to become established on Cocos.

The emergence of numerous species from a common ancestor introduced to new and diverse environments is called **adaptive radiation.** The effects of adaptive radiation are evident in the many types of beaks, specialized for different foods, of Darwin's finches.

A. Long-distance dispersal of plant seeds

Camarhynchus pauper, an insect-eating tree finch

Geospiza magnirostris, a seed-eating ground finch

Pinaroloxias inornata, a warblerlike finch

B. Examples of Darwin's finches

New species can also arise within the same geographic area as the parent species

Not all species arise as a result of geographical isolation. In **sympatric speciation** (Greek *syn,* together, and *patra,* fatherland), reproductive isolation develops and new species arise without geographical isolation. A new species can arise in a single generation if a genetic change produces a reproductive barrier between mutants and the parent population. Sympatric speciation is rare, but it has been especially important in plant evolution.

Many plant species have originated from accidents during cell division that result in extra sets of chromosomes. In this type of sympatric speciation, the new species is **polyploid,** meaning that it has more than two complete sets of chromosomes. Figure A shows how a polyploid zygote can result from a single parent species that is diploid. The key abnormality is that meiosis fails to occur properly during gamete formation; instead, the cells divide by mitosis. Consequently, the chromosome number is not reduced, and diploid, rather than haploid, gametes are produced. If self-fertilization occurs (and it is common in plants), the resulting zygote is tetraploid; that is, it has four of each type of chromosome. This zygote may develop into a mature plant that can reproduce by self-fertilization.

These new tetraploid plants will also be able to breed with diploid plants of the parental type, but the resulting hybrids will be triploid (3n). The triploid zygote comes from the fusion of a diploid (2n) gamete from the tetraploid parent and a haploid (1n) gamete from the diploid parent. Triploid individuals are sterile; they cannot produce normal gametes because the odd number of chromosomes cannot form homologous pairs and separate normally during meiosis. Thus, the creation of the tetraploid plant has been an instantaneous speciation event: A new species, reproductively isolated from its parent species, has been produced in just one generation.

Sympatric speciation by polyploidy was first discovered by Dutch botanist Hugo de Vries. Working in the early 1900s, de Vries studied genetic diversity in evening primroses (Figure B). In one study he started with *Oenothera lamarckiana* (left inset in Figure B), a diploid species of primrose with 14 chromosomes. From this plant he developed, by self-fertilization, a polyploid—a

tetraploid, with 28 chromosomes. De Vries named this new primrose, which could not interbreed with its parent species, *Oenothera gigas,* for its large size (right inset).

Polyploids do not always come from a single parent species. In fact, most polyploid species arise from the hybridization of two parent species. As we see in the next module, the creation of a polyploid species in this way requires the coupling of two accidental events: the hybridization of the two parent species, and a failure of meiosis in the resultant hybrid.

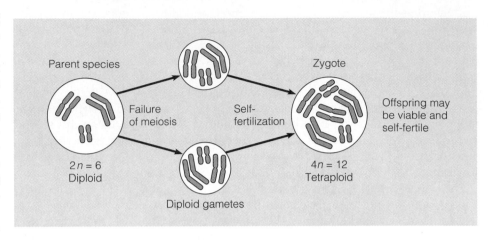

Parent species

Failure of meiosis

Self-fertilization

Zygote

Offspring may be viable and self-fertile

$2n = 6$
Diploid

Diploid gametes

$4n = 12$
Tetraploid

A. Sympatric speciation by polyploid formation

B. Botanist Hugo de Vries with two species of evening primrose

O. lamarckiana

O. gigas

Plant biologists estimate that 25–50% of all plant species are polyploids. Hybridization between two species accounts for most of this polyploidy, perhaps because many of the hybrid polyploids combine the best qualities of their two parent species.

Many of the plants we grow for food are polyploids, including oats, potatoes, bananas, peanuts, barley, plums, apples, sugarcane, coffee, and wheat. Cotton, also a polyploid, remains the source of one of the world's most popular clothing fibers, despite the great numbers of synthetic materials that have been on the market for decades. Cotton cloth is made from the long white plumes that extend from the seeds of the plant.

Wheat, the most widely cultivated plant in the world, occurs as 20 different species of *Triticum*. We know that diploid species of wheat have been cultivated for at least 11,000 years because wheat grains of the diploid species *Triticum monococcum* (with $2n = 14$) have been found in the remains of Middle Eastern farming villages about this old. These diploids have small seed heads and are not highly productive, but some are still cultivated in the Middle East, and many grow wild there.

Our most important wheat species today is bread wheat (*Triticum aestivum*, Figure A), a polyploid with 42 chromosomes. Figure B illustrates how this species evolved; the uppercase letters represent not genes but *sets of chromosomes* that have been traced through the lineage. The process began with hybridization between two diploid wheats, one the domesticated species *T. monococcum* (AA), the other one of several wild species that probably grew as weeds at the edges of cultivated fields (BB). Chromosome sets A and B of the two species would not have been able to pair at meiosis, making the AB hybrid sterile. However, a failure of meiosis in this sterile hybrid and self-fertilization among the resulting gametes produced a new species (AABB) with 28 chromosomes. Today, we know this species as emmer wheat (*T. turgidum*). Varieties of emmer wheat are grown widely in Eurasia and western North America. It is used mainly for making macaroni and other noodle products, because its proteins have a tendency to hold their shape better than bread-wheat proteins.

The final step in the evolution of bread wheat is believed to have occurred in early farming villages on the shores of European lakes over 8000 years ago. At that time, the cultivated emmer wheat, with its 28 chromosomes, hybridized spontaneously with the closely related wild species *T. tauschii* (DD), which has 14 chromosomes. The hybrid (ABD, with 21 chromosomes) was sterile, but a failure of meiosis in this hybrid and self-fertilization doubled the chromosome number to 42. The result was bread wheat, with two each of the three ancestral sets of chromosomes (AABBDD).

A. Bread wheat

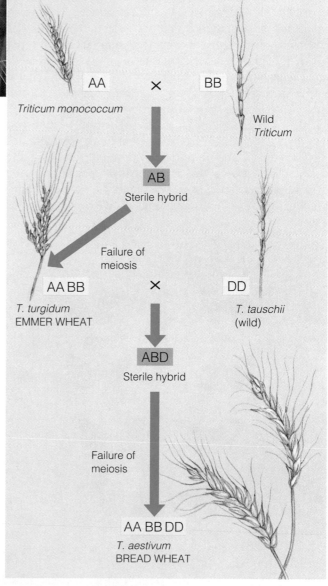

AA × BB

Triticum monococcum

Wild *Triticum*

AB

Sterile hybrid

Failure of meiosis

AA BB × DD

T. turgidum
EMMER WHEAT

T. tauschii
(wild)

ABD

Sterile hybrid

Failure of meiosis

AA BB DD

T. aestivum
BREAD WHEAT

B. The evolution of wheat

15.8 Is the tempo of evolution steady or jumpy?

As we have just seen, when species evolve by polyploidy, evolution occurs suddenly—in a single generation. In contrast, speciation resulting from geographical isolation takes a more gradual course, as gene pools change over many generations. Thus, we might conclude that evolution can occur either in jumpy spurts (such as by polyploidy) or at a slower and steadier tempo.

Evidence of evolution over the millions of years of geological time comes from the fossil record, the chronicle of evolution engraved in the sequence of fossils in layers of rock. Today, there is a lively debate about how to interpret the fossil record. Does it tell us that evolution by natural selection mostly occurs gradually and regularly or that it has a jumpy tempo through geological time?

Figure A illustrates the **gradualist model.** This model is consistent with Darwin's view of the origin of species as an extension of adaptation by natural selection: Populations that have been isolated from common ancestral stock evolve differences gradually as they adapt to their local environments. The figure indicates how eight hypothetical species of butterflies might evolve from a single ancestral species. The narrow base of the arrow on the left indicates that this lineage began as a small isolated population and then diverged gradually from its parent lineage. In both lineages, there is a smooth transition from one species to the next, and evolution occurs as an ancestral species gradually turns into a new species. In both lineages, species change little by little as unique adaptations evolve over long spans of time.

The mechanism underlying the gradualist model is an extension of the processes of microevolution: Changes in allele frequencies in gene pools can lead to the divergence of species. In other words, big changes (speciations) occur by the accumulation of many small ones.

Many evolutionary biologists since Darwin's time, and even Darwin himself, have been struck by the failure of the fossil record to conform to the gradualist model. Few sequences of fossils have ever been found that represent gradual transitions of species. Instead, fossil species usually appear suddenly in a layer of rocks, and many persist essentially unchanged for the whole time they exist on Earth, finally disappearing from the record of the rocks as suddenly as they appeared. Perhaps anticipating a nongradualist model for evolution, Darwin wrote:

> Although each species must have passed through numerous transitional stages, it is probable that the periods during which each underwent modification, though many and long as measured by years, have been short in comparison with the periods during which each remained in an unchanged condition.

In the 1960s and 1970s, several evolutionary biologists developed a nongradualist model known as **punctuated equilibrium,** which is more consistent than gradualism with the fossil record. Figure B illustrates this alternative model, indicating how the evolution of our hypothetical butterflies might occur in spurts instead of being gradual.

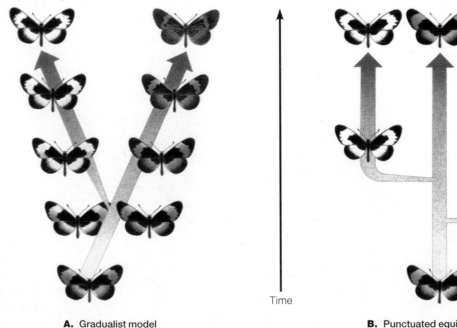

Time

A. Gradualist model   **B.** Punctuated equilibrium model

The narrow bases of the branching arrows indicate small isolated populations that diverge from the parental lineage. The horizontal sections of these arrows indicate that new species change abruptly when they first diverge from a parental lineage. The vertical sections indicate that species change little, if at all, for the rest of their existence. (Notice that the butterfly species represented by the central vertical line, the ancestral lineage, has not changed at all.) The punctuated equilibrium model holds that evolution's tempo has been mainly jumpy, with abrupt episodes of speciation punctuating long periods of little change, or equilibrium.

Is it possible that most species evolve abruptly? In fact, for a small population isolated in a challenging new environment, both genetic drift and natural selection can significantly change the gene pool in just a few hundred to a few thousand generations—a short period in geological time.

The large photograph in Figure C shows the results of a change in climate that led to geographical isolation and rather abrupt speciation. About 50,000 years ago, what is now the Death Valley region of California and Nevada had a wet climate and an extensive system of interconnected lakes and rivers. A drying trend began about 10,000 years ago, and by 4000 years ago, the region had become a desert. Today all that is left of the network of lakes and rivers are isolated springs scattered in the desert, mostly in deep clefts between rocky walls.

The springs vary greatly in water temperature and salinity. Some of them are home to small animals called pupfishes, classified as species of *Cyprinodon*. In some cases, a single spring is home to a species of pupfish adapted to that pool and found nowhere else in the world. Apparently, these desert-pool fishes evolved from a single ancestral species whose range was broken up when the region became arid, confining a number of small populations. Indeed, by either genetic drift or natural selection—and in just a few thousand years—the isolated populations evolved into separate species, each adapted to its home spring. The photograph at the upper right shows ecologists counting pupfish of the species that inhabit a spring called Devil's Hole; the species is fittingly named *C. diabolis* (shown in the small photo).

Can we really consider speciation that takes several thousand years "abrupt"? The fossil record suggests that successful species last for a few million years, on average. Let's say that a particular species survived for 5 million years, but that most of the changes in its body features (the changes involved in speciation) occurred during the first 50,000 years of its existence. In this case, speciation took up only 1% of the lifetime of the species. In the fossil record, the species might ap-

pear abruptly in rocks of a certain age and then continue to appear unchanged in rocks spanning millions of years, before disappearing.

The punctuated equilibrium model helps explain why there are so few transitional species in the fossil record. Nonetheless, some gradualists argue that punctuated equilibrium is an illusion based on only certain types of changes in species. They point out that fossils only give us impressions of the external anatomy and hard parts (the fossilizable skeletons) of extinct species. Changes in internal anatomy, body functions, and behavior would go undetected. Also, population geneticists point out that many of the effects of microevolution occur at the molecular level, without overtly affecting an organism's external body form. Gradualists also note that there is much we do not yet know about the fossil record.

Whatever the outcome of these debates, the punctuated equilibrium model has stimulated research and catalyzed new interest in fossil studies. There will always be debate about aspects of evolutionary theory, but this in no way implies any disagreement about the reality of evolution. Biologists agree that evolution occurs, and debate about its mechanisms only leads to a better understanding of the process. Arguing about evolutionary mechanisms is like arguing about different theories of gravity: We know that objects keep right on falling, even as the debate goes on.

C. Geographical isolation and speciation in Death Valley

15.9 Evolutionary biologist Ernst Mayr connects Darwinism with the modern era

Ernst Mayr is one of the foremost evolutionary biologists of the twentieth century. He was one of the main architects of the modern synthesis, the melding of population genetics with Darwin's theory of natural selection. Now in his nineties and Professor Emeritus at Harvard University, Mayr has a unique grasp of the sweep of ideas that extend to us from Darwin's era. His latest book, *One Long Argument*, published by Harvard University Press in 1991, analyzes Darwin's theories from this broad perspective. In a 1992 interview, Mayr commented:

> I've found in the literature that there are seven or eight different meanings of the word Darwinism for different people. In Darwin's own time, for instance, Darwinism meant evolution without supernatural causation—nothing more, nothing less. Today, Darwinism means the theory of natural selection.

Like Darwin, Mayr spent his early twenties as a field naturalist. In our interview, he talked about some of his early studies on speciation in New Guinea and the Solomon Islands:

> Just as Wallace had found in the Indonesian Islands, and Darwin in the Galapagos, islands are the best place to demonstrate speciation. There is no place in the world in which . . . speciation is as well demonstrated as in the changes from island to island in the Solomon Islands.

Mayr has written extensively about speciation and how it is often tied to geographic isolation on islands—of all sorts. In the interview, he commented:

> Each mountain range in New Guinea has its [own] species of birds; the mountains are like islands in the sky. Geographic speciation takes place on these mountaintops just as it takes place on islands in the ocean.

Extending the island concept more broadly:

> One of the things that I followed up . . . was to see if geographic separation also caused speciation on continents. Indeed, wherever there are barriers, due to water,

mountains, or vegetational changes, speciation can occur.

One of Mayr's most important contributions was developing the idea that founder populations, those few individuals that establish new colonies "beyond the existing species border," have played a key role in the formation of many, and perhaps most, species. As Mayr sees it, the founder principle explains why the fossil record seems inconsistent with the idea of gradualism:

> [Because] the really important evolutionary changes probably . . . came out of founder populations, [we] find little evidence for the gradual origin of new species. [Instead,] a lineage changes somewhat, but not very drastically, and then suddenly a new species originates somewhere else, next to it. Once a species has become widespread and successful, it no longer changes very much [and] may continue [unchanged] for many millions of years.

We asked Professor Mayr if he thought the punctuated equilibrium model threatens the gradualist Darwinian view. His response:

> A mistake people make is thinking that if something evolves very rapidly, it is no longer Darwinian gradual evolution. But as long as the evolution occurs at the populational level and not at the level of individuals, then it is gradual evolution, occurring over many generations. Some paleontologists call the origin of a species relatively sudden if it takes place during 1% of the total life span of the species. But that 1% is 50,000 years to 100,000 years in a species living 5 or 10 million years. Hardly anyone else would call that sudden.

Finally, we asked Mayr what he would most like to tell Darwin if he had the chance. Noting that Darwin died before anyone but Mendel understood genetics, Mayr responded:

> I would just convey to him what we know about genetic variation in populations, because that's the area where he was most puzzled and where he was most anxious to learn more—and never succeeded.

Chapter Review

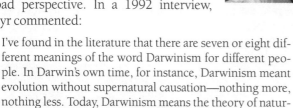

Begin your review by rereading the module headings and scanning the figures before proceeding to the Chapter Summary and questions.

Chapter Summary

Introduction–15.1 We generally distinguish between kinds, or species, of plants and animals on the basis of their appearance. Linnaeus also used physical appearance to identify species when he established the binomial system for naming organisms. Though

modern biologists usually distinguish species this way, both similarities between different species and diversity within a species can be confusing.

15.1, 15.3 The biological species concept is one way of defining a species. A biological species is a population or group of populations whose members can interbreed and produce fertile offspring. Prezygotic and postzygotic reproductive barriers prevent individuals of different species from interbreeding. Prezygotic barriers are habits or mechanisms that prevent mating or fertilization. Postzygotic barriers operate after fertilization; they result in inviability or sterility of hybrids or their offspring.

15.1–15.2 The biological species concept is not applicable in every case. It cannot be applied to fossils, asexual organisms, or organisms in isolated populations having little opportunity to interbreed. Limited interbreeding between populations makes it difficult to define species, and such situations may illustrate speciation, the formation of new species.

15.4–15.5 When a population is cut off from its parent stock, it may veer off on its own evolutionary course. Species evolution that occurs in this manner is called allopatric speciation. Often the initial block to gene flow is a geographical barrier such as a canyon or mountain range. Cut off from other populations, a small splinter population is more likely than a large one to have its gene pool changed by selection, genetic drift, or unique mutations. Small populations sometimes colonize and adapt to islands, making them natural laboratories for evolution. For example, the Galapagos Islands have become populated by plants and animals from South America. On these islands, repeated isolation and adaptation have resulted in adaptive radiation of 13 species of Darwin's finches from a common ancestral population.

15.6–15.7 In sympatric speciation, relatively common in plants, new species may arise without geographical isolation. Failure of meiosis can produce diploid gametes. Self-fertilization can then produce a tetraploid zygote, which may develop into a plant that can reproduce by self-fertilization. Because of its polyploid set of chromosomes, this plant is an instant new species, isolated from its parent. Many plants, including many food plants, are polyploid. They are often the products of hybridization, followed by meiotic failure and self-fertilization. Two such episodes appear to have occurred in the evolution of modern bread wheat.

15.8–15.9 According to the gradualist model of the origin of species, new species evolve by the gradual accumulation of small changes brought about by natural selection. But few gradual transitions are found in the fossil record; new species appear suddenly and may persist unchanged for their entire history. The punctuated equilibrium model suggests that speciation occurs in spurts. Rapid change might occur when an isolated population diverges from the ancestral stock, but through most of its history, a species changes little. Scientists continue to debate these models. However, they are debating not whether evolution takes place, but rather the relative importance of different evolutionary mechanisms.

Testing Your Knowledge

Multiple Choice

1. Biologists have found more than 500 species of fruit flies on the various Hawaiian Islands, all apparently descended from a single ancestor species. This example illustrates
   a. polyploidy
   b. temporal isolation
   c. adaptive radiation
   d. hybrid breakdown
   e. meiotic failure

2. Bird guides once listed the myrtle warbler and Audubon's warbler as distinct species, but recent books show them as eastern and western forms of a single species: the yellow-rumped warbler. Apparently, it has been found that the two kinds of warblers
   a. live in the same areas
   b. successfully interbreed
   c. are almost identical in appearance
   d. are merging to form a single species
   e. live in different places

3. Which of the following is an example of a postzygotic reproductive barrier?

   a. One *Ceanothus* shrub lives on acid soil, another on basic soil.
   b. Mallard and pintail ducks mate at different times of year.
   c. Two species of leopard frogs have different mating songs.
   d. Hybrid offspring of two species of jimsonweeds always die.
   e. Pollen of one kind of tobacco cannot fertilize another kind.

4. Species of horseshoe crabs and cockroaches have existed virtually unchanged for millions of years. Which of the following models of speciation best explains these "living fossils"? (*Explain your answer.*)
   a. gradualism
   b. polyploidy
   c. punctuated equilibrium
   d. sympatric speciation
   e. adaptive radiation

5. A small, isolated population is more likely to undergo speciation than a large one, because a small population
   a. is more affected by genetic drift and natural selection
   b. is more susceptible to gene flow
   c. contains a greater amount of genetic diversity
   d. is more subject to errors during meiosis
   e. is more likely to survive in a new environment

Describing, Comparing, and Explaining

1. Explain how each of the following makes it difficult to clearly define a species: diversity within a species, geographically isolated populations, asexual reproduction, fossil organisms.

2. The mating of a horse and a donkey produces a mule. Does this mean that horses and donkeys are the same species? Why or why not?

3. Explain why islands are "living laboratories for speciation," and describe an example.

Thinking Critically

1. Cultivated American cotton plants have a total of 52 chromosomes ($2n = 52$). In each cell, there are 13 pairs of large chromosomes and 13 pairs of smaller chromosomes. Old World cotton plants have 26 chromosomes ($2n = 26$), all large. Wild American cotton plants have 26 chromosomes, all small. Explain how cultivated American cotton probably originated.

2. The yellow-bellied marmot (*Marmota flaviventris*) is a large rodent that lives, among other places, in alpine meadows in the Cascade Mountains of the Pacific Northwest. The Olympic marmot (*Marmota olympus*) lives in the Olympic Mountains to the west, separated from the yellow-bellied marmot by the forested lowlands around Puget Sound. Experts think the two species are closely related. Describe how an ancestral population may have given rise to both species. (*Hint:* During the ice ages, the habitat favored by marmots occurred at lower elevations.)

Science, Technology, and Society

In 1991, the U.S. government listed the Snake River sockeye salmon as an endangered species. Thousands of sockeyes used to spawn in Redfish Lake, Idaho, but in 1990 not a single fish completed the 1400-km journey from the Pacific Ocean to the spawning grounds. The salmon are impeded on their journey up the Columbia and Snake rivers by hydroelectric dams, and young fish going downstream are killed passing through the dams. The Endangered Species Act prohibits any disturbance of a listed species, and it grants protection not just to whole biological species, but also to "distinct population segments." There are other sockeye populations in other tributaries of this system and in other river systems—some spawning successfully, others declining. A number of biological questions need to be addressed, because proposed changes in operation of the dams could have a great economic impact. In your opinion, what would constitute a "distinct" salmon population? How distinct would it have to be to warrant protection? How would you try to find out whether a salmon population meets your criteria? Is it necessary to preserve local populations if the species as a whole is in no danger of extinction? Why or why not?

Tracing Evolutionary History

Some 80 million years ago, when this duck-billed dinosaur laid her eggs, the North American continent was split in half by a shallow inland sea, now called the Western Interior Seaway. West of the seaway, land that is now part of eastern Montana and the Dakotas rose gently toward the newly formed Rocky Mountains, creating a coastal plain some 400 kilometers wide. Today, this region is arid and sparsely vegetated, with hot summers and cold winters. Eighty million years ago, it was densely vegetated and subtropical—a dinosaur's paradise, with warm rivers meandering down from the Rockies, carrying silt to seaside deltas.

Duck-billed dinosaurs, also called hadrosaurs (Greek for "stout lizards"), were large; the ones shown here were about 7 meters long. Hadrosaurs walked on all fours but could also stand upright on their hind legs. Fossilized imprints of their tracks indicate that these dinosaurs were numerous, often traveled in herds, and spent much of their time eating plants in or near water. Fossils also indicate that hadrosaurs had well-developed ears and were quite vocal. Their ducklike beaks may have served as resonating chambers for sounds produced in their throats, and they may have kept track of their young and other herd members by loud calls. Their roars may have pierced the tropical air of their lush habitat, the way the bellows of modern alligators echo through the wetlands of our Gulf coast states today.

In the late 1970s, researchers made some startling discoveries on the dry plains east of the Rocky Mountains in Montana and Alberta, Canada. Near the small town of Choteau, Montana, they began unearthing entire dinosaur nesting grounds, with many large mud nests (each about 2 meters wide and 1 meter deep) and fossil skeletons of hadrosaurs. Apparently, a large number of the reptiles had nested near a river, and a flash flood had devastated the area, covering everything with silt. Their findings led the researchers to name the species of hadrosaur pictured here *Maiasaura peeblesorum* (Greek *maia*, good mother, and *sauros*, lizard; "of Peebles," the owners of the land where the fossils were found).

The "good mother lizard" had a whole new story to tell about dinosaurs. Until its nesting grounds were unearthed, dinosaur eggs and nests were almost unheard of. A few had been found in the 1920s in the Gobi Desert in East Asia, but North America seemed to lack them altogether. Earlier prevailing opinion held that all dinosaurs, like many reptiles of today, laid their eggs, covered them with sand, and then promptly abandoned them. But some of the *Maiasaura* nests contained fossilized skeletons of well-developed young up to three months old. *Maiasaura* apparently remained with its nest and cared for its young for at least this much time. Some of the nests also contained small bits of fossilized eggshells, another indication of parental care. Today, the nests of some birds that nurture their young contain similar shell fragments; the developing young pulverize the shell fragments as they move around in the nest. In contrast, we see much larger shell fragments in the nests of birds, such as ducks, whose hatchlings leave the nest almost immediately.

Fossil finds like the *Maiasaura* nests have generated a lively debate about dinosaurs. Why were some groups, such as the hadrosaurs, so numerous and successful for millions of years? Was it because they lived in large groups and cared for their young? Hadrosaurs were prey to many carnivorous dinosaurs. Neither the young nor the adults were armored, and the structure of their skeleton indicates that they were not fast runners. Adults and active young could seek safety in the water when attacked, and their island nesting sites undoubtedly afforded some protection. But, perhaps more often, a herd or a large nesting group of adults discouraged predators by swinging their heavy tails and stamping their feet. This group behavior, coupled with parental feeding and care of the young, may have been key to their success.

Fossils have much to tell us about the life histories of extinct organisms. They also have much to tell us about the evolutionary history of life, which is the main subject of this chapter. Let's begin by focusing on fossils themselves.

Fossils form in a variety of ways

A. A paleontologist with dinosaur fossils

Scientists who study fossils, the preserved imprints or remnants of ancient organisms, are called **paleontologists** (from the Greek *palaio-,* old, *ont-,* being, and *logos,* discourse). The paleontologist in Figure A is recovering parts of a dinosaur skeleton at Utah's Dinosaur National Monument. The photographs below illustrate several fossils, each of which formed in a somewhat different way.

The organic substances of a dead organism usually decay rapidly, but hard parts of an animal that are rich in minerals, such as the bones and teeth of dinosaurs and the shells of clams or snails, may remain as fossils. The fossilized skull in Figure B is from one of our early relatives, *Australopithecus africanus,* who lived some 2.5 million years ago in Africa.

Some fossils do retain organic material. The leaf on the left in Figure C is about 40 million years old. It is a thin film pressed in rock, still greenish with remnants of its chlorophyll and well enough preserved that biologists can analyze its molecular and cellular structure. In rare instances, an entire organism, including its soft parts, is fossilized. This can happen only if the individual is buried in a medium that prevents bacteria and fungi from decomposing the corpse. The insect at the right in Figure C got stuck in the resin of a tree about 40 million years ago. The resin hardened into amber (fossilized resin), preserving the insect intact. Today, amber is prized for its yellow or brownish-red sheen and is often used in jewelry.

B. A fossilized skull

C. Fossilized organic material

Sometimes the remains of dead organisms are actually turned into stone by a process called petrification. Petrification occurs when minerals dissolved in groundwater seep into the tissues of dead organisms and replace organic matter. The petrified trees in Figure D stood about 190 million years ago in what is now a desert in eastern Arizona.

D. Petrified trees

The fossils resembling wrinkled butterfly wings in Figure E are imprints of the shells of animals called brachiopods that lived about 375 million years ago. When the animals died, they were buried in mud on the seafloor. When they decayed, the mud remained, forming empty molds (the indented forms you see in the photograph). Some of the molds then filled with water and dissolved minerals, and casts of the brachiopods (the raised forms in the photo) took shape as the minerals settled out and hardened. Brachiopods, which feed on organic particles suspended in seawater, were a dominant marine group with many thousands of species when these fossils formed. Today, only about 300 species remain.

E. Fossilized imprints: molds (right) and casts (left)

Sedimentary rocks are a rich source of fossils 16.2

Sedimentary rocks (from the Latin *sedere*, to settle) form from layers of minerals that settle out of water. Sand and silt eroded from the land are carried by rivers to seas and swamps, where the particles settle to the bottom. Over many millions of years, deposits pile up and compress the older sediments below into rock—sand into sandstone, and mud into shale.

Sedimentary rocks often contain fossils. When aquatic organisms die, they settle along with the sediments and may leave imprints in the rocks, such as those in Figure 16.1E (above). Many terrestrial organisms are swept into swamps and the seas. Some remain in place when they die and are covered by windblown silt. Footprints in mud or volcanic ash may also be covered. When the sea level rises or lakes and swamps fill low-lying areas, sediments cover terrestrial deposits, and many of the organisms leave fossils.

When a region is submerged, the rate of sedimentation and the types of particles that settle may vary with time. As a result of these variations in sedimentation, sedimentary rock forms in layers, or strata, as shown in the photograph here. Each stratum may bear a unique set of fossils representing a local sampling of the organisms that lived when the sedi-

Strata of sedimentary rock

ment was deposited. Younger strata are on top of older ones; thus, the position of fossils in the strata reveals their relative age. For instance, paleontologists know that *Maiasaura* and other hadrosaurs were among the last of the dinosaurs because fossils of these reptiles occur in some of the youngest sedimentary strata from the age of dinosaurs.

The fossil record chronicles macroevolution

The order in which fossils appear in rock strata, called the fossil record, is a catalog of evolutionary history. Sedimentary rocks in one area provide only a local glimpse of evolution, but by studying fossils in strata from many locations, paleontologists can trace **macroevolution,** the main events in the evolutionary history of life on Earth. By compiling a list of the organisms present at widely separated sites, a paleontologist obtains a broad picture of the kinds of organisms that existed when the stratum formed. Additional information comes from fossils present in different kinds of strata that were deposited at the same time in geological history.

The Geological Time Line

Era	Period	Epoch	Millions of Years Ago	Some Important Events in the History of Life
CENOZOIC	Quaternary	Recent	0.01	Historic time
		Pleistocene	1.8	Ice ages; humans appear
	Tertiary	Pliocene	5	Apelike ancestors of humans appear
		Miocene		Continued radiation of mammals and angiosperms
		Oligocene	24	Origins of many primate groups, including apes
		Eocene	38	Angiosperm dominance increases; origins of most modern mammalian orders
		Paleocene	54	Major radiation of mammals, birds, and pollinating insects
MESOZOIC	Cretaceous		65	Flowering plants (angiosperms) appear; dinosaurs become extinct at end of period
	Jurassic		144	Gymnosperms continue as dominant plants; dinosaurs dominant
	Triassic		213	Cone-bearing plants (gymnosperms) dominate landscape; first dinosaurs, mammals, and birds
PALEOZOIC	Permian		248	Radiation of reptiles; origins of mammal-like reptiles and most modern orders of insects; extinction of many marine invertebrates
	Carboniferous		286	Extensive forests of vascular plants; first seed plants; origin of reptiles; amphibians dominant
	Devonian		360	Diversification of bony fishes; first amphibians and insects
	Silurian		408	Diversity of jawless fishes; first jawed fishes; colonization of land by vascular plants and arthropods
	Ordovician		438	Marine algae abundant
	Cambrian		505	Origin of most invertebrate phyla; first vertebrates (jawless fishes); diverse algae
PRECAMBRIAN			590	
			670	Diverse invertebrate animals, mostly soft-bodied
			700	Origin of first animals
			1500	Oldest eukaryotic fossils
			2500	Oxygen begins accumulating in atmosphere
			3500	Oldest definite fossils known (prokaryotes)
			4600	Approximate time of origin of Earth

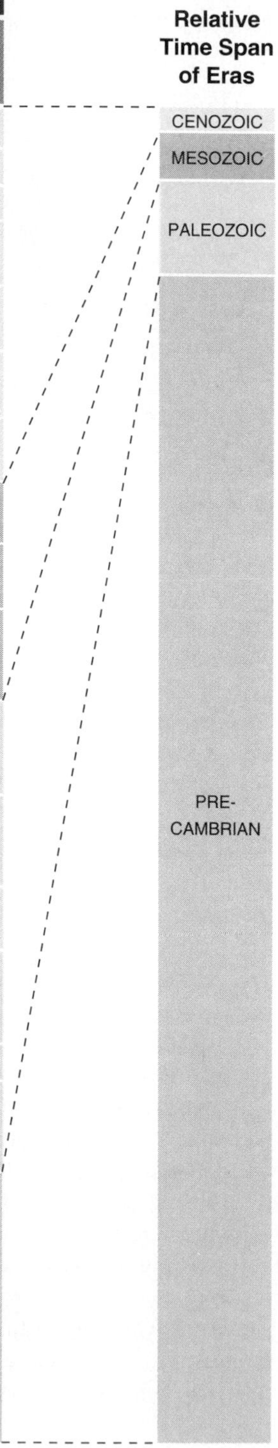

Relative Time Span of Eras

CENOZOIC
MESOZOIC
PALEOZOIC
PRE-CAMBRIAN

Since the mid-1800s, paleontologists throughout the world have been tracing the main events in the history of life, using evidence from the sequence of fossils at many different sites. Their results are summarized in the geological time line on the facing page. The table provides an overview of macroevolution, measured in millons of years. Notice that the time line is divided into major groupings called eras and that the diagram to the right of the table shows the relative lengths of the eras. Also, notice that most eras are subdivided into periods. The boundaries between the eras mark major transitions in the forms of life fossilized in the rocks; this is also true to some extent for periods.

Let's start with the oldest era, at the bottom of the time line. Rocks of most of the Precambrian era have undergone extensive change over time, and much of their fossil contents are no longer visible. Nonetheless, paleontologists have found some fossil-rich Precambrian strata and have pieced together the sequence of major events in life's history indicated at the bottom of the table. The earliest discovered fossils, dating from 3.5 billion years ago, represent microorganisms only. Strata from the late Precambrian (some 670 million years ago) bear highly diverse fossils, including ones of jellyfish, corals, and worms that resemble certain modern species, as well as other animals that seem to have no counterparts on Earth today.

Dating from about 590 million years ago, rocks of the Paleozoic ("ancient animal") era contain fossils of lineages that gave rise to modern organisms, as well as many lineages that have become extinct. During the early Paleozoic, virtually all life was aquatic, but by about 400 million years ago, plants and animals were well established on land.

Following the Paleozoic era was the Mesozoic ("middle animal") era, also known as the age of reptiles. More accurately, it should be called the age of dinosaurs and cone-bearing plants (gymnosperms), because both of these groups of organisms dominated the Mesozoic landscape. The Mesozoic also saw the beginnings of mammals, birds, and flowering plants (angiosperms). By the end of the Mesozoic, all the dinosaur lineages except one, which gave rise to birds, had become extinct, and the stage was set for an explosive period of evolution of mammals, birds, and angiosperms.

The Cenozoic ("recent animal") era began with the downfall of the dinosaurs some 65 million years ago and continues today. Because much more is known about the Cenozoic than about earlier eras, our table subdivides the two Cenozoic periods into finer intervals called epochs. Our own species, *Homo sapiens*, appeared on the scene only about 100,000–200,000 years ago, during the Pleistocene (Ice Age) epoch. Thus, our tenure on the planet is only a tiny postscript to the immense saga of geological time.

The actual ages of rocks and fossils mark geological time　16.4

The record of the rocks chronicles the *relative* ages of fossils; it tells us the order in which groups of species present in a sequence of strata evolved. However, the sequence alone does not tell the *actual* ages in years of the embedded fossils. Imagine yourself peeling the layers of wallpaper from the walls of an old house that has been inhabited by many owners. You can determine the sequence in which the papers were applied but not the year that each layer was added. Likewise, you cannot determine when evolutionary events occurred without finding out the actual ages of fossils.

Geologists and paleontologists use several techniques to determine the ages of rocks and the fossils they contain. The method most often used, called **radioactive dating,** is based on the fact that living organisms contain certain atomic isotopes in certain ratios. For instance, living organisms have the same constant ratio of carbon-14 (^{14}C), a radioactive isotope, to carbon-12 (^{12}C), a stable isotope, as does Earth's atmosphere. However, when an organism dies, its ratio of ^{14}C to ^{12}C starts to drop, because ^{14}C decays to other chemical elements, and the organism no longer obtains any ^{14}C from the atmosphere. Each radioactive isotope has a fixed rate of decay, known as its half-life. For example, ^{14}C has a half-life

of 5600 years, meaning that half the ^{14}C in a specimen decays in about 5600 years, half the remaining ^{14}C decays in the next 5600 years, and so on, until all the ^{14}C is gone. Knowing both the half-life of a radioactive isotope and the ratio of radioactive to stable isotope in a fossil enables us to tell how old the fossil is. For instance, if a fossil has a ^{14}C-to-^{12}C ratio half that of the atmosphere, it is about 5600 years old; a fossil with one-fourth the atmosphere's ratio is about 11,200 years old.

The ^{14}C-to-^{12}C ratio is reliable for dating fossils less than about 50,000 years old. To date older fossils, paleontologists use radioactive isotopes with longer half-lives. For instance, potassium-40, a radioactive isotope with a half-life of 1.3 billion years, can be used to date rocks and fossils hundreds of millions of years old. Radioactive dating has an error factor of plus or minus about 10%.

The dates you see on the geological time line on the opposite page were established by dating rocks and fossils. Notice that the geological periods span unequal intervals of time. The boundaries between periods mark distinct changes in the species composition of the sedimentary rocks. In the next module, we examine some of the geological processes that can produce these distinct changes.

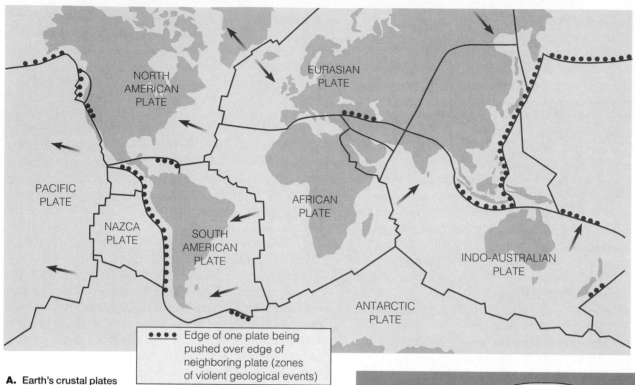

A. Earth's crustal plates

●●●● Edge of one plate being pushed over edge of neighboring plate (zones of violent geological events)

In 1912, German meteorologist Alfred Wegener proposed the hypothesis of **continental drift.** Wegener postulated that all land on Earth was once one great mass, which broke up into continents that drifted like rafts to their present positions. Wegener believed that the shapes of our modern continents, like pieces of a jigsaw puzzle, reflect their former positions in the original supercontinent. Like many ideas generated before their time, Wegener's were not taken seriously for many years. Until the 1960s, the general belief was that the continents have always been fixed in their present positions.

In recent decades, geologists, paleontologists, and biologists have accumulated overwhelming support for the concept of continental drift. We now know that the continents and seafloors form a thin outer layer of planet Earth, called the crust, and that under the crust is a mass of hot material called the mantle. Furthermore, the crust is divided into giant, irregularly shaped plates (outlined in red in Figure A). Because the mantle circulates constantly, the crustal plates move about slowly but incessantly—they literally float—on the underlying mantle. As a result of plate movements, world geography changes constantly, for unless two landmasses are embedded in the same crustal plate (India and Australia, for instance), their positions relative to each other do not remain the same. North America and Eurasia, for example, are presently drifting apart at a rate of about 2 cm per year. Throughout geological time, continental

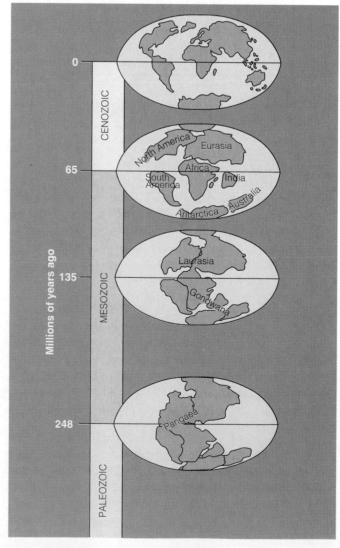

B. Continental drift

movements have greatly influenced the distribution of organisms around the world, and continental drift explains much of the history of life.

Two chapters in the continuing saga of continental drift seem to have been especially significant in their influence on life. The first occurred about 250 million years ago, near the end of the Paleozoic era, when plate movements brought all the land masses back together into the supercontinent Wegener had originally proposed. Shown at the bottom in Figure B, this supercontinent is called **Pangaea,** meaning "all land." Imagine some of the possible effects on life as massive continents came together. Species that had been evolving in isolation came together and competed. When the landmasses fused, the total amount of shoreline and shallow coastal areas was reduced. Then, as now, most marine species inhabited shallow waters, and many of these organisms probably died out as their habitats shrank. The formation of Pangaea also would have altered terrestrial environments, because ocean currents, which affect climates on land, would have changed drastically. Moreover, with Pangaea's reduced coastline, less land area would have been exposed to the ocean's moderating effect on air temperatures. As a result, much of Pangaea may have had a harsh climate.

Overall, the fossil record indicates that the formation of Pangaea reshaped biological diversity, causing great numbers of extinctions. These in turn provided new opportunities for organisms that survived the crisis.

The second dramatic chapter in continental drift began about 180 million years ago, during the early Mesozoic era. Pangaea started to break apart again, causing colossal changes for life forms of the time. As the continents drifted apart, each became a separate evolutionary arena—a huge island on which organisms evolved in isolation from their previous neighbors. At first, Pangaea split into northern and southern landmasses, which we call Laurasia and Gondwana, respectively. This split was more or less complete by about 135 million years ago, as shown in Figure B. By the end of the Mesozoic era (and the Cretaceous period), some 65 million years ago, the modern continents were beginning to take shape. Then, just 10 million years ago, India collided with Eurasia, and the slow, steady crunching of the Indian and Eurasian plates formed the Himalayas, the tallest and youngest of Earth's major mountain ranges. (Regions where two plates crunch together are marked by red dots in Figure A.)

The pattern of continental separation solves many puzzles. One example is the distribution of a group of ancient vertebrates called lungfishes (Figure C). Today, there are six species of lungfish in the world, four in Africa, and one each in Australia and South America (orange areas in Figure D). What is the evolutionary history of these fishes? As the triangles in Figure D indicate, fossil lungfishes have been found on all continents except Antarctica. This widespread fossil record indicates that lungfishes evolved when Pangaea was intact.

C. An African lungfish

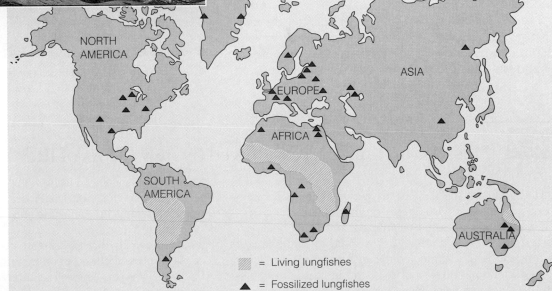

D. Lungfish distribution, a result of continental drift

⌧ = Living lungfishes

▲ = Fossilized lungfishes

16.6 Tectonic trauma imperils local life

Geologists call the forces within the Earth that cause movements of the crust **plate tectonics**. These forces affect life in many ways. Not only do they move continents and cause mountain ranges to rise, they also produce volcanoes and earthquakes. The boundaries of crustal plates are hotspots of geological activity. California's earthquake of October 1989 resulted from movement along the infamous San Andreas fault, which is part of the border where the Pacific plate and the American plate grind together and gradually slide past each other (Figure A). Earthquakes can also occur in the interiors of crustal plates, but these movements ultimately result from forces exerted at the edges of plates.

Volcanoes can also cause tremendous devastation. Mount Saint Helens, the volcano in Washington state that has been active in recent years, is near another plate boundary. Figure B illustrates the eruption of Mount Saint Helens in 1980.

Volcanoes at sea can produce islands, such as the Galapagos, but life on volcanic islands often hangs in a precarious balance. The same volcanic activity that creates an oceanic island may destroy life that evolves there. In 1883, fiery pumice from a volcano covered the small island of Krakatoa, near the boundary between the Indo-Australian plate and the Eurasian plate. Before the eruption, Krakatoa was covered with a dense tropical rain forest. Afterward, it was virtually devoid of life. Despite the devastation, life from neighboring islands began recolonizing Krakatoa soon after its surface cooled. Within 50 years after the eruption, tropical forest with a great diversity of plants and animals again covered the island.

Plate tectonics has played a role in many of the major changes that characterize evolutionary history. As we see in the next module, extraterrestrial objects may also be very important.

A. The San Andreas fault, a boundary between two crustal plates

B. A volcanic eruption at Mount Saint Helens

16.7 Mass extinctions were followed by new forms of life

At the end of the Cretaceous period, about 65 million years ago, the world lost an enormous number of species—more than half of its marine animals, and many lineages of terrestrial plants and animals. For some 150 million years before, dinosaurs dominated the land and air. Then, in less than 10 million years—a brief period in geological time—all the dinosaurs were gone. Scientists have been debating

what happened for more than a decade. The question mark in the panorama on the facing page symbolizes the debate.

What do we know about this unusual time? The fossil record shows us that the climate cooled late in the Cretaceous and that shallow seas were receding from continental lowlands. We also know that many plants needed by dinosaurs for food died out first. Perhaps most telling of all, the sediments deposited at the time contain a thin layer of clay rich in iridium, an element very rare on Earth but common in meteorites. Many paleontologists conclude that the iridium layer is the result of fallout from a huge cloud of dust that billowed into the atmosphere when a large meteorite or asteroid hit Earth. The cloud would have blocked light and disturbed climate severely for months, perhaps killing off many plant species.

The asteroid hypothesis has many supporters, and a large asteroid crater that dates from the late Cretaceous has been found under the Caribbean Sea. Many scientists believe the impact that produced that crater could have caused global climate change and mass extinctions. Other researchers propose that climatic changes due to continental drift could have caused the extinctions, whether an asteroid collided with Earth or not. Still others point to evidence in the fossil record in India indicating that, in the late Cretaceous, massive volcanic activity released particles into the atmosphere, blocking sunlight and thereby contributing to the cooling. The various hypotheses are not mutually exclusive, and researchers continue to debate the extent to which each contributed to the extinctions.

Extinction is inevitable in a changing world. A species may become extinct because its habitat has been destroyed or because of unfavorable climatic changes. Extinctions occur all the time, but extinction rates have not been steady.

There have been six distinct periods of mass extinction in the last 600 million years. These were times when environmental changes on a global scale were so rapid and disruptive that most species were swept away. During these times, losses escalated to nearly six times the average rate.

Of all the mass extinctions, the ones marking the ends of the Cretaceous and the Permian periods have been the most intensively studied. The Permian extinctions, at about the time the continents merged to form Pangaea, claimed over 90% of the species of marine animals and took a tremendous toll on terrestrial life as well. Whatever their causes, mass extinctions affect biological diversity profoundly.

But there is a creative side to the destruction. Each massive dip in species diversity has been followed by an explosive *increase* in diversity. Extinctions seem to have provided the surviving organisms with new environmental opportunities. Consider the case of our own class of animals: mammals. Mammals existed for at least 75 million years before undergoing an explosive increase in diversity just after the Cretaceous. Their rise to prominence was undoubtedly associated with the void left by the extinction of the dinosaurs. The world might be a very different place today if a few families of dinosaurs had escaped the Cretaceous extinctions, or if none of the mammals that lived in the Cretaceous had survived.

Animals and plants before and after the Cretaceous extinctions

Key adaptations may allow species to survive and proliferate after mass extinctions

Geologists and paleontologists have found evidence that each of the six periods of mass extinction in the past 600 million years was followed by a virtual explosion of organisms that had previously been much less prevalent. What allowed some species to survive and proliferate when great numbers of others died out? There is no pat answer. Chance undoubtedly played a major role in determining whether organisms survived or became extinct, and survival does not imply that one species is somehow better than another. It is likely that certain features—key adaptations—allowed surviving species to withstand the great changes of the mass extinction periods. These same features may have played a major role in helping survivors adapt further and proliferate after many other species died out. For example, hair and the ability to nurse young on milk are unique mammalian features that evolved long before the Cretaceous extinctions. So are feathers in birds. These key adaptations probably helped mammals and birds adapt to an environment in which the dinosaurs could not survive.

How do key adaptations arise? One way is by the gradual refinement of existing structures for new functions. Evolutionary biologists use the term **preadaptation** for a structure that evolved in one context and later was adapted for another function. This term suggests that a structure can become adapted to alternative functions; it does not mean that a structure somehow evolves in anticipation of future use. Indeed, natural selection can only result in the improvement of a structure in the context of its current function.

The plants in the photographs here illustrate the concept of preadaptation. They are members of a large group (about 2000 species) called bromeliads, common throughout tropical and subtropical America. A few bromeliads thrive in arid soils, the kind of environment that ancestral bromeliads inhabited. Most bromeliads, however, are "air plants," growing on other plant surfaces, often high off the ground where they do not have access to the soil.

Being arid-adapted, the ancestors of bromeliads had a number of preadaptations that helped them spread into habitats lacking soil. We see evidence of these preadaptations in modern bromeliads. Consider, for example, the pineapple (*Ananas,* Figure A), which grows in arid soil. The bases of its leaves form a catch-basin that holds rain water, and the pineapple's roots absorb water from the catch-basin. Furthermore, pineapple leaves have tiny hairlike projections, called trichomes, that help reduce water loss from the plant. Among the air plants, the vase plant (*Aechmea,* Figure B) also has a catch-basin, but its roots serve only to anchor it to tree branches. Trichomes on its leaves absorb water and dissolved nutrients from the catch-basin. The spread of bromeliads into aerial environments seems to have hinged largely on the presence of catch-basins and trichomes in the ancestral plants and on the adaptation of these features for life without soil. Still requiring soil, the pineapple may be reminiscent of an ancestral bromeliad.

A. Pineapple plant

Another way in which key adaptations can arise is by changes that affect an organism's development. Such changes may result from mutation of only a few genes. The development of an organism depends on regulatory genes that coordinate the activities of other genes that code for protein synthesis. Because each regulatory gene may influence many coding genes, a change in just a single regulatory gene may have a significant effect on an organism's development. Many species that differ significantly in development and body form, such as the human and the chimpanzee, are genetically quite similar. In cases like this, the two species may have diverged rather rapidly as changes in a few regulatory genes produced major changes in their ancestors' development and adult structure.

B. Vase plant

Catch-basins, an adaptation for dry environments in bromeliads

Delayed maturity was a key novelty in human evolution

A subtle change in a species' developmental program can have profound effects. For instance, a slowing of the development of some organs relative to others may produce a very different kind of organism. Figure A is a photograph of an axolotl, a salamander that illustrates a phenomenon called **neoteny,** the retention of juvenile body features in the adult. The axolotl grows to full size and reproduces without losing its external gills, a juvenile feature.

Neoteny has also been important in human evolution. Humans and chimpanzees are much more alike as fetuses than they are as adults, as the skulls in Figure B show. In the fetuses of both species, the skulls are rounded and the jaws are small, making the face rather flat and rounded. As development proceeds, uneven bone growth makes the chimpanzee skull sharply angular, with heavy brow ridges and massive jaws. The adult chimpanzee has much greater jaw strength than we have, and its teeth

A. Axolotl, a neotenous salamander

are proportionately larger. In contrast, the adult human has a skull with decidedly rounded, more fetuslike contours. The jaw is more pronounced than in the fetus, but otherwise the contours of the skull do not change significantly. Put another way, our skull is neotenous; it retains fetal features even after we are sexually mature.

Our large, neotenous skull is one of our most distinctive features. Our large, complex brain is another. Our brain develops to its unparalleled size and complexity during childhood, a period of development unique to humans. All mammals have a period of infant dependency, when the young are fed milk and require parental protection. After weaning, most mammals mature rapidly to adulthood. Apes, including chimpanzees, have a longer period of infancy than most mammals, but only humans have a true *childhood*, a prolonged period between infancy and adolescence when we remain dependent on parental care. The main function of childhood may be in providing more time to learn from adults.

In an essay entitled "Mickey Mouse Meets Konrad Lorenz" (*Natural History*, May 1979), noted evolutionary biologist Stephen Jay Gould contends that there is a connection between our neotenous physical traits and our unusually long period of juvenile dependency. Gould suggests that our juvenile physical traits—rounded head with large forehead, bulging cheeks, and small chin—may be visual clues that make adults feel affectionate, and thus caring and pro-

tective during our childhood. Gould uses the cartoon character Mickey Mouse to illustrate his points. Early Mickey Mouse renditions were not nearly as popular as later ones (Figure C). Gould contends that the present-day Mickey Mouse is successful because he elicits affectionate, parental responses. Compared to earlier versions, he is more youthful: His head and eyes are larger relative to his body, he is higher browed, and his legs are shorter.

Whether or not the Mickey Mouse analogy is appropriate, long retention of juvenile features, coupled with a prolonged period of dependency, are hallmarks of our species. Genetic changes that led to neoteny were probably key events in the divergence of our ancestors from the lineage we share with the great apes.

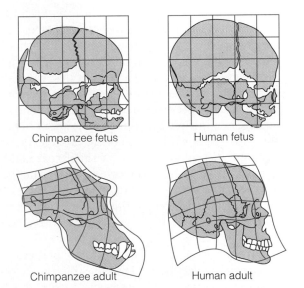

Chimpanzee fetus

Human fetus

Chimpanzee adult

Human adult

B. Chimpanzee and human skulls compared

C. The "evolution" of neoteny in Mickey Mouse

What accounts for evolutionary trends?

The fossil record reveals trends in the evolution of many species and lineages. For instance, two trends in the human lineage were toward a more complex brain and a neotenous life cycle. During the Mesozoic era, there was an overall trend in reptilian evolution toward largeness, a trend that produced the large dinosaurs. A similar trend toward larger body size occurred in certain groups of mammals, such as whales and elephants, during the Cenozoic era.

How can we explain evolutionary trends? Figure A illustrates two evolutionary trends that occurred over 30 million years in a family of extinct mammals called titanotheres (Greek for "giant wild beasts"). One trend was a progressive increase in body size, and the other was the development of horns. Titanotheres, distant relatives of horses and rhinoceroses, were among the dominant herbivores on the plains of North America and Eurasia during the early Cenozoic era. The fossil record shows that the earliest titanotheres, such as *Eotitanops*, were about the size of a large dog and lacked horns. The last of the titanotheres, *Megacerops,* stood nearly 2.5 m at the shoulder and had large horns that it may have used in courtship jousts.

Figure B diagrams how a trend such as increasing body size or horn length might occur in a lineage. Each of the horizontal parts of the branching tree represents the evolution of a new species from a parental lineage. The length of a horizontal segment shows how much larger (right-branching) or smaller (left-branching) a new species was than its immediate ancestor. The vertical segments represent the length of time each species existed, and the dead ends represent species extinctions. In this model, species smaller than their ancestors evolved as often as larger species. (There are 14 speciations leading to smaller size and 14 leading to larger size.) Nevertheless, there was an overall trend toward larger size because there was a higher extinction rate on the left than on the right. Most of the leftward lineages died out, while the rightward ones gave rise to other species.

Thus, Figure B shows that *unequal survival* of species can produce an evolutionary trend. The same trend could occur if there were equal survival but *unequal speciation,* that is, if more new species arose on the right side of Figure B than on the left. One of the current debates in evolutionary biology is about the relative importance of unequal survival and unequal speciation in the macroevolutionary trends that we see in the fossil record.

It is important to recognize that the existence of an evolutionary trend does not imply that the trend is preordained or unchangeable. Evolution is a response to interactions between organisms and their current environment. If conditions change, an evolutionary trend may cease or even reverse itself.

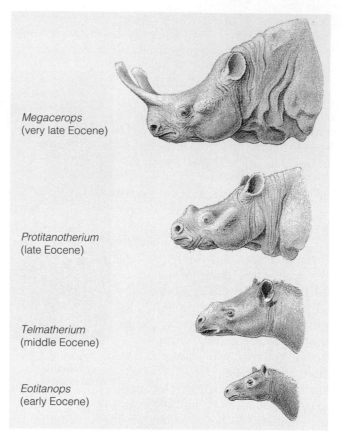

Megacerops
(very late Eocene)

Protitanotherium
(late Eocene)

Telmatherium
(middle Eocene)

Eotitanops
(early Eocene)

A. Trends in the evolution of titanotheres

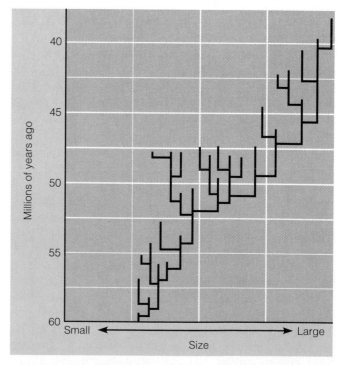

B. A model for an evolutionary trend occurring by the unequal survival of species

The evolutionary history of a group of organisms is called **phylogeny** (from the Greek *phylon*, tribe, and *genesis*, origin). Biologists traditionally represent the genealogies of organisms as **phylogenetic trees**, diagrams that trace evolutionary relationships as best they can be determined. The more we know about an organism and its relatives, the more accurately we can portray its phylogeny. We can be much less certain about the evolutionary history of a fossil group such as the titanotheres than we can about a modern group of organisms whose body structure, biochemistry, behavior, and reproductive biology have been studied. Nonetheless, even the best phylogenetic tree is not gospel; it merely represents the most likely phylogeny based on the available evidence.

Below is a phylogenetic tree proposed for guenons, a group of small (3–7 kg) monkeys that inhabit eastern Africa. There are 25 known species, all members of the genus *Cercopithecus* and all probably descended from one ancestor. This tree, tracing the evolution of 12 arboreal (tree-dwelling) guenon species, is based on a comparison of blood proteins. A very similar tree could be drawn on the basis of facial markings, vocal calls, and geographical distribution.

Each branching point on a phylogenetic tree has meaning. For example, the fork in the main trunk of this tree shows that there are two fundamentally distinct groups of guenons and that these two groups diverged soon after the genus *Cercopithecus* arose. Placement of de Brazza's monkey near the base of the right branch suggests that this species is more closely related to ancestral guenons than are any other species on the tree. On the left branch, six species fit into two groups of three. The branching pattern indicates that the greater spot-nosed monkey is more like the ancestor of the two groups than is either the blue or Syke's monkey. The three monkeys in the other group are about equally distinct from their immediate ancestor.

We will return to phylogenetic trees, but, before we do, we need to examine the connection between phylogeny and the classification system used in biology.

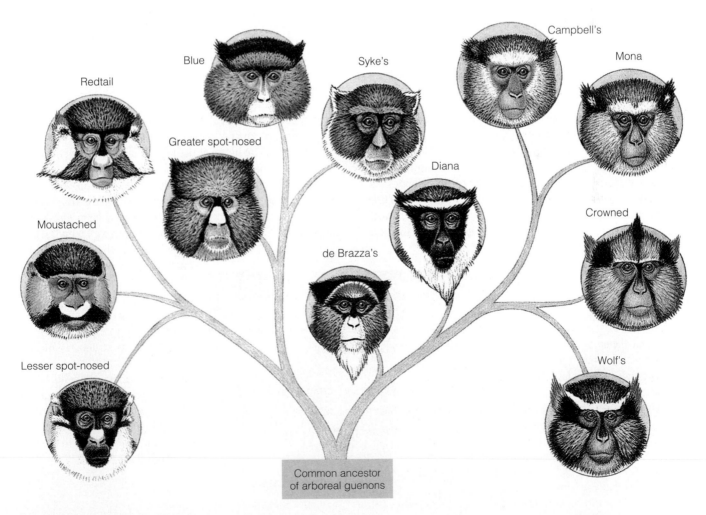

A phylogenetic tree for guenon monkeys

16.12 Systematists classify organisms by phylogeny

Reconstructing phylogeny is part of the science of **systematics,** the study of biological diversity and its classification. Systematists employ taxonomy, the identification and classification of species, and they attempt to arrange organisms in categories that reflect phylogeny.

Assigning scientific names to species is an essential part of systematics. Common names, such as monkey, fruit fly, crayfish, and garden pea, may work well in everyday communication, but they can be ambiguous because there are many species of each of these kinds of organisms. The system of scientific names developed by Linnaeus in the eighteenth century remains in use today. In fact, we still use many of the more than 11,000 names that Linnaeus originated.

As we have seen, Linnaeus's system assigns to each species a two-part latinized name, or **binomial.** The first part of a binomial is the **genus** (plural, *genera*) to which the species belongs. The guenon monkeys illustrated on the previous page, for instance, are all members of the genus *Cercopithecus* (a latinized name derived from the Greek words *kerkos,* tailed, and *pithekos,* monkey). The second part of a binomial refers to one **species** within the genus. For example, the scientific name for the "blue" guenon species is *Cercopithecus mitis* (*mitis* is Latin for harmless).

In addition to defining and naming species, a major objective of systematics is to group species into broader taxonomic categories, using a system also devised by Linnaeus. Beyond the grouping of species within genera, the Linnaean system extends to progressively broader categories of classification. It places similar genera in the same **family,** groups families into **orders,** orders into **classes,** classes into **phyla** (singular, *phylum*), and phyla into **kingdoms.**

The small table here illustrates the taxonomy of the domestic cat. The genus *Felis,* which includes the domestic cat (*Felis silvestris*) and several closely related wild cats, is grouped with the genus *Panthera* in the cat family, Felidae (*Panthera* includes the tiger, leopard, jaguar, and African lion). Family Felidae belongs to the order Carnivora, which also includes the family Canidae (the dog family, including the domestic dog, wolf, and fox) and several other families. Order Carnivora is grouped with many other orders in the class Mammalia, the mammals. Class Mammalia is one of several classes belonging to the phylum Chordata in the kingdom Animalia. Animals with backbones are grouped into a "subphylum" of the chordates called Vertebrata. Each taxonomic level is more comprehensive than the one below. Not all mammals are cats, but all members of the family Felidae belong to the order Carnivora and the class Mammalia.

Each proper name in the taxonomic hierarchy—family Felidae, order Carnivora, or class Mammalia, for instance—is called a **taxon** (plural, *taxa*). Defining taxa usually requires judgment calls; it is often difficult to determine the dividing lines between taxa even at the species level. As we saw in Module 15.1, some species are clearly separated from all others because their individuals cannot produce fertile offspring with individuals of other species, but it is often difficult to tell if a group of organisms constitutes a species. Many species do not reproduce sexually, and, for those that do, we may not know enough to be certain about their identity.

Identifying species often requires judgment calls, but classifying species into higher taxa *always* does. Systematists must weigh all the available evidence about which taxon fits where in the classification system, and decisions about classification often involve heated debate. One systematist may value fine distinctions and favor a relatively large number of taxa for each category above species, whereas another may stress similarities and propose a minimal number of taxa. Even after a classification system is agreed upon, it is subject to updating when new information about the organisms appears.

Ever since Darwin, systematics has had a goal beyond simple organization: to have classification reflect the evolutionary connections among species. As a systematist classifies species in groups subordinate to other groups in the taxonomic hierarchy, the final product takes on the branching pattern of a phylogenetic tree. The figure at the top of the next page shows the pattern for some of the taxa in the order Carnivora; as is conventional, the "highest," or most inclusive, taxa are at the bottom. Systematists judge from current evidence that species classified in each genus shown here evolved from a common ancestor that was also a member of that genus. Likewise, they judge that all the genera in any one family arose from a member of that family, and that all the families in the order Carnivora arose from an ancestral carnivoran. Systematists use many kinds of evidence, such as structural and developmental features, molecular data, and behavioral traits of organisms, to make decisions about classification and its relationship to phylogeny. We look at two major lines of evidence in the next two modules.

Classification of the Domestic Cat	
Category	**Domestic Cat**
Kingdom	Animalia (animals)
Phylum	Chordata (chordates)
(Subphylum)	Vertebrata (vertebrates)
Class	Mammalia (mammals)
Order	Carnivora (carnivorans)
Family	Felidae (cats)
Genus	*Felis* ⎫
Specific name	*silvestris* ⎬ (domestic cat species)

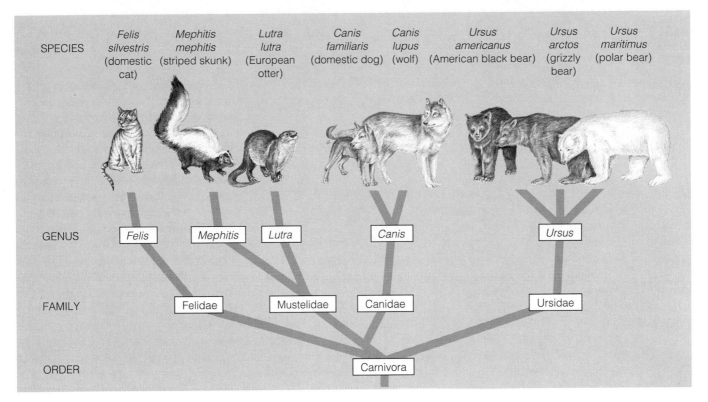

SPECIES: *Felis silvestris* (domestic cat) | *Mephitis mephitis* (striped skunk) | *Lutra lutra* (European otter) | *Canis familiaris* (domestic dog) | *Canis lupus* (wolf) | *Ursus americanus* (American black bear) | *Ursus arctos* (grizzly bear) | *Ursus maritimus* (polar bear)

GENUS: *Felis* | *Mephitis* | *Lutra* | *Canis* | *Ursus*

FAMILY: Felidae | Mustelidae | Canidae | Ursidae

ORDER: Carnivora

The relationship of classification to phylogeny for some carnivorans

Homology indicates common ancestry, but analogy does not 16.13

One of the best sources of information about phylogenetic relationships are homologous structures. As we discussed in Module 14.2, homologous structures may look different and function very differently in different species, but they exhibit fundamental similarities because they evolved from the same structure in a common ancestor. Among the vertebrates, for instance, the whale limb is adapted for steering in the water; the bat wing is adapted for flight. Nonetheless, there is a basic similarity in the bones supporting these two structures (see Figure 14.2A).

A systematist always searches for homologies between species, as these are often keys for determining phylogenetic relationships. Generally, the greater the number of homologous structures between two species, the more closely the species are related.

The search for homologies is not without pitfalls, however, for not all likenesses are inherited from a common ancestor. Evolution often produces *non*homologous similarities among organisms. In a process called **convergent evolution,** species from different evolutionary branches may come to resemble one another if they live in very similar environments. In such cases, natural selection may result in body structures and even whole organisms that look very similar but that are *analogous*, rather than homologous. For example, the two plants shown here look remarkably similar, although they are not closely related. On the left is the ocotillo, which is common on the coastal desert of Baja California. On the right is the allauidia, which grows in desert areas in Madagascar. These two plants are on widely separated lineages and have evolved in isolation for millions of years. Nonetheless, as the plants became adapted to similar environments, analogous equipment evolved, including short, water-retentive leaves and cactuslike thorns. The task for the systematist is to distinguish homologies, which indicate common ancestry, from analogies, which do not.

Analogous structures, resulting from convergent evolution

Molecular biology has provided systematists with powerful new techniques for establishing evolutionary relationships. As we saw in Module 10.5, the characteristics of an organism are ultimately determined by the sequence of nucleotides in its DNA. The reason is that the nucleotide sequences code for sequences of amino acids in proteins, which give an organism its unique characteristics. By comparing the DNA and proteins from different organisms, systematists can obtain quantitative information about differences or similarities among species. For example, the phylogenetic tree for the family Ursidae (bears) and the family Procyonidae (raccoons) shown in Figure A was constructed from comparisons of DNA and blood proteins of the species shown.

Bears and raccoons are closely related mammals, but systematists can also use DNA and protein analyses to assess relationships between groups of organisms that are so phylogenetically distant that structural similarities are absent—human beings and yeasts, for instance.

Among the methods researchers can use to compare proteins, **amino acid sequencing**—determining the sequence of amino acids in a polypeptide—is the most precise. A close match in the amino acid sequences of comparable proteins from different species indicates that the genes that program those proteins evolved from a common gene, in a shared ancestor. Accordingly, the degree of similarity in amino acid sequences indicates the degree of phylogenetic relationship between different species. For example, the amino acid sequence of cytochrome *c*, an electron transport protein common to all aerobic organisms, has been determined for a wide variety of species ranging from bacteria to plants and animals. The sequences for humans and chimpanzees match perfectly for all 104 amino acid positions along the cytochrome *c* polypeptide chain, and this version of cytochrome *c* differs by just one amino acid from the cytochrome *c* in the rhesus monkey. Humans, chimpanzees, and rhesus monkeys all belong to the same mammalian order, Primates.

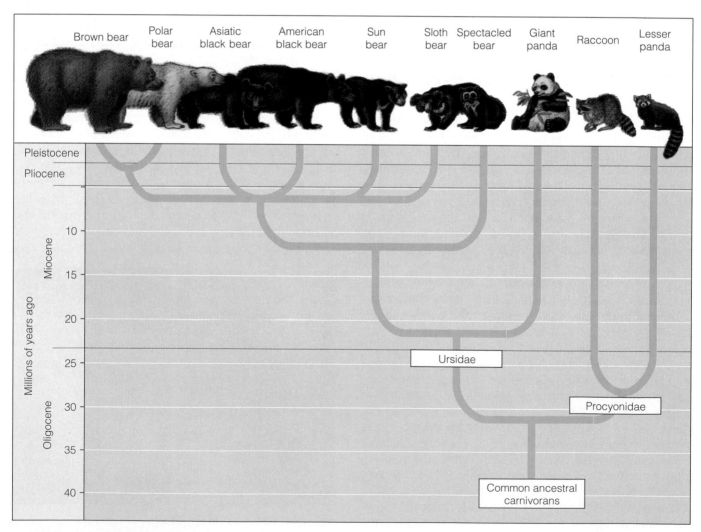

A. A phylogenetic tree based on molecular data

Differences between the cytochromes of these species and more distantly related vertebrates are greater. For example, human cytochrome *c* differs from that of the dog by 13 amino acids, from rattlesnake cytochrome *c* by 20 amino acids, and from tuna cytochrome *c* by 31 amino acids. Phylogenetic trees based on cytochrome *c* are generally consistent with those based on comparative anatomy and the fossil record.

Proteins evolve at different rates, but the rate of evolution for any one protein seems to be nearly constant over time. Compared to many proteins, cytochrome *c* seems to change very slowly and consequently shows relatively few amino acid differences among closely related organisms. Proteins that change more rapidly—for instance, certain albumins in the blood—vary more from one species to the next, and they are more useful in assessing relationships among closely related organisms, such as the carnivorans depicted in Figure A.

The most direct way to determine how closely two species are related is by comparing their DNA. Several methods are available for comparing DNA from different organisms. **DNA sequencing** is the most direct and precise method. After using recombinant-DNA techniques to prepare comparable DNA segments from two species, the researcher determines the nucleotide sequences of the segments. Comparison of the sequences can reveal ultimate similarities or differences between the two species—exactly how similar or different their genes are. Sequencing is the most tedious method of determining DNA similarities. It is not practical at present to sequence more than selected segments of DNA from an organism's genome. (As we mentioned in Chapter 13, sequencing of the entire genome of even a single eukaryotic organism has not yet been achieved.) Systematists will find greater use for this method as the technology for rapidly sequencing DNA improves.

The DNA segments compared using DNA sequencing and other methods involving recombinant DNA technology are usually only a few thousand nucleotides long. Yet the DNA derived from cell nuclei is much longer than this, and it can be unwieldy. Consequently, many researchers use DNA extracted from mitochondria. Mitochondrial DNA is relatively small and more manageable. Moreover, mitochondrial DNA seems to mutate about ten times faster than nuclear DNA, making it possible to sort out genetic relationships among populations of the same species.

Another method of comparing the DNA of different species, **DNA-DNA hybridization,** measures the extent of hydrogen bonding between single strands of DNA obtained from two or more species. Figure B shows how this method works in comparing pieces of DNA from three hypothetical species. ① After DNA is extracted from cells of a species, it is cut up and then ② heated to separate the complementary strands. ③ Single-stranded DNA from two species is then mixed and cooled to re-form double-stranded DNA. In the example illustrated, DNA from species 1 is mixed with DNA from species 2 and from species 3. How tightly a strand of DNA of one species binds to a strand from another species depends on the de-

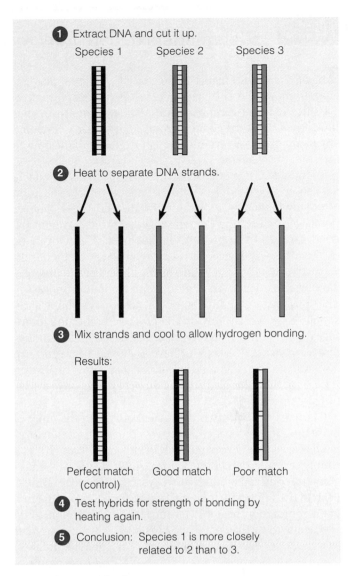

B. DNA-DNA hybridization

gree of similarity. The more alike two DNA molecules are, the greater the number of complementary nucleotides they have, and consequently the more hydrogen bonds will form between the strands in a hybrid molecule. ④ Heat is used to test for complementarity between strands. The higher the temperature (the greater the heat energy) required to separate paired strands, the better the match. More to the point, the higher the temperature needed to separate hybrid DNA strands, the closer the evolutionary connection between species. In Figure B, the hybridization test indicates that ⑤ species 1 is more closely related to species 2 than to species 3.

Systematists are using DNA comparisons more and more to help resolve difficult taxonomic questions. For instance, are the giant panda and lesser panda bears or are they members of the raccoon family? The phylogenetic tree in Figure A reflects evidence from DNA-DNA hybridization that the giant panda is more closely related to bears than to raccoons and that the lesser panda is a member of the raccoon family.

Systematists debate how to interpret phylogenetic data

The modern systematist evaluates all available data, including molecular and anatomical comparisons, for use in classification and reconstruction of phylogeny. No matter how much we learn about a group of organisms, however, systematics will always involve controversy about how to interpret the available data.

Today, there is a heady debate about how the available data should be used, a debate that affects the classification of our own species. As we saw in Module 16.11, branch points of the lineages on a phylogenetic tree highlight the relative times of origin of different taxa. A phylogenetic tree may also indicate how much a particular taxon has diverged from its ancestors and how different two living taxa have become since branching from a common ancestor. The question being debated is whether classification should be based entirely on the order of branching in lineages, or whether it should take into account the amount of divergence that seems to have occurred between taxa. The debate divides modern taxonomy into two main schools of thought: cladistic taxonomy and classical evolutionary taxonomy.

Cladistic taxonomy is concerned only with the order of branching in phylogenetic lineages. The cladistic phylogenetic tree, called a cladogram (Greek *klados,* branch), is a simple series of branches. Figure A indicates how a cladogram is constructed using five vertebrates as examples. Each of the four branch points (numbered dots) on the diagram represents the most recent ancestor common to all species beyond that point. For instance, all the mammals share a more recent common ancestor (branch point 2) than the ancestor common to both the lizard—a reptile—and the mammals (point 1). Like all phylogenetic trees, cladograms indicate the relative ages of lineages, not the

relative ages of any living species. For instance, this cladogram does not imply that a particular lizard species alive today evolved before any of these mammals.

How are the branch points on a cladogram determined? Each branch is defined by homologous features that are unique to the species on that branch. Every species has two types of homologous features, called primitive characters and derived characters. For any group of species, **primitive characters** are those that already existed in the ancestor common to all. For instance, according to fossil evidence, the remote ancestor common to all the vertebrates in Figure A was five-toed, and hence this feature is termed a primitive character.

The sharing of a primitive character tells us nothing about the pattern of evolutionary branching from a common ancestor. Thus, we cannot use the presence of five separate toes to divide the species in Figure A among evolutionary branches. Instead, the branch points are defined by **derived characters,** or homologous features unique to the lineage that arises at each point. For example, hair and mammary glands are two derived characters that define the first branch point on the cladogram below. Since lizards do not have these features and all the mammals do, hair and mammary glands place all the mammals on one branch and the lizard on the other. The second branch point is defined by a number of derived skeletal features, including tooth structure. In the example here, the cheek teeth (premolars and molars) in the seal, lion, and house cat are adapted for shearing or crushing meat. By contrast, cheek teeth in horses have surfaces adapted for grinding plant material. The third branch point (between the seal and the two cats) is defined by a number of features, including the presence of retractable claws in the lion and the domestic cat. The house cat and lion are closely related, but they diverge at

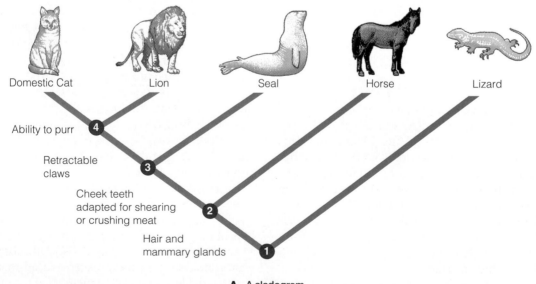

Domestic Cat   Lion   Seal   Horse   Lizard

Ability to purr   ④

Retractable claws   ③

Cheek teeth adapted for shearing or crushing meat   ②

Hair and mammary glands   ①

A. A cladogram

the fourth branch point. Among other distinguishing features is the lion's ability to roar but not to purr; the domestic cat can purr but cannot roar.

Cladistic taxonomy is becoming more and more popular with researchers. Its straightforward approach is an attempt to make systematics less arbitrary and to make classification match evolutionary history more closely. Strict application of cladistics does produce some taxonomic surprises. For instance, a strict cladist would do away with the familiar taxa Class Aves (birds) and Class Reptilia (reptiles), as we conventionally know them, because cladistic analyses and the fossil record indicate that birds are closer relatives of one group of reptiles, the crocodiles, than are lizards and snakes.

The logical method used to construct cladograms is not limited to cladistic taxonomy. It also underlies the approach of **classical evolutionary taxonomy.** However, in contrast to cladistics, the classical approach also takes into account the apparent degree of divergence of taxa. Phylogenetic trees and classification based on the classical approach reflect branching sequences *and* the apparent degree of divergence among the branches. For instance, in classifying birds and reptiles, classical taxonomists acknowledge that birds are more closely related to crocodiles than lizards are. Nevertheless, they combine lizards and crocodiles in the class Reptilia and assign birds to their own class (Aves) because, in their judgment, classification should emphasize the uniqueness of birds. In this view, flight ability in birds and all the adaptations that make it possible were major evolutionary changes, breakthroughs that allowed birds to diverge so far from any reptiles that they should be classified in their own, nonreptilian, group.

Figure B compares the classical evolutionary approach with that of cladistics in interpreting phylogenetic data about humans and our closest living relatives, the great apes. Biochemically, the human, chimpanzee, and gorilla are very similar; in fact, molecular studies show that the chimpanzee and the gorilla both resemble humans more than

they resemble orangutans. Both the classical view and the cladistic view reflect the molecular data. However, in classifying the four species, the classical approach places all the apes in the family Pongidae, separate from the Hominidae. This view emphasizes the overall similarities of the great apes and the uniqueness of the human body, mind, and behavior. By contrast, the cladistic view ignores the apparent distinctiveness of humans and ties classification of the four species strictly to the evolutionary branch points defined by the molecular data. This strict cladistic approach places the human, chimpanzee, and gorilla in the family Hominidae and the orangutan in the family Pongidae.

How should the great apes and our own species be classified? Should birds be placed in their own class? There is no consensus on these issues. The classification used in this book and most others is based on the classical approach, which predates cladistics and can be traced back to Darwin. Classical systematists argue that the subjectivity of their approach is necessary, because degrees of divergence between taxa, such as between birds and reptiles and between humans and apes, are important reflections of the evolutionary process and must be considered in taxonomy.

In being concerned only with branching sequences, cladists have a relatively simple (and more scientifically objective) approach to classification. By contrast, classical taxonomists often have to make subjective judgments about the importance of the degree of divergence. Cladistic taxonomists contend that their more objective approach makes classification more clearly reflect phylogeny. Cladistics has revitalized our thinking about many important issues in systematics, and it will continue to do so. Meanwhile, it is likely that many classical taxa, such as class Aves and family Hominidae (limited to humans), however subjective, will continue to be used. Debate is an essential ingredient in systematics, as it is in all areas of science. The debate over cladistics and classical systematics continues to invigorate the science of classification and phylogeny.

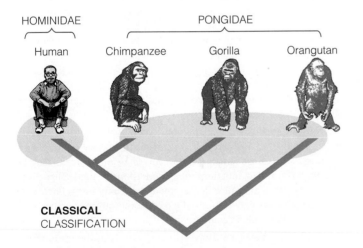

CLASSICAL CLASSIFICATION

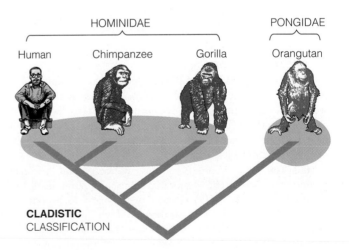

CLADISTIC CLASSIFICATION

B. Two ways to classify humans and their closest living relatives

16.16 Biologists arrange life into five kingdoms

In the next chapter, we begin examining the enormous diversity of organisms that have populated Earth since life first arose 3.5 billion years ago. As we do so, systematics will help us trace the evolutionary relationships among the organisms. As we have seen, the kingdom is the most inclusive taxonomic category. Many schemes for classifying organisms into kingdoms have been proposed over the years. A two-kingdom system that divided all organisms into plants and animals was first proposed by Linnaeus and was popular for over 200 years. But it was beset with problems. Where do bacteria fit in such a system? As prokaryotes, can they be considered members of the plant kingdom? And what about the fungi? In 1969, American ecologist Robert H. Whittaker argued effectively for a five-kingdom system. Whittaker's five kingdoms are the ones we introduced in Chapter 1: Monera, Protista, Plantae, Fungi, and Animalia. The five-kingdom system is commonly accepted today, and we use it in this book. However, there is a vigorous new debate about the number of kingdoms and their evolutionary relationships, and we touch on it when we discuss the bacteria in Chapter 17.

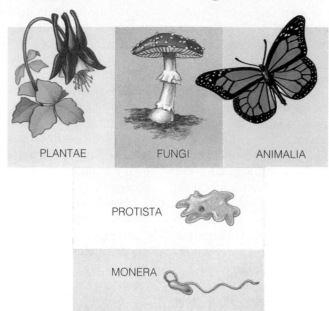

The five-kingdom system recognizes the two fundamentally different types of cell, prokaryotic and eukaryotic, and sets the prokaryotes apart from all eukaryotes by placing them in their own kingdom, Monera. Organisms of the other four kingdoms all consist of eukaryotic cells. Kingdoms Plantae, Fungi, and Animalia consist of multicellular eukaryotes that generally differ in structure, development, and modes of nutrition. Plants make their own food by photosynthesis. Fungi live by decomposing the remains of other organisms and absorbing small organic molecules. Most animals live by ingesting food and digesting it within their bodies.

As we will see, kingdom Protista is something of a grab bag, containing all eukaryotes that do not fit the definitions of plants, fungi, or animals. Most protists are unicellular, but in the version of the five-kingdom scheme used in this text, kingdom Protista also includes relatively simple multicellular organisms that are believed to be direct descendants of unicellular protists. We focus on the kingdoms Monera and Protista in the next chapter.

Chapter Review

Begin your review by rereading the module headings and scanning the figures before proceeding to the Chapter Summary and questions.

Chapter Summary

Introduction–16.2, 16.4 Fossils—remnants or imprints of ancient organisms—record the history of life. They are found mainly in sedimentary rock strata, which are formed from particles that settle out of water. The sequence of strata indicates relative age. Radioactive dating, which measures the decay of radioactive isotopes, can gauge the actual ages of fossils.

16.3 The fossil record chronicles macroevolution, the main events in the history of life. Major transitions in life forms mark separate geological eras; smaller changes divide eras into periods. Life began in the Precambrian era, at the end of which animals first appeared. During the Paleozoic era, plants and animals diversified and moved onto land. During the Mesozoic era, reptiles and cone-bearing plants dominated the land. Extinction of the dinosaurs marked the beginning of the Cenozoic era, during which mammals and flowering plants proliferated.

16.5–16.7 Continental drift and mass extinctions have helped shape the history of life. Movement of Earth's crustal plates brought

the continents together near the end of the Permian period, forming the supercontinent Pangaea. This continental merger altered coastlines and climate, and triggered extinctions. Early in the Mesozoic era, separation of the continents caused the isolation and diversification of organisms. At the end of the Cretaceous period, many life forms disappeared, including the dinosaurs. This mass extinction may have been hastened by the impact of an asteroid or an increase in volcanic activity. Every mass extinction reduced the diversity of life, but each was followed by a rebound in diversity. Mammals filled the void left by the dinosaurs.

16.8–16.9 Distinctive adaptations, such as the hair and milk of mammals, may enable particular organisms to prosper after mass extinctions. Preadaptations that have evolved in one context may be able to perform new functions when conditions change. Changes in regulatory genes may alter development and body form. For example, such changes may result in the retention of juvenile characteristics in the adult, called neoteny.

16.10 The fossil record reveals many trends in the evolution of species and lineages, such as the gradual increase in size in the evolution of dinosaurs. Trends may reflect unequal speciation or unequal survival of species on a branching evolutionary tree.

16.11–16.12 Phylogeny is the evolutionary history of a group of organisms. Reconstructing phylogeny is part of systematics, the study of biological diversity and classification. Taxonomists assign a two-part name, or binomial, to each species. The first name, the genus, covers a group of related species. The second name refers to a species within the genus. Genera are grouped into progressively larger categories: family, order, class, phylum, and kingdom. The species is the most clear-cut category. Taxonomists often differ about particular placement in larger categories, but they strive to ensure that categories reflect evolutionary relationships.

16.13–16.14 Homologous structures and molecular biology provide clues about ancestry. Similarity in the sequence of amino acids or nucleotides indicates phylogenetic relationships among species. DNA-DNA hybridization measures relatedness by testing the hydrogen bonding between DNA strands from different species.

16.15 Systematists differ on how to interpret phylogenetic data. Cladistic taxonomy considers only the sequence of branching in constructing phylogenetic trees. Classical taxonomy weighs both branching sequence and the degree of evolutionary divergence between groups of organisms. For example, classical taxonomy emphasizes the uniqueness of humans and places us in a family separate from the great apes. But molecular data indicate that humans, chimps, and gorillas are more closely related to each other than to the orangutan, and cladistic taxonomy would reclassify these species accordingly, placing humans, chimps, and gorillas in the same family.

16.16 Life is classified into five kingdoms. The prokaryotes—bacteria—make up the kingdom Monera. Plants, animals, and fungi are multicellular eukaryotes that differ in structure, development, and modes of nutrition and thus are grouped into separate kingdoms. Kingdom Protista consists of all the remaining eukaryotes, mostly unicellular forms.

Testing Your Knowledge

Multiple Choice

1. Many species of plants and animals adapted to desert conditions probably did not arise there. Their success in living in deserts could be due to
   a. neoteny
   b. convergent evolution
   c. preadaptation
   d. phylogeny
   e. mass extinction

2. Mass extinctions that occurred in the past
   a. cut the number of species to the few survivors left today
   b. resulted mainly from the separation of the continents
   c. occurred regularly, about every million years
   d. were followed by diversification of the survivors
   e. wiped out land animals, but had little effect on marine life

3. The animals and plants of India are almost completely different from the species in nearby Southeast Asia. Why might this be true?
   a. They have become separated by convergent evolution.
   b. The climates of the two regions are completely different.
   c. India is in the process of separating from the rest of Asia.
   d. Life in India was wiped out by ancient volcanic eruptions.
   e. India was a separate continent until relatively recently.

4. Two worms in the same class must also be grouped in the same
   a. order
   b. phylum
   c. genus
   d. family
   e. species

5. Suppose you double-checked the monkey phylogenetic tree in Module 16.11, using DNA-DNA hybridization. Which of the following hybrid DNAs would you expect to separate at the lowest temperature? (*Explain your answer.*)
   a. Blue and moustached
   b. Diana and Syke's
   c. de Brazza's and crowned
   d. Mona and Campbell's
   e. Redtail and lesser spot-nosed

True/False (*Change false statements to make them true.*)

1. Bacteria are in their own kingdom because they are single-celled.

2. A protein that evolves slowly would be more useful than a rapidly evolving one in determining the relationships among species thought to be closely related.

3. A human hand and a cat's paw are homologous structures.

4. There are four kingdoms of eukaryotic organisms.

5. Biologists agree that unequal speciation or survival of species can produce evolutionary trends such as an increase in size.

Describing, Comparing, and Explaining

1. Briefly describe the major events in the history of life that marked the ends of the Precambrian, Paleozoic, and Mesozoic eras.

2. Explain how the amino acid sequences of proteins can be used to trace evolutionary relationships among organisms.

3. Discuss various hypotheses that have been suggested to explain the Cretaceous mass extinction that ended the reign of the dinosaurs. What kind of evidence supports each hypothesis?

Thinking Critically

1. Fossils suggest that birds and crocodiles are more closely related to each other than either of them is to lizards and turtles. Why do you think this might be a problem for taxonomists?

2. Biologists have discovered spectacular genetic changes called homeotic mutations. Some fruit flies with these mutations develop extra sets of wings or legs. What might be the normal functions of the genes altered by these mutations? Why do you think evolutionists are particularly interested in them?

3. Scientists have calculated that a particular piece of rock contained about 12 g of radioactive potassium-40 when it was formed. It now contains 3 g of potassium-40. The half-life of potassium-40 is 1.3 billion years. About how old is the rock?

Science, Technology, and Society

Experts estimate that human activities cause the extinction of hundreds of species every year. The natural "background" rate of extinction is thought to be a few species per year. As we continue to alter the environment, especially by destroying tropical rain forests, the resulting mass extinction will probably rival the end of the Cretaceous period, when perhaps half the species on Earth disappeared, including the dinosaurs. Many scientists and environmentalists are alarmed at this prospect, which they see as a catastrophe of unprecedented proportions. Other people are not as worried, because life has endured numerous mass extinctions in the past and has always bounced back. Some environmentalists respond that the situation is different this time. Is the current mass extinction different? Why or why not? What might be the eventual consequences for most species? For the survivors? For humans?

The Origin and Evolution of Microbial Life: Prokaryotes and Protists

17

Cruising in shallow water off western Australia, an undersea research vessel glides over an array of what look like pockmarked boulders. Are they lava rocks from an ancient volcano? Mounds of coral? Massive sponges? Their surfaces are greenish and sticky. They seem rigid, and an edge of one shows evidence of layering. Back in the laboratory, a look at a sample through a microscope shows that bacteria coat the mounds. Most of the bacteria turn out to be cyanobacteria, which are photosynthetic and commonly grow in layered mats.

The photograph on this page shows part of a bacterial mat that was growing in a warm lagoon in Baja California. The reddish-brown layers are bands of sediment that the bacteria accumulated. Coating the top of the mat where they are exposed to sunlight, sticky cyanobacteria concentrate sand grains and other fine particles from the seawater. While a sediment layer builds up on them, the bacteria keep migrating to the surface and growing over it. A layered mat builds up as this process is repeated over and over. The large Australian mounds formed in this way over thousands of years, hardening as their older sediment layers solidified.

The cyanobacteria that produce mats and mounds are descended from some of the oldest organisms known—photosynthetic prokaryotes that dominated Earth some 3 billion years ago. The fossil record indicates that at that time, greenish lawns of cyanobacteria covered virtually every wet, sunlit surface on the planet. Animals that might have grazed the bacterial lawns had not evolved yet, and the unchecked growth of ancient prokaryotes built up huge expanses of thick, layered mats. Many of the mats eventually became fossilized, and the layers of sediment became sedimentary rocks. Geologists call fossil bacterial mats **stromatolites** (from the Greek *stroma*, bed, and *lithos*, rock).

Stromatolites occur throughout the fossil record. Their prominence in rocks around 2.5 billion years old marks the time when ancient cyanobacteria were so numerous that their photosynthesis produced enough oxygen (O_2) to make the atmosphere aerobic, setting the stage for the evolution of all aerobic life.

Photosynthetic prokaryotes dominated Earth from nearly 3 billion years ago to about 1 billion years ago and then declined significantly. Their direct descendants, the cyanobacteria and other photosynthetic prokaryotes of today, remain abundant in freshwater lakes and ponds and in shallow oceans, but they only form thick mats or mounds where there is little animal life to eat them. Thus, great expanses of stromatolites are a thing of the past. Nevertheless, the grand, global legacy of ancient cyanobacteria—the aerobic atmosphere—remains with us today. And the aerobic atmosphere is but one illustration of the profound effects that living organisms can have on the environment.

Tracing the roles played by various organisms in the history of life on Earth is one of two main objectives in this and the next two chapters. Our other goal is to introduce you to the diversity of life, starting in this chapter with kingdom Monera (the bacteria) and kingdom Protista (the unicellular eukaryotes and their direct multicellular descendants). We begin with a look at Earth before life existed here.

A section through a cyanobacterial mat

A. The violent birth of Surtsey Island, near Iceland, in 1963

The ocean seems aflame, as lightning bolts flash and molten rock spews from a submerged volcano. The scene in Figure A is reminiscent of the conditions on the early Earth before life arose. This particular scene accompanied the birth in 1963 of the island of Surtsey in the North Atlantic. Soon after the island formed, terrestrial life from Iceland and North America began colonizing it. Since then, Surtsey and its inhabitants have been evolving together, demonstrating on a small scale the inseparable history of planet Earth and its living organisms.

Biological and geological history have been closely intertwined since life began. As we saw in Chapter 16, the formation and subsequent breakup of the supercontinent Pangaea had a tremendous effect on the diversity and geographical distribution of life. Conversely, life has changed the planet it inhabits, sometimes profoundly. The photosynthetic prokaryotes that first released oxygen to the air completely altered Earth's atmosphere. Much more recently, the emergence of *Homo sapiens* has changed the land, water, and air on a scale and at a rate unprecedented for a single species.

Earth is one of nine planets orbiting the sun, which is one of billions of stars in the Milky Way. The Milky Way, in turn, is one of billions of galaxies in the universe. Gazing at stars gives us a look back in time. The star closest to our sun is 4 light years—40 trillion kilometers—away; we see the light it emitted 4 years ago. Some stars are so far away that even if they burned out millions of years ago, we would still see them in the sky tonight. And there are new stars that are invisible because their light has not yet reached Earth.

The universe has not always been so spread out. Apparently, before the universe existed in its present form, all of its matter was concentrated in one mass. The mass seems to have blown apart with a "big bang" sometime between 10 and 20 billion years ago and to have been expanding ever since. Our solar system probably arose from a swirling cloud of dust. Most of the dust condensed in the center to form the sun, but some matter was left orbiting around the infant sun in several concentric rings. Within each orbit, kernels of matter had enough gravity to draw neighboring dust and ice together, forming the planets, including Earth.

B. A section through a fossilized stromatolite

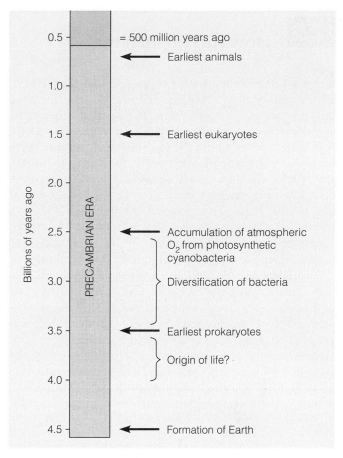

C. Major episodes in the early history of life

Earth probably began about 4.6 billion years ago as a cold world. Later on, heat generated by the impact of meteorites, radioactive decay, and tight compaction by gravity thawed Earth and eventually turned it into a molten mass. The mass then sorted into layers of varying densities, with most of the nickel and iron sinking to the center and forming a core. Less dense material became concentrated in a mantle surrounding the core, and the least dense material settled on the surface, eventually solidifying into a thin crust.

The first atmosphere was probably composed mostly of hot hydrogen gas (H_2). Because the gravity of Earth was not strong enough to hold such small molecules, the H_2 soon escaped into space. Volcanoes and other vents through the crust belched gases that formed a new atmosphere. Analyses of gases vented by modern volcanoes have led scientists to speculate that this second early atmosphere consisted mostly of water vapor (H_2O), carbon monoxide (CO), carbon dioxide (CO_2), nitrogen (N_2), methane (CH_4), and ammonia (NH_3). The first seas were created by torrential rains that began when the Earth had cooled enough for water in the atmosphere to condense. Not only was the atmosphere of the young Earth very different from the one we know today, but lightning, volcanic activity, and ultraviolet radiation were much more intense. It was in such an environment that life began.

Fossil evidence indicates that nonliving matter evolved into primitive prokaryotic organisms within a few hundred million years after Earth's crust solidified. We would guess from the relatively simple structure of the prokaryotic cell (compared with that of the eukaryotic cell) that early organisms were primitive bacteria, and the fossil record supports this idea. Fossils that appear to be prokaryotes about the size of some bacteria living today have been discovered in southern Africa in a rock formation called the Fig Tree Chert, which is 3.4 billion years old. Evidence of even more ancient prokaryotic life

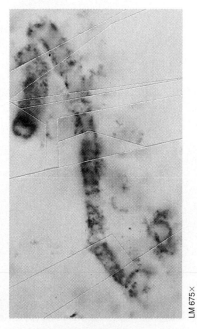

D. A fossilized prokaryote

occurs in stromatolites. Notice how similar the fossilized bacterial mats in Figure B are to the living one we saw on page 313. The stromatolite in Figure B is from western Australia. Dating from the Precambrian period about 3.5 billion years ago, it is among the oldest evidence of life found thus far. The timeline in Figure C puts these most ancient of fossils in historical perspective.

Figure D shows a fossilized prokaryote found in a Precambrian stromatolite. It is a filament formed by a chain of bacterial cells. Fossils like this are common in stromatolites and probably represent some of the vast populations of photosynthetic prokaryotes that produced the atmosphere's oxygen.

Photosynthetic prokaryotes are the simplest kinds of organisms we know that are autotrophic (produce their own food molecules). However, photosynthesis is not a simple metabolic process, so it is unlikely that photosynthetic bacteria were the first forms of life on Earth. In fact, the evidence that photosynthetic prokaryotes thrived 3.5 billion years ago is strong support for the hypothesis that life in a simpler form arose much earlier, perhaps as early as 4 billion years ago.

How did life originate?

From the time of the ancient Greeks until well into the nineteenth century, it was common "knowledge" that life arose from nonliving matter all the time. Many people believed, for instance, that flies came from rotting meat, fish from ocean mud, frogs and mice from wet soil, and microorganisms from broth. Experiments performed in the 1600s showed that relatively large organisms, such as insects, cannot arise spontaneously from nonliving matter. However, debate about how microscopic organisms arise continued until the 1860s. In 1862, the great French scientist Louis Pasteur confirmed what many others had suspected: All life today, including microbes, arises only by the reproduction of preexisting life.

Pasteur ended the argument over spontaneous generation of modern organisms, but he did not address the question of how life arose in the first place. The question remains with us today, although we have come a long way toward answering it in the last half century. We can be fairly certain that the first organisms came into being between about 4.1 billion years ago, when Earth's crust began to solidify, and 3.5 billion years ago, when the planet was inhabited by prokaryotes complex enough to build stromatolites. But what events led to the origin of life?

Most biologists subscribe to the hypothesis that the earliest life was much simpler than anything alive today, and that life did first develop from nonliving materials. The earliest lifelike entities may have been aggregates of molecules with a particular arrangement that made simple metabolism and self-replication possible. Because living organisms consist of polymers formed of small organic molecules (monomers), the origin and accumulation of small organic molecules must have been the earliest chemical steps preceding the origin of life. Some scientists propose that meteorites and comets seeded the planet with monomers, but the prevailing opinion is that the first organic molecules arose mainly from inorganic materials on the early Earth. The formation of polymers, including proteins and nucleic acids, from organic monomers would have been a second stage preceding the origin of life. Once this occurred, the polymers and monomers may have formed aggregates having chemical characteristics different from their surroundings. Heredity may also have originated during this stage. In the next four modules, we examine several stages in chemical evolution, the process by which the molecules of life may have originated on the early Earth.

TALKING ABOUT SCIENCE

Stanley Miller's experiments showed that organic molecules could have arisen on a lifeless Earth

In 1953, when Stanley Miller was a 23-year-old graduate student at the University of Chicago, he performed some experiments that would attract global attention. Miller was the first to show that amino acids and other organic molecules could have been generated on a lifeless Earth.

Miller's experiments were a test of a hypothesis about the origin of life developed in the 1920s by biochemist A. I. Oparin of Russia and geneticist J. B. S. Haldane of England. Oparin and Haldane proposed that the conditions on the early Earth could have generated a collection of organic molecules that in turn could have given rise to the first living organisms. They reasoned that present-day conditions on Earth do not allow the spontaneous synthesis of organic compounds simply because the atmosphere is rich in oxygen. Oxygen is corrosive:

A. Stanley Miller in his laboratory

It is a strong oxidizing agent and tends to disrupt chemical bonds by extracting electrons from them. However, before the early prokaryotes added oxygen to the air, Earth probably had a reducing atmosphere instead of an oxidizing one. A reducing environment tends to *add* electrons to molecules. Thus, an early reducing atmosphere could have caused simple molecules to combine to form more complex ones. As Miller told us a few years ago,

Oparin proposed that the primitive atmosphere contained the gases methane, ammonia, hydrogen, and water, and that chemical reactions in that primitive atmosphere produced the first organic molecules. That hypothesis had a good deal of appeal, but without the experiments, it was talked about but not very well accepted.

Construction of complex molecules from simple ones also requires energy, and Miller reasoned that there were abundant energy sources in the environment of the early Earth. Besides lightning discharges, ultraviolet radiation probably reached Earth's surface with much greater intensity than it does today. Evidence exists that young suns emit more ultraviolet radiation than older suns. Also, in our modern atmosphere, a layer of ozone (O_3) screens out most ultraviolet radiation. Ozone, which forms from ordinary oxygen (O_2), would not have been present before photosynthesis arose.

The apparatus shown with Miller in Figure A on the preceding page is similar to the one he used to test the Oparin-Haldane hypothesis. Figure B shows the main components of Miller's original apparatus and how they simulated conditions on the early Earth. A flask of warmed water simulated the primeval sea. The "atmosphere" consisted of a mixture of water vapor, H_2, CH_4, and NH_3. Electrodes discharged sparks into the gas mixture to mimic lightning. Below the spark chamber, a glass jacket called a condenser surrounded the apparatus. Filled with cold water, the condenser cooled and condensed the water vapor in the gas mixture, causing "rain," along with any dissolved compounds, to fall back into the miniature sea. As material circulated through the apparatus, the solution in the flask slowly changed color. As Miller described it,

> The first time I did the experiment, it turned red. Very dramatic! And then, after it turned red, it got more yellow and then brown as the sparking went on.

After the experiment proceeded for a week, Miller found a variety of organic compounds in the solution, including some of the amino acids that make up the proteins of organisms:

> The surprise was that we . . . got mainly organic compounds of biological significance. And the amino acids were formed, not in trace quantities, but abundantly! The experiment went beyond our wildest hopes.

Miller's early experiments stimulated a great deal of interest and research on the prebiotic (before-life) origin of organic compounds. Since the 1950s, Miller and other researchers using modifications of Miller's apparatus have made most of the 20 amino acids commonly found in organisms, as well as sugars, lipids, the nitrogenous bases present in nucleotides of DNA and RNA, and even ATP. These laboratory studies support the idea that many of the organic molecules that make up living organisms could have formed before life itself arose on the early Earth.

Stanley Miller, who is now professor of chemistry at the University of California, San Diego, has adhered to the hypothesis that the early atmosphere was the original source of organic molecules. Recently, however, this idea has been criticized. Some scientists claim that the Earth never had an atmosphere of methane and ammonia. Miller comments:

> There is considerable controversy about the composition of the primitive atmosphere. We have been doing experiments using carbon monoxide and carbon dioxide [in place of methane]. You can make organic material with these kinds of atmospheres . . . if there is molecular hydrogen around.

What if the early atmosphere, whatever its composition, was not the source of prebiotic organic compounds? A few biologists have suggested that the first organic molecules on Earth arrived with meteorites. Miller responds:

> Some organic substances certainly came in this way. [However] I think that all but a few percent of the synthesis was done on the Earth.

We asked Dr. Miller in what sort of place he thought life actually began on Earth.

> The usual assumption is that it began in the ocean. But you can legitimately propose that some of the processes occurred in different areas. For example, some of the polymerization reactions that made larger organic molecules probably occurred on beaches that had dried out and heated up, and some may have occurred in hot springs. But the oceans form the bulk of the area where organic reactions could take place, and I think most of the chemistry took place there.

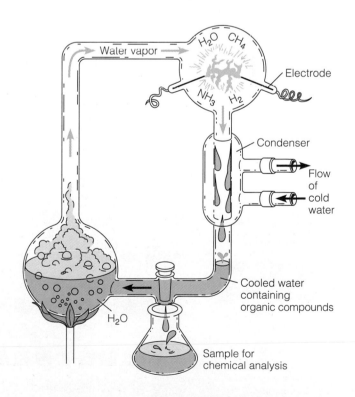

B. The synthesis of organic molecules in Miller's apparatus

17.4 The first polymers may have formed on hot rocks or clay

After small organic molecules formed on the prebiotic Earth, the second major chemical step before life arose must have been polymerization, the formation of organic polymers from monomers.

Polymers such as nucleic acids and proteins are synthesized by dehydration reactions that release a water molecule for each monomer added to the chain (see Module 3.3). In the living cell, specific enzymes catalyze these reactions. Polymerization also occurs in laboratory situations without enzymes, for example, when dilute solutions of organic monomers are dripped onto hot sand, clay, or rock. The heat vaporizes the water in the solutions and concentrates the monomers on the underlying substance. Some of the monomers then spontaneously bond together in chains, forming polymers.

Using this method, biochemist Sidney Fox, at the University of Miami, has made polypeptides that he calls "proteinoids." On the early Earth, in a similar fashion, raindrops or waves may have splashed dilute solutions of organic monomers onto fresh lava or other hot rocks, and then rinsed proteinoids and other polymers back into the sea.

Clay surfaces may have been especially important as early polymerization sites. Even cool clay concentrates amino acids and other organic monomers from dilute solutions, because the monomers bind to electrically charged sites on the clay particles. Such binding sites could have brought monomers close together. Clay also contains metal atoms, such as iron and zinc, that can function as catalysts. On the primitive Earth, metal atoms might have catalyzed the first dehydration reactions that linked monomers together to form polymers.

17.5 The first genes may have been RNA

The formation of polymers on the early Earth set the stage for the origin of early forms of life. But which polymers were most important? Stanley Miller gives us a hint: "The essential difference between life and nonlife is replication. There are other differences, but this is the essential one."

Miller's message is that nucleic acids, the only biological polymers that can replicate and store genetic information, were the essential polymers. Today's cells store their genetic information as DNA, transcribe the information into RNA, and then translate RNA messages into specific enzymes and other proteins. As we have seen in earlier chapters, this DNA → RNA → protein assembly system is extremely intricate. Most likely, it emerged gradually through a series of refinements to much simpler processes. What were the first genes probably like?

One hypothesis is that the first genes were short strands of RNA that replicated themselves, perhaps on clay surfaces. A number of laboratory experiments support this idea. Short RNA molecules have been produced from nucleotide monomers in test tubes, without cells or enzymes. Also, if RNA is added to a test-tube solution containing a supply of free RNA monomers, short new RNA molecules complementary to parts of the starting RNA sometimes assemble. Furthermore, in recent years, scientists have discovered that some RNA can act as an enzyme, catalyzing RNA splicing (see Module 11.9) and even RNA polymerization of a limited sort. So it's not hard to imagine a scenario like the one in the diagram here: A mixture of RNA nucleotide monomers spontaneously joins to form the first, small genes. Then, some of these RNA genes catalyze their own replication.

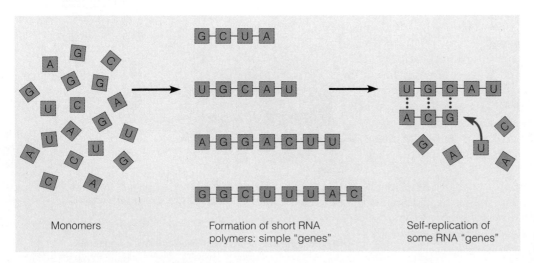

Monomers

Formation of short RNA polymers: simple "genes"

Self-replication of some RNA "genes"

A hypothesis for the origin of the first genes

Molecular cooperatives enclosed by membranes probably preceded the first real cells

If self-replicating RNA did form in an environment without life—and it is important to recognize that this is only a hypothesis—the early RNA molecules were still a far cry from a living cell. Being alive requires a great number of complex organic molecules, and the molecules must interact and cooperate in precise ways. Put another way, life depends on intricate metabolic machinery that derives from the cooperation of many complex organic molecules. It is likely that some amount of molecular cooperation preceded life's origin.

The earliest form of molecular cooperation may have involved a primitive form of translation of simple RNA genes into polypeptides, a form of translation that did not use ribosomes or tRNA. Suppose a strand of RNA acted as a rough template by weakly binding to amino acids and holding a few of them together long enough for them to be linked. Zinc or some other metal in the vicinity may have acted as a catalyst for such a chemical reaction. If the order of nucleotides in the RNA influenced the order of amino acids, the resulting polypeptide would be a rough translation of the RNA gene. If, in turn, the polypeptide behaved as an enzyme for helping the RNA molecule replicate (Figure A), then reciprocal molecular cooperation between nucleic acids and polypeptides would have begun.

Let's suppose that cooperative associations between RNA and polypeptides became common on the prebiotic Earth. We might imagine an "RNA world" where molecular "co-ops" often existed in aquatic environments, especially in small puddles or films of water on clay surfaces. These same sites may have harbored large numbers of other molecules, such as the proteinoids mentioned in Module 17.4. Experiments show that in a watery environment, proteinoids tend to self-assemble into fluid-filled droplets called microspheres (Figure B).

Microspheres are not alive, but they display some of the properties of living cells. They have a selectively permeable membranelike covering, can grow by absorbing other proteinoid molecules from their surroundings, divide when they reach a certain size, and swell or shrink osmotically when placed in solutions of different salt concentrations.

On the early Earth, microspheres might have formed, and certain of them might have taken in some of the RNA–polypeptide co-ops in their surroundings (Figure C). Once a membrane enclosed one of the co-ops, the stage was set for another giant leap toward life. A co-op housed in a membrane is isolated from other co-ops in its surroundings; it is therefore the sole beneficiary of its own protein products. Any membrane-bounded co-op that grew and replicated more efficiently than others would have been favored by natural selection. In other words, the molecular co-ops could have begun evolving in a Darwinian sense.

We have now plotted a scenario whereby local concentrations of organic chemicals on the prebiotic Earth may have given rise to cooperative associations of molecules bounded by membranes. Within these membranes, the molecular co-ops would have developed the ability to replicate and to carry out essential chemical reactions—a primitive metabolism. But millions of years probably passed before the first living cells appeared on Earth. During this long period, the molecular co-ops would have evolved into complex metabolic machines containing DNA and capable of efficiently using a variety of raw materials from the environment. Even the simplest cell would have been vastly more complex than any molecular co-op we have described. Whatever the evolutionary steps that led to the first cellular life, we can be fairly certain that the earliest cells were prokaryotes, as we discuss next.

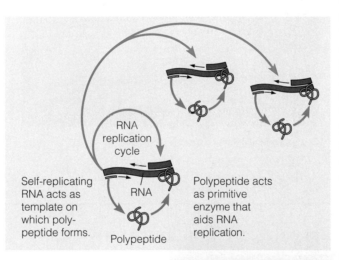

RNA replication cycle

Self-replicating RNA acts as template on which polypeptide forms.

RNA

Polypeptide

Polypeptide acts as primitive enzyme that aids RNA replication.

A. The beginnings of cooperation among macromolecules, in the absence of membranes

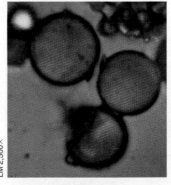

LM 2,500×

B. Microspheres formed by proteinoids

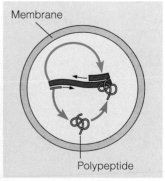

Membrane

Polypeptide

C. Macromolecular cooperation within a membrane-bounded compartment

17.7 Prokaryotes have inhabited Earth for billions of years

The fossil record shows that prokaryotes, or bacteria, making up the **kingdom Monera**, were abundant 3.5 billion years ago. They evolved all alone on Earth for about the next 2 billion years. Today, bacteria are found wherever there is life, and they outnumber all eukaryotes combined. More bacteria inhabit a handful of fertile soil or the mouth or skin of a human than the total number of people who have ever lived. Bacteria also thrive in habitats too cold, too hot, too salty, too acidic, or too alkaline for any eukaryote. You can get an idea of the size of prokaryotes from the figure here, a colorized scanning electron micrograph of the point of a pin (purple), covered with numerous bacteria (orange). Most prokaryotic cells have diameters in the range of 1–10 micrometers (μm), much smaller than most eukaryotic cells (10–100 μm).

Despite their small size, bacteria have an immense impact on our world. We hear most about a few species that cause serious illnesses. During the fourteenth century, Black Death—bubonic plague, a bacterial disease—spread across Europe, killing an estimated 25% of the human population. Tuberculosis, cholera, many sexually transmitted diseases, and certain types of food poisoning are some other human problems caused by bacteria. Bacteria also cause many kinds of diseases in other animals and in plants. We focus on bacterial diseases in Module 17.15.

Far more common than harmful bacteria are those that are benign or beneficial. We have bacteria in our intestines that provide us with important vitamins, and other species found in our mouth help prevent harmful fungi from growing there. Essential to all life on Earth are bacteria that decompose dead organisms. Found in soil and at the bottom of lakes, rivers, and oceans, these bacteria return vital chemical elements to the environment in the form of inorganic compounds that can be used by plants, which in turn feed animals. If bacterial decomposers were to disappear, the chemical cycles that sustain life would halt, and the other four kingdoms of life would also be doomed. In contrast, prokaryotic life would undoubtedly persist in the absence of eukaryotes, as it once did for billions of years.

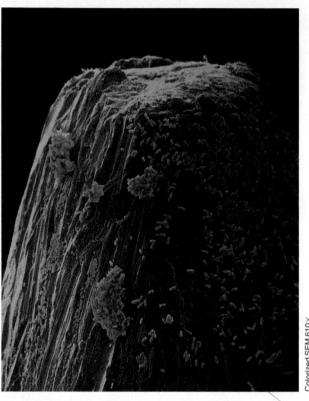

Colorized SEM 610×

Bacteria on the point of a pin

17.8 Review: Prokaryotes are fundamentally different from eukaryotes

Before discussing prokaryotes further, let's review the fundamental differences between prokaryotes and eukaryotes. First, the prokaryotic cell lacks a nucleus and other organelles enclosed by membranes. Membrane-bounded organelles are the hallmark of eukaryotic cells. Second, most of the DNA of a typical bacterial cell exists as a single, large molecule in the form of a ring. In contrast, the eukaryotic cell has its DNA packaged in a number of separate chromosomes, which are elaborate structures containing many protein molecules. You can review the basic structure of a prokaryotic cell in Figure 4.4.

Another unique feature of prokaryotes is that they do not carry out sexual reproduction as we usually think of it. When bacteria transfer genetic material from one cell to another (see Module 12.1), the amount of DNA transferred varies. This process is very different from the precise distribution of parental chromosomes to zygotes that occurs in the eukaryotic process of fertilization.

Bacteria carry out neither mitosis nor meiosis. Growing cells synthesize DNA almost constantly instead of at specific intervals, as in the closely regulated cell cycle of eukaryotes, and bacteria reproduce by the form of cell division called binary fission (see Module 8.3). A single bacterium in a favorable environment will divide repeatedly to yield a clone of offspring. When we speak of bacterial "growth," we actually mean the multiplication of cells—population growth rather than the enlargement of individual cells. In the photograph at the right, each spot you see in the culture dish is a colony, a clone of millions of bacterial cells. Bacteria multiply exponentially: One cell divides to form 2, 2 cells form 4, 4 form 8, and so on, the numbers doubling with each generation. For most bacteria, a generation lasts several hours, but some reproduce as often as every 10–20 minutes in the laboratory. Rates like these could give rise to colonies weighing hundreds of thousands of kilograms in a matter of days. This enormous growth potential is checked by environmental factors such as nutrient limitations and the buildup of toxic metabolic wastes from the bacteria themselves.

Yet another distinctive feature of prokaryotes is the cell surface. Nearly all prokaryotes have a cell wall outside their plasma membrane. As in plants, the wall maintains the cell's shape and provides physical protection. But a bacterial cell wall is made of materials very different from those of a plant cell wall. Instead of cellulose, the wall of most bacteria contains a unique material called **peptidoglycan**, which

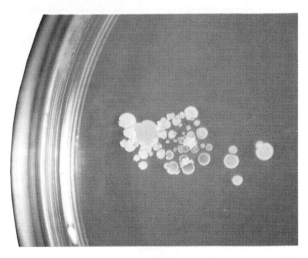

Bacterial colonies in a culture dish

is a polymer of sugars cross-linked by short polypeptides. One giant peptidoglycan molecule surrounds each bacterial cell. Many antibiotics, such as penicillin, inhibit bacterial growth by blocking the synthesis of peptidoglycan. Because eukaryotic cells do not make peptidoglycan, antibiotics can be used to cure many bacterial diseases without harming the patient.

Bacteria come in a variety of shapes 17.9

Determining cell shape by microscopic examination is an important step in identifying bacteria. The micrographs below show the three most common bacterial cell shapes. Spherical bacteria are called **cocci** (from the Greek word meaning berries). Cocci (singular, *coccus*) that occur in clusters, like the ones in Figure A, are called staphylococci (from the Greek *staphyle,* cluster of grapes). Other cocci occur in chains; they are called streptococci (from the Greek *streptos,* twisted). The bacterium that causes strep throat in humans is a streptococcus.

Figure B shows rod-shaped bacteria, which are called **bacilli** (singular, *bacillus*). Most bacilli occur singly, but some species occur in pairs (diplobacilli) and in chains

(streptobacilli). The species shown here, which is common in fertile soil, has a solitary form.

A third group of bacteria are curved or spiral-shaped. Some bacteria in this category resemble commas and are called *vibrios*. Others have a helical shape, like a corkscrew. Helical bacteria that are relatively short and rigid are called *spirilla*; those with longer, more flexible cells are called *spirochetes* (Figure C). The bacterium that causes syphilis, for example, is a spirochete. Spirochetes include some giants by prokaryotic standards—cells 0.5 mm long (though too thin to be seen without a microscope).

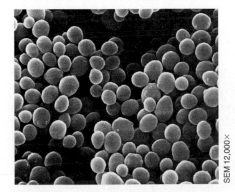

A. Cocci

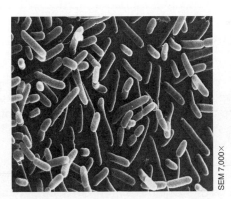

B. Bacilli

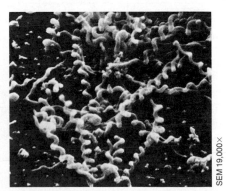

C. Spirochetes

17.10 Bacteria obtain nourishment in a variety of ways

When classifying diverse organisms, biologists often use the phrase "mode of nutrition" to describe how an organism obtains two main resources: energy and carbon (for synthesizing organic compounds). As a group, bacteria exhibit more nutritional diversity than all eukaryotes combined.

Many bacteria are **autotrophs** ("self-feeders"), making their own organic compounds from inorganic sources. As shown in the table below, autotrophs obtain their carbon from carbon dioxide (CO_2). They get their energy from sunlight or from inorganic chemicals, such as hydrogen sulfide (H_2S), elemental sulfur (S), or ammonia (NH_3). Those, like the cyanobacteria, that harness sunlight for energy do so by photosynthesis. Cyanobacteria, such as *Chroococcus* (Figure A), use H_2O as a source of electrons for photosynthesis and produce O_2 as a waste product, just like plants. Other groups of photosynthetic bacteria use inorganic molecules *other* than water (H_2S, for example) as a source of electrons.

Most bacteria are **heterotrophs** ("other-feeders"), meaning they obtain their carbon from organic compounds. Some heterotrophic bacteria obtain their energy from sunlight. By far the largest group of bacteria, however, are nutritionally similar to animals in that they obtain both energy and carbon from organic molecules. Called **chemoheterotrophs,** these bacteria are so diverse that almost any organic molecule can serve as food for some species. Many species, such as *Escherichia coli,* a resident of the human intestine, thrive on sugars. Figure B shows a culture of *E. coli* grown with only the sugar glucose as an organic nutrient. This bacterium can also thrive on many other organic nutrients.

Chemoheterotrophs dominate the bacterial world today. As we discuss in the next module, this may have been true since the dawn of life.

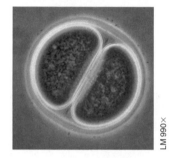

LM 990×

A. *Chroococcus,* a photosynthesizing cyanobacterium

Nutritional Classification of Organisms		
Nutritional Type	**Energy Source**	**Carbon Source**
Autotroph	Sunlight (photosynthesizers)	CO_2
	Inorganic chemicals	CO_2
Heterotroph	Sunlight	Organic compounds
	Organic compounds (chemoheterotrophs)	Organic compounds

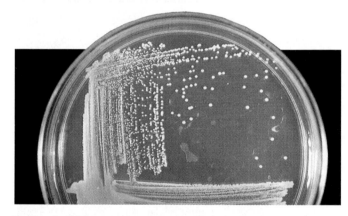

B. *E. coli* colonies grown on glucose as the sole organic nutrient

17.11 The first life forms were probably chemoheterotrophs

With bacterial nutrition in mind, let's return to our discussion of the origin of life. Earlier, we proposed that the first prokaryotes evolved from membrane-bounded molecular co-ops. These co-ops had a simple form of metabolism; they were chemoheterotrophs that absorbed from their surroundings the molecules they needed for synthesizing organic compounds. Because absorbing nutrients from the environment is chemically simpler than synthesizing them from scratch, it is likely that the first prokaryotes were chemoheterotrophs.

The figure on the facing page shows how an early form of energy metabolism may have evolved. Stage ①: Living in the complex soup of organic molecules from which it evolved, an early prokaryotic chemoheterotroph probably obtained all the energy and carbon it needed by absorbing organic compounds from its surroundings. Among the small organic molecules in its environment may have been ATP, which could have been a major source of energy for the cell. The fact that all modern forms of life use ATP as their main energy currency makes it likely that prokaryotes started using this molecule very early.

Stage ② in the diagram carries the chemoheterotroph through a second phase in the evolution of metabolism. Life may have faced its first energy crisis as early prokaryotes began using up the ATP they depended on for fuel. Some of the prokaryotes probably evolved the ability to synthesize enzymes that could regenerate ATP from ADP using energy extracted from other organic nutrients that were still available. The result may have been the step-by-step evolution of glycolysis, the metabolic pathway that uses the

energy released in the breakdown of glucose to generate ATP. (For a review of glycolysis and ATP synthesis, see Chapter 6.)

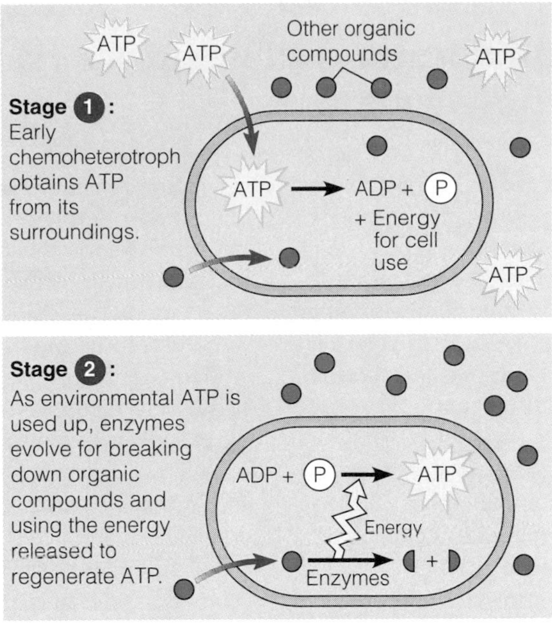

The fact that glycolysis is the only metabolic pathway common to nearly all living organisms suggests its great antiquity. Because glycolysis does not require O_2, it could have evolved in an environment lacking this molecule, like the environment of the ancient Earth. Fermentation, which is essentially a minor expansion of glycolysis, probably became a common way of life on the anaerobic Earth. In the next module, we examine a group of prokaryotes alive today whose modes of nutrition may resemble those of the first cells.

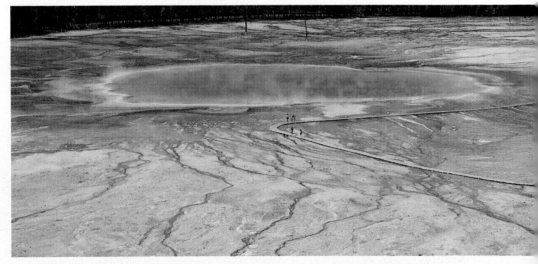

Stage ❶:
Early chemoheterotroph obtains ATP from its surroundings.

Other organic compounds

ATP → ADP + ⓟ
+ Energy for cell use

Stage ❷:
As environmental ATP is used up, enzymes evolve for breaking down organic compounds and using the energy released to regenerate ATP.

ADP + ⓟ → ATP
Energy
◖ + ◗
Enzymes

Hypothesis for the early evolution of energy metabolism

Archaebacteria are fundamentally different from all other bacteria

The name **archaebacteria** comes from the Greek *archaios,* meaning "ancient," and most biologists believe that this group of prokaryotes diverged from other bacteria in very ancient times. Archaebacteria differ from all other prokaryotes in several fundamental ways. For instance, their tRNA and rRNA molecules have base sequences different from those found in all other bacteria. Also, their plasma membranes have unique lipids, and their cell walls lack peptidoglycans. Because they lack peptidoglycans, their growth is not inhibited by penicillin.

Certain species of contemporary archaebacteria, called methanogens, may be metabolically quite similar to early prokaryotes. Methanogens are chemoheterotrophs that live by fermentation in anaerobic environments and give off methane as a waste product. Some methanogens thrive in our intestines, while others live in stagnant swamps. You may have seen methane gas bubbling up from a swamp; it is called marsh gas.

Archaebacteria typically live where few other organisms can survive. A group called the halophiles ("salt lovers") thrive in extremely salty places, such as the Great Salt Lake in Utah. Another group, the thermoacidophiles ("heat-and-acid lovers"), thrive in hot springs,

where temperatures may top 80°C (176°F) and the pH may be as low as 2. The photograph here shows the Grand Prismatic Pool in Yellowstone National Park, where thermoacidophiles thrive and give the pool's edge its greenish color. A thermoacidophile named *Sulfolobus* that lives here can obtain energy by oxidizing sulfur, a mode of nutrition that some biologists think predated chemoheterotrophic nutrition.

Archaebacteria are so different from other bacteria in cellular and metabolic features that some biologists propose classifying them in a separate kingdom. However, in this text we include the archaebacteria as a distinctive group within the kingdom Monera.

An acidic hot pool at Yellowstone containing archaebacteria

Diverse structural features allow eubacteria to thrive in various environments

Today, all living prokaryotes except the archaebacteria are classified as **eubacteria,** a huge group found literally everywhere and upon which all other forms of life depend. As a group, eubacteria exhibit all four modes of nutrition described in Module 17.10. Chemically, they share a number of common features, such as peptidoglycans in their cell walls and the same types of RNA molecules. In this module, we discuss some of the structural features that help eubacteria thrive in their environments.

Many eubacteria (and archaebacteria, too) are equipped with flagella, which enable them to move about. Bacteria with flagella can move toward more favorable places or away from less favorable ones. Flagella may be scattered over the entire cell surface or concentrated at one or both ends of the cell. Entirely different in structure from the flagellum of eukaryotic cells (described in Module 4.18), the bacterial flagellum is a naked protein structure that lacks microtubules. It is attached to the cell surface by a system of rotating rings anchored in the plasma membrane and cell wall. The rings give the flagellum a propellerlike rotary movement, as shown in Figure A. The flagellated bacterium in the photograph is *Proteus,* an especially fast swimmer.

Shorter and thinner than flagella are the bacterial appendages called pili (singular, pilus). Pili are not visible in the Figure A photo but show up clearly in the more highly magnified photograph of another bacterium in Figure B. Pili help bacteria stick to each other and to surfaces, such as rocks in flowing streams or the lining of human intestines. As we mentioned in Module 12.1, special "sex pili" are required for initiating bacterial "mating" (conjugation).

Many eubacteria are capable of surviving extremely harsh conditions. Figure C (top of the facing page) shows an example of such an organism, *Bacillus anthracis,* the bacterium that produces the deadly disease called anthrax in cattle, sheep, and humans. There are actually two cells here, one inside the other. The outer cell produced the specialized inner cell, called an **endospore.** The endospore has a thick, protective coat, its cytoplasm is dehydrated, and it does not metabolize. Under harsh conditions, the outer cell may disintegrate, but the endospore survives all sorts of trauma, including lack of water and nutrients, extreme heat or cold, and most poisons. When the environment becomes more hospitable, the endospore absorbs water and resumes growth.

Eubacteria of several genera can form endospores, and some endospores can remain dormant for centuries. Not even boiling water kills most of these resistant cells. To sterilize laboratory equipment, microbiologists use an appliance called an autoclave, a pressure cooker that kills endospores by heating to a temperature of 121°C (250°F) with high-pressure steam. The food-canning industry uses similar methods to kill endospores of dangerous eubacteria such as *Clostridium botulinum,* which produces the potentially fatal disease botulism.

The mass of branching cell chains (filaments) in Figure D is a structural feature unique to the bacterial group called **actinomycetes.** These eubacteria are very common in soil, where they break down organic substances. The filaments allow the organism to bridge dry gaps between soil particles. Actinomycetes were once mistaken for fungi, which

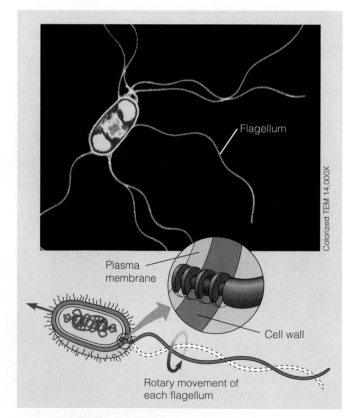

Colorized TEM 14,000X

Flagellum

Plasma membrane

Cell wall

Rotary movement of each flagellum

A. Bacterial flagella

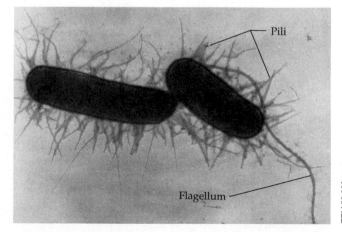

TEM 50,000×

Pili

Flagellum

B. Pili

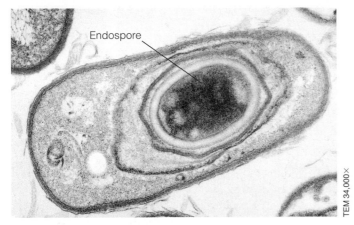

C. An endospore within a bacterial cell

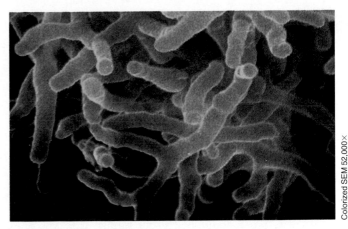

D. Filaments of an actinomycete

also grow in branching filaments; this similarity explains their name, which is from the Greek for "ray fungus." The actinomycete in Figure D is of the genus *Streptomyces*, a common soil bacterium. *Streptomyces* secretes the antibiotic streptomycin and a number of other antibiotics, which inhibit the growth of competing microbes. Pharmaceutical companies use various species of actinomycetes to produce commercial quantities of many antibiotics.

Branching chains of cells are unusual, but many other bacteria grow in unbranched chains. Next, we look at a filamentous cyanobacterium and the effect it can have on a polluted lake.

Cyanobacteria sometimes "bloom" in aquatic environments — 17.14

The larger lake pictured in Figure A below is undergoing a population explosion—often called a "bloom"—of **cyanobacteria.** The blue-green color results from the presence of trillions of cyanobacterial cells. Figure B is a micrograph of *Anabaena,* the predominant cyanobacterium in the lake.

Cyanobacteria make up one group of photosynthetic eubacteria. They are common in lakes, ponds, and tropical oceans. Blooms of a reddish species give the Red Sea its name and, along with other microorganisms, often tint waters of the Gulf of California. Extensive blooms of cyanobacteria in a lake usually indicate polluted water conditions. In the case shown here, the water was loaded with organic wastes from agricultural runoff.

This lake's condition may be reminiscent of the age of cyanobacteria, a time when these prokaryotes dominated Earth. At that time, some 3.0–1.5 billion years ago, cyanobacteria gave Earth its first greenish coat, generated the stromatolites we discussed earlier, and made the atmosphere aerobic. The molecular machinery for photosynthesis that is housed in cyanobacteria today may be much like that which first added O_2 to Earth's atmosphere.

A. Bloom of cyanobacteria in a lake

B. *Anabaena*

Some bacteria cause disease

All organisms, humans included, are almost constantly exposed to bacteria, some of which are potentially harmful. In fact, most of us are well most of the time only because our body defenses check the growth of bacterial **pathogens,** disease-causing agents. Occasionally, the balance shifts in favor of a pathogen, and we become ill. Even some of the bacteria that are normal residents of the human body can make us ill when our defenses have been weakened by poor nutrition or by a viral infection.

Bacteria cause about half of all human diseases. Some species injure us by destroying tissues, but most bacteria cause disease by producing poisons. Bacterial poisons are of two types: exotoxins and endotoxins. **Exotoxins,** toxic proteins secreted by bacterial cells, include some of the most potent poisons known. A single gram of the exotoxin that causes botulism, for instance, could kill a million people.

The culture dish in Figure A contains golden yellow colonies of the bacterial species *Staphylococcus aureus,* another exotoxin producer. (The name *aureus* means golden in Latin.) *S. aureus* is a common and usually harmless inhabitant of our skin surface. If it enters the body through a cut or other wound, however, or if swallowed in contaminated food, it can cause serious diseases. One variety of *S. aureus* produces exotoxins that cause layers of skin to slough off; another can cause food poisoning with vomiting and severe diarrhea; yet another can produce the potentially deadly toxic shock syndrome.

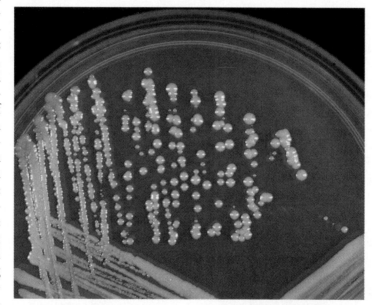

A. An exotoxin producer, *Staphylococcus aureus*

In contrast to exotoxins, **endotoxins** are not cell secretions, but components of the cell walls of certain bacteria. Endotoxins are glycolipids, large molecular complexes of polysaccharides and lipids. All endotoxins induce the same general symptoms: fever, aches, and sometimes a dangerous drop in blood pressure (shock). The severity of symptoms varies with the species of bacterium and with the condition of the host. Species of *Salmonella,* for example, produce endotoxins that cause food poisoning and typhoid fever.

In the last 100 years, following the nineteenth-century discovery that "germs" cause disease, the incidence of bacterial diseases has declined, particularly in developed nations. Sanitation is generally the most effective way to prevent bacterial diseases, and the installation of water-treatment and sewage systems continues to be a priority of public health workers throughout the world. In the past few decades, antibiotics have been discovered that can cure most bacterial diseases.

In addition to sanitation and antibiotics, a third tool in our defense against bacterial diseases is education. A case in point is **Lyme disease,** which is currently on the increase in the United States. First reported in the U.S. in 1969, Lyme disease is caused by *Borrelia burgdorferi,* a bacterium carried by ticks that live on deer and field mice. As shown in Figure B, Lyme disease usually starts out as a red rash that clears in the center, resembling a bull's-eye around the site of a tick bite. Antibiotics can cure the disease if administered within about a month after exposure or when the skin rash appears. If not treated, this infection can produce heart disease, debilitating arthritis, and nervous disorders. The best tactic against Lyme disease right now is public education about precautions for avoiding tick bites and the importance of seeking medical treatment if a rash develops. Wearing light-colored clothing and using a chemical repellant can reduce contact with ticks.

Tick that carries the Lyme-disease bacterium

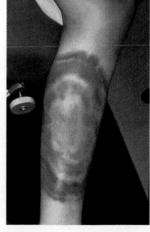

"Bull's-eye" rash

B. Lyme disease, a bacterial disease transmitted by ticks

Discovering what species of bacterium causes a particular disease is a key step in developing ways to prevent or cure it. The diagnosis of bacterial diseases has been on firm scientific footing for over 100 years, thanks to the work of German physician Robert Koch (Figure A). In 1876, Koch discovered rod-shaped bacteria, which we now call *Bacillus anthracis*, in the blood of cattle suffering from anthrax. Koch used a set of diagnostic criteria—now called **Koch's postulates**—to prove that the bacteria were the cause of the disease.

Koch's postulates are as follows:

1. The same, specific pathogen must be identified in each animal (host) that has the disease.

2. The pathogen must be isolated from a host and grown in a pure culture, one in which no other kinds of cells are present.

3. The original disease must be produced in experimental hosts that are inoculated with the pathogen from the pure culture.

4. The same pathogen must be isolated from the experimental hosts after the disease develops in them.

Figure B outlines the procedure that microbiologists use to apply Koch's postulates. Like all scientific work, it hinges on the *repeatability of results*. It is the repeated production of a specific disease by a bacterium and the cultivation of the same bacterium from a number of experimentally infected hosts that confirm the identity of the pathogen.

Koch's postulates have been used to identify most bacterial pathogens, but there are some exceptions. For example, no one has yet been able to grow the bacterium *Treponema pallidum* in pure culture, but strong circumstantial evidence leaves no doubt that this microorganism causes syphilis.

Microbiologists have used Koch's criteria to identify pathogens outside the bacterial world—viruses, fungi, protists, and parasitic worms—but frequently they must modify the procedures. Viruses, for instance, reproduce only inside host cells, and protists may require other living cells for food. Koch's postulates are nevertheless the basis for determining the causative agent of most infectious diseases.

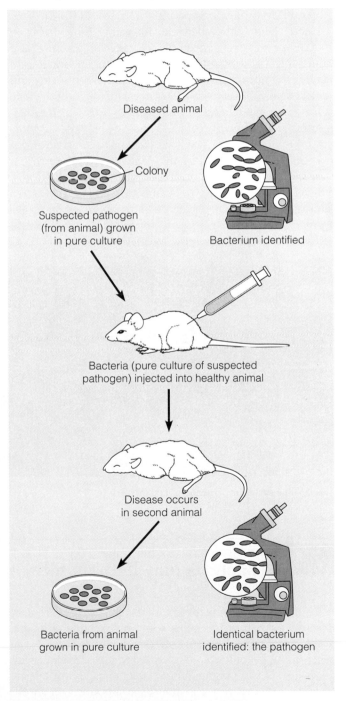

B. Procedure for demonstrating Koch's postulates

A. Robert Koch studying bacterial diseases in Africa in 1906

Chapter 17   The Origin and Evolution of Microbial Life   **327**

Chemical cycles in our environment depend on bacteria

Despite the attention they demand, bacteria that cause disease are in the minority. Far more common are species that are vital to our well-being, either directly or indirectly. All life depends on the cycling of chemical elements between organisms and the nonliving components of our environment (as we describe in detail in Chapter 36). Bacteria are indispensable components of chemical cycles. For example, cyanobacteria not only restore oxygen to the atmosphere, they also convert nitrogen gas (N_2) in the atmosphere into nitrogen compounds (nitrates and nitrites) in soil and water. Other bacteria, including those living in nodules on the roots of bean plants and other legumes, also contribute large amounts of nitrogen compounds to soil. In fact, all the nitrogen that plants use to make proteins and nucleic acids comes from bacterial metabolism in the soil. In turn, animals get their nitrogen compounds from plants.

Another vital function of bacteria is the breakdown of organic wastes and dead organisms. Bacteria decompose organic matter and, in the process, return elements to the environment in inorganic forms available for use by other organisms. If it were not for decomposers, organic wastes and dead organisms would literally pile up on Earth, and the atoms composing them, such as carbon, nitrogen, and oxygen, would not become available for reuse by later generations.

Bacterial decomposers are also the mainstays of our sewage-treatment facilities. In a modern treatment plant, raw sewage is first passed through a series of screens and shredders, and solid matter is allowed to settle out from the liquid waste. This solid matter, called sludge, is then gradually added to a culture of anaerobic bacteria. The bacteria decompose the organic matter in the sludge, converting it to material that can be used as landfill or fertilizer, after chemical sterilization.

Liquid wastes are treated separately from the sludge. In the figure below, you can see a trickling filter system, one type of mechanism for treating liquid wastes. The long horizontal pipes rotate slowly, spraying liquid wastes through the air onto a thick bed of rocks, the filter. Aerobic bacteria and fungi growing on the rocks digest much of the organic matter in the waste. Outflow from the rock bed is chemically sterilized and then released, usually into a river or ocean. In a few sewage systems, such as one at Lake Tahoe, between California and Nevada, the outflow is further purified to remove most organic substances and certain inorganic chemicals. Systems like this one release virtually no contaminants into the environment, but their high cost prevents most cities from building them.

Trickling filter system at a sewage-treatment plant

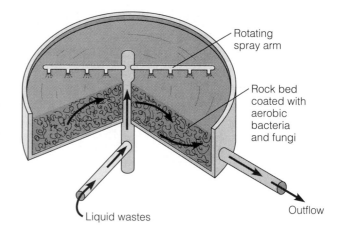

Rotating spray arm

Rock bed coated with aerobic bacteria and fungi

Liquid wastes

Outflow

Bacteria may help us solve some environmental problems

We are just beginning to explore the great potential bacteria have for helping us solve some of our environmental problems. Certain bacteria that occur naturally on ocean beaches can decompose petroleum and are useful in cleaning up oil spills. In Figure A on the facing page, workers are spraying nitrogen and phosphorus compounds on an oil-soaked beach in Alaska. The chemicals act as fertilizers,

stimulating the growth of "oil-eating" bacteria. Although there is some danger of water pollution from the fertilizers, this technique is the most rapid and inexpensive way yet devised to restore oil-fouled beaches.

Other bacteria may help us clean up old mining sites. The water that drains from mines is highly acidic and is also laced with poisons, often compounds of arsenic, copper, zinc,

A. Treatment of an oil spill in Alaska

B. Toxic drainage water from a copper mine in Montana

and the heavy metals lead, mercury, and cadmium. Contamination of our soils and groundwater by these toxic substances poses a widespread threat, and cleaning up the mess is extremely expensive.

The photograph in Figure B shows the remains of a half-mile-deep open-pit mine known as the Berkeley Pit, in Butte, Montana. The mine has not operated since 1983, when its high-quality copper ore was used up. Water in the pit is highly acidic and loaded with heavy metal ions. It threatens groundwater, soils, adjacent rivers, and communities for hundreds of kilometers downstream. The Environmental Protection Agency (EPA) considers the Berkeley Pit the nation's most expensive cleanup site, and the EPA has already spent over $50 million just to initiate waste removal here.

Although there are no simple solutions to environmental problems like this one, bacteria may be able to help. Species of the bacterium *Thiobacillus* thrive in the acidic waters that drain from mines. These bacteria obtain energy from oxidizing minerals and can absorb and accumulate toxic metal ions. The bacteria can be seeded into the contaminated water, or the water can be passed through filters containing the bacteria. The use of filters makes it possible to recover metals from the water, so that the bacteria are actually mining the toxic waste water. (Some mining companies already use one species of *Thiobacillus* to extract copper from low-grade ores.) Water in the Berkeley Pit carries added dividends. Besides copper, zinc, and mercury, it carries gold and silver compounds that also can be extracted by bacteria. These valuable metals may provide the incentive to make the bacterial cleanup of mining sites a financially attractive industry.

Summary: Bacteria are at the foundation of life on Earth 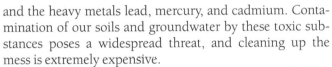 17.19

Summarizing the vital roles of prokaryotes on Earth, we could simply say that all life depends on bacteria. This is true today and has always been so. Another truism is that environmental change and biological evolution are closely connected. Both have been episodic, marked by what were, in essence, revolutions that opened up new ways of life. Bacteria have been at the center of these revolutions and have had a greater impact on the environment and on evolution than all other forms of life combined. Only bacteria have produced an environmental change as radical and significant to the evolution of life as making the atmosphere aerobic. Bacteria not only created an oxygen revolution, they were the first organisms to tolerate the corrosive effects of atmospheric oxygen and the first to use oxygen to metabolize organic molecules. If the oxygen revolution had not occurred, it is likely that Earth would still be populated only by anaerobic prokaryotes.

Cyanobacteria continue to play a major role in maintaining the atmospheric reservoir of oxygen and in removing carbon dioxide from the atmosphere. The role of cyanobacteria and other prokaryotes in converting atmospheric nitrogen into chemical forms that plants can use is essential to all life. Indeed, the cycling of nitrogen between the atmosphere and living organisms would stop altogether without bacteria.

In brief, bacteria are at the foundation of life in both an environmental and an evolutionary sense. In the next module, we plot a scenario illustrating the essential role of prokaryotes in yet another monumental event—the formation of the first eukaryotic cells.

17.20 The eukaryotic cell probably originated as a community of bacteria

The fossil record indicates that eukaryotes evolved from prokaryotes between 1.5 and 1 billion years ago, but how did this seminal event occur? Two hypotheses have been proposed. One is that eukaryotic cells evolved by specialization of internal membranes in ancestral prokaryotes. According to this model, all the membrane-bounded organelles in the eukaryotic cell, including the nucleus, mitochondria, and chloroplasts, were derived by the infolding of a prokaryote's plasma membrane. Figure A suggests how the nuclear envelope and endoplasmic reticulum may have developed, in this model.

Another hypothesis is that eukaryotic cells started out as cooperative communities of prokaryotic cells. This model is called the **endosymbiotic hypothesis** (from the Greek meaning "living together within"), and it focuses on the origins of chloroplasts and mitochondria.

A **symbiotic relationship** is a close association between two or more species. The endosymbiotic hypothesis proposes that chloroplasts and mitochondria evolved from small prokaryotes that established residence within other, larger prokaryotes (Figure B). The ancestors of mitochondria may have been small heterotrophic bacteria that were able to use O_2 to release energy from organic molecules by cellular respiration. At some point, an ancestral host cell lacking this ability may have ingested some of these aerobic bacteria. If some of the smaller cells were indigestible, they might have remained alive and continued to perform respiration in the host cell. In a similar way, small photosynthetic prokaryotes ancestral to chloroplasts may have come to live inside a larger host cell.

It's not hard to imagine how the relationship between the engulfed bacteria and the host cell might have become mutually beneficial. In both cases, the small cells may have grown increasingly dependent on the host cell for water and inorganic ions needed to make food molecules. Likewise, the host cell may have derived increasing amounts of food from the small photosynthetic cells and increasing amounts of ATP from the small aerobes. As the cells in these prokaryotic communes became more interdependent, they may have become truly a single organism, its parts inseparable.

A considerable amount of circumstantial evidence supports the endosymbiotic hypothesis. For instance, present-day mi-

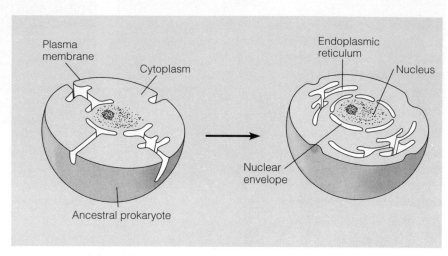

A. Infolding of prokaryotic plasma membrane giving rise to eukaryotic organelles

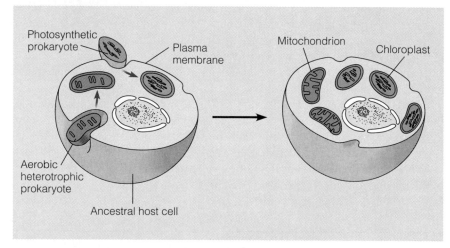

B. Endosymbiotic bacteria giving rise to mitochondria and chloroplasts

tochondria and chloroplasts are similar to bacteria in a number of ways. They contain small amounts of DNA, RNA, and ribosomes, all of which resemble their counterparts in prokaryotes more than those in eukaryotes. These components enable chloroplasts and mitochondria to exhibit some autonomy in their activities. These organelles transcribe and translate their DNA into proteins, making some of their own enzymes. They also replicate their own DNA and reproduce within the cell by a process resembling the binary fission of bacteria.

The endosymbiotic hypothesis can also explain how chloroplasts and mitochondria came to be bounded by two membranes. As Figure B indicates, the inner membranes of these organelles could have been derived from the plasma membranes of the engulfed bacteria, and their outer membranes could have come from the infolded plasma membranes of the original host cells. In fact, the inner membranes of mitochondria and chloroplasts have several enzymes and electron transport molecules that resemble those found in the plasma membranes of modern prokaryotes, perhaps a result of an endosymbiotic origin.

The endosymbiotic hypothesis is an attractive model for the evolution of mitochondria and chloroplasts, and it is accepted as such by most biologists. But not all components of the eukaryotic cell may have evolved this way. Most likely, the nucleus and other organelles of the endomembrane system (see Module 4.7) arose, as indicated in Figure A, by modification of the host cell. Thus, the model in Figure A seems to account for some of the unique features of eukaryotic cells, and the endosymbiotic model seems to better explain the origin of others.

Unicellular eukaryotes and their direct multicellular descendants are called protists

Having presented models for the evolution of the eukaryotic cell, we now leave the prokaryotes and their kingdom Monera and enter a much more complex part of the living world. The photograph below—a drop of pond water viewed with the light microscope—illustrates a variety of **protists**, members of the **kingdom Protista.** Protists are a diverse group of mostly unicellular eukaryotes. Some, called algae, synthesize their own food by photosynthesis. Others, called protozoans, are heterotrophic, eating bacteria, other protists, or organic matter suspended in the water. Also included in the kingdom Protista are a number of colonial and multicellular eukaryotes whose immediate ancestors were unicellular.

Almost any aquatic environment is home to great numbers of protists. Most species, including those shown here, are aerobic. Some protists, however, are anaerobic, living in mud at the bottom of lakes and stagnant ponds or thriving in the digestive tracts of animals, including humans.

As eukaryotes, protists are more complicated than any prokaryotes. Their cells have a true nucleus (containing chromosomes), mitochondria, and other organelles characteristic of eukaryotic cells. Flagella and cilia on protistan cells have a 9 + 2 pattern of microtubules, another typical eukaryotic trait (see Module 4.18).

Kingdom Protista occupies a pivotal position in the history of life: The earliest protists were ancestral to all plants, fungi, and animals, as well as to all modern protists.

Because most protists are unicellular, as a kingdom they are justifiably considered to be the simplest eukaryotes. But the kingdom Protista includes some species whose cells are the most elaborate in the world. This level of cellular complexity is not really surprising, for each unicellular protist is a complete eukaryotic organism analogous to an entire animal or plant.

In the next five modules, we illustrate some of the diversity in four major groups of protists: the protozoans, the slime molds, the unicellular algae, and the multicellular algae, commonly called seaweeds. Multicellular algae are included in the kingdom Protista because they are direct descendants of unicellular algae.

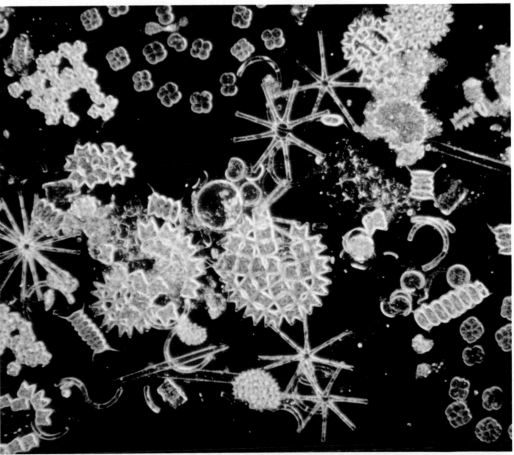

LM 143×

Protozoans are protists that ingest their food

Protists that live primarily by ingesting food, a heterotrophic mode of nutrition that is animal-like, are called **protozoans** (from the Greek *protos,* first, and *zoion,* animal). Protozoans thrive in all types of aquatic environments, including the tiny water droplets in wet moss and soil, and the watery environment inside animals. Most species eat bacteria or other protozoans. Species that live as parasites on animals, though in the minority, cause some of the world's most harmful human diseases. The organisms featured in this module represent the four most common groups of protozoans: amoebas, flagellates, apicomplexans, and ciliates.

Amoebas are characterized by great flexibility and the absence of permanent locomotor organelles. Most species move and feed by means of **pseudopodia** (singular, *pseudopodium*), which are temporary extensions of the cell. The individual in Figure A is ingesting another, much smaller protozoan for food. The amoeba's pseudopodia arch around the prey, engulfing it into a food vacuole (see Module 4.11). Amoebas can assume virtually any shape, as they creep over rocks, sticks, or mud at the bottom of a pond or ocean.

Flagellates are protozoans that move by means of one or more flagella. Most species are free-living (nonparasitic), but we picture a parasitic one here because of its importance in human disease. The protozoans that appear purple in Figure B are *Trypanosoma,* a flagellate that lives in the bloodstream of vertebrate animals. The disk-shaped cells in this micrograph are human red blood cells. Trypanosomes obtain their nutrients from the host blood. The ones shown here cause sleeping sickness, a debilitating disease common in parts of Africa. Living in the bloodstream, trypanosomes are under constant attack by the host's immune system. Trypanosomes escape being killed by their host's defenses by being quick-change artists. They alter the molecular structure of their coats frequently, thus preventing immunity from developing in the host.

Apicomplexans, another large group of protozoans, are all parasitic, and some cause serious human diseases. As seen with the electron microscope, one end (the *apex*) of the infectious cell of these parasites contains a *complex* of organelles specialized for penetrating host cells and tissues—thus the name apicomplexan. The micrograph in Figure C shows *Plasmodium,* the apicomplexan that causes malaria. This parasite actually enters red blood cells, feeding on them from within and eventually destroying them. Spread by mosquitos, malaria is one of the most debilitating and widespread human diseases. Each year in the tropics, more than 200 million people become infected, and at least a million die in Africa alone.

A fourth group of protozoans, the **ciliates,** use cilia to move and feed and are common in all types of aquatic environments. Nearly all ciliates are free-living, including the very common freshwater protist *Paramecium* (Figure D, left, top of the next page). Ciliates are unique in having two types

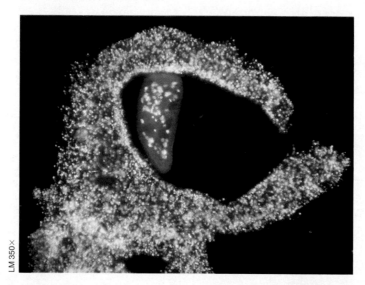

LM 350×

A. An amoeba ingesting a smaller protozoan

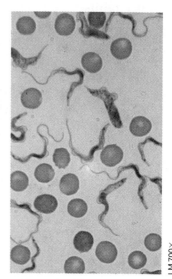

LM 700×

B. A flagellate: *Trypanosoma*

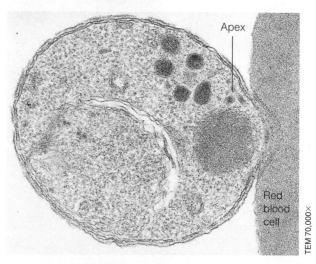

Apex

Red blood cell

TEM 70,000×

C. An apicomplexan: *Plasmodium*

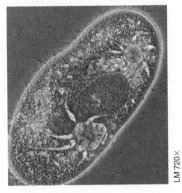

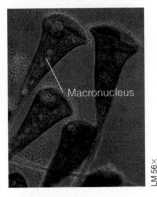

Macronucleus

LM 720× LM 56×

D. Ciliates: *Paramecium* (left) and *Stentor* (right)

of nuclei: a single, large macronucleus, which controls everyday activities, and from one to as many as 80 tiny micronuclei, which function in sexual reproduction. You can see the macronucleus both in *Paramecium* and in the striking, bluish ciliate *Stentor* (Figure D, right). The macronucleus of *Stentor* resembles a string of beads. Like *Paramecium*, *Stentor* is common in freshwater ponds.

Protozoans are highly diverse protists. Indeed, they show us how structurally varied single-celled life can be. It is likely that the amoebas, flagellates, apicomplexans, and ciliates represent several evolutionary lineages that evolved from ancestral eukaryotes.

Cellular slime molds have multicellular stages 17.23

In contrast to protozoans, protists called **cellular slime molds** lead a dual existence; they have both unicellular and multicellular life stages. Cellular slime molds are common on rotting logs and other decaying organic matter.

The micrographs here show three stages in the life cycle of a typical cellular slime mold, *Dictyostelium*. Most of the time, this organism exists as solitary amoeboid cells, creeping about by pseudopodial movement and engulfing bacteria as they go. The top picture shows two amoeboid cells. The small dark rods are bacteria; the bacteria inside the slime-mold cells are being digested within food vacuoles.

When bacteria are plentiful, amoeboid slime-mold cells multiply by mitotic cell division but remain solitary. However, when bacteria are in short supply, the amoeboid cells swarm together, forming a colony. The center picture shows a sluglike colony of *Dictyostelium*. Cells in the colony secrete a cellulose covering coated with a slimy sheath. After wandering around for a short time, the colony develops into a multicellular reproductive structure, shown in the bottom picture.

Dictyostelium is easily cultured in the laboratory, and its relatively simple structure, compared to most multicellular eukaryotes, makes it an attractive research organism. Because the amoeboid cells in the sluglike colony differentiate when they form the reproductive structure, *Dictyostelium* is a useful model for researchers studying the genetic mechanisms and chemical changes underlying cellular differentiation.

Cellular slime molds were formerly classified as fungi because their colonial stage superficially resembles a mold. However, these organisms have virtually nothing in common with true molds or any other fungi. Since all their life stages are composed of amoeboid cells or their derivatives, it is likely that the closest living relatives of cellular slime molds are the single-celled amoebae.

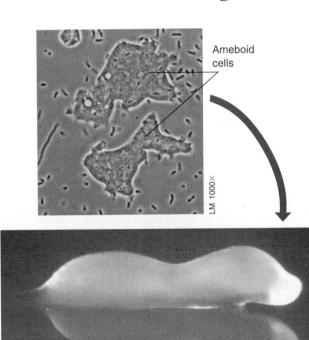

Ameboid cells

LM 1000×

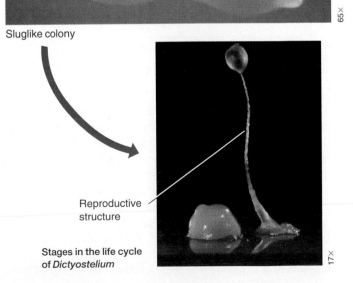

Sluglike colony

65×

Reproductive structure

Stages in the life cycle of *Dictyostelium*

17×

17.24 Plasmodial slime molds have brightly colored stages with many nuclei

The yellow, branching growth on the dead log in Figure A is a protist called a **plasmodial slime mold.** Plasmodial slime molds are common almost everywhere there is moist, decaying organic matter. They exist as single cells and as organisms that grow to several centimeters in diameter, like the one shown here.

Large and branching as it is, the organism in Figure A is not multicellular; it is a single mass of cytoplasm, undivided by membranes and containing many nuclei. Called a **syncytium** (from the Greek *syn,* together, and *kytos,* cell), this is an amoeboid life stage that extends pseudopodia for feeding on bacteria, yeasts, and bits of dead organic matter. Its weblike form is an adaptation that enlarges the organism's surface area, increasing its contact with food, water, and oxygen. Within the fine channels of the syncytium, cytoplasm streams first one way and then the other, in pulsing flows that are beautiful to watch with a microscope. The cytoplasmic streaming probably helps distribute nutrients and oxygen.

As long as a plasmodial slime mold has ample food and water, it usually stays in the syncytial stage. When food and water are in short supply, however, the syncytium stops growing and differentiates into reproductive structures, such as the yellow ones in Figure B. When conditions again become favorable, the life cycle continues with amoeboid cells or flagellated cells emerging from the reproductive structures.

Like the cellular slime molds, plasmodial slime molds were formerly classified in the kingdom Fungi. They are clearly not fungi, and having amoeboid and flagellated cells indicates an affiliation with protozoans. However, plasmodial slime molds are probably not very closely related to any other living organisms.

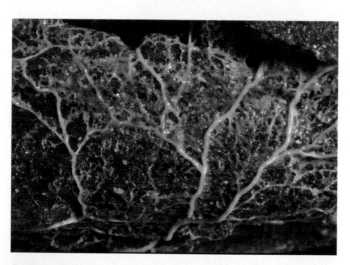

A. Syncytial stage of a plasmodial slime mold, *Physarum*

B. Reproductive stage of *Physarum*

17.25 Photosynthetic protists are called algae

Whereas the protozoans and slime molds obtain their organic food molecules from other organisms, almost all **algae** are plantlike in that they synthesize their own food molecules from carbon dioxide and water.

Photosynthetic algae—the vast majority of algae—have chloroplasts containing the pigment chlorophyll *a,* the same type of chlorophyll plants have. In addition, certain heterotrophic protists are regarded as algae because many of their features suggest a close evolutionary relationship with photosynthetic algae. Many algae are unicellular, others live in colonies, and, as we will see in the next module, still others are multicellular. In this module, we focus on three groups of the unicellular and colonial types: dinoflagellates, diatoms, and green algae. The first two groups are noteworthy because they are numerous and widespread, the third because it was of pivotal importance in the history of life.

Dinoflagellates (Figure A on the facing page) are unicellular, photosynthetic algae that are very common in marine and freshwater environments. Each dinoflagellate species has a characteristic shape reinforced by plates made of cellulose. The beating of two flagella in perpendicular grooves in the cellulose plates produces the spinning

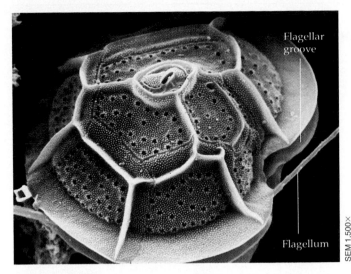

A. A dinoflagellate

SEM 1,500×

Flagellar groove

Flagellum

movement for which these organisms are named (Greek *dinos,* whirling). Dinoflagellate blooms, episodes of explosive population growth, occasionally cause red tides in warm coastal waters. Toxins produced by these protists can kill fish on a large scale and are deadly to humans as well. The dinoflagellate shown in Figure A is *Gonyaulax tamarensis,* a species that has produced deadly red tides along the New England coast.

Diatoms (Figure B) are unicellular, photosynthetic algae with a unique, glassy cell wall containing silica, the mineral actually used to make glass. The cell wall consists of two halves that fit together like the bottom and lid of a shoe box. Both freshwater and marine environments are rich in diatoms; a bucket of water scooped from the surface of the sea or from almost any pond or lake may have millions of these micro-

scopic algae. The cells store their food reserves in the form of an oil, which also provides buoyancy, keeping diatoms floating near the surface, in the sunlight. Massive accumulations of fossilized diatoms make up thick sediments known as diatomaceous earth, which is mined for use as both a filtering medium and an abrasive.

The photographs in Figure C show two types of **green algae.** *Chlamydomonas* is a unicellular alga common in freshwater lakes and ponds. Surrounded by a cellulose wall, it is propelled through the water by two flagella. (Cells with two flagella are said to be biflagellated.) *Volvox* is a colonial green alga, also common in fresh water. Each *Volvox* colony is a hollow ball composed of hundreds or thousands of biflagellated cells. The large colonies shown here will eventually release the small green and red daughter colonies within them. Each of the cells in a *Volvox* colony closely resembles the biflagellated cell of *Chlamydomonas.* It is likely that complex colonial protists such as *Volvox* evolved from unicellular green algae that were structurally similar to *Chlamydomonas.* The cells of *Volvox* and *Chlamydomonas* also resemble the biflagellated gametes of many multicellular algae and certain plants. This feature and others common to green algae and plants—cellulose cell walls, starch as a food-storage compound, and chloroplasts that are virtually the same—lead most biologists to believe that ancient green algae gave rise to the first plants. In the next module, we look at the most complicated of the modern algae, the multicellular forms.

Chlamydomonas

TEM 2,000×

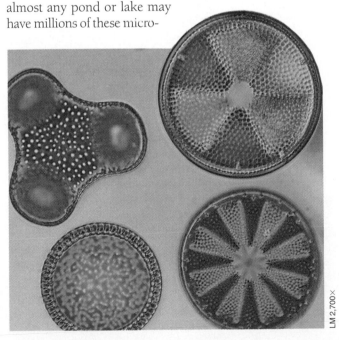

B. Diatoms

LM 2,700×

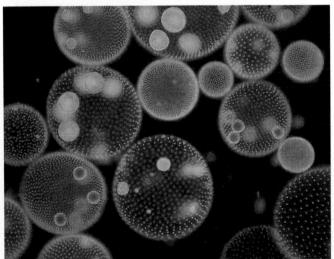

Volvox colonies

C. Green algae

LM 260×

A. Brown alga: a kelp "forest"

B. Red alga: coralline type, on a coral reef

We use the word "seaweeds" here to refer to marine algae that have large multicellular bodies. These organisms can be so large and complex that some biologists consider them to be plants. However, even the most complex seaweeds lack true stems, leaves, roots, and the internal tubes (vascular systems) that transport nutrients and water in most plants. Because seaweeds lack these structures and because they show a clear phylogenetic relationship to certain unicellular algae, we regard the seaweeds as members of the kingdom Protista.

Seaweeds grow on rocky beaches and just offshore beyond the zone of pounding surf. There are three groups of seaweeds, the largest and most complex of which is the **brown algae**. Figure A (above) is a photograph of an underwater forest of brown algae called **kelp,** off the coast of California. Anchored to the seafloor by rootlike structures called holdfasts, kelp may grow to heights of 100 m. Kelp forests are common in cool water. Analogous to terrestrial forests, they support the growth of many other species, including other protists and numerous animals. Thousands of fishes, sea lions, sea otters, and gray whales regularly use kelp forests as their feeding grounds.

The warm coastal waters of the tropics are home to the majority of species of a second group of seaweeds, the **red algae**. Red algae are typically soft-bodied, but some, called the **coralline algae,** have cell walls encrusted with hard, chalky deposits. Coralline algae, such as the one in Figure B, are common on coral reefs (as the name coralline implies), and their hard parts are important in reef-building.

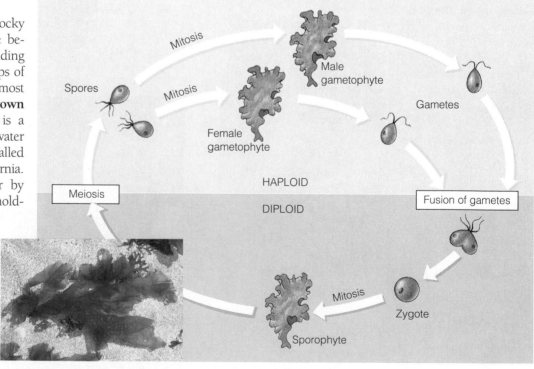

C. Multicellular green alga: *Ulva* (sea lettuce) and its life cycle

The third group of seaweeds, the multicellular **green algae**, is represented here by *Ulva*, the sea lettuce (Figure C on the preceding page). *Ulva* is harvested for human food in many countries. Its life cycle, the sequence of life stages leading from the adults of one generation to those of the next generation, follows a pattern called **alternation of generations**. Because this pattern occurs in a number of multicellular algae and in all plants, we highlight it in Figure C. Notice that a haploid generation (pale orange background here and in the life cycles we study in Chapter 18) is called the **gametophyte** generation. It alternates with a diploid generation (gray background) called the **sporophyte** generation. In *Ulva*, the gametophyte and sporophyte organisms are identical in appearance; both look like the one in the photograph, although they differ in chromosome number. The haploid gametophyte produces gametes by mitosis, and fusion of the gametes begins the sporophyte generation. In turn, cells in the sporophyte undergo meiosis and produce haploid, flagellated spores. The life cycle is completed when a spore settles to the bottom of the ocean and develops into a gametophyte.

Alternation of generations in *Ulva* may resemble a stage in the evolution of plants from green algae. As we will see in Chapter 18, however, the gametophyte of a plant looks physically quite different from the sporophyte.

Multicellular life may have evolved from colonial protists

Multicellular organisms—seaweeds, plants, animals, and most fungi—are fundamentally different from unicellular organisms. In a unicellular organism, all of life's activities occur within a single cell. In contrast, a multicellular organism has different kinds of cells that perform different activities and are dependent on each other. For example, some cells give the organism its shape, while others make or procure food, transport materials, or provide movement.

Multicellularity probably evolved on many separate occasions in the kingdom Protista, with today's seaweeds, plants, animals, and fungi arising from different kinds of unicellular protists. The majority of seaweeds, some plants, and most animals have some flagellated cells, and it is likely that their ancestors were flagellated as well. Fungi, which we study in Chapter 18, lack flagella and probably derived from nonflagellated ancestors.

The most widely held view is that the links between multicellular organisms and their unicellular ancestors were loose colonial aggregates of interconnected cells. The figure below suggests how a unicellular protist with flagellated cells may have formed colonies that eventually gave rise to multicellular organisms. ① An ancestral colony may have formed, as colonial protists do today, when a cell divided and its offspring remained attached to one another. ② Next, the cells in the colony may have become somewhat specialized and interdependent, with different cell types becoming more and more efficient at performing specific, limited tasks. Cells that retained a flagellum may have become specialized for locomotion, while others that lost their flagellum could have assumed functions such as ingesting or synthesizing food. ③ Later on, additional specialization among the cells in the colony may have led to distinctions between sex cells (gametes) and nonreproductive cells (somatic cells).

We see specialization and cooperation among cells today in several colonial protists, such as the green alga *Volvox* in Figure 17.25C. *Volvox* produces gametes, which depend on somatic cells while developing. Cells in truly multicellular organisms, as we know them today, are specialized for many more nonreproductive functions, including feeding, waste disposal, gas exchange, and protection, to name a few. Evolution of the division of labor to this extent involved many additional steps in somatic cell specialization.

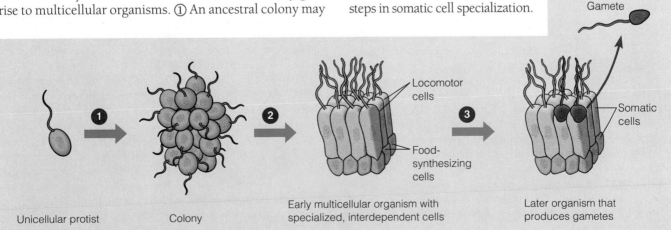

A model for the evolution of a multicellular organism from a unicellular protist

Unicellular protist — Colony — Early multicellular organism with specialized, interdependent cells — Later organism that produces gametes

Locomotor cells / Food-synthesizing cells / Somatic cells / Gamete

Multicellular life has diversified over hundreds of millions of years

The timeline here summarizes the evidence from the fossil record for some key stages in the evolution of multicellular algae, plants, animals, and fungi. Notice at the bottom that the oldest fossils of multicellular organisms, dating from about 700 million years ago, during the Precambrian era, were red algae and animals (mostly organisms resembling corals, jellyfishes, and worms). Although they have not yet been found in the fossil record, other kinds of multicellular algae probably also existed at that time. A period of mass extinction separated the Precambrian from the Paleozoic era, and multicellular life again flourished soon thereafter. By about 500 million years ago, diverse animals, fungi, and multicellular algae populated aquatic environments.

Until about 500 million years ago, all life was aquatic. About that time, the move onto land began, probably as certain green algae living in the company of fungi along the edges of lakes gave rise to primitive plants. In the next chapter, we trace the long evolutionary movement of plants onto land and their diversification there. After that, we pick up the threads of animal evolution in Chapter 19.

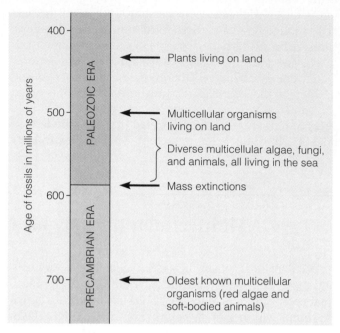

Fossil record of early multicellular life

Chapter Review

Begin your review by rereading the module headings and Module 17.19 and scanning the figures before proceeding to the Chapter Summary and questions.

Chapter Summary

Introduction–17.1 Biological and geological history are closely intertwined. Planet Earth formed some 4.6 billion years ago. The early atmosphere probably contained H_2O, CO, CO_2, N_2, CH_4, and NH_3, but no O_2. Volcanic activity, lightning, and UV radiation were intense. Fossils of bacteria date back 3.5 billion years, but life may have developed from nonliving materials as long as 4 billion years ago.

17.2–17.6 Organic monomers must have appeared first, probably when inorganic chemicals were energized by lightning or ultraviolet radiation. Simulations of such conditions have produced amino acids, sugars, nucleotide bases, and ATP. These monomers could have polymerized on hot rocks or clay, forming polypeptides and short nucleic acids. The first genes may have been RNA molecules that catalyzed their own replication. These molecules could have acted as templates for the formation of polypeptides, which in turn assisted in RNA replication. Protective and selective membranes may have helped these molecular co-ops reproduce and carry out rudimentary metabolism. Natural selection would favor the most efficient co-ops, and they might have eventually evolved into the first prokaryotic cells.

17.7–17.11 Prokaryotes, or bacteria, make up the kingdom Monera. Prokaryotic cells lack nuclei and other membrane-bounded organelles. The bacterial chromosome is a single, circular DNA molecule. Bacteria reproduce rapidly by binary fission and can swap pieces of DNA in a simple form of sex. Bacteria come in a variety of shapes, most commonly spheres (cocci), rods (bacilli), and curves or spirals (spirilla). They also vary in mode of nutrition. Autotrophs obtain carbon from CO_2 and usually obtain energy from sunlight. Heterotrophs obtain carbon from organic compounds. Most bacteria are chemoheterotrophs, which obtain both carbon and energy from organic compounds. Early bacteria may have simply absorbed ATP. As ATP became scarce, glycolysis could have evolved, allowing bacteria to tap energy from organic molecules to generate ATP in an anaerobic environment. Bacteria are still abundant wherever life exists.

17.12–17.16 Archaebacteria are bacteria of ancient origin that are common in anaerobic swamps, salt lakes, and acidic hot springs. All other prokaryotes are classified as eubacteria. Many eubacteria are propelled by rotating flagella. Fibrous pili help some cling to surfaces. Some produce tough endospores that can resist environmental extremes. Cyanobacteria are photosynthetic eubacteria and may "bloom" in polluted water. Ancient cyanobacteria transformed Earth by releasing oxygen and creating an aerobic atmosphere. Pathogenic bacteria, which are identified by criteria known as Koch's postulates, can cause disease by producing exotoxins or endotoxins.

17.17–17.19 Many bacteria are important in decomposition and chemical cycles. (We exploit them in sewage treatment and may use them to clean up oil spills and toxic wastes.) Bacteria are the foundations of all life, in both an environmental and an evolutionary sense.

17.20–17.21 Fossils indicate that eukaryotic cells appeared between 1.5 and 1 billion years ago. Evidence suggests that their organelles, particularly chloroplasts and mitochondria, evolved from small symbiotic prokaryotes that took up residence inside larger prokaryotic cells. Protists are considered to be the simplest eukaryotes, and early protists were the ancestors of plants, animals, and fungi.

17.22–17.28 Protozoans are unicellular protists that ingest their food. They include amoebas, flagellates, apicomplexans, and ciliates.

Most protozoans live freely in water or moist soil, but some live in humans and other animals and cause disease. Cellular and plasmodial slime molds are two groups of protists that live on decaying organic matter. Algae, including the unicellular dinoflagellates, diatoms, and green algae, are photosynthetic protists. Seaweeds—brown algae, red algae, and multicellular green algae—are multicellular. But because they seem closely related to unicellular algae and lack the structural specializations of plants, they are usually classified as protists. The life cycles of many seaweeds involve the alternation of haploid gametophyte and diploid sporophyte generations. Multicellular organisms probably evolved by gradual specialization of the cells of colonial protists.

Testing Your Knowledge

Multiple Choice

1. Ancient cyanobacteria, found in fossil stromatolites, were very important in the history of life because they
   a. were probably the first living things to exist on Earth
   b. produced the oxygen in the atmosphere
   c. are the oldest known archaebacteria
   d. were the first multicellular organisms
   e. extracted heat from the atmosphere, cooling Earth

2. You set your time machine for 3 billion years ago and push the start button. When the dust clears, you look out the window. Which of the following describes what you would probably see?
   a. plants and animals very different from those alive today
   b. a cloud of gas and dust in space
   c. green scum in the water
   d. land and water sterile and devoid of life
   e. an endless expanse of red-hot molten rock

3. In terms of nutrition, autotrophs are to heterotrophs as (Explain your answer.)
   a. algae are to protozoans
   b. archaebacteria are to eubacteria
   c. protozoans are to algae
   d. kelp are to diatoms
   e. pathogenic bacteria are to harmless bacteria

4. The bacteria that cause tetanus can be killed only by prolonged heating at temperatures considerably above boiling. This suggests that tetanus bacteria
   a. have cell walls containing peptidoglycan
   b. protect themselves by secreting antibiotics
   c. secrete endotoxins
   d. are autotrophic
   e. produce endospores

5. Glycolysis is the only metabolic pathway common to nearly all organisms. To scientists, this suggests that it
   a. evolved many times during the history of life
   b. was first seen in early eukaryotes
   c. first appeared early in the history of life
   d. must be very complex
   e. appeared rather recently in the evolution of life

True/False (Change false statements to make them true.)

1. Most eubacteria live in anaerobic, hot, salty, acidic environments.

2. Amoebas, flagellates, and ciliates are all examples of algae.

3. Many bacteria are motile.

4. *Bacillus thuringiensis* is a spherical bacterium.

5. Most bacteria are pathogenic.

6. Cellular slime molds are unicellular part of the time and multicellular part of the time.

Describing, Comparing, and Explaining

1. Describe the conditions probably present on Earth at the time life is thought to have originated. What was the temperature like? What chemicals were present in the atmosphere? Were there bodies of water? What sources of energy were present?

2. How do most biologists think that the mitochondria and chloroplasts of eukaryotic cells originated? What is the evidence for this idea?

3. *Chlamydomonas* is a unicellular green alga. How does it differ from a photosynthetic bacterium, which is also single-celled? How does it differ from a protozoan, such as an amoeba? How does it differ from larger green algae, such as sea lettuce (*Ulva*)?

4. Biologists think that cooperation between simple genes (perhaps made of RNA) and polypeptides may have been an important step in the origin of life. What may have been the nature of this cooperation? In other words, what might RNA and polypeptides have done for each other?

Thinking Critically

1. You have discovered what you think is a new disease of domestic cats. The cats develop a fever and bumps on the skin. Examining the blood of a sick cat under a microscope, you observe some unfamiliar comma-shaped bacteria. What steps should you follow to determine whether these bacteria cause the disease?

2. Imagine you are on a team designing a moon base that will be self-contained and self-sustaining. Once supplied with building materials, equipment, and organisms from Earth, the base will be expected to function indefinitely. One of the members of your team has suggested that everything sent to the base be chemically treated or irradiated so that no bacteria of any kind are present. Do you think this is a good idea? Why or why not?

3. Under suitable conditions, some bacteria are capable of reproducing by binary fission once every half hour. Suppose you placed a single bacterium in a culture medium and it (and its descendants) proceeded to reproduce at this rate. How many bacteria would be present after 2 hours? After a total of 12 hours? (This is easier to figure out with a calculator.) What determines how long this kind of increase can continue?

4. Billions of years ago, life apparently arose from inorganic chemicals, but all life today apparently arises only by the reproduction of preexisting life. If life could come from nonlife on the ancient Earth, why do you think this does not continue to happen today?

Science, Technology, and Society

The burning of fossil fuels is increasing the concentration of carbon dioxide in the atmosphere, and many experts think that this will intensify the greenhouse effect, trap more heat, and lead to a warming of global climate. By photosynthesis, diatoms and other microscopic algae in the oceans use enormous quantities of atmospheric carbon dioxide. To photosynthesize, these algae need minute quantities of iron, and experts suspect that a shortage of iron may limit photosynthesis, especially in the Antarctic Ocean. Looking at global warming on the one hand and algal growth on the other, oceanographer John Martin suggests that one way to reduce carbon dioxide buildup and global warming might be to fertilize the ocean with iron. This would stimulate algal growth, which would remove carbon dioxide from the air. A single supertanker of iron dust, spread over a wide enough area, could affect atmospheric CO_2 buildup significantly. Do you think this would be worth a try? Why or why not? Do you see any reasons to be cautious?

We tend to take our orange juice for granted, but it is no small feat for citrus growers to produce it at a reasonable cost. Orange groves like the one pictured at the left are found in Florida, Texas, and California. An enormous investment, the trees take 3–7 years to start producing fruit and require a rich supply of fertilizer. They are also vulnerable to freezing and to a long list of pathogenic bacteria, insects, and, especially, fungi.

Fungi are not always harmful to plants. There is another kind of association between fungi and plants in which the fungus plays the role of vital benefactor. You can see an example of this relationship in the micrograph on this page, which shows a fungus growing inside a piece of a small onion root. Everything that is darkly stained in the micrograph is part of the fungus: a dense network of thin strands and a number of round vesicles, which are used for lipid storage. (In order to make the fungus more visible, the sample was chemically treated to destroy the root cells.) This fungus is called a mycorrhizal fungus; together, the root and the fungus form an intimate, mutually beneficial association, called a **mycorrhiza** (pronounced MY-koh-RY-za; the word means "fungus root"). For their part, mycorrhizal fungi absorb phosphorus and other essential minerals from the soil, and these nutrients are then available to the plant. The sugars produced by the plant nourish the fungi.

Citrus trees can also have mycorrhizae, and the fungi may offer a way to cut the high economic and environmental costs of producing citrus fruits. Mycorrhizae can make a tree more resistant to disease, thereby reducing the need for pesticides that kill disease-causing organisms. Also, by enhancing a tree's uptake of nutrients, mycorrhizae can reduce, or even eliminate, the need for fertilizers. Unfortunately, the conditions in a typical citrus grove undermine the growth of mycorrhizae. Citrus growers use fungus-killing chemicals (fungicides) to control fungi that cause disease, and the fungicides poison the mycorrhizal fungi as well. As a result, the grower loses the benefits of mycorrhizae and must apply expensive fertilizers. The environment also suffers, because fungicides harm many kinds of organisms, and excess fertilizers can pollute streams, lakes, and groundwater.

There is no immediate solution to this dilemma. The citrus industry has relied on fungicides and fertilizers for many years, and few growers are willing to risk replacing the use of chemicals with the cultivation of mycorrhizae. One solution would be for researchers to find a way to control disease without hurting beneficial fungi.

Cultivated citrus groves are unnatural in lacking mycorrhizae. In nature, nearly all plants have mycorrhizae. In fact, mycorrhizae appear in fossils of the oldest known plants, suggesting that these beneficial relationships with fungi may have been essential for plant evolution. Some 400 million years ago, when plants first appeared, soil would have been poor in nutrients, and early plants or the algae that gave rise to them probably would have had difficulty obtaining enough phosphorus or other nutrients without the help of fungi. Thus, as the first plants evolved from green algae and adapted to land, symbiotic fungi probably played an important role.

The colonization of land by plants was a major event in the history of life. Plants transformed the landscape, creating new environmental opportunities for protists and prokaryotes, and making it possible for herbivorous animals and their predators to evolve on land. This chapter continues our account of the evolution of life's diversity, focusing on plants and fungi.

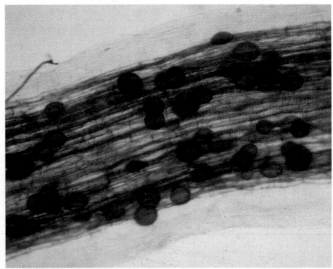

LM 473×

Mycorrhizal fungus in an onion root

What distinguishes a plant from an alga?

For many years, biologists classified the algae and plants in a single kingdom. Why do we now classify plants in their own kingdom, the **kingdom Plantae**, separate from the algal protists?

Figure A illustrates the major differences between a plant and a multicellular green seaweed, the type of algal protist most similar to plants. Most of the distinctions arise from the fact that the alga is adapted for aquatic life and the plant is adapted for terrestrial life.

Living in water is very different from living on land. A seaweed is supported by the surrounding water. Its holdfast anchors it to the seafloor, but most algae have no supporting tissues of their own. The entire algal body has direct access to water and to O_2 and minerals dissolved in the water. Because almost none of the organism is underground, almost all of it receives light and can photosynthesize.

The aquatic ancestors of plants changed drastically as they became adapted to the unique problems of living on land. In contrast to an alga, a plant must hold itself up, because air provides no support. Also, a plant must obtain water and dissolved minerals from the soil and have all photosynthetic machinery above the ground. A typical plant has discrete organs—roots, stems, and leaves—that help solve these problems. (The holdfast and other parts of a seaweed are only analogous, not homologous, to the organs of a plant.) Plant roots provide anchorage and absorb water and nutrients from the soil. In most plants, as we have noted, mycorrhizae greatly enhance nutrient absorption. The stems and leaves of a plant contain rigid support elements and photosynthetic chloroplasts. A waxy **cuticle** covering the stems and leaves helps retain water and prevent drying out in the air. Gas exchange cannot occur across the waxy surfaces, but carbon dioxide and oxygen diffuse across the leaf surfaces through microscopic pores called **stomata** (singular, *stoma*).

The evolution of stems, leaves, and roots solved some of the problems of life on land. For instance, water and minerals had to be conducted upward from roots to leaves, and sugars produced in the leaves had to be distributed throughout the plant. These problems were solved in most plant lineages by the evolution of an efficient transport system—the **vascular system**, a network of narrow tubes that extends throughout the plant body. The photograph of part

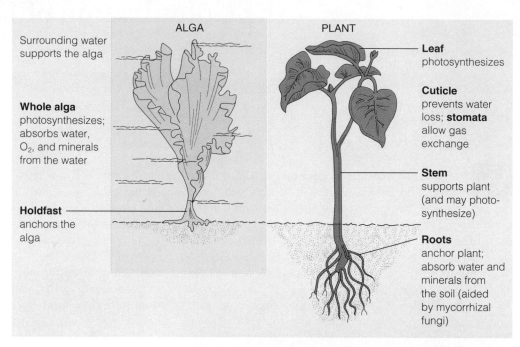

A. A multicellular alga and a plant compared

ALGA

Surrounding water supports the alga

Whole alga photosynthesizes; absorbs water, O_2, and minerals from the water

Holdfast anchors the alga

PLANT

Leaf photosynthesizes

Cuticle prevents water loss; **stomata** allow gas exchange

Stem supports plant (and may photosynthesize)

Roots anchor plant; absorb water and minerals from the soil (aided by mycorrhizal fungi)

B. The network of veins in a leaf

of an aspen leaf in Figure B shows the extensive network of veins, which are fine branches of the vascular system. The vascular system consists of two types of tissue. One, called **xylem**, is made of dead cells forming microscopic water pipes that convey water and minerals up from the roots. The other, called **phloem**, consists of living cells that distribute sugars throughout the plant.

One of the main differences between algae and plants is their mode of reproduction. For an alga, the surrounding water provides a means of dispersal for gametes and offspring. Plants, on the other hand, are faced with the problem of keeping their gametes, zygotes, and developing embryos from drying out in the air. A plant produces its gametes in special structures called gametangia. A **gametangium** (Greek *gamos*, marriage, and *angeion*, vessel) is a multicellular structure with protective jackets of nonreproductive cells that keep the gametes moist. As we will see when we examine some plant life cycles later in the chapter, the egg remains in the female gametangium and is fertilized there. Either the sperm swim to the egg through water, or sperm-producing cells are conveyed close to the egg, where the sperm are produced. The zygote develops into an embryo inside the female gametangium. Most plants release their offspring to the wind or rely on animals, such as seed-eating birds and mammals, for dispersal.

We can see that plants *are* quite different from algae—different enough, in fact, to be classified in their own king-dom. It is important to realize, however, that although classification is a useful tool, it is not a perfect mirror of nature's evolutionary process. Evolution is a continuum, and no matter how much we try to make classification reflect this fact, the very act of classifying an organism tends to set it off from related, even if distinctly different, organisms in other groups. Let's examine the evolutionary connection between green algae and plants.

Plants probably evolved from green algae

Despite their differences, plants and green algae (see Modules 17.25 and 17.26) have a number of homologous features—for instance, chloroplasts and a particular combination of photosynthetic pigments. Also, most green algae and plants have cell walls made of cellulose, both store carbohydrates in the form of starch, and during cell division in both, vesicles derived from the Golgi apparatus form a cell plate that divides the cytoplasm. Because of these and other homologies, biologists generally agree that plants evolved from green algae.

We do not know which green algae gave rise to plants because the soft bodies of algae have left few fossils. The algal ancestors of plants may have carpeted the moist fringes of lakes or coastal salt marshes as early as 500 million years ago. At that time, the continents were relatively flat and may have been subject to periodic flooding and draining. Natural selection would have favored algae that could survive periods when they were not submerged. Eventually, some species might have accumulated adaptations that enabled them to live permanently above the water line. The modern green alga *Coleochaete*, shown in Figure A, may resemble a very early plant ancestor. It grows at the edges of lakes, and its disklike, multicellular colonies resemble plants in having jacketed zygotes.

Whatever organisms first colonized land, early plant life would have thrived in the new environment. Bright sunlight was virtually limitless on land, and at first there were no herbivorous (plant-eating) animals. Among the earliest terrestrial organisms known were simple plants called *Cooksonia*, which have been found as 415-million-year-old fossils. Growing along the shores of lakes, *Cooksonia* (Figure B) had a branched, upright stem containing primitive vascular tissues, though it lacked leaves. The tips of some of its branches bore a bulbous structure called a **sporangium** (plural, *sporangia*) that produced reproductive cells called spores. A plant **spore** is a haploid cell that can develop into a multicellular adult without fusing with another cell.

Cooksonia was a truly primitive plant. Slightly more complex plants flourished about 400 million years ago. They still lacked leaves but branched more profusely, had more complex sporangia, and were more firmly anchored in the soil than *Cooksonia*. By 375 million years ago, plants with well-developed leaves and roots were numerous and diverse.

LM 473×

A. *Coleochaete,* a modern alga that may resemble a plant ancestor

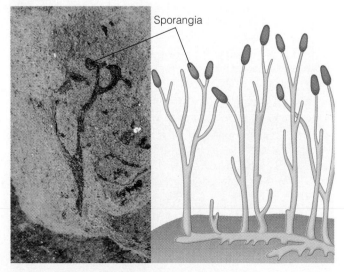

Sporangia

B. *Cooksonia,* one of the earliest plants (fossil at left)

18.3 Plant diversity provides clues to the evolutionary history of the plant kingdom

The phylogenetic tree in Figure A highlights some of the major events in the history of the plant kingdom. The diversity of modern plants, represented by the four major groups of plants listed across the top of the tree, lets us paint a broad picture of plant evolution.

An initial period of plant evolution saw the origin of plants from aquatic ancestors, probably green algae. As the geological timeline on the left of the figure indicates, two distinct lineages arose from ancestral plants about 425 million years ago. One of these lineages gave rise to modern plants called **bryophytes,** a group that includes the mosses. Bryophytes resemble other plants in having gametangia and a cuticle, but they are different in that they lack the vascular tissues xylem and phloem. Without these tissues, they also lack rigid internal support, because the cells of vascular tissue have rigid cell walls that provide most of the support for upright stems. A mat of moss, like the one shown in Figure B, actually consists of many plants growing in a tight pack, holding one another up. The mat is spongy and can retain water. Each moss plant has many small stems and leaves and grips the underlying ground with elongated cells or rootlike extensions from cells. Mosses and other bryophytes also have flagellated sperm, which closely resemble those of some green algae. The sperm must swim to the eggs to fertilize them, so fertilization requires the plant to be covered with a film of water.

Back on our phylogenetic tree, the other lineage of early plants, the **vascular plants,** have xylem and phloem as well as gametangia. Their vascular tissues provide strong support, enabling stems to stand upright on land. Ancestral

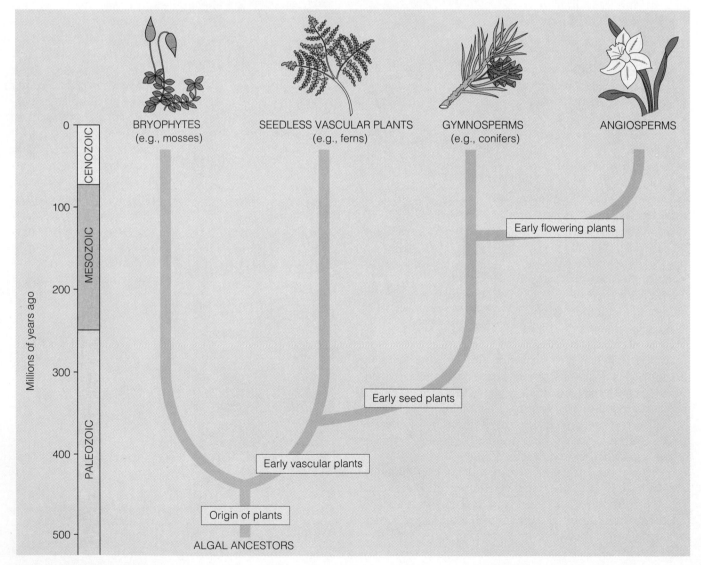

A. Highlights of plant evolution

B. Moss, a bryophyte

vascular plants may have resembled *Cooksonia*, which we discussed in Module 18.2. As indicated on the phylogenetic tree, the vascular plant lineage split in two about 360 million years ago. Plants in one of these lineages have seeds, protective packaging for their embryos. Seeds did not evolve in the other lineage.

Ferns belong to the seedless vascular plant lineage. A fern has well-developed roots and rigid stems. In many species, the leaves, commonly called fronds, sprout from stems that grow along the ground (Figure C; the "fiddleheads" in the inset are young fronds ready to unfurl). Ferns are common in shady areas in temperate forests, but they are most diverse in the tropics. In some of the tropical species, called tree ferns, upright stems grow several meters tall. In common with mosses, ferns have flagellated sperm that require moisture for fertilization.

The other branch of the phylogenetic tree that arose from the same ancestor as the ferns led to a great variety of plants that produce seeds. Today, the seed plant lineage accounts for nearly 90% of the approximately 265,000 species of living plants.

Several key adaptations underlie the enormous success of seed plants. First, they make seeds, which are survival packets for life on land. A **seed** consists of an embryo packaged with a food supply within a protective covering. Second, seed plants do not require a water layer for fertilization. Instead of producing

sperm that swim to the eggs, they produce pollen, a vehicle that transfers nonflagellated sperm-forming cells to the female parts of the plant. Pollen is carried passively by wind or animals, and the transfer of pollen to the female is called **pollination.** As we will see, fertilization occurs sometime after pollination.

Among the earliest seed plants were the **gymnosperms** (Greek *gymnos,* naked, and *sperma,* seed). The seed of a gymnosperm is said to be naked because it is not contained in a fruit. Actually, a gymnosperm seed has a thin protective covering that is part of the seed itself. Gymnosperms coexisted with ferns and other seedless plants in great forests that dominated the landscape for more than 200 million years. Today, the conifers—pine, spruce, fir, and many other kinds of trees with seed-bearing cones and needlelike leaves—are the largest group of gymnosperms.

The most recent major branching point in plant evolution was a split in the seed plant lineage. About 130 million years ago, the flowering plants, or **angiosperms** (Greek *angeion,* vessel, and *sperma,* seed) diverged from the gymnosperm lineage. Flowers are complex reproductive structures that develop seeds within protective chambers. The great majority of modern plants—some 235,000 species—are angiosperms.

In summary, four key adaptations for life on land mark the main lineages of the plant kingdom. (1) Gametangia, which protect gametes and zygotes from drying out, are present in all plants. The earliest species of plants probably had gametangia. (2) Vascular tissues (phloem and xylem) mark a lineage that gave rise to most modern plants. Vascular tissues transport water and nutrients throughout the plant body and also provide support for upright stems. (3) Seeds appeared in a lineage that dominates the plant kingdom today. (4) Flowers mark the angiosperm lineage, which is the dominant group of seed plants. As we will see in the next several modules, the life cycles of modern plants reveal additional details about plant evolution.

C. Ferns, seedless vascular plants

Haploid and diploid generations alternate in plant life cycles

Plants have life cycles very different from ours. Each of us is a diploid individual; the only haploid stages in the human life cycle, as for nearly all animals, are sperm and eggs. By contrast, plants have alternating generations: Diploid (2n) individuals called sporophytes and haploid (1n) individuals called gametophytes generate each other in the life cycle. Starting at the top of the diagram here, you can see that haploid male and female gametophytes produce gametes by mitosis. Fertilization results in a diploid zygote. The zygote undergoes mitosis and develops into the diploid sporophyte. The sporophyte produces haploid spores by meiosis. Completing one life cycle, a spore develops by mitosis into a gametophyte.

Alternation of generations has the important advantage of providing two chances to produce large numbers of offspring (zygotes and spores). Many kinds of algae undergo alternation of generations; we saw one example, the seaweed *Ulva*, in Figure 17.26C. Alternation of generations most

likely evolved in aquatic organisms, and plants seem to have inherited it from their algal ancestors. As plants evolved on land, generations continued to alternate between haploid and diploid, but with some major changes. In seaweeds, both gametes and spores are flagellated and disperse by swimming. But on land, flagellated stages are a handicap because they require water. Accordingly, only some modern plants (such as mosses and ferns) have flagellated sperm, and none have flagellated eggs or spores. Plant egg cells are retained and fertilized in gametangia on the gametophyte. Spores remain on the sporophyte or are adapted for dispersal in the air.

In the next two modules, we see two different ways in which alternation of generations occurs in plants. In mosses, the gametophyte is said to be dominant because it is larger, and the sporophyte depends on it for nourishment. In nearly all other plants, the sporophyte is dominant, with the gametophyte usually depending on it for nourishment.

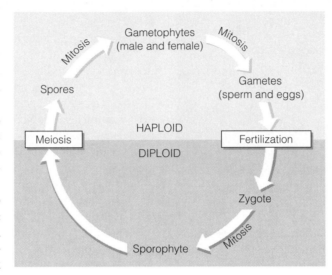

Mosses have a dominant gametophyte   (See text at top of facing page.)

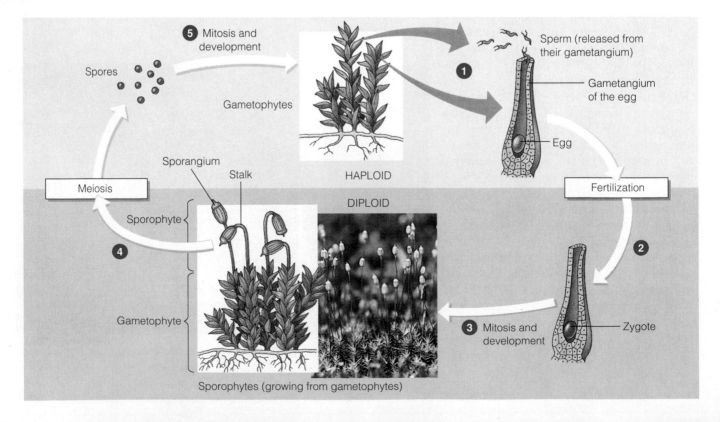

Sporophytes (growing from gametophytes)

In a moss, most of the green, cushiony growth we see consists of gametophytes. ① Gametes develop in gametangia on the gametophytes. The flagellated sperm require a film of water in which to swim to the egg, which stays in its gametangium on the gametophyte. After fertilization, ② the zygote remains in the gametangium. ③ There it divides by mitosis, and develops into a sporophyte. Each sporophyte remains attached to a gametophyte, as you can see in the photograph, and ④ meiosis occurs in the sporangia at the tips of the sporophyte stalks. Haploid spores resulting from meiosis are released. Later, ⑤ they undergo mitosis and develop into gametophytes, completing one life cycle.

Ferns, like most plants, have a dominant sporophyte

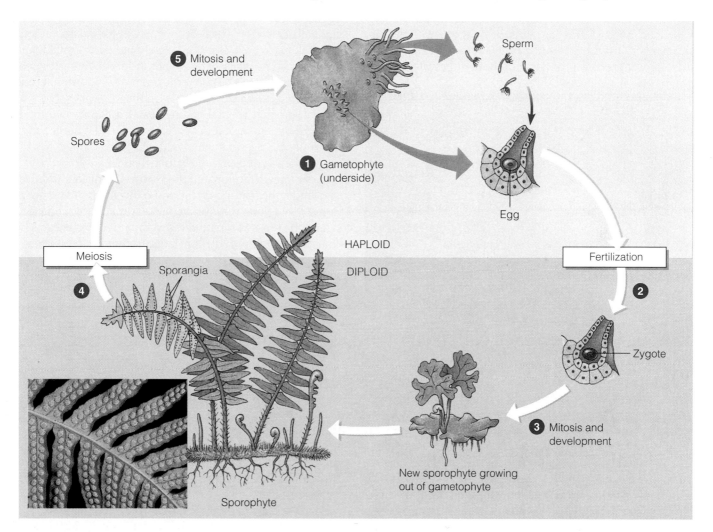

The life cycle of a fern illustrates a dominant sporophyte generation. In fact, all we usually see of a fern is the sporophyte. But let's start the fern life cycle with the gametophyte. ① Fern gametophytes often have a distinctive heartlike shape (top of figure), but they are quite small (about 0.5 cm across) and inconspicuous. Like mosses, ferns have flagellated sperm that require moisture to reach the egg. ② The zygote remains on the gametophyte, where ③ it develops into the sporophyte. ④ Cells in sporangia undergo meiosis, producing haploid spores. ⑤ These spores, in turn, develop into gametophytes by mitosis.

Today, about 95% of all plants, including all seed plants, have a dominant sporophyte generation in their life cycle. As seed plants evolved, their dominant sporophyte became adapted to house all the plant's reproductive stages (including eggs, sperm, spores, zygotes, and embryos). The evolution of pollen, produced by the dominant sporophyte, was a key step in the adaptation of seed plants to dry land. By providing transport for sperm-producing cells, pollen makes it possible for sperm to reach and fertilize the eggs without being immersed in water.

We will resume this story in Module 18.9. But first, let's glance back to a time in plant history before seed plants rose to dominance, a time when ferns and other seedless plants covered much of the land surface.

18.7 Seedless plants formed vast "coal forests"

Ferns and other seedless plants have a long history. During the Carboniferous period (about 285–360 million years ago), vast forests of them grew in swampy areas that covered much of what is now Eurasia and North America. At that time, these continents were close to the equator and had tropical climates. The painting below, based on fossil evidence, reconstructs one of those great seedless forests. Most of the large trees with straight trunks are seedless plants called lycopods. On the far right, the dark tree with numerous feathery branches is a horsetail, another seedless plant. Tree ferns were also prominent, although they are not fea-

tured in this picture. Animals, including the giant dragonfly near the horsetail, also thrived in the swamp forests.

The tropical swamp forests of the Carboniferous period generated great quantities of organic matter. As the plants died, they fell into stagnant wetlands and did not decay completely. Their remains formed thick deposits of organic rubble called peat. Later, the swamps were covered by seawater, marine sediments covered the peat, and pressure and heat gradually converted the peat to coal. Coal is black sedimentary rock made up of fossilized plant material. It formed during several geological periods, but the most extensive coal beds formed from Carboniferous deposits. (Carboniferous comes from the Latin, *carbo,* coal, and *fer-,* bearing.)

Coal, oil, and natural gas are **fossil fuels,** fuels formed from the remains of extinct organisms. Fossil fuels generate much of our electricity. As we deplete our oil and gas reserves, the use of coal is likely to increase.

"Coal forests" dominated the North American and Eurasian landscapes until near the end of the Carboniferous period. At that time, the world climate turned drier and colder, and the vast swamps and forests began to disappear. The climatic change provided an opportunity for seed plants, which can complete their life cycles on dry land and withstand long, harsh winters.

18.8 Gymnosperms called conifers replaced the swamp forests

Of the earliest seed plants, the most successful were the gymnosperms, and several kinds grew along with the seedless plants in the Carboniferous swamps. You can see one gymnosperm in Figure 18.7, the low-growing shrub with large fernlike leaves on the bottom left. Called a seed fern, although only its leaves resembled ferns, this type of gymnosperm did not survive beyond the Carboniferous. The group of gymnosperms that prevailed after the swamps dried up and that is still dominant among gymnosperms today is the **conifers**—naked-seed plants that produce cones. (Conifer means cone-bearing.)

Conifers have been widespread for the past 250 million years, thriving in regions with dry, cool climates and short

growing seasons. The pine tree on the facing page is a typical conifer. Adapted to harsh winter storms, its branches readily shed snow; when wet, heavy snow does stick to them, they usually bend rather than break. The tree's needlelike leaves resist drying because they have little surface area for evaporation; a thick cuticle covering the leaf surface also helps retain water. Because the tree keeps its leaves throughout the year, it can start photosynthesizing as soon as the short growing season begins.

Let's now return to our evolutionary story and see how the pine tree life cycle fits with the major trends in plant history: dominance of the sporophyte generation and protection of the delicate reproductive stages.

18.9 A pine tree is a sporophyte with tiny gametophytes in its cones

Pines and other conifers illustrate how drastically the relative roles of the haploid and diploid generations changed as plants evolved on land. A pine tree itself is a sporophyte; the gametophyte generation consists of microscopic stages that grow inside the tree's cones.

Cones are a significant adaptation to land, for they harbor all of a conifer tree's reproductive structures. These are the same structures we described earlier for the mosses and ferns: diploid sporangia, which produce haploid spores by meiosis; haploid female and male gametophytes; gametes, which are produced by the gametophytes; and zygotes, resulting from fertilization.

A pine tree bears two types of cones. The hard, woody ones we usually notice are female cones (step ① in the figure below, to the left of the tree). The female cone has many hard, radiating scales, each bearing a pair of **ovules.** An ovule starts out as a sporangium and a covering, or integument. ② Male cones are generally much smaller than female cones; they are also soft and short-lived. Each scale on a male cone produces many sporangia, each of which makes numerous spores. Male gametophytes, or **pollen grains,** develop from the spores. When male cones are mature, the scales open and release a cloud of pollen (millions of microscopic grains). You may have seen yellowish conifer pollen covering car tops and windshields or floating on ponds in the spring.

Carried by the wind and independent of water, pollen grains house the cells that will become sperm. ③ Pollination occurs when a pollen grain lands on and enters an ovule. After pollination, meiosis occurs in the ovule, and ④ a haploid spore cell begins developing into the female gametophyte. Not until months later do eggs appear within the female gametophyte. It also takes months for sperm to develop in the pollen grain. ⑤ A tiny tube grows out of the pollen grain and eventually releases a sperm into the egg. Fertilization does not occur until more than a year after pollination.

⑥ Following fertilization, the zygote develops into a sporophyte embryo, and the whole ovule transforms into the seed. The seed contains the embryo's food supply (the remains of the female gametophyte) and has a tough seed coat (the ovule's integument). In a typical pine, seeds are shed from the cones about two years after pollination. ⑦ The seed falls to the ground, or is dispersed by wind or animals, and when conditions are favorable, it germinates (the embryo starts growing). Eventually, the embryo grows into a tree.

In summary, all the reproductive stages of conifers are housed in cones borne on sporophytes. The ovule is a key adaptation—a protective device for all the female stages in the life cycle, as well as the site of pollination, fertilization, and embryonic development. The ovule becomes the seed, a major factor in the success of the conifers and flowering plants on land.

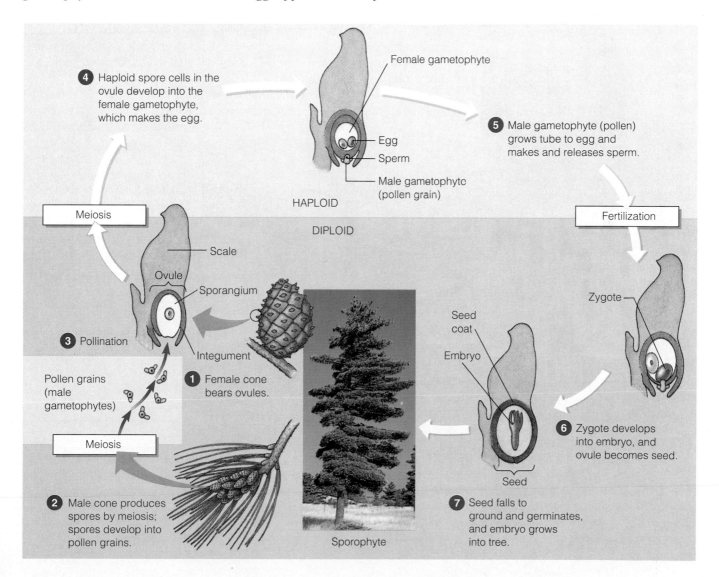

4 Haploid spore cells in the ovule develop into the female gametophyte, which makes the egg.

Female gametophyte

Egg
Sperm
Male gametophyte (pollen grain)

HAPLOID

5 Male gametophyte (pollen) grows tube to egg and makes and releases sperm.

Meiosis

DIPLOID

Fertilization

Scale

Ovule

Sporangium

Zygote

3 Pollination

Integument

Seed coat

Embryo

Pollen grains (male gametophytes)

1 Female cone bears ovules.

Meiosis

6 Zygote develops into embryo, and ovule becomes seed.

Seed

2 Male cone produces spores by meiosis; spores develop into pollen grains.

Sporophyte

7 Seed falls to ground and germinates, and embryo grows into tree.

Vast coniferous forests are threatened by our demand for wood and paper

Attesting to the success of the conifers, a broad band of coniferous forests covers much of northern Eurasia and North America and extends southward in mountainous regions. The photograph in Figure A shows a conifer forest in northern Idaho in autumn. The green trees are mostly pines and firs. The yellow trees are larches, which are unusual among conifers in that they shed their leaves in the fall.

Today, about 90 million acres in the United States, mostly in the western states and Alaska, are designated National Forest lands. Some of these areas are set aside as wilderness; most are managed by the U.S. Forest Service for multiple uses, including lumbering, grazing, mining, public recreation, and wildlife habitat. Balancing these diverse, often conflicting, uses while maintaining forests for future generations is a formidable challenge.

Coniferous forests provide much of our lumber. They are also harvested for wood pulp used to make paper. Currently, our demand for wood and paper is so great that clear-cut areas like the one in Figure B are becoming commonplace. Literally hundreds of thousands of acres of coniferous forests are being cut each year. Some areas are being replanted, and trees will eventually repopulate them, but the rate of cutting vastly exceeds the rate of regrowth.

At the same time that our demand for forest products is increasing, more people are becoming dedicated to conserving forests. The 90 million acres of National Forest may sound like a lot, but as conservationists point out, over 90% of the forest has been cut in many areas.

To those who depend on employment in the lumber industry to support their families, forest conservation may be largely a frustration—an attempt to lock up natural resources we need and to limit an industry that can be an important part of our economy. The debate between those who depend on the commercial use of timber and others who fear we are doing irreparable damage to our forest lands continues to escalate. Meanwhile, coniferous forests are being destroyed at a rate that will eventually put the logger out of work.

There seem to be two possible solutions. One is to increase our efforts to recycle paper and lumber. The other is to reduce consumption, especially of paper. The average person in the United States now consumes nearly 50 times more paper than the average person in less developed nations. And by the year 2000, the demand for forest products in the United States is expected to be double what it was in the 1970s. Reducing paper waste and increasing our recycling efforts could help reverse that trend.

A. A forest of conifers

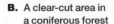

B. A clear-cut area in a coniferous forest

Angiosperms dominate the plant kingdom today

The photograph of the coniferous forest above could give us a somewhat distorted view of today's plant world. Conifers do dominate northern lands, but it is the angiosperms, or flowering plants, that dominate most other land areas. Today, worldwide, there are about 400 times more species of angiosperms than gymnosperms, and nearly 80% of all plants are angiosperms. Whereas gymnosperms supply most of our lumber and paper, angiosperms supply nearly all our food and much of our fiber for textiles. Cereal grains, including wheat, corn, oats, and barley, are flowering plants, as are citrus and other fruit trees, garden vegetables, cotton, and flax. Fine hardwoods from flowering plants, such as oak, cherry, and walnut trees, supplement the lumber we get from conifers.

Several unique adaptations account for the success of angiosperms. The leaves of most species are broad and flat, a shape that makes them very effective collectors of solar energy. Angiosperms also have vascular tissues with thicker and therefore stronger cell walls than gymnosperms. Above all, it is the flower—the trademark of angiosperms—that accounts for the unparalleled success of these plants.

The flower is the centerpiece of angiosperm reproduction

No organisms make a showier display of their sex life than angiosperms. From roses to dandelions, flowers expose a plant's male and female parts and are the sites of pollination and fertilization. They also generate fruits, which contain the angiosperm's seeds. Figure A shows a tulip flower.

Figure B shows the anatomy of a flower. Although the generalized flower drawn here looks different from the tulip (as do most flowers), these and all other kinds of flowers have a common basic anatomy. A **flower** is actually a short stem with four sets of modified leaves called sepals, petals, stamens, and carpels. At the bottom of the flower are the **sepals,** which are usually green. They enclose the flower before it opens (think of a rosebud). Above the sepals are the **pet-**

A. Close-up view of a tulip

als, which are usually the most striking part of the flower and are usually important in attracting animal pollinators. The actual reproductive structures are the stamens and one or more carpels. The **stamens** are the flower's male parts, and there may be only a few or many. Each stamen consists of a stalk bearing a sac called an **anther,** in which pollen grains develop. The female **carpel** consists of a stalk with an ovary at the base and a sticky tip known as the **stigma,** which traps pollen. The **ovary** is a protective chamber containing one or more ovules, in which the eggs develop. As we see in the next module, a seed develops from each ovule, and the fruit develops from the ovary.

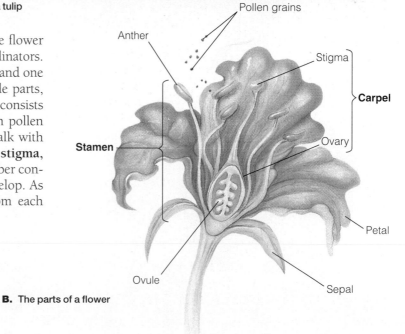

B. The parts of a flower

The angiosperm plant is a sporophyte with gametophytes in its flowers

In broad outline, the angiosperm life cycle resembles that of a gymnosperm (see Module 18.9). The plant we see is a sporophyte, and the tiny gametophyte generation lives on it. In contrast to a gymnosperm, whose gametophytes grow in its cones and whose seeds are naked, an angiosperm has its gametophytes in its flowers and its seeds packaged inside fruits.

The figure below, illustrating the life cycle of a typical flowering plant, highlights features that have been especially important in angiosperm evolution. (We will discuss

these, and the unique angiosperm adaptation called double fertilization, in more detail in Modules 31.10–31.14.) Starting at the "Meiosis" box on the left side of the figure, ① meiosis occurring in the anthers of the flower leads to the male gametophytes, or pollen grains. ② Meiosis in the ovules leads to the female gametophytes, each of which produces an egg. ③ Pollination occurs when a pollen grain, carried by the wind or an animal, lands on the stigma. As in gymnosperms, a tube grows from the pollen grain to an egg, and a sperm fertilizes the egg, creating a ④ zygote.

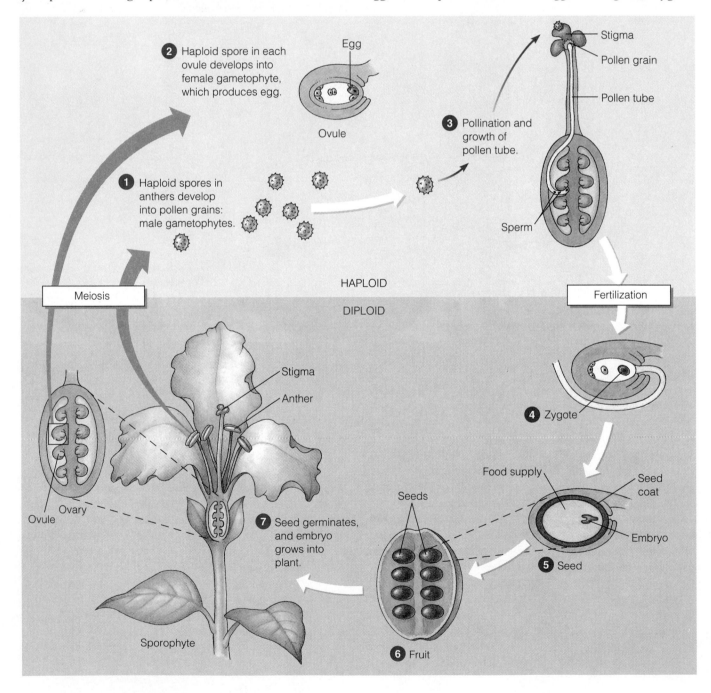

Also as in gymnosperms, ⑤ a seed develops from each ovule. Each seed consists of an embryo (a new sporophyte) surrounded by a store of food and a seed coat. While the seeds develop, ⑥ the ovary's wall thickens, forming the fruit that encloses the seeds. Fruits help disperse the seeds. When conditions are favorable, ⑦ the seed germinates, and the embryo grows into a mature sporophyte, completing the life cycle.

Besides the packaging of angiosperm seeds inside fruit, several other features of angiosperm life have enhanced the success of these plants. One is the evolution of mutually dependent relationships with animals, which carry pollen more reliably than the wind. Another is the ability to reproduce rapidly. Fertilization in angiosperms usually occurs about 12 hours after pollination, making it possible for the plant to produce seeds in only a few days or weeks. As we mentioned in Module 18.9, a typical gymnosperm usually takes well over a year to produce seeds. Rapid seed production is advantageous, particularly in environments such as deserts, where growing seasons are extremely short.

The structure of a fruit reflects its function in seed dispersal

A. Dandelion fruit dispersed by the wind

B. Cockleburs (fruit) carried by animal fur

A fruit, the ripened ovary of a flower, is a special adaptation that helps disperse seeds. Some angiosperms depend on wind for dispersal. For example, the dandelion fruit (Figure A) acts like a kite, carrying a seed away from the parent plant on wind currents. Some other angiosperms produce fruits that hitch a free ride on animals. The two dark cockleburs attached to the fur of the donkey in Figure B are fruits that may be carried miles before they open and release their seeds.

Many angiosperms produce fleshy, edible fruits that are attractive to animals as food. When the mouse in Figure C eats a berry, it digests the fleshy part of the fruit, but most of the tough seeds pass unharmed through its digestive tract. The mouse may then deposit the seeds, along with a supply of natural fertilizer, some distance from where it ate the fruit. Many types of garden produce, including tomatoes, squash, and melons, as well as strawberries, apples, cherries, and oranges, are edible fruits.

The fruit of flowering plants usually develops and ripens quickly, so the seeds can be produced and dispersed in a single growing season. The dispersal of seeds in fruits is one of the main reasons angiosperms are so numerous and widespread.

C. Edible fruit

Interactions with animals have profoundly influenced angiosperm evolution

Flowering plants and land animals have had mutually beneficial relationships throughout their evolutionary history. Most angiosperms depend on insects, birds, or mammals for pollination and seed dispersal. And most land animals depend on angiosperms for food. These mutual dependencies tend to improve the reproductive success of both the plants and the animals and, thus, are favored by natural selection. Let's examine some cases of pollination.

Many angiosperms produce flowers that attract pollinators that rely entirely on the flowers' nectar and pollen for food. Nectar is a high-energy fluid that is of use to the plant only for attracting pollinators. The color and fragrance of a flower are usually keyed to a pollinator's sense of sight and smell. Many flowers also have markings that attract pollinators, leading them past pollen-bearing organs on the way to nectar. For example, flowers that are pollinated by bees often have markings that reflect ultraviolet light. Such markings are invisible to us, but vivid to bees. The bee in Figure A is harvesting nectar and pollen from a scotch broom flower. The flower parts arched over the insect are the pollen-bearing stamens. Some of the pollen the bee picks up here will rub off onto the female parts of the next flower it visits.

Flowers pollinated by birds are usually red or pink, colors to which bird eyes are especially sensitive. The shape of the flower may also be important. Flowers that depend largely on hummingbirds, for example, typically have their nectar located deep in a floral tube, where only the long, thin beak and tongue of the bird are likely to reach. As a hummingbird (Figure B) flies among flowers in search of nectar, its feathers and beak pick up pollen from the anthers of the flowers. It will deposit the pollen in other flowers of the same shape, and so probably of the same species, as it continues the search.

Insects and birds are active mainly during the day. Some flowering plants, however, depend on nocturnal pollinators, such as bats. These plants typically have large, light-colored, highly scented flowers that can easily be found at night. Taken at night, the photograph in Figure C shows a bat approaching the large, white flower of a baobab tree in tropical Africa. While the bat eats part of the flower, the yellow anthers dust its hair with pollen, which it passes on as it visits other flowers.

To a large extent, flowering plants are as diverse and successful as they are today because they have close connections with other organisms. As we discussed in the chapter's introduction, these connections include ones with fungi.

A. A bee picking up pollen as it feeds on nectar

B. A hummingbird attracted to a red flower that fits the shape of its long, thin beak

C. A bat, a nighttime pollinator

If plants had not evolved mutually beneficial relationships with fungi, they might never have been able to colonize land. Mycorrhizae, the associations of plant roots with fungi that we discussed in the chapter's introduction, helped make the colonization of land possible. Conversely, a connection with plants probably also helped fungi make the move onto land. Fungi are heterotrophic; like animals, they cannot make their own food molecules and must obtain them from other organisms. Before plants colonized land and began stocking soil with organic molecules that heterotrophs could use for food, fungi probably thrived only in aquatic environments. The first fungi on dry land may have been the mycorrhizal partners of the first land plants.

A. A fungal parasite, black stem rust, growing on a wheat leaf

Today, fungi are found virtually everywhere. They abound in the soil as well as in all types of aquatic environments. Not all fungi have beneficial associations with plants. Some are parasites, obtaining their nutrients at the expense of plants or other organisms. The dark streaks and rust-colored spots on the wheat leaf in Figure A are growths of black stem rust, a fungus that infects a number of agriculturally important grains and is highly destructive.

The micrograph in Figure B illustrates another kind of fungus, one that catches and eats small animals. The dark, snakelike form in the picture is an animal called a roundworm, which lives in soil. It has been snared by a predatory fungus (the thread and hoops in the picture). The hoops constrict around the worm in a fraction of a second. The fungus then penetrates its prey and digests it.

Many fungi are decomposers of dead organisms. Fungi that decompose (decay) organic matter are essential to all forms of life because they restock the environment with inorganic nutrients essential for plant growth. The whitish growth on the rotting log in Figure C is a fungus that digests wood. If this and other fungi and bacteria in a forest suddenly stopped decomposing for just a few years, leaves, logs, feces, and dead animals would pile up on the forest floor. Plants and the animals they feed would starve because elements taken from the soil would not be returned. Eventually, the forest and all its inhabitants would die.

Fortunately, fungi are among the most adaptable of all living organisms. For example, the white rot fungus in Figure C can decompose not only wood but also many toxic pollutants, including the pesticide DDT and certain chemicals that cause cancer. Fungal decomposers can usually thrive anywhere there is organic matter.

SEM 680×

B. A fungal predator trapping a roundworm

C. A fungal decomposer, white rot fungus, breaking down a dead log

Fungi absorb food after digesting it outside their bodies

Fungi are classified in their own kingdom, the **kingdom Fungi**. These organisms were formerly thought to be plants that lack chlorophyll, but about the only trait fungi share with plants is that most of them grow in the ground. Just what are fungi, and what makes them so distinctive that we put them into their own kingdom?

In brief, **fungi** are heterotrophic eukaryotes that digest their food externally and absorb the small nutrient molecules that result. Fungi have body structures and modes of reproduction unlike those of any other organism. We

concentrate on fungal structure in this module and on fungal reproduction in Module 18.18.

Most fungi, including the molds and mushrooms, are multicellular. Yeasts are unicellular fungi, but their simple structure probably evolved from multicellular ancestors. A typical fungus consists of a netlike mass of filaments called **hyphae** (singular, *hypha*). Hyphae start as single filaments and then branch repeatedly, forming a network known as a **mycelium** (plural, *mycelia*), illustrated in Figure A. Figure B is a photograph of the mycelium of a mold growing on decaying leaves. Hyphae actually grow into the cells of the leaf and digest the cytoplasm. The mushrooms in Figure C, solid as they seem, are actually made of tightly packed hyphae. A mushroom is just an above-ground reproductive structure attached to a much more extensive underground mycelium.

Figure D is a micrograph of hyphae in a culture of a bread mold. In this mold, the hyphae are syncytial, consisting of threads of cytoplasm undivided by membranes and containing many nuclei (not visible here). In many other fungi, the hyphae consist of chains of cells. In either case, the hyphae are surrounded by a plasma membrane covered by a cell wall. Unlike plants, which have cellulose cell walls, most fungi have cell walls made of chitin, a strong, flexible polymer of a nitrogen-containing sugar.

Most fungi are nonmotile; they do not move about in search of food or mates. And, unlike animals, most protists, and many plants, fungi do not have flagellated or amoeboid cells at any stage in their life cycles. But the mycelium makes up for the lack of mobility by being able to grow at a phenomenal rate. A fungus concentrates its resources on extending its hyphae around and through food sources. Because its hyphae grow longer without getting thicker, the fungus develops a huge surface from which it can secrete digestive enzymes and through which it can absorb food. A mycelium can add as much as a kilometer of hyphae each day. A mushroom can grow to its full size in a single night. The tips of the hyphae are just the right size and shape to invade plant cells or grow between them. Whether a fungus is a decomposer infiltrating the remains of a dead plant, a parasite infecting a living plant, or a mutualist that participates in mycorrhizae, the hyphae secrete enzymes that digest plant cell walls and then grow into them.

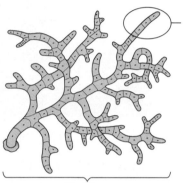

A. A mycelium, made of numerous hyphae

B. Mycelium of a mold growing on decaying leaves

C. Mushrooms, reproductive structures made of packed hyphae

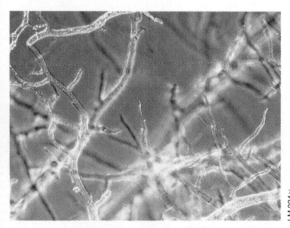

LM 224×

D. Syncytial hyphae (continuous cytoplasm with many nuclei)

Many fungi have three distinct phases in their life cycle

Fungal life cycles range from simple to complex. For example, many yeasts reproduce only by mitotic cell division. By contrast, mushrooms and many other kinds of fungi have three distinct phases in their life cycles. As illustrated below, they have diploid and haploid phases, and they also have a unique third phase, called the **dikaryotic phase**, in which cells contain two nuclei.

Let's follow the life cycle of a mushroom, starting at the top left. ① The mushroom itself is called a **fruiting body.** ② Numerous zygotes develop in specialized cells on the underside of the fruiting body's cap. The zygotes are the only diploid stage in the life cycle. Each is simply a diploid nucleus resulting from fertilization, which in this case is the fusion of the two haploid nuclei present in a cap cell (there are no sperm or eggs). Then, without going through any mitotic divisions, each zygote undergoes meiosis, and haploid spores are formed. ③ The fruiting body releases enormous numbers of these spores. Carried by wind, water, or animals, the spores may happen to land on moist matter that can serve as food. ④ There they germinate and grow into haploid mycelia.

Much of a mushroom's sex life occurs underground. The haploid mycelia are of discrete kinds, called **mating types.** The different mating types contain genetically distinct nuclei, and only certain types are sexually compatible. ⑤ The dikaryotic stage begins when hyphae of two mycelia of compatible types (indicated here by red and yellow nuclei) grow together. The hyphae fuse, but the nuclei do not. You see the result, a dikaryotic mycelium, in ⑥. Note that each of its cells contains two *genetically different* nuclei.

The fruiting body is an extension of the dikaryotic mycelium, and cells making up the fruiting body are dikaryotic. The dikaryotic phase ends and a new diploid stage begins when haploid nuclei in the cap fuse to form the zygotes.

Much of the success of fungi is due to their reproductive capacity. Dikaryotic fruiting bodies lead to the formation of genetically different kinds of mycelia. In this way, the fungus retains genetic variability and its capacity to adapt to the changing conditions of its environment. A single round of the three-phased life cycle of a mushroom also increases the number of organisms enormously. A massive boost in numbers occurs when the haploid spores form. As mycelia develop from the spores (step 4 in the life cycle) and from the dikaryotic hyphae (step 6), the fungus spreads extensively through new soil areas. Dikaryotic mycelia often last for years, branching widely as they decompose organic matter in soil, rotting logs, or leaf litter. The dikaryotic mycelia of some mushrooms are among the world's oldest and largest organisms. One growing in forest soil in northern Michigan extends over about 30 acres, and scientists estimate that it began growing from a single spore about 1500 years ago.

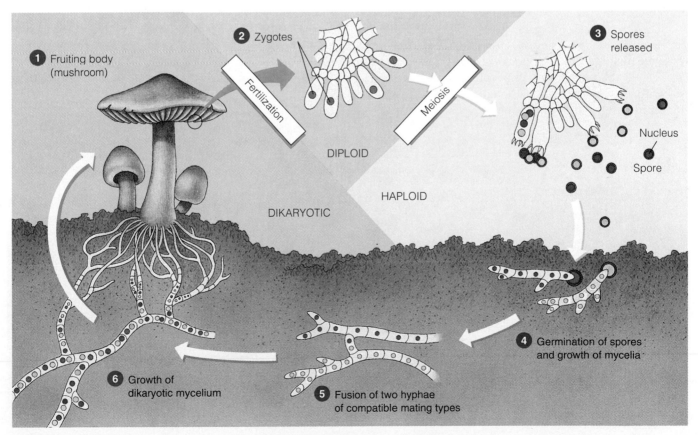

① Fruiting body (mushroom)
② Zygotes
③ Spores released
Nucleus
Spore
DIPLOID
HAPLOID
DIKARYOTIC
Fertilization
Meiosis
④ Germination of spores and growth of mycelia
⑤ Fusion of two hyphae of compatible mating types
⑥ Growth of dikaryotic mycelium

Lichens consist of fungi living mutualistically with photosynthetic organisms

Mutually beneficial connections are a recurring theme in both the plant and the fungal kingdoms. The bright yellow and red forms that stripe the surface of the rock outcropping in Figure A are another example of a mutual relationship involving fungi. These lichens superficially resemble mosses or other simple plants, but they are not plants at all—nor are they even individual organisms. **Lichens** are associations of millions of green algae or cyanobacteria held in a tangled network of fungal hyphae (Figure B). Not all lichens are as vividly colored as those in Figure A. Bright colors in lichens usually result from pigments in the hyphal walls of the fungi.

The lichen association is still not completely understood. The fungus is known to receive food from its photosynthetic partners, and the algae or cyanobacteria receive housing, water, and certain minerals from the fungus. But much remains to be learned about exactly what molecules are exchanged and how the exchanges occur. A few of the algae that live in lichens can grow independently, but lichen fungi cannot survive on their own.

Lichens reproduce asexually by fragmenting or, as shown in Figure B, by dispersing tiny airborne units containing both the fungus and some photosynthetic cells. The fungus and alga or bacterium can also reproduce independently, either sexually or asexually, although they must reassociate for the fungus to survive in nature.

A. Lichens growing on rock

Whatever the precise nature of the lichen mutualism, it gives the two organisms the ability to survive in habitats that are inhospitable to either organism alone. Lichens are rugged and able to live where there is little or no soil. As a result, they are important pioneers on new land. Lichens grow into tiny rock crevices, adding to the forces that erode hard surfaces and paving the way for future plant growth. Some lichens can tolerate severe cold, and carpets of them

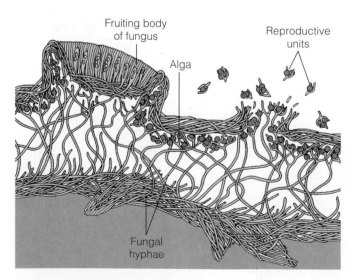

B. Asexual reproduction of a lichen

Fruiting body
of fungus

Reproductive
units

Alga

Fungal
hyphae

C. A caribou eating lichens in the arctic tundra

cover the arctic tundra. The caribou in Figure C eats lichens and is especially dependent on them in its winter feeding grounds in Alaska.

Lichens can also withstand severe drought. They are opportunists, growing in spurts when conditions are favorable. When it rains, a lichen quickly absorbs water and photosynthesizes at a rapid rate. In dry air, it dehydrates and photosynthesis may stop, but the lichen remains alive more or less indefinitely. Some lichens are thousands of years old, rivaling the oldest plants and fungi as the oldest organisms on Earth.

As tough as lichens are, many do not stand up very well against air pollution. Because they get most of their minerals from the air, in the form of dust or compounds dissolved in raindrops, lichens are very sensitive to airborne pollutants such as sulfur dioxide. The death of lichens may be a sign that air quality in an area is deteriorating.

Parasitic fungi harm plants and animals

About 100,000 species of fungi are known to science. Approximately one-third of these are mutualists, in either lichens or mycorrhizae. About another one-third are decomposers living in soil and rotting plant material. The rest—some 30,000 species—are parasitic, mostly in or on plants. In some cases, fungi that infect plants have literally changed landscapes. The dead tree in Figure A is an American elm killed by the parasitic fungus that causes Dutch elm disease. The fungus evolved with European species of elm trees, and it is relatively harmless to them. But it is deadly to American elms. Accidentally introduced into the United States on logs sent from Europe just after World War I, the fungus was carried from tree to tree by bark beetles. Since then, it has destroyed elm trees all across North America. The spread of the Dutch elm fungus continues despite decades of research and attempts to control it.

Fungi are a serious problem as agricultural pests. Species called smuts and rusts (see Figure 18.16A) are common on grain crops and cause tremendous economic losses each year. The ear of corn shown in Figure B is infected with a widespread fungal pathogen called corn smut. The grayish growths are called galls. Analogous to the fruiting body of a mushroom, a gall is made up of dikaryotic hyphae that grow into a developing corn kernel and eventually displace it. When a gall matures, it breaks open and releases thousands of blackish spores.

In parts of Central America, the smutted ears are cooked and eaten as a delicacy, but generally corn smut is regarded as a scourge. Fortunately, certain genetic strains of corn are resistant to it.

The seed heads of many other kinds of grain, including rye, wheat, and oats, are sometimes infected with fungal growths called ergots. If ergot-infested grain is milled into flour and consumed, poisons from the ergots can cause gangrene, nervous spasms, burning sensations, hallucinations, temporary insanity, and death. One epidemic in Europe in 944 A.D. killed more than 40,000 people. Several kinds of toxins have been isolated from ergots. One, called lysergic acid, is the raw material from which the hallucinogenic drug LSD is made. Certain others are medicinal when administered in small doses. An ergot compound is useful in treating high blood pressure, for example.

Animals are much less susceptible to parasitic fungi than plants are. Only about 50 species of fungi are known to be parasitic in humans and other animals. However, their effects are significant enough to make us take them seriously. Among the diseases that fungi cause in humans are yeast infections of the lungs, some of which can be fatal, and of the vagina. Another fungal parasite produces a skin disease called ringworm, so named because it appears as a circular red area on the skin. The ringworm fungus can infect virtually any skin surface. Most commonly, it attacks the feet, where it causes intense itching and sometimes blisters. This condition, known as athlete's foot, is highly contagious but can be treated with various fungicidal ointments.

A. Dutch elm disease

B. Corn smut

It would not be fair to fungi to end our discussion with an account of diseases. Far more important are the benefits we derive from these interesting eukaryotes. On a global scale, we depend on them as decomposers and recyclers of organic matter.

Fungi also have a number of practical uses for humans. Most of us have eaten mushrooms, although we may not have realized that we were ingesting the fruiting bodies of subterranean fungi. And mushrooms are not the only fungi we eat. The distinctive flavors of certain kinds of cheeses, including Roquefort and blue cheese, come from the fungi used to ripen them. Perhaps the fungi most prized by gourmets are truffles, the fruiting bodies of certain mycorrhizal fungi associated with tree roots. Truffle hunters traditionally used pigs, which are attracted to the smell of truffles, to locate the underground fungi. Today, dogs are used more commonly than pigs for locating truffles because dogs do not try to eat the fungi when they find them. The photograph here shows a Swiss market that features truffles (*tartufi*, in Italian), as well as several varieties of mushrooms (right foreground) and other fungi.

More important in food production are unicellular fungi, the yeasts. As discussed in Chapter 6, yeasts are used in baking, brewing, and winemaking. And fungi are medically valuable as well. Like the bacteria called actinomycetes (see Module 17.13), some fungi produce antibiotics that are used to treat bacterial diseases. In fact, the first antibiotic discovered was penicillin, which is made by a common mold. As producers of antibiotics and food, as decomposers, and as mutualistic partners in mycorrhizae and lichens, fungi are vital contributors to the living world.

Fungi are the third group of eukaryotes we have surveyed so far. In the next chapter, we examine the most diverse group, the animals.

Edible fungi in a Swiss open-air market

Chapter Review

Begin your review by rereading the module headings and scanning the figures before proceeding to the Chapter Summary and questions.

Chapter Summary

Introduction–18.1 The movement of plants and fungi onto land was a major event in the history of life. Plants and green algae share many traits, but plants changed extensively as they moved onto land. Their roots, stems, and leaves function in absorption, support, and photosynthesis. Stems and leaves are covered by a cuticle, and leaves have stomata. Most plants have vascular systems, and special jacketed organs (gametangia) protect their gametes and embryos.

18.3–18.5 Plants probably evolved from green algae and soon split into two major lineages. One lineage gave rise to bryophytes, the group that lacks the vascular tissues characteristic of the other main plant lineage. Bryophytes include the mosses, which, not having the support of vascular tissues, form a low, spongy mat. Like all plants, mosses have a life cycle with alternation of generations. A moss mat consists of haploid gametophytes that produce eggs and swimming sperm. The eggs and sperm unite to develop into the less conspicuous diploid sporophytes. Meiosis in the sporophyte produces haploid spores, which grow into gametophytes.

18.3, 18.6–18.7 Ferns and seed plants have vascular tissues, well-developed roots, and rigid stems. The sporophyte dominates their life cycles. Cells of the sporophyte fern frond undergo meiosis to form spores, which develop into gametophytes. The gametophytes produce eggs and sperm. Swimming sperm fertilize the eggs, and the zygotes develop into sporophytes. Seedless plants once dominated ancient forests; their remains formed coal.

18.3, 18.8–18.10 A major step in plant evolution was the appearance of seed plants—gymnosperms and angiosperms—which have pollen grains for producing and transporting sperm and protect their embryos in seeds. Most gymnosperms are conifers such as pines. The pine tree is a sporophyte; tiny gametophytes grow in its cones. Sporangia in male cones make spores that develop into male gametophytes, the pollen grains. Sporangia called ovules in female cones produce female gametophytes. A sperm from the pollen grain fertilizes an egg in the female gametophyte. The zygote becomes a sporophyte embryo, and the ovule becomes a seed, with stored food and a protective coat. The conifer life cycle takes about two years. Coniferous forests are important sources of timber and paper.

18.3, 18.11–18.15 Most plants are angiosperms, or flowering plants. An angiosperm is a sporophyte with gametophytes in its flowers. The angiosperm life cycle is similar to that of conifers, except that it is much more rapid, and angiosperm seeds are protected and dispersed in fruits that develop from ovaries. Interactions with animals influenced the evolution of flowers and fruits. Angiosperms are an important food source for animals, while animals aid plants in pollination and seed dispersal.

18.16–18.18 Plants probably moved onto land along with mycorrhizal fungi, which help plants absorb water and nutrients. Fungi are heterotrophic eukaryotes that digest their food externally and absorb the nutrients. A fungus usually consists of a mass of threadlike hyphae, forming a network called a mycelium. Most fungi cannot move, but they can grow around and through their food at a rapid rate. Fungal spores germinate to form haploid hyphae. In some fungi, such as mushrooms, the fusion of hyphae results in the unique dikaryotic phase of the life cycle, in which each cell contains two unfused nuclei. The dikaryotic mycelium forms a fruiting body—the mushroom—where fertilization occurs. The diploid zygotes then undergo meiosis, producing a new generation of spores.

18.19 Lichens are associations of algae or cyanobacteria with a network of fungal hyphae. The fungus receives food in exchange for housing, water, and minerals. Lichens survive in hostile environments. They cover rocks and frozen tundra soil, and they are pioneers on new land.

18.20–18.21 Fungi are important in many ways. Many are mutualists in lichens or mycorrhizae. Others are parasites, causing diseases like corn smut and athlete's foot. Many fungi are important in the decomposition of organic matter and nutrient recycling. Fungi are also commercially important as food, in baking and beer and wine production, and in the production of antibiotics.

Testing Your Knowledge

Multiple Choice

1. Angiosperms are different from all other plants because only they have
   a. a vascular system
   b. flowers
   c. a life cycle that involves alternation of generations
   d. seeds
   e. a sporophyte phase

2. Which of the following would you expect to see in a mushroom?
   a. swimming, flagellated sperm
   b. cells containing two nuclei
   c. vascular tissues
   d. ovules
   e. cell walls made of cellulose

3. Which of the following produce eggs and sperm? (*Explain your answer.*)
   a. the fruiting bodies of a fungus
   b. fern sporophytes
   c. moss gametophytes
   d. the anthers of a flower
   e. moss sporangia

4. The eggs of nonflowering seed plants are fertilized within ovules, and the ovules then develop into
   a. seeds
   b. spores
   c. gametophytes
   d. fruit
   e. sporophytes

5. The diploid sporophyte stage is dominant in the life cycles of all of the following except
   a. a pine tree
   b. a dandelion
   c. a rose bush
   d. a fern
   e. a moss

6. Under a microscope, a piece of a mushroom would look most like
   a. jelly
   b. a tangle of string
   c. grains of sugar or salt
   d. a piece of glass
   e. foam

True/False (*Change false statements to make them true.*)
1. Nearly all fungi are harmful parasites.

2. Lichens can only survive in moist, fertile environments.

3. Cucumbers, tomatoes, and pumpkins are all fruits.

4. A pine tree is the sporophyte phase of the pine life cycle.

5. The sporophyte stage of a plant life cycle starts with meiosis.

6. Fungi engulf food particles and then digest them.

7. Ferns lack vascular tissues.

Describing, Comparing, and Explaining

1. Explain how the relationship between a mycorrhizal fungus and an orange tree is similar to the relationship between the fungus and alga making up a lichen.

2. Compare a seed plant with an alga in terms of adaptations for life on land versus life in the water.

3. How do animals help flowering plants reproduce? What do the animals get in return?

4. Why are fungi and plants classified in different kingdoms?

Thinking Critically

1. Mosses and ferns are among the most widely distributed plants. Some species live in moist habitats throughout the Northern Hemisphere. In addition, mosses and ferns are usually among the first organisms to colonize distant islands that have been scoured by tidal waves or scorched by volcanic eruptions. What is it about their life cycles that makes mosses and ferns such good travelers?

2. As described in Chapter 16, there have been several periods of mass extinction during the history of life—for example, the extinction of the dinosaurs at the end of the Cretaceous period. Fossils indicate that plants were much less severely affected than animals at these times. What features of their life cycles may have enabled plant species to weather such disasters?

3. Many fungi produce antibiotics, such as penicillin, which are valuable in medicine. But of what value might the antibiotics be to the fungi that produce them? Similarly, fungi often produce compounds with unpleasant tastes and odors as they digest their food. What might be the value of these chemicals to the fungi? How might production of antibiotics and odors have evolved?

Science, Technology, and Society

1. Much of the conifer forest in the U.S. Pacific Northwest has been clearcut; less than 10% of the original ancient forest, dominated by giant firs and hemlocks, remains. There is no law protecting endangered habitats, so in order to protect the northern spotted owl, which lives only in oldgrowth conifers, conservationists have sued to stop logging under the Endangered Species Act. The lawsuits have halted logging in many National Forest areas. Lumber companies buy trees from National Forests, loggers work there, and the economies of many small communities depend on logging. The reduction in timber supply has driven up the cost of lumber. Imagine you have been named by the President of the United States to deal with this situation. What are the opposing issues? What would you suggest to resolve this conflict, and how would you defend your policy?

2. American chestnut trees once made up more than 25% of the hardwood forests of the eastern United States. These trees were wiped out by a fungus that had been accidentally introduced on imported Asian chestnuts, which are not severely affected. More recently, a fungus has killed large numbers of dogwood trees from New York to Georgia; some experts suspect the parasite was accidentally introduced from elsewhere. Dutch elm disease is a similar case (see Module 18.20). Why are plants particularly vulnerable to fungi imported from other regions? What kinds of human activities might contribute to the spread of plant diseases? Do you think the introduction of plant pathogens like the chestnut-blight fungus are more or less likely to occur in the future? Why?

The Evolution of Animal Diversity   19

Of some 1.5 million species of organisms known to science, over two-thirds are animals. As a group, animals thrive in nearly all environments on Earth. Most species of animals live on land, but the greatest number of animal *phyla* are found in the sea, where animals undoubtedly arose. A diver on a tropical coral reef like the one at the left could see more different phyla in half an hour than most people see in a lifetime.

The photograph at the left was taken on the Great Barrier Reef, along the northeast coast of Australia. To experience even a fraction of the teeming life in these waters, a diver has to stop swimming and focus on small areas one at a time. Schools of fishes explode out of caves. A startled moray eel wriggles back into a crevice. Other brightly colored fishes nibble on the coral surface. A red sea star, an orange one, and a bright blue one appear in one visual field. A cactuslike cluster of purple sponges is seen nearby.

The main structure of a coral reef results from the activities of only two of the many kinds of organisms that live there, coralline algae (see Module 17.26) and coral animals. These reef builders occupy only the outer surface of a reef, which consists of dense layers of limestone built up over many generations. Coralline algae add to the reef when they die, and the limestone in their cell walls accumulates on the reef surface. The coral animals build skeletal support for themselves by secreting a layer of limestone. Gradually, over many years, as one generation of algae and coral animals is replaced by another, layer upon layer of limestone gradually build up, and a reef forms.

The photograph at the right gives you an impression of what coral animals look like. They live in interconnected colonies, with the individuals, each called a polyp, housed in pockets formed of their limestone skeletons. When feeding, the polyps (each only about 3 mm in diameter) extend the whitish tentacles you see here. Mucus on the tentacles traps fine organic food particles suspended in the seawater. When disturbed, the polyps pull back into the coral skeleton.

Coral animals also obtain some of their food from dinoflagellates, photosynthetic protists that live in the polyp tissues. In turn, the dinoflagellates make use of some of the coral animals' waste products, such as carbon dioxide and nitrogen compounds. Coral animals can survive without their dinoflagellates, but they cannot form reefs without them. Apparently, the metabolic exchange between the animal and the protist makes it possible for the animal to produce enough limestone to build the reef.

In a coral reef, we see many such mutually beneficial connections among organisms. We saw another example of such a relationship in Chapter 18's introduction, where we discussed mycorrhizae, beneficial associations between plants and fungi. In Chapter 18, we also described lichens, associations between fungi and algae or cyanobacteria (see Module 18.19). And we saw that mutually beneficial interactions between flowering plants and pollinating animals account for much of plant diversity. The association between coral animals and dinoflagellates also contributes to diversity, helping create reef habitats for a host of animals. A diver on a reef might actually see animals representing as many as 27 of some 30 phyla in the **kingdom Animalia.**

In this chapter, we look at nine major animal phyla, those that contain the greatest number of species and that are the most abundant and widespread. Along the way, we will give special attention to the major milestones in animal evolution.

Coral polyps extending tentacles

What is an animal?

We know when we see one—or do we? What about the coral animals on the previous page or the sea star in the figure here? Can we put into words why we call these organisms animals? We can distinguish animals from organisms in the other four kingdoms very generally by saying that animals are eukaryotic, multicellular heterotrophs that lack cell walls. Let's start by examining that statement more closely.

First, being *eukaryotic* groups animals with protists, fungi, and plants, and separates them from bacteria (kingdom Monera). Second, being *multicellular* separates animals, fungi, and plants from most protists, which are unicellular. Third, being *heterotrophic* separates animals and fungi from plants and plantlike protists (the algae), which are photosynthetic. Finally, *lacking cell walls* distinguishes animals from plants, algae, and fungi. So we have four features that, taken together, distinguish animals from other organisms.

Animals also have several other distinctive features. For example, most of them take food molecules into their bodies and digest them there. The majority of animals also have a digestive tract, an internal cavity or tube where hydrolytic enzymes are secreted and digestion occurs.

Other important features show up in the life cycles of animals. The figure here shows the life cycle of a sea star. Except for the haploid eggs and sperm, all stages are diploid, as is true for most animals. ① Male and female adult animals make haploid gametes by meiosis, and ② an egg and a sperm fuse to produce a zygote. The zygote then undergoes mitosis and starts developing, going through a series of stages that are characteristic of the animal kingdom. ③ Early mitotic divisions lead to an embryonic stage called a **blastula,** common to all animals. Typically, the blastula consists of a single layer of cells surrounding a hollow cavity. ④ Later, in the sea star and many other animals, one side of the blastula folds inward, forming an embryonic stage called a **gastrula.** In its early stages, the gastrula looks like an indented blastula. ⑤ The gastrula develops into a saclike embryo with an opening at one end. At this point, the gastrula has an outer cell layer and an inner cell layer that results from infolding. Eventually, the outer layer develops into the animal's epidermis and nervous system, and the inner layer forms the digestive tract. Still later in development, in most animals, a third layer forms between the other two and develops into most of the other internal organs.

Following the gastrula, many animals develop directly into adults. But others, including the sea star, ⑥ develop into one or more larval stages first. A **larva** is an immature individual that looks very different from an adult. The larva ⑦ undergoes a complete change of body form, called **metamorphosis,** in becoming an adult— a mature animal capable of reproducing sexually.

Thus, we have a number of characteristics that answer the question, What is an animal? To our earlier description, we can add that a typical animal ingests its food and digests it in an internal digestive tract; is diploid; and has unique embryonic stages, including a blastula, a gastrula with distinct layers of cells, and, in many species, a larva. How might an organism with these distinctive, animal-like features have evolved?

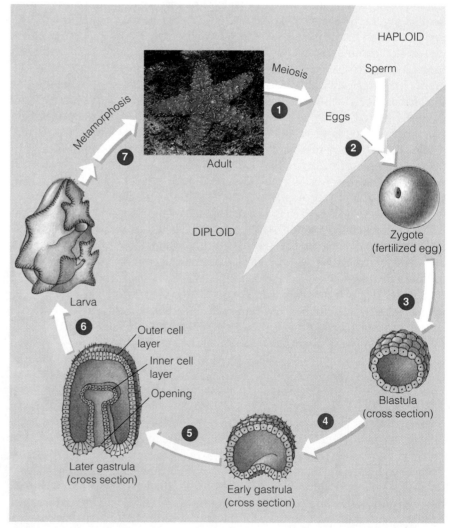

The life cycle of a sea star

The animal kingdom probably originated from colonial protists

Biologists have speculated about the origins of animal diversity ever since Darwin's time. Fossils of the oldest known animals have been found in rocks from the late Precambrian era, about 700 million years ago. Diverse and complex, many of these ancient animals seem to have belonged to phyla that no longer exist. The modern animal phyla—the basic types of animals we see today—apparently evolved during the Cambrian period, which began about 600 million years ago. Neither the Cambrian fossils nor those from the Precambrian tell us much about the origin of the animal kingdom: All fossils found so far are the remains of animals whose body structure was too complex to have been the first animals. So let's speculate a bit.

Animals, like fungi and plants, are multicellular organisms. They probably evolved from protists that lived as colonies of cells. The figure below shows a sequence of stages that may have led to the first animal. ① The earliest colonial aggregates may have consisted of only a few cells, all of which were flagellated and virtually identical. Colonies form when cells divide but do not separate, and early colonies in the animal lineage may have grown larger by simply adding cells. At some point, aggregates probably appeared that consisted of hundreds or thousands of cells. ② Such colonies may have been hollow spheres—floating aggregates of heterotrophic cells—that ingested organic particles suspended in the water. ③ Eventually, cells in the colony may have undergone a division of labor, resulting in some adapted for reproduction and others having somatic (nonreproductive) functions, such as locomotion and feeding. This simple way of dividing up the tasks of staying alive involves a degree of specialization—into two types of cells, reproductive and somatic.

A colony of cells, even one with some division of labor, is still a long way from a multicellular organism. The cells of truly multicellular organisms are highly specialized and interdependent. In addition, multiple layers of cells are a fundamental feature of animals, as we saw in Module 19.1. If and when a hollow colony became animal-like, it first had to fill with cells, and the cells had to form layers. Stage ④ in the figure shows how a simple multicellular organism with cell layers might have evolved from a hollow colony of heterotrophic cells. Layers may have developed as cells on one side of the colony folded inward, the way they do in the gastrula we discussed in Module 19.1. Originally, the infolding may have provided a temporary digestive cavity, a region where cells were specialized for feeding and digestion.

Eventually, as shown in stage ⑤, the infolding may have almost completely eliminated the original hollow space, producing an organism with two cell layers. The main advantage of the layers is to allow further division of labor among the cells. The outer flagellated cells would have provided locomotion and some protection, while the inner cells could have become specialized for reproduction or feeding. It is probably appropriate to call organisms with this level of organization primitive animals, or at least "protoanimals" (organisms that gave rise to animals). Our protoanimal was probably a crawler rather than a floater. Even today, most animals live and feed on some type of surface, and the bottom of an ocean seems a likely candidate for a protoanimal's domain. With its specialized cells and a simple digestive tract formed by infolding, the protoanimal shown here could have fed on organic matter on the seafloor.

What protoanimals and their forerunners really looked like remains a mystery. The two main stages we have proposed—the hollow colony and the "infolded" stage—are modeled after the blastula and gastrula stages in animal development. Indeed, these embryonic stages may resemble organisms that were early ancestors of animals.

We now leave the hypothetical world of protoanimals and begin looking at some real organisms. We turn first to animals with relatively simple body organization and then proceed to more complicated ones. We'll focus on the major landmarks in animal evolution, as reflected in body features of animals living today.

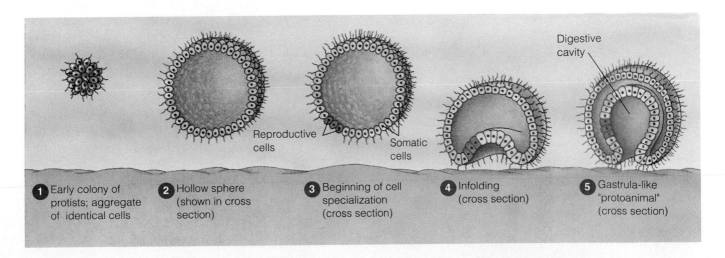

① Early colony of protists; aggregate of identical cells

② Hollow sphere (shown in cross section)

Reproductive cells

③ Beginning of cell specialization (cross section)

Somatic cells

④ Infolding (cross section)

⑤ Gastrula-like "protoanimal" (cross section)

Digestive cavity

Sponges have relatively simple, porous bodies

There are about 5000 species of **sponges** (phylum **Porifera**). Most are marine, but about 150 species live in fresh water. Sponges do not move from place to place. They live singly or in clusters formed by budding. Most species have great powers of regeneration and often reproduce by this means. If a sponge body is fragmented into several pieces, each piece can often develop into a whole new sponge.

Figure A shows two individuals of the genus *Scypha,* a small marine sponge measuring only about 1–3 cm high. Certain other sponges reach heights of 2 m or spread irregularly over large areas on seafloors and lake bottoms. Vase-shaped or cylindrical sponges like *Scypha* have **radial symmetry.** This means that the body parts are arranged like pieces of a pie around an imaginary central axis. As Figure B shows, any imaginary slice passing longitudinally through the central axis of *Scypha* will divide it into mirror images.

Sponges are among the simplest animals. *Scypha,* for example, resembles a simple tube open at one end and perforated with holes. (Porifera means pore-bearers in Latin.) The entire body of *Scypha* and most other sponges consists of three loosely associated layers of cells, which you can see in Figure C. An outside layer of flattened cells, marked by numerous pores (pink), provides protection. A middle body layer has skeletal elements (yellow) composed of either a flexible protein called spongin or mineral-containing particles. A third cell layer, consisting of flagellated cells

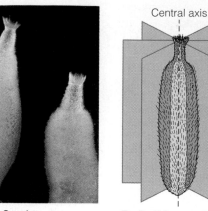

A. *Scypha*

B. Radial symmetry

Central axis

called choanocytes (purple), lines a large central cavity. A **choanocyte,** also called a collar cell, has a collar-like ring surrounding the base of its flagellum. Aided by the flagella on choanocytes, water flows through the pores into the central cavity and then out of the sponge through its one large opening. More complex sponges have folded body walls and branching water canals, rather than a single central cavity.

Sponges feed by collecting food particles (mostly bacteria) from water that streams through their porous bodies. Their choanocytes trap food particles in mucus on their collars and then engulf the particles by phagocytosis (see Module 5.19). Other cells, called **amoebocytes,** are constantly on the move within the sponge body. They pick up food packaged in food vacuoles from the choanocytes, digest it, and carry the nutrients to other cells. Amoebocytes are the "do-all" cells of sponges. Moving about by means of pseudopodia, they digest and distribute food, transport oxygen, dispose of wastes, and manufacture skeletal elements. They can even change into other cell types.

What about the origin of the phylum Porifera? Figure D is a drawing of a colonial protist called a **choanoflagellate,** an organism that lives at the bottom of ponds and shallow seas. Each colony consists of a small cluster of flagellated collar cells attached to a stalk that is anchored to a rock or submerged stick. Sponge choanocytes are remarkably similar to choanoflagellate cells. Actually, sponges have more in common with choanoflagellates than with other animals.

Sponges lack a digestive tract and digest their food inside their cells like choanoflagellates and other protists—and unlike most animals. Also, sponges have no gastrula stage, their three cell layers are not homologous with the body layers of other animals, and they lack nerves and muscles. For these reasons, many biologists believe that sponges arose from ancestors similar to choanoflagellates on a lineage separate from that of the rest of the animal kingdom.

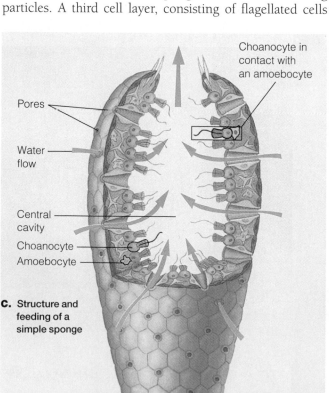

Choanocyte in contact with an amoebocyte

Pores

Water flow

Central cavity

Choanocyte

Amoebocyte

C. Structure and feeding of a simple sponge

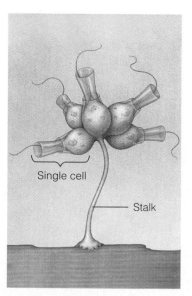

Single cell

Stalk

D. A choanoflagellate colony

Cnidarians are radial animals with stinging threads

Radial symmetry, common in sponges, is a hallmark of the phylum **Cnidaria**: the hydras, jellyfishes, sea anemones, and corals. Cnidarians can have two kinds of body forms, both of which are radially symmetrical. *Hydra*, a common organism in freshwater ponds and lakes, has a cylindrical body with arms, called tentacles, projecting from one end (Figure A). *Hydra's* body form is a **polyp**, like the individual coral animals we saw earlier.

The other type of cnidarian body is the **medusa**,

A. Polyp body form: a hydra

B. Medusa body form: a jellyfish

exemplified by the marine jellyfish in Figure B. Whereas polyps are mostly stationary, medusas move freely about in the water. They are shaped like umbrellas with fringes of tentacles around the lower edge. This type of radial body is different from the cylindrical form of *Hydra*, but any cut through the center of the umbrella from the top to the bottom will divide it into mirror images. Some cnidarians exist only as medusas, others (such as *Hydra*) only as polyps, and still others have both a medusa and a polyp in their life cycles.

Cnidarians are carnivores that use their tentacles to capture small animals and protists and to push the prey into their mouths. In a polyp, the mouth is located on the top of the body, at the hub of the radiating tentacles. In a medusa, the mouth is in the center of the undersurface of the umbrella. In both polyp and medusa, the mouth leads into a digestive compartment called the **gastrovascular cavity** (Greek *gaster*, belly, and Latin *vas*, vessel). Undigested food and other wastes exit through the mouth; there is no anus, and therefore the digestive system is said to be incomplete. The cavity also circulates fluid that services internal cells (hence the "vascular" in gastrovascular). Fluid in the cavity provides body support, much like water in a balloon; to a large extent, this fluid gives a cnidarian its shape.

If we were to list two traits that, taken together, mark an animal as a cnidarian, one would be radial symmetry. The other would be the specialized cells, called cnidocytes, for which the phylum is named. **Cnidocytes** (meaning "stinger cells"), found on the surface of the tentacles of polyps and medusas, function in defense and prey capture. Each cnidocyte contains a fine thread coiled with-

in a capsule (Figure C). When discharged, the thread can sting or entangle prey. *Hydra* uses these threads to capture prey that swim close to its tentacles; certain large marine cnidarians use their stinging threads to catch fish.

Cnidarians have several features that are absent in sponges but present in nearly all other animals. One is a digestive cavity; another is a gastrula stage in development; and a third is the presence of tissues, organized groups of similar cells adapted to perform a specific function. *Hydra*, for example, has muscle tissue that moves its tentacles and shortens its whole body when disturbed. Nervous tissue enables the animal to respond to stimuli and coordinates its muscle activity.

As a group, cnidarians are what we might call "tissue animals." Although sea anemones and certain jellyfishes have organs (several kinds of tissues working together to perform specific functions), most vital functions in cnidarians are performed by individual tissues. In contrast, the rest of the animals we will discuss have most of their tissues organized into organs, and many of their organs organized into functional units called organ systems.

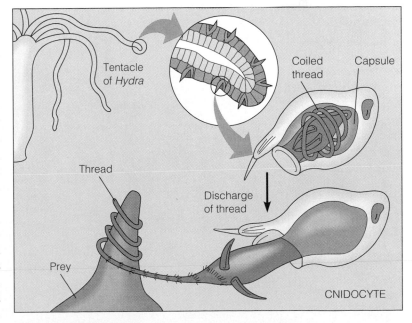

C. Cnidocyte action

19.5 Most animals are bilaterally symmetrical

In contrast to radially symmetrical animals such as cnidarians, most animals are bilaterally symmetrical. **Bilateral symmetry** (Latin *bi-*, double, and *latus*, side) means that an animal can be divided equally by a single cut, as shown in the figure below, and has mirror-image right and left sides. A bilaterally symmetrical animal, such as this crayfish, has a distinct head, or **anterior**, end, and tail, or **posterior**, end. It also has a back (**dorsal**) surface, a bottom (**ventral**) surface, and two side (**lateral**) surfaces. The head is a prominent part of a bilaterally symmetrical animal. It houses its main sensory structures (such as eyes), its brain, and usually its mouth. The brain and sensory structures are organs that, along with nerves that branch throughout the body, form an organ system—the nervous system.

Bilateral animals are fundamentally different from radial ones. A radial animal lacks a head or any forward orientation of its body and typically spends most of its time sitting on the seafloor or drifting about in water currents. In contrast, most bilaterally symmetrical animals are quite active and travel head-first through the environment. Their eyes and other sense organs are up front on the head, where they contact the environment first and help the animal respond appropriately. Humans are bilateral, but because we walk on two feet instead of four, we move ventral-surface-first instead of head-first.

Bilateral symmetry and a head end are prerequisites for the forward movement typical of animals. Most animals crawl, walk, run, burrow, swim, or fly in a head-first direction. The evolution of bilateral symmetry, a head end, and forward movement were important milestones in the history of the animal kingdom.

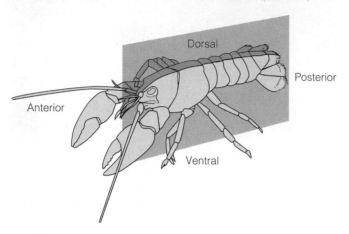

19.6 Flatworms are the simplest bilateral animals

Flatworms, of the phylum **Platyhelminthes** (Greek *platys*, flat, and *helmis*, worm), are leaflike or ribbonlike animals, ranging in length from about 1 mm to 20 m. They are bilaterally symmetrical, but with bodies that are unusually simple for bilateral animals. In common with cnidarians, most flatworms have an incomplete digestive tract (no anus). Most bilateral animals, by contrast, have a complete digestive tract with both a mouth and an anus. Also in common with cnidarians, the digestive cavity is the only space inside the flatworm body. Typical bilateral animals have another space, called the body cavity, between the digestive tract and the body wall. Except for the space *within* their digestive tract, flatworms lack an internal cavity. (We discuss body cavities and their significance in Module 19.7).

There are three major groups of flatworms. The worm called a planarian (Figure A) represents a group called the **free-living** (nonparasitic) **flatworms**. The planarian has a head with two large eyespots and a flap at each side that detects chemicals in the water. Dense clusters of nervous tissue form a simple brain, and a pair of nerve cords connect the brain with small nerves that branch throughout the body. Together, the sensory and nervous structures constitute a nervous system.

The digestive tract of a planarian is highly branched. Its single opening, the mouth, is located not on the head but on the ventral surface. When the animal feeds, a muscular tube projects through the mouth (as shown in the figure) and pulls food in. Planarians live on the undersurfaces of rocks in freshwater ponds and streams. Using cilia on their ventral surface, they crawl about in search of food. They also have muscles that enable them to twist and turn.

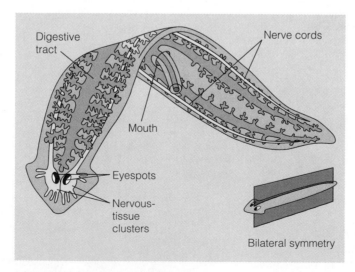

A. A free-living flatworm, the planarian

A second group of flatworms, the **flukes**, are parasites. The photograph in Figure B shows a male and female blood fluke (*Schistosoma*). The female spends much of her time in a groove running the length of the body of the larger male. In this position, the worms copulate frequently, and a single pair can produce over a thousand eggs a day. Both the female and the male have suckers that attach to the inside of the blood vessels near the host's intestines. Blood flukes infect humans and cause a severe, long-lasting disease called schistosomiasis (blood fluke disease). Blood fluke disease is widespread in Africa, Southeast Asia, and South America. It afflicts some 250 million people in 70 countries, causing severe abdominal pain, anemia, and dysentery.

Most flukes, including *Schistosoma*, have a complex life cycle that includes reproduction in more than one host. As outlined in the Figure B diagram, ① blood flukes living in a human host reproduce sexually, and fertilized eggs pass out in the host's feces. If an egg lands in a pond or stream, ② a ciliated larva hatches and ③ can enter a snail, the next host. ④ Asexual reproduction in the snail eventually produces ⑤ other larvae that can infect humans. ⑥ A person becomes infected when these larvae penetrate the skin.

Tapeworms, which are also parasitic, make up the third group of flatworms. Adult tapeworms inhabit the digestive tracts of vertebrate animals, including reptiles, birds, and mammals. In contrast to planarians and flukes, most tapeworms have a very long,

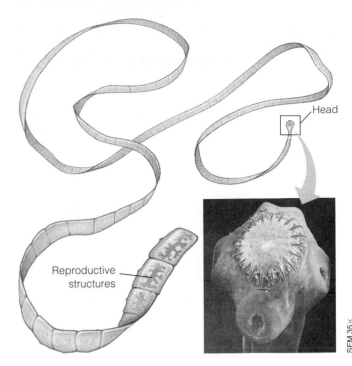

C. A tapeworm, a parasitic flatworm

ribbonlike body with repeated parts. They also differ from other flatworms in not having any digestive tract at all. Living in partially digested food in the intestines of their hosts, they simply absorb nutrients across their body surface. As the drawing in Figure C above shows, the head is the smallest part of the tapeworm body and is armed with suckers and teeth that grasp the host. Behind the worm's head, a short neck generates the repeated parts of the worm. The youngest part is just behind the neck, and the oldest is at the posterior end. The repeated parts are filled with both male and female reproductive structures. Full of ripe eggs, those at the posterior end break off and pass out of the host's body in feces.

Like parasitic flukes, tapeworms have a complex life cycle, usually involving more than one host. Most species benefit from the predator-prey relationships of their hosts. A prey species—a sheep or a rabbit, for example—may become infected by eating grass contaminated with tapeworm eggs. Larval tapeworms develop in these hosts, and a predator—a coyote or a dog, for instance—becomes infected when it eats an infected prey animal. The adult tapeworms develop in the predator's intestine.

Several kinds of tapeworms infect humans. We can be infected, for example, by a large tapeworm called *Taeniarhynchus* by eating rare beef infected with the worm's larvae. The larvae are microscopic, but the adults can reach lengths of 20 m in the human intestine. An orally administered drug called niclosamide kills the adult worms.

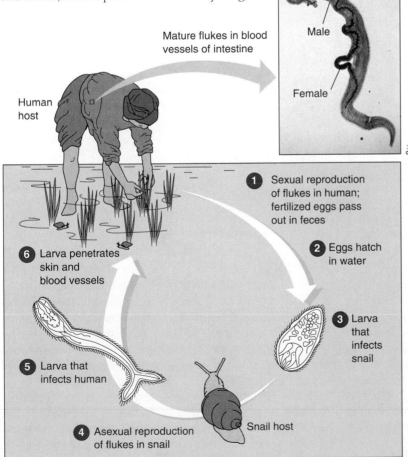

B. A fluke (*Schistosoma*) and its life cycle

Mature flukes in blood vessels of intestine

Male

Female

Human host

1. Sexual reproduction of flukes in human; fertilized eggs pass out in feces
2. Eggs hatch in water
3. Larva that infects snail
4. Asexual reproduction of flukes in snail
5. Larva that infects human
6. Larva penetrates skin and blood vessels

Snail host

Most animals have a body cavity

The evolution of a **body cavity**, a fluid-filled space between the digestive tract and the body wall, was significant in animal history. Sponges, cnidarians, and flatworms lack a body cavity, but nearly all other animals have one.

The figures at the right compare the internal structure of three animals. In all three cross sections, the colors indicate the same tissue layers: the outside covering (blue), a middle region (pink), and the lining of the digestive tract (yellow). These tissue layers form from the cell layers of the gastrula stage during embryonic development (see Figure 19.1).

The cross section through a flatworm (Figure A) reveals a body that is solid—filled with cells of the three tissue layers—except for the cavity of the digestive tract. The other two animals on this page have a body cavity. The roundworm (Figure B) has a body cavity called a pseudocoelom. A **pseudocoelom** (Greek *pseudes,* false, and *koilos,* hollow) is an internal space in direct contact with the wall of the digestive tract. The outer edge of the pseudocoelom contacts a muscle layer that is part of the body wall.

The third animal, an earthworm (Figure C), has a more complex body than either the flatworm or the roundworm. Its body cavity, called a **coelom**, is completely lined by the middle tissue layer (shown in pink). This tissue layer extends from the body wall and wraps around the digestive tract. In effect, this middle tissue layer suspends the digestive tract and other internal organs from the body wall.

There are many advantages to having a body cavity. The cavity makes the animal more flexible and better able to crawl and burrow. It also allows the internal organs to grow and move independently of the outer body wall. The fluid in the body cavity cushions the internal organs, helping prevent internal injury when the animal receives a sharp blow or is pinched. In an animal with a hard skeleton—especially one with internal bones like ours—even mild exercise could harm internal organs if it were not for the coelom and its cushioning fluid. In soft-bodied animals such as earthworms, fluid in the body cavity functions as a watery skeleton against which muscles in the body wall can exert force to move the body. The fluid may also help circulate nutrients and oxygen throughout the body and assist in waste disposal. Amoeboid cells in the fluid also assist with these functions.

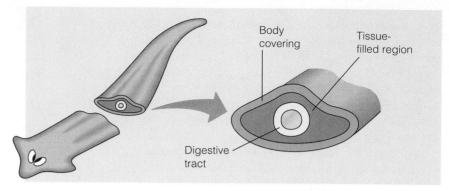

A. Flatworm

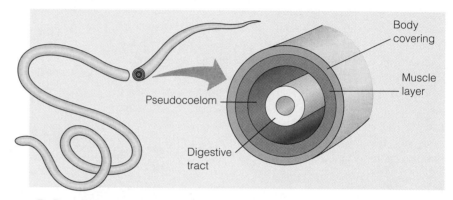

B. Roundworm

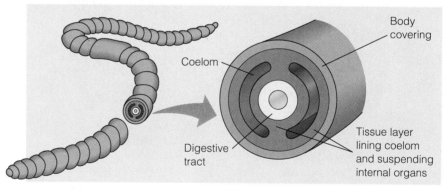

C. Earthworm

There is a connection between an animal's shape and size and the presence or absence of a body cavity. For instance, the small size and thinness of a flatworm's solid body is an adaptation that allows nutrients and oxygen to diffuse into all body cells and wastes to diffuse out. Most flatworms are only a few millimeters long, and even long ones such as tapeworms are very thin.

Flatworms are a widespread, successful group. Considering the many advantages of body cavities, however, it's not surprising that most members of the animal kingdom, an assemblage of over a million species, have some type of body cavity.

Roundworms, also called nematodes, make up the phylum **Nematoda**. These are cylindrical worms with a finely tapered tail; the head is more blunt, as you can see at the right end of the worm in Figure A. The nematode body is covered by a tough, nonliving skin, or **cuticle**, that resists drying and crushing. The roundworm in Figure A has a transparent cuticle, and you can see some of its internal organs.

Nematodes have a pseudocoelom. They also have a complete digestive tract, extending as a straight tube from a mouth at the tip of the head to an anus near the tip of the tail. Food travels only one way through the system, and regions are specialized for certain functions. In animals with a complete digestive tract, the anterior region of the tract churns and mixes food with enzymes, while the posterior region absorbs nutrients and disposes of wastes. This division of labor allows each part of the digestive tract to be highly efficient at its particular function.

Nematodes are among the most numerous of all animals in both number of species and number of individuals. Nematodes live virtually every place there is rotting organic matter, and these worms are important decomposers in soil and on the bottom of lakes and oceans. Other nematodes thrive as parasites in the moist tissues of plants and in the body fluids and tissues of animals. About 90,000 species are known, and it is likely that at least ten times that number actually exist.

Little is known about most free-living nematodes. A notable exception is the soil-dwelling species *Caenorhabditis elegans*, an important research organism and one of the best-understood animals. An adult *C. elegans* consists of only about 1000 cells—in contrast to the human body, which consists of some 60 trillion cells. Because of the simplicity of this worm, re-searchers have been able to trace the lineage of individual cells in the adult back to parts of the zygote. These studies are contributing to our understanding of how genes control animal development.

Though free-living nematodes are the most abundant, many species of roundworms are serious agricultural pests that attack the roots of plants or parasitize animals. Humans are host to at least 50 species of roundworms, including a number of disease-causing organisms that cause major health problems. Among these are hook-worms, nematodes that attach to the intestinal wall and suck blood. Humans, dogs, cats, and many other mammals are susceptible to hookworms. Nematodes called heartworms are deadly to dogs. Spread by mosquitoes and also infectious to humans, heartworms seem to be on the increase in the United States.

One of the most notorious roundworms is *Trichinella spiralis*, which causes a disease called trichinosis in a wide variety of mammals, including humans. People usually acquire the worms by eating undercooked pork containing the juvenile worms. You can see some of these sausage-shaped juveniles in the piece of muscle shown in Figure B. Trichinosis causes severe nausea and sometimes death when large numbers of the worms penetrate heart muscle. Cooking meat until it is no longer pink kills the worms.

We might expect that as numerous and widespread an animal group as the nematodes would include a great diversity of body form. In fact, the opposite is true. Most species of nematodes look very much alike: cylindrical, with a blunt head and a tapered tail. In sharp contrast, animals in the phylum Mollusca, which we examine next, exhibit enormous diversity in body form.

LM 228×

A. A free-living roundworm

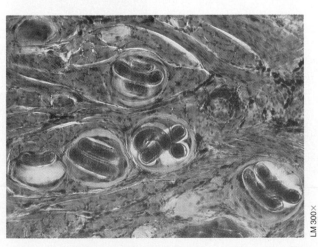

LM 300×

B. A parasitic roundworm (*Trichinella*)

Diverse mollusks are variations on a common body plan

Snails, slugs, oysters, clams, octopuses, and squids are just a few of the great variety of animals known as **mollusks**. Phylum **Mollusca** comprises more than 100,000 species. Most mollusks have soft bodies protected by a hard shell; their phylum name comes from the Latin *molluscus*, meaning soft.

It may seem that animals as different as squids and clams could not belong in the same phylum, but these and other mollusks have inherited several common features from their ancestors. Figure A illustrates the basic body plan of a mollusk. Two hallmarks of the phylum are the muscular **foot** (gray in the drawing) that functions in locomotion and the **mantle** (purple), an outgrowth of the body surface that drapes over the animal. A key molluskan feature, the mantle produces the shell in mollusks such as clams and snails. It also functions in respiration, waste disposal, and sensory reception. In addition, the mantle creates another distinctive feature, the mantle cavity. In many mollusks, the mantle cavity houses a gill, which you can see in Figure A. The gill extracts oxygen dissolved in the water and may also dispose of fluid wastes.

Figure A shows yet another body feature found in many mollusks—a unique rasping organ called a **radula**, which is used to scrape up food. In a snail, for example, the radula extends from the mouth and slides back and forth like a backhoe, scraping and scooping algae off rocks.

Mollusks are the first animals we have discussed in this chapter that have a coelom (introduced in Module 19.7).

Their coelom (light blue in Figure A) consists of three small cavities: one around the heart, one around the reproductive organs, and one that forms part of the kidney. Mollusks also are the first animals we have examined that have a true circulatory system—an organ system that distributes materials, such as nutrients, water, and oxygen, throughout the body. Only the heart and main vessels extending from it are shown in the diagram, but the circulatory system branches throughout the body and has taken over the functions of a large coelom.

Evolution has modified the basic body features in different ways in different groups of mollusks. The three most diverse groups are the gastropods (including snails and slugs), bivalves (such as clams, scallops, and oysters), and cephalopods (including squids and octopuses).

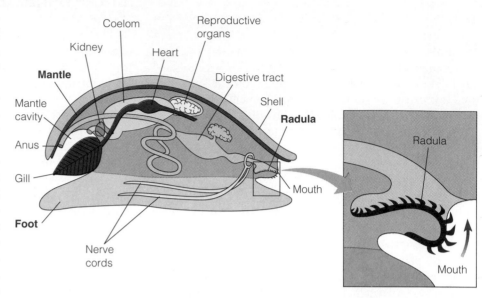

A. The general body plan of a mollusk

B. A terrestrial gastropod: a land snail

C. A marine gastropod: a sea slug

Gastropods (Greek *gaster,* belly, and *pous,* foot), the largest group of mollusks, with over 75,000 species, are adapted to live in fresh water, salt water, and terrestrial environments; in fact, they are the only mollusks that live on land. Terrestrial slugs and snails like the one in Figure B lack the gills typical of aquatic mollusks; instead, their mantle cavity has evolved into a lung that breathes air. Most gastropods are marine, and the group includes some of the most colorful animals in the sea. The gastropod shown in Figure C is called a sea slug. Sea slugs are unusual mollusks in that they lack a mantle, mantle cavity, and shell. The long projections on this species serve as gills, obtaining oxygen and disposing of wastes.

Bivalves (Latin *bi-,* double, and *valva,* "leaf of a folding door"), including numerous species of clams, oysters, mussels, and scallops, have shells divided into two halves that are hinged together. Most bivalves are sedentary, living in sand or mud. They use their muscular foot for digging and anchoring, and most have mucus-coated gills that trap fine food particles suspended in the water. The scallop in Figure D is an unusual bivalve in that it sits on the ocean floor rather than digging into it. Notice the many bluish eyes peering out between the two halves of the hinged shell. The eyes are set into the fringed edges of the animal's mantle. When they detect movement nearby, or if a predator touches one of the long projections of the mantle shown in the photograph, the scallop can jet a short distance away by clapping its valves together and squirting water out of its mantle cavity.

Cephalopods (Greek *kephale,* head, and *pous,* foot) generally differ from gastropods and sedentary bivalves in being built for speed and agility. A few cephalopods have large, heavy shells, but in most the shell is small and internal (as in squids) or missing altogether (as in octopuses). Cephalopods are marine predators, using beaklike jaws and a radula to crush or rip prey apart. Their mouth is at the center of their foot, which is drawn out into several long tentacles for catching and holding prey.

The squid in Figure E (about 20 cm long) ranks with fishes as a fast, streamlined predator. It darts about by drawing water into its mantle cavity and then firing a jet of water back out. Squids have a large, complex brain, rivaling that of fishes, and their eyes (one on each side of the body) are among the most complex sense organs in the animal kingdom. Each squid eye has a lens that focuses light and a retina on which clear images form.

All cephalopods have large brains and sophisticated sense organs, and these contribute to their being successful, mobile predators. The octopus in Figure F (about 30 cm long) lives on the seafloor, where it scurries about in search of crabs and other food. Its brain is larger and more complex, proportionate to body size, than that of any other animals except the vertebrates. Octopuses are highly intelligent and have shown remarkable learning abilities in laboratory experiments.

D. A bivalve: a scallop

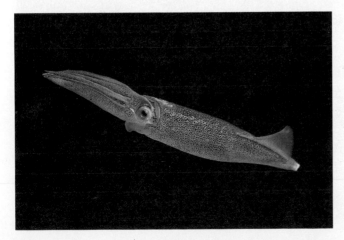

E. A cephalopod with an internal shell: a squid

F. A cephalopod without a shell: an octopus

Most animals have segmented bodies

The phyla we will now discuss illustrate another body feature that had profound influence on animal history—body **segmentation,** the subdivision of the body along its length into a series of repeated parts (segments). This feature played a central role in the evolution of many complex animals.

Segmentation is an obvious feature of an animal like an earthworm (Figure A), in which the segments are marked off externally by grooved rings. Internally, the coelom is partitioned by walls (only two are fully shown here). The nervous system (yellow) includes a ventral nerve cord with a dense cluster of nerve cells in each segment. Excretory organs (green), which dispose of fluid wastes, are also repeated in each segment. The digestive tract, however, is not segmented; it penetrates the segment walls and runs the length of the animal. The main channels of the circulatory system—a dorsal blood vessel and a ventral blood vessel—are also unsegmented. But they are connected by segmental vessels, including five pairs of accessory hearts near the anterior end. The main heart is simply the anterior region of the dorsal blood vessel.

The dragonfly (Figure B) is also segmented, though less uniformly than the earthworm. Its segments are most pronounced in its abdomen; its head and mid-region (thorax) are each formed from several fused segments. Each pair of its six walking legs and each pair of its four wings emerge from a body segment in the thorax.

Segmentation also occurs in the human body (Figure C). We have a backbone formed of a repeated series of bones called vertebrae, and muscles associated with our vertebrae

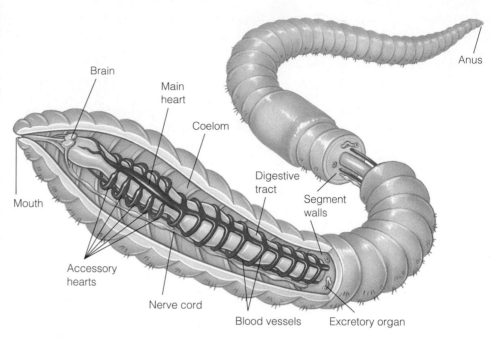

A. Segmentation in an earthworm

are segmented. We also have segmented abdominal muscles, clearly visible in body builders as "stomach ripples."

A segmented body is advantageous in many ways. It allows greater body flexibility and mobility, and it probably evolved as an adaptation that aids in movement. The earthworm uses its flexible, segmented body to crawl over wet surfaces and burrow rapidly into the soil. In the dragonfly, segmentation occurs in both its tough external skeleton and the muscles that move its head, thorax, and abdomen. Also, the segments provide flexibility for flying, perching, mating, and laying eggs.

B. Dragonfly segmentation

C. Segmentation in the human body

Phylum **Annelida** (Latin *anellus*, ring), comprising earthworms and other segmented worms, is named for its characteristic ringlike body parts. Except for a distinct head and tail, a typical **annelid** consists of body segments that are all very similar.

There are about 15,000 annelid species, ranging in length from less than 1 mm to a 3-m-long giant earthworm that lives in Australia. Annelids live in the sea, in most freshwater habitats, and in damp soil. Some aquatic annelids swim in pursuit of food, but most are bottom-dwelling scavengers that burrow in sand and mud.

Earthworms are one of three large groups of annelids. An earthworm eats its way through the soil, extracting nutrients as soil passes through its digestive tube. Undigested material passes out the anus at the posterior end of the worm. Farmers value earthworms because the animals actually till the soil and because earthworm feces improve the soil's texture. Darwin estimated that a single acre of British farmland had about 50,000 earthworms, producing 18 tons of feces per year.

The largest group of annelids are **polychaetes** (Greek *polys*, many, and *chaeta*, hair). Figure A shows a polychaete called a sandworm (about 15 cm long), which lives on the seafloor. Segmental appendages and hard bristles that project from them help the worm wriggle about in search of small invertebrates to eat. The appendages also increase the animal's surface area for taking up oxygen and disposing of wastes. Most polychaetes are marine, although a few inhabit estuaries and fresh water. Many marine species live in tubes and extend feathery appendages that trap suspended food particles. Tube-dwellers usually build their tubes out of sticky proteins secreted by glands near their mouth. Some, such as the Christmas tree worm (Figure B) bore their tubes in the limestone of coral reefs. The brightly colored spire you see in the photograph is made up of feeding appendages extending from the worm's head. The funnel-like projection on the right is a stopper that seals off the tube when the worm withdraws.

Leeches make up a third large group of annelids (Figure C). Leeches are notorious for their blood-sucking habits. However, most species are free-living carnivores that eat small invertebrates such as snails and insects. The majority of leeches inhabit fresh water, but a few terrestrial species inhabit moist vegetation in the tropics.

Until this century, blood-sucking leeches were frequently used by physicians for bloodletting, removing what was considered "bad blood" from sick patients. Some leeches have razorlike jaws that cut through the skin, and they secrete saliva containing a strong anesthetic and an anticoagulant into the wound. The anesthetic makes the bite virtually painless, and the anticoagulant keeps the blood from clotting. Leech anticoagulant is now being produced commercially by genetic engineering and may find wide use in human medicine. Tests show that it prevents blood clots that can cause heart attacks.

Leeches are still occasionally used to remove blood from bruised tissues and to help relieve swelling in fingers or toes that have been sewn back on after accidents. Blood tends to accumulate and cause swelling in a reattached finger or toe until small veins have a chance to grow back into it. Leeches are applied to remove the excess blood.

A. A sandworm, a polychaete that lives on the seafloor

B. A Christmas tree worm, a tube-building polychaete

C. A leech

Arthropods are the most numerous and widespread of all animals

Nearly a million types of segmented animals—including crayfish, lobsters, crabs, barnacles, spiders, ticks, and insects—are members of the phylum **Arthropoda**. It is estimated that the arthropod population of the world numbers about a billion billion (10^{18}) individuals! In terms of species diversity, geographical distribution, and sheer numbers, Arthropoda is the most successful phylum of animals that has ever existed.

Arthropods probably evolved from annelids or from a segmented ancestor of annelids. A number of fossils from the Cambrian period (some 550 million years ago) seem intermediate between annelids and arthropods, and may represent an evolutionary link between these two types of segmented animals.

Arthropods are equipped with jointed appendages, for which the phylum is named (Greek *arthron,* joint, and *pous,* foot). As indicated in the drawing of a lobster in Figure A, the appendages are variously adapted for walking, swimming, feeding, sensory reception, and defense. The arthropod body, including the appendages, is covered by a hard external skeleton, the **exoskeleton.** The exoskeleton consists of layers of chitin, a polysaccharide, mixed with proteins. It protects the animal and provides points of attachment for the muscles that move the appendages. The exoskeleton is thick around the head, where its main function is to house and protect the brain. It is paper-thin and flexible in many other locations, such as the joints of the legs. To grow, an arthropod must periodically shed its old exoskeleton and secrete a larger one, a process called **molting.**

In contrast to annelids, which have similar segments throughout the body, the body of most arthropods is formed of several distinct groups of segments. The lobster, for example, has three groups: head, thorax, and abdomen. (Actually, as shown, the exoskeleton of the head and thorax are partly fused, forming the "cephalothorax.") Each of the segment groups is specialized for a different function. The head bears sensory antennae, eyes, and jointed mouthparts underneath. The thorax bears a pair of defensive appendages (the pincers) and four pairs of legs for walking. The abdomen has swimming appendages.

Figures B–E illustrate representatives of four major groups of arthropods. Figure B shows a number of horseshoe crabs. The **horseshoe crab** is a "living fossil"; that is, it has survived with little change for hundreds of millions of years. It is the only surviving member of a group of spider-

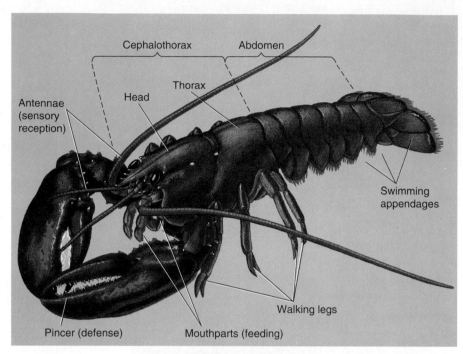

A. Structure of an arthropod, a lobster

B. Horseshoe crabs

like arthropods that were abundant in the sea some 300 million years ago. Horseshoe crabs are common on the Atlantic and Gulf coasts of the United States.

The closest living relatives of horseshoe crabs are the scorpions, spiders, ticks, and mites, collectively called **arachnids,** a second major arthropod group. Most arachnids live on land. Scorpions (Figure C, left) are nocturnal hunters. Their ancestors were among the first terrestrial carnivores, preying on herbivorous arthropods that fed, in turn, on the early land plants. Scorpions have a large pair of pincers (analogous to those of lobsters) for defense and the capture of prey. The tip of the tail bears a poisonous stinger. Scorpions eat mainly insects and spiders and will

A scorpion

A garden spider

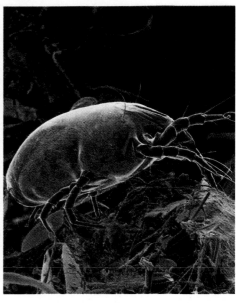

SEM 120×

A dust mite

C. Arachnids

attack people only when prodded or stepped on. Only a few species are dangerous to humans.

Spiders, a diverse group of arachnids, are usually active during the daytime, hunting insects or trapping them in webs. The common garden spider (Figure C, center) spins a complex orb web (described in the introduction to Chapter 3).

Mites make up another large group of arachnids. On the right in Figure C is a micrograph of a house-dust mite, a ubiquitous scavenger in our homes. Thousands of these microscopic animals can thrive in a few square centimeters of carpet or in one of the dust balls that form under a bed. Dust mites do not carry infectious diseases, but many people are allergic to them.

A third major group of arthropods, the **crustaceans,** are nearly all aquatic. Lobsters and crayfish are in this group, along with numerous crabs, shrimps, and barnacles. Barnacles (Figure D) are sedentary marine crustaceans that live in a limestone shell. Their jointed appendages project from the shell and capture small invertebrates and organic particles suspended in seawater. Each of the appendages and the main body of the barnacle are covered with a chitinous exoskeleton that is separate from the surrounding shell.

Millipedes and centipedes make up a fourth group of arthropods. They have similar segments over most of the body and superficially resemble annelids; however, their jointed legs identify them as arthropods. **Millipedes** (Figure E) are wormlike landlubbers that eat decaying plant matter. They have two pairs of short legs per body segment. **Centipedes** are terrestrial carnivores, with a pair of poison claws used in defense and to paralyze prey, such as cockroaches and flies. Each of their body segments bears a single pair of long legs.

The four groups of arthropods illustrated in this module account for about 170,000 living species. We turn next to a fifth major group of arthropods, the insects, whose numbers dwarf all other groups combined.

D. Crustaceans: barnacles

E. A millipede

Insects are the most diverse group of organisms

The total number of insect species is greater than the total of all other species combined. About a million insect species are known today, and researchers estimate that at least twice this many exist (mostly in tropical forests) but have not yet been discovered. Insects have been prominent on land for the last 400 million years. They have been much less successful in aquatic environments; there are only about 20,000 species in freshwater habitats and far fewer in the sea.

Despite their great diversity, insects have a number of common features. Like the grasshopper in Figure A, most have a three-part body, consisting of a head, a thorax, and an abdomen. The head usually bears a pair of sensory antennae and a pair of eyes. Several pairs of mouthparts are adapted for particular kinds of eating—for example, for biting and chewing plant material in grasshoppers; for lapping up fluids in houseflies; and for piercing skin and sucking blood in mosquitoes and other biting flies. Most adult insects have three pairs of legs and one or two pairs of wings, all borne on the thorax. Insects are the only animals other than bats and birds that have wings, and the ability to fly has been a major factor in their evolutionary success.

Systematists classify insects into about 26 orders, mostly on the basis of wing and mouthpart structure. The drawings in Figures A–G illustrate representative insects in seven of the most common orders.

A. ORDER ORTHOPTERA. The grasshopper represents this group, which contains about 20,000 species. Other orthopterans are the crickets, katydids, locusts, cockroaches, walking sticks, and praying mantises. These insects have biting and chewing mouthparts, and most species are herbivorous. Among the carnivorous species are the praying mantises, which use their forelegs to grip prey. Some orthopterans lack wings, but most have two pairs: a pair of forewings and a pair of hindwings. The forewings are often thickened.

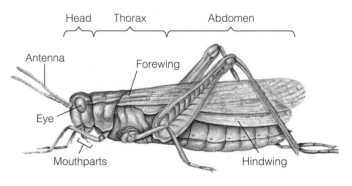

A. Insect anatomy, as seen in a grasshopper

B. ORDER ODONATA. This order includes about 5000 species of dragonflies and damselflies. These insects have two pairs of similar wings. They have biting mouthparts and are carnivorous, often catching and eating other insects on the wing. The larvae of larger species sometimes eat tadpoles and small fishes.

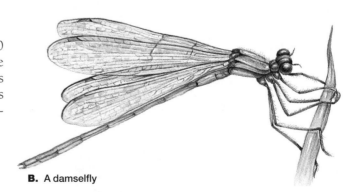

B. A damselfly

C. ORDER HEMIPTERA. Often called the true bugs, hemipterans (about 55,000 species) include bedbugs, plant bugs, stinkbugs, and water striders. They have piercing, sucking mouthparts, and most species feed on plant sap. A few, such as bedbugs, feed on blood. The true bugs have two pairs of wings, and the front half of each forewing is thickened and leathery. Water striders walk on water by taking advantage of surface tension (see Module 2.11).

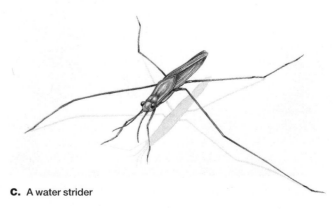

C. A water strider

D. ORDER COLEOPTERA. Beetles make up the largest order in the animal kingdom. There are about 500,000 species known worldwide and some 30,000 species in the United States. Beetles occur almost everywhere, from high mountains to the seashore, and their habitats are extremely varied. Forests, streams, ponds, soil, dung, carrion, and plant material all support their quota of species. Beetles have biting and chewing mouthparts, and most species are either carnivorous or herbivorous. Beetles vary in length from less than 1 mm to about 12 cm. They have two pairs of wings, but only the hindwings function in flight. The forewings are hardened and thickened as protective covers for the hindwings.

D. A ground beetle

E. ORDER LEPIDOPTERA. These are the moths and butterflies (about 140,000 species). They have two pairs of wings, with the hind pair smaller. Typically, the wings and body are covered by scales—the dust you find on your fingers after holding a butterfly. The mouthparts form a long drinking tube that is adapted for drinking nectar from flowers. The tube is coiled under the head when not in use. When the insect drinks nectar, the tube is uncoiled and extended deep into a flower.

E. A hawk moth

F. ORDER DIPTERA. Dipterans (about 80,000 species) are the flies, including fruit flies, houseflies, gnats, and mosquitoes. Flies have a single pair of wings; instead of hindwings, they have small, club-shaped organs called halteres, which function in maintaining balance during flight. Mosquitoes have piercing, sucking mouthparts, and the females suck blood. Most other dipterans have lapping mouthparts and feed on nectar or other liquids. Mosquitoes and other bloodsuckers may transmit disease, such as malaria or African sleeping sickness, to their animal hosts.

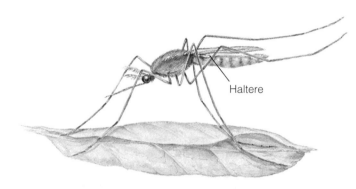

Haltere

F. A mosquito

G. ORDER HYMENOPTERA. The hymenopterans (about 90,000 species) are the ants, bees, and wasps. They have two pairs of wings, both used in flight. The hindwings are smaller and are hooked to the rear of the forewings. This arrangement improves flight efficiency by making the wings behave as single functional units. Generally, the thorax and abdomen are separated by a narrow waist, which, along with the four translucent wings, makes it easy to identify an insect as a member of this order. With chewing and sucking mouthparts, some hymenopterans are herbivorous; others are carnivorous, eating other insects. Many hymenopterans display complex behavior, including social organization (see Modules 37.1 and 37.19).

G. A paper wasp

Echinoderms have spiny skin, an endoskeleton, and a water vascular system for movement

Annelids and arthropods represent one large evolutionary branch of the animal kingdom. **Echinoderms,** such as sea stars and sea urchins, represent a very different evolutionary branch.

Echinoderms are all marine. They lack body segments, and most are radially symmetrical as adults. Both the external and the internal parts of a sea star, for instance, radiate from the center like spokes of a wheel. In contrast to the adult, the larval stage of echinoderms is bilaterally symmetrical. This tells us that echinoderms are not closely related to cnidarians or other animals that never show bilateral symmetry.

The phylum name **Echinodermata** is derived from the Greek words meaning "like a sea urchin's skin," and most echinoderms do have a rough or spiny surface. The spininess of a sea star or sea urchin comes from hard spines or plates embedded under the skin. The spines and plates are actually components of a hard internal skeleton, the **endoskeleton.**

Unique to echinoderms is the **water vascular system,** a network of water-filled canals that branch into extensions called tube feet (Figure A). A sea star pulls itself slowly over the seafloor using its suctionlike tube feet. Its mouth is centrally located on its undersurface. When a sea star encounters an oyster or clam, its favorite food, it grips the mollusk's shell with its tube feet and positions its mouth next to the narrow opening between the two valves of the shell. The sea star then pushes its stomach out through its mouth. The stomach enters the mollusk through its shell opening and proceeds to digest the soft parts of the prey (Figure B).

Sea stars and other echinoderms have strong powers of regeneration. In an attempt to limit losses of clams and oysters, commercial growers have sometimes hacked up sea stars—only to find that the sea star population increased because some of the fragments regenerated their lost parts and became whole new animals.

In contrast to sea stars, sea urchins are spherical and have no arms (Figure C). They do have five rows of tube feet that project through tiny holes in the animal's globelike case. If you look carefully, you can see the long, threadlike tube feet projecting among the spines of this purple sea urchin (both spines and tube feet are purple). Sea urchins move by pulling with their tube feet. They also have muscles that pivot their spines, and some species can walk on their spines. In contrast to sea stars, which are carnivorous, most sea urchins eat algae.

Though echinoderms have many unique features, we see evidence of their relation to other animals in their embryonic development. In sea stars, for instance, the digestive tract and coelom develop from the same parts of the gastrula as they do in animals called lancelets, which are members of the phylum Chordata. We examine lancelets next, followed by a survey of the other major groups of chordates. We return to the subject of echinoderm-chordate relationships when we summarize animal phylogeny in Module 19.23.

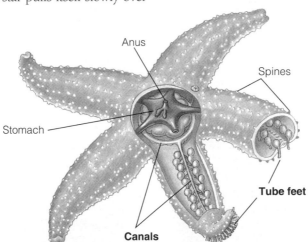

Anus

Spines

Stomach

Tube feet

Canals

A. The water vascular system of a sea star (canals and tube feet)

B. A sea star feeding on a clam

C. A sea urchin

Our own phylum, Chordata, is distinguished by four features

Four distinctive features appear in the embryos, and sometimes in the adults, of animals in the phylum **Chordata**. These four chordate hallmarks are: (1) a **dorsal, hollow nerve cord**; (2) a **notochord,** a flexible, longitudinal rod located between the digestive tract and the nerve cord; (3) **gill structures** (gill slits and structures supporting them) in the pharynx, the region of the digestive tube just behind the mouth; and (4) a **post-anal tail** (a tail posterior to the anus). Together, these features identify an animal as a chordate. (We discuss these features in human embryos in Module 27.17.)

The most diverse chordates are the **vertebrates,** animals with a segmented backbone. All other animals, including two groups of chordates, lack a backbone and are called **invertebrates.** We discuss vertebrates in Modules 19.16–19.22. But first, let's take a look at the invertebrate chordates, the lancelets and the tunicates.

Lancelets are small (5–15 cm) bladelike chordates that live in marine sands. They illustrate the four chordate features more clearly than any other members of the phylum. As shown in the drawing in Figure A, the notochord, dorsal nerve cord, numerous gill slits, and post-anal tail are all prominent. Lancelets also have segmental muscles, which flex the body from side to side, producing slow swimming movements.

A lancelet eats fine organic particles suspended in seawater. When feeding, it sits vertically in sand with its head sticking out. Water passes into its pharynx, where food particles become trapped in mucus around the gill slits. The particles remain in the digestive tract and are digested, while the water flows out through the gill slits into a large cavity in the animal. From here, the water passes back into the ocean via an opening in front of the anus. A similar feeding mechanism is believed to have been used by chordate ancestors.

Tunicates, another group of marine chordates, also feed on suspended particles (Figure B). Unlike lancelets, adult tunicates are stationary and look more like small sacs than anything we usually think of as chordates. Tunicates often adhere to rocks and boats, and they are common on coral reefs. The adults have no trace of a notochord, nerve cord, or tail, but they do have a large gill apparatus. Each of the small whitish lines inside the adult animal in the photograph surrounds a gill slit. Seawater enters through the large opening at the top of the animal. As in lancelets, the gill apparatus traps suspended particles and retains them in the digestive tract. Water passes out of the gill slits and exits through an opening on the side of the body.

In a sequel to *The Origin of Species,* Darwin remarked about the brilliant tints and elegant stripes of tunicates. He also embraced an idea that was gaining popularity at the time—that these animals have something to tell us about early chordate evolution. The tunicate larva (Figure B drawing) supports this idea. Very different from the adult, the larva is a free-swimming, tadpolelike organism that exhibits all four chordate trademarks. Darwin and several of his contemporaries thought that the tunicate larva was similar to an early stage in chordate evolution. Might an ancestral chordate living 500 million years ago have looked something like this? Today, many biologists think that an animal resembling the tunicate larva was a common ancestor of all chordates, including the vertebrates.

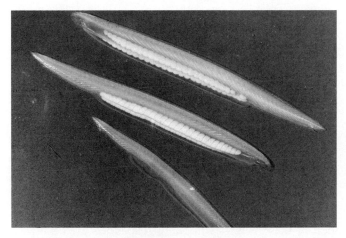

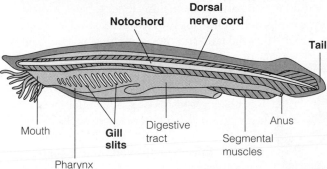

A. Lancelets

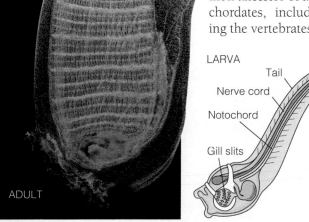

B. Tunicates

19.16 A skull and a backbone are hallmarks of vertebrates

Vertebrates (Latin *vertebra*, joint) comprise most of the phylum Chordata. The cat skeleton in this figure illustrates two distinguishing features of vertebrates: a **skull**, and a **backbone** composed of a series of segmented units called **vertebrae** (singular, *vertebra*). These skeletal elements enclose the main parts of the nervous system. The skull forms a case for the brain, and the vertebrae enclose the nerve cord. In addition to the vertebrae and skull, most vertebrates have skeletal parts supporting their body appendages (legs or fins).

The vertebrate skeleton, an endoskeleton, is made of either flexible cartilage or a combination of hard bone and cartilage, as in the cat. Bone and cartilage are mostly nonliving material, but they contain living cells that secrete the nonliving material. Because of its living cells, the endoskeleton can grow with the animal; it is unlike the arthropod's nonliving exoskeleton, which must be shed periodically. We illustrate the major groups of vertebrates (agnathans, fishes, amphibians, reptiles, birds, and mammals) in the next six modules.

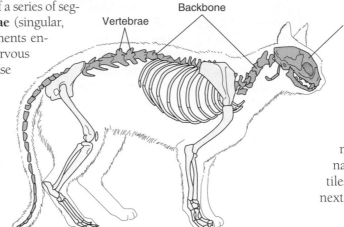

19.17 Most vertebrates have hinged jaws

Just seeing the mouth of a sea lamprey (Figure A, below) almost tells us what it can do. It bores a hole in the side of a fish and sucks its victim's blood. Introduced into the Great Lakes in the early 1900s, these voracious vertebrates multiplied rapidly, decimating fish populations as they spread. By mid-century, sea lampreys, along with an increase in commercial fishing and water pollution, had destroyed virtually all the large fish in the lakes. Since the 1960s, streams that flow into the lakes have been treated with a chemical that reduces lamprey numbers, and fish populations have been recovering.

Lampreys, about 60 species worldwide, range from about 75 mm to 1 m long. They belong to a group of primitive vertebrates called **agnathans** (Greek *a-*, without, and *gnathos*, jaw), which are distinctly different from all other vertebrates, including the fishes. Though agnathans are superficially fishlike, they lack paired fins, and their mouth lacks jaws. Fishes, amphibians, reptiles, birds, and mammals—the vast majority of living vertebrates—have jaws that consist of two parts held together by a hinge. In contrast, the lamprey's toothed, sucking disk is an unhinged, circular outgrowth of the mouth. Lampreys are more closely related to extinct jawless vertebrates than to any other living vertebrate. Larval lampreys, perhaps like those ancestors, trap suspended particles with their gills.

Where did the hinged jaws of vertebrates come from? They evolved by modification of skeletal supports of the gill slits. The first part of Figure B shows the skeletal rods supporting the gills in a hypothetical ancestor. The main function of these gills was trapping suspended food particles. The other two parts show changes that probably occurred as jaws evolved. By following the red and green structures, you can see that the jaws and their supports evolved from two pairs of skeletal rods located between gill slits that were near the mouth. Similar events occur during embryonic development in fishes today.

The oldest fossils of jawed vertebrates, an extinct group of fishes, appear in rocks about 450 million years old. Jaws were of paramount importance in vertebrate evolution, enabling vertebrates to catch and eat a wide variety of prey, instead of feeding on suspended particles.

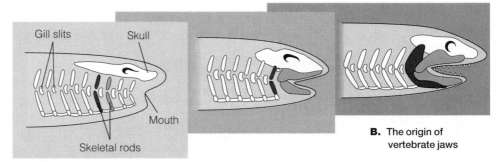

B. The origin of vertebrate jaws

A. The mouth of a lamprey (no jaws)

A. A cartilaginous fish: a shark

Fishes, the first jawed vertebrates, have been numerous and diverse for well over 400 million years. Today, they make up about 30,000 of the 45,000 species of vertebrates. Besides having hinged jaws, a fish has gills, which extract oxygen from water, and paired forefins and hindfins, which help maneuver the body when swimming. Nearly all fishes are carnivores, thriving virtually everywhere in the watery two-thirds of Earth's surface. There are two major groups of living fishes: the cartilaginous fishes (sharks, rays, and skates) and the bony fishes, a group that includes the familiar trout and goldfish.

Cartilaginous fishes have a flexible skeleton made of cartilage. Most sharks, the blacktip reef shark in Figure A, for example, are adept predators—fast swimmers with a streamlined body, acute senses, and powerful jaws. A shark does not have keen eyesight, but its sense of smell is very acute, and on its head it has special electrosensors, organs that can detect minute electrical fields produced by muscle contractions in nearby animals. Sharks also have a **lateral line system,** a row of sensory organs running along each side of the body. Sensitive to changes in water pressure, the lateral line system enables a shark to detect minor vibrations caused by animals swimming in its neighborhood.

There are fewer than 1000 living species of cartilaginous fishes, nearly all of them marine. Far more diverse and abundant in the seas and nearly every freshwater habitat are the bony fishes (some 29,000 species). The bluefin tuna and the sea horse in Figure B are two of the great variety of bony fishes. The bluefin reaches about 5 m in length

and weighs as much as 900 kg (almost 2000 lbs)—among the largest of all bony fishes. In sharp contrast, the sea horse species shown here is only about 10 cm long and weighs only a few grams.

Bony fishes have a stiff skeleton reinforced by hard calcium salts. They also have a lateral line system, a keen sense of smell, and excellent eyesight. Figure C highlights the diagnostic features of a bony fish (the labels in **bold** type). On each side of the head, a protective flap called the **operculum** (plural, *operculi*) covers a chamber housing the gills. Movement of the operculum allows the fish to breathe without swimming. (By contrast, sharks lack operculi and must swim to pass water over their gills.) Bony fishes also have a specialized organ that helps keep them buoyant—the **swim bladder,** a gas-filled sac. Some bony fishes have a connection between the swim bladder and the digestive tract that enables them to gulp air and extract oxygen from it when the dissolved oxygen level in the water gets too low. As we see next, this ability played a significant role in the invasion of land by vertebrates.

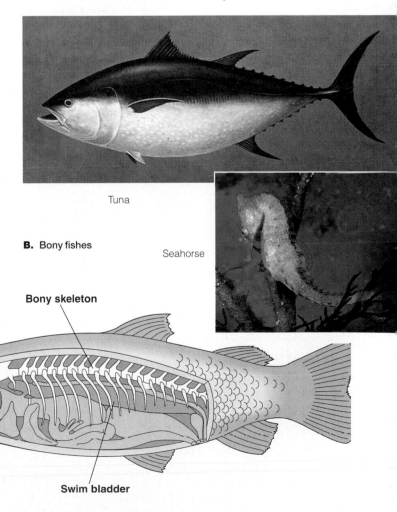

Tuna

B. Bony fishes

Seahorse

Bony skeleton

Operculum

Gills

Swim bladder

C. Diagnostic features of a bony fish

Amphibians were the first land vertebrates

In Greek, the word *amphibios* means "living a double life," and most **amphibians** exhibit a mixture of aquatic and terrestrial adaptations. Most species are tied to water because their eggs dry out quickly in the air. The frog in Figure A spends much of its time on land, but it lays its eggs in water. An egg develops into a larva called a tadpole (Figure B), a legless aquatic scavenger with gills, a lateral line system resembling that of fishes, and a long finned tail. In changing into a frog, the tadpole undergoes a radical metamorphosis. When a young frog crawls onto shore and begins life as a terrestrial hunter, it has four legs, air-breathing lungs instead of gills, a pair of external eardrums, and no lateral line system. Because of metamorphosis, amphibians truly live a double life.

The amphibians of today (frogs, toads, and salamanders) account for only about 8% of all existing vertebrate species. Historically, amphibians are of major importance, for they were the first terrestrial vertebrates.

Figure C illustrates the changes in skeletal structures that took place as vertebrates began to move onto land. The

A. An adult frog

drawing on the left shows a type of bony fish that could have given rise to the first land vertebrates. We know from fossils that such fishes lived in freshwater lakes and rivers some 400 million years ago. This is called a **lobe-finned fish**; it had strong, muscular fins supported by bones that resembled those we see in land vertebrates. With their strong fins, these fishes were able to wriggle over mud flats on drying lakeshores and riverbeds. They also had saclike lungs with connections to their digestive tract. Their lungs enabled them to breathe air during periods of drought or when lakes or rivers became stagnant.

The drawing on the right in Figure C shows an early amphibian that may have evolved from lobe-finned fishes. This creature, which lived some 350 million years ago, had limbs much like those of modern amphibians. With legs instead of fins, it could have walked on land rather than just wriggling around in mud the way its lobe-finned ancestors did.

The early amphibians encountered favorable habitats, the lush coal forests of the Carboniferous period (see Module 18.7). There they found a cornucopia of food—a great diversity of insects and other invertebrates that preceded them onto land. As a result, amphibians became so widespread and diverse that the Carboniferous period is sometimes called the Age of Amphibians. With the decline of the coal forests about 300 million years ago, the early amphibians also waned. Many lineages became extinct, some gave rise to modern amphibians, and others gave rise to the reptiles.

B. A tadpole, a frog larva

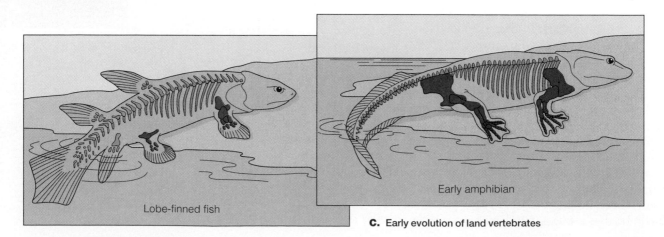

Lobe-finned fish

Early amphibian

C. Early evolution of land vertebrates

Reptiles have more terrestrial adaptations than amphibians

Reptiles (snakes, lizards, turtles, and crocodilians) have several adaptations for terrestrial living not generally found in amphibians. Reptilian skin, covered with scales water-proofed with the tough protein keratin, keeps the body from drying out. Reptiles also have eggs covered with parchmentlike shells that retain water. Unlike most amphibians, which have to lay eggs in a pool of water, a reptile such as the bull snake in Figure A can lay them in a rotten log or in sandy soil under leaves. Inside each egg, the embryo develops within a protective, fluid-filled sac called the amnion. The developing embryo is nourished by yolk until it hatches as a juvenile, ready to move about freely and feed itself. The evolution of this self-contained, drought-resistant **amniotic egg** enabled reptiles to complete their life cycles on land and sever ties with their aquatic origins. As we saw in Module 18.3, the seed played a similar role in the evolution of plants.

Reptiles in existence today are sometimes called cold-blooded animals because they do not use their metabolism to control body temperature. Nonetheless, reptiles do have ways to raise or lower their body temperature. The horned lizard of our southwestern deserts (Figure B) commonly warms up in the morning by sitting on warm rocks. Its skin keeps it from drying out as it basks in the sun. If the lizard starts getting too hot, it seeks shade. Because reptiles warm up by absorbing external heat rather than by generating much of their own, they are said to be **ectothermic** (Greek *ektos*, outside, and *therme*, heat), a term more appropriate than cold-blooded.

Like the amphibians from which they evolved, reptiles were once much more prominent than they are today. Following the decline of amphibians, reptilian lineages expanded rapidly, creating a dynasty that lasted until about 65 million years ago. During that Age of Reptiles, dinosaurs, the most diverse group, included the largest animals ever to inhabit land. Some were "gentle giants" like the large dinosaur in Figure C, lumbering about while browsing vegetation. Others, like the 3-m-long *Deinonychus* (Greek for "terrible claw"), were voracious carnivores that ran on two legs. Probably a pack hunter, *Deinonychus* may have used its sickle-shaped claws to slash at large prey. Unlike modern reptiles, *Deinonychus* and many other dinosaurs may have been **endothermic**, using heat generated by metabolism to maintain a warm, constant body temperature.

The Age of Reptiles began to wane about 70 million years ago. As we discussed in Module 16.7, periods of mass extinction killed off all but one of the dinosaur lineages by about 65 million years ago. That lineage survives today as the birds.

A. A bull snake laying eggs

B. A horned lizard

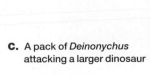

C. A pack of *Deinonychus* attacking a larger dinosaur

Birds share many features with their reptilian ancestors

A **bird** may seem very unlike a reptile, but there are more similarities than differences between these two groups of vertebrates. Fossil evidence indicates that birds evolved from a lineage of two-legged dinosaurs about 150–200 million years ago. Birds evidently inherited amniotic eggs, scales on their legs, toenails containing keratin, and their general body form from their reptilian ancestors. Even feathers, a hallmark of birds, were derived from reptilian scales.

Figure A is an artist's reconstruction based on a 150-million-year-old fossil clearly identifiable as a bird. Called *Archaeopteryx* (Greek *archaios*, ancient, and *pteryx*, wing), this animal lived near tropical lagoons in central Europe. Feather imprints in its fossils identify the archaeopteryx as a bird; otherwise, it was more like some small bipedal dinosaurs that were its contemporaries.

Feathers shape bird wings into airfoils, providing lift and maneuverability in the air (see Figure 30.1E). With flight feathers similar to those of modern birds, an archaeopteryx could have flown by flapping its wings. However, it had a fairly heavy body and could not have been a strong flyer. It may have relied mainly on gliding from trees, perhaps the earliest mode of flying in birds. Recently, bird fossils dating from about 125 million years ago were discovered in Spain and China. These birds had flight adaptations intermediate between the archaeopteryx and modern birds, indicating that they may have been on a direct lineage that led to modern birds.

A. Birds of the extinct genus *Archaeopteryx*

Almost every part of the body of most birds is adapted to enhance flight. Many features help reduce weight for flight: Modern birds lack teeth, their tail is supported by only a few small vertebrae, their wings lack claws, their feathers have hollow shafts, and some of their bones are hollow and contain air sacs. Providing power for flight are large breast (flight) muscles, which are anchored to a keel-like breastbone. Most of what we call white meat on a turkey or chicken are the flight muscles.

Flying requires a great amount of energy, and modern birds have a high rate of metabolism. Birds also are endothermic, with their feathers and body fat helping maintain body temperature by providing insulation. Supporting their high metabolic rate, birds have a highly efficient circulatory system, and their lungs are even more efficient at extracting oxygen from the air than the lungs of mammals.

With their wings driven by powerful flight muscles, modern birds are masterly flyers. Some species, such as the turkey vulture in Figure B, have wings adapted for soaring on air currents, and they flap their wings only occasionally. Others, such as hummingbirds, are adapted for maneuvering but must flap almost continuously to stay aloft. Flight ability is typical of birds, but there are a few flightless species, including the ostrich and the emu of Australia (Figure C). Modern flightless birds evolved from flying ancestors.

B. A turkey vulture

C. An emu, a flightless bird

Mammals evolved from reptiles about 225 million years ago, long before there were any dinosaurs. During the peak of the Age of Reptiles, there were a number of mouse-sized, nocturnal mammals, which lived on a diet of insects. With the downfall of the dinosaurs, mammalian lineages exploded into the vacant habitats. Most mammals are terrestrial, but there are nearly 1000 species of winged mammals, the bats, and about 80 species of totally aquatic forms, the dolphins, porpoises, and whales. The blue whale, an endangered species, grows to lengths of nearly 30 m and is the largest animal that has ever existed.

Like birds, mammals are endothermic, with a high rate of metabolism. Two features—hair and mammary glands that produce milk to nourish young—are mammalian hallmarks. Like feathers, hair is made of keratin. Its main function is to insulate the body and help maintain a warm, constant body temperature. There are three major groups of mammals: the monotremes, the marsupials, and the placentals.

The duck-billed platypus (Figure A) is one of only three existing species of **monotremes,** the egg-laying mammals. The platypus lives along rivers in eastern Australia and on the nearby island of Tasmania. It eats mainly small shrimps and aquatic insects. The female usually lays two eggs and incubates them in a leaf nest. After hatching, the young nurse by licking up milk secreted onto the mother's fur.

Marsupials, the so-called pouched mammals, give birth to tiny, embryonic offspring that complete development while attached to nipples on the abdomen of the mother. The nursing young are usually housed in an external pouch, called the **marsupium,** on the mother's abdomen. Of about 250 species of marsupials, only one, the American opossum (Figure B), occurs in the Western Hemisphere north of Mexico. Most marsupials live in Australia, on neighboring islands, and in Central and South America. Australia has been a marsupial sanctuary for much of the past 60 million years, and marsupials have diversified extensively there, filling terrestrial habitats that on other continents are occupied by placental mammals.

Placentals are mammals whose embryos are nurtured inside the mother by an organ called the placenta. They make up almost 95% of the 4500 species of living mammals. The placenta consists of both embryonic and maternal tissues, and it joins the embryo to the mother within the mother's uterus. The embryo is nurtured by maternal blood that flows close to the embryonic blood system in the placenta. The placenta is the reddish portion of the afterbirth still clinging to the newborn zebra in Figure C. The large silvery membrane is the amniotic sac, which contained the developing embryo in a bath of protective amniotic fluid.

Humans are placental mammals that belong to the order Primates, along with monkeys and apes. We discussed our phylogenetic relationships with other primates in Module 16.15, and we will examine human evolution in more detail in the final chapter of this book.

A. Monotreme: a duck-billed platypus

B. Marsupial: an American opossum with young

C. Placental: a zebra and newborn

19.23

A phylogenetic tree gives animal diversity an evolutionary perspective

With the mammals, we conclude our journey through the animal kingdom. We have looked at examples of an immense diversity of animals—representatives of about one-third of the known animal phyla. The table at the right recaps the phyla we discussed and key animals within those phyla. The phylogenetic tree on the facing page summarizes the evolutionary threads that tie all this diversity together.

At the bottom of the tree are the colonial protists that probably gave rise to the earliest animals. The position of phylum Porifera (the sponges) on the lowest branch of the tree reflects the relatively simple body structure of sponges and their marked similarity to protists. Sponges consist of aggregrates of only a few types of cells, and they lack tissues. Also, they have larvae that are unlike those of any other group of animals. Whereas many biologists contend that all animals arose from a common stock of protists, others postulate that because sponges are so different from other animals, they must have arisen from a separate stock of colonial protists.

Above the sponge lineage, the main trunk of the animal kingdom represents ancestral animals whose bodies consisted of specialized cells organized into tissues—a more complex level of organization than sponges have. These "tissue-level" animals gave rise to two distinct lineages, represented by the second branching point on the phylogenetic tree. In one lineage, radial symmetry evolved as the adult body form. Radial animals of today include the hydras, jellyfishes, sea anemones, and corals of phylum Cnidaria. The other lineage, the mainstream of the animal kingdom, gave rise to the bilateral animals—those that swim, crawl, run, or fly in a head-first direction.

Bilateral animals with the least complex body construction are the flatworms (phylum Platyhelminthes). Judging from their simplicity, flatworms diverged from the main trunk of bilateral animals soon after bilateral symmetry evolved. Recall that a typical flatworm, in common with cnidarians, lacks a body cavity and has an incomplete digestive tract.

As we discussed in Module 19.7, body cavities played a signficant role in animal history. Above the flatworm side-branch on the tree, bilateral animal phyla divide on the basis of the type of body cavity they have. A body cavity called a pseudocoelom makes the roundworms (phylum Nematoda) fundamentally different from other phyla, and they are placed on another side-branch of the tree.

After the nematode branch, the main trunk of the phylogenetic tree continues with animals that have a coelom, a body cavity lined with a sheet of tissue that wraps around the digestive tract and other internal organs. The coelom was a key adaptation, for it provided a way to stabilize body organs while still allowing them to move within the body.

Some Major Animal Phyla: A Review	
Phylum	**Representative Animals**
Porifera	Sponges
Cnidaria	Hydras, jellyfishes, sea anemones, corals
Platyhelminthes	Flatworms (planarians, flukes, tapeworms)
Nematoda	Roundworms
Mollusca	Snails, slugs, bivalves (such as clams), squids, octopuses
Annelida	Segmented worms (earthworms, polychaetes, leeches)
Arthropoda	Horseshoe crabs, arachnids such as spiders, crustaceans such as lobsters, insects, centipedes, millipedes
Echinodermata	Sea stars, sea urchins
Chordata	Lancelets, tunicates, lampreys, fishes, amphibians, reptiles, birds, mammals

An animal's embryological development often provides clues to its evolutionary history (phylogeny). For example, animals with a coelom sort out into two large groups, according to the way their body cavity develops. These two groups are represented on the upper trunk of the tree by the split into two limbs. On one limb, the phyla Echinodermata and Chordata have a coelom that forms from hollow outgrowths of the digestive tube of the early embryo. On the other limb (phyla Arthropoda, Annelida, and Mollusca), the coelom develops from solid masses of cells that form between the digestive tube and the embryonic body wall. Other embryological evidence supports the distinction between these two groups of animal phyla.

We have yet to mention body segmentation, which we discussed earlier as one of the major evolutionary developments in the animal kingdom (see Module 19.10). Segmentation was an important evolutionary development affecting three major phyla: Chordata, Arthropoda, and Annelida. Embryological evidence places the echinoderms in the same lineage as the chordates, but echinoderms are not segmented. This indicates that segmentation as we see it in the body muscles and skeleton of vertebrates did not evolve until after the echinoderms and chordates diverged. Accordingly, the echinoderms are placed on a branch of the phylogenetic tree separate from the chordates. Similarly, on the part of the tree that has animals with coeloms formed from cell masses, there are two phyla with segmentation (annelids and arthropods) and one without (mollusks). Segmentation in annelids and arthropods most likely evolved after mollusks diverged as a phylum. Thus, segmentation arose twice, as two completely independent events in the animal kingdom. Chordate segmentation evolved in an immediate ancestor of the chordates, and segmentation in annelids and arthropods arose independently in the annelid-arthropod lineage.

Our phylogenetic tree linking the major phyla of animals is based on what appear to be fundamental structural features in adult animals and on patterns of embryological development. Built on this circumstantial evidence, the tree's branching is only a hypothetical replica of animal history. Like other phylogenetic trees, this one serves mainly to stimulate research and discussion, and it is subject to revision as new information is acquired. Whether or not some of its branch points are revised, however, the tree's overall message remains the same: The animal kingdom's great diversity arose through the common process of evolution, and all animals exhibit features reminiscent of their evolutionary history.

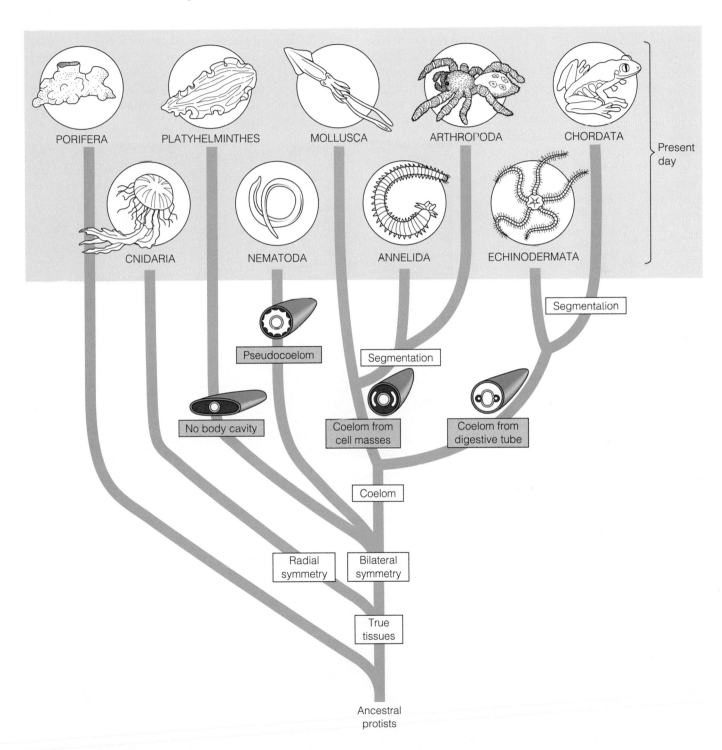

Phylogenetic tree of the animal kingdom

In this photograph, we see an aerial view of a portion of the Great Barrier Reef, which parallels the northeast Australian coast for nearly 2000 km. This reef is a major feature of our planet's surface, visible from satellites orbiting hundreds of kilometers above the Earth. Built mainly by tiny coral animals and red algae, the reef is a showplace of life's diversity. It is also a dramatic measure of life's capacity to alter Earth's surface. For all of the billions of years it has existed, life has had a profound impact on the globe. From the time of the earliest life forms to the bacteria, protists, fungi, plants, and animals of today, every part of Earth's surface has been extensively changed by living organisms.

Biological change and change in the physical environment are closely connected, and to a large extent the word "change" summarizes this unit on evolution and diversity. We have seen that species change in becoming adapted by natural selection to a changing environment. Natural selection is an editing mechanism, shaping gene pools by selecting phenotypes best suited to the environment while eliminating others. Evolution—descent with modification, as Darwin put it—is thus a restructuring process, changing species and fashioning new ones from preexisting ones as environments change and gene pools become isolated. The connection between biological change and environmental change has been crucially important in the history of life and of planet Earth.

We have documented the role of change in shaping the vast diversity of life. We have also chronicled the role of chance. Chance has affected the evolutionary process in the generation of genetic diversity through mutation. Chance has also played a role at every major milestone in the history of life. Before life began, over 3.5 billion years ago, the chance union of certain small organic molecules ignited a chain of events that led to the first genes. Much later—about 65 million years ago—a chance collision between Earth and an asteroid may have caused mass extinctions. What are the odds for and against asteroid collisions? How would the evolution of biological diversity have been different if the dinosaurs had not become extinct? One of the great wonders of our existence and of life itself is that it has all arisen through a combination of evolutionary processes and chance events.

The Great Barrier Reef

Begin your review by rereading the module headings and Module 19.23 and scanning the figures before proceeding to the Chapter Summary and questions.

Chapter Summary

Introduction–19.2   Over two-thirds of all known organisms are animals—eukaryotic, multicellular heterotrophs that lack cell walls and digest their food internally. Most animals are diploid except for haploid eggs and sperm. The animal zygote grows into a hollow blastula, which folds inward to form a gastrula. The animal may go through a larval stage, which metamorphoses into an adult. Animals probably evolved from colonial protists, whose cells gradually became more specialized and layered.

19.3   Of about 30 phyla of animals, sponges are among the simplest. Many sponges are radially symmetrical; their parts are arranged around a central axis. A lining of flagellated choanocytes creates a current and collects food particles from the water passing though the porous, saclike sponge body. Many biologists think sponges arose on a lineage separate from other animals.

19.4   Cnidarians are the simplest animals with tissues. Cnidarians exist in two radially symmetrical body forms: cylindrical polyps (such as coral animals) and umbrella-shaped medusas (jellyfishes). Cnidocytes on their tentacles sting prey, and the tentacles, controlled by nerves, push the food through the mouth into a gastrovascular cavity where it is digested and distributed.

19.5–19.6   Most animals are bilaterally symmetrical, having mirror-image right and left sides. In contrast to sponges and cnidarians, a bilaterally symmetrical animal has a head with sensory structures, and it moves head-first through its environment. Flatworms, such as the free-living planarian, are the simplest bilaterally symmetrical animals. The planarian has a simple nervous system consisting of a brain and sense organs in its head, nerve cords, and branching nerves. As in cnidarians, its mouth is the only opening for its gastrovascular cavity. Flukes and tapeworms are parasitic flatworms with complex life cycles.

19.7–19.8 Most animals have a body cavity, a fluid-filled space between the digestive tract and body wall. The body cavity aids in movement, cushions internal organs, and may stiffen the body and help in circulation. Roundworms (nematodes) have a pseudocoelom, a body cavity in contact with the digestive tract. Like most animals, they also possess a complete digestive tract, a tube with both a mouth and an anus. Many nematodes are free-living; others are plant or animal parasites.

19.9 Mollusks are a large and diverse phylum that includes gastropods (such as snails and slugs), bivalves (such as clams and scallops), and cephalopods (such as octopuses and squids). All have a muscular foot and a mantle, which secretes the shell. Mollusks have a coelom, a body cavity lined by a sheet of tissue that encloses the digestive tract. They also have a true circulatory system. Many mollusks feed with a rasping radula.

19.10–19.11 The segmented bodies of annelids, such as earthworms, give them added mobility for burrowing and swimming. An earthworm eats its way through soil. Polychaetes, the largest group of annelids, search for prey on the seafloor, or live in tubes and filter food particles from the water. Their appendages aid in locomotion and oxygen uptake. Most leeches, the third group of annelids, are free-living carnivores, but some suck blood.

19.12–19.13 Arthropods are segmented animals with exoskeletons and jointed appendages. In terms of numbers, distribution, and diversity, they are the most successful phylum of animals. There are several groups of arthropods. Horseshoe crabs are ancient marine arthropods. Their closest relatives are the arachnids: scorpions, spiders, ticks, and mites. Most arachnids live on land, and most are carnivorous. Crustaceans, such as lobsters, crabs, and barnacles, are nearly all aquatic. Another group is the millipedes and centipedes. Insects are the most numerous and successful arthropods. They have a three-part body (consisting of head, thorax, and abdomen) and three pairs of legs, and most have wings. Insects are grouped into about 26 orders, primarily on the basis of their wings and mouthparts. The largest orders are beetles, moths and butterflies, ants and their relatives, and flies.

19.14 Sea stars and sea urchins are echinoderms, radially symmetrical marine creatures with spiny skins and endoskeletons. Their unique water vascular system moves an array of suction-cuplike tube feet. Sea stars are carnivores, and sea urchins eat algae.

19.15–19.18 Chordates are segmented animals with a dorsal hollow nerve cord, a stiff notochord, gill structures behind the mouth, and a post-anal tail. The simplest chordates are lancelets and tunicates, marine animals that feed on suspended particles. Most chordates are vertebrates, with endoskeletons that include a skull and a backbone composed of vertebrae. All vertebrates except agnathans such as lampreys have hinged jaws and paired fins or legs. These include two kinds of fishes: cartilaginous fishes, such as sharks, and the more diverse bony fishes, such as tuna and trout. Bony fishes have more mobile fins, operculi that move water over the gills, and a buoyant swim bladder.

19.19–19.20 Amphibians, represented today by frogs, toads, and salamanders, were the first terrestrial vertebrates. Their limbs allow them to move on land, but amphibian embryos and larvae still must develop in water. Waterproof scales and a shelled amniotic egg

enable reptiles (snakes, lizards, turtles, crocodiles, and extinct dinosaurs) to live on dry land. Modern reptiles are ectothermic, warming their bodies by absorbing heat from their environment. Some dinosaurs were probably endothermic, using metabolic heat to maintain a warm, constant temperature.

19.21 Birds share many characteristics with their reptilian ancestors, including amniotic eggs, scales, and overall body form. Birds have wings, feathers, endothermic metabolism, and many other adaptations (such as hollow bones and a highly efficient circulatory system) related to flight.

19.22 Mammals are also descended from reptiles. Like birds, they are endothermic. Two unique characteristics of mammals are hair, which insulates their bodies, and mammary glands, which produce milk that nourishes their young. A few mammals, the monotremes, lay eggs, but most give birth to live young. The tiny offspring of marsupials, such as opossums, complete their development attached to their mothers' nipples, usually in a pouch. Most mammals, from horses to humans, are placentals. The young of placental mammals develop inside the body of the mother, nurtured by an organ called the placenta.

19.23 A phylogenetic tree traces the evolutionary relationships among the animal phyla. These relationships are based on fundamental structural features of adult animals and patterns of embryological development. Phylogenetic trees reflect the effects of chance events and evolutionary processes on the history of life.

Testing Your Knowledge

Multiple Choice

1. Jon found an organism in a pond, and he thinks it's a freshwater sponge. His friend Liz thinks it looks more like an aquatic fungus. How can they decide whether it is an animal or a fungus?

   a. see if it can swim
   b. figure out whether it is autotrophic or heterotrophic
   c. see if it is a eukaryote or prokaryote
   d. look for cell walls under a microscope
   e. determine whether it is unicellular or multicellular

2. Reptiles are much more extensively adapted to life on land than amphibians because reptiles

   a. have a complete digestive tract
   b. lay shelled eggs
   c. are endothermic
   d. have legs
   e. go through a larval stage

3. Which of the following most clearly demonstrates the evolutionary relationship between annelids and arthropods?

   a. a complete digestive tract
   b. an exoskeleton
   c. body segmentation
   d. jointed appendages
   e. radial symmetry

4. A small group of worms classified in the phylum Gnathostomulida (not discussed in this chapter) live between sand grains on ocean beaches. These animals are unsegmented and lack a body cavity, but some have a complete digestive tract. These characteristics suggest that gnathostomulid worms might branch from the main trunk of the animal phylogenetic tree between

   a. flatworms and nematodes
   b. annelids and arthropods
   c. mollusks and annelids
   d. cnidarians and flatworms
   e. nematodes and annelids

5. A lamprey, a shark, a lizard, and a rabbit share all the following characteristics except

    a. gill structures in the embryo or adult
    b. vertebrae
    c. hinged jaws
    d. a dorsal hollow nerve cord
    e. a post-anal tail

6. Which of the following categories includes the largest number of species? (*Explain your answer.*)

    a. invertebrates
    b. chordates
    c. arthropods
    d. insects
    e. vertebrates

True/False (*Change false statements to make them true.*)

1. Echinoderms capture prey with their stinging tentacles.

2. Cephalopods may be the most intelligent animals without backbones.

3. Mammals and birds both evolved from reptiles.

4. Nematodes are cylindrical, unsegmented worms.

5. A sea urchin's water vascular system transports food to its cells.

6. Mammals are separated into three major groups on the basis of diet.

7. Eggs and sperm are the only haploid stage in most animal life cycles.

8. Fossils show how animals evolved from colonial protists.

Matching

1. Include the vertebrates	**a.** annelids
2. Medusa and polyp body forms	**b.** nematodes
3. The simplest bilateral animals	**c.** sponges
4. May have a separate origin from other animals	**d.** arthropods
5. Earthworms, polychaetes, and leeches	**e.** flatworms
6. Largest phylum of all	**f.** cnidarians
7. Closest relatives of chordates	**g.** mollusks
8. Body cavity is a pseudocoelom	**h.** echinoderms
9. Have a muscular foot and a mantle	**i.** chordates

Describing, Comparing, and Explaining

1. Birds are a lot like their reptile ancestors, but their bodies are highly modified for flight. What characteristics do birds share with reptiles? What are their adaptations for flight?

2. Compare the structure of a planarian (a flatworm) and an earthworm with regard to the following: digestive tract, body cavity, and segmentation.

3. Name two phyla of animals that are radially symmetrical and two that are bilaterally symmetrical. How do the overall lifestyles of radial and bilateral animals differ?

4. One of the key characteristics of arthropods is their jointed appendages. Describe four functions of these appendages, in four different arthropods.

Thinking Critically

1. Construction of a dam and irrigation canals in an African country has allowed farmers to increase the amount of food they can grow. In the past, crops were planted only after spring floods; the fields were too dry the rest of the year. Now fields can be watered year-round. Improvement in crop yield has had an unexpected cost— a tremendous increase in the incidence of schistosomiasis, or blood fluke disease. Look at the blood fluke life cycle in Figure 19.6B, and imagine that your Peace Corps assignment is to help local health officials control the disease. Why do you think the irrigation project increased the incidence of schistosomiasis? It is difficult and expensive to control the disease with drugs. Suggest three other methods that could be tried to prevent people from becoming infected.

2. A marine biologist has dredged up a previously unknown animal from the seafloor. Describe some of the important characteristics she should look at to determine the animal phylum to which the creature should be assigned.

3. On the basis of relationships described in the text, sketch a phylogenetic tree like the one in Figure 19.23 for the seven major groups of vertebrates: mammals, reptiles, birds, amphibians, bony fishes, cartilaginous fishes, and agnathans. Use the ancestral group for the trunk of the tree. Show which groups branched from the ancestral group and which groups branched from them. (*Hint:* Bony and cartilaginous fishes are separate branches that probably arose at about the same time.)

Science, Technology, and Society

1. Coral reefs harbor a greater diversity of animals than any other environment in the sea. Australia's Great Barrier Reef has been protected as a marine reserve and is a mecca for scientists and nature enthusiasts. In other parts of the world, such as Indonesia and the Philippines, coral reefs are in danger. Many reefs have been depleted of fish, and runoff from the shore has covered coral heads with sediment. Nearly all the changes in the reefs can be traced back to human activities. What kinds of activities do you think might be contributing to the decline of the reefs? What are some reasons to be concerned about this decline? Do you think the situation is likely to improve or worsen in the future? Why? What might the local people do to halt the decline? Should the developed countries help? Why or why not?

2. There are 90 million acres in the U.S. National Wildlife Refuge System, representing most of the habitats within the United States and protecting thousands of animal species. Many refuges permit logging, grazing, farming, mineral exploration, and recreational activities. Each year, 400,000 animals are legally taken from refuges by hunters and trappers. There has recently been pressure on Congress to permit oil and gas exploration in the Arctic National Wildlife Refuge of Alaska. What kinds of human activities do you think should be permitted on wildlife refuges, what kinds not permitted, and under what circumstances? Should refuges be sanctuaries where animals are affected as little as possible by humans? Give reasons for your answers.

3. A lot of money and attention have been directed at saving a handful of endangered animals, such as the black-footed ferret and California condor. Millions of dollars have been spent in efforts to maintain and breed these two species— more than have been spent on all endangered invertebrates, reptiles, and plants put together. Condors and ferrets have recently been returned to the wild, but it is questionable whether they can withstand the human encroachments that endangered them in the first place. Should we try to save every endangered species? Why or why not? What criteria should guide us in deciding which ones to save first? What are some methods we might use to save them?

Unit 4
Animals:
Form and Function

Unifying Concepts of Animal Structure and Function

20

In the frigid South Atlantic off the coast of Antarctica lies tiny Bird Island, part of which you can see at the left. The bird, a wandering albatross, is here for just a brief time. Its scientific name, *Diomedea exulans*, means "exiled warrior" in Greek, and, indeed, the albatross spends most of its time far from land, alone on the open ocean.

Only a few birds can live at sea, but the albatross is a model of fitness for this environment. It even drinks seawater, which is much saltier than its body fluids. Few animals can tolerate such salty liquid, because the salt draws water out of their tissues, and they become severely dehydrated. An albatross can thrive on salt water because it has special salt-excreting glands in its nostrils. The glands dispose of excess salts, allowing the bird to eat salty fish and squid and drink all it needs without ever visiting land.

Along with its remarkable salt tolerance, the wandering albatross has extraordinary flight abilities specially suited to its seafaring life. Its wing span is greater than that of any other bird—about 3.6 meters (nearly 12 feet)! Most birds have to flap their wings to stay up in the air very long, but the albatross's long wings provide so much lift that the bird can stay aloft for hours by gliding almost effortlessly up and down on wind currents. Albatrosses frequent areas where steady winds blow over the ocean. The wandering albatross spends most of its time in an area of high winds and rough seas, between about 40° and 50° latitude in the Southern Hemisphere. The winds carry it from west to east, and in a year's time, an albatross may circle the globe, seeing little land except New Zealand and the tip of South America.

The drawing below illustrates how an albatross uses the energy of ocean winds to stay aloft. Because of the friction between air and water, wind speed is lowest where the air meets the ocean. On the open sea, where there are no obstructions, the wind speed regularly increases from the surface up to an altitude of about 15 meters. An albatross takes full advantage of these conditions. Almost effortlessly, it glides downward in the same direction as the wind, picking up momentum as it descends. Near the ocean surface, it turns sharply into the wind and uses its downward momentum to soar upward, the way a bicyclist uses downhill momentum to coast partway up the next hill. As the bird rises, the increasing wind helps lift it back up to its starting altitude. An albatross can glide up and down like this over and over—with just a few turns of its body—as long as the wind holds.

The albatross's special flight ability and salt tolerance are functions that result from special structural adaptations of its body. The connection between structure and function is a basic concept of biology, and this chapter is an introduction to how the concept applies to animals.

Function fits structure in the animal body

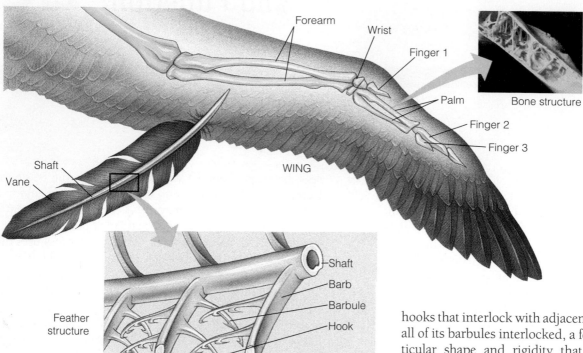

Forearm
Wrist
Finger 1
Palm
Bone structure
Finger 2
Finger 3
Shaft
Vane
WING
Shaft
Barb
Barbule
Hook
Feather
structure

We can approach the study of an animal in several ways. An anatomist studying the albatross, for instance, would focus on structure, perhaps on the shape and size of the wing or the anatomy of its muscles, bones, feathers, and skin. A physiologist (a biologist who studies the function of body parts) might study the mechanics of flight or the action of muscles in holding the wings out for hours on end. Despite their different approaches, the anatomist and the physiologist are both working toward a better understanding of the connection between structure and function—that is, how structural adaptations give the albatross its remarkable ability to exploit the energy of ocean winds.

Let's look at some of the parts of an albatross and see how they illustrate the correlation between structure and function. Take the feathers of the wing, for example. Feathers give the wing its shape and insulate it against the chill of the strong sea wind, without adding much weight to the body. They remain dry because they are lightly coated with oil, and they also trap air, which helps the bird float on the water.

The functions of feathers result from their unique structure. Produced by special pits in the bird's skin, feathers consist entirely of dead material, mainly the protein keratin. A flight feather, like the one enlarged in the figure above, has a hollow keratin shaft that provides trunklike support with minimum weight. A row of small flat rods called barbs extend from both sides of the shaft, forming the vane of the feather. Still finer rods called barbules extend from the sides of the barbs. Each barbule has tiny

hooks that interlock with adjacent barbules. With all of its barbules interlocked, a feather has a particular shape and rigidity that support flight. When barbules become detached, these features are lost, and the bird's flight ability is impaired. When a bird preens, it draws its feathers through its beak, hooking the barbules back together in a zipperlike manner.

The muscles and bones in a bird wing also demonstrate the relationship between structure and function. Muscles provide power, and the bones provide support for flight. The flight muscles are situated on the breast and around the base of the wings, keeping most of the weight off the wings and helping the bird maintain balance in flight.

The bones in a bird wing are homologous with those in the human arm, but the number of bones in the wing diminished as birds evolved. The bird wing has only three fingers (numbered in the figure), and only the middle one (finger 2) has a complete set of bones. The bird's wrist and palm also have fewer bones than ours do. This adaptation helps make the wing lighter but less mobile than the human arm. The reduced mobility stabilizes the wing and makes it better able to function as a unit in flight. The photograph of the cut-open bone at the upper right of the figure illustrates another significant adaptation in the bird skeleton. Many bones are hollow but reinforced internally with trusses similar to those used in airplane wings. This structure provides maximum strength with minimum weight, the ideal combination for flight.

The ability to fly thus results from special structural features of bones, muscles, feathers, and skin—arranged in a particular way to form the bird wing. It is one illustration of the universal theme of a specific function correlated with, and emerging from, specific structures.

Structure in the living world is organized at several hierarchical levels. For instance, at one level, a feather consists of molecules of keratin. At another level, resulting from different arrangements of the keratin molecules, a feather consists of a shaft, barbs, and barbules. At still another level, the way the barbs and barbules fit together forms the feather's vane.

The same hierarchy that exists in an albatross exists in all animal bodies, including our own. The figure here illustrates structural hierarchy in a zebra. Part 1 shows a single muscle cell in the zebra's heart. This cell's main function is to contract and relax in a coordinated way with the other heart muscle cells. Each heart muscle cell is branched, so that it can connect with several other similar cells. The multiple connections allow the cells to cooperate, coordinating their contractions, so the heart can pump blood effectively. Many thousands of these cells make up a tissue (part 2), the second level of structure and function. Heart muscle tissue is the main component of the heart walls.

Part 3, the heart itself, illustrates the organ level of the hierarchy. As an organ, the heart is made up of several types of tissue, including muscle tissue, nervous tissue, connective tissue, and other types of tissue. Part 4 shows the cardiovascular system, the organ system of which the heart is a part. Organ systems have multiple parts. The other parts of the cardiovascular system are the blood vessels: arteries, veins, and capillaries. All the parts combine to carry out the function of this organ system, transporting blood throughout the body.

In part 5, the zebra itself forms the final level of this hierarchy: the organism—in this case, a large mammal. The whole animal consists of a number of organ systems, each specialized for certain functions and all functioning together as an integrated, cooperative unit.

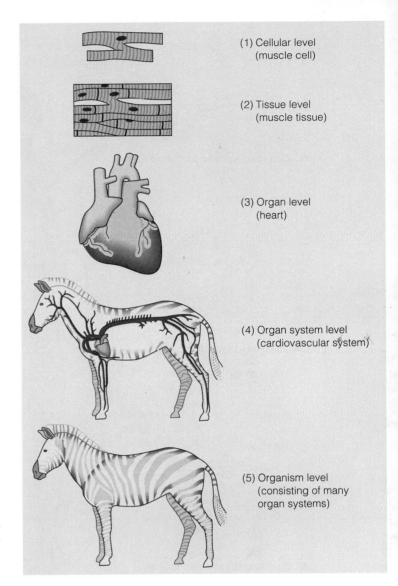

(1) Cellular level (muscle cell)

(2) Tissue level (muscle tissue)

(3) Organ level (heart)

(4) Organ system level (cardiovascular system)

(5) Organism level (consisting of many organ systems)

Tissues are groups of cells **20.3**
with a common structure and function

A **tissue** is a cooperative unit of many very similar cells that perform a specific function. Most of the cells in multicellular organisms, including animals, are organized into tissues. The cells composing a tissue are specialized; they have a particular structure that enables them to perform a specific task. As we saw in the last module, each of the heart's muscle cells has branches that connect to several other muscle cells. The branches help the cells contract in a coordinated manner.

The term *tissue* is from a Latin word meaning "weave," and some tissues resemble woven cloth in that they consist of a meshwork of nonliving fibers surrounding living cells. Other tissues are held together by a sticky glue that coats the cells or by special junctions between adjacent plasma membranes (see Module 4.19). An animal has four major categories of tissue: epithelial tissue, connective tissue, muscle tissue, and nervous tissue. We examine each of these separately in the next four modules.

Epithelial tissue covers and lines the body

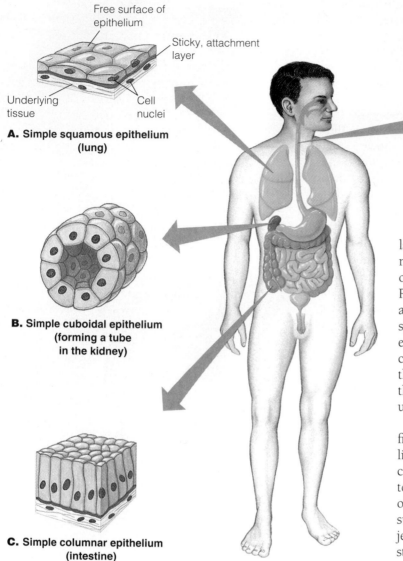

Free surface of epithelium

Sticky, attachment layer

Underlying tissue

Cell nuclei

A. Simple squamous epithelium (lung)

B. Simple cuboidal epithelium (forming a tube in the kidney)

C. Simple columnar epithelium (intestine)

D. Stratified squamous epithelium (esophagus)

layers. The shape of the cells may be squamous (like floor tiles), cuboidal (like dice), or columnar (like bricks or cones on end). Parts A, B, and C of the figure here show examples of simple epithelia with the three cell shapes; part D shows stratified squamous epithelium. In each case, the light purple color identifies the epithelium itself, white the underlying tissue, and dark blue the thin, sticky layer that attaches the epithelium to the underlying tissue.

The structure of each type of epithelium fits its function. Stratified squamous epithelium regenerates rapidly by division of the cells at its attached surface. New cells move toward the free surface as older cells slough off. Stratified squamous epithelium is well suited for covering and lining surfaces subject to abrasion. The esophagus, for instance, can be abraded by rough food. Our epidermis is also stratified squamous epithelium, with a thick layer of dead cells at the free surface. In contrast, simple squamous epithelium is thin and leaky, suitable for exchanging materials by diffusion. We find it in such places as the linings of our lungs and blood vessels.

Cuboidal epithelium and columnar epithelium both have cells with a relatively large amount of cytoplasm, where secretory products are made, and a large surface area, where substances are secreted or absorbed. Lining our digestive tract and air tubes, these cells form a smooth, moist epithelium called a **mucous membrane.** The mucous membrane of our air tubes helps keep our lungs clean by trapping dust, pollen, and other particles in its secretions. The beating of cilia on this mucous membrane then sweeps the mucus-trapped materials upward and out of the breathing passageways.

Epithelial tissue, also called **epithelium,** occurs as sheets of closely packed cells that cover body surfaces and line internal organs. The epidermis, the outer portion of our skin, is one example. Other epithelial tissues line our lungs, the tubular passageways where urine forms in our kidneys, and the organs of our digestive tract, including the esophagus, stomach, and intestines. One side of an epithelium—the "free" surface—forms the actual lining of the passageway, while the other side is anchored to underlying tissues. The tightly knit cells in an epithelium form a barrier and, in some cases, an exchange surface between underlying tissues and the air or fluid in a passageway or hollow organ.

Epithelial tissues are named according to the number of cell layers they have and according to the shape of most of the cells composing them. A simple epithelium has a single layer of cells, whereas a stratified epithelium has multiple

Connective tissue binds and supports other tissues

In contrast to epithelium, **connective tissue** consists of a sparse population of cells scattered through a nonliving substance called a matrix (plural, matrices). The cells synthesize the matrix, which is usually a web of fibers embedded in a liquid, jelly, or solid.

There are six major types of connective tissue, and the figure below illustrates one example of each. The most common type in the human body is called **loose connective tissue** (part A), because its matrix is a loose weave of fibers. Many of the fibers consist of the strong, ropelike protein collagen. Loose connective tissue serves mainly as a binding and packing material, holding other tissues and organs in place. In the figure, we show loose connective tissue lying directly under the skin, where it helps bind the skin to underlying muscles.

Adipose tissue (B) contains fat, which pads and insulates the body and stores energy. Each adipose cell contains a large fat droplet that swells when fat is stored and shrinks when fat is used as fuel.

Blood (C) is a connective tissue with a fluid rather than a solid matrix. The blood matrix, called plasma, consists of water, salts, and dissolved proteins. Red and white blood cells are suspended in the plasma. Blood functions mainly in transporting substances from one part of the body to another, and in immunity.

The other three types of connective tissue have dense matrices. **Fibrous connective tissue** (D) has a matrix of densely packed parallel bundles of collagen fibers. It forms tendons, which attach muscles to bone, and ligaments, which join bones together. **Cartilage** (E), a connective tissue that forms a strong but flexible skeletal material, contains an abundance of collagen fibers embedded in a rubbery matrix. Cartilage commonly surrounds the ends of bones, where it forms a smooth, flexible surface. It also supports the nose and the ears, and it forms the cushioning discs between our vertebrae. **Bone** (F), a rigid connective tissue, has a matrix of collagen fibers embedded in calcium salts. This combination makes bone hard without being brittle. As shown here, bones may contain repeating circular units of matrix, each with a hollow central canal. Blood vessels and nerves enter the bone through the canals and keep the bone cells alive. Because bone contains living cells, it can grow with the animal.

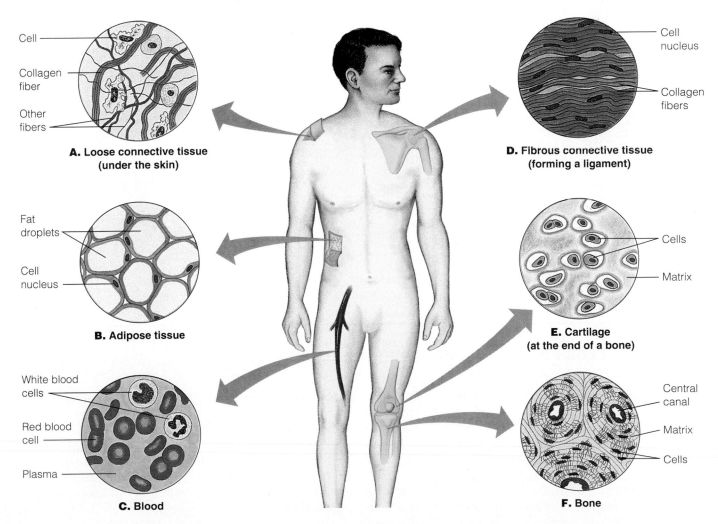

A. Loose connective tissue
(under the skin)

Cell
Collagen fiber
Other fibers

B. Adipose tissue

Fat droplets
Cell nucleus

C. Blood

White blood cells
Red blood cell
Plasma

D. Fibrous connective tissue
(forming a ligament)

Cell nucleus
Collagen fibers

E. Cartilage
(at the end of a bone)

Cells
Matrix

F. Bone

Central canal
Matrix
Cells

20.6 Muscle tissue functions in movement

Muscle tissue, which consists of bundles of long cells called muscle fibers, is the most abundant tissue in a typical animal. Albatrosses, zebras, humans, and all other vertebrates have three types of muscle tissue: skeletal muscle, cardiac muscle, and smooth muscle.

Skeletal muscle is attached to bones by tendons. It is responsible for the voluntary movements of the body. Skeletal muscles are called "voluntary" because an animal can generally contract them at will. As you can see in Figure A below, a skeletal muscle fiber is packed with strands that have alternating light and dark bands. Because of these bands, skeletal muscle is said to be striated, or striped. The bands occur in repeated groups, each of which is a structural and functional unit of muscle contraction. Adults have a fixed number of skeletal muscle cells. Exercise does not increase the number of our muscle cells; it simply enlarges those already present.

Cardiac muscle (Figure B) forms the contractile tissue of the heart. It is striated like skeletal muscle, but its cells are branched. Also, the ends of the cells mesh tightly together, forming relay structures that carry the signals to contract from cell to cell during the heartbeat.

Smooth muscle (Figure C) gets its name from its lack of striations. This type of muscle is found in the walls of the digestive tract, urinary bladder, arteries, and other internal organs. The cells (fibers) are shaped like spindles. They contract more slowly than skeletal muscles, but they can sustain contractions for a longer period of time.

Smooth muscles and cardiac muscles are mostly involuntary; in contrast to skeletal muscles, they are not generally subject to conscious control. We can decide to use our skeletal muscles to step forward or raise a hand, but our smooth muscles churn our stomach and our cardiac muscles pump blood without our conscious command.

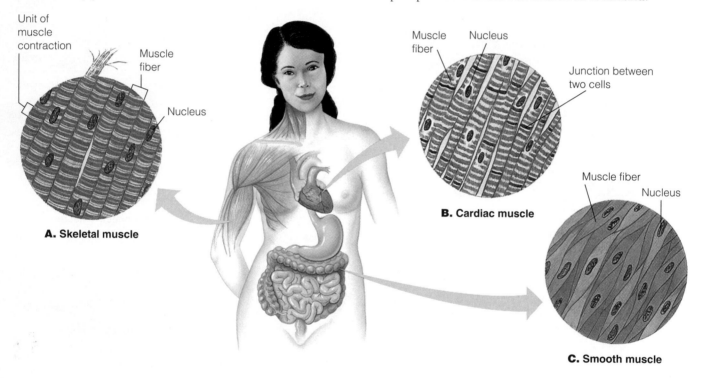

Unit of muscle contraction

Muscle fiber

Nucleus

A. Skeletal muscle

Muscle fiber

Nucleus

Junction between two cells

B. Cardiac muscle

Muscle fiber

Nucleus

C. Smooth muscle

20.7 Nervous tissue forms a communication network

An animal's survival depends on its ability to respond appropriately to stimuli from its environment. **Nervous tissue** forms a communication and coordination system within the body. It senses stimuli and transmits signals from one part of the animal to another, making it possible for all of the body's parts to function as a coordinated whole.

The structural and functional unit of nervous tissue is the nerve cell, or **neuron**, which is uniquely specialized to con-

duct nerve signals. The micrograph at the top of the next page shows three neurons. Each neuron consists of a cell body (containing the cell's nucleus) and several slender extensions. One type of extension, called a dendrite, conveys signals toward the cell body; another type, the axon, transmits signals away from the cell body. Some axons, such as those extending from our brain into our lower abdomen, are half a meter or more in length.

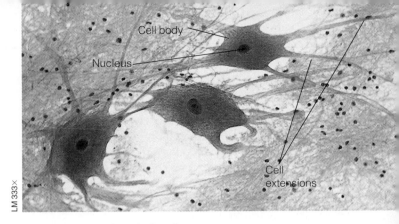

Nerve signals generally travel from one part of the body to another via several neurons. Transmission of a signal from neuron to neuron is usually brought about by chemicals that diffuse from one cell to the next.

Nervous tissue is not made up entirely of neurons. It also contains many cells that support the neurons. Some of these cells help nourish the neurons. Others surround and insulate axons and dendrites, promoting faster transmission of signals.

Having examined the four main kinds of tissue, we turn next to organs.

Several neurons in the spinal cord

Multiple tissues are arranged into organs 20.8

Virtually all animals except sponges, which have a very simple body construction, have organs. An **organ** consists of several tissues adapted to perform specific functions as a group. The heart, for example, while mostly muscle, also has epithelial, connective, and nervous tissues. Epithelial tissue lining the heart chambers prevents leakage and provides a smooth surface over which blood can flow with little friction. Connective tissue makes the heart elastic and strengthens its walls and valves. Nerve cells direct the rhythmic contractions of cardiac muscles.

Another organ, the stomach, consists mainly of three types of tissue. As you can see in the figure below, the tissues are arranged in multiple layers. The lumen, or space, within the stomach is lined by a thick, columnar epithelium that secretes mucus and digestive juices. Surrounding this layer is a zone of connective tissue that contains nerves and blood vessels (shown in red). Three layers of smooth muscle surround the connective tissue, and they, in turn, are surrounded by an outer layer of connective tissue.

An organ represents a higher level of structure than the tissues composing it, and it performs functions that none of its component tissues can carry out alone. These organ-level functions emerge from the cooperative interaction of tissues. Cooperative interaction is a basic feature at all levels in the structural hierarchy of animals.

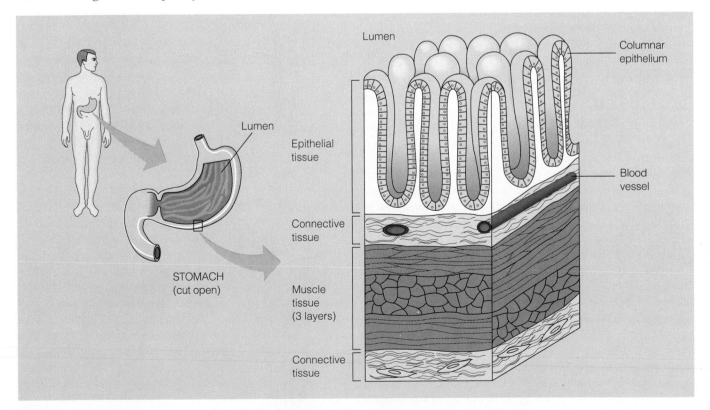

Tissue layers of the stomach wall

The body is a cooperative of organ systems

A level of organization still higher than an organ is an **organ system,** a group of several organs that work together to perform a vital body function. There are twelve major organ systems in vertebrates. The figures on these two pages introduce the main parts of these systems, using the human as an example. We examine each system in detail in the chapters of this unit.

Figure A shows the main components of the **digestive system,** which ingests food and breaks it down into smaller chemical units. Food enters the mouth and travels via the esophagus to the stomach. Digestion occurs mainly in the stomach and small intestine. Nutrient molecules and some water are absorbed into the bloodstream through the walls of the small intestine. The large intestine absorbs additional water and compacts indigestible material into feces, which leave the body through the anus.

The liver is the largest organ in the body and has multiple functions. As part of the digestive system, it discharges bile, which aids in fat digestion, into the small intestine. Among its other functions, the liver carries out metabolic reactions involving carbohydrates and lipids, produces a number of important blood proteins, and removes toxins and worn-out cells from the blood.

The **respiratory system** (Figure B) is the body's assemblage of organs for exchanging gases with the environment. It supplies the blood with oxygen (O_2) and disposes of carbon dioxide (CO_2), a waste product of cellular metabolism. Air enters and exits the system through the nose and mouth, passing through the larynx (voicebox) into the trachea (windpipe). The trachea is a single, large air tube that branches into two smaller tubes called bronchi (singular, bronchus), which go to the lungs. Oxygen diffuses into the blood, and carbon dioxide diffuses out through air sacs in the lungs.

The **cardiovascular system** (Figure C) consists of the heart, which pumps blood, and the blood vessels that transport it. The blood supplies nutrients and O_2 to body cells. It also carries CO_2 to the lungs and other wastes from body cells to other disposal sites, such as the kidneys.

Figure D illustrates the lymphatic system and the immune system, which share some structures and work closely together. The **lymphatic system** is a network of fine vessels interspersed with knotlike masses of tissues called lymph nodes. The lymph vessels supplement the work of the cardiovascular system, and together the cardiovascular system and lymphatic system are sometimes called the **circulatory system.** Lymph vessels pick up fluid that leaks through blood vessels into tissue spaces and return it to the blood, keeping the blood volume constant. Lymph nodes house and dispense specialized white blood cells called lymphocytes, which are components of the immune system. The **immune system** protects the body by attacking foreign substances and infectious microbes. Products of the immune

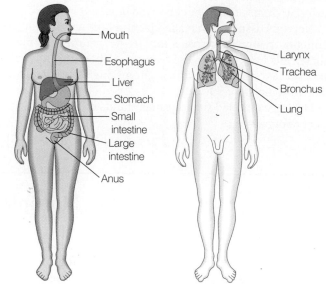

A. Digestive system   **B.** Respiratory system

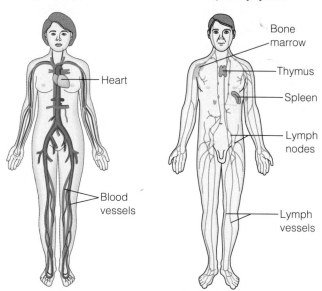

C. Cardiovascular system   **D.** Lymphatic and immune systems

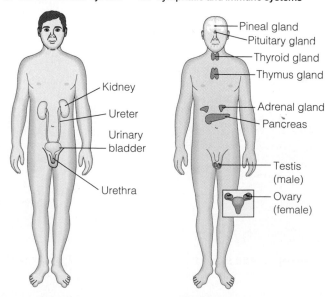

E. Excretory system   **F.** Endocrine system

system are white blood cells, including lymphocytes, and specialized cell secretions called antibodies. These products are transported throughout the body in the bloodstream and in the lymphatic vessels. The thymus, bone marrow, and spleen also play roles in the immune system.

The body's main waste-disposal system is the **excretory system** (Figure E). The kidneys remove nitrogen-containing waste products of cellular metabolism from the blood. In urine, these wastes pass through the ureters to the bladder for temporary storage; they finally leave the body via the urethra. The kidneys also have the vital function of regulating the osmotic balance of the blood.

A number of organs in the body produce chemicals, called hormones, that regulate the activity of organ systems. These organs are called endocrine glands; as a group, they constitute the **endocrine system** (Figure F). The endocrine glands discharge hormones into the blood, and the blood transports the hormones to the parts of the body they act on. Hormones regulate such activities as digestion, metabolism, growth, reproduction, heart rate, and water balance. Some of the organs that are part of the endocrine system serve double duty. The pancreas, for instance, produces hormones that regulate the amount of sugar in the blood, and nonendocrine tissues in the pancreas produce juices that aid in digestion. Likewise, the ovaries and testes, which produce sex hormones, also produce gametes.

The gamete-producing portions of the ovaries and testes are part of the female and male **reproductive systems** (Figure G). Whereas all other organ systems are essential to the survival of the individual organism, an animal can

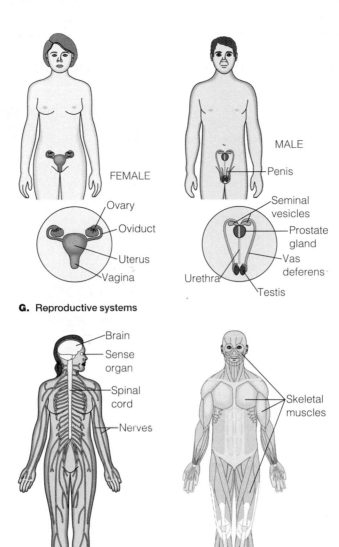

G. Reproductive systems

FEMALE
- Ovary
- Oviduct
- Uterus
- Vagina

MALE
- Penis
- Seminal vesicles
- Prostate gland
- Vas deferens
- Urethra
- Testis

Brain
Sense organ
Spinal cord
Nerves

H. Nervous system

Skeletal muscles

I. Muscular system

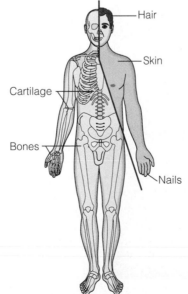

Hair
Skin
Cartilage
Bones
Nails

J. Skeletal and integumentary systems

live without its reproductive system. These systems help the species, rather than the individual, survive. In the female, the ovaries produce egg cells and release them into the oviducts (fallopian tubes), where they may be fertilized. A fertilized egg develops into an embryo in the uterus. The vagina accommodates the male's penis during sexual intercourse and acts as a birth canal for an infant. In the male, the testes produce sperm, and the other organs shown in the figure help keep the sperm viable and convey it into the female's body.

The **nervous system** (Figure H) works together with the endocrine system to coordinate body activities. The brain receives information from the sense organs, such as the eyes. In response, it sends signals to muscles or glands via the spinal cord and nerves. The nervous system also responds to internal information from the body itself.

The **muscular system** (Figure I) consists of all the skeletal muscles in the body. Skeletal muscles can move parts of the body because they are attached to rigid bones or cartilage structures. The muscular system enables us to move about, to manipulate our environment, and to change our facial expressions.

Figure J illustrates the main parts of the skeletal system and the integumentary system (the body covering). The main function of the **skeletal system** is to provide body support, but it also has a protective function. The skull houses and protects the brain, and the rib cage protects the lungs and heart. The **integumentary system** consists of the skin and its derivatives, the hair and nails. Its major function is to protect the internal body parts from mechanical injury, infection, and drying out.

New imaging technology reveals the inner body

Among the most exciting new developments in medical technology are techniques that allow physicians to "see" the organs and organ systems we have just surveyed, without resorting to surgery. We mentioned one of these techniques—ultrasound—in Module 13.11. Some of the others are new versions of X-ray technology.

X-rays, discovered in 1895, were the first means of producing a photographic image of internal organs and the only imaging method available until the 1950s. X-rays are a type of high-energy radiation (see Module 7.6). They pass readily through soft tissues, such as skin, nerves, and muscle. The photographic film on which an X-ray image is recorded is placed behind the body, and what show up most distinctly are the shadows of hard structures that block the rays—bones, cartilage, and dense tumors, for instance.

Conventional X-rays are used routinely to check for broken bones, torn cartilage, and tooth cavities. However, there are three major problems with using X-rays more extensively. One obvious shortcoming is their failure to make soft tissues clearly visible. In addition, the standard X-ray

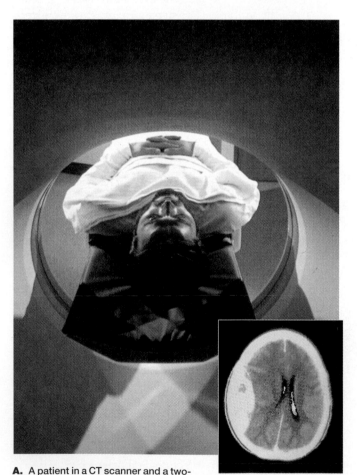

A. A patient in a CT scanner and a two-dimensional CT image of a brain

B. A three-dimensional CT image of a pelvis

technique produces only a flat, two-dimensional image, with anatomical structures often confusingly overlapped. Finally, the X-rays themselves, in large enough doses, can cause cancer.

Modern versions of the standard X-ray technique have overcome some of these disadvantages. They use much lower doses of radiation and also reveal more details of soft tissues. An exciting extension of the standard technique has conquered the overlapping-image problem. This newer X-ray method is called **computed tomography (CT)**, a computer-assisted technique that produces images of a series of thin cross sections through the body. ("Tomography" comes from the Greek words *tomos,* meaning slice, and *graphe,* drawing.) The patient is slowly moved through a doughnut-shaped CT machine (Figure A), as the X-ray source circles around the body, illuminating successive sections from many angles. The CT scanner's computer then produces high-resolution video images of the cross sections, which can be studied individually or combined into various three-dimensional views.

CT scans can detect small differences between normal and abnormal tissues in many organs, but they are especially useful for evaluating brain problems—an area where conventional X-ray procedures are of little help. In the CT image shown in Figure A, you can see a brain hemorrhage (pale oval at far left), the result of a ruptured blood vessel. This CT image is two-dimensional; it shows only a single, thin slice of the brain.

A promising new use for the three-dimensional capability of CT is in the field of reconstructive surgery. Surgeons can study lifelike images of a patient's bones, for example, before operating. Figure B shows a three-dimensional CT image of the dislocated right hip (left side of photo) of a young woman. In this case, the surgeon used the information in the CT scan to plan an artificial hip implant that fit perfectly.

A completely different technique, **magnetic resonance imaging (MRI),** uses no X-rays or any other high-energy radiation. Instead, MRI takes advantage of the behavior of the hydrogen atoms in water molecules. The nuclei of hydrogen atoms are usually oriented in random directions, but in a magnetic field they align in the same direction. MRI uses powerful magnets to align the hydrogen nuclei, then knocks the nuclei out of alignment with a brief pulse of radio waves. Still under the magnets' influence, the hydrogen atoms immediately spring back into alignment, giving out faint radio signals of their own. These signals are picked up by the MRI scanner and translated by computer into an image.

Since water is a major component of all of our soft tissues, MRI visualizes them well. At the same time, dense structures such as bone, which contains little water, are nearly invisible to MRI. These qualities make MRI particularly good for detecting problems in nervous tissue that is surrounded by bone. For example, Figure C shows a tumor (artificially colored red) in the spinal cord of a 5-year-old girl who had lost the use of her legs. With this information, her physician was able to remove the tumor safely, restoring the girl's ability to walk.

Positron-emission tomography (PET) is an imaging technology that differs from both CT and MRI in its ability to yield information about metabolic processes at specific locations in the body. In preparation for a PET scan, the patient is injected with a biological molecule—glucose, for example—labeled with a radioactive isotope (see Module 2.4). The isotope is not dangerous because it is used only in small

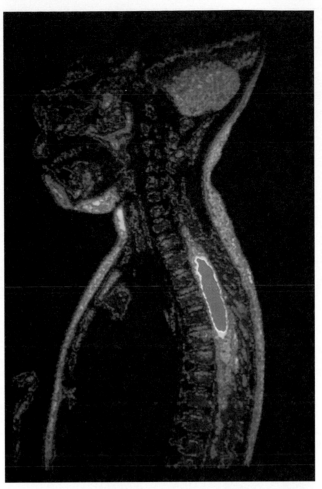

C. An MRI scan showing a spinal tumor

quantities. As it decays, it gives off enough high-energy radiation (gamma rays) to be detected. Metabolically active cells will take up relatively large amounts of the labeled glucose from the blood, and will therefore emit relatively large amounts of radiation.

PET is proving most valuable for measuring the metabolic activity of various parts of the brain. This technique is providing insights into brain activity in people affected by illnesses such as schizophrenia and Alzheimer's disease, and in stroke patients. (A stroke is a loss of brain function caused by the blockage or rupture of a blood vessel.) Equally exciting is the use of PET to learn about the healthy brain. Figure D, for example, shows PET scans of a brain during four different kinds of mental activity. These scans clearly reveal which regions of the brain are most active during each activity.

Because of the high cost of the necessary machinery, PET is still primarily a research tool. However, some less expensive versions are in clinical use, and we will probably see wider use of PET in the future.

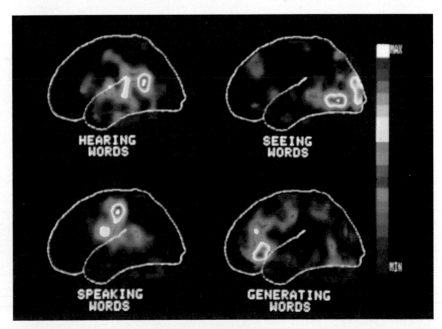

D. PET scans of a brain engaged in different mental activities

An animal's body must allow exchange with the environment

Although animals are covered with protective skin and they are chemically distinct from their environment, they are not closed systems. They cannot survive unless they can exchange materials with their environment. Only when molecules are dissolved in water will they diffuse across a cell's plasma membrane—a requirement that applies to oxygen, nutrients, and metabolic wastes such as carbon dioxide. Consequently, every living cell in the body must be bathed in an aqueous fluid.

The freshwater invertebrate animal *Hydra* has a body wall only two cell layers thick (Figure A). The outside layer is in direct contact with the environment. The inner layer contacts fluid in the animal's saclike gastrovascular cavity. The gastrovascular cavity opens directly to the outside via the mouth, and water flushes in and out of it, bathing the inner layer of cells. As indicated by the arrows in the figure, materials diffuse back and forth between the cells, the hydra's surroundings, and the gastrovascular cavity. With this arrangement, virtually every cell in the animal's body has some of its plasma membrane exposed to an aqueous environment with which it can directly exchange materials. In *Hydra*'s case, each cell has enough exposed surface area to service its entire volume of cytoplasm by direct diffusion and active transport.

The saclike body of a hydra or a paper-thin one like the flatworms we discussed in Module 19.6 works well for animals with a simple body structure. However, most animals have an outer surface that is relatively small compared with the animal's overall volume. As an extreme example,

the surface-to-volume ratio (see Module 4.3) of a whale is millions of times smaller than that of a hydra. Still, every cell in the whale's body must be bathed in fluid, have access to oxygen and nutrients, and be able to dispose of its wastes. How is all this accomplished?

Instead of relying on the general body surface, most animals have specialized surfaces for exchanging materials with the environment. Figure B is a schematic model illustrating four of the organ systems of an animal with a structurally complex body and a large internal surface area that services it. Each system has a specialized internal exchange surface. We have placed the circulatory system in the middle because of its central role in transporting substances among the other three systems. The blue arrows indicate exchange of materials between the circulatory system and the other systems.

Actually, direct exchange does not occur between the blood and the cells making up tissues and organs. At each exchange surface (in Figure B, wherever blue arrows pass into or out of the circulatory system), the body cells are bathed in a solution called **interstitial fluid** (see the circular enlargement). Materials are exchanged between the blood and the interstitial fluid and between the interstitial fluid and the body cells. In other words, to get from the blood to body cells or vice versa, materials must pass through interstitial fluid.

The digestive system, especially the intestine, of this model animal has an expanded surface area resulting from folds and projections of its cells and tissues. Finely branched vessels of the circulatory system (not shown

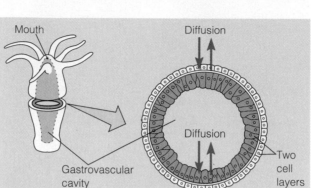

A. Direct exchange between the environment and the cells of a structurally simple animal (*Hydra*)

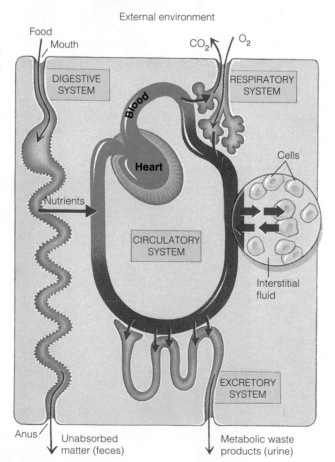

B. Indirect exchange between the environment and the cells of a complex animal

here) form an exchange network with the digestive surfaces. This system of exchange from the cells of the intestine to the surrounding interstitial fluid to the blood is so effective that enough nutrients diffuse from it into the circulatory system to support the rest of the cells in the body.

The folded tubes of the excretory system are equally effective. Enmeshed in tiny blood vessels, they extract metabolic wastes that the blood brings from cells all over the body. The wastes diffuse out of the blood into the excretory tubes and pass out of the body in the urine.

The respiratory system also includes an enormous internal surface area with a vast number of fine blood vessels. Figure C shows a model of the interior of the human lungs. The yellow branches represent a multitude of tiny air sacs, and the red branches represent the finely branched blood vessels that convey blood to the lungs from the heart.

Figure C highlights a basic concept in animal biology: Any animal with a complex body—one with most of its cells out of direct contact with the outside environment—must have internal structures that provide enough surface area to service those cells.

C. The interior surface of the human lungs

Animals regulate their internal environment   20.12

A. An albatross in snow

Over a century ago, French physiologist Claude Bernard recognized that two environments are important to an animal. The external environment surrounds the animal; the internal environment is where its cells actually live. The internal environment of vertebrates is the interstitial fluid that fills the spaces around the cells. Many animals can maintain relatively constant conditions in their internal environment. A hydra is powerless to affect the temperature of the fluid that soaks its cells, but it does regulate the salt and water content of its cells. Our own bodies maintain salt and water balance and also keep our internal fluids at about 37°C (98.6°F). A bird like the albatross also maintains salt and water balance and temperature (about 40°C; 104°F), even in winter when the temperature can drop to −50°C (−58°F). The bird uses energy from its food to generate body heat, and it has a thick, insulating coat of down feathers (Figure A).

Today, Bernard's concept of the constant internal environment is incorporated into the broader principle of **homeostasis,** which means "a steady state." Figure B illustrates this principle, using a pink box to represent an animal. Outside the box, the large double-headed arrow indicates that conditions such as air temperature may fluctuate widely in the animal's external environment. The small double-headed arrow stands for the smaller fluctuations in the animal's internal environment, which are regulated by the animal's control systems (purple box). For example, birds and mammals have a control system that keeps body temperature within a narrow range, despite wide fluctuations in the external environment. Control systems also regulate factors, such as salt concentration, that may not fluctuate much outside the body but that must be maintained at a different level inside.

Though more or less constant, the internal environment still fluctuates slightly. Homeostasis is a dynamic state, an interplay between outside forces that tend to change the internal environment and internal control mechanisms that oppose such changes. An animal's homeostatic control systems maintain internal conditions within a range where life's metabolic processes can occur.

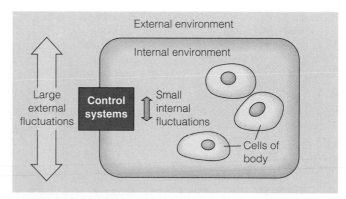

B. A model of homeostasis

20.13 Homeostasis depends on feedback control

Most of the known control mechanisms of homeostasis are based on **negative feedback,** which we discussed in relation to cellular metabolism in Module 5.9. Figure A below shows a mechanical example of negative-feedback control. A thermostat is the control center for regulating the temperature of a room. When the room temperature falls below a set point, such as 20°C (68°F), a sensor (thermometer) turns on the thermostat's switch, which sends a signal to turn on the heater. Heat is produced, warming the room. When the sensor detects that the temperature is above the set point, the thermostat switches the heater off. Physiologists would say that the sensor is triggered by a *stimulus* (room temperature, in this case) and would call the heater an *effector.* An effector produces a *response* (here, heat).

This mechanism is called negative feedback because a change in one condition (in this case, temperature) triggers the control mechanism to *counteract* further change in the same direction. Negative feedback prevents small changes from becoming too large.

The topic of homeostatic control brings us back to the connection between structure and function. Like all biological functions, homeostatic control results from the interactions of specialized structures. For example, our body temperature actually fluctuates between about 36.1°C and 37.8°C (97°F–100°F). As Figure B indicates, our brain has a thermostatic control center that is sensitive to slight changes in the temperature of our blood. When the temperature goes above the set point of 37°C (top half of figure), the control center sends signals to two sets of struc-

tures in the skin: sweat glands and a dense network of blood vessels. As a result, evaporative cooling occurs as sweating increases, and the blood vessels dilate and fill with warm blood. Heat radiates from the skin until the blood cools back to the set point. When this happens, the control center turns off its signals to the skin.

The same structures are involved when blood temperature drops below the set point (bottom half of figure). Signals from the control center shut off the sweat glands and constrict the skin's blood vessels. Blood is then shunted to deeper tissues, reducing heat loss from the skin until the blood temperature returns to the set point and again turns off the control center. If the body does not warm up, shivering—involuntary contractions of skeletal muscles—may occur. These muscle contractions generate a lot of heat, and body temperature increases as a result. We will see other examples of homeostatic control and negative feedback as we examine each of the body's organ systems in detail in the chapters of this unit.

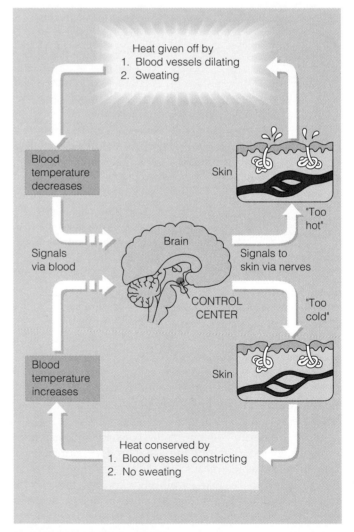

A. Control of room temperature

B. Control of body temperature

Begin your review by rereading the module headings and scanning the figures before proceeding to the Chapter Summary and questions.

Chapter Summary

Introduction–20.2 Structure and function are correlated at each level in the structural hierarchy of an animal's body: Each kind of cell performs specific functions. Groups of cooperating cells form tissues. Tissues form organs, and organs cooperate in organ systems. The whole animal consists of a number of systems, each specialized for certain functions.

20.3–20.7 An animal has four main kinds of tissues: epithelial, connective, muscle, and nervous. Epithelial tissue occurs as sheets of closely packed cells that cover surfaces and line internal organs. Connective tissue binds and supports other tissues. Skeletal muscle, attached to the skeleton, is responsible for voluntary body movements; cardiac muscle pumps blood; and smooth muscle moves the walls of internal organs such as the stomach and arteries. Nervous tissue includes neurons and supporting cells. The branching neurons carry nerve impulses that control and coordinate body activities.

20.8–20.9 The four tissue types cooperate to form organs. Organs, in turn, work together in systems. The integumentary system covers and protects the body. Skeletal and muscular systems support and move it. The digestive and respiratory systems gather food and oxygen, and the cardiovascular system, aided by the lymphatic system, transports them. The excretory system disposes of certain wastes, while the immune system protects the body from infection. The nervous and endocrine systems control and coordinate body functions. The reproductive system perpetuates the species.

20.10 New technologies enable us to see body organs and systems without surgery. Ordinary X-rays produce flat images of hard tissues, but soft tissues do not show up, and high doses are dangerous. In computerized tomography (CT), a low-energy X-ray beam scans sections of the body, combining images to create three-dimensional views that also show soft tissues. Magnetic resonance imaging (MRI) uses magnets and radio waves to produce images of soft tissues. Positron-emission tomography (PET) uses molecules labeled with radioactive isotopes to reveal metabolic activity.

20.11–20.13 An animal must exchange materials with its environment and respond to environmental changes. Small animals with simple body construction have enough surface to meet their cells' needs, but larger, more complex animals have specialized structures that increase surface area for exchanging materials with the environment. Animals regulate their internal environment in response to changes in external conditions. Homeostasis—an internal steady state—depends on control by negative feedback. Control systems sense change and trigger responses that counteract further change in the same direction. Feedback mechanisms keep fluctuations in internal conditions, such as temperature or salt balance, within the narrow range compatible with life.

Testing Your Knowledge

Multiple Choice

1. Which of the following body systems primarily regulate the activities of the other systems?
   a. cardiovascular and muscular systems
   b. nervous and endocrine systems
   c. lymphatic and integumentary systems
   d. endocrine and lymphatic systems
   e. integumentary and nervous systems

2. Every cell in the human body is in contact with an internal environment consisting of
   a. blood
   b. connective tissue
   c. interstitial fluid
   d. matrix
   e. mucous membranes

3. Which of the following best illustrates homeostasis? (*Explain your answer.*)
   a. Most adult human beings are between 5 and 6 feet tall.
   b. The lungs and intestines have large surface areas for exchange.
   c. When blood salt concentration goes up, the kidney expels more salt.
   d. All the cells of the body are about the same size.
   e. When oxygen in the blood decreases, you may feel light-headed.

Matching (*Terms in the right-hand column may be used more than once.*)

1. Closely packed cells covering a surface
2. Neurons
3. Adipose tissue, blood, and cartilage
4. May be simple or stratified
5. Scattered cells embedded in nonliving matrix
6. Senses stimuli and transmits signals
7. Long cells called fibers
8. Cells may be squamous, cuboidal, or columnar
9. Skeletal, cardiac, or smooth

   a. connective tissue
   b. muscle tissue
   c. nervous tissue
   d. epithelial tissue

Describing, Comparing, and Explaining

1. Starting with a muscle cell in the wall of your stomach, briefly describe your body's hierarchy of structural levels.

2. Briefly explain how the structure of each of the following tissues is well suited to its function: fibrous connective tissue in a ligament, stratified squamous epithelium in the skin, neurons in the brain, simple squamous epithelium lining the lung, bone in the skull.

3. Describe ways in which the bodies of large, complex animals are structured for exchange of materials with the environment. Why can some smaller creatures get along without such structural features?

Thinking Critically

A hormone called glucagon is involved in homeostatic control of the amount of sugar in your blood. When the blood sugar level drops, glucagon production increases, and the hormone signals cells to respond to the change. If this works like other homeostatic control systems, what do you suppose glucagon causes to happen in the body? What do you suppose happens to glucagon production when the blood sugar level rises?

Science, Technology, and Society

Physicians, hospitals, and clinics in the United States have eagerly embraced the use of CT and MRI scanners. These machines are quite expensive, costing millions of dollars apiece, and a single scan may cost a patient $1000. In the U.S., even many medium-sized hospitals are equipped with the new technology. For example, as of 1992, there were 13 MRI scanners in Portland, Oregon, a city of fewer than 1 million people. By contrast, in Canada there were only 15 MRI scanners in a country with a population of about 27 million. What are some possible reasons for the different approaches in the U.S. and Canada? What are the advantages and disadvantages of having a scanner in every community?

Whales are the largest animals in the world. Few other species, living or extinct, even approach their great size. The humpback whale, shown in the pictures here, is a medium-sized member of the whale clan. It can be 16 meters (53 ft) long and weigh up to 65,000 kg (72 tons), about as much as 70 midsize cars.

It takes an enormous amount of food to support a 72-ton animal. Humpback whales eat small fishes and crustaceans called krill. The painting at the left shows a remarkable technique they often use to corral food organisms before gulping them in. Beginning about 20 meters below the ocean surface, a humpback swims slowly in an upward spiral, blowing air bubbles as it goes. The rising bubbles form a cylindrical screen, or "bubble net." Krill and fish inside the bubble net swim away from the bubbles and become concentrated in the center of the cylinder. The whale then surges up through the center of the net with its mouth open, harvesting the catch in one giant gulp.

Humpback whales are filter feeders, meaning they strain their food from seawater. Instead of teeth, these giants have an array of brushlike plates called baleen on each side of their upper jaw. You can see the white, comblike baleen in the open mouth of the whale in the photograph at the right. The baleen is used to sift food from the ocean. To start feeding, a humpback whale opens its mouth, expands its throat, and takes a huge gulp of seawater. When its mouth closes, the water squeezes out through spaces in the baleen, and a mass of food is trapped in the mouth. The food is then swallowed whole, passing into the stomach, where digestion begins. The humpback's stomach can hold about half a ton of food at a time, and in a typical day, the animal's digestive system will process as much as 2 tons of krill and fish.

The humpback and most other large whales are endangered species, having been hunted almost to extinction for meat and whale oil by the 1960s. Today, most nations honor an international ban on whaling, and some species are showing signs of recovery. Humpbacks still roam the Atlantic and Pacific oceans. They feed in polar regions during summer months and migrate to warmer oceans to breed when temperatures begin to fall. The photograph below was taken during summer in Glacier Bay, Alaska. Food is so abundant there that humpbacks harvest much more energy than they burn each day. Much of the excess is stored as a thick layer of fat, or blubber, just under their skin. After a summer of feasting, humpback whales leave Glacier Bay and head south to breeding and calving grounds off the Hawaiian Islands, some 6000 km (3600 mi) away. Living off body fat, they eat little, if at all, until they return to Alaskan waters eight months later.

In about four months, a humpback whale eats, digests, and stores as fat enough food to keep its 72-ton body active for an entire year—a remarkable feat, and a fitting introduction to this chapter on animal nutrition and digestion. We will return to the whale as we examine the diverse ways that animals obtain and process nutrients.

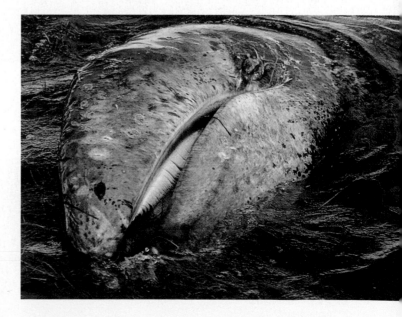

Animals ingest their food in a variety of ways

Animal diets vary enormously, and so do methods of feeding. Certain parasites—tapeworms, for instance—are absorptive feeders; lacking a mouth or digestive tract, they absorb nutrients through their body surface. In contrast, the majority of animals, including the great whales, are ingestive feeders; they eat (ingest) living or dead organisms, either plants or animals or both, through a mouth. We will concentrate on ingestive feeders for the rest of this chapter.

Animals that ingest *both* plants and animals are called **omnivores** (from the Latin *omnis,* all, and *-vorus,* devouring). We humans are omnivores, as are crows, cockroaches, and raccoons. In contrast, plant-eaters, such as cattle, deer, gorillas, and a vast array of aquatic species that graze on algae—sea urchins, for instance—are called **herbivores** (Latin *herba,* green crop). **Carnivores** (Latin *carne,* flesh),

such as lions, sharks, hawks, spiders, and snakes, eat other animals. Figure A illustrates both an herbivore, the impala, and a carnivore, the oxpecker clinging to the side of the impala's head. The oxpecker plucks blood-engorged ticks and other parasites from the impala's skin and fur.

Ingestive feeders use several different mechanisms to obtain their food. **Filter feeders,** such as the humpback whale, actively screen small organisms or food particles from the water. Clams, oysters, and scallops (Figure B) are **suspension feeders.** A film of mucus on their gills traps tiny morsels, which are then swept along to the mouth by beating cilia.

Substrate feeders are ingestors that live in or on their food source and eat their way through the food. Figure C shows a caterpillar eating its way through the soft green tissue inside an oak leaf. The dark spots are a trail of feces that the caterpillar leaves in its wake. Earthworms are also substrate feeders. They eat their way through the soil, digesting partially decayed organic material as they go. In doing so, they help aerate the soil, making it more suitable for plants.

Fluid feeders obtain food by sucking nutrient-rich fluids from a living host, either a plant or an animal. Aphids, for example, tap into the sugary sap in plants. Mosquitoes, ticks, and blood-suckers pierce animals with

A. An herbivore (impala) and a carnivore (oxpecker)

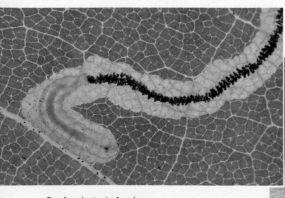

C. A substrate feeder (caterpillar)

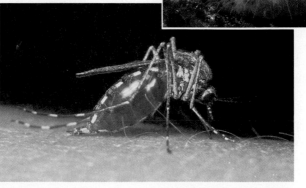

B. A suspension feeder (scallop)

D. A fluid feeder (mosquito)

needlelike mouthparts. The female mosquito in Figure D at the bottom of the previous page has just filled her abdomen with a meal of human blood. (Only female mosquitoes suck blood; males live on plant nectar.)

Rather than filtering food from water, eating their way through a substrate, or sucking fluids, most animals ingest relatively large pieces of food. They use equipment such as tentacles, pincers, claws, poisonous fangs, or jaws and teeth to kill their prey, to tear off pieces of meat or vegetation, or to take mouthfuls of animal or plant products. The scene in Figure E shows that humans clearly fit into this category.

E. Ingestion of mouthfuls of food (by an omnivore)

Food processing occurs in four stages

So far we have discussed what animals eat and how they feed. As shown in the figure below, ① **ingestion,** the act of eating, is only the first of four main stages of food processing. ② **Digestion,** the second stage, is the act of breaking food down into molecules small enough for the body to absorb. Most of the organic matter in food consists of proteins, fats, and carbohydrates—all large polymers (multi-unit molecules made up of small monomers). Animals cannot use these materials directly for two reasons. First, as macromolecules, these polymers are too large to pass through plasma membranes and enter the cells. Second, an animal needs monomers to make the polymers of its own body. Most of the polymers in food (for instance, the proteins in a lima bean we might eat) are different from those making up the animal's body. Therefore, an animal has to break the food polymers into monomers and then use the monomers to make its own brand of polymers.

All organisms use the same monomers. For instance, humpback whales, humans, and lima bean plants all make their proteins from the same 20 kinds of amino acids. Digestion in an animal breaks the macromolecules in food into their component monomers. Proteins are split into amino acids, polysaccharides and disaccharides

are split into simple sugars, fats are split into glycerol and fatty acids, and nucleic acids are split into nucleotides.

As the figure shows, digestion typically occurs in two phases. First, food is mechanically broken into smaller pieces. In a baleen whale, this occurs in the stomach. In animals with teeth, the process of chewing breaks large chunks of food into smaller ones. The second phase of digestion is the chemical breakdown process called hydrolysis. Catalyzed by specific enzymes, hydrolysis breaks chemical bonds in the food polymers by adding water to them (see Module 3.3). In the process, polymers are broken down into monomers, which are small molecules.

The last two stages of food processing occur after digestion. ③ In the third stage, **absorption,** the cells lining the digestive tract take up (absorb) small nutrient molecules. From here the molecules travel in the blood to other body cells, where they are incorporated into the cells or broken down further to provide energy. In a humpback whale, as in almost any animal that eats much more than its body immediately uses, many of the nutrient molecules are converted into fat for storage. ④ In the fourth and last stage of food processing, **elimination,** undigested wastes pass out of the digestive tract.

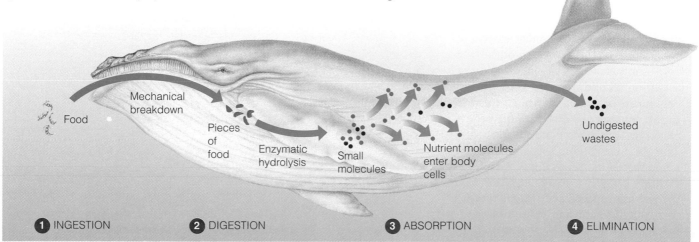

Mechanical breakdown

Food

Pieces of food

Enzymatic hydrolysis

Small molecules

Nutrient molecules enter body cells

Undigested wastes

① INGESTION ② DIGESTION ③ ABSORPTION ④ ELIMINATION

Digestion occurs in specialized compartments

To process food, an animal's body must provide an environment that favors the action of digestive enzymes. In addition, that environment must be contained in some type of compartment where the enzymes will not attack the organism's own macromolecules.

Even single-celled organisms have digestive compartments. An amoeba, for instance, has food vacuoles in which the cell can digest food without the hydrolytic enzymes mixing with its own cytoplasm. Among the animals, sponges are like amoebas in carrying out all of their digestion within their cells (see Module 19.3). In contrast, most other animals have a specific compartment within the body, but outside of cells, in which at least some digestion occurs.

As we saw in Chapter 19, relatively simple animals such as *Hydra* have a **gastrovascular cavity**, a digestive compartment with a single opening, the **mouth**. The gastrovascular cavity functions in both digestion and distribution of nutrients throughout the body. A hydra's gastrovascular cavity enables it to ingest prey much larger than any of its cells could take in directly. In contrast, a sponge, which lacks a digestive compartment, can eat only bacteria and organic particles small enough for its cells to engulf.

Figure A illustrates the main food processing events in a hydra. A hydra is a carnivore that stings its prey (here, a small crustacean called *Daphnia*). It then uses its tentacles to stuff the food into its mouth. Once the prey is in the gastrovascular cavity, cells lining the cavity secrete hydrolytic enzymes (represented by red dots in the figure). Flagella on the cells keep the food mixed with the enzymes, and hydrolysis breaks down the soft tissues of the prey into tiny particles. Once the pieces are small enough, the cells lining the gastrovascular cavity engulf them into food vacuoles, where additional hydrolytic enzymes complete the digestion of the food into simple nutrient molecules. After the hydra has digested its meal, undigested materials remaining in the gastrovascular cavity are eliminated through the mouth.

In contrast to the gastrovascular cavity, a second type of digestive compartment, called an **alimentary canal**, consists of a tube running between two openings, a mouth and an anus. Because food moves in one direction from mouth to anus, the alimentary canal can be adapted along the way into specialized regions that carry out digestion and absorption of nutrients in sequence. For example,

food ingested through the mouth usually passes into a **pharynx**, or throat, and then into a channel called the **esophagus**.

Depending on the species, the esophagus may channel food to a crop, a gizzard, or a stomach; many animals have all three of these organs. A **crop** is a pouchlike organ in which food is usually softened and stored temporarily. ("Crop" comes from the Greek word for "bulge.")

Stomachs and **gizzards** are also pouchlike, but they are more muscular than crops. They actively churn and grind the food. Gizzards often contain teeth or grit to assist in grinding. Chemical digestion and nutrient absorption occur mainly in the **intestine**, the region of the digestive tract between the stomach or gizzard and the anus. The **anus** is the opening through which undigested wastes are expelled.

The regions of the alimentary canal vary according to the type of food they process. Figure B on the facing page illustrates the alimentary canals of animals with three different types of diet. The earthworm is an omnivorous substrate feeder. As it burrows through the ground, its muscular pharynx sucks food—actually, soil—into its mouth. The food passes through the esophagus and is stored and moistened in the crop. The muscular gizzard retains small bits of sand and gravel, which pulverize the food. Organic matter in the food is chemically digested and nutrients are absorbed in the intestine. As the enlargement to the right of the worm shows, the intestine is not a simple cylinder. The intestinal wall has a large inward fold, which increases the intestine's inside surface area (reddish crescent in the cross section). As a result, a greater number of cells are ex-

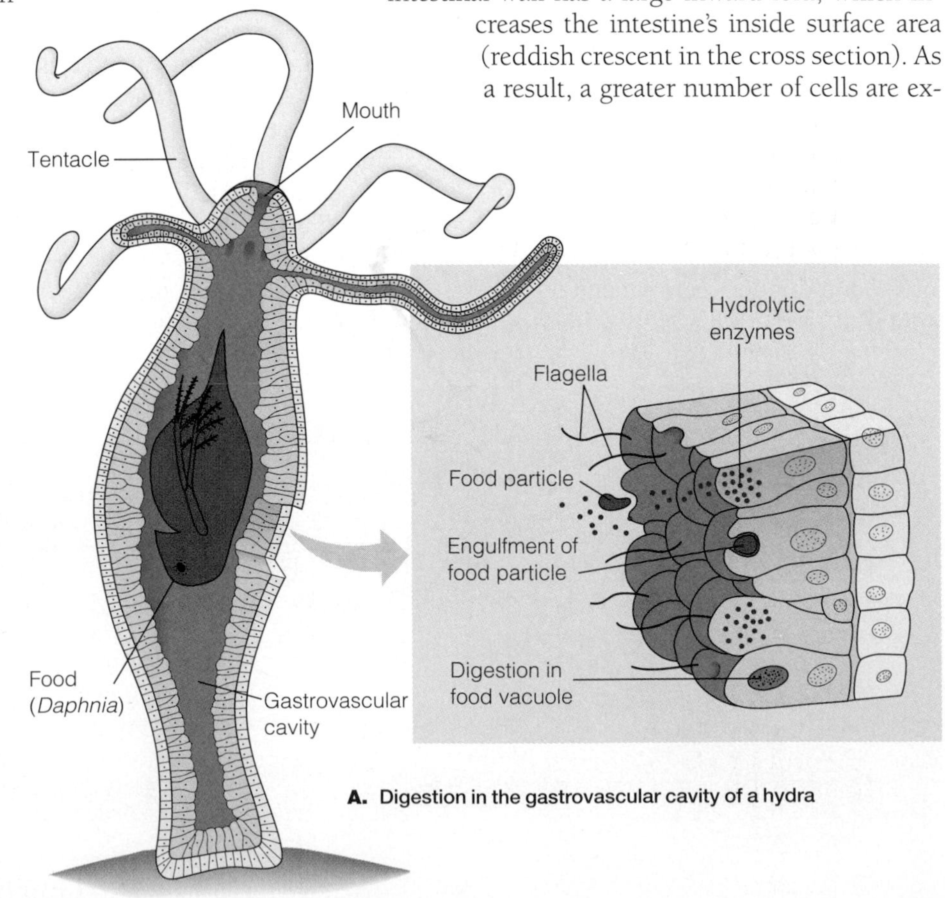

Mouth

Tentacle

Hydrolytic enzymes

Flagella

Food particle

Engulfment of food particle

Digestion in food vacuole

Food (Daphnia)

Gastrovascular cavity

A. Digestion in the gastrovascular cavity of a hydra

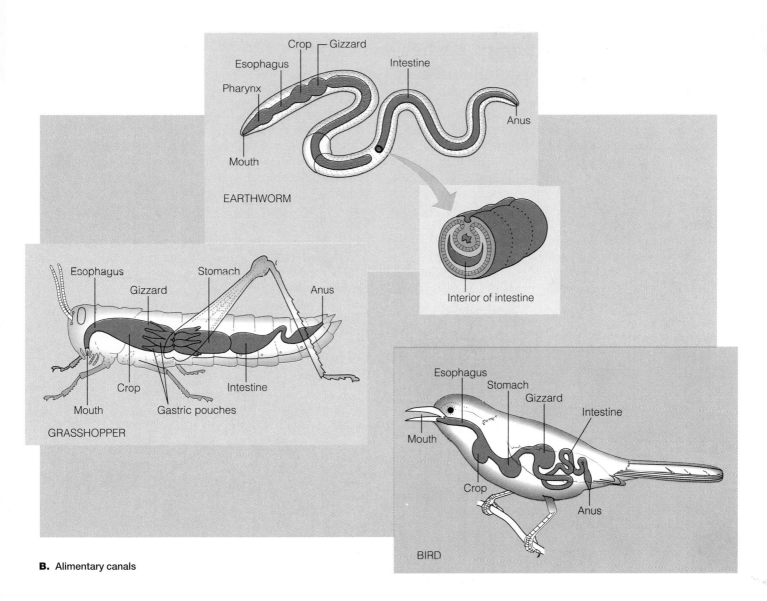

B. Alimentary canals

posed to the food passing through. Without this increased surface area, the intestine would be much less effective in digesting food and absorbing nutrients and water. The undigestible material in the soil that an earthworm eats is eliminated through the anus. This is mostly sand, fine gravel, and other inorganic matter.

The grasshopper is an herbivore. It has jawlike mouthparts that cut and chew plant leaves into small pieces. Like the earthworm, it also has a crop where food is stored and moistened, and a gizzard where hard teeth reduce the food to a pulverized mass. From the gizzard, food passes into the stomach, where most chemical digestion occurs. Nutrients are absorbed in the stomach and also in a cluster of gastric pouches, which extend from the stomach. The short intestine functions mainly to absorb water and compact the wastes, which are expelled through the anus.

The alimentary tract of a typical bird (bottom drawing) consists of the same major organs as those in the earthworm and grasshopper. Birds eat a variety of foods. Hawks and owls, for instance, are carnivores, catching and eating

mice, small birds, snakes, and insects. Many other birds, such as robins and chickens, are omnivores, eating worms, insects, nuts, and fruits. Most birds have a crop, which enables them to eat a large amount of food quickly. Lacking teeth, birds swallow food whole into the crop. The crop stores and softens the food, then passes it into the stomach. Mechanical and chemical digestion begin in the stomach and continue in the gizzard. Many birds eat gravel, as you may see them doing along roadsides. The gravel collects in the gizzard and helps pulverize tough plant fibers and hard parts of insects. (If you've ever kept a parakeet or other seed-eating bird, you know they must have fine gravel in their diet.) After food passes through the gizzard, chemical digestion is completed in the bird's intestine. Nutrients and water are absorbed through the intestinal wall, and wastes pass out through the anus.

Like earthworms, robins, and chickens, humans are omnivorous. We turn to the main features of the human digestive system next.

The human digestive system consists of an alimentary canal and accessory glands

As an introduction to our own digestive system, the drawing below provides an overview of the human alimentary canal and the digestive glands associated with it. The main parts of the canal are the mouth, oral cavity, tongue, pharynx, esophagus, stomach, small intestine, large intestine, rectum, and anus. The digestive glands—the salivary glands, pancreas, and liver—are labeled in blue on the figure. They secrete digestive juices that enter the alimentary canal through ducts. Secretions from the liver are stored in the gallbladder before they are released into the intestine.

Once food is swallowed, muscles propel it through the alimentary canal by **peristalsis**, rhythmic waves of contraction of smooth muscles in the walls of the digestive tract. In only 5–10 seconds, food passes from the pharynx down the esophagus and into the stomach. Constriction at the base of the esophagus keeps food in the stomach.

A muscular ring, called the **pyloric sphincter**, regulates the passage of food out of the stomach and into the small intestine. The sphincter works like a drawstring, closing off the tube and keeping food in the stomach long enough for stomach acids and enzymes to begin digestion. The final steps of digestion, and nutrient absorption, occur in the small intestine over a period of 5–6 hours. Undigested material passes through the large intestine, where water is taken into the body from the remains of the food and digestive juices, and feces are compacted.

In the next several modules, we follow a snack through the alimentary canal, to see in more detail what happens to the food in each of the processing stations along the way. Let's start when you walk into a snack bar for a slice of cheese pizza and a soft drink.

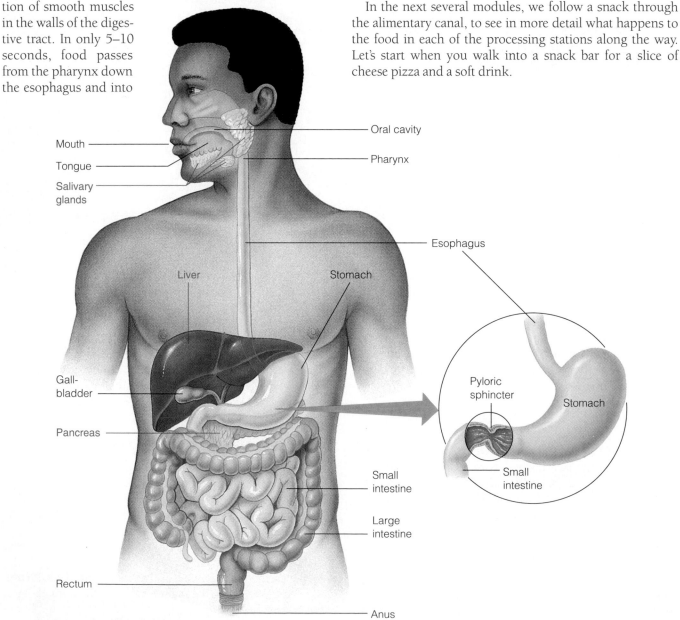

In the snack bar, your salivary glands may start delivering saliva through ducts to the oral cavity even before you place your order. This is a response to the sight or smell of food, or to your usual eating time. In a typical day, our salivary glands secrete over a liter of saliva.

Saliva contains several substances important in food processing. A slippery glycoprotein protects the soft lining of the mouth and lubricates the solid food for easier swallowing. Buffers neutralize food acids, such as those in soft drinks and tomato sauce, helping prevent tooth decay. Antibacterial agents kill many potentially harmful bacteria that may enter the mouth with food. Saliva also contains a digestive enzyme that begins hydrolyzing starch, a major ingredient of your pizza crust.

Mechanical and chemical digestion begin in the oral cavity, as we chew our food. Chewing cuts, smashes, and grinds solid food, making it easier to swallow and exposing more food surface to digestive enzymes. As the drawing at the right shows, we have four kinds of teeth. Starting at the front and proceeding backward on each side of the upper or lower jaw, there are two bladelike incisors. These you use for biting. Behind the incisors, a single pointed canine tooth helps you tear loose a bite of your pizza. Next come two premolars and three molars, which grind and crush the

morsel. (The third molar, a "wisdom tooth," does not appear in some people.)

Also prominent in the oral cavity is the tongue, a muscular organ covered with taste buds. Besides enabling you to taste your meal, the tongue manipulates food and helps shape it into a ball called a bolus. In swallowing, the tongue pushes the bolus to the back of the oral cavity and into the pharynx.

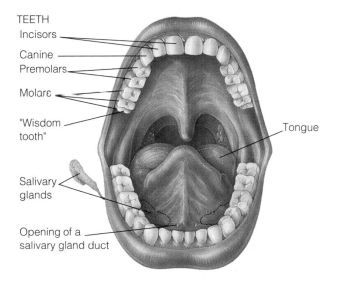

TEETH
Incisors
Canine
Premolars
Molars
"Wisdom tooth"
Salivary glands
Opening of a salivary gland duct
Tongue

Openings into both our esophagus and our **trachea** (windpipe) are in the pharynx. Most of the time, as shown on the far left in the figure below, the esophageal opening is closed off by a sphincter (blue arrows), and the trachea is open for breathing. This situation changes when you start to swallow some of the pizza you've just finished chewing. As you do, a bolus of the food enters the pharynx, triggering the swallowing reflex, shown in the center drawing. The esophageal sphincter relaxes and allows the bolus to

enter the esophagus. At the same time, the larynx (voicebox) moves upward and tips the epiglottis (a flap of cartilage) over the tracheal opening. In this position, the epiglottis prevents food from passing into the windpipe. You can see this motion in the bobbing of your larynx (also called your Adam's apple) during swallowing. After the bolus has entered the esophagus, the larynx moves downward, and the breathing passage reopens (right drawing). The esophageal sphincter contracts above the bolus.

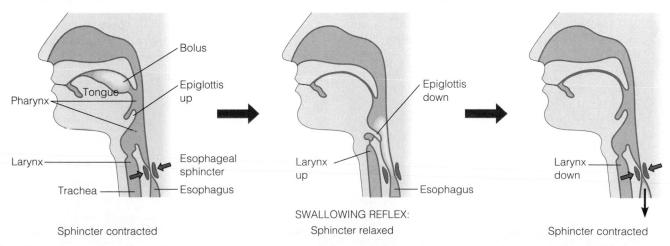

Pharynx
Tongue
Bolus
Epiglottis up
Larynx
Trachea
Esophageal sphincter
Esophagus
Sphincter contracted

Larynx up
Epiglottis down
Esophagus
SWALLOWING REFLEX:
Sphincter relaxed

Larynx down
Sphincter contracted

21.7 The esophagus squeezes food along to the stomach

The **esophagus** is a muscular tube that conveys food boluses from the pharynx to the stomach. The esophageal muscles are arranged in two layers. One muscle layer is circular, running around the esophagus (blue in the figure here); the other is longitudinal, running the length of the esophagus (yellow). Contraction of the circular layer constricts the esophageal passageway. Contraction of the longitudinal layer shortens the esophagus.

The diagrams show how wavelike contractions—**peristalsis**—of the circular and longitudinal muscles squeeze a bolus toward the stomach. As food is swallowed, circular muscles above the bolus contract (blue arrows), pushing the bolus downward. At the same time, longitudinal muscles below the bolus contract (yellow arrows), shortening the pas-

sageway ahead of the bolus. These contractions continue in waves until the bolus enters the stomach.

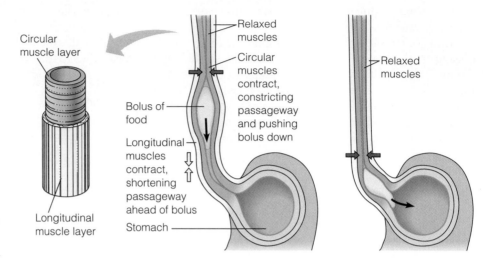

Muscle layers of the esophagus and their roles in peristalsis

21.8 The stomach stores food and breaks it down with acid and enzymes

Having a stomach is the main reason we do not need to eat constantly. Our stomach is highly elastic and can stretch to accommodate about 2 liters of food and drink, usually enough to satisfy our body's needs for many hours.

Some chemical digestion occurs in the stomach. The stomach secretes a digestive, or gastric, juice, made up of mucus, enzymes, and strong acid. The interior surface of the stomach wall is highly folded, and, as the figure on the facing page shows, it is dotted with pits leading down into tubular **gastric glands.** The gastric glands have three types of cells that secrete different components of the gastric juice. Mucous cells (dark pink) secrete mucus, which lubricates and protects the cells lining the stomach; parietal cells (gold) secrete hydrochloric acid (HCl); and chief cells (reddish-tan) secrete pepsinogen, an inactive form of the digestive enzyme pepsin.

The diagram on the far right indicates how pepsinogen, HCl, and pepsin interact during digestion in the stomach. In step ①, pepsinogen and HCl are secreted into the interior of the gastric gland. ② Next, the HCl converts pepsinogen to pepsin. ③ Pepsin itself then activates more pepsinogen, starting a chain reaction. Pepsin begins the chemical digestion of proteins—those in the cheese on your pizza, for instance. It splits the polypeptide chains of the proteins into

smaller polypeptides. This action primes the proteins for further digestion, which will occur in the small intestine.

What prevents gastric juice from digesting away the stomach lining? Secreting pepsin in the inactive form of pepsinogen helps protect the cells of the gastric glands, and mucus helps protect the stomach lining from both pepsin and acid. Still, the epithelium is constantly eroded. Mitosis must generate enough new cells to replace the stomach lining completely about every three days.

Cells in our gastric glands do not secrete gastric juice constantly. Their activity is regulated by a combination of nerve signals and hormones. When you see, smell, or taste food, a signal from your brain to your stomach stimulates your gastric glands to secrete gastric juice. Once you have food in your stomach, substances in the food stimulate cells in the stomach wall to release the hormone **gastrin** into the circulatory system. Gastrin circulates in the bloodstream, returning to the stomach wall. When it arrives there, it stimulates further secretion of gastric juice. Thus, an initial burst of gastric secretion at mealtime triggers more secretion, adding gastric juice to the food for some time. A negative-feedback mechanism like the one we described in Module 20.13 inhibits the secretion of gastric juice when the stomach contents become too acidic. The acid inhibits the release of gastrin, and, with less gastrin in the blood, the gastric glands secrete less gastric juice.

Contraction of muscles in the stomach wall aids chemical digestion. The active stomach churns the food with the gastric juice, forming a mixture called **acid chyme.** Most of the time, the stomach is closed off at both ends. The opening between the esophagus and the stomach is closed except when a bolus driven downward by peristalsis arrives there. Acid chyme is thus kept from flowing backward into

the esophagus. Occasional backflow of acid chyme into the lower end of the esophagus causes the feeling we call heartburn. During vomiting, peristalsis reverses direction and drives the stomach contents upward into the oral cavity. Between the stomach and the small intestine, the pyloric sphincter helps regulate the passage of acid chyme from the stomach into the small intestine. With the acid chyme leaving the stomach only a squirt at a time, the stomach takes about 2–6 hours to empty after a meal.

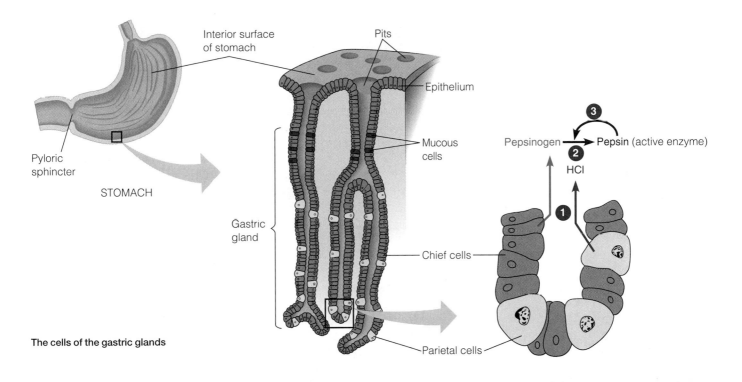

The cells of the gastric glands

Ulcers can form if gastric juice digests the stomach lining   21.9

A stomachful of digestive juice laced with strong acid lets us digest a diverse array of foods. At the same time, these chemicals, acidic enough to dissolve steel, can be harmful. Though the stomach normally makes enough mucus to protect itself from the corrosive effect of digestive juice, this mechanism is not foolproof. When it fails, what we call gastric ulcers can develop in the stomach wall. The symptoms are usually a gnawing pain in the upper abdomen, which may come and go for several hours after eating.

A **gastric ulcer** is an open sore that appears when pepsin and hydrochloric acid destroy the stomach lining faster than it can regenerate. This can happen when the cells lining the stomach produce too much pepsin and/or acid, or when too little protective mucus is made. Tobacco, alcohol, caffeine, large doses of aspirin, and even table salt can irritate the stomach lining and make it more prone to ulcers. People who are under constant stress may also be prone to ulcers, and recent studies suggest that certain bacterial infections also contribute.

An early step in the growth of a stomach ulcer is a mild case of stomach irritation called gastritis. If the irritating conditions continue, a full-fledged ulcer may develop. Eventually, the stomach wall may erode to the point that it actually has a hole in it. This hole can lead to inflammation, infection within the abdomen, or life-threatening internal bleeding.

Fortunately for ulcer sufferers, treatment is available. Self-help measures include eating frequent small meals, reducing the intake of irritants, and taking antacids. Prescription drugs can also reduce the stomach's acid production. If a stomach ulcer does not heal after several months of dietary change and/or drug treatment, surgery to remove the ulcerous portion of the stomach may be advised.

When digesting food leaves the stomach, it is accompanied by gastric juices. Therefore, the first section of the small intestine—the duodenum—is also susceptible to ulcers. Duodenal ulcers usually respond readily to dietary changes or drugs.

The small intestine is the major organ of chemical digestion and nutrient absorption

Returning to our journey through the digestive tract, what is the status of your meal as it passes out of the stomach into the small intestine? The food has been mechanically reduced to smaller pieces and mixed with liquid; it now resembles a thick soup. Chemically, starch digestion began in the mouth, and protein breakdown began in the stomach. Aside from this, virtually all chemical digestion of the original macromolecules in the pizza and soft drink occurs in the small intestine. Nutrients are also absorbed into the blood from the small intestine. With a length of over 6 m, the **small intestine** is the longest organ of the alimentary canal. (Its name is based not on its length but on its diameter, which is only about 2.5 cm; the large intestine is much shorter but has twice the diameter.)

Two large glandular organs, the pancreas and the liver, contribute to digestion in the small intestine (Figure A). The **pancreas** produces digestive enzymes and an alkaline solution rich in bicarbonate. The alkaline solution neutralizes acid chyme as it enters the small intestine. The **liver** performs a wide variety of functions, including the production of bile. **Bile** contains no digestive enzymes, but bile salts dissolved in it make fats more susceptible to enzyme attack. The **gallbladder** stores bile until it is needed in the small intestine. The first 25 cm or so of the small intestine is called the **duodenum.** This is where the acid chyme squirted from the stomach mixes with bile from the gallbladder and digestive enzymes from the pancreas and from the wall of the intestine itself.

The table below summarizes the processes of enzymatic digestion that occur in the duodenum. All four types of

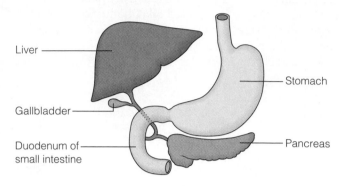

A. The relation of other organs to the small intestine

macromolecules (carbohydrates, proteins, nucleic acids, and fats) are digested. As we discuss the digestion of each, the table will help you keep track of the enzymes involved.

The digestion of carbohydrates begun in the oral cavity is completed in the small intestine. An enzyme called pancreatic amylase hydrolyzes starch (a polysaccharide) into the disaccharide maltose. The enzyme maltase then splits maltose into the monosaccharide glucose. Maltase is one of a family of enzymes, each specific for the hydrolysis of a different disaccharide. Another enzyme, sucrase, hydrolyzes table sugar (sucrose), and lactase digests milk sugar (lactose, common in milk and cheese). Children have much more lactase than adults. Some adults lack lactase altogether, and if they drink milk in quantity, they develop cramps and diarrhea because they cannot digest the lactose.

The small intestine also completes the digestion of proteins begun in the stomach. The pancreas and the duode-

Enzymatic Digestion in the Small Intestine

CARBOHYDRATES					
	Pancreatic amylase			Maltase Sucrase Lactase (etc.)	
Starch (a polysaccharide)	⟶	Maltose (a disaccharide)	Disaccharides	⟶	Monosaccharides
PROTEINS					
	Trypsin Chymotrypsin			Aminopeptidase Carboxypeptidase Dipeptidase	
Polypeptides	⟶	Smaller polypeptides	Small polypeptides and dipeptides	⟶	Amino acids
NUCLEIC ACIDS					
	Nucleases			Other enzymes	
DNA and RNA	⟶	Nucleotides	Nucleotides	⟶	Nitrogenous bases, sugars, and phosphates
FATS					
	Bile salts			Lipase	
Fat globules	⟶	Fat droplets (emulsified)	Fat droplets	⟶	Fatty acids and glycerol

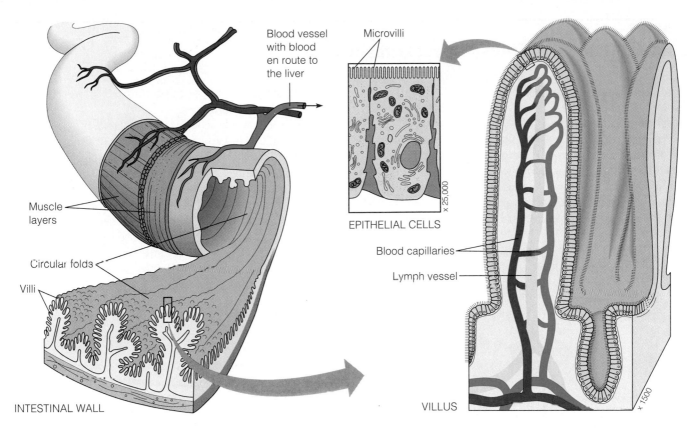

Blood vessel with blood en route to the liver

Muscle layers

Circular folds

Villi

INTESTINAL WALL

Microvilli

EPITHELIAL CELLS

×25,000

Blood capillaries

Lymph vessel

VILLUS

×1500

B. Structure of the small intestine

num secrete a group of enzymes that completely dismantles polypeptides into their component amino acids. Two hydrolytic enzymes, trypsin and chymotrypsin, break polypeptides into shorter chains than those resulting from pepsin digestion. Two other enzymes, carboxypeptidase and aminopeptidase, split off one amino acid at a time, working from the ends of the polypeptides. A fourth type of enzyme, dipeptidase, hydrolyzes fragments only two or three amino acids long. Working together, this enzyme team digests proteins much faster than any single enzyme could.

Another team of enzymes, the nucleases, hydrolyzes the nucleic acids in food. Nucleases from the pancreas split DNA and RNA (which would be present in any meat or vegetables on your pizza) into their component nucleotides. The nucleotides are then broken down into nitrogenous bases, sugars, and phosphates by other enzymes produced by the duodenal cells.

In contrast to starch and proteins, nearly all the fat in your pizza remains completely undigested until it reaches the duodenum. Hydrolysis of fats is a special problem because fats are insoluble in water. First, bile salts from the gallbladder coat tiny fat droplets and keep them separate from one another, a process called emulsification. When there are many small droplets, a large surface area of fat is exposed to lipase, an enzyme that breaks fat molecules down into fatty acids and glycerol.

By the time peristalsis has moved the chyme mixture through the duodenum, chemical digestion of our meal is just about complete. The remaining regions of the small intestine are adapted for the absorption of nutrients.

Structurally, the small intestine is well suited for the task of absorbing nutrients. It has a huge surface area—roughly 600 m² , about the size of a baseball diamond. As Figure B indicates, the extensive surface area results from folds and projections at several levels. Around the wall of the small intestine are large circular folds with numerous small, fingerlike projections called **villi** (singular, *villus*). If we look at the epithelial cells of a villus with an electron microscope, we see many tiny surface projections, called **microvilli.** The microvilli extend into the interior of the intestine. Notice that a small lymph vessel and a network of capillaries (microscopic blood vessels, shown in red and blue) penetrate the core of each villus. Nutrients pass first across the intestinal epithelium and then through the thin walls of the capillaries or lymph vessel. Some nutrients simply diffuse from the digested food into the epithelial cells and then into the blood or lymph. Others are pumped against concentration gradients by the membranes of epithelial cells.

The capillaries that drain nutrients away from the villi converge into larger blood vessels and eventually a main vessel (not shown) that leads directly to the liver. The liver thus gets first access to nutrients absorbed from a meal. The liver converts many of the nutrients into new substances that the body needs. One of its main functions is to remove excess glucose from the blood and convert it to glycogen (a polysaccharide), which is stored in liver cells. From the liver, blood travels to the heart, which pumps the blood and the nutrients it contains to all parts of the body. Your pizza and soft drink are now on their way to being incorporated into your body.

21.11 The large intestine reclaims water

The **large intestine**, or **colon**, is about 1.5 m long and 5 cm in diameter. As the enlargement in the figure shows, it joins the small intestine at a T-shaped junction, where a sphincter controls the passage of unabsorbed food material out of the small intestine. One arm of the T is a blind pouch called the **cecum.** The **appendix,** a small fingerlike extension of the cecum, contains a mass of white blood cells that make a minor contribution to immunity. Despite this role, the appendix itself is prone to infection (appendicitis). If this occurs, the appendix can be surgically removed without weakening the immune system.

The colon's main function is to absorb water from the alimentary canal. Altogether, about 7 liters of fluid enter the canal each day as the solvent of the various digestive juices. About 90% of this water is absorbed back into the blood and tissue fluids, with the small intestine reclaiming much of it and the colon finishing the job. As the water is absorbed, the remains of the digested food become more solid as they are conveyed along the colon by peristalsis. These waste products of digestion, the **feces,** consist mainly of undigestible plant fibers (cellulose in the crust of your pizza, for instance) and bacteria that normally live in the colon. Some of our colon bacteria, such as *E. coli,* produce important vitamins, including biotin, folic acid, several B vitamins, and vitamin K. These vitamins are absorbed into the bloodstream through the colon.

The terminal portion of the colon is the **rectum,** where the feces are stored until they can be eliminated. Strong contractions of the colon create the urge to defecate. Two rectal sphincters, one voluntary and the other involuntary, regulate the opening of the anus.

If the lining of the colon is irritated—by a viral or bacterial infection, for instance—the colon is less effective in reclaiming water, and diarrhea may result. The opposite problem, constipation, occurs when peristalsis moves the feces along too slowly; the colon reabsorbs too much water, and the feces become too compacted. Constipation often results from a diet that does not include enough plant fiber or from a lack of exercise.

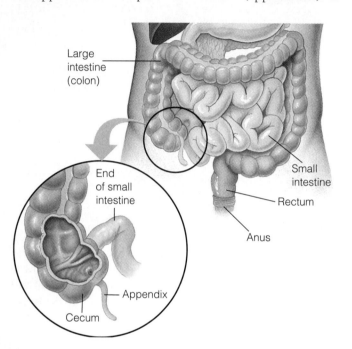

Large intestine (colon)

End of small intestine

Small intestine

Rectum

Anus

Appendix

Cecum

21.12 Vertebrate digestive systems reflect diet

In following a meal through our own alimentary canal, we have seen the basic plan of the vertebrate digestive system. As a group, the vertebrates exhibit many variations on this basic plan. In every case, the structure and function of the digestive system are keyed to the kind of food the animal eats.

A glance at the length of an animal's digestive tract tells us something about its diet. In general, herbivores and omnivores have longer alimentary canals, relative to their overall body size, than carnivores. A longer canal provides the extra time it takes to extract nutrients from vegetation, which is more difficult to digest than meat because of the cell walls in plant material. A longer canal also provides more surface area for the absorption of nutrients, which are usually less concentrated in vegetation than in meat. A model case is the frog, which changes its diet when it transforms from a tadpole to an adult. Tadpoles eat mainly algae and organic matter suspended in the water and in mud at the bottom of ponds; adult frogs are carnivores, eating mainly insects. As Figure A (on facing page) shows, a tadpole has a coiled intestine that is long relative to its body size. When a tadpole transforms into an adult, the rest of its body grows more than its intestine, leaving the adult frog with an intestine that is shorter relative to its overall size.

Herbivorous mammals typically have very long alimentary canals. Most of them also have special chambers in the canal that house teeming numbers of microbes, both bacteria and protozoans. The mammals themselves cannot digest cellulose in the plants they eat. The bacteria and protozoans convert the cellulose to simple sugars and other essential nutrients, and the mammals then obtain these nutrients by digesting the microbes.

Many herbivorous mammals—horses and elephants, for example—house cellulose-digesting microbes in the colon and in a large cecum, the pouch where the small and large intestines connect. Some of the nutrients produced by the microbes are absorbed in the cecum and colon. Most of the

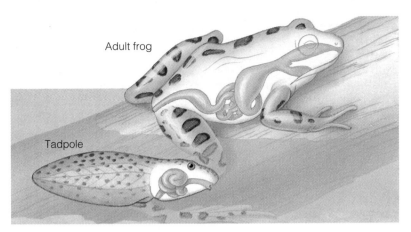

A. The alimentary canal in a tadpole and an adult frog

B. A rabbit eating soft fecal pellets

nutrients, however, are lost in the feces because they do not go through the small intestine, the main site of absorption. Rabbits compensate for this loss by passing food through their digestive system twice and producing two kinds of feces. The first time through the alimentary canal, cellulose in the food is digested by microbes in the cecum. Much of the digested material is then eliminated as soft fecal pellets. The rabbit eats these pellets (Figure B), and its small intestine then absorbs the nutrients that would otherwise have been lost. The leftovers of food that has passed through the digestive tract a second time are eliminated as hard fecal pellets, which are not reingested. Fecal consumption is common in mammals. Many rodents that live in the desert conserve water by eating their first round of feces.

Ruminant mammals, such as cattle, sheep, deer, and goats, have a much more elaborate system for cellulose digestion. The stomach of a ruminant has four chambers, which we show in pink in Figure C. The arrows in Figure C indicate the pathway of food through these chambers. When the cow first chews and swallows a mouthful of grass, the food enters the rumen and the reticulum (green arrows). Bacteria in the rumen and reticulum immediately go to work on the cellulose-

rich meal, and the cow helps by periodically regurgitating and rechewing her food (red arrows). This rumination, or "chewing the cud," softens and helps break down plant fibers, making them more accessible to bacterial digestion.

As indicated by the purple arrows, the cow swallows her cud into the omasum, where water is absorbed. The cud finally passes to the abomasum (black arrows), where the cow's own enzymes complete digestion. Here, the cow obtains many of her nutrients by digesting the bacteria along with nutrients they produce. The bacteria reproduce so rapidly that their numbers remain stable despite this constant loss. With its bacteria and multistage food-processing system, a ruminant harvests considerably more energy and nutrients from the cellulose in hay or grass than a nonruminant herbivore like a horse or an elephant can.

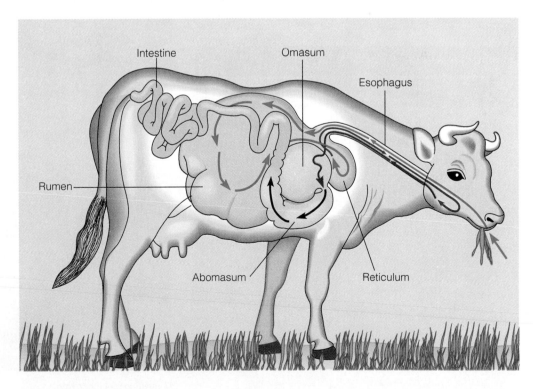

C. Ruminant digestion

21.13 A healthful diet satisfies three needs

All animals—whether herbivores like cows, carnivores like cats, or omnivores like humans—have the same basic nutritional needs. All animals must obtain (1) fuel for cellular metabolism; (2) organic raw materials needed to make the animal's own molecules; and (3) essential nutrients, or substances the animal cannot make for itself from any raw material but must obtain in prefabricated form from food. The different digestive systems we have been examining represent diverse evolutionary adaptations that meet these three basic needs. Let's turn next to the need for fuel, paying particular attention to humans.

21.14 Food energy powers the body

Like all organisms, animals are powered by the energy in ATP. Cellular metabolism generates ATP by oxidizing the small nutrient molecules (monomers) digested from the macromolecules in food. Usually cells use carbohydrates and fats as fuel sources, although when these substances are in short supply, they will use proteins. The energy content of food is measured in **kilocalories** (1 **kcal**=1000 calories). (The "calorie" in popular use is a kilocalorie.)

Cellular metabolism must continuously drive several processes in order for an animal to remain alive. These include breathing, the beating of the heart, and, in birds and mammals, the maintenance of body temperature. The number of kilocalories a resting animal requires to fuel these essential processes for a given time is called the **basal metabolic rate (BMR)**. The BMR for adult humans averages 1300–1800 kcal per day. This is about equivalent to the daily energy consumption of a 100-watt light bulb. But this is only a basal (base) rate—the amount of energy we "burn" lying motionless. Any activity, even working quietly at your desk, consumes kilocalories in addition to the BMR. The more strenuous the activity, the greater the energy de-

mand. The table at the lower left gives you an idea of the amount of activity it takes for a 150-lb (68-kg) person to use up the kilocalories contained in several common foods. The numbers in the table are above and beyond the BMR.

The photograph below shows an apparatus that estimates a person's metabolic rate at various activity levels. The apparatus actually measures the amount of oxygen consumed as the body cells oxidize food in a given time period. For every liter of O_2 consumed, cellular respiration liberates about 4.83 kcal of energy from food molecules. If, for instance, your body used 16 L of oxygen in an hour (about the rate at which you consume oxygen while washing dishes), your metabolic rate would be 77.28 kcal/hr (that is, 16 L/hr × 4.83 kcal/L).

Food–Exercise Energy Equivalents

	Jogging	Swimming	Walking
Speed	9 min/mi	30 min/mi	20 min/mi
Kcal/kg/min	0.173	0.132	0.039
Big Mac® 560 kcal	47 min	1 hr, 3 min	3 hr, 31 min
Cheese pizza (1 slice) 450 kcal	38 min	50 min	2 hr, 50 min
Coca Cola Classic® (10 oz) 144 kcal	12 min	16 min	54 min
Whole wheat bread (1 slice) 86 kcal	7 min	10 min	32 min

These data are for a person weighing 68kg (150 lb).

Measuring metabolic rate

What happens when we take in more kilocalories than we consume in meeting our energy requirements? Rather than discarding the extra energy, our cells store it in various forms. Our liver and muscles store energy in the form of glycogen, a polymer of glucose molecules. Most of us can store enough glycogen to supply about a day's worth of basal metabolism. If our glycogen stores are full and we take in more kilocalories than we expend, our cells store excess energy as fat. This happens even if our diet contains little fat, because the liver converts excess carbohydrates or proteins into fat.

Body fat and fad diets 21.15

In our culture, excess fat is considered unattractive. Surveys indicate that, at any one time, about half of American women and a fourth of American men are dieting to reduce their weight. A majority of cultures in the world today see fat differently; they tend to equate a well-rounded body—even a plump one—with beauty and prosperity. In fact, the ability to maintain fat reserves in the body was an advantage in early human history, and it still is in societies where food is available only sporadically.

Whatever our cultural perspective, one thing is clear: Fat is an essential component of the human body. It helps insulate us against extreme temperature changes. And a moderate amount of body fat seems to be correlated with a healthy immune system. Extremely thin people tend to have lower levels of vitamin A and beta-carotene in their blood, which may make them more susceptible to certain forms of cancer. In general, a healthy woman has between 22% and 26% of her body weight in fat. The ideal amount for men is between 15% and 19%. When we are overweight—more precisely, "overfat"—our body fat may be 20% or more above these ideal amounts. Some people have a particularly hard time keeping their weight down; recent studies have shown that the efficiency with which excess kilocalories are stored in fat is genetically determined.

Being overweight is more than a cosmetic or cultural problem. It can increase our chances of developing certain serious diseases (such as heart disease) and decrease our life span. Our response to excess fat is often to go on a diet—that is, to limit our food intake drastically for a short period of time. Fad diets like the first three in the table below are designed to take pounds off fast. Such diets rarely succeed. Whatever weight is lost in fad dieting is usually quickly regained once the diet ends and we return to our usual eating habits. Fad diets are not only ineffective in the long run, but, as the table indicates, they can also be harmful.

For twentieth-century Americans, the best approach to weight control is to make long-term changes in behavior. These changes are usually a combination of increased exercise and a diet that is restricted, but balanced and satisfying. As indicated in the bottom row of the table, a balanced diet provides a minimum of 1200 kcal per day and includes adequate amounts of all essential nutrients. Such a diet would meet the **Recommended Dietary Allowances (RDAs)** for all nutrients. The RDAs are recommendations for daily nutrient intake established by the U.S. National Academy of Sciences. The RDA for each food substance varies according to age, sex, and weight. A simplified standard, called the U.S. RDA, is used in nutritional labeling; it is set high enough to be appropriate for almost all healthy people from age 4 through adult. A diet that meets these standards, along with regular aerobic exercise, can trim the body gradually and keep extra fat off without harmful side effects.

Diet Type	Health Effects and Potential Problems
Extremely low-carbohydrate diets Less than 100 g of carbohydrates per day.	Initial loss of weight is primarily water. Problems may include fatigue and headaches. May predispose the dieter to subsequent bingeing.
Extremely low-fat diets Less than 20% of kilocalories from fat. Elimination of most or all animal protein sources and all fats, nuts, and seeds.	May be inadequate in essential fatty acids, protein, and certain minerals. May decrease absorption of fat-soluble vitamins. Low-fat diets tend to be unsatisfying.
Formula diets Based on formulated or packaged products. Many are very low in kilocalories.	If very low in kilocalories (less than 800 kcal per day), may result in loss of body protein and may cause dry skin, thinning hair, constipation, and salt imbalance. Usually weight is regained when diet is ended.
Balanced diet of 1200 kcal or more	If carefully chosen, such a diet can meet the RDAs for all nutrients. Weight loss is usually 1–2 pounds per week; dieter may become discouraged.

21.16 Nine amino acids are essential nutrients

Speaking of diets, is it possible for a person to be a true vegetarian—that is, obtain all necessary nutrients by eating only plant material, without animal products of any kind, including eggs, milk, or cheese? The answer is yes, and many people do so, but they have to know how to get all the essential nutrients.

Given sources of organic carbon (such as sugar) and nitrogen (such as amino acids from the digestion of protein), the human body can make a great variety of organic molecules for its own use. For instance, it can make a number of different amino acids using the nitrogen from a single type of amino acid in food. Nonetheless, there are some substances that the body needs but cannot make on its own, and we must obtain these substances from food.

Adult humans cannot make nine of the 20 kinds of amino acids needed to synthesize proteins. These nine, known as the **essential amino acids,** must be obtained in prefabricated form in the diet. A diet that is low in one or more of the essential amino acids results in protein deficiency, a serious type of malnutrition. Malnutrition due *solely* to protein deficiency is very rare, but protein deficiency accompanied by an inadequate intake of other energy-yielding foods is fairly widespread in developing countries. The victims are usually children, who, if they survive infancy, are likely to be retarded mentally and underdeveloped physically.

The simplest way to get all the essential amino acids is to eat meat and animal by-products such as eggs, milk, and cheese. The proteins in these products are said to be complete, meaning they provide all the essential amino acids in the proportions needed by the human body. In contrast, most plant proteins are incomplete, or deficient in one or more essential amino acids. People may become vegetarians by choice or, more commonly, because they simply cannot afford to buy animal protein; animal protein is usually more expensive than plant protein. Nutritional problems can result when people have to rely on a single type of plant food—just corn, rice, or wheat, for instance. When they do, they stand a good chance of becoming protein-deficient.

The key to being a healthy vegetarian is to eat a variety of plant foods that together supply sufficient quantities of all the essential amino acids. Simply by eating a combination of beans and corn, for example, vegetarians can get all nine essential amino acids (see the figure below). Because the body cannot store amino acids, a deficiency of a single essential amino acid limits the use of other amino acids and retards protein synthesis. Most societies have, by trial and error, developed balanced diets that prevent protein deficiency. The Mexican staple of corn tortillas and beans is one example.

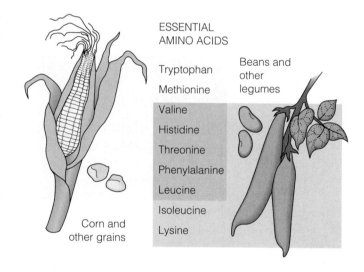

ESSENTIAL
AMINO ACIDS

Tryptophan

Methionine

Beans and other legumes

Valine

Histidine

Threonine

Phenylalanine

Leucine

Isoleucine

Lysine

Corn and other grains

21.17 A healthful diet includes 13 vitamins

Few nutritional topics are as controversial today as vitamins. How much of each vitamin do we need? Do we need vitamin supplements? Some nutritionists argue that vitamin supplements are a waste of money for healthy people who eat a balanced diet. But do we pay enough attention to eating a balanced diet? Are there dangers in taking too much of a supplement? Recent research on vitamins C and E indicates that doses higher than the RDAs are probably harmless and may provide some protection against cancer. What do we really know about vitamins?

A **vitamin** is an organic nutrient that is essential but required in much smaller quantities than the essential amino acids. Most vitamins serve as coenzymes or parts of coenzymes; they have catalytic functions and are used over and over in metabolic reactions (see Module 5.7). The table on the facing page lists 13 vitamins that are essential in the human diet. Even though needed in only tiny amounts, these substances are absolutely necessary, and deficiencies can cause serious problems, as indicated in the far-right column of the table. Vitamins in excess can also be dangerous; symptoms of extreme excess are given in red type.

Vitamins are grouped into two categories: water-soluble and fat-soluble. Water-soluble vitamins include the B complex (the first 8 vitamins in the table), which are important in cellular metabolism. Vitamin B-1, B-6, folic acid, B-12, and biotin are coenzymes used in the metabolism of sugars, amino acids, nucleic acids, and fats. Vitamin B-2, niacin, and pantothenic acid form parts of the coenzymes central to cellular respiration—NAD$^+$, NADP$^+$, FAD, and coenzyme A. Vitamin C, also water-soluble, seems to play many roles, including that of an antioxidant, preventing cell damage caused by certain toxic ions produced by cellular respiration. Except in extreme cases, consuming more of the water-soluble vitamins than our bodies need is not harmful because the excess passes out in urine and feces.

Fat-soluble vitamins play a variety of important roles. Vitamin A is part of the visual pigment of the eye and helps maintain epithelial ztissues. Vitamin D aids in calcium and phosphorus absorption and bone growth. The functions of vitamin E are not yet well understood, but it seems to protect the phospholipids in cell membranes from oxidation. Vita-

min K is required for blood clotting. Unlike the water-soluble vitamins, excessive amounts of the fat-soluble vitamins do not readily leave the body but instead are deposited in body fat. As a result, overdoses of at least some may accumulate and cause toxic effects. Excess vitamin A, for instance, can cause liver and bone damage, among other things.

The vitamin debate is mainly about dosage. Many nutritionists believe that the vitamin RDAs are sufficient. Health-food enthusiasts argue that the RDAs are set too low for at least some vitamins. Debate is especially intense over optimal doses of vitamins C and E. We can say with some certainty that people who eat a balanced diet are unlikely to develop the symptoms of vitamin deficiency listed in the table. We can say with more certainty that we still have much to learn about the biological roles of vitamins.

Vitamin	Major Dietary Sources	Functions in the Body	Symptoms of Deficiency or Extreme Excess*
Water-Soluble Vitamins			
Vitamin B-1 (thiamine)	Pork, legumes, peanuts, whole grains	Coenzyme used in removing CO_2 from organic compounds	Beriberi (nerve disorders, emaciation, anemia)
Vitamin B-2 (riboflavin)	Dairy products, meats, enriched grains, vegetables	Component of coenzyme FAD	Skin lesions such as cracks at corners of mouth
Niacin	Nuts, meats, grains	Component of coenzymes NAD$^+$ and NADP$^+$	Skin and gastrointestinal lesions, nervous disorders Flushing of face and hands, liver damage
Vitamin B-6 (pyridoxine)	Meats, vegetables, whole grains	Coenzyme used in amino acid metabolism	Irritability, convulsions, muscular twitching, anemia Unstable gait, numb feet, poor coordination
Pantothenic acid	Most foods: meats, dairy products, whole grains, etc.	Component of coenzyme A	Fatigue, numbness, tingling of hands and feet
Folic acid (folacin)	Green vegetables, oranges, nuts, legumes, whole grains	Coenzyme in nucleic acid and amino acid metabolism	Anemia, gastrointestinal problems Masks deficiency of vitamin B-12
Vitamin B-12	Meats, eggs, dairy products	Coenzyme in nucleic acid metabolism; needed for maturation of red blood cells	Anemia; nervous system disorders
Biotin	Legumes, other vegetables, meats	Coenzyme in synthesis of fat, glycogen, and amino acids	Scaly skin inflammation; neuromuscular disorders
Vitamin C (ascorbic acid)	Fruits and vegetables, especially citrus fruits, broccoli, cabbage, tomatoes, green peppers	Used in collagen synthesis (e.g., for bone, cartilage, gums); antioxidant; aids in detoxification; improves iron absorption	Scurvy (degeneration of skin, teeth, blood vessels), weakness, delayed wound healing, impaired immunity Gastrointestinal upset
Fat-Soluble Vitamins			
Vitamin A	Deep green and orange vegetables and fruits, dairy products	Component of visual pigments; needed for maintenance of epithelial tissues	Vision problems; dry, scaling skin Headache, irritability, vomiting, hair loss, blurred vision, liver and bone damage
Vitamin D	Dairy products, egg yolk (also made in human skin in presence of sunlight)	Aids in absorption and use of calcium and phosphorus, promotes bone growth	Rickets (bone deformities) in children; bone softening in adults Brain, cardiovascular, and kidney damage
Vitamin E (tocopherol)	Vegetable oils, nuts, seeds	Antioxidant; prevents damage to lipids of cell membranes and probably other molecules	None well documented
Vitamin K	Green vegetables, tea	Important in blood clotting	Defective blood clotting Liver damage and anemia

*Symptoms of extreme excess are given in red type.

Essential minerals are required for many body functions

In addition to amino acids and vitamins, which are organic nutrients, a number of minerals are essential for body function. In the area of nutrition, **minerals** are chemical elements *other than* carbon, hydrogen, oxygen, and nitrogen, the four staples of organic compounds. We must acquire the essential minerals listed in the table below from our diet, and some of the major dietary sources are listed.

Along with other vertebrates, we require relatively large amounts of calcium and phosphorus to construct and maintain our skeleton. Calcium is also necessary for the normal functioning of nerves and muscles, and phosphorus is an ingredient of ATP and nucleic acids. The third mineral in the table, sulfur, is a necessary component of

several amino acids. Iron is a component of several electron-carrier molecules that function in cellular respiration. It is also a component of hemoglobin, the oxygen-carrying protein of red blood cells. Small amounts of fluorine help maintain bones and teeth. Magnesium, manganese, zinc, copper, cobalt, selenium, and molybdenum are components of various enzymes. Vertebrates need iodine to make a hormone called thyroxin, which regulates metabolic rate.

Sodium, potassium, and chlorine are important in nerve function and help maintain the osmotic balance of cells. A common source of sodium and chlorine is sodium chloride, which we know as table salt. Because salt is often in limited supply in natural environments, many herbivores such as

Mineral	Dietary Sources	Functions in the Body	Symptoms of Deficiency*
Calcium (Ca)	Dairy products, dark green vegetables, legumes	Bone and tooth formation, blood clotting, nerve and muscle function	Stunted growth, possibly loss of bone mass
Phosphorus (P)	Dairy products, meats, grains	Bone and tooth formation, acid–base balance, nucleotide synthesis	Weakness, loss of minerals from bone, calcium loss
Sulfur (S)	Proteins from many sources	Component of certain amino acids	Symptoms of protein deficiency
Potassium (K)	Meats, dairy products, many fruits and vegetables, grains	Acid–base balance, water balance, nerve function	Muscular weakness, paralysis
Chlorine (Cl)	Table salt	Acid–base balance, gastric juice	Muscle cramps, reduced appetite
Sodium (Na)	Table salt	Acid–base balance, water balance, nerve function	Muscle cramps, reduced appetite
Magnesium (Mg)	Whole grains, green leafy vegetables	Component of certain enzymes	Nervous system disturbances
Iron (Fe)	Meats, eggs, legumes, whole grains, green leafy vegetables	Component of hemoglobin and of electron-carriers in energy metabolism	Iron-deficiency anemia, weakness, impaired immunity
Fluorine (F)	Drinking water, tea, seafood	Maintenance of tooth (and probably bone) structure	Higher frequency of tooth decay
Zinc (Zn)	Meats, seafood, grains	Component of certain digestive enzymes and other proteins	Growth failure, scaly skin inflammation, reproductive failure, impaired immunity
Copper (Cu)	Seafood, nuts, legumes, organ meats	Component of enzymes in iron metabolism	Anemia, bone and cardiovascular changes
Manganese (Mn)	Nuts, grains, vegetables, fruits, tea	Component of certain enzymes	Abnormal bone and cartilage
Iodine (I)	Seafood, dairy products, iodized salt	Component of thyroid hormones	Goiter (enlarged thyroid)
Cobalt (Co)	Meats and dairy products	Component of vitamin B-12	None, except as B-12 deficiency
Selenium (Se)	Seafood, meats, whole grains	Component of enzyme; functions in close association with vitamin E	Muscle pain, maybe heart muscle deterioration
Chromium (Cr)	Brewer's yeast, liver, seafood, meats, some vegetables	Involved in glucose and energy metabolism	Impaired glucose metabolism
Molybdenum (Mo)	Legumes, grains, some vegetables	Component of certain enzymes	Disorder in excretion of nitrogen-containing compounds

*All of these minerals are also harmful when consumed in extreme excess.

deer and elk do not get enough salt from plants. To satisfy their needs, they regularly visit salt licks, places where the soil or rocks have a high salt content. Sometimes you even see deer and rabbits licking the salt left over from winter de-icing efforts on the gravel along the sides of highways. In contrast to many other animals, humans often ingest far more salt than they need. In the U. S., the average person eats enough salt to provide about twenty times the required amount of sodium. Too much sodium may cause high blood pressure, and low-sodium foods are becoming increasingly popular.

What do food labels tell us? 21.19

Have you ever found yourself sitting at the breakfast table reading the label on a box of cereal or loaf of bread? It can be interesting to compare the way a product is presented on the front of the package or in its advertising with the actual nutritional information on the side or back of the package. For instance, so-called fiber-enriched white bread may have less fiber than an unenriched whole-grain product. Cheese, chips, and yogurt, for instance, are often labeled "low fat" or "lite." This merely means that they have less fat than the same products without these labels—their actual fat content may be surprisingly high. About 20 supposedly "lite" potato chips, for instance, contain about 7 g of fat, compared to about 10 g in 20 regular chips. You only have to eat about 29 "lite" chips to take in slightly more fat than you would get from 20 regular ones.

Recently, public pressure has led to regulations requiring more informative food labels. Two types of information are given on food labels, as shown in the example here. One is simply a list of ingredients. The ingredients must be listed in order from the greatest amount (by weight) to the least. Though the exact amounts are usually trade secrets, this part of the label can give you an idea of the relative amount of each ingredient. Ingredient lists are becoming easier to understand; identifying niacin as a B vitamin is one example. Still, it is useful to know something about food chemistry. For instance, honey and raisin juice are essentially sugars, as are corn syrup and dextrose (in fact, all chemicals ending in -ose). It is also useful to know that 1 g of fat contains twice the kilocalories as 1 g of carbohydrate or protein.

The second type of information on food labels is nutritional information. This part of the label gives the amounts of nutrients in the food by category and the relative levels of nutrients supplied in one serving of the food. The categories of nutrients listed include protein, carbohydrate, and fat, and often sodium (for the benefit of people who are limiting their salt intake). The relative levels of nutrients are expressed as percentages of the U.S. RDAs. When the nutritional information states that a serving (one slice) of the bread provides 4% of the U.S. RDA of protein, it means that it provides *at least* 4% of the U.S. RDA.

In the future, more types of packaged food will be required to specify the amounts of cholesterol, saturated and unsaturated fats, simple carbohydrates (sugars), complex carbohydrates (chiefly starches), and dietary fiber (undigestible complex carbohydrates, mainly cellulose). Emphasizing unsaturated fats and complex carbohydrates in our diet reduces our risk of developing certain diseases, including heart disease and cancer, as we discuss in the next module.

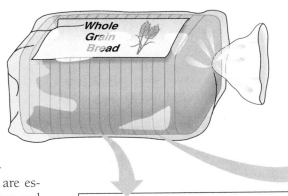

INGREDIENTS

Whole cracked wheat, unbleached enriched wheat four (flour, malted barley flour, niacin [a B vitamin], iron, thiamine. [B-1], riboflavin [B-2]), water, honey, raisin juice, salt, butter, wheat gluten, partially hydrogenated soy bean oil, wheat bran, wheat germ, vinegar, lecithin.

NUTRITIONAL INFORMATION
(Per Serving)

Serving size	1 slice
	(approx. 34 g)
Servings per package	20
Calories	90
Protein	3 g
Carbohydrate	17 g
Fat	1 g
Sodium	180 mg
Total dietary fiber	2 g

PERCENTAGES OF
U.S. RECOMMENDED
DAILY ALLOWANCES
(U.S. RDA)

Protein	4%
Vitamin A	*
Vitamin C	*
Thiamine	6%
Riboflavin	2%
Niacin	4%
Calcium	*
Iron	4%

* Contains less than 2% of the U.S. RDA for these nutrients.

21.20 Diet influences cardiovascular disease and cancer

The figure at the right shows some of the major risk factors associated with cardiovascular disease. Though a few of these factors are unavoidable, we can influence several others through our behavior. Diet is an example of a behavioral factor that affects cardiovascular health in more than one way. A diet rich in saturated fats is linked to both high blood cholesterol and high blood pressure, which in turn are linked to cardiovascular disease.

Cholesterol travels through the body in blood lipoproteins, which are particles made up of thousands of molecules of cholesterol and other lipids, and one or more protein molecules. As we saw in Module 5.20, most blood cholesterol is carried by **low-density lipoproteins (LDLs)**. High LDL levels in the blood correlate with a tendency to develop blocked blood vessels and consequent heart attacks. In contrast, **high-density lipoproteins**, or **HDLs**, another form of cholesterol carrier, probably decrease the risk of vessel blockage. Many researchers believe that reducing LDLs, while maintaining or increasing HDLs, is the best way to reduce the cholesterol risk for cardiovascular disease. Exercise tends to increase the HDL level, while smoking lowers it. A diet high in saturated fats tends to increase LDL levels.

Most animal fats, including those in eggs, lard, and butter, are saturated. Fish oil, on the other hand, and most vegetable oils—for instance, corn, soybean, and olive oil—are unsaturated. When we eat these unsaturated oils in place of saturated fats, our total blood cholesterol level tends to decrease. Fish oil may also reduce blood pressure by helping the body dispose of excess sodium.

As discussed in Module 11.17, diet also seems to be involved in some forms of cancer. Here again, dietary fat is a significant risk factor. Research suggests a link between total fat intake and the incidence of breast, colon, and prostate cancers. And many researchers recommend increasing our fiber intake while reducing fat consumption. The table at the right lists dietary guidelines for reducing cancer risk.

* * *

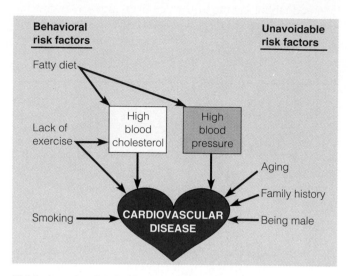

Risk factors associated with cardiovascular disease

Dietary Guidelines for Reducing Cancer Risk

- Limit fat calories to no more than 30% of daily total calories.
- Eat plenty of high-fiber foods such as whole-grain breads and cereals, fruits, and vegetables.
- Eat foods rich in vitamins A and C daily. Include vegetables from the mustard family, such as broccoli and cabbage.
- To avoid possible carcinogens, minimize consumption of cured and smoked foods, such as hot dogs, salami, and bacon. Avoid moldy foods. Avoid charred foods.
- If you drink alcoholic beverages, limit yourself to a maximum of two drinks per day.

In this chapter, we have seen that a sound diet supplies enough raw materials to make all the macromolecules we need, the proper amounts of prefabricated essential nutrients, and enough kilocalories to satisfy our energy needs. Energy remains a theme in Chapter 22, as we look at how the body obtains the oxygen it needs to harvest energy from food molecules.

Chapter Review

Begin your review by rereading the module headings and scanning the figures before proceeding to the Chapter Summary and questions.

Chapter Summary

21.1, 21.13 Animals require nutrients from food for energy and for building cellular materials. Some animals are herbivores (plant-eaters), some are carnivores (meat-eaters), and some are omnivores (eaters of both plants and other animals). Most animals ingest chunks of food, but some are filter or suspension feeders (sifting particles from water), some are substrate feeders (living in or on their food source), and some are fluid feeders (sucking liquids).

21.2–21.3 Food is ingested, digested, and absorbed, and undigested wastes are eliminated. Protozoans such as amoebas digest food particles in food vacuoles. Most animals have a specialized digestive tract. In relatively simple animals like *Hydra*, the digestive compartment is a gastrovascular cavity, a sac with a single opening. In most animals, the digestive compartment is an alimentary canal, a tube running from mouth to anus, divided into specialized regions that carry out food processing sequentially.

21.4–21.7 The human digestive tract consists of an alimentary canal and accessory glands. Digestion begins in the oral cavity. The teeth break up food, saliva moistens it, and salivary enzymes begin the hydrolysis of starch. The tongue pushes the chewed food into the pharynx, and the swallowing reflex moves it into the esophagus. The rhythmic muscle contractions of peristalsis squeeze food toward the stomach and through the rest of the alimentary canal. Ringlike sphincter muscles regulate its passage.

21.8–21.9 The stomach stores food, and its muscular walls mix food with acidic gastric juice. Pepsin in gastric juice begins the hydrolysis of protein. If the gastric glands secrete too much acid or pepsin, or not enough mucus, a gastric ulcer can result.

21.10–21.11 Most digestion and nutrient absorption occur in the small intestine. Alkaline pancreatic juice neutralizes stomach acid, and its enzymes digest polysaccharides, proteins, nucleic acids, and fats. Bile, made in the liver and stored in the gallbladder, emulsifies fat droplets for attack by pancreatic enzymes. Enzymes from the walls of the small intestine complete the digestion of many nutrients. The lining of the small intestine is folded and covered with tiny fingerlike villi that increase its absorptive surface. Nutrients pass through the epithelium of the villi and into the blood, which flows to the liver. The liver can store nutrients and convert them to other substances the body can use. Undigested material passes to the large intestine, or colon, where water is absorbed and feces are produced.

21.12 The digestive systems of vertebrates reflect their diet. Herbivores and omnivores, which eat plant material that is more difficult to digest and has less concentrated nutrients than meat, generally have longer alimentary canals. Some mammals house cellulose-digesting microbes in the colon or cecum, and some reingest their feces to recover nutrients. Ruminants such as cows process food in a four-chambered stomach.

21.13 An animal's diet provides fuel for cellular metabolism, raw materials for making the body's own molecules, and essential nutrients that the body cannot make itself. Cellular metabolism generates ATP by oxidizing nutrient molecules.

21.14 The energy a resting animal requires each day to stay alive—its basal metabolic rate (BMR)—amounts to 1300–1800 kcal per day for human adults. More energy is required for an active life. Excess energy is stored as fat.

21.16–21.18 There are some substances that the body cannot make and must obtain from food. The nine essential amino acids are easily obtained from animal protein and are also available from the proper combination of plant foods. Thirteen vitamins are essential in the human diet. Most function as coenzymes. Essential minerals are elements other than carbon, hydrogen, oxygen, and nitrogen that play a variety of roles in the body.

21.15, 21.20 Diet choice can reduce the risk of cardiovascular disease and cancer. Fad diets are often ineffective and can be harmful; a balanced diet and exercise are the best way to maintain health.

Testing Your Knowledge

Multiple Choice

1. Which of the following sequences correctly traces the passage of food through the human digestive tract?

   a. pharynx, esophagus, stomach, small intestine, large intestine
   b. esophagus, stomach, small intestine, large intestine, pharynx
   c. esophagus, pharynx, stomach, large intestine, small intestine
   d. pharynx, stomach, esophagus, small intestine, large intestine
   e. pharynx, esophagus, stomach, large intestine, small intestine

2. Which of the following is a fluid-feeding herbivore?

   a. aphid   **d.** earthworm
   b. clam   **e.** spider
   c. whale

3. Which of the following is considered an essential nutrient in the human diet?

   a. pepsin   **d.** fat
   b. glucose   **e.** vitamin A
   c. starch

4. A human requires hundreds of grams of carbohydrates every day. The daily requirement for most vitamins is in the milligram range. Why are vitamins needed in such small quantities? (*Explain your answer.*)

   a. Vitamins are not very important in metabolism.
   b. The energy content of vitamins is so great that you don't need much.
   c. The body can store large quantities of most vitamins.
   d. Vitamins are reusable.
   e. Every cell needs carbohydrates, but only a few need vitamins.

True/False (*Change false statements to make them true.*)

1. Villi propel food through the small intestine.

2. In humans, most digestion occurs in the stomach. *sm intestine*

3. Basal metabolic rate is a measure of the energy needed to stay alive.

4. To lose weight, it is best to go on a low-carbohydrate diet.

5. It is desirable to have high levels of low-density lipoproteins (LDLs). *H*

6. If the diet contains certain amino acids, the body can make others.

7. Most nutrient molecules are absorbed in the large intestine. *sm*

8. Bile aids in the digestion of proteins. *lipids*

Describing, Comparing, and Explaining

1. A peanut butter and jelly sandwich contains carbohydrates, proteins, and fats. Describe what happens to the sandwich when you eat it. Discuss ingestion, digestion, absorption, and elimination.

2. Explain how the secretion of gastric juice is controlled. What causes the gastric glands to secrete it? What halts its secretion?

3. Briefly describe three herbivore adaptations related to a plant diet.

Thinking Critically

1. Most large, complex animals, from worms to mammals, have alimentary canals with a mouth and anus. Explain why digestion and absorption in an alimentary canal might be a more efficient way for a large animal to process food than processing it in a gastrovascular cavity.

2. Some essential mineral nutrients, such as selenium and molybdenum, are required in the human diet in miniscule amounts, thousandths of a milligram. What kinds of experiments would you have to carry out to prove that one of these minerals is required by humans? Why might this be difficult?

3. Using the information in the table and text of Module 21.14, calculate how many liters of oxygen you would consume in liberating the energy from one slice of cheese pizza.

Science, Technology, and Society

1. According to recent surveys, 75% of Americans age 18–35 think they are fat. At any given time, 30 million women and 18 million men are on diets. Only 25% are actually medically overweight. Other data show that more than twice as many high school girls as boys see themselves as overweight, and 45% of underweight women see themselves as fat. Why do you think there is a difference between men and women regarding the perception of their weight? What do you think causes people to be so concerned about their weight? To the extent that unwarranted concern over weight is a problem, what can be done about it?

2. In recent years, numerous claims and counterclaims have been made regarding the dangers and values of certain foods. For example, it has been stated that large doses of certain vitamins can retard aging and prevent colds, that oat bran reduces blood cholesterol, that the artificial sweetener saccharin can cause cancer, and that sugar contributes to hyperactivity in children. Can you think of other examples? Have you modified your eating habits on the basis of such information? Why or why not? How should a person evaluate whether or not such nutritional claims are valid?

Respiration: The Exchange of Gases 22

Imagine yourself on a jetliner. The pilot announces that you have reached an altitude of 30,000 feet. You glance out the window and see a flock of geese just below the plane. Your cabin is warm and pressurized. Outside, it is far below zero, and the air is so thin you would pass out almost instantly from lack of oxygen (O_2). But the geese are doing quite well out there.

If you were ever to fly over the Himalaya Mountains between India and China, you might actually see geese this high up. Twice a year, the bar-headed goose migrates over the Himalayas, sometimes at nearly 30,000 feet, en route between its winter quarters in India and its summer breeding grounds in northern Russia.

Closer to home, it is not unusual to see migrating geese and ducks, such as the ones in the photograph at the left, flying as high as 20,000 feet. This altitude is still far out of the comfort range for most humans. Warm clothes insulate our bodies almost as effectively as a bird's feathers, but most of us start feeling light-headed and lethargic at about 10,000 feet and pass out from a lack of oxygen at about 17,000 feet.

How do geese and ducks manage to fly long distances and stay alive at such great heights? One factor is the efficiency of bird lungs, which can extract considerably more oxygen from a given quantity of air than our lungs can. (We will compare bird and human lungs later in this chapter.) Birds such as the mallard and the bar-headed goose also have hemoglobin with a very high affinity for oxygen, picking it up in the lungs and carrying it to body tissues. They also have a very large number of capillaries (tiny blood vessels) that carry oxygen-rich blood to their flight muscles. The flight muscles themselves contain a protein called myoglobin that picks up oxygen from the blood and keeps a ready supply of it in the muscle. All these adaptations combine to allow high-flying birds to obtain enough oxygen even when the air is very thin.

Throughout the world, most people live well below 10,000 feet and are helpless at very high elevations without an oxygen mask. But there are permanent villages as high as

19,000 feet in the Himalayas and nearly that high in the Andes. Like the Nepalese Sherpa shown here, most people born at such heights are small of stature and have unusually large lungs and a large heart. In addition, their blood carries an unusually large amount of oxygen, because they have more hemoglobin and more red blood cells than most of us.

Even if you live at sea level, chances are good that you can condition yourself for fairly high altitudes. If you drive from sea level up into the mountains, your body starts adjusting immediately: Your heart pumps faster, and your capillaries may increase in diameter. If you stay in the mountains more than a few days, your heart slows down, as other changes begin to compensate for the thin air. In a matter of weeks, the rate and depth of your breathing increase, bringing more air into the lungs. At the same time, your body may actually develop more capillaries, and your hemoglobin level and red blood cell count may go up, enabling your blood to carry more oxygen. After rigorous, long-term training, a few mountain climbers have been able to survive for a short time without oxygen masks on the top of the world's highest peak, Mount Everest, at 29,028 feet.

Our study of cellular respiration in Chapter 6 showed why animals require O_2. Without a continuous supply of this gas, most animals cannot obtain enough energy from their food to survive. Without O_2, the metabolic machinery that releases energy from food molecules shuts down or runs far too slowly to keep a goose or human awake, much less flying or running. Energy metabolism also generates waste—notably, the gas carbon dioxide (CO_2). Thus, cells must dispose of CO_2 as well as obtain O_2. The process of **gas exchange**, the interchange of O_2 and CO_2 between an animal and its environment, is the subject of this chapter.

Gas exchange involves breathing, the transport of gases, and the servicing of tissue cells

Gas exchange makes it possible for animals to put to work the food molecules the digestive system provides. The figure here presents an overview of three phases of gas exchange, all of which we examine in detail in this chapter. ① Breathing is one phase of the gas-exchange process. When a bird or mammal inhales air, O_2 passes into cells lining the inside of its lungs. As it exhales, CO_2 passes from lung cells into the environment.

② A second phase of gas exchange is the transport of gases by the circulatory system. It begins with the diffusion of O_2 from the lung cells into blood vessels. The O_2 attaches to hemoglobin and is carried in the blood from the lungs to the body's other tissues. At the same time, the blood also conveys CO_2 from the tissues to the lungs.

③ A third phase of gas exchange occurs in the body tissues, where the cells take in O_2 from the blood and give up CO_2 to the blood. Giving up CO_2 and gaining O_2 makes it possible for the body cells to obtain energy from food molecules by cellular respiration. The O_2 helps a cell "burn" fuel—that is, release energy from the food molecules the body has digested and absorbed. As we saw in Module 6.4, O_2 is a key ingredient in the cell's energy-harvesting operation because it tends to pull electrons away from organic molecules. The CO_2 the cell gives up to the blood is the main waste product that results when food molecules are split apart as the cell harvests energy-rich electrons from them.

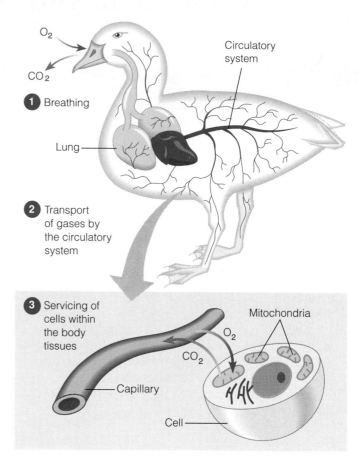

① Breathing

Circulatory system

Lung

② Transport of gases by the circulatory system

③ Servicing of cells within the body tissues

Mitochondria

O_2

CO_2

Capillary

Cell

Animals exchange O_2 and CO_2 across moist body surfaces

The part of an animal where O_2 diffuses across cell membranes into the animal and where CO_2 diffuses into the outside environment is called the **respiratory surface.** (In this context, the word "respiratory" refers to the process of breathing, not to cellular respiration.) Oxygen and carbon dioxide in gaseous form cannot cross membranes, but individual molecules of O_2 and CO_2 can diffuse through membranes if the molecules are first dissolved in water. This tells us that a respiratory surface must be wet in order to function. It must also be extensive enough to take up sufficient O_2 for every cell in the body and to dispose of all waste CO_2. Usually a single layer of moist cells lines the entire respiratory surface. Being thin as well as moist, the layer allows O_2 to diffuse rapidly into the circulatory system or directly into body tissues and allows CO_2 to diffuse out.

The four figures on the facing page illustrate, in simplified form, four types of animal respiratory organs. In each case, the circle represents a cross section of the animal's body in

the region of the respiratory surface. The yellow color, outlined by brown, represents the respiratory surface. The enlargements in the boxes show a portion of the respiratory surface in the process of exchanging O_2 and CO_2.

Some animals use their entire outer skin as a gas-exchange organ. The earthworm in Figure A is an example. Notice in the cross-sectional diagram that its whole body surface is yellow; there are no specialized gas-exchange surfaces. Oxygen diffuses into a dense net of thin-walled capillaries lying just beneath the skin. Earthworms and other "skin-breathers" must live in damp places or in water because their whole body surface has to stay moist. Skin-breathing animals are generally small, and many are long and thin or flattened—the flatworms we saw in Module 19.6, for example. Small size or flatness provides a high ratio of respiratory surface to body volume, ensuring that all body cells will receive enough O_2 and be able to dispose of their waste CO_2.

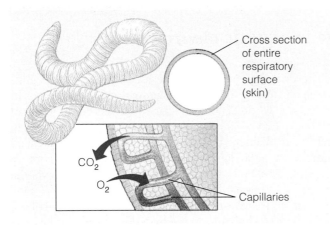

Cross section of entire respiratory surface (skin)

CO_2

O_2

Capillaries

A. Entire outer skin

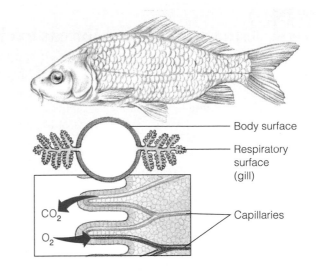

Body surface

Respiratory surface (gill)

CO_2

O_2

Capillaries

B. Gills

In most animals, either the skin surface is not extensive enough to exchange gases for the whole body, or it cannot be kept moist. Consequently, certain parts of the body have become adapted as respiratory surfaces. Gills have evolved in most aquatic animals. Lungs or an internal system of gas-exchange tubes have evolved in most terrestrial animals. Gills, lungs, and internal tubes all have extensive surfaces for gas exchange, as shown in Figures B, C, and D.

Gills are featherlike extensions of parts of the body surface specialized for gas exchange. A fish (Figure B) has a set of gills on each side of its head. As indicated in the enlargement, O_2 diffuses across the gill surfaces into capillaries, and CO_2 diffuses in the opposite direction, out of the capillaries. Since the respiratory surfaces of aquatic animals extend into the surrounding water, keeping the surface moist is not a problem. Gills are generally unsuitable for an animal living on land because the wet surfaces tend to stick together and collapse when exposed to air. Nevertheless, a number of invertebrates with gills live on land at least part of the time. In such animals (certain crabs, for example), the gills are housed in special chambers that keep the respiratory surfaces moist and prevent the gills from collapsing.

In most terrestrial animals, the respiratory surfaces are folded into the body rather than projecting from it. The infolded surfaces open to the air only through narrow tubes, an arrangement that helps retain the moisture that keeps the respiratory surfaces wet.

Insects have an extensive system of internal tubes called **tracheae** (singular, *trachea*) (Figure C). As we will see in Module 22.5, tracheae branch throughout the body, exchanging gases directly with body cells and requiring no assistance from the circulatory system.

The majority of terrestrial vertebrates have **lungs** (Figure D), which are internal sacs lined with a moist epithelium. As the diagram indicates, the inner surfaces of the lungs branch extensively, forming a large respiratory surface. Gases are carried between the lungs and the body cells by the circulatory system.

We examine gills, tracheae, and lungs more closely in the next several modules.

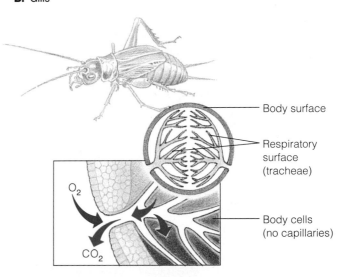

Body surface

Respiratory surface (tracheae)

O_2

CO_2

Body cells (no capillaries)

C. Tracheae

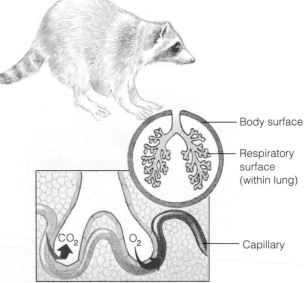

Body surface

Respiratory surface (within lung)

CO_2

O_2

Capillary

D. Lungs

Chapter 22   Respiration: The Exchange of Gases   **435**

Gills are adapted for gas exchange in aquatic environments

Oceans, lakes, and other bodies of water contain O_2 in the form of dissolved gas. The gills of fishes and many invertebrate animals, including lobsters and clams, tap this source of O_2. In fishes, the respiratory surface area of the gills is much greater than the whole remaining body surface. The gills are full of minute blood vessels covered by only one or a few layers of cells; the vessels are so narrow that red blood cells must pass through them in single file. As a result, every red blood cell comes in close contact with oxygen dissolved in the surrounding water.

The chief advantage of exchanging gases in water is that the animal does not have to spend energy to keep its respiratory surface wet. On the other hand, the concentration of available oxygen (dissolved O_2) in water is only about 3–5% what it is in the air, and the warmer and saltier the water, the less dissolved O_2 it holds. Thus, gills—especially those of large, active animals—must be very efficient to obtain enough oxygen from water.

The drawings here show the architecture of fish gills, which are among the most efficient gas-exchange organs in the aquatic world. There are four gill arches on each side of the body. Two rows of gill filaments project from each gill arch. Each filament bears many platelike structures called lamellae (singular, lamella), which are the actual respiratory surfaces. The blue arrows in the drawing represent the route of water flowing over these surfaces.

What you can't see in the drawing are the breathing movements (inhalation and exhalation) that ventilate the gills of the fish. We use the term **ventilation** to refer to any mechanism that increases contact between the surrounding water or air and the respiratory surface (gills or lungs). A fish inhales water by opening its mouth. As it does this, the gill coverings (opercula) on the sides of its body close tightly over the gills. The animal exhales water by closing its mouth and pumping water from its mouth cavity over its gills and out the sides of its body. The gill coverings open during exhalation, allowing the water to escape. These ventilatory movements enhance the gills' gas-exchange efficiency by keeping the water around the gills from stagnating and becoming depleted of oxygen.

In the circular enlargement at the bottom of the figure, notice that the direction of water flow over the fish gills (blue arrows) is opposite that of the blood flow (black arrows) in the gill lamellae. Because of the way the blood vessels are arranged, blood that is low in oxygen (oxygen-poor blood from the body, shown in blue here) arrives at one side of a lamella. Oxygen-rich blood (blood laden with oxygen; red here) then leaves the gills from the other side of the lamella. The opposite flows of blood and water here constitute countercurrent flow, which, as we see in the next module, further increases the efficiency of the gills.

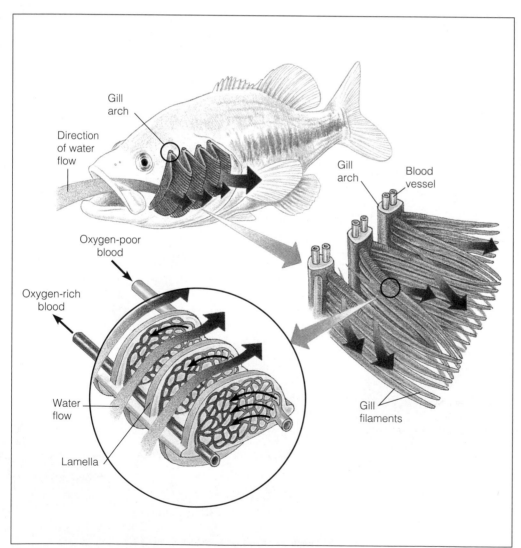

Gill arch

Direction of water flow

Oxygen-poor blood

Oxygen-rich blood

Water flow

Lamella

Gill arch

Blood vessel

Gill filaments

Countercurrent flow in the gills enhances O₂ transfer

Countercurrent exchange is the transfer of something from a fluid moving in one direction to another fluid moving in the opposite direction. The name comes from the fact that the two fluids are moving *counter* to each other. As the two fluids pass each other, their opposite flows maintain a concentration gradient that enhances transfer of the substance. Let's see how this principle works in a fish gill.

Figure 22.3 showed us that water flows over the surface of a lamella in one direction, while blood is flowing in the opposite direction within the lamella. In the diagram here, the changing colors of the large arrows indicate the changing amount of O_2 dissolved in each fluid: the dark-

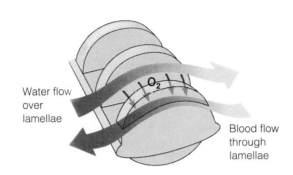

Water flow over lamellae

O₂

Blood flow through lamellae

er the color, the higher the concentration of O_2. The small red arrows represent O_2 diffusing from the water into the blood. Notice that as the blood flows through a lamella and picks up more and more O_2, the blood comes in contact with water that has more and more O_2 available. As a result, a concentration gradient is maintained that favors the transfer of O_2 from the water to the blood along the entire surface of the lamella. Because of this, O_2 diffuses from the water into the blood all along the lamella. So efficient is this countercurrent exchange mechanism that fish gills can remove more than 80% of the oxygen dissolved in the water flowing through them.

The tracheal system of insects provides direct exchange between the air and the body cells

Unlike water-dwellers, land animals exchange gases by breathing air. There are two big advantages to doing so. For one thing, a given volume of air contains much more O_2 than an equal volume of water. Also, air is much lighter and easier to move than water. Thus, a terrestrial animal expends much less energy than an aquatic animal ventilating its respiratory surface.

The tracheal system of insects minimizes energy expense in gas exchange. Insect tracheae are different from the vertebrate trachea (windpipe). Whereas the trachea in a mammal or bird is a single straight tube, the tracheae in an insect are tubes that branch throughout the body, as Figure A indicates. The tubes convey air directly to the body cells, and no energy is spent in moving gases via the circulatory system.

Figure A also shows that an insect's tracheal system includes a number of enlargements called air sacs. The air

sacs work like bellows. When muscles surrounding them contract, air is forced out of the system through openings on the body surface; when the muscles relax, air is drawn in.

The main problem facing any air-breathing animal is the drying effect of the atmosphere. Figure B indicates how the tracheal system is adapted to minimize water loss. Air passes inward through smaller and smaller tubes. The narrowest tubes extend to nearly every cell in the insect's body. The tiny tips of the tracheae are closed and contain fluid (dark blue) that dissolves O_2 and CO_2. The O_2 diffuses into the body cells from the tracheal fluid, and CO_2 diffuses from the cells to the fluid and then out through the tracheae. With the moist respiratory surface in tiny tubes deep in the body, very little water can evaporate from the respiratory system.

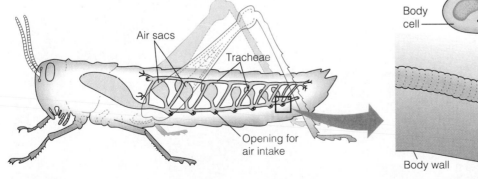

Air sacs

Tracheae

Opening for air intake

A. The tracheal system of an insect

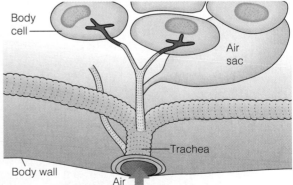

Body cell

Air sac

Trachea

Body wall

Air

B. Tracheal connections to body cells

Terrestrial vertebrates have lungs

Reptiles, birds, mammals, and many amphibians exchange gases in lungs. In contrast to the tracheae of insects, lungs are restricted to one location in the body. Therefore, the circulatory system must transport gases between the lungs and the rest of the body.

Amphibians have small lungs (some lack lungs altogether) and rely heavily on diffusion of gases across other body surfaces. The moist skin of frogs, for example, supplements gas exchange in the lungs. Compared to the lungs of amphibians, those of reptiles, birds, and mammals are very large and contain an enormous respiratory surface relative to the animal's body size. The total respiratory surface of human lungs, for instance, is about equal to the surface area of a tennis court. Most reptiles and all birds and mammals rely entirely on their lungs for gas exchange. Turtles are an exception; their rigid shell restricts breathing movements, and they supplement lung breathing with gas exchange across moist surfaces in their mouth and anus.

Figure A shows the organs of the human respiratory system (along with the esophagus and heart, for orientation). Our lungs reside in the chest cavity, which is bounded at the bottom by a thin sheet of muscle called the **diaphragm.** Air is conveyed to our lungs by a system of branching tubes. The tubes are narrow, minimizing loss of water by evaporation from the moist respiratory surface in the lungs.

Air usually enters our respiratory system through the nostrils. It is filtered by hairs, and warmed, humidified, and sampled for odors as it flows through a maze of spaces in the nasal cavity. Air can also be drawn in through the mouth, but mouth breathing does not allow the air to be processed by the nasal cavity. From the nasal cavity or mouth, air passes to the **pharynx,** where the paths for air and food cross. As we saw in Module 21.6, the air passage in the pharynx is open for breathing except when we swallow.

From the pharynx, air is inhaled into the **larynx** (voicebox). When we exhale, the outgoing air rushes by a pair of **vocal cords** in the larynx, and we can produce sounds by voluntarily tensing muscles in the voicebox, stretching the cords and making them vibrate. We produce high-pitched sounds when our vocal cords are short and tense and therefore vibrating very fast. When the cords are less tense, they vibrate slowly and produce low-pitched sounds.

From the larynx, inhaled air passes toward the lungs through the **trachea,** or windpipe. Rings of cartilage maintain the shape of the trachea, much as metal rings keep the hose of a vacuum cleaner from collapsing. The trachea forks into two **bronchi** (singular, *bronchus*), one leading to each lung. Within the lung, the bronchus branches repeatedly into finer and finer tubes called **bronchioles.**

Though gas exchange occurs only in the lungs, our respiratory system is lined by a moist epithelium from the nasal cavity and mouth all the way to the lungs. The epithelium lining the trachea and all but the smallest bronchioles are covered by cilia and a thin film of mucus. The cilia and mucus are the system's cleaning elements. The mucus traps

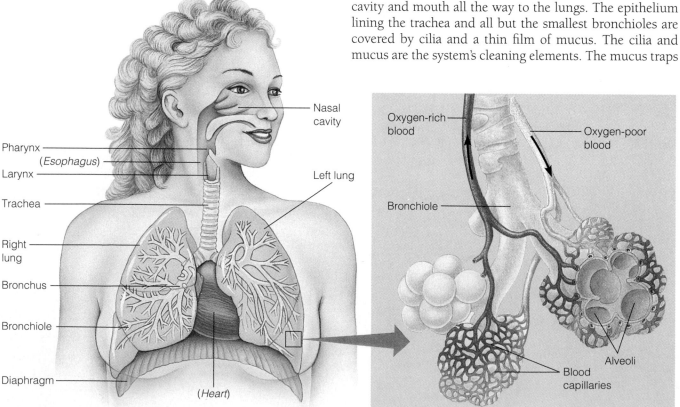

A. The human respiratory system

B. The structure of alveoli

dust, pollen, and other contaminants, and the beating cilia move the mucus upward to the pharynx, where it is usually swallowed.

As Figure B shows, the bronchioles dead-end in grapelike clusters of air sacs called **alveoli** (singular, *alveolus*). Each of our lungs contains millions of these tiny sacs. Figure C is an electron micrograph of lung tissue. Each of the spaces in the micrograph is an alveolus. Although you can't see them here, the inner surface of each alveolus is lined with a

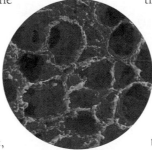

C. An electron micrograph of human alveoli
Colorized SEM 168×

thin layer of epithelial cells that form the respiratory surface, where gases are actually exchanged. The O_2 in inhaled air dissolves in a film of moisture on the epithelial cells. It then diffuses across the epithelium and into a web of blood capillaries that surrounds each alveolus (shown in Figure B). The CO_2 diffuses the opposite way—from the capillaries, across the epithelium of the alveolus, into the air space of the alveolus, and finally out in the exhaled air.   ·

Smoking is one of the deadliest assaults on our respiratory system

The epithelial tissue lining our respiratory system is extremely delicate. Its only protection is the mucus covering the cells and the beating cilia that sweep dirt particles off their surfaces. Virtually everywhere today, the air we breathe exposes the cells in our respiratory system to chemicals that they are not adapted to tolerate. A breath of air in a polluted city may contain thousands of chemicals, many of which are potentially harmful. Air pollutants such as sulfur dioxide, carbon monoxide, and ozone are associated with serious respiratory diseases. Asbestos, once widely used as building insulation, can cause lung cancer. When buildings are torn down or renovated, asbestos must be removed by professionals or sealed over with paint to prevent fine particles from becoming airborne. Another air pollutant, radon, is a naturally occurring radioactive gas emitted from soil and bedrock. Entering homes in some parts of the United States through basement walls and floors, radon may cause lung cancer.

One of the worst sources of toxic air pollutants is cigarette smoke. A single drag on a cigarette exposes a person to over 4000 different chemicals, many of which are known to be harmful and potentially deadly. Every year in the United States, smoking kills about 350,000 people, more than all the deaths caused by traffic accidents or illegal drug abuse. Every cigarette a person smokes probably takes about five minutes away from his or her life expectancy.

Components of cigarette smoke are known to irritate the cells lining the bronchi, destroying their cilia. This interferes with the normal cleansing mechanism of the respiratory system, allowing more smoke and the dirt particles and foreign chemicals in inhaled air to reach the delicate alveoli. Frequent coughing—common to heavy smokers—is the

only way the system can clean itself. A disease known as emphysema can develop as cigarette smoke makes the thin-walled alveoli brittle. Many of the alveoli eventually rupture, reducing the lungs' capacity for gas exchange. Breathlessness and constant fatigue result, as the body is forced to spend more and more energy just breathing. The heart is also forced to work harder, and emphysema often leads to heart disease.

Lung cancer is the deadliest legacy of smoking, almost always killing its victims. The left photograph below shows a cutaway view of a pair of healthy human lungs (and heart). The photograph on the right shows the lungs of a smoker with cancer. The lungs are black from the long-term buildup of smoke particles—except where pale cancerous tumors appear. A person who smokes cigarettes is at least ten times more likely to develop lung cancer than a nonsmoker. And increasing evidence indicates that nonsmokers exposed to secondary cigarette smoke in the environment are also at risk. Efforts are being made to reduce the well-documented dangers of smoking. Some of these measures are banning cigarette ads on TV, prohibiting smoking on airplanes, and creating nonsmoking areas in restaurants.

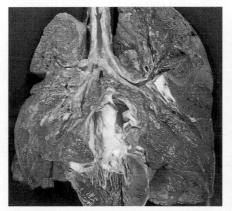

Healthy lungs

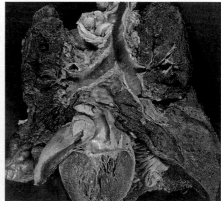

Cancerous lungs

Breathing ventilates the lungs

Breathing is the alternation of inhalation and exhalation. It supplies our lungs with O_2-rich air and expels the CO_2-rich gas mixture that remains after O_2 and CO_2 are exchanged in the alveoli. Like all mammals, we ventilate our lungs by sucking air into the alveoli and then pushing it back out.

Figure A shows the changes that occur in our rib cage, chest cavity, and lungs during breathing. During inhalation (left diagram), both the rib cage and chest cavity expand, and the lungs follow suit. You can feel what happens if you touch your rib cage and inhale deeply. The ribs move upward and the rib cage expands as muscles between the ribs contract. Meanwhile, the diaphragm contracts, moving downward and expanding the chest cavity as it goes.

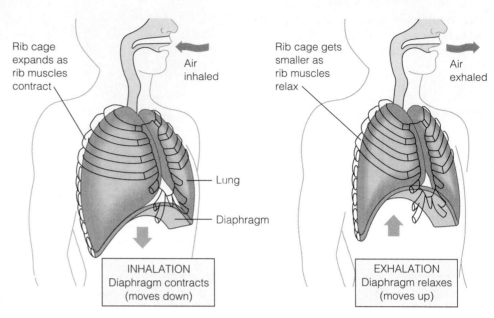

A. How a human breathes

The lungs expand and fill with air during inhalation because the enlargement of the rib cage and chest cavity reduces the air pressure in the alveoli to less than that of the atmosphere. Flowing from a region of higher pressure to one of lower pressure, air rushes through the nostrils and down the breathing tubes to the alveoli.

The right-hand diagram in Figure A shows exhalation. The rib muscles and diaphragm both relax, decreasing the volume of the rib cage and chest cavity and forcing air out of the system. Notice that the diaphragm is enlarged and has a domelike shape when relaxed.

Each year, a human adult may take between 4 million and 10 million breaths. The volume of air in each breath is about 500 mL when we breathe quietly. The maximum volume of air that we can inhale and exhale—as in strenuous exercise—is called **vital capacity.** It averages about 3500 mL and 4800 mL for college-age females and males, respectively. (Women tend to have smaller rib cages and lungs.) Vital capacity depends heavily on the resilience (springiness) of the lungs. The lungs actually hold more air than the vital capacity. Because the alveoli do not completely collapse, a residual volume of "dead" air remains in the lungs even after we blow out as much air as we can. As lungs lose resilience with age or as the result of disease, such as emphysema, our residual volume increases at the expense of vital capacity.

As mentioned in this chapter's introduction, the gas-exchange system of birds is different from ours. Let's return now to the bird to see some of the adaptations that make its respiratory system a more efficient one.

In contrast to the in-and-out flow of air in the human alveoli, the bird's system maintains a one-way flow of air through the lungs. Birds have several large air sacs in addition to their lungs. As the simplified diagrams in Figure B indicate, the air sacs' function is to ventilate the respiratory surface in the lungs. During inhalation, both sets of air sacs expand. The posterior sacs fill with fresh air (blue) from the

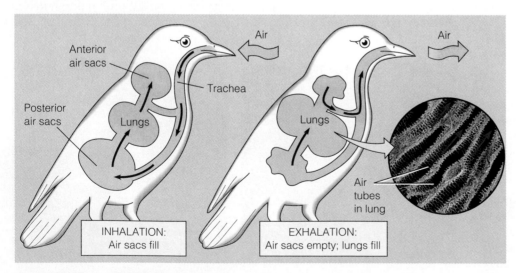

B. The respiratory system of a bird

outside, while the anterior sacs fill with stale air (gray) from the lungs. During exhalation, both sets of air sacs deflate, forcing air from the posterior sacs into the lungs, and air from the anterior sacs out of the system via the trachea.

The electron micrograph in the circular inset shows another key feature in the bird. Instead of alveoli, bird lungs contain tiny, parallel tubes (each about 0.5 mm across). Air passes one-way through the tubes (red arrows), and gases are exchanged across the walls of the tubes. Because of the one-way flow of air, there is no dead air (residual volume) in the bird lung. As a result, birds can extract about 5% more oxygen from a volume of inhaled air than we can.

Breathing is automatically controlled

What controls our breathing? We obviously have some conscious control over it because we can voluntarily hold our breath and voluntarily breathe faster and deeper. Most of the time, however, automatic control centers in our brain regulate our breathing movements. Automatic control is essential, for it ensures coordination between the respiratory system and other body systems, especially the circulatory system.

Our **breathing control centers** (represented by the blue circles in the drawing below) are located in parts of the brain called the pons and the medulla oblongata (medulla, for short). Nerves from the medulla's control center (solid yellow arrows) signal the diaphragm and rib muscles to contract, making us inhale. These nerves send out about 10–14 signals per minute when we are at rest. Between nerve signals, the muscles relax, and we exhale. The control center in the pons smooths out the basic rhythm of inhalation and exhalation set by the medulla.

The medulla's control center also regulates how much CO_2 our alveoli dispose of. It does so by monitoring the pH of the blood. Blood pH starts to drop when the amount of carbon dioxide in the blood increases (blue arrow). When we exercise vigorously, for instance, our metabolic rate goes up and our body cells generate more CO_2 as a waste product. The CO_2 goes into the blood where it reacts with water, forming carbonic acid. The acid lowers the pH of the blood slightly. When this happens, the medulla senses the pH drop, and its breathing control centers increase the breathing rate and depth. As a result, more CO_2 is eliminated in the exhaled air, and blood pH returns to normal.

When you were a kid, did you ever try making yourself dizzy by **hyperventilating,** rapidly taking several deep breaths? Hyperventilating demonstrates the action of your breathing control centers, but it's hard on your body. Deep, rapid breathing can purge the blood of so much CO_2 that the control centers temporarily cease to send signals to the rib muscles and diaphragm. Breathing actually stops until the CO_2 level increases enough to switch the breathing centers back on.

Our breathing control centers respond directly to CO_2, levels, but they usually do not respond directly to oxygen levels. So how is the level of blood *oxygen* controlled? Since the same process that consumes O_2, cellular respiration, also produces CO_2, a drop in pH (rise in CO_2) in the blood is generally a good indication of a drop in blood oxygen.

Thus, by responding to lowered pH, the breathing control centers also control blood oxygen level. Also, some of our large arteries have O_2 sensors, and, when the O_2 level in the blood is severely depressed, these sensors signal the control centers via nerves (dashed yellow arrow in the diagram) to increase the rate and depth of breathing. This response may occur, for example, at high altitudes, where the air is so thin that we cannot get enough O_2 by breathing normally.

By responding quickly to the signals they receive from the blood and the arterial sensors, our breathing control centers keep the rate and depth of our breathing in tune with the changing demands of the body. The breathing centers are only effective, however, if their regulatory activity is coordinated with the activity of the circulatory system. During exercise, the rate at which our heart beats and the amount of blood it pumps with each beat must be increased to match the increased breathing rate. We examine the role of the circulatory system in gas exchange in more detail in the next module.

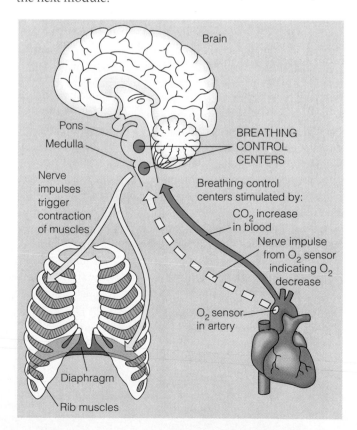

Brain

Pons

Medulla

Nerve impulses trigger contraction of muscles

Diaphragm

Rib muscles

BREATHING CONTROL CENTERS

Breathing control centers stimulated by:

CO_2 increase in blood

Nerve impulse from O_2 sensor indicating O_2 decrease

O_2 sensor in artery

Blood transports the respiratory gases, with hemoglobin carrying the oxygen

We have yet to focus on how O_2 gets from our lungs to all the other tissues in our body, and on how CO_2 travels from the tissues to the alveoli. To do so, we must jump ahead a bit, to the subject of Chapter 23, and look at the basic organization of our circulatory system.

Figure A shows the main components of the human circulatory system and their roles in gas exchange. Let's start with the heart, in the middle of the diagram. One side of the heart handles oxygen-poor blood (colored blue). The other side handles oxygen-rich blood (red). At the lower left in the diagram, oxygen-poor blood returns to the heart from body tissues. The heart pumps this blood to the alveolar capillaries in the lungs. At the top of the diagram, gases are being exchanged between the air in the alveolar spaces and the blood in the capillaries. Off to the right, blood leaves the alveolar capillaries, having gained O_2 and having lost CO_2. This oxygen-rich blood returns to the heart, and the heart pumps it out to the body tissues.

The small red and blue bars in Figure A indicate the relative amounts of O_2 and CO_2 at each critical point where gases are exchanged. At the top, for instance, the bars indicate that there is less O_2 and more CO_2 in the exhaled air (right side) than in the inhaled air (left side). The exhaled air contains O_2 because the respiratory system is not 100% efficient at extracting it.

The red and blue bars also help explain how gases are exchanged. At the bottom of the diagram, for instance, oxygen moves out of the blood into the tissue cells because, like all substances, it diffuses from a region where it is more concentrated to a region where it is less concentrated. The tissue cells maintain the concentration gradient by consuming O_2 in cellular respiration. Meanwhile, as O_2 enters the tissue cells, CO_2, produced as a waste product of cellular respiration, diffuses down its own concentration gradient out of the cells and into the capillaries. Diffusion also accounts for gas exchange in the alveoli.

Oxygen is not very soluble in water, and the blood actually transports very little of it as dissolved O_2. Most of the O_2 in blood is carried by **hemoglobin** in the red blood cells. As indicated in Figure B, a hemoglobin molecule consists of four polypeptide chains (of two different types, distinguished by the two shades of purple). Attached to each polypeptide is a chemical group called a heme (blue-green), at the center of which is an iron atom (black). Each iron atom can carry an O_2 molecule. Thus, a hemoglobin molecule can carry up to four molecules of oxygen. Hemoglobin loads up with oxygen in the lungs, retains it while traveling in the bloodstream, and then unloads it in the body tissues.

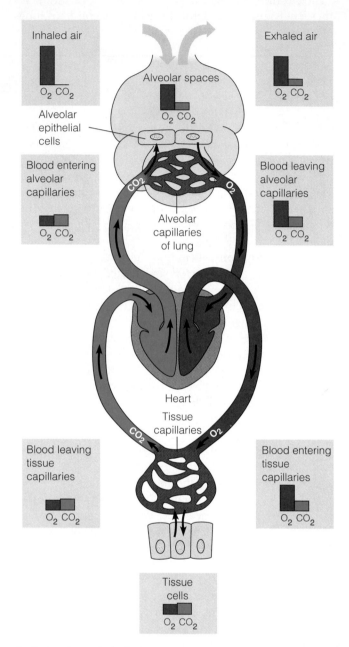

A. Gas exchange in the body

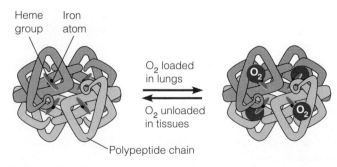

B. Hemoglobin loading and unloading of O_2

Hemoglobin is a multipurpose molecule. It transports oxygen, loading and unloading it when diffusion allows, and it also helps the blood transport carbon dioxide. A third important function of hemoglobin is helping to buffer the blood—that is, prevent changes in pH that a cell could not survive.

The diagrams below illustrate how CO$_2$ is carried in the blood. As shown in Figure A, when CO$_2$ leaves a tissue cell, it first diffuses into the interstitial fluid (light blue) and then across the wall of a capillary into the blood fluid (plasma, shown in yellow). A small amount of the CO$_2$ remains dissolved in the plasma, but most of it enters the red blood cells, where some of it combines with hemoglobin. The rest reacts with water molecules, forming carbonic acid (H$_2$CO$_3$). The H$_2$CO$_3$ forms more readily inside the red blood cells than in the plasma because the cells contain an enzyme that hastens the reaction.

As shown in the center of the red blood cell in Figure A, H$_2$CO$_3$ does not remain intact. It breaks apart into a hydrogen ion (H$^+$) and a bicarbonate ion (HCO$_3^-$). Hemoglobin acts as a buffer by picking up most of the H$^+$ ions, thus preventing them from acidifying the blood. Most of the bicarbonate ions diffuse out of the red blood cell into the blood plasma and are carried in this form to the lungs.

Figure B shows what happens to the bicarbonate ions and CO$_2$ as the blood flows through the lungs. The processes in Figure A are reversed. Carbonic acid forms when bicarbonate combines with H$^+$ ions given up by hemoglobin. The carbonic acid is then converted back into CO$_2$ and water. Finally, the CO$_2$ diffuses from the blood into the alveoli, and out of the body in exhaled air.

Carbon dioxide is a waste gas, but as we see here, the body puts CO$_2$ to good use before disposing of it. The CO$_2$ built into the bicarbonate ions in the blood is an important part of the blood-buffering system. If the pH of the plasma drops (that is, if the concentration of H$^+$ increases), the bicarbonate ions remove H$^+$ ions by combining with them to form carbonic acid. When the pH rises, the carbonic acid releases H$^+$ back into the blood.

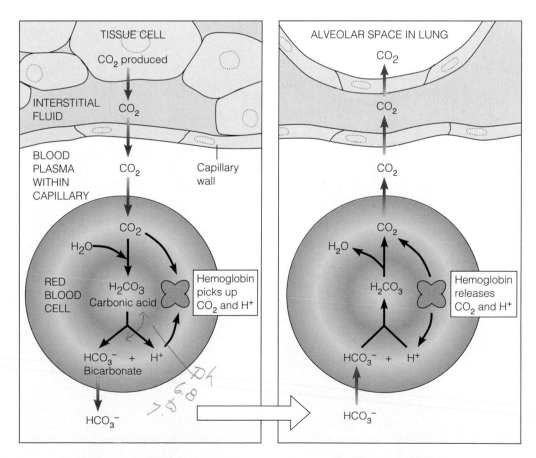

A. CO$_2$ transport in the body tissues

B. CO$_2$ transport in the lungs

The human fetus exchanges gases with the mother's bloodstream

The illustration here shows a human fetus inside the mother's uterus. The fetus literally swims in a protective watery bath, the amniotic fluid. The lungs are full of fluid and are nonfunctional. How does the fetus exchange gases with the outside world? The answer lies in the function of the placenta, a composite organ that includes tissues from both the fetus and the mother. A large net of capillaries fans out into the placenta from blood vessels in the umbilical cord of the fetus. These capillaries exchange gases with the maternal blood that circulates in the placenta, and the maternal circulatory system carries the gases to and from the mother's lungs. Enhancing gas exchange in the placenta, the maternal and fetal blood flow in opposite directions, providing countercurrent exchange, as we saw in fish gills in Module 22.4. Also aiding O_2 uptake by the fetus is the fetal hemoglobin, a special type that attracts O_2 more strongly than the mother's hemoglobin.

What happens when a baby is born? Suddenly placental gas exchange ceases, and the baby's lungs must begin to work. Carbon dioxide in the fetal blood acts as a signal. As soon as CO_2 stops diffusing from the fetus into the placenta, a CO_2 rise in the fetal blood causes a drop in the blood pH. The pH drop stimulates the breathing control centers in the infant's brain, and the newborn takes its first breath.

A human birth and the radical changes in gas-exchange mechanisms that accompany it are extraordinary events. Resulting from millions of years of evolutionary adaptation, these events are on a par with the remarkable flight ability of the bar-headed goose we discussed in the chapter's introduction. For a goose to breathe the thin air and fly great distances high above Earth, or for a human baby to switch almost instantly from living in water and exchanging gases with maternal blood to breathing air directly, requires truly remarkable adaptations in the organism's respiratory system. Also required are adaptations of the circulatory system, which, as we have seen, supports the respiratory system in its gas-exchange function. We turn to circulatory systems themselves in Chapter 23.

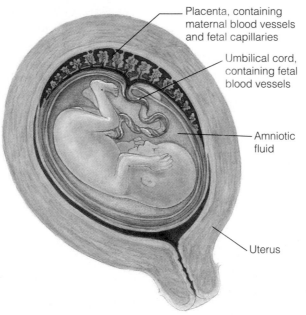

Placenta, containing maternal blood vessels and fetal capillaries

Umbilical cord, containing fetal blood vessels

Amniotic fluid

Uterus

Chapter Review

Begin your review by rereading the module headings and Module 22.1 and scanning the figures before proceeding to the Chapter Summary and questions.

Chapter Summary

Introduction–22.2 Because energy metabolism requires oxygen and produces carbon dioxide, an animal must exchange these gases with the environment. Gas exchange occurs by diffusion through a thin, moist respiratory surface. Some animals, like the earthworm, use their entire skin as a gas-exchange organ. In most animals, specialized body parts carry out gas exchange.

22.3–22.4 Gills are extensions of the body that absorb O_2 dissolved in water. In a fish, gill filaments bear numerous platelike lamellae, which are packed with blood vessels. Blood flows through the lamellae in a direction opposite to water flow; this countercurrent maintains a concentration gradient that maximizes the uptake of O_2.

22.5 Land animals exchange gases by breathing air. Air contains more oxygen and is easier to move than water, but it can dry respiratory surfaces. In insects, a network of tracheal tubes carries out gas exchange. The movements of air sacs pump air in and out, and O_2 diffuses from the tracheae into cells.

22.6 Terrestrial vertebrates have lungs. In humans and other mammals, air enters through the nasal cavity and passes through the pharynx and larynx (the site of the vocal cords) into the trachea (windpipe). The trachea forks to form two bronchi, each of which branches into numerous bronchioles. The bronchioles end in clusters of tiny sacs called alveoli, which form the respiratory surface of the lungs. Oxygen diffuses through the thin walls of the alveoli into the blood.

22.7 Mucus and cilia in the respiratory passages protect the lungs, but pollutants can destroy or overcome these protections. Smoking kills thousands of people each year.

22.8 Breathing supplies the lungs with O_2-rich air and expels CO_2-rich air. During inhalation, muscles expand the rib cage, and the diaphragm contracts and moves downward. The chest cavity expands, reducing air pressure in the alveoli. Air rushes through the respiratory passages into the lungs. During exhalation, the rib muscles and diaphragm relax, forcing air out. Vital capacity is the maximum volume of air we can inhale and exhale, but our lungs hold more air than this amount. Diseases such as emphysema increase residual dead air at the expense of vital capacity.

22.9 Breathing control centers in the pons and medulla of the brain keep breathing in tune with body needs. During exercise, the CO_2 level in the blood rises, lowering blood pH. The medulla senses this drop and increases the rate and depth of breathing, eliminating

CO_2 and returning blood pH to normal. Since CO_2 is produced by the same process that uses oxygen, pH changes reflect O_2 levels; the control centers do not normally respond directly to a lack of oxygen.

22.10 The heart pumps oxygen-poor blood to the lungs, where it picks up O_2 and drops off CO_2. Then it pumps the oxygen-rich blood to body cells, where it drops off O_2 and picks up CO_2. At the tissues, O_2 diffuses down its concentration gradient to the cells, and CO_2 diffuses down its concentration gradient into the blood. In the lungs, the gradients are reversed; CO_2 leaves the blood, and O_2 enters from the air.

22.10–22.11 The protein hemoglobin in red blood cells carries most of the oxygen in the blood. It loads up with oxygen in the lungs and unloads it in the tissues. Hemoglobin also helps buffer the pH of blood and carries some CO_2.

22.11 Most CO_2 in the blood combines with water to form carbonic acid. In turn, carbonic acid breaks down to form bicarbonate ions and H^+ ions, which help buffer blood pH. Most CO_2 is transported to the lungs in the form of bicarbonate ions.

22.12 A human fetus depends on the placenta for gas exchange. A network of capillaries exchanges O_2 and CO_2 with maternal blood that carries gases to and from the mother's lungs. At birth, increasing CO_2 in the fetal blood stimulates the fetus's breathing control center to initiate breathing.

Testing Your Knowledge

Multiple Choice

1. When you exhale, air passes through the respiratory structures in which of the following sequences?
    a. alveolus, trachea, bronchus, bronchiole, larynx, pharynx, nasal cavity
    b. alveolus, bronchiole, bronchus, trachea, pharynx, larynx, nasal cavity
    c. alveolus, bronchiole, bronchus, trachea, larynx, pharynx, nasal cavity
    d. alveolus, bronchus, bronchiole, trachea, larynx, pharynx, nasal cavity
    e. alveolus, bronchiole, bronchus, larynx, trachea, pharynx, nasal cavity

2. Countercurrent exchange in the gills of a fish
    a. speeds up the flow of water through the gills
    b. maintains a concentration gradient that enhances diffusion
    c. enables the fish to obtain oxygen while swimming backward
    d. means that blood and water flow in the same direction
    e. interferes with the efficient absorption of oxygen

3. When you inhale, the diaphragm
    a. relaxes and moves upward
    b. relaxes and moves downward
    c. contracts and moves upward
    d. contracts and moves downward
    e. is not involved in the breathing movements

4. In _____ , oxygen diffuses directly from the air through a moist surface to cells, without being carried by the blood.
    a. an ant   d. a sparrow
    b. a whale   e. a mouse
    c. an earthworm

5. Which answer ranks these locations in order of decreasing oxygen concentration (highest to lowest)? (*Explain your answer.*)
    a. tissue cells, inhaled air, blood leaving lungs
    b. inhaled air, blood leaving lungs, tissue cells
    c. blood leaving lungs, inhaled air, tissue cells
    d. inhaled air, tissue cells, blood leaving lungs
    e. tissue cells, blood leaving lungs, inhaled air

Describing, Comparing, and Explaining

1. What are two advantages of breathing air, compared to obtaining dissolved oxygen from water? What is a comparative disadvantage of breathing air?

2. Trace the path of an oxygen molecule from the air to a muscle cell in your arm, naming all the structures involved along the way.

3. Explain how the structure and arrangement of alveoli are well suited to their function in gas exchange.

4. Compare and contrast the structure and function of human lungs and bird lungs.

Thinking Critically

1. Researchers have discovered that under normal conditions, hemoglobin does not unload all its oxygen in body tissues, but rather keeps some in reserve. An increase in blood temperature or a drop in pH (increase in acidity) will cause hemoglobin to release some of this extra oxygen. How might this phenomenon help exercising muscles?

2. An increase in breathing rate is one symptom of poisoning by acidic substances such as aspirin. Explain how an aspirin overdose would cause breathing rate to increase, and how this increase would help the victim.

3. Briefly explain how each of the following might interfere with gas exchange:
    a. An earthworm's skin dries when exposed to air above the ground.
    b. Iron deficiency can cause a decrease in the number of red blood cells.
    c. Cystic fibrosis is an inherited disease in which the surfaces of the alveoli become coated with thick, sticky mucus.
    d. A farmer sprays thick oil on insects attacking fruit trees.
    e. Carbon monoxide in cigarette smoke takes the place of oxygen in hemoglobin molecules.
    f. As the water in a fish pond warms up in summer, its dissolved O_2 concentration drops.

Science, Technology, and Society

1. In 1990, the tobacco industry briefly succeeded in having a respected scientist, Dr. David Burns, removed from a U.S. Environmental Protection Agency advisory panel reviewing a report on the effects of passive smoking. Scientists agreed that Burns was a leading authority in the field, having worked on several studies and reports on the effects of smoking on health. The Tobacco Institute, an industry group, protested his selection, stating, "Frankly, we are mystified how an individual with Dr. Burns's long and intense involvement with the antismoking movement can be expected to contribute to a reasonable, objective, examination" of the EPA report. Do you agree that this is a reason for dismissal from the panel of experts? Six of the members of the same panel came under fire because they had conducted research financed by the tobacco industry. Would they be expected to be reasonable and objective? How should experts be selected?

2. The mineral asbestos is known to cause lung cancer in workers who have used it in building construction. There is asbestos in pipe insulation, fireproofing, tile, and acoustical plaster in many older buildings. Many experts think the presence of asbestos presents a cancer risk to the general public, so millions of dollars have been spent during the last decade to remove it. Environmental tobacco smoke (passive smoking) is estimated to represent 200–400 times the cancer risk of low-level asbestos exposure, but to date there has been no large-scale effort to protect the public from smoke. Why do you think the effort to protect the public from these two pollutants has not been proportional to their risks?

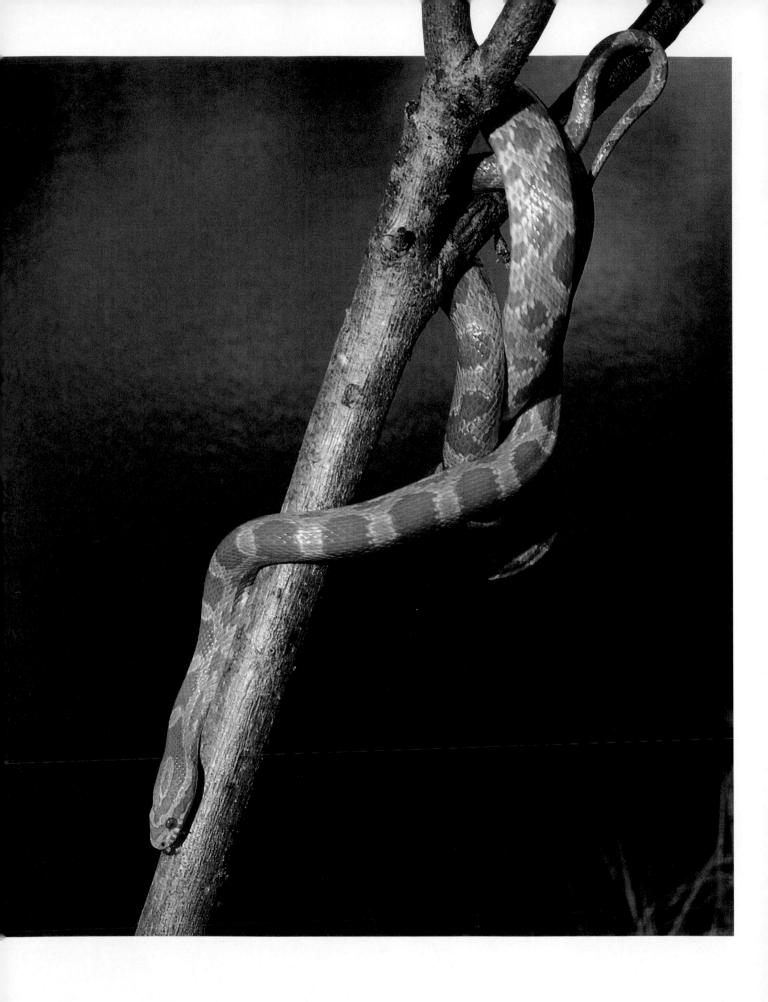

ew animals seem less alike than those on these two pages. At the left is a corn snake, found throughout much of the U.S. This nonpoisonous predator eats mostly rats and mice, but it can also climb trees and dine on bird eggs. The giraffe (below) lives in central Africa. It is an herbivore, using its long neck to browse trees.

Despite their differences, snakes and giraffes have many features in common. As land vertebrates, they both have a backbone, lungs, and a circulatory system that carries oxygen, carbon dioxide, and other materials. They also have something in common with *all* animals that live on land— every part of their body is subject to the persistent, unwavering force of gravity.

Gravity does not greatly affect aquatic animals because their bodies are supported by water, but it has profound effects on terrestrial species. Our own bodies show signs of constant, long-term exposure to gravity as the skin on our face begins to sag with age. As is generally true for vertically oriented animals, our circulatory system is strongly affected by gravity. Gravity tends to pull blood downward into the lower parts of the body. The pull of gravity is a problem for us because we stand upright on two legs; the giraffe is affected because of its long neck and legs; the corn snake faces the problem when it climbs a tree after bird eggs. If gravity prevailed, blood would accumulate in our legs and in the giraffe's, and when a snake climbed upward, its long, thin body would bulge like a balloon at the tail.

What are the solutions to these problems? Mammals, including humans and giraffes, have very strong hearts that keep blood circulating despite gravity's pull. The blood flow slows as it travels uphill through the veins back to the heart, but muscle contractions help it on its way. As we walk or run, our leg muscles squeeze the veins and force the blood upward toward the heart. Our veins also have valves that allow the blood to flow in only one direction, preventing it from flowing back down the legs. Giraffes and humans are also endowed with tight skin and abundant connective tissue in the legs. These adaptations help keep blood vessels there from enlarging. In the same way, the tight suits worn by fighter pilots counteract the g-forces pulling on their blood during high-speed maneuvers.

Despite the safeguards that help prevent blood from pooling in our legs, gravity sometimes wins. If we don't exercise adequately or if we stand too long in one position, our circulation may slow down, and blood can pool in our leg veins. This is often a problem for people who work standing up. It can also be a problem for a giraffe standing for long periods of time in a zoo. Over time, reduced circulation weakens the leg veins, causing them to stretch and enlarge. The valves in the veins also weaken and become less effective at preventing the backflow of blood. As a result, some people develop varicose veins in their legs— veins just under the skin that become swollen and unsightly because they have too little support from other tissues.

The corn snake does not have valves in its veins, but its blood still circulates, even when it is climbing. Because its heart is located close to its head, the snake's brain receives enough blood even when it is vertical. Also, its thin, tight body helps prevent blood pooling, and blood vessels in its tail constrict when it climbs. After a climb, a corn snake wriggles vigorously. This motion contracts muscles all over its body, squeezing veins and increasing circulation.

The ability to keep the blood circulating despite the pull of gravity is a matter of life or death for most animals. All organisms must exchange materials with their environment and distribute materials within their bodies. Most animals have a system of internal transport—a **circulatory system**—that transports oxygen and carbon dioxide, distributes nutrients to body cells, and conveys wastes to specific sites for disposal. This chapter focuses on the structure, function, and evolution of circulatory systems.

23.1 The circulatory system associates intimately with all body tissues

A circulatory system is necessary in any animal whose body is too large or too complex for vital chemicals to reach all its parts by diffusion alone. Diffusion is inadequate for transporting chemicals over distances greater than a few cell widths—far less than the distance oxygen must travel between our lungs and brain, or the distance nutrients must go between our small intestine and the muscles in our arms or legs. Without our circulatory system, nearly every cell in the body would quickly die from lack of oxygen.

To be effective, the circulatory system must have an intimate connection with the tissues. In the human body, for instance, the heart pumps blood that has just been oxygenated in the lungs through a system of blood vessels into microscopic vessels called **capillaries**. The capillaries form an intricate net among the tissue cells, such that no substance has to diffuse far to enter or leave a cell. The micrograph in Figure A illustrates the relationship between capillaries and our body tissues. This particular capillary supplies freshly oxygenated, nutrient-rich blood to smooth muscle cells. (The oval structures outside the capillary, each a little bigger than a red blood cell, are nuclei of the muscle cells; see Module 20.6.) Notice that red blood cells pass single-file through the capillary. Each red blood cell comes close enough to the surrounding tissue that O_2 and nutrients can diffuse out of it into the muscle cells.

In Figure B, the downward arrows show the route that molecules take in diffusing from the blood into tissue cells. As we discussed in Module 20.11, materials are not exchanged directly between the blood and the body cells. Each body cell is immersed in a watery interstitial fluid. Molecules such as O_2 (red dots in the figure) diffuse first out of a capillary into the interstitial fluid, and then from the interstitial fluid into a tissue cell.

The circulatory system has several other major functions, besides transporting O_2 and nutrients. It conveys metabolic wastes to waste-disposal organs—CO_2 to the lungs, and a variety of other metabolic wastes to the kidneys. The upward arrows in Figure B represent the diffusion of waste molecules out of a tissue cell into the capillary.

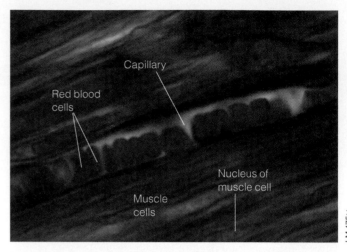

A. A capillary passing through muscle tissue

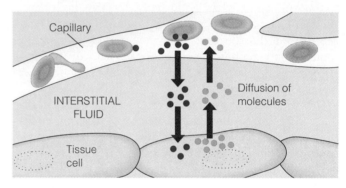

B. Diffusion between blood and tissue cells

The circulatory system also plays two key roles in maintaining a constant internal environment (homeostasis). First, by exchanging molecules with the interstitial fluid, the circulatory system helps control the makeup of the environment in which the tissue cells live. Second, the circulatory system helps control the makeup of the blood by continuously moving it through organs, such as the liver and kidneys, that regulate the blood's contents.

23.2 Several types of internal transport have evolved in animals

Not all animals have a circulatory system like ours. Cnidarians such as hydras and jellyfishes, for example, have no true circulatory system—that is, no organ system specifically for internal transport. As we saw in Module 21.3, the body wall of a hydra is only two or three cells thick, so all the cells can exchange materials directly with the water surrounding the animal or with the water in the gastrovascular cavity. Water is drawn in through the mouth,

circulates in the gastrovascular cavity, and then passes back out through the mouth. Hydras have no blood.

The jellyfish in Figure A (facing page) has an elaborate gastrovascular cavity (gold), with branches radiating to and from a circular canal. Cells lining all these branches bear flagella, whose beating helps circulate the gastrovascular fluid. Digestion also occurs in the gastrovascular cavity and in the cells lining it. Only these cells have direct access to

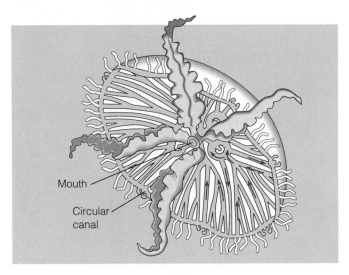

A. A gastrovascular cavity in a Jellyfish

Mouth

Circular canal

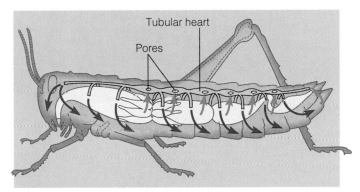

Tubular heart

Pores

B. An open circulatory system in a grasshopper

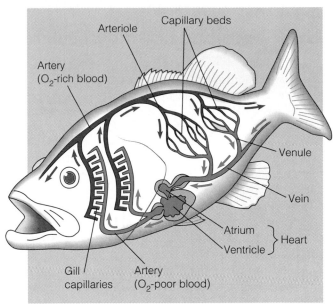

Arteriole

Capillary beds

Artery (O_2-rich blood)

Venule

Vein

Atrium

Heart

Ventricle

Gill capillaries

Artery (O_2-poor blood)

C. A closed circulatory system in a fish

nutrients, but the nutrients have only a short distance to diffuse to other body cells.

A gastrovascular cavity provides adequate internal transport for animals whose body cells are not arranged in numerous layers, but it is not adequate for animals with thick, multiple layers of cells. Such animals have a true circulatory *system* containing a specialized circulatory fluid, **blood.**

Two basic types of circulatory systems have evolved in animals. Many invertebrates—spiders, crayfish, and grasshoppers, for example—have what is called an **open circulatory system.** The system is termed "open" because blood is pumped through open-ended vessels and flows out among the cells; there is no separate interstitial fluid. In the grasshopper (Figure B), pumping of the tubular heart drives the blood into the head and the rest of the body (red arrows). Nutrients diffuse from the blood directly into the body cells, as contractions of body muscles move the blood toward the tail. When the heart relaxes, blood returns to it (blue arrows) through several pores. Each pore has a valve that closes when the heart contracts, preventing backflow of the blood.

Vertebrates, including humans, have a **closed circulatory system,** often called a **cardiovascular system** because it consists of a heart and a network of tubelike vessels (Greek *kardia*, heart, and Latin *vas*, vessel). The blood is

confined to the vessels, which keep it distinct from the interstitial fluid. There are three kinds of vessels in a closed circulatory system. **Arteries** carry blood away from the heart to organs throughout the body, **veins** return blood to the heart, and **capillaries** convey blood between arteries and veins within each organ. Arteries and veins are distinguished by the *direction* in which they convey blood, not by the quality of the blood they contain. Although most arteries convey oxygen-rich blood and most veins carry blood depleted of oxygen (O_2-poor blood), there are important exceptions. For example, we have two arteries, called pulmonary arteries, that carry O_2-poor blood from our heart to our lungs; and we have four pulmonary veins that carry freshly oxygenated blood from the lungs to the heart.

The cardiovascular system of a fish illustrates the main features of a closed circulatory system. Figure C shows a simplified version. (A bony fish actually has four gills on each side of its head, instead of just two as shown, and there are hundreds of thousands of gill capillaries.) The heart has two main chambers. The **atrium** (plural, *atria*) receives blood from the veins, and the **ventricle** pumps blood to the gills via large arteries. As in all the closed circulatory system figures in this chapter, red represents O_2-rich blood and blue represents O_2-poor blood. After passing through the gill capillaries, the O_2-rich blood flows into other large arteries that carry it to all other parts of the body. The large arteries branch into **arterioles,** small vessels that give rise to capillaries. Networks of capillaries called **capillary beds** infiltrate every organ and tissue in the body. The thin walls of the capillaries allow chemical exchange between the blood and the interstitial fluid. The capillaries converge into **venules,** which in turn converge into veins that return blood to the heart.

Vertebrate cardiovascular systems reflect evolution

The colonization of land by vertebrates was a momentous episode in the history of life, opening vast new opportunities for this group of animals. As aquatic vertebrates became adapted for terrestrial life, nearly all of their organ systems underwent major changes. One of the most drastic evolutionary changes was the switch from gill breathing to lung breathing, and this switch was accompanied by equally drastic changes in the cardiovascular system.

The diagrams at the right indicate the general features of the cardiovascular system of a fish and a mammal. Other terrestrial vertebrates—amphibians, reptiles, and birds—also have lungs and a cardiovascular system adapted for life on land. By comparing a fish and a mammal, we see the major cardiovascular differences between vertebrates with gills and those with lungs.

As indicated in Figure A, a fish has a single circuit of blood flow. Its heart receives and pumps only O_2-poor blood. After leaving the heart, the blood first passes through a capillary bed in the gills, where it picks up oxygen. In passing through the many capillaries in the gills, the blood slows down considerably, but it is helped on its way to the other organs by the animal's swimming movements. In the other organs, it passes through other beds of capillaries, known as systemic capillaries, before returning to the heart.

Terrestrial vertebrates have a more complex cardiovascular system, which accommodates air breathing. The mammalian heart (Figure B) has four chambers: two atria (A) and two ventricles (V). Notice that the right side of the heart (from the animal's viewpoint) is on the left in the diagram, and the left side of the heart is on the right. (It is customary to draw the system this way, as though the heart is in a body facing you from the page.) Terrestrial vertebrates also have two blood circuits instead of the single one in the fish. The **pulmonary circuit** carries blood between the heart and the gas-exchange tissues in the lungs, and the **systemic circuit** carries blood between the heart and the rest of the body. With this double circulation, only the right side of the heart handles blood that has been depleted of oxygen. The right atrium receives this blood from body tissues, and the right ventricle pumps it to capillary beds in the lungs.

Completely separate from the right side of the mammalian heart, the left side acts as a second pump, handling only O_2-rich blood. As indicated in Figure B, the left atrium receives blood from the lungs, and the left ventricle pumps it out to body organs via the systemic circuit. Thus, the mammalian heart is actually two pumps in one; the system accommodates lung breathing, while keeping O_2-rich blood flowing rapidly to body organs. Blood returns to the mammalian heart just after being oxygenated in the lungs, and the left ventricle gives this blood a strong boost that propels it rapidly out to the systemic capillaries. Rapid delivery of O_2-rich blood to the body tissue supports the high metabolic rate characteristic

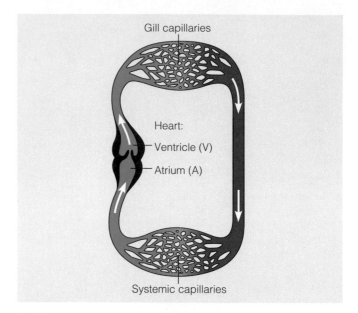

A. Fish

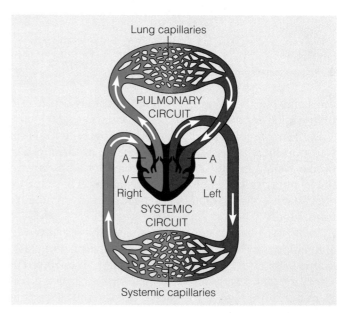

B. Mammal

of many land vertebrates, especially birds and mammals, which are endothermic.

The differences between the fish and mammalian cardiovascular systems are representative of the kinds of changes that occurred as terrestrial vertebrates evolved from their fishlike aquatic ancestors. The change from gills to lungs was accompanied by the evolution of a single-circuit cardiovascular system into a dual system with a double-pump heart. Let's now take a look at the mammalian system in more detail.

The human heart and cardiovascular system typify those of mammals. Your heart (Figure A at the right) is about the size of a clenched fist. It is enclosed in a sac just under your breastbone. The heart is formed mostly of cardiac muscle tissue. Its thin-walled atria collect blood returning to the heart and pump it only the short distance into the ventricles. The thicker-walled ventricles pump blood to all other body organs. The valves in the heart regulate the direction of blood flow, as we will see in Module 23.6.

Let's follow the blood through the entire circulatory system, shown diagrammatically in Figure B. Beginning with the pulmonary (lung) circuit, ① the right ventricle pumps blood to the lungs via ② two **pulmonary arteries.** As the blood flows through ③ capillaries in the lungs, it unloads CO_2 and loads up on O_2. The O_2-rich blood then flows back to the left atrium via the **pulmonary veins** (actually two from each lung, though only one per lung is shown here). Next, ④ the left atrium pumps the O_2-rich blood into ⑤ the left ventricle.

In Figure A, you can see clearly that the walls of the left ventricle are thicker than those of the right ventricle. The powerful muscles in the left ventricle pump blood to all body organs (including the lungs) through the systemic circuit. As Figure B shows, oxygen-rich blood leaves the left ventricle through ⑥ the **aorta.** The aorta is our largest blood vessel, with a diameter of roughly 2.5 cm, about the same as that of a quarter. Several large arteries branch from the aorta (only one is shown in Figure B) and lead to ⑦ the upper body. The aorta then curves down, behind the heart, and more arteries branching from it supply blood to ⑧ abdominal organs and the lower body. For simplicity, Figure B does not show the individual organs, but within each one, arteries carry the blood into arterioles, which in turn convey it to capillaries. The capillaries rejoin as venules, which carry the blood back into veins. Oxygen-poor blood from the upper body and head is channeled into a large vein called ⑨ the **superior vena cava.** Another large vein, ⑩ the **inferior vena cava,** drains blood from the lower body. The two venae cavae empty their blood into ⑪ the right atrium. As the right atrium pumps the blood into the right ventricle, we complete our journey.

Now that we have surveyed the cardiovascular system as a whole, let's take a closer look at the structure and function of its parts, first the vessels and then the heart.

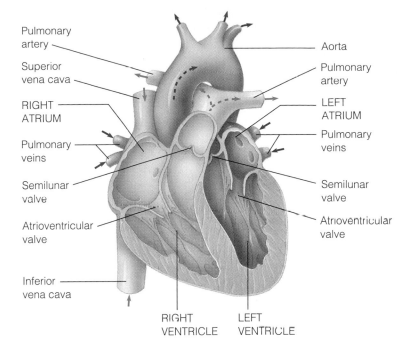

A. The human heart

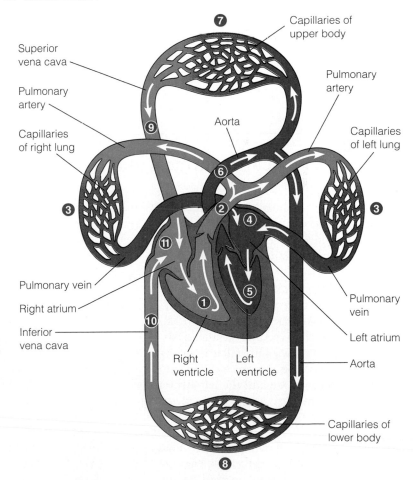

B. The path of blood flow through the cardiovascular system

23.5 The structure of blood vessels matches their functions

This drawing illustrates the structures of the different kinds of blood vessels and how the vessels are connected. Look first at the capillaries (center), which form fine branching networks where materials are exchanged between the blood and tissue cells. Appropriate to this function, capillaries have very thin walls formed of a single layer of epithelial cells, which is wrapped in a thin layer of fibrous material. The inner surface of the capillary is smooth and keeps the blood cells from being abraded as they tumble through the system.

Arteries, arterioles, veins, and venules have walls that are thicker than those of capillaries. The walls have the same epithelium as capillaries, but they are reinforced by two other tissue layers, both of which are thicker and sturdier in the arteries than in the veins. The middle layer, mainly smooth muscle, allows arteries and some veins to regulate blood flow by constricting. The thick muscle layer in the large arteries near the heart also allows these vessels to withstand surges of blood carrying the full force of the heartbeat. An outer layer of connective tissue is elastic and enables the vessels to stretch. Many of the veins have valves, flaps of tissue projecting toward the heart. The valves prevent backflow, permitting blood to flow only toward the heart.

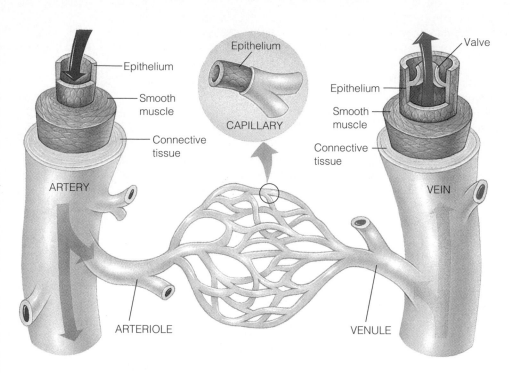

23.6 The heart contracts and relaxes rhythmically

The heart is the hub of the circulatory system. In a continuous, rhythmic cycle, it passively fills with blood from the large veins and then actively contracts, propelling the blood throughout the body. Its alternating relaxations and contractions make up the **cardiac cycle.**

As shown in this diagram, when the heart is relaxed, in the phase called ① **diastole,** blood flows into all four of its chambers. Blood enters the right atrium from the venae cavae and the left atrium from the pulmonary veins. Also during diastole, the valves between the atria and the ventricles (atrioventricular, or AV, valves) are open, allowing blood to flow from the atria into the ventricles. Diastole lasts about 0.4 sec, long enough for the ventricles to nearly fill with blood.

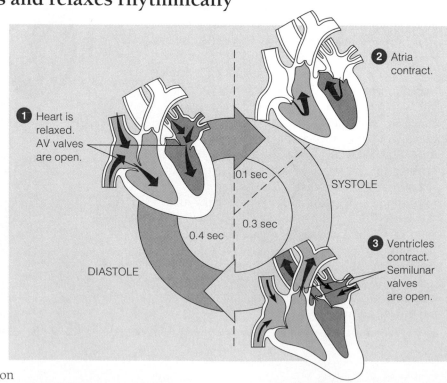

The contraction phase of the heart cycle is called **systole.** ② Systole begins with a very brief (0.1-sec) contraction of the atria that completely fills the ventricles with blood. This is the only time in the cardiac cycle when the atria contract. Finally, ③ the ventricles contract for about 0.3 sec. The force of their contraction closes the AV valves and opens the valves at the exit from each ventricle, the semilunar valves. With its semilunar valves open, the heart pumps O_2-poor blood to the lungs via the pulmonary artery and O_2-rich blood to the rest of the body through the aorta. Blood also flows into the atria during the second part of systole, as the small arrows in part 3 indicate.

The volume of blood per minute that the left ventricle pumps into the aorta is called **cardiac output.** This volume is equal to the amount of blood pumped by the left ventricle each time it contracts (about 75 mL per beat for the average person) times the heart rate. An average person at rest might have a heart rate of about 70 beats per minute. At this rate, the cardiac output would be 70 x 75 = 5.25 L/min.

Your heart rate and cardiac output will vary, depending on your level of activity and other factors. Both will increase, for instance, when you consume stimulants such as caffeine.

At different times in its cycle, the heart makes distinctive sounds, which you can hear if you press your ear against the breastbone of a friend. The sound pattern is "lub-dupp, lub-dupp." The "lub" sound comes from the contraction of the ventricles and the discharge of the blood against the AV valves. The "dupp" comes from the discharge of the blood against the semilunar valves.

A trained ear can also detect the sound of a heart murmur, indicating a defect in one or more of the heart valves. A murmur sounds like a hiss, and it occurs when a stream of blood squirts backward through a valve. Some people are born with murmurs, while others have their valves damaged by infection (from rheumatic fever, for instance). Most cases of heart murmur are not serious, and those that are can be corrected by replacement of the damaged valves with artificial ones or with valves taken from an organ donor.

The pacemaker sets the tempo of the heartbeat 23.7

What we call our heartbeat results from the contraction of cardiac muscles in the walls of the atria and ventricles. A specialized region of cardiac muscle called the **pacemaker,** or **SA (sinoatrial) node,** maintains the heart's pumping rhythm by setting the rate at which the heart contracts.

As indicated in the figure below, the pacemaker is situated in the wall of the right atrium. ① When the pacemaker contracts, it generates signals (gold color) much like those produced by nerve cells. ② The signals spread quickly through both atria, making them contract in unison. The signals also pass to a relay point called the **AV (atrioventricular) node,** in the wall between the right atrium and right ventricle. Here the signals are delayed about 0.1 second before passing to the ventricles. The delay ensures that the atria will contract first and empty completely before the ventricles contract. ③ Specialized muscle fibers (orange) relay the signals from the AV node to the cardiac muscles of the ventricles. There the signals trigger the strong contrac-

tions that drive the blood out of the heart to the lungs (from the right ventricle) and into the aorta (from the left ventricle).

So we see that the tempo of our heartbeat is set within the heart itself—by the pacemaker and a relay station called the AV node. In certain kinds of heart disease, this self-pacing system fails to maintain a normal heart rhythm. The remedy is an **artificial pacemaker,** a tiny electronic device surgically implanted near the AV node. Artificial pacemakers emit electrical signals that trigger normal heart muscle contractions.

Actually, there is more to the heart rate story than the pacemaker/AV node system. The brain also exerts some influence on heart rate. When we exercise or become excited, for example, cardiovascular control centers in our brain send nerve signals to both the pacemaker and the AV node, making them increase the heart rate. In contrast, when we are asleep or depressed, the brain's control centers decrease the activity of the pacemaker and AV node and the tempo of the heartbeat. Thus, the brain's control centers provide a way for the heart to respond to stimuli from our surroundings—something the pacemaker and AV node cannot do.

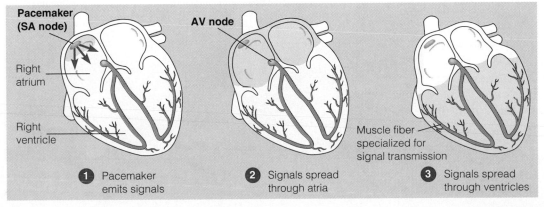

Right atrium

Right ventricle

Pacemaker (SA node)

AV node

Muscle fiber specialized for signal transmission

1 Pacemaker emits signals

2 Signals spread through atria

3 Signals spread through ventricles

What is a heart attack?

Diseases of the heart and blood vessels—cardiovascular diseases—are the leading cause of death in the United States today. Every year, several hundred thousand Americans die from heart attacks. A **heart attack** is the death of cardiac muscle cells and the resulting failure of the heart to deliver enough blood to the rest of the body. Cardiac muscle cells are nourished and supplied with oxygen by blood vessels called **coronary arteries.** Shown in red in Figure A, the coronary arteries branch from the aorta just as it emerges from the heart. If these arteries become clogged, the heart muscle dies from lack of oxygen. A blockage at the point indicated in Figure A, for instance, would cut off the blood supply to the part of the heart muscle shown in blue.

What causes blockage of the coronary arteries? Sometimes, a gradual narrowing of the channel within these arteries results from atherosclerosis, the buildup of lipids on the inside of the vessels. In such cases, a person may feel occasional chest pains, a condition known as angina pectoris (Latin *angere,* to choke, and *pector,* breast). Angina is most likely to occur when the heart is laboring hard because of physical or emotional stress. Angina is a signal that part of the heart is not receiving a sufficient supply of oxygen, and that a heart attack could occur in the future.

Many heart attacks occur without warning. A blood clot may completely block a coronary artery, or atherosclerosis may reach a critical level, causing massive damage to the heart muscle. All of a sudden, the person feels a heavy squeezing ache or discomfort in the center of the chest. The pain may radiate to the shoulder, arm, neck, or jaw. Other symptoms may include sweating, nausea, shortness of breath, and dizziness or fainting.

When heart muscles die, they are not replaced, because cardiac muscle cells do not divide. When a person survives a heart attack, scar tissue (a type of connective tissue) grows into the areas where the heart muscle has died. This is the body's way of repairing the damaged areas. However, the scar tissue cannot contract, as cardiac muscle does. As a result, the damaged heart is permanently weakened, and its ability to pump blood throughout the body, including to its own coronary arteries, may be seriously impaired.

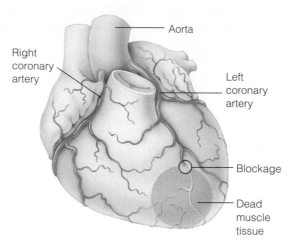

A. Heart attack

Corrective surgeries, such as coronary bypasses, heart transplants, and artificial hearts, offer some relief for heart disease patients. Bypass surgery is the most common type of surgery for blockage of coronary arteries. In this procedure, blood vessels removed from a patient's legs are sewed into the heart as replacements for clogged arteries. Newer, less radical techniques include balloon angioplasty, in which a tube with a deflated balloon attached to one end is inserted in a blocked artery. When the tube runs into the blockage, the physician inflates the balloon, which opens the vessel by squeezing the blocking tissue against the vessel wall, where it sticks. Laser beams are also used to destroy arterial blocks. Unfortunately, surgery of any kind only treats the disease symptoms, not the underlying causes, and heart disease often develops again, even after these expensive treatments.

On the positive side, over the past two decades, the death rate from cardiovascular disease in the U.S. has declined by more than 25%. Health education and early diagnosis seem to have been the main reasons for this decrease. In recent years, many people have become more conscious of their health; as a group, Americans are smoking less, exercising more, watching their diet, and having regular medical checkups. Early diagnosis and treatment of problems such as high blood pressure may prevent a deadly heart attack. Figure B shows another important factor in preventing many deaths from heart attacks: Paramedics, skilled emergency-aid specialists, help save thousands of lives every year.

B. Paramedics aiding a heart attack victim

Blood pressure is the force that blood exerts against the walls of our blood vessels. Created by the beating of the heart, blood pressure is the main force driving the blood from the heart through the arteries and arterioles to the capillary beds. When the ventricles contract, blood is forced into the arteries faster than it can flow into the arterioles. This stretches the elastic walls of the arteries. You can feel this effect of blood pressure when you measure your heart rate by taking your pulse. The **pulse** is the rhythmic stretching of the arteries caused by the pressure of blood forced through the arteries by the powerful contractions of the ventricles during systole. What you actually feel is an artery bulging with each heartbeat. Between heartbeats, the arteries snap back because the pressure is reduced when the heart relaxes.

Blood pressure depends partly on cardiac output (the volume of blood per minute the left ventricle pumps into the aorta) and partly on the resistance to blood flow imposed by the blood vessels. As Figure A indicates, blood pressure (expressed in millimeters of mercury, mm Hg) and the blood's velocity (rate of flow, expressed in centimeters per second, cm/sec) are highest in the aorta and change little as the blood passes through the other arteries. However, blood pressure and velocity both decline abruptly as the blood enters the arterioles. The pressure drop results mainly from a resistance to blood flow caused by friction between the blood and the inner walls of the arterioles. The resistance is high because the blood contacts a large amount of wall surface in the numerous, tiny arterioles. By lowering the blood pressure, the high resistance eliminates pressure peaks in the arterioles. This is why you can feel your pulse only in your arteries.

Friction also reduces the velocity of the blood in the arterioles. But velocity declines mainly because of the structural arrangement indicated at the top of Figure A. The total, combined width of all the openings into a set of arterioles is much greater than the width of the artery that feeds blood into them. If there were only one arteriole per artery, the blood would flow faster through the arteriole, the way water does when you pinch a garden hose. However, there are many arterioles per artery, so the effect is like letting go of a pinched hose: As you increase the width of the opening, the flow rate goes down. In the circulatory system, the combined width of the openings of the arterioles is analogous to the larger opening in the hose.

The overall result of the decline in velocity and pressure in the arterioles is a steady, leisurely flow of blood in the capillaries. The gentle flow allows the exchange of substances between the blood and the interstitial fluid.

By the time the blood reaches the veins, its pressure has dropped to near zero. The blood has encountered so much resistance as it passes through the millions of tiny arterioles

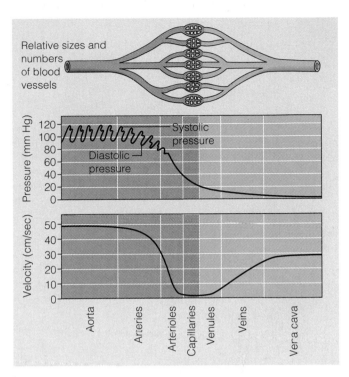

A. Blood pressure and velocity in the various blood vessels

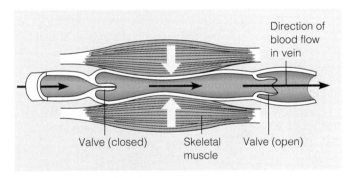

B. Blood flow in a vein

and capillaries that the force from the pumping heart no longer propels it. How, then, does blood return to the heart, especially when it must travel up from the legs, against gravity? As we discussed in the chapter's introduction, the veins of mammals such as humans or giraffes (and corn snakes, as well) are sandwiched between skeletal muscles (Figure B). Consequently, whenever the body moves, the muscles pinch the veins and squeeze blood along toward the heart. The large veins of mammals have valves that allow the blood to flow only toward the heart. Breathing also helps return blood to the heart. When we inhale, the change in pressure within our chest cavity causes the large veins near our heart to expand and fill.

Measuring blood pressure can reveal cardiovascular problems

When we talk about a person's blood pressure, we are referring to the pressure exerted by the blood against the walls of the arteries. A typical blood pressure reading for an average young adult is 120/70, as indicated in part ① of the figure below. The number 120 represents the pressure in millimeters of mercury during systole, and 70 represents the pressure during diastole. A blood pressure of 120 is a force that can support a column of mercury 120 mm high.

Blood pressure is usually measured with an instrument called a sphygmomanometer, an inflatable cuff attached to a pressure gauge. As shown in part ② of the figure, the cuff is wrapped around the upper arm. By squeezing the rubber bulb, the examiner inflates the cuff until its pressure closes the main artery in the arm. When this occurs, the pressure exerted by the cuff is greater than the blood pressure in the main artery (120 in this case, though the examiner does not yet know it). Since the blood cannot flow past the cuff, there is no pulse below the cuff.

Part ③ shows how systolic pressure is actually measured. While listening with a stethoscope for a pulse in the arm below the cuff, the examiner gradually deflates the cuff. The systolic pressure (120) is recorded when the examiner hears the first sounds of blood spurting through the constricted artery—a mild tapping.

In part ④, diastolic pressure is measured by deflating the cuff further and releasing pressure on the artery. At first, as

the blood moves more rapidly through the artery, the sounds become louder. When the artery is no longer constricted, the sounds cease. At this point, the examiner records the diastolic pressure (70, in our example), the pressure remaining in the artery when the heart is relaxed between beats.

Blood pressure in a typical healthy adult ranges between 110 and 140 mm Hg systolic and between 70 and 90 mm Hg diastolic. Although individual differences occur, pressures significantly below or above these levels may indicate cardiovascular problems. A person with a systolic blood pressure persistently below 100 mm Hg is considered to have **low blood pressure.** This may be no cause for concern, and may even suggest that the person will live a long, healthy life. However, some cases of low blood pressure are associated with poor nutrition or disorders of the thyroid or adrenal glands.

High blood pressure, or hypertension, is a type of cardiovascular disorder and an important cause of heart disease. **Hypertension** is defined as a persistent blood pressure of 140/90 or higher; the higher the values, the greater the risk of serious cardiovascular disease. Elevated pressure caused by reduced flexibility or partial blockage of the arteries makes the heart work harder, because it has to pump against greater resistance in the arteries. As a result, the heart muscle may tend to wear out prematurely. Hyperten-

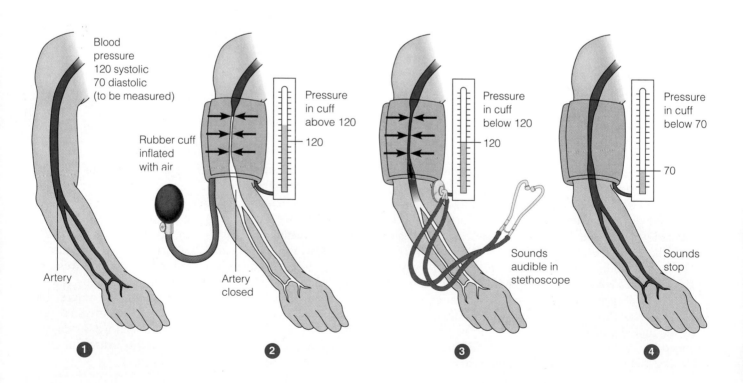

Blood pressure
120 systolic
70 diastolic
(to be measured)

Rubber cuff
inflated
with air

Artery

Pressure
in cuff
above 120

120

Artery
closed

1

2

Pressure
in cuff
below 120

120

Sounds
audible in
stethoscope

3

Pressure
in cuff
below 70

70

Sounds
stop

4

sion also may contribute to arterial blockage by causing small tears in the epithelium that lines the blood vessels and increasing the tendency of atherosclerotic deposits to form.

Hypertension is sometimes called the "silent killer" because many people with the disease experience no symp-toms until tragedy strikes. Even those who know they have high blood pressure may tend to ignore their condition and not seek medical treatment. Fortunately, hypertension can usually be controlled by diet, exercise, and medication.

Muscle cells control the distribution of blood

Because the heart beats constantly and each tissue has many capillaries, every part of the body is supplied with blood at all times. Nonetheless, at any given time, only about 5–10% of the body's capillaries have blood flowing through them. Capillaries in a few organs, such as the brain and heart, usually carry a full load of blood, but in many other sites, the blood supply varies from time to time.

Two mechanisms control the distribution of blood to capillaries of the various organs; both depend on muscle cells. Figure A (above) shows one mechanism, which involves contraction and relaxation of the smooth muscle layer of the arteriole wall. ① When the muscle cells in this layer relax, the arteriole opens (dilates), allowing blood to enter the capillaries. ② Contraction of the muscle layer, under the influence of nerves and hormones, constricts the arteriole, decreasing blood flow through it to a capillary bed.

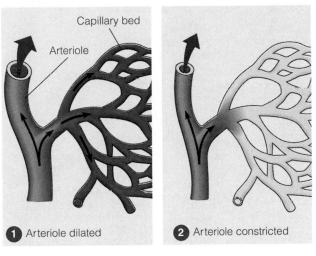

A. Control of capillary blood flow by arteriole constriction

① Arteriole dilated ② Arteriole constricted

The other controlling mechanism is illustrated in Figure B (below). Notice that in both parts of the figure there is a capillary called a thoroughfare channel, through which blood streams directly from arteriole to venule. This channel is always open. Capillaries branching off from thoroughfare channels form the bulk of the capillary bed. Passage of blood into these branching capillaries is regulated by rings of smooth muscle called precapillary sphincters. As you can see in Figure B, ① blood flows through a capillary bed when its precapillary sphincters are relaxed. ② It bypasses the bed when the sphincters are contracted. After a meal, for instance, precapillary sphincters in the wall of the digestive tract let a larger quantity of blood pass through the capillary beds there than when food is not being digested. During strenuous exercise, many of the capillaries in the digestive tract are closed off, and blood is supplied more generously to skeletal muscles.

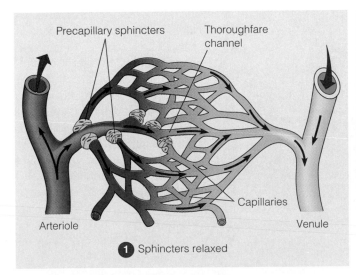

① Sphincters relaxed

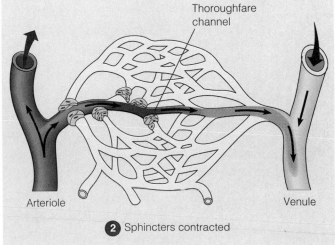

② Sphincters contracted

B. Control of capillary blood flow by precapillary sphincters

Capillaries allow the transfer of substances through their walls

Capillaries are the only blood vessels with walls thin enough to permit substances to cross between the blood and the interstitial fluid that bathes the body cells. The transfer of materials between the blood and interstitial fluid is the most important function of the circulatory system, so let's take a closer look at it.

Figure A, at the right, shows a cross section of a capillary that serves skeletal muscle cells. The capillary wall (tan in the drawing) consists of overlapping epithelial cells that enclose a lumen, or space, that contains the blood. The nucleus you see here belongs to one of the two cells making up this portion of the capillary. (The other cell's nucleus does not appear in this particular cross section.) The light area around the capillary is a space containing interstitial fluid.

Exchange of substances between the blood and the interstitial fluid occurs in several ways. Some substances simply diffuse through the epithelial cells of the capillary wall or are carried in endocytotic vesicles across these cells (see Module 5.19). In addition, the capillary wall is leaky—there are narrow clefts between the epithelial cells making up the wall. Through these clefts, water and small solutes, such as sugars, salts, and oxygen, pass freely. Blood cells and dissolved proteins remain inside the capillary because they are too large to pass.

In addition to the passive diffusion of substances out of the capillary, active forces also push fluid through the leaky capillary wall. The diagram in Figure B (below) shows part of a capillary with blood flowing from its arterial end (near an arteriole) to its venous end (near a venule). The blue arrows represent active forces driving fluid into or out of the capillary. One such force is blood pressure, which tends to push fluid outward. Another is osmotic pressure, a force that tends to draw fluid inward because the blood has a

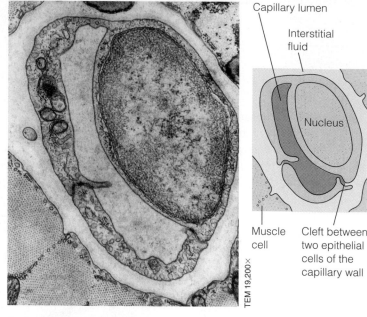

A. A capillary in cross section

TEM 19,200×

Capillary lumen

Interstitial fluid

Nucleus

Muscle cell

Cleft between two epithelial cells of the capillary wall

higher concentration of solutes than the interstitial fluid. Proteins dissolved in the blood account for much of this high solute concentration. (To review the principles of osmosis, see Module 5.15.)

The direction of fluid movement into or out of the capillary at any point depends on the difference between the blood pressure and the osmotic pressure. At the arterial end of the capillary, the blood pressure exceeds the osmotic pressure (as indicated by the relative widths of the blue arrows). Thus, there is a net pressure forcing fluid outward, and more fluid moves out of the capillary than in.

At the venous end of the capillary, the situation is reversed. The blood pressure drops so much in the capillary bed that the osmotic pressure outweighs it, and fluid reenters the capillary.

About 99% of the fluid that leaves the blood at the arterial end of a capillary bed reenters the capillaries at the venous end. The remaining 1% is eventually returned to the blood by the vessels of the lymphatic system, which we discuss in Module 24.5.

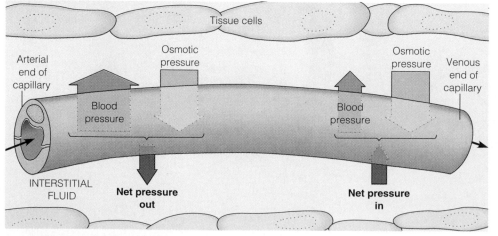

B. The movement of fluid into and out of a capillary

Blood consists of cells suspended in plasma

Now that we have examined the structure and function of the heart and blood vessels, let's focus on the composition of blood itself. In an average adult human, the circulatory system contains 4–6 liters of blood. Blood consists of several types of cellular elements suspended in a liquid called **plasma.** When a blood sample is taken, as shown below, the cellular elements, which make up about 45% of the blood, can be separated from the plasma by spinning the sample in a centrifuge. Red and white blood cells and small cell pieces called platelets settle to the bottom of the centrifuge tube, underneath the transparent, straw-colored plasma. **Platelets** are bits of cytoplasm pinched off from large cells in the bone marrow. As we will see in Module 23.16, platelets are important in blood clotting.

Plasma, making up just over half the volume of the blood, is about 90% water. The other 10% is made up of dissolved salts, proteins, and various other substances being transported by the blood. Salts are dissolved in the plasma as inorganic ions. The dissolved ions have several functions, such as maintaining the osmotic balance between the blood and the interstitial fluid and keeping the pH of the blood at about 7.4. Inorganic ions also help regulate the permeability of cell membranes.

Plasma proteins work together with the salts in maintaining osmotic balance and pH. The protein fibrinogen functions with the platelets in blood clotting, as we will see. Another group of plasma proteins, the immunoglobulins, are important in body defense (immunity), which we discuss in Chapter 24.

What about the blood cells shown below? We discuss their functions in the next two modules.

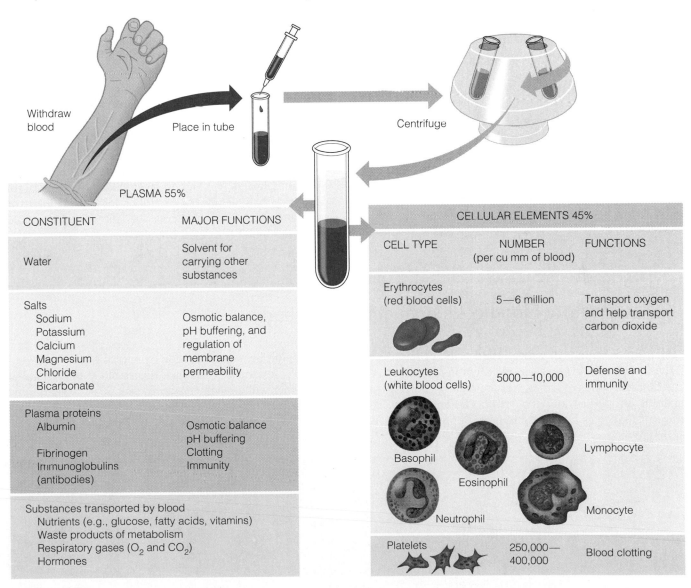

23.14 Red blood cells transport oxygen

Red blood cells, also called **erythrocytes,** are by far the most numerous blood cells. There are about 25 trillion of these tiny cells in the average person's bloodstream. Our red blood cells and those of other mammals lack nuclei, an unusual characteristic for living cells. The nuclei are lost as the cells develop.

The structure of a red blood cell suits its main function, which is to carry oxygen. The micrograph here shows that human red blood cells are biconcave disks, thinner in the center than at the sides. A biconcave disk has more surface area for gas exchange than a flat disk or a sphere has. Red blood cells are small relative to other cells—white blood cells, for instance—and their small size also gives them greater total surface area for gas exchange.

As small as a red blood cell is, it contains about 250 million molecules of hemoglobin. As red blood cells pass through the capillary beds of lungs or gills, oxygen diffuses into the red blood cells, where it binds to the iron in hemoglobin. This process is reversed in the capillaries of the systemic circuit, where the hemoglobin unloads its cargo of oxygen to the tissues (see Module 22.10).

Red blood cells are formed in bone marrow. Their production is controlled by a negative-feedback mechanism that is sensitive to the amount of oxygen reaching the tissues via the blood. If the tissues are not receiving enough oxygen, the kidneys secrete a hormone called erythropoi-

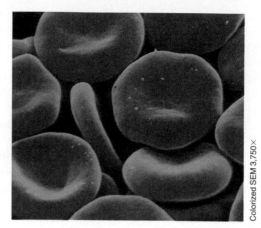

Colorized SEM 3,750×

etin that stimulates the bone marrow to produce more red blood cells. On the other hand, if the tissues are receiving more oxygen than they can use, the kidneys stop secreting the hormone, and erythrocyte production slows.

On average, red blood cells circulate in the blood for three or four months before starting to wear out. Worn-out red blood cells are broken down in the liver, where enzymes digest the hemoglobin and use its amino acids to make other proteins. Much of the iron returns to the bone marrow, where it is used in making more red cells.

Maintaining adequate amounts of iron, hemoglobin, and red blood cells is essential to normal body function. Someone with an abnormally low amount of hemoglobin or a low number of red blood cells has a condition called **anemia.** An anemic person feels constantly tired and run down, and is often susceptible to infections, because the body cells do not get enough oxygen. Anemia can result from a variety of factors, including excessive blood loss, vitamin or mineral deficiencies, and bone marrow cancer. Iron deficiency is the most common cause. A person may not get enough iron in the diet, or the digestive tract may not absorb enough of it. Women are more likely to develop iron deficiency than men because of blood loss during menstruation. Pregnant women generally benefit from iron supplements to support the developing fetus and placenta.

23.15 White blood cells help defend the body

There are five kinds of **white blood cells,** or **leukocytes.** As you can see below in these micrographs of stained cells, the different kinds of leukocytes are distinguished by their staining properties and the shape of their nuclei. As a group, leukocytes fight infections and prevent cancer cells from growing.

Basophils help fight infection by releasing chemicals——for example, histamine. Histamine dilates blood vessels and allows other white blood cells to move out of capillaries into surrounding tissues. Two of the most common white blood cells that move into body tissues are neutrophils and monocytes. These are **phagocytes;** they "eat" bacteria, foreign pro-

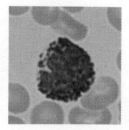
Basophil

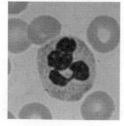
Neutrophil

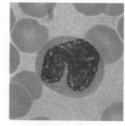
Monocyte

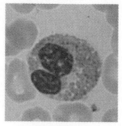
Eosinophil

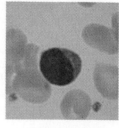
Lymphocyte

LM 1,350×

teins that enter the body through wounds, and the debris from other body cells that have died. The blood transports large numbers of neutrophils and monocytes to sites of injury, where the cells combat bacterial infections and help the tissue heal by removing debris.

Less is known about a fourth type of leukocyte, the eosinophil. Eosinophils are phagocytic and seem to be important in fighting infections by parasitic protozoans and worms. They may also help reduce allergy attacks.

The fifth type of leukocyte, the lymphocyte, is the key cell in

immunity. Some lymphocytes produce antibodies, the plasma proteins that react against foreign substances. Others defend the body against viral infections and destroy cancerous cells.

As a group, white blood cells actually spend most of their time outside the circulatory system, moving through interstitial fluid, where most of the battles against infection are waged. There are also great numbers of white cells in the lymphatic system. Like red blood cells, leukocytes arise in bone marrow. Their numbers increase whenever the body is fighting an infection.

Blood clots plug leaks when blood vessels are injured 23.16

We all have cuts and scrapes from time to time, yet we don't bleed to death because our blood contains self-sealing materials that plug leaks in our vessels. The sealants—platelets and the plasma protein **fibrinogen**—are always present in our blood. They are activated to form a clot when a blood vessel is injured.

Figure A shows the stages of the clotting process. ① It begins when the epithelium (tan) lining a blood vessel is damaged and connective tissue in the wall of the vessel is exposed to blood. Platelets (white) respond immediately. Those in the vicinity adhere to the exposed connective tissue and release a substance that makes nearby platelets sticky. ② Soon a cluster of sticky platelets forms a plug that seals minor breaks in the blood vessel.

A platelet plug provides emergency protection against blood loss; a tiny wound, such as from a pinprick, might require nothing more.

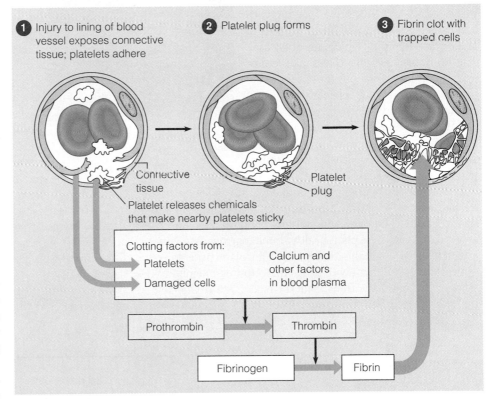

A. The blood-clotting process

However, when damage is more severe—an open cut, for instance—a chain of reactions is set off that culminates in the formation of a more complex plug called a fibrin clot. First, as shown in the yellow box in Figure A, clotting factors released by platelets and damaged cells mix with other factors in the plasma. The mixture activates a protein called prothrombin, converting it into the enzyme thrombin. The thrombin then converts fibrinogen into a threadlike protein called **fibrin**. As shown in part ③ of the figure, threads of fibrin trap blood cells. In this way, the injured vessel is sealed until connective tissue forms a permanent patch in it. Figure B is a micrograph of a fibrin clot.

The clotting mechanism is so important that any defect in it can be life-threatening. In the inherited disease hemophilia, excessive, sometimes fatal, bleeding occurs from

even minor cuts and bruises. Another type of defect can lead to a blood clot in the *absence* of injury; such a clot is called a thrombus. If a thrombus blocks one of the coronary arteries, a heart attack occurs.

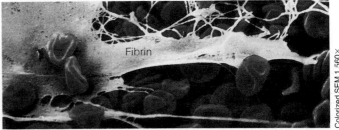

B. A fibrin clot

Purified stem cells offer a potential cure for leukemia and other blood cell diseases

Our white blood cells form a main line of defense against infections and may eliminate cancer cells that start to grow in the body. This cellular defense system is effective, but unfortunately it is not foolproof. As we saw in Module 10.19, the AIDS virus can overwhelm our body defenses by infecting and killing white blood cells. In addition, the very tissues that produce our defensive cells may become cancerous themselves. **Leukemia** is cancer of the bone marrow cells that produce white blood cells (leukocytes). A person with leukemia has an unusually high number of leukocytes, many of which are abnormal. The bone marrow cells that make leukocytes may also crowd out those that make red blood cells and platelets. Without enough red blood cells to oxygenate tissues and with an impaired clotting mechanism, a person with leukemia suffers from severe anemia. Leukemia is usually fatal unless treated, and not all cases respond to current treatments.

Irving Weissman in his laboratory at Stanford University

The standard treatments for leukemia are radiation and chemotherapy (see Module 8.11); both treatments slow the spread of cancer by destroying rapidly dividing cells. Another approach is to transplant healthy bone marrow tissue into a patient from a suitable donor, but transplant patients require lifelong treatment with drugs that suppress their body's tendency to reject the transplanted tissue. Such drugs have the side effect of suppressing immunity to infections. To avoid the transplant-rejection problem, patients are sometimes treated with their own bone marrow. Marrow from the patient is removed, processed to remove most or all cancerous cells, and then injected back into the patient.

A sophisticated variation of this approach may soon be available, one that could be used to treat other serious problems as well. Immunologist Irving Weissman and co-workers have developed a method for isolating the type of bone marrow cell—aptly called a **stem cell**—that gives rise to all of our blood cells, both red and white. These stem cells are very rare, only about one in 2000 marrow cells, but they are very powerful. In tests performed in mice, the injection of as few as 30 stem cells can completely repopulate an animal's blood and immune system.

Purified stem cells might also be used to treat genetic diseases of the blood, such as sickle-cell anemia, if some of the stem cells were genetically engineered before reinjection. Even AIDS, which is caused by a virus (HIV) that affects lymphocytes, might conceivably be fought by stem cells. If HIV-resistance genes could be inserted into stem cells, the lymphocytes and other white blood cells arising from those stem cells might be resistant to HIV.

We pursue the diverse roles of white blood cells in more detail when we study the human immune system in Chapter 24.

Chapter Review

Begin your review by rereading the module headings and scanning the figures before proceeding to the Chapter Summary and questions.

Chapter Summary

Introduction Most animals have a circulatory system that transports oxygen and nutrients to cells, removes carbon dioxide and other wastes, and regulates the makeup of the interstitial fluid in which the cells live.

23.1–23.5 Several means of internal transport have evolved. In hydras and jellyfish, the gastrovascular cavity functions in both digestion and internal transport. Most animals have either an open or a closed circulatory system. In open systems, a heart pumps blood through open-ended vessels into spaces among the cells. In closed systems, blood circulates within a network of vessels. The heart pumps blood through arteries to capillaries, and the blood returns to the heart via veins. The fish heart pumps blood through a single loop, first through capillaries in the gills, then through capillaries in other organs, and finally back to the heart. The human cardiovascular system, like that of other land vertebrates, accommodates air breathing. The right atrium and ventricle pump O_2-poor blood from the tissues through the pulmonary circuit, to the capillaries of the lungs and back to the left atrium. The left atrium and ventricle pump this O_2-rich blood around the systemic circuit, to the capillaries in the tissues and back.

23.6–23.7 During heart relaxation (diastole), blood flows from the veins into all four heart chambers. During contraction (systole), the atria first push blood into the ventricles, then stronger contractions of the ventricles propel blood into the pulmonary artery and aorta. Heart valves prevent backflow of blood. The pacemaker (SA node), a specialized bit of muscle in the wall of the right atrium, initiates each beat by triggering contraction of the atria. The AV node then relays these signals to the ventricles. Control centers in the brain adjust heart rate to body needs.

23.8 A heart attack is damage to cardiac muscle that occurs when a coronary artery feeding the heart is blocked, usually by lipid deposits or a blood clot. Heart disease can be treated by surgery. Treatment of high blood pressure, regular checkups, and lifestyle changes can help prevent heart disease.

23.9–23.10 Blood pressure, the force blood exerts on vessel walls, depends on cardiac output and the resistance of vessels. Pressure is highest in the arteries, and it drops as blood flows into arterioles and capillaries. Blood flows the slowest in the capillaries. Blood pressure drops nearly to zero by the time the blood reaches the veins, but muscle contractions, breathing, and one-way valves keep blood moving back to the heart.

23.11–23.12 The exchange of substances between the blood and the interstitial fluid is the most important function of the circulatory system. Constriction of arterioles and sphincters regulates capillary blood flow. Substances can leak through clefts in the capillary walls or pass through the thin walls by diffusion. In addition, blood pressure forces fluid out through the walls of a capillary at the arterial end, and osmotic pressure draws fluid in at the venous end.

23.13–23.17 Blood consists of cells in a fluid plasma. Plasma consists of water carrying various salts, proteins, nutrients, wastes, gases, and hormones. There are three types of cellular elements in blood. Red blood cells (erythrocytes) are the most numerous. They contain hemoglobin, which enables them to transport oxygen. White blood cells (leukocytes) function both inside and outside the circulatory system to fight infections and cancer. When a blood vessel is damaged, platelets (fragments of bone marrow cells) help trigger the conversion of soluble fibrinogen to an insoluble fibrin clot that plugs the leak. Stem cells in bone marrow give rise to all blood cells. Purified stem cells might be used to treat blood disorders.

Testing Your Knowledge

Multiple Choice

1. Which of the following has an open circulatory system?
   a. trout
   b. monkey
   c. ant
   d. frog
   e. jellyfish

2. Blood pressure is highest in _____, and blood moves most slowly in _____.
   a. veins . . . capillaries
   b. arteries . . . capillaries
   c. veins . . . arteries
   d. capillaries . . . arteries
   e. arteries . . . veins

3. When the doctor listened to Janet's heart, he heard "lub-hisss," "lub-hiss" instead of the normal "lub-dupp" sounds. The hiss is most likely due to (Explain your answer.)
   a. a clogged coronary artery
   b. a defective atrioventricular (AV) valve
   c. a damaged pacemaker
   d. a defective semilunar valve
   e. high blood pressure

4. Which of the following is the biggest difference between your cardiovascular system and the cardiovascular system of a fish?
   a. In a fish, blood is oxygenated by passing through a capillary bed.
   b. Your heart has two chambers; a fish heart has four.
   c. Your circulation has two circuits; fish circulation has one circuit.
   d. Your heart chambers are called atria and ventricles.
   e. Yours is a closed system; the fish's is an open system.

5. Paul's blood pressure is 125/80. The 125 indicates _____, and the 80 indicates _____.
   a. pressure in the left ventricle . . . pressure in the right ventricle
   b. arterial pressure . . . heart rate
   c. pressure during ventricular contraction . . . pressure during heart relaxation
   d. systemic circuit pressure . . . pulmonary circuit pressure
   e. pressure in the arteries . . . pressure in the veins

6. Which of the following *initiates* the process of blood clotting?
   a. damage to the lining of a blood vessel
   b. exposure of blood to the air
   c. conversion of fibrinogen to fibrin
   d. attraction of leukocytes to a site of infection
   e. conversion of fibrin to fibrinogen

True/False (*Change false statements to make them true.*)

1. The walls of arteries are thicker than the walls of veins.
2. Arteries carry blood toward the heart.
3. The conversion of fibrinogen to fibrin is a key step in blood clotting.
4. Red blood cells defend the body against infection.
5. Arterioles can regulate blood flow into capillaries.
6. The arteries of the systemic circuit branch from the aorta.
7. Stem cells regulate the heartbeat.

Describing, Comparing, and Explaining

1. Describe the shapes, sizes, typical numbers, and functions of red blood cells, white blood cells, and platelets.

2. Why can you feel your pulse in arteries but not in veins? If there is no pulse in your veins, what pushes the blood in veins back to the heart?

3. Imagine a drop of blood in a pulmonary vein. Trace its path through the heart and around the body, returning to the pulmonary vein. Name, in order, the heart chambers and types of vessels the blood passes through on its circulatory journey.

4. Explain how the structure of capillaries relates to their function of exchange of substances with the surrounding interstitial fluid.

Thinking Critically

1. The volume of blood that the left ventricle pumps into the aorta each minute is called the cardiac output. A swimmer's heart rate is 150 beats per minute, and 100 mL of blood is pumped by the left ventricle with each beat. What is the swimmer's cardiac output?

2. Juan has a disease in which damaged kidneys allow some of his normal plasma proteins to be removed from the blood. How might this condition affect the osmotic pressure of blood in capillaries, compared to the surrounding interstitial fluid? One of the symptoms of this kidney malfunction is an accumulation of excess interstitial fluid, which causes Juan's arms and legs to swell. Can you explain why this occurs?

Science, Technology, and Society

1. The incidence of cardiovascular disease is much lower in the Mediterranean countries (such as Spain, France, Italy, and Greece) than in North America. Suggest several alternative hypotheses that might explain the difference. How would you test your hypotheses?

2. Recently, a 19-year-old woman received a bone marrow transplant from her 1-year-old sister. The woman was suffering from a deadly form of leukemia and was almost certain to die without a transplant. Their parents had decided to have another child in a final attempt to provide their daughter with a matching donor. Although the ethics of the parents' decision were criticized, doctors reported that this situation is not uncommon. (Another couple conceived three children in an unsuccessful effort to provide a kidney donor.) In your opinion, is it acceptable to have a child in order to provide an organ or tissue donation? Why or why not?

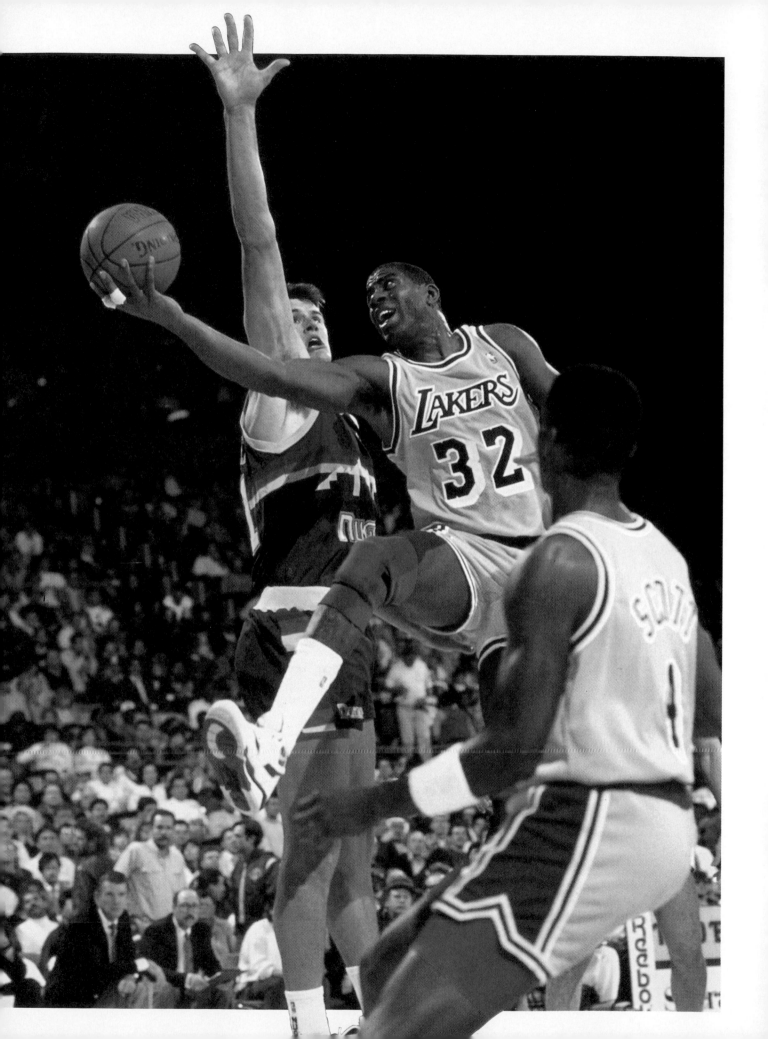

The Immune System 24

First identified in the United States in 1981, AIDS (acquired immune deficiency syndrome) is now epidemic throughout much of the world. Today, about one million Americans have tested positive for the AIDS virus. One of those is Earvin (Magic) Johnson, basketball superstar for twelve years with the Los Angeles Lakers. About a month after he was diagnosed, Johnson told of his initial shock in a *Sports Illustrated* interview (November 18, 1991):

> "[On] October 25 [1991], as I walked into the offices of [my physician] Dr. Michael Mellman, I was more curious than worried. I had been rejected for a life-insurance policy. . . ."
>
> "Earvin, sit down. I have your test results," Dr. Mellman said. "You're HIV-positive. You have the AIDS virus."
>
> "Suddenly I felt sick. I was numb. In shock. And, yes, I was scared. Dr. Mellman quickly told me that I didn't have AIDS, that I was only infected with the virus that could someday lead to the disease. But I didn't really hear him. Like almost everyone else who has not paid attention to the growing AIDS epidemic in the U.S. and the rest of the world, I didn't know the difference between the virus and the disease. While my ears heard HIV-positive, my mind heard AIDS."

As we discussed in Module 10.18, HIV, or human immunodeficiency virus, attacks the immune system, the body's main line of defense against infectious diseases and cancer. HIV enters the cells of the immune system and may lurk there for years, safe from the body's attempts to fight back. As Magic Johnson points out, testing positive for HIV is different from having AIDS, although, tragically, it appears that nearly all people with HIV will eventually develop AIDS.

Once HIV starts to multiply, it kills its host cells, infects other cells, and eventually destroys the body's ability to fight even the mildest infections. It may take about ten years for AIDS to develop after the initial HIV infection. In someone with AIDS, the immune system may be totally undermined, and diseases that it ordinarily fights off become deadly. Most AIDS patients die from other infectious diseases or from certain types of cancer, usually within three years of an AIDS diagnosis.

The AIDS virus is transmitted mainly in blood and semen. Most often, it enters the body through imperceptible wounds during sexual contact or via needles contaminated with infected blood. Routine screening of donated blood has greatly reduced the risk from blood transfusions, but the sharing of needles to inject drugs remains a major source of infection. Unfortunately, many people still think of AIDS as a disease confined to drug addicts and homosexuals. However, the rate of new infections is rising among heterosexual people who do not inject drugs, including teenagers and college students. In the U.S. alone, over 1,000,000 people are known to be infected with HIV, as of 1993. The clear implication is that HIV is transmitted sexually, regardless of sexual preference. In fact, anyone who has sex with a person who has had any chance of exposure to HIV in the past 20 years risks exposure to the virus.

Soon after being diagnosed as HIV-positive, Magic Johnson resigned from the Lakers. He made a brief comeback in 1992, starring in the Olympics and rejoining the Lakers for a short time, but resigned again the same year. Magic now spends much of his time working as an advocate for safer sex practices. He believes he acquired HIV and may have infected many others during the years before he was married, when he was promiscuous. In his own words, "I could easily have avoided being infected at all. All I had to do was wear condoms." Condoms do, in fact, minimize the AIDS risk, but they do not eliminate it. For a sexually active person, using condoms and having only one, nonpromiscuous partner are the best defenses against the AIDS virus.

The AIDS story demonstrates how much we depend on our body's built-in defense system. As we see in this chapter, our immune system is a very *specific* defense system, in that it recognizes an invader, such as a virus, and then produces large numbers of cells that combat that particular agent. In a healthy individual, this specific defense backs up several mechanisms of resistance termed *nonspecific* because they do not distinguish one infectious agent from another. Following a brief look at our nonspecific defenses in the first two modules, we concentrate on the mechanisms of immune defense. At the end of the chapter, we return to AIDS and examine how the insidious HIV defeats our body defenses.

Nonspecific defenses against infection include the skin, phagocytes, and antimicrobial proteins

The body's first line of defense against intruders consists of several nonspecific obstacles to infection. Our first defense is our skin. The skin's outer layer is a tough physical barrier of dead cells that most bacteria and viruses cannot penetrate. Our skin also has chemical defenses. For example, acids in sweat and oils secreted by glands in the skin inhibit the growth of many microorganisms. Sweat, saliva, and tears also contain lysozyme, an enzyme that attacks the cell walls of many bacteria.

Two organ systems that open to the external environment—the digestive and respiratory systems—are also guarded by nonspecific defenses. Stomach acids kill most bacteria swallowed with food. Guarding the respiratory route, the mesh of hairs in our nostrils filters incoming air, and mucus in our respiratory tubes traps most microbes and dirt that get past the nasal filter. Cilia on cells lining the tubes sweep the mucus upward and out of the system.

Microbes that do penetrate the skin or enter the tissues of the digestive or respiratory system are soon confronted by nonspecific defensive cells. These are all classified as white blood cells (see Module 23.15), although they are found in interstitial fluid as well as in blood vessels. **Neutrophils** and **monocytes** are phagocytic white blood cells; they engulf bacteria and viruses in infected tissues. **Macrophages** ("big eaters") are large phagocytic cells that develop from monocytes. Macrophages wander actively in the interstitial fluid, eating any bacteria and virus-infected cells they encounter. In Figure A, a macrophage is using multiple pseudopodia to snare bacteria. Other protective cells, called **natural killer cells**, attack cancer cells and infected body cells, especially those harboring viruses.

Other nonspecific defenses include antimicrobial proteins that either attack microorganisms directly or impede their reproduction. Especially important are interferons and the complement proteins.

Interferons are proteins produced by virus-infected cells that help other cells resist viruses. Figure B shows how the interferon mechanism seems to work. ① The virus infects a cell and ② turns on interferon genes in the cell's nucleus, ③ causing the cell to make interferon. The infected cell then dies, but ④ its interferon molecules may diffuse to neighboring healthy cells, ⑤ stimulating them to produce other proteins that inhibit viral reproduction. The defense is not virus-specific; interferon made in response to one virus confers resistance to unrelated viruses. The resistance is short-term and seems to be most effective against viruses that cause influenza and the common cold.

Interferons have been in the news fairly regularly since they were discovered in the late 1950s. The body makes these protective proteins in very small quantities, but recombinant DNA technology has made it possible to produce amounts large enough for use in treating viral infections. Also, in some experimental studies, interferons have been useful in treating certain cancers.

Another type of antimicrobial protein, **complement**, is actually a group of over 20 proteins. Complement is named for its cooperation (complementation) with other defense mechanisms. The complement proteins circulate in the blood in inactive form. Once activated, they coat the surfaces of microbes, making them more susceptible to engulfment by macrophages. Complement proteins also cut lethal holes in the membranes of pathogens, and they amplify another nonspecific defense mechanism, the inflammatory response.

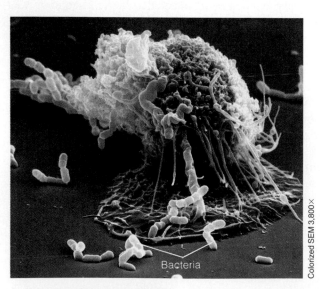

A. Phagocytosis by a macrophage

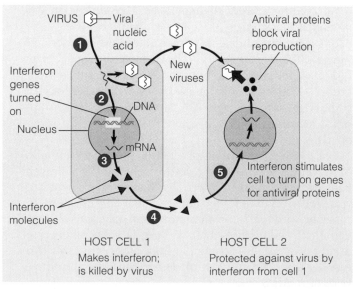

B. Interferon action

The inflammatory response mobilizes nonspecific defense forces

The **inflammatory response** is a major component of our nonspecific defense system. Any damage to tissue, whether caused by an infectious microorganism or by physical injury, even just a scratch or an insect bite, triggers this response. You can see outward signs of the inflammatory response if you watch your skin react to a mosquito bite. The bite area becomes red, swollen, and warmer than the surrounding area. This reaction is inflammation, which literally means "setting on fire."

The figure below shows the chain of events that make up the inflammatory response, in a case where a pin has broken the skin and infected it with bacteria. ① The first thing that happens when a tissue is injured is that the damaged cells release chemical alarm signals such as **histamine.** ② The chemicals spark the mobilization of various defenses. Histamine, for instance, induces neighboring blood vessels to dilate and become leakier. Blood flow to the damaged area increases, and blood plasma passes out of the leaky vessels into the interstitial fluid of the affected tissues. Other chemicals attract phagocytes and other white blood cells to the area. Squeezing between the cells of the blood vessel wall, these white cells (yellow in the figure) migrate out of the blood into the tissue spaces. The local increase in blood flow, fluid, and cells produces the redness, heat, and swelling characteristic of inflammation.

The major results of the inflammatory response are to disinfect and clean injured tissues. ③ The white blood cells mustered into the area engulf bacteria and the remains of any body cells killed by them or by the physical injury. Many of the white cells die in the process, and their remains are also engulfed and digested. The pus that often accumulates at the site of an injury or infection consists mainly of dead white cells and fluid that has leaked from the capillaries during the inflammatory response.

The inflammatory response also helps prevent the spread of infection to surrounding tissues. Clotting proteins (see Module 23.16) present in blood plasma pass into the interstitial fluid during inflammation. Along with platelets, these substances form local clots that help seal off the infected region and allow repair of the damaged tissue to begin. Healing is now under way.

The inflammatory response may be localized, as we have just described, or widespread (systemic). Sometimes microorganisms such as bacteria or protozoans get into the blood or release toxins that are carried throughout the body in the bloodstream. The body may react with one or several inflammatory weapons. For instance, the number of white blood cells circulating in the blood may increase. Another response is fever, an abnormally high body temperature. Toxins themselves may trigger the fever, or certain white blood cells may release compounds that set the body's thermostat at a higher temperature. A very high fever is dangerous, but a moderate one may contribute to defense by stimulating phagocytosis and inhibiting the growth of many kinds of microorganisms.

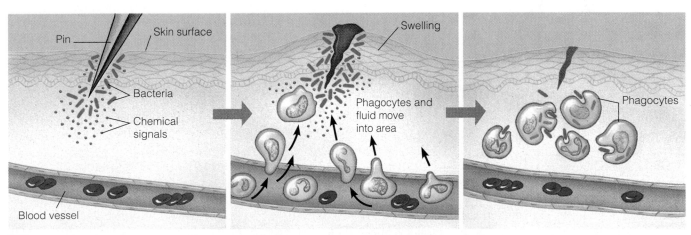

① Tissue injury; release of chemical signals such as histamine

② Dilation and increased leakiness of local blood vessels; migration of phagocytes to the area

③ Phagocytes (macrophages and neutrophils) consume bacteria and cell debris; tissue heals

The inflammatory response

24.3 The immune response counters specific invaders

With our nonspecific defense system warding off so many infectious agents, why do we need an immune system? Our nonspecific defenses are not effective against every microorganism we encounter. The **immune system** is another line of defense, one that recognizes and defends against specific kinds of invasive microbes and cancer cells. Our immune system often acts more effectively than nonspecific defenses, and it can amplify certain nonspecific responses, such as inflammation and the complement reactions.

Whereas our nonspecific defenses are always ready to fight a variety of infections, the immune response must be primed by the presence of a foreign substance, called an antigen. When our immune system detects an antigen, it responds with an increase in the number of cells that either attack the invader directly or produce defensive proteins called antibodies. The defensive cells and antibodies produced against that antigen are usually ineffective against any other foreign substance.

An **antigen** is defined as a molecule that elicits an immune response. (The word "antigen" is a contraction of "*antibody-generating*," a reference to the fact that the foreign agent provokes the immune response.) Antigens include certain molecules on the surfaces of viruses, bacteria, mold spores, pollen, and house dust, as well as molecules on the cell surfaces of transplanted organs. An **antibody** is a plasma protein that attaches to one particular kind of antigen and helps counter its effects.

The immune system is extremely specific, and it has a remarkable "memory." It can "remember" antigens it has encountered before and react against them more promptly and vigorously on second and subsequent exposures. For example, if a person gets rubella (German measles), the immune system remembers certain molecules on the virus that causes this disease. The person is then immune to re-infection because the body will recognize and destroy the rubella virus before it can produce symptoms of illness. Thus, the immune response, unlike our nonspecific defenses, is adaptive; exposure to a particular foreign agent enhances future response to that same agent.

The term **immunity**, in the context of our immune system, means resistance to *specific* invaders. Immunity can be acquired by having an infection; it can also be achieved by the procedure known as **vaccination.** Immunity may be accomplished by presenting the immune system with a **vaccine** composed of a harmless variant of a disease-causing microbe. The vaccine stimulates the immune system to mount defenses against this variant, defenses that will also be effective against the actual pathogen because it has similar antigens. Once we have been successfully vaccinated, our immune system will respond quickly if it is exposed to the microbe. Vaccination has been particularly effective in combating viral diseases, including smallpox, polio, mumps, and measles. Researchers are trying to develop a vaccine against AIDS, but so far results are not encouraging.

Whether antigens enter the body naturally (if you catch the flu) or artificially (if you get a flu shot), the immunity achieved is called **active immunity,** because the body is stimulated to produce antibodies in its own defense. It is also possible to acquire **passive immunity.** For example, antibodies pass from a pregnant woman's bloodstream to that of her fetus, and travelers often get a shot containing antibodies to several disease organisms (rather than a vaccine based on antigens). Both the fetus and the travelers have acquired antibodies passively. This type of immunity is temporary because the person's immune system has not been stimulated by antigens, and the introduced antibodies will remain effective in the body for only a few weeks or months.

24.4 Lymphocytes mount a dual defense

Lymphocytes, white blood cells that spend most of their time in the tissues and organs of the lymphatic system, produce the immune response. Like all blood cells, lymphocytes originate from stem cells in the bone marrow (see Module 23.17). Stem cells give rise to two distinct kinds of immature lymphocytes. As you can see in the top part of the figure on the facing page, one kind of lymphocyte (left) is carried via the blood from the bone marrow to the thymus, a gland in the upper chest region. There the lymphocytes become specialized as T lymphocytes, or **T cells.** The other kind of lymphocyte (right) continues developing in the bone marrow and becomes specialized as a B lymphocyte, or **B cell.**

The B cells and T cells of our immune system mount a dual defense. The B cells secrete antibodies, and since antibodies become dissolved in the blood, the B-cell component of the immune system is called **humoral immunity.** (The blood and other body fluids were formerly called "humors.") The humoral system defends primarily against bacteria and viruses present in body fluids. Humoral immunity can be passively transferred by injecting the plasma from an immune individual into a nonimmune individual.

The second component of our immune system, produced by the T cells, is called **cell-mediated immunity.** In contrast to humoral immunity, cell-mediated immunity cannot be transferred passively with plasma. Cell-mediated immunity can be passively transferred only by giving actual T cells from an immune individual to a nonimmune one. T cells circulate in the blood and attack bacteria and viruses that have already infected (entered) body cells. The T cells also work against infections caused by fungi and protozoans and are thought to be important in protecting the body from its own cells if they become cancerous. T cells also function indirectly by promoting phagocytosis by other white blood cells and helping the B cells produce antibodies.

T cells and B cells develop in the thymus and bone marrow, respectively (center of the diagram below). Certain genes in a T cell are turned on, and the cell synthesizes a specific protein and builds it into its plasma membrane. Sticking out from the cell's surface, these protein molecules act as specific receptors, capable of binding one type of antigen. Each B cell also synthesizes specific receptor proteins on its surface. Its receptors are essentially molecules of a specific antibody. Once it has developed its surface proteins, each B cell and T cell is competent at recognizing one specific antigen and mounting an immune response against it. One cell may recognize an antigen on the mumps virus, for instance, while another detects a particular antigen on a tetanus-causing bacterium.

An enormous number of different kinds of B cells and T cells develop. Researchers estimate that each of us has between 100 million and 100 billion different kinds—enough to recognize and bind virtually any kind of antigen we would ever encounter. A small population of each kind of lymphocyte lies in wait in our bodies, genetically programmed to recognize and respond to a specific antigen. A key feature of our immune system is its preparedness for an almost unlimited variety of potentially harmful antigens.

The bottom part of the diagram shows where the T cells and B cells go after they have developed their surface receptors. From the thymus and bone marrow, they move via the blood to the lymph nodes, spleen, and other parts of the lymphatic system. Here they are able to confront infectious agents that penetrate the body's outer defenses. As we see in the next module, capillaries of the lymphatic system extend into all the tissues of the body. Consequently, bacteria or viruses infecting nearly any part of the body eventually turn up in the lymphatic fluid. When an individual B cell or T cell actually encounters the specific antigen it is programmed to recognize, it differentiates further and becomes a fully mature component of the immune system.

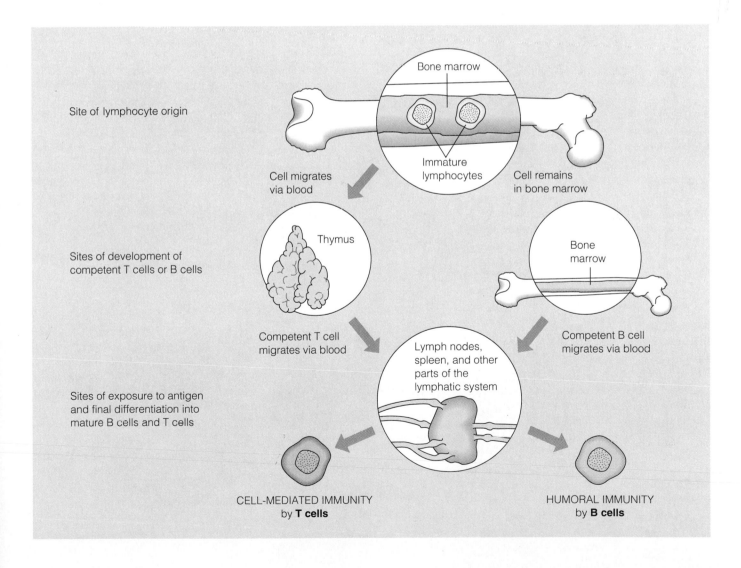

Site of lymphocyte origin

Bone marrow

Immature lymphocytes

Cell migrates via blood

Cell remains in bone marrow

Sites of development of competent T cells or B cells

Thymus

Bone marrow

Competent T cell migrates via blood

Competent B cell migrates via blood

Sites of exposure to antigen and final differentiation into mature B cells and T cells

Lymph nodes, spleen, and other parts of the lymphatic system

CELL-MEDIATED IMMUNITY by **T cells**

HUMORAL IMMUNITY by **B cells**

The lymphatic system becomes a crucial battleground during infection

The lymphatic system, illustrated in Figure A, circulates a fluid called **lymph**, which is similar to interstitial fluid but contains less oxygen and fewer nutrients. The **lymphatic system** consists of a branching network of lymphatic vessels with numerous lymph nodes (saclike organs densely packed with lymphocytes), as well as the thymus, tonsils (including the adenoids), appendix, spleen, and bone marrow. The system has two main functions: to return tissue fluid to the circulatory system, and to fight infection.

As we noted in Module 23.12, about 1% of the fluid that enters the tissue spaces from the blood in a capillary bed does not reenter the blood capillaries. Instead, this small amount of fluid is returned to the blood via the vessels of the lymphatic system. The enlargement in Figure B shows a branched lymphatic vessel in the process of taking up fluid from tissue spaces in the skin. As shown here, fluid enters the lymphatic system by diffusing into tiny, dead-end lymphatic capillaries that are intermingled among the blood capillaries.

Lymph drains from the lymphatic capillaries into larger and larger lymphatic vessels. It reenters the circulatory system via two large lymphatic vessels, which fuse with veins in the shoulders. These large lymphatic vessels are the thoracic duct and the right lymphatic duct, which you can see in Figure A. As Figure B indicates, the lymphatic vessels resemble veins in having valves that prevent the backflow of fluid toward the capillaries. Also like veins, lymphatic vessels depend mainly on the movement of skeletal muscles to squeeze their fluid along. The black arrows in Figures B and C indicate the flow of lymph.

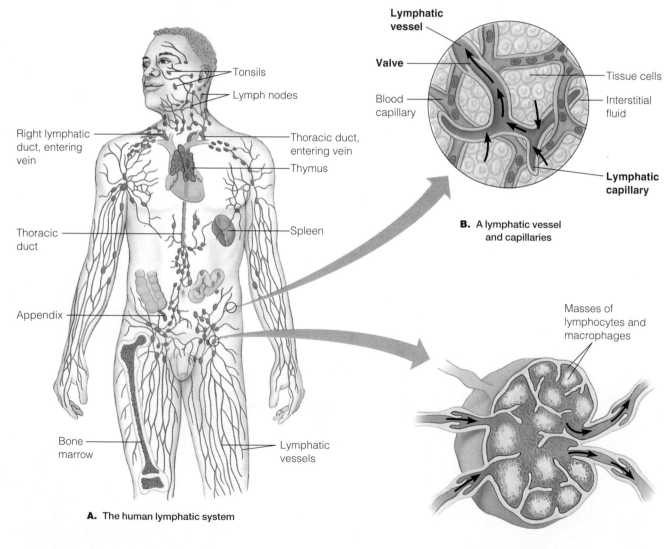

Tonsils

Lymph nodes

Right lymphatic duct, entering vein

Thoracic duct, entering vein

Thymus

Thoracic duct

Spleen

Appendix

Bone marrow

Lymphatic vessels

A. The human lymphatic system

Lymphatic vessel

Valve

Blood capillary

Tissue cells

Interstitial fluid

Lymphatic capillary

B. A lymphatic vessel and capillaries

Masses of lymphocytes and macrophages

C. A lymph node

Lymph nodes and the other lymphatic organs shown in Figure A perform the defensive function of the lymphatic system. As shown in Figure D for a lymph node, these organs are densely packed with lymphocytes (smaller cells) and macrophages (larger cells). Lymph flows through the lymphatic organs, and the macrophages ingest any infectious microorganisms in the fluid. The macrophages also play a key role in activating the lymphocytes, as we will see in Module 24.13.

When the body is fighting an infection, the lymphatic system becomes a major battleground. Foreign invaders swept into the system from infection sites in the body contact and activate lymphocytes. An activated lymphocyte multiplies and forms large numbers of activated B or T cells. The B cells produce antibodies, which are carried in the lymph and bloodstream to sites of infection. T cells also travel in the lymph and blood and actively fight infectious agents wherever they occur in the body. You can see evidence of your body's cellular response to an infection when your lymph nodes become swollen and tender as a result of producing large numbers of defensive lymphocytes. Swollen lymph nodes—often referred to as "swollen glands"—commonly occur in the neck.

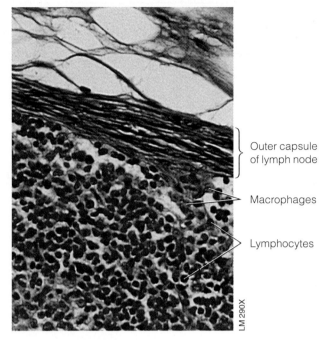

D. Lymphocytes and macrophages within a lymph node

Outer capsule of lymph node

Macrophages

Lymphocytes

LM 290X

24.6

Antigens have specific regions where antibodies bind to them

As we indicated earlier, antigens are molecules that elicit the immune response. Usually antigens do not belong to the host animal. Most antigens are proteins or large polysaccharides on the surfaces of viruses or foreign cells. Common examples are protein-coat molecules of viruses, parts of the capsules and cell walls of bacteria, and macromolecules on the surface cells of other kinds of organisms, such as protozoans and parasitic worms. Sometimes a particular microbe is called an antigen, but this usage is misleading because the microbe will almost always have several kinds of antigenic molecules. Blood cells or tissue cells from other individuals (of the same species or a different species) can also provide antigenic molecules.

Our immune system can recognize millions, perhaps billions, of different antigens, as noted in Module 24.4. In fact, we have so many different kinds of B and T cells that many of them never encoun-

ter the antigen they would recognize, and simply remain idle throughout our lives. As shown in the figure here, antibodies usually identify localized regions, called antigenic determinants, on the surface of an antigen molecule. An antigen-binding site on the antibody molecule recognizes an antigenic determinant by the fact that the binding site and antigenic determinant have complementary shapes, like an enzyme and substrate, or a lock and key. An antigen usually has several different determinants (there are three in the diagram here), so different antibodies (two, in this case) can bind to the same antigen. A single antigen molecule may stimulate the immune system to make several distinct antibodies against it. Notice that each antibody molecule has two identical antigen-binding sites. We'll return to antibody structure in Module 24.10.

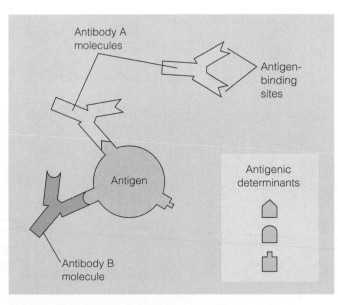

Antibody A molecules

Antigen-binding sites

Antigen

Antigenic determinants

Antibody B molecule

The binding of antibodies to antigenic determinants

Clonal selection musters defensive forces against specific antigens

The immune system's ability to defend against an almost infinite variety of antigens depends on a process called **clonal selection**. At first, an antigen introduced into the body activates only a tiny fraction of lymphocytes. These "selected" cells then proliferate, forming a clone of cells (a population of genetically identical cells) that are specific for the stimulating antigen.

The cluster of cells at the top in the figure below represents a diverse population of B cells in a lymph node. Each B cell is unique because it has its own specific type of antigen receptor embedded in its surface. The cells have their receptors in place *before* they ever encounter an antigen. The surface of each cell has multiple copies of its particular receptor (actually thousands, though only a few are shown here). Because these are B cells, the receptors are antibody molecules.

Once an antigen enters the body and is swept into the lymph node, it binds with the receptors that fit it. Binding of the antigen activates the lymphocyte. The figure shows only one cell (with purple receptors) being activated. In real life, a larger number of cells would respond—but still only a small fraction of the cells in the body's lymph nodes. Other lymphocytes, without the appropriate binding sites, would not be affected.

The rest of the figure shows what happens after antigen molecules bind with the specific receptors on one of the B cells. Primed by the interaction with the antigen, the selected cell grows, divides, and differentiates further. The result is a clone of **effector cells** specialized for defending against the very antigen that triggered the response. The effector cells shown here secrete specific antibodies. (These effector cells have large amounts of endoplasmic reticulum, a characteristic of cells actively synthesizing and secreting proteins.) The same kind of clonal selection mechanism operates on T cells to produce the mature ones that carry out cell-mediated immunity.

Thus, we see that the versatility of the immune system—its ability to defend against a virtually unlimited variety of invaders—depends on a great diversity of lymphocytes with different antigen receptors. It does not depend on cells' having the flexibility to change their antigenic specificity on demand.

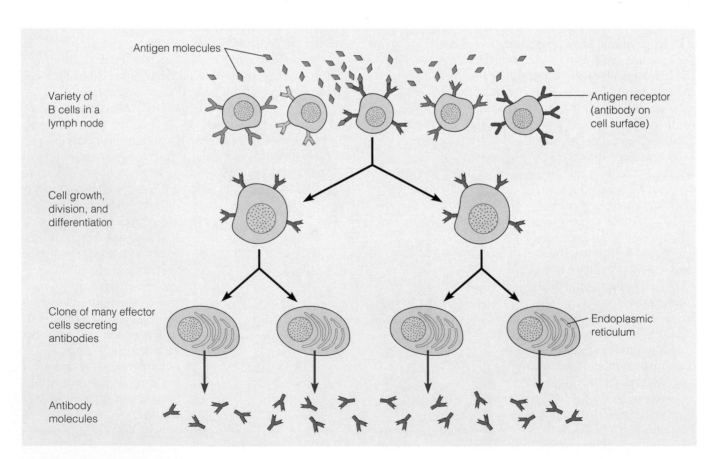

Clonal selection operating on B cells

Now that we have seen how clonal selection works, we can see how it fits into the rest of the immune system's response to an antigen. It takes two exposures to an antigen for the immune system to mount its full response. As indicated by the blue curve in Figure A, the two exposures trigger two distinct phases of immune response. The initial phase, called the **primary immune response**, occurs when lymphocytes are first exposed to an antigen and form a clone of effector cells. On the far left of the graph, you can see that the primary response does not start right away; it usually takes several days for the lymphocytes to become activated by an antigen (X) and form clones of effector cells. When the effector cell clone forms, antibodies start showing up in the blood, as the graph shows.

After the primary immune response, a second exposure to the *same* antigen elicits a faster and stronger response, called the **secondary immune response**. In the case of humoral immunity (the antibody response shown here), the secondary response produces very high levels of antibodies that are often more effective against the antigen than those produced during the primary response. As the blue curve shows, the secondary response also lasts much longer than the primary response.

The red curve in Figure A illustrates the specificity of the immune response. If the body is exposed to a different antigen (Y), even after it has already responded to antigen X, it responds with another primary response, this one directed against antigen Y. The response to Y is not enhanced by the response to X.

Figure B outlines the cellular events that produce the primary and secondary immune responses. Each exposure to the antigen triggers clonal selection, and two clones of lymphocytes result. The cells of each clone, while all identical genetically, fall into two sets. One set consists of effector cells; the other set consists of cells called **memory cells**, which differ from effector cells in both appearance and function. During the primary response, the effector cells of the clone combat the antigen, by producing antibodies if they are B cells. These effector cells usually survive only a few days. In contrast, the memory cells of the clone may last for decades. They remain in the lymph nodes, ready to be activated by a second exposure to the antigen. When this happens, the memory cells initiate the secondary immune response: They multiply quickly, producing a large clone of lymphocytes that mount the secondary response. Like the first clone, the second clone includes effector cells that actually produce the antibodies of the secondary response and memory cells capable of responding to future exposures to the antigen. In some cases, memory cells seem to confer lifetime immunity, as they may in such childhood diseases as mumps, measles, and polio.

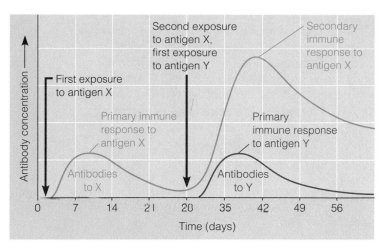

A. Immunological memory

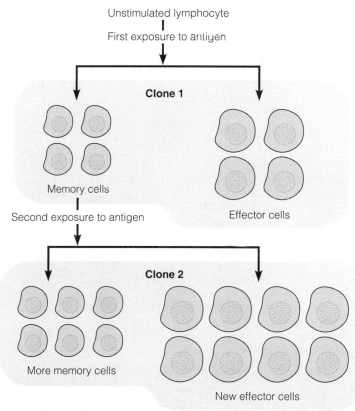

B. The cellular basis of immunological memory

Although we have focused on humoral immunity (B cells) in this module, clonal selection, effector cells, and memory cells are features of cell-mediated immunity (T cells) as well. In the next four modules, we discuss humoral immunity further. After that, we will focus on how the cell-mediated arm of the immune system helps defend the body against pathogens.

Overview: B cells are the warriors of humoral immunity

By connecting the concepts of clonal selection and immunological memory, we can obtain an overview of the "defensive machine" we call humoral immunity. Continuing our military analogy, B cells are the "warriors" of this machine.

Starting with the first encounter between an antigen and a group of B cells, the figure below summarizes the primary and secondary responses in humoral immunity. Recall that a B cell is first "selected" by the antigen. In the first step of this process, surface receptors (specific antibodies) on the B cell bind with antigen molecules (actually antigenic determinants on the antigen's surface). This binding triggers the growth, division, and further differentiation of the selected cell. The resulting clone contains many effector B cells, called **plasma cells,** and a smaller number of memory B

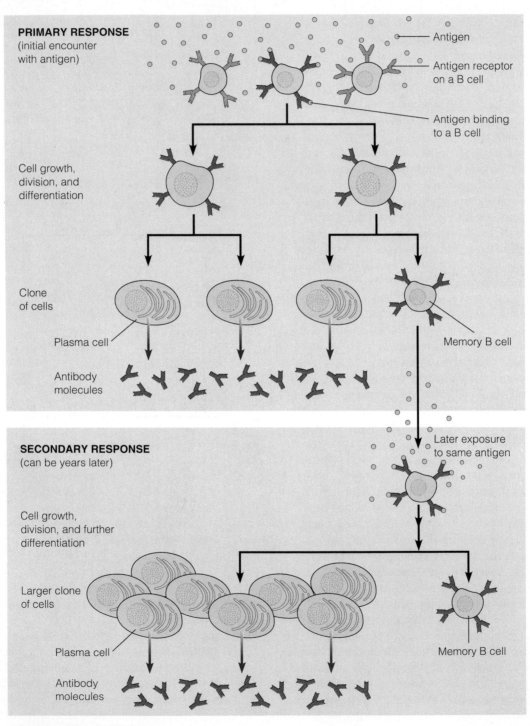

An overview of humoral immunity

cells (only a few of the two types of cells are shown in the figure). The plasma cells secrete antibody molecules, completing the primary response. In producing antibodies that can actually combat antigen molecules, plasma cells are the "front-line warriors" of humoral defense. They may secrete as many as 2000 antibody molecules per second for their brief lifetime of four or five days. These antibodies circulate in the blood and lymphatic fluid, binding to antigens and contributing to their destruction. Then the plasma cells of the primary immune response die out, as the primary response subsides.

In a military sense, we could consider memory B cells the "reservists" of humoral immunity. Rather than actively combating antigens, they await future exposure to antigens. The secondary response in humoral immunity can occur only if these cells contact the same antigen that triggered their production, perhaps not until years later. If this contact does occur, the events in the bottom part of the figure on the preceding page are set in motion. The memory cells bind antigens and are stimulated to quickly produce large new clones of cells. The cloning process is the same as in the primary response, but it occurs more rapidly and yields more plasma cells. As a result, antibody levels in the blood and lymph are much higher than during the primary response.

Antibodies are the weapons of humoral immunity

Antibodies are proteins that serve as molecular weapons. We have been using Y-shaped symbols to represent these molecules, and the Y shape actually does resemble their shape. Figure A is a computer graphic of a single antibody molecule. Figure B is a simplified diagram explaining its structure. Each antibody molecule is made up of four polypeptide chains, two "heavy" chains and two "light" chains. In both figures, the regions of reddish color represent the fairly long heavy chains of amino acids that give the molecule its Y shape. Bonds (black lines in Figure B) at the fork of the Y hold these chains together. The two green regions in each figure are shorter chains of amino acids, the light chains. Each of the light chains is bonded to one of the heavy chains. As the computer graphic indicates, the bonded chains intertwine.

As we have seen, an antibody molecule has two functions in humoral immunity: to recognize and bind to a certain antigen, and to assist in destroying the antigen it recognizes. The structure of an antibody allows it to perform these functions. Notice in Figure B that each of the four chains of the molecule has a constant (C) region and a variable (V) region. A pair of V regions forms an antigen-binding site at each tip of the Y; thus, the V regions perform the antibody's recognition-and-binding function. A huge variety in the shapes of the binding sites arises from a similarly large variety in the amino acid sequences in the V regions, hence the term "variable." This structural variety gives the humoral immune system the ability to react to virtually any kind of antigen.

The rest of the antibody molecule—the constant regions—helps destroy and eliminate antigens. Antibodies with different kinds of C regions are grouped into different classes. Humans and other mammals have five major classes of antibodies, and each of the five classes has a particular role in humoral immunity. For instance, one type functions mainly in the primary immune response; another comes into play during the secondary response. Next, we look at how antibody–antigen binding leads to the destruction of invading microbes.

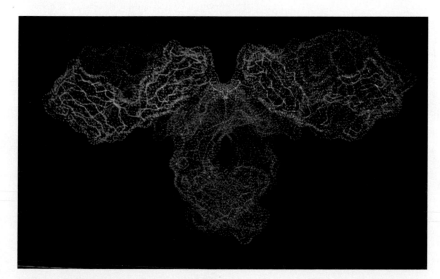

A. Computer graphic of an antibody molecule

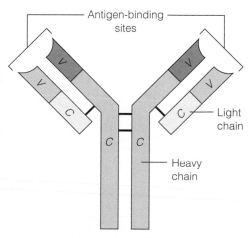

B. Antibody structure

The actual role of an antibody in destroying invading cells or molecules is to mark the invaders. An antibody marks an antigen by combining with it to form an antigen–antibody complex. Weak chemical bonds between antigen molecules and the antigen-binding sites on antibody molecules hold the complex together.

As the figure below illustrates, it is the binding of antibodies to antigens that actually triggers mechanisms to neutralize or destroy an invader. Such a mechanism is called an effector mechanism. Several effector mechanisms are depicted in the figure. In neutralization, the binding of antibodies physically blocks harmful antigens, making them harmless. For example, antibodies may bind to the surface molecules a virus uses to attach to a host cell. Antibodies may also bind to toxin molecules on bacterial cells. Phagocytes such as macrophages then dispose of the complexes.

Another effector mechanism is the agglutination (clumping together) of bacteria or other foreign cells. Because each antibody molecule has at least two binding sites, antibodies can hold a clump of invading cells together.

Agglutination makes the cells easy for phagocytes to capture.

A third effector mechanism, precipitation, is similar to agglutination, except that the antibody molecules link *dissolved* antigen molecules together. This makes the antigen molecules precipitate out of solution as solids. The precipitated antigens, like clumps of agglutinated cells, are easily engulfed by phagocytes.

One of the most important effector mechanisms in humoral immunity is the activation of complement proteins (see Module 24.1) by antigen–antibody complexes. Activated complement proteins (green in the diagram) can attach to a foreign cell and open holes in its plasma membrane, causing cell lysis (rupture).

Taken as a whole, this figure illustrates a fundamental concept of immunity: All effector mechanisms involve a *specific* recognition and attack phase followed by a *nonspecific* destruction phase. Thus, the antibodies of humoral immunity, which identify and bind to foreign invaders, work with nonspecific defenses, such as phagocytes and complement, to form a complete defense system.

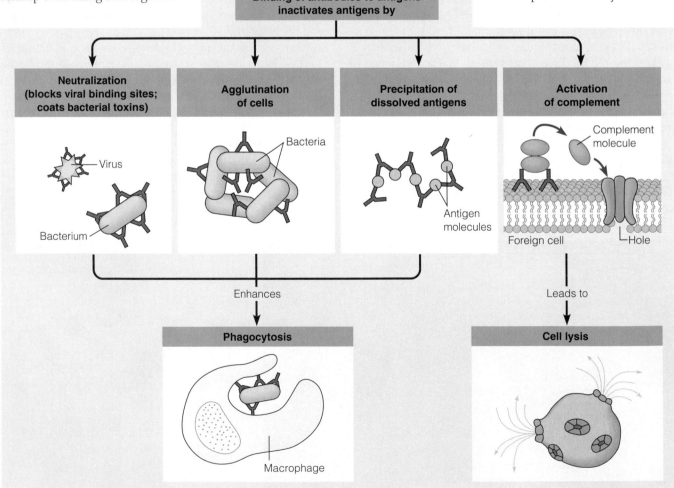

Binding of antibodies to antigens inactivates antigens by

Neutralization (blocks viral binding sites; coats bacterial toxins)
Virus
Bacterium

Agglutination of cells
Bacteria

Precipitation of dissolved antigens
Antigen molecules

Activation of complement
Complement molecule
Foreign cell
Hole

Enhances

Leads to

Phagocytosis
Macrophage

Cell lysis

Effector mechanisms of humoral immunity

Monoclonal antibodies are powerful tools in the lab and clinic

Because of their ability to tag specific molecules or cells, antibodies have been widely used in biological research and clinical testing. In the original procedure for preparing antibodies, a small sample of antigen was injected into a rabbit or mouse. In response to the antigen, the animal would produce antibodies, which could be collected directly from its blood. However, because the antigen usually had many different antigenic determinants, the result would be a mixture of different antibodies rather than a pure preparation of one. And to make a large amount of antibody, more than one animal had to be used.

Such problems sharply limited the use of antibodies until the late 1970s, when a technique for making **monoclonal antibodies** was developed. The term "monoclonal" means that all the cells producing the antibodies are descendants of a single cell; thus, they all produce identical antibody molecules. Monoclonal antibodies are harvested from cell cultures rather than from animals.

As shown in Figure A, the trick to making monoclonal antibodies is the fusion of two cells to form a hybrid cell with a combination of desirable properties. First, an animal is injected with the antigen that will provoke the desired antibody. At the same time, cancerous tumor cells, which can multiply indefinitely, are grown in a culture. The scientist then fuses a tumor cell with a normal antibody-producing B cell from the animal. As we saw earlier, a single such B cell makes antibody to only one antigenic determinant. Fusing the two cells yields a hybrid cell with the desirable features of each of the parent cells. The hybrid cell makes a single type of antibody molecule and is able to multiply indefinitely in a laboratory dish. Thus, large amounts of identical antibody molecules are easily prepared.

The ability to make mono-clonal antibodies has spawned a new industry. A common area of application is medical diagnosis. For example, monoclonal antibodies may bind with bacteria that cause a sexually transmitted disease, or with a hormone that indicates pregnancy. If the antibodies have been labeled (by a dye, for instance) for easy detection, they will reveal the presence of the bacteria or hormone. This procedure is the basis for a sensitive home pregnancy test. The photograph in Figure B shows a woman testing a sample of her urine for a hormone produced during early pregnancy. If the hormone is present, the monoclonal antibodies will bind with it, producing a color change in the sample.

Monoclonal antibodies also have potential use in the treatment of certain diseases, including cancer. For instance, a monoclonal antibody specific to an antigen on a cancer cell can be combined with a drug that kills cells. Such an antibody–drug combination should act like a molecular guided missile. When injected into a cancer patient, the specificity of the antibody should target it to cancer cells, and the drug should kill the cells. This approach to cancer treatment is currently being tested, with encouraging results in several cases of bone marrow cancer.

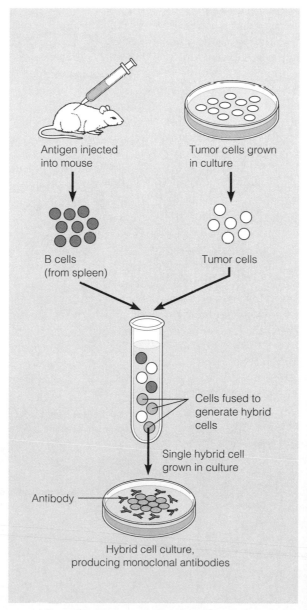

A. The procedure for making monoclonal antibodies

B. A home pregnancy test

The antibodies of humoral immunity make up one task force of the body's defense network. The humoral defense system identifies and helps destroy invaders that are in our blood, lymph, or interstitial fluid—in other words, outside our body cells. But many invaders, including all viruses, enter cells and reproduce there. It is the cell-mediated immunity produced by T cells that battles pathogens that have already entered body cells.

Unlike B cells, which can respond to free antigens present in body fluids, T cells can respond only to antigens present on the surfaces of the body's own cells. There are three main kinds of T cells. **Cytotoxic T cells** actually attack body cells that are infected with pathogenic microbes. The other two types of T cells, suppressor T cells and helper T cells, have regulatory functions in immunity. **Suppressor T cells** inhibit the activity of other T cells as well as B cells; they are responsible for terminating immune activities after an infection has been eliminated. **Helper T cells** are fundamentally important in all aspects of immunity. They play a central role in activating other types of T cells, and they even help activate B cells to produce antibodies.

Helper T cells interact with other white blood cells that function as **antigen-presenting cells (APCs)**. In studying this cellular interaction, researchers found that cell-mediated immunity works by a lock-and-key mechanism similar to the antibody–antigen interaction in humoral immunity. All of cell-mediated immunity and much of humoral immunity depend on the precise interaction of APCs and helper T cells.

The function of an APC is to *present* a foreign antigen to a helper T cell. A typical APC is a macrophage. After ingesting a foreign substance or microbe, the APC attaches antigenic portions of molecules from the ingested invader to its surface. As shown in Figure A, a special protein molecule belonging to the APC, which we will call the **self protein** (because it belongs to the body itself), holds the foreign antigen (a **nonself molecule**) and displays its antigenic determinants. Each of us has a unique set of self proteins, which serve as identity markers for our body cells. Our helper T cells can recognize only the *combination* of a self protein (green in Figure A) and a foreign antigen (orange) displayed on an APC. This double-recognition system used by helper T cells is analogous to the system banks use for safe-deposit boxes: Opening your box requires the use of the banker's key along with your specific key.

The ability of a helper T cell to recognize and be activated by unique

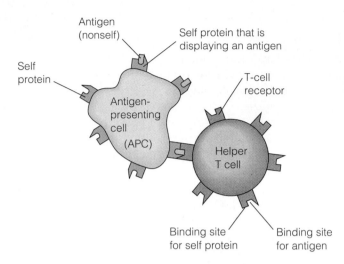

A. A helper T cell interacting with an APC

self–nonself complexes on APCs depends on T-cell receptors, specific proteins embedded in the T cell's own plasma membrane. As shown on the helper T cell in Figure A, a T-cell receptor molecule (purple) has two binding sites: an antigen-binding site that recognizes a nonself protein (antigen) and a self protein binding site. The two binding sites enable a T-cell receptor to recognize the overall shape of a self–nonself complex on an APC. Cell-mediated immunity is highly specific because each helper T cell can bind with only one kind of self–nonself complex on an APC.

What happens as a result of the interaction between a helper T cell and an APC? After this key event occurs, both cells secrete chemical signals that amplify the immune response (Figure B). The APC secretes a protein (interleukin-1, green arrow) that helps activate the helper T cell. Interleukin-1 makes the helper T cell secrete a second protein (interleukin-2, dark blue arrow) that has three major ef-

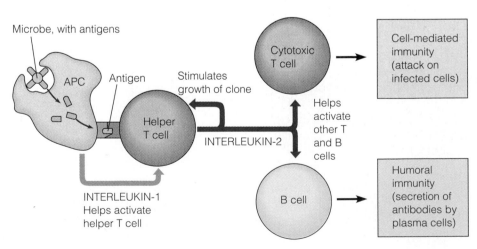

B. The central role of helper T cells

fects. One effect is to make the helper T cell grow and divide, producing a clone. Cells of the clone all have the same receptors keyed to the specific combination of self and nonself molecules on the original APCs. The clone includes memory cells as well as active helper T cells. The active helper T cells amplify the defenses against the antigens at hand. The other effects of interleukin-2 are to help activate the cytotoxic T cells of cell-mediated immunity and to help activate the B cells of humoral immunity.

Cytotoxic T cells are the only T cells that actually kill other cells. Their targets are infected body cells. Once activated, cytotoxic T cells identify their targets in the same way that helper T cells identify APCs. A target cell has fragments of foreign (nonself) antigens—molecules belonging to the viruses or bacteria infecting it—attached to self proteins on its surface (Figure C). These self proteins (light green) are different from those that display antigens on APCs, so the cytotoxic T cells do not attack APCs. Like a helper T cell, a cytotoxic T cell carries receptors (purple) that can bind with a self–nonself complex on the target cell.

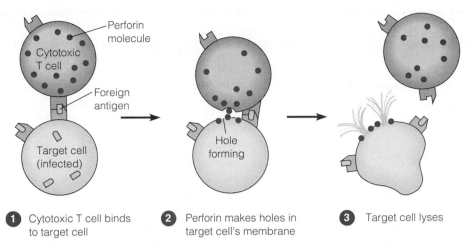

1 Cytotoxic T cell binds to target cell

2 Perforin makes holes in target cell's membrane

3 Target cell lyses

C. How a cytotoxic T cell kills a target cell

The self–nonself complex on an infected body cell is like a red flag to cytotoxic T cells that have matching receptors. As shown in Figure C, ① the cytotoxic T cell binds to the target cell and synthesizes a protein called **perforin.** ② Perforin is discharged and attaches to the target cell's membrane, making holes in it. ③ Cell lysis results. This is similar to one of the actions of complement protein, although complement usually perforates *microbial* membranes.

Cytotoxic T cells may prevent cancer

As we discussed in Module 11.14, cancer is a variety of diseases affecting cells of many different types in the body. Regardless of causes and origins, however, all cancers involve changes in normal body cells, and some of these changes take place on the outer membrane surfaces of the cells. Such changes may alter surface molecules in such a way that the immune system identifies the cancer cells as foreign intruders and destroys them.

Research on the role of the immune system in combating cancer has not yet produced definitive results. However, the fact that people with immune deficiencies are often more susceptible to cancer has led to the assumption that the immune system plays a watchdog role against at least some forms of cancer. The micrograph below shows a troop of cytotoxic T cells (pale green) attacking a tumor cell in a laboratory culture. This same phenomenon may occur in the body; cytotoxic T cells may attack and kill cancer cells whenever they appear. Why this built-in surveillance system sometimes fails, allowing tumors to develop, is still largely a mystery. Some scientists suggest that tumors develop when cancer cells shed surface molecules that mark them as foreign, or when cancer cells secrete chemicals that suppress the immune system. Studies of immune suppression are an important part of the cancer research effort.

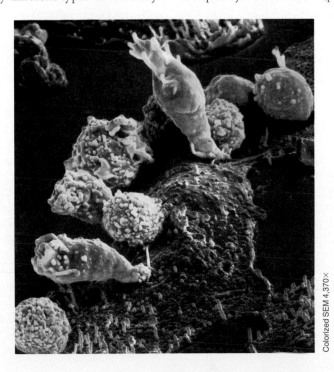

Colorized SEM 4,370×

Cytotoxic T cells attacking a cancer cell

24.15 The immune system depends on our molecular fingerprints

As we have seen, our immune system's ability to recognize the body's own molecules—that is, to distinguish *self* from *nonself*—enables it to battle foreign molecules and cells without harming healthy body cells. The self molecules are the keys to this ability. Each person's cells have a particular collection of self proteins that provide the molecular "fingerprints" recognized by the immune system.

Each of us has two sets of self proteins on the surfaces of our cells. Class I proteins occur on all cells in the body. Class II proteins are found only on B cells and on macrophages, cells that can function as antigen-presenting cells. Both sets are unique to the individual in which they are found, marking the body cells as "off-limits" to the immune system; our lymphocytes do not attack these molecules.

Our immune system not only distinguishes our own cells from invading microbes, it also can tell our body cells from those of other people. Genes determine the specific structure of self molecules, and each of us has about 20 genes that encode these proteins. Because there are 50 or more alleles in the human population for each of the 20 genes, it is virtually impossible for any two people (except identical twins) to have matching sets of self proteins.

The immune system's ability to recognize foreign antigens does not always work in our favor. For example, when an organ from one person is transplanted into another person, the recipient's immune system recognizes the donor's cells as foreign and attacks them. To minimize the chance of this rejection, medical workers try to find a donor whose self proteins match the recipient's as closely as possible. Also, various drugs are used to suppress the immune response against the transplant. Unfortunately, these drugs may also interfere with the ability of the immune system to fight infections. A few drugs, such as cyclosporine, have the advantage of suppressing only cell-mediated responses; they do not cripple humoral immunity.

Two new approaches hold promise for avoiding transplant rejection. In one approach, monoclonal antibodies would be used as "guided missiles" to target and eliminate the T cells that attack transplants. A second approach would be to find a way to use isolated stem cells to establish a new immune system that would recognize the transplanted organ as self.

24.16 Immunological failure causes several diseases

Overall, our immune system is highly effective, protecting our bodies against a vast array of potentially harmful invaders. When the immune system malfunctions or fails, a number of serious diseases can result.

Several **autoimmune diseases**—including insulin-dependent diabetes (see Module 26.9), rheumatoid arthritis, systemic lupus erythematosus, and rheumatic fever—result when the immune system goes awry and turns against the body's own molecules. In rheumatoid arthritis, for instance, some of the patient's own antibodies form molecular complexes with one another and with complement. The molecular complexes tend to accumulate in the joints, where they cause painful inflammation and damage to bone and cartilage.

People with systemic lupus erythematosus (lupus) develop immune reactions against parts of their own cells, particularly nucleic acids. Lupus is not well understood, but deposits of antibody–nucleic acid complexes accumulate in, and inflame, the joints and kidneys. Drugs that suppress the immune system and the inflammatory response can relieve lupus symptoms, but such drugs may also damage the kidneys.

Rheumatic fever seems to result from another kind of autoimmunity. Rheumatic fever is always preceded by a strep infection, caused by streptococci bacteria. The immune system responds with antibodies that fight off the infection. However, in people with self proteins that closely resemble one or more of the streptococcal antigens, the antibodies also react with some of the person's own tissues, producing rheumatic fever. The symptoms include inflammation of the joints and damage to heart muscles and valves.

In contrast to autoimmune diseases are a variety of defects called **immunodeficiency diseases.** Immunodeficient people lack one or more of the components of the immune system and, as a result, are susceptible to infections that would ordinarily not cause a problem. Immunodeficiency can have a number of causes. For instance, Hodgkin's disease, a type of cancer that affects the lymphocytes, can depress the immune system. Radiation therapy and the drug treatments used against many cancers can have the same effect. In the rare congenital disease called severe combined immunodeficiency (SCID), both T cells and B cells are absent or inactive. People with SCID are extremely sensitive to even minor infections. Until recently, their only hope for survival was to live behind protective barriers that kept out all pathogens, or to receive and retain a bone marrow transplant. Cases of SCID involving a single faulty gene may soon be routinely treated with gene therapy, as we discussed in Module 13.17. In the future, SCID may be treated with purified stem cells.

In addition to autoimmune and immunodeficiency diseases, physical and emotional stress may also weaken the immune system. Research indicates, for example, that people who are generally content tend to have a higher ratio of helper T cells to suppressor T cells than those who admit to being unhappy. In one study, students were examined just after a vacation and then again during exams. Their natural killer cells were less effective, and they produced less interferon during exams.

How might stress affect the immune system? A link between our nervous and immune systems may be the answer. Nerve fibers penetrate the organs that produce lymphocytes. And lymphocytes have receptors for chemical signals secreted by nerve cells. Might chemical signals from our nerve cells enhance our immunity when we are relaxed and happy? Research on the role of stress in immunity may yield some answers in the near future.

Allergies are overreactions to certain environmental antigens

Allergies, another type of immune system disorder, are abnormal sensitivities to antigens in our surroundings. Antigens that cause allergies are called **allergens.** Protein molecules on pollen grains, on the surface of tiny mites that live in house dust, and in animal dander are common allergens. Many people who are allergic to cats and dogs are actually allergic to proteins in the animal's saliva. They become sensitized to salivary proteins deposited on the fur when the animal licks itself. Allergic reactions typically occur very rapidly and in response to tiny amounts of an allergen. A person allergic to cat or dog saliva, for instance, may react to a few molecules of the allergen in a matter of minutes. Allergic reactions can occur in many parts of the body, including the nasal passages, bronchi, digestive tract, and skin. Symptoms may include sneezing, coughing, wheezing, upset stomach, and itching.

The symptoms of an allergy result from a two-stage reaction sequence outlined in the figure below. The first stage, called sensitization, occurs when a person is first exposed to an allergen—pollen, for example. B cells make a special class of antibodies (one of the five classes in mammals) in response to allergens. Some of these antibodies attach to receptor proteins on the surface of **mast cells,** normal body cells that produce hista-

mine and other chemicals that trigger the inflammatory response (see Module 24.2). In our example, the affected mast cells are in the nose.

The second stage of an allergic response begins when the person is exposed to the same allergen at a later time. The allergen binds to the antibodies attached to mast cells, causing the cells to release histamine and other inflammatory chemicals. Histamine triggers the allergic symptoms. As in inflammation, it causes blood vessels to dilate and leak fluid. Histamine also elicits other symptoms, such as nasal irritation, itchy skin, and tears. **Antihistamines** are drugs that interfere with histamine's action and give temporary relief from an allergy.

Allergies range from seasonal nuisances to severe, life-threatening responses. **Anaphylactic shock** is an especially dangerous type of allergic reaction. Some people are extremely sensitive to certain allergens, such as the venom from a bee sting. Any contact with these allergens makes their mast cells release inflammatory chemicals very suddenly. As a result, their blood vessels dilate abruptly, causing a precipitous drop in blood pressure (shock), which is potentially fatal. Fortunately, anaphylactic shock can be counteracted with injections of the hormone epinephrine.

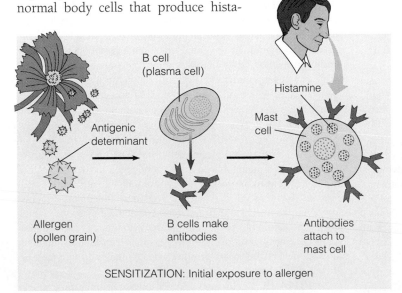

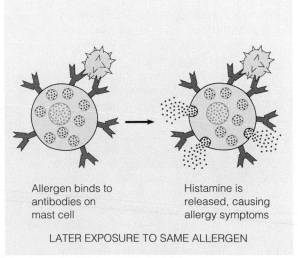

The two stages of an allergic reaction

Almost 300,000 living Americans have been diagnosed with AIDS. Worldwide, as many as 10–12 million may be infected with the AIDS virus, HIV. The World Health Organization estimates that by the year 2000, 30–40 million people will carry HIV.

HIV is deadly because it destroys the immune system, leaving the body defenseless against most invaders. HIV can infect a variety of cells, but it seems to have a preference for helper T cells—the cells that activate other T cells and B cells as well. When HIV depletes the body of helper T cells, the immune system cannot carry out either the cell-mediated or the humoral response. Death usually results, not from AIDS itself but from another infectious agent or from cancer. We described the course of cell infection by HIV in Module 10.18. If you turn back to Figure 10.19, you will see a micrograph showing HIV attacking a helper T cell.

At present, AIDS is incurable, and the AIDS virus continues to evade a worldwide effort to control its transmission and deadly effects. A drug now in use called AZT (azidothymidine) may postpone the development of AIDS in a person infected with HIV. Unfortunately, AZT does not prevent AIDS from developing or from causing death. Other drugs now being tested may keep HIV from binding to the receptors on T cells, or they may prevent the virus from reproducing. Simultaneous treatment with several drugs offers some promise. It may also be possible to make antibody–drug complexes that home in on HIV-infected T cells and destroy them (see Module 24.12). Other approaches that may bear fruit in the future include injecting patients with HIV-resistant stem cells that could give rise to HIV-resistant T cells.

What about vaccines against HIV? Vaccines made from killed, whole AIDS virus elicit an immune response to HIV and are currently being tested in asymptomatic patients and in people with early symptoms. Other experimental vaccines consisting of HIV surface proteins are being made by recombinant DNA techniques. Some of these produce HIV-specific immune responses (antibodies and T cells) that may prove capable of killing the AIDS virus. Nevertheless, so far the development of vaccines has not progressed beyond the testing stage.

Time is a big problem in the fight against AIDS. Drugs and vaccines must undergo exhaustive tests for efficacy and for toxic side effects before they can be released for general use. While hearing the outcry of AIDS patients and activists, some experts predict that effective drugs and vaccines may not see widespread use until the turn of the century. At present, and probably for the next several years, education remains our most effective weapon against AIDS. The World Health Organization concurs with Magic Johnson's campaign for safer sex. The use of condoms could save millions of lives. For more information on AIDS and its mode of transmission, anyone may call the National AIDS Hotline, 1-800-342-AIDS.

Chapter Review

Begin your review by rereading the module headings and scanning the figures before proceeding to the Chapter Summary and questions.

Chapter Summary

Introduction–24.2 The body has two main defenses against invaders. Our first line of defense consists of nonspecific obstacles to infection. The skin blocks microbes. Skin secretions, stomach acids, interferons, and protein complement slow or kill them. White blood cells attack them. Tissue damage triggers the inflammatory response, which disinfects tissues and limits infection.

24.3–24.4 The second line of defense is the immune system, which counters specific invaders by responding to foreign substances called antigens. Infection or vaccination triggers active immunity. The immune system then reacts to antigens and "remembers" an invader. We can also temporarily acquire passive immunity. Two kinds of lymphocytes—B cells and T cells—carry out the immune response. As each lymphocyte develops, it makes membrane receptors able to bind to one kind of antigen. Thus, millions of kinds of B cells and T cells wait to confront invaders in the blood and the lymphatic system.

24.5 The lymphatic system is a network of lymphatic vessels, lymph nodes, and other organs. Lymph vessels collect fluid and return it to the blood. Lymph nodes house lymphocytes and macrophages that respond to infection.

24.6–24.8 When a foreign antigen enters the body, it activates only lymphocytes with matching receptors, a process called clonal selection. The selected cells multiply into a clone of short-lived effector cells, specialized for defending against the antigen that triggered the response. At the same time, some lymphocytes become long-lived memory cells. These developments are the primary immune response. Subsequent exposure to the antigen activates the memory cells, which mount a more rapid and massive secondary immune response. Memory cells may confer lifelong immunity.

24.9–24.12 B cells are responsible for humoral immunity, which defends against microbes in body fluids. Antigens trigger B cells to differentiate into plasma cells, which secrete antibodies. An antibody is a protein molecule that binds to a particular antigen and helps destroy or eliminate it. Antibodies block harmful antigens on microbes, clump bacteria together, and precipitate dissolved antigens. They also activate complement proteins, which kill bacteria and enhance inflammation. Because they can tag molecules and cells, antibodies are used for research and diagnosis. Monoclonal antibodies are produced by fusing B cells with easy-to-grow tumor cells.

24.13–24.15 T cells carry out cell-mediated immunity, attacking pathogens inside cells. First, an antigen-presenting cell (APC), such as a macrophage, displays an antigen (a nonself molecule) and one of the body's own self proteins to a helper T cell. The helper cell's receptor recognizes the self–nonself complex, and the activated cell differentiates into a clone. Then the helper T cells activate cytotoxic T cells with the same receptors. Cytotoxic T cells identify body cells with both self proteins and foreign antigens, bind to the infected cells, and destroy them. (When the threat has passed, suppressor T cells terminate immune activities.) Cytotoxic T cells also attack cancer cells, whose antigens may be altered by the disease, and transplanted organs, whose cells lack the body's unique "fingerprint" of self proteins.

24.16–24.17 The immune system sometimes fails. In autoimmune diseases, the system turns against the body's own molecules. In immunodeficiency diseases, immune components are lacking, and infections recur. Allergies are abnormal sensitivities to antigens (allergens) in the surroundings. Physical and emotional stress weaken the immune system.

24.18 Millions throughout the world are infected with HIV, the AIDS virus. HIV attacks helper T cells, crippling both humoral and cell-mediated immunity. Death usually results from infections or cancer. So far, AIDS is incurable, but drugs and vaccines offer some hope. AIDS education and safer sex could save many lives.

Testing Your Knowledge

Multiple Choice

1. Foreign molecules that evoke an immune response are called
   - **a.** pathogens
   - **b.** antibodies
   - **c.** lymphocytes
   - **d.** histamines
   - **e.** antigens

2. Which of the following is *not* part of the body's nonspecific defense system?
   - **a.** natural killer cells
   - **b.** antibodies
   - **c.** interferons
   - **d.** complement
   - **e.** inflammation

3. Which of the following best describes the difference in the way B cells and cytotoxic T cells deal with invaders?
   - **a.** B cells confer active immunity; T cells confer passive immunity.
   - **b.** B cells send out antibodies to attack; T cells themselves do the attacking.
   - **c.** T cells handle the primary immune response; B cells handle the secondary response.
   - **d.** B cells are responsible for cell-mediated immunity; T cells are responsible for humoral immunity.
   - **e.** B cells attack the first time the invader is present; T cells attack subsequent times.

4. The antigen-binding sites of an antibody molecule are formed from the molecule's variable regions. Why are these regions called variable?
   - **a.** They can change their shapes on command to fit different antigens.
   - **b.** They change their shapes when they bind to an antigen.
   - **c.** Their specific shapes are unimportant.
   - **d.** They can be different shapes on different antibody molecules.
   - **e.** Their sizes vary considerably from one antibody to another.

5. Researchers suspect that cytotoxic T cells are usually able to find and attack cancer cells because
   - **a.** cancer changes the surfaces of cancerous cells
   - **b.** suppressor T cells help them
   - **c.** cancer is a bacterial infection
   - **d.** cancer cells release antibodies into the blood
   - **e.** cancer is an autoimmune disease

6. The structural relationship between an antigen and antibody is like the relationship between a (*Explain your answer.*)
   - **a.** bullet and gun
   - **b.** hand and glove
   - **c.** car and driver
   - **d.** mother and child
   - **e.** hammer and nail

Matching

1. Attacks infected body cells
2. Carries out humoral immunity
3. Triggers allergy symptoms
4. Nonspecific white blood cell
5. General name for a B or T cell
6. Carries out the secondary immune response
7. Cell most commonly attacked by HIV
8. Inhibits the immune response

   - **a.** suppressor T cell
   - **b.** lymphocyte
   - **c.** cytotoxic T cell
   - **d.** helper T cell
   - **e.** mast cell
   - **f.** neutrophil
   - **g.** B cell
   - **h.** memory cell

Describing, Comparing, and Explaining

1. Briefly describe four different methods (effector mechanisms) used by antibodies to neutralize or destroy invading microorganisms.

2. Were your lymphocytes "born" with the ability to make antibodies against the chickenpox virus, or did the cells have to "learn" how to make them? Explain.

3. Describe (a) how AIDS is transmitted and (b) how immune system cells in an infected person are affected by HIV. Why is AIDS particularly deadly, compared to other viral diseases?

4. What is inflammation, and how is it triggered? How does it protect the body?

Thinking Critically

1. Influenza (flu) viruses have a particular tendency to undergo genetic mutations and rearrangements that produce viruses with new combinations of surface antigens. How and why would this affect the ability of the immune system to defend against the flu?

2. As settlers, miners, and construction workers move into the Amazonian rain forest in increasing numbers, the native South Americans who live there are exposed to new diseases. It has recently been reported that large numbers of Yanomamo Indians in Brazil have been dying of measles and mumps. These diseases usually are much less harmful to individuals of European ancestry. Why do you think these diseases are so much more serious to the Indians? (Think about clonal selection and the past histories of the two groups of people.)

3. Many allergy medications contain antihistamines. How might these substances work to stop the runny nose and itchy eyes of an allergy victim?

Science, Technology, and Society

Concern is increasing over the rising rate of HIV infection among teenagers. In San Francisco, for example, the incidence of HIV in teenagers is estimated to be doubling every 16–18 months. Schools in some large cities have instituted programs to make condoms available to students, along with advice and counseling about safer sex. These plans have divided school boards and communities. Some citizens and church groups are opposed to giving condoms to students, because it might appear to encourage sexual activity. Many school and public-health officials view the situation differently. New York City Schools Chancellor Joseph Fernandez says, "This is not an issue of morality. It is a matter of life and death." All agree that the spread of AIDS is a serious problem. The heart of the controversy seems to be whether the schools should take such a direct role in this part of student life. What are the reasons that support school distribution of condoms? What are the reasons for opposing such plans? What do you think the school's role should be?

Control of the Internal Environment

One of the most confusing notions about animals is that they are either "cold-blooded" or "warm-blooded." These terms are deeply entrenched, and when we think about the body temperature of an invertebrate, a fish, an amphibian, or a reptile, we typically think "cold-blooded." When we think about warm-blooded animals, we think only of mammals or birds. But many tropical fishes and invertebrates have body temperatures as warm as those of any mammal. How do we classify them? And what about the fish shown at the left, the great white shark? Is it cold-blooded? It cruises through cool and warm oceans alike, maintaining a warm body temperature wherever it goes. The great white's body temperature stays fairly constant. But so does the body of many "cold-blooded" fishes and invertebrates that live in tropical or polar seas where water temperatures stay about the same year-round.

Instead of calling animals cold-blooded or warm-blooded, biologists prefer the terms ectothermic and endothermic, referring to the main source of body heat instead of the animal's body temperature. An **ectotherm** is an animal that warms itself *mainly* by absorbing heat from its surroundings. An **endotherm** derives *most* of its body heat from its own metabolism. In general, endothermic animals maintain a fairly constant body temperature, whereas an ectotherm's body temperature usually fluctuates with the environmental temperature. The distinction between ectotherms and endotherms often blurs, however, and, as we will see in this chapter, many animals are not strictly ectotherms or endotherms.

Most fishes are ectothermic, but there are exceptions. The great white shark is an endotherm; it has adaptations that conserve metabolic heat produced by its muscles, keeping it warm—often warmer than the surrounding water. We examine these adaptations in Module 25.2.

The great white shark is a large predator, reaching 6–7 meters in length. Attracted by blood, it is one of the most dangerous of all sharks and is the species usually depicted in films and popular novels about shark attacks. Despite its reputation as a "man-eater," however, most of its attacks on humans seem to be in defense of its territory or because it mistakes a diver for its preferred prey, a seal or sea lion. Its warm muscles give the great white shark the extra power and speed it often needs to catch fast-swimming marine mammals.

The great white shark's ability to maintain its body temperature is extraordinary for a fish, but it is a poor second to most mammals, especially the one pictured below. The polar bear is an endotherm that maintains a consistently warm body temperature despite spending most of its time in subzero weather. Its dense fur and thick body fat provide such superb insulation that the bear can plunge into the Arctic Sea among floating icebergs and swim long distances after seals, its favorite prey. Out of the water, the animal's hairs not only minimize heat loss, they also help its skin absorb heat. At a distance, the hairs appear white, but they are actually transparent and beam sunlight to the skin. The bear's skin itself is black, maximizing heat absorption.

Animals survive fluctuations in the external environment because they have homeostatic ("steady-state") control mechanisms, which allow only a narrow range of fluctuation in the fluid environment that bathes their cells. This chapter examines three elements of homeostasis: **thermoregulation,** the maintenance of internal temperature within a range that allows cells to function efficiently; **osmoregulation,** control of the gain and loss of water and dissolved solutes; and **excretion,** the disposal of nitrogen-containing wastes. Let's look at thermoregulation first.

The polar bear, a master of thermoregulation

25.1 Heat is gained or lost in four ways

In regulating its internal temperature, an animal's body must adjust to heat gained from or lost to the environment by four physical processes. An animal in actual contact with an object, such as a warm rock, can exchange heat with the object by direct transfer, a process called conduction. Heat is always conducted from a body of higher temperature to one of lower temperature. The lizard in the drawing at the right is elevating its body temperature using heat conducted from the rock, a common practice among reptiles.

In contrast to conduction, convection is the transfer of heat by the movement of air or liquid past a body surface. In the figure, a breeze removes heat (orange arrow) from the lizard's tail. We experience the same effect when a fan brings comfort on a hot, still day.

Radiation, a third means of heat exchange, is the emission of electromagnetic energy. Radiation can transfer heat between objects that are not in direct contact, as when an animal absorbs heat (yellow arrows) from the sun. The yellow arrow emerging from the lizard's back indicates that the animal radiates some of its own heat into the external environment.

The fourth type of heat transfer results from evaporation. This is the loss of heat from the surface of a liquid that is transforming into a gas. A lizard has dry skin but may lose some heat as moisture evaporates from its nostrils (blue arrow). Evaporative cooling of the human body is increased greatly by sweating.

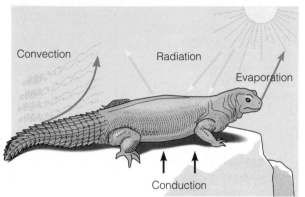

25.2 Thermoregulation depends on both heat production and heat gain or loss

Different animals are adapted to different external temperature ranges, and there is an optimal range for each species. Within the optimal range, endotherms and many ectotherms maintain a fairly constant internal temperature as the external temperature fluctuates. They may do so in two ways: (1) Endotherms and even some ectotherms may alter their rate of heat production, and (2) both endotherms and ectotherms may change their rate of heat gain or loss by conduction, convection, radiation, or evaporation.

In cold weather, hormonal changes tend to boost the metabolic rate of birds and mammals, increasing their rate of heat production. For birds and mammals, simply moving around more or shivering increases heat produced by the skeletal muscles as a metabolic by-product. Many ectotherms use a similar mechanism when environmental temperatures become extreme. Honeybees, for instance, survive cold winters by clustering together and shivering in their hive. Each bee in the cluster in Figure A shivers almost constantly when it is cold. The collective metabolic activity of all the bees generates enough heat to keep the cluster alive, although a few bees on the outside usually succumb to the cold. In using metabolic heat to warm up, honeybees show that the distinction between ectotherms and endotherms often blurs.

Several kinds of adaptations adjust the rate of an animal's heat loss or gain. Heat loss in mammals is often regulated by coat thickness—a thin coat of fur in summer and a thick, insulating coat in winter. Mammals also have muscles in the skin that raise their hairs in the cold. The raised hairs increase the fur's insulating power by trapping a layer of air next to the warm skin. (Although humans have relatively little hair, we do have muscles that raise our hair in the cold. As a result, we get goose bumps, a vestige from our furry ancestors.)

Heat loss can also be altered by the amount of blood flowing to the skin. In a bird or mammal, nerves signal surface blood vessels to constrict or dilate, depending on the external temperature. When the vessels are constricted, less blood flows from the warm body core to the body surface, decreasing the rate of heat loss to the environment. Conversely, dilation of surface blood vessels increases the rate of heat loss. Heat loss can be further increased by evaporative cooling, such as sweating or panting.

The circulatory system plays other roles in thermoregulation. Figure B illustrates adaptations of the circulatory sys-

A. Heat production by bees

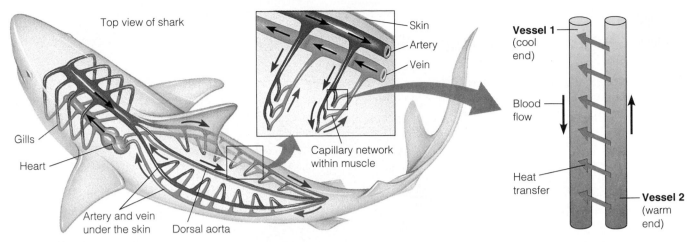

Top view of shark

Skin
Artery
Vein

Capillary network
within muscle

Gills

Heart

Artery and vein
under the skin

Dorsal aorta

B. Thermoregulation in the great white shark

Vessel 1
(cool
end)

Blood
flow

Heat
transfer

Vessel 2
(warm
end)

C. Countercurrent heat exchange

tem that conserve heat in the endothermic great white shark. Like other fishes, this shark loses some heat to the surrounding water when its blood passes through its gills. In most fishes, the dorsal aorta carries most of the blood from the gills directly inward along the core of the body. As a result, the typical fish's body temperature is about the same as its environment. In the great white shark, however, most of the blood from the gills flows into two large arteries that lie just under the skin. The enlargement in Figure B shows a segment of one of the large arteries and of a large vein that parallels it. A network of capillaries (small red vessels) carries cool blood inward from the artery. Other capillaries (blue) carry warm blood from the inner body to the vein. The two sets of capillaries parallel each other, forming what is called a **countercurrent heat exchanger.**

Figure C shows how a countercurrent heat exchanger works to prevent the loss of much of the metabolic heat generated by the shark's muscles. (It may be useful to compare this system with the countercurrent mechanism discussed in Module 22.4.) Warm and cold blood flow in opposite (countercurrent) directions in two adjacent blood

vessels. In Figure C, vessel 1 carries cool blood inward from the surface of the shark, and vessel 2 carries warmed blood outward toward the surface. Because of the countercurrent flow, heat passes from the warmer blood in vessel 2 to the cooler blood in vessel 1 along the whole length of the vessels. In the shark, the warm, outflowing blood loses its heat to the cooler, inflowing blood in the capillary network. As a result, much of the metabolic heat from the muscles is retained in the body instead of being lost, as it would if warmer blood reached the surface. The shark is endothermic because its muscles generate a large amount of heat and because its countercurrent heat exchanger conserves much of that heat.

Countercurrent heat exchange is important in controlling heat loss in many animals. Many birds, for instance, have heat-retaining countercurrent systems in their legs. Seals, sea lions, and whales have them in their flippers. Heat loss is minimal, even when the animal is swimming in frigid water or standing on ice, because heat carried by blood flowing outward from the animal's core is picked up by blood flowing inward.

Behavior often affects body temperature `25.3`

Behavior is often important in thermoregulation. Many animals increase or decrease body heat by simply relocating. Some species migrate to a more suitable climate. Oth-

ers bask in the sun or huddle together when it is cold, and find cool, damp areas or burrow when it is hot. Bathing brings immediate relief from the heat by convection and continues to cool the surface for some time by evaporation. Elephants often seek relief from tropical heat by bathing and by flapping their ears to cool their blood. Packed with blood vessels, their ears can radiate a lot of heat. Dressing for the weather is a thermoregulatory behavior unique to humans.

By seeking warmer or cooler locations, many ectotherms, especially reptiles, keep their body temperature quite stable. If a sunny spot is too warm, for instance, a lizard will alternate between the sun and the shade, or it will turn in a direction that exposes less surface area to the sun.

25.4 Torpor saves energy

With the ability to control their internal temperature, endothermic animals can generally remain active in more severe weather conditions than ectotherms. On the other hand, endothermy consumes a lot of energy. Endotherms spend much more of their food energy on heat production than ectotherms do. When food supplies are low and environmental temperatures are extreme, many birds and mammals temporarily abandon endothermy, thereby expending less energy on thermoregulation.

Torpor is a state of reduced activity in which body temperature and metabolic rate decrease and the heart and respiratory system slow down. Many endothermic animals have short periods of torpor, usually at times when they cannot feed. Most bats, for instance, go into torpor when they are inactive, most often during daylight hours. Hummingbirds feed during the day and often undergo torpor on cold nights.

Bat in a state of torpor

Sleep, and the slight drop in body temperature that accompanies it, may be left over from a more pronounced daily torpor in ancestral species.

Hibernation (Latin *hibernus,* winter) is a type of long-term torpor by some animals in cold weather. Many ground squirrels, for example, hibernate in a grass-lined burrow. With a reduced metabolic rate, they live on energy stored in body fat when food was abundant. Some ground squirrels also have a summer torpor, called **aestivation** (Latin *aestas,* summer). Aestivation allows an animal to survive long periods of high environmental temperatures and reduced food and water supplies.

Many animals can tolerate wide fluctuations in their body temperature, but no animal can withstand much change in the relative amounts of dissolved solutes and water in its internal fluids. We now turn to homeostasis in body fluids.

25.5 Osmoregulation: All animals balance the uptake and loss of water and dissolved solutes

The metabolic reactions on which life depends require a precise balance of water and dissolved solutes. Among the solutes whose concentration must be regulated are a variety of amino acids, proteins, and dissolved ions such as Na^+, Cl^-, K^+, Ca^{2+}, and HCO_3^-. As we discussed in Module 5.15, osmosis occurs whenever two solutions separated by a membrane differ in total solute concentration. There is a net movement of water from the hypotonic solution to the hypertonic one until the solute concentrations are equal on both sides of the membrane. Whether an animal inhabits land, fresh water, or salt water, its cells cannot survive a *net* water gain or loss that significantly alters the concentration of its dissolved solutes. Animal cells swell and burst if there is a net uptake of water, or they shrivel and die if there is a net loss of water.

As terrestrial animals, we acquire most of our water in our food and drink. Like other animals, we lose water by urinating and defecating. We also lose it by evaporation as we breathe and perspire. Evaporation is unimportant for aquatic animals. Water continuously enters and leaves their bodies by osmosis.

What prevents an aquatic animal from having a net gain or loss of water? Some aquatic animals that live in the sea have body fluids with a solute concentration equal to that of seawater. Called **osmoconformers,** such animals do not undergo a net gain or loss of water because equal amounts of water move back and forth between two solutions with equal solute concentrations. The sea, where animals first evolved, is the only environment that supports osmoconformers. The total solute concentration in jellyfish, scallops, lobsters, and most other marine invertebrates conforms with that of seawater, and thus these animals do not expend energy regulating their water content. However, the concentration of certain ions in their body fluids—especially the fluid within their cells—is different from that of seawater. For example, in order for their cell membranes to function properly, the concentration of potassium ions (K^+) must be higher within their cells than in either their interstitial fluid or in seawater. Because it takes energy to actively transport ions into cells, even an osmoconformer expends some energy to maintain its ion concentrations.

All freshwater animals, all land animals, and most marine vertebrates—whales, seals, sea birds, and most fishes—have body fluids whose solute concentration is different from that of their environment. Therefore, they must use energy in controlling water loss or gain. Such animals are called **osmoregulators.**

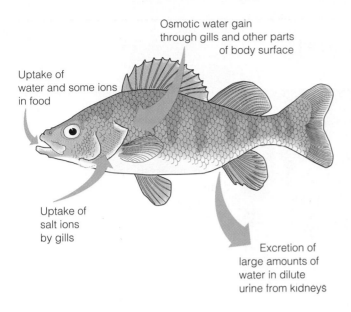

Osmotic water gain
through gills and other parts
of body surface

Uptake of
water and some ions
in food

Gain of water and salt
ions from food
and by drinking
seawater

Osmotic water loss
through gills and other parts
of body surface

Uptake of
salt ions
by gills

Excretion of
large amounts of
water in dilute
urine from kidneys

Excretion of
salt ions
from gills

Excretion of salt ions
and small amounts
of water in scanty
urine from kidneys

A. Osmoregulation in a freshwater fish

B. Osmoregulation in a saltwater fish

The freshwater fish in Figure A (a yellow perch) has a very different internal solute concentration than its environment. The concentration of solutes in its internal fluids is much higher than that of fresh water, creating an osmotic problem for the animal. Because water flows by osmosis from a solution with a lesser solute concentration to one with a greater solute concentration, the fish constantly takes in water from its surroundings. A freshwater fish gains water through its body surface, especially through its gills, and also in its food. It does not drink water except with its food. The freshwater fish also loses some solutes in its **urine**, the waste material produced by its excretory system.

It takes the work of three organ systems to achieve the proper water and solute balance in a freshwater fish. The animal's digestive system takes up ions from the food. Its gills (respiratory system) also take up salt ions, especially Na^+ and Cl^-. The fish's kidneys (excretory system) work constantly to produce large amounts of dilute urine—that is, urine with a much lower solute concentration than that of the animal's internal fluids. By excreting dilute urine, the fish disposes of excess water and conserves solutes.

The saltwater fish in Figure B (a cod) has osmoregulatory problems opposite those of its freshwater relatives. Its internal fluids must be lower in total solutes than seawater. A saltwater fish loses water by osmosis through its body surfaces but compensates for the loss by drinking seawater and pumping excess salt ions, such as Na^+ and Cl^-, out through its gills. It also saves water and disposes of some salts by producing small amounts of urine with the same solute concentration as its body fluids.

Most fishes have little tolerance for changes in the solute concentration of their surroundings. The yellow perch and the cod shown here, for instance, are restricted to fresh water and the ocean, respectively. In contrast, a few fishes, such as salmon, can migrate between seawater and fresh water. Salmon have remarkably adaptable osmoregulatory

mechanisms. While in the ocean, they drink seawater and excrete excess salts from their gills, osmoregulating like a cod. When salmon move into fresh water to spawn, their osmoregulatory mechanism switches to the freshwater mode of the yellow perch; they cease drinking, and their gills take up salts from the dilute environment.

What about land animals? They are osmoregulators, but they are not surrounded by water; therefore, they cannot directly exchange water with the environment by osmosis. A terrestrial animal gains water by drinking and by eating moist foods, and it constantly loses water as it breathes and disposes of wastes. Its osmoregulatory situation resembles that of a marine fish like the cod; its paramount problem is losing water and becoming dehydrated. In fact, dehydration is such a severe problem on land that it may largely explain why only two groups of animals, arthropods and vertebrates, have colonized land with great success.

What is it about arthropods and vertebrates that gives them an edge against the dehydration threat? Insects, the prevalent arthropods on land, have tough exoskeletons impregnated with waterproof wax. Most terrestrial vertebrates, including humans, have an outer skin formed of multiple layers of dead, water-resistant cells. Also key to survival on land are adaptations that protect fertilized eggs and developing embryos from drying out. Many insects lay their eggs in humid or moist areas, and the eggs of many species are surrounded by a tough, watertight shell that keeps the developing embryo moist. Likewise, the embryos of terrestrial vertebrates—reptiles, birds, and mammals—develop in a water-filled sac surrounded by protective membranes. Behavioral adaptations can also contribute to water conservation. For instance, many desert mammals and arthropods spend much of their time in moist burrows and venture out only at night. And, as we will see in Module 25.11, the kidney also plays a major role in conserving water.

25.6 Sweating can produce serious water loss

Despite the cold day, a cross-country skier is sweating heavily. Sweating helps her body thermoregulate, but it can also cause osmoregulatory problems. Knowing that replacing lost water is essential to preventing dehydration, the skier stops in the middle of a race to drink.

Water loss is the main problem with sweating, although serious salt losses can also occur in extreme cases. Most of the salt we lose in sweat is NaCl (dissolved as the ions Na^+ and Cl^-). Sweat itself is much less concentrated than our body fluids; a volume of sweat contains only about one-third as many Na^+ and Cl^- ions as an equal volume of body fluid. As a result, sweating causes serious water loss well before ion losses become a problem.

Rehydration

The way to prevent dehydration from sweating is simply to drink water. Plain water is absorbed faster than beverages that contain sugar or dissolved ions (electrolytes). It is usually not necessary to take salt tablets or drink fluids that contain dissolved ions. In fact, if a person who has been sweating heavily takes salt tablets before water, the salts in the digestive tract tend to extract water from the body tissues by osmosis and dehydrate the body even more. This is why patients who are severely dehydrated are first given hypotonic fluids (solutions less concentrated than body fluids). Salt supplements are given later, if necessary.

25.7 Many freshwater animals face seasonal dehydration

When any animal loses more water than it takes in over a period of time, it becomes dehydrated. Even some aquatic animals—for instance, those that inhabit ponds or small lakes—are at risk because their habitats tend to dry up seasonally. At even greater risk are many small invertebrates that live in films of water in the soil or on plant surfaces. The moss-covered roof in Figure A is a seasonal home to a great number of tiny animals. During wet months, the roof teems with insects and aquatic invertebrates that live in droplets of water on the moist, green moss. When the moss dries up, most of the animals lay eggs with tough waterproof coverings and die. But there are exceptions.

The micrograph in Figure B shows a tiny (less than 1 mm long) invertebrate called a tardigrade, or water bear. Tardi-grades thrive in droplets of water on mosses and other moist plants and are often common on moss-covered roofs. When active, a tardigrade lumbers around on its stumpy legs, feeding on plant juices. When its habitat dries up, the tardigrade has the remarkable ability to lose more than 95% of its body water and survive in a dried-out, dormant state for decades! When a dried-out water bear is rehydrated, it can begin walking around again within minutes.

Biologists are just beginning to learn how this type of dehydration works. The problem for the animal is that proteins and cell membranes can be damaged by dehydration. Many protein molecules are fragile and tend to break down when a cell loses water. Cell membranes that are fluid when intact tend to collapse and stick together when they dry out. Studies of nematodes, which can withstand dehydration, have shown that sugar molecules seem to replace the water required for keeping proteins and cell membranes intact, protecting these cellular structures from the effects of dehydration.

A. A moss-covered roof, a wet or dry habitat

B. A tardigrade

SEM 180×

Waste disposal is as important to homeostasis as water and solute balance. Metabolism produces a number of toxic by-products, particularly the nitrogenous (nitrogen-containing) wastes that result from the breakdown of proteins and nucleic acids. An animal must dispose of these substances or be poisoned by them. Excretion is the disposal of metabolic wastes.

The form of an animal's nitrogenous wastes depends on the animal's evolutionary history and its habitat. As the diagram below shows, most aquatic animals dispose of their nitrogenous wastes as ammonia, which simply diffuses into the surrounding water. Small, soft-bodied invertebrates, such as planarian worms, excrete ammonia across their whole body surface. Fish excrete it mainly from the gills. Among the most toxic of all metabolic by-products, ammonia (NH_3) forms from the amino groups ($-NH_2$) removed from proteins and nucleic acids. Ammonia is highly toxic because it tends to raise the pH of body fluids and interfere with membrane transport functions. It is, in fact, too toxic to be stored in the body, but it is highly soluble in water and diffuses rapidly across cell membranes. If an animal is surrounded by water, ammonia readily diffuses out of cells and out of the body.

Ammonia excretion works only for aquatic animals. A terrestrial animal cannot dispose of ammonia fast enough to avoid poisoning because ammonia does not diffuse rapidly into the air. As indicated in the figure, land animals convert the amino groups from nitrogen-containing compounds into less toxic compounds such as urea or uric acid. These substances can be safely transported and stored in the body and then released periodically from the excretory system. The disadvantage of excreting urea or uric acid is that the animal must use energy to produce these compounds from amino groups. In contrast, no energy is expended when ammonia forms from an amino group.

Two groups of terrestrial vertebrates, mammals and amphibians, and some fishes excrete the organic molecule urea. Urea is highly soluble in water. It is also some 100,000 times less toxic than ammonia, so it can be held in a concentrated solution in the body and disposed of with relatively little water loss. Many ani-

mals that excrete urea also dispose of some nitrogenous wastes as ammonia. Moreover, some animals can switch between these waste products, depending on environmental conditions. Certain toads, for example, excrete ammonia (thus conserving energy) when in water, but they primarily excrete urea when on land.

Urea can be stored in a concentrated solution, but it still takes water to dispose of it. By contrast, land animals that excrete uric acid (birds, insects, many reptiles, and snails) avoid the water-loss problem almost completely. As you can see in the figure, uric acid is a considerably more complex molecule than either urea or ammonia. Uric acid is also thousands of times less soluble in water than either ammonia or urea. Uric acid can be excreted as small, solid crystals. In most cases, it is excreted as a paste or dry powder, combined with the feces. (The white material in bird droppings is mostly uric acid.) An animal must expend more energy to excrete uric acid than urea, but the higher energy cost is offset by the great savings in body water.

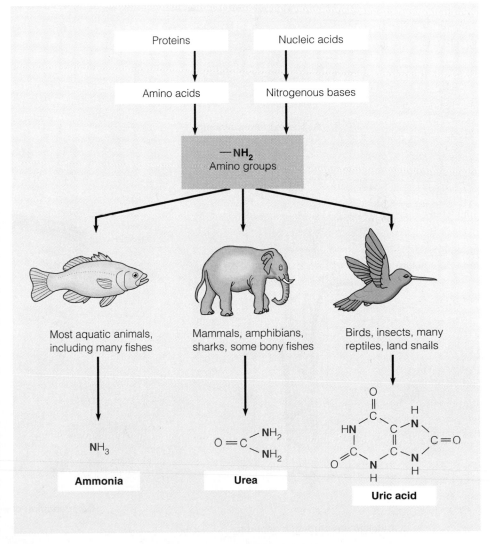

The excretory system plays several major roles in homeostasis

The problems of nitrogenous waste disposal and water–solute balance go hand in hand. Survival in any environment requires close coordination between excretion and water and salt needs. In humans, the excretory system plays a central role in homeostasis, forming and excreting urine while regulating the amount of water and salts in the body fluids.

Our two kidneys are the main processing centers of the excretory system. Each is a compact organ, about the size of your fist, nearly filled with about 80 km of fine tubes (tubules) and an intricate network of blood capillaries. The human body contains only about 5 liters of blood, but since this blood circulates repeatedly, about 1100–2000 L pass through the capillaries in our kidneys every day. As the blood circulates, our kidney tubules extract daily about 180 L (45 gal) of fluid, called **filtrate,** consisting of water, urea, and a number of valuable solutes, including Na^+, K^+, Cl^-, bicarbonate (HCO_3^-), glucose, and amino acids. If we lost all of the filtrate as urine, we would dehydrate rapidly. But our kidneys refine the filtrate, concentrating the urea and returning most of the water and solutes to the blood. In a typical day, we excrete only about 1.5 L of urine.

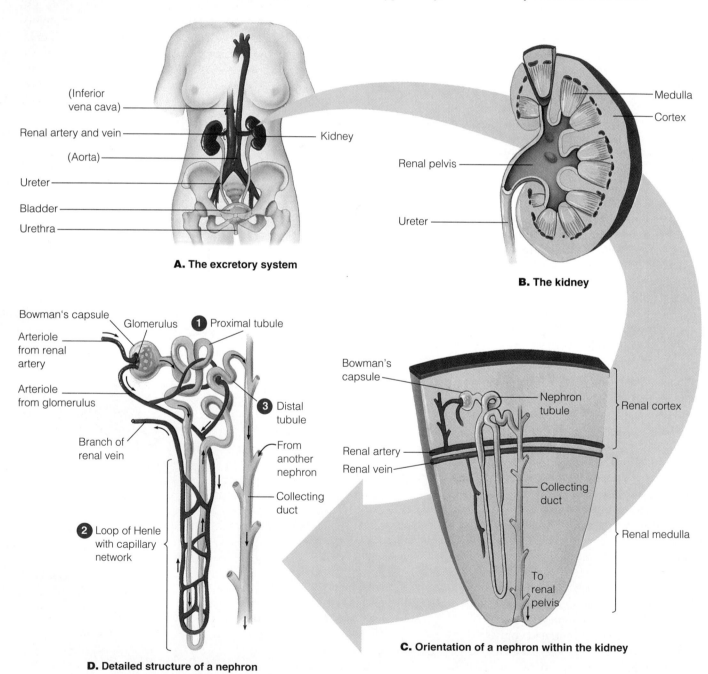

A. The excretory system

(Inferior vena cava)
Renal artery and vein
(Aorta)
Kidney
Ureter
Bladder
Urethra

B. The kidney

Medulla
Cortex
Renal pelvis
Ureter

D. Detailed structure of a nephron

Bowman's capsule
Glomerulus
❶ Proximal tubule
Arteriole from renal artery
Arteriole from glomerulus
Branch of renal vein
❸ Distal tubule
From another nephron
Collecting duct
❷ Loop of Henle with capillary network

C. Orientation of a nephron within the kidney

Bowman's capsule
Nephron tubule
Renal cortex
Renal artery
Renal vein
Collecting duct
Renal medulla
To renal pelvis

The drawings on the opposite page illustrate the "plumbing" plan and the blood supply of the human excretory system. Starting with the whole system in Figure A, blood to be filtered enters each kidney via a **renal artery,** shown in red; blood that has been filtered leaves the kidney in the **renal vein,** shown in blue. (The word "renal" comes from the Latin *renes,* kidney.) Urine leaves each kidney through a duct called a **ureter** and passes into the **urinary bladder.** The bladder empties periodically during urination. Urine actually leaves the body through a tube called the **urethra,** which empties near the female vagina or through the male penis.

As indicated in Figure B, the kidney has two main regions, the cortex (outer layer) and the medulla (inner region). From the medulla, the urine flows into a chamber called the renal pelvis, and from there into the ureter.

Each of our kidneys contains thousands of tiny functional units called **nephrons,** one of which is diagrammed in Figure C. Each nephron consists of a single long tubule and its associated blood vessels. Performing the kidney's functions in miniature, each nephron extracts a tiny amount of filtrate from the blood and then refines the filtrate into a much smaller quantity of urine. The nephron extends through the kidney's cortex and medulla. The receiving end of the nephron is a cup-shaped swelling, called **Bowman's capsule.** At the nephron's other end is the **collecting duct,** which carries urine to the renal pelvis.

Figure D shows a nephron in more detail, along with its blood vessels. Bowman's capsule envelops a ball of capillaries called the **glomerulus** (plural, *glomeruli*). Together, the glomerulus and Bowman's capsule make up the blood-filtering unit of the nephron. Here, blood pressure forces water and solutes from the blood in the glomerular capillaries across the wall of Bowman's capsule and into the nephron tubule. This process creates the filtrate, leaving blood cells and macromolecules behind in the capillaries.

Downstream from Bowman's capsule, the rest of the nephron is a filtrate refinery. There are three sections of the tubule: ① the **proximal tubule** (in the cortex); ② the **loop of Henle,** a long hairpin loop carrying filtrate down into the medulla and then back toward the cortex; and ③ the **distal tubule** (called distal because it is the most distant from Bowman's capsule). The distal tubule empties its filtrate into a collecting duct (right side of Figure D), which receives filtrate from many nephrons. By the time the filtrate passes through a collecting duct, it is actually urine. From the kidney's many collecting ducts, urine passes into the renal pelvis, and then into the ureter.

The intricate association between blood vessels and the nephron is the key to nephron function. As shown in Figure D, the nephron has two distinct networks of capillaries. A small network forms the glomerulus, which is actually a finely divided portion of an arteriole that branches from the renal artery. Leaving the glomerulus, the arteriole re-forms and carries blood to the second capillary network, which surrounds the rest of the nephron tubule. This second network functions with its tubule in refining the filtrate. Notice how some of the vessels in this network are arranged around the loop of Henle. Blood is conveyed via these capillaries into the medulla along one side of the loop, then back along the other side of the loop. Leaving the nephron, the capillaries converge to form a small branch of the renal vein.

With the structure of a nephron in mind, we focus next on what actually happens as our excretory system filters blood, refines the filtrate, and excretes urine.

Overview: The key functions of the excretory system are filtration, reabsorption, secretion, and excretion

Our excretory system produces and disposes of urine in four major processes, diagrammed below. First, during **filtration,** water and virtually all other molecules small enough to be forced through the capillary wall enter the nephron tubule from the glomerulus.

After filtration, two processes simultaneously refine the filtrate. In **reabsorption,** water and valuable solutes, including glucose, salts, and amino acids, are reclaimed from the filtrate and returned to the blood. During **secretion,** certain substances are removed from the blood and added to the filtrate. When there is an excess of K^+ or H^+ in the blood, for example, these ions are transported into the cells of the nephron tubule, which secrete them into the filtrate. In removing excess H^+ from the blood, secretion keeps the blood from becoming acidic. Secretion also eliminates certain drugs and toxic substances from the blood.

Finally, in **excretion,** urine, the product of filtration, reabsorption, and secretion, passes from the kidneys to the outside via the ureters, urinary bladder, and urethra.

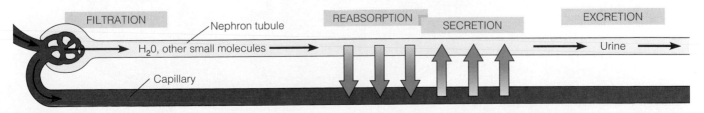

From blood filtrate to urine: a closer look

Let's take a closer look at how a single nephron and collecting duct in the kidney form a filtrate from the blood and then refine the filtrate into urine. As we have seen, the filtrate forced into a Bowman's capsule by blood pressure initially consists of a large amount of water and a number of valuable solutes, along with waste molecules.

The colored arrows in the figure below indicate where the refining processes of reabsorption and secretion occur along the length of the nephron tubule. For simplicity, we have not illustrated the capillary network that surrounds the nephron. The gray area surrounding the nephron and collecting duct represents interstitial fluid. Small molecules and ions travel between the tubule and the capillaries by passing through the interstitial fluid during reabsorption and secretion.

The red and pink arrows pointing out of the tubule represent reabsorption. The blue arrows pointing into the tubule represent secretion. The red and blue arrows both indicate active transport; the pink arrows indicate passive transport. The intensity of the gray color corresponds to the concentration of solutes in the interstitial fluid surrounding the tubule: The concentration is lowest in the cortex of the kid-

ney and is progressively higher toward the inner medulla. We will see that it is by maintaining this solute gradient that the kidney is able to extract and save most of the water from the filtrate. All along the tubule, wherever you see water passing out of the filtrate into the interstitial fluid, it moves by osmosis. It does so because the solute concentration of the interstitial fluid exceeds that of the filtrate.

For our discussion, it is convenient to separate the activities of the proximal and distal tubules, which are located in the cortex of the kidney, from the activities of the loop of Henle and collecting duct, which are situated mainly in the medulla. As the figure indicates, the proximal and distal tubules reabsorb NaCl and H_2O from the filtrate. They are also the sites where secretion of excess H^+ ions into the filtrate helps regulate the blood's pH. The proximal tubule also reabsorbs nutrients such as glucose and amino acids. The distal tubule further buffers the blood against pH changes by reabsorbing bicarbonate ions (HCO_3^-). In the same area, poisons, such as ammonia, and drugs, such as penicillin, may be secreted into the filtrate.

In contrast to the varied functions of the proximal and distal tubules, the loop of Henle and the collecting duct

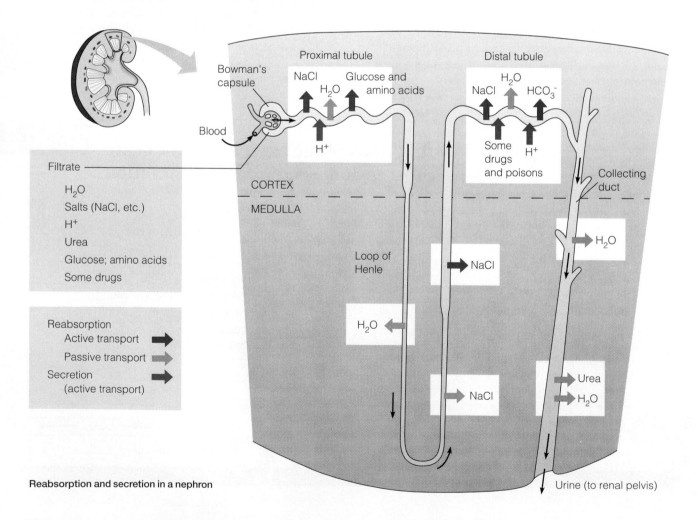

Reabsorption and secretion in a nephron

have one major function—water reabsorption. As the diagram shows, NaCl and some urea are also reabsorbed in these parts; however, their removal from the filtrate serves mainly to increase water reabsorption. Reabsorption of NaCl and urea maintains the concentration gradient of solutes in the interstitial fluid.

Notice that the loop of Henle carries the filtrate deep into the medulla and then back to the cortex. As the filtrate passes into the medulla, water leaves the tubule because the interstitial fluid there has a higher solute concentration than the filtrate. As soon as the water passes into the interstitial fluid, it is carried away by nearby blood capillaries. This is essential, since the water would otherwise dilute the interstitial fluid surrounding the loop, destroying the concentration gradient.

Just after the filtrate rounds the hairpin turn in the loop of Henle, water reabsorption stops, because the tubule there is impermeable to water. As the filtrate moves through the medulla back toward the cortex, NaCl is reabsorbed from it. At first, this reabsorption occurs passively, because the concentration of NaCl is higher in the filtrate than in the interstitial fluid. (The high solute concentration around the turn of the loop of Henle is largely due to urea, not to NaCl.) Active transport then takes over, and even more NaCl is reabsorbed as the filtrate nears the distal tubule. Because of this active transport, the NaCl concentration of the interstitial fluid surrounding the distal tubule and the first part of the collecting duct remains high. Therefore, more water leaves the filtrate in these regions.

Final refinement of the filtrate occurs in the collecting duct in the medulla, and urea is the main substance involved. Until the collecting duct enters the medulla, urea has remained in the filtrate because the nephron is impermeable to it. As a result, the filtrate's concentration of urea has become quite high—in fact, higher than that of the interstitial fluid. In the medulla, the collecting duct is permeable to urea, and some urea leaks out of it into the interstitial fluid. Deep in the medulla, the urea makes the interstitial fluid so concentrated that even more water is reabsorbed just before the finished urine passes into the renal pelvis.

So we see that the nephron can reabsorb and thus save much of the water that filters into it from the blood. Water conservation is a major part of body fluid homeostasis for a land animal, but our nephrons do much more than just save water. Under hormonal control, they maintain a precise balance between water and solutes in our body fluids. When the solute concentration of the body fluids rises above a set point, hormones signal the tubules to reabsorb much of the water from the filtrate. As a result, we excrete concentrated urine. Conversely, when we drink too much water, the nephrons maintain homeostasis by becoming less permeable to water, and we excrete dilute urine.

Kidney dialysis can be a lifesaver **25.12**

Knowing how the nephron works helps us understand how some of its functions can be performed artificially when the kidneys are damaged. Although kidney failure is not common, it means certain death from the buildup of toxic wastes in the blood unless treatment is available.

This figure illustrates a type of artificial kidney, called a dialysis machine. **Dialysis** means "separation" in Greek, and, like the nephrons of the kidney, the machine sorts small molecules of the blood, keeping some and discarding others. The patient's blood is pumped from an artery through a series of tubes that operate like nephrons. The tubes are made of a selectively permeable membrane, which allows only water and certain solutes to pass through it, and they are immersed in a dialyzing solution much like the interstitial fluid that bathes the nephrons. As the blood circulates through the dialysis tubing, urea and excess salts diffuse out of it (instead of leaving by filtration and secretion, as in the nephron). Needed substances, such as bicarbonate ions, diffuse from the dialyzing solution into the blood (reabsorption). The machine continually discards the used dialyzing solution as wastes build up in it (excretion).

Dialysis may be called for whenever the kidneys are damaged but can be repaired or will be replaced by transplant. Such cases include impairment of the nephrons from physical injury, from bacterial or viral infections, or from chemical poisoning. Unfortunately, dialysis machines are not always available when needed, and not everyone can afford them when they are.

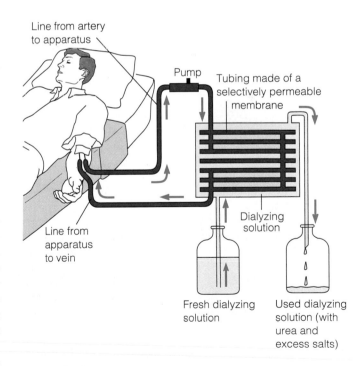

Line from artery to apparatus

Pump

Tubing made of a selectively permeable membrane

Line from apparatus to vein

Dialyzing solution

Fresh dialyzing solution

Used dialyzing solution (with urea and excess salts)

A discussion of homeostasis would not be complete without mention of the liver, which performs more functions than any other organ in the body. In addition to its roles in digestion (see Module 21.10), the liver supports the vital activities of the kidney. Specifically, the liver prepares nitrogenous wastes for disposal by synthesizing urea from ammonia. It also helps the kidney get rid of other potentially poisonous chemicals, such as alcohol and other drugs. The liver converts these into inactive products that the kidney can filter from the blood and excrete in the urine.

How can one organ do so much? Part of the explanation is that the liver's cells have the metabolic machinery to chemically alter a great number of substances. Another consideration is the liver's strategic location in the body—between the intestines and the heart. As indicated in the figure at the right, the intestinal capillaries all converge to form a single blood vessel, the **hepatic portal vessel**, which conveys blood directly to the liver. Nutrients absorbed by the intestine, as well as any harmful substances picked up by the intestinal capillaries, pass to the liver. Thus, the liver has a chance to modify substances absorbed by the digestive tract before the blood carries these materials to the heart for distribution to the rest of the body. Blood travels from the liver to the heart via the hepatic vein and the inferior vena cava.

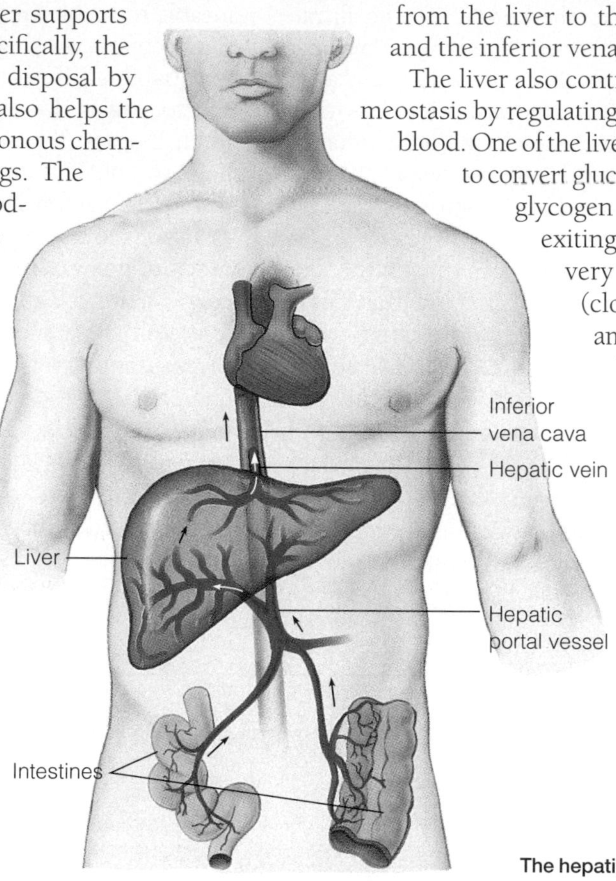

The liver also contributes in a major way to homeostasis by regulating the amount of glucose in the blood. One of the liver's most important functions is to convert glucose into glycogen and store the glycogen for use at a later time. Blood exiting from the liver usually has a very low glucose concentration (close to 0.1%). In balancing the amount of glucose carried in the blood with the amount of glycogen it stores, the liver exercises crucial control over body metabolism.

The liver's diverse roles in homeostasis and its interaction with other organs make the point that homeostasis requires the concerted action of several body systems. The coordination of all the body's regulatory systems by chemicals called hormones is the subject of Chapter 26.

Inferior vena cava

Hepatic vein

Liver

Hepatic portal vessel

Intestines

The hepatic portal system

Chapter Review

Begin your review by rereading the module headings and scanning the figures before proceeding to the Chapter Summary and questions.

Chapter Summary

Introduction Animals have homeostatic control mechanisms that compensate for fluctuations in the external environment. Thermoregulation, osmoregulation, and excretion are three elements of homeostasis.

25.1–25.2 Thermoregulation—the maintenance of body temperature within a tolerable range—depends on heat production, gain, and loss. Endotherms maintain body temperature mainly by using metabolic heat, ectotherms primarily by absorbing heat from their surroundings. Both endotherms and ectotherms regulate heat gain or loss by conduction, convection, radiation, or evaporation.

25.2 Fur and feathers help the body retain heat. Blood flow to the skin also affects heat loss. In a countercurrent heat exchanger, blood from the core of the body warms cooler blood returning from the skin or limbs, conserving body heat.

25.3–25.4 Behavior also affects thermoregulation. Animals may bask in the sun, seek shade, bathe, burrow, huddle, or migrate. Tor-por, which includes hibernation in cold weather and aestivation in warm weather, is a state of reduced activity that saves energy.

25.5–25.7 Osmoregulation is control of the gain and loss of water and dissolved solutes. Many marine animals are osmoconformers; their body fluids have the same concentration of solutes as seawater. Osmoregulators control water and solute concentrations. A freshwater fish gains water by osmosis but tends to lose solutes; it osmoregulates by excreting water but conserving solutes. A saltwater fish loses water by osmosis; it drinks seawater and excretes excess salts. Land animals gain water by drinking and eating but lose water and solutes by evaporation and waste disposal; their waterproof skin, kidneys, and behavior conserve water.

25.8 Excretion is the disposal of metabolic wastes, mostly nitrogenous by-products of protein and nucleic acid breakdown. Ammonia is poisonous but soluble, and it is easily disposed of by aquatic animals. Urea is less toxic and easier for many land animals to store and excrete. Some land animals save water by excreting uric acid, which is virtually insoluble.

25.9 The excretory system expels wastes and regulates water and salt balance. Every day, the human kidneys extract 180 L of filtrate from the blood, nearly all of which is reabsorbed. The remainder, about 1.5 L of urine, leaves the kidney via the ureters, is stored in the urinary bladder, and expelled through the urethra.

25.9–25.11 Millions of nephrons carry out the functions of the kidneys, in four steps: filtration, reabsorption, secretion, and excretion. Each nephron consists of blood vessels and a folded tubule. In the first step, filtration, blood pressure forces water and many small solute molecules from the blood and into the nephron tubule. Then valuable solutes, such as glucose and salts, are reabsorbed into the blood. This is accomplished mainly through the active transport of solutes, with water following passively via osmosis. Active and passive transport of salt and urea in parts of the tubule make the interstitial fluid highly concentrated, enhancing the osmotic reabsorption of water. In a third step, hydrogen ions, poisons such as ammonia, and drugs are secreted via active transport from blood to filtrate. The product of these processes—urine—is then excreted.

25.12 Kidney dialysis sometimes is used to compensate for kidney failure. A dialysis machine performs the functions of the kidney by removing wastes from the blood and maintaining its solute concentration.

25.13 The liver performs many homoeostatic functions. It assists the kidneys by making urea from ammonia and breaking down toxic chemicals. Blood from the intestines flows through the liver before distribution to the rest of the body, allowing the liver to adjust the blood's chemical content.

Testing Your Knowledge

Multiple Choice

1. The main difference between endotherms and ectotherms is
   a. how they conserve water
   b. whether they are warm or cold
   c. where they get most of their body heat
   d. whether they live in a warm or cold environment
   e. whether they live on land or in the water

2. In each nephron of the kidney, the glomerulus and Bowman's capsule
   a. filter the blood and capture the filtrate
   b. reabsorb water into the blood
   c. break down harmful toxins and poisons
   d. reabsorb salts and nutrients
   e. refine and concentrate the urine for excretion

3. As filtrate passes through the loop of Henle, salt is removed and concentrated in the interstitial fluid of the kidney medulla. Because of this high salt concentration, the nephron is able to
   a. excrete the maximum amount of salt
   b. neutralize toxins that might accumulate in the kidney
   c. control the pH of the interstitial fluid
   d. excrete a large amount of water
   e. reabsorb water most efficiently

4. Birds and insects excrete uric acid, while mammals and amphibians excrete mainly urea. What is the chief advantage of uric acid over urea as a waste product?
   a. Uric acid is more soluble in water.
   b. Uric acid is a much simpler molecule.
   c. It takes less energy to make uric acid.
   d. Less water is lost excreting uric acid.
   e. More solutes are lost excreting uric acid.

True/False (*Change false statements to make them true.*)

1. The body temperature of a fish is always equal to the temperature of its environment.

2. Nitrogenous wastes are produced when the body breaks down proteins.

3. An ectotherm generates much of its own body heat from metabolism.

4. After leaving the kidney, urine passes out of the body through the urethra, the urinary bladder, and then the ureter.

5. After a meal, the liver removes excess glucose from the blood and stores it for later use.

6. Reabsorption and secretion of substances by the kidney tubules help regulate blood pH.

Matching

Match each of the following components of blood with what happens to it as the blood is processed by the kidney.

1. Water
2. Glucose
3. Plasma protein
4. Hydrogen ion (H^+)
5. Red blood cell
6. Urea

a. Remains in blood
b. Passes into filtrate; mostly reabsorbed
c. Passes into filtrate; partially reabsorbed; excreted in urine
d. Passes into filtrate; also secreted; excreted in urine

Describing, Comparing, and Explaining

1. Explain why biologists prefer to use the the terms "endothermic" and "ectothermic" rather than "warm blooded" and "cold-blooded."

2. Choose an animal, and describe four ways in which it regulates heat gain and loss.

3. Compare the problems of water and salt regulation faced by a salmon when it is swimming in the ocean and when it migrates into fresh water to spawn.

Thinking Critically

1. Describe what you might do on a hot day to increase heat loss from your body by (a) evaporation, (b) conduction, and (c) convection. Similarly, describe what you might do on a cold day to reduce heat loss by (a) radiation, (b) convection, and (c) conduction.

2. Compare the makeup and concentration of the blood, the filtrate produced in the kidney, and the urine finally excreted by the body. In what ways do they differ? Is the kidney simply a "blood filter"? What else does the kidney do?

3. Compare the countercurrent heat exchanger with the countercurrent mechanism that enables a fish to maximize oxygen absorption from water. Where are the two mechanisms found? What fluids are flowing in opposite directions in each case? How do the opposing flows increase the efficiency of heat or gas exchange?

Science, Technology, and Society

The kidneys remove many drugs from the blood, and these substances show up in the urine. Some employers require a urine drug test at the time of hiring and/or at intervals during the term of employment. An employee who fails a drug test can lose his or her job. Why do some employers feel that drug testing is necessary? Why do some individuals oppose drug testing? Do you think that passing a drug test is a valid criterion for employment? If so, for what types of jobs? Would you take a drug test to get or keep a job? Why or why not?

The call of the male in the spring breeding season—a distinctive, deep-voiced "jug o' rum" or "br' wum"—can carry nearly half a mile. Up to about 20 cm (8 inches) long, the bullfrog is the biggest frog in North America. A bullfrog eats mainly insects, but a large adult like the female at the left will occasionally take a small bird or snake. A frog locates prey mainly by sight, and when an insect gets within range, the bullfrog's tongue flips out and snatches it so fast that your eye can hardly follow.

Eyesight plays a key role in a bullfrog's everyday life. Frogs actually have three eyes: the two large ones that bulge from the head and a tiny one you can't see, between the other two. The third eye, called the frontal organ, is embedded in the skin. What can an eye covered with skin do? Scientists do not yet have a complete answer, but the frontal organ and an adjacent structure—an outgrowth of the brain called the pineal gland—are known to detect light and may help time the frog's daily activities. The pineal gland secretes a chemical called melatonin into the blood. The more light the pineal and frontal organ detect, the less melatonin the pineal puts out. The amount of melatonin in the blood varies over a 24-hour period, with peaks at night and lows during the day. Regular daily changes in the blood level of melatonin, cued to the amount of sunlight and thus the time of day, seem to regulate a frog's daily activities. During the breeding season, for instance, male frogs are silent during most of the day but call through much of the night. Changes in melatonin level at nightfall and daybreak may control the timing of this behavior.

Does this have any connection to humans? We do not have a third eye like a bullfrog, but we do have a pineal gland, and it secretes melatonin. The level of melatonin in our blood also changes according to a daily cycle, most of us having a peak about 2:00 A.M. Scientists hypothesize that melatonin helps make us sleepy at night and may maintain daily rhythms in some of our body activities. As in most other mammals, our melatonin level also seems to follow a seasonal cycle, with highs during the short days of late fall and early winter and lows in the spring and early summer. In other mammals, the seasonal cycle triggers seasonal ac-

tivities such as mating. The role of the seasonal melatonin cycle in humans is not known, but mounting evidence suggests that abnormal cycles can cause a condition called seasonal affective disorder (SAD).

A person with SAD is beset with severe, seasonal mood swings. The short winter days bring extreme irritability, depression, and anxiety, often with a craving for sweets, a tendency to eat ravenously between meals, and a resultant weight gain. SAD sufferers may go to bed early and still feel exhausted when they arise nine or ten hours later. Their condition is likely to worsen if they move north in the winter, but if they move south, they improve. SAD symptoms usually disappear in the spring.

It is not yet clear whether melatonin abnormalities cause SAD. But SAD patients do have unusual daily melatonin cycles, with peaks often occurring at midnight or dawn instead of around 2:00 A.M. Also, SAD patients become more depressed if they are given oral doses of melatonin. In contrast, their symptoms are relieved and the melatonin in their blood decreases if they are exposed to high-intensity light (about ten times normal room lighting) for an hour or more each day.

SAD seems to be a rare disorder, yet its symptoms are not very different from what we call "winter blues" or "cabin fever." Is it just our imagination when we feel depressed, drowsy, and irritable during the winter months? Or do some of us have a mild melatonin imbalance? Studies of winter moodiness in people living in northern latitudes (above 40°N) indicate that most people feel their worst at the same time of year that patients with SAD have their severest problems.

Much remains to be learned about the normal functions of melatonin. We know that it is an important regulator of daily and seasonal activities, but we do not yet know how it works. Melatonin is one of many chemical signals that regulate body activities. In this chapter, we examine the structure of such chemicals and how they make our organ systems function cooperatively within our bodies. We also examine how the organs that produce chemical signals are themselves regulated.

Chemical signals coordinate body functions

Animals rely on many kinds of chemical signals to regulate their body activities. One type, exemplified by melatonin, is called a **hormone** (from the Greek *hormon,* to excite). Hormones are produced and secreted mainly by organs called **endocrine glands.** Figure A outlines the action of a hormone-secreting cell in an endocrine gland. Secretory vesicles in the endocrine cell are full of the hormone (blue). Endocrine glands do not have ducts to transport their secretions; the endocrine cell secretes hormone molecules directly into the circulatory system. The hormones travel in the blood, often to quite distant **target cells,** cells that respond to the regulatory signals.

Figure B shows a second type of hormone-secreting cell. This **neurosecretory cell** is a specialized nerve cell that, in addition to conducting nerve signals, makes and secretes hormones. Like endocrine cells, neurosecretory cells release hormones into the blood for transport to target cells.

Collectively, all hormone-secreting cells constitute the **endocrine system,** the main chemical-regulating system of the body. Because hormones are carried in the blood, they reach all parts of the body, and the endocrine system is especially important in controlling whole-body activities. For example, hormones govern our metabolic rate, growth, maturation, and reproduction. In many cases, a single hormone molecule can dramatically alter a target cell's metabolism by turning on the production of a number of enzymes. Thus, a tiny amount of a hormone can govern the activity of enormous numbers of target cells in a variety of organs.

The endocrine system often collaborates with the body's other major coordinating system, the nervous system. The nervous system transmits signals via nerve cells. When a nerve signal reaches the end of a nerve cell, it triggers the secretion of molecules called neurotransmitters (Figure C). **Neurotransmitters** are chemical messengers that carry information from one nerve cell to another, or from a nerve cell to another kind of cell that will react, such as a muscle cell. Thus, the nervous system, like the endocrine system, relies on chemical messengers. Unlike hormones, however, most neurotransmitters do not travel in the bloodstream.

Why does an animal need two kinds of regulatory systems? Timing is part of the answer. In many cases, the endocrine system takes minutes, hours, or even days to act, partly because of the time it takes for hormones to be made and carried in the blood to all their target organs. And some hormones act by stimulating their target cells to synthesize new proteins, adding even more time to the process. In contrast to the relative slowness of the endocrine system, the nervous system provides split-second control. The flick of a frog's tongue catching a fly and the jerk of your hand away from a flame result from high-speed nerve signals, not from hormonal control. In the case of your hand, a single nerve cell carries a signal that conveys the sensation of burning to your spinal cord. A second nerve cell conveys a signal into the arm muscle that actually pulls away. A chemical signal is involved: A neurotransmitter carries the signal from the first nerve cell to the second one. But the distance it travels is tiny, and little time is consumed.

A neurotransmitter is a type of chemical messenger called a local regulator. A **local regulator** is secreted into the interstitial fluid and causes changes in cells very near the point of secretion—in contrast to a hormone, whose target cells typically are distant from the gland cells that secrete it. Local regulators of another type, called **prostaglandins,** are made by nearly all cells and have a wide variety of functions. For example, prostaglandins secreted by cells of the placenta cause the nearby muscles of the uterus to contract, helping induce labor during childbirth. We discuss local regulators further in Chapter 28. For the rest of this chapter, we'll explore the endocrine system and its hormones.

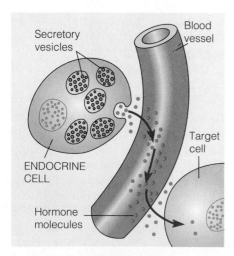

A. Hormone from an endocrine gland

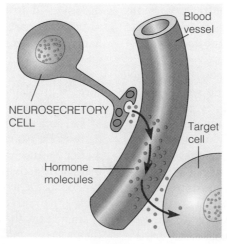

B. Hormone from a neurosecretory cell

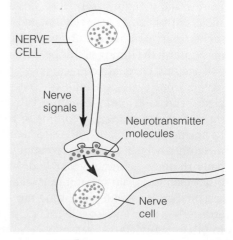

C. Neurotransmitter

There are over 50 known hormones in the human body. Each hormone is carried in the blood and may contact all the tissues in the body, but only those cells with specific receptors for a particular hormone are affected by that hormone. Despite the diversity of hormones, there are only two general mechanisms by which hormones trigger changes in target cells.

The steroid hormones, including our sex hormones (see Table 26.3), use the mechanism modeled in Figure A. **Steroid hormones** are lipids made from cholesterol (see Module 3.9). As lipids, ① they can diffuse through the phospholipid plasma membrane of their target cells. Once inside a target cell, a steroid hormone ② enters the nucleus and ③ binds to a specific receptor protein. ④ The hormone–receptor complex then attaches to certain sites on the cell's DNA, ⑤ activating transcription of specific genes, which are ⑥ translated into new proteins. In other target cells, an identical hormone–receptor complex may activate different genes. As a result, different types of target cells can respond differently to the same steroid hormone.

Nonsteroid hormones are all synthesized from amino acids, and there are three main classes of these substances. The amine hormones are modified versions of single amino acids. The peptide hormones are short chains of amino acids (as few as three). The protein hormones are made of polypeptides having as many as 200 amino acids. (For examples of amine, peptide, and protein hormones, see Table 26.3.)

Most nonsteroid hormones have the same basic mode of action. Our current understanding of their action stems from the pioneering work of American physiologist Earl W. Sutherland in the 1950s. Sutherland studied the effects of epinephrine on liver cells and muscle cells. Epinephrine (also known as adrenaline) is an amine hormone. It is called the "fight-or-flight" hormone because it prepares the body for sudden action. Among other effects, it stimulates the breakdown of the energy-storage molecule glycogen in liver cells. Glycogen breakdown yields glucose, which provides body cells with a ready supply of energy.

Sutherland found that epinephrine binds to a specific receptor protein in the plasma membrane of a liver cell, and that the binding is correlated with an increase of a compound called **cAMP** (cyclic adenosine monophosphate) within the liver cell. It became clear that epinephrine signals the liver cell to synthesize cAMP and that cAMP then makes the cell break down glycogen. A hormone that has this effect on a target cell is called a first messenger. As indicated in Figure B, ① the hormone binds to a membrane receptor protein, ② setting off a series of reactions that activate an enzyme. ③ The enzyme, in turn, converts ATP to cAMP. Once it is formed, ④ cAMP serves as a **second messenger**, in this case triggering the breakdown of glycogen. The cAMP molecule is one of several types of molecules known to function as second messengers for hormones. In other types of cells, cAMP triggers different reactions. Thus, a single nonsteroid hormone can produce different responses in different target cells, as can a steroid hormone.

We can summarize the two mechanisms of hormone action by saying that most steroid hormones interact with the DNA in the nucleus, stimulating the synthesis of new proteins by their target cell. In contrast, the nonsteroid hormones never enter the cell, but produce their effects through a second messenger.

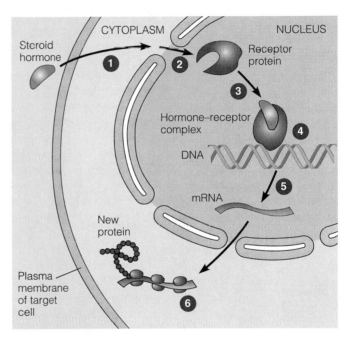

A. Steroid hormone action

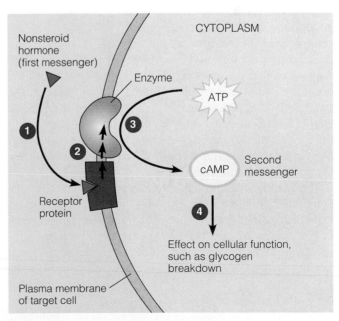

B. Nonsteroid hormone action

Overview: The vertebrate endocrine system

The vertebrate endocrine system consists of more than a dozen glands. Some of these, such as the thyroid and the pituitary gland, are endocrine specialists; their sole or main function is secreting hormones into the blood. Several other glands have both endocrine and nonendocrine functions. The pancreas, for example, contains endocrine cells that secrete three hormones that influence the level of glucose in the blood. The pancreas also has nonendocrine cells that secrete digestive enzymes into the intestine via specialized ducts (see Module 21.10).

The figure here shows the locations of the major endocrine glands in a human. The table on the facing page summarizes the actions of the main hormones they produce, as well as how the glands themselves are regulated. The table provides an overview of the human endocrine system, and you may wish to refer back to it as we focus on the individual glands and their hormones in later modules. For now, you may find it helpful to note several general features that are shared by the endocrine systems of all vertebrates. For one thing, notice the distribution of the four chemical classes of hormone (steroids, amines, peptides, and proteins) in the table. Only the sex organs and the cortex of the adrenal gland produce steroids. Most of the endocrine glands produce nonsteroid hormones and achieve their control over body processes via the second-messenger mechanism we discussed in Module 26.2.

The hormones in the table have a wide range of targets. Some, like the sex hormones, which promote male and female characteristics, affect most of the tissues of the body. Growth hormone, produced by the pituitary, promotes protein synthesis in virtually every tissue. Other hormones, such as glucagon from the pancreas, have only a few kinds of target cell (liver and fat cells for glucagon). Some hormones have other endocrine glands as their targets. For example, the pituitary gland produces thyroid-stimulating hormone, which stimulates activity of the thyroid gland.

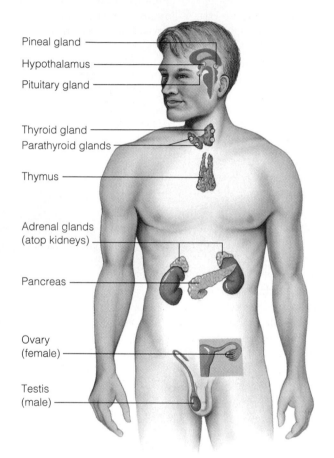

Pineal gland
Hypothalamus
Pituitary gland
Thyroid gland
Parathyroid glands
Thymus
Adrenal glands (atop kidneys)
Pancreas
Ovary (female)
Testis (male)

The major endocrine glands in humans

The close association of the endocrine system and the nervous system is apparent in both the drawing and the table. For example, the hypothalamus, which is part of the brain, secretes many hormones that regulate other endocrine glands, especially the pituitary. We will explore the structural and functional connections between the endocrine system and nervous system in more depth when we focus on the hypothalamus in the next module.

The only glands we will not be discussing in later modules are the pineal gland and the thymus. Much remains to be learned about both of these organs. The **pineal gland** is part of the brain—actually, an outgrowth of it. As we mentioned in the chapter's introduction, the pineal gland secretes the hormone melatonin, which links environmental light conditions and certain activities with daily or seasonal rhythms. We know the most about melatonin's function in mammals that breed during certain seasons. For example, in sheep and deer that breed in the fall, when days are short, elevated melatonin in the blood cues reproductive activity. In contrast, in mammals that breed in the spring, longer days and less melatonin in the blood are the reproductive cues. We do not yet know what specific effects melatonin has on the body cells that produce these rhythms.

The **thymus gland** lies under the breastbone in humans and is quite large during childhood. Before the 1960s, it was generally thought that this gland was as nonessential as the appendix, and when the thymus seemed enlarged, physicians routinely removed portions of it. It is now known that the thymus plays an important role in the immune system. It secretes several important hormones, including a peptide that stimulates the development and differentiation of T cells (see Module 24.4). At puberty, when the immune system is well established, the thymus begins to shrink, and it virtually disappears by adulthood.

Major Vertebrate Endocrine Glands and Some of Their Hormones

Gland	Hormone	Chemical Class	Representative Actions	Regulated by
Pineal body	Melatonin	Amine	Involved in rhythmic activities (daily and seasonal)	Light/dark cycles
Hypothalamus	Hormones released by the posterior pituitary and hormones that regulate the anterior pituitary (see below)			
Pituitary gland				
Posterior lobe (releases hormones made by hypothalamus)	Oxytocin	Peptide	Stimulates contraction of uterus and mammary gland cells	Nervous system
	Antidiuretic hormone (ADH)	Peptide	Promotes retention of water by kidneys	Water/salt balance
Anterior lobe	Growth hormone (GH)	Protein	Stimulates growth (especially bones) and metabolic functions	Hypothalamic hormones
	Prolactin (PRL)	Protein	Stimulates milk production	Hypothalamic hormones
	Follicle-stimulating hormone (FSH)	Protein	Stimulates production of ova and sperm	Hypothalamic hormones
	Luteinizing hormone (LH)	Protein	Stimulates ovaries and testes	Hypothalamic hormones
	Thyroid-stimulating hormone (TSH)	Protein	Stimulates thyroid gland	Thyroxine in blood; hypothalamic hormones
	Adrenocorticotropic hormone (ACTH)	Protein	Stimulates adrenal cortex to secrete glucocorticoids	Glucocorticoids; hypothalamic hormones
Thyroid gland	Thyroxine (T_4) and triiodothyronine (T_3)	Amine	Stimulate and maintain metabolic processes	TSH
	Calcitonin	Peptide	Lowers blood calcium level	Calcium in blood
Parathyroid glands	Parathyroid hormone (PTH)	Peptide	Raises blood calcium level	Calcium in blood
Thymus	Thymosin	Peptide	Stimulates T cells	Not known
Adrenal glands				
Adrenal medulla	Epinephrine and norepinephrine	Amines	Increase blood glucose; increase metabolic activities; constrict certain blood vessels	Nervous system
Adrenal cortex	Glucocorticoids	Steroids	Increase blood glucose	ACTH
	Mineralocorticoids	Steroids	Promote reabsorption of Na^+ and excretion of K^+ in kidneys	K^+ (potassium) in blood
Pancreas	Insulin	Protein	Lowers blood glucose	Glucose in blood
	Glucagon	Protein	Raises blood glucose	Glucose in blood
Testes	Androgens	Steroids	Support sperm formation; development and maintenance of male secondary sex characteristics	FSH and LH
Ovaries	Estrogens	Steroids	Stimulate uterine lining growth; development and maintenance of female secondary sex characteristics	FSH and LH
	Progesterone	Steroid	Promotes uterine lining growth	FSH and LH

The hypothalamus, closely tied to the pituitary, connects the nervous and endocrine systems

The distinction between the endocrine system and the nervous system often blurs, especially when we consider the diverse roles of the hypothalamus and its intricate association with the pituitary gland. As part of the brain (Figure A), the hypothalamus receives information from nerves about the internal condition of the body and about the external environment. It then responds to these conditions by sending out appropriate nervous or endocrine signals. The **hypothalamus** is the master control center of the endocrine system. Its endocrine signals directly control the pituitary gland, which in turn secretes hormones that influence numerous body functions (see Table 26.3).

As Figure A shows, the pituitary gland consists of two distinct parts: a posterior lobe and an anterior lobe, both situated in a pocket of skull bone just under the hypothalamus. The posterior lobe, or **posterior pituitary**, is composed of nervous tissue and is actually an extension of the hypothalamus. It stores and secretes hormones made in the hypothalamus.

The anterior lobe, or **anterior pituitary**, is composed of non-nervous, glandular tissue. Unlike the posterior pituitary, it synthesizes its own hormones, several of which control the activity of other endocrine glands. The hypothalamus exerts control over the anterior pituitary by secreting two kinds of hormones into the blood. **Releasing hormones** make the anterior pituitary secrete its hormones. **Inhibiting hormones** make the anterior pituitary stop secreting hormones.

Figure B is a diagram of how the hypothalamus operates through the anterior pituitary to regulate the activity of another endocrine organ, the thyroid gland. The hypothalamus secretes a releasing hormone known as TRH (thyroid-releasing hormone). In turn, TRH makes the anterior pituitary secrete TSH (thyroid-stimulating hormone). Under the influence of TSH, the thyroid secretes the hormone thyroxine into the blood. Thyroxine increases the metabolic rate of most body cells, warming the body as a result.

Figure B also shows how the TRH-TSH-thyroxine system is regulated, keeping the hormones at homeostatic levels. The hypothalamus takes its cues from the environment; for instance, cold temperatures generally trigger an increase in TRH secretion. As the red arrow indicates, a negative-feedback mechanism regulates TSH secretion by the pituitary. As thyroxine increases in the blood, it inhibits TSH secretion, probably by making the pituitary cells insensitive to the effects of TRH. Negative feedback is important throughout the endocrine system, as we'll see in this chapter.

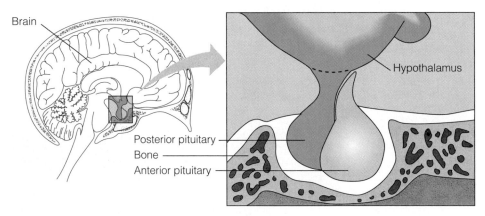

A. Location of hypothalamus and pituitary

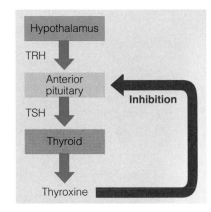

B. Control of thyroxine secretion

The hypothalamus and pituitary have multiple endocrine functions

The TRH-TSH-thyroxine model illustrates a regulatory hierarchy characteristic of the vertebrate endocrine system. The hypothalamus exerts master control over the system, using the pituitary to relay directives to other glands and as a regulatory center for feedback control.

The two figures on the facing page emphasize the structural and functional connections between the hypothalamus and the pituitary. As Figure A indicates, a set of neurosecretory cells extend from the hypothalamus into the posterior pituitary. These cells synthesize the hormones **oxy-**

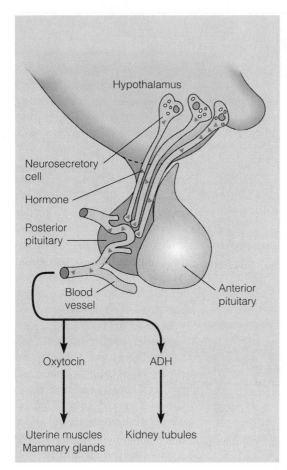

A. Hormones of the posterior pituitary

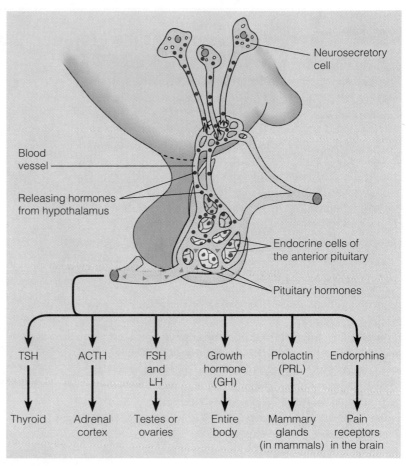

B. Hormones of the anterior pituitary

tocin and **antidiuretic hormone (ADH).** The hormones are channeled along the neurosecretory cells into the posterior pituitary. When released into the blood from the posterior pituitary, oxytocin induces contraction of the uterine muscles during childbirth and causes the mammary glands to eject milk during nursing. ADH helps the kidney tubules reabsorb water, thus decreasing urine volume when the body needs to retain water. When the body has too much water, the hypothalamus responds to negative feedback, slowing the release of ADH from the posterior pituitary.

Figure B illustrates a second set of neurosecretory cells in the hypothalamus. These cells secrete releasing and inhibiting hormones that control the anterior pituitary. A system of small blood vessels carries these hormones from the hypothalamus to the anterior pituitary. In response to hypothalamic-releasing hormones, the anterior pituitary synthesizes and releases many different peptide and protein hormones, which influence a broad range of body activities. Like **thyroid-stimulating hormone (TSH), adrenocorticotropic hormone (ACTH), follicle-stimulating hormone (FSH),** and **luteinizing hormone (LH)** activate other endocrine glands. Feedback mechanisms control the secretion of all these hormones by the anterior pituitary.

Of all the pituitary secretions, none has a broader effect than the protein called **growth hormone (GH).** GH promotes protein synthesis and the use of body fat for energy

metabolism in a wide variety of target cells. In young mammals, GH promotes the development and enlargement of all parts of the body. When the pituitary produces too much GH in a young person, gigantism can result, while too little GH during development can lead to dwarfism. One of the most dramatic achievements of genetic engineering has been the artificial production of human growth hormone, which can be used to treat pituitary dwarfism and may reverse some of the symptoms of aging (see Module 11.18).

Another anterior pituitary hormone, **prolactin (PRL),** produces very different effects in different species. In mammals, it stimulates mammary glands to produce milk; in birds, it controls fat metabolism and reproduction; in amphibians, it regulates larval development; and in freshwater fishes, it regulates salt and water balance. These diverse effects suggest that prolactin is an ancient hormone whose functions diversified during vertebrate evolution.

The **endorphins,** another kind of anterior pituitary hormone, are sometimes called the body's natural painkillers, or "natural opiates." These chemical signals have a pain-inhibiting effect on the nervous system similar to that of the drug morphine. They are produced by the brain and by the anterior pituitary. It has been suggested that the so-called "runner's high" results partly from the release of endorphins when stress and pain in the body reach critical levels.

26.6 The thyroid regulates development and metabolism

Our **thyroid gland** is located just under the voicebox. Thyroid hormones affect virtually all the tissues of vertebrate animals.

The thyroid produces two very similar amine hormones, both of which contain the element iodine. One of these, thyroxine, is often called T_4 because it contains four iodine atoms; the other, triiodothyronine, is called T_3 because it contains three iodine atoms.

T_3 and T_4 have essentially the same effects on their target cells. One of their crucial roles is in development and maturation. In bullfrogs, for example, they trigger the profound reorganization of body tissues that occurs as a tadpole—a strictly aquatic organism—transforms into an adult frog, which spends much of its time on land. Thyroid hormones are equally important in mammals, especially in bone and nerve cell development. In humans, a congenital thyroid deficiency known as cretinism results in retarded skeletal growth and poor mental development.

The thyroid hormones continue to play vital roles during adulthood. For example, T_3 and T_4 help maintain normal blood pressure, heart rate, muscle tone, digestion, and reproductive functions. Throughout the body, these hormones tend to increase the rate of oxygen consumption and cellular metabolism. Too much or too little thyroid hormone in the blood can result in serious metabolic disorders. An excess of T_3 and T_4 in the blood (*hyper*thyroidism) can make a person overheat, sweat profusely, become irritable, develop high blood pressure, and lose weight. Conversely, insufficient amounts of T_3 and T_4 (*hypo*thyroidism) can cause weight gain, lethargy, and intolerance to cold. Fortunately, hypothyroidism and hyperthyroidism are uncommon disorders, and both can be successfully treated.

Hypothyroidism can result from a defective thyroid gland or from dietary disorders. Figure A illustrates a severe condition called **goiter,** an enlargement of the thyroid that can result if dietary hypothyroidism is not treated. Goiter can occur when there is not enough iodine in the diet. In such cases, the thyroid gland cannot synthesize adequate amounts of its T_3 and T_4 hormones. The lack of T_3 and T_4 interrupts the normal feedback loop that controls thyroid activity (Figure B). As a result, the blood never carries enough of the T_3 and T_4 hormones to shut off secretion of thyroid-stimulating hormone (TSH) by the pituitary. The thyroid enlarges because TSH continues to stimulate it.

Goiter can be prevented simply by including iodine in the diet. Seawater is a rich source of iodine, and goiter rarely occurs in people living near the seacoast, where the soil is iodine-rich and a lot of seafood is consumed. Goiter has also been reduced in many nations by the incorporation of iodine into table salt. Unfortunately, goiter still affects many thousands of people in developing nations.

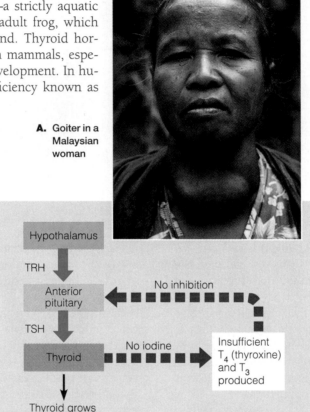

A. Goiter in a Malaysian woman

Hypothalamus

TRH

Anterior pituitary

No inhibition

TSH

No iodine

Thyroid

Insufficient T_4 (thyroxine) and T_3 produced

Thyroid grows to form goiter

B. Why a goiter forms

26.7 Hormones from the thyroid and parathyroids maintain calcium homeostasis

An appropriate level of calcium in the blood and interstitial fluid is essential for many body functions. Without calcium, nerve signals cannot be transmitted from cell to cell, muscles cannot function properly, blood cannot clot, and cells cannot transport molecules across their membranes. The thyroid and parathyroid glands function in homeostasis of calcium ions (Ca^{2+}), keeping the concentration of the ions within a narrow range (about 9–11 mg per 100 mL of blood).

There are four **parathyroid glands,** all embedded in the surface of the thyroid. Two peptide hormones, **calcitonin** from the thyroid gland and **parathyroid hormone (PTH),** secreted by the parathyroids, regulate the blood calcium level. Calcitonin and PTH are said to be **antagonistic hormones** because they have opposite effects. Calcitonin lowers the calcium level in the blood, whereas PTH raises it. As the figure below indicates, these two antagonistic hormones operate through a feedback system. The system operates with a set point at approximately 10 mg of Ca^{2+} per 100 mL of blood. To read the diagram, start with the yellow box on the left, and follow the arrows to the top part of the figure. You can see that a rise in the blood Ca^{2+} level above 10 mg/100 mL induces the thyroid gland to secrete calcitonin. Calcitonin, in turn, has three effects: (1) It causes more Ca^{2+} to be deposited in the bones, (2) it makes the intestines absorb less Ca^{2+} from the diet, and (3) it makes the kidneys reabsorb less Ca^{2+} as they form urine. The result is a lower Ca^{2+} level in the blood.

Starting from the yellow box on the right, now follow the bottom part of the diagram to see how PTH from the parathyroid glands reverses all three of calcitonin's effects. When the blood Ca^{2+} level drops below 10mg/100 mL of blood, the parathyroids release PTH into the blood. PTH stimulates the release of calcium ions from bone and increases Ca^{2+} uptake by the kidneys and intestines. The diagram also indicates the important role that vitamin D plays in calcium homeostasis. Synthesized in an inactive form by skin exposed to sunlight, vitamin D is carried in the blood to the liver and kidneys, where it is converted to an active form. The active form enables PTH to increase calcium uptake by the intestine.

In summary, a sensitive balancing system maintains calcium homeostasis. The system depends on feedback control involving two antagonistic hormones. Failure of the system can have far-reaching effects in the body. For example, a shortage of PTH causes the blood level of calcium to drop dramatically, leading to convulsive contractions of the skeletal muscles. If unchecked, this condition, known as tetany, is fatal.

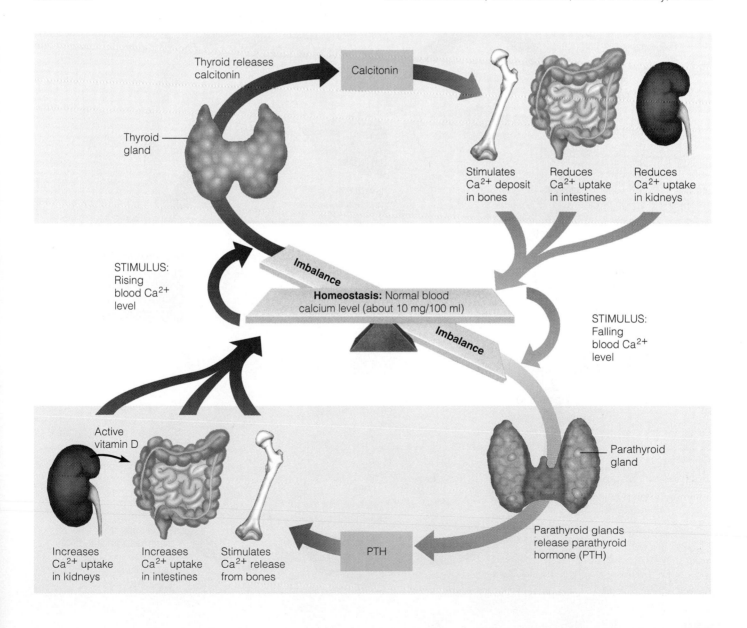

The pancreas controls blood glucose level

The **pancreas** produces two hormones that regulate blood glucose and also control the synthesis of proteins and fats in many body cells. One of the hormones is **insulin,** probably the most familiar of all the hormones. Much of what we know about this hormone stems from study of the disease called diabetes mellitus, which results from insulin deficiency or abnormal function.

Insulin is a protein hormone produced by specialized clusters of pancreatic cells known as **islet cells.** There are several distinct types of islet cells, but only those known as beta cells synthesize and secrete insulin. Islet cells of another type, called alpha cells, secrete a different hormone, a peptide called **glucagon.**

As shown in the figure below, insulin and glucagon are antagonists, countering each other in a feedback circuit similar to the one that regulates blood calcium. The concentration of glucose in the blood determines how much insulin or glucagon is secreted into the blood. Follow the top half of the diagram to see what happens when the glucose concentration of the blood rises above the set point of about 90 mg/100 mL, as it does shortly after eating a carbohydrate-rich meal. The rising blood glucose level (yellow box on the left) stimulates the beta cells in the pancreas to secrete more insulin. The insulin makes the body cells take up more glucose from the blood, thereby decreasing the blood glucose level. Liver cells take up much of the glucose and use it to

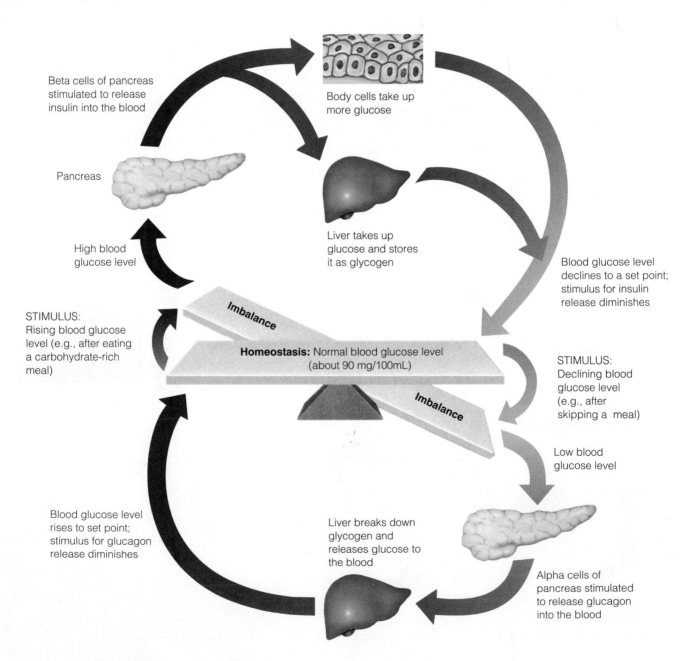

Beta cells of pancreas stimulated to release insulin into the blood

Body cells take up more glucose

Pancreas

Liver takes up glucose and stores it as glycogen

High blood glucose level

Blood glucose level declines to a set point; stimulus for insulin release diminishes

STIMULUS:
Rising blood glucose level (e.g., after eating a carbohydrate-rich meal)

Imbalance

Homeostasis: Normal blood glucose level (about 90 mg/100mL)

Imbalance

STIMULUS:
Declining blood glucose level (e.g., after skipping a meal)

Low blood glucose level

Blood glucose level rises to set point; stimulus for glucagon release diminishes

Liver breaks down glycogen and releases glucose to the blood

Alpha cells of pancreas stimulated to release glucagon into the blood form glycogen, which the liver stores. Insulin also stimulates cells to metabolize the glucose for immediate energy use, for storage of energy in fats, or for the synthesis of proteins. When the blood glucose level returns to the set point, the beta cells lose their stimulus to secrete insulin.

If you follow the bottom half of the diagram, you can see that when the blood glucose level dips below the set point (yellow box on the right), as it may between meals or during heavy exercise, the pancreatic alpha cells respond by secreting glucagon. Glucagon makes liver cells break glycogen down into glucose and release the glucose to the blood. (It also makes fat cells release into the blood fatty acids, an alternative energy source.) Then, when the blood glucose level returns to the set point, the alpha cells lose their stimulus to produce glucagon.

In the next module, we see what can happen when this delicately balanced control system fails.

Diabetes is a common endocrine disorder

Diabetes mellitus, a serious hormonal disease in which the body cells are unable to absorb glucose from the blood, affects as many as five in every 100 people in the United States. This disease occurs when there is not enough insulin in the blood or when the body cells do not respond normally to blood insulin. In either case, the cells cannot obtain enough glucose from the blood, and thus, starved for fuel, they are forced to burn the body's supply of fats and proteins. Meanwhile, since the digestive system can continue to absorb glucose from the diet, the glucose concentration in the blood can become extremely high—so high, in fact, that glucose is excreted and can be detected in the urine. (Normally, the kidney leaves no glucose in the urine.) There are treatments for diabetes mellitus—insulin supplements and/or special diets—but no cure. Every year some 350,000 Americans die from the disease or from its complications, which include severe dehydration, cardiovascular and kidney disease, nerve damage, and gangrene.

There are two types of diabetes mellitus. Type I (insulin-dependent) diabetes results from destruction of the pancreatic beta cells, either by a viral infection, by faulty genes that make the immune system destroy the cells, or by both. In any case, the pancreas does not produce enough insulin, and glucose builds up in the blood. Type I diabetes often develops before the age of 15. Patients require regular supplements of insulin, and most take the hormone by direct injection. The insulin they use is now made commercially by genetically engineered bacteria. Insulin-dependent diabetics also may benefit from a diet high in soluble fiber (found in foods such as apples and oatmeal).

A second type of diabetes mellitus develops even though the pancreatic beta cells are functioning normally and there is plenty of insulin in the blood. Type II diabetes occurs because the body cells fail to respond adequately to insulin. The disease seems to be inherited and may result from genes that code for malfunctional insulin receptors on the cells. Type II diabetes accounts for about 90% of the cases in the United States. It is almost always associated with obesity and often does not show up until a person is over 40. The disease is controlled by diet to control sugar in take and reduce weight; recommended diets are high in soluble fiber and low in fat and sodium. An oral drug that reduces the blood glucose level is also available.

How is diabetes detected? The early signs of either type of diabetes are a lack of energy, a craving for sweets, frequent urination, and persistent thirst. A combination of these symptoms and a family history of diabetes indicate that a person should be tested for the disease. The diagnostic test for diabetes is a glucose-tolerance test, in which the person swallows a sugar solution, then a physician measures the blood glucose level at prescribed time intervals. In the graph below, you can see the differences between the glucose-tolerance values recorded for a normal person and those for a diabetic.

Diabetes is not the only disease that can result from problems with insulin. Some people have hyperactive beta cells that put too much insulin into the blood when they eat sugar. As a result, their blood glucose level can drop well below normal. This condition, called **hypoglycemia**, usually occurs two to four hours after a meal and may be accompanied by hunger, weakness, sweating, and nervousness. In severe cases, when the brain receives inadequate amounts of glucose, a person may develop convulsions, become unconscious, and even die. Hypoglycemia is not common, and most forms of it can be controlled by reducing sugar intake and eating more frequently, in smaller amounts.

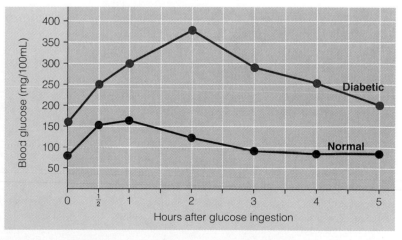

Glucose-tolerance tests

The adrenal glands mobilize responses to stress

The human body has two **adrenal glands** sitting atop the kidneys. As you can see at the top left of the figure below, each of these organs is actually two glands in one: a central portion called the **adrenal medulla** and an outer portion called the **adrenal cortex.** Though the cells they contain and the hormones they produce are different, both the medulla and the cortex secrete hormones that enable the body to respond to stress.

The adrenal medulla produces the "fight-or-flight" hormones, which ensure a rapid, short-term response to stress. You've probably felt your heart beat faster and your skin develop goose bumps when sensing danger or approaching a stressful situation, like speaking in public. Positive emotions—extreme pleasure, for instance—can produce the same effects. These reactions are triggered by two amine hormones secreted by the adrenal medulla, **epinephrine** (adrenaline) and **norepinephrine** (noradrenaline).

Stressful stimuli, whether negative or positive, activate certain nerve cells in the hypothalamus. As indicated on the left side of the diagram, these cells send signals to nerve cells in the spinal cord. The spinal cord cells extend to the adrenal medulla and stimulate it to release epinephrine and norepinephrine into the blood. Norepinephrine and epinephrine have somewhat different effects on tissues, but both contribute to the short-term stress response. Both hormones stimulate liver cells to release glucose, thus making more energy available for cellular fuel. They also prepare the body for action by increasing the blood pressure, the breathing rate, and the metabolic rate. In addition, epinephrine and norepinephrine change blood-flow patterns, making some organs more active and others less so. For example, epinephrine dilates blood vessels in the brain and skeletal muscles, thus increasing alertness and the muscles' ability to react to stress. At the same time, epinephrine and norepinephrine constrict blood vessels elsewhere, thereby reducing activities that are not immediately involved in the stress response, such as digestion. The short-term stress response is rapid and usually subsides shortly after we first encounter stress.

In contrast to hormones from the adrenal medulla, those secreted by the adrenal cortex can provide a slower, more long-term response to stress. Also in contrast to the adrenal

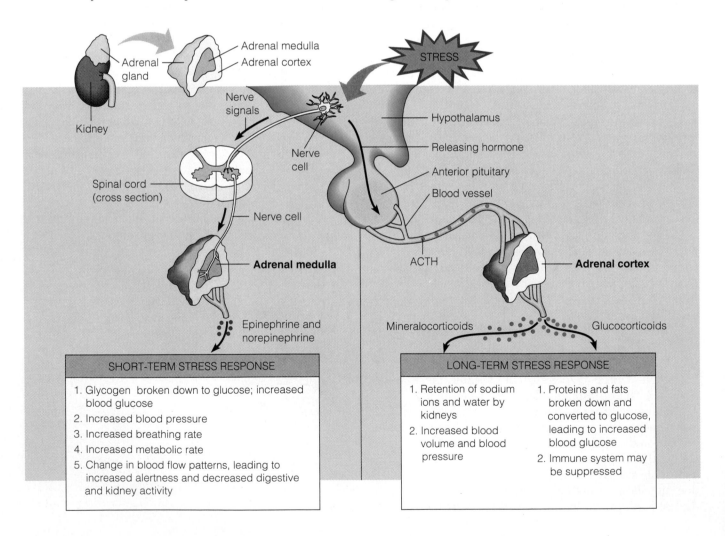

SHORT-TERM STRESS RESPONSE
1. Glycogen broken down to glucose; increased blood glucose
2. Increased blood pressure
3. Increased breathing rate
4. Increased metabolic rate
5. Change in blood flow patterns, leading to increased alertness and decreased digestive and kidney activity

LONG-TERM STRESS RESPONSE	
1. Retention of sodium ions and water by kidneys	1. Proteins and fats broken down and converted to glucose, leading to increased blood glucose
2. Increased blood volume and blood pressure	2. Immune system may be suppressed

medulla, the adrenal cortex responds to endocrine signals—chemical signals in the blood—rather than to nerve cell signals. As the right side of the diagram indicates, the hypothalamus secretes a releasing hormone that stimulates target cells in the anterior pituitary to secrete the hormone **ACTH**. In turn, ACTH stimulates cells of the adrenal cortex to synthesize and secrete a family of steroid hormones called **corticosteroids.** The two main types in humans are the mineralocorticoids and the glucocorticoids. Both are essential to homeostasis, helping the body function normally, whether or not it is stressed.

Mineralocorticoids have their main effects on salt and water balance. One of these hormones makes the kidney reabsorb sodium ions and water, with the overall effect of increasing the volume of the blood and raising blood pressure as a response to prolonged stress.

The **glucocorticoids** promote the synthesis of glucose from noncarbohydrates such as proteins and fats. This makes more glucose available in the blood as cellular fuel in response to stress.

Very high levels of glucocorticoids in the blood can suppress the body's defense system, including the inflammatory response that occurs at the site of an infection. For this reason, physicians may use glucocorticoids to treat diseases in which excessive inflammation is a problem. The glucocorticoid cortisone, for example, was once regarded as a miracle drug for treating serious inflammatory conditions such as arthritis. Cortisone and other glucocorticoids can relieve swelling and pain from inflammation, but by suppressing immunity, they can also make a person highly susceptible to infection. We discuss some other dangers of glucocorticoids next.

Glucocorticoids offer relief from pain, but not without serious risks

26.11

Pain is often part of a professional athlete's life, and few are better acquainted with it than Bill Walton, former basketball superstar for UCLA, the Boston Celtics, and the Portland Trail Blazers. Walton was born with high arches and a malformed left foot. Running or jumping usually hurt, but for years he accepted the pain as part of his heavy workouts.

In 1977, Walton, a 6'11", 225-lb center, had been with the Blazers three years and led them to the 1977 National Basketball Association championship. Walton's stardom with the Blazers was all too brief, however, for the steady pounding up and down the basketball court and the heavy impact of rebounding eventually made his pain intolerable. Midway through the 1978 season, he was sidelined with painful injuries. Following the team physician's advice, he started taking glucocorticoids and other painkillers so he could stay active. During the '78 playoffs, primed with oral doses of dexamethasone, a glucocorticoid, and several other drugs, he played in two games. Though limping badly, Walton scored 27 points and got 22 rebounds. The morning after the second game, X-rays showed he had been playing with a fractured bone in the arch of his left foot. Walton's superstar days were over. Amid a storm of media attention, he made several comebacks in the 1980s, but he was

Bill Walton (right), who used a glucocorticoid for foot pain

never his former self on the basketball court. Nevertheless, in recognition of his early triumphs, he was elected to the Basketball Hall of Fame in 1993.

Physicians often prescribe glucocorticoids to relieve pain from athletic injuries, and the oral use of these drugs is not uncommon. Potentially, glucocorticoids are very dangerous; taking them by mouth for more than five days can depress the activity of the adrenal glands and may cause psychological changes. It is safer, but still potentially dangerous, to inject a glucocorticoid at the site of an injury. With this treatment, the pain usually subsides, but its underlying cause remains. Masking the pain covers up the pain's message—that tissue is damaged and may get worse if not allowed to heal. If an athlete exercises an injured site before the tissue has recovered, the added stress can cause more serious damage.

Bill Walton's case was complicated. It never was firmly established that glucocorticoids worsened his condition, because he had been playing with foot pain and may have seriously injured his foot before he started using painkillers. One physician contends that Walton fractured the same bone in his left foot four times during his basketball career. An important outcome of Walton's plight was the widespread attention he drew, making more people aware of the potential dangers of painkilling drugs.

The glucocorticoids and mineralocorticoids secreted by the adrenal cortex are steroid hormones. Some other steroids, the sex hormones, affect growth and development and also regulate reproductive cycles and sexual behavior. The **gonads,** or sex glands (ovaries in the female and testes in the male), secrete sex hormones, in addition to producing gametes.

There are three major categories of sex hormones: androgens, estrogens, and progestins. Both females and males have all three types, but in different proportions. Females have a high ratio of estrogens to androgens. The **estrogens** maintain the female reproductive system and promote the development of such female features as the generally smaller body size, higher-pitched voice, breasts, and wider hips of women. **Progestins,** at least in mammals, are primarily involved in preparing the uterus to support the developing embryo.

In general, **androgens** stimulate the development and maintenance of the male reproductive system. Males have a high ratio of androgens to estrogens, their main androgen being **testosterone.** Androgens produced by male embryos early in development stimulate the embryo to develop into a male rather than a female. High concentrations of andro-

Male elephant seals

gens trigger the development of male characteristics: in humans, for instance, a lower-pitched voice, facial hair, and large skeletal muscles. (In Module 3.10, we discussed the powerful and potentially harmful effects of anabolic steroids, chemical relatives of testosterone, which athletes sometimes use to build heavy muscles.) Androgens have somewhat different effects in different animals. In bullfrogs, the effects include both large body size and the tendency to call loudly in the spring. In elephant seals (see photo), male androgens produce bodies weighing 2 tons and more, an inflatable enlargement of the nasal cavity, and aggressive behavior toward other males.

As with hormone production by the thyroid gland and the adrenal cortex, the synthesis of sex hormones by the gonads is regulated by the hypothalamus and anterior pituitary. In response to a releasing factor from the hypothalamus, the anterior pituitary secretes follicle-stimulating hormone (FSH) and luteinizing hormone (LH), which stimulate the ovaries or testes to synthesize and secrete the sex hormones. We examine the complex effects of these hormones when we focus on human reproduction in the next chapter.

Chapter Review

Begin your review by rereading the module headings and Module 26.3, and scanning the figures before proceeding to the Chapter Summary and questions.

Chapter Summary

26.1   Endocrine glands and neurosecretory cells secrete hormones, chemical signals that are carried by the blood and cause specific changes in target cells. The hormone-secreting cells make up the endocrine system, which works with the nervous system in regulating body activities.

26.2   Hormones trigger changes in target cells by two general mechanisms. Steroid hormones, which are lipids, enter target cells and bind to receptor proteins, interact with DNA, and trigger protein synthesis. Nonsteroid hormones, which are amines, peptides, or proteins, attach to receptor proteins on target cell membranes. This triggers the formation of a second messenger, such as cAMP, which causes changes inside the target cell. The effects of a hormone on different cells depend on their receptors and the specific reactions triggered inside the cells.

26.3   The vertebrate endocrine system consists of more than a dozen glands, secreting more than 50 hormones. Some glands are

specialized for hormone secretion only; some also do other jobs. Only the sex glands and adrenal cortex secrete steroids; the remaining glands make other types of hormones. Some hormones have a very narrow range of targets and effects; others have numerous effects on many targets. So far, scientists do not fully understand the functions of the pineal gland and the thymus gland.

26.4–26.5   The hypothalamus is the master control center of the endocrine system. It regulates the pituitary gland, which secretes hormones that influence many organs and activities. Neurosecretory cells from the hypothalamus make the hormones oxytocin and ADH and transmit nerve signals that trigger their release from the posterior pituitary. Releasing and inhibiting hormones secreted by the hypothalamus control the secretion of TSH, ACTH, GH, and other hormones from the anterior pituitary. The brain and anterior pituitary also produce endorphins, the body's natural painkillers. Blood hormone levels exert negative-feedback control over the secretion of many hormones.

26.6–26.7 T_4 and T_3 from the thyroid gland regulate development and metabolism. Thyroid imbalance can cause cretinism, metabolic disorders, and goiter. Blood calcium level is regulated by the balance between calcitonin from the thyroid and parathyroid hormone from the parathyroid glands.

26.8–26.9 The pancreas secretes two hormones, insulin and glucagon, that control the blood glucose level. When blood glucose increases, insulin signals cells to use and store it. When blood glucose drops, glucagon causes cells to release stored glucose. Diabetes mellitus results from a lack of insulin or a failure of cells to respond to it.

26.10–26.11 The adrenal glands release hormones in response to stress. Nerve signals from the hypothalamus stimulate the adrenal medulla to secrete epinephrine and norepinephrine, which quickly trigger the fight-or-flight response. ACTH from the pituitary causes the adrenal cortex to secrete glucocorticoids and mineralocorticoids, which boost blood pressure and energy in response to long-term stress. Glucocorticoids such as cortisone relieve inflammation and pain, but they can suppress immunity and mask injury.

26.12 Estrogens, progestins, and androgens are steroid sex hormones produced by the ovaries in females and the testes in males. Estrogens and progestins stimulate the development of female characteristics and maintain the female reproductive system. Androgens such as testosterone trigger the development of male characteristics. The secretion of sex hormones is controlled by the hypothalamus and pituitary.

Testing Your Knowledge

Multiple Choice

1. Which of the following controls the activity of all the others?
   - **a.** thyroid gland
   - **b.** pituitary gland
   - **c.** adrenal cortex
   - **d.** hypothalamus
   - **e.** ovaries

2. The pancreas secretes insulin in response to
   - **a.** an increase in body temperature
   - **b.** changing cycles of light and dark
   - **c.** a decrease in blood glucose
   - **d.** a hormone secreted by the anterior pituitary
   - **e.** an increase in blood glucose

3. Which of the following hormones have antagonistic (opposing) effects?
   - **a.** parathyroid hormone and calcitonin
   - **b.** insulin and thyroxine
   - **c.** growth hormone and epinephrine
   - **d.** ACTH and cortisone
   - **e.** epinephrine and norepinephrine

4. The body is able to maintain a relatively constant level of thyroxine in the blood because (*Explain your answer.*)
   - **a.** thyroxine stimulates the pituitary to secrete thyroid-stimulating hormone (TSH)
   - **b.** thyroxine inhibits secretion of TSH from the pituitary
   - **c.** TSH-releasing hormone (TRH) inhibits secretion of thyroxine by the thyroid gland
   - **d.** thyroxine stimulates the hypothalamus to secrete TRH
   - **e.** thyroxine stimulates the pituitary to secrete TRH

5. Which of the following hormones has the broadest range of targets?
   - **a.** ADH
   - **b.** prolactin
   - **c.** TSH
   - **d.** epinephrine
   - **e.** ACTH

6. Several kinds of chemicals act as hormones in the human body. Which of the following is *not one* of them?
   - **a.** proteins
   - **b.** steroids
   - **c.** amines
   - **d.** peptides
   - **e.** carbohydrates

True/False (*Change false statements to make them true.*)

1. Fast-acting hormones like epinephrine generally exert their effects by entering target cells and interacting with DNA.

2. An increase in endorphins blocks pain.

3. An iodine deficiency might interfere with the production of thyroxine.

4. Some cells can conduct nerve signals and secrete hormones.

5. Hormones from the adrenal cortex control salt and water balance.

6. Most endocrine glands produce steroid hormones.

Matching

Match each hormone (left-hand column) with its effect on target cells (middle column) and the gland where it is produced (right-hand column).

1. thyroxine	a. lowers blood glucose	p. pineal gland
2. insulin	b. stimulates ovaries	q. testes
3. PTH	c. triggers "fight or flight"	r. parathyroid gland
4. epinephrine	d. promotes male traits	s. adrenal medulla
5. melatonin	e. regulates metabolism	t. posterior pituitary
6. ADH	f. related to daily rhythm	u. pancreas
7. androgen	g. raises blood calcium level	v. anterior pituitary
8. FSH	h. boosts water retention	w. thyroid gland

Describing, Comparing, and Explaining

1. Explain how the hypothalamus controls body functions through its action on the pituitary gland. How do control of the anterior and posterior pituitary differ?

2. Some hormones, such as insulin, have almost instant effects on target cells. Others, like the sex hormones, may take hours or even days to work. Explain this difference in terms of the two mechanisms of hormone action on target cells.

3. Explain why type I diabetes must be treated with injections of insulin, while type II diabetes is controlled in other ways.

4. Explain how the same hormone might have different effects on two different target cells and no effect on a third type of cell.

Thinking Critically

When a person takes cortisone as a medication over a long period, the body's natural secretion of this hormone drops. Assuming that the level of cortisone is regulated like other hormones, explain why this drop occurs.

Science, Technology, and Society

A low rate of secretion of growth hormone (GH) causes pituitary dwarfism. Until recently, victims of pituitary dwarfism ended up being abnormally short. Growth hormone produced by recombinant DNA technology has enabled hundreds of children who suffer from pituitary dwarfism to grow normally and reach a stature within the normal range. So far, the long-term side effects of this use of GH are unknown. Now that the hormone is readily available and relatively inexpensive, some parents who are afraid their normal children are not growing fast enough want to use GH to make them grow faster and taller. Why would parents want to do this? Should the child's wishes be taken into account? Are there reasons to hesitate about treating a child with growth hormone, or are the potential benefits worth the risk?

Reproduction and Embryonic Development 27

Are these lizards mating? In a sense, they are, although both of them are females. There are no males in this species, the desert-grassland whiptail (*Cnemidophorus uniparens*). The two individuals here are involved in a complex ritual that primes the one on the bottom to lay her eggs. There is no copulation, but the female on top behaves much like a male in other species of whiptail lizards. She grabs her "mate" by the neck, mounts her, and wraps her tail around her abdomen. If these lizards find mates again a few weeks later, their roles will reverse. The one on top here will be on the bottom, and vice versa.

Desert-grassland whiptails inhabit dry prairies and deserts of the southwestern United States and northern Mexico. About 25 cm (10 in) long, they are active predators, darting quickly about in search of insects. Research indicates that this unusual species arose from a single female. And DNA-sequencing studies find that the ancestral female was a hybrid of two still-existing species having both male and female individuals. The ancestral female would have reproduced without having her eggs fertilized, and she would have passed this ability on to all of her offspring. "Mating" by today's desert-grassland females seems to be an evolutionary leftover—a ritual derived from one or both of the ancestral species. Despite the lack of copulation and fertilization, the ritual has an important result: After mating, a female produces about three times as many eggs as she would if she did not mate. Apparently, desert-grassland whiptails still require the sexual stimuli their ancestors did to ensure maximum reproductive success.

Desert-grassland whiptails are truly unusual in the way they reproduce. However, they are not unusual at all in the way their embryo develops. The diagram here shows a whiptail embryo developing within an egg. Nourished by a large supply of yolk, it undergoes mitotic cell division. The embryo's body tissues take form as its cells differentiate—that is, take on the specific characteristics of various types of tissue. Body structures such as the whiptail's legs, head, and tail take form under the direction of master control

genes like the ones we discussed in Module 11.13. These same basic processes occur during the development of all animal species. Also, the four membranes associated with the whiptail embryo—the chorion, amnion, allantois, and yolk sac—occur in all reptiles, birds, and mammals. Supporting the embryo and keeping it (and its food, the yolk) from drying out, these membranes are hallmarks of all terrestrial vertebrates except amphibians.

This chapter surveys animal reproduction and development. Following a brief introduction to the diverse ways that animals reproduce, we focus on the reproductive system of our own species. We examine how human eggs and sperm form, and then we return briefly to the subject of hormones, to see how they affect our reproductive activities. In the second half of the chapter, we discuss the processes of fertilization and embryonic development in vertebrates and conclude with several modules on human embryonic development and birth.

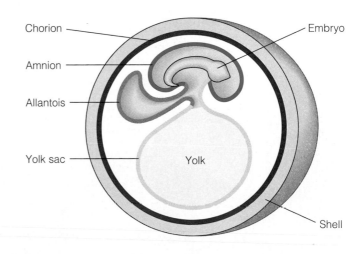

An embryo developing within an egg

Sexual and asexual reproduction are both common

An individual animal has a finite life span. In contrast, a species transcends the finite life spans of its individual members by **reproduction**, the creation of new individuals from existing ones. Animals reproduce in a great variety of ways.

Asexual reproduction is the creation of offspring whose genes all come from one parent without the fusion of egg and sperm. The desert-grassland whiptail lizard, for example, reproduces only asexually; a female produces eggs that, without being fertilized, develop into a clone of female offspring.

Many invertebrates reproduce asexually by **budding**, splitting off new individuals from existing ones (see Module 8.12). The sea anemone in the center of Figure A, below, is undergoing **fission**, another means of asexual reproduction. In fission, one individual separates into two or more individuals of about equal size. The offspring of budding and of fission are genetic copies of the parent.

Asexual reproduction can also occur by **fragmentation**, the breaking of the parent body into several pieces. For an animal to reproduce this way, fragmentation must be accompanied by **regeneration**, the regrowth of body parts from pieces of an animal. Reproduction—an increase in the number of individuals—occurs if two or more pieces of a parent body regenerate into complete adults. In some animals, the entire parent body fragments, and the pieces develop into a clone of new individuals. In certain others, only parts of the parent body break off. Sea stars have remarkable powers of regeneration and asexual reproduction. If a sea star loses one of its arms, for example, it will regenerate a new one in a matter of weeks. If a lost arm has a

piece of the animal's central disk attached to it, reproduction can occur—that is, the arm can develop into a whole new sea star.

Asexual reproduction has a number of advantages. For one thing, it allows animals that do not move from place to place or that live in isolation to produce offspring without finding mates. Indeed, the desert-grassland whiptail survived as a species because its single female ancestor was capable of reproducing asexually. Another advantage of asexual reproduction is that it allows an animal to produce many offspring quickly; no time or energy is lost in gamete production or fertilization. Asexual reproduction perpetuates a particular genotype precisely and rapidly. Therefore, it can be an effective way for animals that are genetically well suited to a particular environment to quickly expand their populations and exploit available resources.

A potential disadvantage of asexual reproduction is that it produces genetically uniform populations. Genetically similar individuals may thrive in one particular environment, but if the environment changes and becomes less favorable to survival, all individuals may be affected equally, and the entire population may die out.

In contrast to asexual reproduction, **sexual reproduction** is the creation of offspring by the fusion of two haploid (n) sex cells, or **gametes**, to form a diploid (2n) **zygote**. The male gamete, the **sperm**, is generally a small cell that moves by means of a flagellum. The female gamete, the **ovum** (unfertilized egg), is usually a relatively large cell that is not self-propelled. The zygote and the new individual it develops into contain a unique combination of genes carried from the parents via the egg and sperm.

Unlike asexual reproduction, sexual reproduction increases genetic variability among offspring. As we discussed in Modules 8.17 and 8.19, meiosis and random fertilization can generate enormous genetic variation. The variability produced by the reshuffling of genes in sexual reproduction may provide greater adaptability to changing environments. In theory, when an environment changes suddenly or drastically, there is a better chance that some of the variant offspring will survive and reproduce than if all offspring are genetically very similar.

Many animals can reproduce both sexually and asexually, benefiting from both modes. The microscopic animal in Figure B on the facing page is a rotifer. Rotifers abound in freshwater ponds and lakes, where

A. Asexual reproduction of a sea anemone by fission

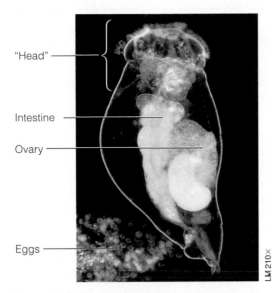

"Head"

Intestine

Ovary

Eggs

LM 210×

B. A rotifer laying eggs (side view)

Hermes and Aphrodite, fused with a woman, forming a single, bisexual individual.)

In some species, such as the majority of tapeworms, hermaphrodites can fertilize their own eggs. However, mating must occur in many other hermaphroditic animals. When hermaphrodites mate (for example, the two earthworms in Figure C), each animal serves as both male and female, donating and receiving sperm. For hermaphrodites, every individual encountered is a potential mate, and many more offspring can be produced from a mating than if only one individual's eggs were fertilized.

The mechanics of fertilization play an important part in sexual reproduction. Many aquatic invertebrates, and most fishes and amphibians, have **external fertilization.** The parents discharge their gametes into the water where fertilization occurs, often without the male and female even making physical contact. Timing is crucial because the eggs must be ripe for fertilization when sperm contact them. For many species—certain clams that live in freshwater rivers and lakes, for instance—environmental cues such as temperature and day length cause both males and females to release gametes all at once. Males or females may also emit a chemical signal as they release their gametes. The signal triggers gamete release in members of the opposite sex. Most fishes and amphibians with external fertilization have specific courtship rituals that trigger simultaneous gamete release in the same vicinity by the female and male. An example of such a ritual is the clasping of a female toad by a male (Figure D; the sperm are too small to see in the photo).

their diet consists mainly of algae, bacteria, and protozoans. Most rotifers reproduce asexually when there is ample food and when water temperatures are favorable for rapid growth and development. The female in the photograph is laying unfertilized eggs produced by mitosis. They will hatch almost immediately into a clone of new females. Asexual reproduction usually continues until cold temperatures signal the approach of winter, or until the food supply dwindles or a pond starts to dry up. The rotifers then reproduce sexually, producing a generation of genetically varied individuals. The fertilized eggs of the sexual generation have a thick shell and can withstand harsh conditions such as freezing and drying.

Although sexual reproduction has advantages, it presents a problem for nonmobile animals and for those that live solitary lives: How to find a mate? One solution that has evolved is **hermaphroditism.** Each hermaphroditic individual has both female and male reproductive systems. (The term comes from the Greek myth in which Hermaphroditus, son of the gods

C. Earthworms mating

D. Toads in an embrace that triggers the release of eggs and sperm

Eggs

In contrast to external fertilization, **internal fertilization** occurs when sperm are deposited in or close to the female reproductive tract, and gametes unite within the female's body. Nearly all terrestrial animals have internal fertilization, which is an adaptation that protects developing eggs from excessive heat and drying. Internal fertilization usually requires **copulation,** or sexual intercourse. It also requires complex reproductive systems, including copulatory organs and receptacles for storing sperm and transporting them to the eggs. For examples of these complex structures, we turn next to the human female and male.

27.2 Reproductive anatomy of the human female

The drawings in this and the next module illustrate the structures of the human female and male reproductive systems. Both sexes have a pair of gonads (ovaries or testes) where the gametes are produced, a system of ducts that house and conduct the gametes, and structures that facilitate copulation.

As illustrated in Figure A, below, a woman's **ovaries** are almond-shaped; each is about an inch long. They contain many **follicles** (the small, round structures of different sizes that you can see within the ovary on the right). Each follicle contains a single developing egg cell surrounded by one or more layers of follicle cells that nourish and protect the developing cell. In addtion to producing egg cells, the ovaries produce hormones, as we saw in Chapter 26. Specifically, the follicle cells produce the female sex hormone estrogen.

A woman is born with between 40,000 and 400,000 follicles, but only several hundred will release egg cells during her reproductive years. Starting at puberty and continuing until menopause, one follicle (or rarely two or more) matures and releases its egg cell about every 28 days. An egg cell is ejected from the follicle in the process called **ovulation,** shown in the photograph in Figure B, at the right. The orangish mass below the ejected egg cell is part of the ovary.

After ovulation, the remaining follicular tissue grows within the ovary to form a solid mass called a **corpus luteum** (Latin for "yellow body"); you can see one on the right side of Figure A. The corpus luteum secretes progesterone, the hormone that helps maintain the uterine lining during pregnancy, and additional estrogen. If the egg is not fertilized, the corpus luteum degenerates, and a new follicle matures during the next cycle. We discuss ovulation and female hormonal cycles further in later modules.

Notice in Figure A that each ovary lies next to the opening of an **oviduct,** also called a fallopian tube. The oviduct opening resembles a funnel fringed with fingerlike projections. The projections touch the surface of the ovary, but

B. Ovulation

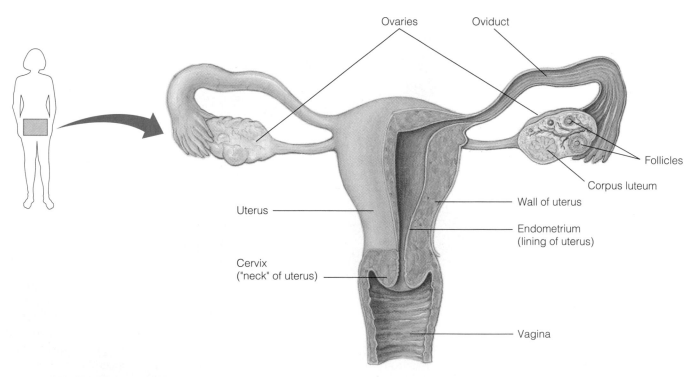

A. Front view of female reproductive anatomy (upper portion)

518 Unit 4 Animals: Form and Function

the ovary is actually separated from the opening of the oviduct by a tiny space. When ovulation occurs, the egg cell passes across the space and into the oviduct, where cilia sweep it toward the uterus. Fertilization usually occurs in the upper third of the oviduct. The resulting zygote starts to divide, thus becoming an embryo, as it moves along within the oviduct.

The **uterus,** commonly called the womb, is the actual site of pregnancy. The uterus is less than 4 inches long in a woman who has never been pregnant, but during pregnancy, it can expand to accommodate a baby weighing 4 kg (8.8 lb) or more. The uterus has a thick muscular wall, and its inner lining, the **endometrium,** is richly supplied with blood vessels. The embryo implants (digests a place for itself) in the endometrium, and development is completed there. The term **embryo** is used for the stage in development from the first division of the zygote until body structures begin to appear, about the ninth week in humans. From the ninth week until birth, a developing human is called a **fetus.**

The uterus is the *normal* site of pregnancy. However, in about one out of 100 pregnancies, the embryo implants somewhere else, resulting in an **ectopic pregnancy** (Greek *ektopos,* "out of place"). Most ectopic pregnancies occur in the oviduct and are called tubal pregnancies. Ectopic pregnancies require surgical removal; otherwise, they can rupture surrounding tissues, causing severe bleeding and even death.

The narrow neck of the uterus is the **cervix,** which opens into the vagina. The **vagina** is a thin-walled, but strong, muscular chamber that serves as the birth canal through which the baby is expelled. The vagina also accommodates the male's penis and is a repository for sperm during copulation.

You can see more features of female reproductive anatomy in Figure C, a side view. Notice that the vagina opens to the outside just behind the opening of the urethra, the tube through which urine is excreted. A pair of skin folds, the **labia minora,** border the openings, and a pair of thick, fatty ridges, the **labia majora,** protect the entire genital region. Until sexual intercourse or vigorous physical activity ruptures it, a thin membrane called the **hymen** partly covers the vaginal opening; the hymen has no known function. **Bartholin's glands,** near the vaginal opening, secrete lubricating fluid during sexual arousal, as does the vaginal lining.

Several female reproductive structures are important in sexual arousal, and stimulation of them can produce highly pleasurable sensations. The vagina, labia minora, and a structure called the **clitoris** all engorge with blood and enlarge during sexual activity. The sole function of the clitoris is sexual arousal. It consists of a short shaft supporting a rounded **glans,** or head, covered by a small hood of skin called the **prepuce.** In Figure C, blue highlights the spongy tissue within the clitoris that fills with blood during arousal. The clitoris, especially the glans, has an enormous number of nerve endings and is very sensitive to touch. Accompanied by other arousing stimuli, gentle stimulation of the glans can often trigger orgasm. We discuss the human sexual response in more detail in Module 27.6.

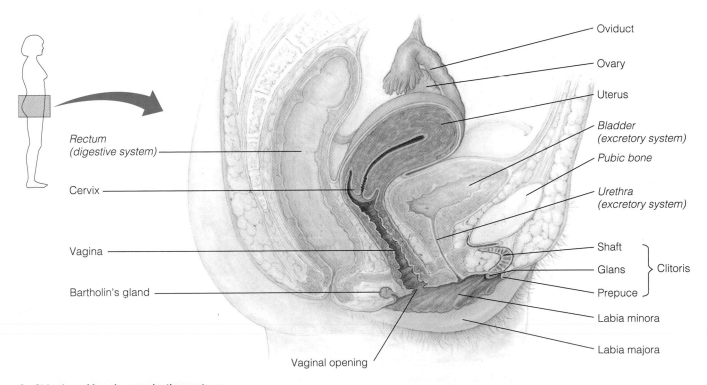

C. Side view of female reproductive anatomy

Reproductive anatomy of the human male

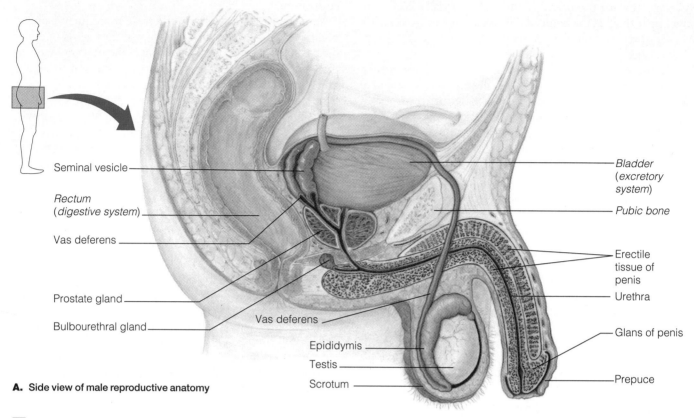

Seminal vesicle

Rectum
(*digestive system*)

Vas deferens

Prostate gland

Bulbourethral gland

Vas deferens

Epididymis

Testis

Scrotum

Bladder
(*excretory
system*)

Pubic bone

Erectile
tissue of
penis

Urethra

Glans of penis

Prepuce

A. Side view of male reproductive anatomy

Figures A and B present two views of the male reproductive system. As we did for the female, we will describe the path the gametes follow. In the male, we track the sperm from the gonads to the outside of the body.

The male gonads, or **testes** (singular, *testis*), are both housed outside the abdominal cavity in a sac called the **scrotum.** Sperm cannot develop at human body temperature, but having the testes in the scrotum keeps the sperm-forming cells cool enough to function normally.

Hormones from the anterior pituitary control the production of sperm and the secretion of androgens (mainly testosterone) by the testes. Follicle-stimulating hormone (FSH) increases sperm production, while luteinizing hormone (LH) promotes the secretion of androgens. In turn, a releasing hormone from the hypothalamus regulates the release of LH and FSH, and negative feedback by androgens controls the secretion of both the releasing hormone and LH. The testes produce sperm continuously from puberty well into old age. Hundreds of millions are in production every day.

From each testis, sperm pass into the **epididymis**, a coiled tube that stores the sperm while they become fully mature—a process that occurs over three to four weeks. Mature sperm leave the epididymis during **ejaculation**, the

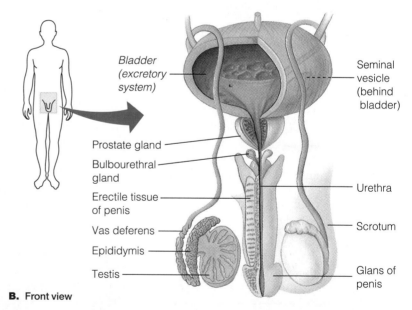

Bladder
(*excretory
system*)

Prostate gland

Bulbourethral
gland

Erectile tissue
of penis

Vas deferens

Epididymis

Testis

Seminal
vesicle
(behind
bladder)

Urethra

Scrotum

Glans of
penis

B. Front view

expulsion of sperm-containing fluid from the penis. At that time, muscular contractions propel the sperm from the epididymis through another duct called the **vas deferens.** The vas deferens passes upward into the abdomen and loops around the urinary bladder. Behind the bladder, it joins the vas deferens from the other testis. A common duct formed by the two joins the urethra. The urethra conveys both urine and sperm to the outside through the penis, although

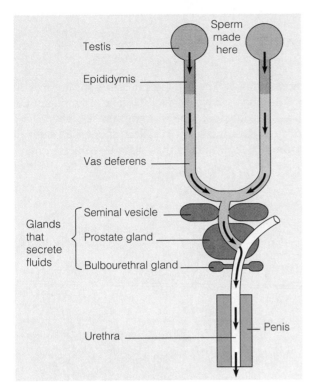

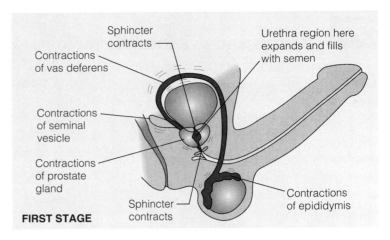

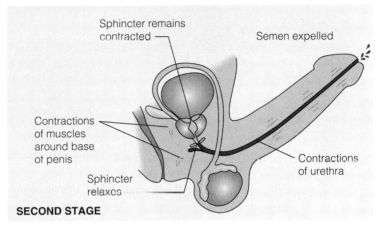

C. Flowchart summarizing the sperm path

FIRST STAGE

SECOND STAGE

D. The two stages of ejaculation

not at the same time. Thus, unlike the female, the male has a connection between the reproductive and excretory systems.

In addition to the testes and ducts, the male reproductive system contains three sets of glands: the seminal vesicles, the prostate gland, and the bulbourethral glands. The two **seminal vesicles** secrete a thick, clear fluid that lubricates and nourishes the sperm. The **prostate gland** secretes a milky, alkaline fluid that balances the acidity of any traces of urine in the urethra and helps protect the sperm from the natural acidity of the vagina. The two **bulbourethral glands** secrete only a few drops of fluid into the urethra, during sexual arousal. The fluid may help lubricate the urethra, helping sperm move through it.

The path taken by the sperm and the order in which the three kinds of glands add their secretions to it are summarized in the flowchart in Figure C. Together the sperm and the secretions of the glands make up **semen**, the fluid that is discharged (ejaculated) from the penis during orgasm. About 5 milliliters (1 teaspoonful) of semen are discharged during a typical ejaculation. About 95% of the fluid consists of glandular secretions. The other 5% is made up of 200 million to 500 million sperm, only one of which may fertilize an egg cell. The other sperm seem to play an accessory role in fertilization, probably causing chemical changes in the environment of the female reproductive tract needed for fertilization to occur.

The human **penis** consists mainly of tissue that can fill with blood to cause an erection during sexual arousal. The erectile tissue is shown in blue in Figures A and B. Erection is essential for insertion of the penis into the vagina. Like the

clitoris, the penis consists of a shaft that supports the glans, or head. The glans is richly supplied with nerve endings and is highly sensitive to stimulation. As in the female, a fold of skin called the prepuce, or foreskin, covers the glans. Circumcision, the surgical removal of the prepuce, is commonly performed for religious or health reasons. However, scientific studies have not proved that circumcision has any effect on a man's health or on that of his sexual partner.

Figure D illustrates the process of ejaculation and summarizes what we have said about the production of semen and its expulsion. Ejaculation occurs in two stages. At the peak of sexual arousal, muscles in the epididymis, seminal vesicles, prostate gland, and vas deferens contract (upper drawing). These contractions force secretions from the glands into the vas deferens and propel sperm from the epididymis. At the same time, a sphincter muscle contracts, preventing urine from leaking into the urethra from the bladder. Another sphincter contracts and closes off the entrance of the urethra into the penis. The section of the urethra between the two sphincters fills with semen and expands. In the second stage of ejaculation, the expulsion stage (lower drawing), the sphincter at the base of the penis relaxes, admitting semen into the penis. Simultaneously, a series of strong muscle contractions around the base of the penis and along the urethra expels the semen during orgasm.

Sperm and ova form by meiosis

Both sperm and ova are haploid and develop by meiosis from diploid cells in the gonads. Before we turn to the formation of gametes, you may want to review Modules 8.12 and 8.13 as background for our discussion.

Spermatogenesis, the formation of sperm cells, takes about 65–75 days in the human male. Figure A outlines sper-matogenesis. Though the diploid chromosome number in humans is 46 (that is, $2n = 46$), for the sake of simplicity, the figure illustrates only four diploid chromsomes ($2n = 4$).

Sperm develop in the testes in coiled tubes called **seminiferous tubules**. Diploid cells that begin the process are located near the outer wall of

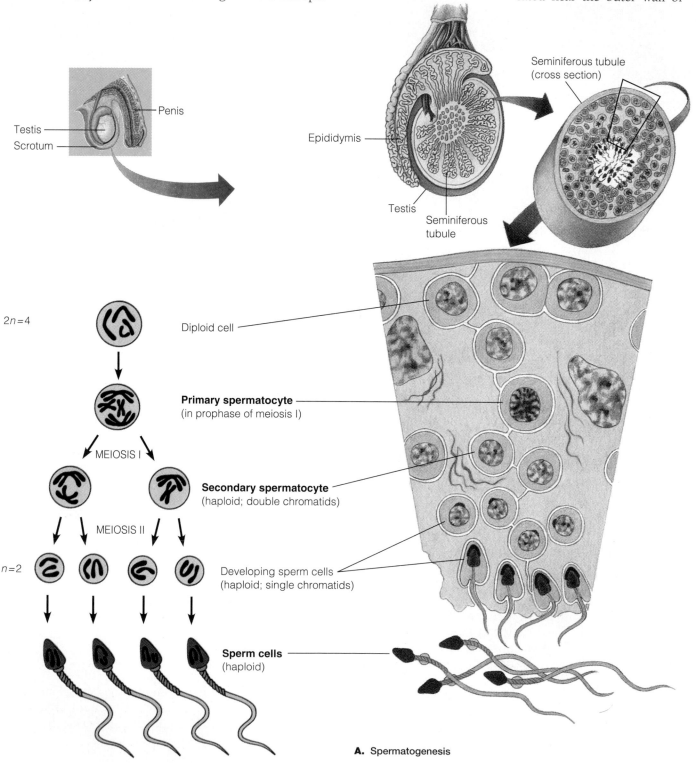

Penis

Testis

Scrotum

Epididymis

Testis

Seminiferous tubule

Seminiferous tubule (cross section)

$2n=4$

Diploid cell

Primary spermatocyte
(in prophase of meiosis I)

MEIOSIS I

Secondary spermatocyte
(haploid; double chromatids)

MEIOSIS II

$n=2$

Developing sperm cells
(haploid; single chromatids)

Sperm cells
(haploid)

A. Spermatogenesis

the tubules (at the top of the enlarged wedge of tissue in Figure A). These cells multiply constantly by mitosis, and each day about 3 million of them develop into **primary spermatocytes,** the cells that undergo meiosis. Meiosis I of a primary spermatocyte produces two **secondary spermatocytes,** each with the haploid number of chromosomes (*n*). The chromosomes are still in their duplicated state, each consisting of two identical chromatids. Meiosis II then forms four cells, each with the haploid number of single-chromatid chromosomes. A sperm cell develops from each of these haploid cells and is gradually pushed toward the center of the seminiferous tubule. From there it passes into the epididymis, where it matures and becomes motile. Sperm is stored in the epididymis until ejaculation.

Figures B and C show **oogenesis,** the development of an ovum, which occurs in the ovary. As we indicated in Module 27.3, a female's ovary at birth contains all the follicles (ovum-forming bodies) she will ever have. Oogenesis actually begins prior to birth, when a diploid cell in each developing follicle begins meiosis. At birth, each follicle contains a dormant **primary oocyte,** a diploid cell that is resting in prophase of meiosis I. A primary oocyte can be hormonally triggered to develop into an ovum. After puberty, about every 28 days, FSH (follicle-stimulating hormone) from the pituitary stimulates one of the dormant follicles to develop. The follicle enlarges, and the primary oocyte completes meiosis I. In the female, the division of the cytoplasm in meiosis I is unequal, with a single **secondary oocyte** receiving almost all of it. The smaller of the two daughter cells, called the **first polar body,** receives almost no cytoplasm.

The secondary oocyte is the stage released by the ovary during ovulation. It enters the oviduct, and if a sperm cell penetrates it, the secondary oocyte undergoes meosis II. Meiosis II yields a **second polar body** and the actual ovum. The haploid nucleus of the ovum can then fuse with the haploid nucleus of the sperm cell, producing a zygote.

As Figure B indicates, the first polar body also undergoes meiosis II, forming two cells. These and the second polar body receive virtually no cytoplasm and degenerate soon after the ovum forms. Polar body formation allows the ovum to acquire nearly all the cytoplasm and thus the bulk of the nutrients contained in the original diploid cell.

Figure C is a cutaway view of an ovary. The series of follicles here represent the changes one follicle undergoes over time; the arrows indicate the sequence. An actual ovary would have thousands of dormant follicles, each containing a primary oocyte. Usually, only one follicle has a dividing oocyte at any one time, and as it develops, that follicle stays in one place in the ovary. Meiosis I occurs as the follicle matures. About the time the secondary oocyte forms, the pituitary hormone LH (luteinizing hormone) triggers ovulation, the expulsion of the secondary oocyte from the follicle. The ruptured follicle then develops into a corpus

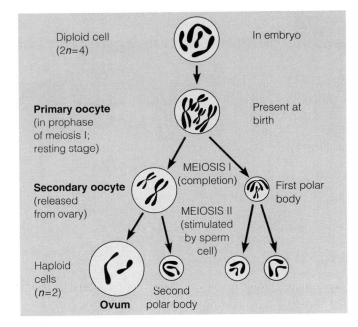

B. Meiosis in oogenesis

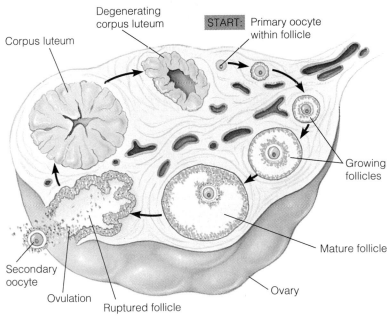

C. Development of an ovarian follicle

luteum. Unless fertilization occurs, the corpus luteum degenerates before another follicle starts to develop.

Oogenesis and spermatogenesis are alike in that they produce haploid gametes. However, these two processes differ in three important ways. First, only one ovum results from each diploid parent cell that undergoes oogenesis, whereas four sperm cells result from each parent cell that undergoes spermatogenesis. Second, an ovary at birth contains all the primary oocytes it will ever have, whereas the testes produce new primary spermatocytes throughout the male's reproductive years. Third, oogenesis is not completed without stimulation from a sperm cell, whereas spermatogenesis produces mature sperm in an uninterrupted sequence.

Hormones synchronize cyclic changes in the ovary and uterus

Oogenesis is one part of a female mammal's reproductive cycle, a recurring sequence of events that produces gametes, makes them available for fertilization, and, as we will see, prepares the body for pregnancy. In our discussion of oogenesis in Module 27.4, we described the cyclic events that occur about every 28 days in the human ovary; this **ovarian cycle** is represented in part 3 of the figure on the facing page. As the figure indicates, hormonal messages synchronize the ovarian cycle with related events in the uterus called the **menstrual cycle** (part 5). The hormone story (parts 1, 2, and 4) is complex and involves intricate feedback controls. Therefore, we'll need to move up and down the figure to follow the actions of the hormones. At the end of this module, you will be able to use the figure from top to bottom to review the female reproductive cycle in chronological order.

We begin our discussion with the straightforward, structural events of the ovarian and menstrual cycles. For simplicity, we have divided the ovarian cycle (part 3 of the figure) into two phases separated by ovulation: the pre-ovulatory phase, when a follicle is growing and a secondary oocyte is developing, and the post-ovulatory phase, after the follicle has become a corpus luteum.

Events in the menstrual (or uterine) cycle (part 5) occur in tune with the ovarian cycle. By convention, the first day of a woman's "period" is designated as day 1 of the menstrual cycle. Uterine bleeding, called **menstruation**, usually persists for 3 to 5 days. Notice that this corresponds to the beginning of the pre-ovulatory phase of the ovarian cycle. During menstruation, the endometrium (inner lining of the uterus) breaks down and leaves the body through the vagina. The menstrual discharge consists of blood, small clusters of endometrial cells, and mucus. After menstruation, the endometrium regrows. It continues to thicken through the time of ovulation, reaching a maximum at about 20 to 25 days. If an embryo has not implanted in the uterine lining by this time, menstruation begins again, marking the start of the next ovarian and menstrual cycles.

Now let's look at the hormones that regulate the ovarian and menstrual cycles. The ebb and flow of the five hormones listed in the table on this page synchronize the growth of the follicle and ovulation with the preparation of the uterine lining for possible implantation of an embryo. A releasing hormone from the hypothalamus of the brain regulates secretion of the two pituitary hormones FSH and LH. The blood levels of FSH, LH, and two other hormones—estrogen and progesterone—coincide with specific events in the ovarian and menstrual cycles.

Hormones of the Ovarian and Menstrual Cycles		
Hormone	**Secreted by**	**Major Roles**
Releasing hormone	Hypothalamus	Regulates secretion of LH and FSH by pituitary
FSH	Pituitary	Stimulates growth of ovarian follicle
LH	Pituitary	Stimulates growth of ovarian follicle and production of secondary oocyte; promotes ovulation; promotes development of corpus luteum and secretion of hormones
Estrogen	Ovarian follicle	Low levels inhibit pituitary; high levels stimulate hypothalamus; promotes endometrium
Estrogen and progesterone	Corpus luteum	Maintain endometrium; high levels inhibit hypothalamus and pituitary; sharp drops promote menstruation

The ovarian cycle is closely tuned to the changing levels of all five hormones. The ovarian cycle begins when rising blood levels of the releasing hormone from the hypothalamus stimulate the anterior pituitary to increase its output of FSH and LH (part 1 of the figure). The graph in part 2 indicates that the pituitary releases relatively small quantities of LH and FSH during most of the pre-ovulatory phase of the ovarian cycle. True to its name, FSH stimulates the growth of an ovarian follicle. In turn, the follicle secretes estrogen. Early in the pre-ovulatory phase, the follicle is small (part 3) and secretes relatively little estrogen (part 4). As the follicle grows, it secretes more and more estrogen, and the rising but still relatively low level of estrogen exerts negative feedback on the pituitary. This keeps the blood levels of FSH and LH low for most of the pre-ovulatory phase (part 2). As the time of ovulation approaches, hormone levels change drastically, with estrogen reaching a critical peak (part 4) just before ovulation. This high level of estrogen exerts positive feedback on the hypothalamus, which then makes the pituitary secrete bursts of FSH and LH. By comparing parts 2 and 4 of the figure, you can see this response as the peaks in FSH and LH that occur just after the estrogen peak.

The role of FSH after a follicle matures is unknown, but the massive surge of LH has pronounced effects. It stimulates the completion of meiosis, transforming the primary oocyte in the follicle into a secondary oocyte. It also signals enzymes to rupture the follicle, allowing ovulation to occur, and triggers development of the corpus luteum from the ruptured follicle (hence its name, luteinizing hormone). LH also promotes the secretion of progesterone and estrogen by the corpus luteum. In part 4 of the figure, you can see the progesterone peak and the second (lower and wider) estrogen peak after ovulation.

High levels of estrogen and progesterone in the blood following ovulation exert a strong influence on both ovary and uterus. The combination of the two hormones exerts negative feedback on the hypothalamus and pituitary, producing the marked declines in FSH and LH levels shown on the right side of the graph in part 2. The declines of FSH and LH prevent follicles from developing and ovulation from occurring during the post-ovulatory phase. Also, the LH drop is followed by the gradual degeneration of the corpus luteum. Near the end of the post-ovulatory phase, unless an embryo has implanted in the uterus, the corpus luteum stops secreting estrogen and progesterone. As the blood levels of these hormones drop, the hypothalamus once again can stimulate the pituitary to secrete more FSH and LH, and a new cycle begins.

Hormonal regulation of the menstrual cycle is simpler than that of the ovarian cycle. The menstrual cycle (part 5) is directly controlled by the levels of only estrogen and progesterone. You can see the effects of these hormones by comparing parts 4 and 5 of the figure. Starting at about day 5 of the cycle, the endometrium thickens in response to the increasing levels of estrogen and, later, progesterone. When the levels of these hormones drop (right side of part 4), the endometrium begins to slough off. Menstrual bleeding begins soon thereafter, on day 1 of a new cycle.

To this point, we have described what happens in the human ovary and uterus in the absence of fertilization. As we'll see later, the ovarian and menstrual cycles are put on hold if fertilization and pregnancy occur. Early in pregnancy, the developing embryo, implanted in the endometrium, releases a hormone that acts like LH. The hormone maintains the corpus luteum, which continues to secrete progesterone and estrogen, keeping the endometrium intact. We will return to the hormones of pregnancy in Modules 27.16 and 27.17.

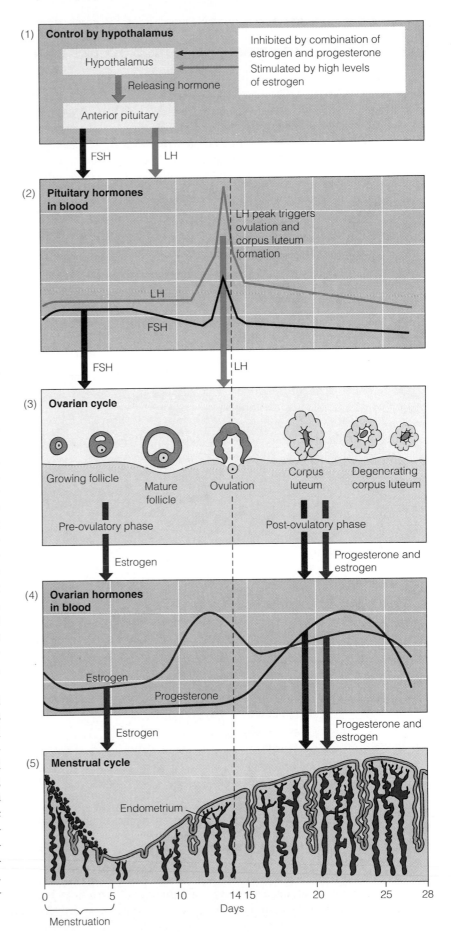

27.6 The human sexual response occurs in four phases

Most female mammals are receptive to males only on certain days—in many species, for only a brief period once or a few times a year. A female deer or bear, for example, will mate only during a few weeks in the autumn. During specific mating times, a female is said to be in estrus, meaning she is at her peak of sexual readiness. This is the only time she ovulates and the only time her uterus is primed for implantation.

Humans and several other primates are unusual in having no distinct mating periods; females are potentially receptive to males throughout the year. In humans, the sexual behavior called "making love" may have evolved as a way to strengthen the bond between mates, as well as to promote the union of egg and sperm. Our sexuality is emotional as well as physical, and we have highly varied sexual expression.

The physical events of the human sexual response occur in a sequence of four phases. During the excitement phase, sexual passion builds, and the penis and clitoris become erect, as do the testes, labia, and nipples. The vagina secretes lubricating fluid, and muscles tighten in the arms and legs. These responses continue during the plateau phase, which is marked by increases in breathing and heart rates. **Orgasm** follows, characterized by rhythmic contractions of the reproductive structures, extreme pleasure for both partners, and ejaculation by the male. The resolution phase reverses the previous responses; the structures return to normal size, muscles relax, and passion subsides.

27.7 Sexual activity can transmit disease

Speaking to the widespread fear of AIDS, one advertisement for condoms reads, "I enjoy sex, but I'm not willing to die for it." We discussed the importance of safer sex as a deterrent to AIDS in Chapter 24's introduction. AIDS is only one of many **sexually transmitted diseases (STDs)**, contagious diseases spread by sexual contact. Latex condoms can usually prevent STD spread. Viral STDs are not curable, but the bacterial, protozoal, and fungal STDs generally are. Usually both partners must be treated to prevent reinfection.

STDs are epidemic throughout the world. Many of these diseases cause long-term problems if they are not treated. A condition called pelvic inflammatory disease (PID) is a common secondary result of STDs caused by bacterial infections in women. In PID, bacteria spread from the vagina into the uterus, oviducts, and ovaries, and may cause acute pain, scarring, and sterility. The table here lists most of the STDs common in the U.S.

STDs Common in the United States

Disease	Microbial Agent	Major Symptoms and Effects	Treatment	Number of Cases (1992)
Chlamydial infections	*Chlamydia trachomatis* (bacterium)	Genital discharge, itching, and/or painful urination; often no symptoms in women; PID	Antibiotics	4,000,000
Gonorrhea	*Neisseria gonorrhoeae* (bacterium)	Genital discharge; painful urination; sometimes no symptoms in women; PID	Antibiotics	501,000
Syphilis	*Treponema pallidum* (bacterium)	Ulcer (chancre) on genitals in early stages; spreads throughout body and can be deadly if not treated	Antibiotics can cure in early stages	113,000
Genital herpes (see Module 10.17)	Herpes simplex virus type 2, occasionally type 1	Recurring symptoms: small blisters on genitals, painful urination, skin inflammation; linked to cervical cancer; miscarriage, birth defects	Acyclovir can shorten duration of symptoms in recurrences	31,000,000
AIDS and HIV infection	HIV (virus)	See Module 24.18		1,000,000 carry virus
Trichomoniasis	*Trichomonas vaginalis* (protozoan)	Vaginal irritation, itching, and discharge; usually no symptoms in men	Antiprotozoal drugs	3,000,000
Candidiasis (yeast infections)	*Candida albicans* (fungus)	Similar to symptoms of trichomoniasis; frequently acquired nonsexually	Antifungal drugs	Unknown

27.8 Contraception prevents unwanted pregnancy

Contraception is the prevention of pregnancy. A variety of contraceptive methods are available and used throughout the world today, but only complete abstinence (no intercourse) is totally effective. Other forms of contraception, which have varying degrees of effectiveness, work in one of three ways: (1) preventing the release of gametes from the gonads, (2) preventing fertilization, or (3) preventing the embryo from implanting.

Contraceptive Methods

Method	Pregnancies/100 Women/Year* Used Correctly	Typically
Prevents Release of Gametes		
Birth control pill (combination)	0.1	3
Prevents Fertilization		
Vasectomy	0.1	0.15
Tubal ligation	0.2	0.4
Progestin minipill or implant	0.5	3
Rhythm	1–9	20
Withdrawal	4	18
Condom	2	12
Diaphragm and spermicide	6	18
Cervical cap and spermicide	6	18
Sponge and spermicide	6–9	18–28
Spermicide alone	3	21
Prevents Implantation		
Intrauterine device (IUD)	1–2	3

*Without contraception, about 85 pregnancies would occur.

Methods that prevent the release of gametes are highly effective, as the table above indicates. Birth control pills, now used by over 60 million women worldwide, have been available since the 1960s. The most widely used pills today are combinations of a synthetic estrogen and a synthetic progesteronelike hormone called progestin. "The pill" prevents ovulation and keeps follicles from developing. Its potential side effects include a decreased risk of ovarian and endometrial cancers and an increased risk of cardiovascular problems, especially in women who smoke or have a history of such problems. In rare cases, long-term use of the pill may increase the risk of liver tumors. All things considered, taking the combination pill is far less risky than pregnancy. Death rates associated with pregnancy are about twice those from using the pill.

A second type of birth control pill, called the minipill, contains only progestin. Slightly less effective than the combination pill, the minipill prevents fertilization by altering a woman's cervical mucus so that it blocks sperm from entering the uterus. Norplant®, a time-release capsule that is implanted under a woman's skin, releases progestin into the blood and is effective for five years.

Sterilization prevents conception permanently. In **vasectomy,** in men, a doctor cuts a section out of each vas deferens to prevent sperm from entering the urethra. In **tubal ligation,** in women, a doctor cuts a short section out of each oviduct (and may tie, or ligate, the remaining ends) to prevent eggs from reaching the uterus. Preliminary research has suggested a link between vasectomy and increased risk of prostate cancer, but both forms of sterilization are still considered relatively safe and free from side effects. They are difficult to reverse, however, and should be considered permanent.

The effectiveness of other methods that prevent fertilization depends on how they are used. Temporary abstinence, often called the **rhythm method,** depends on refraining from intercourse during the few days before and after ovulation, when conception is most likely. The rhythm method is generally not reliable because it is difficult to predict or detect the time of ovulation. Equally unreliable is **withdrawal** of the penis from the vagina before ejaculation. Even before ejaculation, sperm may be present in the penis and may be deposited in the vagina.

Barrier methods are more effective in preventing sperm and egg from meeting. The condom is a thin sheath, usually made of latex, that fits over the penis to collect the semen. The diaphragm is a dome-shaped rubber cap that covers the cervix and is inserted before intercourse; it stops the sperm from moving into the uterus and oviducts. The cervical cap, which fits closely over the cervix, and the contraceptive sponge, which is inserted into the vagina, also work as barriers. However, to achieve the maximum effectiveness, diaphragms, cervical caps, and sponges must be used with **spermicides,** sperm-killing chemicals in the form of a cream, jelly, or foam. Spermicides alone are less effective.

For preventing implantation, intrauterine devices (IUDs) can be used. IUDs are small, usually plastic objects that a doctor inserts into the uterine cavity. They probably work by irritating the endometrium. Unfortunately, they may cause bleeding, uterine infection, or uterine perforation.

Contraception in all its forms is a uniquely human invention that allows us to enjoy sexual intimacy without reproducing. Let's now rejoin our main subject and follow what happens in humans and other animals when an egg and sperm actually meet.

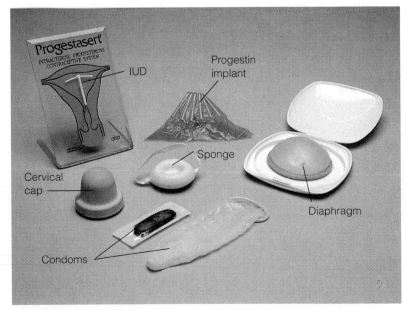

Some contraceptive devices

Fertilization results in a zygote and triggers embryonic development

Embryonic development begins with **fertilization**, the union of a sperm and an egg to form a diploid zygote. Fertilization introduces the sperm's haploid set of chromosomes into the egg and also activates the egg by triggering metabolic changes that start embryonic development.

Figure A is a micrograph of an unfertilized human egg almost covered by sperm. Of all these sperm, only a single one will enter and fertilize the egg. All the other sperm—the ones shown here and millions more that were ejaculated with them—will die. The sperm that penetrates the egg gains a chance to have its unique set of genes combine with those of the egg and contribute to the next generation.

Figure B illustrates the structure of a mature human sperm. Here is another case of form fitting function. The sperm's streamlined shape is an adaptation for swimming through fluids in the vagina, uterus, and oviduct of the female. The sperm cell's thick head contains a haploid nucleus and is tipped with a membrane-bounded sac, the **acrosome**, which lies just inside the plasma membrane. The acrosome contains enzymes that help the sperm penetrate the egg. The neck and middle piece of the sperm contain a long, spiral mitochondrion. The sperm absorbs high-energy nutrients, especially the sugar fructose, from the semen. Thus fueled, its mitochondrion provides ATP for movement of the tail, which is actually a flagellum. By the time a sperm has reached the egg, it has consumed much of the energy available to it. But a successful sperm will have enough energy left to enter the egg and deposit its nucleus in the egg's cytoplasm.

Figure C illustrates the sequence of events in fertilization. This diagram is based on fertilization in sea urchins, on which a great deal of research has been done. Similar processes occur in other animals, including humans. The diagram traces one sperm through the successive activities of fertilization. Notice that, to reach the egg nucleus, the sperm nucleus must pass through three barriers: the egg's jelly coat (yellow), a middle region of glycoproteins called the vitelline layer (pink), and the egg cell's plasma membrane (black line).

Let's follow the steps shown in the figure. As a sperm ① approaches and then ② contacts the jelly coat of the egg, the acrosome in the sperm head releases a cloud of enzyme molecules that digest a cavity in the jelly. When the sperm head reaches the vitelline layer, ③ species-specific protein

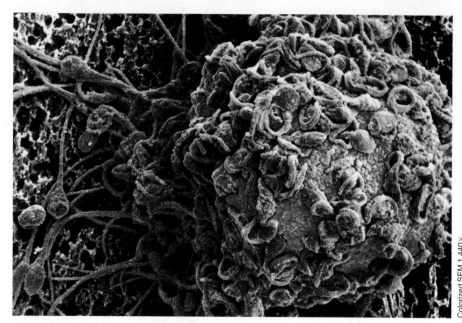

A. A human egg cell surrounded by sperm

Colorized SEM 1,440×

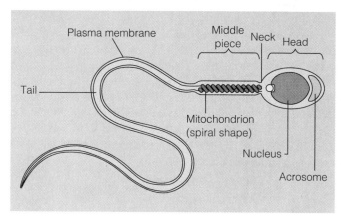

B. Structure of a human sperm cell

molecules on its surface bind with specific receptor proteins on the vitelline layer. The specific binding between the proteins of the sperm and egg ensures that sperm of other species cannot fertilize the egg. This specificity is especially important when fertilization is external, because the sperm of other species may be present in the water. After the specific binding occurs, the sperm proceeds through the vitelline layer, and ④ the sperm's plasma membrane fuses with that of the egg. Fusion of the two membranes makes it possible for ⑤ the sperm nucleus to enter the egg.

Fusion of the sperm and egg plasma membranes triggers a number of important changes in the egg. Two such changes prevent other sperm from entering the egg. Less than a second after the membranes fuse, the entire egg plasma membrane becomes impenetrable to other sperm cells. Shortly thereafter, ⑥ the vitelline layer hardens and separates from the plasma membrane. The space quickly fills with water, and the vitelline layer becomes the so-called **fertilization membrane,** another barrier impenetrable to sperm. If these events did not occur and an egg were fertilized by more than one sperm, the resulting zygote nucleus would contain too many chromosomes, and the zygote could not develop normally.

Membrane fusion also triggers a burst of metabolic activity in the egg. In preparation for the enormous growth and development that will follow fertilization, the egg's metabolic machinery suddenly gears up from virtual dormancy. At the same time, ⑦ the egg and sperm nuclei fuse, producing the diploid nucleus of the zygote. In the next module, we begin to trace the development of the zygote into a whole new animal.

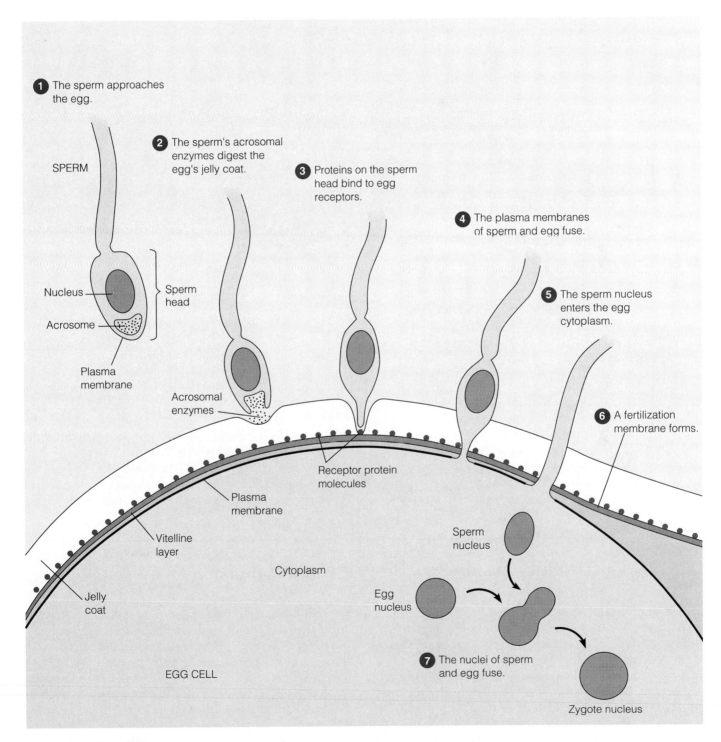

c. The process of fertilization

Cleavage produces a ball of cells from the zygote

An animal is made up of many thousands, millions, even trillions of cells organized into complex tissues and organs. The transformation to this multicellular state from a zygote is truly phenomenal. Order and precision are required at every step. They are clearly displayed in the first two major phases of development: cleavage and gastrulation. We focus on cleavage in this module and gastrulation in the next.

Cleavage is a rapid succession of cell divisions that produces a ball of cells—a multicellular embryo—from the zygote. In most animals, the embryo does not feed or grow larger during cleavage. Nutrients stored in the original egg cell nourish the dividing cells, and the successive cell divisions merely partition the zygote into many smaller cells.

The figure here illustrates cleavage in a sea urchin. As the first three steps show, the number of cells doubles with each cleavage division. In a sea urchin, a doubling occurs about every 20 minutes, and the whole cleavage process takes about 3 hours to produce a solid ball of cells. Notice that each cell in the ball is much smaller than the zygote. As cleavage continues, a fluid-filled cavity called the **blastocoel** forms in the center of the embryo. At the completion of cleavage, there is a large cavity surrounded by one or more layers of cells. This hollow ball of cells is called the **blastula.**

Cleavage makes two very important contributions to early development. By decreasing the size of the embryo's cells, it increases the surface-to-volume ratio of each cell, thereby enhancing oxygen uptake and other important exchanges with the environment. Cleavage also partitions the embryo into developmental regions. As developmental biologists are beginning to discover, the cytoplasm of the zygote contains a variety of chemicals that control where specific parts of the embryo will develop.

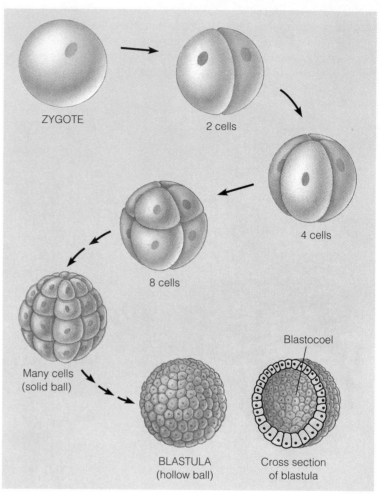

ZYGOTE

2 cells

4 cells

8 cells

Many cells (solid ball)

Blastocoel

BLASTULA (hollow ball)

Cross section of blastula

After cleavage, these chemicals are localized in particular groups of cells, where they activate the genes that direct the formation of specific parts of the animal.

Cleavage creates a multicellular embryo, the blastula, from a single-celled zygote. Gastrulation, the next phase of development, is mainly an organizing process.

Gastrulation produces a three-layered embryo

Gastrulation, the second major phase of development, adds more cells to the embryo, but more importantly, it sorts all the cells into distinct cell layers. In the process, the embryo is transformed from a hollow ball of cells—the blastula—into a three-layered stage called the **gastrula.**

The three layers produced in gastrulation are embryonic tissues called **ectoderm, endoderm,** and **mesoderm.** The ectoderm forms the outer layer (skin) of the gastrula. The endoderm forms an embryonic digestive tract. And the mesoderm partly fills the space between the ectoderm and

the endoderm. Eventually, these three cell layers develop into all the parts of the adult animal. For instance, our nervous system and the outer layer (epidermis) of our skin come from ectoderm; the innermost lining of our digestive tract arises from endoderm; and most other organs and tissues, such as the kidney, heart, muscles, and the inner layer of our skin (dermis), develop from mesoderm.

The mechanics of gastrulation vary somewhat depending on the species. We have chosen the frog, a vertebrate that has long been a favorite of researchers, to demonstrate how gastrulation produces the three cell layers. The figure below takes us from a blastula at the top to a three-layered gastrula at the bottom. The diagrams in the left column show an external view of gastrulation; each drawing represents a multicellular embryo, and the arrows indicate cell movements.

The cutaway drawings on the right reveal the internal structures that develop as gastrulation occurs. The timing of these events varies with the species and the temperature of the lake or pond in which the frog develops. In many frogs, cleavage and gastrulation together take about 15–20 hours.

1 **The blastula.** Formed by cleavage, the frog blastula is a partially hollow ball of unequally sized cells. As the cross section shows, the cells toward one end, called the animal pole, are smaller than those near the opposite end, the vegetal pole. The cells near the vegetal pole are larger because they contain yolk granules, which make them divide at a slower rate than those of the animal pole. The three colors on the blastula indicate regions of cells that will give rise to the primary cell layers: ectoderm (blue), endoderm (yellow), and mesoderm (magenta). (Notice that each layer may be more than one cell thick.) In real embryos, these regions have been identified by dyeing the cells with harmless stains and observing where the cells go as development proceeds. A glance ahead through drawings 2–4 will show you that, as gastrulation occurs, the cells that form mesoderm and endoderm move from the surface to the inside of the embryo.

2 **Blastopore formation.** Gastrulation begins when a small groove, called the **blastopore,** appears on one side of the blastula. The blastopore is the place where cells that will form endoderm and mesoderm move inward from the surface. While these cells are moving, other cells, which will form ectoderm, spread over the surface of the embryo.

3 **Cell migration to form layers.** The beginnings of the three layers can now be seen in the cross section. Migrating endodermal cells (yellow) have produced a simple digestive cavity called the **archenteron.** The advancing endoderm and the archenteron have filled some of the space formerly occupied by the blastocoel. Cells that will form the mesoderm (magenta) are located between the endoderm and the ectoderm (blue).

4 **Completion of gastrulation.** Gastrulation is completed when the embryo is three-layered. Ectoderm covers the surface except for a cluster of endodermal cells called the **yolk plug.** The yolk plug marks the site of the blastopore and of the future anus. At this stage, the endoderm and its archenteron have replaced the blastocoel. Mesoderm forms a layer between the ectoderm and the endoderm.

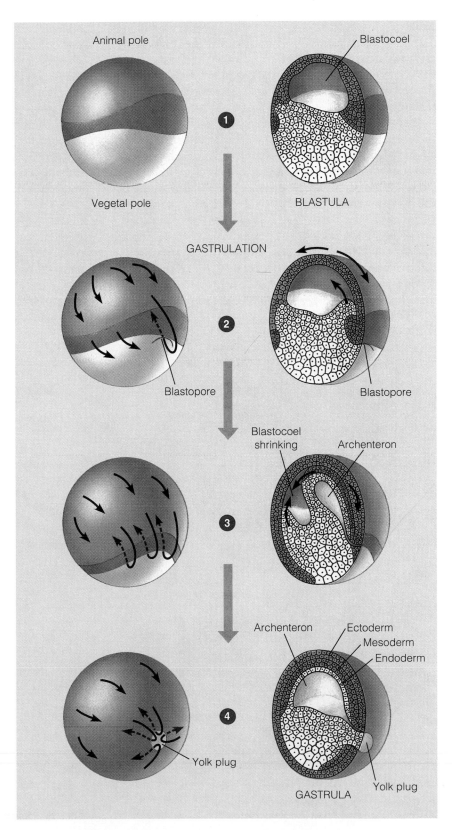

27.12 Organs start to form after gastrulation

In organizing the embryo into three layers, gastrulation sets the stage for the shaping of an animal. Once the ectoderm, endoderm, and mesoderm form, cells in each layer begin to differentiate into tissues and embryonic organs. The cutaway drawing in Figure A shows the developmental structures that appear in a frog embryo a few hours after the completion of gastrulation. The orientation drawing at the upper left of the figure indicates a corresponding cut through an adult frog.

We see two structures in the embryo in Figure A that were not present at the gastrula stage described in Module 27.11. An organ called the notochord has developed in the mesoderm, and a structure that will become the hollow nerve cord is beginning to form in the ectoderm. Recall that the notochord and hollow nerve cord are hallmarks of the chordates.

The notochord is visible in cross section in the drawing in Figure A. It forms from mesoderm just above the archenteron. Made of a cartilagelike substance, the **notochord** extends for most of the embryo's length and provides support for other developing tissues. Later in development, the no-

tochord will function as a core around which mesodermal cells gather and form the frog's backbone.

You can also see in Figure A the beginnings of the frog's hollow nerve cord. The area shown in green in the cutaway drawing is a thickened region of ectoderm called the neural plate. From it arises a pair of pronounced ectodermal ridges, called neural folds, visible in both the drawing and the micrograph below it. If you now look at the series of diagrams in Figure B, you will see what happens as the neural folds and neural plate develop further. The neural plate rolls up and forms the neural tube, which then sinks beneath the surface of the embryo and is covered by an outer layer of ectoderm. The **neural tube** is destined to become the brain and spinal cord.

Figure C shows a later frog embryo (about 12 hours older than the one in Figure A), in which the neural tube has formed. Notice in the drawing that the neural tube lies directly above the notochord. The relative positions of the neural tube, notochord, and archenteron give us a preview of the basic body plan of a frog. The spinal cord will lie within extensions of the dorsal surface of the backbone

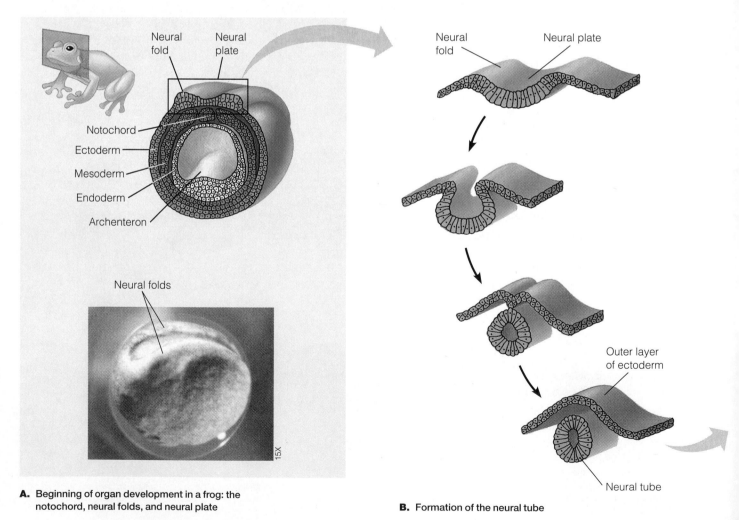

A. Beginning of organ development in a frog: the notochord, neural folds, and neural plate

B. Formation of the neural tube

(which will replace the notochord), and the digestive tract will be ventral to the backbone. We see this same arrangement of organs in all vertebrates.

Figure C shows several other fundamental changes. In the micrograph, which is a side view, you can see that the embryo is more elongated than the one in Figure A. You can also see the beginnings of an eye and a tail (called the tail bud). Part of the ectoderm has been removed to reveal a series of internal ridges called somites. The **somites** are blocks of mesoderm that will give rise to segmental structures, such as the vertebrae and associated muscles of the backbone. In the cross-sectional drawing, you can see that the mesoderm next to the somites is developing a hollow space—the body cavity, or **coelom.** Body parts that are segmented (constructed of repeating units) and a coelom are basic features of chordates (see Module 19.15).

In this and the previous module, we have observed the sequence of changes that occur as an animal begins to take shape. To summarize, the key phases in embryonic development are cleavage (which creates a multicellular animal from a zygote), gastrulation (which organizes the embryo into three discrete layers), and organ formation (differentiation of the three embryonic tissue layers into embryonic organs). These same three phases take place in nearly all animals.

Derivatives of the Three Embryonic Tissue Layers	
Embryonic Layer	**Organs and Tissues in the Adult**
Ectoderm	Epidermis of skin and its derivatives; epithelial lining of mouth and rectum; sense receptors in epidermis; nervous system; adrenal medulla
Endoderm	Epithelial lining of digestive tract (except mouth and rectum); epithelial lining of respiratory system; liver; pancreas; thyroid; parathyroids; thymus; lining of urinary bladder
Mesoderm	Notochord (in animals retaining it as adults); skeletal system; muscular system; circulatory system; excretory system; reproductive system (except germ cells, which differentiate during cleavage); dermis of skin; lining of body cavity; adrenal cortex

If we followed a frog's development beyond the stage represented in Figure C, within a few hours we would be able to monitor muscular responses and a heartbeat, and see a set of gills with blood circulating in them. A long tail fin would grow from the tail bud. The timing of the later stages in frog development varies enormously, but in many species, by 5–8 days after development begins, we would see all the body tissues and organs of a tadpole emerge from cells of the ectoderm, mesoderm, and endoderm. Eventually, the structures of the tadpole (Figure D) would transform into the tissues and organs of an adult frog. The table above lists the major organs and tissues that arise in frogs, and other vertebrates, from each of the three main embryonic tissue layers.

Watching embryos develop helps us appreciate the enormous changes that occur as one tiny cell, the zygote, gives rise to a highly structured, many-celled animal. Your own body, for instance, is a complex organization of some 60 trillion cells, all of which arose from a zygote smaller than the period at the end of this sentence. Discovering how this incredibly intricate arrangement is achieved is one of the greatest challenges of biology. By manipulating embryos in a variety of experiments, developmental biologists have begun to work out the cellular mechanisms that underlie development. We examine some of these mechanisms in the next three modules.

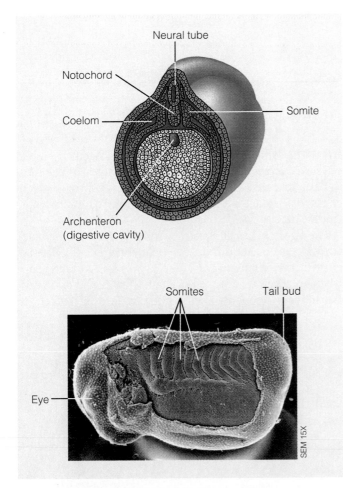

C. An embryo with completed neural tube, somites, and coelom

D. A tadpole

27.13 Cell shape changes, cell migration, and cell aggregation are key processes in early development

The transformation of an animal from a blastula to a gastrula, and many later events in development, result largely from changes in the shapes and locations of embryonic cells—what we might call cellular animation. Figure A shows how two changes in cell shape bring about the formation of the neural tube (see Figure 27.12B). Cells of the ectoderm fold inward by first elongating and then becoming wedge-shaped. The result of these two processes is a tube of ectoderm—the start of the brain and spinal cord. Cell elongation and wedge formation also cause the infolding that occurs at the blastopore when gastrulation begins.

The micrograph in Figure B illustrates another key process in early development: cell migration. The photo

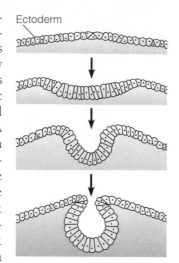

Ectoderm

A. Changes in cellular shape in neural tube formation

shows a cluster of ectodermal cells migrating within a gastrula; you can see pseudopodia, the fingerlike cellular extensions by which the cells move. Many embryonic cells migrate to specific destinations by following chemical trails, which may be secreted by cells near their destination. Some of the cells shown here, for example, are probably following a trail toward the surface of the embryo, where they will give rise to skin cells.

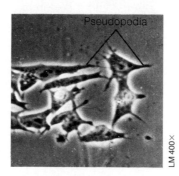

Pseudopodia

LM 400×

B. Migrating cells from an early embryo

Once a migrating cell reaches its destination, specific proteins on its surface allow it to recognize similar cells. The similar cells then aggregate, or clump together, and secrete proteins that glue them in place. Through the process of **differentiation**, the cells then take on the characteristics of a particular tissue.

27.14 Embryonic induction initiates organ formation

When a cell differentiates, certain of its genes are turned on and expressed, while others remain inactive (see Module 11.6). But what determines the particular genes that are expressed—and thus the particular kind of cell that results? Scientists do not yet have a complete answer.

We do know, however, that the result of differentiation is often determined by **embryonic induction,** a mechanism in which one group of cells influences the development of another group of cells. Induction plays a major role in the early development of tissues and organs from ectoderm,

endoderm, and mesoderm. Induction can occur by actual physical contact between two groups of cells or by chemical signals. In either case, the effect is to switch on a set of genes that make the receiving cells differentiate into a specific tissue.

The figure below illustrates a series of inductions (black arrows) that occur during the differentiation of cells that form the vertebrate eye. ① The eye begins to take shape when an outgrowth of the developing brain, the optic vesicle, induces ectoderm on the body surface to thicken. In

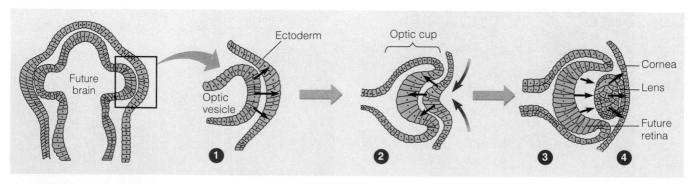

Induction during eye development

turn, ② the thickened ectoderm induces the optic vesicle to form a cup. Then, as the optic cup enlarges and begins to form the retina (the eye's light detector), ③ the cup induces the thickened ectoderm to indent and start forming the lens (which focuses light). Finally, ④ the developing lens induces development of the cornea (the eye's transparent outer covering). Experimenters can make this same series of events occur almost anywhere on the head by surgically implanting

the optic vesicle there. In other words, they can make an eye form where there would normally be skin. Thus, it is clear that the optic vesicle cells initiate eye formation.

Induction plays a role in the early development of virtually all organs and tissues. Learning more about the physical and chemical signals responsible for induction is one of the main goals of modern research in developmental biology.

Pattern formation organizes the animal body

Forming the parts of an eye is one thing, but what about the development of an entire region of the body? An arm and a leg, for instance, have the same kinds of tissues—muscle, connective tissue, cartilage, and skin—but these tissues are arranged somewhat differently in the two limbs. What directs the formation of major body parts?

The shaping of an animal's major parts involves **pattern formation,** the emergence of a body form with specialized organs and tissues all in the right places. Research indicates that the master control genes we discussed in Module 11.13 respond to chemical signals that tell a cell where it is relative to other cells in the embryo. These positional signals determine which master control genes will be expressed and, consequently, which body parts will form.

Figure A indicates how positional signals affect the development of the limbs of vertebrates. Vertebrate limbs develop from embryonic structures called limb buds. Bird wings, for example, develop from the two anterior limb buds. In

order for a wing to form properly, each embryonic wing cell must receive signals specifying its position in three dimensions: How close is it to the embryo's main axis, to the anterior or posterior edge of the developing wing, and to the dorsal or ventral surface of the embryo? Only with this information will the cell's genes direct synthesis of the proteins needed for normal differentiation in that cell's specific location.

Experiments have revealed that vertebrate limbs have zones of cells that provide positional information to other cells via chemical signals. Researchers have located one such pattern-forming zone on the posterior surface of the wing-forming limb buds of birds. Cells nearest the zone—presumably those exposed to the highest concentration of chemical signals from it—develop into posterior wing structures; cells farthest from the zone form anterior structures. As indicated in Figure B, if a block of cells from this zone is removed from one bird embryo (the donor) and grafted onto the anterior part of the limb bud of another embryo (the host), the host will develop additional wing structures—almost a double wing.

A major goal of research on pattern formation and other developmental studies is to learn how the one-dimensional information encoded in the nucleotide sequence of a zygote's DNA directs the development of the three-dimensional form of an animal. In the next two modules, we'll see the results of this process as we watch an individual of our own species take shape.

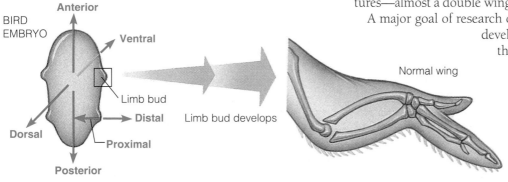

A. Normal development of a wing

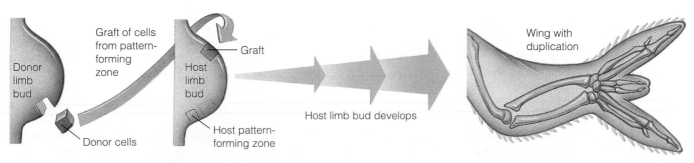

B. Experimental evidence for a pattern-forming zone

Human development: the first month of pregnancy

Pregnancy, or **gestation**, is the condition of carrying developing young within the female reproductive tract. It begins at conception, the fertilization of the egg by a sperm cell, and continues until the birth of the baby. The total time of development from conception to birth is called the gestation period. In humans, the gestation period averages 266 days (38 weeks) from conception. This is a long time compared to some other mammals; mice, for instance, have a gestation period of only about one month. At the other extreme, elephants have a 22-month gestation period.

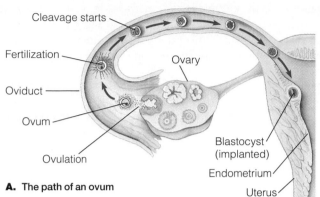

A. The path of an ovum

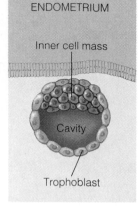

B. Blastocyst (5 days after conception)

The figures on this and the next page illustrate the changes that occur during the first month of human development. The insets at the lower left of Figures C–F show the actual sizes of the embryo at each stage.

Figure A takes us back to ovulation, with the arrows indicating the route of an egg cell, or ovum. Gestation begins when the ovum is fertilized in the oviduct. Cleavage begins about 24 hours later and continues as the embryo moves down the oviduct toward the uterus. By the fourth or fifth day after conception, the embryo has reached the uterus, and cleavage has produced about 100 cells. The embryo is now a hollow ball of cells called a **blastocyst** (the mammalian equivalent of the frog blastula we saw in Module 27.10).

The human blastocyst (Figure B) has a fluid-filled cavity, an inner cell mass that will actually form the baby, and an outer layer of cells called the **trophoblast**. The trophoblast secretes enzymes that enable the blastocyst to implant in the endometrium, the uterine lining (gray in all the figures).

The blastocyst starts to implant in the uterus about a week after conception. In Figure C, you can see extensions of the trophoblast spreading into the endometrium; these are multiplying cells. The trophoblast cells eventually form part of the **placenta**, the organ that provides nourishment and oxygen to the embryo and helps dispose of its metabolic wastes. As we'll see, the placenta consists of both embryonic and maternal tissues.

The actual embryo in Figure C is the two-layered plate of cells (one layer blue, the other yellow) called the **embryonic disc.** The blue cells of the embryonic disc will form the ectoderm, endoderm, and mesoderm of the embryo; the yellow cells will form the yolk sac. In Figure D, gastrulation has started, and mesoderm cells are appearing. By about 16 days after conception (Figure E), gastrulation has produced all three embryonic layers, from which the entire human body will develop.

The human embryo, like that of the lizard in the chapter's introduction—and, in fact, like the embryos of all other reptiles, birds, and mammals—develops with a set of four supportive structures, the **extraembryonic membranes.**

Called the chorion, amnion, allantois, and yolk sac, these structures form a life-support system for the developing embryo. The extraembryonic structures are already starting to form in the 9-day embryo shown in Figure D; they stand out more clearly in Figure E.

The outermost extraembryonic membrane, the **chorion,** becomes the embryo's part of the placenta. Cells in the chorion secrete a hormone called **human chorionic gonadotropin (HCG),** which maintains the corpus luteum of the ovary during the first three months of pregnancy. In turn, the corpus luteum continues to secrete estrogen and progesterone into the mother's blood. Without these hormones, menstruation would occur, and the embryo would abort spontaneously.

You can trace the development of the chorion from the time of implantation; it arises from the trophoblast cells colored dark-orange in Figure C. In Figure D, the chorion shows outgrowths on its outer surface. In Figure E, the chorion's outgrowths, now called **chorionic villi,** are larger and contain mesoderm from the embryo. In Figure F, the chorionic villi contain embryonic blood vessels formed from the mesoderm. By this stage, the placenta is fully developed. Starting with the chorion and extending outward, the placenta is a composite organ consisting of chorionic villi closely associated with the blood vessels of the mother's endometrium. The villi are actually bathed in tiny pools of maternal blood (purple in the figure). There is no direct contact between the mother's blood and the embryo's blood. However, the chorionic villi absorb nutrients and oxygen from the mother's blood, and these materials are passed to the embryo via the chorionic blood vessels colored red. The blue chorionic vessels carry wastes away from the embryo. The wastes diffuse into the mother's bloodstream and are excreted by her kidneys.

The placenta takes care of the embryo's every need. It even allows protective antibodies to pass from the mother to the fetus. Depending on what is circulating in the mother's bloodstream, however, the placenta can also be a source of trouble. A number of viruses—the German measles virus

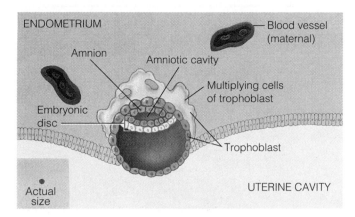

C. Implantation (7 days)

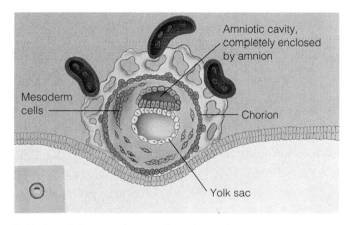

D. Embryonic layers and extraembryonic membranes starting to form (9 days)

and the AIDS virus, for example—can pass across the placenta. German measles can cause serious birth defects; HIV-infected babies usually die of AIDS within a few years. Many harmful chemical substances freely cross the placenta, including the addictive drugs cocaine, heroin, morphine, and amphetamines. Most prescription drugs cross the placenta, and certain ones, such as the antibiotic tetracycline, can cause birth defects. Smoking and drinking alcohol during pregnancy can seriously raise the chances of miscarriage and birth defects. Passage of alcohol across the placenta can cause a set of birth defects called **fetal alcohol syndrome,** which includes mental retardation.

Returning to the other extraembryonic membranes, we see in Figure C that the **amnion** starts forming from ectoderm soon after implantation. It first appears as a dome of cells surrounding a cavity above the embryonic disc. The dome and cavity expand (Figures D and E) and eventually enclose the embryo (Figure F). The amnion protects the embryo. Its cavity is filled with fluid, which absorbs shocks and prevents the embryo from drying out. The amnion usually breaks just before childbirth, and the amniotic fluid (commonly called water) leaves the mother's body through her vagina.

As you can tell from Figure D, the **yolk sac** develops from part of the embryonic disc. In humans and most other mammals, it contains no yolk, but it is given the same name as the homologous structure in other vertebrates. In a bird or reptile

egg, the yolk sac contains a large mass of yolk. Isolated within a shelled egg outside the mother's body, a developing bird or reptile obtains nourishment from the yolk rather than from a placenta. In humans and other mammals, the yolk sac, which remains small, has other important functions. It produces the embryo's first blood cells and its first germ cells, those that will give rise to the gamete-forming cells in the gonads.

You can see the fourth extraembryonic membrane, the **allantois,** developing as an extension of the yolk sac in Figure E. In birds and reptiles, the allantois expands around the embryo and becomes important in waste disposal. In humans and other mammals, the allantois remains small and forms part of the umbilical cord—the lifeline between the embryo and the placenta. It also forms part of the embryo's urinary bladder.

The embryo shown in Figure F is about one month old. In the next module, photographs illustrate this stage and the rest of human development in the uterus.

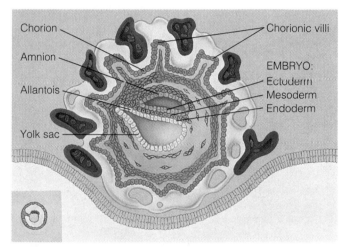

E. Three-layered embryo and four extraembryonic membranes (16 days)

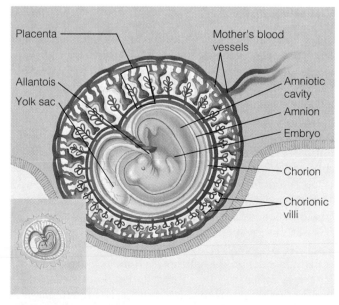

F. Placenta formed (31 days)

Human development from conception to birth is divided into three trimesters

For convenience in studying human development, we divide the period from conception to birth into three **trimesters** of about three months each. The first trimester is the time of most radical change.

FIRST TRIMESTER

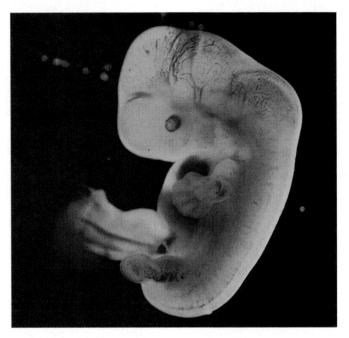

A. 5 weeks

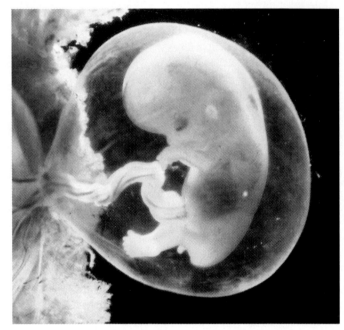

B. 9 weeks

The photograph in Figure A shows a human embryo about one month after fertilization. In that brief time, this highly organized multicellular embryo has developed from a single cell. Not shown here are the extraembryonic membranes that surround the embryo or most of the umbilical cord that attaches it to its life-support organ, the placenta. A month-old human embryo is about 7 mm (0.28 in) long and has a number of features in common with the somite stage of a frog embryo (see Figure 27.12C). The embryo has a notochord and a coelom, both formed from mesoderm. Its brain and spinal cord have begun to take shape from a tube of ectoderm, as in the frog. The human embryo also has four stumpy limb buds, a short tail, and elements of gill pouches. The gill pouches appear during embryonic development in all chordates; in land vertebrates, they eventually develop into parts of the throat and middle ear. Overall, a month-old human embryo is similar to other vertebrates at the somite stage of development.

Figure B shows a developing human, now called a fetus, about 9 weeks after fertilization. The large pinkish structure on the left is the placenta, attached to the fetus by the umbilical cord. The clear sac around the fetus is the amnion. By this time, the fetus is decidedly human, rather than generally vertebrate. It is about 5.5 cm (2.2 in) long and has all of its organs and major body parts, including a disproportionately large head. The somites have developed into the segmental muscles and the bones of the back and ribs. The limb buds have become tiny arms and legs with fingers and toes.

Beginning at about 9 weeks, the fetus can move its arms and legs, turn its head, frown, and make sucking motions with its lips. By the end of the first trimester, the fetus looks like a miniature human being, although its head is still oversized for the rest of the body. The sex of the fetus is usually evident at this time.

SECOND TRIMESTER

The main developmental changes during the second and third trimesters involve an increase in size and general refinement of the human features—nothing as dramatic as the changes of the first trimester. The photograph in Figure C shows a fetus at 14 weeks, 2 weeks into the second trimester. The fetus is now about 6 cm (2.4 in) long. During the second trimester, the placenta takes over the task of maintaining itself by secreting progesterone, rather than receiving it from the corpus luteum. At the same time, the placenta stops secreting HCG, and the corpus luteum, no longer needed to maintain pregnancy, degenerates.

At 20 weeks (Figure D), well into the second trimester, the fetus is about 19 cm (7.6 in) long, weighs about half a kilogram (1 lb), and has the face of an infant, complete with eyebrows and eyelashes. Its arms, legs, fingers, and toes have lengthened. It also has fingernails and toenails, and is covered with fine hair. By this time, the fetal heartbeat is readily detected, and the fetus usually is quite active. The mother's abdomen has become markedly enlarged, and she may often feel her baby move. Because of the limited space in the uterus, the fetus flexes forward into the so-called fetal position. By the end of the second trimester (about 30 weeks), the fetus's eyes are open, and its teeth are forming.

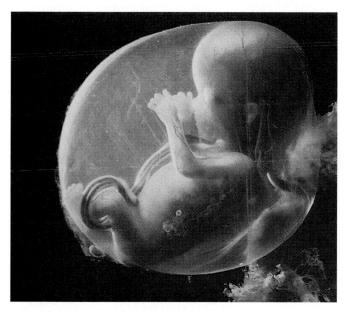

C. 14 weeks

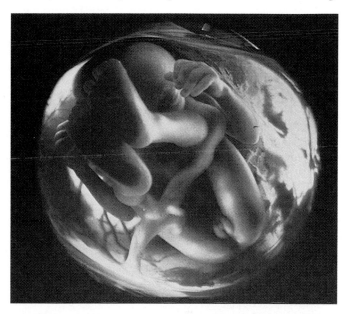

D. 20 weeks

THIRD TRIMESTER

The third trimester (24 weeks to birth, Figure E) is a time of rapid growth, as the fetus gains the strength it will need to survive outside the protective environment of the uterus. Babies born prematurely—as early as the 24th week—may survive, but they require special medical care after birth. During the third trimester, the fetus's circulatory system and respiratory system undergo changes that will allow the switch to air breathing (see Module 22.12). The fetus also gains the ability to maintain its own temperature, its bones begin to harden, and its muscles thicken. It also loses much of its fine body hair, except on its head. The head itself changes its proportions. The fetus becomes less active, as it fills the space in the uterus. At birth, a typical baby is about 50 cm (20 in) long and weighs 2.7–4.5 kg (6–10 lb).

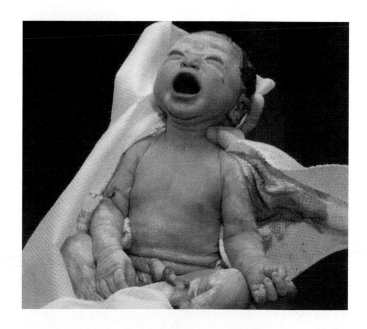

E. At birth

Childbirth is hormonally induced and occurs in three stages

The birth of a child is brought about by a series of strong, rhythmic contractions of the uterus, commonly called **labor.** As illustrated in Figure A, hormones play a role in inducing labor. One of the hormones, estrogen, reaches its highest level in the mother's blood during the last weeks of pregnancy. A key effect of all this estrogen is to trigger the formation of numerous oxytocin receptors on the uterus. Cells of the fetus produce the hormone oxytocin, and later in pregnancy, the pituitary gland produces it in increasing amounts. Oxytocin is a powerful stimulant for the smooth muscles in the wall of the uterus, causing them to contract. It also stimulates the placenta to make prostaglandins, local tissue regulators that also stimulate uterine muscle cells, making them contract even more.

The hormonal induction of labor involves positive-feedback control. In this case, oxytocin and prostaglandins cause uterine contractions that in turn stimulate the release of more and more oxytocin and prostaglandins. The result is climactic—the intense muscle contractions that propel a baby from the womb.

Figure B shows the three stages of labor. As the process begins, the cervix (neck of the uterus) gradually opens, or dilates. ① The first stage of childbirth, dilation, is the time from the onset of labor until the cervix reaches its full dilation of about 10 cm. Dilation is the longest stage of labor, lasting 6–12 hours or even considerably longer.

② The period from full dilation of the cervix to delivery of the infant is called the expulsion stage. Strong uterine contractions, lasting about 1 minute each, occur every 2–3 minutes, and the mother feels an increasing urge to push or bear down with her abdominal muscles. Within a period of 20 minutes to an hour or so, the infant is forced down and out of the uterus and vagina. An attending physician or midwife clamps and cuts the umbilical cord after the baby is expelled.

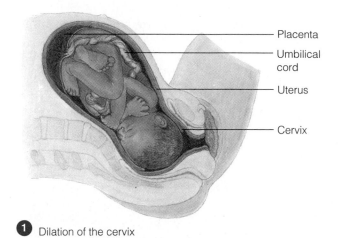

- Placenta
- Umbilical cord
- Uterus
- Cervix

① Dilation of the cervix

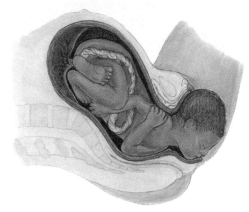

② Expulsion: delivery of the infant

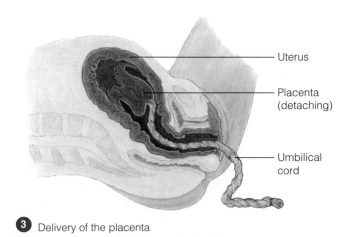

- Uterus
- Placenta (detaching)
- Umbilical cord

③ Delivery of the placenta

B. The three stages of labor

ESTROGEN — from ovaries

OXYTOCIN — from fetus and pituitary

Induces oxytocin receptors on uterus

Stimulates uterus to contract

Stimulates placenta to make PROSTAGLANDINS

Stimulate more contractions of uterus

Positive feedback

A. The hormonal induction of labor

③ The final stage of labor is the delivery of the placenta, which is usually accomplished within 15 minutes after the birth of the baby.

Hormones continue to play important roles after delivery. Decreasing levels of progesterone and estrogen allow the uterus to start returning to its prepregnancy state. Less progesterone in the maternal blood also allows the pituitary hormone prolactin to promote milk production by the mammary glands. About 2–3 days after birth, the mother begins to secrete milk under the direct influence of both oxytocin and prolactin.

Reproductive technology increases reproductive options 27.19

Today, most couples have a choice about whether or not, and when, to have children. The choice is not available, however, to millions of couples who are unable to conceive children because of one or more physical problems. More often than not, the man is infertile; his testes may not produce enough sperm, or his sperm may not be vigorous enough to swim through the uterus and vagina and fertilize an egg. Female infertility can result from a failure to ovulate or from a blockage in the oviducts, preventing egg and sperm from meeting. Sexually transmitted diseases often produce scar tissue that blocks the oviducts. Some women also have antibodies that immobilize sperm in the uterus.

Reproductive technology can solve a number of infertility problems. Hormone therapy will sometimes increase sperm or egg production. Surgery can correct certain disorders, such as blocked oviducts. For some couples, it is simply a matter of timing sexual activity to maximize the chance for fertilization to occur. But for as many as half the couples who seek medical help, treatment is not effective or the cause of infertility remains unknown.

A breakthrough in reproductive technology has been **in vitro fertilization.** In the first step of this procedure, a woman whose oviducts are blocked has ova surgically removed from her ovaries. The ova are then mixed with sperm in culture dishes (in vitro means "in glass" in Latin). The photograph here shows a fertilized human ovum that is undergoing cleavage in vitro. About 2 days after fertilization, when it has divided to form about 8 cells, it is carefully inserted into the woman's uterus and allowed to implant. Embryos can also be frozen in advance to be used later if the first attempt is aborted.

In vitro fertilization is now performed in major medical centers throughout the world. It is very expensive—usually $5000 or more per attempt—but hundreds of children have been conceived this way. In no case has there been any evidence of abnormalities.

A spin-off of in vitro fertilization is surrogate motherhood. A woman may be able to conceive but unable to carry a fetus. The blastocyst may fail to implant in her uterus, or she may have repeated miscarriages. In such cases, a couple may produce an embryo by in vitro fertiliza-

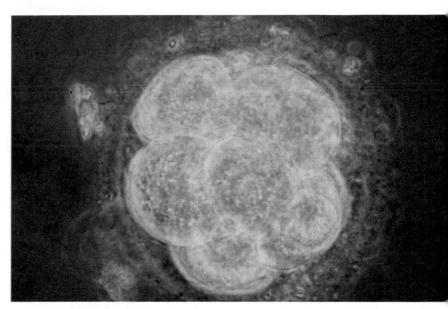

A human embryo in culture

tion and then enter into a legal contract with another woman to become a surrogate mother—to have the embryo implanted in her uterus and carry it through to birth. This method has worked in a number of cases, but serious ethical and legal problems can arise if a surrogate mother changes her mind and wants to keep the baby she has carried for 9 months. There is also the question of who is at fault if the child is born with genetic defects. In the United States, about one-third of the states have laws restricting surrogate motherhood.

Surrogate motherhood is one of many important social issues that center on human reproduction. In coming years, we can expect debate over reproductive issues to intensify. Where do you stand on these issues?

* * *

In this chapter, we have considered the structural and functional bases of animal reproduction and some of the mechanics of embryonic development. We have watched a single-celled product of sexual reproduction, the zygote, become transformed into a new organism, complete with all organ systems. One of the first of those organ systems to develop is the nervous system. In the next chapter, we see how the nervous system functions together with the endocrine system to regulate virtually all body activities.

Chapter Review

Begin your review by rereading the module headings and scanning the figures before proceeding to the Chapter Summary and questions.

Chapter Summary

27.1 Reproduction is required for the continuity of a species. In asexual reproduction, one parent produces offspring by budding, fragmentation, or development from unfertilized eggs. Sexual reproduction involves the fusion of gametes from two parents. Asexual reproduction enables a single individual to produce many offspring rapidly. Sexual reproduction increases the variability of offspring, which may enhance reproductive success in changing environments.

27.2 In humans, the reproductive system consists of a pair of ovaries in females or testes in males, ducts that carry gametes, and structures for copulation. A woman's ovaries contain follicles that nurture eggs and produce sex hormones. Oviducts convey eggs to the uterus, where a fertilized egg develops. The uterus opens into the vagina, which receives the penis during intercourse and forms the birth canal.

27.3 A man's testes produce sperm, which are expelled through ducts during ejaculation. Several glands, including the prostate, contribute to the formation of fluid that carries, nourishes, and protects sperm. This fluid and the sperm constitute semen.

27.4 Spermatogenesis and oogenesis produce sperm and ova, respectively. Primary spermatocytes are made continuously in the testes; each diploid cell undergoes meiosis to form four haploid sperm. A woman's ovaries contain her lifetime supply of primary oocytes at birth. Each month, one matures to form a secondary oocyte, which, if penetrated by a sperm cell, becomes the haploid ovum.

27.5 Hormones synchronize cyclic changes in the ovaries and uterus. Approximately every 28 days, the hypothalamus signals the anterior pituitary to secrete FSH and LH, which trigger the growth of a follicle and ovulation, the release of an egg. The follicle secretes estrogen; after ovulation, the follicle becomes the corpus luteum, which adds progesterone. These two hormones stimulate the endometrium (the uterine lining) to thicken, which prepares the uterus for implantation. They also inhibit the hypothalamus, reducing FSH and LH secretion. If the egg is not fertilized, the drop in LH shuts down the corpus luteum and its hormones. This triggers menstruation, the breakdown of the endometrium. The hypothalamus and pituitary then stimulate another follicle and start a new cycle. If fertilization occurs, a hormone from the embryo maintains the uterine lining and prevents menstruation.

27.6–27.8 Human sexual behavior promotes the union of egg and sperm and may promote bonding between mates. The human sexual response occurs in four phases: excitement, plateau, orgasm, and resolution. Sexual intercourse may carry risks of sexually transmitted diseases and unwanted pregnancy. Contraception prevents pregnancy by blocking the release of gametes, preventing fertilization, or preventing implantation.

27.9 Fertilization initiates embryonic development. When a sperm reaches the egg, its enzymes pierce the egg's coat. Sperm proteins bind to egg receptor proteins, sperm and egg plasma membranes fuse, and their nuclei unite. Changes in the egg membrane prevent entry of additional sperm, and the fertilized egg, now called a zygote, boosts its metabolic activity.

27.10–27.12 Cleavage is a rapid series of cell divisions that turns a zygote into a hollow ball of cells called a blastula. Next, in gastrulation, cells migrate inward and form a rudimentary digestive cavity. The resulting gastrula has three layers of cells: ectoderm, en-

doderm, and mesoderm. After gastrulation, each of these embryonic tissue layers gives rise to specific organ systems. The notochord develops above the digestive cavity. The cell layer above the notochord rolls up to become the neural tube, which becomes the brain and spinal cord.

27.13–27.15 In accomplishing these changes, cells in an embryo elongate and contract, migrate, and aggregate at particular sites. Cells then begin to differentiate, or specialize for different functions. In a process called induction, adjacent cell layers may influence each other's differentiation via chemical signals. Pattern formation, the emergence of all the parts of a structure in the correct relative positions, seems to involve the response of genes to spatial variations of chemicals in the embryo.

27.16 Human development begins with fertilization in the oviduct. Cleavage produces a blastocyst, whose inner cell mass becomes the embryo. The blastocyst's outer layer, the trophoblast, implants in the uterine wall. Gastrulation occurs, organs develop from the three embryonic layers, and the embryo becomes surrounded by four extraembryonic membranes. The embryo floats in the fluid-filled amnion, while the chorion forms the embryo's part of the placenta. The placenta's chorionic villi absorb food and oxygen from the mother's blood. Viruses and drugs may also pass through the placenta and harm the embryo.

27.17 Human embryonic development is divided into three trimesters of about three months each. The most rapid change occurs during the first trimester. By 9 weeks, all organs are formed, and the embryo is called a fetus. The second and third trimesters are a time of growth and preparation for birth.

27.18 Hormonal changes bring on birth. Estrogen makes the uterus more sensitive to oxytocin, which, along with prostaglandins, acts to initiate labor. The cervix dilates, the baby is expelled by strong muscular contractions, and the placenta follows. Oxytocin and prolactin then stimulate milk secretion.

27.19 In vitro fertilization and surrogate motherhood give hope to infertile couples, but these new technologies have raised ethical and legal questions.

Testing Your Knowledge

Multiple Choice

1. After a sperm penetrates an egg, a fertilization membrane forms. This membrane
   a. secretes important hormones
   b. enables the fertilized egg to implant in the wall of the uterus
   c. prevents more than one sperm from entering the egg
   d. attracts additional sperm to the egg
   e. activates the egg for embryonic development

2. In an experiment, a researcher colored a bit of tissue on the outside of a frog gastrula with a fluorescent dye. The embryo developed normally, but when the tadpole was placed under an ultraviolet light, which of the following glowed bright orange? (*Explain your answer.*)
   a. the heart
   b. the bones
   c. the brain
   d. the stomach
   e. the liver

3. How does a zygote differ from an ovum?
   a. A zygote has more chromosomes.
   b. A zygote is smaller.
   c. A zygote consists of more than one cell.

d. A zygote is much larger.

e. A zygote divides by meiosis.

4. Which of the following traces the path of sperm out of the body of a human male?

a. epididymis, seminiferous tubule, vas deferens, urethra

b. seminiferous tubule, vas deferens, epididymis, urethra

c. epididymis, seminiferous tubule, urethra, vas deferens

d. seminiferous tubule, epididymis, vas deferens, urethra

e. seminiferous tubule, epididymis, urethra, vas deferens

5. A woman had several miscarriages. Her doctor suspected that a hormonal insufficiency was causing the lining of the uterus to break down, as it does during menstruation, terminating her pregnancies. Treatment with which of the following might help her remain pregnant?

a. oxytocin
d. luteinizing hormone

b. follicle-stimulating hormone
e. prolactin

c. testosterone

6. Which of the following most reduces the chances of both conception and the spread of sexually transmitted diseases?

a. condom

b. birth control pill

c. diaphragm

d. intrauterine device

e. withdrawal

Matching

1. Turns into the corpus luteum
2. Female gonad
3. Site of spermatogenesis
4. Site of fertilization in humans
5. Human gestation occurs here
6. Sperm duct
7. Secretes seminal fluid
8. Lining of uterus

a. vas deferens
b. prostate gland
c. endometrium
d. testis
e. follicle
f. uterus
g. ovary
h. oviduct

Describing, Comparing, and Explaining

1. Some animals, such as rotifers and aphids, are able to alternate between sexual and asexual reproduction. Under what conditions might it be advantageous to reproduce sexually? Asexually?

2. The graph below plots the rise and fall of pituitary and ovarian hormones during the monthly ovarian cycle. Identify each hormone (A–D) and the reproductive events with which each one is associated (P–S).

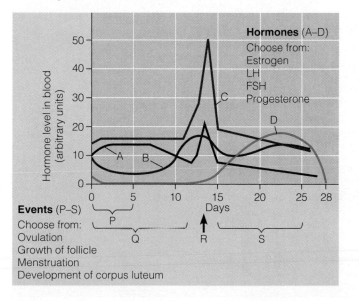

3. Compare sperm formation (spermatogenesis) with egg formation (oogenesis). In what ways are the processes similar? In what ways are they different?

4. The embryos of reptiles, birds, and mammals have systems of extraembryonic membranes. What are the functions of these membranes, and how do fish and frog embryos survive without them?

Thinking Critically

1. As a frog embryo develops, the neural tube forms from ectoderm along what will be the frog's back, directly above the notochord. To study this process, a researcher carefully extracted a bit of notochord tissue and inserted it underneath the ectoderm where the belly of the frog would normally develop. What can the researcher hope to learn from this experiment? What are the possible outcomes? What experimental control would you suggest?

2. Sally was a heavy drinker, but a few weeks after she found out she was pregnant, she stopped. Unfortunately, her baby was born suffering from fetal alcohol syndrome, a complex of birth defects resulting from the action of alcohol on the embryo. In a few sentences, explain why drugs do the most harm if they are present, as in Sally's case, during the earliest weeks of pregnancy.

3. The birth control pill contains a combination of synthetic estrogen and progesterone. How would this hormone combination prevent conception? How is it similar to what is present when a woman is pregnant?

4. In an embryo, nerve cells grow out from the spinal cord and form connections with the muscles they will eventually control. What mechanisms, described in this chapter, might explain how these cells "know" where to go and which cells to connect with?

Science, Technology, and Society

1. New techniques for sorting sperm, combined with in vitro fertilization or artificial insemination, allow a couple to choose their baby's sex. Would you want to try such techniques if you had the chance? Can you foresee problems if these procedures become widely available?

2. Nerve cells transplanted from aborted fetuses can relieve the symptoms of Parkinson's disease, a brain disorder. Fetal tissue transplants might also be used to treat epilepsy, diabetes, Alzheimer's disease, and spinal cord injuries. Why might tissues from a fetus be particularly useful for replacing diseased or damaged cells? Since 1989, there has been controversy over whether the U.S. government should allow fetal tissues from induced abortions to be used in transplant research. Opponents would allow only tissues from miscarriages to be used. Why would most researchers prefer to use tissues from aborted fetuses? Why do some people oppose this? What is your position on this issue? Why?

3. In the use of the contraceptive called Norplant, several small tubes are implanted under the skin. The tubes gradually release progestin, one of the hormones in birth control pills, for as long as five years. Why do you think this is considered by many to be a major advance in contraceptive technology? In what situations do you think it would be most useful? Shortly after this device was introduced, a California woman was convicted of child abuse and sentenced to a year in prison and three years' probation. The judge ordered the woman to have Norplant implanted as a condition of her probation. Is this a fair or effective response to the woman's crime? How else might this new technology be used or abused?

4. New technology has made it possible for doctors to save a small percentage of babies born 16 weeks prematurely. A baby born this early weighs just over a pound and faces months of care in an intensive-care nursery. The cost for care may be hundreds of thousands of dollars per infant. Some people wonder whether such a huge technological and personnel investment should be devoted to such a small number of babies. They feel that limited resources might better be directed at providing prenatal care that could prevent many premature births. What do you think? Why?

Squids are superbly equipped predators of the open sea. They drift and glide—watching, waiting—then suddenly dart after a fish with knifelike precision. In a chase, a squid can reach almost 40 km/hr (about 25 mi/hr) in a matter of seconds, faster than any other invertebrate and most fish. Suction cups on a squid's tentacles and arms grasp prey, and beaklike jaws at the base of the tentacles quickly chop the food into bite-sized pieces.

The squid in this photograph (*Loligo pealei*) is only about 20 cm (8 in) long. Often pursued by sharks or other large fish, it can twist and quickly turn out of the way. It can also shoot a cloud of black ink into the water to divert attackers, and it can change color almost instantly to blend with the background. The squid changes color by changing its spots, which are due to pigment-containing cells in the skin. When tiny muscles attached to the cells contract, the cells stretch, spreading the pigment out so that the spots enlarge and darken the squid's skin. When the muscles relax, the pigment cells shrink, and the skin pales as the pigmented areas become smaller.

A squid's abilities to attack or escape depend on a sophisticated nervous system (colored gold in the drawing on this page). A squid has a large brain for its body size. Nerves carry sensory signals into the brain from the eyes and other sense organs. The brain quickly sorts the information and sends out commands, via other nerves, to the muscles. As indicated in the drawing, large nerves run forward from the brain into the tentacles and arms, and even larger nerves course downward into the main body. Some of the nerves carrying commands to the muscles contain unusual-ly large nerve cells with long, thick extensions called giant fibers. The giant fibers can conduct nerve signals at great speed, from the brain to the tip of the tail in about a hundredth of a second. Command signals sent through these big nerve cells give the squid the quick reflexes it needs to catch fast-swimming prey and escape predators.

The discovery of the squid's giant fibers, some 50 years ago, set the stage for a revolution in research on animal nervous systems. The fibers are easy to remove and keep alive for study in the laboratory. The photograph below was taken with a light microscope. It shows a small piece of a giant fiber with a glass electrode inserted into it. A researcher can use the electrode to make direct measurements of tiny electrical changes that occur in the fiber when it is stimulated. From such research has come much of our knowledge about how nerve cells, including our own, conduct signals.

To survive and reproduce, a squid or any other animal must respond appropriately to environmental stimuli. Most animals have two coordinating systems, the endocrine (hormone) system, which we discussed in Chapter 26, and the nervous system. The two systems often cooperate in triggering responses to stimuli. In a squid, for instance, hormones sometimes assist the nervous system in altering skin color. In our own bodies, an interplay of hormones and nerve signals produces our fight-or-flight response to threats and stress. The nervous system alone, however, provides the pinpoint control underlying delicate body movements, as when a squid suddenly changes direction in the water, or when your eyes move back and forth between words and pictures as you read this chapter.

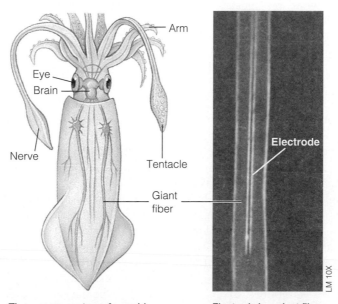

Arm
Eye
Brain
Nerve
Tentacle
Giant fiber
Electrode
LM 10X

The nervous system of a squid   **Electrode in a giant fiber**

28.1 Nervous systems receive sensory input, interpret it, and send out appropriate commands

Nervous systems are the most intricately organized data-processing systems on Earth. A cubic centimeter of your brain, for instance, contains several million nerve cells, each of which may communicate with thousands of other nerve cells in data-processing networks that make the most elaborate computer circuit boards look primitive. These living networks enable us to learn and remember, and they control our every perception and movement.

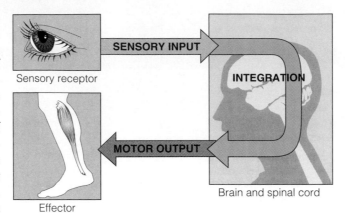

SENSORY INPUT

INTEGRATION

MOTOR OUTPUT

Sensory receptor

Effector

Brain and spinal cord

The diagram here shows the basic organization of a nervous system. The system has three interconnected functions: sensory input, integration, and motor output. Sensory input is triggered by stimuli, such as light or touch. **Sensory input** is the conduction of signals from sensory receptors, such as the light-detecting cells of the eye, to processing centers, the brain and (in vertebrates) the spinal cord. **Integration** is the interpretation of the sensory signals within processing centers. **Motor output** is the conduction of signals from a processing center to effector cells, such as the muscle cells that actually perform many of the body's responses to stimuli. Motor commands may also control secretory cells in the glands; thus, gland cells can also be effectors. In any case, information in the form of signals passes from receptors to processing centers, and from there to effectors. The signals travel along pathways of cells called neurons.

28.2 Neurons are the functional units of nervous systems

Nervous systems consist of two major types of cells: **neurons** (nerve cells), which are specialized for carrying signals from one location in the body to another, and **supporting cells,** which protect, insulate, reinforce, and assist neurons. Supporting cells outnumber neurons by as many as 50 to 1, but only neurons carry signals.

Neurons come in a variety of types, but most of them share some common features, which you can see in the figure at the top of the facing page. The figure depicts a neuron like those that carry motor commands from your spinal cord to your skeletal muscles. The neuron has a large **cell body** housing its nucleus and most of its cytoplasmic organelles. Two types of fibers extend from the cell body. Cells of one type, called **dendrites** (Greek *dendron,* tree), are often short (as on this neuron), numerous, and highly branched. Dendrites convey signals *toward* the cell body. The signals may come from a sensory cell or from a neighboring neuron. The second type of neuron fiber is the **axon** (Greek *axon,* axle), which on many neurons is a single fiber. The axon conducts signals *away from* the cell body—that is, to another neuron or to an effector such as a muscle cell. Many axons are long. Certain ones in your leg, for instance, stretch from the lower part of your spinal cord all the way to muscles in your toes. The giant fibers of a squid are also axons.

In many animals, including the vertebrates, axons that convey signals very rapidly are enclosed along most of their length by a thick insulating material. In vertebrates, the insulating material is called the **myelin sheath.** As shown in the figure, the sheath resembles a chain of oblong beads strung along the axon. Each bead is actually a supporting cell called a Schwann cell. The myelin sheath is essentially a chain of Schwann cells, each wrapped many times around the axon. If you roll a pencil up in several layers of aluminum foil, then pinch the foil at regular intervals and cut through the pinched layers down to the pencil, you will have something that resembles a myelinated axon. The foil represents the multilayered myelin sheath, and the cuts represent tiny spaces between adjacent Schwann cells, places where there is no myelin sheath.

On a real axon, the spaces between Schwann cells are called **nodes of Ranvier,** and they are the only points on the axon where signals can be transmitted. Everywhere else, the myelin sheath insulates the axon, preventing signals from passing along it. Therefore, when a signal travels along a myelinated axon, it jumps from node to node. By jumping along the axon, the signal travels much faster than it could if it had to take the long route along the whole length of the axon. In the human nervous system, signals can travel about 100 m/sec (over 300 mi/hr), which means that a command from your brain can make your fingers move in just a few milliseconds (thousandths of a second). Without myelin sheaths, the signals could go only about 5 m/sec.

The debilitating disease called multiple sclerosis (MS) demonstrates the importance of our myelin sheaths. MS

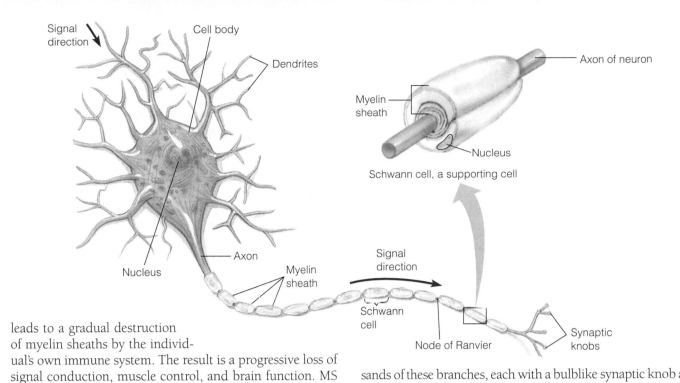

leads to a gradual destruction of myelin sheaths by the individual's own immune system. The result is a progressive loss of signal conduction, muscle control, and brain function. MS is not yet curable, but drugs that suppress the immune system can relieve symptoms and slow its progress.

Returning to the figure above, you can see that the axon ends in a cluster of branches. A typical axon has hundreds or thousands of these branches, each with a bulblike synaptic knob at the very end. As we will see later, the **synaptic knobs** relay signals to another neuron or to an effector such as a muscle cell. With the basic structure of a neuron in mind, we can now focus on how neurons are arranged in the nervous system.

Nervous systems have central and peripheral divisions

The diagram below shows the relationship between neurons and nervous system structure and function. With few exceptions, nervous systems have two divisions. The division called the **central nervous system (CNS)** consists of the brain and, in vertebrates, the spinal cord. The other division, the **peripheral nervous system (PNS),** is made up mostly of communication lines, called nerves, that carry signals into and out of the CNS. A **nerve** is a ropelike bundle of neuron fibers (axons and dendrites) tightly wrapped in connective tissue. The PNS also has **ganglia** (singular, *ganglion*), which are clusters of cell bodies belonging to the neurons making up the nerves. For simplicity, our diagram shows only one neuron in each of the nerves and only one cell body in the ganglion.

There are three functional types of neurons, corresponding to the three main functions of the nervous system. **Sensory neurons** function in sensory input. They communicate information about the environment—stimuli—from sensory receptors such as eyes, ears, and skin surface into the CNS. The dendrites of many sensory neurons extend from a sensory receptor to a cell body in a ganglion. A sensory cell's axon carries signals into the CNS. **Interneurons,** a second type of nerve cell, are entirely within the CNS. Interneurons integrate data obtained from one or more sensory neurons, and then relay appropriate signals to **motor neurons,** the third type of nerve cell. A motor neuron's axon conveys signals from the CNS to effector cells (in this case, muscle cells), which then produce an appropriate response.

With its wirelike, interconnecting nerves and central switchboard of interneurons, the nervous system resembles a complex telephone network. Telephone lines carry electrical signals. What kind of signals traverse the "wires" of the nervous system? To find out, we must examine a neuron and its function in more detail.

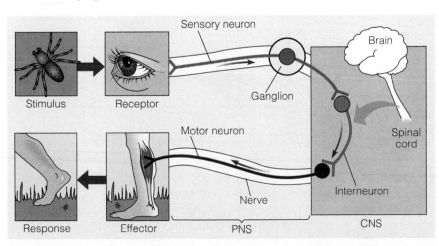

A neuron maintains a resting potential across its membrane

To understand nerve signals and how they are transmitted along a neuron, we must first study a resting neuron, one that is not transmitting a signal.

Functionally, a resting neuron resembles a flashlight battery. Both contain potential energy, energy that can be put to work (see Module 5.2). The battery's potential energy can be used to produce light, the neuron's to send signals from one part of the body to another. The neuron's potential energy resides in an electrical charge difference across its plasma membrane. The cytoplasm just inside the membrane is negative in charge, and the fluid just outside the membrane is positive. Since opposite charges tend to move toward each other, a membrane stores energy, much as a battery does, by holding opposite charges apart. As you can see in Figure A, the strength (voltage) of this stored energy in a neuron can be measured with microelectrodes connected to a voltmeter. The voltage across the plasma membrane of a resting neuron is called the **resting potential.** Cells have a negative resting potential; that is, the inside of the cell is negative relative to the outside. A neuron has a resting potential of about -70 millivolts (mV), about 5% of the voltage in a size AA flashlight battery.

What causes the resting potential? The answer lies with the membrane itself (Figure B). The membrane keeps dissolved proteins and other large organic molecules inside the cell. Most of these molecules are negatively charged. Also, the membrane has channels and pumps, made of proteins, that regulate the passage of inorganic ions, especially sodium (Na^+) and potassium (K^+). A resting membrane allows much more K^+ than Na^+ to diffuse across. In Figure B, you can see that Na^+ (pink) is more concentrated outside the cell than inside; the Na^+ channels allow very little to diffuse in. But K^+ (blue), which is more concentrated inside, can freely flow out. As the positively charged K^+ ions diffuse out, they leave behind an excess of negative charge. Together, the negative molecules trapped inside the cell and the diffusion of K^+ out of the cell create most of the resting potential. (For simplicity, Figure B omits Cl^- and other negative inorganic ions on both sides of the membrane.)

The **sodium–potassium (Na^+–K^+) pump** is another membrane component that helps maintain the resting potential. By actively transporting Na^+ out of the cell and K^+ in, these pumps keep the concentration of Na^+ low in the cell and K^+ high. The pumps also contribute slightly to the loss of positive charge from the cell by moving more Na^+ out than K^+ in.

As we see next, stimuli that trigger nerve signals act by triggering a release of the resting potential energy.

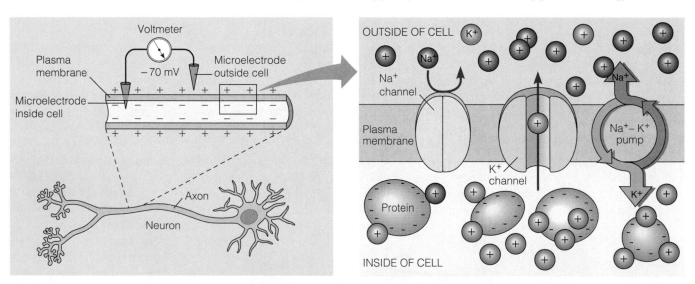

A. Measuring a neuron's resting potential

B. How the resting potential is generated

A nerve signal originates as a change in the resting potential

Turning a flashlight on uses the energy stored in a battery to create light. In a similar way, stimulating a neuron's plasma membrane can trigger the release and use of the resting potential energy to generate a nerve signal. A **stimulus** is any factor that causes a nerve signal to be generated—for instance, an electric shock, physical pressure on a cell, or a sudden temperature change.

The discovery of squid giant fibers (axons) gave researchers their first chance to study how stimuli trigger signals in a living neuron. From microelectrode studies with squid neurons, British biologists A. L. Hodgkin and A. F. Huxley worked out the details of nerve signal transmission in the 1940s. Their findings, summarized in the figure below, apply to neurons in all animals.

Graphing electrical changes in neuron membranes was the first step in discovering how nerve signals are generated, so let's begin by looking at these changes. The multi-colored line on the graph traces the electrical changes that make up the **action potential,** the technical name for a nerve signal. All the changes indicated by the graph occur at the place on the membrane where a stimulus is applied. The graph records electrical events over time (in milliseconds) at that particular place. ① The graph starts out at −70 mV, the membrane's resting potential. ② The stimulus is applied, at time 0, and in 2–3 msec, the voltage rises from −70 mV to what is called the **threshold potential** (−50 mV, in this case). The difference between the threshold potential and the resting potential is the minimum change in the membrane's voltage that must occur to generate the action potential. ③ The threshold potential triggers the action potential, the steep upswing (red) on the graph, which reaches a peak of about +35 mV. The entire change from −70mV to +35 mV occurs within 3–4 msec of stimulation. ④ The voltage then drops back down, undershoots the resting potential, and finally returns to it.

Once they had recorded the electrical changes of the action potential, Hodgkin and Huxley wanted to find out what caused the changes. They found that specific ion movements coincide with, and in fact cause,

the electrical changes. Their conclusions have since been strengthened by studies of the membrane channels used by Na$^+$ and K$^+$ to cross the membrane. These channels have voltage-sensitive gates that open and close and actually control the movements of the ions.

If you look at the diagrams surrounding the graph, you'll see the Na$^+$ and K$^+$ movements that are the basis of the action potential. In ①, at the lower left, the resting membrane is positively charged on the outside, and the cytoplasm just inside the membrane is negatively charged. ② The stimulus applied at time 0 triggers the opening of voltage-sensitive Na$^+$ gates in the membrane. At first, only a few gates open, and a tiny amount of Na$^+$ enters the axon. Just this tiny change, however, makes the inside surface of the membrane slightly less negative than before, and as a result, the membrane's voltage begins to change. If the stimulus is strong enough, a sufficient number of Na$^+$ gates open to change the voltage to the threshold level. ③ Once the threshold is reached, the increasing positive charge inside the membrane is all it takes to trigger the opening of more and more Na$^+$ gates. As more Na$^+$ moves in, the voltage soars to its peak. ④ As soon as the peak is reached, the whole process starts to reverse. The peak voltage itself triggers closing and inactivation of the Na$^+$ gates. Na$^+$ stops moving in, while the K$^+$ gates open, allowing K$^+$ to diffuse rapidly out. These changes produce the downswing on the graph and a very brief undershoot of the resting potential. The membrane then returns to its resting potential, less than 7 msec after the stimulus occurred.

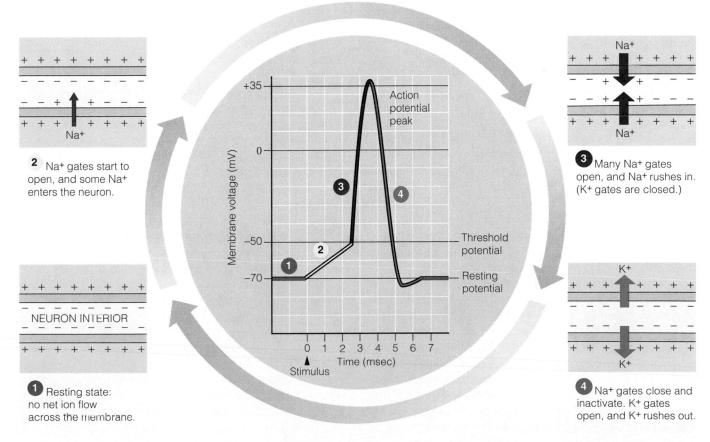

2 Na+ gates start to open, and some Na+ enters the neuron.

1 Resting state: no net ion flow across the membrane.

NEURON INTERIOR

3 Many Na+ gates open, and Na+ rushes in. (K+ gates are closed.)

4 Na+ gates close and inactivate. K+ gates open, and K+ rushes out.

Action potential peak

Threshold potential

Resting potential

Membrane voltage (mV)

Time (msec)

Stimulus

An action potential is a localized electrical event—a change from a neuron's resting potential at a specific point. To function as a signal, this local event must travel along the neuron. You can follow this movement in the figure at the right. A nerve signal starts out as one action potential, generated on the axon near the cell body of the neuron. The effect of this action potential is like tipping the first of a row of standing dominoes: The first domino does not travel along the row, but its fall is relayed to the end of the row, one domino at a time.

The numbered parts of the figure show the changes that occur in part of an axon at three successive times, as a nerve signal passes from left to right. As we saw in Module 28.5, all the ion movements associated with a particular action potential occur at one place on the axon. Let's first focus on the axon region on the far left. ① When this region of the axon (pink here) has its Na^+ gates open, Na^+ rushes inward (magenta arrows), and an action potential is generated. This corresponds to the upswing of the curve on the graph in Figure 28.5. ② When that same region has its K^+ gates open, K^+ diffuses out of the axon (blue arrows); at this time, its Na^+ gates are closed and inactivated, and the action potential is subsiding. On a graph, we would see the downswing of the action potential initiated in part 1. ③ A short time later, we would see no signs of an action potential at this (far-left) spot, because the axon membrane here has returned to its resting potential.

Now let's see how these events lead to the "domino effect" of a nerve signal. In part 1 of the figure, the pink arrows pointing sideways within the axon indicate local spreading of the electrical changes caused by the inflowing Na^+ ions associated with the first action potential. These changes trigger the opening of Na^+ gates in the membrane just ahead of the action potential. As a result, a second action potential is generated, as indicated by the pink region in part 2. In the same way, a third action potential is generated

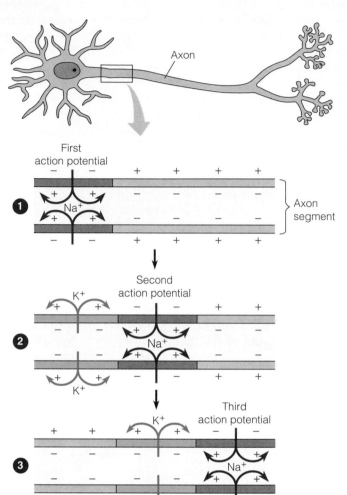

in part 3, and, each action potential generates another all the way down the axon.

You might be wondering why action potentials are propagated in only one direction along the axon (left to right in the figure). As the pink arrows indicate, local electrical changes do spread in both directions in the axon. However, these changes cannot open Na^+ gates and generate an action potential when the Na^+ gates are inactivated. Thus, an action potential cannot be generated in the regions where K^+ is leaving the axon (blue in the figure).

So we see that a nerve signal, also known as an action potential, propagates itself in one direction by the electrical changes it produces in the neuron membrane. An action potential is a bit of coded information that can travel from one end of a neuron to another. What does this tell us about how a nervous system actually works? If you rap your finger on a desk, for instance, the contact is a stimulus that triggers action potentials in the tips of sensory neurons in your skin. The action potentials regenerate, carrying the information (that your finger has hit a hard object) into your central nervous system.

Action potentials are *all-or-none* phenomena; that is, they do not vary in size with the strength of the stimulus that triggers them. How do they relay different intensities of information to your central nervous system? If you rap your finger hard against the desk, your central nervous system receives many more action potentials per millisecond than after a soft tap. It is the *frequency* of action potentials that changes with the intensity of stimuli.

Once your central nervous system receives information in the form of action potentials, it can process the information and formulate a response to it. The nervous system's ability to respond depends on the sensory neurons' passing their signals to other neurons in the CNS. Our next step is to see how signals pass from one neuron to another.

One of the key components of the nervous system is the **synapse**, the junction, or relay point, between two neurons. When an action potential arrives at the end of one neuron's axon, the information coded in the action potential passes to a receiving neuron across the synapse.

Synapses are either electrical or chemical. In an electrical synapse, action potentials themselves pass from one neuron to the next. The receiving neuron is stimulated quickly and always at the same level (same frequency of action potentials) as the transmitting neuron. Lobsters, crayfish, and many fishes can flip their tails with lightning speed because the neurons that carry signals for these movements communicate by electrical synapses. In the human body, electrical synapses are common in the heart and digestive tract, where nerve signals maintain steady, rhythmic muscle contractions. In contrast, chemical synapses are prevalent in most other organs, including skeletal muscles, and in the central nervous system, where signaling among neurons is complex and varied.

Unlike an electrical synapse, a chemical synapse has a narrow gap, called the **synaptic cleft**, separating a synaptic knob of the transmitting neuron from the receiving neuron. The cleft prevents the action potential in the transmitting neuron from spreading directly to the receiving neuron. Instead, the action potential is first converted to a chemical signal at the synapse. The chemical signal, consisting of molecules of **neurotransmitter** (see Module 26.1), then may generate an action potential in the receiving cell.

Now let's follow the sequence of events that occur at a chemical synapse in the figure to the right. The neurotransmitter is contained in vesicles in the synaptic knob of the transmitting neuron. Following the numbered sequence, ① an action potential (red arrow) arrives at the synaptic knob. ② The action potential triggers chemical changes that make neurotransmitter vesicles fuse with the plasma membrane of the transmitting cell. ③ The fused vesicles release their neurotransmitter molecules (green) into the synaptic cleft. ④ The released neurotransmitter molecules diffuse across the cleft and bind to receptor molecules on the receiving cell's plasma membrane. ⑤ The binding of neurotransmitter to receptor opens chemical-sensitive ion channels in the receiving cell's membrane. With the channels open, ions can diffuse into the receiving cell and trigger new action potentials. ⑥ The neurotransmitter is broken down by an enzyme, and the ion channels close. Step 6 ensures that the neurotransmitter's effect on the receiving cell is brief and precise.

You can review the aspects of the nervous system we have covered so far by thinking about the cellular events occurring right now in your own nervous system. Action potentials carrying coded information about the words on this page are streaming along sensory neurons from your eyes to your brain. Arriving at synapses with receiving cells (interneurons in the brain), the action potentials are trig-

gering the release of neurotransmitters at the ends of the sensory neurons. The neurotransmitters are diffusing across synaptic clefts and triggering changes in some of your interneurons—changes that lead to integration of the signals and ultimately to a determination of what the signals mean (in this case, the meaning of words and sentences). When you finish this module, motor neurons in your brain will send out action potentials to muscle cells in your fingers, telling them to contract in just the right way to turn the page.

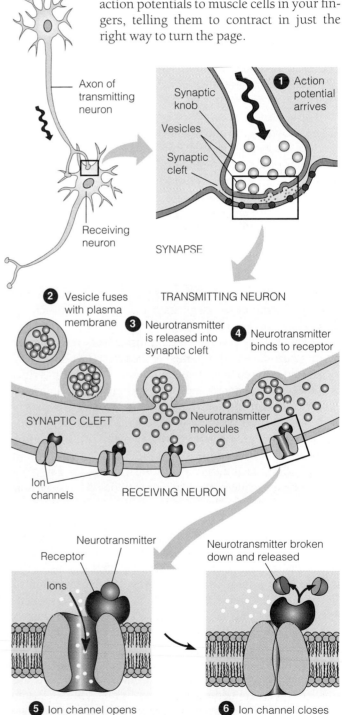

Axon of transmitting neuron

Receiving neuron

1 Action potential arrives

Synaptic knob

Vesicles

Synaptic cleft

SYNAPSE

TRANSMITTING NEURON

2 Vesicle fuses with plasma membrane

3 Neurotransmitter is released into synaptic cleft

4 Neurotransmitter binds to receptor

SYNAPTIC CLEFT

Neurotransmitter molecules

Ion channels

RECEIVING NEURON

Neurotransmitter

Receptor

Ions

5 Ion channel opens

Neurotransmitter broken down and released

6 Ion channel closes

Neurotransmitters excite or inhibit neurons

What do neurotransmitters do to receiving neurons? It all depends on the kind of membrane channels they open. Neurotransmitters that open Na^+ channels, for instance, may trigger action potentials in the receiving cell. Such neurotransmitters are said to be excitatory. In contrast, many neurotransmitters are inhibitory. These open membrane channels for ions (specifically Cl^-) that *decrease* the receiving cell's tendency to develop action potentials. The excitatory and inhibitory effects of neurotransmitters are the basis of information processing in the nervous system.

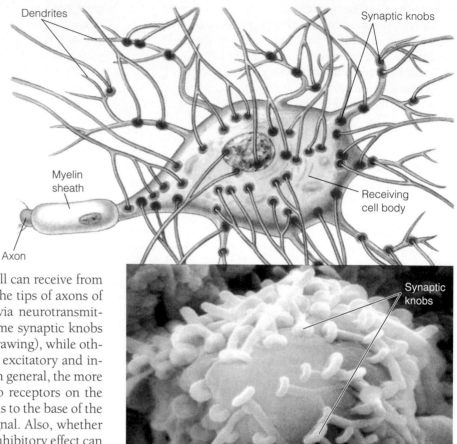

The drawing and micrograph here give you an idea of the many inputs a cell can receive from other neurons. The synaptic knobs are the tips of axons of other neurons, delivering information via neurotransmitters to the receiving cell. Notice that some synaptic knobs deliver excitatory signals (green in the drawing), while others deliver inhibitory signals (red). Both excitatory and inhibitory signals can vary in magnitude. In general, the more neurotransmitter molecules that bind to receptors on the receiving cell and the closer the synapse is to the base of the receiving cell's axon, the stronger the signal. Also, whether a neurotransmitter has an excitatory or inhibitory effect can depend on the receiving neuron.

A receiving neuron may obtain input from hundreds of other neurons via thousands of synaptic knobs—many more than you see here. The membrane of a neuron is thus like a tiny circuit board, receiving and processing information in the form of neurotransmitter molecules. The nervous system relies on these living circuit boards to perform all of its integrating activities and to make precise, appropriate responses to stimuli.

Depending on the information other cells send it, a receiving neuron's membrane may or may not generate action potentials. At any particular instant, the inhibitory signals it receives may cancel out the excitatory signals. If the excitatory signals are collectively strong enough to overpower the inhibitory ones, the receiving cell may then send out action potentials to other cells. The neuron passes signals to the other cells at a frequency that represents a **summation** of all the information it has received. Each new receiving cell, in turn, processes this information along with other inputs it obtains. The key to this complex information-processing system is the variability in the effects that neurotransmitters produce at chemical synapses.

What kinds of chemicals are neurotransmitters? Most are small, nitrogen-containing organic molecules. One such substance is the neurotransmitter **acetylcholine**, which slows our heart rate and makes our skeletal muscles contract. Several other nitrogen-containing neurotransmitters—epinephrine, norepinephrine, serotonin, dopamine, and the natural painkillers called endorphins—are also hormones (see Chapter 26). As neurotransmitters, epinephrine and norepinephrine increase the heart rate during stress. Serotonin and dopamine are active in the brain, where they affect sleep, mood, attention, and learning. The degenerative disorder Parkinson's disease is associated with a lack of dopamine in the brain. Certain psychoactive drugs, including LSD and mescaline, apparently produce their hallucinatory effects by binding with serotonin and dopamine receptors in the brain.

Three other neurotransmitters—glutamate, glycine, and GABA (gamma aminobutyric acid)—are actually amino acids. Glycine and GABA are inhibitors; glutamate is an excitatory neurotransmitter, very common in the brain and spinal cord. Research on neurotransmitters is a fast-moving area in biology. Studies have recently found that ATP (the energy currency molecule) serves as a neurotransmitter in the brain. Other studies show that our neurons employ some potentially toxic gases, notably nitric oxide (NO) and probably carbon monoxide (CO), as neurotransmitters. A vigorous research effort focuses on how neurons use gases such as these as signals.

Many stimulants and depressants act on chemical synapses

28.9

Knowing about neurotransmitters helps us understand some of the effects on our nervous system of substances we call stimulants and depressants. Stimulants are chemicals that increase the activity of the central nervous system, often by altering the effect of neurotransmitters at synapses. For instance, cocaine and amphetamines increase the normal stimulatory effect of the neurotransmitter norepinephrine. The most commonly used stimulant is caffeine, found in coffee, tea, chocolate, many soft drinks, and a number of over-the-counter drugs. (The related compounds theophylline and theobromine are also found in many caffeine food sources.) Caffeine, used in nearly all cultures, generally raises our heart rate and breathing rate and keeps us awake by countering the effects of inhibitory signals on synapses in our central nervous system. As a result, many of us cannot drink more than one or two cups of coffee at a sitting without becoming restless. The average American consumes several hundred milligrams of caffeine a day, mostly from the sources listed in the table above. Although research has yet to prove that average consumption is harm-

Caffeine Content of Foods*	
Brewed coffee (6 oz)	47–211 mg
Instant coffee (6 oz)	35–140 mg
Black tea (6 oz)	34–60 mg
Cola drink (12 oz)	30–58 mg
Green tea (6 oz)	11–43 mg
Cocoa (6 oz)	2–8 mg
Milk chocolate candy (1 oz)	1–15 mg
Decaffeinated coffee (6 oz)	1–10 mg

*The ranges in caffeine content are due to variations in starting materials and preparation methods.

ful, several studies indicate that consuming more than 600 mg of caffeine a day can cause stomach problems, depression, and sleeplessness.

In contrast to stimulants, depressants often inhibit signal transmission at synapses. Among the depressant drugs, barbiturates and the tranquilizers Valium and Librium intensify the effect of the neurotransmitter GABA. Although often thought of as a stimulant, alcohol is, in fact, a strong depressant. Alcohol's precise effect on the nervous system is not yet known, but it seems to increase GABA's inhibitory effect. As people become dependent on alcohol, their nervous system produces less and less GABA.

More commonly used than any other depressant, alcohol can be a very dangerous drug. Habitual heavy drinking (three or more drinks a day at least three times a week) can cause liver and brain damage and may be associated with cardiovascular disease and several forms of cancer. When a woman drinks during pregnancy, alcohol enters the fetus's bloodstream and can cause serious birth defects, as we discussed in Module 27.16.

Depression is linked to neurotransmitter imbalance

28.10

Tiny amounts of neurotransmitters underlie all of our moods, from ecstasy to despair, and the balance is a very delicate one. Norepinephrine, dopamine, serotonin, and the endorphins are our main "feel-good" neurotransmitters. The link between neurotransmitters and the mood disease called depression first came to light with the discovery that some people taking a drug called reserpine for high blood pressure suffered severe depression. Research showed that reserpine's main effect is to decrease the amount of norepinephrine at synapses. Besides reducing blood pressure, a reduced amount of norepinephrine in the brain seems to produce depression.

Another line of evidence for a connection between neurotransmitters and depression has come from studies of antidepressants, drugs used to treat depression. In recent years, research has shown that a group of antidepressants called tricyclics increase levels of serotonin and norepinephrine in the brain. It has also been shown that the anti-

depressants called MAO (monoamine oxidase) inhibitors decrease the activity of enzymes that break down certain neurotransmitters, including norepinephrine.

Unfortunately for the sufferers of major depression, its mechanisms seem to be diverse and complex, and difficult to treat. A variety of neurotransmitters are involved in mood, and, as we have seen, a variety of membrane receptors determine how the neurotransmitters affect receiving neurons. As a result, any particular case of depression might result from several different factors. Depression may result from inadequate or excessive neurotransmitter production or breakdown, or a problem with the production or function of receptor molecules. In addition, nonneurotransmitter hormones may be involved, as in the case of seasonal affective disorder (see the introduction to Chapter 26). A thorough understanding of depression, as well as other mental disorders, will only come with further knowledge of the basic biochemistry of the brain.

28.11　Diverse nervous systems have evolved in the animal kingdom

Thus far, we have concentrated on the cellular mechanisms that are fundamental to the nervous system. There is remarkable uniformity throughout the animal kingdom in the way nerve cells function. However, as we will now see, there is great variety in how nervous systems as a whole are organized. The cnidarian *Hydra* (Figure A) has one of the simplest types of nervous system. It has what is called a **nerve net**, a web-

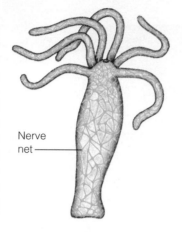

A. *Hydra* nervous system

like system of neurons extending throughout the body. A hydra has no head or brain, and its nerve net has no central or peripheral divisions. It can glide slowly or somersault from place to place, but much of the time it is stationary, attached to submerged plant stems or rocks. The nerve net is quite adequate for a hydra's headless, radially symmetrical body and limited activity.

Most animals are bilaterally symmetrical, with a head and tail and a tendency to move head-first through the environment. The head—most often the first part of the animal to encounter new stimuli—is usually equipped with sense organs and a brain. Flatworms (Figure B) are the simplest animals that show two evolutionary hallmarks of bilateral symmetry: **cephalization**, or concentration of the nervous system at the head end, and **centralization**, the presence of a central nervous system (CNS) distinct from a peripheral nervous system (PNS). The planarian worm in Figure B has a small brain composed of ganglia (masses of nerve cell bodies) and two parallel nerve cords (bundles of axons and dendrites). These elements constitute the worm's CNS. Other, smaller nerves make up the PNS.

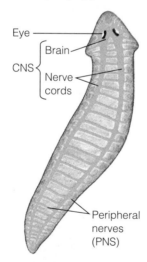

B. Flatworm nervous system

Other invertebrates show greater cephalization and centralization of the nervous system than flatworms. They exhibit the basic pattern of a CNS composed of a brain and one or more nerve cords, but their brains are typically larger and more complex, and the nerve cords are more distinctly set off from peripheral nerves. The insect shown in Figure C, for example, has a brain composed of several

fused ganglia and a ventral nerve cord with a ganglion in each body segment. Each of these ganglia directs the activity of muscles in its segment of the body.

The vertebrate nervous system (Figure D) is highly centralized and cephalized. As we have seen, the CNS consists of a prominent brain, enclosed in a skull, and a **spinal cord**, which lies inside a bony vertebral column, or spine. The spinal cord receives sensory information from the skin and muscles and sends out motor com-

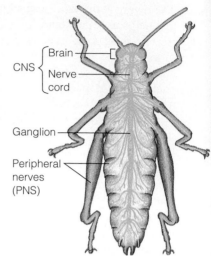

C. Insect nervous system

mands for movement. The brain is the system's master control center. It includes steady-state (homeostatic) centers that keep all body organs functioning smoothly; sensory centers that integrate information from the eyes, ears, nose, and other sense organs; and (in humans, at least) centers of

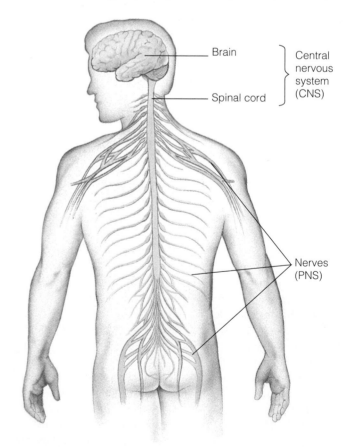

D. Vertebrate nervous system (back view)

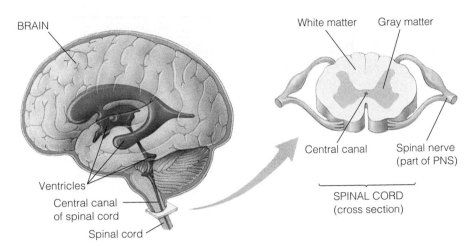

E. Fluid-filled spaces of the vertebrate CNS

called **ventricles** in the brain are continuous with the **central canal** of the spinal cord. The ventricles and central canal contain **cerebrospinal fluid** (blue). This fluid, which also surrounds the brain and spinal cord, cushions the CNS and supplies it with nutrients, hormones, and white blood cells.

The cross section of the spinal cord shows that the CNS has two distinct areas. **White matter** consists of myelinated axons and dendrites; **gray matter** is mainly nerve cell bodies. In the brain, most of the gray matter is cerebral cortex, which is the integration center for sophisticated brain functions such as problem solving. The PNS in vertebrates consists of all the ganglia and nerves outside the brain and spinal cord. We'll examine the peripheral nervous system more closely next and then return to the brain.

emotions and intellect. The brain also exerts master control over the spinal cord and sends out its own motor commands to voluntary muscles. The brain and spinal cord both contain hollow regions (Figure E). Fluid-filled spaces

---

The vertebrate peripheral nervous system is a functional hierarchy

The ganglia and nerves of the peripheral nervous system of vertebrates form a vast communication network. Nerves that carry signals to or from the brain are called **cranial nerves.** Nerves that convey signals to or from the spinal cord are called **spinal nerves.** Your eyes, nose, and ears, for instance, are serviced by branches of your cranial nerves. The muscles and skin of your arms and legs contain branches of your spinal nerves. Every moment, your sensory neurons transmit enormous amounts of information to your CNS; simultaneously, motor neurons relay commands from the CNS to effectors (muscles and glands). All spinal nerves and most cranial nerves contain both sensory neurons and motor neurons. Thousands of incoming and outgoing signals pass each other within the same nerves all the time.

The figure at the right shows the functional hierarchy of the PNS—the organization that allows the PNS to perform its complex tasks. The body's sensory neurons form the **sensory division** (blue) of the PNS. The sensory division has two sets of neurons. One set brings in information about the outside environment (from the eyes, ears, and other external sense organs). The other set supplies the CNS with information about the body itself—for example, data about the acidity of the blood from sensors in large arteries.

The body's motor neurons make up the second major division of the PNS, the **motor division** (purple). Within the

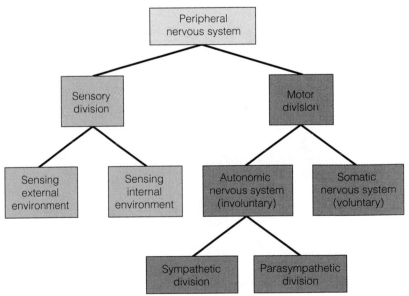

motor division, neurons of the so-called **somatic nervous system** carry signals to skeletal muscles, mainly in response to external stimuli. When you touch a hot stove, for instance, these neurons carry commands to your arm to pull away. The somatic nervous system is said to be voluntary because many of the actions it produces are under conscious control. In contrast, the motor neurons of the **autonomic** (self-governing) **nervous system** are generally involuntary. We discuss the varied functions of the autonomic nervous system next.

Opposing actions of sympathetic and parasympathetic neurons regulate the internal environment

Our autonomic nervous system consists of two sets of neurons with opposing effects on most body organs. One set, called the **parasympathetic division,** primes the body for digesting food and resting—activities that gain and conserve energy for the body. A sample of the effects of parasympathetic signals appear on the left in the figure below. These include stimulation of digestive organs such as the salivary glands, stomach, and pancreas; decrease of the heart rate; and narrowing of the bronchi, which correlates with a decreased breathing rate.

The other set of neurons, the **sympathetic division,** tends to have the opposite effect, preparing the body for in-

tense, energy-consuming activities, such as fighting, fleeing, or competing in a strenuous game. You see some of the effects of sympathetic signals on the right side of the figure. The digestive organs are inhibited, the bronchi are relaxed so that more air can pass through them, the heart rate is increased, the liver releases the high-energy compound glucose into the blood, and the adrenal glands secrete the fight-or-flight hormones epinephrine and norepinephrine.

Fight-or-flight and relaxation are opposite extremes. Our bodies usually operate at intermediate levels, with most of our organs receiving both sympathetic and parasympathetic signals (the liver and adrenal medulla are exceptions). The

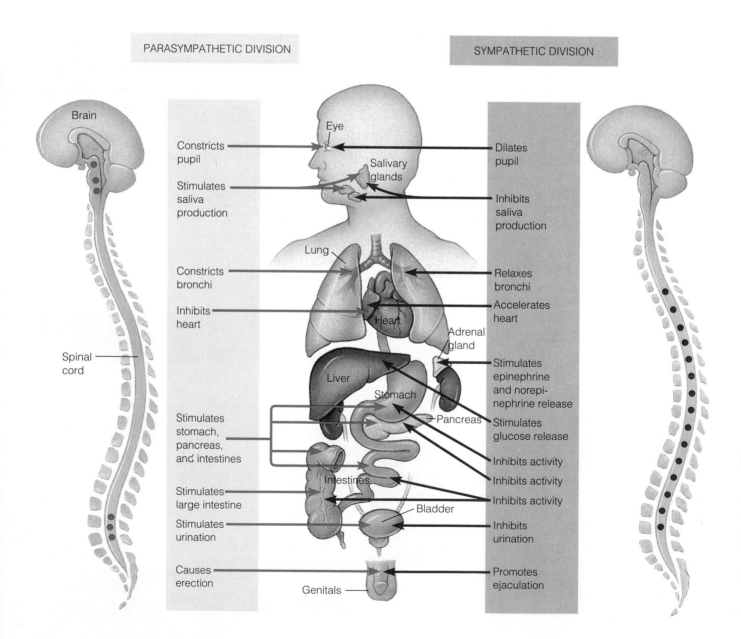

opposing signals adjust an organ's activity to a suitable level. Right now, for instance, your salivary glands are probably secreting just enough saliva to keep your mouth moist. The glands are receiving parasympathetic signals that tend to increase their activity, but these signals are counterbalanced by sympathetic signals with the opposite effect. The situation changes when you think about eating your favorite snack. Your thoughts then trigger an increase in parasympathetic signals without changing sympathetic signals. You will notice more saliva in your mouth as the parasympathetic signals overpower the sympathetic ones.

Sympathetic and parasympathetic neurons emerge from different regions of the CNS, and they use different neurotransmitters. As the green dots in the figure indicate, neurons of the parasympathetic system emerge from the brain and the lower part of the spinal cord. Most parasympathetic neurons produce their effects by releasing the neurotransmitter acetylcholine at synapses with target organs. In contrast, neurons of the sympathetic system emerge from the middle regions of the spinal cord (red dots). Most sympathetic neurons release the neurotransmitter norepinephrine at target organs.

Carrying command signals, the sympathetic and parasympathetic neurons constitute lower levels of the nervous system's hierarchy. In the next several modules, we take a closer look at the higher levels of the hierarchy. The components of the CNS, the spinal cord and the brain, are the sites where command signals originate.

The spinal cord produces the simplest behaviors in vertebrates 28.14

The spinal cord, shown below in cross section, has two principal functions: It carries information to and from the brain, and it formulates simple responses, called reflexes, to certain kinds of stimuli. A **reflex** is an unconscious response to a specific stimulus. The knee-jerk reflex, as depicted in Figure A, is an example of the simplest type of reflex, involving only two neurons. Doctors often test this reflex to determine whether nerves that control leg movements are functioning normally. A tap with a rubber mallet, as shown, stimulates sensory receptors in the thigh muscles. Unless there is damage to nerves in the thigh, the receptors detect the blow and generate action potentials in sensory neurons. Action potentials in each sensory neuron then pass up the thigh to the spinal cord. In the spinal cord, the sensory neuron synapses directly with a motor neuron. If the signal is strong enough, it generates an action potential in the motor neuron. When an action potential reaches the thigh muscle, it makes it contract, jerking the lower leg forward.

Most reflexes are more complex than the knee jerk and include one or more interneurons in the spinal cord between the sensory neuron and the motor neuron. The response to sharp pain is one example (Figure B). In this case, an interneuron in the spinal cord receives sensory signals from a sensory neuron, integrates the signals, and sends a command to withdraw via a motor neuron out to the arm muscles. This reflex can be modified by input from the brain. Acting on the interneuron, signals from the brain may increase or decrease the reflex or even override it. This explains why we can consciously modify our responses to pain.

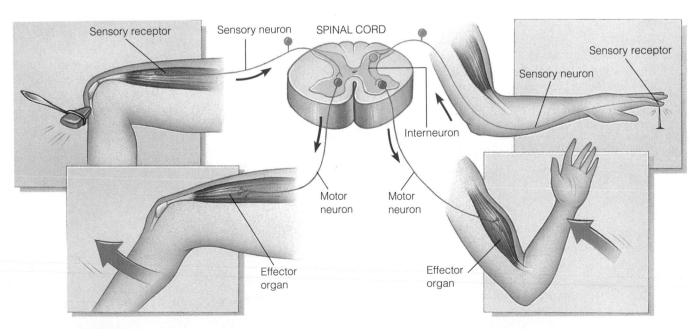

A. The knee-jerk reflex

B. A reflex response to a sharp pain

Complex behavior evolved with the increasing complexity of the vertebrate brain

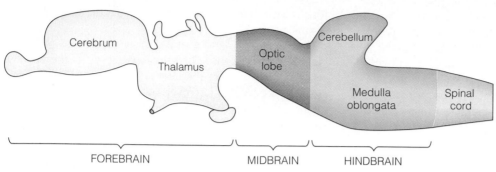

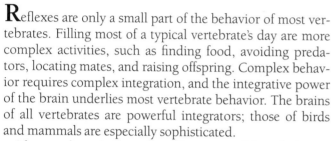

A. The ancestral vertebrate brain

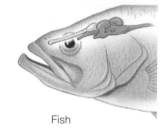

Fish

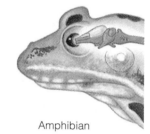

Amphibian

Reptile

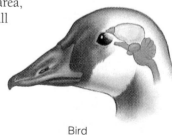

Bird

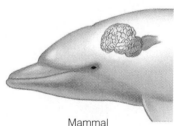

Mammal

Reflexes are only a small part of the behavior of most vertebrates. Filling most of a typical vertebrate's day are more complex activities, such as finding food, avoiding predators, locating mates, and raising offspring. Complex behavior requires complex integration, and the integrative power of the brain underlies most vertebrate behavior. The brains of all vertebrates are powerful integrators; those of birds and mammals are especially sophisticated.

The vertebrate brain evolved from a set of three bulges at the anterior end of the spinal cord. These three ancestral regions, called the **forebrain, midbrain,** and **hindbrain** (Figure A), still appear during embryonic development in all vertebrates. As the vertebrate brain evolved, three major trends altered the three ancestral brain regions. First, the relative size of the brain increased in certain evolutionary lineages. Birds and mammals have bigger brains relative to body size than fishes, amphibians, and reptiles.

A second trend in brain evolution was the subdivision of the three original regions. We see evidence of this today when we watch the brain of a vertebrate developing from its embryonic form into its adult form. The forebrain, midbrain, and hindbrain gradually become subdivided into regions that assume specific responsibilities. We will take a closer look at these regions and their functions when we focus on the human brain in the next module.

The third trend in the evolution of the vertebrate brain was the increasing integrative power of the forebrain. Evolution of the most complex vertebrate behavior paralleled evolution of the **cerebrum,** the dominant part of the forebrain. As shown in Figure B, the cerebrums (gold areas) of birds and mammals are much larger relative to the other parts of the brain (blue areas) than the cerebrums of other vertebrates. (The porpoise is drawn at a smaller scale than the other animals.) The larger cerebrum is directly correlated with the more sophisticated behavior of birds and mammals.

Notice in Figure B that the drawing of the porpoise's cerebrum is marked with lines. These represent folds that increase the surface area of the cerebrum. Because the nerve cell bodies are in the cortex, or outer layer, of the cerebrum, the amount of surface area of the brain is more important in determining performance than the overall volume. Porpoises and primates have much larger and more complex cerebral cortices than any other vertebrates. Although less than 5 mm thick, the highly folded human cerebral cortex occupies over 80% of the total brain mass. This is the largest brain surface area, relative to body size, of all animals. The cerebral cortex of the porpoise comes in second. Porpoises have complex social interactions and communicate using a large repertory of sounds. They also have the ability to locate objects, such as prey, using sound echoes. Much of their cerebral cortex may be devoted to processing information about their sound-oriented world.

B. Vertebrate brains compared

The structure of a living supercomputer: the human brain

Composed of about 100 billion intricately organized neurons, with a much larger number of supporting cells, the human brain is more powerful than the most sophisticated computer. Let's take a look at its overall structure.

In Figure A, we see that the three ancestral brain regions have evolved considerably from their original state. The parts of the hindbrain are shown in shades of blue. The **medulla oblongata** and the **pons** of the hindbrain house all the sensory and motor neurons passing between the spinal cord and the forebrain. Thus, one of the main functions of the medulla and pons is conducting information. They also control breathing rate, heart rate, and digestion, and help coordinate whole-body movements, such as walking.

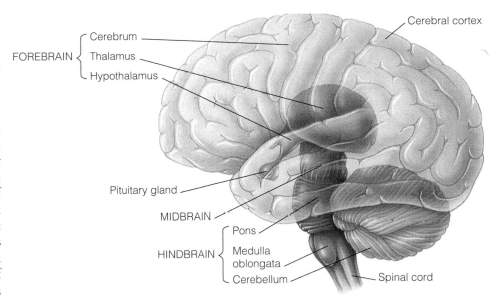

A. Main parts of the human brain

The third part of the hindbrain, the **cerebellum**, is mainly a planning center for coordinating body movement. The cerebellum receives information from sensory receptors in muscles. At the same time, it receives signals from the cerebrum outlining changes that are needed in body movements. The cerebellum evaluates all these inputs and, in a few milliseconds, sends a plan for smooth, coordinated movements back to the cerebrum. When you step off a curb, for instance, your cerebellum evaluates your body position automatically and relays to your cerebrum a plan for smoothly getting the rest of your body to follow. The cerebrum then fine-tunes the plan and sends appropriate commands to the muscles.

The midbrain (violet) also relays sensory information to the cerebrum and integrates several types of sensory information, such as sound and, in nonmammalian vertebrates, sight. In mammals, vision is integrated mainly in the forebrain, with the midbrain coordinating eye reflexes. Part of the midbrain helps regulate sleep and arousal.

The most sophisticated neural processing occurs in the forebrain. Its three main integrating centers are the thalamus (red), the hypothalamus (orange), and the cerebrum (gold). The **thalamus** contains most of the cell bodies of neurons that relay information to the cerebral cortex. Before sending information to the cerebrum, the thalamus sorts data into categories (all the touch signals from a hand, for instance). It also suppresses some signals and enhances others. Thus, the thalamus exerts some control over what information gets from sense organs and other parts of the brain to the cerebrum.

The **hypothalamus** is small, but it is very important for regulating homeostasis. In Module 26.4, we saw that it con-trols the pituitary gland and the secretion of many hormones. The hypothalamus also regulates body temperature, blood pressure, hunger, thirst, sexual urges, and fight-or-flight responses, and it helps us experience emotions such as rage and pleasure. A "pleasure center" in the hypothalamus could also be called an ad-

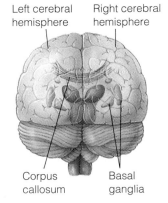

B. Rear view of the brain

diction center, for it is strongly affected by certain addictive drugs. Cocaine, for example, stimulates this pleasure center, producing a short-term (5–20 min) high, often followed by depression. Cocaine addiction may involve chemical changes in the pleasure center and elsewhere in the hypothalamus.

The **cerebrum**, the largest and most sophisticated brain center, consists of right and left **cerebral hemispheres** (Figure B). In placental mammals, a thick band of nerve fibers called the **corpus callosum** connects the hemispheres and allows them to process information together. Under the corpus callosum, clusters of nerve cell bodies referred to as the **basal ganglia** (green) form another part of the cerebrum. These are important in motor coordination. If they are damaged, a person may become passive and immobile. Degeneration of neurons entering the basal ganglia occurs in Parkinson's disease.

The largest part of the cerebrum, the part of the brain that has changed the most during evolution, is the cerebral cortex, the subject of the next module.

28.17 The cerebral cortex is a mosaic of specialized, interactive regions

The human **cerebral cortex** is a highly folded sheet of gray matter containing some 10 billion neurons and hundreds of billions of synapses. Its intricate neural circuitry produces our most distinctive human traits: all of our reasoning and mathematical abilities, language skills, powers of imagination, artistic talents, and personality traits. Assembling information it receives from our eyes, ears, nose, taste buds, and touch sensors, the cortex also creates our sensory perceptions—what we are actually aware of when we see, hear, smell, taste, or touch.

Discovering how the cerebral cortex works is a monumental task. What we know so far comes mainly from studies of the effects of tumors, strokes (death of brain cells due to blocked blood vessels), and accidental damage to parts of the cortex. The effects of electrical stimulation of the cortex performed during surgery have also advanced our knowledge. One thing the cortex lacks are cells that detect pain; thus, after anesthetizing the scalp, a neurosurgeon can operate on the cerebrum with the patient awake. Various parts of the cortex can then be stimulated with microelectrodes that produce a harmless electrical current, and a researcher can obtain information about the effects by simply questioning the patient. Brain researchers are also using PET scans and other imaging techniques we discussed in Module 20.10 to find out which parts of the cortex are most active when the brain is performing specific tasks.

The surface of the cerebral cortex, for both the right and the left hemispheres, has four discrete lobes (represented by different colors in the drawing below). Researchers have identified a number of functional areas within each lobe. The figure identifies the main functional areas in the brain's left cerebral hemisphere. Two of these areas are located where the frontal and parietal lobes meet. The area called the motor cortex, at the rear edge of the frontal lobe, functions mainly in sending commands to skeletal muscles, signaling appropriate responses to sensory stimuli. Next to the motor cortex, in the parietal lobe, lies the somatosensory cortex. This functional area receives and partially integrates signals from touch, pain, pressure, and temperature receptors throughout the body. The cerebral cortex also has centers that receive and begin processing sensory information concerned with vision, hearing, taste, and smell. Each of these centers, as well as the somatosensory cortex, cooperates with an adjacent area, called an association area. In ways researchers are just beginning to understand, a complicated interchange of signals among receiving centers and association areas produces our sensory perceptions.

Making up most of our cerebral cortex, the association areas are the sites of higher mental activities—roughly, what we call thinking. A large association area in the frontal lobe uses varied inputs from many other areas of the brain to evaluate consequences, make considered judgments, and plan for the future. Language results from some extremely complex interactions among several association areas. For instance, the parietal lobe of the cortex has association areas used for reading and speech. These areas obtain visual information (the appearance of words on a page) from the vision centers. Then, if the words are to be spoken aloud, they arrange the information into speech patterns and tell another speech center, in the frontal lobe, how to make the motor cortex move the tongue, lips, and other muscles to form words. When we hear words, the parietal areas perform similar functions using information from auditory centers of the cortex. You can see the locations of the various language centers in the PET scans in Figure 20.10.

What we have said about association areas so far applies to the left hemisphere of the cerebral cortex. As we see in the next module, the association areas in the right hemisphere have somewhat different functions.

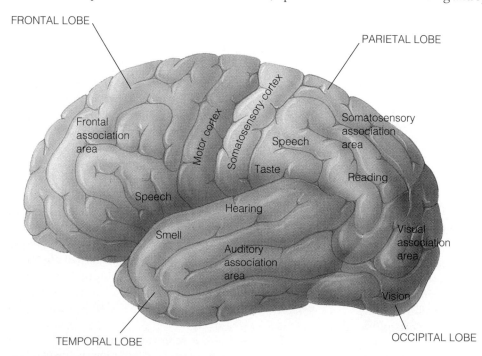

FRONTAL LOBE

PARIETAL LOBE

Frontal association area

Motor cortex

Somatosensory cortex

Speech

Somatosensory association area

Taste

Reading

Speech

Hearing

Smell

Visual association area

Auditory association area

Vision

TEMPORAL LOBE

OCCIPITAL LOBE

Functional areas of the cerebrum's left hemisphere

Roger Sperry discovered that the right and left cerebral hemispheres function differently

Our two cerebral hemispheres look alike superficially. Their primary motor and sensory areas also function in the same way. However, the left and right association areas function so differently that we could almost say we have two brains in one. The left hemisphere houses our language centers, as we have just seen; it also has association areas for logic and mathematical abilities. In contrast, the right hemisphere lacks language, logic, and math centers, but has association areas that underlie our imagination, spatial perceptions, artistic and musical abilities, and emotions.

Much of what we know about the differing functions of our right and left cerebral hemispheres stems from the work of Nobel Prize winner Roger Sperry (above). Starting in the 1960s, Sperry and his colleagues performed extensive studies of "split-brain" patients, people who had a severed corpus callosum, the mass of neuron fibers that pass information back and forth between the right and left hemispheres (see Figure 28.16B). Most of Sperry's subjects were patients whose corpus callosum had been surgically cut in a now-outmoded treatment for severe epileptic seizures. The technique was drastic, but the patients improved rapidly and did not seem to suffer from personality changes or loss of intelligence. However, as Roger Sperry was to show, cutting masses of neurons does cause problems.

Sperry's first subject was a veteran whose war wounds had left him with uncontrollable seizures. Sperry wrote in "Left-Brain, Right-Brain" (*Saturday Review*, August 9, 1975):

> Upon recovery from the [corpus callosum] surgery, he was free of seizures and seemed quite normal in his everyday behavior. . . . But his surface normality seemed to overlie some startling changes in his inner mental makeup—a suspicion we were to confirm in studies of additional subjects who had had the same operation.

The figure here illustrates what Sperry was referring to. As shown in panel 1, a split-brain patient holding a key in the left hand, with both eyes open, can readily name it

as a "key." If blindfolded, the subject can recognize the key and use it to open a lock, but will be unable to name it (panel 2). The problem results from the fact that the right side of our brain receives much of the information it integrates from sense organs on the left side of the body, and the left side of the brain from the right side of the body. Thus, sensory information about a key in a patient's left hand goes to the right hemisphere (panel 3). But the speech center is in the left hemisphere. Because there is no intact corpus callosum to bring the information about the key from the right hemisphere, the patient cannot name the key.

Summarizing this and other findings, Sperry wrote:

> It is most impressive and compelling to watch a [split-brain] subject solve a given problem like two different people . . . using two quite different strategies—depending on whether he is using his left or his right hemisphere. Both the left and right hemispheres of the brain have been found to have their own specialized forms of intellect. The left is highly verbal and mathematical, performing with analytic, symbolic, computerlike, sequential logic. The right, by contrast, is spatial and mute, performing with a . . . kind of information processing that cannot yet be simulated by computers.

The realization that our right and left hemispheres contain different association areas has led to interesting speculations. One idea is that our brain is actually a fusion of two very distinct parts: a right "intuitive" brain and a left "rational" brain. A related idea is that, in each person, either the right brain or the left brain is dominant—that artists, for instance, are right-brain dominant, whereas mathematicians and engineers are left-brain dominant. These hypotheses remain largely unsubstantiated, and they leave open the question of what might be the evolutionary advantages of having different functional areas on the right and left sides of our brain.

In the remaining modules of this chapter, we examine several other topics of active brain research—arousal and sleep, emotions, memory, and learning.

(1) Can name key

(2) Cannot name key in left hand

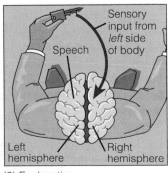

(3) Explanation

A split-brain patient demonstrating the different functions of the left and right hemispheres

28.19 The brain is active during sleep

As anyone who has sat through a lecture on a warm spring day knows, attentiveness and mental alertness vary from moment to moment. Arousal is a state of awareness of the outside world. Its counterpart is sleep, a state in which we continue to receive external stimuli but are not conscious of them.

Researchers study the patterns of electrical activity in the brain during arousal and sleep by a technique called electroencephalography. Electrical contacts placed on the head (Figure A) detect signals from action potentials in brain neurons. The wires lead to a recording device that converts streams of such electrical signals, called brain waves, to a graph called an **electroencephalogram**, or EEG (Figure B). In general, the less mental activity taking place, the more regular the graph appears. The top part of Figure B shows the fairly regular, slow brain waves (known as alpha waves) recorded when a person is awake but quiet, with closed eyes. The middle recording, made when a person was solving a complex mental problem, consists of faster, more irregular waves (beta waves).

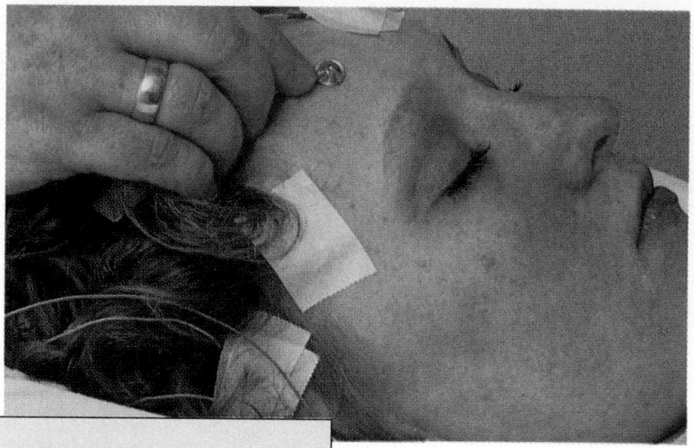

The bottom part, showing a typical sleep cycle, reveals that sleep is a dynamic process. The EEG shows two, alternating patterns: a period of slow, fairly regular brain waves (delta waves), indicating deep sleep, when the brain is largely inactive; and a period called **REM sleep**, when the brain waves are fairly rapid and less regular, more like those of the awake state. During REM (rapid-eye-movement) sleep, the eyes move rapidly under the closed lids. The brain itself is highly active and may consume more oxygen than it does when awake. We have most of our dreams during REM sleep, which typically occurs about six times a night for 5- to 50-minute periods. It has been suggested that dreams help the brain analyze recent experiences and emotional problems or store information in its memory banks, or that dreams are simply the brain's attempt to make sense of random nerve signals. The true function of dreaming is still unknown.

Several parts of the brain regulate sleep, dreams, and arousal. Processing centers in the hindbrain, for example, induce sleep and cause REM sleep and non-REM sleep. The cerebral cortex creates dream images (what we imagine we see or hear, for instance) from signals it receives from the hindbrain. An arousal center in the midbrain fires the action potentials that wake us up. The mechanisms of arousal and sleep are fairly well understood, but the question of what sleep actually does for us remains unanswered.

A. Electrical contacts on the scalp for an EEG

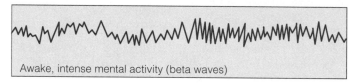

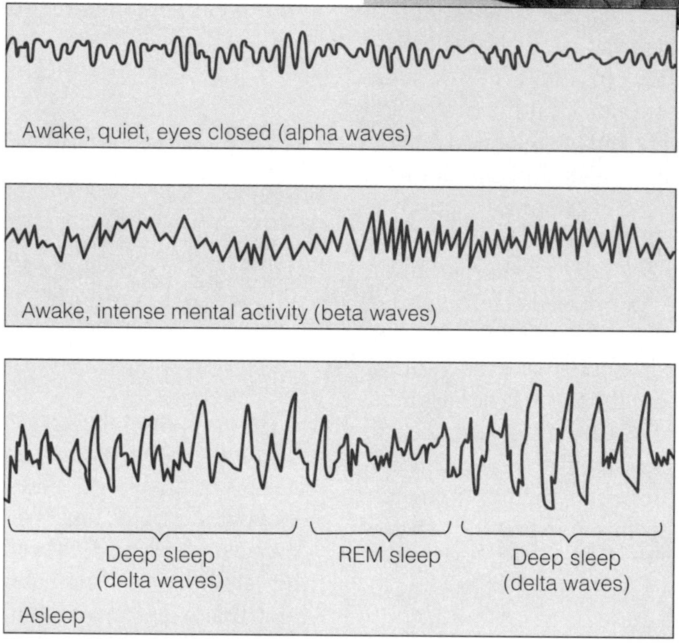

Awake, quiet, eyes closed (alpha waves)

Awake, intense mental activity (beta waves)

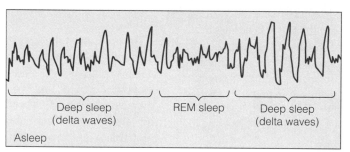

Deep sleep (delta waves)   REM sleep   Deep sleep (delta waves)

Asleep

B. Some EEG printouts

28.20 The brain's limbic system underlies emotions and memory

What enables us to have emotions—to feel joy, sadness, or anger? What gives us the ability to remember past experiences? These questions have intrigued philosophers and biologists for generations, and we are still far from complete answers. What we have so far is a partial map of parts of the brain that generate emotions and serve as our memory banks. This map comes mainly from studies of brain-damaged people and animals.

Emotions and memory arise in our **limbic system,** a functional unit of several integrating centers and interconnecting neuron tracts in our forebrain. As you can see in the figure below, the limbic system (colored gold) includes the thalamus and a ring around it formed by parts of the hypothalamus and inner portions of the cerebrum. Two of its components, the amygdala and the hippocampus, function with another part of the cerebral cortex, the frontal cortex, in memory processing and retrieval.

We sense our limbic system's role in both emotion and memory when certain odors bring back "scent memories," odor-related emotional experiences of our past. Have you ever had a particular smell suddenly make you nostalgic for something that happened when you were a child? You can see the structural basis for scent memories in the figure. Signals from your nose enter your brain through the olfactory bulb (lower left), which is part of the limbic system. Thus, a specific scent can immediately trigger activity in the memory and emotion centers of the limbic system. Any other sensory memories (visual, touch, taste, or sound) associated with the scent probably come from other centers in the cortex.

Memory, which is essential for learning, is the ability to store and retrieve information related to previous experiences. The computer this chapter was typed on can hold about 5 million bits (about 2500 pages) of information. The human brain, which may be the most complex structure in the universe, probably stores and retrieves billions of bits of information during a person's lifetime.

We have more than one kind of memory. **Short-term memory,** as the name implies, lasts only a short time—usually only a few minutes. It is short-term memory that allows you to dial a phone number just after looking it up. If you call the number frequently, happen to associate it with some intense emotion, or try to memorize it, the number may become stored in **long-term memory.** You will be able to recall it weeks after you originally looked it up, or even longer. Short-term memory may result from the ongoing electrical activity of action potentials in the brain's memory centers. Long-term memory involves structural or chemical changes at synapses. Our use of long-term memory requires **working memory,** the ability to retrieve stored information and apply it to current situations.

By experimenting with animals, studying amnesia (memory loss) in humans, and using brain imaging techniques, scientists have mapped some of the major brain pathways involved in memory. The following is a possible pathway by which your brain might memorize a telephone number (the circled numbers correspond to the figure):

① When you read a phone number, signals from your eyes first go into the vision centers at the back of your cerebral cortex. There the signals are integrated.

② The vision centers send action potentials representing the phone number to the amygdala and the hippocampus.

③ From there, the information is relayed to the thalamus.

④ Then the information travels to other parts of the forebrain, including the prefrontal cortex.

⑤ The pathway is completed and the information filed in long-term memory when signals return to the vision centers in the cortex.

What do the amygdala, hippocampus, and prefrontal cortex actually do? The answer is far from complete. The amygdala seems to act as a type of memory filter, somehow determining what information will be stored and what will not. The amygdala may label information to be saved by tying it to an event or emotion of the moment. For example, a telephone number that you associate with affection for a close friend may be labeled with that feeling before being relayed to long-term memory centers. This may explain why we remember information better when we are particularly interested in it or when it is associated with an intense emotion or unusual event.

The hippocampus, which is part of the cerebral cortex, seems to be a major center for long-term memory. Chemical changes at synapses in the hippocampus store at least some types of information for weeks before relaying them to other parts of the cortex.

The prefrontal cortex seems to be the center of working memory. Somehow it accesses long-term memory banks and works with other brain centers in using memories to modify behavior.

Unraveling more about how neurons in these brain centers construct emotions and store, retrieve, and use memories will be one of the truly exciting areas of future research.

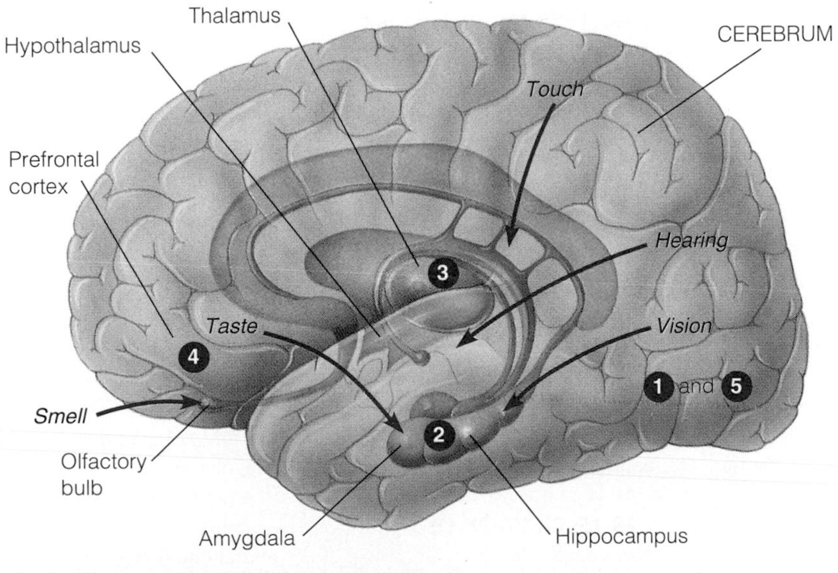

The limbic system (gold areas) and a possible pathway for establishing a visual memory (numbers)

28.21 Studies of learning in invertebrates bridge the gap between neurons and behavior

We began this chapter with a discussion of how the squid, a member of the phylum Mollusca, has contributed to our understanding of nervous system function. The photograph here shows a different mollusk, the sea hare *Aplysia*, a type of marine snail; it is about 30 cm (1 ft) long. The sea hare's entire nervous system consists of only about 20,000 neurons (compared to about 100 billion neurons in the human brain). Despite its simplicity, *Aplysia* displays some simple forms of learning, and its particular combination of simplicity and learning ability makes it a very useful experimental animal.

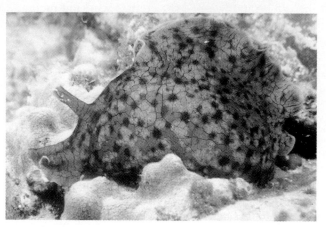

The sea hare *Aplysia*

Aplysia has a number of simple reflexes, and it can be taught to modify some of them. For example, it ordinarily withdraws quickly when prodded by a blunt instrument, but it will learn to ignore repeated mild prodding. This type of behavioral change involves only a few neurons, and re-

searchers can use it to study the cellular events that underlie learning in general.

Some of the most interesting studies of *Aplysia* have to do with the chemical basis of learning. The animal's withdrawal reflex, for instance, is strengthened by repeated prods that are coupled with mild electrical stimuli. As the reflex gets stronger, the synaptic knobs of the sensory neuron that detects the stimuli release more neurotransmitter, which, in turn, stimulates the motor neuron that causes the reflex.

Studies of relatively simple nervous systems, like that of the sea hare, are part of a rapidly expanding research effort that aims to work out the biological basis of learning, memory, problem solving, feeling, sleep, and virtually everything else our brain does so well for us. In the next chapter, we examine another aspect of nervous systems—how sense organs gather information about the environment.

Chapter Review

Begin your review by rereading the module headings and Module 28.1 and scanning the figures before proceeding to the Chapter Summary and questions.

Chapter Summary

Introduction–28.3 The nervous system enables an animal to respond quickly and precisely to environmental changes. The functional units of the nervous system are neurons, cells that carry signals. Three kinds of neurons illustrate the three main functions of the nervous system. Sensory neurons conduct signals from receptors to the brain and spinal cord. Interneurons integrate this information and transmit it to motor neurons. Motor neurons, in turn, conduct signals to effectors, muscles or glands that carry out responses.

28.4–28.6 At rest, a neuron's plasma membrane has an electrical voltage called the resting potential. A stimulus alters the permeability of a portion of the membrane, allowing ions to pass through and changing the membrane's voltage. A nerve signal, called the action potential, is an all-or-none change in the membrane voltage, from the resting potential to a maximum level and back to the resting potential. Action potentials are self-propagated in a one-way chain-reaction sequence along a neuron.

28.7–28.10 At most synapses—junctions between neurons—the transmitting cell secretes a chemical neurotransmitter, which crosses the synaptic cleft and binds to the receiving cell. Some neurotransmitters excite the receiving cell; others inhibit, by decreasing

the receiving cell's ability to develop action potentials. A cell may receive differing signals from many neurons; the summation of excitation and inhibition determines whether or not it will transmit nerve signals. Hallucinogens, stimulants, and depressants often act at synapses, and there seems to be a link between depression and brain neurotransmitters.

28.11–28.13 The vertebrate nervous system shows a high degree of cephalization, which is concentration in the head end, and centralization, which is the presence of a central nervous system (CNS). A CNS consists of a brain and spinal cord. The nerves and ganglia outside the CNS make up the peripheral nervous system (PNS). The PNS has sensory and motor functions, and its motor functions can be further subdivided. The somatic nervous system of the PNS exerts mostly voluntary control over the skeletal muscles. The autonomic nervous system of the PNS exerts mostly involuntary control over the internal organs.

28.14–28.16 The spinal cord carries information to and from the brain and controls many reflexes, which are simple, involuntary responses. More complex behaviors are controlled by the brain. The parts of the human brain specialize in different tasks. Various subdivisions of the hindbrain, midbrain, and lower forebrain conduct information to and from higher brain centers, regulate homeostatic functions, keep track of body position, and sort sensory information.

28.15, 28.17–28.18 In humans, the cerebrum is the largest and most complex brain center. Most of the neural circuitry is located in the outer cerebral cortex of the two cerebral hemispheres. The somatosensory cortex and centers for vision, hearing, taste, and smell integrate sensory information. The motor cortex directs responses. Association areas, concerned with higher mental activities such as reasoning and language, make up most of the cerebrum. The right and left cerebral hemispheres specialize in different mental tasks.

28.19–28.21 Several parts of the brain function in arousal, sleep, emotions, and memory. The brain is active during sleep, but the functions of sleep and dreams are not known. Our emotions appear to depend on interaction between the cortex and the limbic system, a collection of brain centers in the cerebrum and lower forebrain. Some of these centers are also involved in short-term memory, a first step in sorting information for long-term memory in the cortex. Research suggests that learning and memory involve chemical changes at synapses.

Testing Your Knowledge

Multiple Choice

1. Joe accidentally touched a hot pan. His arm jerked back, and an instant later, he felt the burning pain. How would you explain that his arm moved before he felt the pain?

    a. His limbic system blocked the pain momentarily, but the important pain signals eventually got through.

    b. His response was a spinal cord reflex that occurred before the pain signals got to the brain.

    c. It took a while for his brain to search long-term memory and figure out what was going on.

    d. The autonomic nervous system responded to the danger because the brain was too busy to react quickly.

    e. This scenario is not actually possible. The brain must register pain before a person can react.

2. Suppose a researcher carefully removes the myelin sheath from the axon of a neuron. The most likely result would be that the altered neuron would

    a. no longer be able to produce neurotransmitters

    b. be much more sensitive to stimuli

    c. be much less sensitive to stimuli

    d. transmit nerve signals more rapidly

    e. transmit nerve signals more slowly

3. The folds of the cerebrum increase its surface area. This is important in the brain's performance because it

    a. maximizes the number of cell bodies that process information

    b. prevents a "short circuit" between adjacent areas

    c. allows the cerebrum to absorb more oxygen

    d. increases myelin, which speeds up nerve signals

    e. more effectively protects the cerebrum from damage

4. Anesthetics block pain by blocking the transmission of nerve signals. Which of these three chemicals might work as anesthetics? (*Explain your answer.*)

    I. a chemical that prevents the opening of sodium gates in membranes

    II. a chemical that inhibits the enzymes that degrade neurotransmitter

    III. a chemical that blocks neurotransmitter receptors

    a. I   **c.** III   **e.** I and III

    b. II   **d.** II and III

5. A man suffered a head injury in a car accident. He subsequently suffered from uncontrollable mood changes, emotional outbursts, and a curious form of amnesia. Although he could remember incidents before the accident in detail, he was not able to form new memories. Which part of the brain did his doctor conclude had been damaged?

    a. cerebellum   **c.** limbic system   **e.** hypothalamus

    b. medulla oblongata   **d.** corpus callosum

Matching

1. Controls breathing and heart rate
2. Regulates temperature, hunger, thirst
3. Connects the cerebral hemispheres
4. Assists the cerebrum in coordinating movements
5. Handles language, judgment, and reasoning
6. Carries information to and from the brain

    a. cerebellum
    b. cerebral cortex
    c. spinal cord
    d. corpus callosum
    e. hypothalamus
    f. medulla oblongata

Describing, Comparing, and Explaining

1. As you hold this book, nerve signals are generated in nerve endings in your fingertips and sent to your brain. Once a touch has caused an action potential at one end of a neuron, what causes the nerve signal to move from that point along the length of the neuron to the other end? What is the nerve signal, exactly? Why can't the signal go backward? How is the nerve signal transmitted from one neuron to the next across a synapse? Write a short paragraph that answers these questions.

2. Suppose you decide to turn to the next page of this book. By what route does your brain receive the nerve signals telling it that your fingers are securely grasping the corner of the page? What part of your brain makes the decision to initiate the muscle movements required to turn the page? What part of your brain helps make the movements smooth and coordinated? How do the nerve signals get to the muscles of your arm and hand?

Thinking Critically

1. Movement of a nerve signal along a neuron is most like which of the following? (a) the flow of water through a pipe, (b) the movement of a car along a highway, (c) the burning of a dynamite fuse. In a sentence or two, explain why. In what ways is a nerve signal different from your choice?

2. Using microelectrodes, a researcher recorded nerve signals in four neurons in the brain of a snail, referred to as A, B, C, and D in the table below. A, B, and C all can transmit signals to D. In three experiments, the animal was stimulated in different ways. The numbers of nerve signals transmitted per second by each of the cells is recorded in the table. Write a short paragraph explaining the different results of the three experiments.

	Signals/sec			
	A	B	C	D
Experiment #1	50	0	40	30
Experiment #2	50	0	60	45
Experiment #3	50	30	60	0

Science, Technology, and Society

1. Elaine suffered a serious head injury in an accident. She lapsed into a coma, and her EEG indicated that her cerebral cortex had stopped functioning. Her lower brain centers continued to control activites such as heartbeat and breathing. Her family asked that she be taken off all medication and tube feeding so she could "die with dignity." Legal authorities refused to allow the hospital to withdraw treatment. Should the family's wishes be respected? Should the woman's life be maintained at all costs? Under what circumstances do you think life support should be withdrawn from a severely brain-damaged person? How would you want this to be handled if you were such a patient?

2. Alcohol's depressant effects on the nervous system cloud judgment and slow reflexes. Alcohol consumption is a factor in most fatal traffic accidents in the United States. What are some other impacts of alcohol abuse on society? What are some of the responses of people and society to alcohol abuse? Do you think this is primarily an individual or societal problem? Do you think our responses to alcohol abuse are appropriate and proportional to the seriousness of the problem?

A fatal encounter between a grizzly bear and a migrating salmon in a glacial stream in Alaska: Salmon are fast and agile, but easy prey for a bear in shallow water. Despite the grizzly's rather poor eyesight and great size (up to about 450 kg, or nearly 1000 pounds), it has lightning reflexes and can snatch up a large fish with little effort.

What brought these animals together at this particular moment? The grizzly learned as a cub to head for the salmon streams in late autumn. It followed its mother there and learned how to fish by watching her. The bear's acute sense of smell helped it find the stream. For the salmon, becoming a meal for the bear suddenly ends a remarkable odyssey. After nearly eight years at sea, it was returning to reproduce in the stream where its life began. The salmon has a keen sense of smell, and odors led it to this particular stream among thousands of possibilities.

In the fall, a female salmon lays her eggs in a shallow depression on the bottom of a small, fast-flowing stream. A male covers the eggs with sperm, and both parents die soon thereafter. The following spring, larval salmon hatch and begin drifting downstream toward the ocean, starting a journey that will take them far out to sea. While at sea, they feed in large schools with salmon from other river systems. When they are 4 to 6 years old and sexually mature, the salmon segregate into groups of common geographic origin and start migrating back toward the river from which they emerged as juveniles.

Research on how salmon find their way to their home stream has been going on since the early 1950s. The fish seem to navigate to the mouth of their home river system by using the angle of the sun for reference. Once there, however, their sense of smell takes over.

The water that flows from each stream into a river seems to carry a unique scent resulting from the mixture of plants and soils in the area. The scent of its home stream apparently becomes fixed in the memory of a young salmon before it migrates to the sea. When a mature salmon arrives in the vicinity of its home river system, it swims along the coast until it detects the faint odor matching the scent memory in its brain. In response to perhaps only a few molecules of its "home chemicals," it enters that river and begins its upstream journey.

A salmon has a nostril on each side of its head. In the photograph below, you can see that each nostril has two openings. As the fish swims, water flows in one opening and out the other, and highly sensitive cells in the nostril detect chemicals dissolved in the water. When researchers blocked the nostrils of migrating salmon with cotton, the fish lost their homing ability. They still migrated, but could not make the correct choices at forks of streams.

As a salmon swims upstream, its nostrils follow a scent trail in the water. The closer it gets to its home stream, the stronger the scent becomes. More and more of the scent molecules stimulate sensory cells in its nostrils, and the cells send more and more signals to the fish's brain, telling it that it is on the right track.

Sensory information gathered by sensory receptors and processed by the brain guide salmon and grizzly bears to specific stream sites. This chapter focuses on sensory structures and how they gather information. To begin, we examine the distinction between information gathering and information processing—the difference between sensing chemicals in the air or water, for instance, and recognizing them as odors.

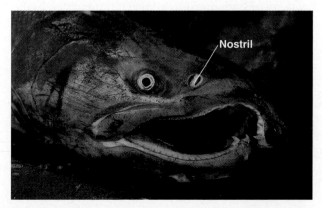

Nostril

29.1 Sensory inputs become sensations and perceptions in the brain

Sensory receptor cells are tuned to the condition of the external world and the internal organs. They detect stimuli, such as chemicals, light, tension in a muscle, sounds, electricity, cold, heat, and touch, and send information to the central nervous system. The sensory cells in a bear's nose, for instance, detect chemicals in the air and send reports about the chemicals to the brain. The reports take the form of action potentials, the same signals used throughout the nervous system (see Module 28.5). The sensory receptor's job is completed when it triggers action potentials that go to the central nervous system.

What happens to sensory information in the central nervous system—our brain, for example? When action potentials reach the brain, we may experience a **sensation**, an awareness of sensory stimuli. Sensations result when the brain integrates new information. If you detect the scent of a carnation, for instance, and you have never had any experience that would connect the scent with a flower (or with a substance that smells like one), you simply have a sensation; your brain integrates sensory signals from your nose and makes you aware of a pleasant odor.

In most cases, the brain does much more than form sensations. It integrates the new information with other information and forms a **perception**, a meaningful interpretation or conscious understanding of sensory data. In the case of a carnation, for example, if you connect the scent with the flower, either because you see the flower and smell it or because you have seen one before, your brain has formed a perception.

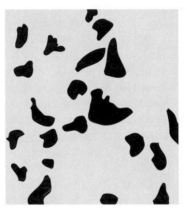

The figure below may further demonstrate the difference between sensation and perception for you. What do you see when you first look at the figure? If you see only some black splotches and blue space, you have developed a sensation—in this case, a more-or-less meaningless interpretation of some sensory information. What if we say that the figure shows a person riding a horse? With this clue, the brain forms a perception; it converts the sensation into a meaningful image. The figure may not have worked this way for you. If you saw the horse and rider right away, you experienced both a sensation and a perception. Your brain integrated the new information with some of its stored data; perhaps you've had a lot of experience with horses or have seen this figure before.

What does the brain actually do with sensory information in creating a perception? As we discussed in Chapter 28, researchers using brain imaging techniques are beginning to find out. The perception of a sweet-smelling flower, for instance, results from communication among neurons arranged in an extremely complex circuitry within the vision and odor centers of our cerebral cortex. The vision centers analyze and connect information on the flower's color, form, and any detectable motion. Neuronal communication involving the visual centers, odor centers, and perhaps the memory banks creates the overall perception.

Perceptions are the product of a continuum of information processing, beginning with the detection of stimuli by sensory receptors. For the rest of this chapter, we concentrate on sensory receptors and how they function.

29.2 Sensory receptor cells convert stimuli into electrical energy

Sensory organs such as your eyes or the taste buds on your tongue contain specialized receptor cells that detect stimuli. The receptor cells in your eyes, for instance, detect light energy; those in your taste buds detect chemicals dissolved in saliva, such as salt or sugar.

What exactly do we mean when we say that a receptor cell detects a stimulus—a molecule of sugar or a photon of light, for instance? Stimulus detection means that the cell converts one type of signal (the stimulus) into an electrical signal. This conversion, called **sensory transduction**, occurs in the plasma membrane of the receptor cell.

Figure A shows sensory transduction occurring when receptor cells in a taste bud detect sugar molecules. ① The molecules first enter the taste bud, where ② they bind to

specific protein molecules in a receptor cell membrane. The binding changes the membrane permeability by causing ion channels in the membrane to open. Positively charged ions then flow into the cell from the surrounding fluid and alter the membrane voltage (the electrical potential energy stored by the membrane; see Module 28.5), raising it to a higher level called the receptor potential. The **receptor potential** is the electrical signal produced by sensory transduction. In contrast to action potentials, receptor potentials vary; the stronger the stimulus, the stronger they are.

Once a receptor cell converts a stimulus into a receptor potential, it sends signals representing the stimulus toward the central nervous system. In our taste bud example, ③ each receptor cell forms a synapse with a sensory neuron.

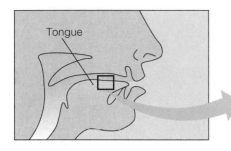

Tongue

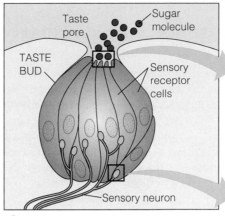

Sugar molecule

Taste pore

TASTE BUD

Sensory receptor cells

Sensory neuron

1 Taste bud anatomy

A. Sensory transduction at a taste bud

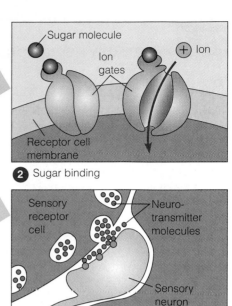

Sugar molecule

Ion gates

(+) Ion

Receptor cell membrane

2 Sugar binding

Sensory receptor cell

Neuro-transmitter molecules

Sensory neuron

3 Synapse

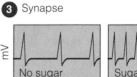

mV

No sugar

Sugar present

4 Action potentials

This is a chemical synapse just like the one between neurons described in Module 28.7. Notice that the neurotransmitter molecules diffuse across the synapse from the receptor cell to the sensory neuron. The receptor cell constantly secretes neurotransmitter at a set rate, which triggers a steady stream of action potentials in the sensory neuron. **4** The graph on the left represents the rate at which the sensory neuron triggers action potentials when the taste bud is not detecting any sugar. The right-hand graph shows what happens when there are enough sugar molecules to trigger a strong receptor potential. This receptor potential makes the receptor cell release more neurotransmitter than usual, enough to increase the rate of action potential generation in the sensory neuron. This change in the rate of action potentials signals the brain that the receptor cell detects the stimulus.

Thus we see that sensory receptors transduce (convert) stimuli into electrical signals and trigger action potentials that go into the central nervous system for processing. Since action potentials are the same no matter where or how they are produced, how do they communicate different information, like a sweet taste instead of a salty one? In Figure B, the taste bud on the left has receptor molecules only for sugar, the one on the right only for salt. The sensory neurons from the salt-detecting taste bud synapse with interneurons in the brain that are different from those contacted by neurons from the sugar-detecting taste bud. The brain distinguishes stimulus types (in this case, salt from sugar) by which interneurons are stimulated.

The graphs in Figure B indicate how action potentials communicate information about the *intensity* of stimuli (for example, very sweet or less sweet). In each case, the graph on the left represents the rate at which the sensory neurons in the taste bud transmit action potentials when the receptor cells are not stimulated. The right-hand graphs show that the rate of transmission depends on the intensity of the stimulus. The stronger the stimulus, the more frequently the receptor transmits action potentials to the brain. The brain interprets the intensity of the stimulus from the rate at which it receives action potentials. It gains additional information about stimulus intensity by keeping track of how many sensory neurons it receives signals from.

There is an important qualification to what we have just said about stimulus intensity. Have you ever noticed how an odor that is strong at first seems to fade with time, even when you know the substance is still there? The same effect helps you adjust to a hot or cold shower and enables you to wear clothes without being constantly aware of them. The effect is called **sensory adaptation,** the tendency of sensory receptor cells to become less sensitive when they are stimulated repeatedly. When receptors become less sensitive, they trigger fewer action potentials, and the brain may lose its awareness of stimuli as a result. Sensory adaptation keeps the body from reacting to normal background stimuli. Without it, our nervous system would become overloaded with useless information.

This overview of sensory transduction, transmission, and adaptation explains how sensory receptors work in general. Now let's begin looking at the various kinds of receptors.

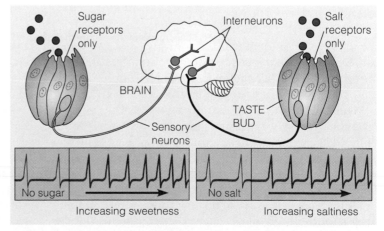

Sugar receptors only

Interneurons

Salt receptors only

BRAIN

Sensory neurons

TASTE BUD

No sugar — Increasing sweetness

No salt — Increasing saltiness

B. How action potentials represent different taste sensations

Based on the type of signals to which they respond, we can group sensory receptors into five general categories: pain receptors, thermoreceptors, mechanoreceptors, chemoreceptors, and electromagnetic receptors.

Figure A, showing a section of human skin, reveals why the surface of our body is sensitive to such a variety of stimuli. Our skin contains sensory receptors falling into three of the five categories: pain receptors, thermoreceptors (sensors for both heat and cold), and mechanoreceptors (sensors for touch and pressure). Each of these receptors has a small nerve fiber made up of several neurons that transduce stimuli and also send action potentials to the central nervous system. In other words, each neuron in the receptor serves as both a receptor cell and a sensory neuron. The branched ends of some of the nerve fibers are wrapped in one or more layers of connective tissue (green areas in the figure), but in the pain and touch receptors just under the skin surface and around the base of the hair, the ends of the nerve fibers are naked.

Probably all animals have **pain receptors,** although we cannot say what nonhuman perceptions of pain are like. Pain is important because it often indicates danger and usually makes an animal withdraw to safety. Pain can also make us aware of injury or disease.

Thermoreceptors in the skin detect either heat or cold. Other temperature sensors located deep in the body monitor the temperature of the blood. The hypothalamus in the brain is the body's major thermostat. Receiving action potentials from both surface and deep sensors, the hypothalamus keeps a mammal's or bird's body temperature within a narrow range (see Module 20.13).

Mechanoreceptors are the most diverse of the sensory receptors. Different types are stimulated by various forms of mechanical energy, such as touch and pressure, stretching of muscles, motion, and sound. All these forces produce their effects by bending or stretching the plasma membrane of a receptor cell. When the membrane changes shape, it becomes more permeable to positive ions, and if the stimulus is strong enough, the mechanical energy of the stimulus is transduced into a receptor potential.

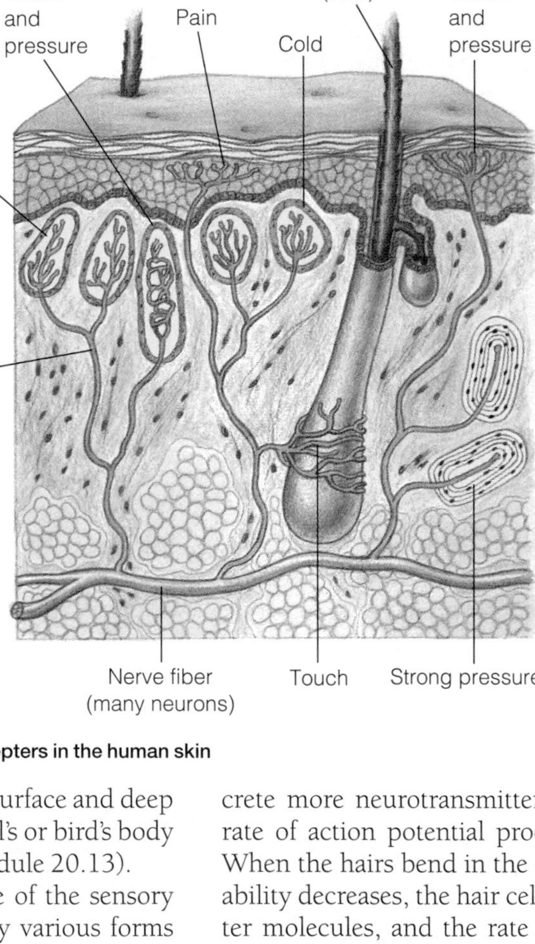

At the top of Figure A are two types of mechanoreceptors that detect both touch and pressure. Both types are near the skin surface and transduce very slight inputs of mechanical energy into action potentials. A third type of pressure sensor, lying deeper in the skin, is stimulated by strong pressure. A fourth type of mechanoreceptor is the touch receptor you see wound around the base of the hair. These receptors are sensitive to the hair's movements. Similar touch receptors at the base of the stout whiskers on a cat or a bear are extremely sensitive; they enable such animals to detect close objects by touch when hunting at night. A fifth type of mechanoreceptor (not shown) are the stretch receptors found in our skeletal muscles. Sensitive to changes in muscle length, **stretch receptors** monitor the position of body parts.

A variety of mechanoreceptors collectively called **hair cells** detect sound waves and other forms of movement in air or water. The "hairs" on these sensors are either specialized types of cilia or cellular projections called microvilli. The sensory hairs project from the surface of a receptor cell into either the external environment, such as the water surrounding a fish, or an internal fluid-filled compartment, such as our inner ear. Figure B (top of the facing page) shows how hair cells work. When fluid moves, bending the hairs in one direction, the hairs stretch the cell membrane, increasing its permeability to positively charged ions. This makes the hair cell secrete more neurotransmitter molecules and increases the rate of action potential production by a sensory neuron. When the hairs bend in the opposite direction, ion permeability decreases, the hair cell releases fewer neurotransmitter molecules, and the rate of action potential generation decreases. We'll see later that hair cells are important in both hearing and balance.

Chemoreceptors include the sensory cells in our nose and taste buds, which are attuned to chemicals in the external environment, as well as some internal receptors that detect chemicals in the internal environment. Internal chemoreceptors include sensors in some of our arteries that

A. Sensory receptors in the human skin

Labels: Touch and pressure · Pain · (Hair) · Touch and pressure · Cold · Heat · Nerve fiber (several receptor cells/ sensory neurons) · Nerve fiber (many neurons) · Touch · Strong pressure can detect changes in the amount of CO_2 in the blood. In all types of chemoreceptors, a receptor cell develops receptor potentials in response to chemicals dissolved in fluid (for instance, blood, saliva, the fluid coating the inside surface of the nose, or the water surrounding a fish).

Figure C (below) shows one of the most sensitive and specific chemoreceptors in the animal kingdom—the antennae of the male silkworm moth *Bombyx mori*. The antennae are covered with more than 50,000 tiny bristles (visible in the lower photo, a micrograph). Each bristle is a chemoreceptor that is highly sensitive to a sex attractant called bombykol, produced by the female moth. A male begins to respond to the female

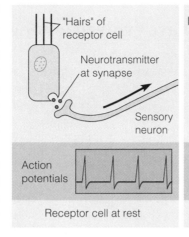

B. Mechanoreception by a hair cell

when as few as 50 of his bristles detect one bombykol molecule per second.

Electromagnetic receptors are sensitive to energy of various wavelengths, which takes such forms as electricity, magnetism, and light. Certain fishes, for instance, locate prey by sensing minute amounts of electricity produced by the prey's muscle activity. The so-called electric fishes of Africa and Australia discharge electric currents into the water. Electroreceptors in the fish's skin detect nearby obstacles and animals by the disturbances they produce in the current.

Little is known about the sense of magnetism, but there is strong evidence that many species can detect the Earth's magnetic fields and use this information when migrating. The heads of certain fishes, mammals, and birds contain the mineral magnetite, which is strongly attracted by magnets. One such species is the bobolink, a bird that breeds in the northern United States and winters in South America. Researchers have found that, when migrating, the bobolink responds to changes in the local magnetic field, but they do not yet know if its magnetite is part of a magnetic sensor. Magnetite has recently been found in human brains, but so far its role there is unknown.

Photoreceptors, including eyes, are probably the most common type of electromagnetic receptor. Photoreceptors detect the electromagnetic energy we call light, which may be in the visible, infrared, or ultraviolet part of the electromagnetic spectrum (see Module 7.6). The rattlesnake in Figure D has two kinds of photoreceptors. Its prominent eyes detect visible light. Below its eyes are two receptors extremely sensitive to infrared radiation, a form of heat. These receptors detect the body heat of its preferred prey, small mammals and birds. The receptors can detect the infrared radiation emitted by a warm mouse a meter away. The snake moves its head from side to side until the radiation is detected equally by the two receptors, indicating that the mouse is straight ahead. We examine photoreceptors in more detail in the next several modules.

SEM 80×

C. Chemoreceptors on insect antennae

D. Electromagnetic receptors in a snake

Three different types of eyes have evolved among invertebrates

Three main types of photoreceptors have evolved in the animal kingdom, and we find examples of all three among the invertebrates. The simplest type of photoreceptor is the **eye cup**, such as that of a planarian worm. As shown in Figure A, the eye cup is made up of a cup-like cluster of dark-colored cells that partially shield adjacent photoreceptor cells. Pigment molecules in the photoreceptor cells absorb light and cause the light energy to be transduced, and action potentials are sent through the cells to the brain. The photoreceptors do not provide data that the brain can use to form images. Rather, depending on the position of the animal and which part of the photoreceptors are shielded from the light, the cells simply detect the intensity and direction of light. Stimulated by action potentials from the photoreceptor cells, the animal's brain can determine which way the worm should crawl to escape bright light and find a dark hiding place.

Two other types of eyes have lenses that focus light and form images. A large number of invertebrates, including crayfish, crabs, and nearly all insects, have compound eyes. A **compound eye** consists of many tiny light detectors, or facets. You can see the facets making up the two compound eyes of a fly in Figure B. Each facet has its own cornea (a transparent covering) and a lens, which focuses light onto several photoreceptor cells. Every facet in the compound eye picks up light from a tiny portion of the field of view. Assembled from thousands of facets, the image a compound eye forms is actually a mosaic of many parts. The photograph of a bed of zinnias in Figure B resembles the image created by a compound eye. An insect's brain may smooth and sharpen the image when it integrates the visual information.

Compound eyes are extremely acute motion detectors, an important advantage for insects and other small animals that are constantly threatened by predators. The compound eyes of most insects also provide excellent color vision. Some species, such as honeybees, can even see ultraviolet light (invisible to humans), which helps them locate certain nectar-bearing flowers.

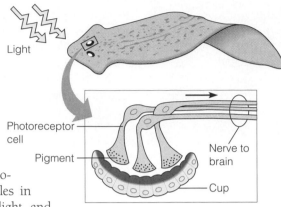

A. The eye cup of a planarian worm

The third type of invertebrate eye, the **single-lens eye**, works on a principle similar to that of a camera. The eye of a squid, for example, has structures that perform the same functions as the parts of the simple box camera shown in Figure C. Light enters the eye or the camera through a small opening, called the pupil in the eye and the aperture in the camera. The diameter of the pupil is changed by an adjustable iris, analogous to a camera's shutter. Behind the pupil, the single lens focuses light onto the retina, which consists of light-transducing receptor cells. In the camera, the retina is replaced by photosensitive film. The single lens produces one uninterrupted image, in contrast to the composite image formed by the compound eye. Like fine-grain film, the retina of a single-lens eye can provide very sharp images. Human eyes are also of this type, as we discuss next.

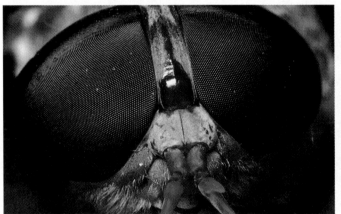

B. Compound eyes of a fly (left) and a mosaic image (right)

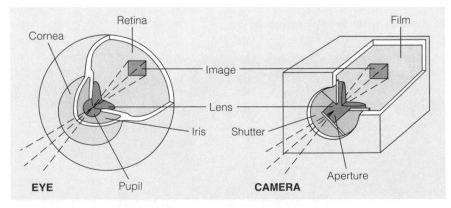

C. A single-lens eye, compared with a camera

The vertebrate eye is like the eye of a squid in that it has a single lens and is cameralike, but it evolved independently and differs in several details from any invertebrate photoreceptor. Our eyes are remarkable sense organs, able to detect a multitude of colors, form images of objects miles away, and respond to minute amounts of light energy.

The human eyeball, illustrated here, consists of a tough, whitish layer of connective tissue called the **sclera** surrounding a thin, pigmented layer called the **choroid**. The transparent **cornea**, which lets light into the front of the eye, is part of the sclera. The choroid at the front of the eye forms the **iris**, which gives the eye its color. Muscles in the iris regulate the size of the **pupil**, the opening in the center of the iris that lets light into the interior of the eye. After going through the pupil, light passes through the disklike **lens**, which is held in position by ligaments. As in the squid, the lens focuses images onto the **retina**, which is a layer just inside the choroid. Photoreceptor cells of the retina transduce light energy, and action potentials pass via sensory neurons in the optic nerve to the visual centers of the brain. Because there are no photoreceptor cells in the optic nerve, the place where the optic nerve passes through the back of the eye is a **blind spot**. We cannot detect light that is focused on the blind spot, but because we have two eyes with overlapping fields of view, we perceive uninterrupted images.

Two chambers make up the bulk of the eye. The large chamber behind the lens is filled with jellylike **vitreous humor**. The much smaller chamber in front of the lens is full of a liquid similar to blood plasma, called **aqueous humor**. The humors help maintain the shape of the eyeball. In addition, the aqueous humor circulates through its chamber. Secreted by capillaries, this fluid supplies nutrients and oxygen to the lens, iris, and cornea and carries off wastes. It reenters the blood through tiny ducts near the iris. Blockage of these ducts can cause glaucoma, increased pressure inside the eye that may lead to blindness by compressing the retina.

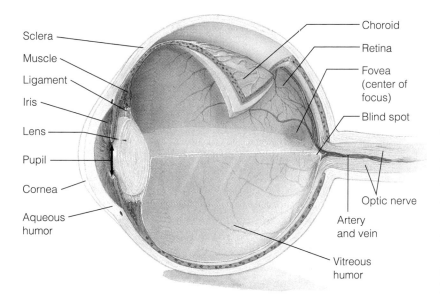

Sclera
Muscle
Ligament
Iris
Lens
Pupil
Cornea
Aqueous humor

Choroid
Retina
Fovea (center of focus)
Blind spot
Optic nerve
Artery and vein
Vitreous humor

To focus, a lens changes position or shape **29.6**

In almost every type of eye, the lens focuses light onto a retina by bending light rays. Focusing can occur in two ways. The lens may be rigid, as in many fishes, and focus by moving back or forth, as you might focus on an object using a magnifying glass. Or, as in the mammalian eye, the lens may be elastic and focus by changing shape. The thicker the lens, the more sharply it bends light.

The shape of the mammalian lens is controlled by the muscles attached to the choroid. When the eye focuses on a nearby object, these muscles contract. This makes the ligaments that suspend the lens slacken. With this reduced tension, the elastic lens becomes thicker and rounder, as shown in the left diagram below; this change is called **accommodation**. When the eye focuses on a distant object, the muscles controlling the lens relax, putting tension on the ligaments and flattening the lens, as shown in the right diagram.

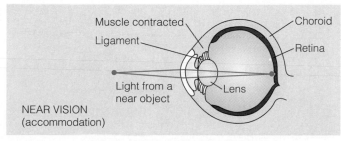

Muscle contracted
Ligament
Choroid
Retina
Light from a near object
Lens
NEAR VISION (accommodation)

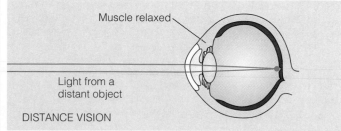

Muscle relaxed
Light from a distant object
DISTANCE VISION

How do corrective lenses work?

When you have your vision tested, you are asked to read letters on a special chart. The chart measures your **visual acuity**, the ability of your eyes to distinguish fine detail. The examiner asks you to read a line of letters sized for legibility at a distance of 20 feet, using one eye at a time. If you can do this, you have so-called normal (20/20) acuity in each eye. This means that from a distance of 20 feet, each of your eyes can read the chart's line of letters designated for 20 feet.

Suppose you find out that your visual acuity is 20/10. This is actually better than normal; it means that you can read letters from a distance of 20 feet that a person with 20/20 vision can only read at 10 feet. On the other hand, someone with 20/50 acuity has worse than normal vision. He or she must approach to a distance of 20 feet to read what a person with normal acuity can read at 50 feet. Visual acuity tests can show that there is a vision problem, but they cannot determine the cause.

Three of the most common visual problems are near-sightedness, farsightedness, and astigmatism. All three are focusing problems, easily corrected with artificial lenses. Nearsighted people cannot focus well on distant objects, although they can see well at short distances. A nearsighted eyeball (Figure A) is longer than normal. The lens cannot flatten enough to compensate, and it focuses distant objects in front of the retina, instead of on it. As shown by the right-hand drawing in Figure A, **nearsightedness** (also known as myopia) is corrected by glasses or contact lenses that are thinner in the middle than at the outside edge. The corrective lenses make the light rays from distant objects diverge slightly as they enter the eye. The focal point formed by the lens in the eye then falls directly on the retina.

Farsightedness (also known as hyperopia) is the opposite of nearsightedness. It occurs when the eyeball is shorter than normal, and the focal point of the lens is behind the retina (left part of Figure B). Farsighted people see distant objects normally, but they can't focus at short distances. Corrective lenses that are thicker in the middle than at the outside edge compensate for farsightedness by making light rays from nearby objects converge slightly before they enter the eye. Another type of farsightedness, called presbyopia (Greek for "old eye") develops with age. Beginning around the mid-forties, the lens of the eye becomes less elastic. As a result, the lens gradually loses its ability to focus on nearby objects, and reading without glasses becomes difficult.

Astigmatism is blurred vision caused by a misshapen lens or cornea. Any such distortion makes light rays converge unevenly and not focus at any one point on the retina. Astigmatism is corrected by lenses that are asymmetrical in a way that compensates for the asymmetry in the eye.

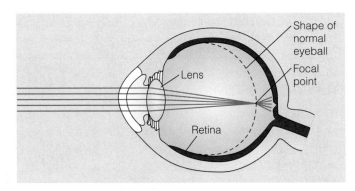

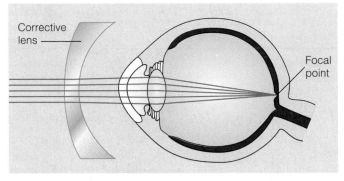

A. A nearsighted eye (eyeball too long)

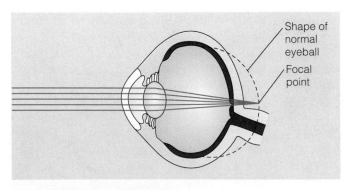

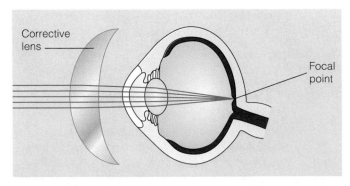

B. A farsighted eye (eyeball too short)

Our photoreceptor cells are rods and cones

Built into the human retina are about 130 million photoreceptor cells called rods and cones because of their shapes (Figure A below). Over twenty times more numerous than cones and also more sensitive to light, **rods** enable us to see in dim light at night, though only in shades of gray. **Cones** are stimulated by bright light and can distinguish color, but they do not function in night vision.

In humans, rods are found in greatest density at the outer edges of the retina, and are completely absent from the **fovea,** the eye's center of focus (Figure B; see also Figure 29.5). If you face directly toward a dim star in the night sky, the star is harder to see than if you look at it at an angle. Viewing it at an angle makes your lens focus the starlight onto the parts of the retina with the most rods. By contrast, you achieve your sharpest day vision by looking straight at the object of interest. This is because cones are densest (about 150,000 per mm^2) in the fovea. Some birds have over ten times more cones in their foveas than we do, which enables such species as hawks to spot small prey from high in the air.

How do rods and cones detect light? As Figure A shows, each rod and cone contains an array of membranous discs containing light-absorbing visual pigments. Rods contain a visual pigment called **rhodopsin,** which functions by absorbing dim light. Cones contain visual pigments called **photopsins,** which absorb bright, colored light. We have three types of cone cells, each containing a different type of photopsin. These cells are called blue cones, green cones, and yellow cones, referring to the colors absorbed best by their photopsin. All three types of cones actually absorb a wide range of colors, and together they can detect virtually any color in the visible spectrum (between ultraviolet and infrared). We can perceive an almost limitless number of colors because the light from each particular color triggers a unique pattern of stimulation in the millions of cones in our retina. Color blindness results from a deficiency in one or more types of cones.

Like all receptor cells, rods and cones are stimulus transducers. When rhodopsin and photopsin absorb light, they change chemically, and the change alters the permeability of the cell's membrane. The resulting receptor potentials trigger a complex integration process that actually begins in the retina. Notice in Figure B that the rods and cones have their tips embedded in the back of the retina (pink cells). Light must pass through several layers of neurons in the retina before reaching the pigments in the rods and cones. Visual information transduced by the rods and cones passes in the opposite direction (black arrows), from the photoreceptor cells through the network of neurons. Notice the numerous synapses between the photoreceptor cells and the neurons and among the neurons themselves. Integration in this maze of synapses helps sharpen images and increases the contrast between their light and dark areas. Action potentials carry the partly integrated information into the brain via the optic nerve. Sensations and perceptions of the images result from further integration in vision centers of the cerebral cortex.

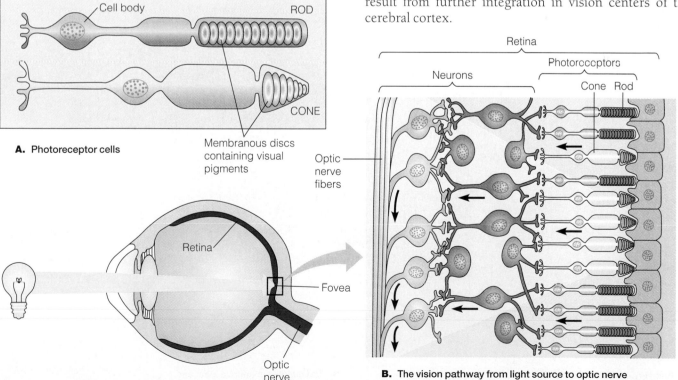

A. Photoreceptor cells
Cell body
ROD
CONE
Membranous discs containing visual pigments

Retina
Optic nerve
Fovea

B. The vision pathway from light source to optic nerve
Retina
Neurons
Photoreceptors
Cone Rod
Optic nerve fibers

The ear converts air pressure waves into action potentials that are perceived as sound

The human ear is really two separate organs, one for hearing and the other for maintaining balance. We look at the structure and function of our hearing organ in this module and then turn to our sense of balance in Module 29.10. Both organs operate on the same basic principle, the stimulation of cilialike projections on hair cells in fluid-filled canals.

The ear is complex, and it helps to learn its basic structure before studying how it functions. The ear is composed of three regions: the outer ear, the middle ear, and the inner ear (Figure A). The **outer ear** consists of the flaplike **pinna**—the structure we commonly refer to as our "ear"—and the **auditory canal**. The pinna and the auditory canal collect sound waves and channel them to the **eardrum**, a sheet of tissue that separates the outer ear from the **middle ear** (Figure B). When sound waves strike the eardrum, it vibrates and passes the sound waves to three small bones: the hammer, anvil, and stirrup. From the stirrup, the vibrations pass into the **inner ear** through the **oval window**, a membrane-covered hole in the skull bone. The **Eustachian tube** conducts air between the middle ear and the back of the throat, ensuring that air pressure is kept equal on either side of the eardrum. The tube is what enables you to move air in or out to equalize pressure ("pop" your ears) when changing altitude rapidly in an airplane or car.

The inner ear consists of several channels in the bones of the skull. The channels contain fluid that is set in motion by sound waves or by movements of the head. One of these channels, the **cochlea** (Latin for snail), is a long, coiled tube that contains what is actually the hearing organ. The cross-sectional view of the cochlea in Figure C shows that inside it are three fluid-filled canals. Our hearing organ, the **organ of Corti**, is a long, thin spiral within the middle canal. As you can see in the enlargement, the organ of Corti consists

of a **basilar membrane** (the floor of the middle canal), an array of hair cells embedded in the basilar membrane, and a shelf of tissue that projects over the hair cells from the wall of the middle canal. The hair cells are the receptor cells of the ear. Notice that they project into the fluid in the middle canal, and that the tips of most are embedded in the overlying shelf. Sensory neurons at the base of the hair cells carry action potentials from the organ of Corti into the brain via the auditory nerve.

Now let's see how the parts of the ear function in hearing. As indicated in Figure D (top of the facing page), a vibrating object, such as a plucked guitar string, creates pressure waves in the surrounding air, represented by the up-and-down waves in the figure. Collected by the pinna and auditory canal of the outer ear, these waves make your eardrum vibrate with the same frequency as the sound. The frequency, measured in hertz (Hz), is the number of vibrations per second (1 Hz is equal to one vibration per second).

From the eardrum, the vibrations pass through the hammer, anvil, and stirrup in the middle ear. These small bones amplify the vibrations (heighten the sound waves) and transmit them to the oval window between the middle ear and inner ear. Vibrations of the oval window then produce pressure waves in the fluid within the cochlea. The vibrations first pass from the oval window into the fluid in the upper canal of the cochlea. Pressure waves travel through the upper canal to the tip of the cochlea, at the coil's center. The pressure waves then enter the lower canal and gradually fade away.

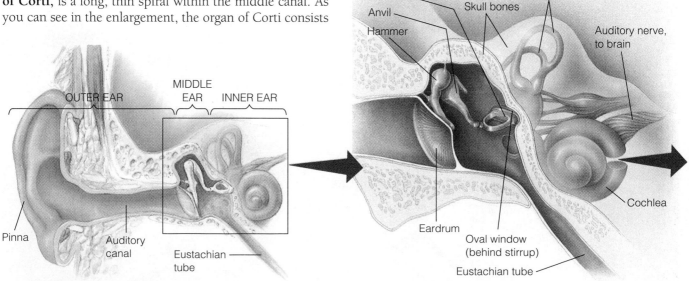

A. An overview of the human ear

B. The middle ear and the inner ear

As a pressure wave passes through the cochlea's upper canal, it pushes downward on the middle canal, making the basilar membrane vibrate. As the membrane vibrates, it alternately presses the hair cells into the overlying shelf and draws them away. This distorts the plasma membrane of each hair cell, making it more permeable to sodium ions.

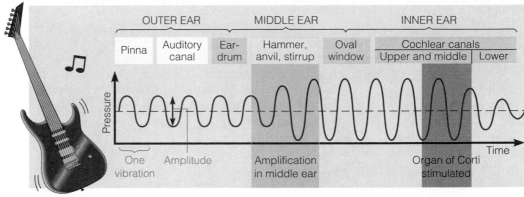
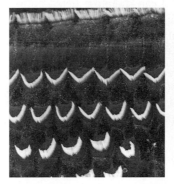

D. The route of sound waves through the ear

The hair cell develops a receptor potential and releases more neurotransmitter molecules at its synapse with a sensory neuron. In turn, the sensory neuron sends more action potentials to the brain.

The brain perceives a sound as an increase in the frequency of action potentials it obtains from the auditory nerve. But how is the quality (volume and pitch) of the sound determined? The higher the volume (loudness) of sound, the higher the amplitude of the pressure wave it generates. In the ear, the higher the amplitude, the more vigorous the vibrations of fluid in the cochlea, the more pronounced the bending of the hair cells, and the more action potentials generated in the sensory neurons. The loudness of sound is measured in decibels (dB). The decibel scale for human hearing ranges from 0 to 120 dB, the loudest we can hear without intolerable pain.

The pitch of a sound depends on the frequency of the sound waves. High-pitched sounds, such as high notes sung by a soprano, generate high-frequency waves. Low-pitched sounds, like low notes sung by a bass, generate low-frequency waves. How does the cochlea distinguish sounds of different pitch? The key is that the basilar membrane is not uniform along its spiraling length. The end near the oval window is relatively narrow and stiff, while the other end, near the tip of the cochlea, is wider and more flexible. Each region of the basilar membrane is most sensi-

SEM 975×

E. Healthy (left) and damaged (right) hair cells in an organ of Corti

tive to a particular frequency of vibration, and the region vibrating most vigorously at any instant sends the most action potentials to auditory centers in the brain. The brain interprets the information and gives us a sensation of pitch. Young people with healthy ears can hear pitches in the range of 20–20,000 Hz. Dogs can hear sounds as high as 40,000 Hz, and bats can emit and hear clicking sounds as high-pitched as 75,000 Hz. (A bat detects moths and other flying insects by sending out high-pitched clicks and plotting the position of its prey from the echoes.)

Few parts of our anatomy are more delicate than the organ of Corti. Quiet sounds make its hairs wave gently against the overlying shelf. But loud sounds can damage or destroy hair cells, causing hearing loss. The left micrograph in Figure E (above) shows normal hairs in the organ of Corti of a mouse. The right micrograph shows an area with damage similar to what could result from extended exposure to a very loud noise. Amplified rock music often exceeds 90 dB, an intensity that will cause hearing loss if experienced often.

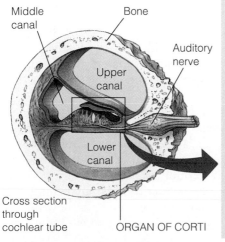

C. The organ of Corti, within the cochlea

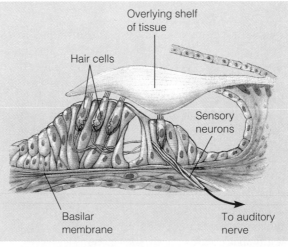

29.10 The inner ear houses our organ of balance

We have two sets of balance, or equilibrium, receptors, one on each side of the skull, in the inner ear. Each set lies next to the cochlea in five fluid-filled structures, which you can see on the left in the figure below: three semicircular canals and two chambers, the utricle and the saccule. All the equilibrium structures operate on the same principle, by the bending of hairs on hair cells, much like the way the organ of Corti detects sound.

The three **semicircular canals** detect changes in the head's rate of movement. As shown below, the canals are arranged in three perpendicular planes and can therefore detect movement in all directions. A swelling at the base of each semicircular canal contains a cluster of hair cells with their hairs projecting into a gelatinous mass called a cupula (shown in the enlargements). When you rotate your head in any direction, the thick, sticky fluid in the canals moves more slowly than your head. Consequently, the fluid presses against the cupula, bending the hairs. The faster you rotate your head, the greater

the pressure and the higher the frequency of action potentials sent to the brain. If you rotate your head at a constant slow speed, the fluid in the canals can move with the head, and the pressure on the cupula is reduced. But if you rotate your head fast and then stop suddenly, the fluid continues to move after the head stops; as a result, you may feel dizzy.

Clusters of hair cells in the **utricle** and **saccule** detect the position of the head with respect to gravity. When you move your head, tiny limestone crystals in a gelatinous layer coating the hairs are pulled by gravity, and their movement bends the hair cells in a specific direction. This alters the rate at which action potentials are sent to the brain. The brain determines the new position of the head by interpreting the altered flow of action potentials.

The equilibrium receptors provide data the brain needs to determine the position and movement of the head. Using this information, the brain develops commands that make the skeletal muscles balance the body.

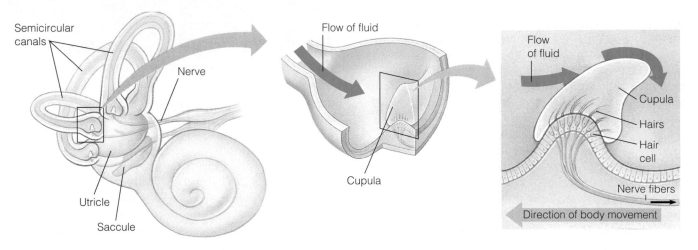

29.11 What causes motion sickness?

Boating, flying, or even riding in a car can make us dizzy and nauseous, a condition called motion sickness. Some people start feeling ill just from thinking about getting on a boat or plane. Many others get sick only during storms at sea or during "rough air" in flight. The cause of motion sickness is not known, but it seems to result from the brain's receiving signals from equilibrium receptors in the inner ear that conflict with signals from a different set of receptors, usually the ones in the eyes. When a susceptible person is inside a moving ship, for example, signals from the equilibrium receptors indicate, correctly, that the body is moving (in relation to the environment outside the ship). In conflict with these signals, the eyes may tell the brain that the body is in a stationary environment, the

cabin. Somehow the conflicting signals make the person feel ill. Sometimes, closing the eyes or looking straight ahead relieves symptoms. Many sufferers of motion sickness take a sedative such as Dramamine® to relieve their symptoms.

Motion sickness can be a severe problem for astronauts, and the National Aeronautics and Space Administration (NASA) actively conducts research on the problem. One of NASA's most interesting findings is that some people can learn to consciously control body functions, such as the vomiting reflex, that are ordinarily under involuntary control. Astronauts receive intensive training in how to exert "mind over body" when zero gravity starts to induce motion sickness.

Our senses of smell and taste depend on receptor cells that detect chemicals in the environment. Chemoreceptor cells in our nose detect airborne molecules; those in our taste buds detect molecules in food. In both cases, a cell responds to a group of chemically related molecules, not just to one kind of molecule. In the nose, for example, each type of receptor cell detects one of about fifty general types of odor (such as spicy, musky, or putrid). Research indicates that a particular odor triggers a specific level of stimulation in all the receptor cells of one type. The brain perceives the odor of cinnamon, for instance, when it receives a specific pattern of action potentials from the spice receptors; it perceives the odor of cloves when it receives another pattern from the same receptor cells.

Figure A illustrates the mechanics of smell in a human. The olfactory chemoreceptor cells are sensory neurons that extend into the upper portion of our nasal cavity. Notice the cilia extending from the tips of the receptor cells into the mucus that coats the nasal cavity. When you smell an odor, molecules (represented by the blue dots in the figure) have entered your nose, dissolved in the mucus, and bound to receptor molecules on the cilia. The binding triggers receptor potentials, which alter the rate of action potentials passing into the brain. Integration of the signals in the brain results in an odor perception.

Many animals, such as the salmon and grizzly bear in the chapter's introduction, rely heavily on their sense of smell for survival. Odors often provide more information than visual images about food, the presence of mates, or danger. In contrast, humans often pay more attention to sights and sounds than to smells and usually notice odors only when they are especially pleasant or unpleasant. The smell of decay, for instance, warns us of food spoilage. The sense of smell was more important to our ancestors, whose survival may have depended on the ability to smell prey, edible plants, and danger such as wildfires.

Our sense of taste depends on taste buds on the tongue (see Module 29.2). We have four types of taste buds, detecting the categories sweet, sour, salty, and bitter, and they are arranged on the

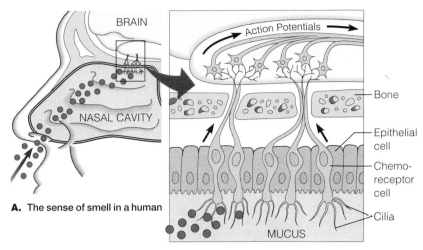

A. The sense of smell in a human

Bone

Epithelial cell

Chemoreceptor cell

Cilia

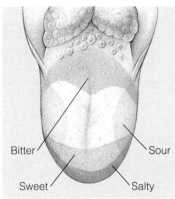

Bitter

Sour

Sweet

Salty

B. Taste buds on the human tongue

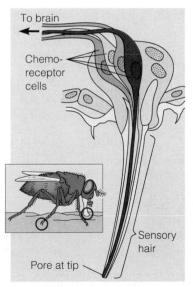

To brain

Chemoreceptor cells

Sensory hair

Pore at tip

C. Taste in a fly

tongue as shown in Figure B. Each type of taste bud can be stimulated by a broad range of chemicals in its category. Sweetness receptors, for example, detect several kinds of sugar molecules, such as sucrose and fructose. When you taste something, your brain receives a variety of taste inputs, and the flavors you perceive usually result from a combination of the four taste categories in varying proportions.

Imagine tasting with your hands and feet, instead of with your tongue. Insects do just that. They have chemoreceptors in sensory hairs on their feet and can taste food simply by stepping in it. Some, like the fly, also have liplike mouthparts covered with sensory hairs. As shown in Figure C, the fly's sensory hairs each contain four chemoreceptor cells that extend to a pore. Like the cells in our taste buds, each of the fly's taste cells detects a category of chemicals and responds to a broad range of them. The fly's brain probably receives signals from two or more types of receptor cells for any food the insect touches.

Probably all animals can detect certain chemicals in their surroundings, and the underlying mechanism of chemoreception is similar in all species that have been studied. Receptor cells detect groups of chemicals, and the brain develops odor or taste sensations by integrating data received from a variety of chemoreceptors.

Review: The central nervous system couples stimulus with response

In this chapter and the previous one, we have focused on information gathering and processing. Sensory receptors provide an animal's nervous system with vital data that enable the animal to avoid danger, communicate with others of its kind, find food and mates, and maintain homeostasis—in short, to survive.

A grizzly bear catching a salmon helps us summarize the sequence of information flow in an animal. A bear sees a flash in the stream. Within milliseconds, photoreceptor cells in the bear's retinas transduce the light energy focused on them by the lens, and action potentials representing a glimpse of the salmon enter the brain. As the salmon swims close, movement of the water stimulates touch receptors at the base of the bear's whiskers. Action potentials triggered by these receptors reach the brain at the same time as those from the eyes. Before the salmon can swim out of reach, a vast network of neurons in the grizzly's brain, with thousands of synapses, integrates all the information and sends out command signals, again in the form of action potentials. The commands go out via motor neurons to muscles in the bear's paws, neck, and jaws. The bear lunges and grabs its meal.

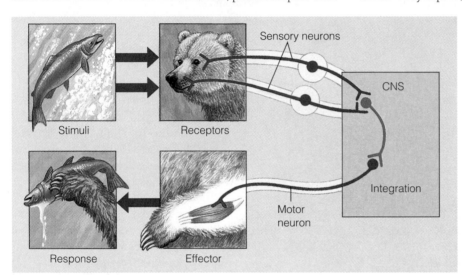

The nervous system links stimulus reception with response. It takes in information coded as action potentials, integrates it, plans a response, and sends out action potentials commanding an appropriate action. In doing so, the nervous system couples the various forms of stimulus signals to body response—in the case of the bear, to split-second muscle contractions. In the next chapter, we see how muscles carry out the commands they receive.

Chapter Review

Begin your review by rereading the module headings and Module 29.13 and scanning the figures before proceeding to the Chapter Summary and questions.

Chapter Summary

Introduction–29.2 Sensory receptors inform an animal about conditions inside and outside its body. Receptors convert stimuli into electrical energy, a process called sensory transduction. Action potentials representing the stimuli are transmitted via sensory neurons to the central nervous system for processing. The brain distinguishes different types of stimuli. The strength of the stimulus alters the rate of action potential transmission. Sensory neurons become less sensitive when stimulated repeatedly, a phenomenon known as sensory adaptation. The awareness of sensory stimuli is called sensation. Perception is the brain's full integration of sensory data.

29.3 There are five categories of sensory receptors. Pain receptors sense dangerous stimuli. Thermoreceptors detect heat and cold. Mechanoreceptors respond to mechanical energy, and chemoreceptors to chemicals in the external environment or body fluids. Electromagnetic receptors respond to electricity, magnetism, and light.

29.3–29.8 The most common types of electromagnetic receptors are photoreceptors, which sense light. The photoreceptors of flatworms are simple eye cups that sense the intensity and direction of light. The compound eyes of insects consist of many lenses that together produce a mosaic image. Vertebrates have single-lens eyes. In the human eye, the cornea, lens, and fluid in the eyeball focus light on photoreceptor cells—rods and cones—in the retina. Rods contain the visual pigment rhodopsin and function in dim light. Cones contain photopsin, which enables us to see color in full light. The human lens changes shape to bring objects at different distances into sharp focus. Nearsightedness results when the eyeball is longer than normal, and the lens cannot flatten enough to compensate. Farsightedness occurs when the eyeball is shorter than normal, and the focal point is behind the retina.

29.9 The human ear functions in hearing and balance. The outer ear channels sound waves to the eardrum, which passes the vibrations to a chain of bones. The bones transmit the vibrations to fluid in the cochlea. The waves generated in the cochlear fluid move hair cells (mechanoreceptors) of the organ of Corti against an overlying shelf of tissue. Movement of the hair cells triggers nerve signals to the brain. Louder sounds cause greater movement and more action potentials; sounds of different pitches stimulate hair cells in different parts of the cochlea.

29.10–29.11 The utricle and saccule, near the cochlea, contain clusters of hair cells that are bent by gravity. Fluid in the three semicircular canals also bends hair cells when the head moves. Changes in neural output from these hair cells enable the brain to sense body position and movement. Conflicting signals from the inner ear and eyes may cause motion sickness.

29.12 The senses of smell and taste depend on chemoreceptors, which send nerve signals to the brain when specific molecules bind to them. In humans, olfactory receptors line the upper part of the nasal cavity. Taste receptors are in taste buds on the tongue. Olfactory receptors respond to chemicals that produce many different smell sensations, but there are only four basic types of taste receptors, responding to chemicals that taste sweet, sour, salty, and bitter.

Testing Your Knowledge

Multiple Choice

1. Mr. Johnson was becoming slightly deaf. To test his hearing, his doctor held a vibrating tuning fork tightly against the back of Mr. Johnson's skull. This sent vibrations through the bones of the skull, setting the fluid in the cochlea in motion. Mr. Johnson could hear the tuning fork this way, but not when it was held away from the skull a few inches from his ear. The problem was probably in the (*Explain your answer.*)

   a. auditory center in Mr. Johnson's brain
   b. auditory nerve leading to the brain
   c. hair cells in the cochlea
   d. bones of the middle ear
   e. fluid of the cochlea

2. Which of the following correctly traces the path of light into your eye?

   a. lens, cornea, pupil, retina
   b. cornea, pupil, lens, retina
   c. cornea, lens, pupil, retina
   d. lens, pupil, cornea, retina
   e. pupil, cornea, lens, retina

3. You turn on the shower and check the water temperature with your hand. Your brain is able to monitor increasing temperature because the heat receptors in your skin

   a. transmit action potentials to different brain regions as the water warms up
   b. transmit bigger action potentials to the brain as the water warms up
   c. gradually adapt to the stimulus
   d. transmit action potentials at a greater rate as the water warms up
   e. integrate temperature information

4. If you look away from this book and focus your eyes on a distant object, the eye muscles _____ and the lenses _____ to focus images on the retinas.

   a. relax . . . flatten
   b. relax . . . become more rounded
   c. contract . . . flatten
   d. contract . . . become more rounded

5. Which of the following receptors are *not* present in human skin?

   a. thermoreceptors
   b. chemoreceptors
   c. touch receptors
   d. pressure receptors
   e. pain receptors

6. Jim had his eyes tested and found that he has 20/40 vision. This means

   a. the muscles in his iris accommodate too slowly
   b. he is farsighted
   c. the vision in his left eye is normal, but his right eye is defective
   d. he can see at 40 feet what a person with normal vision can see at 20 feet
   e. he can see at 20 feet what a person with normal vision can see at 40 feet

True/False (*Change false statements to make them true.*)

1. Sensory adaptation is the tendency for receptors to gradually become more sensitive to stimuli.

2. Between the eyes of some snakes are receptors sensitive to infrared radiation.

3. Cone cells function in bright light and can detect colors.

4. A salmon finds its way up its home stream by sensing the Earth's magnetic field.

5. The compound eyes of insects are good at detecting motion.

6. The semicircular canals sense body position, and the saccule and utricle sense movement.

7. Hair cells of the ear are examples of electromagnetic receptors.

8. The organ of Corti is the site of receptor cells for vision.

Describing, Comparing, and Explaining

1. Ann passed the New Morning Bakery and was drawn inside by the scent of freshly baked cinnamon rolls. Describe how the sensory receptors in her nasal cavity converted the scent stimulus into action potentials and sent them to her brain.

2. Listen for a moment to the sounds around you. How are the sound waves converted to action potentials in your ears? How does your brain determine the volume and pitch of the sounds?

3. As you read these words, the lenses of your eyes project images of the letters on your retinas. There the photoreceptors respond to the patterns of light and dark, and transmit nerve signals to the brain. The brain then interprets the words. In this example, what is the difference between sensation and perception?

Thinking Critically

1. In what ways might you expect the eyes of a nocturnal animal, such as an owl, to be different from your eyes?

2. Sensory organs tend to come in pairs. We have two eyes and two ears. Similarly, a planarian worm has two eye cups, a rattlesnake has two infrared receptors, and a butterfly has two antennae. What is the advantage of paired sensory receptors? In other words, how are two ears better than one?

3. Sea turtles bury their eggs on the beach above the high-tide line. When the baby turtles hatch, they dig their way to the surface of the sand and quickly head straight for the water. How do you think the turtles know which way to go? Outline an experiment to test your hypothesis.

Science, Technology, and Society

1. Have you ever felt your ears ringing after listening to loud music from a stereo or at a concert? Can this music be loud enough to permanently impair your hearing? Do you think people are aware of the possible danger of prolonged exposure to loud music? Should anything be done to warn or protect them? If you think so, what action would you suggest? What effect might warnings have?

2. Radial keratotomy is a surgical procedure for correcting nearsightedness that was pioneered in Russia. Small incisions are made around the edge of the cornea and the cornea flattens out, allowing the eye to focus better on distant objects. Many people who have had this operation no longer need to wear glasses. There has been controversy over use of radial keratotomy in the United States. Although many people are helped by this procedure, in some cases the cornea is damaged and vision is actually made worse. Do you think surgery is justified to correct a condition like nearsightedness, which most people would consider simply an inconvenience? Would you undergo this operation, knowing there is some risk?

How Animals Move 30

Ants rival humans in living just about any place on land except where there is permanent snow cover. Literally trillions of these insects walk the planet, and they are almost always on the move, engaged in some activity. Strenuous labor and almost constant body movement are two of their hallmarks.

The ants in this photograph are leaf-cutters that live in huge underground nests in tropical rain forests. A colony of these insects, often well over a million individuals, can strip a large tree of its leaves and flowers in a single night. The two ants shown here are part of a "leaf parade" of workers—all females—returning home after a successful nocturnal raid high in the trees. Up on a high branch, a worker uses her razor-sharp mouthparts to slice up a leaf, pick up a piece, and carry it back to her nest. You can't see them in this photograph, but other, much smaller workers often ride on the leaves and guard the leaf-carriers. They, too, are active, gnashing their mouthparts at parasitic flies that try to lay eggs on the load-bearers.

Adult leaf-cutter ants eat mostly plant sap, but their young eat fungi that grow on the leaves the adults cut. A leaf-cutter nest, often 4–5 meters deep and 7–8 meters in diameter at the soil surface, is a subterranean farm—a maze of tunnels and chambers in which the ants cultivate a particular species of fungus for food. Returning from a leaf-cutting trip, the loaded workers descend into a tunnel and drop their cuttings off in a brood chamber. Other workers chew up the leaves and start new fungal growth by placing bits of live fungus on them. Still other members of the colony "weed" undesirable fungi out of the fungus gardens, cover the fungal growths with chemicals that inhibit competing fungi, harvest the food, feed it to the young, and carry refuse out of the nest. The work is nonstop.

Over a four- to five-year period, a colony of leaf-cutters may excavate nearly 50 tons of forest soil in constructing and renovating a nest. It will also cut, haul, and process many tons of forest leaves and pile up large refuse heaps in keeping the nest clean. Constant activity in service of the colony is the life of every worker ant, and the ant body is a model of strength and mobility. Typical of insects, an ant's skeleton is a stiff outer coat called an exoskeleton. The insect exoskeleton is made of chitin, an unusually strong, durable polysaccharide. Chitin threads are embedded in a matrix of protein, forming a material analogous to fiberglass.

As shown below, each of the six legs of an ant consists of several exoskeletal pieces held together at flexible joints. The joints give the ant its mobility. Each exoskeletal leg piece is a tube that is virtually unbreakable by any force the insect confronts in hauling a bit of soil or a piece of leaf. The muscles in an ant's legs are tiny but powerful, and because the load of an ant's body is distributed over six legs, each leg uses only a tiny fraction of its muscle power in supporting the ant's own weight. The rest is used to power movements and perform useful work.

Movement is one of the most distinctive features of animals. Whether an animal walks or runs on two legs or six, swims, crawls, flies, or sits in one place and only moves its mouthparts, the interplay of three organ systems provides its movement. The nervous system plays a key role in issuing commands to the muscular system. The muscular system exerts the force that actually makes an animal or its parts move. The skeletal system is essential because the force exerted by muscles produces movement only when the force is applied against a firm structure, the skeleton. In this chapter, we focus on skeletons, muscles, and the movement their interactions produce.

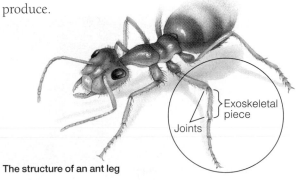

Joints

Exoskeletal piece

The structure of an ant leg

Diverse means of animal locomotion have evolved

Animal movement is extremely diverse. Many animals stay in one place and only move certain parts of their bodies. For example, a sponge's only movements are the opening and closing of cellular pores on its surface and the beating of flagella, drawing suspended food particles in through the pores. Most animals exhibit more movement than sponges, and animals that move about in search of food may spend much of their time and energy doing so. Active travel from place to place, also called **locomotion,** is our focus in this chapter. Locomotion in all its forms requires an animal to overcome two forces that tend to keep it stationary: friction and gravity. The relative importance of these two forces varies, depending on the environment.

SWIMMING Gravity is not much of a problem for a swimming animal, because water supports much or all of the animal's weight. On the other hand, overcoming friction is more difficult for a swimmer, because water is dense and offers considerable resistance to a body moving through it.

Many different modes of swimming have evolved. Many insects, for example, swim the way we do, using their legs as oars to push against the water. Squids and some jellyfish are jet-propelled, taking in water and squirting it out in bursts. Fishes swim by moving their body and tail from side to side (Figure A). Whales and other aquatic mammals move their body and tail from top to bottom. A sleek, streamlined shape, like that of many fishes, is an adaptation that aids rapid swimming.

LOCOMOTION ON LAND: HOPPING, WALKING, RUNNING, AND CRAWLING The problems of locomotion on land are more or less the opposite of those in the water. Air offers very little resistance to an animal moving through it. However, air provides little support for an animal's body, and a land animal must be able to support itself and overcome the force of gravity. When a

A. Fish swimming

B. Kangaroos, hopping and at rest

C. A dog walking

land animal walks, runs, or hops, its leg muscles expend energy both to propel it and to keep it from falling down. To move on land, powerful muscles and strong skeletal support are more important than a streamlined shape.

The kangaroo travels mainly by hopping (Figure B). Large muscles in its hind legs generate a lot of power. Muscles and tendons (which connect muscle to bone) in the legs also momentarily store energy when the kangaroo lands—somewhat like the spring on a pogo stick. The higher the jump, the tighter the spring coils when a pogo stick lands, and the greater the tension in the muscles and tendons when a kangaroo lands. In both cases, the stored energy is available for the next jump. For the kangaroo, the tension in its legs is a cost-free energy boost that reduces the total amount of energy the animal expends to travel. The pogo stick analogy applies to hundreds of thousands of animal species. The legs of an ant or a dog, for instance, retain some spring during walking or running, although less than those of a hopping kangaroo.

The kangaroo illustrates a solution to another problem of terrestrial life: maintaining balance. The animal can sit upright with relatively little energy cost because it has three parts of its body touching the ground simultaneously, its two hind legs and the base of its tail. Similar to a camera tripod, this arrangement stabilizes the upright body. The same principle applies to the dog in Figure C. When walking, a dog keeps three feet on the ground at all times (in this case, both hind feet and the right front foot). Bipedal (two-footed) animals such as birds and humans are less stable on land, but keep part of at least one foot on the ground when walking. When an animal is running, all of its feet may be off the ground momentarily. At running speeds, momentum, more than foot contact, stabilizes the body's position, just as a moving bicycle stays upright.

A crawling animal such as a snake or an earthworm faces very different problems. Because much of the animal's body is in contact with the ground, friction offers considerable resistance to movement. Many snakes crawl rapidly by undulating the entire body from side to side. In the process, the snake's body pushes against the ground, and this drives it forward. Snakes also have large, movable scales on their underside that assist in locomotion. Boa constrictors and pythons, for instance, creep forward in a straight line, driven by the leglike action of their belly scales. Muscles lift the scales away from the ground, tilt them forward, and then push them backward against the ground. The backward push drives the snake forward.

Earthworms crawl by peristalsis, a type of movement produced by rhythmic waves of muscle contractions passing from head to tail. (In Module 21.7, we saw how peristalsis squeezes food through our digestive tract.) To move by peristalsis, an animal needs a set of muscles that elongate the body, another set that shortens it, and a way to anchor itself to the ground. As illustrated in Figure D, the contraction of circular muscles constricts and elongates certain regions (groups of body segments) of a crawling earthworm, while longitudinal muscles shorten and thicken other regions. In position ①, segments at the head of the worm are short and thick (longitudinal muscles contracted) and anchored to the ground by bristles. Just behind the head, a group of segments is thin and elongated (circular muscles contracted), with bristles held away from the ground. In position ②, the head has moved forward because circular muscles in the head segments have contracted. Segments just behind the

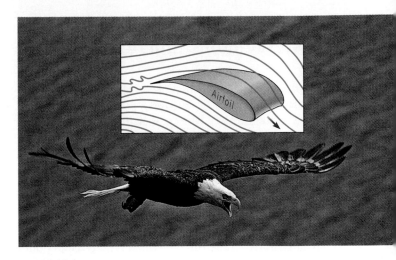

E. A bird flying

head and near the tail are now thick and anchored, thus preventing the head from slipping backward. In position ③, the head segments are thick again and anchored to the ground in their new position, well ahead of their starting point. The rear segments of the worm now release their hold on the ground and are pulled forward.

FLYING   Many phyla of animals include species that crawl, walk, or run, and almost all phyla include swimmers. But flying has evolved in only a few animal groups: insects, reptiles, birds, and, among the mammals, bats. A large group of flying reptiles died out millions of years ago, leaving birds and bats as the only flying vertebrates.

For an animal to become airborne, its wings must develop enough lift to completely overcome the downward pull of gravity. The key to flight is in the shape of wings. All types of wings, including those of airplanes, are airfoils—structures whose shape alters air currents in a way that creates lift. As Figure E shows, an airfoil has a leading edge that is thicker than the trailing edge. It also has an upper surface that is somewhat convex and a lower surface that is flattened or concave. This shape makes the air passing over the wing travel farther than the air passing under the wing. As a result, air molecules are spaced farther apart above the wing than under it. Thus the air pressure underneath is greater, and this greater pressure lifts the wing.

All types of animal movement, including the diverse forms of locomotion, have certain underlying similarities. At the cellular level, every form of movement is based on one of two basic contractile systems, microtubules and microfilaments, both of which involve protein strands moving against one another. The movements of cilia and flagella result from the bending of microtubules, as we discussed in Module 4.18. Microfilaments play a major role in amoeboid movement and are the contractile elements of muscle cells. Later in this chapter, we will look at the contraction of muscles and how it translates into movement when the muscles work against a firm skeleton. First, let's look at skeletons.

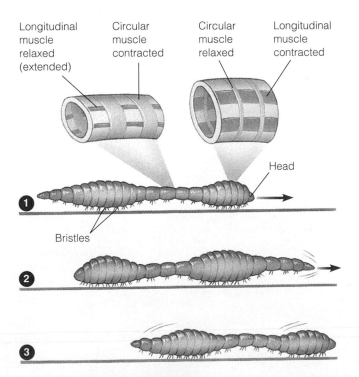

Longitudinal muscle relaxed (extended)

Circular muscle contracted

Circular muscle relaxed

Longitudinal muscle contracted

Head

Bristles

D. An earthworm crawling, by peristalsis

Skeletons function in support, movement, and protection

A skeleton is multifunctional. An animal could not move without its skeleton, and most land animals would sag from their own weight if they had no skeleton to support them. Even an animal in water would be a formless mass with no skeletal framework to maintain its shape. Skeletons also may protect an animal's soft parts. For example, the vertebrate skull protects the brain, and the ribs form a cage around the heart and lungs. There are three main types of skeletons: hydrostatic skeletons, exoskeletons, and endoskeletons. All three types have multiple functions.

A **hydrostatic skeleton** consists of fluid held under pressure in a closed body compartment. This is very different from the more familiar skeletons made of hard materials. Nonetheless, a hydrostatic skeleton helps protect other body parts, cushioning them from shocks. It also gives the body shape and provides support for muscle action, as does a hard skeleton.

Earthworms have a fluid-filled internal cavity—their body cavity, or coelom (see Module 19.7). As a segmented animal, the earthworm has its coelom divided into separate compartments. The fluid in this compartmented coelom functions as a hydrostatic skeleton, and the action of circular and longitudinal muscles working against the hydrostatic skeleton produces the peristaltic movement described in Module 30.1.

Cnidarians, such as *Hydra,* also have a hydrostatic skeleton. A hydra holds fluid in its gastrovascular cavity and can alter its body shape drastically by contracting muscles in its body wall. When a hydra closes its mouth and the muscles encircling its body wall constrict, it elongates and its tentacles extend, as shown on the left in Figure A. Because the muscles cannot compress the water in the gastrovascular cavity, constricting the cavity forces the water to elongate the animal, somewhat like squeezing a water-filled balloon. A hydra often sits in this position for hours, waiting for prey to swim by. If it is disturbed, its mouth opens, allowing water to flow out, and longitudinal muscles in its body wall contract, shortening the body (Figure A, right photograph).

A. The hydrostatic skeleton of a hydra in two states

Most animals with hydrostatic skeletons are soft and flexible. Its hydrostatic skeleton allows a hydra to extend its body and spread out its tentacles, as well as expand its body around ingested prey. A similar hydrostatic skeleton also helps an earthworm burrow through soil and permits many tube-dwelling animals (see Figure 19.11B) to expand their bodies for feeding and gas exchange when out of their tubes, and then quickly squeeze back into their tubes when threatened. Hydrostatic skeletons work well for many aquatic animals and for terrestrial animals that crawl or burrow by peristalsis. However, a hydrostatic skeleton cannot support the forms of terrestrial locomotion in which an animal's body is held off the ground, such as walking.

A great variety of aquatic and terrestrial animals have **exoskeletons.** In insects and other arthropods (see Module 19.12), muscles attached to knobs and plates on the inner surfaces of the exoskeleton move the jointed body parts. At the joints of legs, the skeleton is thin and flexible, allowing for a wide variety of body movements.

The armorlike protection, support, and flexibility provided by the exoskeleton have helped assure the great evolutionary success of the arthropods. The arthropod exoskeleton has only one major drawback: It does not grow with the animal. It must be shed (molted) and replaced by a larger exoskeleton at intervals to allow for the animal's growth. Depending on the species, most insects molt from four to eight times before reaching adult size. A few insect species and certain other arthropods, such as lobsters and crabs, molt at intervals throughout life.

In Figure B, you see a crab in the process of molting. Its old shell (the dark one on the right) split open when the

B. The exoskeleton of an arthropod: a crab molting

crab outgrew it. An arthropod is never without an exoskeleton of some sort. A newly molted crab, for instance, has a new, soft, elastic exoskeleton, which formed under the old one. Soon after molting, the crab expands its body by gulping air or water. Its new exoskeleton then hardens in the expanded position, and the animal has room for further growth. Until its exoskeleton hardens, an arthropod is very susceptible to predation; besides being weakly armored, it is usually less mobile, because the soft exoskeleton cannot support the full action of its muscles.

The shells of mollusks such as clams (Figure C) are also exoskeletons, but unlike the chitinous arthropod exoskeleton, mollusk shells are made of a mineral, calcium carbonate. The mantle, a sheetlike extension of the animal's body wall, secretes the shell (see Module 19.9). As a mollusk grows, it does not molt; rather, it enlarges the diameter of its shell by adding to its outer edge.

An **endoskeleton** consists of hard supporting elements situated among the soft tissues of an animal. Sponges, for example, are reinforced by hard spicules consisting of inorganic material such as calcium salts or silica, or by softer fibers made of protein. Sea stars, sea urchins, and most other echinoderms have an endoskeleton of hard plates beneath their skin. In living

C. The exoskeleton of a mollusk: a clam

sea urchins about all you see are movable spines, which are attached to the endoskeleton by muscles (Figure D, left). A dead urchin with its spines removed reveals the plates that form a rigid case (right photo).

Vertebrates have endoskeletons consisting of cartilage or a combination of cartilage and bone. One major lineage of vertebrates, the sharks, have entirely cartilaginous skeletons. In Figure E, you see the more common vertebrate condition. Bone makes up most of the adult frog skeleton, as it does in most fishes and in land vertebrates. Cartilage (blue in the figure) remains in the frog skeleton, and in the skeletons of most other vertebrates, mainly in areas where flexibility is needed. Our own skeleton is one example, as we see next.

D. A sea urchin and its endoskeleton (right)

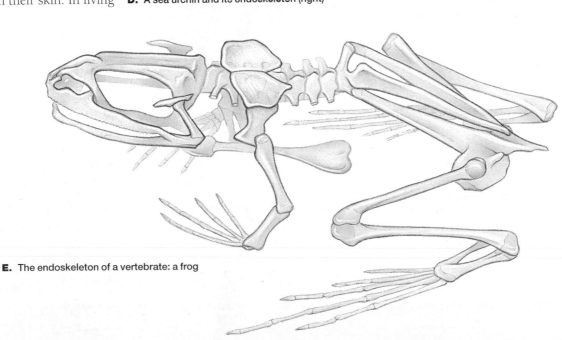

E. The endoskeleton of a vertebrate: a frog

The human skeleton is a unique variation on an ancient theme

In contrast to the frog skeleton, which supports an animal that sits on all fours and hops on its hind legs, the human skeleton supports an upright body that sits on its hindquarters and walks or runs on two legs. Despite the differences, the skeletons of these and other vertebrates have a number of similarities. For instance, all vertebrates have an axial skeleton (green in Figure A) supporting the axis, or central trunk, of the body. The **axial skeleton** consists of the skull,

surrounding and protecting the brain; the backbone (vertebral column), enclosing the spinal cord; and, in most vertebrates, a rib cage around the lungs and heart.

Most vertebrates also have appendages (arms, legs, wings, fins) and an appendicular skeleton (gold in Figure A) supporting the appendages. In a land vertebrate, the **appendicular skeleton** is made up of the bones of the forelimbs and hind limbs, the shoulder girdle, and the pelvic girdle. Bones of the girdles provide a base of support for the arm and leg bones. Notice in the human that three bones support each arm: the humerus in the upper arm, and the radius and ulna in the lower arm. Corresponding bones in the leg are the femur in the thigh, and the tibia and fibula in the lower leg. These same limb bones are found in most land vertebrates, although in the frog and certain others, the bones in the lower forelimb and lower hind limb are fused.

The skeletal features the human body has in common with other vertebrates stem from an ancient pattern, a basic skeletal model that probably originated in a group of fishes ancestral to land vertebrates. Hundreds of millions of years of adaptation have reshaped the model in a great variety of ways, but the main components remain. The skeleton of each species of land vertebrate, including the human, is a unique variation on the ancestral theme.

What are some of the distinctive features of the human skeleton? Our distant ancestors were quadrupedal (four-footed), and virtually every part of the skeleton changed drastically as upright posture and bipedalism evolved in the human lineage. Figure B contrasts the human skeleton (side view) and that of a quadrupedal primate, a baboon. Housing our large brain, our skull is large and flat-faced; its rounded part is the largest brain case relative to body size in the animal kingdom. In humans, the skull is balanced atop the backbone (red), whereas in quadrupeds, it is attached to the leading edge of the backbone. Our backbone is S-shaped, which helps balance the body in the vertical plane. In contrast, the baboon's is

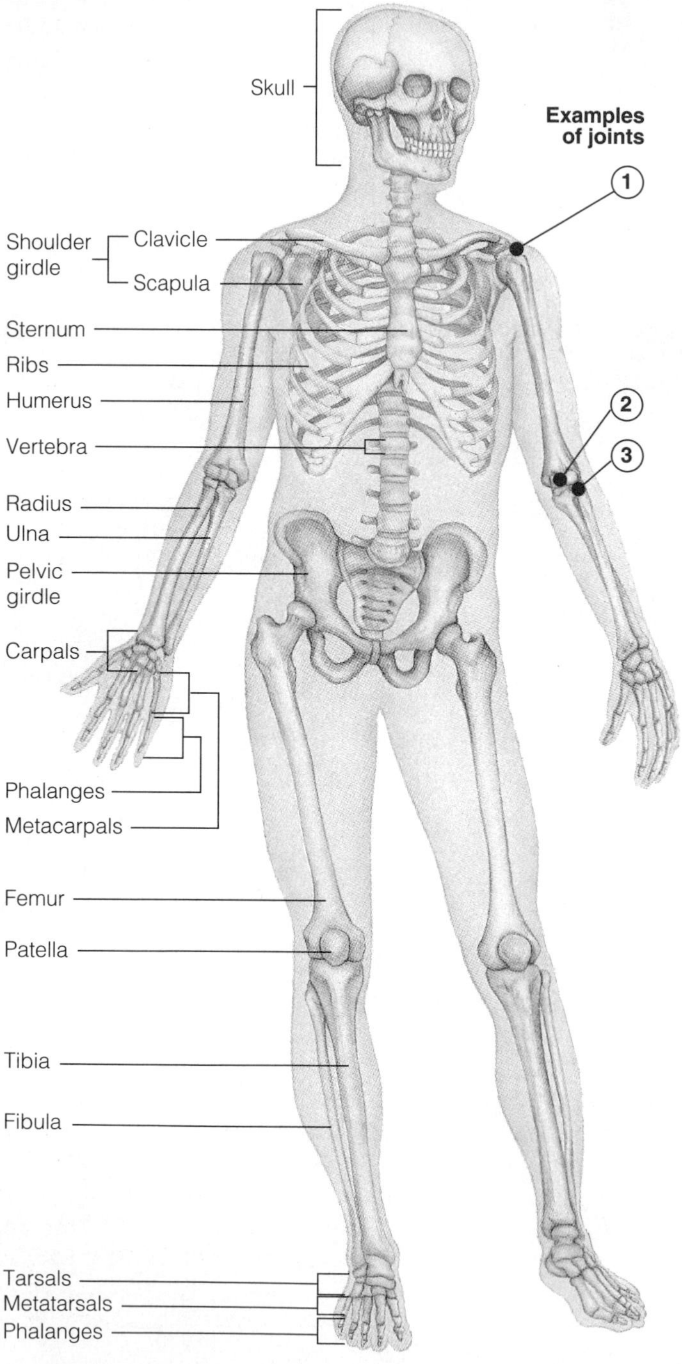

Skull

Examples of joints

①

Shoulder girdle ⎡ **Clavicle**
⎣ **Scapula**

②
③

Sternum

Ribs

Humerus

Vertebra

Radius

Ulna

Pelvic girdle

Carpals

Phalanges

Metacarpals

Femur

Patella

Tibia

Fibula

Tarsals
Metatarsals
Phalanges

A. The human skeleton

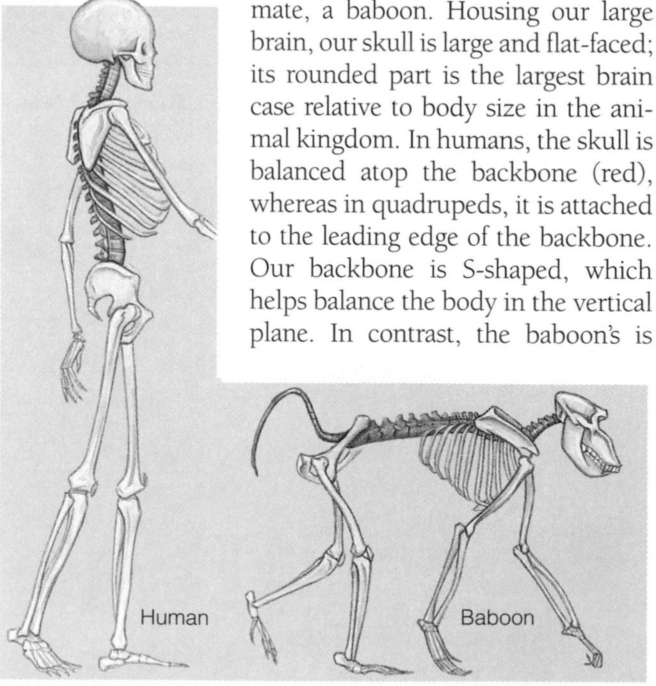

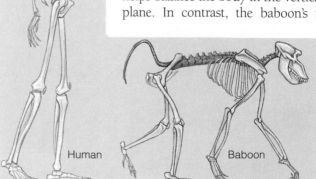

B. Bipedal and quadrupedal primate skeletons compared

Human   Baboon

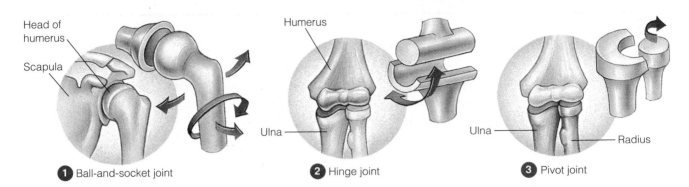

Head of humerus

Scapula

1 Ball-and-socket joint

Humerus

Ulna

2 Hinge joint

Ulna

Radius

3 Pivot joint

C. Three kinds of joints

arched horizontally. Our pelvic girdle is shorter and rounder and oriented more vertically than the quadruped's. The bones of our hands and feet are also different from the baboon's. Free of locomotor functions, the human hand is adapted for strong gripping and precise manipulation. Our two feet are specialized for supporting the entire body and for bipedal walking and running.

Much of the versatility of the vertebrate skeleton comes from its movable joints (Figure C, and the numbered locations in Figure A). We have ① **ball-and-socket joints** where the humerus joins to the shoulder girdle and where the femur joins to the pelvic girdle. These joints enable us to rotate our arms and legs and move them in several planes. Two other kinds of joints provide flexibility at the

elbow and knee. Shown here in the arm, ② a **hinge joint** between the humerus and the head of the ulna permits movement in a single plane. ③ A **pivot joint** allows us to rotate the forearm at the elbow. Hinge and pivot joints between the bones in our wrists and hands enable us to make precise manipulations.

Our skeleton is a showpiece of functional form. Many animals outperform us in specific activities—flying, swimming, running, digging. But the human body and the skeleton that supports it are unique in their versatility. We can swim, walk, run, crawl, jump, and burrow, as well as manipulate objects with our hands. Indeed, our success on the planet is due partly to the diverse movements our skeleton makes possible.

Skeletal disorders afflict millions 30.4

Nothing produced by evolution's restructuring process is perfect, and the human skeleton, despite its many adaptive advantages, is no exception. Our distant ancestors were quadrupedal and had an arched backbone similar to that of the baboon in Figure 30.3B. An arched backbone easily carries the weight of the ribs and organs suspended more-or-less evenly under it, much as a suspension bridge bears weight. In contrast, our vertical backbone, restructured from the arched condition, bears weight unevenly. Our internal organs hang down along our backbone, and our lower back (the lower part of the S curve) bears much of the load. This strains the lower back, especially when we bend over or lift heavy objects. Our tendency to have lower back problems is not surprising

Another common skeletal disorder, **arthritis**—inflammation of joints—affects one out of every seven people in the U.S. One form of arthritis seems to occur as part of aging. The joints become stiff and sore and often swell, as the cartilage between the bones wears down. Sometimes the bones thicken at the joints, producing crunching noises when they rub together and restricting movement. This form of arthritis is irreversible but not crippling in most cases, and moderate exercise, rest, and over-the-counter pain relievers usually alleviate it.

A much more serious form of arthritis, rheumatoid arthritis, is an autoimmune disease. The joints become highly inflamed, and their tissues may be destroyed by the body's immune system. Rheumatoid arthritis usually begins between ages 30 and 40 and affects more women than men. It may be triggered by a microbial infection, stress, or genetic factors. Anti-inflammatory drugs help relieve symptoms, but there is no cure. In some cases, patients are fitted with artificial joints.

Osteoporosis, another serious bone disorder, seems to be related to hormonal changes that accompany aging. Affecting about 20 million Americans, it is most common in women after menopause, when estrogen levels drop. Estrogen contributes to normal bone maintenance, and with lowered production of the hormone, bones may become thinner, more porous, and easily broken. Insufficient exercise, an inadequate intake of protein and calcium (for maintaining bone mass), and diabetes mellitus may also contribute to the disease. Calcium and sometimes estrogen are prescribed as treatment, but many researchers feel that the best approach is prevention. They encourage young women to maximize their calcium intake and exercise regularly in order to build bone mass.

Bones are complex living organs

Familiar expressions can seldom be taken literally, and "dry as a bone" is a good example. Bones are actually complex organs consisting of several kinds of moist, living tissues, amply supplied with blood. You can get a sense of some of a bone's complexity from these drawings of a human tibia (one of the lower leg bones). A sheet of fibrous connective tissue, shown in pink (most visible in the enlargement on the lower right), covers most of the outside surface. This tissue is able to form new bone in case of a fracture. Replacing the connective tissue at either end of the bone, a thin sheet of cartilage (blue), also living tissue, forms a cushionlike surface for joints. The bone itself contains living cells that secrete a surrounding material, or matrix. Bone matrix consists of flexible fibers of the protein collagen embedded in hard calcium salts (see Figure 20.5). Analogous to the steel rods in reinforced concrete, the collagen fibers resist cracking; analogous to the mineral substance of concrete, the calcium salts resist compression.

The shaft of this long bone is made of compact bone, so called because it has a dense matrix. Notice that the compact bone surrounds a central cavity. The central cavity contains **yellow bone marrow**, which is mostly stored fat brought into the bone by the blood. The ends, or heads, of the bone have an outer layer of compact bone and an inner layer of spongy bone, so named because it is honeycombed with small cavities. The cavities contain **red bone marrow** (not shown in the figure), a specialized tissue that produces our blood cells.

Like all living tissues, tissues in a bone require servicing. Blood vessels course through channels in the bone, transporting nutrients and regulatory hormones to its cells. Nerves (not shown) paralleling the blood vessels help regulate the traffic of materials between the bone and the blood.

As we see in the next module, the diverse living tissues in bones are key to bone growth.

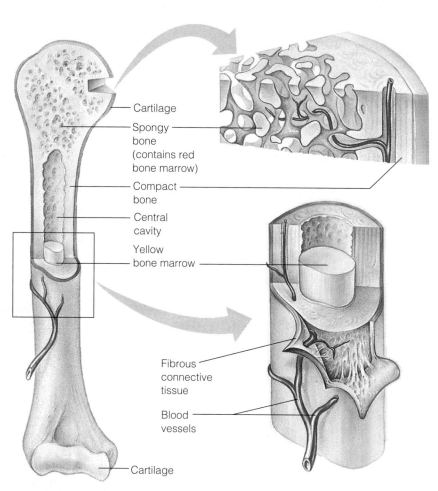

Cartilage
Spongy bone (contains red bone marrow)
Compact bone
Central cavity
Yellow bone marrow
Fibrous connective tissue
Blood vessels
Cartilage

The structure of a leg bone

Bone growth is a major feature of human development

Our living endoskeleton is with us from about a month after conception. It starts out as fibrous connective tissue and cartilage, with actual bone starting to form when the embryo is about six weeks old. The X-ray photograph in Figure A shows the developing skeleton in a 12-week-old human fetus. The bones that form most of the skull develop from sheets of fibrous connective tissue. In contrast, long bones, such as those in the legs and arms, develop from shafts of cartilage, each surrounded by a sheet of connective tissue.

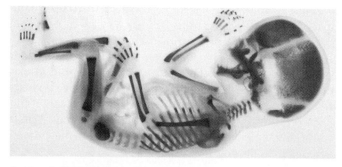

A. The skeleton of a 12-week-old fetus

Figure B illustrates how a long bone develops. Part ① shows an early embryonic stage. The bone originates as a shaft of cartilage surrounded by a sheet of connective tissue (not shown). A ring of bone then forms around the shaft. Bone also starts to replace cartilage at the center of the shaft. ② In the fetus, the bone grows in length and thickness as blood vessels penetrate the center of the shaft, and the yellow marrow cavity begins to form. Other blood vessels penetrate the cartilage near the ends of the shaft, and new bone formation also begins there.

Part ③ shows the growing bone of a child. The bone grows in overall width as bone cells produce new bone along the outer edge (green arrows); at the same time, bone on the inner surface (purple arrows) is broken down, and the marrow cavity increases in volume. The bone grows in length as long as cartilage continues to grow in the areas indicated in dark blue. Simultaneously, new bone replaces the cartilage in the areas colored orange.

Skeletal growth stops at about age 18 in women and about age 21 in men. Until those times, a complex interaction of hormones from the pituitary gland, thyroid gland, and gonads maintains cartilage growth and keeps the bones growing in proper proportion. The body reaches its adult size and bones stop growing when cells involved in bone growth begin to respond differently to the hormones.

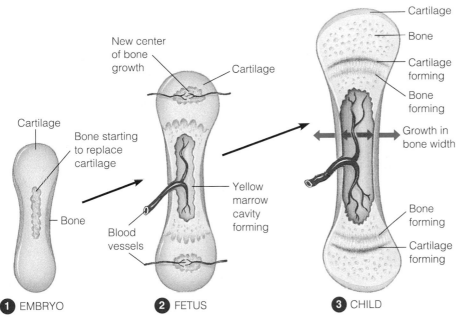

B. Development of a bone

① EMBRYO   ② FETUS   ③ CHILD

The skeleton and muscles interact in movement   30.7

We now focus on how an animal's skeleton interacts with its muscles in producing movement. Starting from a resting position, a muscle can *only* contract, or shorten, under its own power; to extend, it must be pulled by the action of another muscle. If we had only one muscle in our arm, for example, we could not move it back and forth or rotate it. The ability to move an arm in opposite directions requires that muscles be attached to the arm bones in antagonistic pairs—that is, two muscles working against each other.

The figure here illustrates how two muscles interact with parts of the skeleton and with each other to produce certain movements of the human arm. Muscles are connected to bones by **tendons** made of fibrous connective tissue. As you can see, one end of the biceps muscle is attached by a tendon to bones of the shoulder. The other end is attached to one of the bones in the forearm. As shown on the left, contraction of the biceps muscle raises the forearm. The triceps muscle is the biceps' antagonist. The upper end of the triceps attaches to the shoulder, while its lower end attaches to the elbow. As shown on the right, contraction of the triceps lowers the forearm, extending the biceps in the process.

All animals—even very small ones like the ants in the chapter's introduction—have antagonistic pairs of muscles that apply opposite forces against parts of their skeleton. Next, let's see how the structure of a muscle explains its ability to contract.

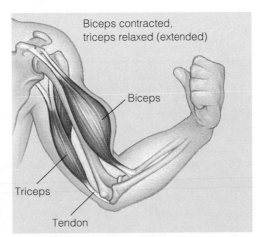

Biceps contracted, triceps relaxed (extended)

Biceps

Triceps

Tendon

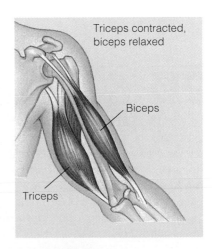

Triceps contracted, biceps relaxed

Biceps

Triceps

30.8 Each muscle cell has its own contractile apparatus

We looked briefly at the various types of muscle tissue in Module 20.6. **Skeletal muscle**, which is attached to the skeleton and produces body movements, is made up of a hierarchy of smaller and smaller parallel strands. As indicated at the top in this figure, a muscle consists of bundles of parallel muscle fibers. Each muscle fiber is a single cell with many nuclei.

Further down in the drawing, notice that each muscle fiber is itself a bundle of smaller **myofibrils.** Skeletal muscle is also called striated (striped) muscle because the myofibrils exhibit alternating light and dark bands when viewed with a light microscope. A myofibril consists of repeating units called **sarcomeres.** Structurally, a sarcomere is the region between two dark, narrow lines, called Z lines, in the myofibril. Functionally, the sarcomere is the contractile apparatus in a myofibril—the muscle fiber's fundamental unit of action.

The micrograph and the diagram below it reveal the structure of a sarcomere in more detail. They show that a myofibril is composed of regular arrangements of two kinds of filaments: thin filaments, colored blue in the diagram, and thick filaments, colored magenta. Biochemical analyses have shown that a **thin filament** consists of a double strand of the protein actin and one strand of a regulatory protein, coiled around each other. Each **thick filament** consists of a number of parallel strands of the protein myosin. The broad, dark band taking up most of the sarcomere is where the thick filaments are, as you can see in the bottom diagram; they are interspersed with thin filaments extending almost to the center of the sarcomere. In contrast, the light bands have only thin filaments. The Z lines consist of proteins that connect adjacent thin filaments.

This specific arrangement of repeating units of thin and thick filaments is directly connected to the mechanics of muscle contraction. In the next module, we see how this structure contributes to the functioning of a muscle cell, and hence how an entire muscle contracts.

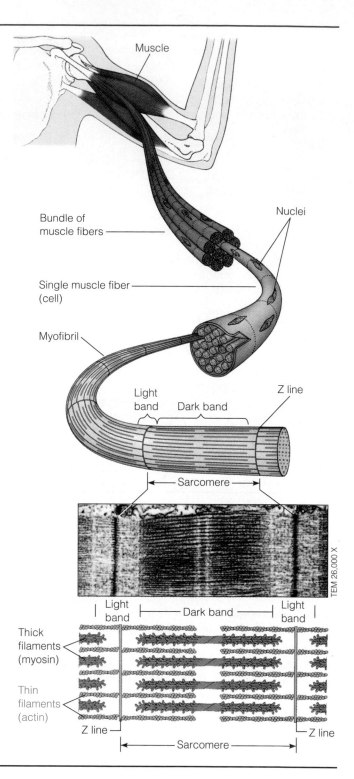

30.9 A muscle contracts when thin filaments slide across thick filaments

In the 1950s, British Nobel laureate A. F. Huxley and other researchers proposed what has become known as the **sliding filament model** of muscle contraction. This model explains the relationship between the structure of a sarcomere and its function. According to the model, a sarcomere contracts (shortens) when its thin filaments slide across its thick filaments. Figure A (facing page), a simplified diagram of Huxley's model, shows a sarcomere in a relaxed muscle, in a contracting one, and in a fully contracted one. Notice in the contracting sarcomere that the Z lines and the thin filaments (blue) are closer together horizontally than in the relaxed one. And when the muscle is fully contract-

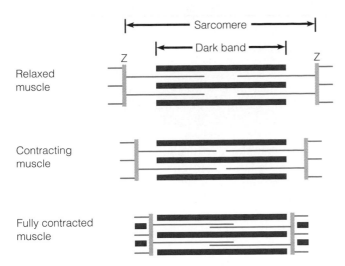

A. The sliding filament model

ed, the thin filaments overlap in the middle of the sarcomere. Contraction only shortens the sarcomere; it does not change the lengths of the thick and thin filaments. A whole muscle can contract to about half of its resting length when all its sarcomeres shorten.

What makes the thin filaments slide when a sarcomere contracts? Following Huxley's lead, researchers have sought answers at the molecular level. The key events are energy-consuming interactions between the myosin molecules of the thick filaments and the actin of the thin filaments. Electron micrographs show that parts of myosin molecules, called heads, bind with specific sites on the thin filaments. Energy for sliding comes from ATP. Also essential for muscle contraction are calcium ions (Ca^{2+}), which trigger the initial events in sliding.

Figure B indicates how sliding seems to work. ① ATP binds to a myosin head, causing the head to detach from a binding site (dark blue) on actin. (Only two of many binding sites are indicated.) ② Next, energy is made available for contraction when the ATP is broken down (hydrolyzed) to ADP and ⓟ (phosphate), which remain bound to the myosin head. The head gains some of the energy, and its position changes as a result. In its new position, the myosin head is cocked like a pistol ready to fire (actually, ready to bind with another site on the actin molecule). ③ Then Ca^{2+} enters the picture. Acting like a molecular trigger, it opens up a binding site on the actin molecule, making it possible for the myosin head to bind to the actin. ④ The molecular event that actually causes sliding is called the power stroke. The myosin head bends when ADP and ⓟ are released from it. The bending pulls the thin filament toward the center of the sarcomere in the direction of the blue arrow. After the power stroke, more ATP binds with the myosin head, and the whole process repeats. On the next power stroke, the myosin head attaches to another binding site upstream along the thin filament. This cycle—detach, straighten out (cock), attach, bend—occurs again and again in a contracting muscle. Though we show only one myosin head in the figure, a typical thick filament has about 350 heads, each of which can bind and unbind to a

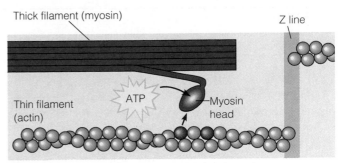

① ATP binds to a myosin head, which is released from an actin filament.

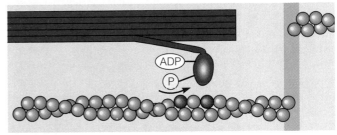

② Hydrolysis of ATP cocks the myosin head.

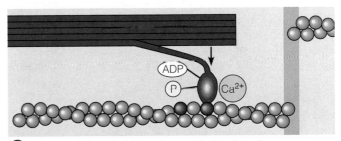

③ The myosin head attaches to an actin binding site, with the help of calcium.

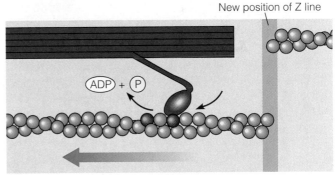

④ The power stroke slides the actin (thin) filament.

B. The mechanism of filament sliding

thin filament about five times per second. Preventing the filaments from backsliding during contraction, some myosin heads hold the thin filaments in position, while others are reaching for new binding sites. The process continues until the muscle is fully contracted, or until it stops contracting.

The sliding filament model explains how the sarcomeres of skeletal muscles contract. Next, let's return to an earlier topic and see how the nervous system controls muscle contraction.

Motor neurons stimulate muscle contraction

Sarcomeres do not contract on their own. They must be stimulated to contract by motor neurons. In humans and other vertebrates, each muscle fiber is stimulated by one motor neuron. However, a typical motor neuron can stimulate more than one muscle fiber, because each neuron has many branches. In the example shown in Figure A, you see two so-called **motor units,** each consisting of a neuron and all the muscle fibers it controls (two or three, in this case). A motor neuron has its dendrites and cell bodies in the central nervous system (here, the spinal cord). Its axon extends out to synapses, called **neuromuscular junctions,** with the muscle fibers. When a motor neuron sends out an action potential, its synaptic knobs release the neurotransmitter acetylcholine. Acetylcholine diffuses across the neuromuscular junctions to the muscle fibers, making all the fibers of the motor unit contract simultaneously.

The organization of individual neurons and muscle cells into motor units is the key to the action of whole muscles. We know that we can vary the amount of force our muscles develop—an arm wrestler, for example, may change the amount of force developed by the biceps and triceps several times in the course of a match. The ability to do so depends mainly on the nature of motor units. Each motor neuron in a large muscle like the biceps may serve several hundred fibers scattered throughout the muscle. Stimulation of the muscle by a single motor neuron, however, would produce only a weak contraction. More forceful contractions would result when additional motor units were activated. Thus, depending on how many motor units your brain commands to contract, you can apply a small amount of force to lift a fork, or considerably more to lift, say, this textbook. In muscles requiring precise control, such as those controlling eye movements, a motor neuron may control only one fiber.

How does a motor neuron make a muscle fiber contract? The initial events of stimulation are the same as those that occur at a synapse between two neurons in the nervous system (see Module 28.7): The acetylcholine that diffuses across the neuromuscular junction changes the permeability of the muscle fiber's plasma membrane. The change triggers action potentials that sweep across the muscle cell membrane. As indicated in Figure B, the action potentials (black arrows) then pass deep into the muscle cell along membranous tubules that fold inward from the plasma membrane. Inside the cell, the action potentials make the endoplasmic reticulum (ER) release Ca^{2+} (green dots) into the cytoplasm. The Ca^{2+} then triggers the binding of myosin to actin, initiating filament sliding, as we saw in the previous module. When a muscle relaxes, the process reverses: Motor neurons stop sending action potentials to the muscle fibers, the ER pumps Ca^{2+} back out of the cytoplasm, and the sarcomeres stop contracting.

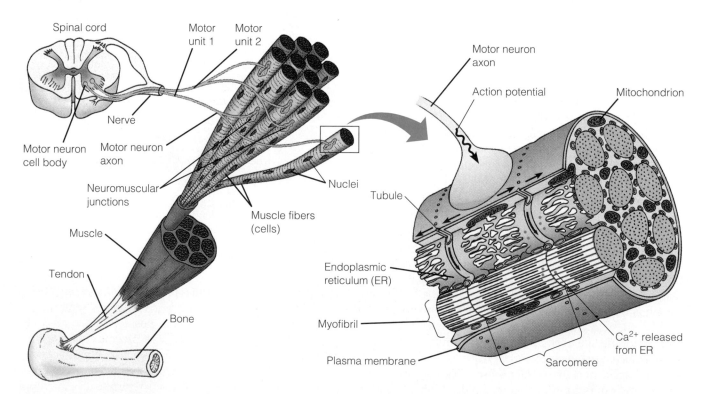

A. The relation between motor neurons and muscle fibers

B. Part of a muscle fiber (cell) at a neuromuscular junction

Probably no one appreciates the precise control that nervous systems exert over contracting muscles more than someone who has tried to build a robot. Robotic arms and hands are becoming common on assembly lines, performing such routine tasks as welding auto body parts together. But even the most advanced robotic machines appear very simple compared to human beings. Even most tasks that we do without thinking, and that seem simple, require finely tuned muscle movements controlled by commands formulated by a network of neurons in our brain and transmitted to muscles by motor neurons.

A robot capable of independent action must have a complex computer for a brain and a locomotor system that mimics that of a complex animal. So far, no computer comes close to the integrating capacity of the human brain, nor has robotic science devised any mechanical system of movement that closely mimics what our muscles can do. The robot in Figure A is a model built by artificial intelligence researcher Stanley Rosenschein of Palo Alto, California. Rosenschein uses the task of buttering a piece of toast as a research model in trying to construct robots that can perform complex jobs requiring precise actions. Buttering toast is no great challenge for the human nervous system or for the manipulative skills of our hands. Our brain is usually so preoccupied with more complex thinking and problem solving that we pay little attention to what it is telling the muscles in our hands to do when we pick up a piece of toast, put it on a plate, decide on a certain amount of butter, turn a knife in just the right way, judge the hardness of the butter, apply a certain pressure to spread it on the toast, and make any necessary readjustments to perform all these tasks efficiently. Doing all this smoothly and

A. A robot that butters toast

independently is a major challenge for a robot, even one that is programmed to do precisely this job.

The difficulty of building robots with humanlike intelligence, movement, and manipulative abilities has led some robotic scientists to simplify their goals. A number of years ago, Rodney Brooks, at the Massachusetts Institute of Technology, became intrigued with the relative simplicity of insect brains and muscle systems and with what insects can do with what they have. As we saw in the chapter's introduction, leaf-cutter ants perform a variety of complex tasks with their tiny brains and small bodies. The challenge Brooks set for his research team was to build small robots with some of the abilities of insects—abilities that could be adapted to perform useful tasks.

The insectlike robot in Figure B, known as Genghis, is the brainchild of Rodney Brooks' robotics laboratory. Genghis has six legs with wires for muscles, and it walks like a clumsy ant. For sensory receptors, it has six infrared-sensing eyes and two mechanical whiskers. These sensors feed information into a brain consisting of only 57 electronic circuits. The brain does little integrating but manages to steer Genghis around objects in its path. Other tiny circuit boards between each pair of legs resemble the ganglia in an insect's ventral nerve cord (see Module 28.11). The artificial ganglia run tiny motors that propel the legs.

Modeled after animals with relatively simple nervous systems, Genghis is incapable of anything resembling human thoughts. Its brain is designed only to receive data from sensors and produce a specific action in response. This eliminates the need for the complex circuitry that it takes to process sensory data. As Brooks points out, it also makes it possible to build a cheap, miniature robot that can still do many useful things. Imagine a swarm of antlike robots constructing computer chips, cleaning the precision parts that focus a space telescope, patrolling crops for pests, or collecting ore samples from the surface of Mars. Sound like a science fiction novel? Brooks and his co-workers are convinced that insectlike robots will be performing routine tasks for us long before we see humanlike robots walking down the street.

B. Genghis on the move

The structure-function theme underlies all the parts and activities of an animal

A. A baseball player in action

The batter drives the ball hard toward center field. Will it get out of the infield? The shortstop dives and robs the batter of a base hit.

Though far removed from the arena of natural selection, a baseball game demonstrates some of the remarkable evolutionary adaptations of the human body. A hit and a catch are both spectacular displays of our nervous system's ability to almost instantly link environmental stimuli with appropriate muscular activity. The batter hits a ball that was traveling over 90 miles per hour. The shortstop's eyes detect the ball as soon as it leaves the bat. In both cases, a supercomputer we call the human brain integrates information about the angle and speed of the ball and then signals multiple muscles to perform a very specific action—to swing a bat at precisely the right moment, or dive and thrust out a hand to meet the ball, again at just the right moment. These remarkable abilities result from adaptations of the human body, derived by natural selection and key to our survival as a species.

No less remarkable are the adaptations that underlie the cooperative work of a leaf-cutter ant colony. Movement is a common denominator of everything the million or so ants in a colony do. Each ant's tiny brain keeps it performing a specific set of movements that make it a contributing member of the colony as a whole. The chores of colony maintenance and fungus gardening involve an assembly line of cooperative legwork and delicate manipulations of mouthparts. Colony defense often takes the form of slashing or crushing invaders by the powerful mouthparts of detachments of soldier ants. Out in a rain forest at night (Figure B), the ants follow a precise route, a scent trail laid down earlier by scouts from their nest. Their senses are so acute that just a milligram of scent chemical would be enough to lead an entire colony of leaf-cutters three times around the world.

* * *

We have now completed our unit on the organ systems of animals. Our overriding theme has been that structure underlies function—that the structural adaptations of a cell, tissue, organ, or organ system determine the job it can perform. We have seen numerous examples of this basic principle, from our early studies of the digestive system to our discussion of the cells of the immune system, and also in our last several chapters on nervous control, sensory receptors, and movement systems. Watching the precise movements of a baseball player, the work of an ant colony, or the outward activity of any whole animal gives us an overview of the structure–function relationship and a reminder of the evolutionary process that generated it. The visible activities of any animal result from a nervous system (composed of cells with a specific structure) directing muscles (capable of contracting because of their cellular structure) to respond to information that sensory receptors (specifically structured cellular detectors) have gathered about the environment. We'll see the structure–function theme emerge again in the context of adaptations to the environment when we take up the study of plants in the next unit.

B. Leaf-cutter ants in motion

Begin your review by rereading the module headings and scanning the figures before proceeding to the Chapter Summary and questions.

Chapter Summary

Introduction–30.1 Movement is one of the most distinctive features of animals. Animals move in a variety of ways. Some aquatic animals, such as sponges, do not move about in search of food, but use body parts to generate currents that bring food to them. Other animals swim, using their limbs as oars, expelling water for jet propulsion, or moving their tails from side to side. Aquatic animals are supported by water, so gravity has little effect on them, but they are slowed by friction. Many are smooth and streamlined.

30.1 Animals that walk, hop, or run on land are not affected as much by friction, but they must be able to support themselves against the force of gravity. Legs stabilize the body and propel the animal forward. Muscles generate the power for movement, and springy legs store energy for each step. Animals that crawl or burrow must overcome friction. They move by side-to-side undulation or head-to-tail waves of muscle contraction. The wings of birds, bats, and flying insects generate lift that overcomes the pull of gravity.

30.2–30.4 All animal movement is based on contractile systems working against a firm skeleton. Skeletons function in support, movement, and the protection of internal organs. Worms and cnidarians have hydrostatic skeletons, which consist of fluid held in closed compartments. These animals can change shape by contracting muscles in their body walls. Exoskeletons are hard external cases, like clam shells and the armorlike skeletons of insects. Vertebrates have endoskeletons composed of cartilage or cartilage and bone, with movable joints. The human endoskeleton consists of an axial portion (skull, vertebrae, and ribs) and an appendicular portion (shoulder girdle, arms, pelvic girdle, and legs). The human skeleton is versatile, but it is also subject to problems such as back pain, arthritis, and osteoporosis.

30.5–30.6 A bone is a living organ containing several kinds of tissues. It is covered with a connective tissue membrane. Cartilage at the ends of the bone cushions the joints. The bone is served by blood vessels and nerves. Bone cells live in a matrix of flexible protein fibers and hard calcium salts. Long bones have a central cavity that stores fat, and spongy bone at their ends that contains red marrow, where blood cells form. Most of the skull bones develop from sheets of connective tissue. Long bones start out as shafts of cartilage, which are gradually replaced by bone tissue. A bone keeps growing until cartilage growth ceases.

30.7–30.9 Muscles pull on bones, which act as levers that produce movements. Antagonistic pairs of muscles produce opposite movements. Muscles perform work only when contracting. A muscle extends when pulled by another muscle. Each muscle cell, or fiber, contains bundles of overlapping thick and thin protein filaments. Calcium ions trigger myosin heads on the thick filaments to pull themselves along the actin molecules of the thin filaments. ATP provides the energy for this movement. The sliding of thick and thin filaments across one another increases their degree of overlap, shortening the muscle fiber. This shortening of muscle fibers produces the pull, or contraction, of a muscle.

30.10 Motor neurons carry action potentials that initiate muscle contraction. A neuron releases the neurotransmitter acetylcholine at a neuromuscular junction, signaling a muscle fiber to contract. A neuron can branch to a number of muscle fibers; the neuron and the muscle fibers it controls constitute a motor unit. The strength of a muscle contraction depends on the number of motor units activated. Motor units vary in size, and the control of small motor units creates precise movements.

30.11 An animal's nervous system connects sensations derived from environmental stimuli to responses carried out by its muscles. Robots can imitate, in a simple way, some of the senses and responses of animals.

Testing Your Knowledge

Multiple Choice

1. A human's internal organs are protected mainly by the
   a. hydrostatic skeleton
   b. motor unit
   c. axial skeleton
   d. exoskeleton
   e. appendicular skeleton

2. When your biceps muscle contracts, your arm bends at the elbow. Contraction of the muscle results from the ———— of thick and thin filaments inside its muscle fibers.
   a. collapse
   b. rotation
   c. shortening
   d. folding
   e. sliding

3. Gravity would have the least effect on the movement of which of the following? (*Explain your answer.*)
   a. a salmon
   b. a human
   c. a snake
   d. a sparrow
   e. a grasshopper

4. Arm and leg muscles are arranged in antagonistic pairs. How does this affect their functioning?
   a. It provides a backup if one of the muscles is injured.
   b. One muscle of the pair pushes while the other pulls.
   c. A single neuron controls both of them.
   d. It allows the muscles to produce opposing movements.
   e. It doubles the strength of contraction.

5. Which of the following bones in the human arm would correspond to the femur in the leg?
   a. radius
   b. tibia
   c. humerus
   d. metacarpal
   e. ulna

6. Which of the following animals is correctly matched with its type of skeleton?
   a. fly—endoskeleton
   b earthworm—exoskeleton
   c. dog—exoskeleton
   d. lobster—exoskeleton
   e. bee—hydrostatic skeleton

7. When a horse is running fast, its body position is stabilized by
   a. side-to-side undulation
   b. energy stored in muscles and tendons
   c. the lift generated by its movement through the air

d. foot contact with the ground

e. its momentum

8. What is the role of acetylcholine in muscle contraction?

   a. It makes up part of the thin filaments inside a muscle fiber.
   b. It provides energy for contraction.
   c. It blocks contraction when the muscle relaxes.
   d. It signals a muscle fiber to contract
   e. It forms the heads of the myosin molecules in the thick filaments inside a muscle fiber.

9. Muscle A and muscle B are the same size, but muscle A is capable of much finer control than muscle B. Which of the following is likely to be true of muscle A? (*Explain your answer.*)

   a. It is controlled by more neurons than muscle B.
   b. It contains fewer motor units than muscle B.
   c. It is controlled by fewer neurons than muscle B.
   d. It has larger sarcomeres than muscle B.
   e. Each of its motor units consists of more cells than the motor units of muscle B.

True/False (*Change false statements to make them true.*)

1. The shoulder joint is a hinge joint.

2. Friction is the main impediment to animals that walk on land.

3. In rheumatoid arthritis, joint tissues are attacked by the body's own immune system.

4. A duck has an endoskeleton.

5. The scapula (shoulder blade) is part of the axial skeleton.

6. Calcium provides the energy for muscle contraction.

7. Thin filaments of a muscle fiber contain actin, and thick filaments contain myosin.

8. A single neuron may signal many muscle fibers to contract at once.

9. Arthritis often accompanies aging.

10. Tendons connect bones together at joints.

11. Blood is manufactured in yellow bone marrow.

12. When its sarcomeres shorten, a muscle fiber contracts.

Describing, Comparing, and Explaining

1. A hawk swoops down, seizes a mouse in its talons, and flies back to its perch. In a few sentences, explain how its wings enable it to overcome the downward pull of gravity as it flies upward.

2. In terms of both numbers of species and numbers of individuals, insects are the most successful land animals. Write a paragraph explaining how their exoskeletons help them live on land. Are there any disadvantages to having an exoskeleton?

3. Describe how you bend your arm, starting with action potentials and ending with the contraction of a muscle. How does a strong contraction differ from a weak contraction?

4. In what ways did the human skeleton change as an upright posture and bipedalism evolved? Describe the changes by comparing the human skeleton and the skeleton of a quadruped like a baboon.

Thinking Critically

1. Drugs are often used to relax muscles during surgery. Which of the following two chemicals do you think would make the best muscle relaxant, and why? Chemical A: Blocks acetylcholine receptors on muscle cells. Chemical B: Floods the cytoplasm of muscle cells with calcium.

2. An earthworm's body consists of a number of fluid-filled compartments, each with its own set of longitudinal and circular muscles. In a different kind of worm, the roundworm, a single fluid-filled cavity occupies the body, and there are only longitudinal muscles that run its entire length. In what ways would you expect the movement of a roundworm to differ from the movement of an earthworm?

3. When a person dies, muscles often become rigid and fixed in position—a condition known as rigor mortis, which often figures importantly in mystery novels. Rigor mortis occurs because muscle cells use up their supply of ATP as they die. The rigor disappears several hours after death because the biological molecules break down. Explain, in terms of the mechanism of contraction described in Module 30.9, why the lack of ATP would cause muscles to become rigid, rather than limp, after death.

Science, Technology, and Society

1. A goal of the recent Americans with Disabilities Act is to allow people with physical limitations to fully participate in and contribute to society. Perhaps you have a disability or know someone with a disability. Imagine that a neuromuscular disease or injury makes it impossible for you to walk. Think about your activities during the last 24 hours. How would your life be different if you had to get around in a wheelchair? What kinds of barriers or obstacles would you encounter? What kinds of changes would have to be made in your activities and surroundings to accommodate your change in mobility?

2. Athletes sometimes take anabolic steroids illegally to increase the buildup of muscle that occurs with training. In young people, whose bones are still growing, steroids also act to speed up the conversion of cartilage to bone. With regard to bone growth, what might be some negative consequences of steroid use by young athletes? Does steroid use seem to you to be worth the risks? Why or why not? As a competing athlete, would you use anabolic steroids if you knew you could get away with it? Why or why not?

3. You may know an elderly person who has broken a bone (often a hip) as a consequence of osteoporosis. To prevent osteoporosis, researchers recommend exercise and maximizing calcium intake beginning in the teens and twenties. Do you think young people think of themselves as future senior citizens? How would you recommend that they be encouraged to develop health habits that might not pay off for 40 or 50 years?

Unit 5
Plants:
Form and Function

Plant Structure, Reproduction, and Development

31

In 1951, author and conservationist Freeman Tilden wrote, "Not a single *Sequoiadendron gigantea* [giant sequoia] should be cut. They represent a unique survival; they belong to our remote and future histories." The giant sequoia at the left, named the General Sherman for the controversial Union commander in the Civil War, is a case in point; it has been growing for about 2500 years, and plant biologists believe it could live for many more.

A unique survivor and a truly awesome representative of the plant kingdom, the General Sherman resides in the Giant Grove in Sequoia-Kings Canyon National Park, in the Sierra Nevada of central California. At 83 m (272 ft) tall, it is over 100 feet taller than Niagara Falls. Its massive trunk, weighing nearly 1400 tons, is about 10 m in diameter at the base; its lowest limb, about 40 m above the ground, is over 2 m in diameter. This is the largest plant on Earth.

The photograph on this page shows the ancient trunk of another giant sequoia. The rings you see in this cross section are called growth rings because each one marks a year in the tree's life. The rings vary in thickness, depending on weather conditions during the growing seasons. About 1700 years old when it was felled in 1917, this tree was a seedling at the time of the Roman Empire. It was over 200 years old when the Saxons invaded England in 449 A.D. By 1492, when Columbus landed in the New World, it had laid down 80% of its growth rings. Having survived California earthquakes, numerous fires, droughts, and storms, most of the giant sequoias were cut down between 1850 and 1900, a time of virtually uncontrolled lumbering. A tree like this one or the General Sherman would yield enough prime boards for nearly 50 four- or five-room houses, and many giant sequoias were cut down for just that purpose. Today, only a few groves remain, nearly all in national and state parks, where they are viewed as living treasures.

Though we are becoming increasingly aware of the importance of environmental conservation, we still tend to think of plants in utilitarian terms. We depend on them for lumber, fabric, paper, landscaping, shade, and most of all

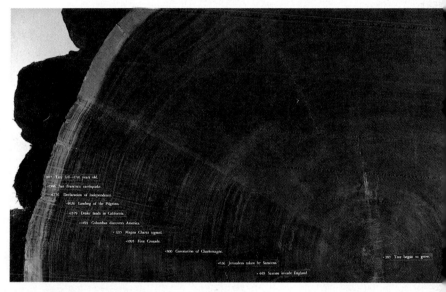

A cross section of the trunk of a giant sequoia

for food. Beyond our immediate needs, plants are vital to the planet's well-being. Virtually all land animals depend on them for food, either eating plants directly or eating other animals that eat plants. Above and below the ground, plants provide food, shelter, and breeding areas for animals, fungi, and microorganisms. Plant roots also prevent soil erosion, and photosynthesis in plant leaves helps reduce carbon dioxide levels in the atmosphere and adds oxygen to the air.

The giant sequoia is one of about 550 living species of conifers, or cone-bearing plants. As we saw in Chapter 18, conifers are gymnosperms, one of two groups of seed plants. Gymnosperms are the naked-seed plants, bearing seeds in cones. Plants of the other group, the angiosperms (flowering plants), produce seeds enclosed in fruits. Because angiosperms make up nearly 90% of the plant kingdom, we concentrate on them in this unit. In this chapter, we continue our form-and-function theme, studying plant structure in relation to how plants reproduce and grow. We begin with a look at one of plant biology's most distinguished personalities.

Plant scientist Katherine Esau, preeminent student of plant structure and function

A. Katherine Esau with sugar beets, about 1924

Ever scornful of retirement, Dr. Katherine Esau has been one of this century's most prolific plant scientists. During a research and teaching career spanning more than 60 years, Dr. Esau published over 150 articles in research journals and six books about plant biology. Two of her texts, *Plant Anatomy* and *Anatomy of Seed Plants,* are classics that have been translated into many languages and used throughout the world.

Born in Ukraine in 1898, Katherine Esau was fascinated by plants at an early age. She entered the Women's Agricultural College in Moscow when she was 17. The Esau family moved to Germany in 1918, and Katherine graduated from the Agricultural College of Berlin in 1922 with a degree in plant breeding. That same year, she emigrated with her parents to the United States.

The Esau family settled in California, and soon thereafter, Katherine was hired as a plant breeder on a sugar-beet seed farm. She describes the photograph in Figure A, taken in a field near Oxnard, California, as "my first professional acquaintance with a sugar beet." Sugar beets played a major role in Esau's career. Her early research led to an understanding of how viruses infect the plant and destroy its tissues. Her later studies of sugar beets and other angiosperms made Dr. Esau an international expert on phloem (the food-conducting tissue in plants). By the time the electron microscope became available to biologists, Dr. Esau was already in her sixties. Nevertheless, she became skilled with the instrument (Figure B) and continued

to make important discoveries of the detailed structure of plants well into her nineties.

For much of her career, Katharine Esau was a professor of botany at the Davis and Santa Barbara campuses of the University of California. Soft-spoken, meticulous, witty, and a strong leader by example, she excelled in both teaching and research. Dr. Esau studied plant cell structure with the goal of understanding how plant parts work, and she remains one of plant biology's strongest advocates of the integrated study of biological structure and function. Among her many accomplishments, Dr. Esau's microscopy studies led to the discovery of plasmodesmata, the open connections between plant cells (see Figure 4.19A). Her research on plasmodesmata in phloem suggested how sugars are transported throughout the plant body. Dr. Esau also discovered that plant viruses are transmitted through plant tissues via plasmodesmata. Her energy and focus lead us into this chapter.

B. Dr. Esau with an electron microscope

There are two major groups of angiosperms: monocots and dicots

Angiosperms have dominated the land for over 100 million years, and there are about 235,000 species of flowering plants living today. Most of our food comes from a few hundred species of flowering plants that have been domesticated. Among these are the root crops, such as sugar beets, potatoes, and carrots; the fruits, such as apples, oranges, squash, nuts, and berries; the legumes, such as peas and beans; and the grains, such as rice, wheat, corn, and barley.

Plant biologists classify angiosperms in two groups, called the monocots and dicots, on the basis of several structural features, illustrated below. The names "monocot" and "dicot" refer to the first leaves that appear on the plant embryo. These embryonic leaves are called seed leaves, or **cotyledons.** A **monocot** embryo has one seed leaf; a **dicot** embryo has two seed leaves.

Monocots (about 65,000 species) include the orchids, bamboos, palms, and lilies, as well as the grains and other grasses. You can see the single cotyledon inside the seed on the top left in the figure.

The leaves, stems, flowers, and roots of monocots are also distinctive. Most monocots have leaves with parallel veins. Monocot stems have vascular tissues (tissues that transport water and nutrients) arranged in a complex array of bundles. The flowers of most monocots have their petals

and other parts, which we will describe later, typically in multiples of three. The roots of monocots form a fibrous system—a mat of threads—that spreads out below the soil surface. With most of their roots in the top few centimeters of soil, monocots, especially grasses, make excellent ground cover that reduces erosion. The roots of a rye-grass plant, for instance, have more than 20 times the surface area of its stems and leaves.

Most angiosperms (about 170,000 species) are dicots. This group includes most shrubs and trees (except for the conifers), as well as the majority of our ornamental plants and many of our food crops.

You can see the two cotyledons of a typical dicot in the seed at the lower left. Dicot leaves have a multibranched network of veins, and dicot stems have vascular bundles arranged in a ring. The dicot flower usually has petals and other parts in multiples of four or five. The large, vertical root of a dicot, known as a taproot, goes deep into the soil, as you know if you've ever tried to pull up a dandelion.

As we saw in the preceding unit on animals, a close look at a structure often reveals its function. Conversely, function provides insight into the logic of a structure. In the modules that follow, we'll begin a detailed look at the correlation between plant structure and function.

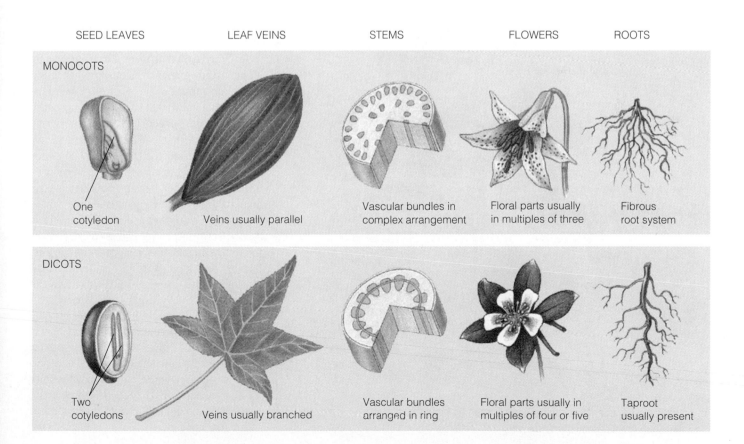

Chapter 31   Plant Structure, Reproduction, and Development   **603**

The plant body consists of roots and shoots

Among the evolutionary adaptations that made it possible for plants to move onto land were the ability to take up water and minerals from the soil, to absorb light and take in carbon dioxide from the air for photosynthesis, and to survive dry conditions. The subterranean roots and aerial shoots of a land plant, such as the generalized flowering plant shown in the drawing here, perform all these vital functions. Neither the root nor the shoot can survive without the other. Lacking chloroplasts and living in the dark, the roots would starve without sugar and other organic nutrients transported from the photosynthetic leaves of the shoot system. Conversely, stems and leaves depend on the water and minerals absorbed by roots.

A plant's **root system** anchors it in the soil, absorbs and transports minerals and water, and stores food. The fibrous root system of a monocot provides broad exposure to soil water and minerals, as well as firm anchorage. The root system of a dicot, with many small secondary roots growing out from one large taproot, is different but also effective at absorbing water and minerals. In both dicots and monocots, most absorption of water and minerals occurs near the root tips through tiny projections called root hairs. A **root hair** is an outgrowth of an epidermal cell (one of the cells forming the outer layer of the root). These microscopic cell extensions vastly increase the absorptive surface area of the root.

The **shoot system** of a plant is made up of stems, leaves, and adaptations for reproduction—flowers, in angiosperms. (We'll return to the angiosperm flower in Module 31.9.) As indicated in the drawing, the **stems** are the parts of the plant that are generally above the ground and that support the leaves and flowers. In the case of a tree, the stems are the trunk and all the branches, including the smallest twigs. A young stem has **nodes,** the points at which leaves are attached, and **internodes,** the portions of the stem between nodes. The **leaves** are the main site of photosynthesis in most plants, although some species have green, photosynthetic stems. A leaf consists of a flattened blade and a stalk, or petiole, which joins the leaf to the stem.

The two types of buds you see in the figure are undeveloped shoots. The apex (tip) of the shoot is usually where growth is concentrated. When a plant stem is growing in length, the **terminal bud,** at the apex of the stem, has developing leaves and a compact series of nodes and internodes. The **axillary buds,** one in each of the angles formed by a leaf and the stem, are usually dormant. In many plants, the terminal bud produces hormones that inhibit growth of the axillary buds, a phenomenon called **apical dominance.** By concentrating resources on growing taller, apical dominance is an evolutionary adaptation that increases the plant's exposure to light. This is especially important where vegetation is dense. However, branching is also important for increasing the exposure of the shoot system to the environment, and under certain conditions, the axillary buds begin growing. Some develop into shoots bearing flowers, and others become nonreproductive branches complete with their own terminal buds, leaves, and axillary buds. In some cases, removing the terminal bud stimulates the growth of axillary buds. This is why fruit trees and houseplants become bushy when they are pruned or "pinched back."

The plant shown here gives us an overview of plant structure, but it by no means represents the enormous diversity of angiosperms. Let's look briefly at some variations on the basic themes of root and stem structure.

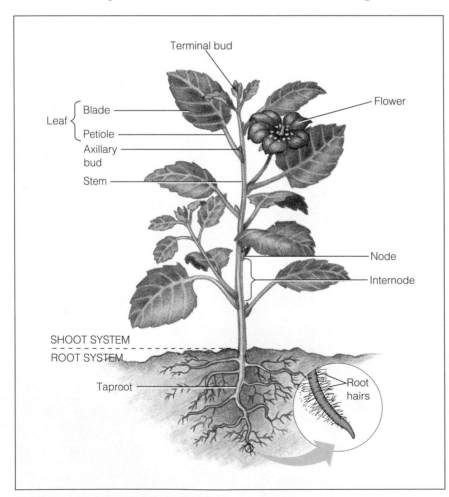

The body plan of a flowering plant (a dicot)

Roots, stems, and leaves come in a variety of sizes and shapes, and they are adapted for a variety of functions. Many are adapted for storing food. The vegetables we call carrots, turnips, sugar beets, and sweet potatoes, for instance, are unusually large taproots that store food in the form of carbohydrates such as starch. The plants use the stored sugars during periods of active growth and when they are producing flowers and fruits. Plant breeders have improved the yields of root crops by selecting varieties, such as the sugar beet plant in Figure A, with very large taproots.

Figure B shows some examples of modified stems. The strawberry plant has a horizontal stem, or runner, that grows along the ground surface. A runner is a means of asexual reproduction; as shown here, a new plant can emerge from its tip. You've seen a different stem modification if you have ever dug up an iris plant; the large, brownish, rootlike structures near the soil surface are actually horizontal stems called **rhizomes**. The rhizomes store food, and they can

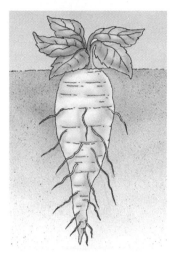

A. The modified root of a sugar beet plant

also spread and form new plants. The white potato plant has rhizomes that end in enlarged structures called **tubers** (the potatoes we eat), where food is stored in the form of starch.

Plant leaves, too, are highly varied. Grasses and most other monocots, for instance, have long leaves without petioles. In sharp contrast, some dicots, such as celery, have enormous petioles—the stalks we eat—which contain a lot of water and stored food. The upper photograph in Figure C shows a modified leaf called a tendril, with its tip coiled around a wire. Tendrils help plants such as this cucumber climb. The spines of the prickly pear cactus (lower photo) are modified leaves that may protect the plant from plant-eating animals. The main part of the cactus is the green stem, which is adapted for photosynthesis and water storage.

So far we have examined plants as we see them with the unaided eye. Next, we begin to dissect the plant and explore its microscopic organization.

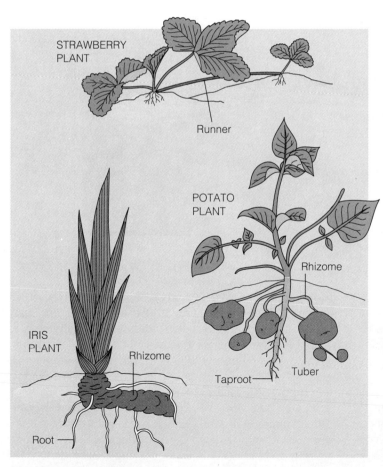

B. Three kinds of modified stems

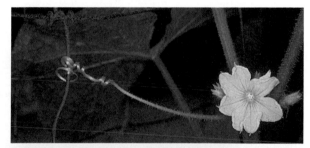

C. Modified leaves: a cucumber tendril (top) and cactus spines (bottom)

Plant cells and tissues are diverse in structure and function

Plant cells are unique in several ways. Many have chloroplasts, which contain the photosynthetic pigment chlorophyll. Mature plant cells like the one in Figure A often have a large central vacuole. Fluid in the vacuole helps maintain the cell's firmness, or turgor, which we discuss in Chapter 32. The single most distinctive feature of plant cells is their cell wall surrounding the plasma membrane. The wall is made mainly of the structural carbohydrate cellulose. Many plant cells, especially those that provide structural support, have a two-part cell wall; a primary cell wall is laid down first, and then a more rigid secondary cell wall is secreted between the plasma membrane and the primary wall. The enlargement on the right in Figure A shows the adjoining cell walls of two cells. The primary walls of adjacent cells in plant tissues are held together by a sticky layer called the middle lamella. Pits, where the cell wall is relatively thin, allow the contents of adjacent cells to lie close together. Plasmodesmata are channels of communication and circulation between adjacent plant cells.

The structure of the cell and the nature of its wall often determine the cell's main functions. This is evident in the six major types of plant cells shown in Figures B–F.

Parenchyma cells (Figure B) are the most abundant type of cell in most plants. They are relatively unspecialized and flexible, with thin primary walls (orange in the drawing) and no secondary walls. Parenchyma cells carry out a variety of functions, such as food storage, photosynthesis, and aerobic respiration. Parenchyma cells come in a variety of shapes, but they are often multisided, like the one in Figure A. Most parenchyma cells remain able to divide and differentiate into other types of plant cells, which they may do during the repair following an injury to a plant.

Collenchyma cells (Figure C) resemble parenchyma cells in lacking secondary walls, but they have thicker primary walls. Their main function is to provide support in parts of the plant that are still growing. Young stems, for instance, often have collenchyma cells just below their surface. These living cells elongate along with the growing stem.

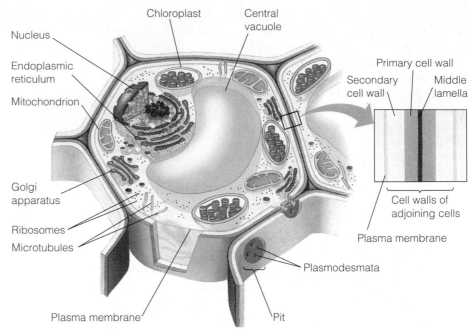

A. Plant cell structure

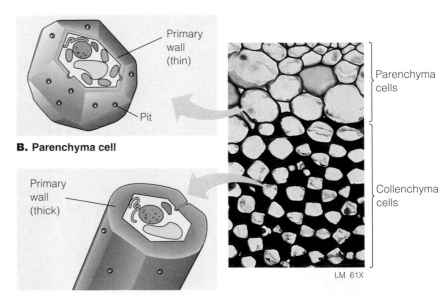

B. Parenchyma cell

C. Collenchyma cell

LM 61X

Sclerenchyma cells have rigid secondary cell walls (yellow in Figure D) hardened with lignin, which is the main chemical component of wood. Mature sclerenchyma cells cannot elongate, and they occur only in regions that have stopped growing in length. When mature, most sclerenchyma cells are dead, their cell walls forming a rigid scaffold that supports the plant.

Figure D shows the two types of sclerenchyma cells. One, called a **fiber**, is long and slender, and usually occurs in bundles. Some plant tissues formed of fiber cells are commercially important; hemp fibers, for example, are used to make rope. The other type of sclerenchyma cell, called a **sclereid**,

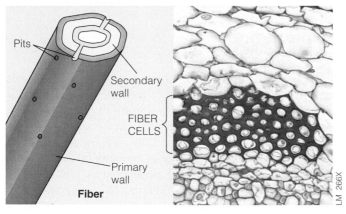

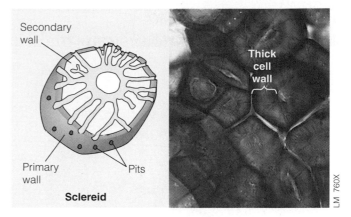

D. Sclerenchyma cells

or stone cell, is shorter than the fiber, and has a thick, irregular, and very hard secondary wall. Nutshells and seed coats owe their hardness to sclereids, and sclereids scattered in the soft tissue of a pear feel gritty when eaten.

Water-conducting cells are of two types, both having rigid, lignin-containing secondary cell walls. As Figure E shows, **tracheids** are long cells with tapered ends. **Vessel elements** are wider, shorter, and less tapered. Chains of tracheids or vessel elements arranged end-to-end form a system of tubes that convey water from the roots to the stems and leaves. The tubes are hollow because when mature, both tracheids and vessel elements are dead, and only their end walls remain. Water passes through pits in the end walls of tracheids and through the open ends of the vessel elements.

Food-conducting cells, also known as **sieve-tube members,** are also arranged end-to-end, forming tubes (Figure F). Unlike water-conducting cells, however, sieve-tube members have thin primary walls and no secondary walls, and they remain alive at maturity. Their end walls, perforated by pores, form **sieve plates,** through which sugars, other compounds, and some mineral ions move between adjacent food-conducting cells. Each sieve-tube member is flanked by at least one **companion cell,** which is connected to the sieve-tube member by numer-

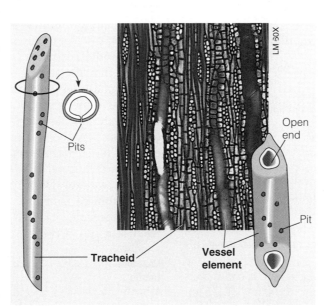

E. Water-conducting cells

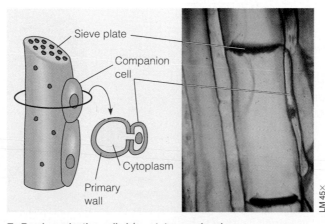

F. Food-conducting cells (sieve-tube members)

ous plasmodesmata. The nucleus and ribosomes of the companion cells may make certain proteins for the sieve-tube member, which has no nucleus or ribosomes of its own.

As in animals, the cells of plants are grouped into tissues with characteristic functions, such as photosynthesis. Simple tissues consist of a single cell type, such as parenchyma, collenchyma, or sclerenchyma. Thus, we can speak of a parenchyma cell, but we can also refer to parenchyma tissue, an association of many parenchyma cells.

A plant tissue composed of more than one type of cell is called a complex tissue. The vascular tissues of plants—tissues that conduct water and food—are complex tissues. Vascular tissue called **xylem** contains water-conducting cells that convey water and dissolved minerals upward from the roots. Vascular tissue called **phloem** contains sieve-tube members that transport sugars made in the leaves to the nonphotosynthetic parts of the plant, and from storage tissues in roots to actively growing tissues. In addition to their water-conducting cells or sieve-tube members, xylem and phloem also contain sclerenchyma cells that provide support and parenchyma cells that store various materials.

Plants, like animals, have several levels of structural organization. We examine the level above tissues, called tissue systems, in the next module.

Three tissue systems make up the plant body

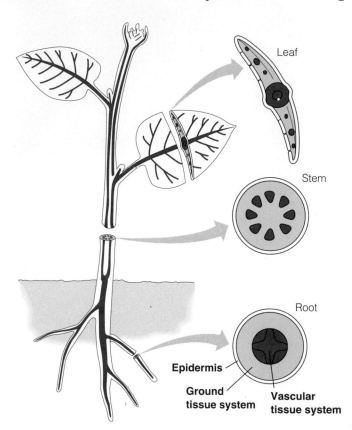

A. The three tissue systems

$\mathbf{A}$ plant's organs—its roots, stems, and leaves—are made up of three **tissue systems:** the epidermis, the vascular tissue system, and the ground tissue system. We examine the tissue systems of young roots and shoots in this module. Later, we will see that the tissue systems are somewhat different in older roots and stems.

As indicated in Figure A, each tissue system is continuous throughout the plant. The **epidermis,** or "skin," of a plant (white in Figure A), covers and protects its leaves, young stems, and young roots. Like our own skin, the epidermis is a plant's first line of defense against physical damage and infectious organisms. On the leaves and on some stems, epidermal cells secrete a waxy coating called the **cuticle,** which helps the plant retain water. The **vascular tissue system** (pink) is continuous throughout the plant. Made up of xylem and phloem, the vascular tissue system transports water and nutrients and also provides support. The **ground tissue system** (aqua) makes up the bulk of a young plant, filling the spaces between the epidermis and vascular tissue systems. The ground tissue system, mainly parenchyma, but usually including some collenchyma and sclerenchyma, has diverse functions, including photosynthesis, storage, and support.

Figure B, a cross section of a young dicot root (from a buttercup), shows what the three tissue systems look like under a microscope. The epidermis is a single layer of tight-

ly packed cells covering the entire root. Water and minerals enter the plant from the soil through these cells. Notice that some of the epidermal cells are starting to form root hairs, which will increase the absorptive surface area of the epidermis. In the center of the root, the vascular tissue system forms a cylinder, with xylem cells radiating from the center like spokes of a wheel, and phloem cells filling in the wedges between the spokes. The ground tissue system of the root forms the **cortex,** which consists mostly of parenchyma tissue. The cells here store food (the purple granules are starch) and take up minerals that have entered the root through the epidermis. The innermost layer of cortex is the **endodermis,** a thin cylinder one cell thick. The endodermis is a selective barrier, determining which substances pass between the rest of the cortex and the vascular tissue. (We discuss how this barrier works in Chapter 32.)

The roots of dicots and monocots are similar, but, as Figure C shows, the young stem of a dicot (sunflower) looks quite different from that of a monocot (corn). Both stems have their vascular tissue system arranged in numerous **vascular bundles.** However, the monocot stem has vascular bundles throughout its ground tissue system, whereas the dicot stem has a distinct ring of vascular bundles and a two-part ground tissue system. The dicot cortex fills the space between the vascular ring and the epidermis. The other part of the dicot ground tissue system, called the **pith,** fills the center of the stem and is often important in food storage.

Figure D illustrates the arrangement of the three tissue systems in a leaf. Leaves are distinctive in having pores, called **stomata** (singular, *stoma*), in their epidermis. You can see the stomata clearly in the micrograph on the right. Each is a gap between two specialized epidermal cells,

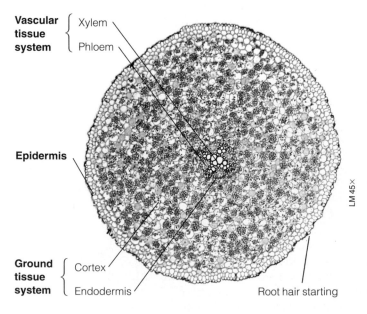

LM 45×

B. Tissue systems in a young root (a dicot)

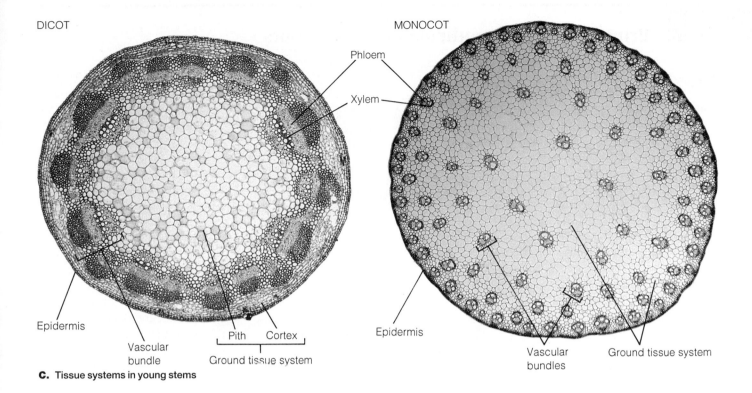

DICOT   MONOCOT

Phloem

Xylem

Epidermis   Epidermis

Vascular   Pith   Cortex   Vascular   Ground tissue system
bundle   bundles
            Ground tissue system

C. Tissue systems in young stems

called guard cells. **Guard cells** regulate the size of the stomata, allowing gas exchange between the surrounding air and the photosynthetic cells inside the leaf. The stomata are also the major avenues of water loss from the plant, as we discuss in Chapter 32.

The ground tissue system of a leaf is called **mesophyll.** Mesophyll consists mainly of parenchyma cells equipped with chloroplasts (green in Figure D) and specialized for photosynthesis. Notice that cells in the lower mesophyll are loosely arranged, with many air spaces. This is the chief site of gas exchange. Air enters through stomata and circulates freely in the spaces among the cells. The stomata are more numerous in the leaf's lower epidermis. This adaptation minimizes water loss, which occurs more rapidly from the sunlit upper surface of the leaf.

The leaf's vascular tissue system is made up of a network of veins. As you can see in the figure, each **vein** is a vascular bundle composed of xylem and phloem surrounded by a sheath of parenchyma cells. The veins' xylem and phloem, continuous with the vascular bundles of the stem, are in close contact with the leaf's photosynthetic tissues. This ensures that those tissues are supplied with water and mineral nutrients from the soil and that sugars made in the leaves are transported throughout the plant. In some plants (the C_4 plants; see Module 7.12), the sheath cells are the sites of the Calvin cycle in photosynthesis.

This completes our survey of basic plant anatomy. In the next two modules, we examine how plants grow in length and thickness.

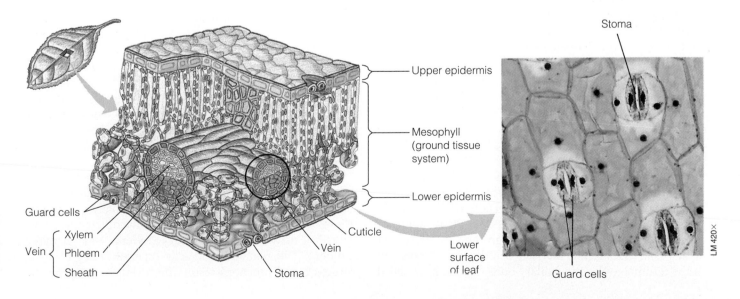

Upper epidermis

Mesophyll
(ground tissue
system)

Lower epidermis

Stoma

Guard cells

Vein { Xylem / Phloem / Sheath }

Cuticle

Vein

Stoma

Lower
surface
of leaf

Guard cells

LM 420×

Primary growth lengthens roots and shoots

Most plants continue to grow as long as they live, a condition known as **indeterminate growth**. Most animals, in contrast, are characterized by **determinate growth**; that is, they cease growing after reaching a certain size. These differences underlie a broader distinction between plants and animals. Most animals *move* through their environment. Plants, in contrast, *grow* through their environment. This indeterminate growth enables plants to increase their exposure to sunlight, air, and soil throughout life.

Indeterminate growth does not mean that plants are immortal. In fact, most plants have a finite life span. Species called **annuals**, for instance, complete their life cycle (the time from when they begin to grow through flowering and seed production to death) in a single year or growing season. Our most important food crops (wheat, corn, and rice, for example) are annuals, as are a great number of wildflowers. **Biennials** complete their life cycle in two years; flowering usually occurs during the second year. Beets and carrots are biennials, but we usually harvest them in their first year and miss seeing their flowers. Plants that live many years, including trees, shrubs, and some grasses, are known as **perennials**.

Some perennial plants are among the oldest organisms alive. The giant sequoias we discussed in the chapter's introduction are ancient (the oldest living one began growing about 3000 years ago), and another conifer, the bristlecone pine, can outlive the sequoias by thousands of years. Older still are clones of certain shrubs and grasses (see Figure 31.14C). It is likely that even the oldest plants could live for many more thousands of years if they could escape disease, fires, floods, climatic change, and human encroachment.

Indeterminate growth in all plants is made possible by tissues called meristems. A **meristem** (from the Greek *meristos,* divided) consists of unspecialized cells that divide and generate new cells and tissues. Plant biologists call these cells "embryonic," although they are present at all stages in a plant's life. Meristems at the tips of roots and in the terminal (top) and axillary buds of shoots are called

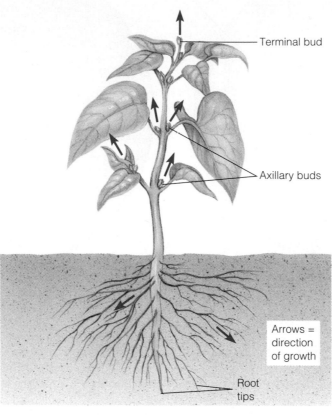

A. Locations of apical meristems, responsible for primary growth

Labels on figure:
Terminal bud
Axillary buds
Arrows = direction of growth
Root tips

apical meristems (Figure A). Cell division in the apical meristems produces the new cells that enable a plant to grow in length.

The lengthwise growth produced by apical meristems is called **primary growth**. Figure B is a longitudinal section through a growing onion root. Primary growth is what pushes this root through the soil. The very tip of the root is the **root cap**, a thimblelike cone of cells that protect the delicate, actively dividing cells of the apical meristem. The root's apical meristem (indicated by the blue oval) has two roles: It replaces cells of the root cap that are scraped away by the soil (downward arrow), and it produces the cells (upward arrow) for primary growth. Its primary growth cells form three concentric cylinders of embryonic tissue. The outermost cylinder will differentiate into the epidermis of the root. The intermediate cylinder—the bulk of the root tip—will develop into the root's cortex. The innermost cylinder will become the vascular tissue.

The apical meristem sustains growth of the root by continuously adding cells to the three embryonic tissue cylinders. However, cell division alone does not make the root elongate. A more important factor is the lengthening of cells. Notice in Figure B that the meristem grades upward into a zone of elongation. Cells in this zone can undergo a tenfold increase in length. They lengthen, rather than expand equally in all directions, mainly because of the parallel arrangement of cellulose fibers in their cell walls. The enlargement diagrams at the left of the root indicate how this works. The cells elongate by taking up water, and as they do, the cellulose fibers (shown in red) separate, somewhat like an expanding accordion. The cells cannot expand greatly in width because the cellulose fibers do not stretch much.

The epidermis, cortex, and vascular cylinder actually take shape in an area known as the zone of differentiation, above the elongation zone. (One zone actually grades into the next.) Cells of the vascular cylinder differentiate into vascular tissues called **primary xylem** and **primary phloem**. Differentiation of cells (specialization of their structure and function) results from differential gene ex-

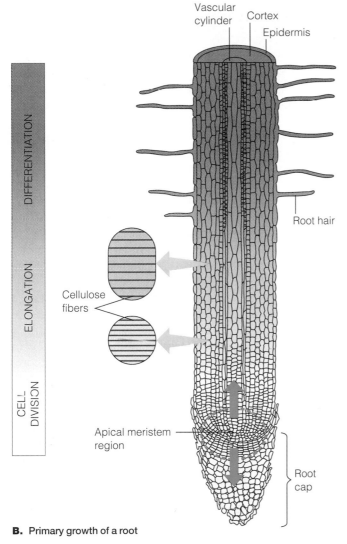

B. Primary growth of a root

Below the zone of elongation in the stem, the cells of the three cylinders differentiate into the three tissue systems.

The diagrams in Figure C show three stages in the growth of a shoot. Stage ① is just like the longitudinal section in the micrograph. At stage ②, the apical meristem has been pushed upward by the elongating cells underneath. As the apical meristem advances upward, some of its cells are left behind, and these become new axillary bud meristems at the base of the leaves. At stage ③, the process has repeated one more time.

Primary growth accounts for a plant's lengthwise growth. The stems and roots of many plants increase in thickness too, and in the next module, we see how this usually happens.

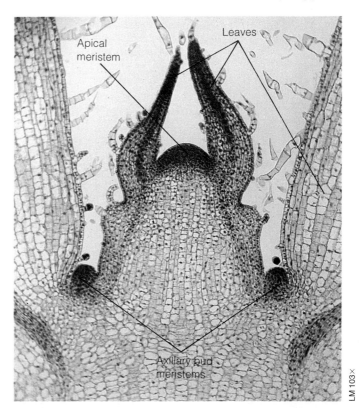

pression, as we discussed in Module 11.4. Cells in the vascular cylinder, for instance, develop into primary xylem or phloem cells because a certain set of genes is turned on, and is therefore expressed as specific proteins, while other genes in these cells are turned off. Genetic control of cellular differentiation is one of the most active areas of plant research. Recently, developmental biologists have discovered that plants, like animals, have homeotic genes (see Module 11.13), master control genes that determine major features of the body plan, such as the placement of sepals and petals on flowers.

The micrograph in Figure C shows a longitudinal section through the tip of a growing shoot; for reference, this is the terminal bud region shown in Figure A, cut lengthwise from the tip of the shoot to just below its uppermost pair of axillary buds. The apical meristem is a dome-shaped mass of dividing cells at the tip of the terminal bud. All the leaves you see have arisen on the flanks of the apical meristem. As in the root, the apical meristem forms three embryonic cylinders in the shoot. Elongation occurs just below this meristem, and the elongating cells push the apical meristem upward, instead of downward as in the root.

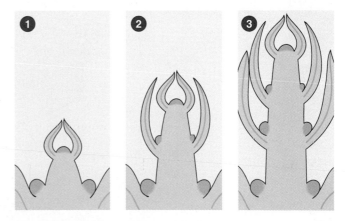

C. Primary growth of a shoot

Secondary growth increases the girth of woody plants

Stems and roots often begin to thicken after their apical meristems have produced the three embryonic tissue cylinders we just described. An increase in a plant's girth is called **secondary growth.** The results of secondary growth are most evident in the woody plants—trees, shrubs, and vines—whose stems last from year to year and consist mainly of thick layers of dead tissue, called wood.

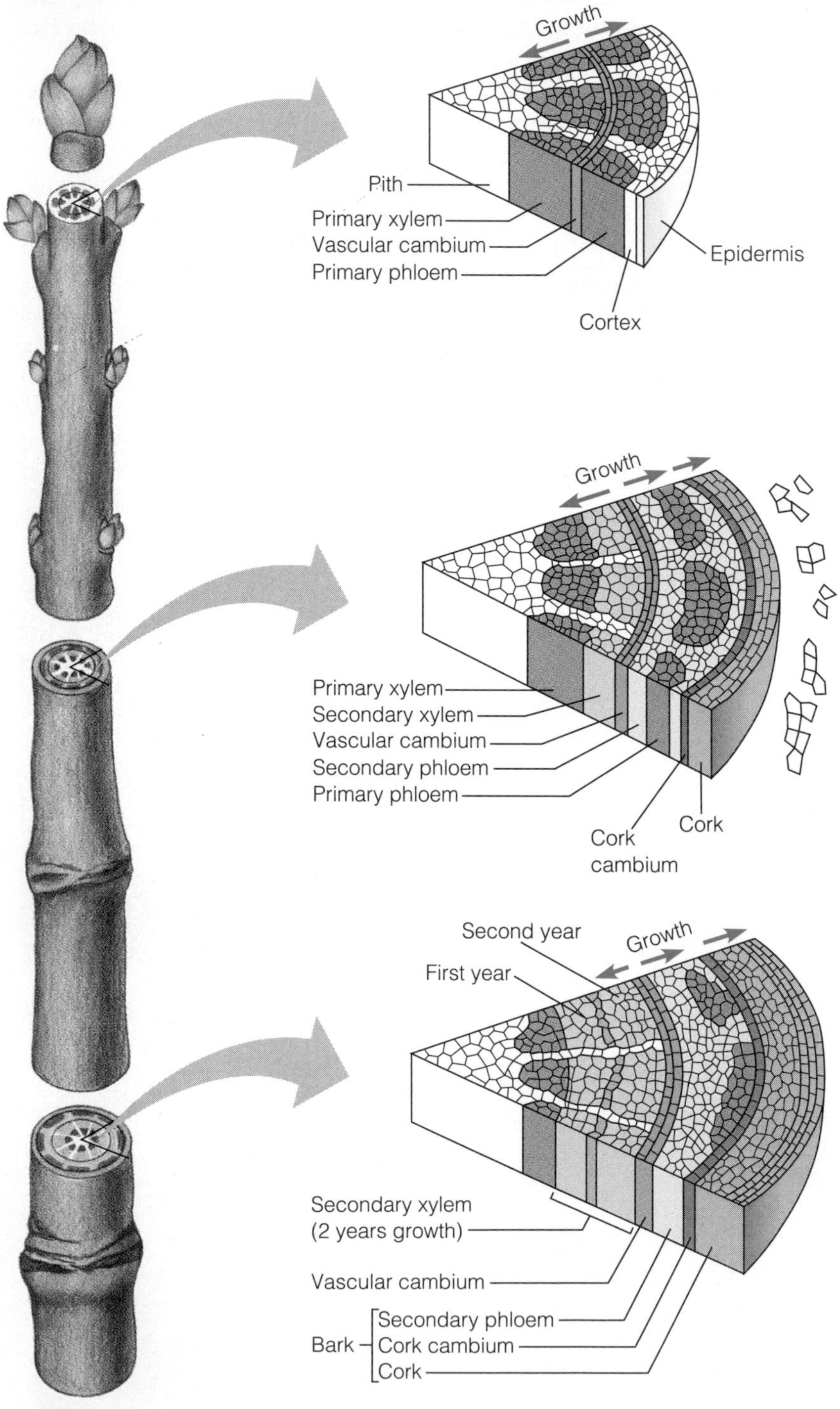

Secondary growth involves cell division in two meristems we have not yet discussed: the vascular cambium and the cork cambium. The **vascular cambium** arises from parenchyma cells in the vascular tissue cylinder. Vascular cambium first appears as a cylinder of actively dividing cells between the primary xylem and primary phloem, as you can see in the pie section at the top of Figure A. This region of the stem is just beginning secondary growth. Except for the vascular cambium, the stem at this stage of growth is virtually the same as one growing only in length—that is, showing only primary growth (see Figure 31.6C). Secondary growth occurs on either side of the vascular cambium, as indicated by the green arrows.

The other two drawings show the results of secondary growth. In the center one, the vascular cambium has given rise to two new tissues. One is the **secondary xylem,** next to the inner surface of the vascular cambium. The other is the **secondary phloem,** just outside the vascular cambium.

By focusing first on the secondary xylem, you can see what causes most of the increase in a stem's thickness. Notice there are *two* layers of secondary xylem in the bottom drawing. The inner layer is the one produced by the vascular cambium during the first year of secondary growth. The outer layer was produced in the second year of secondary growth. This yearly production of a new layer of secondary xylem accounts for most of the growth in thickness of a perennial plant.

Consisting of xylem cells and fibers that have thick walls rich in lignin, the secondary xylem makes up the **wood** of a tree, shrub, or vine. Over the years, a woody stem gets thicker and thicker as its vascular cambium produces layer upon layer of secondary xylem. The annual growth rings, such as those in the trunk of a giant sequoia or the locust tree in Figure B, result from the layering of secondary xylem. The layers are visible as rings because of uneven activity of the vascular cambium during each year. In woody plants that live in temperate regions, such as most of the United States, the vascular cambium becomes dormant each year during winter, and secondary growth is interrupted. When secondary growth resumes in the spring, a cylinder of spring wood forms. Made up of the first new xylem cells to develop, spring wood cells are usually larger and

A. Secondary growth of a woody stem

thinner-walled than those produced later in summer (the summer wood). Each tree ring consists of a cylinder of spring wood surrounded by a cylinder of summer wood.

Now let's return to Figure A and see what happens to the parts of the stem that are *external* to the vascular cambium. Unlike xylem, the external tissues do not accumulate over the years. Instead, they are sloughed off at about the same rate they are produced.

Notice at the top of the figure that the epidermis and cortex, both the result of primary growth, make up the young stem's external covering. When secondary growth begins, the epidermis and cortex start sloughing off. In the center drawing, you see that a new outer layer called **cork** (gray) has replaced the cortex and epidermis, which have sloughed off. Mature cork cells are dead and have thick, waxy walls, which protect the underlying tissues of the stem. Cork is produced by meristematic tissue called the **cork cambium** (dark gray), which first forms from parenchyma cells in the cortex. As the stem thickens and the secondary xylem expands, the cortex and the original cork cambium are pushed outward and sloughed off. You see their remains as the three outermost gray layers in the bottom diagram. By this time, a new cork cambium has developed from parenchyma cells in the secondary phloem.

As indicated in the bottom diagram of Figure A, everything external to the vascular cambium is called **bark:** all of the secondary phloem, the cork cambium, and the cork. The youngest secondary phloem (next to the vascular cambium) functions in sugar transport. The older secondary phloem dies, as does the cork cambium you see here. Pushed outward, these tissues and cork produced by the cork cambium help protect the stem until they, too, are sloughed off as part of the bark. Keeping pace with secondary growth, cork cambium keeps regenerating from the young secondary phloem, and keeps producing a steady supply of cork.

The log on the left in Figure B, from a locust tree, shows the results of several decades of secondary growth. The bulk of a trunk like this is dead tissue. The only living tissues in it are the vascular cambium, the youngest secondary phloem, the cork cambium, and cells in the wood rays, which you can see radiating from the center of the log in the drawing. The **wood rays** consist of parenchyma cells that transport water to the outer living tissues in the trunk. The **heartwood,** in the center of the trunk, consists of older layers of secondary xylem. These cells no longer transport water; they are clogged with resins and other metabolic byproducts that make the heartwood resistant to rotting. The lighter-colored **sapwood** consists of younger secondary xylem that actually conducts water.

Thousands of useful products are made from wood—from construction lumber to fine furniture, musical instruments, paper, insulation, and a long list of chemicals, including turpentine, alcohols, artificial vanilla flavoring, and preservatives. Among the qualities that make wood so useful are a unique combination of strength, hardness, lightness, high insulating properties, durability, and workability. In many cases, there is simply no good substitute for wood. A wooden oboe, for instance, produces far richer sounds than a plastic one. Fence posts made of locust tree wood last much longer in the ground than metal ones. Ball bearings are sometimes made of a very hard wood called lignum vitae. Unlike metal bearings, they require no lubrication, because a natural oil completely penetrates the wood.

In a sense, wood is analogous to the hard skeletons of many land animals. It is an evolutionary adaptation that allows a shrub or tree to remain upright and keep growing year after year on land—sometimes to attain enormous masses and heights, as we saw in the chapter's introduction. In the next few modules, we examine some of the adaptations that allow plants to reproduce on land.

Sapwood

Heartwood

Bark

B. Anatomy of a log

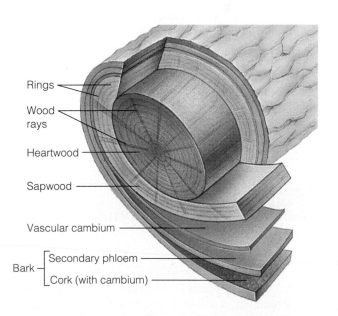

Rings

Wood rays

Heartwood

Sapwood

Vascular cambium

Bark
- Secondary phloem
- Cork (with cambium)

Overview: The sexual life cycle of a flowering plant

It has been said that an oak tree is an acorn's way of making more acorns. Indeed, evolutionary fitness for an oak tree, as for any organism, is measured only by its ability to replace itself with healthy, fertile offspring. Thus, from an evolutionary viewpoint, all the structures and functions of a plant can be interpreted as mechanisms contributing to reproduction.

In flowering plants, the flower houses the reproductive structures (Figure A). A flower is actually a compressed shoot, and its main parts, the sepals, petals, stamens, and carpels, are modified leaves. The **sepals** are usually green and look more like leaves than the other flower parts. Before the flower opens, the sepals enclose and protect the flower bud. The **petals** are often bright and

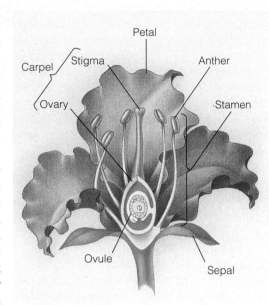

A. Structure of a flower

colorful and advertise the flower to insects and other pollinators.

The stamens and carpel contain the flower's reproductive structures. The **stamens** are the male structures of the flower. At the tip of each stamen is an **anther,** a sac in which pollen grains develop. As we will see in the next module, pollen grains house the cells that develop into sperm.

The **carpel** contains the female parts of the flower. The tip of the carpel, the **stigma,** is the receiving surface for pollen grains brought from other flowers, or from the same flower, by wind or animals. The base of the carpel is the **ovary,** which houses the reproductive structure called the ovule. The **ovule** contains the developing egg and cells that support it.

Figure B shows the life cycle of a generalized angiosperm. Fertilization occurs within the ovule, which then matures into a seed containing the embryo. Meanwhile, the ovary develops into a fruit, which protects the seed and aids in dispersing it. Completing the life cycle, the seed **germinates** (begins to grow) in a suitable habitat, the embryo develops into a seedling, and the seedling grows into a mature plant.

In the next four modules, we focus on key stages in the angiosperm sexual life cycle in more detail. We will see that there are many variations on its basic themes.

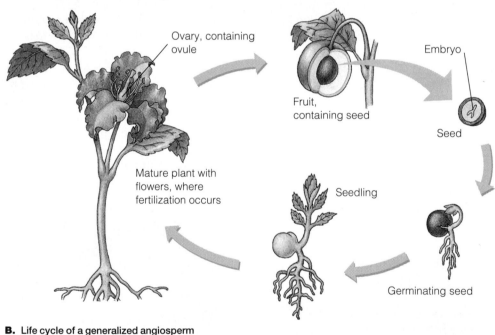

B. Life cycle of a generalized angiosperm

The development of pollen and ovules culminates in fertilization

Recall from Chapter 18 that the life cycles of all plants include alternation of haploid (n) and diploid ($2n$) generations. The roots, stems, leaves, and most of the reproductive structures of a rose plant, an oak tree, or a grass—in fact, all angiosperms and gymnosperms—are diploid. The diploid plant body is called the **sporophyte.** A sporophyte produces special structures, the anthers and ovules in angiosperms, in which cells undergo meiosis. The resulting haploid cells then divide mitotically, and each becomes a multicellular **gametophyte,** the plant's haploid generation.

The gametophyte produces gametes by mitosis. At fertilization, the male and female gametes unite, producing a diploid zygote. The life cycle is completed when the zygote divides by mitosis and develops into a new sporophyte. Without a microscope, all we can see of the angiosperm and gymnosperm life cycles are the sporophytes and occasionally the pollen (male gametophytes) produced by them. In this module, we take a "microscopic look" at the gametophytes of a flowering plant.

We begin with the development of the male gametophyte, the pollen grain. The cells that develop into pollen grains are found in chambers within a flower's anthers (top left in the figure). Each cell first undergoes meiosis, forming four haploid cells called spores. Each spore then divides mitotically, forming two haploid cells, called the tube cell and the generative cell. A thick wall forms around these cells, and the resulting pollen grain is ready for release from the anther.

Moving to the top right of the figure, we can follow the development of the flower parts that form the female gametophyte and eventually the egg. In most species, the ovary of a flower contains several ovules, but only one is shown here. An ovule contains a central cell (pink) surrounded by a protective covering of smaller cells (yellow). The central cell enlarges and undergoes meiosis, producing four haploid cells. Three of these cells usually degenerate, but the surviving one (called the spore) enlarges and divides mitotically, producing a multicellular structure called the **embryo sac** (pink area). Housed in several layers of protective cells (yellow) produced by the sporophyte plant, the embryo sac is the female gametophyte. The sac contains a large central cell with two haploid nuclei. One of its other cells (dark pink in the drawing) is the haploid egg, ready to be fertilized.

The first step leading to fertilization is **pollination** (at the center of the figure), the delivery of pollen to the stigma of a carpel. Most angiosperms are dependent on animals to transfer their pollen. But the pollen of some plants, such as grasses, is windborne, as anyone with allergies knows.

After pollination, the pollen grain germinates on the stigma. Its tube cell gives rise to the pollen tube, which grows downward into the ovary. Meanwhile, the generative cell divides mitotically, forming two sperm. When the pollen tube reaches the base of the ovule, it enters the embryo sac through a pore and discharges both its sperm. One sperm fertilizes the egg, forming the zygote (purple). The other sperm contributes its haploid nucleus to the large central cell of the embryo sac. This cell, now with a triploid (3n) nucleus, will give rise to tissue that nourishes the embryo that develops from the zygote.

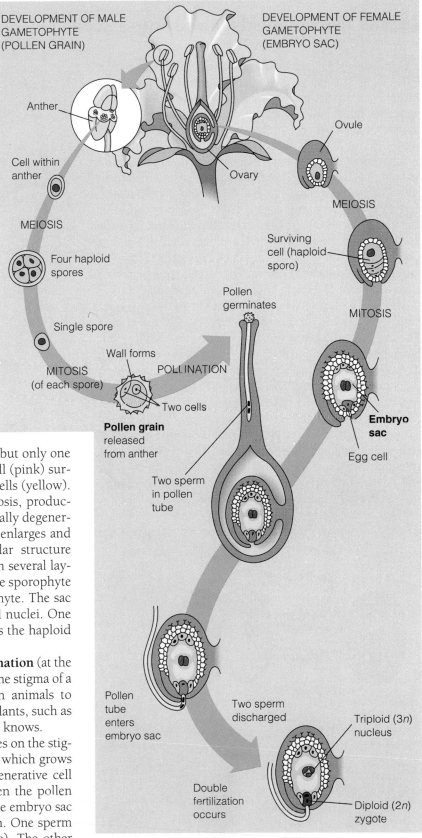

DEVELOPMENT OF MALE GAMETOPHYTE (POLLEN GRAIN)

Anther

Cell within anther

MEIOSIS

Four haploid spores

Single spore

MITOSIS (of each spore)

Wall forms

Two cells

Pollen grain released from anther

DEVELOPMENT OF FEMALE GAMETOPHYTE (EMBRYO SAC)

Ovule

Ovary

MEIOSIS

Surviving cell (haploid sporo)

MITOSIS

Pollen germinates

POLLINATION

Two sperm in pollen tube

Embryo sac

Egg cell

Pollen tube enters embryo sac

Two sperm discharged

Triploid (3n) nucleus

Double fertilization occurs

Diploid (2n) zygote

The formation of both a zygote and a cell with a triploid nucleus is called **double fertilization.** This occurs only in plants, mainly in angiosperms. Next, let's see what happens after double fertilization.

The ovule develops into a seed

After fertilization, the ovule, containing the triploid central cell and the zygote, begins developing into a seed. The triploid cell divides and develops into a nutrient-rich, multicellular mass called the **endosperm.** The endosperm nourishes the embryo until it becomes a self-supporting seedling.

Embryonic development begins when the zygote divides into two cells (Figure A). Both of these cells divide repeatedly. The figure's color coding will help you trace the fate of each cell. The large blue cell divides to form a threadlike anchor that links the embryo to the parent plant. Repeated division of the small purple cell produces a ball of cells that becomes the embryo. The bulges you see on the embryo are the cotyledons starting to form. The plant in this drawing is a dicot; a monocot would have only one cotyledon.

The result of embryonic development in the ovule is a mature seed, which you see at the bottom right in Figure A. The ovule's coat has lost most of its water and has formed a resistant **seed coat** (brown) that encloses the embryo and its food supply, the endosperm (pink). At this time, the embryo stops developing, and the seed becomes dormant; it will not develop further until the seed germinates. **Seed dormancy,** a condition in which growth and development are suspended temporarily, is an important evolutionary adaptation. It allows time for a plant to disperse its seeds and increases the chance that a new generation of plants will begin growing only when environmental conditions, such as temperature and moisture, favor survival.

The dormant embryo consists of a miniature root and shoot, each equipped with an apical meristem. The meri-

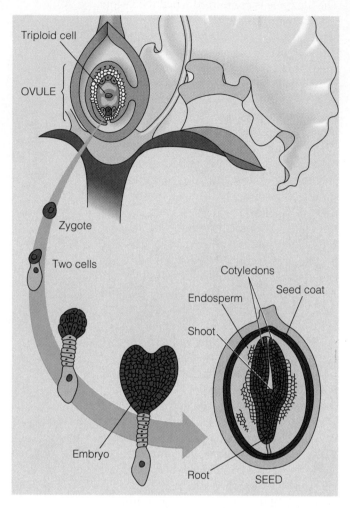

A. Development of a dicot plant embryo

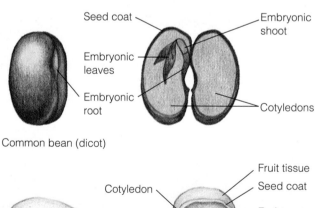

Common bean (dicot)

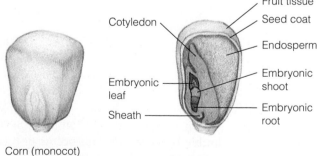

Corn (monocot)

B. Seed structure

stems will produce cells that elongate the embryo when the seed germinates. Also present in the embryo are the three tissue cylinders that will form the epidermis, cortex, and primary vascular tissues.

Figure B shows what two seeds look like when they are split open. In the bean, a dicot, the embryo is an elongated structure with two fleshy cotyledons (yellow). The embryonic root develops just below the point at which the cotyledons are attached to the rest of the embryo. The embryonic shoot, tipped by a pair of miniature embryonic leaves, develops just above the point of attachment. The bean seed contains no endosperm, because its cotyledons absorb the endosperm nutrients as the seed forms. The nutrients pass from the cotyledons to the embryo when it germinates.

The kernel of corn, an example of a monocot, is actually a fruit containing one seed. Everything you see in the drawing is the seed, except the kernel's outermost covering. The covering is the dried tissue of the fruit, tightly bonded to the seed coat. Different from the bean, the corn seed contains a large endosperm and a single cotyledon. The cotyledon absorbs the endosperm's nutrients during germination. Also unlike the bean, the corn's embryonic root and shoot have a protective sheath.

We have now followed the angiosperm life cycle from the flower on the sporophyte plant through the transformation of an ovule into a seed. Fruits, one of the most distinctive features of angiosperms, develop at the same times seeds do, but from the outer parts of the ovary. We study them next.

The ovary develops into a fruit **31.12**

A. Development of a simple fruit, a pea pod

A **fruit** is a thickened ovary, specialized as a vessel that houses and protects seeds and helps disperse them from the parent plant. A corn kernel is a fruit, as is a peach, orange, tomato, cherry, or pea pod.

The photographs in Figure A illustrate the changes in a pea plant leading to pod formation. ① Pea flowers are pollinated by insects. Soon after pollination, ② the flower drops its petals, and hormonal changes make the ovary start growing. The ovary expands tremendously, and its wall thickens, ③ forming the pod, or fruit.

Figure B matches the parts of a pea flower with what they become in the pod. The wall of the ovary becomes the pod. The ovules, within the ovary, develop into the seeds. The small, threadlike structure at the end of the pod is what remains of the upper part of the flower's carpel. The sepals of the flower often stay attached to the base of the green pod. Peas are usually harvested at this stage of fruit development. If the pods are allowed to develop further, they become dry and brownish and will split open, releasing the seeds.

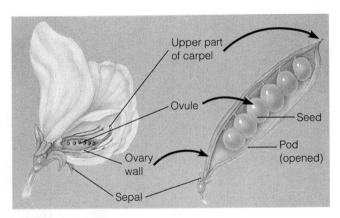

B. The correspondence between flower and fruit in the pea plant

Upper part of carpel
Ovule
Seed
Pod (opened)
Ovary wall
Sepal

Fruits are highly varied, as Figure C illustrates. In some cases, what we commonly call fruits are actually more than developed ovaries. In an apple, for instance, the part we discard, the core, is the thickened ovary, and therefore the true fruit. The soft, fleshy part we eat develops from parts of the flower base that fuse with the ovary. In contrast to the pea pod, which dehydrates when it is fully developed, a fleshy fruit like an apple becomes softer as enzymes weaken the cell walls. It also may change color from green to red, orange, or yellow, and become sweeter as organic acids or starch molecules are converted to sugar. An unripe apple or pear is sour to the taste because of high acid concentrations. Fleshy, edible fruits such as apples serve as enticements to animals that help spread seeds.

Apples, pea pods, and cherries are examples of **simple fruits**—those that develop from a flower with a single carpel and ovary. In contrast, a blackberry is an **aggregate fruit**, because it develops from a flower with many carpels. Each of the small parts of the berry develops from a single ovary. Different yet is a **multiple fruit**, which develops from a group of separate flowers tightly clustered together. When the walls of the many ovaries start to thicken, they fuse together and become incorporated into one fruit. The pineapple is an example of a multiple fruit. Each of its many parts develops from a separate flower.

C. A variety of fruit

Seed germination continues the life cycle

The germination of a seed is often used to symbolize the beginning of life, but as we have seen, the seed already contains a miniature plant, complete with embryonic root and shoot. Thus, at germination, the plant does not begin life but rather resumes the growth and development that was temporarily suspended during seed dormancy.

Germination usually begins when the seed takes up water. This hydration makes the seed expand, rupturing its coat and triggering metabolic changes in the embryo that make it start growing again. Enzymes in the endosperm or cotyledons begin digesting stored nutrients, and the nutrient molecules are transported to the growing regions of the embryo.

The figures below trace germination in a dicot, the garden pea, and a monocot, corn. In Figure A, notice that the embryonic root of a pea emerges first and grows downward from the germinating seed. Next, the embryonic shoot emerges, and a hook forms near its tip. The hook protects the delicate shoot tip by holding it downward, rather than pushing it up through the abrasive soil. As the shoot breaks into the light, its tip is lifted gently out of the soil as the hook straightens. The first foliage leaves then expand from the shoot tip and begin making food by photosynthesis. The cotyledons, depleted of their food reserves by the germinating embryo, remain behind in the soil and decompose.

In corn (Figure B), the embryonic root begins to grow just slightly before the embryonic shoot. A hook does not develop on the shoot. Instead, a protective sheath surrounding the shoot pushes straight upward and breaks through the soil. Then, safe from soil abrasion, the shoot tip grows up through the tunnel provided by the sheath. As in the pea, the corn cotyledon remains in the soil and decomposes.

A germinating seed is a fragile stage in a plant's life cycle. In the wild, only a small fraction of seedlings endure long enough to reproduce. Most plants compensate for the odds against their seedlings by producing enormous numbers of seeds. Many plants also reproduce asexually, as we see next.

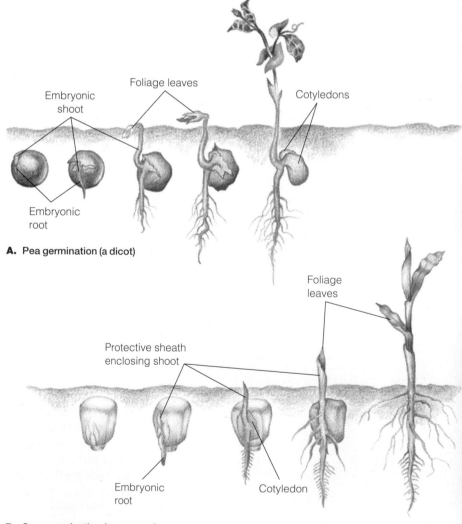

A. Pea germination (a dicot)

Foliage leaves

Embryonic shoot

Cotyledons

Embryonic root

Foliage leaves

Protective sheath enclosing shoot

Embryonic root

Cotyledon

B. Corn germination (a monocot)

Imagine some of your fingers separating from your body, taking up life on their own, and eventually developing into copies of yourself. This would be asexual reproduction, the creation of offspring derived from a single parent without fertilization. The result would be a clone, a population of asexually produced, genetically identical organisms.

As these photographs suggest, asexual reproduction, often called **vegetative reproduction,** is common among plants. Vegetative reproduction involves **fragmentation,** the separation of parts from the parent plant and regeneration of the parts into whole plants. The garlic bulb (Figure A) is actually an underground stem that functions in storage. A single large bulb fragments into several parts, called cloves. Each clove can give rise to a separate plant, as indicated by the green shoots emerging from some of them.

Each of the small trees you see in Figure B is a sprout from the roots of a coast redwood tree, a close relative of the giant sequoia we discussed in the chapter's introduction. Eventually, one or more of these root sprouts may take the place of its parent in the forest.

The ring of plants in Figure C is a clone of creosote bushes growing in the Mojave Desert about 250 km southeast of Los Angeles. Known as the King Clone, this ring makes the oldest sequoias seem youthful. All these bushes came from generations of vegetative reproduction by roots. Apparently, the clone began with a single plant that germinated from a seed about 12,000 years ago. The original plant probably occupied the center of the ring.

Figure D shows a patch of dune grass in Australia. This species and most other grasses can propagate asexually by sprouting shoots and roots from underground runners. A small patch of grass can spread in this way until it covers an acre or more of surface.

Vegetative reproduction is an extension of the capacity of plants to continue growing throughout their lives. Their meristematic tissues can sustain or renew growth indefinitely. In addition, the parenchyma cells throughout a plant can divide and differentiate into the various types of specialized cells.

Asexual reproduction has several potential advantages. For one thing, a parent plant well suited to its environment can clone many copies of itself. Also, early life for vegetative offspring, which are mature fragments of the parent, may be less hazardous than for seedlings. Both asexual reproduction and sexual reproduction have played important roles in the evolutionary adaptation of plant populations to their environments.

A. Cloves of a garlic bulb

B. Sprouts from the roots of coast redwood trees

C. A ring of creosote bushes

D. Dune grass

Chapter 31   Plant Structure, Reproduction, and Development   **619**

Vegetative reproduction is a mainstay of modern agriculture

The ability of plants to reproduce vegetatively provides many opportunities for producing large numbers of plants with minimal effort and expense. For example, most of our fruit trees, ornamental trees and shrubs, and houseplants are asexually propagated from stem or leaf cuttings. A number of other plants, such as strawberries, raspberries, and potatoes, are propagated from pieces of underground stems.

Plants can also be propagated by test-tube methods. The culture tube in Figure A contains an apple plantlet that was grown from a few meristem cells cut from a large tree and cultured on a chemical medium. Using this method, a single plant can be cloned into thousands of copies that will continue to grow when transferred to soil. Orchids and certain pine trees used for mass plantings are commonly propagated this way. Furthermore, test-tube methods allow genetically engineered plant cells to be grown into plants (see Module 12.14). Foreign genes are incorporated into a single parenchyma cell, and the cell is then cultured until it multiplies and develops into a new plantlet.

Figure B shows another new development, a so-called artificial seed. It is a tiny alfalfa plant embryo cloned from a few cells of a parent plant. In the lab, the embryo is packaged with some nutrients in a polysaccharide capsule.

A. Test-tube cloning

B. An artificial seed germinating

Using artificial seeds, a farmer can grow a truly uniform crop, one in which all the plants have the same set of desirable traits and will be ready for harvest at the same time. The main disadvantage of artificial seeds is that they are more expensive than the natural ones produced by plants.

Despite the promising new developments, modern agriculture faces some potentially serious problems. Nearly all of today's crop plants have very little genetic variability, and the use of gene-cloning techniques may compound the problem. Also potentially hazardous is the fact that we grow most crops in **monocultures,** large areas of land with a single plant variety. Given these conditions, plant scientists fear that a small number of diseases could devastate large crop areas. In response, plant breeders are working to maintain "gene banks," storage sites for seeds of many different plant varieties that can be used to breed new hybrids. We'll have more to say about plants and agriculture when we take up the subject of plant nutrition in Chapter 32.

Chapter Review

Begin your review by rereading the module headings and scanning the figures before proceeding to the Chapter Summary and questions.

Chapter Summary

Introduction, 31.2 Plants are important to global environment, and humans depend on them for such essentials as lumber, fabric, paper, and food. The most diverse and familiar plants are the angiosperms, or flowering plants. There are two main types of angiosperms. The monocots include the grains and other grasses, orchids, palms, and lilies. The dicots are a larger group, including most trees, shrubs, ornamentals, and many food crops. Monocots and dicots differ in the number of seed leaves and structure of roots, stems, leaves, and flowers.

31.3–31.4 The body of a plant consists of a root system and a shoot system. Roots anchor the plant, absorb and transport minerals and water, and store food. Tiny root hairs increase the absorptive surface of roots. The shoot system of an angiosperm consists of stems, leaves, and flowers. Leaves are the plant's main photosynthetic organs. At the tip of a stem is a terminal bud, the growth point of the stem. Axillary buds can give rise to branches.

31.5–31.6 There are five major types of plant cells: parenchyma, collenchyma, sclerenchyma (including fiber and sclereid cells),

water-conducting cells (including tracheids and vessel elements), and food-conducting cells (sieve-tube members). They form three tissue systems within roots, stems, and leaves: the epidermis, the vascular tissue system, and the ground tissue system. The epidermis covers and protects the plant. The vascular tissue system consists of bundles of xylem and phloem. The water-conducting cells of xylem carry water and minerals. The food-conducting cells of phloem transport sugars. The ground tissue system consists of spongy parenchyma cells and supportive collenchyma and sclerenchyma cells. The cortex and pith of stems and roots are ground tissues that function mainly in storage. Mesophyll, the ground tissue of a leaf, is where most photosynthesis occurs.

31.7–31.8 Most plants grow as long as they are alive. Growth originates in meristems, areas of unspecialized, dividing cells. Apical meristems at the tips of roots and in the terminal and axillary buds of shoots initiate lengthwise growth by producing new cells. A root or shoot lengthens as the cells elongate and differentiate; this is called primary growth. An increase in a plant's girth, called secondary growth, arises from cell division in a cylindrical meristem called the vascular cambium. The vascular cambium thickens a stem by adding layers of secondary xylem, or wood, next to its inner surface. Outside the vascular cambium, the bark consists of secondary phloem (also produced by the vascular cambium), a meristem called the cork cambium, and protective cork cells produced by the cork cambium. The outer layers of bark are sloughed off as the plant thickens.

31.9–31.12 The plant life cycle alternates between diploid ($2n$) and haploid (n) generations. Most of the body of a plant is a diploid sporophyte. A flower is a reproductive shoot consisting of sepals, petals, stamens, and a carpel. Haploid spores are formed within an ovary at the base of the carpel and in anthers at the tips of stamens. The spores in the anthers give rise to male gametophytes—pollen grains—which produce sperm. The ovary contains ovules, where spores produce the female gametophytes, called embryo sacs. Each embryo sac contains an egg. Pollen grains may be carried to the carpel by wind or animals. A pollen tube grows to the ovary, and sperm pass through it and fertilize both the egg and a second cell. This process, called double fertilization, is unique to plants. After fertilization, the ovule becomes a seed, and the fertilized egg within it divides to become an embryo. The other fertilized cell develops into endosperm, or stored food. Embryo and endosperm are protected by a tough seed coat. At the same time, the ovary is developing into a fruit, which helps protect and disperse the seeds.

31.13 A seed starts to germinate when it takes up water, expands, and bursts its seed coat. Metabolic changes cause the embryo to resume growth and absorb nutrients from endosperm. An embryonic root emerges, and a shoot pushes upward and expands its leaves.

31.14–31.15 Many plants are also able to reproduce asexually, via bulbs, sprouts, or runners. Propagating plants from cuttings or cloning plants from bits of tissue exploits vegetative reproduction, which can increase agricultural productivity but can also reduce genetic diversity.

Testing Your Knowledge

Multiple Choice

1. Which of the following is closest to the center of a woody stem? (*Explain your answer.*)

   a. vascular cambium
   b. young phloem
   c. old phloem
   d. young xylem
   e. old xylem

2. A pea pod is formed from _____. A pea inside the pod formed from _____.

   a. an ovule . . . a carpel
   b. an ovary . . . an ovule
   c. an ovary . . . a pollen grain
   d. an anther . . . an ovule
   e. endosperm . . . an ovary

3. While walking in the woods, you encounter a beautiful and unfamiliar flowering plant. If you want to know whether it is a monocot or dicot, it would *not* help to look at the

   a. number of seed leaves, or cotyledons, present in its seeds
   b. shape of its root system
   c. number of petals in its flowers
   d. arrangement of vascular bundles in its stem
   e. size of the plant

4. In angiosperms, each pollen grain produces two sperm. What do these sperm do?

   a. Each one fertilizes a separate egg cell.
   b. One fertilizes an egg, and the other fertilizes the fruit.
   c. One fertilizes an egg, and the other is kept in reserve.
   d. Both fertilize a single egg cell.
   e. One fertilizes an egg, and the other fertilizes a cell that develops into stored food.

True/False (*Change false statements to make them true.*)

1. The growth points of a stem are called anthers.
2. Collenchyma and sclerenchyma cells are specialized for support.
3. Parenchyma cells are the most abundant cells in most plants.
4. Vascular tissues are clustered at the center of a root.
5. Wood consists mainly of xylem.
6. Annual plants continue to live for many years.
7. The root cap absorbs minerals and water from the soil.

Matching

1. Attracts pollinator
2. Develops into seed
3. Protects flower before it opens
4. Produces sperm
5. Produces pollen
6. Houses ovules

   a. pollen grain
   b. ovule
   c. anther
   d. ovary
   e. sepal
   f. petal

Describing, Comparing, and Explaining

1. The scent of apple blossoms and the buzzing of bees fill an orchard on a warm spring day. Describe the processes by which the pollen carried from flower to flower by the bees results in the apple you might pick in the fall.

2. Name three kinds of vegetative (asexual) reproduction. Explain two advantages vegetative reproduction has over sexual reproduction.

Thinking Critically

1. Study the illustrations in Module 31.6, and notice the arrangement of vascular tissues in a stem and a leaf. Notice the way a leaf joins a stem. Why does it make sense that in the veins of a leaf, the xylem is always on top and the phloem on the bottom?

2. Plant scientists are searching Peru, Mexico, and the Middle East for the wild ancestors of potatoes, corn, and wheat. Why is this search important?

3. What part of a plant do you think you are eating when you consume each of the following? Tomato, onion, celery stalk, peanut, strawberry, lettuce, artichoke, beet.

Science, Technology, and Society

Most of our food comes from only a few hundred species of plants. Tropical forests contain a wealth of plants that are potential new sources of food, as well as sources of medicine and other useful products. The developing nations of the tropics are not able to efficiently study and develop these resources themselves. Under pressure from growing populations and debt, they are cutting their forests for lumber and farmland, and many species are disappearing. The plants of the tropics are a virtually untapped source of useful products, and developed countries are pressuring the tropical countries to protect the forests before even more species are lost. Many people in the developing nations see little incentive to preserve the forests, only to have corporations from industrialized countries profit from new products obtained from them. This issue was at the center of conflict between the developed and developing nations at the 1992 United Nations Earth Summit in Rio de Janeiro. What would you suggest to resolve this conflict? Is there a way to preserve the tropical forests so that both the developed and developing nations will benefit from their abundance?

Plant Nutrition and Transport

32

George Washington Carver (in bow tie) with students at Tuskegee Institute.

Peanuts, peanut butter, and peanut oil are part of our culture. Yet how many consumers know where peanuts come from? Do they grow on trees, shrubs, or nonwoody plants? You can tell from the photograph at the left that the peanut plant is short and nonwoody. But unless you live in the southeastern United States, where large peanut crops are grown, you may not know some of the truly unusual features of this important crop plant.

The peanut is one species in a large group called the legumes (about 10,000 species). Other legumes include peas, beans, lentils, clovers, alfalfa, and locust trees. Most species, including the peanut, bear seeds in pods.

Outside the U.S., peanuts are often called groundnuts, and this is actually a more appropriate name. The plant's scientific name, *Arachis hypogaea* (Greek *arakos,* legume plant, and *hypogaia,* "under earth") is also fitting. Unlike nearly all other legumes, which produce their pods on aerial stems, the peanut bears its fruit—the tough fibrous pods we crack open—about 5 cm *below* the ground. When the plant matures, small yellow flowers appear above the ground at the base of its leaf stalks. After fertilization, each flower's ovary grows a long stalk that points downward and pushes the developing pod into the ground. Notice the stalks attached to the ripe pods in the photograph. The seeds—peanuts—develop in the pods.

The history of the peanut industry in the U.S. is closely tied to the career of George Washington Carver (1864–1943). Carver's life was filled with accomplishments in education, literature, art, and science. Born a slave in strife-torn southwestern Missouri during the Civil War, he rose to become a professor at the Tuskegee Institute in Alabama and one of the world's most respected agricultural researchers.

One of Carver's major contributions to agriculture was to promote the practice of rotating crops. He recognized that by alternating a cotton or corn crop with a legume like peanuts, all the plants grew better, and the soil remained more fertile. Carver also championed the nutritional and economic potential of the peanut plant. In the late 1800s, the economic survival of the South depended on finding crops to replace or rotate with cotton, which had seriously diminished the soil's fertility. Carver's work helped open new markets for peanuts and peanut products. In his laboratory at Tuskegee, he separated peanut meal into its chemical constituents and showed that the meal could be used to make a host of marketable foods, among them a nutritious milk substitute, a rich oil, flour, dye, and cheese.

One of the main reasons peanuts and other legumes are grown so widely is that they are rich in protein; as Carver discovered, as much as 30% of a peanut's dry weight is protein. Small bumps, or nodules, on the roots of the peanut plant (not visible in the photograph) play a large role in protein production. Found on all legumes, root nodules house bacteria that supply nitrogen, needed for protein synthesis, to the plant. With their root nodule bacteria, legumes often thrive in nitrogen-poor soils and actually improve soil fertility by adding nitrogen compounds to it.

Nitrogen is one of several inorganic nutrients that plants must have in order to grow. In this chapter, we see how plants obtain nitrogen and other essential nutrients and how they transport them throughout their roots, stems, and leaves.

Watch a plant grow from a tiny seed, and you can't help wondering where all the mass comes from. Aristotle thought soil provided all the substance for plant growth. As we discussed in Chapter 7, seventeenth century Belgian physician Jan Baptista van Helmont performed an experiment to find out if plants grew by absorbing material from soil. He planted a willow seedling in a pot containing 90.0 kg of soil. After five years, the willow had grown into a tree weighing 76.8 kg, but only 0.06 kg of soil had disappeared from the pot. Van Helmont concluded that the willow had grown mainly from the water he had added regularly. A century later, Stephen Hales, an English botanist, postulated that plants were nourished mostly by air.

As it turns out, there is some truth in all these early ideas about plant nutrition. As indicated in Figure A, a plant's leaves absorb CO_2 from the air; in fact, about 95% of a plant's dry weight is organic (carbon-containing) material built mainly from CO_2. The figure also points out that a plant gets water, inorganic ions (minerals), and some oxygen (O_2) from the soil.

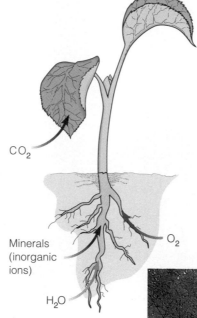

A. The uptake of nutrients by a plant

What happens to the materials a plant takes up from the air and soil? The sugars a plant makes by photosynthesis are composed of the elements carbon, oxygen, and hydrogen. In Chapter 7, we saw that the carbon and oxygen used in photosynthesis come from atmospheric CO_2 and that the hydrogen comes from water molecules. Plant cells use the sugars made by photosynthesis in constructing all the other organic materials they need. The giant trunks of the redwood trees in Figure B, for instance, consist mainly of sugar derivatives.

Plants use cellular respiration to break down some of the sugars they make, obtaining energy from them in a process that consumes O_2. A plant's leaves take up some O_2 from the air, but we do not show this in Figure A because plants are actually net producers of O_2, giving off more of this gas than they use. When water is split during photosynthesis, O_2 gas is produced and released through the leaves. The O_2 being taken up from the soil by the plant's roots in Figure A is actually atmospheric O_2 that

has diffused into the soil; it is used in cellular respiration in the roots themselves.

What does a plant do with the minerals it absorbs from the soil? A look at three elements (nitrogen, phosphorus, and magnesium) that plant roots take up as inorganic ions provides a partial answer. Nitrogen is a component of many plant hormones and coenzymes and of ATP, all nucleic acids, and all proteins. Nitrogen and magnesium are both components of chlorophyll, the plant's key light-absorbing molecule. Phosphorus is a major component of nucleic acids, phospholipids, and ATP.

A plant's ability to move water from its roots to its leaves or to deliver sugars to specific areas of its body are staggering feats of evolutionary engineering. Figure B highlights the distance between the bottom of a tree and its leaves; the roots of a redwood can be over 100 m (330 ft) below the topmost leaves! In the next four modules, we follow the movements of water, dissolved mineral nutrients, and sugar throughout the plant body.

B. Redwood trees, giant products of photosynthesis

With its surface area enormously expanded by thousands of root hairs (Figure A), a plant root has a remarkable ability to extract materials from soil. Recall from Chapter 31 that root hairs are extensions of epidermal cells that form the outside covering of the root. Because of its large root surface area, a plant can absorb enough water and inorganic ions to survive and grow.

All substances that enter a plant root are in solution. In order for water and solutes to be transported from the soil throughout the plant, they must move through the epidermis and cortex of the root, and then into the water-conducting xylem tissue in the central cylinder of the root. (You can see these tissues in the root cross section at the top of Figure B.) Any route the water and solutes take from the soil to the xylem requires that they pass through some of the plasma membranes of the root cells. Since plasma membranes are selectively permeable, only certain solutes can enter the xylem.

You can see two possible routes to the xylem in the bottom part of Figure B. The blue arrows indicate an *intracellular* route. Water and selected solutes cross the cell wall and plasma membrane of an epidermal cell (usually at a root hair). The cells within the root are all interconnected by plasmodesmata (channels through the walls of adjacent cells); there is a continuum of living cytoplasm among the root cells. Therefore, once inside the epidermal cell, the solution can move inward from cell to cell without crossing any other plasma membranes, diffusing through the interconnected cytoplasm all the way into the root's endodermis (yellow). An endodermal cell then discharges the solution into the xylem (gray).

The red arrows indicate an alternative route. This route is *extracellular*; the solution moves inward within the walls of the root cells but does not enter the cytoplasm of the epidermis or cortex cells. The solution crosses no plasma membranes, and there is no selection of solutes until they reach the endodermis. Here, a barrier called the **Casparian strip** stops water and solutes from entering the xylem via cell walls. Shown in black in Figure B, the Casparian strip is a waxy belt that extends through the walls of the endodermal

A. Root hairs of a radish seedling

cells and is continuous from cell to cell. Because of the Casparian strip, water and ions that travel the extracellular (red) route can enter the xylem only by crossing a plasma membrane into an endodermal cell. Ion selection occurs at this membrane instead of in the epidermis, and once the selected solutes and water are in the endodermal cell, they can be discharged into the xylem.

The routes we have followed are only two of many possibilities. Water and solutes can enter the xylem by any combination of the intracellular and extracellular routes. Because of the Casparian strip, however, there are no nonselective routes. Next, we see how water and minerals move upward within the xylem from the roots to the shoots.

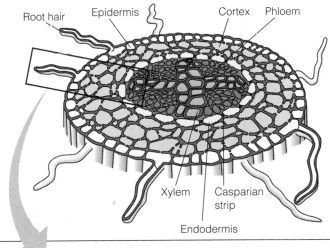

Root hair   Epidermis   Cortex   Phloem

Xylem   Casparian strip

Endodermis

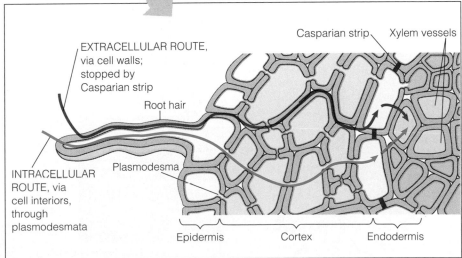

EXTRACELLULAR ROUTE, via cell walls; stopped by Casparian strip

Root hair

INTRACELLULAR ROUTE, via cell interiors, through plasmodesmata

Plasmodesma

Casparian strip   Xylem vessels

Epidermis   Cortex   Endodermis

B. Routes of water and solutes from soil to root xylem

32.3 Transpiration pulls water up xylem vessels

The ability to transport water and dissolved ions upward from the soil is a significant adaptation for a land plant. It enables the plant to supply nutrients to its stems and leaves while growing upward and exposing its leaves to sunlight. But how does a plant transport water and nutrients dissolved in it from its roots to its leaves?

We saw in Chapter 31 that xylem tissue consists of porous cells called tracheids and open-ended ones called vessel elements, both of which form very thin, vertical tubes. The solution of inorganic nutrients in a plant's xylem tissue, called **xylem sap**, flows all the way up from the center of the root to the tips of the leaves in these tubes. What comes to mind when you think about a fluid moving upward in tubes? In humans, for example, blood flows vertically through vessels into the head because the heart pumps it there. Do plants also have some kind of pump that pushes their xylem sap upward? Plant biologists have found that the roots of some plants do exert a slight upward push on xylem sap. The root cells actively pump inorganic ions into the xylem, and the root's endodermis holds the ions there. As ions accumulate

in the xylem, water tends to enter by osmosis, pushing xylem sap upward ahead of it. This force, called **root pressure**, can push xylem sap up a few meters. But the push exerted by root pressure does not account for most of the sap's ascent, and some of the tallest trees, including redwoods and giant sequoias, generate no root pressure at all.

It turns out that xylem sap is mainly *pulled*, rather than pushed, from the roots of a plant to the leaves. Plant biologists have determined that the pulling force is **transpiration**, which is the loss of water from the leaves and other aerial parts of a plant.

The figure here illustrates transpiration and its effect on water movement in a tree. At the top right, water molecules (blue dots) are shown leaving the leaf through a stoma. This occurs as long as the stoma is open. The water diffuses out of the leaves because the concentration of water molecules is higher in the spaces between cells inside the leaf than in the surrounding air. Transpiration can pull xylem sap up the tree because of two special properties of water: cohesion and adhesion. **Cohesion** is the sticking together of molecules of the same kind. In the case of water, hydrogen bonds make the H_2O molecules stick to one another, as the circular inset in the figure shows (see also Module 2.10). The cohering water molecules in the xylem tubes form continuous strings, extending all the way from the leaves down to the roots. In contrast to cohesion, **adhesion** is the sticking together of molecules of different kinds. Water molecules tend to adhere to cellulose molecules in the walls of xylem cells.

What effect does transpiration have on a vertical string of water molecules that tend to adhere to the walls of xylem tubes in a plant? Before a water molecule can leave the leaf, it must break off from the top of the string. In effect, it is pulled off by a large diffusion gradient between the moist interior of the leaf and the surrounding air. Cohesion resists the pulling

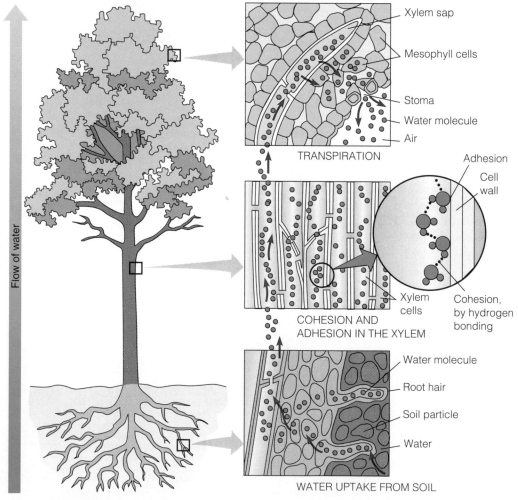

Flow of water

Xylem sap

Mesophyll cells

Stoma

Water molecule

Air

TRANSPIRATION

Adhesion

Cell wall

Xylem cells

Cohesion, by hydrogen bonding

COHESION AND ADHESION IN THE XYLEM

Water molecule

Root hair

Soil particle

Water

WATER UPTAKE FROM SOIL

The flow of water up a tree

force of the diffusion gradient, but it is not strong enough to overcome it. The molecule breaks off, and the opposing forces of cohesion and transpiration put tension on the rest of the molecular string. As long as transpiration continues, the string is kept tense and is pulled upward as one molecule exits the leaf and one right behind it is tugged up into its place. Adhesion of the string of water molecules to the walls of the xylem cells assists the upward movement of the xylem sap by counteracting the downward pull of gravity.

Plant biologists call this explanation for the ascent of xylem sap the **transpiration-adhesion-cohesion mechanism.** We can summarize it as follows: Transpiration exerts a pull that is relayed downward along a string of water molecules held together by cohesion and helped upward by adhesion. It is especially important to the plant that none of its own energy is required to transport xylem sap. A plant's xylem tissue is adapted to use outside forces—cohesion, adhesion, and the evaporating effect of sunlight—to move water and dissolved minerals from its roots to its shoots.

Guard cells control transpiration   32.4

Transpiration actually works both for and against plants. In using the pull of transpiration to move its xylem sap, a plant can lose an astonishing amount of water. Transpiration is usually greatest on days that are sunny, warm, dry, and windy, because these climatic factors increase evaporation. An average-sized maple tree (about 20 m high), for instance, can lose more than 200 L of water an hour during a summer day. As long as water moves up from the soil fast enough to replace the water that is lost, even this amount of transpiration presents no problem. But if the soil dries out and transpiration exceeds the delivery of water to the leaves, the leaves will wilt, and the plant will eventually die.

The leaf stomata, which can open and close, are adaptations that help plants regulate their water content and adjust to changing environmental conditions. As shown in the diagrams below, a pair of guard cells flanks each stoma. The guard cells control the opening of a stoma by changing shape. Guard cells usually keep the stomata open during the day and closed at night. Having the stomata open during the day allows CO_2 to enter the leaf from the atmosphere and thus to keep photosynthesis going when sunlight is available. Keeping the stomata closed at night saves water when there is no light for photosynthesis and therefore no need to take up CO_2.

What actually causes guard cells to change shape and thereby open or close stomata? The figure here illustrates the principle. A stoma opens (left) when its guard cells gain K^+ ions (red dots) and water (blue arrows) from surrounding cells (shown in light green).

The cells actively take up K^+, and water then enters by osmosis. (For a review of osmosis, see Module 5.15.) When the vacuoles in the guard cells gain water, the cells become turgid and swell, buckling outward and increasing the size of the gap (stoma) between them. Conversely, when the guard cells lose K^+, they also lose water by osmosis, become flaccid, and sag together, closing the space between them (right).

Several factors influence the opening and closing of stomata. One is sunlight, which stimulates guard cells to take up K^+ and water, opening the stomata in the morning. A low level of CO_2 in the leaf can have the same effect. A third factor is an internal timing mechanism found in the guard cells. Called a biological clock, the timer triggers ion uptake (stomatal opening) in the morning and ion release (stomatal closing) at night. (We'll return to biological clocks in Module 33.10.) The guard cells also close the stomata during the day if the plant loses water too fast. This response reduces further water loss and may prevent wilting, but it also slows down CO_2 uptake and photosynthesis—one reason that droughts reduce crop yields. Overall, the mechanisms that regulate stomatal opening enable a plant to strike a balance between the need to save water and the need to make sugars.

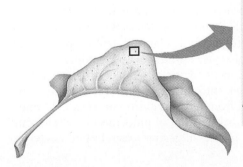

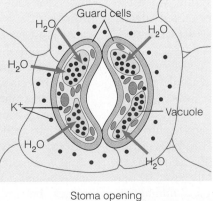

Stoma opening

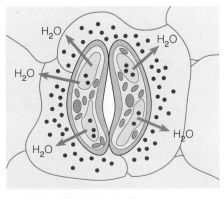

Stoma closing

Phloem transports sugars

A plant has two separate transport systems: xylem, which we have already looked at, and phloem, which moves mainly food molecules, the sugars the plant makes by photosynthesis. As we saw in Module 31.5, phloem contains food-conducting cells called sieve-tube members arranged end-to-end as tubes. The micrograph in Figure A shows the structure of sieve-tube members. You may recall that the end walls between these cells form perforated sieve plates. Since the plasma membranes of adjacent cells do not extend across the perforations, the sugary solution, called **phloem sap,** moves freely from cell to cell. Phloem sap may contain inorganic ions, amino acids, and hormones in transit from one part of the plant to another, but its main solute is usually the disaccharide sucrose.

In contrast to xylem sap, which only flows upward from the roots, phloem sap moves throughout the plant in various directions. A location in a plant where sugar is being produced, either by photosynthesis or by the breakdown of stored starch, is called a **sugar source.** Phloem usually moves sugar out of a sugar source, such as a leaf or green stem, into the nonphotosynthetic parts of the plant. A location in a plant where sugar is stored or consumed is called a **sugar sink.** Growing roots, shoot tips, and fruits are sugar sinks, as are nonphotosynthetic stems and tree trunks. Storage structures such as the taproot of a beet, the tubers of a potato plant, and the bulb of a lily are sugar sinks during the summer, when the plant is stockpiling sugars. In early spring, when the plant renews its growth and consumes its stored sugars, beet roots, tubers, bulbs, and other storage structures become sugar sources, and phloem transports sugar away from them to growing organs. Thus, each food-conducting tube in phloem tissue has a source end and a sink end, but these may change with the season or the developmental stage of the plant.

What causes phloem sap to flow from a sugar source to a sugar sink? Flow rates may be as high as 1 m/hr, which is much too fast to be accounted for by diffusion. (Phloem sap could flow only about 1 m in eight years if it moved by diffusion alone.) Plant biologists have tested a number of hypotheses for phloem sap movement. A model called the **pressure-flow mechanism** is now widely accepted. Figure B on the facing page illustrates how this works, using a beet plant as an example. The gradient of pink in the phloem tube represents both a sugar concentration gradient and a parallel gradient of water (hydrostatic) pressure in the phloem sap.

At the sugar source (leaves, in this example), ① sugar is loaded into a phloem tube by active transport. Sugar loading at the source end raises the solute concentration inside the phloem tube. ② The high solute concentration makes water flow into the tube by osmosis. The inward flux of water raises the water pressure at the source end of the tube.

At the sugar sink (the beet root, in this case), both sugar and water leave the phloem tube. ③ As sugar departs from the phloem, ④ water follows by osmosis. The exit of sugar lowers the sugar concentration in the sink end; the exit of water lowers the hydrostatic pressure in the tube. The building of water pressure at the source end and the reduction of that pressure at the sink end cause water to flow from source to sink—down a gradient of hydrostatic pressure. Since the sugar is dissolved in the water and the sieve plates allow free movement of solutes as well as water, the sugar is carried along from source to sink at the same rate as the water. As indicated on the right side of Figure B, xylem tubes transport the water back from sink to source.

The pressure-flow mechanism explains why phloem sap always flows from a sugar source to a sugar sink, regardless of their locations in the plant. However, the mechanism is somewhat difficult to test because most experimental procedures disrupt the structure and function of the phloem tubes. Some of the most interesting studies have taken advantage of natural phloem probes: insects called aphids, which feed on phloem sap.

The three photographs in Figure C (facing page) show how plant biologists use aphids to study phloem sap. On the left, an aphid feeds by inserting its needlelike mouthpart, called a stylet, into the phloem of a tree branch. The aphid is releasing from its anus a drop of so-called honeydew—actually, a tiny amount of phloem sap lacking some solutes that the insect's digestive tract has removed for

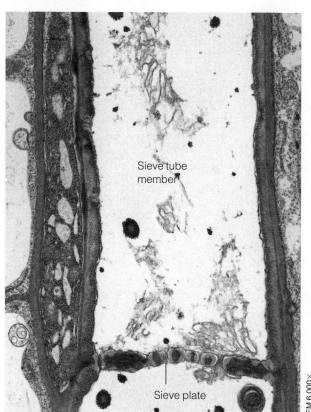

Sieve tube member

Sieve plate

TEM 6,000×

A. Food-conducting cells of phloem

food. The micrograph in the center shows an aphid's stylet inserted into one of the plant's food-conducting cells. The pressure within the phloem force-feeds the aphid, swelling it to several times its original size. While the aphid is feeding, it can be anesthetized and severed from its stylet. The stylet then serves the researcher as a miniature tap that drips phloem sap for hours. The right-hand photograph shows a droplet of phloem sap on the cut end of a stylet. Studies using this technique support the pressure-flow model: The closer the stylet to a sugar source, the faster the sap flows out, and the greater its sugar concentration. This is what we would expect if pressure is generated at the source end of the phloem tube by the active pumping of sugar into the tube.

We now have a broad picture of how a plant transports materials from one part of its body to another. Water and inorganic ions enter from the soil and are distributed by xylem. The xylem sap is pulled upward by transpiration. Carbon dioxide enters the plant through leaf stomata and is converted into sugars in the leaves. A second transport system, phloem, distributes the sugars. Pressure flow drives the phloem sap from leaves and storage sites to other parts of the plant, where the sugars are used or stored.

In Chapter 7, we discussed how plants convert raw materials into organic molecules by photosynthesis. We have yet to say much about the kinds of inorganic nutrients a plant needs and what it does with them. This is the subject of plant nutrition, which we discuss in the next module.

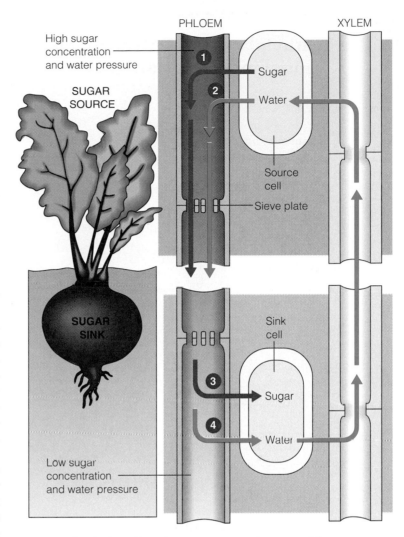

B. Pressure flow in plant phloem from a sugar source to a sugar sink (and the return of water to the source via xylem)

Aphid feeding on a small branch

C. Tapping phloem sap with the help of an aphid

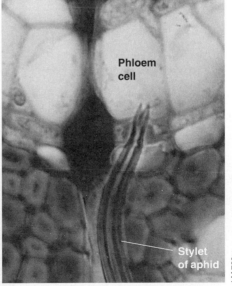

Aphid's stylet inserted into a phloem cell

Severed stylet dripping phloem sap

Healthy plants obtain a complete diet of essential inorganic nutrients

In contrast to animals, which require a complex diet of organic foods, plants survive and grow solely on inorganic substances. The ability of plants to assimilate CO_2 from the air, extract water and inorganic ions from the soil, and synthesize organic food compounds is essential not only to the survival of plants but also to humans and other animals.

A chemical element is considered an essential plant nutrient if the plant must obtain it to complete its life cycle— that is, to grow from a seed and produce another generation of seeds. A method known as hydroponic culture can be used to determine which of the chemical elements are essential nutrients. As shown in Figure A, the roots of a plant

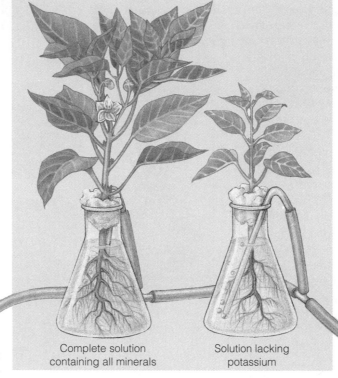

Complete solution containing all minerals

Solution lacking potassium

A. A hydroponic culture experiment

are bathed in solutions of various minerals in known concentrations. Air has to be bubbled into the water to provide the roots with oxygen for cellular respiration. A particular element, such as potassium, can be omitted from the culture medium to test whether it is essential to the plants.

If the element left out of the solution is an essential nutrient, then the incomplete medium will make the plant abnormal in appearance compared to control plants grown on a complete nutrient medium. The most common symptoms of a nutrient deficiency are stunted growth and discolored leaves. Studies like this have helped identify sixteen elements that are essential nutrients in all plants, and a few other elements that are essential to certain groups of plants. Most research has involved crop plants and houseplants; little is known about the nutritional needs of uncultivated plants.

Of the sixteen essential elements, nine are called **macronutrients** because plants require relatively large amounts of them. Six of the nine macronutrients—carbon, oxygen, hydrogen, nitrogen, sulfur, and phosphorus—are the major ingredients of organic compounds. These six elements make up almost 98% of a plant's dry weight. The other three macronutrients—calcium, potassium, and magnesium— make up another 1.5%.

Calcium has several functions. For example, it combines with certain proteins, forming a glue that holds plant cells

together in tissues. It also helps regulate the selective permeability of membranes.

Potassium is a crucial activator of several enzymes. It is also the main solute for osmotic regulation in plants; we saw in Module 32.4 how K^+ movements regulate the opening and closing of guard cells. In addition, K^+ plays a role in elongation during the primary growth of a plant. Elongation results mainly from the swelling of cells due to water uptake. An elongating cell accumulates K^+ and other solutes in its central vacuole, and water then enters the cell by osmosis.

Magnesium is a component of chlorophyll, the photosynthetic pigment. This element is also a cofactor required for the activity of several enzymes. (As we saw in Module 5.7, a cofactor is a substance that cooperates with an enzyme in catalyzing a reaction.)

Elements that plants need in very small amounts are called **micronutrients.** The seven known micronutrients are iron, chlorine, copper, manganese, zinc, molybdenum, and boron. These elements function in the plant mainly as components or cofactors of enzymes. Iron, for example, is a metallic component of cytochromes, proteins that function in the electron transport chains of chloroplasts and mitochondria. Since micronutrients play mainly catalytic roles

B. The effect of nitrogen availability on the growth of corn

(and are therefore used over and over), plants need only minute quantities of these elements. The requirement for molybdenum, for example, is so modest that there is only one atom of this rare element for every 16 million atoms of hydrogen in dried plant material. Yet a deficiency of molybdenum or any other micronutrient can kill a plant.

The quality of soil—that is, the nutrients available to plants growing in it—determines the quality of our own nutrition. The photograph in Figure B on the preceding page shows two corn crops. The plants on the left are growing in soil rich in nitrogen-containing compounds that the plants can use to build proteins. The small, lighter-colored plants on the right are growing in soil deficient in nitrogen. Even if the nitrogen-deficient plants produce grain, the crop will have a lower food value, and its nutrient deficiencies will then be passed on to livestock or human consumers. Maximizing the nutritional value of crops such as corn is one of the goals of research in plant nutrition.

You can diagnose some nutrient deficiencies in your house and garden plants

The symptoms of nutrient deficiency in plants are usually obvious, and although a number of deficiencies produce similar outward signs, it is possible to diagnose some problems quite readily. Many growers make visual diagnoses of their own ailing house or garden plants and then check their conclusions by having samples of soils and plants chemically analyzed at a state or local soils laboratory.

Deficiencies of macronutrients are seen most often, and the photographs here compare a healthy tomato plant (Figure A) with genetically identical plants suffering from three macronutrient deficiencies. Nitrogen shortage is the single most common nutritional problem for plants. Soils are usually not deficient in nitrogen, but they are often deficient in the nitrogen compounds that plants can use, dissolved nitrate ions (NO_3^-) and ammonium ions (NH_4^+). Stunted growth and yellow-green leaves (Figure B) are signs of nitrogen deficiency.

Phosphorus deficiency is the second most common nutritional ailment in plants. As with nitrogen, soils usually contain plenty of phosphorus, but it may not be in the ionic, water-soluble forms—$H_2PO_4^-$ or HPO_4^{2-}—that plants can use. A phosphorus-deficient plant may have green leaves, but its growth rate is markedly reduced, and its new growth is often spindly and brittle. Also, in some plants, such as the one in Figure C, phosphorus deficiency produces a reddish color on the undersides of the leaves.

Figure D illustrates potassium deficiency. Plants take up potassium as K^+ ions dissolved in the soil water. Once again, most potassium compounds in the soil are only slightly soluble in water and therefore not available to plants. The outward signs of a potassium shortage are generally more localized than those of nitrogen and phosphorus deficiencies. For example, the older leaves usually show the most pronounced signs; they often turn yellow and develop dead, brownish tissue at the edges or in scattered spots. Stems and roots are also weakened, leading to stunting.

Once a diagnosis of a nutrient deficiency is made, treating the problem is usually simple. You can choose from a number of fertilizer products for enriching the soil. Some of these consist of the inorganic compounds plants can use directly, such as nitrates or phosphates; others contain organic materials that are broken down to the usable inorganic compounds by microbes in the soil.

A. A healthy plant   **B.** Nitrogen deficiency   **C.** Phosphorus deficiency   **D.** Potassium deficiency

A. Three soil horizons seen in a cotton field

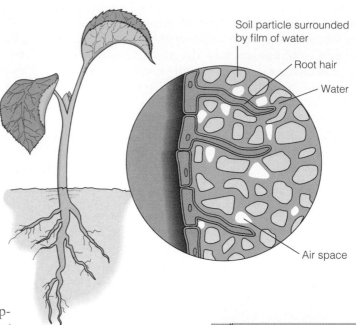

B. A close-up view of root hairs in soil

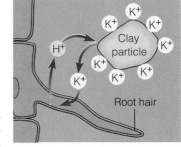

C. Positive-ion exchange

Soil characteristics determine how plants grow in a particular location. Fertile soil is soil that supports abundant plant growth by providing adequate water and dissolved nutrients. It also provides the particular conditions that allow plant roots to absorb their requirements.

Figure A shows a researcher photographing a cross section of the soil in a Tennessee cotton field. You can see three distinct layers, called **soil horizons,** in the cut. The A horizon, the top 8 inches (20 cm) in this case, is the **topsoil.** The topsoil is subject to extensive weathering (freezing, drying, and erosion, for example). Fertile topsoil contains a mixture of fine rock particles and clays, providing an extensive surface area that retains water and inorganic nutrients. It also contains decomposing organic material, called **humus,** and living organisms. Humus is an important source of plant nutrients; it also tends to retain water while keeping the topsoil porous enough for good aeration of the plant roots. Fertile topsoils usually support teeming numbers of bacteria, protozoans, fungi, and small animals such as earthworms, roundworms, and burrowing insects. Along with plant roots, these organisms loosen and aerate the soil and contribute organic matter to the soil as they live and die. Nearly all plants depend on bacteria and fungi in the soil to break down organic matter into inorganic molecules that roots can absorb. Plant roots branch out in the A horizon and usually extend into the next layer, the B horizon.

The soil's B horizon contains many fewer organisms and much less organic matter than the topsoil and is less subject to weathering. Fine clays and nutrients dissolved in soil water drain down from the topsoil and often accumulate in the B horizon. Below the B horizon, the C horizon is composed mainly of slightly weathered rock.

Figure B illustrates the intimate association between a plant's root hairs, soil water, and the tiny particles of topsoil. The root hairs are in direct contact with soil water held in tiny spaces among the particles. The soil water is actually a solution containing dissolved oxygen (O_2) and inorganic ions, many of them plant nutrients. Oxygen diffuses into the water from small air spaces in the soil. The root hairs take up dissolved oxygen, ions, and water from the film of soil water that surrounds them.

A **positive-ion exchange** is a mechanism whereby root hairs take up certain inorganic ions. Positively charged ions, such as calcium (Ca^{2+}), magnesium (Mg^{2+}), and potassium (K^+), adhere by electrical attraction to the negatively charged surfaces of clay particles. This adherence helps prevent positively charged nutrients from draining away during heavy rain or irrigation. In positive-ion exchange (Figure C), the root hairs release hydrogen ions (H^+) into the soil solution. The H^+ ions displace nutrient ions on the clay particle surfaces, and the root hair can then absorb the free ions.

In contrast to positively charged ions, negative ions, such as nitrate (NO_3^-), are usually not bound tightly by soil particles. Unbound ions are readily available to plants, but they tend to drain out of the soil quickly. This is often how soils become nitrogen-deficient.

It may take centuries for a soil to become fertile. The loss of soil fertility is one of our most pressing environmental problems, as we discuss next.

Our survival as a species literally depends on soil, yet the loss of soil by erosion and chemical pollution threaten this vital resource throughout the world. As the human population continues to grow, and more and more land is cultivated, prudent farming practices that conserve soil fertility will become essential to our survival. Three critical aspects of soil conservation are irrigating properly, preventing erosion, and fertilizing.

Irrigation can turn a desert into a garden, but farming in dry regions is a huge drain on water resources. Irrigation also tends to make the soil salty. The whitish deposits on the soil in the photograph in Figure A are salts that were dissolved in irrigation water flooded onto a field. Left behind when the excess water evaporated, the deposits will eventually make the soil too salty for crop plants to tolerate. Instead of flooding fields, modern irrigation often employs perforated pipes that drip water slowly into the soil close to plant roots. This drip irrigation uses less water, allows the plants to absorb most of the water, and reduces water loss from evaporation and drainage.

Preventing erosion—the blowing or washing away of soil—is one of the most important challenges of modern agriculture. Thousands of acres of farmland are lost to water and wind erosion each year in the United States alone. Plowed soil is especially vulnerable. Plowing aerates the soil and buries weeds and crop stubble, but it also exposes the soil to eroding winds and rains. A method known as minimal tillage farming can reduce erosion. Farmers using this method do not plow every year, and they usually rely on herbicides to kill weeds. Unfortunately, the herbicides contribute to chemical pollution of

A. Flood irrigation

the soil. Other ways of reducing soil losses to erosion include planting trees along field edges to prevent wind erosion, and taking special care in hilly terrain. The crops in Figure B are planted in rows that go around, rather than up and down, the hill in the field. This contour tillage helps slow the runoff of water and topsoil after heavy rains.

Fertilizers have probably been applied to crops since prehistoric farmers noticed that grass grew faster and greener where animals had defecated. Today, in developed nations, most farmers use commercially produced fertilizers containing minerals that are either mined or prepared by industrial processes. These fertilizers contain inorganic compounds of nitrogen, phosphorus, and potassium, the three elements that are most commonly deficient in farm soils.

Manure, fishmeal, and compost (decaying plant matter) are referred to as "organic" fertilizers because they are of biological origin. Before the nutrients in these substances can be used by plants, the organic material must be broken down by bacteria and fungi to inorganic nutrients that roots can absorb. In the end, the inorganic ions a plant extracts from the soil are in the same form whether they came from organic fertilizer or from a chemical factory. The difference is that organic fertilizers release nutrients gradually, whereas inorganic fertilizers make them available immediately. Problems arise when fields are overfertilized with inorganic products and excess nutrients are not taken up by plants. The nutrients are not usually retained in the soil, so they can enter and pollute the groundwater, as well as streams and lakes. In contrast, organic fertilizers are retained in the soil because their organic components are relatively insoluble in water.

Commercial, inorganic fertilizers have greatly increased agricultural productivity, and they are used so extensively today that if farmers suddenly stopped using them, widespread famine could result. Agricultural researchers are attempting to develop ways to reduce the amount of inorganic fertilizers applied to croplands while maintaining crop yields. In the next module, we discuss one farmer's use of exclusively organic fertilizers.

B. Planting to prevent soil erosion in a hilly area

32.10 Organic farmer Stephen Moore uses no commercial chemicals

Stephen Moore has a Ph.D. in environmental engineering, was on the faculty at MIT, and spent five years as an environmental consultant. In 1981, Moore changed direction and took over his parents' 60-acre southern California farm, which had been in the family since 1875. As he puts it, "I have a love for getting my hands right in it." With a work force of eight, he grows lemons, limes, avocados, persimmons, and a variety of vegetables, without any pesticides or commercial fertilizers.

Prior to the 1950s, soil fertility on Moore's land was maintained by regular applications of organic matter, mainly animal manure. As a result, the humus content and fertility of the topsoil remained high. But between the 1950s and 1980s, the Moores followed the nationwide trend of using more and more inorganic chemicals and less organic fertilizer. Steve Moore found that, after more than 30 years of treatment with commercial chemicals, the soil he inherited was virtually infertile. He decided to revert to "the old ways," rebuilding soil fertility with compost and animal manure, and by what he calls "growing our own fertilizer." Among his orchard trees and on recently harvested cropland, he raises legumes and grasses; both while growing and when plowed under, these crops help rebuild soil fertility. With healthier soil, there is little trouble with pests. In 1985, Moore stopped using commercial chemicals altogether, and today he takes great pride in running a profitable, completely organic farm.

Organically grown foods are becoming increasingly popular. Many growers are able to market their produce locally, and there is also a system established for national and international distribution. Moore sells most of his vegetables locally, and his avocados and citrus fruits sell regionally on the west coast. The operating expenses of an organic farm are higher than those of conventional farms, and organic foods usually cost more than those produced conventionally. But a growing number of consumers are willing to pay extra for what they believe to be safer, perhaps more nutritious, and often tastier food. Steve Moore thinks of it this way:

> When you buy organically grown products, you benefit from eating nutritious food. You also make an investment in the future, because you are paying the real costs up front. It is cheaper to grow food the conventional way, but there are hidden costs. Pesticides and chemical fertilizers are costly to the environment. They contaminate groundwater, and we pass the bill for that on to future generations.

Moore believes, as do a growing number of agricultural experts, that the future of farming depends on an international effort in sustainable agriculture—growing crops and maintaining healthy, fertile soil at the same time.

32.11 Fungi help most plants absorb nutrients from the soil

Reliance on the soil for nutrients that may be in short supply makes it imperative that plants have a large absorptive surface area on their roots. As we have seen, root hairs add a great deal of surface to plant roots. Most plants gain even more absorptive surface by teaming up with fungi. The micrograph here shows a small root of a eucalyptus tree. The root is covered with a twisted mat of fungal filaments. Together, the roots of this plant and the fungus comprise a mutually beneficial association called a **mycorrhiza** (Greek *mykes*, fungus, and *rhiza*, root).

Mycorrhizae are particularly beneficial adaptations on plants growing in poor soils, but almost all plants are capable of forming this mutualistic relationship if their roots are exposed to appropriate species of fungi. The fungal filaments around the root provide an enormous surface, which absorbs water and inorganic ions, especially phosphate, more rapidly than the root can. Some of the water and ions taken up by the fungus are transferred to the plant. The fungus may also secrete acid that increases the solubility of some minerals in the soil and may convert them to forms that are more readily used by the plant. In turn, the plant's

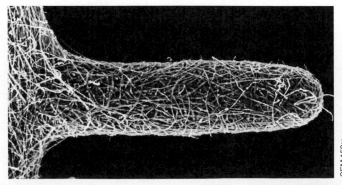

SEM 158×

Mycorrhiza on a eucalyptus root

photosynthetic products nourish the fungus. The fungi may also help protect the plant against certain disease microorganisms common in soil.

As more is learned about mycorrhizal associations, important agricultural applications are likely. As we mentioned in the introduction to Chapter 18, citrus trees require less fertilizer when grown with mycorrhizae. Techniques for supplying root fungi to other crops may eventually enable us to reduce fertilizer use to a greater extent.

The widespread presence of mycorrhizae is a reminder of the theme of connections among living organisms. Despite their ability to make their own food molecules, plants are not independent of other organisms. The fossil record shows that mycorrhizae have been common since plants first evolved. Indeed, the mycorrhizal connection probably altered the entire course of evolution by helping make possible the colonization of land.

The plant world includes parasites and carnivores 32.12

Mycorrhizae are not the only means by which plants gain more nutrients and improve their chances of survival. Some plants have evolved ways of obtaining food from other plants or animals. Figure A shows a parasitic plant called dodder (the yellow-orange threads wound around the green plant). Dodder cannot photosynthesize; it obtains organic molecules from other plant species, using modified roots that tap into the host's vascular tissue.

Figure B shows part of an oak tree parasitized by mistletoe, the plant we may tack above doorways during the Christmas season. There are about 1000 species of mistletoe, and one or more usually occur in areas where there are deciduous trees. All the leaves you see here are mistletoe; the oak has lost its leaves for winter. Mistletoe is photosynthetic, but it supplements its diet by using its roots to siphon sap from the vascular tissue of the host tree. Both dodder and mistletoe may kill their hosts by blocking light or taking too much food from them.

Certain plants are carnivorous, obtaining some of their nutrients, especially nitrogen, from animal tissues. Carnivorous plants grow in bogs where the soil is highly acidic. Organic matter decays so slowly in acidic soils that there is little inorganic nitrogen available for plant roots to take up. Though they are photosynthetic, the sundew and Venus flytrap (Figures C and D) thrive by obtaining their nitrogen from insects.

Few species illustrate the correlation of structure and function more clearly than carnivorous plants. The sundew plant (Figure C) has modified leaves, each bearing many club-shaped hairs. A sticky, sugary se-

A. Dodder growing on a pickleweed

B. Mistletoe growing on an oak

C. A sundew plant trapping a fly

D. A Venus flytrap, with a katydid

cretion at the tips of the hairs attracts insects and traps them. The presence of an insect triggers the hairs to bend and the leaf to cup around its prey. The hairs then secrete digestive enzymes, and the plant absorbs nutrients released as the insect is digested.

The Venus flytrap (Figure D) has V-shaped leaves that close around small insects. As insects enter the open V, they touch sensory hairs that trigger closure of the trap. The leaf then secretes digestive enzymes and absorbs nutrients from the prey.

Using insects as a source of nitrogen is a nutritional adaptation that allows carnivorous plants to thrive in soils where most other plants cannot. Only a few plants are carnivorous, however. Most plants obtain nitrogen with the help of soil bacteria, as we discuss next.

32.13 Most plants depend on bacteria to supply nitrogen

It is ironic that plants often have nitrogen deficiencies, for the atmosphere is nearly 80% nitrogen. Atmospheric nitrogen, however, is gaseous N_2, and plants cannot use nitrogen in that form. In fact, nearly all plants depend on nitrogen supplies in the soil.

For plants to absorb nitrogen from the soil, the nitrogen must first be converted to ammonium ions (NH_4^+) or nitrate ions (NO_3^-). Ammonium and nitrate in the soil are produced from atmospheric N_2 by bacteria. As shown in this figure, certain soil bacteria, called nitrogen-fixing bacteria, convert N_2 into ammonium. This process, called **nitrogen fixation,** is vital to most plants. A second group of bacteria, called ammonifying bacteria, adds to the soil's supply of ammonium by decomposing organic matter (humus).

The dashed red arrow in the diagram indicates that plant roots absorb only a small amount of nitrogen as ammonium. The availability of ammonium to the plant is limited, because it tends to remain bound to clay particles in the soil. Fortunately for plants, a third type of soil bacteria, nitrifying bacteria, converts soil ammonium to nitrate. Plants take up most of their nitrogen in this form. They then convert the nitrate back into ammonium and then into amino acids, which they can use to make proteins and other nitrogen-containing organic molecules.

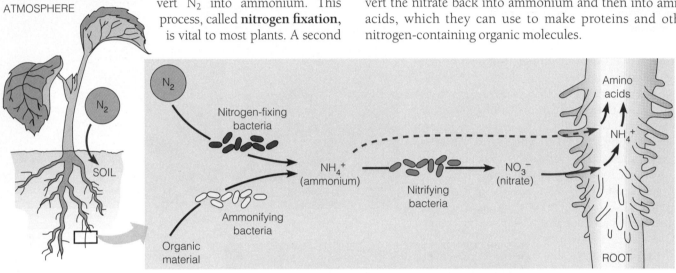

32.14 Legume plants have built-in nitrogen-fixing bacteria

As we discussed in the chapter's introduction, legumes (peanuts, peas, beans, and many other plants that produce their seeds in pods) have a built-in source of fixed nitrogen—bacteria housed in swellings, or nodules, on their roots. The photograph in Figure A shows these structures on the roots of a pea plant. The **nodules** consist of plant cells that contain nitrogen-fixing bacteria. The micrograph at the top of the facing page (Figure B) shows a cross section of one such cell, with vesicles full of bacteria. The nitrogen-fixing bacteria in legume nodules belong to the genus *Rhizobium* (from the Greek *rhiza,* root, and *bios,* life).

The relationship between a legume and its nitrogen-fixing bacteria is mutually beneficial. The plant provides the bacteria with carbohydrates and other organic compounds. The bacteria have enzymes that catalyze the conversion of atmospheric N_2 into ammonium (NH_4^+). When conditions are favorable, root nodule bacteria actually fix so much ni-

A. Root nodules on a pea plant

trogen that the nodules secrete excess ammonium, which increases the fertility of the soil. This is one of the reasons farmers rotate crops, one year planting a nonlegume such as corn, and the following year planting a legume to raise the concentration of fixed nitrogen in the soil. The legume crop is often soybeans or alfalfa. Soybeans are harvested as a crop, and alfalfa may be cut for hay. In either case, the roots remain in the ground and add their fixed nitrogen and other nutrients to it. Instead of being harvested, the legume crop may be plowed under so that it will decompose and add even more fixed nitrogen to the soil.

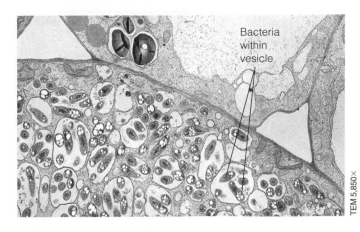

TEM 5,850×

B. Bacteria within a root nodule cell

A major goal of agricultural research is to improve the protein content of crops

The ability of plants to incorporate inorganic nitrogen into proteins and other organic substances has a major impact on human welfare. Either by choice or by economic necessity, the majority of people in the world have a predominantly vegetarian diet. Thus, particularly in the developing countries, people depend mainly on plants as immediate sources of protein. Unfortunately, many plants have a low protein content, and the proteins that are present in them may be deficient in one or more of the amino acids that humans need from their diet. Protein deficiency is one of the most common forms of malnutrition in humans.

Improving the quality and quantity of proteins in crops is a major goal of agricultural researchers. Among some of the most important advances to date are new varieties of corn, wheat, and rice (Figure A). However, many of these "super" varieties have an extraordinary demand for nitrogen, which is usually supplied to them in the form of commercial, inor-

A. Plant researchers with "super" rice

ganic fertilizer. Besides being hazardous to the environment, commercial fertilizers are expensive to produce. Most of the countries that have the greatest need for high-protein crops are the ones least able to afford to grow them.

One of the most promising lines of agricultural research is directed toward improving the output of nitrogen-fixing bacteria that are mutualistic in plants. A negative-feedback mechanism normally regulates the rate at which *Rhizobium,* the root nodule bacterium, fixes nitrogen, forming nitrogen compounds. When the

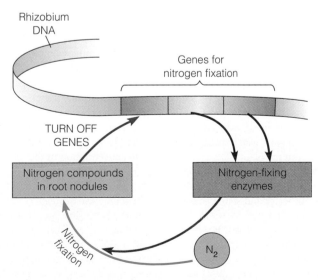

B. Regulation of nitrogen fixation in *Rhizobium* bacteria

quantity of fixed nitrogen in a root nodule reaches a certain level (the green rectangle in Figure B), it switches off the bacterial genes that code for the enzymes needed for nitrogen fixation. Researchers have isolated certain *Rhizobium* mutants that continue to make these enzymes even after nitrogen compounds accumulate. Farmers may someday grow plants colonized by these mutant bacteria, a breakthrough that would increase the protein content of the legume crop and also add more fixed nitrogen to the soil. Genetic engineering also promises additional improvements in crops, as we discuss next.

Genetic engineering could greatly increase crop yields

A technician readies a .22-caliber gun for shooting foreign genes into plant cells. The gun fires plastic bullets containing tiny metal pellets coated with DNA. As indicated in the drawing, the bullet stays in the gun, but the particles are driven into the plant cells. They travel through the cell walls into the cytoplasm, and the foreign DNA becomes integrated into plant cell DNA. An engineered cell can be grown into a whole new plant that will produce proteins encoded by the foreign DNA, along with the proteins of the original parent cell.

Many new varieties of crop plants have already been produced with the gene gun and by an older technique that uses bacterial plasmids for gene transfer (see Module 12.14). Cotton and tobacco plants have been engineered that are resistant to viral attack. Potato plants have been engineered to synthesize their own insecticide, making them resistant to attack by beetles that can destroy whole crops. Tomato plants have been engineered to produce fruit that is slow to spoil. In the future, genetic engineering may also produce crop plants that can synthesize pharmaceuticals, industrial oils, and other useful chemicals.

A major goal of genetic engineering is to create crop plants that provide more nutritious food—for instance, corn, wheat, and other grains that have a full complement of the amino acids humans need to make proteins. Another goal is to engineer varieties of nitrogen-fixing bacteria that are more efficient than naturally occurring varieties at making NH_4^+. Eventually, it may be possible to transplant genes for nitrogen fixation directly into the DNA of nonlegume crop plants.

Genetic engineering holds great potential for increasing agricultural production. There are potential problems, however. Gene-spliced crop plants, containing genes that resist natural diseases, might, for example, escape into the wild and overgrow native species. Engineered crop plants might also hybridize with their wild relatives, creating weeds that grow out of control. Also, there is a concern that new proteins in foods produced by gene-spliced plants could be toxic or cause serious allergies in some people. Governments throughout the world are grappling with how to proceed—whether to promote the agricultural revolution offered by gene splicing, or slow its progress until more information is available about the potential hazards. We will touch on the subject of agricultural productivity again in Chapter 33, which discusses plant hormones.

Using a gene gun

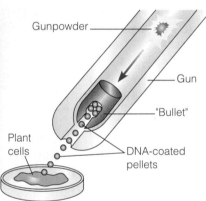
Gunpowder
Gun
"Bullet"
Plant cells
DNA-coated pellets

Chapter Review

Begin your review by rereading the module headings and scanning the figures before proceeding to the Chapter Summary and questions.

Chapter Summary

Introduction–32.1 As a plant grows, its roots absorb water, inorganic nutrients, and oxygen from the soil. Its leaves take carbon dioxide from the air. Xylem and phloem transport water, nutrients, and the products of photosynthesis throughout the plant.

32.2 Root hairs greatly increase a root's absorptive surface. Water and solutes move freely through or between cells of the epidermis and cortex toward the center of the root. All water and solutes must pass through the endodermis before entering the xylem for transport upward. The endodermal cells admit only certain solutes. In some plants, solute transport may raise water pressure in the xylem. This root pressure can push water a short way up the stem.

32.3–32.4 Most of the force that moves water and solutes upward in the xylem comes from transpiration, the evaporation of water from the leaves. Cohesion causes water molecules to stick together, relaying the pull of transpiration along a string of water molecules all the way to the roots. The adhesion of water molecules to xylem cell walls helps counter gravity. These processes are capable of moving xylem sap, consisting of water and dissolved inorganic nutrients, to the top of the tallest tree. Guard cells surrounding stomata in the leaves control transpiration.

32.5 Phloem transports food molecules made by photosynthesis by a pressure-flow mechanism. At a sugar source, such as a leaf, sugar is loaded into a phloem tube by photosynthetic cells. This raises the solute concentration in the tube, and water follows by osmosis, raising the pressure in the tube. As sugar is removed and stored in a sugar sink, such as the root, water follows. The increase in pressure at the sugar source and decrease at the sugar sink causes phloem sap to flow from source to sink. In the same way, sugar stored in roots may be moved upward to developing leaves.

32.6–32.7 A plant's ability to make food depends on the nutrients it obtains from its surroundings. Macronutrients, such as carbon, oxygen, nitrogen, and phosphorus, are required in large amounts, mostly to build organic molecules. Micronutrients, including iron, copper, and zinc, act mainly as cofactors of enzymes. Growing plants in solutions of known composition enables researchers to determine nutrient requirements. Stunting, wilting, and color changes indicate nutrient deficiencies.

32.8–32.10 Soil characteristics determine whether a plant will be able to obtain the nutrients it needs to grow. Fertile soil contains a mixture of small rock and clay particles that hold water and ions. Humus—decaying organic material—holds nutrients, air spaces, and water and supports the growth of organisms that enhance soil fertility. Water-conserving irrigation, erosion control, and the prudent use of herbicides and fertilizers are aspects of good soil management. Organic farming protects the environment, and organically grown foods have become increasingly popular.

32.11–32.12 Relationships with other organisms aid plants in obtaining nutrients. Many plants form mycorrhizae, mutually beneficial associations with fungi. A network of fungal threads increases a plant's absorption of nutrients and water, and the fungus receives some nutrients from the plant. Parasitic plants such as mistletoe siphon sap from host plants. Carnivorous plants obtain some of their nitrogen by digesting insects.

32.13–32.16 Bacteria in the soil recycle nitrogen by decomposing organic matter. Other soil bacteria fix nitrogen from the air, and still others convert it to a form used by plants. Legume plants have a built-in source of nitrogen; nodules in their roots house nitrogen-fixing bacteria. Plants use nitrogen to make proteins, which are important in the human diet, as well as other important organic molecules. Plant scientists are using genetic engineering to develop new food crops for a growing world population.

Testing Your Knowledge

Multiple Choice

1. Houseplants require the smallest amount of which of the following nutrients?

    a. oxygen   **d.** iron
    b. phosphorus   **e.** hydrogen
    c. carbon

2. The clay particles in soil are important because they

    a. are composed of nitrogen needed by plants
    b. allow spaces for air and drainage
    c. fill spaces and keep oxygen out of the soil
    d. are charged and hold ions needed by plants
    e. supply humus needed by plants

3. By trapping insects, carnivorous plants obtain ———, which they need ———.

    a. water . . . because they live in dry soil
    b. nitrogen . . . to make sugar
    c. phosphorus . . . to make protein
    d. sugars . . . because they can't make enough in photosynthesis
    e. nitrogen . . . to make protein

4. A major long-term problem resulting from flood irrigation is the

    a. drowning of crop plants
    b. accumulation of salts in the soil
    c. erosion of fine soil particles
    d. encroachment of water-consuming weeds
    e. excessive cooling of the soil

True/False (Change false statements to make them true.)

1. If a plant gets too hot, guard cells change shape and open the stomata.

2. Lower air pressure in the leaves "sucks" water to the tops of tall trees.

3. Potassium is carried from the roots to the leaves in the xylem.

4. Most of the organic material produced by a plant as it grows comes from materials obtained from the air.

5. Transpiration moves sugar from leaves to roots.

6. Carbon, nitrogen, oxygen, and chlorine are macronutrients.

7. Negatively charged ions such as NO_3^- (nitrate) are easily washed from soil.

Describing, Comparing, and Explaining

1. Explain how guard cells control the rate of water loss from a plant on a hot, dry day. Why is this both helpful and harmful to the plant?

2. Write a short paragraph describing the three ways in which plants depend on bacteria for their supply of nitrogen.

3. Describe the characteristics of good topsoil. What are the roles of fine rock and clay particles, humus, and living organisms in the soil?

Thinking Critically

1. Acid rain is acidic because it contains an excess of hydrogen ions (H^+). One effect of acid rain is to deplete the soil of nutrients such as calcium (Ca^{2+}), potassium (K^+), and magnesium (Mg^+). Why do you think acid rain washes these nutrients from the soil?

2. Researchers have found that, in some situations, the application of nitrogen fertilizer to crops may have to be increased each year. Fertilizer decreases the rate of natural nitrogen fixation that occurs in the soil, and more fertilizer is needed to make up the difference. Explain how this might occur.

3. A tip for making cut flowers last longer without wilting is to cut off the cut ends of the stems under water and then keep them wet, so no air bubbles get into the xylem. Explain why this works.

Science, Technology, and Society

1. About 10% of U.S. cropland is irrigated. Agriculture is by far the biggest user of water in arid western states, including Colorado, Arizona, and California. The populations of these states are growing, and there is an ongoing conflict between cities and farm regions over water. To ensure water supplies for urban growth, cities are purchasing water rights from farmers. This is often the least expensive way for a city to obtain more water, and it is possible for some farmers to make more money selling water than growing crops. Discuss the possible consequences of this trend. Is this the best way to allocate water for all concerned? Why or why not?

2. The first genetically engineered food plant—a tomato that resists spoiling—is now in food stores. Citing health and safety concerns, several prominent chefs have announced they will boycott genetically altered foods. What might be some hazards of eating genetically engineered foods? Should the public be informed that the food they purchase has been genetically altered? Would you have any reservations about eating a tomato genetically engineered to resist spoiling? Why or why not?

3. This chapter discusses several trends in modern agriculture, such as organic farming and the use of genetic engineering to develop improved crops. How might organic farming benefit from advances in genetic engineering, and vice versa? How might organic farming and genetic engineering undermine each other?

Control Systems in Plants 33

The pink-flowered plant in the photograph at the left, a popular houseplant, is a tropical species originally from South America. It is called the sensitive plant because its leaves fold up when touched. Few people who know about it can resist the urge to touch a leaf and watch it react. As much for its fascinating behavior as for its beauty, the sensitive plant has been grown indoors for hundreds of years. In the eighteenth century, Swedish botanist Carolus Linnaeus gave the sensitive plant its scientific name. Linnaeus called it *Mimosa pudica*, which means "modest mimic" in Latin.

The photos below illustrate the difference between an undisturbed and a disturbed *Mimosa*. The undisturbed plant on the left has its leaves extended. *Mimosa* has a main leaf stalk, which branches into four smaller stalks, each of which, in turn, bears numerous small leaflets. When touched, the leaflets fold upward, and the four small stalks move closer together, making the plant look wilted, as shown on the right. If you pinch just a single leaflet, you can watch the effect of the stimulus travel throughout the plant. Within a few seconds, all the plant's leaves appear wilted. This response results from a rapid loss of water from specialized cells in cushionlike thickenings, called pulvini, located at the base of each leaflet, small stalk, and main leaf stalk. The signals that trigger the response travel as action potentials, similar to, but slower than, those in the nervous systems of animals (see Module 28.5). The action potentials trigger cells on one side of the pulvinus to lose water and become flaccid. After a sensitive plant reacts, it takes about 10 minutes for the cells to regain water and become turgid, restoring the plant's original form.

What is the function of this behavior? Scientists have studied *M. pudica* for centuries, but we still can only speculate. In the wild, the plant in-

habits dry, windy areas in the tropics. Perhaps its ability to fold its leaves is an adaptation that reduces surface area and saves water. The wilting response may also make the plant less attractive to herbivores.

The sensitive plant not only responds to touch, it also carries out what plant biologists call sleep movements. At dusk, it folds its leaves in the same way it does when disturbed. You may have noticed that dandelions, morning glories, and many other plants fold their flowers or leaves in the evening and unfold them in the morning. Again, we can only speculate about the significance of this behavior. Charles Darwin thought sleep movements might slow the loss of heat from the leaves after dark. Whatever their function, these movements demonstrate a daily regularity in a plant's activities. In fact, plants as well as animals have internal biological clocks that keep them on a regular daily cycle.

In this chapter, we focus on the ability of plants to sense and respond to their environment and on the control mechanisms that make plant behavior possible. We will see that chemical messengers—hormones—play important roles in many kinds of plant movements and in controlling other activities, including growth, flowering, and fruit production.

Main leaf stalk

The effect of stimulating a sensitive plant, *Mimosa pudica*

Experiments on how plants turn toward light led to the discovery of a plant hormone

A houseplant on a windowsill grows toward light (Figure A). If you rotate the plant, it will soon reorient its growth until its leaves again face the window. The growth of a shoot toward light is called **phototropism** (from the Greek *photos,* light, and *tropos,* turn). Phototropism is an adaptive response, directing growing seedlings and the shoots of mature plants toward the sunlight they need for photosynthesis.

Microscopic observations of plants growing toward light indicate the cellular mechanism that underlies phototropism. Figure B shows a grass seedling curving toward light coming from one side. As the enlargement shows, cells on the darker side of the seedling are larger—actually, they have elongated faster—than those on the brighter side. The different cellular growth rates made the shoot bend toward the light. Other experiments (not illustrated) have shown that the growth rates of these cells do not differ if a seedling is kept in the dark or if it is illuminated uniformly from all sides. In these situations, all the cells elongate at a similar rate, and the seedling grows straight upward.

What causes plant cells in a shoot to grow at different rates? We found in Chapter 26 that hormones help coordinate internal activities, such as growth rates, in animals. We might predict, therefore, that plants also have hormones that regulate growth. The idea that plants have hormones, in fact, emerged from a series of classic experiments on how shoots respond to light.

In 1880, Charles Darwin and his son, Francis, conducted some of the earliest experiments on phototropism and set the stage for the discovery of an important plant hormone. They found that grass seedlings would bend toward light only if the *tips* of their shoots were present. The first five grass plants in Figure C summarize the Darwins' findings. When they removed the tip of a grass shoot, the shoot grew straight up, rather than curving toward the

A. A houseplant growing toward light

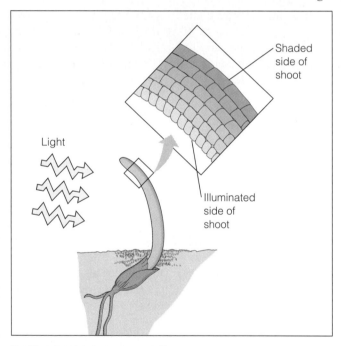

B. Phototropism in a grass seedling

Light

Shaded side of shoot

Illuminated side of shoot

light. The shoot also remained straight when the Darwins placed an opaque cap on its tip. However, the shoot curved normally when they placed a transparent cap on its tip or an opaque shield around its base. The Darwins concluded that the tip of the shoot was responsible for sensing light. They also recognized that the growth response that makes the seedling curve toward light actually occurs below the tip. Therefore, they speculated that some signal was transmitted downward from the tip to the growth region of the shoot.

In 1913, Danish botanist Peter Boysen-Jensen further tested the chemical signal idea of the Darwins; the last two plants in Figure C summarize his findings. In one group of grass seedlings, Boysen-Jensen inserted a block of gelatin between the tip and the lower part of the shoot. The gelatin block prevented cellular contact between the tip and the rest of the shoot but allowed chemicals to diffuse through. The seedlings with gelatin blocks behaved normally, bending toward the light. In a second set of seedlings, Boysen-Jensen inserted a thin piece of the mineral mica under the shoot tip. Mica is an impermeable barrier, and the seedlings with mica had no phototropic response. These experiments supported the idea that the signal for phototropism is a mobile chemical.

In 1926, Dutch botanist Fritz Went modified Boysen-Jensen's techniques and discovered the chemical messenger for phototropism. As shown in Figure D, Went first removed the tips of grass seedlings and placed them on blocks of agar, a gelatinlike material. He reasoned that the chemical messenger (pink in the figure) from the shoot tips should diffuse into the agar, and that the blocks should then be able to substitute for the shoot tips. Went tested the effects of the agar blocks on tipless seedlings. He grew the

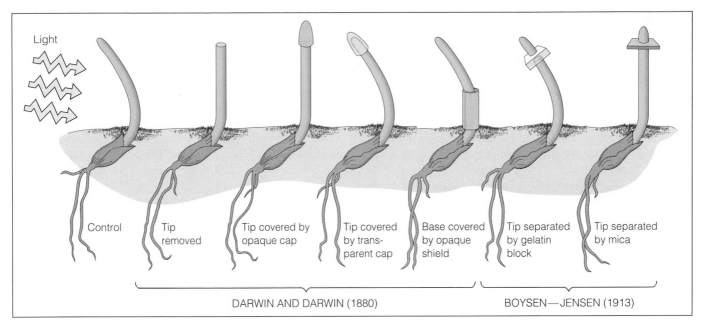

Light

| Control | Tip removed | Tip covered by opaque cap | Tip covered by transparent cap | Base covered by opaque shield | Tip separated by gelatin block | Tip separated by mica |

DARWIN AND DARWIN (1880)　　　　　BOYSEN—JENSEN (1913)

C. Early experiments on phototropism

seedlings in the dark to eliminate the effect of sunlight in order to test the effect of the chemical alone. First, he centered the treated agar blocks on the cut tips of a batch of seedlings. These plants grew straight upward, in contrast to the tipless control seedlings, which did not grow much at all. Went concluded that the agar had absorbed the chemical messenger produced in the shoot tip, and that the chemical passed down into the shoot and stimulated it to grow. He then placed some agar blocks off-center on another batch of tipless seedlings. These seedlings bent away from the side with the chemical-laden agar block, as though growing toward light. Seedlings that received blank agar blocks (whether offset or not) did not grow. Went concluded that when shoots curve toward light, they do so because there is a higher concentration of the growth-promoting chemical on the darker side of the shoot. For this chemical messenger, or hormone, Went chose the name auxin, from the Greek *auxein*, "to increase."

Went's pioneering work sparked an explosion of research on plant growth regulators. Follow-up experiments showed that the tips of growing shoots secrete auxin in equal amounts in the dark

or light. When a shoot is exposed to light from one direction, auxin diffuses from the lighted side to the shaded side of the shoot and stimulates cell growth there. At the same time, the growth rate on the lighted side may decrease because of a reduced auxin level there. In the 1930s, biochemists determined the chemical identity of Went's auxin, finding it to be a small organic molecule that plant cells make from the amino acid tryptophan.

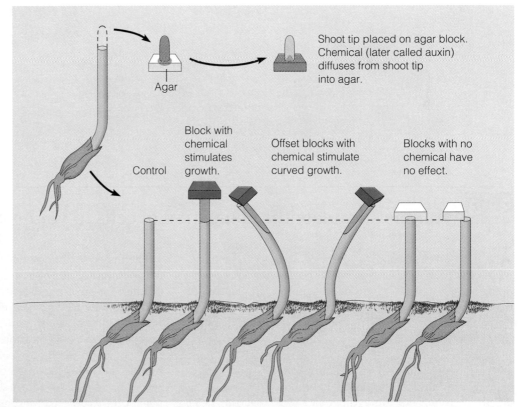

Agar

Shoot tip placed on agar block. Chemical (later called auxin) diffuses from shoot tip into agar.

Control

Block with chemical stimulates growth.

Offset blocks with chemical stimulate curved growth.

Blocks with no chemical have no effect.

D. Went's experiments

33.2 Five classes of hormones regulate plant growth and development

To date, plant biologists have identified five different types of plant hormones. Three of these—the auxins, cytokinins, and gibberellins—are actually hormone *classes*. Each class includes several chemicals with similar structure and function.

Plants produce hormones in very small concentrations, but a minute amount of any of these chemicals can have profound effects on target cells. Just a few molecules of a hormone can redirect the metabolism and development of a plant's cells. Plant hormones may do this by altering the expression of genes, by activating or inhibiting enzymes and other proteins, or by changing the properties of membranes.

As the table below indicates, each type of hormone can produce a variety of effects. Notice that all five types of hormones influence growth, and four of them affect development (cell differentiation). Hormones stimulate growth by signaling target cells to divide or elongate; some of the hormones inhibit growth by depressing cell division or elongation. The effects of a hormone depend on the plant species, the hormone's site of action, the developmental stage of the plant, and the concentration of the hormone. In most situations, no single hormone acts alone. Instead, several plant hormones usually interact in controlling the growth and development of a plant. We will see several examples of these interactions as we focus on each of the plant hormones in the next five modules.

Hormone	Major Functions	Where Produced or Found in Plant
Auxins	Stimulate stem elongation, root growth, differentiation and branching, development of fruit; apical dominance; phototropism and gravitropism (response to gravity)	Meristems of apical buds and young leaves; embryo within seed
Cytokinins	Affect root growth and differentiation; stimulate cell division and growth, germination, and flowering; delay senescence (aging)	Synthesized in roots and transported to other organs
Gibberellins	Promote seed and bud germination, stem elongation, leaf growth; stimulate flowering and development of fruit; affect root growth and differentiation	Meristems of apical buds, roots, and young leaves; embryo
Abscisic acid (ABA)	Inhibits growth; closes stomata during water stress; helps maintain dormancy	Leaves, stems, green fruit
Ethylene	Promotes fruit ripening; opposes some auxin effects; promotes or inhibits growth and development of roots, leaves, flowers, depending on species	Tissues of ripening fruits, nodes of stems, dying leaves

33.3 Auxins stimulate the elongation of cells in young shoots

The term **auxin** is used to describe a class of chemicals whose chief function is to promote the elongation of developing shoots. Several auxins occur naturally in plants, and many others have been synthesized by chemists. The most important naturally occurring auxin is a compound named **indoleacetic acid, or IAA.**

Figure A (on the facing page) shows the effect of IAA on growing pea plants. All the seedlings in this photograph were grown under controlled conditions for the same length of time. The only difference was that the taller seedlings, on the right, were treated with IAA.

The major site of auxin synthesis in a plant is the apical meristem at the tip of a shoot (see Module 31.7). As IAA produced in the tip moves downward, it stimulates growth of the stem by making cells elongate. As the black curve in Figure B shows, IAA promotes cell elongation in stems only over a certain concentration range. Above a certain level (0.9 g of auxin per liter of solution, in this case), it usually inhibits cell elongation in stems. This inhibitory effect probably occurs because a high level of IAA makes the plant cells synthesize another hormone, ethylene, which generally counters the effects of IAA.

A. The effect of auxin (IAA) on pea plants

B. The effect of auxin concentration on cell elongation

The red curve on the graph shows the effect of IAA on root growth. An IAA concentration too low to stimulate shoot cells will cause root cells to elongate. On the other hand, an IAA concentration high enough to make stem cells elongate is in the concentration range that inhibits root cell elongation. These effects of IAA on cell elongation reinforce two points: (1) the same chemical messenger may have different effects at different concentrations in one target cell, and (2) a given concentration of the hormone may have different effects on different target cells.

How do auxins make plant cells elongate? One hypothesis currently being tested suggests that auxins initiate elongation by weakening cell walls. As shown in Figure C, auxins may stimulate certain proteins (purple) in a plant cell's plasma membrane to pump hydrogen ions into the cell wall. The hydrogen ions activate enzymes that break bonds between cellulose molecules in the cell wall. The cell then swells with water and elongates because its weakened wall no longer resists the cell's tendency to take up water osmotically. After this initial elongation, the cell sustains the growth by synthesizing more wall material and cytoplasm.

Auxins produce a number of other effects, in addition to stimulating cell elongation and causing stems and roots to grow in length. These hormones can also trigger the development of vascular tissues and induce cell division in the vascular cambium, thus promoting growth in stem diameter (see Module 31.8). Furthermore, auxins are produced by developing seeds, and, in many plants, auxins in seeds promote the growth of fruit. Some plants will even develop fruits without being fertilized if they are sprayed with auxins. Farmers sometimes produce tomatoes, cucumbers, and eggplants, for example, by spraying the plants with synthetic auxins. The fruits that result are seedless.

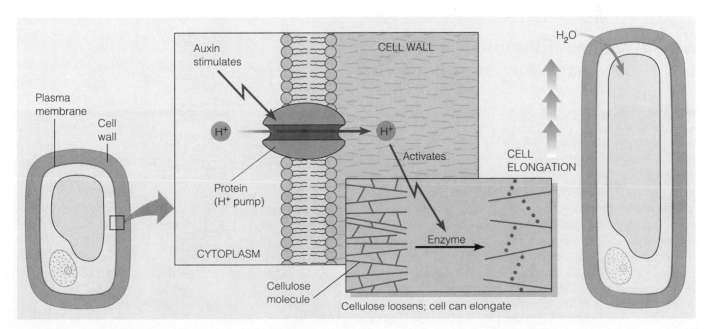

C. A hypothesis to explain how auxin stimulates cell elongation

33.4 Cytokinins stimulate cell division

Cytokinins (from the Greek *kytos,* cell, and *kineo,* to move) are growth regulators that promote cell division, or cytokinesis. A number of cytokinins have been extracted from plants, and several synthetic ones have been made. Natural cytokinins are produced in actively growing tissues, particularly in roots, embryos, and fruits. Cytokinins made in the roots reach target tissues in stems by moving upward in xylem sap.

Plant biologists have found that cytokinins enhance the division, growth, and development of plant cells grown in culture. Cytokinins can also retard the aging of flowers and fruits, and cytokinin sprays are used to keep cut flowers fresh.

In whole plants, the effects of cytokinins are often influenced by the concentration of auxins present. The photographs here show the results of a simple experiment that partially separates the different effects of auxins and cytokinins. Both basil plants pictured are the same age. The one on the left has an intact terminal bud; the one on the right had its terminal bud removed several weeks earlier. In the plant on the left, auxin transported down the stem from the terminal bud promoted lengthwise growth while inhibiting the growth of the axillary buds (the buds that produce side branches). As a result,

the shoot grew in height but did not branch out to the sides very much. In the plant on the right, the lack of a terminal bud eliminated the inhibitory effect of auxin on the axillary buds. This allowed cytokinins transported up from the roots to activate the axillary buds, making the plant grow more branches and become bushy.

Most plants have complex growth patterns, with some lateral growth occurring even when terminal buds are intact. These patterns probably result from the interaction of auxins and cytokinins, with the ratio of the two hormones playing a critical role. Cytokinins entering the shoot system from the roots counter the effects of auxins coming down from the terminal buds. The lower axillary buds on a shoot usually begin to grow before those closer to the terminal bud, reflecting the higher ratio of cytokinins to auxins in the lower parts of the plant.

The antagonistic interaction of auxins and cytokinins may also be one way the plant coordinates the growth of its root and shoot systems. As roots become more extensive and produce more and more cytokinins, the increased level of the cytokinins would signal the shoot system to form more branches. Antagonistic interactions are common among plant hormones, as we will see in the next several modules.

Terminal bud

No terminal bud

The effects of naturally occurring auxins and cytokinins on plant growth

33.5 Gibberellins affect stem elongation and have numerous other effects

Figure A shows two clusters of young rice plants. The cluster on the left is normal; the taller, yellowish plants on the right are infected with a fungus of the genus *Gibberella.* The infected seedlings will not produce grain; in fact, they will topple over and die before they can mature and flower. Rice growers in Asia have suffered crop losses from *Gibberella* for centuries. In Japan, the aberrant growth pattern is called "foolish seedling disease." In the 1920s, Japanese scientists found that the fungus releases a chemical that actually causes the disease. The chemical was named **gibberellin.** Researchers later discovered that gibberellin exists naturally in plants, where it is a growth regulator. Foolish seedling disease occurs when rice plants that have been infected with the *Gibberella* fungus get an overdose of gibberellin.

A. Foolish seedling disease (right), caused by the *Gibberella* fungus

More than 70 different gibberellins are now known, and many of them occur naturally. Synthesized at the tips of both stems and roots, gibberellins produce a wide variety of effects. One of their main effects is to stimulate elongation in stems and leaves. This action generally enhances that of the auxins, although the precise way these hormones interact is not yet known. Also in combination with auxins, gibberellins can influence fruit development, and gibberellin-auxin sprays can make apples, currants, and eggplants develop without fertilization. One of the most widespread uses of gibberellins is in the production of the Thompson variety of seedless grapes. Gibberellins make the grapes grow larger and farther apart in a cluster. The left cluster of grapes in Figure B is untreated; the right one shows the effect of gibberellin treatment.

Gibberellins are also important in seed germination in many plants. Many seeds that require special environmental conditions to germinate, such as exposure to light or cold temperatures, will begin to germinate if they are sprayed with gibberellins. In nature, gibberellins in seeds

B. The effect of gibberellin treatment on grapes

are probably the link between environmental cues and the metabolic processes that renew growth of the embryo. For example, after a grass seed absorbs water, gibberellins released from the embryo support germination by mobilizing nutrients stored in the seed. In some plants, the gibberellins seem to interact antagonistically with another hormone, abscisic acid, which maintains seed dormancy.

Abscisic acid inhibits many plant processes   33.6

There are times, such as the onset of winter or severe drought, when it is adaptive for a plant to become dormant, temporarily ceasing to grow. At such times, the hormone **abscisic acid (ABA)**, produced in buds, inhibits cell division in buds and in the vascular cambium, thus suspending both primary and secondary growth. ABA also signals the buds to form scales that will protect them from harsh conditions. Abscisic acid was named when it was believed that this hormone caused abscission, or the breaking off, of leaves from deciduous trees in autumn. The name has prevailed even though research has never shown that ABA plays a role in leaf drop.

Another time in the life of a plant when it is advantageous to suspend growth is the onset of seed dormancy. Seed dormancy is especially important to annual plants in deserts and semiarid regions because germination without sufficient water for growth would quickly kill the plants. In many plants, ABA seems to act as the growth inhibitor. The seeds of such annuals remain dormant in parched soil until a downpour washes ABA out of the seeds, allowing them to germinate. For example, these dune primroses (white) and desert sunflowers (yellow), photographed in the Anza-Borrego Desert in California, grew from seeds that germinated just after a hard rain.

As we saw in the previous module, gibberellins also participate in seed germination. For many plants, the ratio of ABA to gibberellins, which promote germination, determines whether the seed will remain dormant or germinate. Similarly, the dormancy of buds is controlled more by a balance of these growth regulators than by their absolute concentrations. In apple trees, for example, the ABA concentration is actually higher in growing buds than in dormant buds, but a high level of gibberellins in the growing buds overpowers the inhibitory effect of ABA.

In addition to its role in dormancy, ABA acts as a "stress hormone" in growing plants, helping them cope with adverse conditions. For instance, when a plant is dehydrated, ABA accumulates in the leaves and causes stomata to close. This reduces transpiration and prevents further water loss. However, it also reduces the plant's rate of sugar production by photosynthesis.

The Anza-Borrego Desert blooming after a rain

Early in this century, oranges and grapefruits were ripened for market in sheds equipped with kerosene stoves. Fruit growers thought it was the heat that ripened the fruit, but when they tried newer, cleaner-burning stoves, the fruit did not ripen fast enough. Plant biologists learned later that ripening in the sheds was actually due to **ethylene**, a gaseous by-product of kerosene combustion. We now know that plants produce their own ethylene, which functions as a hormone that triggers a variety of aging responses, including fruit ripening.

Fruit ripening involves the breakdown of cell walls, changes in color (often from green to yellow or red), and sometimes drying. These are aging processes triggered by ethylene, which is produced by all the cells of a plant. As fruits age, they produce more and more ethylene, thus hastening the ripening process. As a gas, ethylene diffuses through the fruit in the air spaces between cells. It also diffuses through the air from fruit to fruit, and, as a result, if one apple in a bin starts to spoil, it really can spoil the whole lot. You can make some fruits ripen faster if you store them in a plastic bag so that ethylene gas can accumulate. Figure A shows the results of a demonstration with bananas. Three unripe bananas were stored in plastic bags: (1) with an ethylene-releasing orange, (2) with a beaker of an ethylene-releasing chemical, and (3) alone (the control). As you can see, the more ethylene present, the riper the banana. On a commercial scale, many kinds of fruit—tomatoes, for instance—are often picked green and then partially ripened in huge storage bins into which ethylene gas is piped—a modern variation on the old storage shed.

In other cases, growers take measures to *retard* the ripening action of natural ethylene. Stored apples are often flushed with CO_2, which inhibits the action of ethylene. Also, gas is circulated around the apples to prevent ethylene from accumulating. In this way, apples picked in autumn can be stored for shipping to grocery stores the following summer.

Like fruit ripening, the changes that occur in deciduous trees each autumn—color changes, drying, and the loss of leaves—are also aging processes. Fall colors result from a combination of new pigments made in autumn and pigments that were already present in the leaf but masked by the dark-green chlorophyll. Leaves lose their green color in autumn because they stop making chlorophyll. Ethylene probably plays a role in fall color changes and in leaf drying, but little is known about its effects. Much more is known about how ethylene promotes the loss of leaves (abscission).

When an autumn leaf falls, the base of the leaf stalk separates from the stem. The separation region is called the abscission layer. As shown in Figure B, the abscission layer consists of a narrow band of small parenchyma cells with thin, weak walls. The leaf drops off when its weight, often helped by wind, splits the abscission layer apart. Notice the layer of protective cells adjacent to the abscission layer. Even before the leaf falls, these cells form a leaf scar on the stem. Dead cells covering the scar help protect the plant from infectious organisms.

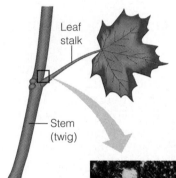

Leaf stalk

Stem (twig)

LM 28×

Stem   Protective layer   Abscission layer   Leaf stalk

B. Abscission layer at the base of a leaf

A. The effect of ethylene on the ripening of bananas

Leaf drop is triggered by environmental stimuli, the shortening days and cooler temperatures of autumn. These stimuli apparently cause a change in the balance of ethylene and auxin—another case of antagonism between plant hormones. The auxin prevents abscission and helps maintain the leaf's metabolism, but as a leaf ages, it produces less and less auxin. Meanwhile, cells in the abscission layer begin producing ethylene. The ethylene primes the abscission layer to split by promoting the synthesis of enzymes that digest cell walls in the layer. Autumn leaf drop is an adaptation that helps keep the tree from drying out in winter. Without its leaves, a tree loses less water by evaporation when its roots cannot soak up water from the frozen ground.

We have now completed our survey of the five types of plant hormones. Before moving on to the topic of plant behavior, let's take a closer look at the practical importance of these chemical regulators.

Plant hormones have many agricultural uses 33.8

Much of what we know about plant hormones—and there is much more to be learned—has a direct application to agriculture. As already mentioned, the control of fruit ripening and the production of seedless fruits are two of several major uses of these chemicals. Plant hormones also allow farmers to control when plants will drop their fruit. For instance, auxins are often used to prevent orange and grapefruit trees from dropping their fruit before they can be picked. The photograph shows a citrus grower in Florida spraying an orange grove with auxins. The quantity of auxin must be carefully monitored because too much of the hormone may stimulate the plant to release more ethylene, making the fruit ripen and drop off sooner.

Large doses of auxins are often used intentionally to promote premature fruit drop. For example, auxins may be sprayed on apple and olive trees to thin the developing fruits; the remaining fruits will grow larger. Ethylene is used to thin peaches and prunes, and it is sometimes sprayed on berries, grapes, and cherries to loosen the fruit so it can be picked by machines.

Using the hormone auxin to prevent early fruit drop

Gibberellins are very useful for producing seedless fruits, as mentioned earlier. They are also important in commercial seed production. A large dose of gibberellins will induce many biennial plants, such as carrots, beets, and cabbage, to flower and produce seeds during their first year of growth. Ordinarily, biennials will not produce seeds until their second year of growth.

Research on plant hormones has had other spin-offs. One of the most widely used herbicides, or weed killers, is 2,4-D, a synthetic auxin that disrupts the normal balance of hormones that regulate plant growth. Because dicots are more sensitive than monocots to this herbicide, 2,4-D can be used to selectively remove dandelions and other "broadleaf" dicot weeds from a lawn or grainfield. By applying herbicides to cropland, a farmer can reduce the amount of tillage required to control weeds, thus reducing soil erosion, fuel consumption, and labor costs.

Modern agriculture relies heavily on the use of synthetic chemicals. Without chemically synthesized herbicides to control weeds and synthetic plant hormones to help grow and preserve fruits, less food would be produced, and food prices could increase considerably. At the same time, there is growing concern that the heavy use of artificial chemicals in food production may pose environmental and health hazards. A chemical called dioxin, for example, is a by-product of 2,4-D synthesis. Though 2,4-D itself does not appear to be toxic to mammals, dioxin causes birth defects, liver disease, and leukemia in laboratory animals. Therefore, dioxin is a serious hazard when it leaks into the environment. Also, many consumers are concerned that foods produced with artificial help may not be as tasty or nutritious as those raised more naturally. At present, however, "natural" foods are relatively expensive to produce. As we discussed in Module 32.10, these issues involve both economics and ethics: Should we continue to produce cheap, plentiful food using artificial chemicals and tolerate the potential problems, or should we put more of our agricultural effort into farming without these potentially harmful substances, recognizing that foods may be less plentiful and more expensive as a result?

Tropisms tune plant growth to the environment

Having discussed the hormones that carry signals within a plant, we now shift our focus to the responses of plants to physical stimuli from the environment. In the chapter's introduction, we discussed the rapid leaf movements and sleep activities of the sensitive plant, *Mimosa*. These dramatic movements result from changes in turgor in specialized cells. Less obvious because they occur so slowly that they can be observed only by time-lapse photography are plant movements called tropisms. **Tropisms** are growth responses that change the shape of a plant or make it grow toward or away from a stimulus. Phototropism, the growth of a plant shoot toward light, is one type of tropism. Two other types are gravitropism, a response to gravity, and thigmotropism, a response to touch.

As we saw in Module 33.1, the mechanism for phototropism is a differential rate of cell elongation on opposite sides of a stem. Cells on the darker side of a stem elongate faster than those on the brighter side because of an unequal distribution of auxin. Experiments have shown that illuminating a grass shoot from one side causes auxin to migrate across the tip from the bright side to the dark side. It is not yet known how light induces this effect, although grass shoot tips contain a light-sensitive pigment that may play a key role.

A plant's growth response to gravity, **gravitropism,** is illustrated by the corn seedlings in Figure A. These seedlings were both germinated in the dark. The one on the left was germinated so that the shoot would grow straight up and the root straight down. The seedling on the right was germinated in the same way, but two days later it was turned on its side so that the shoot and root were horizontal. By the time the photograph was taken, the shoot had turned back upward, exhibiting a negative response to gravity, and the root had turned down, exhibiting positive gravitropism.

Plant biologists have yet to work out how plants tell up from down. One idea is that gravity pulls special organelles containing dense starch grains to the low points of cells. The uneven distribution of organelles may in turn signal the cells to redistribute auxin. The differential concentrations of auxins may then control the direction of growth. Whatever the underlying mechanisms, gravitropism is an important adaptation, making the shoot of a germinating plant grow toward light and the root grow into the soil no matter how the seed lands or is planted in the soil.

Growth movement in response to touch, **thigmotropism** (Greek *thigma*, touch), is illustrated in Figure B by the tendril of a green bean plant coiling around a wire fence. The tendril (actually, a modified leaf) grew straight until it touched the support. Contact then stimulated the differential growth of cells on opposite sides of the tendril, making it coil around the wire. Most vines and other climbing plants have tendrils that respond by coiling and grasping when they touch rigid objects. Thigmotropism enables these plants to use outside objects for support while growing toward sunlight.

Tropisms all have one function in common: They help plant growth stay in tune with the environment. In the next module, we see that plants also have a way of keeping time with their environment.

A. Gravitropism

B. Thigmotropism

Your pulse rate, blood pressure, body temperature, rate of cell division, blood cell count, alertness, urine composition, metabolic rate, sex drive, and responsiveness to drugs all fluctuate rhythmically with the time of day. Plants also display rhythmic behavior; examples include the opening and closing of stomata (see Module 32.4) and the sleep movements of *Mimosa pudica* and many other species.

A biological cycle of about 24 hours is called a **circadian rhythm** (from the Latin *circa*, about, and *dies*, day). A circadian rhythm is not merely a daily response to environmental cycles, such as the rotation of Earth that causes day and night. These rhythms persist even when an organism is sheltered from environmental cues. A *Mimosa*, for example, will exhibit sleep movements at about the same intervals even if it is kept in constant light or constant darkness. In fact, plants and animals continue their daily rhythms when placed in the deepest mine shafts or when orbiting in satellites. Thus, circadian rhythms are programmed to fit normal environmental conditions—that is, day and night cycles—and they occur with or without external stimuli such as sunrise and sunset. All research thus far indicates that circadian rhythms are controlled by internal timekeepers called **biological clocks.**

A biological clock continues to mark time in the absence of environmental cues, but to remain tuned to a period of exactly 24 hours, it requires daily signals from the environment. If an organism is kept in a constant environment, its circadian rhythms will deviate somewhat from a 24-hour period. Consider bean plants, for instance, which have sleep movements similar to those of the sensitive plant. As shown in the photographs, the leaves of a bean plant are held horizontally at noon, but are folded downward at midnight. When the plant is held in constant darkness, its sleep movements change to a cycle of about 26 hours.

Deviation of the cycle from 24 hours does not mean that biological clocks drift erratically. The clocks are still keeping time; they are just not synchronized precisely with the outside world. The light-dark cycle due to Earth's rotation is the most common factor keeping biological clocks synchronized with external light conditions. If the timing of these cues changes, it takes a few days for a clock to be reset. Thus, a plant kept for several days in the dark will be out of phase with plants growing in the normal environment where the sun rises and sets each day. The same sort of thing happens when we cross several time zones in an airplane: When we reach our destination, our internal clock is not synchronized with the clocks on the wall. Moving a plant across several time zones produces a similar response. In the case of either the plant or the human traveler, resetting the clock usually takes several days.

Noon

Midnight

Sleep movements of a bean plant

There are many more questions than answers about biological clocks. We do not yet know exactly what biological clocks are, where in an organism they are found, or how they work. Most scientists agree that the clocks are cellular mechanisms located either in membranes or in the machinery for protein synthesis. Wherever they are and however they work, biological clocks and the circadian rhythms they control are affected little by temperature. In this way they differ from most metabolic processes, which are generally sensitive to temperature changes. Somehow, a biological clock compensates for temperature shifts. This adjustment is essential, for a clock that speeds up or slows down with the rise and fall of outside temperature would be an unreliable timepiece.

In attempting to answer questions about biological clocks, it is essential to distinguish between the clock and the rhythmic processes it controls. You could think of the sleep movements of leaves as the "hands" of the biological clock, but they are not the essence of the clockwork itself. You can restrain the leaves of a bean plant for several hours so that they cannot move. But on release, they will rush to the position appropriate for the time of day. Thus, we can interfere with an organism's rhythmic activity, but its biological clock goes right on ticking off the time.

Plants mark the seasons by measuring photoperiod

A biological clock not only times a plant's everyday activities, it may also influence seasonal events that are important in a plant's life cycle. Flowering, seed germination, and the onset and ending of dormancy are examples of stages in plant development that usually occur at specific times of the year. The environmental stimulus plants most often use to detect the time of year is called **photoperiod,** the relative lengths of day and night. In this module, we see how photoperiod affects flowering in certain plants. In the next module, we will focus on a mechanism that may set the biological clock and help the plant keep track of photoperiod.

Plants whose flowering is triggered by photoperiod fall into two groups. One group, the **short-day plants,** generally flower in late summer, fall, or winter, when light periods shorten. Chrysanthemums and poinsettias are examples of short-day plants. In contrast, **long-day plants,** such as spinach, lettuce, iris, and many cereal grains, usually flower in late spring or early summer, when light periods lengthen. Spinach, for instance, flowers only when daylight lasts at least 14 hours.

In the 1940s, researchers discovered that flowering and other responses to photoperiod are actually controlled by *night* length, not day length. In fact, the so-called short-day plants are actually long-night plants, and the so-called long-day plants are actually short-night plants. Unfortunately, the day-length terms are embedded firmly in the literature of plant biology.

The figure below illustrates the evidence for the night-length effect and also shows the difference between the flowering response of a short-day plant and a long-day plant. The left side of the figure represents short-day plants. The first two bars and corresponding plant drawings show that a short-day plant will not flower until it is exposed to a continuous dark period exceeding a critical length (about 10 hours, in this case). The continuity of darkness is important. The short-day plant will not blossom if the nighttime part of the photoperiod is interrupted by even a flash of light (third bar). (There is no effect if the daytime portion of the photoperiod is broken by a brief exposure to darkness.)

Florists apply this information about short-day plants to bring us flowers out of season. Chrysanthemums, for instance, are short-day plants that normally bloom in the autumn, but their blooming can be stalled until spring by punctuating each long night with a flash of light, thus turning one long night into two short nights.

The right side of the figure demonstrates the effect of night length on a long-day plant. In this case, flowering occurs when the night length is *shorter* than a critical length (less than 10 hours, in this example). As the third bar shows, flowering can be induced in a long-day plant by a flash of light during the night.

Notice that we distinguish long-day from short-day plants not by an *absolute* night length but by whether the critical night length sets a minimum (short-day plants) or maximum (long-day plants) number of hours of uninterrupted darkness required for flowering. The actual length of the critical period varies from species to species.

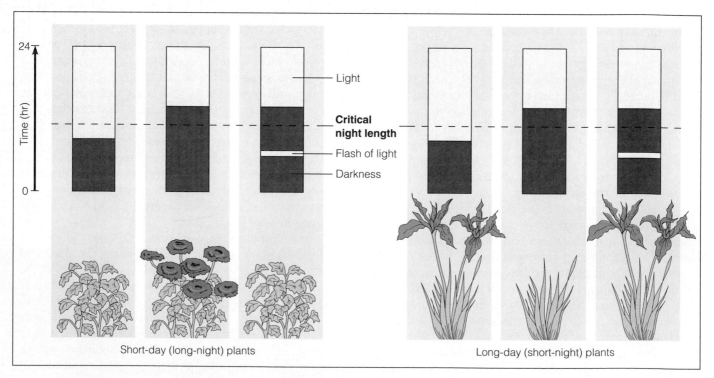

Photoperiodic control of flowering

Phytochrome is a light detector that may help set the biological clock

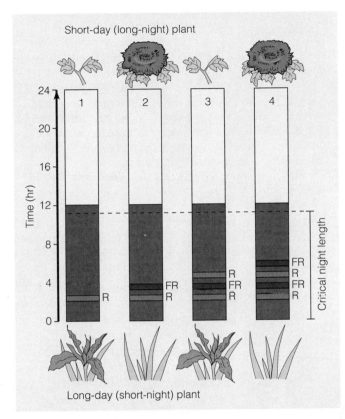

Short-day (long-night) plant

Long-day (short-night) plant

A. The reversible effects of red and far-red light

The discovery that photoperiod (specifically night length) determines seasonal responses of plants poses another question: How does a plant actually measure photoperiod? Much remains to be learned about this, but a pigment named phytochrome is part of the answer. **Phytochrome** is a colored protein that absorbs light.

Before we look at the role of phytochrome, let's see how two different wavelengths of light affect flowering in short-day and long-day plants. Bar 1 in Figure A repeats the effect we saw in the previous module for both short-day and long-day plants that receive a flash of light during their critical dark period. The letter R on the light flash stands for red light with a wavelength of 660 nanometers (nm). Researchers have discovered that this type of light (which is one component of white daylight) is the most effective wavelength for interrupting night length.

The other three bars in Figure A make some additional points about the effects of different wavelengths of light on flowering. FR (far-red) stands for light having a wavelength of about 730 nm. This wavelength is in the far-red part of the spectrum, just barely visible to humans. As bar 2 shows, the shortening of critical night length by a flash of red light (R) can be prevented by a subsequent flash of FR

light: Both types of plants behave as though there is no interruption in the night length. Bars 3 and 4 indicate that no matter how many flashes of light a plant receives, only the wavelength of the last flash will affect the plant's measurement of night length. Thus, the sequence R-FR-R produces the same results as in bar 1, and the sequence R-FR-R-FR yields the same effect as in bar 2.

Now we're ready to consider the role of phytochrome. The light-absorbing portion of this pigment molecule is what actually enables a plant to detect light. Phytochrome alternates between two forms that differ only slightly in structure. One form absorbs red light and the other absorbs far-red light, and this difference accounts for the reversible effects of red and far-red light on a plant.

The two forms of phytochrome are designated P_r (red-absorbing) and P_{fr} (far-red-absorbing). As indicated in Figure B, the conversion from one form to the other is caused by two different wavelengths of light. When the P_r form absorbs red light (660 nm), it is quickly converted to P_{fr}, and when P_{fr} absorbs far-red light (730 nm), it is converted back to P_r. A plant synthesizes phytochrome as P_r, and, if the plant is kept in the dark, the pigment remains in this form. Also, any P_{fr} in a plant reverts to P_r in the dark. The P_{fr} to P_r conversion occurs each day after sunset. At sunrise, the P_r form of phytochrome is rapidly converted to P_{fr} because sunlight is richer in red than in far-red light. Apparently, these phytochrome conversions are cues that set a plant's biological clock. The clock may measure the time between the beginning of the P_{fr} to P_r change (sunset) and the P_r to P_{fr} change (sunrise). In doing so, it monitors photoperiod and becomes synchronized to the actual length of a day.

Found in all species of plants and algae examined to date, the phytochrome system may play a universal role in helping photosynthetic organisms stay synchronized with seasonal changes in light conditions. Also, as indicated in Figure B, its P_{fr} component seems to trigger certain physiological responses, such as seed germination, flowering, and stomatal opening.

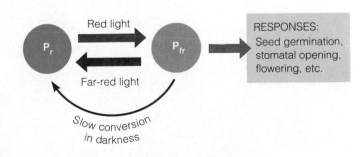

Red light

Far-red light

Slow conversion in darkness

RESPONSES: Seed germination, stomatal opening, flowering, etc.

B. Interconversion of the two forms of phytochrome

33.13

Plant scientist Ruth Satter tells of a life in research and teaching

Photoperiodism and biological clocks provide a fitting finale to this chapter and to our unit on plant biology, for they integrate what we have learned about internal and external stimuli that control plant growth and development. The person who first called the daily sleep movements of plant leaves the "hands of the biological clock" was the plant scientist Dr. Ruth Satter. Until her untimely death in 1989, following a long battle with leukemia, Satter was a leader in research on how organisms measure time and control rhythms.

During her research career at Yale University and the University of Connecticut, Satter made many breakthrough discoveries about biological clocks and seasonal behavior in plants. By monitoring leaf movements, she demonstrated how sunrise and sunset synchronize the plant's internal clock with the external environment. Satter also recognized the importance of basic plant research to agriculture. As she put it in a 1988 interview:

> Although our study of timekeeping processes seems very theoretical, it has a practical application. A large number of processes in many plants occur at close to the same date each year. These include the time at which a plant switches from vegetative growth to flowering. Spinach, for instance, is a plant that grows vegetatively in the spring and flowers during the long days of summer. Knowing this, one can manipulate day length artificially and grow several crops during the year instead of just growing one. Understanding how the plant measures time should have additional practical benefits, possibly [allowing

A. Ruth Satter in 1988

us] to alter the plant's photoperiodic requirements by genetic engineering.

Dr. Satter's career in plant research developed from an early interest in gardening.

> My grandmother stimulated my interest when I was a child. She was a natural experimentalist, and her gardening consisted of taking dried lima beans purchased to make soup and saving some of them to plant out in the yard. Or taking potatoes that had sprouted, and cutting out the sprouts and planting them. I worked at these experiments with her and was fascinated by them.

Starting in college, however, Satter took a detour from plants. At Barnard College, she majored in mathematics and physics, and following graduation, near the end of World War II, she joined a research team working on radar development. After the war, she married and had four children. The inadequate child-care facilities of the time gave her little opportunity to pursue a scientific career while her children were young.

But after a break of about 15 years, Satter returned to school and she entered a doctoral program in biology.

> Going to graduate school was a real privilege and something that I had looked forward to. It added a whole new dimension to my life. I was much older than most of my peers, and . . . some of the younger students were not particularly friendly toward the older students. Returning to school was not an accepted thing to do at that time.

There was also a gender gap, with far fewer women than men in science. In Satter's words:

> A generation or two ago it was very difficult for a woman to have a career in science. Opportunities were very limited, and science in many institutions was a male enclave. The gender gap is much smaller today than it was a decade ago. But there are [still] a few basic problems. Barriers are breaking down, but it will take a while before the break is complete. There are not enough older women in science to act as role models. I think that the situation will continue to improve, resulting in a more comfortable environment for women.

B. Satter with her research group at the University of Connecticut (1987)

Though strongly oriented toward research and training graduate students, Ruth Satter also liked to involve undergraduate students in her work. Her advice to those with a particular interest in some aspect of science:

Go see the faculty members who are working in this area. Visit their laboratories, express your interest, and ask if you can do any work in the laboratory, even if it is washing glassware. While there, you will come into contact with people working in the lab and will have the opportunity to talk with them about their work. Express an interest and an enthusiasm. I think that relatively few faculty members will turn down an enthusiastic student. In fact, they will encourage you to become more involved.

This completes our unit on plant biology. Our central theme has been the connection between structural adaptation and function at the chemical, cellular, tissue, and organ levels. In our next and final unit, we pursue connections at a different level, that of organisms interacting in the environment.

Begin your review by rereading the module headings and scanning the figures before proceeding to the Chapter Summary and questions.

Chapter Summary

Introduction—33.1 Hormones coordinate the activities of plant cells and tissues. The study of plant hormones began with observations of plants bending toward light, a phenomenon called phototropism. This bending involves faster cell growth on the shaded side of the shoot than on the lighted side. Experiments carried out by Charles Darwin and others showed that the tip of a shoot detects light and transmits a signal down to the growing region of the shoot. This signal is a hormone named auxin; it accumulates on the dark side of the shoot and stimulates growth some distance below the tip.

33.2 Auxins are now known to be a class of chemicals, one of five types of plant hormones. These hormones are produced in very small amounts. They affect gene expression, enzyme activity, and membrane permeability. Thus, they shape plant growth, development, and metabolism.

33.3 Auxins are produced in the apical meristems at the tips of shoots. One auxin is IAA. At different concentrations, IAA stimulates or inhibits the elongation of shoots and roots. It may act by weakening cell walls, allowing them to stretch when cells take on water. Auxins also stimulate the development of vascular tissues and cell division in vascular cambium, promoting growth in stem diameter. Farmers use auxins to produce seedless fruits and promote fruit drop. A synthetic auxin called 2,4-D is used to kill weeds. The application of plant hormones increases productivity, but in some cases there are questions about the safety of using such artificial chemicals.

33.4 Cytokinins, produced by growing roots, embryos, and fruits, promote cell division. Cytokinins from roots may act to balance the effects of auxins from apical meristems, causing lower buds to develop into branches. Thus, the ratio of auxins to cytokinins may coordinate the growth of roots and shoots.

33.5 Gibberellins, produced by the tips of stems and roots, stimulate the elongation of stems and leaves and the development of fruit. When applied to dwarf plants, they cause stem elongation. They also induce some plants to flower and produce seeds early, or to develop seedless fruits without fertilization. Gibberellins released from embryos function in some of the early events of seed germination.

33.6 Abscisic acid (ABA) inhibits the germination of seeds. The ratio of ABA to gibberellins often determines whether a seed will remain dormant or germinate. Seeds of many desert plants remain dormant until rain washes away their ABA. During winter or drought, ABA from buds inhibits primary and secondary growth in buds and vascular cambium. ABA also acts as a "stress hormone," causing stomata to close when a plant is dehydrated.

33.7 As fruit cells age, they give off ethylene gas, which hastens ripening. A changing ratio of auxin to ethylene, triggered by shorter, cooler days, probably causes autumn color changes and the loss of leaves from deciduous trees. Fruit growers use ethylene to control ripening.

33.9 Plants sense and respond to environmental changes in a variety of ways, known as tropisms. Phototropism, bending toward light, seems to be caused by the migration of auxin from the light side to the dark side of a stem. A response to gravity, or gravitropism, may be caused by a buildup of special organelles on the low sides of shoots and roots. Thigmotropism, a response to touch, seems to be responsible for the coiling of tendrils and vines around objects.

33.10 An internal biological clock controls sleep movements and other daily cycles in plants. These cycles, called circadian rhythms, have periods of about 24 hours and persist even in the absence of environmental cues. The exact mechanisms and locations of biological clocks are not known.

33.11—33.12 Plants mark the seasons by measuring photoperiod, the relative lengths of night and day. The timing of flowering is one of the seasonal responses to photoperiod. Short-day plants flower when nights exceed a certain critical length; long-day plants when nights are shorter than a critical length. A light-absorbing protein called phytochrome enables plants to monitor photoperiod. The conversion of phytochrome to alternate chemical forms may play an important role in the functioning of a plant's biological clock.

Testing Your Knowledge

Multiple Choice

1. During winter or periods of drought, which of the following plant hormones inhibits growth and seed germination?

    a. ethylene
    b. abscisic acid
    c. gibberellin
    d. auxin
    e. cytokinin

2. A certain short-day plant flowers when days are less than 12 hours long. Which of the following would cause it to flower? *(Explain your answer.)*

    a. a 9-hr night and 15-hr day with 1 min of darkness after 7 hr
    b. an 8-hr day and 16-hr night with a flash of white light after 8 hr
    c. a 13-hr night and 11-hr day with 1 min of darkness after 6 hr
    d. a 12-hr day and 12-hr night with a flash of red light after 6 hr
    e. alternating 4-hour periods of light and darkness

3. Auxins cause a shoot to bend toward light by

   a. causing cells to shrink on the dark side of the shoot
   b. stimulating growth on the dark side of the shoot
   c. causing cells to shrink on the lighted side of the shoot
   d. stimulating growth on the lighted side of the shoot
   e. inhibiting growth on the dark side of the shoot

4. In the autumn, the amount of _____ increases and _____ decreases in fruit and leaf stalks, causing a plant to drop fruit and leaves.

   a. ethylene . . . auxin
   b. gibberellin . . . abscisic acid
   c. cytokinin . . . abscisic acid
   d. auxin . . . ethylene
   e. gibberellin . . . auxin

5. Plant hormones act by affecting the activities of

   a. genes
   b. membranes
   c. enzymes
   d. genes, membranes, and enzymes
   e. genes and enzymes

6. Buds and sprouts often form on tree stumps. Which of the following hormones would you expect to stimulate their formation?

   a. auxin
   b. cytokinins
   c. abscisic acid
   d. ethylene
   e. gibberellins

True/False (*Change false statements to make them true.*)

1. If a plant is exposed to constant light, sleep movements cease.

2. Day length controls flowering and other responses related to photoperiod.

3. Shoots show negative gravitropism.

4. The biological clock is known to be located in terminal buds.

5. It is possible to get some seeds to germinate by spraying them with gibberellins.

6. A hormone that stimulates growth at one concentration may inhibit growth at a different concentration.

7. There are many kinds of auxins and gibberellins that affect plant growth.

Matching

1. Bending of a shoot toward light	**a.** circadian rhythm
2. Where a leaf separates from the stem	**b.** phytochrome
3. Growth response to touch	**c.** photoperiod
4. A cycle with a period of about 24 hours	**d.** sleep movement
5. Pigment that helps control flowering	**e.** abscission layer
6. Relative lengths of night and day	**f.** thigmotropism
7. Growth response to gravity	**g.** phototropism
8. Folding of plant leaves at night	**h.** gravitropism

Describing, Comparing, and Explaining

1. If apples are to be stored for long periods, it is best to keep them in a place with good air circulation. Explain why.

2. Write a short paragraph explaining why a houseplant becomes more bushy if you pinch off the terminal bud.

3. Explain the roles of plant hormones in the dormancy of buds and seeds.

Thinking Critically

1. Jon just started a new job as night watchman at a flower nursery. His boss told him to stay out of a room where chrysanthemums (which are short-day plants) were about to flower. Looking for the rest room, Jon accidentally opened the door to the chrysanthemum room and turned on the lights for a moment, at just about midnight. How might this affect the chrysanthemums? What could Jon do to correct his mistake?

2. Lauren noticed that dandelions in her yard opened each morning and closed each evening. Describe an experiment that would determine whether this daily activity is controlled by an internal biological clock and not just by the presence or absence of light.

3. A plant biologist observed a peculiar pattern when a tropical shrub was attacked by caterpillars. He noticed that a caterpillar might start eating any leaf on a stem. Once that leaf was eaten, it would skip over nearby leaves and attack a leaf some distance away. The researcher found that when a leaf was eaten, nearby leaves started making a chemical that deterred the caterpillars. Simply removing a leaf did not trigger the same change nearby. The biologist suspected that a damaged leaf sent out a hormone that signaled other leaves. What kinds of experiments could he perform to find out whether his guess is correct?

4. Space Shuttle astronauts are now able to perform animal and plant experiments in orbit, free of Earth's gravity and its 24-hour period of rotation. What phenomena discussed in this chapter might be worth investigating in space? Why? Describe how such an experiment might be carried out. What might your experiment demonstrate?

5. You may have noticed that evergreen trees growing in the open often have limbs all the way down to the ground, but the same kind of tree in a dense forest might have a bare trunk, with its lowest limbs 20–30 m above the ground. Can you suggest a hormonal mechanism that might cause lower limbs to die and break off in this situation? Of what value might this be to the tree?

Science, Technology, and Society

1. Imagine the following scenario: A plant scientist has developed a synthetic chemical that mimics the effects of a plant hormone. The chemical can be sprayed on apples before harvest to prevent flaking of the natural wax that is formed on the skin. This makes the apples shinier and gives them a deeper red color. What kinds of questions do you think should be answered before farmers start using this chemical on apples? How might the scientist go about finding answers to these questions?

2. Daminozide, known commercially as Alar, is a synthetic hormone used to keep fruit firm. Because there is some evidence that Alar can cause cancer, its use was recently phased out. There is concern about the health effects of dioxin, a contaminant in the herbicide 2,4-D. Are you concerned about possible harmful chemicals in the food you eat? What are the benefits and costs of using artificial chemicals on food crops? What are the benefits and costs of discontinuing their use? Where can you obtain information about the safety or harmfulness of the herbicides, hormones, and other chemicals used on food? How can you act on this information?

3. Recombinant DNA techniques can be used to alter crop plants to make them more resistant to herbicides such as 2,4-D, so that the herbicides can be used to "weed" crops without harming them. Not everyone thinks this is a good idea. What do you think? Why?

Unit 6
Ecology

The Biosphere: An Introduction to Earth's Diverse Environments

34

I magine finding yourself in a strange, dark environment full of species never before seen by humans—species whose ultimate energy source is not sunlight, but energy from the molten interior of their planet. Sound like science fiction? It actually happened to scientists in the battery-powered deep-sea research vessel *Alvin,* pictured at the left.

Alvin can withstand water pressures that would flatten an ordinary submarine. Accommodating a pilot and two other people, and equipped with spotlights and sampling arms, it has carried scientists down some 2500 m, over a mile deeper than sunlight penetrates. In the late 1970s, diving off the southern tip of Baja California in Mexico, scientists in *Alvin* discovered seafloor life that does not depend on photosynthesis. They found new organisms inhabiting the unique world of hydrothermal vents, sites near the adjoining edges of giant plates of Earth's crust where molten rock and hot gases surge upward from Earth's interior. Figure A shows one such site, where a chimneylike vent, perhaps as much as 30 m high, emits scalding water and hot gases such as hydrogen sulfide. The water is at 350°C but is kept from boiling by immense water pressures. Just a few meters from the vents, the water temperature is only about 2°C. This is truly an environment of extremes.

A variety of animals, including sea anemones, giant clams (30 cm long), shrimps, crabs, a few fishes, and tube worms, thrive near hydrothermal vents. In Figure B, a crab crawls over a dense growth of 3-m-long tube worms. The worms were unknown to science until hydrothermal vents were explored. The bright red head of each worm projects from a rigid tube anchored in the seafloor. These and other vent animals live on energy extracted from chemicals by bacteria, rather than on light energy trapped by plants, algae, or photosynthetic bacteria, as other animals do.

The bacteria in hydrothermal vents also have an unusual existence. Living in crevices among the hot rocks, most of them obtain energy by oxidizing hydrogen sulfide (H_2S) to sulfates (SO_4^{2-}) or elemental sulfur (S). These so-called

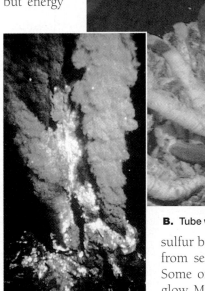

A. A hydrothermal vent

B. Tube worms

sulfur bacteria use the energy to convert CO_2 from seawater into organic food molecules. Some of the vents give off a dull, yellowish glow. Might some of the bacteria use the faint light to make food by photosynthesis? Scientists are eager to find out.

Many of the animals in vent communities obtain nutrients by eating bacteria. Others, such as the giant tube worms, harbor sulfur bacteria within their bodies. The worms absorb sulfur compounds from the water; the bacteria use the compounds as an energy source and make organic food molecules.

The unusual organisms living at hydrothermal vents indicate how diverse life is on our planet. Until the *Alvin* expeditions, no one knew that entire communities of organisms could live on energy from Earth itself. Scientists have since discovered hydrothermal vents at several locations in the Pacific and Atlantic oceans.

The scientific study of the interactions of organisms with their environments—from hydrothermal vents to the more typical solar powered environments we depend on—is called **ecology** (from the Greek *oikos,* home). As we will see in this chapter and those that follow, ecology is a critically important field of biology, with implications for all forms of life on Earth.

Ecologists study how organisms interact with their environments at several levels

Research on giant clams near an ocean vent

The definition of ecology just given sounds straightforward, but it covers an enormously complex area of biology. The interactions between organisms and their environment are two-way. Organisms are affected by their environment, but, by their very presence and activities, they also change the environment, often dramatically. In transforming energy, bacteria change the ocean vent environment and make it habitable for animals. Likewise, plants drastically alter their environments, extracting CO_2 from the atmosphere and adding O_2 to it, in the process providing themselves and animals with food.

Ecologists study environmental interactions at several levels. At the organism level, they may examine how one kind of organism meets the challenges of its environment. An ecologist working at this level might study, for instance, the adaptations of clams to the extreme temperatures around hydrothermal vents.

Another level of study in ecology is the **population,** an interbreeding group of individuals belonging to the same species and living in a particular geographic area. The clams of one species living near a particular ocean vent would constitute a population. One ecologist might study a clam population's rate of growth relative to the temperature of the surrounding water. Someone else, interested in clam evolution, might look at genetic differences among populations from different vents.

A third level, the **community,** consists of all the organisms—that is, all the populations of different species—that inhabit a particular area. For example, all the organisms supported by a particular hydrothermal vent would constitute a community. An ecologist working at this level might focus on interactions among organisms, such as the effect of predation by crabs on tube worms or clams.

The fourth level of ecological study, the **ecosystem,** includes all the life forms existing in a certain area and all the nonliving factors as well. The nonliving factors, or **abiotic** factors, include temperature, forms of energy, gases, water, nutrients, and other chemicals. The organisms making up the community of species in the area are called **biotic** factors. Some critical questions at the ecosystem level concern how chemicals cycle, and how energy flows between organisms and their surroundings. For a vent community, one ecosystem-level question would be: How much of the energy available to them do the giant clams and other animals actually use?

Ecological research at any level employs the usual methods of science: observation, hypothesis, prediction, and testing. However, in contrast to the laboratory, where conditions can be simplified and experiments can be precisely controlled, environments are very complex. As we saw in Module 1.5, testing ecological hypotheses usually requires adapting the idealized model of science to the multiple variables of the environment.

The biosphere is the total of all of Earth's ecosystems

Recalling a view of Earth from space (as in Figure A), Apollo astronaut Rusty Schweickart once remarked: "On that small blue-and-white planet below is everything that means anything to you. National boundaries and human artifacts no longer seem real. Only the biosphere, whole and home of life."

What exactly is the biosphere? The **biosphere** is the global ecosystem—that portion of Earth that is alive, or all of life (at least as we know it) and where it lives. The most complex level in ecology, the biosphere includes the atmosphere to an altitude of only a few kilometers, the land to a soil depth of a few meters, lakes and streams, and the ocean to a depth of

A. Earth as seen from the moon

several kilometers. Isolated in space, the biosphere is self-contained, or closed, except that its photosynthesizers derive energy from sunlight, and it loses heat to space.

Another feature of the biosphere is its patchiness, and we can see this on several levels. On a global scale, we see it in the distribution of continents and oceans. On a regional scale, patchiness occurs in the distribution of deserts, grasslands, forests, lakes, and streams, for example. The aerial view of a wilderness area in Figure B shows patchiness on a local scale. Here we see a mixture of forest, small lakes, a meandering river, and open meadows. If we moved even closer, into any one of these different environments, we would find patchiness on yet a smaller scale. For example, we would find that each lake has several different **habitats** (places where organisms live), each with a characteristic community of organisms. Abiotic factors, especially water depth, temperature, and dissolved O_2, largely determine the kinds of organisms that live in the different lake habitats.

Standing in a wilderness can be misleading; the lakes and streams appear untouched, and the forest seems almost boundless. Views from space are more sobering, for they

B. Patchiness of the environment in the Alaskan wilderness

show planet Earth as only a small sphere in the vastness of the universe. Unfortunately, we humans tend to treat the biosphere as an unlimited resource for our own consumption.

Environmental problems reveal the limits of the biosphere 34.3

Our current awareness of the biosphere's limits stems mainly from the 1960s, a time of growing disillusionment with environmental practices of the past. In the 1950s, technology seemed poised to free humankind from several age-old bonds. New chemical fertilizers and pesticides, for example, showed great promise for increasing agricultural productivity and eliminating insect-borne diseases. Fertilizers were applied extensively, and pests were attacked by the use of massive aerial sprays. The immediate results were astonishing: Increases in farm productivity allowed developed nations such as the United States to grow surplus food and market it overseas, and the worldwide incidence of malaria and several other insect-borne diseases was markedly reduced. DDT and other chemicals were hailed as miracle weapons with potential use anywhere insects caused problems.

Our enthusiasm for chemical fertilizers and pesticides began to wane as some of the side effects of DDT and other widely used poisons began appearing in the late 1950s. One of the first to perceive the global dangers of pesticide abuse was the late Rachel Carson. Much of our current environmental awareness stems from her book *Silent Spring*, published in 1962. Her warn-

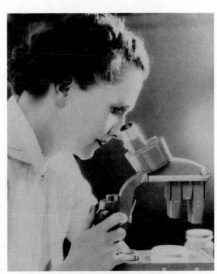

"The 'control of nature' is a phrase conceived in arrogance, born of the Neanderthal age of biology and philosophy, when it was supposed that nature exists for the convenience of man."

Rachel Carson, *Silent Spring*

ings were underscored when, shortly thereafter, scientists reported that DDT was threatening the survival of predatory birds and was showing up in human milk. Another serious problem to arise was genetic resistance to pesticides, evolving in an increasing number of pest populations. At the same time, some of the ill effects of other technological developments began to be widely publicized. By the early 1970s, disillusionment with chemicals and a realization that our finite biosphere could not tolerate unlimited exploitation had developed into widespread concern about environmental problems.

Today, it's clear that no part of the biosphere is untouched by human activities, and many people are concerned about the abusive impact of human populations and technology. The misuse of natural resources, localized famine aggravated by land misuse and expanding population, the growing list of species extinguished or endangered by loss of habitat, and the poisoning of soil and streams with toxic wastes—these are just a few of the problems that we have created and must solve. We will examine some of our environmental problems in Chapter 38. But first, let's examine the biosphere and the major concepts of ecology in more detail.

Abiotic factors are important in the biosphere

The biosphere is finite, but it is also extremely diverse. Understanding its structure and dynamics can help us understand our environmental dilemma. A variety of physical and chemical factors affect the organisms living in single ecosystems and in the biosphere as a whole. Solar energy, water, temperature, soil, oxygen, fire, and wind are some of the most important abiotic factors.

Solar energy powers all ecosystems except hydrothermal vent systems, which, as we have seen, are driven by energy from Earth's interior. In aquatic environments where sunlight reaches, the availability of light has a significant effect on the growth and distribution of plants and algae. Because water itself and the microorganisms in it absorb light and keep it from penetrating very far, most photosynthesis occurs near the surface of a body of water. In terrestrial environments, light is often not the most important factor limiting plant growth. In many forests, however, shading by trees creates intense competition for light at ground level.

Water, a second abiotic factor, is essential to all life. Aquatic organisms have a seemingly unlimited supply of water, but they face problems of water balance if their own solute concentration does not match that of their surroundings. As we saw in Module 25.5, aquatic organisms confront very different solute concentrations in the sea than in freshwater lakes and streams. For a terrestrial organism, the main water problem is the threat of drying out. Therefore, many land species have water-tight coverings that reduce water loss. Many also have kidneys that save water by excreting very concentrated urine.

Temperature is an important abiotic factor because of its effect on metabolism. Few organisms can maintain a sufficiently active metabolism at temperatures close to 0°C, and temperatures above 50°C destroy the enzymes of most organisms. Extraordinary adaptations enable some species to live outside this temperature range. For example, some of the frogs and turtles living in the northern U.S. and Canada can freeze during winter months and still survive. Bacteria living in hydrothermal vents and hot springs have enzymes that function optimally at extremely high temperatures. Mammals and birds can remain considerably warmer than their surroundings and can be active in a fairly wide range of temperatures, but even these animals function best at certain temperatures.

Soil—its structure, pH, and inorganic nutrients—is an important environmental factor limiting the distribution of plants and thus the animals that feed on them (see Module 32.8). Variations in soil account for many of the differences we see in the kinds of plants growing in different ecosystems.

Oxygen is plentiful in the atmosphere and rarely limits the rate of cellular respiration in terrestrial organisms. In contrast, oxygen (as dissolved O_2) is often in short supply for aquatic organisms.

Fires and other catastrophic events, such as hurricanes and volcanic eruptions, are infrequent and highly unpredictable in most ecosystems. However, fire recurs frequently in such ecosystems as grasslands and dry forests. The photograph below was taken in 1989, only a short time after extensive fires swept through Yellowstone National Park. The area was already being colonized by small plants that can take advantage of nutrients released from the trees that burned.

Wind is an important abiotic factor for several reasons. Some organisms—for example, the bacteria, protists, and many insects that live on snow-covered mountain peaks—depend on nutrients blown to them by winds. Local wind damage often creates openings in forests, contributing to patchiness in ecosystems. Wind also increases an organism's rate of water loss by evaporation. The consequent increase in evaporative cooling can be advantageous on a hot summer day, but it can cause dangerous wind chill in the winter.

In the next module, we examine the interaction of one animal species with the abiotic factors of its environment. We also look at some of the biotic factors that act with the abiotic ones in the animal's ecosystem.

The aftermath of fire in Yellowstone National Park

Organisms are adapted to abiotic and biotic factors by natural selection

An organism's ability to survive and reproduce in a particular environment is a result of natural selection, as we discussed in Chapter 14. By eliminating the least fit individuals in populations, environmental forces help shape species to the mix of abiotic and biotic factors that they encounter.

The presence of a species in a particular place can come about in two ways: The species may evolve in that location, or it may disperse to that location and be able to survive once it is there. The pronghorn antelope, pictured at the right, evolved on the open plains and shrub deserts of North America. It is not found elsewhere and is not closely related to the numerous species of antelopes in Africa. Taking the pronghorn as an example, let's see how some of its unique adaptations fit the environmental conditions in which it evolved.

First, what about the major abiotic factors? The pronghorn's habitat is arid, windswept, and subject to extreme temperature fluctuations both daily and seasonally. The pronghorn is superbly adapted to these conditions. If you drive through Wyoming or parts of Colorado in the winter, you will see herds of these animals foraging in the open when temperatures are well below 0°C. The pronghorn has a thick coat made of hollow hairs that trap air and use it as insulation. Water is rarely a problem for a pronghorn because it can live on the moisture it obtains from vegetation.

What about the pronghorn's adaptations to the biotic components of its habitat? The pronghorn's main foods are short, coarse grasses and woody shrubs, and its teeth are adapted for biting and chewing these plants. Also, like a cow, it has a stomach containing cellulose-digesting bacteria. As the pronghorn eats plants, the bacteria digest cellulose, and the antelope obtains most of its nutrients from the bacteria. As the pronghorn evolved, it became adapted to predation by wolves, coyotes, and cougars. The pronghorn's main adaptations to escape predators are great speed and endurance. Capable of sprinting about 95 km/hr on flat ground, it is one of the fastest mammals. An adult pronghorn can also keep up a pace of about 65 km/hr for at least 30 minutes—a definite advantage when being chased by long-distance runners such as wolves. Other adaptations that help the pronghorn foil predators include its tan-and-white coat, which camouflages the animal on the open plains, and its keen eyes, which can detect movement at great distances. The pronghorn also derives protection from living in herds. When one pronghorn starts to run, its white rump patch seems to alert other herd members to danger.

Pronghorn antelopes

Organisms vary a great deal in their ability to tolerate fluctuations and long-term changes in their environments. The pronghorn has survived significant environmental changes. During the pioneering days of the American West, its numbers were seriously reduced by human hunting. Unlike the bison, elk, wolf, and cougar (all of which were extinguished from the plains), wild populations of the pronghorn survived. Today, the species is common in several western states, where it competes mainly with domestic cattle and sheep for food and is hunted mainly by humans and coyotes.

The pronghorn is a highly successful, herbivorous, running mammal of open country. In a very different environment, such as a wooded area where predators would be more easily hidden by vegetation and could stalk them at close range, the antelope's adaptations for escaping predators might not be as effective as they are on the open plains. This suggests that an organism can usually tolerate environmental fluctuations only within the set of conditions to which it is adapted. Outside that set, the organism may not survive long enough to reproduce. Thus, in adapting populations to local environmental conditions, natural selection may limit the distribution of organisms. The absence of the pronghorn outside North America, however, does not necessarily imply that the species could not survive elsewhere; it may only mean they were never able to disperse beyond this region.

Regional climate influences the distribution of biological communities

When we ask what determines whether a particular organism or community of organisms lives in a certain area, the climate of the region—especially temperature and rainfall—is often a large part of the answer. A number of interacting factors determine climate on both global and regional scales.

Figure A shows that, because of its curvature, Earth receives an uneven distribution of solar energy. The sun's rays strike equatorial areas most directly. Thus, any particular area of land or ocean there absorbs more heat than do comparable areas in the more northern or southern latitudes. The uneven heating of Earth's surface is a major factor driving air movements and water currents.

The different seasons of the year result from the permanent tilt of the planet on its axis as it orbits the sun. As Figure B shows, the globe's position relative to the sun changes through the year. The Northern Hemisphere, for instance, is tipped most toward the sun in June, creating the longest days of the year and summer in that hemisphere; at the same time, days are short and it's winter in the Southern Hemisphere. Conversely, the Southern Hemisphere is tipped furthest toward the sun in December, creating summer there and winter in the Northern Hemisphere.

When air is warmed, it rises and tends to absorb moisture; when air is cooled, it falls and tends to lose moisture. Figure

A. Uneven heating of Earth's surface

C shows some of the effects of uneven warming on global wind patterns and rainfall. Red arrows indicate air movements. Heated by the direct rays of the sun, air at the equator rises, creating an area of calm or of very light winds known as the **doldrums.** As warm equatorial air rises, it cools, forms clouds, and drops rain. Warm temperatures throughout the year and heavy rainfall largely explain why rain forests are concentrated near the equator.

After losing their moisture over equatorial zones, high-altitude air masses spread away from the equator until they cool and descend again at latitudes of about 30° north and south. Many of the world's great deserts—the Sahara in North Africa and the Arabian on the Arabian peninsula, for example—are centered at these latitudes because of the dry air they receive from the equator. As the dry air descends, some of it spreads back toward the equator. This movement creates the **trade winds,** which dominate the **tropics** (latitudes between 23.5° north and south). As the air moves back toward the equator, it warms and picks up moisture until it is uplifted again.

Latitudes between the tropics and the Arctic Circle in the north and the Antarctic Circle in the south are called **temperate zones.** Generally, these regions have much milder climates than the tropics or the polar regions. Notice in Figure C that some of the descending dry air heads into the latitudes

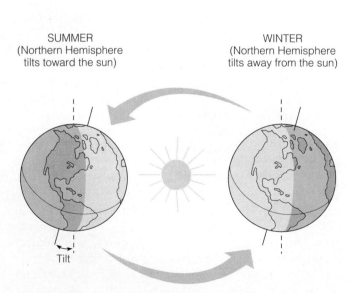

B. How Earth's tilt causes the seasons in the Northern Hemisphere

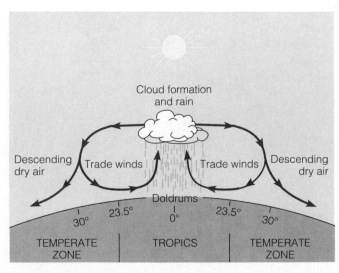

C. How uneven heating causes rain and winds

above 30°. At first these air masses pick up moisture, but they tend to drop it as they cool at higher latitudes. This is why the north and south temperate zones, especially latitudes between about 40° and 60°, tend to be moist. Deserts and arid regions in higher latitudes usually result from mountain ranges that cut off the flow of moist air, as we'll see.

Figure D shows the major global air movements, called the prevailing winds. **Prevailing winds** result from the combined effects of the rising and falling of air masses (red arrows at the left side of the globe) and Earth's rotation (gray arrows). Because Earth is spherical, its surface moves faster at the equator than at other latitudes. In the tropics, Earth's rapidly moving surface deflects vertically circulating air, making the trade winds blow from the northeast in the Northern Hemisphere and from the southeast in the Southern Hemisphere. In temperate zones, the slower-moving surface produces the **westerlies,** winds that blow from west to east.

A combination of the prevailing winds, the planet's rotation, and the locations and shapes of the continents creates **ocean currents,** riverlike flow patterns in the oceans. Carrying warm or cold water, ocean currents have a profound effect on regional climates. For instance, the Gulf Stream circulates warm water northward from the Gulf of Mexico and makes the climate on the west coast of Great Britain warmer than many areas farther south. In contrast, the northward flowing Humboldt Current cools the west coast of South America. Figure E depicts the major ocean currents. Red arrows indicate warm currents; blue arrows, cold ones.

Landforms can also affect local climate. Figure F presents one example, the effect of mountains on rainfall. This drawing represents major landforms across the state of Washington, but mountain ranges cause similar effects elsewhere. Washington is a temperate area in which the prevailing winds are westerlies. As moist air moves in off the Pacific Ocean and encounters the westernmost mountain range (the relatively low Coast Range), it rises and drops a large amount of rain. The biological community in this wet region is sometimes called a temperate rain forest. Some of the world's tallest trees, the Douglas firs, thrive here. Farther inland, precipitation increases again as the air moves up

and over higher mountains (the Cascade Range). On the eastern side of the Cascades, there is little precipitation; as a result of this "rain shadow," eastern Washington is nearly a desert.

Rain forests and deserts are among the world's major biological communities. Just as rain forests appear where there is abundant precipitation, and deserts where dry air descends over land, the appearance of other types of biological communities can be explained by regional climates. We'll see this clearly when we survey the biosphere's major terrestrial communities. First, let's take a brief look at aquatic ecosystems.

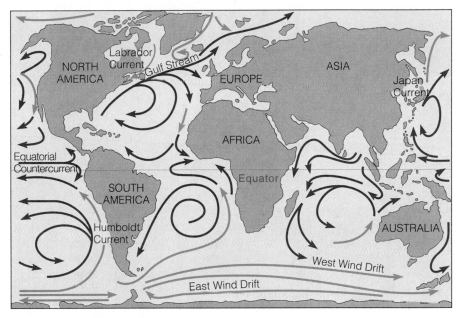

D. Prevailing wind patterns

E. Ocean currents

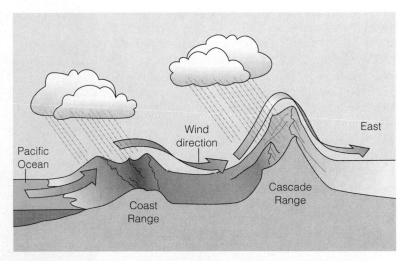

F. How mountains affect rainfall (Washington state)

Oceans occupy most of Earth's surface

A. An estuary in Maryland

Life originated in the sea and evolved there for almost 3 billion years before plants and animals began moving onto land. Covering about 75% of the planet's surface, oceans have always had an enormous impact on the biosphere. Their evaporation provides most of Earth's rainfall, and ocean temperatures have a major effect on climate and wind patterns. Photosynthesis by marine algae supplies a substantial portion of the biosphere's oxygen.

Figure A shows an **estuary,** an area where fresh water merges with seawater. This particular estuary is part of the Chesapeake Bay in Maryland. Several freshwater rivers empty into this bay, which merges with the Atlantic Ocean near Norfolk, Virginia. The saltiness of estuaries ranges from nearly that of fresh water to that of the ocean. Estuaries are among the most productive environments on Earth. Oysters, crabs, and many fishes live in estuaries or reproduce in them.

The shallow zone where estuarine or sea water meets land is called the **intertidal zone.** This area is often flooded by high tides and then left dry during low tides, about every 12 hours. Intertidal zones vary from salt marshes (the green area next to the mudflats in Figure A) to wave-splashed sandy or rocky beaches. All of the intertidal zone is a type of **wetland,** an ecosystem that is intermediate between an aquatic ecosystem and a terrestrial one. Most wetlands have soil that is saturated with water, either permanently or periodically. The highly productive wetlands of the intertidal zone

are home to many sedentary organisms such as algae, barnacles, mussels, sea stars, and sea anemones. Intertidal organisms attach to rocks or vegetation, or burrow into mud or sand, and are thus prevented from being washed away. Figure B shows some of the diverse organisms in a tidepool on the coast of central California. A tidepool is a small body of water that remains in a rock or sand depression in the intertidal zone during low tide.

The intertidal zone is one of several oceanic zones (Figure C, at the top of the facing page). Abiotic conditions often dictate the kinds of communities these zones support. The ocean water itself, called the **pelagic zone** (Greek *pelagos,* sea), supports communities dominated by highly motile animals such as fishes, squids, and marine mammals, including whales and dolphins. Diverse algae and cyanobacteria, collectively called **phytoplankton** (Greek *phyton,* plant, and *plankton,* wandering) drift passively in the pelagic zone. Phytoplankton are the ocean's main photosynthesizers, making most of the organic food molecules on which other ocean-dwellers depend. **Zooplankton** are animals that drift in the pelagic zone either because they are too small to resist ocean currents or because they don't swim. Zooplankton eat phytoplankton and, in turn, are consumed by other animals, including fishes. The seafloor is called the **benthic zone** (Greek *benthos,* "depth of the sea"). Depending on depth and light penetration, the benthic community consists of attached algae, fungi, bacteria, sponges, burrowing worms, sea anemones, clams, crabs, and fishes.

B. Intertidal zone organisms

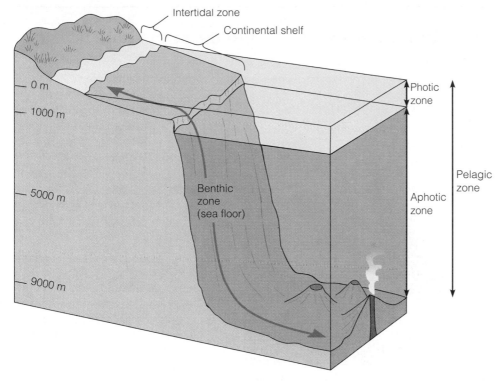

C. Oceanic zones

Labels in figure: Intertidal zone; Continental shelf; 0 m; 1000 m; 5000 m; 9000 m; Benthic zone (sea floor); Photic zone; Pelagic zone; Aphotic zone

continental shelves. Coral reefs are also common around oceanic islands. And some islands, called atolls, consist of a coral reef surrounding a lagoon (Figure D).

As Figure C indicates, marine biologists often group the illuminated regions of the benthic and pelagic communities together, calling them the photic zone. The **photic zone** is a relatively small portion of ocean water and bottom into which light penetrates and in which photosynthesis occurs. Underlying the photic zone is a vast, dark region called the **aphotic zone.** This is the most extensive part of the biosphere. Without light, there are no photosynthetic organisms, but life is still diverse in the aphotic zone. Many kinds of bacteria, invertebrates, such as sea urchins and polychaete worms, and fishes scavenge the remains of dead algae and animals that sink from the lighted waters above. The hydrothermal vent communities, powered by chemical energy rather than sunlight, are densely populated, as we discussed in the chapter's introduction.

The availability of light and inorganic nutrients has a huge impact on the distribution of organisms in the ocean. Water and particles suspended in it absorb light, and sunlight does not penetrate very far into the sea. On the submerged parts of continents called continental shelves, the pelagic and benthic communities usually receive ample light, and nutrients from the seafloor circulate in the shallow water. Estuaries and coral reefs (described in the introduction to Chapter 19), the richest communities in the ocean, are found on

Until fairly recently, many people viewed the ocean as a bountiful, virtually limitless resource, and we have harvested the ocean heavily and used it as a dumping ground for wastes. Estuaries and intertidal wetlands have been especially abused, with few undisturbed areas remaining, and many totally replaced by commercial and residential developments on landfill. We are now seeing the effects of our disregard for marine communities, as seafood is becoming less plentiful, the result of overharvesting and pollution; whales are in danger of extinction, mainly from overhunting; and oil and other pollutants foul coastal areas. Laws in many countries, including the U.S., now prohibit whaling and the disposal of sewage and other wastes at sea. Many countries are also taking steps to restore and conserve estuaries and other intertidal wetlands.

D. A coral atoll in the South Pacific

Freshwater communities populate lakes, ponds, rivers, and streams

A. A freshwater wetland in Massachusetts

The major difference between seawater and fresh water is salinity. As discussed in Module 25.5, the scarcity of dissolved ions in fresh water has a profound effect on organisms living there. Light also has a significant impact on freshwater communities, as it does in the ocean.

In all but the smallest lake or pond, there is usually a distinct photic (lighted) zone and an aphotic zone. Phytoplankton grow in the photic zone, and rooted plants often inhabit shallow waters, as in the marshy wetland in Figure A. There is often a significant "rain" of dead organisms sinking to the bottom, and large populations of bacteria and other microbes in the benthic (bottom-dwelling) community decompose this material. Respiration by bacteria also removes oxygen from water near the bottom, and in some lakes, this makes benthic areas unsuitable for any organisms except anaerobes.

Temperature also has a profound effect on freshwater communities, especially in temperate areas. During the summer, lakes often have a distinct upper layer of water that has been warmed by the sun. The warm-water layer is less dense than underlying, cooler water and does not mix with it. Fishes often spend much of their time in the deep, cool waters of a lake, unless oxygen levels there become depleted by decomposers.

Nitrogen and phosphorus are often limiting nutrients that determine the amount of phytoplankton growth in a

B. A stream in Georgia

lake or pond. When there are temperature layers in a lake, for instance, nutrients released by decomposers can become trapped near the bottom, out of reach of the phytoplankton. During the summer months, this may limit the growth of algae and photosynthesis in the photic zone. As winter approaches, the surface water becomes denser as it cools; it then tends to mix with the deeper water, allowing nutrients to return to the surface where phytoplankton can again use them. Seasonal mixing also restores oxygen to the depths.

Today, many lakes and ponds are affected by large inputs of nitrogen and phosphorus from sewage and runoff from fertilized lawns and agricultural fields. These nutrients often produce blooms, or population explosions of algae. Heavy algal growth reduces light penetration into the water, and, when the algae die and decompose, a pond or lake can suffer serious oxygen depletion.

Rivers and streams generally support quite different communities of organisms than lakes and ponds. A river or a stream changes greatly between its source (perhaps a spring or snowmelt) and the point at which it empties into a lake or the ocean. Near a source, the water is usually cold, low in nutrients, and clear (Figure B). The channel is often narrow, with a swift current that does not allow much silt to accumulate on the bottom. The current also inhibits the growth of phytoplankton; most of the organisms found here are supported by the photosynthesis of algae attached to rocks or organic material (such as leaves) carried into the stream from the surrounding land. The most abundant benthic animals are usually insects that eat algae, leaves, or one another. Trout are often the predominant fishes, locating their food, including insects, mainly by sight in the clear water.

Downstream, a river or stream often widens and slows. Marshes, ponds, and other wetlands are common in downstream areas. There, the water is usually warmer and may be murkier because of sediments and phytoplankton suspended in it. Worms and insects that burrow into mud are often abundant, as are waterfowl, frogs, and catfish and other fishes that find food more by scent and taste than by sight.

The major terrestrial communities are called biomes

The map below introduces the nine major types of biological communities, called **biomes,** that cover the land surface of Earth. The distribution of these terrestrial communities largely depends on climate, with temperature and rainfall often the key factors determining the kind of biome that exists in a particular region. Many of the biomes are named for their predominant vegetation, but each is also characterized by animals adapted to that particular environment. A grassland, for instance, is more likely than a mountain biome to be populated by large herds of grazing animals like pronghorn antelope.

There is no strict method of defining a biome, and through the years, different ecologists have recognized and organized biomes in different ways. Each biome is a *type* of community, not a specific assemblage of certain species. As you can see on the map, some of the biomes extend over massive geographic areas. The taiga, for example, extends in a broad band across North America, Europe, and Asia. In contrast, some biomes, such as deserts, have a more scattered distribution. The same type of biome may occur in two widely separated areas if the climate in the areas is similar. The assemblages of species in widely separated biomes,

such as Africa's Sahara Desert and eastern Asia's Gobi Desert, are different, but the species in both are adapted to desert conditions. Widely separated biomes may look alike because of convergence, the evolution of similar traits in independently evolved species that live in similar environments (see Module 16.13).

Biomes also tend to grade into each other, and within each biome, there is local variation, giving the vegetation a patchy, rather than uniform, appearance. For example, snowfall may break branches and small trees, causing openings in the coniferous taiga. These openings allow deciduous trees such as aspen and birch to grow. Today, in many areas, the natural biomes are broken up by human activity. Most of the eastern United States, for example, was once a temperate deciduous forest, but human activity has eliminated all but a small percentage of that forest. In fact, many ecologists recognize an "urban biome" and an "agricultural biome" scattered over much of the Earth, areas where the earlier biomes have been drastically altered.

We now begin a more detailed survey of the major biomes. To help you locate the biomes, we include with each an orientation map color-coded to match the map here.

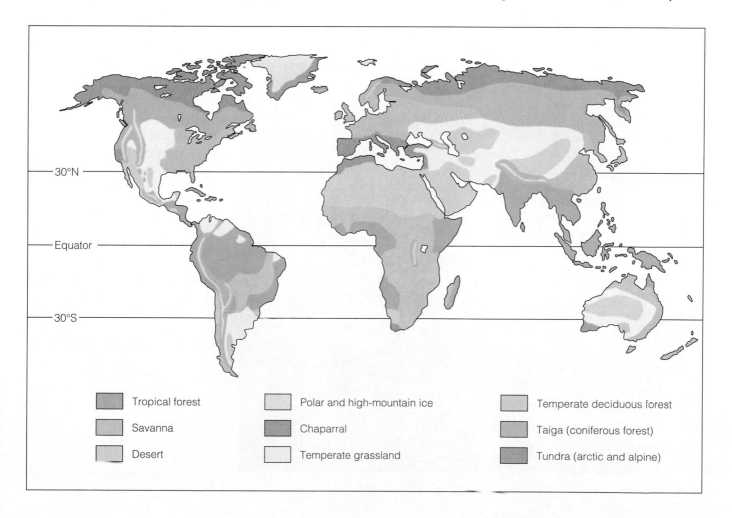

▦ Tropical forest	▦ Polar and high-mountain ice	▦ Temperate deciduous forest
▦ Savanna	▦ Chaparral	▦ Taiga (coniferous forest)
▦ Desert	▦ Temperate grassland	▦ Tundra (arctic and alpine)

Tropical forests cluster near the equator

Tropical forests occur in equatorial areas where the temperature is warm and days are 11–12 hours long year-round. Rainfall in these areas is quite variable, and this, rather than temperature or photoperiod, generally determines the kind of vegetation that grows in a particular kind of tropical forest. There are three major types of tropical forests. Tropical thorn forests are common in equatorial lowlands, such as parts of eastern Africa and northwestern India, where rainfall is scarce. These areas have prolonged dry seasons, and the plants found there are a mixture of thorny shrubs and trees, and nonwoody plants that retain water for long periods. In areas with distinct wet and dry seasons, such as central West Africa and much of India and Southeast Asia, tropical deciduous forests dominate. Tropical deciduous trees and shrubs drop their leaves during the long dry season and releaf only during the following heavy rains or monsoons. Tropical rain forests are found in very humid equatorial areas, such as Indonesia and the Amazon River basin in South America, where rainfall is abundant (greater than 250 cm per year) and the dry season lasts no more than a few months.

The tropical rain forest, such as the luxuriant area in Costa Rica shown below, is the most complex of all biomes, harboring more species than any other community in the world. Up to 300 species of trees, many of them evergreen angiosperms 50–60 m tall, can be found in a single hectare (2.5 acres). Because of the density of large trees, the rain forest often has a closed canopy, with little light reaching the forest floor. Where an opening does occur, perhaps because of a fallen tree, other trees and large, woody vines known as lianas grow rapidly. Many of the animals that live in tropical rain forests are tree-dwellers; monkeys, birds, insects, snakes, bats, and frogs find food and shelter many meters above the ground. The soils of tropical rain forests are typically poor, because high temperatures and rainfall lead to rapid decomposition and recycling rather than to a buildup of organic material. At any given time, almost all of the nutrients are incorporated in living organisms.

Human impact on the tropical rain forest is currently a source of great concern. It is a common practice to clear the forest for lumber or simply burn it, farm the land for a few years, and then abandon it. Mining has also devastated large tracts of rain forest. Once stripped, the tropical rain forest recovers very slowly. The destruction of these forests is now proceeding at an alarming rate. More than half are already gone, and the loss is much more than aesthetic: The devastation of tropical rain forests may cause large-scale changes in world climate, as well as large-scale destruction of species (see Module 38.14). With international financial support, a number of countries, including Costa Rica, Belize, Mexico, Venezuela, and Brazil, have begun preserving some of their rain forests. In the next module, we hear from a scientist who is in the thick of the rainforest conflict.

Tropical forests (all types)

Tropical rain forest

U.S. Forest Service scientist Ariel Lugo shares his views on tropical forest issues

Dr. Ariel Lugo is a tropical ecologist with a firm grasp of the problems and controversies of tropical forests. After an undergraduate education and a master's degree at the University of Puerto Rico, he earned his Ph.D. at the University of North Carolina and served as a biology professor at the University of Florida. Since 1979, Dr. Lugo has been a research scientist with the U.S. Forest Service in his native Puerto Rico. Dr. Lugo understands the global community's concern about the current rate of destruction of tropical forests and the loss of species in them. He tempers these concerns with those of local people who inhabit and use tropical forests. In a recent interview, we asked Dr. Lugo to summarize the situation as he sees it.

Each sector of society sees the forest in a different way. First of all, millions of people live in tropical forests, and they clearly have a plan for the forest that provides them with their survival. But a conservationist who is flying in an airplane over the forest sees all that green and says, "Ah, here's where I'm going to protect the biodiversity of the world. We're not going to touch this forest, because this is going to balance what we've been doing to the rest of the world." So for them, you want to leave the tropical forest alone. But if in the plane you have a timber baron, that person sees in this piece of forest one of the last of the Earth's untouched stores of timber. And you could say, if you were a timber baron, "Ah, I want to make money by cutting those trees and selling them." But there might be other people with different plans—the government, for example. The government sees the vast expanses of tropical forest as the future of the country. The United States made its future by cutting its forests. For the government, it's an area to be developed; the forest provides space for new cities for new economic development.

Ariel Lugo in the Caribbean National Forest

What do you see as the main reasons for tropical forest destruction?

Lumber is really a minor factor in deforestation. The major [one] is people have to cut the forest down and plant their crops. The main reason for deforestation is to satisfy human needs for food. We have to admit that sometimes you need to deforest—we cannot be such purists as to say that we can't cut any forest.

Often, when tropical forests are cut, the land is farmed for only a short time before it becomes infertile and is abandoned. Is there an alternative to this destructive practice?

To me, it is critical that when we convert land from one use to another, we make sure that the use we're after is sustainable. Sustainable productivity means that you benefit from what the land can produce without being so greedy as to wipe out this productivity in the process. How can we manage ecosystems within our own generation, get the benefits of the production, and then pass these ecosystems on with the same capacity so that the next generation is not handicapped by a reduction of productivity? The level to which we're going to achieve this depends on culture. I think different cultures have different expectations, different needs, and so some cultures push nature more to the limit than others.

What are the consequences of pushing tropical rain forests to the limit?

When deforestation occurs in an unplanned, uncontrolled way, then you lose species and productivity and degrade your soil and water resources. These regional effects fragment the landscape. And when you fragment the landscape, you disconnect nature—you lose the value of ecosystems working in synchrony. When you add up all the deforestation, you get into the global impact. Tropical forests are so huge that they help regulate climate, they help regulate the cycles of nutrients and water and gases, and so on. The main thing I want to tell you about the consequences of deforestation is just the waste. If we just wantonly destroy, then we are wasting a resource that could be used to improve the well-being of people.

What needs to be done?

The challenge is to harmonize or integrate all the legitimate claims on the forest. Now, who's going to put it all together? Who's going to organize development such that you have indigenous people, and you have conservation of the resource, and you maintain climatic balances, and you develop some areas, and you have agriculture, and you get fuel wood—that requires management in my view, and it requires a lot of compromise, and a lot of hard thinking, and a lot of good will and political will.

Savannas are grasslands with scattered trees

This photograph, taken in Kenya, shows a typical **savanna,** a biome dominated by grasses and scattered trees. Extensive savannas (from the Spanish *sabana,* meadow) cover wide

areas of the tropics in central South America, central and South Africa, and parts of Australia. Scattered savannas also occur in temperate North America (roughly in a band from Minnesota to eastern Texas), where the temperate forests of the eastern states merge with the grasslands of the west.

Savannas are simple in structure compared to tropical forests. Frequent fires, caused by lightning or human activity, and grazing animals inhibit further invasion by trees. Fires and grazers also maintain the small-growth form of grasses and nonwoody, broad-leaved dicot plants that grow with them. Grasses are wind-pollinated, but the dicots produce showy flowers that attract insect pollinators, often in great numbers, in the summer.

The savanna is home to many of the world's large herbivores and their predators. African savannas are home to giraffes, zebras, and many species of antelope, as well as to baboons, lions, and cheetahs. Several species of kangaroos are the dominant herbivores of Australian savannas. Throughout much of our North American savannas, farms replace areas once inhabited by bison, deer, black bear, coyotes, and wolves.

Burrowing animals, including mice, moles, gophers, snakes, ground squirrels, worms, and many arthropods, are also common on savannas. Grasses grow rapidly, providing a good food source, but they do not provide much cover. As a result, nest sites and shelters are primarily on or under the ground.

Deserts are defined by their dryness

Daytime temperatures in **deserts** sometimes get as high as 54°C, and nights often drop to freezing. However, it is sparse rainfall (less than 30 cm per year), rather than extreme temperatures, that makes an area a desert. The driest deserts are in central Australia and in the central Sahara in Africa, where the average annual rainfall is less than 2 cm and there is no rain at all during some years.

As we discussed in Module 34.6, large tracts of desert occur in two regions of descending dry air, centered around the 30° north and 30° south latitudes. In the Southern Hemisphere, these include the Kalahari in Africa and much of central Australia; in the Northern Hemisphere, they include the Sahara, the Arabian Desert, large areas of Mexico, and much of the southwestern United States. The photograph at the top of the facing page was taken in the Sonoran

Desert in southern Arizona. Large deserts also occur at higher latitudes, in the rain shadows of mountains; these include much of central Asia east of the Caucasus Mountains, southern Argentina east of the Andes, and much of California and Nevada east of the Sierra Nevada.

Unlike other biomes, many deserts are actually growing in size. The process of **desertification,** the conversion of other biomes—especially savannas—to deserts, is a significant environmental problem. In central Africa, for example, a burgeoning human population, overgrazing, and dryland farming are converting large areas of savanna to desert.

The cycles of growth and reproduction in the desert are keyed to rainfall. The driest deserts have no perennial vegetation at all, but in less arid regions the dominant plants are scattered shrubs, often interspersed with cacti, as in the

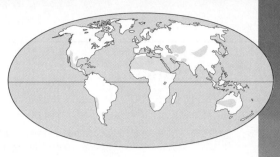

Sonoran Desert. Desert plants typically produce great numbers of seeds, which may remain dormant until a heavy rain triggers germination. Periods of rainfall (often in late winter) may produce spectacular blooms of annual plants.

Like desert plants, desert animals are adapted to drought and extreme temperatures. Many live in burrows and are active only during the cooler nights, and most have special adaptations enabling them to conserve water. Seed-eaters such as ants, many birds, and rodents are common in deserts. Lizards and snakes are important predators of the seed-eaters.

Spiny shrubs dominate the chaparral · 34.14

Chaparral (the Spanish word for "place of evergreen scrub oaks") is a region of dense, spiny shrubs with tough, evergreen leaves. The climate of chaparral areas results mainly from cool ocean currents circulating offshore, which usually produce mild, rainy winters and long, hot, dry summers.

First described in the Mediterranean region, chaparral vegetation is also found in coastal areas of Chile, southwestern Africa, southwestern Australia, and California. This photograph was taken in chaparral just south of San Francisco. In addition to the perennial shrubs that dominate chaparral, annual plants are also commonly seen, especially during the wet winter and spring months.

Chaparral vegetation is adapted to periodic fires, most often caused by lightning; in fact, the vegetation requires occasional fires for long-term maintenance. Many of the shrubs have root systems adapted to fire; the roots do not burn, and the plants regenerate quickly, using nutrients released by the fires. In addition, many chaparral plant species produce seeds that will germinate only after a hot fire. Others reproduce asexually without reliance on seeds.

Animals characteristic of the chaparral are browsers such as deer, fruit-eating birds, and seed-eating rodents, as well as lizards and snakes.

34.15 Temperate grasslands include the North American prairie

Temperate grasslands have some of the characteristics of tropical savannas, but they are mostly treeless and are found in regions of relatively cold winter temperatures. Temperate grasslands include the areas known as pampas in Argentina and Uruguay, steppes in Asia, and prairies in central North America. The keys to the persistence of most grasslands are seasonal drought, occasional fires, and grazing by large mammals, all of which prevent woody shrubs and trees from invading and becoming established.

Grasslands expanded in range following the retreat of the glaciers after the last ice age. Coupled with this expansion was the proliferation of large grazing mammals. The bison and pronghorn of North America, the gazelles and zebras of the African veldt, and the wild horses and sheep of the Asian steppes are some examples. Grasslands also support many species of burrowing animals, including herbivorous rodents, and carnivores such as badgers, skunks, and foxes.

The amount of annual rainfall influences the height of grassland vegetation. The photograph above was taken on

the relatively dry, short-grass prairie of western South Dakota. Tall-grass prairie occurs in wetter areas, such as eastern Kansas. Whereas the vegetation of short-grass prairie usually grows no more than about 1 m high, tall-grass vegetation often attains heights of over 2 m.

Little remains of North American prairies today. Most of the region is intensively farmed, and it is one of the most productive agricultural regions in the world.

34.16 Deciduous trees dominate temperate forests

Temperate deciduous forests grow in latitudes between about 35° and 50°, regions where there is sufficient moisture to support the growth of large trees. This includes most of the eastern United States, most of central Europe, and parts of eastern Asia and Australia. Temperate deciduous forests, such as this dense woodland in northern Michigan, are characterized by broad-leaved, deciduous trees. The dominant trees in the photograph are maples, but the species composition in temperate deciduous forests varies widely around the world; oak, hickory, beech, and birch dominate in many other areas.

Temperatures in temperate deciduous forests range from very cold in the winter to hot in the summer ($-30°C$ to $+30°C$). Precipitation is relatively high and usually evenly distributed throughout the year. These forests usually have a five- to six-month growing season and a distinct annual rhythm, in which the trees drop leaves and become dormant in late autumn, then produce new leaves each spring. Leaf drop seems to be an adaptation that conserves water (see Module 33.7).

The temperate deciduous forest is more open than the tropical rain forest and not as tall or as diverse. Nonetheless, temperate forests have a considerable variety and abundance of food and habitats, and they support a rich diversity of animal life. A great variety of insects and spiders, for example, live in the soil or leaf litter, or feed on the leaves and branches of small trees and shrubs. The temperate deciduous forest is also home to whitetail deer, many species of birds, small mammals, and, where not eliminated by humans, bobcats, foxes, black bears, and mountain lions.

Humans have dramatically altered temperate deciduous forests by logging and by clearing land for agriculture and urban development. Only scattered remnants of the original forests remain today.

Conifers prevail in the taiga

The **taiga**, also known as coniferous or boreal (northern) forest, extends in a broad band across North America and Eurasia, reaching to the southern border of the arctic tundra. The scene at the right shows a fir forest in Banff National Park in Canada.

Taiga (from the Russian word for mountain) is also found at higher elevations in more temperate latitudes, as in many mountainous regions in western North America. The taiga is characterized by harsh winters and short summers that can occasionally be warm. There may be considerable precipitation, mostly in the form of snow. The soil is thin and acidic. It forms slowly because of the low temperatures and the waxy covering of conifer needles, which decompose slowly.

Snow may accumulate to several meters each winter in the taiga, and this has important ecological consequences. The snow usually falls before the coldest temperatures occur, and it insulates the soil. The snow cover keeps the soil from freezing to such depths that it would never thaw out during the short taiga summers.

A typical taiga has only a few species of conifers—spruce, pine, fir, and hemlock—often in dense stands that inhibit undergrowth. Most conifer branches bend and resist breaking even when heavily laden with snow, but severe storms still break limbs and cause trees to fall, yielding patches of increased light on the forest floor. This often results in a higher diversity of plants in the damaged area. Scattered clusters of deciduous trees, such as birch, willow, aspen, and alder, are common where conifers have fallen or were destroyed by lightning-caused fires.

Animals living in the taiga are, like the prevailing conifers, adapted for the cold winters. Mice and many other small mammals remain active all winter in snow tunnels at ground level, where they forage on old vegetation. Squirrels and many birds feed extensively on conifer seeds. Large browsers include deer, moose, elk, snowshoe hares, beavers, and porcupines. Predators in the taiga include grizzly bears, wolves, lynxes, and wolverines.

At the northernmost limits of plant growth, and at high altitudes just below areas covered permanently with ice and snow, is the **tundra** (from the Russian word for "marshy plain"). Plant forms in the tundra are dwarf woody shrubs, grasses, mosses, and lichens. The arctic tundra encircles the North Pole, extending southward to the coniferous forests. Alpine tundras are found above the treeline on high mountains, even in the tropics. For example, tundra is found in the Andes Mountains in Ecuador, where the elevation is high enough to produce a very cold climate.

This photograph shows the arctic tundra in central Alaska in the autumn. The climate here is often extremely cold, with little light for long periods of time. During the brief, warm summers, when there is nearly constant daylight, plants grow quickly and flower in a rapid burst.

The arctic tundra is characterized by **permafrost**, continuously frozen ground. Permafrost underlies about 80% of Alaska, and almost half of Canada, Scandinavia, and Russia. The depth of the permafrost ranges from a few meters to nearly 1500 m (5000 ft) in northern Siberia. Only the upper part of tundra soil, from a few centimeters to several meters deep, thaws in the summer, and then only for a brief time. The permafrost prevents the roots of plants from penetrating very far into the soil. The permafrost and extremely cold winter air temperatures explain the absence of trees. The arctic tundra may receive as little precipitation as some deserts. But poor drainage, due to the permafrost, and slow evaporation, because of the low temperatures, keep the soil continually saturated.

Animals of the tundra withstand the cold by having good insulation that retains heat. Large herbivores of the tundra include musk oxen and caribou. The principal smaller animals are lemmings and a few predators such as the arctic fox and snowy owl. Many species of the tundra, especially birds, are migratory, using the tundra as a summer breeding ground. During the brief growing season, clouds of mosquitoes often fill the tundra air. Populations of many tundra insects and rodents expand rapidly in the summer, and then decline abruptly when the warm season ends. We'll learn more about the ups and downs of biological populations in Chapter 35.

Chapter Review

Begin your review by rereading the module headings and scanning the figures, before proceeding to the Chapter Summary and questions.

Chapter Summary

Introduction–34.1 Ecology is the scientific study of the interactions of organisms with their environments. Ecologists ask questions on several levels: How does an individual organism interact with its environment? What are the dynamics of its population? How do populations interact in a community? How do energy and chemicals flow within an ecosystem?

34.2–34.5 The global ecosystem is called the biosphere: all life on Earth and where it lives. Abiotic factors, including solar energy, temperature, water, soil, wind, and catastrophes both natural and human-caused, shape the biosphere's diverse habitats. Natural selection adapts organisms to these factors, as well as to biotic factors such as predation and competition.

34.6 The distribution of life is shaped by climate, mainly variations in temperature and rainfall. Most climatic variations are due to the uneven heating of Earth's surface. Warm, moist, rising air produces rain, which waters the tropical rain forests. Descending dry air creates deserts at about 30° north and south of the equator. The Earth's rotation deflects moving air, creating the prevailing winds. The winds, the Earth's rotation, and the arrangement of the continents direct ocean currents that warm or cool coastal areas. Landforms such as mountains can affect local climate.

34.7 The oceans cover about 75% of Earth's surface. Light and the availability of nutrients are the major factors shaping communities in the sea. Ocean water is called the pelagic zone, and the ocean bottom is called the benthic zone; the area where the sea meets the land is the intertidal zone. Intertidal wetlands, areas permanently or periodically saturated with water, include estuaries, salt marshes, and tidepools.

34.8 Freshwater communities include lakes, ponds, rivers, and streams. Light, temperature, and the availability of nutrients and dissolved oxygen shape lake and pond communities. Temperature, nutrients, currents, and water clarity vary from the source of a river to its mouth, and river communities vary accordingly.

34.9–34.14 Climatic differences shape the major types of biological communities, called biomes, that cover Earth's land surface. Several kinds of tropical forests occur in the warm, moist belt along the equator. The tropical rain forest is the most diverse community on Earth, but it is endangered by human encroachment. Drier tropical areas and some nontropical areas are characterized by the savanna, a grassland with scattered trees. The savanna is the home of many large herbivores and their predators. Deserts are drier still, and they are growing because of the misuse of surrounding land. The chaparral biome is a shrubland with cool, rainy winters and dry, hot summers, when fires often occur.

34.15–34.16 Temperate grasslands are found in the interiors of the continents where winters are cold. Drought, fires, and grazing animals prevent trees from growing. In temperate areas where rainfall is more abundant, there are forests of deciduous trees. In North America, both the temperate grassland and the deciduous forest biomes have been drastically altered by agriculture and urban development.

34.17–34.18 The taiga is the coniferous forest biome of the far north and high mountains. It occurs where there are short summers and long, snowy winters. Between the taiga and the permanently frozen polar regions lies the arctic tundra, a treeless plain where it is too cold for large trees to survive. Alpine tundra occurs above the tree line on high mountains. The vegetation of the tundra includes shrubs, grasses, mosses, and lichens.

Testing Your Knowledge

Multiple Choice

1. Changes in the seasons are caused by
   a. the tilt of Earth's axis toward or away from the sun
   b. yearly cycles of temperature and rainfall
   c. variation in the distance between Earth and the sun
   d. an annual cycle in the sun's energy output
   e. the periodic buildup of heat energy at the equator

2. What makes the Gobi Desert of Asia a desert?
   a. The growing season there is very short.
   b. Its vegetation is sparse.
   c. It is hot.
   d. Temperatures vary little from summer to winter.
   e. It is dry.

3. Andrea was a passenger on a plane that flew over temperate deciduous forest, then grassland and desert, finally landing at an airport in chaparral. The route of Andrea's flight was between
   a. New York and Denver
   b. Philadelphia and Los Angeles
   c. Denver and Los Angeles
   d. Washington, D.C. and Phoenix
   e. Seattle and Washington, D.C.

4. Which of the following sea creatures might be described as a pelagic animal of the aphotic zone?
   a. a coral reef fish
   b. a giant clam near a deep-sea hydrothermal vent
   c. an intertidal snail
   d. a deep-sea squid
   e. a harbor seal

True/False (*Change false statements to make them true.*)

1. A community consists of the individuals of a particular species living in a particular area.

2. The tundra looks marshy because of heavy rain and snow.

3. Light is an important abiotic factor affecting aquatic life.

4. Zooplankton produce most of the food in the sea.

5. The availability of oxygen limits the numbers and distribution of many terrestrial organisms.

6. Much of the ocean's benthic zone is in the aphotic zone.

Matching

1. Ground permanently frozen
2. Deciduous trees such as hickory and birch
3. Mediterranean climate
4. Spruce, fir, pine, and hemlock trees
5. Home of zebras, wildebeest, and lions
6. The steppes, pampas, and plains
7. The most complex and diverse biome
8. Dry tropical lowlands with trees

   a. tropical rain forest
   b. savanna
   c. temperate forest
   d. temperate grassland
   e. chaparral
   f. tundra
   g. tropical thorn forest
   h. taiga

Describing, Comparing, and Explaining

1. Explain how the following factors change from the source of a river to its mouth: nutrient content, current, sediments, temperature, oxygen content, food sources.

2. Choose any animal or plant in your geographic area, and write a paragraph describing how it is adapted to abiotic and biotic factors in its environment.

3. What climatic conditions allow tropical rain forests to grow along the equator in places such as Brazil and Southeast Asia, but create deserts like the Sahara 30° north and south of the equator?

Thinking Critically

1. The pronghorn of the American plains looks and acts like the antelopes of the African savanna. But the pronghorn is really the only survivor of a separate family of mammals not closely related to the African antelopes. How do you think this animal came to be so much like the "real" antelopes?

2. Estuaries make up only a tiny fraction of the marine environment, but they are important because many marine organisms, such as fishes and shrimps, migrate to estuaries to reproduce. What is an estuary? What characteristics of an estuary would make it a good "nursery"?

Science, Technology, and Society

Near Lawrence, Kansas, there was a rare patch of the original North American temperate grassland that had never been converted to farming. It was home to numerous native grasses, annual plants, and grassland animals. Among the species present were two endangered plants. Environmental activists thought the area should be set aside as a nature preserve, and they started to raise money to save it. In 1990, the owner of the land plowed it, stating that he did not want to be told what he could do with his property. He was within his legal rights, since there are no federal laws protecting endangered plants on private land. What issues and values are in conflict in this situation? How could the story have had a more satisfactory ending for all concerned?

Population Dynamics 35

City-dwellers often detour around buildings and trees where starlings roost. Ranchers and farmers watch dense flocks of these birds devour grain from fields and feedlots. Yet if you lived in North America a little over a century ago, you would not have seen this bird. Once restricted to Europe and Asia, the European starling is now an abundant and destructive pest in North America, eastern Australia, New Zealand, and South Africa. Omnivorous, aggressive, and tenacious, starlings often replace native species. They oust adult woodpeckers, bluebirds, and swallows from nesting sites, and pull nestlings of these and other species out of nests to make room for their own offspring.

During the 1800s and early 1900s, introducing foreign species of animals and plants to North America was a popular, unregulated activity. Many people belonged to "acclimatization societies," whose purpose was to bring in species from other lands. Private groups and state game agencies imported the ring-necked pheasant, an Asian native, for sport hunting. Civic authorities in over 100 cities introduced the now-widespread house sparrow for "aesthetic reasons" and to control pest insects. A citizens' group introduced the starling as part of a campaign to bring all the birds mentioned in Shakespeare's works to the New World. In *King Henry the Fourth*, a character named Hotspur alludes to the starling's ability to mimic human speech: "I'll have a starling shall be taught to speak nothing but 'Mortimer.'" Who would have thought this line from a play would trigger an environmental calamity 300 years after it was written?

Shakespeare enthusiasts released about 120 starlings in New York's Central Park in 1890. New Yorkers cheered as a breeding pair built a nest under the eaves of the American Museum of Natural History. From that foothold, starlings spread rapidly throughout the U.S. and Canada, as shown on the map at the right. In less than a century, the North American starling population increased to about 100 million. The United States now spends millions of dollars each year trying to control starlings. Mass trappings, hunting, electri-

fied wires on buildings, fireworks, chemical repellants, and poisons have all proved ineffective in controlling the birds.

The starling population in North America has some features in common with the global human population. Both are expanding and are virtually uncontrolled. Both are also harming other species—the starling threatens other bird species; the human population threatens much of the biosphere. In this chapter, we discuss why neither of these populations can grow indefinitely. But when will they stop, and what will stop them?

Population dynamics, the subject of this chapter, is concerned with changes in population size and the factors that regulate populations over time. In the first few modules, we look at how ecologists study populations and some of the major factors that control populations in nature. Then, we return to the topic of uncontrolled populations and see how the principles of population dynamics apply to the global human population.

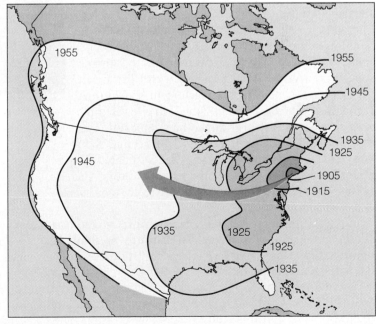

The spread of starlings across North America

35.1 Populations are defined in several ways

The starling population of North America and the global human population are both very large. Most of our knowledge of population dynamics comes from studies of much smaller groups that are confined by more restricted geographical boundaries—for instance, a population of moose in a certain mountain valley, or a population of algae in a lake. Biologists generally define a **population** as an interbreeding group of individuals of a particular species that are more-or-less isolated from other such groups. A somewhat less rigid definition considers a population to be a group of individuals of a species that use common resources and are regulated by the same natural phenomena, such as temperature, water and food supply, and predation.

A researcher must define a population by geographical boundaries appropriate to the questions being asked. For example, a population biologist studying the contribution of asexual reproduction to the population growth of sea anemones might define a population as all the anemones of one species in a tidepool. Another researcher studying the effects of hunting on deer might define a population as all the deer within a particular state. Yet another researcher, attempting to determine which segment of the human population will be most affected by the AIDS epidemic, might study the HIV infection rate of the human population in one nation or throughout the world. Regardless of the scale, two important characteristics of any population are its density and its dispersion patterns. We discuss these characteristics next.

35.2 Density and dispersion patterns are important population variables

Population density is the number of individuals of a species per unit area or volume—the number of oak trees per km^2 in a forest, for example, or the number of earthworms per m^3 in the forest's soil.

How do we measure population density? In rare cases, it is possible to actually count all individuals within the boundaries of the population. As a hypothetical example, we could count the total number of oak trees (say, 200) in a forest covering 50 km^2. The population density would be the total number of trees divided by the area, or 4/km^2.

In most cases, it is impractical or impossible to count all individuals in a population. Instead, ecologists use a variety of sampling techniques to estimate population densities. For example, they might base an estimate of the density of alligators in the Florida Everglades on a count of individuals in a few sample plots of one km^2 each. The larger the number and size of sample plots, the more accurate the estimates. In some cases, population densities are estimated not by counts of organisms but by indirect indicators, such as number of bird nests or rodent burrows, or signs such as animal droppings or tracks.

Another sampling technique commonly used to estimate wildlife populations is the **mark-recapture method.** In Figure A, researchers are sampling a population of mouse-like rodents called meadow voles. Box traps are set within the boundaries of the population; captured voles are tagged ("marked") and then released. After a certain amount of time (two weeks, in this study), traps are set again. This time, both marked and unmarked voles will be captured. The proportion of marked to unmarked individuals gives

A. The mark-recapture method

an estimate of the size of the entire population. The following equation gives an estimate of the number of individuals (*N*) in the population:

$$N = \frac{\text{Marked individuals} \times \text{Total catch second time}}{\text{Recaptured marked individuals}}$$

For example, suppose the researcher captures, tags, and releases 50 voles. Two weeks later, 100 voles are captured. If 5 of this second catch are voles that have been recaptured, we would estimate the entire population of voles as N = 50 x 100 divided by 5 = 1000 voles. This method assumes that a marked individual has the same chance of being trapped as an unmarked individual. This is not always a safe assumption, however. An animal that has been trapped before may be wary of the traps, or, having learned that traps contain food, may have deliberately returned for more. Still, the mark-recapture method is a useful tool for estimating population density.

The **dispersion pattern** of a population refers to the way individuals are spaced within their area. The **clumped** pattern, in which individuals are aggregated in patches, is the most common in nature. Clumping often results from the unequal distribution of resources in the environment. For instance, the cottonwood trees in Figure B are clumped along a stream channel in patches of moist and sandy soil. Clumping of animals is often associated with uneven food distribution or with mating or other social behavior. For instance, mosquitoes often swarm in great numbers, thereby increasing their chances for mating.

B. Clumped dispersion of cottonwood trees

The **uniform,** or even, pattern of dispersion often results from interactions among individuals of a population. For instance, creosote bushes in the desert (Figure C, left) tend to be uniformly spaced because their roots compete for water and dissolved nutrients. Animals often exhibit uniform dispersion as a result of social interactions. Examples are birds nesting in large numbers on small islands, or people living in a housing development (Figure C, right).

Some populations exhibit both clumped and uniform dispersion patterns simultaneously. If you look at the entire population, you usually see clumps of individuals. For instance, if you studied dispersion patterns of the human population of the state of Pennsylvania, you would find most of the population clumped in cities. Within each clump, however, individuals or family groups might be more-or-less uniformly dispersed, as they are in housing developments.

In the third, or **random,** type of dispersion, individuals in a population are spaced in a patternless, unpredictable way. Random dispersion is rare. Clams living in a coastal mudflat, for instance, might be randomly dispersed at times of the year when they are not breeding, when resources are plentiful and do not affect their distribution. Clams might also be randomly dispersed if a great number of factors—food, shelter, predators, and dissolved oxygen, for example—are affecting them in conflicting ways with chaotic results. Researchers often test for clumped or uniform dispersion patterns by using statistical methods to compare a population's dispersal with mathematical models of random dispersal.

Estimates of population density and dispersion patterns are both important in analyzing population dynamics. They enable researchers to compare and contrast growth or stability of populations occupying different geographic areas. Later in this chapter, we will explore some of the factors that can alter population density and dispersion. In the next module, we turn our attention to how populations grow.

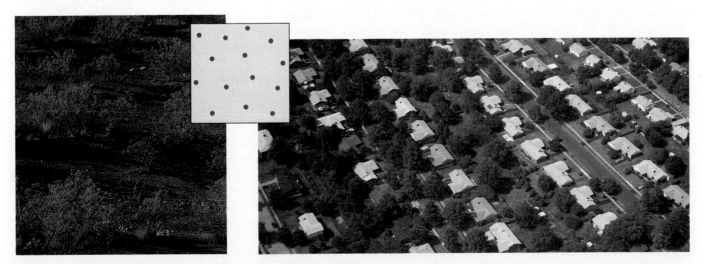

C. Uniform dispersion of creosote bushes (left) and human habitations (right)

Idealized models help us understand population growth

No other organisms reproduce faster than bacteria. Some bacteria can divide every 20 minutes under ideal laboratory conditions. Thus, starting with a single bacterium, after 20 minutes, there are two bacteria; after 40 minutes, there are four; after 60 minutes, eight—and so on, as shown in the table in Figure A. If this rate of population growth could continue for only a day and a half—a mere 36 hours—there would be bacteria enough to form a layer a foot deep over the entire planet!

Time	Number of Cells	
0 minutes	1	$= 2^0$
20	2	$= 2^1$
40	4	$= 2^2$
60	8	$= 2^3$
80	16	$= 2^4$
100	32	$= 2^5$
120 (= 2 hours)	64	$= 2^6$
3 hours	512	$= 2^9$
4 hours	4096	$= 2^{12}$
8 hours	16,777,216	$= 2^{24}$
24 hours	68,719,476,746	$= 2^{72}$

A. Exponential growth of bacteria

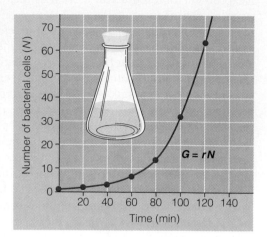

$G = rN$

THE EXPONENTIAL GROWTH MODEL The rate of expansion of a population under ideal conditions is called exponential growth when the whole population multiplies by a constant factor during constant time intervals (the generation time). For the bacterial population in Figure A, the constant factor is 2, and the generation time is 20 minutes. The progression for bacterial growth—2, 4, 8,16, etc.—is the number 2 raised to a successively higher power (exponent) each generation (that is, $2^1 = 2$; $2^2 = 4$; $2^3 = 8$; $2^4 = 16$; etc.).

Suppose you have a summer job working in a microbiology research lab, and you are asked to monitor the growth of a bacterial population. You would sample the population by counting the number of bacterial cells (N) at regular intervals of time (t). You could then plot the numbers you obtained for N against t. The graph in Figure A shows the type of curve you would obtain if you plotted the number of cells in a bacterial population that was expanding exponentially (and connected the points on the graph with a smooth curve).

The simple equation $G = rN$ describes the J-shaped curve, which is typical of exponential growth. The letter G stands for the growth rate of the population, N stands for the population size (the number of individuals in the population), and r stands for the **intrinsic rate of increase,** an organism's inherent capacity to reproduce. The value for r depends on the kind of organism, but it remains constant for any population expanding without limits. We can obtain a rough estimate of the value of r by subtracting the death rate (the number of individuals dying in a given unit of time) from the birth rate (the number of individuals born in that same unit of time).

Overall, the exponential equation tells us that, with r being constant, the rate at which a population grows depends on the number of individuals already in the population. The bigger the value of N, the faster the population increases. As time goes by, N gets bigger faster and faster. On a

graph, the lower part of the J results from the relatively slow growth when the population is small. The steep, upper part of the J results from N's being large.

The **exponential growth model** gives an idealized picture of the unregulated growth of a population. For bacteria, unregulated growth means there is no restriction on the abilities of the cells to live, grow, and reproduce. Given a few days of unregulated growth, bacteria would smother every other living thing. Obviously, long periods of exponential increases are not common in the real world, or life could not continue on Earth.

POPULATION-LIMITING FACTORS AND THE LOGISTIC GROWTH MODEL In nature, a population may grow exponentially for a while, but eventually, some environmental factor or factors will limit its growth. Population size then levels off, or it may even stop. Environmental factors that restrict population growth are called **population-limiting factors.**

You can see the effect of population-limiting factors in the graph in Figure B (at the top of the facing page), which illustrates the growth of a population of fur seals on St. Paul Island, off the coast of Alaska. (For simplicity, only the mated bulls were counted. Each has a harem of a number of females, as shown in the photograph.) Before 1925, the seal population on the island remained low because of uncontrolled hunting, although it changed from year to year. After hunting was controlled, the population increased rapidly until about 1935, when it leveled off and began fluctuating around a population size of about 10,000 bull seals. At this point, a number of population-limiting factors, including some hunting and the amount of space suitable for breeding, restricted population growth.

The fur seal growth curve resembles the **logistic growth model,** a description of idealized population growth that is slowed by limiting factors. Figure C compares the logistic growth model (red) with the exponential growth model (blue). As you can see, the logistic curve is J-shaped at first, but gradually levels off to resemble a lazy S more than a J.

The equation for logistic growth is more complicated than the exponential equation, because it describes the effect of population-limiting factors as well as population growth. As shown in Figure C, the logistic equation is the exponential equation modified by the term $(K–N)/K$. This term represents the overall effect of population-limiting factors.

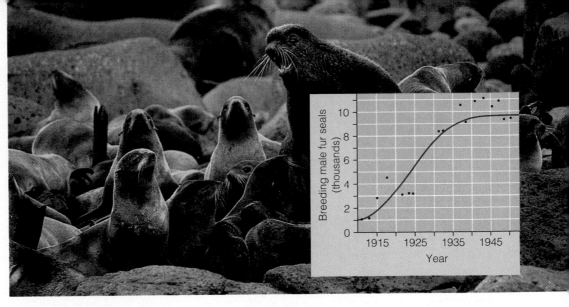

B. Growth of a population of fur seals

The logistic equation is actually simpler than it may appear. Notice that the only new letter in the equation is K, which stands for carrying capacity. **Carrying capacity** is the number of individuals in a population that the environment can just maintain ("carry") with no net increase or decrease. For the fur seal population on St. Paul Island, for instance, K is about 10,000 mated males. The value of K varies, depending on species and habitat. K might be considerably less than 10,000 for a fur seal population on an island with fewer breeding sites.

Let's see how the term $(K–N)/K$ works in producing the S-shaped logistic curve. When the population first starts growing, N is close to zero—very small compared to the carrying capacity K. At this time, N has little effect on the term $(K–N)/K$; in fact, the term nearly equals K/K, or 1. When this is the case, population growth G is close to $rN(1)$, or just rN—that is, exponential growth. However, as the population increases and N gets close to carrying capacity, it has a large effect on the term $(K–N)/K$. In fact, the term becomes an increasingly small fraction. And the value rN is multiplied by that fraction, slowing down the population growth more and more. At carrying capacity, the population is as big as it can theoretically get in its environment; at this point, $N = K$, $(K–N)/K = 0$, and the population growth rate (G) is zero.

Thus, the term $(K–N)/K$ accounts for the leveling off of the logistic growth curve. To summarize: When the population is still very small, the term has little effect, and the logistic curve is very close to the J-shaped curve. As the

population takes off, N makes the term $(K–N)/K$ and the growth rate smaller and smaller. The curve then gets farther and farther away from the J-shaped curve and becomes S shaped.

What does the logistic growth model suggest to us about real populations in nature? The model predicts that a population's growth rate will be small when the population size is either small or large, and highest when the population is at an intermediate level relative to the carrying capacity. At a low population level, resources are abundant, and the population is able to grow nearly exponentially. At this point, however, the increase is small because N is small. In contrast, at a high population level, population-limiting factors strongly oppose the population's potential to increase. In nature, there might be less food available per individual or fewer breeding sites, nest sites, or shelters. What actually happens is that limiting factors make the birth rate decrease, the death rate increase, or both. Eventually, the population stabilizes at the carrying capacity (K), when the birth rate equals the death rate.

It is important to realize that both the logistic growth model and the exponential growth model are mathematical ideals. No natural populations fit either one perfectly. Overall, these models are useful starting points for studying population growth. Ecologists use them to predict how populations will grow in certain environments and as a basis for constructing more complex models. The models have stimulated many experiments and discussions that have led to a greater understanding of populations in nature. In the next module, we take a closer look at some of the factors that limit the growth of natural populations.

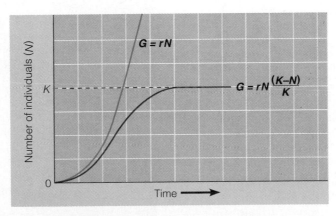

C. Logistic growth and exponential growth compared

Density-dependent and density-independent factors limit population growth

Two general types of factors limit population growth. In one case, limits on population growth are related to the density of the population itself. In the other, population density has nothing to do with population growth.

Population-limiting factors whose effects depend on population density are called **density-dependent factors.** Such factors affect a greater percentage of individuals in a population as the number of individuals increases. Limited food supply and the buildup of poisonous wastes are examples of density-dependent factors. Such factors depress a population's growth rate by increasing the death rate, decreasing the birth rate, or both. We often see density-dependent regulation in laboratory populations. For example, when a pair of fruit flies is placed in a jar with a limited amount of food, the population grows logistically. After a rapid increase, population growth levels off, as the flies become so numerous that they outstrip their limited food supply (Figure A). Each individual in a large population has a smaller share of the limited food than it would in a small population. Also, the more flies, the more concentrated poisonous wastes become in the jar.

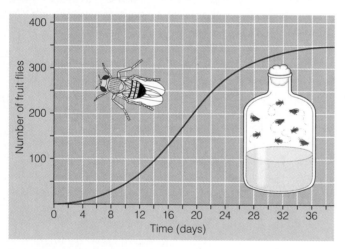

A. Logistic growth of a laboratory population of fruit flies

Laboratory populations are one thing; natural populations are another. We do not often see clear-cut cases of density-dependent factors regulating populations in nature. To test whether such factors are operating, it is necessary to change the density of individuals in the population while keeping other factors constant. This is sometimes done in managing game populations. For instance, state agencies often allow hunters to reduce populations of white-tail deer to levels that keep the animals from permanently damaging the plants they use for food. White-tail deer are browsers, preferring the highly nutritious parts of woody shrubs—

young stems, leaves, and buds. When deer populations are kept low and high-quality food is therefore abundant, a high percentage of females become pregnant and bear offspring; in fact, many of them produce twins (Figure B). On the other hand, when populations are high and food quality is poor, many females fail to ovulate, the spontaneous abortion rate is usually higher, and twinning is rare. We can conclude, therefore, that food supply is a density-dependent factor regulating white-tail deer populations.

Population-limiting factors whose occurrence is not affected by population density are called **density-independent factors.** Included are such forces as climate and weather. A freeze in the fall, for example, may kill a certain percentage of insects in a population. The date and severity of the first freeze obviously are not affected by the density of the insect population. Density-independent factors such as a killing frost affect the same *percentage* of individuals regardless of population size. (In larger populations, of course, greater numbers will die.)

In many natural populations, density-independent factors limit population size well before resources or other density-dependent factors become important. In such cases, the population may decline suddenly. If we look at the growth curve of such a population, we see something like exponential growth followed by a rapid decline, rather than a leveling off. Figure C (on the facing page) shows this effect for a population of aphids, insects that feed on the phloem sap of plants. These and many other insects often show virtually exponential growth in the spring and then rapid die-offs when it becomes hot and dry in the summer. A few individuals may remain, and these may allow population growth to resume again if favorable conditions return. Some insect populations—many mosquitoes and

B. A deer with twin fawns, born when population density was low

grasshoppers, for instance—will die off entirely, leaving only eggs, which will initiate population growth the following year. In addition to seasonal changes in the weather, environmental factors, such as fire, floods, storms, and habitat disruption by human activity, can affect populations in a density-independent manner.

Over the long term, most populations are probably regulated by a mixture of density-independent and density-dependent factors. Many populations remain fairly stable in size and are presumably close to a carrying capacity that is determined by density-dependent factors. In addition, however, many show short-term fluctuations due to density-independent factors. In some cases, the distinction between density-dependent and density-independent factors is not clear. In the case of white-tail deer, for example, in very cold, snowy areas, many individuals may starve to death. The severity of this effect is related to the harshness of the winter; cold temperatures increase energy requirements (and therefore the need for food), while deeper snow makes it harder to find food. But the severity of the effect is also density-dependent, because the larger the population, the less food available per individual.

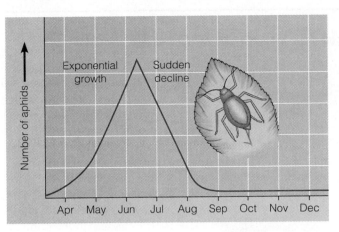

C. The effect of a density-independent factor on aphid population size

Some populations have "boom-and-bust" cycles

The graph below illustrates long-term changes in two populations, the snowshoe hare and the lynx, that fluctuate cyclically. The lynx is one of the main predators of the snowshoe hare in the taiga biome of North America. About every 10 years, both hare and lynx populations have a rapid increase (a "boom") followed by a sharp decline (a "bust").

What causes boom-and-bust cycles? Since ups and downs in the two populations seem to almost match each other on the graph, does this mean that changes in one directly affect the other? In other words, does predation by the lynx make the hare population fluctuate, and do the ups and downs of the hare population cause the changes in the lynx population?

For the lynx and many other predators, the availability of prey often determines population changes. Thus, the 10-year cycles in the lynx population probably result at least in part from the 10-year cycles in the hare population. For the hare, however, researchers have discovered that populations will fluctuate markedly about every 10 years whether lynx are present or not. Thus, we cannot conclude from the graph that lynx predation causes the cyclical changes in the hare population. In fact, rather than being produced by the lynx, the periodic crashes in the hare population may be associated with changes in the hare's own food source. When the hares nibble on certain plants, the plants produce chemicals that repel them. Thus, as a hare population expands, the food it needs may actually become unpalatable.

In addition to altered food supply and predation, several other explanations have been proposed for animal population cycles. One hypothesis, based on laboratory studies of mice and other small rodents, is that stress from crowding may alter hormonal balance and reduce fertility. The causes of cycles probably vary among species and perhaps among populations of the same species.

Population cycles of the snowshoe hare and the lynx

35.6 Evolution shapes life histories

An organism's **life history** is the series of events from birth through reproduction to death. Examples of life history traits are the age at which reproduction first occurs, the number of offspring, and the amount of parental care given to offspring. All such traits influence the growth rate of a population. As much as the body features of an organism, they are shaped by evolution operating through natural selection.

A. Guppies from two different environments

Let's look at the effect of one environmental factor, predation, in shaping a life history through natural selection. Figure A illustrates typical mature individuals from two wild populations of the guppy, a popular home-aquarium fish. Guppies vary in color and body form, but they are all one species. On the Island of Trinidad in the West Indies, certain guppy populations live where predators called killifish eat mainly small, immature guppies. Other guppy populations live where larger fish, called cichlids, eat mostly mature, large-bodied guppies. Where preyed upon by cichlids, guppies tend to be smaller, mature earlier, and produce more offspring each time they give birth than those in areas without cichlids. Thus, guppy populations differ in certain life history traits depending on the kind of predators in their environment. If the differences between the populations result from natural selection, the life history traits should be heritable. To test for heritability, researchers raised guppies from both populations in the laboratory without predators. The two populations retained their life history differences when followed through several generations, indicating that the differences were indeed inherited.

In a follow-up experiment, researchers introduced guppies from a cichlid habitat into a guppy-free area where there were predators that ate small, immature guppies. The scientists predicted that the new kind of predation would bring about changes in the guppy life history traits. So they were not surprised when, within two years, they began to find female guppies maturing later and producing fewer and larger offspring than those from a control site with cichlids. A laboratory test without predators showed that the changes in the population were heritable. These studies demonstrate not only that life history traits are shaped by natural selection, but that questions about evolution can be tested by field experiments.

In nature, every population has a particular life history adapted to its environment. Some populations, usually of small-bodied species, exhibit an **opportunistic life history.** Individuals reproduce when young, and they produce many offspring; the population tends to grow exponentially when conditions are favorable. Such a population typically lives in an unpredictable environment and is controlled by density-independent factors such as the weather. For example, the dandelion (Figure B) and many other annual weeds grow quickly in open, disturbed areas, producing a large number of seeds in a brief time when the weather is favorable. Although most of the seeds will not produce mature plants, their large number and ability to disperse to new habitats ensure that at least some will grow and eventually produce seeds themselves. For such species, natural selection has fostered an emphasis on the quantity of reproduction rather than on individual survivorship.

In contrast, some populations, mostly larger-bodied

B. Seeds of a dandelion, an opportunistic species

C. Seeds of a coconut palm, an equilibrial species

species, exhibit an **equilibrial life history.** Individuals usually mature later and produce few offspring but care for their young. The population size may be quite stable, held near the carrying capacity by density-dependent factors. Natural selection has resulted in the production of better-endowed offspring that can become established in the well-adapted population into which they are born. The life histories of many large terrestrial vertebrates fit this model. Among polar bears, for instance, a female has only one or two offspring every three years, but the cubs remain in her protective custody for over two years. In the plant kingdom, the coconut palm tree shown in Figure C (on the facing page) also fits the equilibrial model. Compared to most other trees, it produces relatively few, very large seeds. A large endosperm (the coconut "milk" and "meat") provides nutrients for the embryo; this is a plant's version of parental care.

For most populations, these two life history categories are only hypothetical models, starting points for studying the complex interplay of the forces of natural selection in nature. Most populations probably fall between the opportunistic and equilibrial extremes.

Life tables track mortality and survivorship in populations

Life Table for the U.S. Population in 1980				
Age Interval	Number Living at Start of Age Interval (N)	Number Dying During Interval (D)	Mortality (Death Rate) During Interval (D/N)	Chance of Surviving Interval (1 − D/N)
0–10	10,000,000	121,678	0.012	0.988
10–20	9,878,322	124,163	0.013	0.987
20–30	9,754,159	174,161	0.018	0.982
30–40	9,579,998	202,773	0.021	0.979
40–50	9,377,225	410,607	0.044	0.956
50–60	8,966,618	882,352	0.098	0.902
60–70	8,084,266	1,810,106	0.224	0.776
70–80	6,274,160	2,999,619	0.478	0.522
80–90	3,274,541	2,628,753	0.803	0.197
90–100	645,788	645,788	1.000	0.000

The principles of population growth have broad application. When the life insurance industry was established about a century ago, insurance companies took notice of some of the early scientific studies of human populations. Needing to determine how long, on average, an individual of a given age could be expected to live, they began using what are called **life tables.** The life table above was compiled in 1980 from a sample of 10 million U.S. citizens. Using this table, an insurance agent would predict that a 22-year-old has about a 0.982 (98.2%) chance of surviving to age 30. Population ecologists have constructed life tables for populations of many other species of animals, as well as for some plants.

A graph like the one below makes the data in a life table easy to comprehend at a glance. These so-called survivorship curves vary with the species and often with populations within species. By using a percentage scale instead of actual ages on the horizontal axis, we can compare species with widely varying life spans on the same graph. The curve for the human population tells us that most people die in older age intervals, as we see in the last column of the life table. Species that exhibit this Type I curve—whales and elephants, for example, as well as humans—usually produce few offspring but give them good care. In general, populations with a Type I survivorship curve have life history traits that fit the equilibrial life history model. In contrast, a Type III curve indicates high death rates for the very young, and then a period when death rates are much lower for those few individuals who survive to a certain age. Species with this type of survivorship curve usually produce very large numbers of offspring but provide little or no care for them. An oyster, for instance, may release millions of eggs, but most offspring die as larvae from predation or other causes. Populations with a Type III survivorship curve generally fit the opportunistic model. A Type II curve is intermediate, with mortality more constant over the life span. This type of survivorship has been observed in several invertebrates, including *Hydra,* and in certain rodents, such as the gray squirrel.

Life tables and survivorship curves allow us to compare populations on the basis of individual life spans. Another way to compare populations is by age structure, as we see next.

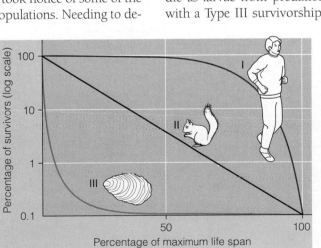

Three types of survivorship curves

A population's age structure indicates its future growth trend

The **age structure** of a population is the proportion of individuals in different age groups. By studying age structure, we get an idea of how a population will probably grow in the future. A typical population has three main age groups: prereproductive, reproductive, and postreproductive. Figure A compares the age structure for the human population in the United States in three selected years. In each diagram, the total colored area represents 100%, or the total population. Within this total, each of the three different colors represents the fraction of the population for one of the three main age groups. And within each age group, each of the horizontal bars represents the population in a five-year age group. The areas to the left of the vertical center line represent the percentages of males in each age group in the population; females are represented on the right side of the line. In 1990, for instance, males aged 15–19 made up just about 3.6% of the total U.S. population; females aged 15–19 made up a slightly smaller percentage. Notice that the percentages on each side of the vertical line go only to 6. This is because no five-year age group (either female or male) makes up more than 6% of the total population. In other years or other countries, the bars might extend further.

In 1960, the U.S. population was expanding rapidly, and that expansion is explained by the shape of the year's age-structure diagram. At that time, the highest percentage of individuals were of reproductive age or younger. The hatched bars represent the "baby boom" that lasted for about two decades after the end of World War II. Notice how this group moves upward on the age-structure diagrams over the years. By 1975, the population bulge due to the post-World War II "boomers" had moved up to the age groups 10–29. And by 1990, the oldest of this generation were in their 40s. The age-structure diagram for 1990 is much less of a pyramid than the ones for the earlier years are.

Figure B (on the facing page) compares populations for one year (1990) in two countries that have very different age structures, Sweden and Mexico. Except for a rather small effect of the post–World War II baby boom (which you can see in the bars for the 40–49 year olds in the left-hand diagram), Sweden has had a stable population for several decades, with individuals distributed fairly evenly through all age classes. This is because, for many years, the birth rate has closely matched the death rate in Sweden. In sharp contrast, Mexico is experiencing a population explosion; over 75% of its individuals are in the reproductive ages or younger. With this bottom-heavy age structure, Mexico has a large number of young people who are likely to reproduce in the near future. In Figure C, children stand out in the typical Mexican street scene (right), in contrast to the situation in Stockholm, Sweden (left).

Another important factor in Mexico's population growth is that the death rate, especially for the prereproductive and reproductive age groups, has declined significantly in the last 25 years. The combination of a large, young population and a low death rate will sustain Mexico's explosive popula-

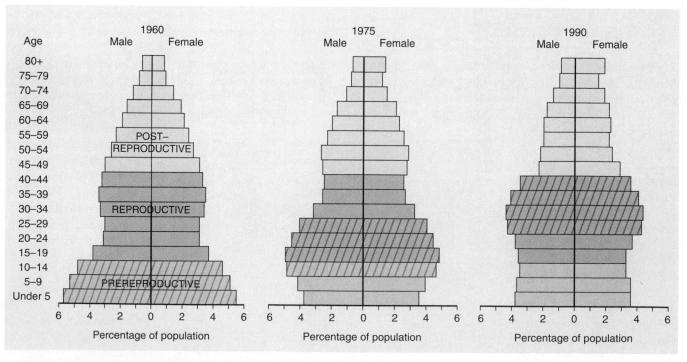

A. The age structure of the U.S. population in 1960, 1975, and 1990

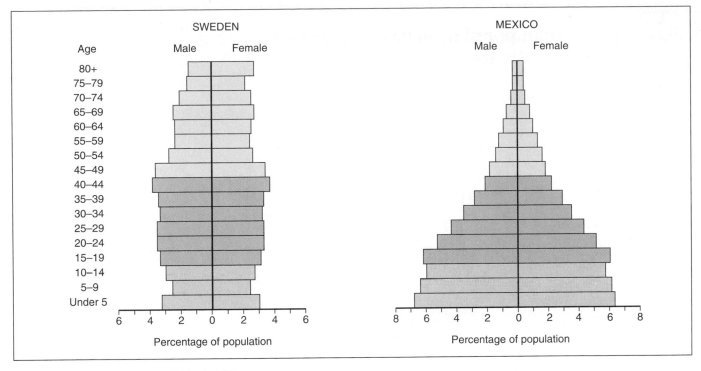

SWEDEN
Age   Male   Female

80+
75–79
70–74
65–69
60–64
55–59
50–54
45–49
40–44
35–39
30–34
25–29
20–24
15–19
10–14
5–9
Under 5

6   4   2   0   2   4   6
Percentage of population

MEXICO
Male   Female

8   6   4   2   0   2   4   6   8
Percentage of population

B. Age structures of Sweden and Mexico in 1990

tion growth. In general, a population with a large percentage of young reproductive or prereproductive individuals will grow more rapidly than a population whose age groups are nearly balanced.

Age-structure diagrams not only reveal a population's growth trends, they also indicate social conditions. Mexico and Sweden, for instance, will continue to have very different age distributions in the foreseeable future. They will also have very different social problems. For Sweden, the age-structure diagram tells us that a decreasing number of working-age people—mostly those of college age today—will soon be supporting an increasing number of retired people. The age structure of the United States is generally similar to Sweden's, although less stable. (Immigration now accounts for much of the growth in the U.S. population.) Programs such as Social Security and Medicare, which are crucial in supporting most older Americans, will become severely strained as the baby boomers retire.

While Sweden and the U.S. must plan for more health-care facilities for their aging populations, this will not be Mexico's most seri-

ous problem. Instead, Mexico will probably have an increasing number of working-age people who are unemployed. It will also have an increasing need for infant-care services and schools as the population rapidly grows.

We began this chapter by describing how a population of starlings grew quickly after a few individuals were introduced into North America. Starlings in North America and humans on a global scale are unusual in having their growth rates relatively unaffected by population-limiting factors at the present time. The global human population is essentially exploding, as we see in the next module.

C. Street scenes in Stockholm (left) and Mexico City (right)

Chapter 35   Population Dynamics   **689**

The human population has been growing exponentially for centuries

Worldwide, human population growth is a mosaic of various rates of growth in different countries. Some developed countries, such as Sweden, have virtually stable populations because birth rates and death rates balance. In sharp contrast to Sweden, most developing nations have burgeoning populations in which birth rates greatly exceed death rates. Partly as a result of such unchecked growth, many people in such countries face serious housing, water, and food shortages, and severe pollution problems. Figure A shows the stifling, unsanitary conditions in a slum that has arisen within a garbage dump in Manila. Currently one of the largest cities in the world, Manila has well over 10 million people. Its problems are not unique, however, and scenes much like this are common in a number of the world's large cities.

What about the human population as a whole? As Figure B shows, it has grown almost continuously for thousands of years. Growth was relatively slow for most of those years, with a slight dip in the fourteenth century, when the bubonic plague, or Black Death, wiped out about a fourth of the population of Europe. By about 1650, approximately 500 million people inhabited the world. Between 1650 and 1850, the population doubled to 1 billion; it doubled again to 2 billion between 1850 and 1930, and doubled still again by 1975 to more than 4 billion. In 1993, the population numbered about 5.6 billion people. At the current rate of increase, it takes less than three years for world population to add the equivalent of another United States (about 260 million). If the present growth persists, there will be about 8 billion people on Earth by the year 2020.

The growth of human populations, like that of other species, depends on birth rates and death rates. When agricultural societies replaced a lifestyle of hunting and gathering about 10,000 years ago, birth rates increased and death rates decreased, leading to the steady, but relatively slow, rate of population growth you see in the graph. Since the mid-1700s, when the Industrial Revolution began in England, the global human population has grown exponentially. This rate of growth has resulted mainly from falling death rates, es-

A. A slum in Manila, Philippines

B. The history of human population growth

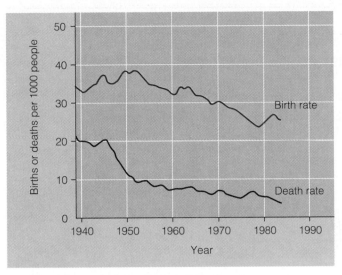

C. Changes in birth and death rates in Sri Lanka

D. Population growth in Sri Lanka

pecially infant mortality, throughout most of the world. Improved nutrition, better medical care, and sanitation—all *technological,* rather than biological, attributes of our species—have contributed to an increased percentage of newborns that survive long enough to leave offspring of their own. The effect of decreasing mortality on population growth is compounded in most developing countries, which tend to have relatively high birth rates. For example, the birth rate in Sri Lanka has decreased fairly steadily for nearly 50 years, but it has never declined enough to offset the drop in the death rate (Figure C). Thus, the overall effect is a growing population (Figures D and E).

For hundreds of years, agricultural and industrial technology have allowed us to exploit environmental resources more and more efficiently and to inhibit the mechanisms that would normally control our population growth. Put another way, technology has increased Earth's carrying capacity for people, allowing our long-term exponential growth. It is unlikely that any other population of large animals has ever sustained the level of growth depicted in Figure B for so long a time.

What is the future of the human population? It is difficult to predict the future size of any population accurately, and the human population is no exception. What we do know is that no population can continue to grow indefinitely. Exactly what the world's carrying capacity for humans is and under what circumstances we will approach it are topics of great concern and debate. Has our population already exceeded the total number that Earth can sustain, or have we not yet reached it? Ideally, human populations would reach carrying capacity smoothly and then level off. But if the population has already exceeded carrying capacity, a sharp decline might occur in the future. If the population fluctuates around carrying capacity, it might mean periods of increase would be followed by mass death, as has occurred during plagues and famines.

Barring some worldwide calamity, it is likely that the human population will continue to grow well into the next century. In 1993, the rate of growth worldwide was about 1.6% per year. As we saw earlier, scientists estimate that by the year 2020, the world population will be about 8 billion, about 1.5 times what it is today.

Whatever happens, we know that the human population will stop growing when birth rates and death rates are equal. A unique feature of human reproduction is that it can be deliberately controlled by voluntary contraception or government sanctions. Social change also affects birth rates. For example, in developed countries, many women are delaying marriage and reproduction, perhaps because of better opportunities for employment and advanced education. It seems more desirable for population control to result from a decrease in the birth rate by social changes or individual choice than by an increase in the death rate.

E. Schoolchildren in Sri Lanka

What does the future hold?
Ask a pessimist and an optimist

In 1968, Paul Ehrlich, Biology Professor at Stanford University and an authority on population growth, startled a complacent world into thinking about the dangers of over-population with his book *The Population Bomb*. Since then, Ehrlich has been a leading spokesman for population control, maintaining that human population growth is the single biggest problem facing humanity and the biosphere. Pointing out that the world is still gaining over 85 million people every year, Ehrlich is pessimistic about our future.

Julian Simon, economics professor at the University of Maryland, challenges Ehrlich and other "doomsdayers." Simon maintains an optimistic view of the world today, and sees even brighter days in the future, resulting from the application of technology. His book *The Ultimate Resource*, published in 1981, hails human ingenuity as the great problem solver.

Simon's views are outrageous to Ehrlich and others who are convinced that we are dangerously depleting natural resources, that there are limits to how much food can be produced, and that the planet's carrying capacity may already be surpassed. To Simon, however, there is no real evidence that such ideas are true. He maintains that none of Ehrlich's predictions of impending disaster from overpopulation have come to pass, and asserts that the more people there are, the more great minds there will be to solve problems.

Paul Ehrlich   Julian Simon

To Ehrlich, the main question is: What will be the effect on the biosphere of supporting billions more people in the near future? He points to the problems of acid rain, global warming, depletion of the ozone layer in the atmosphere, and the fact that these problems worsen as the global population increases exponentially.

Simon answers all doomsday concerns with optimism. He believes that if global warming or other problems become significant, we will find a way to solve them and avoid widespread ecological calamity.

Is it unwarranted paranoia to maintain that population growth is a serious problem? As Simon points out, more people are better fed, better clothed, and live longer today than ever before. Or is it naive to think that human ingenuity and technology will continue to solve global problems? Ehrlich points out that technology itself has created some of the problems we face—for example, the buildup of toxic pesticides in the environment. What if we *could* continue to raise the planet's carrying capacity, molding its environments to support billions more people? What effect would this have on our own quality of life and on all the other species on the planet? We will consider these important questions in light of our current global problems later in this unit.

Chapter Review

Begin your review by rereading the module headings and scanning the figures, before proceeding to the Chapter Summary and questions.

Chapter Summary

Introduction–35.1 Population dynamics is concerned with changes in population size and the factors that regulate populations over time. A population may be defined in a variety of ways, including criteria of reproductive isolation or a common resource base, or by geographical boundaries appropriate to research criteria.

35.2 Density and dispersion pattern are two important characteristics used to analyze and compare populations. Population density is the number of individuals in a given area or volume. It is sometimes possible to count all the individuals in a population, but density is usually estimated by sampling. Dispersion pattern is the pattern of spacing, which may be clumped, uniform, or random.

35.3 Idealized models describe two kinds of population growth. Exponential growth is the accelerating increase that occurs during a time when growth is unregulated. In nature, unlimited exponential growth can occur only for short periods. Logistic growth is slowed by population-limiting factors and tends to level off at carrying capacity, which is the number of individuals the environment can support.

35.4–35.5 The logistic growth model predicts that population-limiting factors have a stronger effect when the population is larger. Examples of these density-dependent factors are limited food supply and the buildup of toxic wastes. Density-independent factors limit many natural populations. These factors, such as weather, fire, and floods, affect the same percentage of individuals regardless of population density. Most populations are probably regulated by a mixture of density-dependent and density-independent factors. Some populations go through boom-and-bust cycles of growth and decline.

35.6 Natural selection shapes a species' life history, the series of events from birth through reproduction to death. There are two hypothetical extremes. Opportunistic populations produce many off-spring and grow exponentially in unpredictable environments. Equilibrial populations raise few offspring and maintain relatively stable populations in stable environments.

35.7 Life tables and survivorship curves predict an individual's statistical chance of dying or surviving during each interval in its life span. The three types of survivorship—living a full life span and dying of old age, reproducing and dying young, or maintaining a constant rate of mortality through the life span—reflect important species differences in life history.

35.8 The age structure of a population—its proportion of prereproductive, reproductive, and postreproductive individuals—affects its future growth. A human population with a large percentage of prereproductive and reproductive individuals (as in developing nations) will grow more rapidly and have different societal needs than a population whose age groups are nearly balanced (as in developed nations).

35.9–35.10 The human population as a whole is growing exponentially, having doubled three times in the last three centuries. It now stands at 5.6 billion and may reach 8 billion by the year 2020. Most of the increase is due to improved health and technology, which have decreased the death rate. The human population faces an uncertain future. Although some are optimistic about our ability to expand Earth's carrying capacity, others are concerned that our increase in numbers may damage the biosphere beyond repair. Eventually, births and deaths will balance; our behavior will probably determine whether this results from a decrease in birth rates or an increase in death rates.

Testing Your Knowledge

Multiple Choice

1. Which of the following shows the effects of a density-dependent limiting factor?
   a. A forest fire kills all the pine trees in a patch of forest.
   b. Early rainfall triggers the explosion of a locust population.
   c. Drought decimates the wheat crop.
   d. Silt from logging kills half the young salmon in a stream.
   e. Voles multiply, and foxes switch to voles as a food source.

2. A population that is growing logistically
   a. grows fastest when density is lowest
   b. has a high intrinsic rate of increase
   c. grows fastest at an intermediate population density
   d. grows fastest as it approaches carrying capacity
   e. is always slowed by density-independent factors

3. Pine trees in a forest tend to shade and kill pine seedlings that sprout nearby. This causes the pine trees to (*Explain your answer.*)
   a. increase exponentially
   b. grow in a clumped pattern
   c. grow in a uniform pattern
   d. exceed their carrying capacity
   e. grow in a random pattern

4. To figure out the human population density of your community, you would need to know the number of people living there and
   a. the land area in which they live
   b. the birth rate of the population
   c. whether population growth is logistic or exponential
   d. the dispersion pattern of the population
   e. the carrying capacity

5. Exponential growth of the human population since the beginning of the Industrial Revolution appears to be a result of
   a. migration to thinly settled regions of the globe
   b. better nutrition's boosting the birth rate
   c. a drop in the death rate due to better health care
   d. the concentration of humans in cities
   e. social changes that make it desirable to have more children

True/False (*Change false statements to make them true.*)

1. Growth of the human population has accelerated in the last century.

2. Because the resources they use are not evenly distributed, many plants and animals show a random pattern of dispersion.

3. Weather and climate are important density-independent factors limiting populations.

4. A population with an opportunistic life history is likely to be regulated mainly by density-independent factors.

Describing, Comparing, and Explaining

1. Compare exponential and logistic population growth. Under what conditions might each occur? What might limit growth?

2. Is carrying capacity a characteristic of a species, its environment, or both? Explain.

3. What is survivorship? What does a survivorship curve show? Explain what the three survivorship curves in Figure 35.7 tell us about humans, squirrels, and oysters.

Thinking Critically

1. The mountain gorilla, spotted owl, giant panda, snow leopard, and California condor are all endangered by human encroachment on their environments. Another thing these animals have in common is that they all have equilibrial life histories. Why might they be more easily endangered than animals with opportunistic life histories?

2. Beavers are released in an effort to repopulate a valley where they were trapped to extinction many years ago. Many of their young are killed by predators, but each original pair produces an average of three offspring that survive to maturity. (This means, in effect, 1.5 offspring per parent per generation.) If this rate of growth continues, roughly how many beavers will there be in the sixth generation for each original pair in the first generation? Graph the number of beavers against generation, and compare to the curves in Figure 35.3C. What kind of growth is seen in the beaver population? What would the growth curve look like if the beavers averaged 1.1 offspring per parent per generation? Do numbers have to double each generation for exponential growth to occur?

3. A fisheries biologist studied a population of cichlids in a small (120 hectares) African lake. He found that the fish lived only in scattered reed beds that accounted for one-fourth the area of the lake. He caught 185 fish, tagged them, and released them. Two days later, he netted 208 fish and found that 35 of them were tagged. How many cichlids are in the lake? What is the density of the cichlid population? What is the pattern of dispersion of the fish? How might the pattern of dispersion affect your interpretation of their density?

Science, Technology, and Society

1. Many people regard the rapid population growth of developing countries as our most serious environmental problem. Others feel that growth of developed countries, though slower, is actually a greater threat to the environment. What kinds of environmental problems result from population growth in (a) developing countries, and (b) developed countries? Which do you think is the greater threat? Why?

2. During the 1980s, domestic political pressures caused the United States to terminate its support of worldwide United Nations population planning programs. Although these programs emphasized prevention of pregnancy, U.S. policy was changed so that funds were denied to any organization that informed women that abortion was one of their options. What effects might the U.S. action have on population planning (education, contraception, abortion, etc.) and population growth in developing countries? Do you feel the action was justified? Why or why not?

A 4-millimeter-long wasp called *Apanteles glomeratus* stabs through the skin of a caterpillar and lays her eggs (photograph at the left). The caterpillar, a larva of the cabbage white butterfly (*Pieris rapae*, below left), is doomed. It will be destroyed from within, as the wasp larvae hatch and nourish themselves on its internal organs. We benefit from the wasp's behavior, for *Pieris* caterpillars eat cabbages and broccoli and are abundant agricultural pests.

Apanteles wasps help control cabbage butterfly populations, but they never come close to eliminating them. There are always fewer *Apanteles* than *Pieris* caterpillars, for *Apanteles* has problems of its own. It is not the only wasp that can inject its eggs into other insects. Other tiny wasps, called ichneumons (below center), not only have this ability, they can also detect when a *Pieris* caterpillar contains *Apanteles* larvae. When a female ichneumon finds one of these caterpillars, she pierces it and deposits her eggs inside the *Apanteles* larvae. And the story may go on. Yet another wasp, a chalcid (below right), may lay its eggs inside the ichneumon larvae. When this happens, the hapless caterpillar houses a three-step food chain: chalcids eating ichneumons eating *Apanteles* eating the caterpillar. Usually, only the chalcids will emerge from the dead husk of the caterpillar.

Though he was unaware of this unusual food chain, eighteenth-century English satirist Jonathan Swift wrote:

> So, Nat'ralists observe, a Flea
> Hath smaller Fleas that on him prey,
> And these have smaller Fleas to bite 'em,
> And so proceed *ad infinitum.*

Swift's flea analogy was meant to ridicule picky literary critics, but his lines paint a good picture of the complex relationships between *Pieris* caterpillars and wasps.

Connections among organisms is the major topic of this chapter. In the first few modules, we see that a biological community derives its structure from the interactions and interdependence of the organisms living within it. Later, we find that an ecosystem functions as a result of the complex interactions between a community and its physical environment. When we get to ecosystems, we take another look at Jonathan Swift's poem; although it works as satire, its last line can't be taken literally for an ecosystem, because no food chain, including the extensive one in a *Pieris* caterpillar, can be infinite.

A cabbage white butterfly (*Pieris rapae*)

An ichneumon wasp

A chalcid wasp

A community is all the organisms inhabiting a particular place

In the previous chapter, we saw that a population is a group of interacting individuals of a particular species. We now move one step up the hierarchy of nature to the level of the community. A **biological community,** such as the forest shown here, is an assemblage of all the organisms living together and potentially interacting in a particular area. The wild turkeys in the photograph interact with many other species, such as oak trees whose acorns they eat.

Just as a population has certain characteristics, such as density and dispersion pattern, a community has its own set of properties. Its defining characteristics are its diversity, its prevalent form of vegetation, its stability, and its trophic structure.

The **diversity** of a community—the variety of different kinds of organisms that make it up—has two components. One is species richness, or the total number of different species in the community. The other is the relative abundance of the different species. For example, imagine two communities, each with 100 individuals distributed among four different species (A, B, C, and D) as follows:

Community 1: 25A, 25B, 25C, 25D
Community 2: 97A, 1B, 1C, 1D

The species richness is the same for both communities, because they both contain four species, but the relative abundance is very different. Suppose these communities were two forests. You would easily notice the four different types of trees in Community 1, but without looking carefully, you might see only the abundant species A in the second forest. You could get a true picture of diversity in these communities only by carefully examining both species richness and relative abundance.

The second property of a community, its **prevalent form of vegetation,** applies mainly to terrestrial situations. As we saw in Chapter 34 for the major world communities called biomes, terrestrial communities (unlike marine communities) are distinguished by their dominant plants. For example, deciduous trees prevail in temperate deciduous forests, whereas coniferous trees prevail in the taiga. When we look at a biome, we see not only which plants are dominant, but also how the plants are arranged, or "structured." For instance, the forest shown here has a pronounced vertical structure: The treetops form a top layer, or canopy, under which there is a subcanopy of lower branches, and small shrubs and herbs carpet the forest floor. Thus, the prevalent form of a community's vegetation has two components: the kinds of dominant plants and their particular structure. The types and structural features of plants largely determine the kinds of animals that live in a community.

The third property of a community, **stability,** refers to the community's ability to resist change and return to its original species composition after being disturbed. Stability

Wild turkeys in a forest community

depends on both the type of community and the nature of disturbances. For example, a forest dominated by cedar and hemlock trees is a highly stable community in that it may last for thousands of years with little change in species composition. However, after a fire that kills the dominant species, a cedar/hemlock forest might seem less stable than, say, a grassland, because it will probably take much longer for the forest to return to its original species composition.

The fourth property of a community is its **trophic structure** (from the Greek *trophe,* nourishment), or the feeding relationships among the various species making up the community. A community's trophic structure determines the passage of energy and nutrients from plants and other photosynthetic organisms to herbivores and then to carnivores. For example, as we saw in the chapter's introduction, the energy in a cabbage leaf eaten by a *Pieris* caterpillar may end up supporting one, two, or three kinds of carnivorous wasp larvae.

With the four main properties of a community in mind, we turn next to the forces that tie species populations together into communities. These forces—actually, the interactions between the species themselves—are of three main types: competition, predation, and symbiosis. In discussing these interactions, we'll see that they are all influenced by evolution through natural selection. We begin with a look at competition.

When two populations (for example, two species of birds in a forest) both require a limited resource (nest sites, for instance), individuals of the two species compete for the resource. The contest, called **interspecific competition**, can inhibit the growth of both populations. Sometimes the competition even eliminates one of the populations from a community. Interspecific competition may play a major role in structuring a community.

In 1934, Russian ecologist G. F. Gause studied the effects of interspecific competition in laboratory experiments with two closely related species of protists, *Paramecium aurelia* and *P. caudatum* (Figure A). Gause cultured the protists under stable conditions with a constant amount of bacteria added every day as food. The top graph shows what he found when he grew the two species in separate cultures. Each population grew rapidly and then leveled off at what was apparently the carrying capacity of the culture (see Module 35.3). In contrast, the bottom graph shows what happened when Gause cultured the two species together. Here, *P. aurelia* apparently had a competitive edge in obtaining food, and *P. caudatum* was driven to extinction in the culture. Gause concluded that two species so similar that they compete for the same limiting resources cannot coexist in the same place. One will use the resources more efficiently and thus reproduce more rapidly. Even a slight reproductive advantage will eventually lead to local elimination of the inferior competitor. His ideas were termed the **competitive exclusion principle.** Later experiments on several species of animals and plants have reinforced the principle.

The competitive exclusion principle applies to what is called a population's niche. Broadly defined, a **niche** is the way a population "fits into" its environment—the sum total of the population's use of the biotic and abiotic resources of its environment. For example, the temperature range within which the population lives, the time of day individuals feed, and the type of food they consume are all part of its niche. Combining the niche concept with the competitive exclusion principle, we can predict that populations of two species cannot coexist in a community if their niches are identical.

Field studies of the interactions between populations of two species of barnacles that attach to intertidal rocks on the North Atlantic coast indicate how interspecific competition and the niche can affect the structure of a community (Figure B). Both types of barnacles, *Balanus* and *Chthamalus,* grow on rocks that are exposed during low tide; when immersed at high tide, they feed on organic particles suspended in the water. Both types are attached as adults, but have free-swimming larvae that may settle and begin to develop on virtually any rock surface. Attachment sites on rocks, the amount of exposure to seawater and air, and interactions with other species are some of the aspects of each barnacle's niche.

Chthamalus (shown in brown) occupies the upper parts of the rocks, which are out of the water longer during low tides. *Balanus* (gray) fails to survive as high on the rocks as *Chthamalus,* apparently because *Balanus* dries out in the air. When experimenters removed *Balanus* from the lower rocks, *Chthamalus* spread lower, colonizing the unoccupied rocks. However, when both species colonize the same rock, *Balanus* eventually displaces *Chthamalus* on the lower part of the rock. Researchers conclude that the upper limit of *Balanus'* distribution is set mainly by the availability of water, whereas the lower limit of *Chthamalus'* distribution is set by competition.

There are actually more questions than answers about the role of competition in natural communities. It is hard to determine which facets of the niche are critical when species seem to be competing. And it's even harder to assess what has happened in the evolutionary past. When competitive exclusion has occurred, has it actually determined how many species coexist in a community? Did "outcompeted" populations become locally extinct or migrate out of the area? Or were they altered by natural selection, becoming able to use a different set of resources? These are some of the fascinating but difficult questions confronting researchers in community ecology.

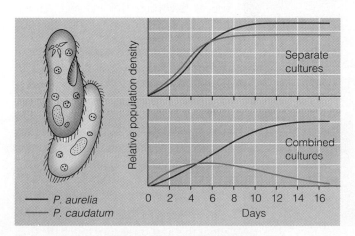

A. Competition in laboratory populations of *Paramecium*

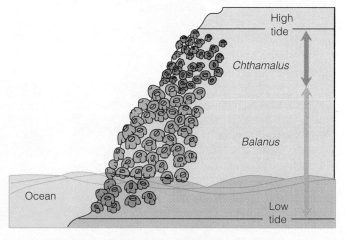

B. The niche concept illustrated by two species of barnacles

Predation leads to diverse adaptations in both predator and prey

In **predation**, an interaction where one species eats another, the consumer is called a **predator** and the food species is known as the **prey**. We will use these terms not only for cases of animals eating other animals, but also for plant-herbivore interactions, where the plant is prey. Thus, the *Pieris* caterpillar in the chapter's introduction is a predator of cabbage and broccoli.

No species is entirely free from predation, especially when young, and mechanisms of defense against being eaten have evolved in every species. Some species grow too big to be consumed by certain predators; adult elephants, for instance, are generally safe from attack by lions and other large cats. Other species can flee or hide, and many plant and animal species have protective armor, spines, or external glands that produce noxious chemicals. These and other antipredator defense mechanisms evolve through natural selection, as predators and prey interact. Adaptations that favor the predator, such as speed and agility, claws, fangs, and ambush tactics, counter the defense systems of prey. Some predator-prey interactions illustrate the concept of **coevolution**, a series of reciprocal adaptations in two species. Coevolution occurs when a change in one species acts as a new selective force on another species, and counteradaptation of the second species, in turn, affects selection of individuals in the first species.

Figure A illustrates an example of coevolution of an herbivorous insect (the caterpillar of the butterfly *Heliconius*, top left) and a plant (the passionflower *Passiflora*, a tropical vine). *Passiflora* produces toxic chemicals that protect its leaves from most herbivorous insects, but *Heliconius* caterpillars have digestive enzymes that break down the toxins. As a result, *Heliconius* gains access to a food source that few other insects can eat. These poison-resistant insects seem to be a strong selection force favoring the survival of *Passiflora* plants that have additional defenses. For instance, the leaves of some species of *Passiflora* produce yellow sugar deposits that look like *Heliconius* eggs. You can see two eggs in the top right photograph of Figure A and two of the sugar deposits in the bottom photo. Female *Heliconius* butterflies avoid laying their eggs on leaves that already have eggs. Presumably, this is an adaptation that ensures an adequate food supply for offspring, since only a few caterpillars will hatch and feed on any one leaf. Because the butterfly often mistakes the yellow sugar deposits for eggs, *Passiflora* species with the deposits are more likely to escape predation.

Because plants cannot run away from herbivores, chemical toxins, often in combination with various kinds of antipredator spines and thorns, are their main arsenals against being eaten to extinction. Among such chemical weapons are the poison strychnine, produced by a tropical vine called *Strychnos toxifera;* morphine, from the opium poppy; nicotine, produced by the tobacco plant; mescaline, from peyote cactus; tannins, from a variety of plant species; and many substances we use as flavorings (for instance, cinnamon, cloves, and mints) but that are toxic to predators.

Animal defenses against predators are extremely diverse. Mechanical defenses, such as the porcupine's sharp quills, may be the most obvious, but chemical defenses are also widespread. Animals with effective chemical defenses are often brightly colored, a warning to predators. The vivid markings of the poison-arrow frog (Figure B), an inhabitant of rain forests in Costa Rica, warn of deadly alkaloids in the frog's skin; predators learn about this as soon as they touch the frog.

Eggs

Sugar deposits

A. Coevolution: *Heliconius* and the passionflower vine

B. Chemical defenses: the poison-arrow frog   **C.** Camouflage: a canyon tree frog on granite

In some parts of South America, human hunters in the rain forest tip their arrows with poisons from similar frogs to bring down large mammals.

Camouflage is an especially common type of defense in the animal kingdom. Figure C shows that a canyon tree frog *(Hyla arenicolor)*, common in the southwestern U.S., becomes almost invisible on a granite background.

Mimicry is another effective defense mechanism. The insect on the bottom in Figure D is a honeybee, which has a stinger armed with a toxic chemical. The insect on the top is a nonstinging flower fly. Many predators learn to avoid the harmful bee and may not attack the harmless fly because it looks like a bee. This type of mimicry, in which a palatable species mimics an unpalatable one, is called **Batesian mimicry.** For this type of mimicry to be effective, the mimic must be considerably less abundant than the species it copies; otherwise, predators would learn that, for example, yellow insects that look like bees are good rather than bad to eat.

Figure E illustrates another kind of mimicry, called **Müllerian mimicry,** in which two unpalatable species that inhabit the same community mimic each other. On the bottom is a

D. Batesian mimicry: a flower fly (top)
and a honeybee (bottom)

E. Müllerian mimicry: a cuckoo bee (top)
and a yellow jacket (bottom)

type of wasp called a yellow jacket; on the top is a species of bee called a cuckoo bee. Both species have stingers that release toxic chemicals. Presumably, both gain an adaptive advantage beyond their own defenses because predators will learn more quickly to avoid any prey with this appearance.

36.4 Predation can maintain diversity in a community

In simple laboratory experiments where a single predator species is kept with a single prey species having no refuge, the predator may devour all the prey and then itself perish from starvation. In Figure A, the creature on the left in the micrograph is a predatory protist called *Didinium*. It is shown here eating a *Paramecium*. The graph shows *Didinium's* effect on a population of *Paramecium* when the two are grown together in a test tube. At first, the *Paramecium* population expands rapidly, but *Didinium* soon kills all the *Paramecium* and then dies out itself. If this scenario occurred often in nature, predators would reduce the diversity of species in communities. However, predators rarely drive their prey to extinction, for several reasons. For one thing, natural communities are complex, with many species; predators themselves are often preyed upon, limiting their numbers. Also, a predator may be able to switch to an alternative food source when the population of one prey species dwindles. And the defensive mechanisms of prey play a major role in keeping prey populations from extinction.

Several studies have shown that predator-prey relationships can actually help maintain community diversity rather than reduce it. Experiments by American ecologist Robert Paine in the 1960s were among the first to provide evidence. Paine removed the dominant predator, a sea star of the genus *Pisaster* (Figure B), from experimental areas within the intertidal zone of the Washington coast. The result was that *Pisaster's* main prey, a mussel of the genus *Mytilus,* outcompeted many of the other shoreline organisms (barnacles and snails, for instance) for the important resource of space on the rocks. The number of species dropped from fifteen to eight.

The experiments of Paine and others have generated the concept of keystone predators. A **keystone predator** is a species, such as *Pisaster,* that reduces the density of the strongest competitors in a community. In doing so, the predator helps maintain species diversity by preventing competitive exclusion of weaker competitors. In Paine's study, predation by *Pisaster* was a key factor in maintaining populations of at least seven other species.

SEM 480×

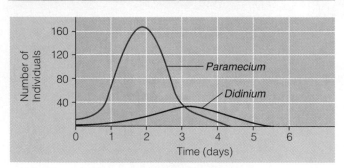

A. Predator-prey dynamics in a laboratory experiment

B. *Pisaster* sea star, a keystone predator, eating a mussel

36.5 Symbiotic relationships help structure communities

A **symbiotic relationship** is an interaction between two or more species in which one species lives in or on another species. There are three main types of symbiotic relationships: parasitism, commensalism, and mutualism. All three can be important to community structure.

Parasitism (Greek, *para,* near, and *sitos,* food) is a kind of predator-prey relationship in which one organism, the parasite, derives its food at the expense of its symbiotic associate, the host. Parasites are usually smaller than their hosts. An example of a parasite is a tapeworm that lives inside the in-

A. Rabbits in Australia before a parasite was introduced

relationship, and today, the virus that was originally introduced has only a mild effect on the rabbit population. Pest managers continue to control the rabbits by introducing other deadly strains of the myxoma virus.

In contrast to parasitism, in **commensalism** (Latin, *com,* together, and *mensa,* table), one partner benefits without significantly affecting the other. Few cases of absolute commensalism probably exist, because it is unlikely that one of the partners will be completely unaffected. Commensal associations sometimes involve one species' obtaining food that is inadvertently exposed by another. For instance, several kinds of birds feed on insects flushed out of the grass by grazing cattle. It is difficult to imagine how this could affect the cattle, but the relationship may help or hinder them in some way not yet recognized.

testines of a larger animal and absorbs nutrients from its host. Natural selection favors the parasites that are best able to find and feed on hosts. At the same time, defensive abilities of the hosts are also selected for. For instance, plants make chemicals toxic to fungal and bacterial parasites, along with ones toxic to predatory animals (sometimes they are the same chemicals). In vertebrates, the immune system provides a multipronged defense against internal parasites.

At times, it is actually possible to watch the effects of natural selection in host-parasite relationships. Scenes like the one shown in Figure A were common in Australia in the 1940s. The continent was overrun by hundreds of millions of European rabbits, whose population had exploded from just twelve pairs imported a century earlier. The rabbits destroyed huge

B. Mutualism between an acacia tree and ants

The third type of symbiosis, **mutualism** (Latin, *mutualis,* reciprocal) benefits both partners in the relationship. We have discussed several mutualistic associations in previous chapters—for instance, the legume plants and their nitrogen-fixing bacteria, and the interactions between flowering plants and their pollinators. Figure B illustrates another case of mutualism. This is part of a branch of a bull's horn acacia tree, which grows in Central and South America. The tree provides room and board for ants of the genus *Pseudomyrmex.* The ants live in large, hollow thorns and eat sugar secreted by the tree. As you can see in the photograph, they also eat the yellow structures at the tips of leaflets; these are protein-rich swellings that seem to have no function for the tree except to attract ants. The ants benefit the host tree by attacking virtually anything that touches

expanses of Australia and threatened the sheep and cattle industries. In 1950, myxoma virus, a parasite that infects rabbits, was deliberately introduced into Australia to control the rabbit population. Spread rapidly by mosquitoes, the virus devastated the rabbit population. The virus was less deadly to the offspring of surviving rabbits, however, and it caused less and less harm over the years. Apparently, genotypes in the rabbit population were selected that were better able to resist the parasite. Meanwhile, the deadliest strains of the virus perished with their hosts as natural selection favored strains that could infect hosts but not kill them. Thus, natural selection stabilized this host-parasite

it. They sting other insects and large herbivores and even clip surrounding vegetation that grows near the tree. When the ants are removed, the trees die, probably because herbivores damage them so much that they are unable to compete with surrounding vegetation for light and growing space.

The complex interplay of species in symbiotic relationships highlights an important point about communities: They are structured by a web of diverse connections among organisms. In the next module, we see what can happen when community structure is disrupted.

Outside disturbances can radically alter community structure

Communities may change drastically after a flood, fire, glacial advance or retreat, or volcanic eruption strips away their vegetation. The disturbed area may be colonized by a variety of species. Later, these may be replaced as yet other species colonize the area. Such a transition in the species composition of a community is called **ecological succession.**

Today, no single force has a greater impact on community change and succession than human disturbance. Logging and clearing for agriculture have reduced large tracts of mature forests to small patches of disconnected woodlots in many parts of the United States and throughout Europe. Vast areas that were once grasslands in the midwestern U.S. have been converted to croplands. In the Southern Hemisphere, tropical rain forests are quickly disappearing as a result of clear-cutting for lumber and pastureland. In parts of Africa, centuries of overgrazing have contributed to famines through desertification, turning seasonal grasslands into great barren areas.

When human activity or other disturbance removes the vegetation from a community but leaves the soil intact, the area may return to its former state if left alone. For instance, forested areas in the eastern U.S. that were cleared for farming will, if abandoned, undergo succession and eventually return to forest. The earliest plants to recolonize an area are usually herbaceous (nonwoody) species that grow from seeds blown in from neighboring areas or carried in by animals. These plants thrive where there is little competition from other plants. Often within a year, if the area is not burned or heavily grazed, woody shrubs will sprout from windblown or animal-borne seeds, eventually replacing most of the herbaceous species. Still later, forest trees will replace most of the shrubs.

You can find early and intermediate successional stages almost anywhere that forests have been clear-cut and in croplands no longer under cultivation. In fact, much of the U.S. is now a hodgepodge of early successional growth where mature communities once prevailed. If you live in a city, you can find early successional plants in many vacant lots. Most of the plants growing in the vacant lot in Figure A are early colonizers, and some are exotic species brought to North America during colonial times. The trees in the photo are Eurasian "trees of heaven," widely planted in the U.S. in the 1800s and now common in many cities. Exotic species such as this thrive in disturbed areas just like early successional plants native to North America.

So far we have described successional changes that occur in disturbed areas where there is

intact soil. Succession also occurs in areas that have no soil and are essentially barren of life. Examples are new islands formed by volcanoes at sea, strip-mined land, and the clay, sand, and gravel left by retreating glaciers. Figure B illustrates this type of succession in Alaska. Though these photographs are not all from the same area, they illustrate the main stages of succession that follow the retreat of glaciers. The entire sequence takes about 200 years. ① The first photograph shows a retreating glacier in Glacier Bay. ② After the retreat, the landscape is barren and soilless. ③ Lichens and mosses, which grow from windblown spores, are the first large photosynthetic organisms to colonize the barren ground. ④ Soil gradually develops, and dwarf willows, cottonwoods, and alders, seeded from nearby areas, overgrow the lichens and mosses. In turn, ⑤ spruce trees begin to colonize the forest. Finally, ⑥ a spruce and hemlock forest dominates the landscape.

Some communities, such as a spruce and hemlock forest, seem to be end points, that is, mature stages that will persist indefinitely. Ecologists once viewed succession as a sequence that inevitably moved to a permanent final stage, for a particular climate and soil type—what they called a climax community. Today, however, this idea is viewed more critically. In many areas, what appear to be climax communities may not be stable over long periods, and many communities are routinely disturbed and never reach a climax. For instance, fires often sweep prairie grasslands, and without fire, some of these communities would eventually become forests. In this case we might say that forest is the climax community, but that makes little sense if the forest community never develops. In reality, periodic fires stabilize the community at a stage that does not exactly fit the traditional idea of a climax community.

In ecological succession, we see the results of organisms reacting to both biotic and abiotic factors in their environment. We have, therefore, gone past the level of community, a unit of interacting species, to the level of ecosystem. We focus on ecosystems next.

A. Succession in a vacant lot

1 Retreating glacier

2 Barren and soilless landscape

3 Lichens and mosses

4 Dwarf willows, cottonwoods, and alders

5 Colonization by spruce trees

6 Spruce and hemlock forest

B. Succession after the retreat of glaciers

36.7 Energy flow and chemical cycling are the two fundamental processes of ecosystems

The terrarium in the figure below contains an **ecosystem,** a community of organisms interacting with the abiotic factors in their environment. The arrows highlight the two fundamental processes of this and every other ecosystem: energy flow and chemical cycling. **Energy flow** is the passage of energy through the components of the ecosystem. Energy enters the terrarium and most other ecosystems in the form of sunlight. The plants (which are autotrophs, or self-feeders) convert the light energy to chemical energy. Animals (heterotrophs, which feed on other organisms) obtain some of this chemical energy in the form of organic compounds

when they eat the plants. Other heterotrophs, such as bacteria and fungi in the soil, obtain much of the chemical energy when they decompose the dead remains of plants and animals. Every use of the chemical energy by the organisms involves a loss of some energy to the surroundings in the form of heat. Eventually, therefore, the ecosystem would run out of energy if it were not powered by a continuous inflow of new energy from an outside source. For most ecosystems, the sun is the outside energy source. One exception is the hydrothermal vent ecosystem (discussed in Chapter 34's introduction), powered by chemical energy from Earth itself.

In contrast to energy flow, **chemical cycling** involves the circular movement of materials *within* the ecosystem. An ecosystem, especially an artificial one like a terrarium, is more-or-less self-contained in terms of materials. Chemical elements such as carbon and nitrogen are cycled between abiotic components (air, water, and soil) and biotic components (organisms) of the ecosystem. The plants acquire these elements in inorganic form from the air and soil and fix them into organic molecules, some of which animals consume. Microorganisms that break down organic wastes and dead organisms return most of the elements in inorganic form to the soil and air. Some elements are also returned as the inorganic by-products of plant and animal metabolism.

In summary, energy flow and chemical cycling both involve the transfer of substances through the feeding levels of the ecosystem. However, energy flows into and out of the ecosystem, whereas chemicals are recycled within the ecosystem. In the rest of this chapter, we discuss energy flow and chemical cycling in ecosystems.

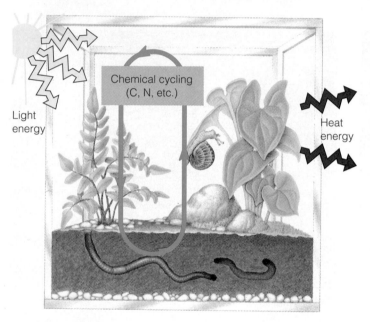

A terrarium ecosystem

36.8 Trophic structure determines ecosystem dynamics

The community of living organisms within every ecosystem has a trophic structure, a pattern of feeding relationships, consisting of several different levels. The trophic structure determines the route that energy takes in flowing through the ecosystem, as well as the pattern of chemical cycling. The sequence of food transfer from trophic level to trophic level is known as a **food chain.** In the figure on the facing page, the trophic levels are arranged vertically, and the names of the levels appear in colored boxes. The arrows connecting the organisms point from the food to the consumer.

The figure compares a terrestrial food chain and an aquatic food chain. Starting at the bottom, the trophic level that supports all others consists of autotrophs, which ecologists call the **producers** in an ecosystem. Most producers

are photosynthetic organisms that use light energy to power the synthesis of organic compounds. Plants are the main producers on land. In water, the producers are mainly photosynthetic protists and cyanobacteria, collectively known as phytoplankton. Multicellular algae and aquatic plants are also important producers in shallow waters.

All organisms in trophic levels above the producers are heterotrophs, or consumers, and all consumers are directly or indirectly dependent on the output of producers. Herbivores, which eat plants, algae, or autotrophic bacteria, are the **primary consumers** of autotrophs and their products. Primary consumers on land include grasshoppers and many other insects, snails, and certain vertebrates, such as grazing mammals and birds that eat seeds and fruit. In aquatic environments, primary consumers include a variety of zooplankton (mainly protists and microscopic animals such as small shrimps) that prey on the phytoplankton.

Above the primary consumers, the trophic levels are made up of carnivores, which eat the consumers from the level below. On land, **secondary consumers** include many small mammals, such as the mouse shown here eating an herbivorous insect, and a great variety of small birds, frogs, and spiders, as well as lions and other large carnivores that eat grazers. In aquatic ecosystems, secondary consumers are mainly small fishes that eat small bottom-dwelling invertebrates and zooplankton.

Higher trophic levels include **tertiary consumers,** such as snakes that eat mice and other secondary consumers. Most ecosystems have secondary and tertiary consumers and, as the figure indicates, some also have a higher level, **quaternary consumers.** These include hawks in terrestrial ecosystems, and killer whales in the marine environment.

Not shown in the figure is another trophic level of consumers called **detritivores,** which derive their energy from **detritus,** the dead material produced by all the trophic levels. Detri-

tus includes animal wastes, plant litter, and all sorts of dead organisms. Most organic matter eventually becomes detritus and is consumed by detritivores. A great variety of animals, often called scavengers, eat detritus. For instance, earthworms, many rodents, and insects eat fallen leaves and other detritus. Other scavengers include crayfish, catfish, and vultures.

An ecosystem's main detritivores are the bacteria and fungi. Enormous numbers of these microorganisms in the soil and in mud at the bottom of lakes and oceans convert (recycle) most of the ecosystem's organic materials to inorganic compounds that plants or phytoplankton can use. The breakdown of organic materials to inorganic ones is called **decomposition.** In a sense, all organisms perform decomposition. In cellular metabolism, they all break down organic material and release inorganic products, such as carbon dioxide and ammonia, to the environment.

Food chains provide an overview of ecosystem structure and function, but they are an oversimplification. Natural ecosystems never have a single, unbranched food chain, as we see in the next module.

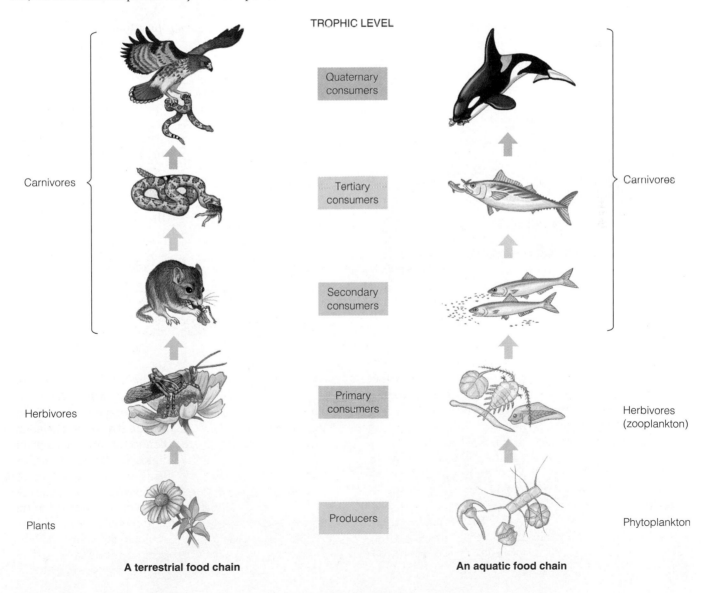

TROPHIC LEVEL

Quaternary consumers

Tertiary consumers

Secondary consumers

Primary consumers

Producers

Carnivores

Herbivores

Plants

Carnivores

Herbivores (zooplankton)

Phytoplankton

A terrestrial food chain

An aquatic food chain

Food chains interconnect, forming food webs

A more realistic view of the trophic structure of an ecosystem than a food chain is a **food web**, a network of interconnecting food chains. This figure shows a simplified example of a food web in a salt marsh. Rooted plants, such as grasses and sedges, are the main producers along the shoreline. In deeper water, a few submerged algae supplement the production of food by phytoplankton.

The tan arrows represent primary consumption. Notice that a consumer may eat more than one type of producer, and several species of primary consumers may feed on the same species of producer. The purple arrows represent secondary consumption. On the far left, the shrew is strictly a secondary consumer, eating insects and earthworms. The mallard duck (to the right of the shrew), however, is both a primary consumer, eating plant seeds, and a secondary consumer, eating insects and other small invertebrates. Likewise, as indicated by the purple and blue arrows pointing to it, the owl is both a secondary and a tertiary consumer, because many of the small mammals and birds it eats are primary as well as secondary consumers. There are no quaternary consumers in this ecosystem, and this is often the case in nature. Finally, the ecosystem also includes detritivores (at the bottom of the figure) that consume dead organic material from all trophic levels.

Though more realistic than a food chain, this figure is still a highly simplified model of the feeding relationships in the ecosystem. An actual food web would involve many more organisms at each trophic level, and virtually all the animals would have a more diverse diet than shown here. Indicating "who eats whom," the arrows in a food web diagram outline an ecosystem's overall structure. The interconnecting arrows show the linkages among the organisms, connections based on the movement of chemical nutrients and energy. In the next module, we take a closer look at energy flow in ecosystems.

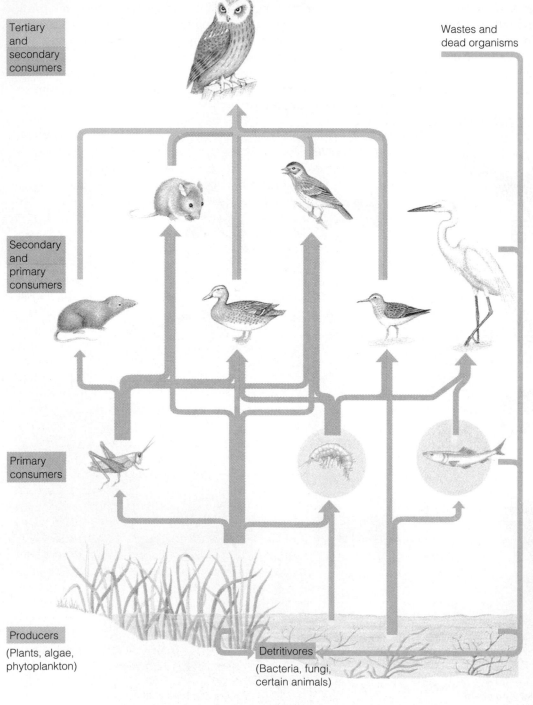

Tertiary
and
secondary
consumers

Wastes and
dead organisms

Secondary
and
primary
consumers

Primary
consumers

Producers

(Plants, algae,
phytoplankton)

Detritivores

(Bacteria, fungi,
certain animals)

Energy supply limits the length of food chains

Each day, planet Earth receives far more solar energy than is necessary to support the biosphere. Most of this energy is absorbed, scattered, or reflected by the atmosphere or by Earth's surface. Of the visible light that reaches leaves, algae, and cyanobacteria, only about 1% is converted to chemical energy by photosynthesis. But on a global scale, this is enough to produce about 170 billion tons of organic material per year in the biosphere.

Ecologists call the amount, or mass, of organic material in an ecosystem the **biomass.** They call the rate at which producers convert solar energy to chemical energy (organic compounds) **primary productivity.** Thus, the primary productivity in the entire biosphere is about 170 billion tons of biomass per year. Because all the consumers in an ecosystem acquire their organic fuels from producers (either directly or via other consumers), primary productivity sets the ecosystem's spending limit for energy. The energy content in 170 billion tons of organic matter is the living world's total yearly energy budget.

Let's see how the world's energy budget gets divided up. The figure here, called an energy pyramid, is typical of the results of numerous studies of energy and productivity in ecosystems. This particular pyramid represents an ecosystem with four trophic levels. Each tier of the pyramid represents a trophic level, and the relative sizes of each tier indicate how much energy flows from one level to the next. The 10,000 kcal of energy in the producer trophic level amounts to 1% of 1 million kcal of sunlight available to the producers. In this generalized pyramid, 10% of the energy available at each trophic level appears at the next higher level. The 90% decline of productivity from one trophic level to the next is a rough average. In actual ecosystems, the decline varies with the species present, generally ranging from as high as 98% to about 80%.

The basic trend you see in the figure, productivity declining significantly with each higher trophic level, holds for all ecosystems. This is partly a consequence of the laws of thermodynamics (see Module 5.2): Energy cannot be created or destroyed, though it can be changed from one form to another (first law of thermodynamics); and all energy transformations (including the metabolic reactions in living organisms) involve the conversion of some energy to heat, which is lost from the system (second law). Thus, not all of the chemical energy stored as biomass by plants can be converted to the chemical energy of consumers. Even if consumers were to eat all the organic matter in the trophic level below them (which, of course, they do not), some energy would still be lost from the ecosystem as heat.

Let's look more closely at energy flow by focusing on a single stage, the transfer of organic matter from producers

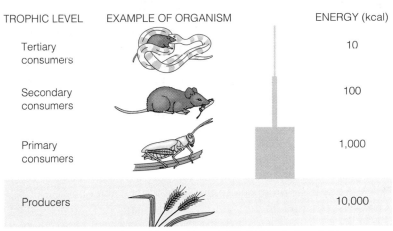

TROPHIC LEVEL	EXAMPLE OF ORGANISM	ENERGY (kcal)
Tertiary consumers		10
Secondary consumers		100
Primary consumers		1,000
Producers		10,000

Energy at each trophic level from 1,000,000 kcal of sunlight

A generalized energy pyramid

to primary consumers (herbivores). In most ecosystems, herbivores manage to eat only a fraction of the plant material produced, and they can't digest all of what they do consume. For example, a grasshopper eating a blade of grass might digest and absorb only about half the organic material it eats, passing the indigestible wastes as feces (which are available to detritivores). Of the organic compounds it does absorb from its food, the grasshopper typically uses about two-thirds as fuel for cellular respiration. In the process of releasing energy that the grasshopper can put to work, cellular respiration degrades the organic compounds to inorganic waste products and heat. Thus, of the compounds absorbed, only the store of chemical energy left over after respiration can add to the biomass of the grasshopper's trophic level, the primary consumer level. Only this amount of energy is available to the next trophic level. If we tracked the fate of organic materials at higher trophic levels, we would find that energy is lost there in similar ways.

An important implication of this stepwise decline of energy in a trophic structure is that the amount of energy available to top-level carnivores is small compared with that available to lower-level consumers. Only about a thousandth of the chemical energy fixed by photosynthesis can flow all the way through a food chain to a tertiary consumer, such as a snake feeding on a mouse. This explains why most food chains are limited to three to five levels; there is simply not enough energy at the very top of an energy pyramid to support another trophic level.

We can now see why the last line of Jonathan Swift's verse quoted in the chapter's introduction—"And so proceed *ad infinitum*"—cannot be taken literally for an ecosystem. The number of levels in a trophic structure is anything but infinite. Instead, as the energy pyramid shows, trophic structures and their component food chains are severely limited by the availability of energy.

36.11 An energy pyramid explains why meat is a luxury for humans

The dynamics of energy flow apply to the human population as much as to other organisms. Like other consumers, we depend entirely on productivity by plants for our food. As omnivores, we eat both plant material and meat. When we eat grain or fruits, we are primary consumers; when we eat beef or other meat from herbivores, we are secondary consumers. When we eat fish like trout and salmon (which eat insects and other small animals), we are tertiary or quaternary consumers.

The energy pyramid on the left below indicates energy flow from primary producers to humans as vegetarians. The energy in the producer trophic level comes from a corn crop. The pyramid on the right illustrates energy flow from the same corn crop, with humans as secondary consumers, eating cattle. These pyramids are generalized models, based on the rough estimate that about 10% of the energy available in a trophic level appears at the next higher trophic level. Thus, the pyramids indicate that the human population has about ten times more energy available to it when people eat grain than when they process the same amount of grain through another trophic level and eat grain-fed beef. Put another way, the pyramids indicate that it takes about ten times more energy to feed the human population when we eat meat than when we eat plants directly.

Actually, the 10% figure is high for energy flow involving cattle and humans. As endotherms, cattle expend a great deal of the energy they take in on heat production—much more than do ectotherms, such as grasshoppers (see Module 25.4). Accounting for the energy loss in heat production, it may actually take closer to 100 times more energy to feed us on cattle (and other mammals and birds) than on plants directly.

Eating meat of any kind is an expensive luxury, both economically and environmentally. In many countries, people cannot afford to buy much meat or their country cannot afford to produce it, and people are vegetarians by necessity. Whenever meat is eaten, producing it requires that more land be cultivated, more water be used for irrigation, and more chemical fertilizers and pesticides be applied to croplands used for growing grain. It is likely that, as the human population expands, meat consumption will become even more of a luxury than it is today.

The laws of thermodynamics and the fact that energy does not cycle within ecosystems explain why the human population has a limited supply of energy available to it. We turn next to the subject of chemical nutrients, all of which differ from energy in that they follow cyclic pathways within ecosystems.

TROPHIC LEVEL

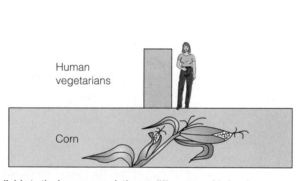
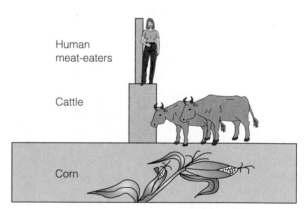

Food energy available to the human population at different trophic levels

---

36.12 Chemicals are recycled between organic matter and abiotic reservoirs

The sun keeps most ecosystems supplied with energy, but there are no extraterrestrial sources of water or the other chemical nutrients essential to life. Life, therefore, depends on the recycling of chemicals. In the next four modules, we look at the cyclic movement of four substances within the biosphere: water, carbon, nitrogen, and phosphorus. In each case, we see that chemicals pass back and forth between organic matter and the abiotic components of ecosystems. We call the part of the ecosystem where a chemical accumulates or is stockpiled outside of living organisms an abiotic reservoir. The main abiotic reservoirs are highlighted in white boxes in the figures. Let's begin with the cycling of water.

Water moves through the biosphere in a global cycle

The figure below illustrates the global water cycle, which is driven by heat from the sun. Three major processes driven by solar heat—precipitation, evaporation, and transpiration from plants—continuously move water between the land, oceans, and the atmosphere. The widths of the blue arrows indicate the relative amounts of water that move to and from the various locations each year. The numbers in parentheses indicate actual amounts of water as billion billion (10^{18}) grams per year. Over the oceans (left side of the figure), evaporation exceeds precipitation by 36×10^{18} grams per year. The result is a net movement of this amount of water vapor in clouds that are carried by winds from the oceans across the land. On land, precipitation exceeds evaporation and transpiration. The excess precipitation forms systems of surface water (such as lakes and rivers) and groundwater, all of which flow back to the sea, completing the water cycle. The water cycle has a global character because there is a large reservoir of water in the atmo-

sphere. Thus, water molecules that have evaporated from the Pacific Ocean, for instance, may appear in a lake or in an animal's body far inland in North America.

Human activity affects the global water cycle in a number of important ways. One of the main sources of atmospheric water is transpiration from the dense vegetation making up tropical rain forests. The destruction of these forests, which is occurring rapidly today, will change the amount of water vapor in the air. This, in turn, will most likely alter local, and perhaps global, weather patterns.

Another change in the water cycle caused by humans results from pumping large amounts of groundwater to the surface to use for irrigation. This practice can increase the rate of evaporation over land, and unless this loss is balanced by increased rainfall over land, groundwater supplies can be depleted. Large areas in the midwestern U.S., the southwestern American desert, parts of California, and areas bordering the Gulf of Mexico currently face this problem.

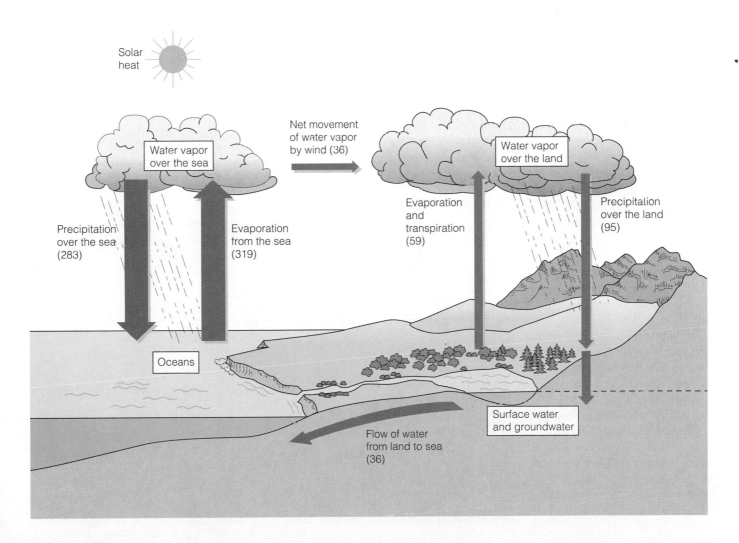

36.14 The carbon cycle depends on photosynthesis and respiration

Like water, the element carbon has an atmospheric reservoir and cycles globally. Moving clockwise from the top of the figure, you see that carbon dioxide (CO_2) from the atmosphere is converted into organic compounds of plants, algae, and cyanobacteria by photosynthesis. Some of this organic material is then eaten by primary consumers, such as rabbits, and serves as the carbon source for these organisms. Higher-level consumers obtain their carbon by eating lower-level consumers. Meanwhile, carbon compounds in detritus—animal wastes, plant litter, and dead organisms of all kinds—are consumed and decomposed by detritivores. Cellular respiration by plants, animals, soil microbes, and other organisms breaks down organic compounds to CO_2, which returns to the atmosphere.

On a global scale, the return of CO_2 to the atmosphere by respiration closely balances its removal by photosynthesis. However, the burning of wood and fossil fuels (coal and petroleum) is steadily increasing the amount of CO_2 in the atmosphere. This may lead to significant environmental problems, such as global warming, as we will discuss in Module 38.13.

36.15 The nitrogen cycle relies heavily on bacteria

The atmosphere contains a huge reservoir of nitrogen; almost 80% of the atmosphere is N_2. Cycling of this nitrogen into and out of ecosystems depends mainly on bacteria, as the figure indicates. Notice the two groups of nitrogen-fixing bacteria on the right side. Nitrogen fixers in legume plants (peas, beans, and alfalfa, for example) and nitrogen fixers in the soil convert atmospheric N_2 to ammonia (NH_3). (In aquatic ecosystems, cyanobacteria are important nitrogen fixers.) Following the arrows toward the lower left, you can see that nitrifying bacteria convert ammonia into nitrates. Nitrates are inorganic compounds containing the nitrate ion NO_3^-, and they are the main source of nitrogen for plants. The amino acids and pro-

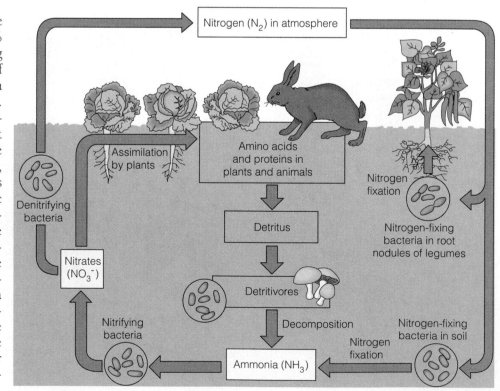

teins made by plants using this nitrogen are then available to consumer organisms. Bacteria and fungi acting as detritivores decompose nitrogen-containing detritus back into ammonia, thus keeping nitrogen moving within the trophic structure. Another group of soil bacteria, the denitrifiers (left side of figure), complete the nitrogen cycle by converting soil nitrates back into atmospheric N_2.

Overall, most of the nitrogen cycling in natural ecosystems involves the inner cycle in the diagram, the thicker arrows linking plants and animals, detritivores, and nitrifying bacteria. The outer cycle (thinner arrows) moves only a tiny fraction of nitrogen into and out of natural ecosystems. However, human activity has altered the nitrogen cycle balance in many areas. Sewage-treatment facilities usually empty large amounts of dissolved inorganic nitrogen compounds into rivers or streams. Farmers routinely apply large amounts of inorganic nitrogen fertilizers, mainly ammonia and nitrates, to croplands. Lawns and golf courses also receive sizable doses of fertilizer. Crop and lawn plants take up some of the nitrogen compounds, and denitrifying bacteria convert some into atmospheric N_2, but chemical fertilizers usually exceed the soil's natural recycling capacity. The excess nitrogen compounds often enter streams, lakes, and groundwater.

In lakes and streams, these nitrogen compounds continue to fertilize, causing heavy growth of algae. Groundwater pollution by nitrogen fertilizers is a serious problem in many agricultural areas, especially in the midwestern United States. Nitrates in drinking water are converted to nitrites, which can be toxic, in the human digestive tract. In Module 32.10, we discussed some alternatives to the extensive use of agricultural fertilizers.

The phosphorus cycle depends on the weathering of rock

In contrast to nitrogen, the element phosphorus has its main abiotic reservoirs in rocks, rather than in the atmosphere. (This is also true of the elements potassium and calcium.) At the center of the figure here, the weathering of rock gradually adds phosphates (compounds containing PO_4^{3-}) to the soil. Plants absorb the dissolved phosphate ions in the soil and build them into organic compounds. Consumers obtain phosphorus in organic form from plants. Decomposers return phosphates to the soil. As indicated at the lower left of the figure, some phosphates also precipitate out of solution at the bottom of deep lakes and oceans. The phosphates in this form may eventually become part of new rocks and will not cycle back into living organisms until geological processes uplift the rocks and expose them to weathering.

Because weathering is generally a slow process, the amount of phosphates available to plants in natural ecosystems is often quite low. Plant growth can, in fact, be limited by the small amount of soluble phosphates in the soil. And in lakes that have not been altered by human activity, a low level of dissolved phosphates often keeps algal growth to a minimum, thereby helping keep the water clear. In many areas, however, excess, rather than limited, phosphates are a problem. Like nitrogen compounds, phosphates are a major component of sewage outflow. They are also used extensively in agricultural fertilizers and are a common ingredient in pesticides. Phosphate pollution of lakes and rivers, like nitrate pollution, leads to heavy algal growth. We discuss some other effects of human alteration of nutrient cycles in the next two modules.

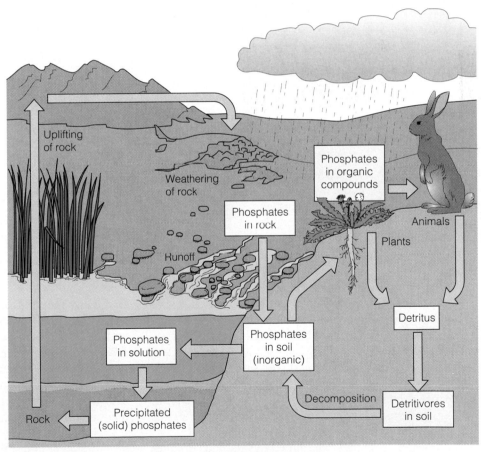

A. The dam at the Hubbard Brook study site

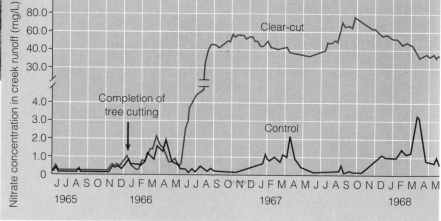

B. Logged watersheds in the Hubbard Brook Forest

C. The loss of nitrate from a clear-cut watershed

The cycling of any chemical in an ecosystem depends on the web of feeding relationships among plants, animals, and microbes. Let's look now at what can happen to nutrient cycling when some key organisms—the dominant plants—are removed.

For the past 35 years, a team of scientists has studied nutrient cycling in a forest ecosystem in two different situations: under natural conditions, and after severe human intrusion. The study site is the Hubbard Brook Experimental Forest in the White Mountains of New Hampshire. It is a nearly mature deciduous forest with several valleys, each drained by a small creek that is a tributary of Hubbard Brook. Thick rock is close to the surface of the soil, and water does not seep into the rock from the soil. As a result, water drains out of each valley only via its creek.

The research team first determined the amounts of water and several key mineral nutrients (for example, nitrate, calcium, and potassium) that normally move in and out of six valleys. Figure A shows a small concrete dam with a V-shaped spillway for water. Such a dam was built across the creek at the bottom of each valley to monitor water and nutrient losses from the stream. Researchers found that when the ecosystem was undisturbed, about 60% of the water

that fell as rain and snow exited through the streams, and the remaining 40% was lost by transpiration from plants and evaporation from the soil. They also found that the flow of mineral nutrients into and out of the ecosystem was nearly balanced, and was relatively small compared with the quantity of nutrients being recycled within the forest itself. During most years, the forest actually gained small amounts of a few nutrients, including nitrates.

In 1966, one of the six valleys was completely logged in an experiment to test the effect of deforestation on nutrient cycling. The clear-cut valley is the white (snow-covered) area just right of center in Figure B. The inflow and outflow of water and minerals in this altered valley were compared with a control valley for three years. Water runoff from the clear-cut valley increased by 30–40% after deforestation, apparently because there were no trees to absorb and transpire water from the soil. Net losses of nutrients from the clear-cut valley were huge. The concentration of calcium (Ca^{2+}) in its creek increased fourfold, for example, and the concentration of potassium (K^+) increased by a factor of 15.

Figure C shows one of the major changes in the clear-cut valley—the pronounced loss of nitrate. The red line traces the concentration of nitrate in runoff from the clear-cut valley from June 1965 through May 1968. The black line indicates nitrate concentrations in a control, an unlogged valley. Notice that nitrate concentrations in the runoff from the clear-cut area began to rise markedly about five months after tree cutting was completed. (The break between 4.0 and 30.0 mg/L simplifies the graph; it represents a one-month period when the rate of nitrate loss increased very rapidly.) Within eight months, the nitrate loss was some 60 times greater in the clear-cut valley than in the control. Without trees to take up and hold nitrate as soil bacteria produce it, this vital nutrient drained rapidly from the ecosystem. In fact, nitrate in the creek reached a level considered unsafe for drinking water.

Similar studies in other experimental forests have affirmed the evidence from Hubbard Brook that clear-cutting severely alters both water drainage and nutrient cycling. The rapid runoff depletes soil nutrients, and, as we see next, it can also pollute areas downstream.

Altered ecosystems trigger changes in other ecosystems 36.18

The Hubbard Brook experiment shows that major change in a terrestrial ecosystem disrupts chemical cycling and moves large amounts of chemical nutrients to other areas. Most of the nutrients lost from deforested lands and from agricultural areas are highly soluble in water and readily pass into aquatic ecosystems.

What do we really mean when we say chemical nutrients can pollute a receiving ecosystem? Let's trace the chain of events that can occur when a freshwater lake, for instance, receives an overload of nutrients from surrounding terrestrial ecosystems. A natural lake typically has a moderate growth of algae and plants and often a rich diversity of fishes and invertebrates. Freshwater lakes change as a result of natural events. Some may gradually undergo **eutrophication,** meaning they become increasingly more productive. As they change, they remain in equilibrium with their surroundings; the input of new nutrients is balanced by the losses from the lake due to outflow and the settling of organic matter to the bottom. The kinds of organisms living in the lake may change, but productivity increases slowly enough that the species diversity usually remains high.

Unfortunately, the natural balance of nutrient cycling in freshwater lakes is easily upset. When a lake receives an excess of mineral nutrients, its entire trophic structure can change very quickly. Today, a lake may receive runoff of inorganic fertilizers from agricultural lands as well as sewage, factory wastes, and animal wastes from pastures and stockyards. The water becomes polluted with these materials—actually, overfertilized—and the lake's photosynthetic organisms multiply rapidly. The result is accelerated eutrophication. Shallow areas may become weed-choked, and algal and bacterial populations often grow explosively. This photograph illustrates the effect of experimental eutrophication of part of a freshwater lake. Researchers used a plastic curtain to separate the part of the lake in the background from the portion in the foreground. They then added inorganic phosphorus to the distant part of the lake, keeping the other part as a control. Within two months, a cyanobacterial bloom changed the color of the fertilized lake to the pale blue you see here.

Experimental eutrophication of part of a lake

Heavy bacterial and algal growth increases oxygen production during the day but greatly reduces oxygen levels at night, when the photosynthesizers respire. As the organisms die and accumulate at the bottom of the lake, the bacteria decomposing them can use up much of the oxygen dissolved in deep waters. When this happens, the lake may lose most of its species diversity, with only the most tolerant organisms surviving. Human-caused eutrophication, for example, wiped out commercially important fish in Lake Erie during the 1950s and 1960s. Since then, tighter regulations on the dumping of wastes into the lake have enabled some fish populations to rebound, but many of the native species of fishes and invertebrates have not recovered. Today, accelerated eutrophication is the most common problem affecting lakes throughout the world. About 60% of the lakes in the U.S. are affected to some degree.

Human disruption of both aquatic and terrestrial ecosystems is a global problem. In the next module, we take a look at measures that some countries are taking to make human development more compatible with the ecosystems they depend on.

Megareserves are an attempt to reverse ecosystem disruption

Today, few, if any, ecosystems remain unaltered by human activities. Accelerated eutrophication reduces species diversity in lakes and rivers because many organisms cannot tolerate the rapid changes in water quality. Large tracts of forest in taiga biomes are still being clear-cut in the U.S., Canada, and Siberia to satisfy demands for lumber and urban development. The current rate of conversion of tropical forests to farmland threatens the survival of thousands of species and may alter global weather patterns.

In an attempt to slow the disruption of ecosystems, a number of countries are setting up what they call megareserves. A **megareserve** is an extensive region of land that includes one or more areas undisturbed by humans. The undisturbed areas are surrounded by lands that have been changed by human activity and are used for economic gain. The key factor to the megareserve concept is the development of a social and economic climate in the surrounding lands that is compatible with ecosystem conservation. These surrounding areas continue to be used to support the human population, but they are protected from extensive alteration. As a result, they serve as a buffer zone, or shield, against further intrusion into the undisturbed areas. As ecologist Daniel Janzen, a leader in tropical conservation, puts it, "The likelihood of long-term survival of a conserved wildland area is directly proportional to the economic health and stability of the society in which that wildland is imbedded."

With international financial and scientific help, the small Central American nation of Costa Rica has become a world leader in establishing megareserves. In exchange for reduction in its international debt, the Costa Rican government has established eight megareserves, called "conservation areas," as shown on the map here. The green areas are national park lands, which remain relatively unchanged by human activity; the yellow areas are buffer zones, privately owned areas where people live and make a living.

Costa Rica is making progress toward managing its megareserves so that the buffer zones provide a steady, lasting supply of forest products, water, and hydroelectric power, and also support sustainable agriculture and tourism. An important goal is to provide a stable economic base for people living there. Destructive practices that are not compatible with long-term ecosystem stability, and from which there is often little local profit, are gradually being discouraged. Such destructive practices include massive logging, large-scale single-crop agriculture, and extensive mining. Costa Rica looks to its megareserve system to maintain at least 80% of its native species. We take a closer look at conservation efforts in the tropics in Chapter 37.

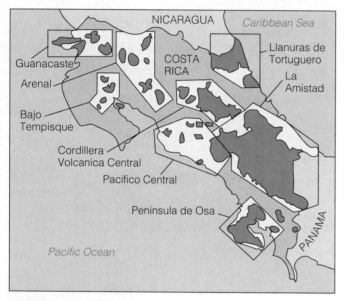

Megareserves in Costa Rica

Begin your review by rereading the module headings and scanning the figures, before proceeding to the Chapter Summary and questions.

Chapter Summary

36.1 All the organisms in a particular area make up a community. A number of factors characterize every community: diversity (variety of species), the prevalent form of vegetation (dominant plants and their structure), stability (the ability to resist change and recover after a disturbance), and trophic structure (feeding relationships).

36.2–36.5 Three types of interactions connect species in communities: interspecific competition, predation, and symbiosis. Two species will compete if they both require the same limited resource. If they have very similar niches, they may not be able to coexist. In some cases, as predators adapt to prey, natural selection also shapes the prey's defenses. This process of reciprocal adaptation is known as coevolution. A keystone predator may maintain community diversity by controlling the strongest competitors among its prey species. Three types of symbiotic relationships exist in communities. In parasitism, a parasite obtains food at the expense of its host. In commensalism, one species benefits while the other is unaffected. In mutualism, both partners benefit.

36.6 Disturbances such as fire or human activities may drastically change a community. A forest clear-cut, a vacant lot, or barren rock are gradually colonized by life in a process called ecological succession. Lichens, mosses, and grasses may invade first, followed by shrubs, and perhaps by trees and other forest species.

36.7–36.9 A community interacts with abiotic factors, forming an ecosystem. Energy flows from the sun, through plants, animals, and decomposers, and is lost as heat. Chemicals are recycled among air, water, soil, and organisms. Food chains and food webs map the flow of energy and nutrients from plants (producers), to herbivores (primary consumers), to carnivores (secondary and higher-level consumers). Detritivores (scavengers, fungi, and bacteria) decompose waste matter and recycle nutrients.

36.10–36.11 Primary productivity is the rate at which producers convert sunlight to chemical energy in organic material. The amount of organic material in an ecosystem is called biomass. An energy pyramid depicts the flow of energy from producers to primary consumers and to higher trophic levels. Some of the energy consumers gain from their food is used in cellular respiration and lost as heat. Only 10% or less of the energy in food is stored at each trophic level and available to the next level. This stepwise energy loss limits most food chains to three to five levels. Because the energy pyramid tapers so sharply, a field of corn or other plant crops can support many more vegetarians than meat-eaters.

36.12–36.18 Ecosystems require daily infusions of solar energy, but nutrients are recycled between organisms and their abiotic environments. Heat from the sun drives the global water cycle of precipitation, evaporation, and transpiration. Carbon is taken from the atmosphere by photosynthesis, used to make organic molecules, and returned to the atmosphere by cellular respiration. Some bacteria in the soil break down organic matter and recycle nitrogen to plants; other bacteria move nitrogen between soil and air. Phosphorus and other soil minerals are also recycled, but are stored primarily in rock. Disturbances such as clear-cutting of forests increases the runoff of water and the loss of soil nutrients. Nutrient runoff may fertilize a lake and cause the heavy growth of algae, choking other aquatic life.

36.19 The alteration of ecosystems by human activities threatens global climate and the existence of thousands of species. To slow the disruption of ecosystems, some nations are establishing megareserves—undisturbed wildlands surrounded by buffer zones of compatible economic development.

Testing Your Knowledge

Multiple Choice

1. When you eat a hamburger, you are a
   a. primary producer
   b. detritivore
   c. secondary consumer
   d. decomposer
   e. primary consumer

2. Which of the following best illustrates ecological succession?
   a. A mouse eats seeds, and an owl eats the mouse.
   b. Decomposition in soil releases nitrogen that plants can use.
   c. Grass grows on a sand dune, then shrubs, and then trees.
   d. Imported pheasants increase, while local quail disappear.
   e. Overgrazing causes a loss of nutrients from soil.

3. A bat locates insect prey in the dark by bouncing high-pitched sounds off them. One species of moth escapes predation by diving to the ground when it hears "sonar" of a particular bat species. This illustrates _____ between the bat and moth. *(Explain your answer.)*
   a. mutualism
   b. competitive exclusion
   c. ecological succession
   d. commensalism
   e. coevolution

4. Local conditions like heavy rainfall or clear-cutting may limit the amount of nitrogen, phosphorus, or calcium available to a particular ecosystem, but the amount of carbon available to the system is seldom a problem. Why?
   a. Organisms do not need very much carbon.
   b. Plants can make their own carbon using water and sunlight.
   c. Plants are much better at absorbing carbon from the soil.
   d. Many nutrients come from the soil, but carbon comes from the air.
   e. Symbiotic bacteria help plants capture carbon.

Describing, Comparing, and Explaining

1. In Southeast Asia, there's an old saying, "There is only one tiger to a hill." Explain in terms of energy flow in ecosystems why big predatory animals such as tigers and sharks are relatively rare.

2. Describe four ways in which humans alter chemical cycles.

Thinking Critically

1. An ecologist studying plants in the desert performed the following experiment. She staked out two identical plots, including a few sagebrush plants and numerous small, annual wildflowers. She found the same five wildflower species in roughly equal numbers on both plots. She then enclosed one of the plots with a fence to keep out kangaroo rats, the most common herbivores of the area. After two years, to her surprise, four of the wildflower species were no longer present in the fenced plot, but one species had increased drastically. The control plot had not changed. Using the principles and terminology of ecology, explain what you think happened.

2. Reread about *Pieris* caterpillars and their wasp parasites in the chapter's introduction. Assume that the species described are the only ones in this food chain, and they are of average productivity. Now make a rough calculation: For every 10,000 kcal of chemical energy in the leaves eaten by *Pieris* caterpillars, about how many calories are stored in the bodies of chalcid larvae? Explain how you arrived at your answer.

3. The carbon cycle is the network of pathways by which carbon atoms cycle and recycle among the atmosphere, soil, and living things. A carbon atom in CO_2 might be used by a dandelion to make a sugar molecule. The sugar molecule might then be consumed by a grasshopper when it eats the dandelion, and be deposited in the soil in an organic molecule in the grasshopper's feces. A bacterium could release the carbon atom back into the air as CO_2 as it decomposes the soil litter. Continue the story, describing ten more places the carbon atom might go. Just for fun, include each of the following: a redwood tree, a peanut butter sandwich, a giant squid, and your own body.

Science, Technology, and Society

1. Sometime in 1986, near Detroit, a foreign freighter pumped out water ballast containing the larvae of European zebra mussels. The mollusks began to multiply wildly, spreading through Lake Erie and entering Lake Ontario. In some places, they now cling to surfaces at densities of $20,000/m^2$. They have blocked the intake pipes of power plants and water-treatment plants, fouled boat hulls, and sunk buoys. We have seen this scenario before, with rabbits in Australia and starlings in North America, among others. What makes this kind of population explosion occur? What might happen to native organisms that suddenly must share the Great Lakes ecosystem with zebra mussels? How would you suggest trying to solve the mussel population problem?

2. By 1935, hunting and trapping had eliminated wolves from the United States outside Alaska. Wolves are now protected, and they have moved south from Canada and become reestablished in the Rocky Mountains and northern Great Lakes. Conservationists would like to speed up this process by reintroducing wolves into Yellowstone National Park, where they were last seen 50 years ago. Local ranchers are opposed to bringing back the wolves; they fear predation on their cattle and sheep. What are some reasons for reestablishing wolves in Yellowstone Park? What effects might the wolves have on the ecological communities in the park? What might be done to mitigate the conflicts between ranching and wolves?

Behavioral Adaptations to the Environment 37

In his 1959 treatise on Mexican wildlife, American naturalist Aldo Leopold wrote:

> The chesty roar of jaguar in the night causes men to edge toward the blaze and draw serapes tighter. It silences the yapping dogs and starts the tethered horses milling. In announcing its mere presence in the blackness of the night, the jaguar puts the animate world on edge.

The jaguar is the largest and strongest member of the cat family in the Americas. An adult male stands nearly 2.5 feet at the shoulder and can weigh up to 300 pounds. Its broad face is shaped by massive jaw muscles that can crush heavy bones.

Jaguars were once common throughout most of South and Central America and ranged through Mexico into the southcentral United States. Today, they are rare almost everywhere except in a few strongholds in the dense rain forests of Central America and Brazil. In 1984, the tiny Caribbean country of Belize established the world's only jaguar refuge, the Cockscomb Forest Jaguar Preserve. The refuge idea grew out of a study of jaguar behavior and ecology carried out in the Cockscomb area in the early 1980s by Alan Rabinowitz, a zoologist with the New York Zoological Society.

Field research in behavior is never easy, but Rabinowitz, in his twenties at the time, weathered monumental difficulties. Nearly all the jaguars he studied were shot within a few months by irate ranchers and farmers who believed the cats were killing livestock. Rabinowitz himself contracted amoebic dysentery, hookworm, and fungal infections, and barely survived the crash of his plane. One of his assistants died from snakebite. Still, Rabinowitz made some significant discoveries. In just two years, aided by the local Mayans, he managed to monitor the activities of six adult jaguars over extended periods. This required trapping the animals, anesthetizing them, and then fitting each one with a radio-transmitter collar (as shown at the right). Using a directional receiver tuned to the transmitter's signals, he could then trace the jaguar's movements in the dense jungle.

Rabinowitz confirmed that jaguars, like most big cats, are solitary hunters able to kill animals twice their size. He also discovered that jaguars shun contact with others of their species, except during the breeding season. Male jaguars in Belize may have overlapping hunting grounds but rarely use the same areas at the same time. A male announces his presence by defecating in open areas, by scratching the ground, and by grunting or low growling. During the breeding season, male and female jaguars pair off, and their behavior changes markedly. The male may lick and caress the female, and after a litter is born, he will often help feed the nursing female and the cubs for days or even weeks.

The study of an animal's behavior is essential to understanding its evolution and ecological interactions. For example, the jaguar's hunting techniques are similar to those of several other species of big cat, such as the mountain lion and leopard. This similarity suggests that basic hunting behavior evolved in a common ancestor of the big cats. From an ecological standpoint, the jaguar's signaling behavior prevents direct confrontations between individuals and increases its chances of having nearly exclusive use of part of the limited hunting grounds available to the population. Its breeding season behavior directly affects its reproductive success. We will return to the jaguar several times as we pursue this chapter's main objective—to illustrate the connections between animal behavior, evolution, and ecology.

Fitting a jaguar with a radio-transmitter collar

Behavioral biology is the study of how animals behave in their natural environments

Animal behavior is broadly defined as externally observable muscular activity triggered by some stimulus, or, more simply, what an animal does when interacting with its natural environment. People have been describing animal activities throughout history, but the science of behavior, or **behavioral biology,** was not set on firm ground until the twentieth century. Nobel laureates Karl von Frisch, Konrad Lorenz, and Niko Tinbergen were among the first experimentalists in behavioral biology. In the early 1900s, Austrian zoologist Karl von Frisch, who pioneered the use of experimental methods in behavior, demonstrated that honeybees have visual senses very different from ours. Bavarian naturalist Konrad Lorenz, often regarded as the founder of

behavioral biology, emphasized the importance of comparing the behavior of various animals. In experiments begun in the 1930s, Lorenz showed that the same stimulus may elicit very different forms of behavior in different animals, including several species of birds, fishes, and invertebrates. Dutch biologist Niko Tinbergen worked closely with Lorenz, concentrating on experimental studies of innate, genetically programmed behavior and on simple forms of learning.

The figure here illustrates a classic Tinbergen experiment. It deals with nesting behavior in an insect called the digger wasp, which builds its nest in a small burrow in the ground. A female wasp will often excavate and care for four or five separate nests, flying to each one daily, cleaning it, and bringing food to the single larva in the nest. To test his prediction that the female digger wasp uses landmarks to keep track of her nests, Tinbergen ① placed a circle of pinecones around a nest opening and waited for the mother wasp to return. When she did, he watched as she tended the nest. When she flew away, he ② moved the pine cones a few feet to one side of the nest opening. The next time the wasp returned, she flew to the center of the pinecone circle instead of to the actual nest opening. This experiment indicated that the wasp did use landmarks, and that she could learn new ones to keep track of her nests. But it also raised another question: Did the wasp respond to the pinecones themselves or to their circular arrangement? To answer this question, Tinbergen ③ arranged the pinecones in a triangle around the nest and made a circle of small stones off to one side of the nest opening. This time, the wasp flew to the stones, demonstrating that she cued in on the arrangement of the landmarks rather than the landmarks themselves.

In asking how a digger wasp locates her nests, Tinbergen's studies approached the wasp's behavior at the level that behavioral biologists call proximate (immediate) causes. A **proximate cause** explains behavior in terms of immediate interactions with the environment. For the digger wasp, the proximate cause of nest-locating behavior is the environmental cue: the arrangement of the landmarks. If we determine the proximate cause of a particular behavior, we have an answer to a "how" question; for the wasp, we have learned how she locates her nest.

But there is more to the wasp's behavior than this. As we have seen throughout this text, biologists also pursue "why" questions, those that can only be answered from an evolutionary perspective. Behavioral biologists call the evolutionary causes of behavior **ultimate causes.** The ultimate cause of the wasp's behavior was not addressed by Tinbergen's experiments. It has to do with natural selection acting on genetically based phenotypic differences: Presumably the fitness (reproductive success) of digger wasps has been enhanced by the female's ability to store information about

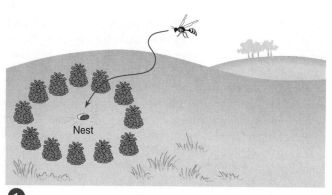

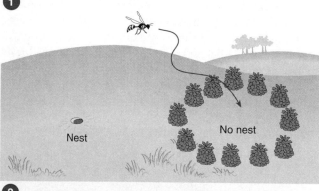

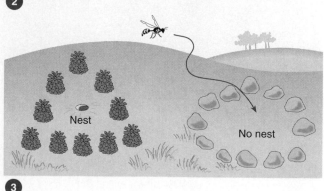

Nest-locating behavior of the digger wasp

nest location and to use that information to find and service her nests. The idea that evolution underlies behavior implies that the capacity for particular behaviors is inherited, for only if behavior stems from an individual's genes can that behavior evolve by natural selection. The search for ultimate causes, an area of biology called **behavioral ecology,** dominates research in behavior today. This chapter draws heavily on research in behavioral ecology.

Genetic programming and experience both contribute to behavior patterns

The relationship between genes and behavior is rarely simple. Some amount of genetic programming and experience seem to be responsible for most kinds of animal behavior. For example, much of the digger wasp's nesting behavior is genetically programmed and not changeable by experience. However, as Tinbergen showed, the wasp's nest-locating behavior can be modified by the experience of seeing different landmarks.

Another example of behavior with a direct genetic component is the gathering of nest material by two closely related species of African parrots, often called lovebirds. A female lovebird builds her nest with thin strips of vegetation that she cuts with her beak. In captivity, lovebirds will use small sheets of paper in place of vegetation. As shown in the top part of the drawing here, the female of one species, Fischer's lovebird, cuts fairly long strips and carries them back to her nest site one at a time in her beak. In contrast, the peach-faced lovebird cuts shorter strips and usually carries several at a time, by tucking them into the feathers of her lower back.

By interbreeding the two species and recording the behavior of hybrids, researchers have confirmed that these behavior patterns are inherited. A hybrid female cuts strips of intermediate length and attempts to carry them using hybrid behavior. For example, she usually tries tucking the strips under her feathers without releasing them from her beak. Eventually, after failing to transport the strips under her feathers, she will learn to carry a strip in her beak. But she will still act out part of the tucking sequence by turning her head to the rear before flying off.

The degree to which behavior is influenced by strict genetic programming or by experience (learning) is an age-old debate. The lovebird example shows that an inherited behavior pattern can be modified, especially if the programmed behavior proves ineffective. On the other hand, it also shows that certain components of the genetically determined behavior (head turning) may be too firmly programmed to change.

What about human behavior? How much is our ability to solve math problems determined by genes and how much by experience? A virtuoso pianist must have extraordinary hand coordination and the intangible ability to make the mechanical instrument resound with the dynamics and emotion of the music. But how much does training contribute to what we think of as inborn talent? And is a capacity for disciplined training itself mainly learned or inherited? Recent evidence indicates that much of our behavior is influenced by genes. However, rarely is behavior *either* completely programmed by genes *or* completely learned. The issue is *the extent to which* a particular behavior is programmed or learned.

Single long strip carried in beak (Fischer's lovebird)

Several short strips tucked under feathers (peach-faced lovebird)

Tucking failure

Strip in beak

Hybrid behavior

Gene-based behavior in lovebirds

Innate behavior often appears as fixed action patterns

Lorenz and Tinbergen were among the first to demonstrate the importance of genetically programmed behavior, or **innate behavior**, in the lives of animals. Many of their studies were concerned with essentially unchangeable behavioral sequences called **fixed action patterns (FAPs)**. Like someone who has memorized a poem or a piece of music but must start over at the beginning if interrupted, an animal can only perform the FAP as a whole. Once an animal initiates a FAP, it usually carries the sequence to completion, even if it receives stimuli indicating that the FAP is no longer appropriate.

Figure A illustrates one of the FAPs that Lorenz and Tinbergen studied in detail. The bird is the graylag goose, a common European species that nests in shallow depressions on the ground. If the goose happens to bump one of her eggs out of the nest, she always retrieves it in the same manner. As indicated in the figure, she stands up, extends her neck, uses her beak and a side-to-side head motion to nudge the egg back, and then sits down on the nest again. If the egg slips away (or is pulled away by an experimenter) while the goose is retrieving it, she stops her side-to-side head motion but still goes through the other motions as though the egg were there. Only after she sits back down on her eggs does she seem to notice that an egg is still outside the nest. Then she begins another retrieval sequence with the egg. If the egg is again pulled away, the goose will still complete the retrieval motion as before. She even performed the sequence when Lorenz and Tinbergen placed a foreign object, such as a small toy or a ball, near the nest.

Fixed action patterns are important in the lives of many species. Figure B (on the facing page) illustrates some key events in the life cycle of the European cuckoo, which lays its eggs in the nests of other species of birds. ① When a female cuckoo (upper left) is ready to lay her eggs, she finds a nest of a suitable host species and waits for the host bird to

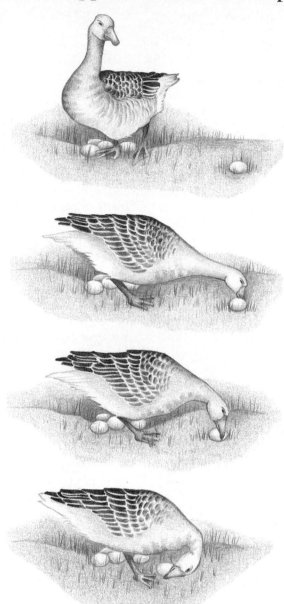

A. A graylag goose retrieving an egg—a FAP

leave the nest unattended. ② She needs only a few seconds to fly to the nest, pick up one of the host's eggs in her beak, and lay one of her own eggs in its place. ③ Mission accomplished, she flies off, abandoning her offspring to the foster parents and eating the stolen egg. When the host bird returns, she usually accepts the cuckoo's egg and incubates it with her own eggs.

The cuckoo's timing is precise, and its egg usually hatches before the host eggs. After a cuckoo hatches, a series of FAPs ensures its survival and dooms the host offspring. First, the hatchling cuckoo, with its eyes not yet open, ejects the unhatched host eggs from the nest (left photograph in Figure B). The process of ejection, which is innate, is a FAP. After ejecting the host's eggs, the young cuckoo will be fed and nurtured by its foster parents. When a hatchling senses that an adult bird is near, it responds with another FAP: It begs for food by raising its head, opening its mouth, and cheeping. In turn, the foster parent responds with yet another FAP: It stuffs food in the gaping mouth. These innate behaviors are replayed over and over, even after the young cuckoo is much larger than the adults (right photo).

In its simplest form, a FAP can be thought of as an innate response to a certain stimulus. A stimulus that triggers, or "releases," a FAP is called a **sign stimulus**. The sign stimulus for the graylag goose egg retrieval is the presence of an egg, or other object, near the nest. For the hatchling European cuckoo disposing of host eggs, it may be the feel of an unhatched egg. For parent birds, it is the chick's gaping mouth.

Humans also have FAPs. For instance, when a human infant is touched on the cheek, the infant responds by turning toward the touch and rooting around for a nipple with its mouth. An infant's smile is also a FAP, induced by an adult's face or even something that vaguely resembles a face, such as two dark spots on a white circle.

Egg-laying bchavior

Ejection of host eggs from nest by cuckoo hatchling Feeding of cuckoo chick by foster mother

B. FAPs in the life of a European cuckoo

In all cases, the specific sign stimulus is a proximate cause of the FAP. Experiments with invertebrates suggest that a sign stimulus activates a specific, genetically programmed nervous pathway that makes appropriate muscles in the animal perform the FAP. There is no evidence to suggest that animals integrate the information in sign stimuli in the brain and then make a decision.

What can we say about FAPs in the context of behavioral ecology? In many cases, we can see a clear advantage of FAPs over learned behavior. FAPs allow an animal to perform an activity correctly the first time, because FAPs do not require time for learning. This advantage can be important for invertebrates whose life spans are so short that

there is little time for learning; and FAPs are common in invertebrates. For birds and mammals, FAPs enable hatchlings and newborns to obtain food from parents without any learning. A newly hatched cuckoo, for instance, has had no chance to learn how to obtain food from its foster parents. Likewise, a newborn jaguar starts nursing without any previous experience in obtaining milk from its mother. In the context of ultimate causes, natural selection seems to have favored behavior that enables offspring to consume their first meal without any previous learning.

Unlike fixed action patterns, which are innate and unchanging, learning involves changes in preexisting behavior. We examine several forms of learning next.

Learning ranges from simple behavioral changes to complex problem solving

Learning is a change in behavior resulting from experience. As this table indicates, there are various forms of learning, ranging from a simple behavioral change in response to a single stimulus, to complex problem solving involving entirely new behavior.

Types of Learning	
Learning Type	**Defining Characteristics**
Habituation	Loss of a response to a stimulus after repeated exposure
Imprinting	Learning that is irreversible and limited to a critical time period in an animal's life; often results in a strong bond between new offspring and parents
Association	Behavioral change resulting from a link between a behavior and a reward or punishment; includes classical conditioning and trial-and-error learning
Imitation	Learning by observing and mimicking others
Innovation	Inventive behavior that arises in response to a new situation without trial and error or imitation; problem solving by reasoning

One of the simplest forms of learning is **habituation**, in which an animal learns not to respond to an unimportant stimulus. There are many examples in invertebrates and vertebrate animals. The cnidarian *Hydra*, for example, contracts when disturbed by a slight touch; it stops responding, however, if disturbed repeatedly by such a stimulus. Similarly, a scarecrow stimulus will usually make birds avoid a tree with ripe fruit for a few days. But the birds soon become habituated to the scarecrow and may even land on it on their way to the fruit tree. Once habituated to a stimulus, an animal still senses the stimulus—its sensory organs detect it—but the animal has learned not to respond to it.

Habituation prevents an animal from wasting time or energy in useless activity. This simple form of learning is highly adaptive, for it allows an animal to concentrate on stimuli that signal food, mates, or real danger and remain unperturbed by a vast number of other stimuli that are irrelevant to its survival and reproduction.

Imprinting is learning that involves both innate behavior and experience

Learning often interacts closely with genetically determined, innate behavior. Some of the most interesting cases involve the phenomenon known as imprinting. **Imprinting** is learning that is limited to a specific time period in an animal's life and that is irreversible. One result of imprinting is the formation of a strong bond between two animals, often a hatchling or other newborn animal and its parent. The specific time during which imprinting occurs is called the **critical period.**

In perhaps his most famous study, Konrad Lorenz used the graylag goose to demonstrate imprinting. He divided a batch of eggs from a nest, leaving some with the mother and putting the rest in an incubator. The young reared by the mother served as the control group. They showed normal behavior, following the mother about as goslings and eventually growing up to mate and interact with other geese. The geese from the artificially incubated eggs formed the experimental group. These geese spent their first few hours after hatching with Lorenz, rather than with their mother. From that day on, they steadfastly followed Lorenz

(Figure A, on the facing page) and showed no recognition of their mother or other adults of their own species. This early imprinting lingered into adulthood: The birds continued to prefer the company of Lorenz and other humans to that of their own species. Some of them even tried to mate with humans.

In other experiments, Lorenz demonstrated that the most important imprinting stimulus for graylag geese was movement of an object (normally the parent bird) away from the hatchlings. The effect of movement was increased if the moving object emitted some sound. The sound did not have to be that of a goose, however; Lorenz found that a box with a ticking clock in it was readily and permanently accepted as a "mother."

The critical period for imprinting varies with the species. Lorenz found it to be the first two days after hatching for the graylag goose. During that time, the hatchlings apparently have no innate sense of "mother" or "I am a goose, you are a goose." Instead, they simply respond to and identify with the first object they encounter that has certain simple

A. Konrad Lorenz with geese imprinted on him

Imprinting also occurs in many adult animals. For instance, just after giving birth, female sheep and many other mammals imprint on the smell of their own offspring. Afterwards, the females will reject all other young of their species.

A special kind of imprinting, called song imprinting, plays an important role for many kinds of birds. In most cases, when we hear a bird sing, it is a male trying to attract a mate or issuing an aggressive warning to would-be competitors. Each species of songbird has its own particular song, although the song varies somewhat among individuals of a population. Researchers study bird songs using instruments that record individual sounds making up the song. Figure B shows two sound tracings, illustrating the song patterns of two male white-crowned sparrows. The upper tracing shows the complete song of the species, sung by a normal wild male (song tracings of other wild males would show only minor differences). The lower tracing illustrates an abnormal sound pattern produced by a male reared in isolation. If during the critical period (10–100 days after hatching) an isolated male hears the song of a normal male of its species, it will imprint on the song and learn to sing it normally. In contrast, if during the critical period an isolated male sparrow hears the song of another species, even a closely related one, the male will not learn that song; he will simply sing an abnormal song like the one in the lower tracing. Thus, it seems that genes determine what song a bird can sing, somehow allowing the bird to imprint only on the particular song sung by its own species.

In summary, imprinting is a simple form of learning that serves a variety of functions in different species. In helping a male bird learn his species' song, imprinting helps prevent potentially harmful confrontations between local rivals. In providing parents and offspring and potential mates with a quick way to identify each other, imprinting helps conserve energy that might be spent in searching. In terms of ultimate causes, the conserved energy can then be put to use in finding food and in rearing young—activities that usually increase evolutionary fitness.

characteristics. Imprinting has both innate and learned components. Its innate component is the ability or tendency to imprint during a critical period. The actual imprinting itself is a form of learning.

Since Lorenz's pioneering work, many other examples of imprinting have been discovered, and not all involve parent–offspring bonding. In some species, new offspring imprint on certain components of their home environment rather than on parents. Newly hatched salmon, for instance, do not receive any parental care but seem to imprint on the complex bouquet of odors unique to the freshwater stream where they hatch. They retain their imprinting into adulthood, and it enables them to find their way back to the stream to spawn after spending a year or more at sea. As we saw in the introduction to Chapter 29, adult salmon recognize the odors of their home stream and swim toward their source from great distances.

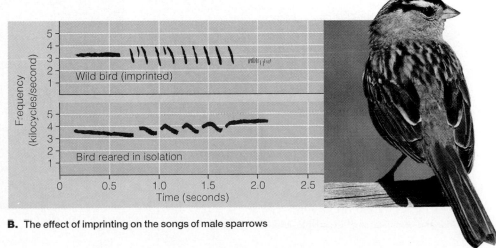

B. The effect of imprinting on the songs of male sparrows

Many animals learn by association and imitation

A. Associative learning by ducks

Association is learning that a particular stimulus or a particular response is linked to a reward or punishment. If you keep a pet, you probably have observed one type of associative learning first-hand. A dog or cat will learn to associate a particular sound or word with some type of punishment or reward. The ducks in the pond pictured in Figure A have learned to associate the presence of people with handouts, and they congregate rapidly whenever someone approaches the shoreline. This type of learning, in which an arbitrary stimulus is associated with a reward or a punishment, is called **classical conditioning.** Eventually, the animal learns to respond to the arbitrary stimulus even in the absence of a reward or punishment.

In natural settings, a much more common form of associative learning is **trial-and-error learning.** In this case, an animal learns to associate one of its own behavioral acts with a reward or punishment. The animal then tends to re-peat the response if it is rewarded or avoid the response if it is punished. For example, predators quickly learn to associate certain kinds of prey with painful experiences. The coyote in Figure B has a face full of quills obtained when it tried attacking a porcupine (right photograph). The porcupine's sharp quills and ability to roll into a quill-covered ball are strong deterrents against many predators. Coyotes, mountain lions, and domestic dogs often learn the hard way to avoid attacking porcupines nose-first. Trial-and-error learning often provides responses that are important to survival.

Another form of learning is **imitation**—learning by observing and mimicking the behavior of others. Imitation is similar to song imprinting in birds, though imitation is not limited to a critical period. Many predators, including cats, coyotes, and wolves, learn some of their basic hunting tactics by observing and imitating their mother. For example, during the two years they stay with their mother, jaguar cubs may observe and learn basic stalking techniques, as well as which prey are easiest to kill. Typically, both trial-and-error learning and imitation play roles as young predators refine their hunting skills. For instance, a young coyote or mountain lion might learn by trial and error to be wary of porcupines. And it might learn by imitation to flip a porcupine onto its back, exposing the unprotected belly.

B. Trial-and-error learning by a coyote (left), shown with quills from a porcupine (right)

Dogs often get into situations like the one in Figure A. Eventually, by trial and error, the animal might make the situation worse by looping its leash around the stump again, or it might actually find a new route to its food bowl. If it did reach the food, it might learn which route was correct and then be able to solve a similar problem more quickly. In any case, a dog does not show an ability to solve problems except by trial and error or by learning from prior experience.

A. A dog demonstrating its *lack* of reasoning ability

Innovation, sometimes called reasoning, is the ability to perform a correct or appropriate behavior on the first attempt without prior experience. Innovation implies an ability to think of possible solutions to a problem and analyze them. We see this type of learning most often in primates. For example, if a chimpanzee is placed in a room with several boxes on the floor and a banana hung high above its head, the chimp will "size up" the situation and stack the boxes in order to reach the food. In Figure B, a chimp is using a stick to extend its reach across the water in order to fetch a desired object. The vast majority of animals do not exhibit this kind of problem-solving ability.

The subject of innovation in problem solving raises the larger question of what goes on in the brain of a nonhuman animal. To what extent is an animal aware of the need to solve a problem? To what extent is it aware of itself and its surroundings in general? It is important to recognize and avoid as much as possible the pitfall of **anthropomorphism,** the tendency to ascribe to other animals human motivations or conscious awareness, or to assume that they experience feelings such as pain or pleasure in the same ways we do. This is not to say that our thoughts and those of other animals are completely dissimilar. It's just that we do not yet know for certain whether, or to what extent, any nonhuman animal has mental experiences similar to ours.

In humans, mental functioning, which includes conscious thinking, awareness of self, judgment, and use of language, is called **cognition** (Latin *cognitio,* knowledge). Because of our ignorance of other animals' mental states, many researchers maintain that questions of cognition in animals other than humans are outside the realm of science. To avoid any trace of anthropomorphism, many researchers focus on behavior patterns that can be described in terms of simple stimuli and responses. In contrast, other researchers, especially those studying chimpanzees and other primates in the wild, find much evidence that cognition is not restricted to humans. In a 1992 book entitled *Animal Minds*, Donald Griffin of Princeton University advanced the view (not shared by many other biologists) that conscious thinking is part of the behavior of many animals. Griffin sees animal cognition arising through the normal process of natural selection and, like so many other major animal functions, having roots extending far back into evolutionary history.

Ultimately, answers to questions about animal thinking may profoundly affect how we interact with other animals and how we view ourselves. Historically, a prevalent view has been that our intellect sets us apart from other animals. There may seem to be a vast difference between human cognition and that of other animals. But is this difference a fundamental biological one, or is it a matter of degree? Are we simply at one end of a continuum of cognitive abilities?

B. A chimpanzee solving a problem by innovation

37.8 An animal's behavior reflects its evolution

As we have seen so far, behavior patterns typically have both innate and learned components. Many animals do complex things, but often, if we compare several individuals of a species, we find that their behavior is more or less the same. This suggests that much of their behavior results from genetically fixed programs. Nonetheless, simple kinds of learning, such as imprinting, frequently modify key aspects of fixed behavior patterns. Also, as we saw in the case of the lovebirds in Module 37.2 and the example of bird song imprinting in Module 37.5, the capacity to learn rests ultimately on some kind of genetic basis.

We have stressed that behavior is an evolutionary adaptation that enhances survival and reproductive success (fitness). In fact, behavior is the result of the fine-tuning of an animal to its environment by natural selection. We have seen evidence of this in every type of behavior we have discussed so far—including the hunting and reproductive behavior of jaguars, nest location by digger wasps, and imprinting and other forms of learning in many animals. In all cases, the behavior patterns involve an interplay between the organism and its environment. In the next several modules, we pay special attention to the ecological role of behavior—that is, its role in enabling an animal to survive in its environment.

37.9 Biological rhythms synchronize behavior with the environment

Animals exhibit a great variety of rhythmic (regularly repeated) behavior patterns. Many mammals—for example, bats, deer, and most cats, including the jaguar—sleep or doze a great deal during the day and feed at dusk and dawn or at night. In contrast, most birds sleep at night and are active during daylight hours. Patterns that are repeated daily, such as sleep/wake cycles in animals and plants, are called **circadian rhythms.** As we saw in plants, internal timers called biological clocks underlie circadian rhythms (see Module 33.10). External cues, especially light/dark cycles, adjust the clocks, keeping rhythms tightly coordinated with the outside world.

How are circadian rhythms studied? The photograph in Figure A shows an animal used for studying activity rhythms, the flying squirrel, a nocturnal inhabitant of North American forests. A squirrel is placed in a cage containing a wheel in which it can run. The cage is connected to a chart recorder, which moves graph paper past a pen at a fixed speed. Whenever the squirrel runs in the wheel, the pen is activated and marks the paper.

The graphs in Figure A trace the activity patterns of two flying squirrels held under different light conditions for 23 days. The longer black bars indicate periods of extended activity. Graph (1) shows the activity pattern of a squirrel

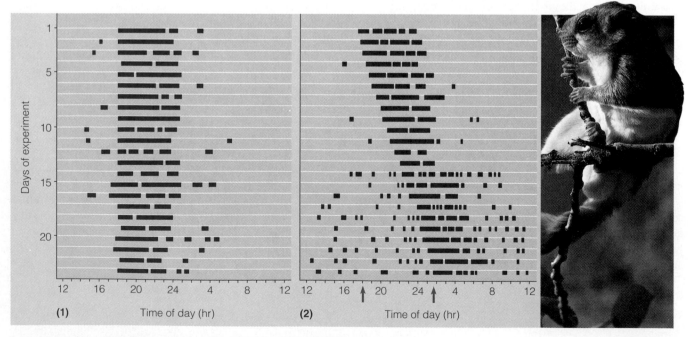

A. Activity rhythms of two flying squirrels under different conditions of light and darkness

exposed to 12 hours of light alternating with 12 hours of darkness, simulating natural conditions. Graph (2) shows the activity pattern of a squirrel held in constant darkness for 23 days.

As you can see, the activity of both squirrels remained rhythmic throughout the recording period, with a distinct period of extended activity every day. The activity rhythm of the squirrel exposed to cycles of 12 hours of light and 12 hours of dark remained virtually unchanged for all 23 days. In contrast, the other squirrel's high activity period shifted slightly each day, and after 23 days was nearly 8 hours out of synchronization with the actual time of day. (The small red arrows indicate when the period of greatest activity began on days 1 and 23.) When the squirrel held in the dark was returned to a regular cycle of 12 hours of light and 12 hours of dark, its activity cycle shifted back to that of the other animal in a few days. (This is not shown on the graph.)

Research with many different species has shown that without environmental cues, biological clocks keep time in a free-running way. In the study shown here, the flying squirrel's clock kept time at 24 hours, 21 minutes. Thus, the activity pattern of the squirrel kept in the dark shifted by 21 minutes each day. In contrast, humans have a biological clock that makes circadian rhythms follow an approximately 25-hour cycle. In humans, flying squirrels, and other organisms that have been studied, environmental cues are needed to keep circadian rhythms synchronized with external conditions. Sunrise and sunset are important cues.

How do researchers study circadian rhythms in humans? In one study, conducted in 1989, Italian interior designer Stefania Follini volunteered to test the effects on her body rhythms of long-term isolation below the ground. Follini spent 131 days alone in a plastic, 3.5-m-by-6-m cubicle 3 m underground in a cave near Carlsbad, New Mexico (Figure B). Temperature in the cubicle was held constant at 21°C. Follini had control of artificial lighting, but the cave itself was totally dark; therefore, without a clock, she had no cues about day or night hours. Researchers monitored a number of physiological factors, including Follini's blood pressure, heart rate, and body temperature. Follini herself kept track of what she thought were days and nights and the passage of time.

During the course of the study, Follini's sleep/wake rhythm followed the typical 25-hour period of the free-running human biological clock. Unexpectedly, however, her blood pressure and heart rate followed a 48-hour to 7-day cycle. When Follini emerged from the cave in June, at the end of the study, she thought it was March. She had also developed a severe calcium deficiency, lost 24 pounds, and stopped menstruating. She did not regain normal body

B. Subterranean living quarters for studying human biorhythms

rhythms until months after she had emerged, and she began menstruating again only after receiving hormone shots. The reasons for these changes are unknown, but researchers think at least some were related to the emotional stress of being isolated from all human contact and normal environmental stimuli for months. Follini's experiences and those of others who have participated in similar projects lead many researchers to be wary of long-term isolation studies. Most studies of human circadian rhythms are now performed in hospitals or research labs, where subjects have contact with other people and are more closely monitored.

It is important that we learn more about biological rhythms, clocks, and the cues that set them, because body rhythms affect our general well-being, work efficiency, and decision-making ability. Working night shifts, keeping irregular hours, and traveling by jet across several time zones often lead to fatigue, reduced job performance, and depression because our internal rhythms can't adjust instantaneously to a different time frame. Researchers studying ways to help the body reset its internal clocks when necessary have had some encouraging results. Specific light/dark regimes can often be used to reset our biological clocks. Also, some people can minimize the symptoms of jet lag by adjusting when they eat, exercise, and sleep.

Although we know that internal clocks time animal rhythms, research has not yet revealed the exact nature of biological clocks. In birds, the clock that times sleep/wake cycles seems to be located in the pineal body (see Module 26.3); researchers have evidence of a similar clock in the hypothalamus of the brain in mammals. Current hypotheses suggest that the clocks are biochemical, perhaps using molecular interactions that occur with regularity.

37.10 Kineses and orientation behavior place animals in favorable environments

Many protists and some animals exhibit a very simple response to certain stimuli: When they sense a stimulus, they merely start or stop moving, change their speed of movement, or turn more or less frequently in a random (nondirected) manner. A random movement in response to a stimulus is called a **kinesis** (plural, *kineses;* Greek for movement). Human body lice, for instance, exhibit kineses; they move randomly but become more active in dry areas and less active in moist ones. The more they move, the greater the chance they will find a moist area, which is more favorable to their survival. Once in a suitable place, their decreased activity tends to keep them there.

In contrast to kineses are directed movements, also called **orientation behavior.** Orientation behavior results in an animal's placing its body in a specific position in reference to an environmental cue. One form of orientation behavior is a taxis (Greek *tasso,* "put in order"). A **taxis** (plural, *taxes*) is a more-or-less automatic movement directed toward (positive) or away from (negative) a stimulus. For example, trout exhibit positive *rheo*taxis (Greek *rheos,* current); they automatically swim in an upstream direction, toward the current. Many animals locate mates by *chemo*taxis. Females of many species, especially insects, emit chemical signals that attract males. The males, able to detect minute quantities of the chemicals, pick up the scent trail and orient to increasing concentrations of the chemical, eventually following it to its source.

A taxis is a more finely tuned response to environmental stimuli than a change in speed or the turning of a random movement. It is, therefore, a more efficient way for an animal to locate a mate or find food. It is not surprising that orientation behavior is much more prevalent than kineses in the animal kingdom. As we see next, orientation behavior is important in animal migration.

37.11 How do migrating animals navigate?

Although many animals reside in one geographical area year-round, many species migrate, often over great distances. **Seasonal migration** is the regular movement of animals from one place to another at particular times of the year. It enables many species to access rich food resources throughout the year and to breed or winter in areas that favor survival.

Figure A shows the migratory route (red line) of one long-distance traveler, the gray whale. During summer, these giant mammals feast on small, bottom-dwelling invertebrates that abound in northern oceans. In the autumn, they leave their northern feeding grounds and begin a long trip south along the North American coastline. Arriving in warm, shallow lagoons off Baja California (Mexico) in the winter months, they breed, and pregnant females give birth to young before migrating back north. The yearly round trip, some 20,000 km, is the longest for any mammal.

Many other species of mammals, as well as numerous birds and some insects, also migrate seasonally. For instance, numerous insect-eating birds winter in the tropics and breed at high latitudes. Their breeding grounds—such as the tundra and taiga of the northern hemisphere—harbor large populations of insects, but only in the summer months. Among the insects themselves, the monarch butterfly has one of the most remarkable seasonal migrations. During winter, these insects festoon certain trees at the western tip of Cuba, in a few mountain valleys of central Mexico, and at a few sites along the California coast (Figure B). In the late summer and fall, all of North America's monarchs fly to these wintering sites. They remain on the trees for 4 to 5 months, not feeding but living off food molecules stored in their tissues. Suitable wintering sites are rare because temperatures must be just cool enough that the monarchs do not metabolize their stored food before spring, and warm enough that they do not freeze. With the onset of spring, monarchs mate at the wintering sites and begin migrating northward. As they arrive at regional destinations, they lay eggs and then die. Two or more generations are produced during the summer, repopulating the United States

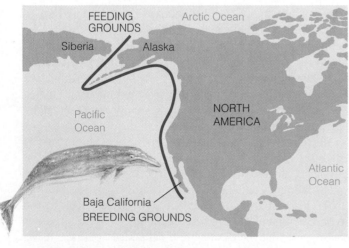

A. The migratory route of the gray whale

B. Monarch butterflies at a wintering site

were placed in funnel-like cages in a planetarium (photograph). Each funnel had an ink pad at its base and was lined with blotting paper. When a bird stepped on the ink pad and then tried to fly in a certain direction, it tracked ink on the paper. The researchers found that wild buntings and those raised in the lab and introduced to the northern sky in a planetarium tracked ink in the direction of the North Star. Birds raised under a sky with a different fixed-location star oriented to that star. Apparently, buntings learn a star map and fix on a stationary star when navigating at night.

Unlike the bunting, many birds can adjust to the apparent movement of the sun or other stars. How they do so is not yet known. Also little understood is how birds continue navigating when the sun or stars are obscured by clouds. There is strong evidence, however, that some birds can orient to Earth's magnetic field. Magnetite, the iron-containing mineral once used by sailors as a primitive compass, is probably involved in sensing the field. This mineral has been found in the heads of pigeons, in the abdomens of bees, and in several species of bacteria. Future research may show that magnetic sensing is a widespread, important orienting mechanism.

and southern Canada. With the approach of fall, the summer's last generation of monarchs flies south to the wintering grounds. Somehow, they are able to migrate as far as 4000 km and end up at a specific site, even though they have never flown the route before.

Researchers are beginning to learn about the kinds of environmental cues that migrators use in navigating. Gray whales, for instance, seem to use coastal landmarks to pilot their way north and south. Migrating south in the autumn, they orient with the North American coastline on their left. Migrating north in the spring, they keep the coast on their right. Whale watchers sometimes see gray whales stick their heads straight up out of the water, perhaps to obtain a visual fix on land. Many birds migrate at night, navigating by the stars the way ancient human sailors did. In contrast, monarch butterflies migrate during the day, resting in trees and bushes at night; they may use the sun as a compass.

Navigating by the sun or stars requires sophisticated orientation mechanisms, including an internal timing device to compensate for the continuous daily movement of Earth relative to celestial objects. Consider what would happen if you started walking one day, orienting yourself by always keeping the sun on your left. In the morning, you would be heading south, but by evening you'd be heading north again, having made a half-circle and gotten nowhere. At night, the stars also shift their apparent position as Earth rotates.

At least one night-migrating bird, the indigo bunting, seems to solve the problem by fixing on the North Star, the one bright star in northern skies that appears almost stationary. Figure C illustrates an experimental setup that was used to study the bunting's navigational mechanism. During the migratory season, wild and laboratory-reared birds

Funnel-shaped cage

Paper

Ink pad

C. An experiment demonstrating star navigation

37.12 Highly evolved feeding behaviors help animals obtain energy efficiently

No aspect of an animal's life affects fitness more than its ability to obtain energy efficiently. Animals feed in a great many ways. Some animals are food "generalists," while others are "specialists." The gull in Figure A is an extreme generalist; it will eat just about anything that is readily available—plant or animal, alive or dead. In sharp contrast, the koala of Australia, an extreme food specialist, eats only the leaves of eucalyptus trees (Figure B).

Most animals have some variety in their diet but are more selective than gulls, even if they are generalists. Often an animal will concentrate on a particular item that is abundant, sometimes to the exclusion of other foods. When an animal does this, it is said to have a **search image** for the favored item. (We often use search images to help us find something more efficiently. For example, if you were looking for a particular package on a kitchen shelf, you would probably scan rapidly, looking for a package of a particular size and color rather than reading labels.) If the favored food item becomes scarce, the animal may develop a search image for a different food item.

Why might an animal be selective in its choice of food? Some researchers propose that the explanation lies in what is called **optimal foraging**, feeding behavior that provides maximal energy gain with minimal energy expense and minimal time spent in foraging (searching for, securing, and eating food). We would expect that natural selection would

A. Feeding generalist, a gull

B. Feeding specialist, a koala

favor animals that forage optimally, but it is difficult to test this hypothesis in nature. Whenever an animal has food choices, there are a number of trade-offs. Consider the bass in Figure C, for example. It can readily consume both minnows and crayfish. If it eats a minnow, it will probably get more usable energy per unit weight (a crayfish has a lot of hard-to-digest exoskeleton), but the minnow is smaller and may be harder to catch. On the other hand, it may take more time to eat a crayfish because of its large claws and tough exoskeleton. Complicating the picture even more, the bass must be alert to other predators while feeding. What would expose it more to a predatory turtle or larger fish, chasing a minnow or mouthing a thrashing crayfish?

In most natural environments, there are so many variables that it is hard to imagine any animal could forage in an absolutely optimal manner. Nonetheless, numerous studies indicate that when prey is plentiful, many different species forage in such a way that their overall energy intake-to-expenditure ratio is high. In the context of behavioral ecology, they tend to forage in a way that maximizes their rate of energy storage, or productivity (see Module 36.10). A bass, for instance, forages efficiently, if not exactly optimally, by switching between minnows, crayfish, aquatic insects, and other invertebrates as conditions, such as a prey's relative sizes, density, and ease of capture, change.

C. A bass eating a crayfish

Armadillo, the preferred prey

Tapir, generally
ignored

D. Alternative prey available
to jaguars in Belize

In another case, Alan Rabinowitz found that jaguars in Belize ate mainly small mammals, especially the abundant and slow-moving armadillo (Figure D, left). An armadillo may weigh only about 5 kg—just a few mouthfuls for a jaguar. It also has hard body armor—no problem for the big cat's jaws, but not digestible. Why would the most powerful cat in North America eat mainly bite-sized prey, especially when it has much larger game available to it? For instance, tapirs (Figure D, right), which may weigh up to 200 kg and are known to be vulnerable to jaguars, also inhabit the jungles of Belize. But Rabinowitz found no evidence of jaguars eating these large mammals in his study area. Most likely, the jaguars concentrate on armadillos because they are abundant and easy to catch. Tapirs, by contrast, run fast, usually into very dense undergrowth, and are not nearly as abundant.

The kangaroo rat (Figure E), an herbivorous rodent found in North American deserts, illustrates the effects of trade-offs in optimal foraging more clearly than either the bass or the jaguar. Foraging at night, the kangaroo rat fills its cheek pouches with high-energy seeds and carries them home to its burrow. Careful studies show that when a choice of seeds is available, the animal picks up seeds that contain more energy

E. A kangaroo rat with a collection of seeds

than most of those it leaves behind. Later, in the safety of its burrow, the rat may be even more selective, eating only the very richest seeds from its cache. Thus, the kangaroo rat does not forage exactly optimally, for it expends energy gathering seeds it does not consume, but it exhibits a healthy compromise: It selects high-energy food in a manner that reduces time spent above the ground, where it is exposed to predators.

Social behavior is an important component of population biology | 37.13

Of the types of behavior we have examined so far, several involve interactions between two or more individuals of a species. Imprinting, for instance, often involves interaction between a parent and an offspring. Many animals migrate and feed in large groups (flocks, packs, herds, or schools). Wolves, for example, usually hunt in a pack consisting of a tightly knit group of family members. Hunting in packs allows them to kill large animals, such as moose or elk, that would be unavailable to a single individual.

Biologists define **social behavior** broadly as any kind of interaction between two or more animals, usually of the

same species. An essential ingredient of all types of social behavior is communication—some means of transferring information between individuals. Because social behavior involves group interactions, it often has a strong effect on the growth and regulation of animal populations. For example, the large prey animals that wolf packs can kill provide more food for each wolf than smaller prey; hence, an opportunity for more wolves to survive in a particular area. As we will see in the next several modules, social behavior may also affect fitness directly by actually determining which animals in a population will produce offspring.

37.14 Rituals involving agonistic behavior often resolve confrontations between competitors

Agonistic behavior (from the Greek *agon,* struggle) includes a variety of threats or actual combat that settles disputes between individuals in a population. Conflicts often arise over limited resources, such as food or mates, especially in dense populations. An agonistic encounter may involve a test of strength or, more commonly, exaggerated posturing and other symbolic displays—rituals—that make the individuals look large or fierce. Eventually, one individual stops threatening and becomes submissive, exhibiting some type of appeasement display—in effect, surrendering.

Because violent combat may injure the victor as well as the vanquished in a way that reduces reproductive fitness, we would predict that natural selection would favor ritualized contests. And, in fact, this is what usually happens in nature. The rattlesnakes pictured

Ritual wrestling by rattlesnakes

here, for example, are rival males wrestling over access to a mate. If they bit each other, both would die from the toxin in their fangs, but this is a pushing, rather than a biting, match. One snake usually tires before the other, and the stronger one pins the loser's head to the ground. In a way, the snakes are like two people who settle a serious argument by arm wrestling instead of resorting to fists or knives. In a typical case, the agonistic ritual appeases the combatants and inhibits further aggressive activity. Once two individuals have settled a dispute by agonistic behavior, future encounters between them usually involve less dispute, with the original loser giving way to the original victor. Often the victor of an agonistic ritual gains first or exclusive access to mates, and so this form of social behavior can directly affect an individual's evolutionary fitness.

37.15 Dominance hierarchies are maintained by agonistic behavior

Many animals live in social groups maintained by agonistic behavior. Chickens are an example. If several hens unfamiliar to one another are put together, they respond by chasing and pecking one another. Eventually, they establish a clear "peck order." The alpha (top-ranked) hen in the peck order (the one on the left in the photograph) is domi-

Chickens exhibiting peck order

nant; she is not pecked by any other hens and can usually drive off all the others by threats rather than actual pecking. The alpha hen also has first access to resources such as food, water, and roosting sites. The beta (second-ranked) hen similarly subdues all others except the alpha, and so on down the line to the omega, or lowest animal. Peck order in chickens is an example of a **dominance hierarchy,** a ranking of individuals one above the other by social interactions.

Dominance hierarchies offer several benefits. Once a hierarchy is established, each animal's status in the group is fixed, often for several months or even years; consequently, rather than fighting with others, group members can concentrate on finding food, watching for predators, locating a mate, or caring for young. Dominance hierarchies are common, especially in vertebrate populations. In a wolf pack, for example, there is a dominance hierarchy among the females, and the hierarchy may control the pack's size. When food is abundant, the alpha female mates and also allows others to do so. When food is scarce, she usually monopolizes males for herself and keeps other females from mating.

Animal behaviorist Jane Goodall discusses dominance hierarchies and cognition in chimpanzees

Chimpanzees are our closest relatives. Chimps and humans are more like each other genetically than either is like other apes. Dr. Jane Goodall (Figure A), one of the world's best-known biologists, has studied these remarkable primates in their natural habitat in East Africa since the early 1960s. Many of her discoveries are described in her seven books, her appearances on National Geographic Society television specials, and her frequent lecture tours. In all of her writings and interviews, Dr. Goodall promotes better understanding of animal behavior, especially of primates. She also works tirelessly to encourage better living conditions for animals in medical research labs and zoos. Despite her crowded schedule, Goodall still spends as much time as possible studying chimpanzees in the wild.

Dominance hierarchies are an integral part of chimpanzee life, as Goodall relates:

> Some male chimpanzees devote much time and effort to improving or maintaining their position in the hierarchy. For the most part, the male uses the impressive charging display, during which he races across the ground, hurls rocks, drags branches, leaps up and shakes the vegetation [see Figure B]. In other words, he makes himself look larger and more dangerous than he may actually be. In this way he can often intimidate a rival without having to risk an actual fight, which could be dangerous for him as well as for his rival. The more frequent, the more vigorous, and the more imaginative his charging display, the more likely it is that he will attain a high social position.

Despite the focus male chimpanzees have on social position, the benefits of high position are not clear. In Goodall's words:

> We don't understand why the chimps devote so much time, effort, and risk [on gaining high rank]. What is the advantage of high rank for a male chimpanzee? What are the evolutionary benefits, the reproductive benefits? A high ranking male has prior access to the best food, but this is not of great benefit since, if the food is short, the chimps typically move about in ones and twos. He can usurp a choice resting place, but he almost never does. He can inhibit other males from copulating with a female in estrus when they are all in a group together, but there is a mechanism in chimp society that enables even a low-ranking male to appropriate a sexually attractive female. All he has to do is persuade her to follow him, away from the other males, to some peripheral area of the community range. It's almost as though humans aren't the only creatures who value high rank for its own sake, and the power that it gives.

And what about female chimpanzees?

> Females have a hierarchy too. . . . The reproductive advantage to the high-ranking female is clear. She can better appropriate choice food items and thus make her milk richer. In addition, her offspring are likely to become high-ranked since she will support them. In the supportive family group situation, all have a better chance of survival.

Studies of chimpanzee behavior can help us understand certain aspects of our own behavior. They can also make us more aware of what we have in common with other species. Jane Goodall's years of chimpanzee research have convinced her that chimpanzees are truly cognitive beings. As she explains:

> Science has been very quick to recognize the incredible similarity in [the anatomy] of the chimpanzee brain and human brain . . . and all the other amazing physiological similarities [between the two species]. So it stands to reason that you would find similarities in the emotions . . . and in certain kinds of behavior and intellect.

A. Jane Goodall with Gohlin, an alpha male

B. Charging display

Territorial behavior parcels space and resources

A. Sunbathers

No one parceled out space to sunbathers on the beach in Figure A. The fact is that people tend to space themselves out like this when close to others, establishing what we might call personal territories.

Many animals exhibit territorial behavior. A **territory** is an area, usually fixed in location, that individuals defend and from which other members of the same species are usually excluded. The size of the territory varies with the species, the function, and the resources available. Territories are typically used for feeding, mating, rearing young, or combinations of these activities.

Another beach scene, Figure B, shows a nesting colony of gannets, seabirds of the Atlantic Ocean. The birds are tightly and evenly spaced on a small island off the coast of Quebec.

Space is at a premium, and the birds maintain tiny nesting territories by agonistic behavior—calling and pecking at each other. Each gannet is literally only a peck away from its closest neighbors. Such small territories are feasible because the gannets don't use them for feeding; they feed in large flocks at sea, where they display little agonistic behavior. In contrast to the small nesting territory of gannets and many other seabirds, most cats, including jaguars, leopards, cheetahs, and even domestic cats, defend much larger territories (often several square kilometers in area), which they use for foraging as well as breeding.

Individuals that have established a territory usually proclaim their territorial rights continually; this is the function of most bird songs, as well as the noisy bellowing of sea lions, the chattering of squirrels, and the defecating in open areas along jungle trails of jaguars. Scent markers are frequently used to signal territories. The male cheetah in Figure C, a resident of Africa's Serengeti National Park, is spraying urine on a large rock. The odor will serve as a chemical "no trespassing" sign. Other males that approach the area will sniff the marked rock and recognize that the urine is not their own. Usually, the intruder will avoid the marked territory and a potentially deadly confrontation with its proprietor. Domestic cats also mark their territories by spray-urinating.

Territorial animals benefit in several ways. They have exclusive access to food supplies and breeding areas within their territories. Also, familiarity with their areas helps them obtain food there and avoid predators. Moreover, they can care for their young without interference from other individuals of the same species. All these factors may increase fitness in territorial species.

B. Gannets at a nesting ground

C. A cheetah spray-urinating

Mating behavior often involves elaborate courtship rituals

Many animals are strongly programmed to view any organism of the same species as a threatening competitor, to be driven away if possible. Even animals that forage and travel in groups maintain a certain distance from their companions. How, then, is mating accomplished? In many species, prospective mates must perform an elaborate court-ship ritual, unique to the species, confirming that the individuals are not threats to each other and that they are of the correct species, sex, and physical condition.

Courtship behavior often involves actions that signal appeasement. For example, the common loon, a water bird that breeds on secluded lakes in the northern United States

①

②

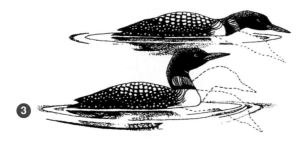

③

④

and Canada, turns its beak away as an appeasement signal. In sharp contrast, a male loon defending his territory often charges at an intruder with his beak pointed straight ahead.

Figure A depicts the courtship and mating of a pair of loons. The birds swim side by side while performing a series of displays, all of which signal appeasement. ① The courting birds frequently turn their heads away from each other, ② dip their beaks in the water, and ③ submerge their heads and necks. Prior to copulation, the male invites the female onto land by ④ turning his head backward with his beak held downward. There, ⑤ they copulate.

Loons represent a large number of species in which the courting pair are more or less isolated from the rest of the population. In certain other species, courtship is a group activity during which members of one or both sexes choose mates from a group of candidates. For instance, sage grouse, chickenlike birds that inhabit high sagebrush plateaus in the western United States, perform mating rituals in large groups. Each day in the early spring, fifty or more males congregate in an open area called an arena. The males strut about, erecting their tail feathers in a bright, fanlike display. Dominant males usually defend a small territory near the center of the arena. Females arrive several weeks after the males, and after watching the males perform (Figure B), a female will select one, and the pair will copulate. Usually, all the females choose dominant males, with the result that only about 10% of the males actually mate.

Is there an advantage to group mating rituals like this? In species that reproduce sexually, an individual's own genes alone do not determine reproductive success; rather, it is the combination of that individual's genes with those of its mate. Researchers hypothesize that there is a connection between a dominant male's display and his evolutionary fitness. We might suppose, therefore, that in choosing a dominant male from several in an arena, a female sage grouse gives her genes the best chance for future survival. Devising ways to test this hypothesis of an ultimate cause is one of the fascinating challenges of research in behavioral ecology.

A. Courtship and mating of the common loon

B. Courtship display by a male sage grouse

Chapter 37   Behavioral Adaptations to the Environment   **735**

Complex social organization hinges on complex signaling

So far, two major concepts emerge from our discussion of social behavior. (1) Social behavior provides organization within populations of animals. For instance, agonistic behavior maintains dominance hierarchies and territoriality, and appeasement behavior is central to courtship and mating. (2) Social behavior and the social organization it provides depend on some form of signaling, or communication, among the participating animals. As we have seen, animals use a variety of signals, including sounds, such as the growl of a male jaguar announcing his presence at night; odors, such as the urine signs left by many cats to mark their territories; visual displays, such as beak-pointing in loons; and touches, such as the caressing of mating jaguars. Sounds and odors have the advantage of carrying long distances. Signals detected by sight or touch often communicate aggression or appeasement between individuals at close range.

In general, the more complex the social organization, the more complex the signaling required to sustain it. In fact, animals with a complex social structure often use more than one type of signal simultaneously. Figure A shows a ring-tailed lemur, a tree-dwelling primate of Madagascar that lives in social groups averaging 15 individuals. Visual displays, scent communication, and vocalizations maintain the dominance hierarchy in the group. The animal shown here is communicating aggression with its prominent tail. Prior to this display, it smeared its tail with odorous secretions from glands on its forelegs. By waving its scented tail over its head, the lemur transmits both visual and chemical signals.

In addition to primates, many other vertebrates exhibit intricate social behavior. However, many of the most complex social systems are found among the invertebrates. The social system of honeybees is a prime example. Often numbering over 50,000 individuals, a honeybee colony has complicated communication needs. For example, a worker bee that has located a good source of food must communicate this information to other workers in the hive, enabling the food to be harvested in quantity. Intrigued with the question of how bees communicate, Karl von Frisch performed several experiments in the 1940s. He put out dishes of scented sugar water as food sources, varying their distance and direction from hives. In addition, he modified the hives so he could see inside them. Von Frisch and several later researchers discovered that a returning worker bee passes some very complex information to other workers using a unique signaling system. The figures on the facing page illustrate von Frisch's hypothesis concerning the bees' communication system.

When a worker returns to a hive, others gather around it (Figure B), and the bee regurgitates some nectar that the others taste. The nectar probably lets the other workers know the type of food that has been found. The worker then performs one of two "dances" that seem to indicate the location of the food. If the source is within 50 m or so of the hive, the bee moves rapidly sideways in tight circles, performing a "round dance" (Figure C). The round dance translates as "food is near" but does not indicate direction. The other workers then leave the hive and begin foraging nearby; tasting the nectar probably helps them identify a scent to fly toward.

A worker returning from a longer distance performs a "waggle dance," instead of a round dance. As shown in part 1 of Figure D, the waggle dance involves a half-circle swing in one direction, followed by a straight run, then a half-circle swing in the other direction. During the straight run, the bee waggles its abdomen vigorously.

The waggle dance may impart a remarkably detailed code. According to von Frisch's hypothesis, it tells other workers both the distance and direction of a food source. The speed at which a worker completes the cycles of the dance indicates distance to the food. A certain high rate, for example, would translate as "about 100 m away," whereas a certain slow rate would mean "about 1000 m away," with intermediate rates corresponding to intermediate distances. The code for direction requires that the waggle dance be performed on a vertical surface in the hive. The angle of the straight run in relation to the vertical is the same as the horizontal angle of the food's location in relation to the position of the sun. Thus, if the dancer runs directly up-

A. A lemur communicating aggression

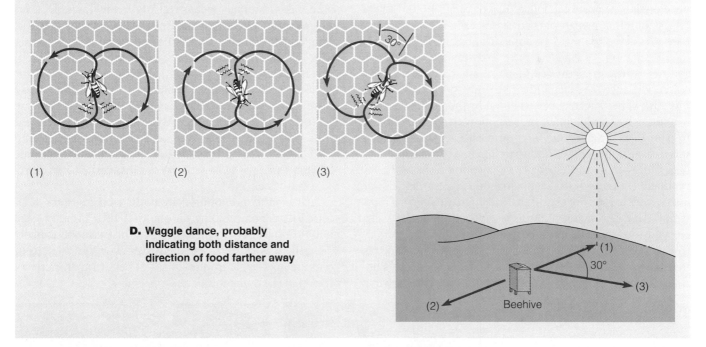

B. Bees clustering around a recently returned worker

C. Round dance, indicating that food is nearby, in an unspecified direction

D. Waggle dance, probably indicating both distance and direction of food farther away

ward (Figure D, part 1), the other workers will fly directly toward the sun when they leave the hive. If the dancer runs directly downward (part 2), the others will fly directly away from the sun. And if the dancer runs at an angle, say 30° to the right of vertical (part 3), the other workers will fly 30° to the right of the horizontal direction of the sun, and so forth. Today, some researchers question the evidence for von Frisch's model and are exploring alternative hypotheses.

Bees also have some kind of innate timekeeping ability that enables them to adjust for the sun's movement. If a rainstorm prevents them from foraging for several hours, they will still fly in the proper direction toward food, even though the sun's angle has changed.

How do bees inside a dark hive follow the dance movements of their fellow workers? Many questions remain to be answered, but several types of communication, including touch, taste, and sound, seem to be involved. Apparently, as the workers swarm closely around a dancer, they detect dance cues by physical contact. The dancers also seem to make different buzzing sounds during the straight run of the waggle dance; researchers think that these sounds correlate with the distance and direction of the food source. Used together, these different signals make it possible for worker bees to communicate the complex information needed to forage as a group and thus gather enough food to supply their large colony.

Altruism can be explained by evolution

All the workers in a honeybee hive are sterile females. They never reproduce, but spend their lives laboring on behalf of the one fertile queen that lays all the eggs in the hive. When a worker stings an intruder in defense of the hive, the worker usually dies. Such behavior, which involves self-sacrifice, is known as **altruism.** It is an important component of social organization in many animal species.

Altruism is often a central feature of cooperative group living. For example, altruism very different from that of honeybees helps the African elephants in Figure A function as a family unit. A typical family group consists of several closely related adult females (often sisters) and their immature offspring. The oldest female leads the group, often selflessly confronting large predators. In fact, all the adults look out for the well-being of the entire group, often assembling in a tight formation when there is a serious threat. Moreover, any lactating female in the family group will allow any infant to nurse.

Altruism is also evident in many species of rodents that live in colonies. The animal in Figure B, a Belding's ground squirrel, is sounding a high-pitched alarm call that alerts other members of the colony to the approach of a predator. By careful observation, researchers have confirmed that when a squirrel issues an alarm call, it gives other members of its colony a chance to retreat to their burrows, but it increases its own risk of being detected and killed by the predator.

You might wonder about the evolutionary advantage of altruism. In being altruistic, a ground squirrel, for example, may sacrifice its own life and therefore its reproductive success. How can altruistic behavior evolve if it reduces the reproductive success of self-sacrificing individuals? One answer is that the frequency of genes for altruism may increase if individuals that benefit from altruistic acts are themselves also carrying those genes. This is most likely to be the case if the altruists and their beneficiaries are related. According to this idea, known as **kin selection,** altruistic behavior evolves because it increases the number of *copies* of a gene common to a group, regardless of which individuals in the group transmit the gene. For instance, if by performing a selfless act, an elephant provides several of her siblings a better chance of reproducing, it is likely that more of her genes (which she shares with siblings) will be passed to the next generation. If some of those genes are involved in altruism, this behavior will evolve. Likewise, worker bees in a hive all share genes with the queen. Their work (or even death) in support of the queen helps ensure that a large number of those genes will survive.

Alarm sounding by ground squirrels is also consistent with kin selection. Usually, only females give alarm calls. Males tend to disperse to other colonies after mating and thus are generally surrounded by nonrelatives. Females tend to stay in the same colony, surrounded by offspring as well as other animals of varying relatedness. Thus, survival of the female's genes is enhanced by alarm sounding, whereas for males, such behavior would not enhance evolutionary fitness.

Kin selection does not explain all types of altruism. In some cases, animals behave altruistically toward others who are not relatives. Jane Goodall has discovered that chimpanzees sometimes save the lives of nonrelatives. Similarly, female dolphins without young will often help unrelated mothers care for their young. In these cases, there can be no immediate enhancement of the altruists' fitness. However, in the future, the current beneficiary may reciprocate—that is, "return the favor"—by performing some other helpful act. Thus, we can explain altruism toward nonrelatives as **reciprocal altruism:** an altruistic act repaid at a later time by the beneficiary (or by another member of the social system).

Altruism and its evolution are hotly debated topics in biology today. Some biologists question how broadly ideas about kin selection and reciprocal altruism can be applied. The debate is especially heated over whether these concepts are applicable to human social behavior, an issue we take up in the next module.

A. A family group of elephants

B. A ground squirrel sounding an alarm

Sociobiology has extended evolutionary theory to human social behavior

In 1975, Edward O. Wilson of Harvard University published a book entitled *Sociobiology: The New Synthesis.* Drawing from numerous studies on vertebrate and invertebrate animals, Wilson promoted a relatively new area of research called **sociobiology,** the study of the evolutionary basis of social behavior. Sociobiologists, including Wilson, support the concept that social behavior evolves, like an animal's anatomical traits, as an expression of genes that have been perpetuated by natural selection. Sociobiologists search for evidence that social activities improve an animal's fitness. The concepts of kin selection and reciprocal altruism, with their focus on how the survival of genes underlies social behavior, are key ideas in sociobiology.

Most of Wilson's book applied evolutionary thinking to social behavior of nonhuman animals. But in the final chapter, Wilson speculated on the evolutionary basis of certain kinds of social behavior in humans. This speculation rekindled an old debate over which most strongly determines human behavior: evolution and genetic programming (nature) or environment (nurture).

Let's examine the nature-versus-nurture controversy in the context of a specific example: avoidance of sibling incest. Nearly all cultures have laws or taboos forbidding sexual relations between brother and sister. Do we humans avoid incest mainly because we are genetically programmed to do so, or mainly because of our socialization (nurturing)? The socialization argument is that avoidance of incest is a learned behavior, that the social stigma attached to incest is based on experience—people who break the taboo have been seen to have more children with birth defects. Someone favoring this side of the debate might also argue that if incest avoidance were chiefly innate, cultural taboos would be unnecessary.

On the other side of the debate, a sociobiologist might point out that incest avoidance is common to diverse human cultures and occurs in many nonhuman animal species that lack what we call culture. This commonality in itself may be evidence that incest avoidance has a strong innate component.

Some sociobiologists have cited the experience of the Israeli communal societies known as kibbutzim (singular, kibbutz) as evidence that an innate avoidance of incest is strong enough to counter cultural influences. From the time of birth, children in traditional kibbutzim usually spent much of their time in large child-care centers like the one in the photograph below. This arrangement gave the kibbutz children a siblinglike social relationship with one another. In the early years of the kibbutzim, parents encouraged their children to marry within their own kibbutz. However, a study of over 5000 people who grew up in these societies showed that marriages between individuals raised together from early childhood were extremely rare, in spite of the wishes of parents and kibbutz leaders. We might conclude, therefore, that people who spend nearly all their time together as children have little sexual attraction for one another as adults. A sociobiologist might argue from this example that humans have an innate resistance to incest—in this case, with nongenetic "siblings"—that overrides cultural pressure encouraging it.

Biologists concerned with human behavior do not take extreme positions on nature versus nurture. Obviously, we are not robots stamped out of a rigid genetic mold. There is also no reason to assume that our behavior is not affected by our genotype. Individuals vary extensively in anatomical features, and we should expect inherent variations in behavior as well. We should also expect that environment (including culture) intervenes in the pathway from genotype to phenotype for behavioral traits as it does for physical traits. Most sociobiologists envision human culture and genetic components of our social behavior to be truly integrated in human nature. Finding out how genes and the environment interact in determining our behavior is central to understanding human nature, and we pursue this topic further in the next module.

A child-care center at an Israeli kibbutz

Culture may reinforce innate social behavior

It is sometimes argued that cultural change is so rapid that it overwhelms any influence of evolution by natural selection on human behavior. However, sociobiologists consider this argument an oversimplification. As an alternative hypothesis, they propose that human culture often enhances or reinforces behavioral traits derived through natural selection. Let's assume, for instance, that incest aversion has an innate component. If most members of a society share the aversion, it is likely to become formalized in laws and taboos. A society that shuns or imprisons members who practice incest lowers the reproductive fitness of incestuous individuals. Thus, the cultural stigma may amplify the evolutionary component of the behavior by acting as an environmental factor in natural selection.

The photograph at the right focuses on one of our most important cultural attributes—education. All cultures are transmitted by the tutoring of the younger generation by the older. Is the passing on of knowledge a purely cultural attribute, or does it have a genetic component? The sociobiologist's view is that the positive reinforcements a culture gives to education are proximate causes that amplify a behavior that ultimately evolved because it increased fitness. The challenge in sociobiology is to uncover evidence of the ultimate causes of education and other cultural traits central to humanity. We trace some of the evolutionary threads of human culture in the final chapter.

Transmission of culture by education: a dance lesson in Indonesia

Chapter Review

Begin your review by rereading the module headings and scanning the figures, before proceeding to the Chapter Summary and questions.

Chapter Summary

37.1 Behavioral biology is the study of what animals do when interacting with their environment. Studying behavior is essential to understanding an animal's evolution and ecological role. Behavior can be interpreted in terms of proximate causes, or immediate interaction with the environment. Behavioral biologists are especially interested in the ultimate causes of behavior, which are evolutionary. Natural selection preserves behaviors that enhance fitness. Behavioral ecology focuses on the evolutionary interpretations of behavior.

37.2–37.3 Some aspects of behavior are innate—genetically programmed—and others are learned. Usually a sign stimulus, such as a baby bird's open mouth, triggers an innate, unchangeable fixed action pattern—in this case, an adult inserting a worm. In such parent–offspring interactions, it is important that behavior be performed correctly without practice.

37.4–37.5 Learning is a change in behavior resulting from experience. Habituation is learning to ignore a stimulus. Imprinting is irreversible learning limited to a critical period in the animal's life. For example, geese imprint on their mother during a short critical period. Imprinting shows that learning can involve innate behavior and experience. It helps animals learn quickly and enhances fitness.

37.6–37.7 Many animals can also learn by association, linking behaviors to reward or punishment, either by trial and error or by imitation. Some animals, such as chimpanzees, can "think through" solutions to problems, a type of learning called innovation. Behavioral biologists disagree about conscious thinking in animals. Are humans unique, or are we simply at one end of a thinking continuum?

37.8–37.9 Many animals, including humans, display rhythmic behavior. Daily (circadian) rhythms, such as sleep, appear to be timed by an internal biological clock. In the absence of environmental cues, these rhythms continue, but they become out of phase with the environment.

37.10–37.11 The simplest animal movements are kineses, changes in the rate of movement that keep animals in favorable environments. More complex orientation behaviors, such as taxes toward or away from stimuli, involve positioning in reference to a cue. Some animals, such as birds and monarch butterflies, undertake long-range migrations. Animals may navigate using the sun, stars, landmarks, or Earth's magnetism.

37.12 Animals are selective and efficient in their food choices. Some species generalize and some specialize. Natural selection has shaped feeding behavior to maximize energy gain and minimize the expenditure of time and energy. This behavior is known as optimal foraging.

37.13–37.19 Social behavior—the interactions among members of a population—influences population dynamics. Agonistic behavior consists of threats and combat that settles disputes. Dominance hierarchies and territories, maintained by agonistic behavior, partition resources. Courtship behaviors reduce aggression and advertise the species, sex, and physical condition of potential mates. Social behavior depends on signaling, in the form of sounds, scents, displays, or touches. Bees, for example, perform dances that communicate the direction and distance of nectar to other members of the colony.

37.20–37.22 Some animals exhibit altruism, endangering themselves to help others. Altruism can be explained in terms of kin selection: An animal can pass on its genes by helping relatives with copies of the same genes. Reciprocal altruism may also occur: A favor may later be repaid by the beneficiary. These are key ideas of sociobiology, the study of the evolutionary basis of social behavior. Most sociobiologists believe that natural selection underlies many human behaviors, and that our culture reinforces these behaviors.

Testing Your Knowledge

Multiple Choice

1. At various times, a behavioral biologist studied squirrels, sparrows, sharks, deer, and caterpillars. Which of the following behaviors do you think he observed least often? (*Explain your answer.*)

   a. agonistic behavior
   b. habituation
   c. imprinting
   d. innovation
   e. orientation behavior

2. Pheasants do not feed their chicks. Immediately after hatching, a pheasant chick starts pecking at seeds and insects on the ground. How might a behavioral ecologist explain the ultimate cause of this behavior?

   a. Pecking is a fixed action pattern.
   b. Pheasants learned to peck, and their offspring inherited this behavior.
   c. Pheasants that pecked survived and reproduced best.
   d. Pecking is a result of imprinting during a critical period.
   e. Pecking is an example of habituation.

3. Which of the following is true of animals that use the sun to navigate? (*Explain your answer.*)

   a. They cannot travel long distances.
   b. Most live in the sea, where there are few landmarks.
   c. They must have accurate biological clocks.
   d. Most migrate in large schools, flocks, or herds.
   e. They more easily travel east and west than north and south.

4. Ants carry dead ants out of the anthill and dump them on a "trash pile." If a live ant is painted with a chemical from dead ants, other ants carry it, kicking and struggling, to the trash pile, until the substance wears off. Which of the following best explains this behavior?

   a. The chemical is a sign stimulus for a fixed action pattern.
   b. The ants have become imprinted on the chemical.
   c. The ants continue the behavior until they become habituated.
   d. The ants can only learn by trial and error.
   e. The chemical triggers a negative taxis.

True/False (*Change false statements to make them true.*)

1. Agonistic behavior is taking a risk to help another individual.

2. A search image helps migrating birds find their way.

3. The ultimate causes of animal behavior have to do with evolution.

4. It is anthropomorphic to say "My cat is always so happy to see me."

5. Biological clocks tend to run a bit fast.

Describing, Comparing, and Explaining

1. Almost all the behaviors of a housefly are innate. What are some advantages and disadvantages to the fly of innate, programmed behaviors, compared with learned behaviors?

2. A wolf pack has both a dominance hierarchy and a territory defended against other wolf packs. What are the benefits to the pack of a dominance hierarchy? What are the benefits of holding a territory?

3. A chorus of frogs fills the air on a spring evening. Frog calls are courtship signals. What are the functions of courtship behaviors? How might a behavioral biologist explain the proximate cause of this behavior? The ultimate cause?

4. Explain how a bird's song is both programmed and learned.

Thinking Critically

1. Explain how you might use your knowledge of associative learning to train a dog to sit, roll over, or stop digging in flower beds.

2. Scientists studying scrub jays found that it is common for "helpers" to assist mated pairs of birds in raising their young. The helpers lack territories and mates of their own. Instead, they help the territory owners gather food for their offspring. What possible advantage might there be for the helpers to engage in this behavior instead of seeking their own territories and mates? Why do you think the scientists tried to find out whether the helpers and the birds they helped were related?

3. Western gulls nest very close together on flat-topped islands or rocks. Herring gulls are different, spacing their nests much farther apart. Small gulls called kittiwakes nest in tiny pockets on the faces of cliffs. Eggs of all three species sometimes roll out of the nest, but they don't go far. Sometimes, chicks wander off. Therefore, it may be advantageous for gulls to be able to recognize their own eggs and young, and some species imprint on their eggs or young. Suggest experiments to determine whether each of these species—the Western gull, herring gull, and kittiwake—is programmed to imprint on its own eggs or young. Which of them do you think will be found to imprint on its eggs? Its chicks? Both eggs and chicks? Neither? Give reasons for your predictions.

Science, Technology, and Society

1. Researchers are very interested in studying identical twins who were raised apart. Among other things, they hope to answer questions about the roles of inheritance and upbringing in human behavior. So far, data suggest that identical twins raised apart are much more alike than researchers would have predicted. They have similar IQs, personalities, mannerisms, habits, and interests. Why do identical twins make such good subjects for this kind of research? What do the results suggest to you?

2. How aware are animals? Do they think and feel the same kinds of things we do? These questions bear on animal rights, a subject much in the news. Many important biological discoveries have come from experiments performed on animals, yet some animal rights activists believe animal experimentation is cruel and should be stopped. They have harassed researchers, even vandalized laboratories and set animals free. Why are animals used in experiments? Are there uses of animals that should be discontinued? What kinds of guidelines should researchers follow in using animals in experiments, and who should establish and enforce the guidelines?

Human Evolution
and Its Ecological Impact

Bedo Jaosolo was 17 years old when the photograph at the left was taken near his home in Madagascar. By then he was recognized internationally as a skilled naturalist and expert guide. Son of a government fisheries manager, Bedo grew up near a small nature reserve called the Analamazoatra in the rain forest of eastern Madagascar. Having spent much of his childhood in the reserve, he had a deep knowledge and appreciation of its diverse animals and plants.

The Analamazoatra reserve was established in 1970 as a refuge for lemurs, a group of primates that occur in nature only in Madagascar. (The small photo below shows a species of lemur called the indri.) As a teenager, Bedo Jaosolo knew the lemurs and their habits as well as anyone. He worked closely with primate researchers who came to the Analamazoatra from around the world and who knew Jaosolo as an authority on the animals of his native rain forest. Jaosolo also introduced tourists, nature writers, and photographers to lemurs and other animals rarely seen and little understood.

Madagascar is a Texas-sized island in the Indian Ocean, about 420 km off the eastern coast of Africa. It is no small feat to be a guide and naturalist there, because with an estimated 200,000 species of plants and animals, Madagascar is among the top five most biologically diverse countries in the world. Severed by plate tectonics from the African continent, the island has been an isolated hotbed of tropical speciation for well over 100 million years. Madagascar is not only diverse; about 80% of its species are found nowhere else on Earth.

As Bedo Jaosolo and the researchers who sought his help knew well, most of Madagascar's species are in serious trouble. People have lived on the island for only about 2000 years, but in that time, Madagascar has lost 80% of its forests and about 50% of its native species. Except for isolated patches such as the

An indri, one species of lemur

Analamazoatra, the rain forest that once covered the eastern side of the island has been cleared for agriculture and living space for a burgeoning human population. Today, Madagascar is home to over 10 million people, most of whom are desperately poor and hardly in a position to be concerned with environmental conservation.

Madagascar's dilemma is not unique, for the needs of an expanding human population are creating environmental problems and threatening biological diversity throughout the world. In this chapter, we trace the roots of our often troubled relationship with the biosphere. We also focus on several of the major problems we now face. In the search for solutions to these problems, the attitude and awareness of people are of utmost importance. At an early age, Bedo Jaosolo acquired an unusual environmental awareness and sensitivity toward nature. Tragically, he died at age 20 on June 18, 1989, apparently killed after an argument at a party. What actually happened has never been resolved. According to one account, Jaosolo was murdered by assailants jealous of the attention and money he was earning from foreign scientists and tourists. In another scenario, he was killed because of his father's efforts to stop poaching and illegal tree-cutting in the Analamazoatra. Or perhaps Jaosolo's death was an accident. Whatever happened, the world lost a gifted young naturalist who understood and cared deeply for his natural surroundings. To conservationist/author David Quammen, who spent time with him in the Analamazoatra, Bedo Jaosolo symbolized "the role to be played by a new generation of . . . biologists and naturalists, a small crucial group of young people who care about nature." The hope of Jaosolo's friends is that his understanding and appreciation of the significance of other species will become increasingly widespread.

743

38.1 Human beings are the most environmentally destructive animals ever to live

The Brazilian city of Rio de Janeiro, part of which we see in the photograph here, is one of the largest cities in the world. The region encompassing Rio de Janeiro and its sister metropolis, São Paulo, is similar in size and growth rate to many large population centers throughout the world. As we saw in Module 35.9, the global human population now totals over 5.5 billion and may reach 8 billion within the next three

Rio de Janeiro

decades. An underlying reason for this population explosion is the human ability to manipulate nature—to change it rapidly and drastically to suit our needs, rather than being largely controlled by it, the way other species are. Our manipulative abilities stem mainly from our intellect and our social evolution. Not only can we develop and use complex tools, we also have a unique capacity to transmit learned information over the generations. Cumulative knowledge, more than any other single factor, gives us the power to control the environment and to expand our numbers to levels unprecedented by any species of large-bodied animal.

If success is measured in numbers and by geographical range, we have surely achieved it. Unfortunately, our success has been purchased at a high price. Altering the environment is tricky business. As we solve a problem in our own interest in a particular place, we often generate other problems in another place or time. Clearing a rain forest, for example, may help Madagascar or Brazil feed more people over the short term, but the loss of forests is creating global problems. Many of our environmental manipulations have become so extensive that we threaten our own existence and that of many other species.

In the next few modules, we examine the evolutionary history of our species, *Homo sapiens* (Latin for "wise man"), and the emergence of our unique ability to dominate nature. After that, we will take a close look at some of the major environmental problems we have created and how we may work toward solving them.

38.2 The human story begins with our primate heritage

Humans, apes, monkeys, and lemurs all belong to the mammalian order Primates. The earliest primates were probably small arboreal (tree-dwelling) mammals that arose when dinosaurs still dominated the planet. Most living primates are arboreal, and the primate body has a number of features that were shaped, through natural selection, by the demands of living in the trees. Although humans have never lived in trees, the human body retains many of the traits that evolved in our arboreal ancestors, and we can learn a great deal about ourselves and our relationship with the environment by studying other primates.

The squirrel-sized slender loris in Figure A illustrates a number of the basic primate features. It has limber shoulder and

A. The slender loris, a prosimian

hip joints, enabling it to climb and move gracefully in the trees. Also, the five digits of its hands and feet are highly mobile; its thumbs and big toes are opposable to the other digits, giving it the ability to hang onto branches and manipulate food. The great sensitivity of the hands and feet to touch also aids in manipulation. Moreover, lorises have a short snout and eyes set close together on the front of the face. The position of the eyes makes their fields of vision overlap, enhancing depth perception, an important trait for arboreal maneuvering. We humans share all these basic primate traits with the slender loris except for opposable big toes.

The slender loris belongs to a group of primates called the **prosimians** (Greek

Baboons

B. Old World monkeys

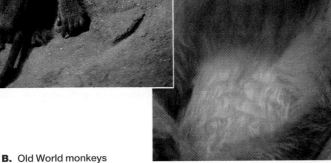

Hanuman langur

Golden lion tamarin

C. A New World monkey

pro, before, and Latin *simia,* ape). Fossil prosimians, the oldest known primates, date back 65 million years, to the period of mass extinctions that saw the end of the dinosaurs (see Module 16.7). The prosimians are one of the mammalian lineages that expanded extensively after the dinosaurs became extinct.

Today, the prosimians are represented by about 35 species, including lorises, bushbabies, and tarsiers of Africa, India, and Southeast Asia and the lemurs of Madagascar. Ranging from the size of a mouse to the 10-kg (22-lb) weight of the colorful indri of Madagascar, lemurs are the most diverse prosimians. Most are agile climbers and leapers that spend nearly all their time in trees. All prosimians live in tropical forests, and nearly all are threatened by habitat destruction. Of about 40 species of lemurs, for instance, 18 have become extinct since humans first colonized Madagascar.

The other group of primates, the **anthropoids** (Greek *anthropos,* man, and *eidos,* form), includes monkeys, apes, and humans. Anthropoids generally have larger brains relative to body size and rely more on eyesight and less on sense of smell than prosimians. The earliest anthropoids probably were monkeylike primates that arose from prosimian ancestors some 40 million years ago in Africa and Asia. Monkeys differ from apes and humans in having a tail and in having forelimbs that are about equal in length to their hind limbs.

Monkeys originated in Africa, spread throughout that continent and much of Asia, and finally reached the Americas. They may have migrated to the Americas on floating logs or via a bridge of land that stretched from Asia to what is now Alaska. The so-called Old World (Africa and Eurasia) and the New World (the Americas) have different types of monkeys, which have been evolving along separate pathways for millions of years.

Old World monkeys include some ground-dwelling species such as the baboons of the African savanna (Figure B, left), as well as many arboreal species, such as the Hanuman langur of the southern Himalayas (Figure B, right). Defining features of Old World monkeys are nostrils that are narrow and close together. Many (baboons, for example) also have a tough seat pad.

New World monkeys, found in Central and South America, are all arboreal. Their nostrils are wide open and far apart, none has a seat pad, and many have a long tail that is prehensile—adapted for grasping tree limbs. (Many Old World monkeys have a tail, but it is not prehensile.) A truly striking New World monkey, the squirrel-sized golden lion tamarin (Figure C), inhabits lowland rain forests of eastern Brazil. With most of its habitat destroyed by housing developments, this species has been reduced to only about 100 individuals in the wild and is the subject of an intense international conservation effort.

Apes are our closest relatives

Apes, including the gibbons, orangutan, gorilla, and chimpanzee, are closely related to humans. In contrast to the monkeys, the apes are all confined to tropical areas (mainly rain forests) in Southeast Asia and Africa. Apes lack a tail and have forelimbs that are longer than their hind limbs. They are chiefly vegetarians, although chimpanzees also eat insects and some vertebrates, such as young antelope, pigs, and monkeys.

Nine species of **gibbons** (Figure A), all found in Southeast Asia, are the only entirely arboreal apes. Gibbons are the smallest, lightest, and most agile of the apes. They are also the only apes that are monogamous, with mated pairs remaining together for life.

The **orangutan** is a shy, solitary species that lives in the rain forests of Sumatra and Borneo. The largest living arboreal mammal, it moves rather slowly through the trees, supporting its heavy-set body with all four limbs (Figure B). In contrast to the gibbons, the orangutan may often venture onto the forest floor.

The **gorilla** (Figure C) is the largest of all primate species, with some males weighing over 200 kg (440 lb). Living in African rain forests, the gorilla can climb trees, but with its massive body, it spends nearly all of its time on the ground. The gorilla is said to be a knuckle walker because when it is on all fours, the knuckles of its hands contact the ground. Gorillas also can stand upright on their hind legs.

Like the gorilla, the **chimpanzee** and a very similar species called the **bonobo** are knuckle walkers. These apes spend as much as a quarter of their time on the ground. Both species inhabit rain forests in central Africa, and the chimpanzee is also found in African savannas. Chimpanzees have been studied extensively, and many aspects of their behavior resemble human behavior. For example, chimpanzees make and use simple tools. The individual in Figure D is using a blade of grass to "fish" for termites. Chimpanzees also raid other social groups of their own species, exhibiting behavior formerly believed to be uniquely human. Researchers have demonstrated repeatedly that chimpanzees can learn human sign language. However, we do not yet know what role symbolic communication plays in the behavior of wild chimpanzees.

One of our most entrenched beliefs is that humans are the only thinking, self-aware beings. The behavior of chimpanzees in front of mirrors, however, challenges this belief. When first introduced to a mirror, a chimpanzee responds the way most other animals do—as if it were seeing another individual of its species. After several days, though, a chimp will begin using a mirror in ways that indicate it has a concept of self. It will inspect its face and other parts of its body that it cannot see without the mirror. It will also make faces at the mirror, using expressions different from those used in communicating with others.

Recent biochemical evidence indicates that the chimpanzee and the gorilla are more closely related to humans than they are to other apes. Humans and chimpanzees are especially closely related; human DNA differs from chimpanzee DNA by less than 3%. Primate re-

A. A gibbon

B. An orangutan

C. A gorilla

D. A chimpanzee

searchers are acutely aware of the special significance of the great apes to us. In the words of chimpanzee authority Jane Goodall, "The most important spin-off of the chimp research is probably the humbling effect it has on us who do the research. We are not, after all, the only aware, reasoning beings on this planet."

The human branch of the primate tree is only a few million years old 38.4

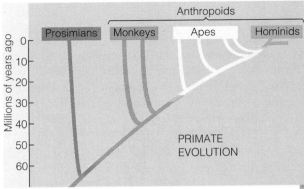

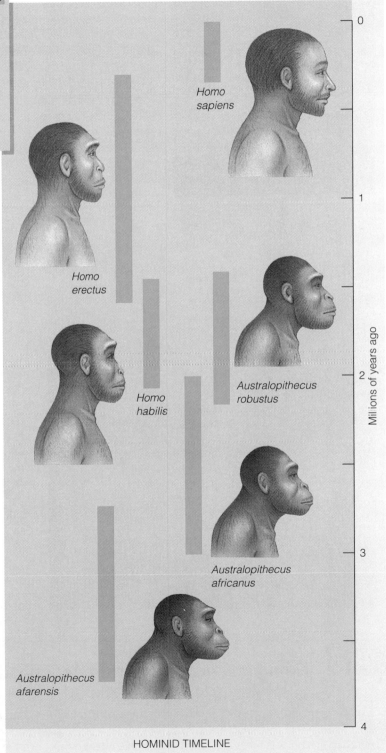

A popular misconception is that humans descended from apes that were just like chimpanzees or gorillas. In truth, apes and **hominids** (species on the human branch of the evolutionary tree) diverged from a common, ape*like* ancestor. The time of divergence is controversial, but most anthropologists believe that it probably occurred only 4–5 million years ago, as indicated in the "tree" above.

The main part of this figure is a rough time line of human evolution. A phylogenetic tree is not used here because the evolutionary connections are still being debated. But we humans clearly have several different species in our lineage, including two species of the genus *Homo* in addition to *Homo sapiens*. The green bars indicate the time during which each species existed (see the time scale on the far right), as judged from the fossil record. The overlaps in the bars represent the coexistence of two or three hominid species. The hominids are drawn with short hair and short beards in order to make their skull shapes more apparent.

The first hominids were a diverse group called **australopithecines** (genus *Australopithecus*, meaning "southern ape"). There were probably at least three australopithecine species, distinguished by skull and tooth anatomy. The earliest and most primitive of these, *A. afarensis*, appeared 3–4 million years ago and became extinct about 2.75 million years ago. Two other species of australopithecines, *A. africanus* and *A. robustus*, overlapped in time with species of *Homo*. Our own species, *Homo sapiens*, is the sole remaining member of the hominid family tree. Several other branches of the tree (the australopithecines and other species of *Homo*) died out. We trace the threads of human evolution in more detail in the next three modules.

HOMINID TIMELINE

The first humans foraged on the African savanna

A. Ancient footprints

Some 3.7 million years ago, several bipedal (upright-walking) human animals of the species *Australopithecus afarensis* left footprints in damp volcanic ash in what is now Tanzania in East Africa (Figure A). The prints fossilized and were discovered by British anthropologist Mary Leakey in 1978. The footprints are part of the strong evidence that upright walking is a very old human trait.

Human evolution began in Africa, where the australopithecines lived in savanna areas (Figure B). One of the most complete fossil skeletons of an australopithecine, an *afarensis* type, dates to about 3 million years ago in East Africa. Nicknamed "Lucy" by her discoverers, the individual was a female, only about 3 feet tall, with a head about the size of a softball. Lucy and her kind may have subsisted on nuts and seeds, bird eggs, and whatever animals they could catch or scavenge from kills made by more efficient predators such as large cats and dogs. They probably spent most of their time foraging on the ground, but they probably took to the trees to escape predators, pick fruits, and steal animal carcasses stored there by leopards and other large cats.

There is a lively debate today about the position of the australopithecines in the human lineage. Some anthropologists believe that at least some of these early hominids were on the lineage that gave rise to *Homo sapiens*. Others argue that the australopithecines were all on a separate branch of the hominid tree, a branch that became extinct. Whatever actually happened, all australopithecine species had become extinct by about 1.5 million years ago. The australopithecines show us that the fundamental human trait of bipedalism evolved millions of years before the other major human trait—an enlarged brain—became evident. As evolutionary biologist Stephen Jay Gould puts it: "Mankind stood up first and got smart later."

Enlargement of the human brain is first evident in fossils from East Africa dating to the latter part of australopithecine times, about 2 million years ago. Skulls have been found with brain capacities intermediate in size between those of the australopithecines and those of *Homo sapiens*. Simple stone tools are sometimes found with the larger-brained fossils, which have been dubbed *Homo habilis* ("handy man"). *Homo habilis* existed on the African savanna with the australopithecines for nearly a million years, probably scavenging, gathering, and hunting in much the same way as its smaller-brained relatives. *H. habilis* may have given rise to another ancestor of ours, called *Homo erectus*.

B. An artist's depiction of an australopithecine family

Homo erectus was our immediate ancestor

Fossils of an extinct hominid called *Homo erectus* ("upright man") range in age from about 1.6 million years to 300,000 years. *H. erectus* was taller than *H. habilis*, had a larger brain, and had a more advanced culture. This species was also the first hominid to spread out of Africa to Eurasia. To survive in the colder climates of the north, *H. erectus* lived in huts or caves, built fires, wore clothes made of animal skins, and designed stone tools that were more elaborate than the tools of *H. habilis*. Spread over much of the Old World, populations of *H. erectus* became regionally diverse. One or more of the *H. erectus* populations probably gave rise to *Homo sapiens*, and we track some of the possible ways that may have occurred in the next module.

When and where did modern humans arise?

The oldest known fossils of *Homo sapiens*, dating back over 300,000 years, are found in Africa. Humans of that time, apparently the earliest descendants of *H. erectus*, were generally heavier boned, with markedly thicker skulls and more pronounced brow ridges, than people living today. Experts often call these early forms of our species "archaic *Homo sapiens*." By about 100,000 years ago, archaic *H. sapiens* had spread throughout the Old World. The archaic group we know most about were the Neanderthals who lived in Europe, the Middle East, and parts of Asia from about 130,000–35,000 years ago. (They are called Neanderthals because their fossils were first found in the Neander Valley of Germany.) The Neanderthals were relatively short and stocky, with a heavily muscled body. They were skilled toolmakers, and they participated in burials and other rituals that required abstract thought. The illustration here shows a reconstruction of a burial scene.

The oldest known fossils of fully modern *Homo sapiens*—humans that if dressed in modern clothes would not stand out in a crowd—date to about 100,000 years ago in Africa. Similar fossils nearly as old have been found in caves in Israel. Apparently, these modern humans coexisted with the archaic types until about 35,000 years ago, when archaic humans disappeared. There is heated debate over what happened to the archaic groups and over the related questions of when and where modern *Homo sapiens* arose.

Some anthropologists argue from fossil evidence that the Neanderthals and other archaic *Homo sapiens* were intermediate between *H. erectus* and modern *H. sapiens*. They also propose that modern *Homo sapiens* arose from the geographically separate archaic populations in Africa, Europe, and Asia. Proponents of this "multiregional hypothesis" contend that the modern races of humankind stem from the regional diversity that evolved over a million years ago in ancestral populations of *H. erectus*. Despite racial differences, however, all living humans are very similar genetically. Multiregionalists believe that interbreeding among neighboring populations accounts for our genetic likeness.

In sharp contrast to the multiregional hypothesis is a proposal that modern *Homo sapiens* arose only in Africa, from a single population of *H. erectus*. According to this "monogenesis hypothesis," the Neanderthals and other archaic peoples were evolutionary dead ends. Proponents of this hypothesis argue that modern humans spread out of Africa about 100,000 years ago and completely replaced the archaic peoples. Most of the support for this idea stems from studies in molecular genetics. Analyses of DNA found in the mitochondria of human cells (mtDNA) indicate that mtDNA is extremely uniform in today's human population. Supporters of the monogenesis hypothesis maintain that such uniformity could only stem from a recent origin of modern humans—specifically, from the single African source. They further argue that human DNA could not be as uniform as it is if our gene pool came from diverse populations of archaic humans. They conclude that the African-derived population of *H. sapiens* replaced the archaic peoples and did not interbreed with them at all. If this view is correct, then the modern races of humans are less than 100,000 years old.

While the debate continues, biochemists are trying to recover DNA samples from fossilized skulls of Neanderthals and other archaic *Homo sapiens*. By comparing fossil DNA with that of living humans, they should be able to tell whether the Neanderthals contributed any genes to the modern population. The continuing studies of human genetics and fossils should fuel a significant expansion of our knowledge of human origins in the coming years.

Neanderthal burial, with widow and child

38.8 | Culture gives us enormous power to change our environment

Three major milestones highlight the evolution of *Homo sapiens*: (1) the evolution of our erect stance, which required major remodeling of the entire skeleton; (2) enlargement of the brain, which paralleled a prolonged period of growth of the skull and its contents after birth (see Module 16.9); and (3) the evolution of a prolonged period of parental care. Humans care for their offspring far longer than any other species, giving young children the chance to learn from the experiences of earlier generations. This is the basis of **culture**—the accumulated knowledge, customs, beliefs, arts, and other human products that are socially transmitted over the genera-

tions. More than any other factor in human history, culture has made *Homo sapiens* a unique force in the history of life on Earth—a species that can defy its physical limitations and alter nature at a rate far exceeding that of biological evolution. Our culture enables us to change the environment to meet our needs. As we discussed in Module 35.9, it has also allowed the human population to grow exponentially, in effect increasing Earth's carrying capacity for people. In the next three modules, we examine three stages of culture, each of which gave humans increasing powers to alter the environment, sometimes with unfortunate consequences.

38.9 | Scavenging-gathering-hunting was the first stage of culture

The first stage of culture, called scavenging-gathering-hunting for its three central activities, began with the earliest australopithecines on the African savannas. It continued to be the way of life for *Homo habilis,* for *H. erectus,* for the Neanderthals, and for modern humans during most of the last 100,000 years. Actually, for most of human existence, people have probably relied more on scavenging, especially stealing fresh kills from other predators, and gathering wild fruits, seeds, and vegetables, than on hunting. Only in the last 50,000 years or so did toolmaking become sophisticated enough to allow hunting to become a major food-producing activity and scavenging to decrease in importance. Gathering and hunting continue successfully to this day in various societies. The photograph here shows a

!Kung bushman, a skilled bow hunter. The !Kung, who live in southwestern Africa, are one of several hunter-gatherer peoples of modern times. (The exclamation point represents a clicking sound used in their language.)

When did human culture begin to create environmental problems? It has been suggested that even scavenging by bands of australopithecines or *Homo habilis* could have created serious problems for other species. Several lion-sized predators, saber-toothed cats, became extinct in Africa some 1.5 million years ago. Could early hominids have become so adept at stealing the kills of these carnivores that they seriously reduced the cats' food supply? Some anthropologists think it very likely.

With the development of tools that could be thrown and cooperative hunting techniques, humans began to have profound effects on certain other species. Fossil records indicate that *Homo sapiens* decimated populations of woolly rhinoceroses and giant deer in Europe. About 50,000 years ago, modern humans reached Australia by boat and may have killed off that continent's giant kangaroos. Nomadic hunters migrated from Asia to the New World via the Bering land bridge about 11,000 years ago. Here, they may have been responsible for driving a number of large animals to extinction.

Hunter-gatherers not only made tools, they also organized communal activities and divided labor. Many groups also developed semipermanent residences near rich hunting grounds and along seacoasts and lakes. Some even grew crops and traded with other groups. These activities ushered in a new stage of culture—the rise of agriculture.

A !Kung bushman hunting with bow and arrow

Agriculture was the second stage of culture

Agriculture developed in Eurasia and the Americas 10,000–15,000 years ago. Some of the earliest sites were in Southeast Asia, where farmers cut and burned small plots of dense rain forest. Typical of tropical forests, the soil was poor, and plots had to be abandoned for new ones every few years. Early Asian farmers domesticated and grew rice and the cereal grain millet.

Other agricultural centers sprang up in the Americas and in the Middle East, in an area called the Fertile Crescent. Extending northeast from the Nile River in Africa and including the modern nations of Israel, Syria, Iraq, and Kuwait, the Fertile Crescent was once covered with forests and grasses and had rich soils. Farmers cleared the forests and exploited the soil for years, growing mostly wheat, without serious loss in fertility. About 5000 years ago, farmers in the area began using primitive plows, like this one of an Egyptian farmer. The plow broke up the thick sod of grasslands and opened vast new areas for farming. With more intensive agriculture, local populations increased, placing an ever-increasing burden on the soil to produce food. Herds of domestic animals overgrazed the land, leaving it exposed to eroding winds and rains, while the crops eventually depleted the soil of nutrients. Today, much of the Fertile Crescent is a desert.

Agriculture changed forever our relationship with the biosphere. In adopting farming as a way of life, *Homo sapiens* took the first big step toward becoming the dominant species on Earth. Along with agriculture came permanent settlements and the first cities. As farming advanced, fewer people were needed to grow a group's food, and many people could specialize in other activities. Agriculture thus paved the way for the development of technology, industry, and the arts—the next great cultural wave, which continues today.

The machine age is the third stage of culture

The Industrial Revolution, which began in England in the eighteenth century, brought a switch from small-scale, hand production of tools and other goods to large-scale machine production. With the machine age came increasing demands for energy sources to fuel the machines—mostly timber and coal at first, then oil, and recently also hydroelectric power and nuclear power. The invention of the tractor and other farm machinery reduced the need for farm laborers, and many people migrated to cities in search of work. As more food was produced and medical advances reduced deaths from diseases, the human population began to grow faster throughout the world (see Module 35.9). Both the human population and technology are now escalating exponentially.

Through the cultural changes from scavenging-gathering-hunting to high-tech societies, humans have not changed biologically in

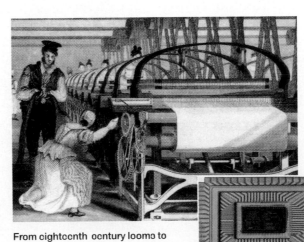

From eighteenth-century looms to twentieth-century computer chips

any significant way. We are probably no more intelligent than our forebears who lived in caves. The same toolmaker who chipped away at stones now fashions microchips for computers. The know-how to build computers and spaceships is stored not in our genes but in what is passed along by parents, teachers, and books. We are presently the most numerous and widespread of all large animals, and everywhere we go, we bring change. As we saw when we discussed biological evolution in Unit 3, there is nothing new about environmental change. What is new is the *speed* of change, for cultural changes outpace biological evolution by orders of magnitude. As we discuss in the following modules, we are changing the world so quickly that many species cannot adapt—in fact, we may be jeopardizing our own existence as well.

Technology and the population explosion compound our impact on the biosphere

The history of life is a story of perpetual change. Periodic global changes have disrupted entire ecosystems, killing off great numbers of species and making way for their replacement by others. The last great global crisis, which occurred about 60 million years ago, marked the end of the Age of Reptiles. As we discussed in Module 16.7, the dinosaurs were swept away after dominating the land and air for millions of years. Global crises are catastrophic; they produce mass extinctions and alter entire ecosystems. But they can also have creative effects. The mass extinction of the dinosaurs cleared the way for mammals and birds to diversify.

Today, entire terrestrial ecosystems, the global atmosphere, the oceans, and many forms of life may once again face a major crisis. This time the cause is not a meteor collision with planet Earth or other physical force of nature, but the impact of a single species, *Homo sapiens,* and our cultural ability to produce rapid change. Technology and cultural change have produced many benefits, including a significant improvement in human health. They also continue to fuel our population explosion; and feeding, clothing, and housing billions of people even at a minimal level strains the biosphere. Even more of a strain are the huge amounts of resources consumed by certain segments of the human population.

Because of our technological advances, those of us in developed nations have become mass consumers. The United States, for example, has less than 5% of the global population, but consumes far more than 5% of the world's resources. On a per capita basis, the average U.S. citizen consumes almost 13 times as much energy as the average person in China, and 35 times the amount consumed by the average person in India (see the table). Overall, the United States consumes more energy than the total population of Central and South America, Africa, India, and China combined—in 1993, some 3.5 billion people compared to the U.S.'s 260 million! The high rate of resource use compounds the danger imposed by the human population explosion.

What are the consequences of overpopulation and our penchant for consuming resources? In future years, if more than a small fraction of the people on Earth assume the high standard of living that the developed countries now enjoy, the resources that sustain the human population, such as soil, water, and fossil fuels, could be depleted. We already see harmful effects virtually everywhere today. Figure A is a scene from a remote ecosystem wounded by human activity. On March 24, 1989, a supertanker ran aground and spilled more than 10 million gallons of crude oil into Prince William Sound in Alaska. This ecological disaster was a shocking reminder that our technological

Energy Consumption per Capita for Selected Countries (1989)		
Country	Population (thousands)	Energy per Person (gigajoules*)
Bangladesh	111,105	2
Nigeria	105,065	5
India	829,932	9
Indonesia	180,856	8
Brazil	146,437	23
China	1,136,397	23
Turkey	54,846	28
Mexico	82,677	51
Japan	123,051	118
Germany	79,050	156
Soviet Union (former)	279,353	191
United States	247,525	295

*1 gigajoule = 239,000 kcal.

A. An oil spill in a remote ecosystem in Alaska

tentacles reach far; as we burn gasoline in Los Angeles, Chicago, or New York, the impact of our demand for oil is felt thousands of miles away.

In a similar way, pollutants emitted into the atmosphere in Seattle, London, Moscow, or just about anywhere may be carried aloft and foul the air thousands of miles away. Acid precipitation, for instance, threatens to destroy lakes far from sources of the sulfur and nitrogen pollutants that cause this problem. (We discussed acid precipitation in detail in Module 2.16.) Pollutants called chlorofluorocarbons (CFCs), emitted into the air from air conditioners, refrigerators, and fire extinguishers, damage the upper atmosphere's ozone layer, which protects us from the harmful ultraviolet rays in sunlight. An international agreement was reached in 1987 to phase out the use of CFCs, but finding alternatives has proved difficult and expensive, and quantities of these chemicals already in the air will continue to deplete the ozone layer—possibly to dangerously low levels—for decades.

Chemical pesticides are another case of the far-reaching effects of human alteration of the biosphere. As we saw in Module 34.3, DDT and other chemical pesticides have helped us grow more food and fight infectious diseases. In the long run, however, many such chemicals tend to accumulate in the body tissues of nontarget animals, including humans. DDT, for instance, is long-lived and readily stored in fat. At least in some species, the effects have been deadly.

Researchers began finding DDT in the fat tissues of a large number of birds and mammals in the early 1960s, after about a decade of widespread use of the chemical. They even found traces in marine mammals in the Arctic, far from any places DDT had been applied. The chemical had been transported and concentrated as it passed through food chains. Because predators consume large numbers of prey, the concentrations of DDT and other substances that persist in living tissue can become increasingly magnified as the substances move through food chains. Because of this process, called **biological magnification**, the top predators in food chains can accumulate DDT and other persistent substances to toxic levels. You can see the effect in Figure B. The tiny dots in the pyramid represent DDT. In this particular food chain, the chemical was magnified by a factor of about 10 million, from 0.000003 parts per million (ppm) in the water to 25 ppm in the osprey.

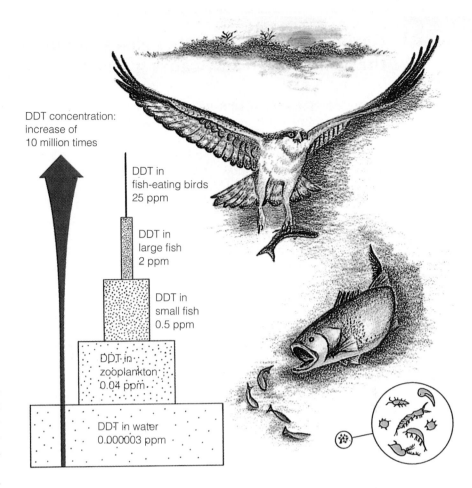

DDT concentration: increase of 10 million times

DDT in fish-eating birds 25 ppm

DDT in large fish 2 ppm

DDT in small fish 0.5 ppm

DDT in zooplankton 0.04 ppm

DDT in water 0.000003 ppm

B. Biological magnification of the pesticide DDT in a food chain

In the 1960s and 1970s, scientists studying predatory birds such as ospreys and eagles found a correlation between high levels of DDT in parent birds and a thinning of eggshells. Young birds died before hatching, producing a marked decline in populations.

The United States banned the use of DDT in 1976, although we still manufactured it and sold it to other countries until 1984. DDT and several close chemical relatives are still used in many developing nations, including Mexico. Despite this continued threat to the biosphere, the osprey and several other predators have made a comeback in the past several decades because the pesticide is no longer used in their immediate environment.

The DDT story represents a common approach to altering the environment to suit our needs. Technology produced a substance that seemed to hold great promise for helping us feed more people and reduce suffering and death. The substance was applied broadly soon after its insecticidal qualities were discovered, before tests of its potential danger to the biosphere were performed. In fact, the biosphere became the testing ground. As we are about to see, this approach may now be putting our life-support systems to the ultimate test.

Carbon dioxide and other gases added to the atmosphere may cause global warming

The rise of agriculture and the Industrial Revolution triggered some profound environmental changes. A potential effect that could alter the entire biosphere is rapid global warming. We introduced global warming in Module 7.13, in the context of the controversy over forest conservation. However, this subject is so important and timely that we need to paint a more complete picture of its causes and potential effects.

Fossil fuels power most of our industries, agricultural equipment, and automobiles, and they heat most of our homes. One of the major side effects of the combustion of fossil fuels is an increase in atmospheric CO_2. There is, of course, nothing odd or new about CO_2 being in the atmosphere. As we discussed in Module 36.14, the addition of CO_2 to the atmosphere by cellular respiration roughly balances the removal of CO_2 from the atmosphere by photosynthesis. What is troubling is the current rate of increase in atmospheric CO_2. When a monitoring station in Hawaii began making very accurate measurements in 1958, the CO_2 concentration was about 315 ppm (Figure A). Today, the concentration of CO_2 in the atmosphere is more than 350 ppm, an increase of over 10% in just the past 36 years.

Carbon dioxide is one of several so-called greenhouse gases—molecules that can trap heat and cause atmospheric warming, often called the **greenhouse effect** (Figure B). CO_2 absorbs infrared radiation and slows its escape from Earth. Other greenhouse gases are methane and nitrous oxide, both of which are also increasing in the atmosphere as a result of fossil-fuel consumption, industry, and agriculture. To a certain degree, the greenhouse effect is beneficial; without the CO_2 that respiration puts into the atmosphere, the average air temperature at Earth's surface would be only about −18°C.

Our current cause for alarm arises from the potential for too much warming. Studies of climatic changes through geological time and mathematical models lead many climatologists to predict that in the next 50–100 years, at the present rate of greenhouse-gas increase, atmospheric temperatures could rise by about 2°–5°C. It is also possible that as the temperature of the atmosphere increases, populations of soil bacteria will increase and in turn produce

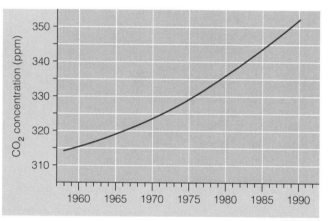

A. The increase of atmospheric CO_2 since 1958

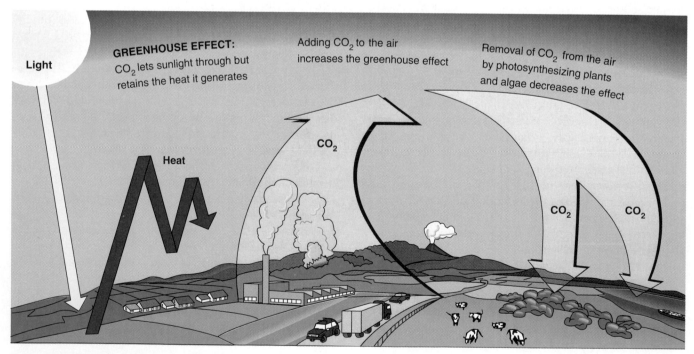

GREENHOUSE EFFECT: CO_2 lets sunlight through but retains the heat it generates

Light

Heat

Adding CO_2 to the air increases the greenhouse effect

Removal of CO_2 from the air by photosynthesizing plants and algae decreases the effect

CO_2

CO_2

CO_2

B. The greenhouse effect and factors influencing it

even more CO_2 and methane. The upswinging (yellow) curve in Figure C shows the range of global warming predictions generated by several computer models. The jagged black line shows the actual increase in global temperatures recorded since about 1850.

What are the possible consequences of this much global warming? As Figure C shows, the world has experienced a temperature rise of about 0.5°C in the past century. Half a degree Celsius sounds minor, and it is, on a local or regional basis. However, on a global scale, an increase of less than 2°C would be enough to melt polar ice and raise sea levels significantly. The United Nation's Intergovernmental Panel on Climate Change has concluded that if the current warming trend continues, sea levels could rise by 1 m or more by the end of the twenty-first century. Unless massive dikes were built, this rise would flood coastal areas, many of which are environmentally sensitive and heavily populated. In addition to flooding, a warming trend might alter patterns of global precipitation. For instance, the grain belts of the central United States and central Asia might become much drier and unable to support the crops currently grown there. Furthermore, forested areas in semiarid zones could lose their trees and become deserts.

At present, the future of greenhouse warming is uncertain. There is no proof that global temperatures will continue to rise or that greenhouse warming will have dire consequences. Although the increases in greenhouse gases and the 0.5°C rise in global temperature are documented, some scientists remain skeptical about whether atmospheric CO_2 and methane have increased mainly from human sources and whether the increases have actually caused the temperature rise. It is also possible that smoke and increased cloud cover from fossil-fuel consumption and deforestation may *decrease* warming by reducing the amount of solar heat that reaches Earth's surface. Another possibility is that marine algae, which take up enormous amounts of CO_2, may help slow greenhouse warming.

Some researchers and government officials say that we need more data, and that we will not be able to reach any conclusions about global warming or what to do about it for another decade or so. Others conclude that immediate steps to slow the warming trend are necessary to prevent catastrophic global change. Professor Wallace Broecker of Columbia University summarizes the situation we now seem to face:

The inhabitants of planet Earth are quietly conducting a gigantic environmental experiment. So vast and so sweeping will be the impacts of this experiment that, were it brought before any responsible council for approval, it would be firmly rejected as having potentially dangerous consequences.

What can be done to lessen the chances of a greenhouse disaster? International cooperation and national and individual action are needed to decrease fossil-fuel consumption and to reduce the destruction of forests and replant many areas throughout the world. Several nations have instituted tree-planting programs. At the 1992 United Nations Earth Summit, leaders of most industrial nations signed a treaty calling for stabilizing CO_2 output by the end of the century. Because fossil-fuel combustion currently powers much of our industry and economic growth, meeting the terms of the treaty will require strong individual efforts and acceptance of some major lifestyle changes. There is much to gain by conserving energy at home, recycling, and reducing personal auto use by walking, bicycling, or using various forms of mass transit. Other strategies include developing solar, wind, and geothermal energy sources to reduce our reliance on fossil fuels. Although such measures are expensive, so are building massive dikes to prevent coastal cities from flooding, or changing a food-producing nation such as the United States to a food-dependent one.

While acknowledging the uncertainties of global warming, British climatologists Philip Jones and Tom Wigley assert:

The longer the world waits to act, the greater will be the climate change that future generations will have to endure. A policy of inaction would be justified only if researchers were sure that the greenhouse effect was negligible.

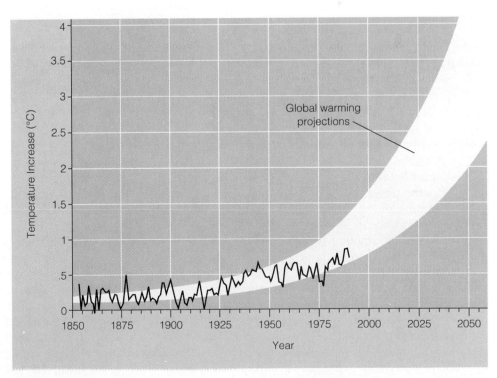

C. Global warming

The consequences of tropical forest destruction show that local activities can change the biosphere

A. Clearing a forest in Guatemala

In Chapter 1, we focused on research in tropical rain forests and saw that scientists have hardly begun to study these ecosystems. Today, extensive areas of rain forests in Central and South America, Southeast Asia, and central Africa are being destroyed in attempts to support national economies. Networks of new roads are being built into the forests, large areas are being logged or simply burned, and the cleared areas are being planted with coffee, rice, or other crops, or used for grazing cattle. Figure A shows forest clearing taking place in Guatemala.

The destruction of tropical rain forests is a global problem for several reasons. These forests play a vital role in maintaining atmospheric gas balance. Their photosynthesis is an important source of atmospheric oxygen and also absorbs large amounts of CO_2 from the atmosphere. When a tract of forest is destroyed, greenhouse warming is boosted because the trees no longer take up CO_2, and burning the trees adds more CO_2 to the atmosphere.

Tropical rain forests also moderate global climate. They receive enormous amounts of precipitation and return much of the moisture to the atmosphere through evaporation and transpiration (see Module 36.13). In fact, much of the world's water cycles continuously between the tropical atmosphere and rain forests. Because so much water is confined to the tropics, certain other latitudes receive only moderate amounts of precipitation—amounts suited for agriculture, for example. When tropical forests are destroyed, much of the water they would have held is carried in the atmosphere to higher latitudes. Those areas may then receive

B. The rosy periwinkle, a source of anticancer drugs

so much rainfall that crops are damaged.

Most authorities agree that the loss of tropical rain forests, in one sense a local phenomenon, is actually one of the most serious problems we *all* face. As we discussed in the chapter's introduction, Madagascar has already lost most of its rain forest and with it many species of animals and plants found nowhere else in the world. Brazil, with more than three times as much rain forest as any other country (about 350 million hectares, or 875 million acres), is currently deforesting a total area nearly the size of Indiana every year. Considering the tropics as a whole, about 17 million hectares are now being destroyed each year. This is an area roughly as big as the state of Washington.

Some progress is being made in tropical forest conservation. Several tropical countries, including Madagascar and Costa Rica (see Module 36.19), with monetary help from developed nations, are beginning to set up biological reserves. Brazil leads the way with what are called **extractive reserves.** An extractive reserve allows people to harvest natural forest products at a rate that does not destroy the ecosystem. The main forest products in Brazil are nuts, fruits, latex (for making rubber), and chicle (chewing gum). Economic profits from these and other natural products have proved to be greater in the long run than the profits from cattle ranching on deforested land.

Tropical rain forests also hold enormous potential for the development of new useful products. A number of medications have been developed from tropical forest plants. For instance, two substances effective against Hodgkin's disease and certain other forms of cancer come from the rosy periwinkle, a flowering plant native to Madagascar (Figure B). Madagascar alone harbors some 8000 species of flowering plants, 80% of which occur only there. Among these unique plants are several species of wild coffee trees, some of which yield beans lacking caffeine (naturally "decaffeinated"). These and many other examples show us that the great biological diversity of tropical forests offers us a storehouse of new drugs, foods, and other useful products.

The rapid loss of species is a warning that the biosphere is in trouble

The key deer, a miniature subspecies of the whitetail deer, is about the size of a German shepherd dog (Figure A). It is found only in the Florida Keys and is a population with a unique gene pool. Key deer were cut off from whitetail populations on the mainland when the sea level rose after the last ice age. Confined to a few islands, key deer have never been numerous and were nearly exterminated by poachers in the early 1900s. Conservation efforts brought the population back from fewer than 50 animals to about 400 in the early 1970s. Key deer are now declining again as the human population in the Florida Keys mushrooms. Housing developments are reducing the habitat, while motorists on new highways have become the main threat to the deer's survival. Now down to some 250 individuals, the deer are dying on the roads faster than they can reproduce.

In many ways, the plight of the key deer illustrates the effect of modern human culture worldwide. We are now presiding over an alarming **biodiversity crisis.** Animal and plant species are being lost at a rate that is unprecedented in the history of life. Some biologists estimate that within the next century, fully half of Earth's current species may pass into extinction. This would reduce diversity to the level of 60 million years ago, after mass extinctions, which occurred over a much longer period of time, reduced species numbers by about 75%.

Thus far, about 1.5 million living species have been discovered and named, and it is likely that there are at least another 2.5 million yet to be found. Tropical forests alone may house 80% of the world's plants and animals. The current rate of deforestation is so great that many tropical species may become extinct before scientists discover them. Deforestation also threatens migratory species. Scientists have documented serious reductions in populations of over half of the more than 200 species of songbirds that winter in Central and South America and migrate to the U.S. and Canada to breed. Destruction of forests in the tropics *and* in North America is thought to be a major factor in the decline.

Habitat loss threatens the greatest number of species, but other factors are also important. Overhunting has eliminated many species and continues to threaten others. Whales, for example, have been hunted to near extinction, and a few countries continue to kill these huge mammals despite a

A. Key deer

B. Black-market ivory confiscated by wildlife officials in Tanzania

nearly worldwide ban on whaling. Many animals and plants have also succumbed to competition from foreign species that humans have introduced. In East Africa, hundreds of species of tropical fishes in Lake Victoria are currently threatened by the nile perch, a large species (up to 200 lb) that was introduced as a sport and food fish.

Often, as for the key deer, a combination of killing and habitat destruction has driven a species over the edge. In the past century, for instance, the Hawaiian Islands have lost half their bird species to overhunting, deforestation, and diseases carried by foreign bird species introduced into the islands. In Africa, elephants and rhinoceroses are being pushed toward extinction by habitat loss and by poachers catering to a black market for ivory and horns (Figure B).

Why should we worry about species losses? First of all, large-scale extinctions of other species may be a warning to us that we are altering the biosphere so fast that we ourselves are threatened. Secondly, we depend on many other species for food, clothing, shelter, oxygen, soil fertility—the list goes on and on. Finally, we might ask ourselves what right we have to destroy other species. Shouldn't we use our intelligence constructively and accept the responsibility of conserving as much of the biosphere and its diversity as possible?

38.16 Responsible environmental decisions require an understanding of the natural world

In numbers, geographical range, and capacity to alter the biosphere, *Homo sapiens* is clearly one of the most successful species ever to inhabit planet Earth. Unfortunately, as we have seen in this chapter, our success is linked with many serious problems and a rate of environmental change unprecedented in the history of life. In short, the fast pace of human cultural change has set us and the rest of the biosphere on a precarious path into the future.

For those of us in countries with relatively stable populations, it is easy to blame environmental problems on na-

A luna moth on a lichen-covered tree

tions with burgeoning numbers of people. But, as we have seen, developed nations extract far more than their share of Earth's resources. We cannot escape our history. The deforestation now occurring in Madagascar, Brazil, and many other tropical countries is very similar to what took place in much of the United States prior to the twentieth century. At that time, a rapidly expanding population with a growing economy and an aim to better feed and clothe its people cut down vast tracts of forests. As a result, farms and cities now occupy most of our once heavily forested eastern seaboard and upper Midwest. Our forebears also decimated the ancient forests of the Pacific Northwest, and the process continues today, as we log our remaining evergreen forests at an unprecedented rate. As Brazilian officials are quick to point out, our environmental history gives us little right to lecture them about the need for conservation.

It is unlikely that human nature will suddenly change drastically—that we will abruptly lose our environmental manipulativeness, or that we could survive if we did. What we must seek instead are ways to be more accommodating with other species and with the biosphere. Fragile connections, such as those sustaining the endangered luna moth and its habitat (photo at left), underlie structure and function in the natural world. Knowledge of these connections and an appreciation of their vulnerability are vital. In the words of biologist E. O. Wilson:

> The drive toward perpetual expansion—or personal freedom—is basic to the human spirit. But to sustain it we need the most delicate, knowing stewardship of the living world that can be devised.

Chapter Review

Begin your review by rereading the module headings and scanning the figures, before proceeding to the Chapter Summary and questions.

Chapter Summary

38.1 The rapidly growing global human population now exceeds 5.5 billion. We are the most numerous, successful, and destructive large animals ever to inhabit the planet. Our success is due to our ability to manipulate nature, but many of our activities now threaten all life on Earth.

38.2 Humans are members of an order of mammals, the primates, that first appeared about 65 million years ago. The first primates lived in trees, and we have inherited some of their characteristics: limber joints, sensitive grasping hands, a short snout, and forward-pointing eyes that enhance depth perception. There are two groups of living primates: the prosimians (such as lorises and lemurs) and the anthropoids (monkeys, apes, and humans).

38.3 Humans are most closely related to the apes, primates that lack tails and have forelimbs equal in length to their hind limbs. The

apes include gibbons, orangutans, gorillas, chimpanzees, and bonobos. We share more than 97% of our genes with chimps, our closest living relatives; our behavior also has some similarities. Most anthropologists think that hominids and chimpanzees diverged from a common apelike ancestor 4–5 million years ago.

38.4–38.6 The human family tree began in Africa. The earliest known hominid (member of the human family) is *Australopithecus afarensis*, who walked upright on the African savanna 3–4 million years ago. There were several branches of the human family tree; one gave rise to *Homo habilis*, which coexisted with australopithecines, had a larger brain, and made simple tools. *Homo habilis* may have given rise to the more advanced *Homo erectus*, who had an even larger brain and a more complex culture, and who spread out of Africa over most of the Old World.

38.7 The oldest fossils of modern humans, *Homo sapiens*, are found in Africa and date back about 300,000 years. By about 100,000 years ago, *Homo sapiens* had spread throughout the Old World. An archaic group of *Homo sapiens*, the stocky and muscular Neanderthals, appeared about 130,000 years ago and died out about 35,000 years ago. Fully modern *Homo sapiens* date back about 100,000 years. Some anthropologists think that modern humans arose from geographically separate populations of *Homo erectus* in Africa, Europe, and Asia, and that Neanderthals and other archaic peoples represent a transitional stage in our evolution. Others believe that *Homo sapiens* arose in Africa and replaced archaic peoples elsewhere.

38.8–38.9 Major milestones in the evolution of *Homo sapiens* are the evolution of an erect stance, enlargement of the brain, and the evolution of a prolonged period of parental care. We have not changed much biologically since modern humans first appeared. But our culture—the accumulated knowledge, customs, beliefs, arts, and other products—has evolved enormously. There have been three main stages in cultural change. The first humans survived by scavenging, gathering, and hunting. Early scavengers and hunters may have depleted populations of some of their prey and some competing carnivores.

38.10–38.11 The second stage in cultural change was the beginning of agriculture 10,000–15,000 years ago, when people settled down, planted crops, and began to domesticate animals. Early farmers in the Fertile Crescent overgrazed the land and depleted the soil, leaving much of the Middle East a desert. The third stage in cultural change, the Industrial Revolution, began in the 1700s. Industrialization brought a change from hand production to energy-intensive, large-scale machine production. Mechanized farming and improved medicine then accelerated the growth of the human population and our impact on the biosphere.

38.12–38.15 The explosive growth of human population and technology continues today. Populations of developing nations are growing the fastest, but the technology and resource consumption of the less-populous developed nations put a much greater strain on the biosphere. Oil spills, acid rain, ozone depletion, and chemical pesticides affect the entire world. Burning of fossil fuels is increasing the amount of CO_2 and other greenhouse gases in the air, which may warm Earth enough to change climate patterns, melt polar ice, and flood coastal regions. Logging and clearing forests, especially tropical rain forests, for farming contribute to global warming, by reducing the uptake of CO_2 by plants (and adding CO_2 to the air when trees are burned). Another serious consequence of tropical deforestation is its threat to global biodiversity. Destruction of the tropical rain forests that are home to perhaps 80% of the world's plants and animals could alter global climate and will deprive us of many potentially useful products. Overhunting, introduced species, and habitat destruction could alter the biosphere to such an extent that we are threatening our own survival.

38.16 Our intelligence enables us to manipulate nature on a global scale, but we have created global environmental problems. Our intelligence can also enable us to better understand the living world and live more harmoniously with nature.

Testing Your Knowledge

Multiple Choice

1. The two major groups of primates are
   a. monkeys and anthropoids
   b. prosimians and apes
   c. monkeys and apes
   d. prosimians and anthropoids
   e. Old World monkeys and New World monkeys

2. Which of the following correctly lists probable ancestors of modern humans from the oldest to the most recent?
   a. *Homo erectus, Australopithecus, Homo habilis*
   b. *Australopithecus, Homo habilis, Homo erectus*
   c. *Australopithecus, Homo erectus, Homo habilis*
   d. *Homo erectus, Homo habilis, Australopithecus*
   e. *Homo habilis, Homo erectus, Australopithecus*

3. Fossils suggest that the first major trait distinguishing human primates from other primates was
   a. forward-facing eyes with depth perception
   b. a larger brain
   c. erect posture
   d. grasping hands
   e. toolmaking

4. Some anthropologists believe that the modern races of *Homo sapiens* evolved from separate populations of *Homo erectus* in different geographic areas. How, then, do proponents of this "multiregional hypothesis" explain the great degree of genetic similarity among all modern humans?
   a. The same mutations occurred in populations in different locations.
   b. There probably was interbreeding among neighboring populations.
   c. All *Homo sapiens* populations were shaped by similar environments.
   d. The ancestral *Homo erectus* originally came from Africa.
   e. Modern races of humans are not at all genetically similar.

5. Ospreys and other top predators in food chains are most severely affected by pesticides such as DDT because
   a. their systems are especially sensitive to chemicals
   b. of their rapid reproductive rates
   c. the pesticides become concentrated in their prey
   d. they cannot store the pesticides in their tissues
   e. they are directly exposed to pesticides in the air

6. Some biologists think that half the species on Earth may be exterminated in the next century. Mass extinctions have happened before. How would this one be different?
   a. This time species are threatened by a change in climate.
   b. This extinction is happening at a much faster rate.
   c. A much larger fraction of Earth's species are likely to be affected.
   d. All life on Earth may eventually be extinguished.
   e. This extinction is much more gradual and difficult to notice.

7. Which of the following consumes the most energy each day? (*Explain your answer.*)
   a. United States
   b. South America
   c. China
   d. India
   e. Africa

True/False (*Change false statements to make them true.*)

1. The development of agriculture was the second stage in cultural change.
2. The lemur and loris are Old World monkeys.
3. Humans probably evolved in Africa.
4. The earliest hominids were probably hunters.
5. There have been significant biological changes in humans since we lived in caves.
6. Earth's human population is more than 5 billion.
7. The biggest threat to biodiversity is habitat destruction.
8. Humans are thought to have evolved from chimpanzees.
9. Carbon dioxide from fossil fuels is destroying the ozone layer.
10. Early scavenger-gatherer-hunters had little effect on the natural environment.

Describing, Comparing, and Explaining

1. What adaptations inherited from our primate ancestors enable humans to make and use tools?
2. In what ways is chimpanzee behavior similar to human behavior?
3. Explain why humans have been able to expand our numbers and distribution to a greater extent than any other animal.
4. Describe the lifestyle of the earliest known hominids.
5. What is the greenhouse effect? How is it important to life on Earth?

6. What are the possible causes and consequences of global warming? Why is international cooperation necessary if we are to solve this problem?

Thinking Critically

1. Anthropologists are interested in locating areas in Africa where fossils 3–10 million years old might be found. Why?

2. Some researchers think that drying and cooling of the climate caused expansion of the African savanna, and that this environment favored upright-walking early humans. Why might bipedalism be advantageous in the savanna? How might an erect posture relate to the evolution of a larger brain?

3. The oceans are a poorly understood factor in global warming. Some researchers are concerned that warming of the oceans might cause the large amounts of CO_2 dissolved in seawater to come out of solution, the way air bubbles form in a glass of tap water when it warms up. Why might this concern them? On the other hand, how might the oceans help reduce global warming?

4. Biologists in the United States are concerned that populations of many migratory songbirds, such as warblers, are declining. The evidence suggests that some of these birds might be victims of pesticides. Most of the pesticides implicated in songbird mortality have not been used in the U.S. since the 1970s. Suggest a hypothesis to explain the current decline in songbird numbers.

Science, Technology, and Society

1. You may have heard that human activities cause the extinction of one species every hour. Such estimates vary widely, because we do not know how many species exist, nor how fast their habitat is being destroyed. You can make your own estimate of the rate of extinction. Start with the number of species thought to exist. To keep things simple, ignore extinction in the temperate latitudes and focus on the 80% of plants and animals that live in the tropical rain forest. Assume that destruction of the forest continues at a rate of 1% per year, so the forest will be gone in 100 years. Assume (optimistically) that half the rainforest species will survive in preserves, forest remnants, and zoos. How many species will disappear in the next century? How many species is that per year? Per day? Recent studies of the rainforest canopy have led some experts, such as Terry Erwin of the Smithsonian Institution, to suggest that there might be as many as 30 million species on Earth. How does starting with this figure change your estimates?

2. One of the reasons the developed countries consume so much energy is that the price of energy does not reflect its real costs. For example, the monetary cost of the 1991 Gulf War would have doubled the price of every barrel of oil imported into the United States that year. What kinds of hidden environmental costs are not reflected in the prices of fossil fuels? How are these costs paid, and by whom? Do you think these costs could or should be figured into the price of oil? How might that be done?

3. Most atmospheric scientists say that global warming is under way, and we need to take action now to avoid drastic environmental change. Some say it is too soon to tell, and we should gather more data before we act. What are the advantages and disadvantages of doing something now to slow global warming? What are the advantages and disadvantages of waiting until more data are available? Is there any way to know for sure whether there is a link between CO_2 and global warming?

4. Until recently, response to environmental problems has been fragmented—an antipollution law here, incentives for recycling there. Meanwhile, the problems of the gap between the rich and poor nations, diminishing resources, and pollution continue to grow. Now people and governments are starting to envision a sustainable society, one in which each generation inherits an adequate supply of natural and economic resources and a relatively stable environment. The Worldwatch Institute, a respected environmental monitoring organization, estimates that we must reach sustainability by the year 2030 to avoid economic and environmental disaster. To get there, we must begin shaping a sustainable society during this decade. In what ways is our present system not sustainable? What might a sustainable society be like? Do you think a sustainable society is an achievable goal? Why or why not? What is the alternative? What might we do to work toward sustainability? What are the major roadblocks in the way of achieving sustainability? How would your life be different in a sustainable society?

Metric Conversion Table

Measurement	Unit and Abbreviation	Metric Equivalent	Approximate Metric-to-English Conversion Factor	Approximate English-to-Metric Conversion Factor
Length	1 kilometer (km)	= 1000 (10^3) meters	1 km = 0.6 mile	1 mile = 1.6 km
	1 meter (m)	= 100 (10^2) centimeters	1 m = 1.1 yards	1 yard = 0.9 m
		= 1000 millimeters	1 m = 3.3 feet	1 foot = 0.3 m
			1 m = 39.4 inches	
	1 centimeter (cm)	= 0.01 (10^{-2}) meter	1 cm = 0.4 inch	1 foot = 30.5 cm
				1 inch = 2.5 cm
	1 millimeter (mm)	= 0.001 (10^{-3}) meter	1 mm = 0.04 inch	
	1 micrometer (μm)	= 10^{-6} meter (10^{-3} mm)		
	1 nanometer (nm)	= 10^{-9} meter (10^{-3} μm)		
	1 angstrom (Å)	= 10^{-10} meter (10^{-4} μm)		
Area	1 hectare (ha)	= 10,000 square meters	1 ha = 2.5 acres	1 acre = 0.4 ha
	1 square meter (m^2)	= 10,000 square centimeters	1 m^2 = 1.2 square yards	1 square yard = 0.8 m^2
			1 m^2 = 10.8 square feet	1 square foot = 0.09 m^2
	1 square centimeter (cm^2)	= 100 square millimeters	1 cm^2 = 0.16 square inch	1 square inch = 6.5 cm^2
Mass	1 metric ton (t)	= 1000 kilograms	1 t = 1.1 tons	1 ton = 0.91 t
	1 kilogram (kg)	= 1000 grams	1 kg = 2.2 pounds	1 pound = 0.45 kg
	1 gram (g)	= 1000 milligrams	1 g = 0.04 ounce	1 ounce = 28.35 g
			1 g = 15.4 grains	
	1 milligram (mg)	= 10^{-3} gram	1 mg = 0.02 grain	
	1 microgram (μg)	= 10^{-6} gram		
Volume (Solids)	1 cubic meter (m^3)	= 1,000,000 cubic centimeters	1 m^3 = 1.3 cubic yards	1 cubic yard = 0.8 m^3
			1 m^3 = 35.3 cubic feet	1 cubic foot = 0.03 m^3
	1 cubic centimeter (cm^3 or cc)	= 10^{-6} cubic meter	1 cm^3 = 0.06 cubic inch	1 cubic inch = 16.4 cm^3
	1 cubic millimeter (mm^3)	= 10^{-9} cubic meter (10^{-3} cubic centimeter)		
Volume (Liquids and Gases)	1 kiloliter (kL or kl)	= 1000 liters	1 kL = 264.2 gallons	1 gallon = 3.79 L
	1 liter (L)	= 1000 milliliters	1 L = 0.26 gallons	1 quart = 0.95 L
			1 L = 1.06 quarts	
	1 milliliter (mL or ml)	= 10^{-3} liter	1 mL = 0.03 fluid ounce	1 quart = 946 mL
		= 1 cubic centimeter	1 mL = approx. $\frac{1}{4}$ teaspoon	1 pint = 473 mL
			1 mL = approx. 15–16 drops	1 fluid ounce = 29.6 mL
				1 teaspoon = approx. 5 mL
Volume (Liquids and Gases)	1 microliter (μl or μL)	= 10^{-6} liter (10^{-3} milliliters)		
Time	1 second (s)	= $\frac{1}{60}$ minute		
	1 millisecond (ms)	= 10^{-3} second		
Temperature	Degrees Celsius (°C)		°F = $\frac{9}{5}$ °C + 32	°C = $\frac{5}{9}$ (°F – 32)

Further Reading

Chapter 1 Introduction

Halle, F. "A Raft Atop the Rain Forest." *National Geographic,* October 1990. More about the research described in Chapter 1's introduction.

Mayr, E. *The Growth of Biological Thought: Diversity, Evolution, and Inheritance.* Cambridge, MA: Harvard University Press, 1982. A classic by one of the greatest evolutionary biologists of the century. (The concept of emergent properties is discussed on pp. 63–64.)

Moore, J. A. *Science as a Way of Knowing: The Foundation of Modern Biology.* Cambridge, MA: Harvard University Press, 1993. A lively account by a famous developmental biologist.

Stolzenburg, W. "Winged Saviors of the Forest." *Nature Conservancy,* March/April 1991. Roberto Roca's study of oilbirds in Venezuela.

Unit One The Life of the Cell

Alberts, B. D. Bray, J. Lewis, M. Raff, K. Roberts, and J. D. Watson. *Molecular Biology of the Cell,* 2nd ed. New York: Garland, 1989. A comprehensive cell biology textbook for advanced students; clearly written and illustrated.

Becker, W. M., and D. W. Deamer. *The World of the Cell,* 2nd ed. Redwood City, CA: Benjamin/Cummings, 1991. A very readable, student-oriented text for undergraduates; excellent explanations of cellular bioenergetics and metabolism.

Mathews, C. K., and K. E. van Holde. *Biochemistry,* 2nd ed. Redwood City, CA: Benjamin/Cummings, 1990. A general biochemistry text, extremely well illustrated.

Stryer, L. *Biochemistry,* 3rd ed. New York: Freeman, 1989. A general biochemistry text, popular through three editions.

Chapter 2 The Chemical Basis of Life

Atkins, P. W. *Molecules.* New York: Scientific American Library, 1987. Beautifully illustrated tour of the world of molecules.

Mohnen, V. A. "The Challenge of Acid Rain." *Scientific American,* August 1988. Causes of the acid precipitation problem and possible solutions.

Chapter 3 The Molecules of Cells

Asimov, I. *The World of Carbon,* 2nd ed. New York: Macmillan, 1962. A primer on organic chemistry by one of America's most popular science writers.

Dushesne, L. C., amd D. W. Larson. "Cellulose and the Evolution of Plant Life." *BioScience,* April 1989. The chemistry and natural history of the most abundant organic molecule in the biosphere.

Flannery, M. C. "Collagen: Complex and Crucial." *The American Biology Teacher,* November/December 1990. The structure and function of our body's most abundant protein.

Vollrath, F. "Spider Webs and Silk." *Scientific American,* March 1992.

Chapter 4 A Tour of the Cell

De Duve, C. *A Guided Tour of the Living Cell.* New York: Scientific American Books, 1986. A beautifully illustrated introduction to the cell by the discoverer of lysosomes.

Symmons, M., A. Prescott, and R. Warn. "The Shifting Scaffolds of the Cell." *New Scientist,* February 18, 1989. The dynamics of the cytoskeleton.

Ezzel, C. "Sticky Situations." *Science News,* June 13, 1992. About the glue that holds animal cells together.

Chapter 5 The Working Cell

Bretscher, M. S. "The Molecules of the Cell Membrane." *Scientific American,* October 1985.

Brown, M. S., and J. L. Goldstein. "How LDL Receptors Influence Cholesterol and Atherosclerosis." *Scientific American,* November 1984.

Koshland, D. E. "Protein Shape and Biological Control." *Scientific American,* October 1973. A discussion of how enzymes are regulated.

Chapter 6 How Cells Harvest Chemical Energy

Angier, N. "A Stupid Cell with All the Answers." *Discover,* November 1986. The important roles that yeast is playing in basic biological research.

Harold, F. M. *The Vital Force: A Study of Bioenergetics.* New York: Freeman, 1986. A challenging introduction to energy and life, and how ATP is made by chemiosmosis.

Chapter 7 Photosynthesis: Using Light to Make Food

Bazzazz, F. A., and E. D. Fajer. "Plant Life in a CO_2-Rich World." *Scientific American,* January 1992. How will increasing atmospheric CO_2 and global warming affect the relative success of C_3 and C_4 plants?

Govindjee, and W. J. Coleman. "How Plants Make Oxygen." *Scientific American,* February 1990.

Unit Two Genetics

Suzuki, D., A. Griffiths, J. Miller, and R. Lewontin. *An Introduction to Genetic Analysis,* 4th ed. New York: Freeman, 1989. A good genetics textbook, including Mendelian, molecular, and population genetics..

Watson, J. D., N. H. Hopkins, J. W. Roberts, J. A. Steitz, and A. M. Weiner. *Molecular Biology of the Gene,* 4th ed. Menlo Park, CA: Benjamin/Cummings, 1987. The classic textbook on genetics at the molecular level.

Chapter 8 The Cellular Basis of Reproduction and Inheritance

Glover, D. M., C. Gonzalez, and J. W. Raff. "The Centrosome." *Scientific American,* June 1993. What is currently known about the role of centrosomes—centrioles plus surrounding material (microtubule organizing center)—in forming the mitotic spindle and cytoskeleton.

McIntosh, J. R., and K. L. McDonald. "The Mitotic Spindle." *Scientific American,* October 1989.

Murray, A. W., and M. W. Kirschner. "What Controls the Cell Cycle." *Scientific American,* March 1991.

See also the cell biology texts listed under the Unit One heading.

Chapter 9 Patterns of Inheritance

Charlesworth, B. "The Evolution of Sex Chromosomes." *Science,* March 1, 1991. How did the *X-Y* mode of sex determination evolve?

Peters, J. A. ed., *Classic Papers in Genetics.* Englewood Cliffs, NJ: Prentice-Hall, 1959. English translations of Mendel's work and other classics.

Chapter 10 Molecular Biology of the Gene

Darnell, J. E., Jr. "RNA." *Scientific American,* October 1985. The role of RNA in protein synthesis and its relationship to DNA.

Judson, H. F. *The Eighth Day of Creation: Makers of the Revolution in Biology.* New York: Simon & Schuster, 1979. The history of molecular biology.

Langone, J. "Emerging Viruses." *Discover,* December 1990.

Watson, J. D. *The Double Helix.* New York: Atheneum, 1968. The brash, controversial best-seller by the codiscoverer of the double helix.

"What Science Knows About AIDS." *Scientific American,* October 1988. An entire issue devoted to AIDS. For more recent information, see Chapter 24.

Chapter 11 The Control of Gene Expression

"Cancer." *Science,* November 22, 1991. A special section of seven articles on basic and applied cancer research.

De Robertis, E., G. Oliver, and C. Wright. "Homeobox Genes and the Vertebrate Body Plan." *Scientific American,* July 1990.

Grunstein, M. "Histones as Regulators of Genes." *Scientific American,* October 1992.

Lowenstein, J. "Genetic Surprises." *Discover,* December 1992. What biologists are learning about the organization of eukaryotic genomes.

Ptashne, M. "How Gene Activators Work." *Scientific American,* January 1989. How regulatory proteins bind to DNA.

Radetsky, P. "Genetic Heretic." *Discover,* November 1990. The story of the discovery that RNA can act as an enzyme in RNA splicing.

Chapter 12 Recombinant DNA Technology

Gilbert, W. "Toward a Paradigm Shift in Biology." *Nature,* January 10, 1991. How the ability to dissect genomes is changing the kinds of questions biologists ask.

Kahn, P. "Germany's Gene Law Begins to Bite." *Science,* January 31, 1992. A case study in attempts to regulate DNA technology.

"The New Harvest: Genetically Engineered Species." *Science,* June 16, 1989. A special issue with seven articles on this topic.

Watson, J. D., J. Tooze, and G. T. Kurtz. *Recombinant DNA: A Short Course,* 2nd ed. New York: Scientific American Books, 1989. The principles and applications of recombinant DNA technology.

Chapter 13 The Human Genome

Beardsley, T. "Diagnosis by DNA." *Scientific American,* October 1992.

"Genome Issue." *Science,* October 2, 1992. Contains five articles on progress in the Human Genome Project. (Look for a similar issue each fall.)

Hall, S. S. "James Watson and the Search for Biology's 'Holy Grail'." *Smithsonian,* February 1990. Watson and the Human Genome Project.

Morell, V. "Huntington's Gene Finally Found." *Science,* April 2, 1993.

Mulligan, R. C. "The Basic Science of Gene Therapy." *Science,* May 14, 1993.

Stine, G. J. *The New Human Genetics.* Dubuque, IA: W. C. Brown, 1989. An introductory human genetics textbook for undergraduates.

Watson, et al. *Recombinant DNA: A Short Course.* See under Chapter 12.

Weiss, R. "Predisposition and Prejudice." *Science News,* January 21, 1989. As more tests to detect "bad" genes are devised, will genetic discrimination emerge?

Unit Three Evolution and The Diversity of Life

Futuyma, E. J. *Evolutionary Biology,* 2nd ed. Sunderland, MA: Sinauer, 1986. An excellent undergraduate text.

Margulis, L., and Schwartz, K. V. *Five Kingdoms: An Illustrated Guide to the Phyla of Life on Earth,* 2nd ed. New York: Freeman, 1987.

Milner, R. *The Encyclopedia of Evolution.* New York: Facts on File, 1990. Many interesting anecdotes.

Chapter 14 How Populations Evolve

Darwin, C. *The Origin of Species* and *The Descent of Man.* New York: Modern Library, 1990. Two classic books in one volume for around $12.

Dawkins, R. *The Blind Watchmaker.* New York: Norton, 1986. How complexity can arise in the absence of design.

Desmond, A., and J. Moore. *Darwin.* New York: Warner, 1992. A recent biography.

Diamond, J. "Founding Fathers and Mothers." *Natural History,* June 1988. The importance of genetic drift in human evolution.

Noonan, D. "Dr. Doolittle's Question." *Discover,* February 1990. How biochemist Russell Doolittle learns about evolution from blood-clotting molecules.

Chapter 15 The Origin of Species

Culotta, E. "How Many Genes Had to Change to Produce Corn?" *Science,* June 28, 1991.

Gould, S. J. "Opus 200." *Natural History,* August 1991. Tracing the origin and impact of the theory of punctuated equilibrium.

Grant, P. R. "Natural Selection and Darwin's Finches." *Scientific American,* October 1991. How a single drought can change a population.

Kluger, K. "Go Fish." *Discover,* March, 1992. Rapid speciation in Lake Victoria.

Rennie, J. "Are Species Specious?" *Scientific American,* November 1991. Problems with the concept of biological species.

Chapter 16 Tracing Evolutionary History

Alvarez, W., F. Asaro, and V. Courtillot. "What Caused the Mass Extinction?" *Scientific American,* October 1990. Two articles on asteroids versus volcanoes as the main cause of Cretaceous extinctions.

Cherfas, J. "Ancient DNA: Still Busy After Death." *Science,* September 20, 1991. Molecular biologists go to work on fossils.

Ezzell, C. "Conserving a Coyote in Wolf's Clothing? Molecular Systematics and Endangered Species." *Science News,* June 15, 1991.

Gould, S. J. "We Are All Monkey's Uncles." *Natural History,* June 1992. Cladists and human classification.

Chapter 17 The Origin and Evolution of Microbial Life: Prokaryotes and Protists

Horgan, J. "In the Beginning. . ." *Scientific American,* February 1991. A discussion of new controversies about the origin of life.

Mann, C. "Lynn Margulis: Science's Unruly Earth Mother." *Science,* April 19, 1991. A profile of the chief advocate of the endosymbiotic theory.

Radetsky, P. "How Did Life Start?" *Discover,* November 1992.

Tortora, G. J., B. R. Funke, and C. L. Case. *Microbiology: An Introduction,* 4th ed. Redwood City, CA: Benjamin/Cummings, 1992. A general text with an emphasis on disease-causing microbes.

Chapter 18 Plants, Fungi, and the Colonization of Land

Alexopoulos, C. J., and C. W. Mims. *Introduction to Mycology,* 3rd ed. New York: Wiley, 1979. A general text.

Gould, S. J. "A Humungous Fungus Among Us." *Natural History,* July 1992. Is a 30-acre fungus the world's largest organism?

Lewington, A. *Plants for People.* New York: Oxford University Press, 1990. The many uses of plant products.

Mauseth, J. *Botany.* Philadelphia: Saunders, 1991. A beautifully illustrated introduction to plants.

Chapter 19 The Evolution of Animal Diversity

Brusca, R. G., and G. J. Brusca. *Invertebrates.* Sunderland, MA: Sinauer Associates, 1990. An evolutionary approach to invertebrate animals.

Carroll, R. C. *Vertebrate Paleontology and Evolution.* New York: Freeman, 1987. An authoritative yet accessible text.

Fischman, J. "One That Got Away." *Discover,* January, 1992. Using molecular systematics to trace our vertebrate roots to fish.

Gould, S. J. *Wonderful Life: The Burgess Shale and the Nature of History.* New York: Norton, 1989. Gould's best-seller about animal evolution.

Mitchell, L. G., J. A. Mutchmor, and W. D. Dolphin. *Zoology.* Menlo Park, CA: Benjamin/Cummings, 1988. An accessible introductory text.

"The Thunder Lizards." *Natural History,* December 1991. Several articles about how dinosaurs lived.

Yoffe, E. "Silence of the Frogs." *The New York Times Magazine,* December 13, 1992. Amphibians are in global decline. What are the causes and ecological implications?

Zimmer, C. "Ruffled Feathers." *Discover,* May 1992. The controversy about the origin of birds.

Unit Four Animals: Form and Function

Eckert, R., and D. Randall. *Animal Physiology: Mechanisms and Adaptations,* 3rd ed. New York: Freeman, 1988. A widely used textbook of comparative animal physiology.

Marieb, E. *Human Anatomy and Physiology,* 2nd ed. Redwood City, CA: Benjamin/Cummings, 1992. A basic text, beautifully illustrated.

Schmidt-Nielsen, K. *Animal Physiology: Adaptations and Environment,* 4th ed. New York: Cambridge University Press, 1990. As its title implies, this comparative text emphasizes adaptations to the environment.

Chapter 20 Unifying Concepts of Animal Structure and Function

Caplan, I. "Cartilage." *Scientific American,* October 1984. The structure and function of an important tissue.

Sochurek, H., and P. Miller. "Medicine's New Vision." *National Geographic,* January 1987. Methods of photographing the human interior.

Chapter 21 Nutrition and Digestion

Christian, J. L., and J. L. Greger. *Nutrition for Living,* 3rd ed. Redwood City, CA: Benjamin/Cummings, 1991. A popular nutrition text.

Diamond, J. "The Athlete's Dilemma." *Discover,* August 1991. Have endurance athletes reached the limits of what metabolism can support?

Sanderson, S., and R. Wasserug. "Suspension Feeding Vertebrates." *Scientific American,* March 1990. An article emphasizing whales.

Chapter 22 Respiration: The Exchange of Gases

Houston, C. S. "Mountain Sickness." *Scientific American,* October 1992. How high altitude interferes with homeostasis.

Stroh, M. "Breathing Lessons." *Science News,* May 9, 1992. The use of computers to develop models for how breathing works.

Chapter 23 Circulation

Golde, D. "The Stem Cell." *Scientific American,* December 1991. The master cell of blood cell production.

Lillywhite, H. B. "Snakes, Blood Circulation, and Gravity." *Scientific American,* December 1988.

Zivin, J., and D. Choi. "Stroke Therapy." *Scientific American,* July 1991. Causes, effects, and experimental treatment of strokes.

Chapter 24 The Immune System

"AIDS: The Unanswered Questions." *Science,* May 28, 1993. A special section devoted to AIDS and HIV.

Anderson, R., and R. May. "Understanding the AIDS Pandemic." *Scientific American,* May 1992.

Robbins, A., and P. Freeman. "Obstacles to Developing Vaccines for the Third World." *Scientific American,* November 1988.

Roitt, I., J. Brostoff, and D. Male. *Immunology,* 2nd ed.. St. Louis: Mosby. A well-illustrated text.

Chapter 25 Control of the Internal Environment

Carey, F. G. "Fishes with Warm Bodies." *Scientific American,* February 1973. How some fish species conserve metabolic heat.

McKenzie, A. "Seeking the Mechanisms of Hibernation." *BioScience,* June 1990.

Smith, H. W. *From Fish to Philosopher.* Boston: Little, Brown, 1953. A classic book on vertebrate evolution as revealed by kidney structure and function.

Chapter 26 Chemical Regulation

Atkinson, M., and N. MacLaren. "What Causes Diabetes?" *Scientific American,* July 1990.

Snyder, S. and D. Bredt. "Biological Roles of Nitric Oxide." *Scientific American,* May 1992. Recent research on a chemical signal with multiple functions.

Snyder, S. H. "The Molecular Basis of Communication Between Cells." *Scientific American,* October 1985. Emphasizes the relationship between the endocrine and nervous systems.

Chapter 27 Reproduction and Embryonic Development

Caldwell, M. "How Does a Single Cell Become a Whole Body?" *Discover,* November 1992.

Crews, D. "Courtship of Unisexual Lizards: A Model for Brain Evolution." *Scientific American,* December 1987. Research on all-female species.

Crooks, R., and K. Baur. *Our Sexuality,* 5th ed. Redwood City, CA: Benjamin/Cummings, 1993. A popular textbook on human sexuality.

Gilbert, S. F. *Developmental Biology,* 3rd ed. Sunderland, MA: Sinauer Associates, 1991. An excellent textbook used in upper-division courses.

Chapter 28 Nervous Systems

Kemp, M. "A Squid for All Seasons." *Discover,* June 1989. An important organism in neurobiology research.

Koch, C. "What Is Consciousness?" *Discover,* November 1992.

Sperry, R. W. "The Great Cerebral Commissure." *Scientific American,* January 1967. An early account of the different functions of the left and right halves of the brain.

Chapter 29 The Senses

Barinaga, M. "The Secret of Saltiness." *Science,* November 1, 1991. The physiology of a basic taste.

Freeman, W. "The Physiology of Perception." *Scientific American,* February 1991. How the brain processes sensory data.

Ramachandran, V. "Blind Spots." *Scientific American,* May 1992. Numerous optical illusions based on the blind spot.

Chapter 30 How Animals Move

Alexander, R. "Muscles Fit for the Job." *New Scientist,* April 15, 1989. How are particular muscles adapted for diverse functions in movement?

Amato, I. "Heeding the Call of the Wild." *Science,* August 30, 1991. Materials scientists study the structure and function of skeletons and other animal products.

Unit Five Plants: Form and Function

Mauseth, J. D. *Botany.* Philadelphia: Saunders, 1991.

Raven, P. H., R. F. Evert, and S. E. Eichhorn. *Biology of Plants,* 5th ed. New York: Worth, 1992.

Taiz, L., and E. Zeiger. *Plant Physiology.* Redwood City, CA: Benjamin/Cummings, 1991.

Chapter 31 Plant Structure, Reproduction, and Development

Bolz, D. M. "A World of Leaves: Familiar Forms and Surprising Twists." *Smithsonian,* April 1985. A delightful article on the adaptations of leaves, with exquisite photographs.

Dale, J. "How Do Leaves Grow?" *BioScience,* June 1992. Researchers are using the techniques of molecular and cell biology to answer questions about plant structure and growth.

"The Trees Told Him So." *Science News,* September 7, 1985. Growth rings may give clues to volcanic eruptions.

Vaughan, D. A., and L. A. Sitch. "Gene Flow from the Jungle to Farmers." *BioScience,* January 1991. Problems and methods of maintaining genetic diversity in crop plants.

Chapter 32 Plant Nutrition and Transport

Beardsley, T. "A Nitrogen Fix for Wheat." *Scientific American,* March 1991. Artificially induced root nodules on nonlegumes.

Gibbons, B. "Do We Treat Our Soil Like Dirt?" *National Geographic,* September 1984.

Reganold, J. P., R. I. Papendick, and J. F. Parr. "Sustainable Agriculture." *Scientific American,* June 1990.

Chapter 33 Control Systems in Plants

Evans, M. L., R. Moore, and K.-H. Hasenstein. "How Roots Respond to Gravity." *Scientific American,* December 1986.

Mores, P. B., and N.-H. Chua. "Light Switches and Plant Genes." *Scientific American,* April 1988. A link between environment and gene expression in plants.

Sussman, M. "Shaking *Arabidopsis thaliana.*" *Science,* May 1, 1992. Cell biology of a plant's response to touch.

Unit Six Ecology

Begon, M., J. L. Harper, and C. R. Townsend. *Ecology,* 2nd ed. New York: Blackwell Scientific, 1990.

Ricklefs, R. E. *Ecology,* 3rd ed. New York: Chiron Press, 1986.

Chapter 34 The Biosphere: An Introduction to Earth's Diverse Environments

Abrahamson, W. G., T. G. Whitham, and P. W. Price. "Fads in Ecology." *BioScience,* May 1989. Changing ideas in a dynamic field.

Burman, A. "Saving Brazil's Savannas." *New Scientist,* March 2, 1991.

Ray, G., and J. Grassle. "Marine Biology Diversity." *BioScience,* July/August 1991. Why we need a program for conserving marine communities.

Chapter 35 Population Dynamics

Ackerman, L., et al. "The Successful Animal." *Science 86,* January/February 1986. Cultural and historical aspects of human population growth.

Daily, G. C., and P. R. Ehrlich. "Population, Sustainability, and Earth's Carrying Capacity." *BioScience,* November 1992.

Savonen, C. "One Salmon, Two Salmon . . . 10,000 Salmon: Counting the Fish in Alaska." *Oceans,* January/February 1985. A population density determination in action.

Chapter 36 Communities and Ecosystems

Beardsley, T. "Desert Dynamics." *Scientific American,* November 1992. An experiment tests the effect of kangaroo rats on desert vegetation.

Culotta, E. "Biological Immigrants Under Fire." *Science,* December 6, 1991. How exotic species can disrupt natural communities.

Oliwenstein, L. "Royal Flush." *Discover,* January 1992. New debate about a classic case of mimicry (monarchs and viceroy butterflies).

Odum, H. T. *Systems Ecology: An Introduction.* New York: Wiley, 1984. A text that emphasizes important aspects of ecosystems.

Chapter 37 Behavioral Adaptations to the Environment

Alcock, J. *Animal Behavior: An Evolutionary Approach,* 4th ed. Sunderland, MA: Sinauer Associates, 1989.

Davies, N., and M. Brooke. "Coevolution of the Cuckoo and Its Hosts." *Scientific American,* January 1991.

Diamond, J. "Sexual Deception." *Discover,* August 1989. Speculations on the evolutionary basis of seduction.

Jolly, A. "A New Science That Sees Animals as Conscious Beings." *Smithsonian,* March 1985.

Lorenz, K. *On Aggression.* New York: Harcourt Brace Jovanovich, 1966. A classic book on competitive social interactions.

Chapter 38 Human Evolution and Its Ecological Impact

Broeker, W. "Global Warming on Trial." *Natural History,* April 1992. How solid is the evidence that Earth is heating up?

Gomez-Pompa, A., and A. Kaus. "Taming the Wilderness Myth." *BioScience,* April 1992. How Western beliefs affect environmental policy.

Hammond, A. L. *World Resources, 1990–1991.* New York: Oxford University Press, 1990. A revised version of this book, prepared by the staff of the World Resources Institute, appears every other year as an updated guide to the global environment. The book is prepared in collaboration with the United Nations Environment Programme and the U. N. Development Programme.

"Managing Planet Earth." *Scientific American,* September 1989. A special issue devoted to the environment.

Tattersall, I. *The Human Odyssey: Four Million Years of Human Evolution.* New York: Prentice-Hall General Reference, 1993. A readable and beautifully illustrated book based on the new Hall of Human Biology and Evolution at the American Museum of Natural History.

Wilson, E. O. *The Diversity of Life.* Cambridge, MA: Harvard University Press, 1992. Why we should be concerned about the extinction of other species.

Chapter 2 Answers

Testing Your Knowledge

Multiple Choice: **1.** b **2.** c **3.** c **4.** e **5.** a (It needs to share 2 more electrons for a full outer shell of 8.) **6.** c **7.** d

True/False: **1.** F (Only salt and water are compounds.) **2.** T **3.** T **4.** T **5.** F (Molecules are farther apart.) **6.** F (The smallest particle is an atom.) **7.** F (7) **8.** F (Most acid precipitation results from burning fossil fuels.) **9.** T **10.** F (Reactants are the starting materials.)

Describing, Comparing, and Explaining

1. For diagram, see Figure 2.10A. Water molecules form hydrogen bonds because they are polar. The unique properties of water that result from hydrogen bonding are cohesion, surface tension, ability to absorb and store large amounts of heat, high boiling point, a solid form (ice) that is less dense than liquid water, and solvent properties.
2. First: It takes energy to break hydrogen bonds, so a large amount of heat is needed to warm water slightly. This resists an increase in body temperature. Also, this allows water to store much heat, and a large heat loss is accompanied by a slight temperature drop. Second: It takes a lot of heat energy to break the hydrogen bonds between a water molecule and its neighbors when it evaporates, so an evaporating water molecule takes much heat away from the body. This is evaporative cooling.
3. A covalent bond forms when atoms complete their outer shells by sharing electrons. Atoms also complete their outer shells by gaining or losing electrons. This leaves the atoms with + and − charges, and the atoms are attracted to each other, forming an ionic bond.
4. An acid is a compound that donates H^+ ions to a solution. A base is a compound that accepts H^+ ions and removes them from solution. Acidity is described by the pH scale, which measures H^+ concentration on a scale of 0 (most acidic) to 14 (most basic).

Thinking Critically

1. Give a mouse sugar or oxygen gas containing a radioactive isotope of oxygen; then see whether the carbon dioxide it exhales is radioactive.
2. Fluorine needs 1 electron for a full outer shell of 8, and if potassium loses 1 electron, its outer shell will have 8. Potassium will lose an electron (becoming a + ion) and fluorine will pick it up (becoming a − ion). The ions will form an ionic bond.
3. In alcohol, most of the heat energy in the marble speeded up alcohol molecules, increasing the temperature of the alcohol. In water, some of the energy in the marble was needed to break the hydrogen bonds between water molecules, leaving less energy to speed up the molecules, so the temperature climbed less.

Science, Technology, and Society

1. *Some issues and questions to consider:* How do pollution credits differ from traditional regulation? How can pollution credits make electricity cheaper? How will pollution credits affect air quality? For whom? Do pollution credits reduce pollution, or just move it around? Are pollution credits a license to pollute? Could environmental groups or citizens bid for pollution credits and "retire" them to improve air quality? Should people have to pay for clean air?
2. *Some issues and questions to consider:* Which is less expensive, power from nuclear power plants or fossil-fuel plants? Does the price of electricity reflect its actual cost, including environmental costs? Which would be more extensive, the environmental effects of acid rain and global warming from fossil-fuel power plants, or the effects of nuclear wastes or potential nuclear accidents? Do you favor development of nuclear energy or fossil-fuel power plants? Which would you prefer to have near you? Do your answers to these last two questions differ? If so, why?
3. *Some issues and questions to consider:* Is it the *kinds* of atoms present in chemical wastes that is important, or the way the atoms are combined to form particular substances? How important is the way the atoms are concentrated and disposed of? Do chemicals produced by human technology differ from naturally occurring substances? How?

Chapter 3 Answers

Testing Your Knowledge

Multiple Choice: **1.** d (monomer and polymer) **2.** c **3.** d **4.** e **5.** a **6.** b **7.** d

True/False: **1.** F (monosaccharides) **2.** T **3.** T **4.** F (type of lipid) **5.** T **6.** F (hydrophobic) **7.** T **8.** F (secondary and tertiary) **9.** T **10.** T

Describing, Comparing, and Explaining

1. Triglycerides—energy storage. Phospholipids—make up membranes. Waxes—make up waterproof coatings. Steroids—are part of cell membranes, act as hormones.
2. Weak bonds that stabilize the three-dimensional structure of a protein are disrupted, and the protein unfolds. Function depends on shape, so if the protein is the wrong shape, it won't function properly.
3. Proteins are made of 20 amino acids arranged in many different sequences into chains of many different lengths. Genes, defined stretches of DNA, dictate the primary sequences of proteins in the cell.
4. Proteins function in structure, contraction, storage, defense, transport, and signaling, and as enzymes (see Module 3.11).

Thinking Critically

1. Pentane isomers are like butane and isobutane (Module 3.1), only one carbon larger.
2. 20 choices for first amino acid, 20 choices for second. 20×20 or $20^2 = 400$ possibilities for 2 amino acids. $20 \times 20 \times 20 \times ... = 20^{129}$ possibilities for a protein 129 amino acids long.
3. This is a hydrolysis reaction, which consumes water. It is essentially the reverse of the diagram in Figure 3.5.

Science, Technology, and Society

1. *Some issues and questions to consider:* How are these chemicals important in agriculture, medicine, and public health? How have they affected humans? Wildlife? Natural vegetation? Are the chemicals themselves harmful, or the way that they are used? What influences have shaped your opinions: The media? Personal experience? Reading? Friends and family? How might the opinion of a villager in a developing country differ from yours?
2. *Some issues and questions to consider:* How will you choose your test subjects? How many should you choose? Will you give them all vitamin C, or just some of them? What criteria will you use to divide the test subjects into groups? What is a control group? Should the subjects know whether they are getting vitamin C or not? Should the experimenters who are giving out the drug and measuring the severity of cold symptoms know which of the subjects are getting vitamin C? What is a "double blind" study? If there is a difference between your groups, how can you be sure it is due to vitamin C?
3. *Some issues and questions to consider:* Is it the responsibility of a corporation to protect the livelihoods of farmers and exporters from

whom they buy a product? Or should these individuals look out for themselves? How might producers protect themselves from economic shocks? What are the roles of the governments of developed and developing countries in this kind of trade? How can consumers make their preferences known? Do the concerns of consumers affect manufacturers?

Chapter 4 Answers

Testing Your Knowledge

Multiple Choice: **1.** e **2.** c (Small cells have a greater ratio of surface area to volume.) **3.** b **4.** d **5.** c

Describing, Comparing, and Explaining

1. Tight junctions form leakproof bonds. Anchoring junctions link cells but allow materials to pass between them. Communicating junctions are channels that allow flow from cell to cell.
2. Both process energy. A chloroplast converts light energy to chemical energy. A mitochondrion converts chemical energy (food) to other chemical energy (ATP).
3. Different conditions and conflicting processes can occur simultaneously in different compartments. Also, there is increased area for membrane-located enzymes that carry out metabolic processes.

Thinking Critically

1. A (chloroplast) and C (cell wall and mitochondrion) must be bean plant; D and F are also eukaryotic, so they must be mouse. By process of elimination, B and E must be bacteria.
2. Cell 1: S = 600 μm^2; V = 1000 μm^3; S/V = 0.6. Cell 2: S = 2400 μm^2; V = 8000 μm^3; S/V = 0.3. The smaller cell has a larger surface in relation to volume for absorbing food and oxygen and excreting waste. Small cells are more efficient.

Science, Technology, and Society

Some issues and questions to consider: Were the cells Moore's property, a gift, or just surplus? Was Moore asked to donate the cells? Was he informed about how the cells might be used? Is it important to ask permission or inform the patient in such a case? How much did the researchers modify the cells? What did they have to do to them to sell the product? Do the researchers and the university have a right to make money from Moore's cells? Is the fact that they saved Moore's life a factor here? Does Moore have the right to sell his cells? Would Moore have been able to sell the cells without the researchers' help?

Chapter 5 Answers

Testing Your Knowledge

Multiple Choice: **1.** d **2.** b **3.** a **4.** c (Only active transport can move solute against a concentration gradient.)

Describing, Comparing, and Explaining

1. Salt water is hypertonic; it has a higher solute concentration than the solution in root cells. Water would therefore leave the roots by osmosis.
2. Heating, pickling, and salting denature enzymes, changing their shapes so they do not fit substrates. Freezing decreases the kinetic energy of molecules, so they lack energy of activation, even in the presence of enzymes.

Thinking Critically

The more enzyme present, the faster the rate of reaction, because it is more likely that enzyme and substrate will meet. The more substrate, the faster the reaction, for the same reason, but only up to a point. An enzyme molecule can work only so fast; once it is saturated (working at top speed), more substrate does not increase the rate.

Science, Technology, and Society

Some issues and questions to consider: Does a woman have a right to work in an unsafe environment, even if it may put her child at risk? What are the rights of the child? Does the company have the right to protect women who are not pregnant? Is the company trying to protect the mother and child, or protect itself from a potential lawsuit? Who is responsible for protecting employees and their children? The employees? The company? The government? Suppose a woman risks lead exposure and bears a retarded child. Who is responsible if, 20 years from now, the child decides to sue?

Chapter 6 Answers

Testing Your Knowledge

Multiple Choice: **1.** c **2.** c (NAD^+ and FAD, which are recycled by electron transport, are limited.) **3.** e **4.** d **5.** c **6.** b **7.** a
True/False: **1.** T **2.** F (Glycolysis takes place in the cytoplasm.) **3.** F (They carry electrons from glycolysis and the Krebs cycle to the electron transport chain.) **4.** T **5.** T **6.** F (Glycolysis does this conversion.) **7.** T **8.** F (Oxygen combines with H from glucose to make H_2O.)

Describing, Comparing, and Explaining

1. Disadvantage: Less ATP is produced (only 2 per glucose molecule versus 36 aerobically). Advantage: No oxygen is needed.
2. Glycolysis is considered the most ancient because it occurs in all living cells.
3. Oxygen picks up electrons and hydrogen atoms produced by the oxidation of glucose, at the end of the electron transport chain. Carbon dioxide results from the breakup of glucose molecules in glycolysis and the Krebs cycle.

Thinking Critically

1. 100 kcal per day is 700 kcal per week. On the basis of the chart in Figure 6.3, walking 4 mi/hr would require 700/231 = about 3 hr; swimming, 1.3 hr; running, 0.8 hr.
2. The amino acids that make up proteins have amino groups containing N atoms. Amino groups must be added to fats or carbohydrates in order to make amino acids to make proteins.
3. 10 NAD^+ and 2 FAD are needed to pick up the electrons and hydrogen atoms from a glucose molecule. NAD^+ and FAD are recycled between electron transport and glycolysis and the Krebs cycle. We need a small additional supply to replace those that are lost or damaged.

Science, Technology, and Society

Some issues and questions to consider: Is your customer aware of the danger? Do you have an obligation to protect the customer, even against her wishes? Does your employer have the right to dismiss you for informing the customer? For refusing to serve the customer? Could you or the restaurant later be held liable for injury to the fetus? Or is the mother responsible for willfully disregarding warnings about drinking?

Chapter 7 Answers

Testing Your Knowledge

Multiple Choice: **1.** d **2.** c **3.** c **4.** b **5.** a **6.** b **7.** e **8.** a **9.** c (NADPH and ATP from light-dependent reactions are used by the Calvin cycle.) **10.** b
True/False: **1.** F (Plants, algae, and some bacteria are able to photosynthesize.) **2.** F (Both herbivores and carnivores would die.) **3.** T

4. T **5.** F (The oxygen comes from CO_2.) **6.** T **7.** F (Cyclic electron flow is the backup method.) **8.** T **9.** F (Most raw materials come from the air.) **10.** F (It reflects and transmits green light.)

Describing, Comparing, and Explaining

1. See Figure 7.5.

2. Young trees would take up CO_2 faster. Harvesting and replanting releases CO_2 stored in old trees (burning and decomposition of wastes or products made from the wood), contributing to the greenhouse effect.

3. In photosynthesis, electrons are from chlorophyll; in cellular respiration, electrons are from organic molecules. In photosynthesis, electron energy is from light; in cellular respiration, electron energy is stored in chemical bonds of organic molecules. In photosynthesis, electrons are finally picked up by $NADP^+$; in cellular respiration, electrons are finally picked up by O_2. In both processes, energy is used to transport H^+ through a membrane. As H^+ flows back, the energy is used to make ATP.

4. Plants can break down the sugar for energy in cellular respiration, or use the sugar as a raw material for making other organic molecules. Excess sugar is stored as starch.

Thinking Critically

1. When a rainforest tree is eaten by an animal, or when the tree or its parts die and decompose, the organic molecules contained in the tree are consumed in cellular respiration in the cells of the animal or decomposer. The same amount of oxygen produced by the tree when the organic molecules were made is consumed in cellular respiration when they are oxidized. The rain forest could produce a surplus of oxygen if the forest were actually growing in size, storing up more energy in photosynthesis than was consumed in cellular respiration.

2. Under the sea, things look greenish because these are the wavelengths that most easily pass through the water without being absorbed. Green light is not absorbed by green algae; this is why they are green. The red pigments in red algae absorb the greenish light under the sea. (This is why they are red, not green.) This enables them to live at great depths, where only greenish light is available.

3. The oxygen atoms in glucose come from CO_2. You could give some plants CO_2 containing the isotope ^{18}O and H_2O containing "normal" oxygen, and see whether the isotope ended up in glucose or O_2 gas. Then use H_2O labeled with ^{18}O, and unlabeled CO^2. Only ^{18}O from CO_2 ends up in glucose.

Science, Technology, and Society

Some issues and questions to consider: What are the risks that we take and costs we must pay if greenhouse warming continues? How certain do we have to be that warming is caused by human activities before we act? Is it possible that greenhouse warming may not be a threat, that it may actually be beneficial? What can we do to reduce CO_2 emissions? What are the risks and costs if we reduce CO_2 emissions and greenhouse warming turns out to not be a real threat? Is it possible that the costs and sacrifices of reducing CO_2 emissions might actually improve our lifestyle?

Chapter 8 Answers

Testing Your Knowledge

Multiple Choice: **1.** c **2.** a **3.** b **4.** e **5.** e (A diploid cell would have an even number of chromosomes; the odd number suggests meiosis I has been completed. Sister chromatids are together only in prophase and metaphase of meiosis II.) **6.** c **7.** b **8.** c

True/False: **1.** T **2.** T **3.** F (Sister chromatids separate at anaphase.) **4.** F (*X* and *Y* are called sex chromosomes.) **5.** F (Homologous pairs split in the first meiotic division.) **6.** F (The haploid phase begins with meiosis.) **7.** T **8.** F (Nondividing cells stay in the G_1 stage.) **9.** F ($n = 23$ in humans.) **10.** T

Describing, Comparing, and Explaining

1. Various orientations of chromosomes at metaphase I of meiosis produce different combinations of chromosomes in gametes. Crossing over during prophase I results in exchange of chromosome segments and new combinations of genes. Random fertilization of eggs by sperm further increases possibilities for variation.

2. Mitosis is a single division that produces two daughter cells that are genetic copies of the parent cell. Meiosis consists of two consecutive divisions that reduce a diploid parent cell to four haploid cells. In animals, mitosis occurs in most body tissues, meiosis only in the testes and ovaries. Sister chromatids of single chromosomes separate in anaphase of mitosis. In anaphase I of meiosis, homologous chromosomes separate.

3. See the photos in Module 8.7. Interphase: Growth; metabolic activity; DNA synthesis. Prophase: Chromosomes shorten and thicken, mitotic spindle forms. Metaphase: Chromosomes line up at the cell equator. Anaphase: Sister chromatids separate and move to the poles. Telophase: Daughter nuclei form around chromosomes; cytokinesis occurs.

4. In culture, cells divide 20–50 times, but only when they are in contact with a surface and when they are not touching. (This is density-dependent inhibition.) Depletion of growth factors may stop division. Controls act at the restriction point in the G_1 stage of the cell cycle. Cancer cells are much less affected by control mechanisms such as density-dependent inhibition, and can grow without contacting a solid surface. When they stop, they do not stop at the restriction point.

5. A ring of microfilaments pinches an animal cell in two, a process called cleavage. In a plant cell, membranous vesicles form a disc called the cell plate at the midline of the parent cell, membranes fuse with the plasma membrane, and a cell wall grows in the space, separating the daughter cells.

Thinking Critically

1. 100,000 genes/23 chromosomes = about 4000 genes/chromosome.

2. A bacterial cell contains a single, relatively small chromosome, which can replicate more easily and quickly. In a bacterium, there is less danger of chromosomes getting tangled, so binary fission is less complex than mitosis. Also, a bacterial cell is smaller, so it takes less material to grow and less time to divide.

3. 1 cm^3 = 1000 mm^3, so 5000 mm^3 of blood contain 5000 × 1000 × 5,000,000 = 25,000,000,000,000 or 2.5 × 10^{13} red blood cells. $\frac{1}{120}$ are replaced each day, so 2.5 × 10^{13}/120 = 2.1 × 10^{11} replaced each day. 24 × 60 × 60 = 86,400 seconds in a day. Therefore, 2.1 × 10^{11}/86,400 = about 2 × 106, or 2 million divisions per second.

4. Each chromosome is on its own in mitosis; replication and splitting of sister chromatids happen independently for each horse or donkey chromosome. Therefore, mitotic divisions, starting with the zygote, are not impaired. In meiosis, however, homologous chromosomes must pair in prophase I. This process of synapsis cannot occur properly because horse and donkey chromosomes do not match in number or content.

Science, Technology, and Society

Some issues and questions to consider: Could it be that less money is spent on prevention because effective prevention is so much cheaper?

Or because prevention has been tried, and it does not work well? Are lifestyle changes the kind of measures that could benefit from a shift in resources? Is prevention an individual matter of avoiding exposure, or a social matter of preventing exposure? How might the answer to this question shape prevention policy? If more money were devoted to prevention, how would this encourage you or others to make lifestyle changes? Would prevention work better for younger or older people? Might older people, already exposed to cancer-causing agents, actually be harmed by a shift of resources to prevention?

Chapter 9 Answers

Testing Your Knowledge

Multiple Choice: **1.** c **2.** b **3.** b **4.** d **5.** d (Neither parent is ruby-eyed, but some offspring are, so it is recessive. Different ratios among male and female offspring show it is sex-linked.) **6.** e **7.** a **8.** e

Describing, Comparing, and Explaining

1. Cross two peas heterozygous for pod color and flower color; both have green pods and purple flowers (genotype *GgPp*). Offspring are produced in the ratio of 9 green purple : 3 green white : 3 yellow purple : 1 yellow white. The ratio from dihybrid cross is the combined 3:1 ratio of two independent monohybrid crosses. Alleles of the parents do not "stick together" when passed to offspring.

2. Fruit flies are small and easy to raise and breed. They have large numbers of offspring, a short generation time, clearly visible traits, few chromosomes, and *X* and *Y* chromosomes like humans.

3. Genes on the single *X* chromosome in males are always expressed because there are no corresponding genes on the *Y* chromosome to mask them. A male needs only one recessive color-blindness allele (from his mother) to show the trait; a girl must inherit the allele from both parents, which is less likely.

4. See Figure 9.16A. The parental gametes are *WS* and *ws*. Recombinant gametes are *Ws* and *wS*, produced by crossing over.

Thinking Critically

1. Height appears to be a quantitative trait, resulting from polygenic inheritance, like human skin color. For example, a tall couple might be *AABbCC* and *AaBBCC*, and produce mostly children with all tall genes. A shorter couple might be *aaBbcc* and *aabbCc*, and produce children with mostly short genes.

2. Start out by breeding the cat to get a population to work with. If the curl allele is recessive, two curl cats can only have curl kittens. If the allele is dominant, curl cats can have "normal" kittens. If the curl allele is sex-linked, ratios will differ in male and female offspring of some crosses. If the curl allele is autosomal, the same ratios will be seen among males and females. To see if a curl cat is true-breeding (homozygous), testcross it with a normal cat. If the curl cat is homozygous, all offspring of a testcross will be curl; if it is if heterozygous, half the offspring will be curl and half normal. The genotype of a true-breeding cat is *CC*. Use only true-breeding curl cats for breeding.

Genetics Problems

1. The brown allele appears to be dominant, the white allele recessive. The brown parent appears to be homozygous dominant, *BB*, and the white mouse is homozygous recessive, *bb*. The F₁ mice are all heterozygous, *Bb*. If two of the F₁ mice are mated, $\frac{3}{4}$ of the F₂ mice will be brown.

2. The best way to find out whether an F₂ mouse is homozygous dominant or heterozygous is to do a testcross: Mate the brown mouse with a white mouse. If the brown mouse is homozygous, all the offspring will be brown. If the brown mouse is heterozygous, you would expect half the offspring to be brown and half to be white.

3. Freckles is dominant, so Tim and Carolyn must both be heterozygous. There is a $\frac{3}{4}$ chance that they will produce a child with freckles, a $\frac{1}{4}$ chance they will produce a child without freckles. The probability that the next two children will have freckles is $\frac{3}{4} \times \frac{3}{4} = \frac{9}{16}$.

4. As in problem 3, both Tim and Carolyn are heterozygous, and Michael is homozygous recessive. The probability of the next child having freckles is $\frac{3}{4}$. The probability of the next child having a straight hairline is $\frac{1}{4}$. The probability that the next child will have freckles and a straight hairline is $\frac{3}{4} \times \frac{1}{4} = \frac{3}{16}$.

5. The genotype of the black short haired parent rabbit is BBSS. The genotype of the brown long-haired parent rabbit is bbss. The F₁ rabbits will all be black and short-haired, BbSs. The F₂ rabbits will be $\frac{9}{16}$ black short-haired, $\frac{3}{16}$ black long-haired, $\frac{3}{16}$ brown short-haired, and $\frac{1}{16}$ brown long-haired.

6. Half their children will be heterozygous and have elevated cholesterol levels. There is a $\frac{1}{4}$ chance that their next child will be homozygous, *hh*, and have an extremely high cholesterol level, like Zoe.

7. The man's genotype is *ii*. His sister's genotype is *IᴬIᴮ*. Their parents must be *Iᴬi* (type A) and *Iᴮi* (type B).

8. If the genes are not linked, the proportions among the offspring are 25% gray red, 25% gray purple, 25% black red, 25% black purple. The actual percentages show that the genes are linked. Recombination frequency is 6%.

9. The recombination frequencies are: black dumpy 36%, purple dumpy 41%; and black purple, 6% (previous question). Since these recombination frequencies reflect distances between the genes, the sequence must be purple-black-dumpy (or dumpy-black-purple).

10. The bristle-shape alleles are sex-linked, carried on the *X* chromosome. Normal bristles is dominant (*F*), and forked is recessive (*f*). The genotype of the female parent is *XᶠXᶠ*. The genotype of the male parent is *XᶠY*. Their female offspring are *XᶠXᶠ*, their male ffspring *XᶠY*.

Science, Technology, and Society

1. *Some issues and questions to consider:* Why do people keep exotic pets? Why do breeders produce them? Do we need these animals, or is keeping them a whim or fancy? Isn't any domestic cat, pigeon, or goldfish a human-made product? How different does an animal have to be from its "wild" kin before it is a grotesque oddity? Does the responsibility for producing these unusual pets lie with the breeder or the fancier who demands them? Is it cruel to breed a goggle-eyed goldfish? A fainting goat? How does the breeder decide which unusual animals to breed and which not to breed? How "different" does an animal have to be for a breeder to refuse to breed it? Should the decision relate to what is best for the animal or what is best for the breeder? Why? How do you decide what is best for the animal?

2. *Some issues and questions to consider:* If the test results are correct, is it possible for Nathan and Diana to have a child with sickle-cell anemia? Could the test results be wrong? How can you find out? If the results are confirmed, what is the most likely explanation? What do you say to the couple? Tell them it is safe to have another child, explain why, and let them draw their own conclusions? What if they ask about their son? Or should you confront the question of their son's paternity head-on? Could you give them the go-ahead without going into the reasons? Is it ethical to withhold information? Is this a realistic dilemma, or is this kind of situation so rare that you need not worry about it? Why?

3. *Some issues and questions to consider:* Do biologists actually see the structures of molecules and cells? What about past evolutionary processes, the origin of life, the physical appearance or behavior of dinosaurs? In other fields of science, what is the evidence for atoms, subatomic particles, the formation of stars, makeup of the interior of Earth, past positions of the continents? How much of science is based on direct observation and how much on "circumstantial" evidence? How clear does evidence have to be before it is acceptable? What is an educated guess in science? Which comes first, the guess or observations and experiments? What prompts a scientist to propose an explanation? Does an object or process have to be seen to be believed, or does the evidence of observations and experiments merely need to be consistent with a proposed explanation? What if more than one explanation can account for the observations? Is it possible to be absolutely sure that an explanation is correct? That it is incorrect? If an explanation has not been disproved, does that make it correct? How valuable is an explanation that is impossible to test through observation or experiment? What is the place of words like "correct," "incorrect," "fact," and "truth" in science? Are some of the "facts" in this textbook "wrong?" Which ones?

Chapter 10 Answers

Testing Your Knowledge

Multiple Choice: **1.** e (Only the phage DNA enters a host cell; T4 DNA determines both DNA and protein.) **2.** e **3.** b **4.** c

Matching: **1.** h **2.** d **3.** f **4.** e **5.** a **6.** b **7.** c **8.** g

Describing, Comparing, and Explaining

1.

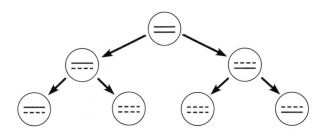

Hydrogen bonds

Sugar (deoxyribose)

Phosphate group

A nucleotide pair

Complementary base pair

2. Ingredients: Original DNA, nucleotides, several enzymes and other proteins, especially DNA polymerases. Steps: Original DNA strands separate, nucleotides line up along each strand according to base-pairing rules, enzymes link nucleotides to form new strands. Product: Two identical DNA molecules, each with one old strand and one new strand.

3. A gene is the polynucleotide sequence with information for making one polypeptide. Each codon—a triplet of bases in DNA or

RNA—codes for one amino acid. Transcription occurs when RNA polymerase produces mRNA from one strand of DNA. A ribosome is the site of translation, or polypeptide synthesis. Each tRNA molecule has an amino acid attached at one end, and a three-base anticodon at the other end. Beginning at the start codon, mRNA moves relative to the ribosome a codon at a time. A tRNA with a complementary anticodon pairs with each codon, adding its amino acid to the polypeptide chain. The amino acids are linked by peptide bonds. Translation stops at a stop codon, and the finished polypeptide is released.

4. See Figure 10.17C. Glycoprotein spikes on the virus attach to receptor proteins on the host cell, enabling the virus envelope to fuse with the cell. Once inside, the viral RNA codes for viral proteins and more viral RNA, both of which make up the new viruses. The viruses leave the cell by surrounding themselves with host membrane. The viruses may damage the cell by interfering with cell metabolism.

Thinking Critically

1.

2. Bacteria reproduce by binary fission, so numbers double and redouble: 1—2—4—8—16—etc. When a virus enters a host cell, some time may pass while viral nucleic acid is replicated, proteins are made, and viruses assemble, but then hundreds of viruses may burst from the cell all at once. They infect other cells and the process is repeated, causing "bursts" of viruses to appear.

3. mRNA: GAUGCGAUCCGCUAACUGA. Amino acids: Met-Arg-Ser-Ala-Asn.

Science, Technology, and Society

Some issues and questions to consider: Is it fair to issue a patent for a gene or gene product that occurs naturally in every human being? Or should a patent only be issued for something new that is invented, rather than found? Suppose another scientist slightly modifies the gene or protein. How different does the gene or protein have to be in order that the patent is not infringed? Might patents encourage secrecy and interfere with the free flow of scientific information? What are the benefits to the holder of a patent? If researchers were not allowed to patent their discoveries, what would be the incentives for doing their research?

Chapter 11 Answers

Testing Your Knowledge

Multiple Choice: **1.** b (Different genes are active in different kinds of cells.) **2.** d **3.** c **4.** b **5.** a **6.** b

Describing, Comparing, and Explaining

1. The nucleus of a differentiated tadpole intestine cell can shape the development of an entire embryo. Salamander cells can dedifferentiate

and regenerate a lost leg. A carrot plant can grow from a single root cell.

2. Homeotic genes are "master genes" that produce homeotic proteins, which in turn may activate or repress the activities of many other genes.

3. An operon is a cluster of genes with related functions, such as producing enzymes that manufacture or break down a particular substance. Some operons are turned on by enzyme substrates, so enzymes are made only when the substrate is present. Other operons are turned off by products, so enzymes for making a product are made only when that product is not present in the environment. Thus, operons enable a bacterium to make only the proteins needed at a particular moment.

4. Apparently, all cancer-causing agents work in essentially the same ways—they somehow turn on oncogenes and/or turn off tumor-suppressor genes. The result is uncontrolled cell division.

Thinking Critically

a. The repressor will not detach from DNA; proteins are not made with or without lactose.

b. The repressor does not block RNA polymerase; proteins are made with or without lactose.

c. The repressor does not block RNA polymerase; proteins are made with or without lactose.

d. RNA polymerase will not transcribe the genes; no proteins are made with or without lactose.

Science, Technology, and Society

Some issues and questions to consider: Is it the responsibility of the non-smoker to protect himself or herself by staying away from tobacco smoke? Or is it the responsibility of the smoker to keep from exposing nonsmokers? Is it the responsibility of employers to protect their employees, or restaurant owners to protect their customers? How might such protection be accomplished? What kinds of rules and laws are likely to be enacted in the wake of the EPA report, and how will they be enforced? Which works better—a rule or an incentive? Could a smoker be punished for child abuse for smoking at home?

Chapter 12 Answers

Testing Your Knowledge

Multiple Choice: **1.** d **2.** e **3.** c **4.** b **5.** c (Bacteria lack the RNA-splicing machinery needed to delete eukaryotic introns.) **6.** b

Describing, Comparing, and Explaining

1. Conjugation: A bacterium replicates part of its DNA and passes it to a recipient cell. Transformation: A bacterium takes up DNA from the surrounding fluid. Transduction: A phage transfers genes from one bacterium to another.

2. Extract plasmids from *E. coli*. Cut plasmids and human DNA with restriction enzyme to produce sticky ends. Join the plasmids and genes with ligase. Allow *E. coli* to take up recombinant plasmids. Bacteria will then replicate plasmids and multiply, producing a clone of *E. coli* with growth hormone genes. Bacteria make growth hormone, which is collected and purified.

3. Medicine: Genes can be used in human gene therapy, to engineer viruses for vaccines, or to produce transgenic lab animals for AIDS research. Proteins can be hormones, enzymes, blood-clotting factor, etc., or can be used in vaccines. Agriculture: Copies of genes can be inserted into plant cells or animal eggs to produce transgenic crop plants or farm animals, or inserted into viruses to make vaccines. Proteins can be used as animal growth hormones.

4. Recombinant organisms might harm or displace other species. They might transfer genes to other species and make them harmful.

Thinking Critically

1. She could start with DNA isolated from liver cells (the entire liver-cell genome) and carry out the procedure outlined in Module 12.6 to produce a collection of recombinant bacterial clones, each carrying a small piece of liver-cell DNA. To find the clone with the desired gene, she could then make a probe of radioactive RNA with a nucleotide sequence complementary to part of the gene: GACCUGACUGU. This probe would bind to the gene, labeling it and identifying the clone that carries it. Alternatively, the biochemist could start with mRNA isolated from liver cells and use it as a template to make DNA (using reverse transcriptase). Cloning this DNA rather than the entire genome would yield a smaller library of genes to be screened—only those active in liver cells. Furthermore, the genes would lack introns, making the desired gene easier to manipulate after isolation.

2. The biggest danger is unknowingly producing a clone of bacteria with an unknown but harmful gene. This danger could be reduced by taking precautions to contain the bacteria and using "crippled" bacteria that can grow only under specialized laboratory conditions, not outside the lab.

3. If the bacteria are grown in a medium containing the antibiotic, only those bacteria that have actually taken up plasmids will survive. This solves the problem of sorting through bacteria that do not contain recombinant DNA to find the few that do.

Science, Technology, and Society

Some issues and questions to consider: What are some of the unknowns in recombinant DNA experiments? Do we know enough to anticipate and deal with possible unforeseen and negative consequences? Do we want this kind of power over evolution? Who should make these decisions? If scientists doing the research were to make the decisions about guidelines, what factors might shape their judgment? What might shape the judgment of business executives in the decision-making process? Does the public have any right to have a say in the direction of scientific research? Does the public know enough about biology to get involved in this decision-making process? Who represents "the public," anyway?

Chapter 13 Answers

Testing Your Knowledge

Multiple Choice: **1.** e **2.** c (This concentrates certain alleles and also makes it easy to get useful family histories.) **3.** b **4.** d **5.** d

Describing, Comparing, and Explaining

1. A dominant allele cannot be carried in a masked form by a heterozygote; every individual with a dominant allele has the disorder. If the disorder is serious enough, it will reduce the survival and reproduction rates of individuals who have it, so it will not be passed on and become widespread. The effects of Huntington's disease do not become apparent until the affected individual is about 40 years old, after he or she may have already had children.

2. See Figures 13.4A and B.

3. Because individuals who have recent common ancestors are likely to carry the same recessive alleles, a mating between them is likely to produce offspring that are homozygous for a harmful recessive trait.

4. To prepare a DNA fingerprint, a sample of DNA is cut up by restriction enzymes. These enzymes cut a specific base sequence, so DNA from different people yields different mixtures of fragments.

The restriction fragments are submitted to electrophoresis, which pulls the fragments through a gel and separates them into bands by size and charge. The pattern of restriction fragment bands differs in DNA fingerprints from different people.

5. To carry out gene therapy, cloned genes can be inserted into cells by engineered retroviruses. The cells are then returned to the patient. One problem is getting the transferred gene to turn on properly and stay turned on. Also, gene therapy cannot be applied to cells that do not multiply, such as nerve cells.

Thinking Critically

1. To control for the age of the father, compare women of various ages who are all married to men of the same age. To see if the age of the father increases the occurrence of Down syndrome, look at men of various ages whose wives are all the same age.

2. One chromosome would suffer a deletion and the other a duplication.

3. Determining the nucleotide sequences is just the first step. Once researchers have written out the DNA "book," they will have to try to figure out what it means—what the nucleotide sequences code for, and how they work.

Genetics Problems

1. $\frac{1}{4}$ (Both parents are carriers of an autosomal recessive allele.)

2. $\frac{9}{16}$ (Probability of normal pigmentation $= \frac{3}{4}$. Probability of being normal for galactosemia trait $= \frac{3}{4}$. Combined probability $= \frac{3}{4} \times \frac{3}{4} = \frac{9}{16}$.)

3. $\frac{1}{2}$ (Marie is normal, so Tim must be heterozygous. Martha is homozygous recessive. Tim will pass the dominant allele to half their children.)

4. Alkaptonuria appears to be caused by a recessive allele, because Ann and Michael are normal but their daughter Carla has the disorder. It is autosomal; if it were X-linked, a father (George or Michael) would have to show the trait in order to pass it on to his daughter. Known genotypes: Arlene, Tom, Wilma, and Carla are homozygous recessive. George, Sam, Ann, Michael, Daniel, and Alan are heterozygous. Sandra, Tina, and Christopher are unknown.

5. $\frac{1}{6}$ (Probability that they will have a CF child $=$ probability that Charles is a carrier $\times$ probability that Elaine is a carrier $\times$ probability that two carriers will have a CF child. Probability that Charles is a carrier $= 1$. Probability that Elaine is a carrier $= \frac{2}{3}$, because her parents had a CF child and must be heterozygous. Probability that two carriers will have a CF child $= \frac{1}{4}$. Combined probability $= 1 \times \frac{2}{3} \times \frac{1}{4} = \frac{1}{6}$.)

6. $\frac{1}{2}$ (Jon is homozygous recessive. If their child is deaf, Christine must be heterozygous.)

7. $\frac{1}{4}$ will be boys suffering from hemophilia. $\frac{1}{4}$ will be female carriers. (The mother is a heterozygous carrier, and the father is normal.)

8. $\frac{1}{2}$ (Jeanne is a heterozygous carrier, and Pierre is normal. Half their sons will have Duchenne muscular dystrophy.)

9. In order for a woman to be color-blind, she must inherit X chromosomes bearing the color-blindness allele from both parents. Her father only has one X chromosome, which he passes on to all his daughters, so he must be color-blind. A male only needs to inherit the color-blindness allele from a carrier mother; both his parents are usually phenotypically normal.

10. It must be caused by a recessive gene because both parents of afflicted boys are normal. It must be sex-linked because it is only seen in boys, who inherit the allele from heterozygous mothers. A girl would have to inherit the allele from both parents to be af-

flicted. A male would himself have to be afflicted to pass on the allele to his daughter. Since afflicted males die young, they do not father children.

Science, Technology, and Society

1. $\frac{1}{2}$ Some issues and questions to consider: Why is being tested for the allele a difficult decision? How would you feel if the test turned out positive? How might a positive result change your relationship to your family, friends, job, and activities?

2. Some issues and questions to consider: What kinds of impacts will gene therapy have on the individuals who are treated? On society? Who will decide what patients and diseases will be treated? What costs will be involved, and who will pay them? How do we draw the line between treating disorders and "improving" the human species?

3. Some issues and questions to consider: Should genetic testing be mandatory or voluntary? Under what circumstances? Why might employers and insurance companies be interested in genetic data? Since genetic characteristics differ among racial groups and between the sexes, might it be used to discriminate? Which of these questions do you think is most important? Which issues are likely to be the most serious in the future?

Chapter 14 Answers

Testing Your Knowledge

Multiple Choice: 1. e 2. c 3. b (Erratic rainfall and differential reproductive success would ensure that a mixture of both forms remained in the population.) 4. b 5. d

Describing, Comparing, and Explaining

1. Your paragraph should mention such evidence as biogeography, fossils, comparative anatomy, comparative embryology, DNA and protein comparisons, examples of .natural selection (peppered moth), and artificial selection.

2. a. There is gene flow between duck populations, which will change the proportions of various alleles in the two gene pools.
   b. This bottleneck may alter the proportions of alleles in the gene pool by chance (genetic drift).
   c. Mutation adds a black gene to the gene pool.
   d. Nearsighted hawks will have less reproductive success than sharp-eyed hawks, and their genes will be less numerous in the next generation.
   e. Mating is nonrandom; males with bright red eyes will have more offspring than less attractive males, and the frequency of the red-eye gene will increase.

3. If $q^2 = 0.002$, then $q = 0.04$ (approximately). Since $p + q = 1$, $p = 1 - q = 0.96$. The proportion of heterozygotes is $2pq = 2 \times 0.96 \times 0.04 = 0.08$. About 8% of African-Americans are carriers.

Thinking Critically

1. The first spraying killed the least resistant mites. Mites with genes that made them resistant to spray were better able to survive and reproduce. Many of their offspring inherited this resistance, and were able to resist the second spraying. The mite population is becoming adapted to the spray—an example of natural selection.

2. The unstriped snails appear to be better adapted. Striped snails made up 47% of the population, but 56% of the meals for the birds. The proportion of unstriped snails will probably increase.

3. Researchers could compare structures of DNA and proteins from the two kinds of bats, lemurs, and perhaps insectivores. If there are fewer differences in nucleotide or amino acid sequence

between megabats and lemurs than between the two kinds of bats, this would suggest that megabats are descended from primates.

Science, Technology, and Society

Some issues and questions to consider: Who should decide curriculum, "experts" or members of the community? Are the two alternatives both scientific ideas? Who judges what is scientific? If it is more fair to consider alternatives, should the door be open to all alternatives? Are constitutional issues (separation of church and state) involved here? Can a teacher be compelled to teach an idea he or she thinks is wrong? Should a student be required to learn an idea he or she thinks is wrong?

Chapter 15 Answers

Testing Your Knowledge

Multiple Choice: **1.** c **2.** b **3.** d **4.** c (According to this model, species appear rapidly and remain unchanged for long periods.) **5.** a

Describing, Comparing, and Explaining

1. Different physical appearance may indicate different species or just differences within a species. Isolated populations may or may not be able to interbreed. There is no way to determine whether fossil or asexual organisms can interbreed.
2. Horses and donkeys are not the same species because the mule is sterile. This is an example of hybrid sterility, a postzygotic reproductive barrier.
3. Organisms that reach an island are geographically isolated from populations on the mainland and other islands. Genetic drift might occur in small island populations, unique mutations might occur, and island populations undergo natural selection and adapt to island conditions. Repeated speciation may occur in island groups; Darwin's finches of the Galapagos Islands are a good example. The Death Valley pupfishes have been similarly isolated in "islands"—small springs in the desert.

Thinking Critically

1. Pollen from an Old World cotton plant (13 large chromosomes) combines with an egg from American wild cotton (13 small chromosomes). Their hybrid offspring has 13 large and 13 small chromosomes. In the hybrid, nondisjunction occurs during meiosis, producing gametes with 13 large chromosomes and 13 small chromosomes. Self-fertilization produces a polyploid plant with 52 chromosomes—13 pairs of large ones and 13 pairs of small ones.
2. During the ice ages, the mountains were covered by ice and snow. A single ancestral marmot population lived in the cold lowlands. As the climate warmed, marmots moved to higher elevations. The Olympic population became geographically isolated from the Cascade population. Different mutations, selection, and genetic drift occurred in each population; eventually, they became reproductively isolated, distinct species. This is an example of allopatric speciation.

Science, Technology, and Society

Some issues and questions to consider: Does "distinct" mean reproductive isolation from other populations? Does it mean a unique gene pool with unique adaptations? What kinds of information would demonstrate unique genes or adaptations? What proportion of the species does (or did) the Redfish Lake population represent? Is it more important to preserve the population if the fish represent the "mainstream" type of sockeye, or if the population is relatively unique? Is the Redfish Lake population a large portion of the total

species or a relatively small portion? Is there value in preserving diverse samples of various sockeye types? How far are we willing to go to preserve distinct populations?

Chapter 16 Answers

Testing Your Knowledge

Multiple Choice: **1.** c **2.** d **3.** e **4.** b **5.** b (Diana and Syke's are on quite separate branches, so they are only distantly related.)
True/False: **1.** F (They are in a separate kingdom because they are prokaryotes.) **2.** F (A rapidly evolving protein would be more useful than a slowly evolving one.) **3.** T **4.** T **5.** F (Biologists disagree on this question.)

Describing, Comparing, and Explaining

1. Precambrian: The appearance of diverse animals. Paleozoic: Radiation of reptiles and insects, extinction of many marine invertebrates. Mesozoic: Mass extinction of dinosaurs and many other organisms.
2. The number of differences in amino acid sequence reflects evolutionary relationship. More similarities means that organisms had a more recent common ancestor and are more closely related.
3. An asteroid or comet may have hit Earth, raising dust, blocking sunlight, cooling the climate, and killing plants. Evidence includes a layer of iridium in rocks and a large crater. Fossils suggest that increased volcanic activity in India may have released dust and cooled the climate. Continental drift may have altered climate and shorelines at the end of the Paleozoic. There is evidence for cooling and change in sea levels.

Thinking Critically

1. This would either separate crocodiles from other reptiles or put birds in the same taxon as reptiles. Both of these changes differ from the way classical taxonomy classifies these animals.
2. The genes affected might be regulatory genes that direct structural genes for legs and wings during development. Such mutations might be responsible for major changes in body structure that might have significant effects on evolution.
3. The rock is about 2.6 billion years old. (If the half-life of potassium-40 is 1.3 billion years, there would be 6 grams left after 1.3 billion years and 3 grams left after another 1.3 billion years.)

Science, Technology, and Society

Some issues and questions to consider: How is the cause of this mass extinction different from the possible causes of previous mass extinctions? How might the rate of this mass extinction compare to the others? Which kinds of organisms are most likely to perish? What kinds are most likely to survive? Just because life bounces back, does that mean we will? Do we have any ethical responsibility to preserve other species?

Chapter 17 Answers

Testing Your Knowledge

Multiple Choice: **1.** b **2.** c **3.** a (Algae are autotrophs; protozoans are heterotrophs.) **4.** e **5.** c **6.** c
True/False: **1.** F (Most archaebacteria live in these environments.) **2.** F (protozoans) **3.** T **4.** F (A bacillus is rod-shaped) **5.** F (Most bacteria are not pathogenic.) **6.** T

Describing, Comparing, and Explaining

1. The early Earth was warmer than it is now, but eventually cool enough for rain to fall and seas to form. The atmosphere contained

water vapor, carbon monoxide, carbon dioxide, nitrogen, methane, and ammonia. There was volcanic heat, lightning, and UV light.

2. Small, free-living prokaryotes were probably engulfed by a larger cell and took up residence inside. A symbiotic relationship developed between the host cell and engulfed cells. DNA, RNA, ribosomes, and inner membranes of mitochondria and chloroplasts are similar to those of bacteria. These organelles make some of their own proteins, replicate their own DNA, and reproduce by a fissionlike process.

3. *Chlamydomonas* is a eukaryotic cell, much more complex than a prokaryotic bacterium. It is autotrophic, while protozoans are heterotrophic. It is unicellular, unlike multicellular sea lettuce.

4. RNA may have acted as a crude template for polypeptide formation. Polypeptides may have aided in RNA replication.

Thinking Critically

1. You could follow Koch's postulates to determine whether the suspect bacteria cause the disease: (1) Confirm the presence of the bacteria in all cats with the disease. (2) Isolate the bacteria, and grow them in pure culture. (3) Inoculate healthy cats with bacteria from the pure culture and see if they get the disease. (4) See if the bacteria have multiplied in the inoculated cats.

2. Bacteria are necessary for life; for example, they are important in waste decomposition and recycling and nutrient cycles.

3. After 2 hours, 16 bacteria would be present. After 12 hours, the number would grow to 16,777,216 bacteria. This kind of growth could continue only until the bacteria run out of food or are inhibited by their own wastes.

4. Conditions on Earth today are quite different from those on the primitive Earth. The ancient Earth may have had a reducing atmosphere containing chemicals different from today's oxidizing atmosphere, which contains abundant oxygen. The ozone layer, formed from oxygen, now shields Earth's surface from UV radiation. If organic chemicals were somehow formed today, the organisms already here would probably consume them.

Science, Technology, and Society

1. *Some issues and questions to consider:* How could we determine beforehand whether the iron treatment would really have the desired effect? Would it just need to be done once, or would it be an ongoing process? Is it a "cure" for the problem, or does it simply treat the "symptoms"? Could the iron treatment have side effects? Do we have the right to experiment on this scale?

Chapter 18 Answers

Testing Your Knowledge

Multiple Choice: **1.** b **2.** b **3.** c (It is the only gametophyte among the possible answers; all the others make spores.) **4.** a **5.** e **6.** b

True/False: **1.** F (About one-third of all fungi are harmful.) **2.** F (Lichens are rugged and can survive with little soil.) **3.** T **4.** T **5.** T **6.** F (Fungi digest food and then absorb nutrients.) **7.** T

Describing, Comparing, and Explaining

1. In both cases, the fungus and photosynthetic organism participate in mutualism. The fungus aids in the absorption of water and nutrients, and the photosynthetic organism provides the fungus with food in the form of organic molecules.

2. The alga is surrounded and supported by water, and has no supporting tissues, vascular system, or special adaptations for obtaining or conserving water. Its whole body is photosynthetic, and its

gametes and embryos are dispersed into the water. The seed plant has tissues that support it against gravity; vascular tissues that carry food and water, and special organs (roots, stems, leaves) that absorb, support, and photosynthesize. It is covered by a waterproof cuticle and has stomata for gas exchange. Its gametes are protected by waterproof jackets, sperm are carried by pollen grains, and embryos are protected and provided for by seeds.

3. Animals carry pollen from flower to flower and thus help fertilize eggs. They also disperse seeds by consuming fruit or carrying fruit that clings to their fur. In return, they get food (nectar, pollen, fruit).

4. Plants are autotrophs. They have chlorophyll and make their own food via photosynthesis. Fungi are heterotrophs that digest food externally and absorb nutrient molecules. There are also many structural differences: The threadlike fungal mycelium is different from the plant body, their cell walls are made of different substances, etc.

Thinking Critically

1. The life cycles of mosses and ferns include a stage when millions of tiny spores are released. These spores can be carried great distances by the wind and germinate wherever and whenever they encounter the proper conditions. Seeds are not as easily and widely dispersed.

2. The seeds and spores of plants can often remain dormant for a long time, enabling the plant to endure environmental change and germinate when the crisis is over.

3. Antibiotics probably kill off bacteria that compete with fungi for food. Similarly, bad tastes and odors deter animal competitors. They are valuable warnings to animals that might eat spoiled food. Those fungi that produce antibiotics and bad-smelling and bad-tasting chemicals would survive and reproduce better than fungi unable to inhibit competitors. Animals that could recognize the smells and tastes also would survive and reproduce better than their competitors. Thus, natural selection would favor fungi that produce the chemicals and, to some extent, the competitors deterred by them.

Science, Technology, and Society

1. *Some issues and questions to consider:* What are the other functions of forest land? How are other uses affected by logging? Must all the trees in an area be clear-cut? Should a particular area have multiple uses, or should different areas be used for different purposes? How much timber do we need? Could we conserve and recycle more? Are government-managed forests subsidizing private industry? Should we protect habitats, as well as species? Aren't trees a renewable resource? Does the rate of regrowth match the rate of harvest? Will the ancient forests grow back? Are jobs and the economy at least as important as the owl?

2. *Some issues and questions to consider:* Why might a tree be more vulnerable to foreign fungi than to fungi in its natural habitat? How have plants and their parasites evolved? How do modern transportation and commerce contribute to the spread of non-native fungal parasites? How might open borders (such as the European community) and free-trade zones affect their spread? How might we safeguard plants from these parasites?

Chapter 19 Answers

Testing Your Knowledge

Multiple Choice: **1.** d **2.** b **3.** c **4.** a **5.** c **6.** a

True/False: **1.** F (Cnidarians use stinging tentacles.) **2.** T **3.** T **4.** T **5.** F (The water vascular system enables the urchin to move.) **6.** F (Mammals are grouped according to mode of reproduction: eggs,

pouch, or placenta.) **7.** T **8.** F (There is little such fossil evidence.)
Matching: **1.** i **2.** f **3.** e **4.** c **5.** a **6.** d **7.** h **8.** b **9.** g

Describing, Comparing, and Explaining

1. Birds share a number of reptile characteristics: amniotic eggs, scales on their legs, beaks and toenails with keratin, and general body form. Their adaptations for flight include feathers, wings, a short tail, bones with air sacs, breast muscles anchored to a keel-like breastbone, a high rate of metabolism, endothermic metabolism, and an efficient circulatory system and lungs.

2. The gastrovascular cavity of a flatworm is an incomplete digestive tract; the worm takes in food and expels waste through the same opening. An earthworm has a complete digestive tract; food travels one-way, and different areas are specialized for different jobs. The flatworm's body is solid and unsegmented. The earthworm has a coelom, allowing its outside and internal organs to grow and move independently. Fluid in the coelom acts as a skeleton. Segmentation of the earthworm allows for greater flexibility and mobility.

3. Cnidarians and most echinoderms are radially symmetrical, while most other animals, such as arthropods and chordates, are bilaterally symmetrical. Most radially symmetrical animals stay in one spot or float passively. Most bilateral animals are more active and move head-first through their environment.

4. For example, the legs of horseshoe crab are used for walking, while the antennae of a grasshopper have a sensory function. Some appendages on the abdomen of a lobster are used for swimming, while the scorpion catches prey with its pincers. (Note that the scorpion stinger and insect wings are not considered jointed appendages.)

Thinking Critically

1. Wet conditions probably increased the survival and reproduction of snails, critical in the fluke life cycle. Possible methods of control include: (a) Better sanitation so human feces with eggs do not enter water, (b) draining fields to kill snails, (c) poisoning snails, and (d) wearing boots to prevent infection when working in the fields.

2. Important characteristics include symmetry, presence and type of body cavity, segmentation, type of digestive tract, type of skeleton, and appendages.

3. Agnathans are the trunk, continuing to the present time. Cartilaginous and body fish are offshoots of the agnathans. Bony fishes gave rise to amphibians, and amphibians gave rise to reptiles. Birds and mammals are separate branches from the reptile branch of the vertebrate tree.

Science, Technology, and Society

1. *Some issues and questions to consider:* How does the decline of the reefs relate to agriculture? Deforestation? Overfishing? Rapid population growth? What value are the reefs to the local people? What is their value as a world biological resource? What might be the consequences if the reefs disappear? What is likely to make the situation worse? What is likely to improve the situation? In what ways might developed countries contribute to this problem? In what ways might developed countries be able to help the local people preserve the reefs? How might developed countries benefit from helping?

2. *Some issues and questions to consider:* Is a refuge a place where animals should be safe from human activities, or should people be allowed to use the refuge's resources? Are there other locations of refuges for these human activities? How do wildlife species benefit from refuges? How do hunters, farmers, trappers, ranchers, etc. benefit from refuges? Of what value to humans are wild animals that are left untouched in wildlife refuges? Is value to humans important, or are the animals valuable for their own sake? What species might be helped by hunting, logging, trapping, or farming? Do wild species need this kind of help?

3. *Some issues and questions to consider:* What criteria should be considered when deciding whether or not we should try to save a species? Value to humans? Ecological role? Nearness to extinction? Likelihood of survival on its own? Can you name species other than the condor or ferret that might be more critical to the health of the ecosystems of which they are a part? Can you name other species that might be easier to save? If we did not expend the money and effort on the condor and ferret, would we expend it saving other creatures? Should we try harder to save a few species that are nearly extinct, or preserve the habitats of many other creatures farther from extinction? Isn't extinction a natural process? Or do we have a special responsibility to intervene in certain cases?

Chapter 20 Answers

Testing Your Knowledge

Multiple Choice: **1.** b **2.** c **3.** c (Expelling salt opposes the increases, thereby maintaining a constant internal environment.)

Matching: **1.** D **2.** C **3.** A **4.** D **5.** A **6.** C **7.** B **8.** D **9.** B

Describing, Comparing, and Explaining

1. Muscle cells cooperate to form smooth muscle tissue, which moves food through the stomach. Muscle, connective, epithelial, and nervous tissue form the stomach, an organ that digests food. The stomach, along with other organs such as the esophagus, intestines, and liver, form the digestive system, which ingests and breaks down food. The human being, an organism, consists of about a dozen systems, each with specialized functions, working together.

2. Fibrous connective tissue cells are embedded in a strong matrix of collagen fibers, which join bones together. Stratified squamous epithelium consists of many cell layers, which protect the body. Neurons are cells with long branches that conduct signals to other cells. Bone cells are surrounded by a matrix containing fibers and calcium salts, forming a hard protective covering around the brain.

3. The surfaces of the intestine, excretory system, and lungs are highly folded and divided, with many blood vessels, increasing their surface area for exchange. Smaller creatures have a greater surface-to-volume ratio, and their cells are closer to the surface, enabling direct exchange between cells and the outside environment.

Thinking Critically

Glucagon causes an increase in blood sugar, pushing blood sugar back toward the set point. When blood sugar rises, glucagon production drops.

Science, Technology, and Society

Some issues and questions to consider: How do the health-care systems of the two countries compare? How do the populations of the two countries and the dispersion of populations compare? Do patients have to wait in line for CT or MRI scans? Do they have to travel long distances? Are CT and MRI scans used mainly for emergencies, or could most patients wait for scans? Are there elements of competition and prestige in having a scanner? How do CT and MRI scans affect the cost of medical care? Are hospitals forced to charge more for other services to pay for scanners? Is there pressure to use the scanners in order to justify their expense?

Chapter 21 Answers

Testing Your Knowledge

Multiple Choice: **1.** a **2.** a **3.** e **4.** d (They generally serve as coenzymes with catalytic functions.)

True/False: **1.** F (Villi absorb nutrients.) **2.** F (Most digestion occurs in the small intestine.) **3.** T **4.** F (A balanced diet with exercise is best) **5.** F (HDLs are desirable.) **6.** T **7.** F (Most nutrients are absorbed in the small intestine.) **8.** F (Bile aids in fat digestion.)

Describing, Comparing, and Explaining

1. You ingest the sandwich one bite at a time. In the oral cavity, chewing begins mechanical digestion, and salivary enzyme action on starch begins chemical digestion. When you swallow, food passes through the pharynx and esophagus to the stomach. Mechanical and chemical digestion continue in the stomach, where pepsin and HCl in gastric juice begin protein digestion. In the small intestine, enzymes from the pancreas and intestinal wall break down starch, protein, and fat to monomers. Bile from the liver and gallbladder emulsifies fat droplets for attack by enzymes. Most nutrients are absorbed into the bloodstream through the walls of the small intestine. In the large intestine, water is absorbed from undigested material, and feces are produced and eliminated.

2. When you eat, nerve impulses from the brain and the hormone gastrin from the stomach lining signal gastric glands in the stomach wall to secrete gastric juice. As the acid level rises in the stomach, secretion of gastrin slows, and glands stop their secretion of gastric juice.

3. Some herbivores have cellulose-digesting microbes in their cecum and colon. Some reingest their feces to extract more nutrients. Ruminants have four-chambered stomachs, and rechew food (cud) as part of a multistage digestive process.

Thinking Critically

1. A gastrovascular cavity is a sac with one opening. Food must be taken in and wastes expelled through the same opening. In an alimentary canal, food moves in a single direction and is processed sequentially, without backing up. An animal with an alimentary canal can eat continuously, and each part of the canal carries out specific digestive steps as food passes through on a one-way trip.

2. To prove a mineral is essential, you could feed experimental subjects a diet without the mineral and look for ill effects. One problem with this is that it might harm the subjects. Also, it is very difficult to remove all traces of a mineral from the diet. Slight contamination or cheating by subjects might supply sufficient amounts. An additional problem is that animal subjects might not have the same needs as humans.

3. A slice of pizza contains 450 kcal. Each liter of oxygen consumed liberates 4.83 kcal. It would take 450/4.83 = 93 liters of oxygen to liberate the energy in the pizza.

Science, Technology, and Society

1. *Some issues and questions to consider:* What are the roles of family, school, advertising, media, and peers with regard to an individual's self-image and perception of weight? Are people, especially young people, getting useful, realistic information about diet, self-image, and health? Why or why not? How might the available information be improved? Will better information be enough? How and why do information and environment affect males and females differently?

2. *Some issues and questions to consider:* Where did you hear these claims? What would you consider a reliable source? What kinds of studies and data backed up the claims? What kinds of studies and data would you find convincing? What does the number of times you hear a diet claim repeated—in the media, from friends, in advertisements—have to do with its validity? What is the motivation behind diet claims? Did individuals making the claims stand to gain in some way from public attention? What would constitute a conflict of interest for an individual making a diet claim? What would you consider to be an unbiased source of information?

Chapter 22 Answers

Testing Your Knowledge

Multiple Choice: **1.** c **2.** b **3.** d **4.** a **5.** b (Oxygen diffuses from the air to blood to the cells because of the gradient of decreasing O_2 concentration.)

Describing, Comparing, and Explaining

1. Advantages of breathing air: It can hold a higher concentration of O_2 than water and is easier to move over the respiratory surface. Disadvantage of breathing air: It dries out the respiratory surface, impeding diffusion.

2. Nasal cavity, pharynx, larynx, trachea, bronchus, bronchiole, alveolus, through wall of alveolus into blood vessel, blood plasma, into red blood cell, attaches to hemoglobin, carried by blood through heart, blood vessel in muscle, dropped off by hemoglobin, out of red blood cell, through blood vessel wall, through plasma membrane into muscle cell.

3. Alveoli are very numerous, maximizing the total area of the respiratory surface. Their moist surface and thin walls (one layer of epithelial cells) allow diffusion of gases. The surfaces of alveoli are covered by networks of capillaries, maximizing the surface for exchange with blood.

4. Air sacs and one-way tubes instead of alveoli promote the one-way flow of air through bird lungs. Fresh air constantly bathes the respiratory surface, allowing birds to extract more oxygen. The in-and-out air flow of human lungs leaves residual air, reducing the extraction of oxygen.

Thinking Critically

1. The higher rate of muscle cell metabolism during exercise causes temperature and CO_2 production of muscle to increase. When blood passes through muscle, the change causes hemoglobin to drop off extra O_2—just what is needed for exercise.

2. The increase in acidity—drop in pH—stimulates the breathing centers to increase the rate of breathing. This causes more CO_2 to be expelled from the lungs, decreasing the blood concentration of carbonic acid and raising pH.

3. **a.** Oxygen will diffuse only through a moist surface.
   b. Red blood cells carry oxygen to tissues.
   c. Oxygen will not diffuse as easily from the air into the blood.
   d. Oil blocks the openings leading to insect tracheae, suffocating insects.
   e. Hemoglobin normally carries oxygen to cells.
   f. The drop in O_2 concentration reduces the O_2 concentration gradient between water and fish blood, reducing the amount of O_2 absorbed through the gills.

Science, Technology, and Society

1 *Some issues and questions to consider:* Does expertise in the area necessarily bias a researcher? Would it be more fair to select scientists who know less about the subject? Should objections to Dr. Burns's

methods, or to his conclusions be more important as selection criteria for the panel? Who selects the panel? Might they be biased? Does the source of research funds make an expert's opinions suspect? What are criteria for awarding funds? Are there strings attached?

2. *Some ideas and questions to consider.* How obvious is it to most people that asbestos causes cancer? That environmental tobacco smoke causes cancer? Which is easier to remove from the environment? What interest groups might conflict over the importance of removing asbestos? Over eliminating environmental tobacco smoke? Who is perceived to be responsible for the asbestos danger? For removing asbestos? For the danger of environmental tobacco smoke? For eliminating environmental tobacco smoke?

Chapter 23 Answers

Testing Your Knowledge

Multiple Choice: **1.** c **2.** b **3.** d (The second sound is the closing of the semilunar valves as the ventricles empty.) **4.** c **5.** c **6.** a

True/False: **1.** T **2.** F (Veins carry blood toward the heart.) **3.** T **4.** F (White blood cells or leukocytes defend the body.) **5.** T **6.** T **7.** F (The SA node or pacemaker regulates heartbeat.)

Describing, Comparing, and Explaining

1. Red blood cells are numerous, medium-sized, biconcave disks that carry oxygen. White blood cells are less numerous, larger, more irregular cells with nuclei. They fight infections and cancer. Platelets are tiny, irregular cell pieces involved in blood clotting.

2. The pulse is a rhythmic stretching of arteries caused by the pressure of blood pushed through the arteries by ventricular contractions. Resistance of the arterioles reduces pressure and smooths out pressure peaks. Squeezing of skeletal muscles on veins, reduced pressure in the chest due to breathing, and one-way valves in veins keep blood moving toward the heart.

3. Pulmonary vein, left atrium, left ventricle, aorta, artery, arteriole, capillary bed (in finger, for example), venule, vein, vena cava, right atrium, right ventricle, pulmonary artery, capillary bed in lung, pulmonary vein.

4. Capillaries are very numerous, producing a large surface area close to body cells. The capillary wall is only one epithelial cell thick. Clefts between epithelial cells allows materials to leak in and out.

Thinking Critically

1. Cardiac output = amount of blood pumped by the left ventricle with each beat × the number of beats per minute. In this example, cardiac output = 100 mL × 150 beats/min = 15000 mL/min or 15 L/min.

2. Proteins are important solutes in blood, accounting for much of the osmotic pressure that draws fluid into the blood. If protein concentration is reduced, the inward pull of osmotic pressure will fail to balance the outward push of blood pressure. The net pressure at the arterial end of a capillary forces fluid out into the interstitial fluid. Net pressure at the venous end is not great enough to draw fluid back in.

Science, Technology, and Society

1. *Some issues and questions to consider:* What are some factors known to contribute to cardiovascular disease? Which of these might differ between the Mediterranean and North America? How might lifestyle, genes, and environmental influences affect these factors? Can you isolate one influence or factor in order to test its effects alone? How? Can you test these factors in the laboratory? How? Can you apply laboratory findings to humans with confidence?

Why? How might North Americans have to change their lives to benefit from your findings?

2. *Some issues and questions to consider:* Is it ethical to have a child in order to save the life of another? Is it right to conceive a child as a means to an end—to produce a tissue or organ? Is this a less acceptable reason than most reasons that parents have for bearing children? Do parents even need a reason for conceiving a child? Do doctors find this acceptable? If they do, why do we seldom hear about it, and why did this story make the front page? Do parents have the right to make decisions like this for their young children? How will the donor (and recipient) feel about this when the donor is old enough to know what happened?

Chapter 24 Answers

Testing Your Knowledge

Multiple Choice: **1.** e **2.** b **3.** b **4.** d **5.** a **6.** b. (Antigen-binding sites on an antibody fit a specific antigen.)

Matching: **1.** c **2.** g **3.** e **4.** f **5.** b **6.** h **7.** d **8.** a

Describing, Comparing, and Explaining

1. Neutralization: Antibodies bind to antigens, masking them, and phagocytes dispose of them. Agglutination: Antibodies clump invading cells together. Precipitation: Antibodies cross-link dissolved antigens and cause them to precipitate out of solution. Activation of complement system: Complement coats and punctures invading cells, and enhances phagocyte activity and inflammation.

2. You were born with a small number of lymphocytes preprogrammed to make antibodies against the chickenpox virus. These cells were not activated until you were first exposed to the virus. As part of the primary immune response, memory cells were produced that were capable of responding much more quickly the next time you were exposed.

3. AIDS is mainly transmitted in blood and semen. It enters the body through slight wounds during sexual contact or via needles contaminated with infected blood. AIDS is deadly because it infects helper T cells, crippling immunity and leaving the body vulnerable to other infections and cancer.

4. Inflammation is triggered by tissue injury. Damaged cells release histamine and other chemicals, which cause nearby blood vessels to dilate and become leakier. Blood plasma leaves vessels, and phagocytes are attracted to the site of injury. An increase in blood flow, fluid accumulation, and increased cell population cause redness, heat, and swelling. Inflammation disinfects and cleans the area and seals off the spread of infection from the injured area.

Thinking Critically

1. Each time the flu virus changes its antigens, a different population of lymphocytes is activated to defend the body. This primary immune response to a new invader is slower and probably starts with fewer cells than the secondary immune response carried out by memory cells. Since the flu virus can be different each time, memory cells are less able to respond to it. The body is always one step behind, immune to the last flu virus but not the next one.

2. There are probably two reasons the Indians are more vulnerable. First: Among Europeans, individuals born without the particular kind of lymphocytes able to respond to measles probably died out centuries ago. Because the Indians weren't exposed to measles,

there would have been no such natural selection among them. Though some of the present-day Indians may have lymphocytes capable of responding to measles, many do not, and they die of the disease. Second: Most Europeans contract these infections (or are immunized) as children. The diseases are more serious in adults.

3. Histamine, from mast cells, trigger allergy symptoms. They cause blood vessels to dilate and leak fluid, producing watery eyes and runny nose. They also cause nasal irritation, sneezing, itchiness, and tears. An antihistamine, as its name implies, blocks histamines and reduces these symptoms.

Science, Technology, and Society

Some issues and questions to consider: How important is it to protect students from AIDS? Is this a function of schools? Do schools serve other such "noneducational" needs? Do parents or citizens' and church groups—on either side of the issue—have a say in this, or is it a matter between the school and the student? Does distribution of condoms condone or sanction sexual activity or promiscuity? Is a school legally liable if a school-issued condom fails to protect a student? Are there alternative measures, such as education, that might be as effective for slowing the spread of AIDS?

Chapter 25 Answers

Testing Your Knowledge

Multiple Choice: **1.** c **2.** a **3.** e **4.** c (The freshwater fish gains water by osmosis.) **5.** d

Matching: **1.** b **2.** b **3.** a **4.** d **5.** a **6.** c

True/False: **1.** F (Body temperature may be higher.) **2.** T **3.** F (An endotherm generates body heat from metabolism.) **4.** F (The order is urete, bladder, urethra.) **5.** T **6.** T

Describing, Comparing and Explaining

1. Many "cold-blooded" animals that live in warm places are as warm as "warm-blooded" ones; "warm-blooded" animals may be quite cold when hibernating. Biologists prefer to refer to the source of heat—metabolism or surroundings—rather than to temperature.
2. Hormonal changes, shivering, and sitting in the sun might enable a squirrel to increase heat gain during the winter. Thicker fur, burrowing, and even hibernation might enable it to decrease heat loss. The opposite would prevail in summer.
3. In salt water, the fish loses water by osmosis. It drinks salt water and disposes of salts through its gills. Its kidneys conserve water and excrete some salts. In fresh water, it gains water by osmosis. Its kidneys excrete a lot of urine, so it loses some salts. Its gills and digestive tract take up salts to replenish those lost.

Thinking Critically

1. (a) Sweat, or splash water on your face; (b) take a dip in a cold pool; (c) sit in front of a fan. (a) (b) (c) Cover up with warm, insulating clothing.
2. Filtrate is similar to blood, but blood contains blood cells and proteins, which are too large to pass through the "filter". Useful solutes, such as glucose, salts, and amino acids, as well as water, are mostly reabsorbed. Urea is mostly not reabsorbed. Some substances, such as H^+ ions, are added to the filtrate by secretion. Urine is more concentrated than filtrate, containing more urea and H^+, but less salt and other usable substances than filtrate. The kidney maintains homeostasis through selective reabsorption and secretion, adjusting blood composition in response to changes in the internal environment.

3. Countercurrent heat exchangers are sets of parallel blood vessels found in animals that live in cold environments. A countercurrent mechanism in fish gills extracts oxygen from water. In the heat exchanger, cool blood from the skin and warm blood from the body interior flow in opposite directions. In the gill, water and blood flow in opposite directions. In the heat exchanger, as blood loses heat, it comes into contact with progressively colder blood, maximizing heat transfer. In the gill, as blood picks up oxygen, it comes into contact with water containing higher and higher concentrations of oxygen, maximizing the diffusion of oxygen.

Science, Technology, and Society

Some issues and questions to consider: Could drug use endanger the safety of the employee or others? Is drug testing relevant to jobs where safety is not a factor? Is drug testing an invasion of privacy, interfering in the private life of employees? Is an employer justified in banning drug use off the job if it does not affect safety or ability to do the job? Do the same criteria apply to employers requiring the test? Could an employer use a drug test to regulate other employee behavior that is legal, such as smoking?

Chapter 26 Answers

Testing Your Knowledge

Multiple Choice: **1.** d **2.** e **3.** a **4.** b (Negative feedback: When thyroxine increases, it inhibits TSH, which reduces thyroxine secretion.) **5.** d **6.** e

True/False: **1.** F (Slower-acting steroid hormones behave this way.) **2.** T **3.** T **4.** T **5.** T **6.** F (Only the gonads and adrenal cortex produce steroid hormones.)

Matching: **1.** e, w **2.** a, u **3.** g, r **4.** c, s **5.** f, p **6.** h, t **7.** d, q **8.** b, v

Describing, Comparing, and Explaining

1. The hypothalamus secretes releasing hormones and inhibiting hormones, which are carried by the blood to the anterior pituitary. In response to these signals from the hypothalamus, the anterior pituitary increases or decreases its secretion of a variety of hormones that directly affect body activities or influence other glands. Neurosecretory cells that extend from the hypothalamus into the posterior pituitary produce and secrete hormones into the blood in response to nerve impulses from the hypothalamus.
2. Slower-acting steroid hormones, such as sex hormones, enter target cells and bind to a receptor protein. This complex attaches to DNA, activating genes and causing a cell to make new proteins. It takes time for these proteins to build up and affect the target cell. Epinephrine binds to a receptor in the target cell membrane, triggering the formation of a second messenger, such as cAMP, which immediately activates cell activities.
3. Type I diabetes is caused by a shortage of insulin, which must be replaced by injection. In type II diabetes, insulin is present in normal quantities, but cells are not able to respond to it normally.
4. Only cells with the proper receptors will respond to a hormone. For a steroid hormone, the presence (or absence) and types of receptor proteins inside the cell determine the hormone's effect. For a nonsteroid hormone, the types of receptors on the cell's plasma membrane are key, and the second messenger may have different effects inside different cells.

Thinking Critically

As the cortisone level increases, the presence of the hormone inhibits cortisone secretion. Cortisone medication must have the same effect.

Science, Technology, and Society

Some issues and questions to consider: Why are parents concerned about their childrens' height? Just because the hormone is available, does this mean that parents should request its use? What is the role of the physician in this situation? Would this kind of use be worth the risk? Is a child able to make an informed decision? How might the child feel in later years if the hormone has unintended side effects? How can people be prevented from misusing a legal drug? How are the distribution and use of these kinds of medications regulated?

Chapter 27 Answers

Testing Your Knowledge

Multiple Choice: **1.** c **2.** c (The outer layer in a gastrula is the ecto-derm; of the choices given, only the brain develops from ectoderm.) **3.** a **4.** d **5.** d **6.** a

Matching: **1.** e **2.** g **3.** d **4.** h **5.** f **6.** a **7.** b **8.** c

Describing, Comparing, and Explaining

1. Asexual reproduction is advantageous in a stable, favorable envi-ronment; it allows an isolated animal to produce many offspring quickly and precisely. Sexual reproduction increases variability; it enhances success in a variable environment.
2. A. FSH. B. estrogen. C. LH. D. progesterone. P. menstruation. Q. pre-ovulatory phase—follicle develops. R. ovulation. S. post-ovu-latory phase—corpus luteum develops.
3. Both produce haploid gametes. Spermatogenesis produces four small sperm; oogenesis produces one large ovum. The ovary con-tains all the primary oocytes at birth, while testes can keep mak-ing primary spermatocytes throughout life. Oogenesis is not com-plete until fertilization, but sperm mature without eggs.
4. The extraembryonic membranes provide a moist environment for the embryos of land animals, and enable the embryos to ab-sorb food and oxygen and dispose of wastes. Such membranes are not needed when an embryo is surrounded by water.

Thinking Critically

1. The researcher might find out whether chemicals from the noto-chord stimulate the nearby ectoderm to become the neural tube, a process called induction. Transplanted notochord tissue might cause ectoderm anywhere in the embryo to become neural tissue. Control: Transplant non-notochord tissue under the ectoderm of the belly area.
2. The most rapid and drastic developmental changes—blastula for-mation, implantation, gastrulation, formation of extraembryonic membranes, shaping of major organs and systems—occur during the first nine weeks of gestation.
3. The hormones inhibit FSH and LH secretion and block follicle de-velopment and ovulation. The same hormones are produced by the corpus luteum when a woman is pregnant.
4. The nerve cells may follow chemical trails to the muscle cells and identify and attach to them by means of specific surface proteins.

Science, Technology, and Society

1. *Some issues and questions to consider:* Why would a person want to choose the sex of a baby? Personal preference? Family name? Family or peer pressure? Possibility of genetic disease? If widely available, will this procedure change the male/female ratio? Might allowing such choice divert resources from more important medical areas?
2. *Some issues and questions to consider:* What causes a miscarriage? Why might tissues from miscarriages be unsuitable for trans-

plants? How would tissues from miscarriages be obtained? Could using tissues from aborted fetuses encourage the practice of abor-tion? Could using tissues from aborted fetuses affect a woman's decision about whether or not to have an abortion? If tissue from aborted fetuses can help patients live longer, more productive lives, can we, in good conscience, block or abandon this line of re-search?

3. *Some issues and questions to consider:* How is this device similar to the birth control pill? How is it different? Under what circum-stances might this device be useful? Should the state have any role in reproductive decisions, or is this an area of individual choice? If you concur with the California decision, where would you draw the line on state intervention in reproductive issues?
4. *Some issues and questions to consider:* If it is technologically possible to save a baby, does that mean we have to do it? What would bring "the greatest good to the greatest number" of babies? If we divert-ed resources away from premature babies, would they really be used for prenatal care? Who pays for the treatment of premature babies? Prenatal care?

Chapter 28 Answers

Testing Your Knowledge

Multiple Choice: **1.** b **2.** e **3.** a **4.** e (Both I and III would prevent ac-tion potentials from occurring; II could actually increase the genera-tion of action potentials.) **5.** c

Matching: **1.** f **2.** e **3.** d **4.** a **5.** b **6.** c

Describing, Comparing, and Explaining

1. At the point where the action potential is triggered, sodium ions rush into the neuron. They diffuse laterally and cause sodium gates to open in the adjacent part of the membrane, triggering another action potential. The moving wave of action potentials, each triggering the next, is a moving nerve signal. Behind the action potential, sodium gates are temporarily inactivated, so the action potential can only go forward. At a synapse, the transmit-ting cell releases a chemical neurotransmitter, which binds to receptors on the receiving cell and triggers a nerve signal in the receiving cell.
2. Fingertips to brain: sensory neuron to spinal cord, interneurons to brain. The motor area of the cerebral cortex makes the decision, and the cerebellum helps coordinate movements. Brain to mus-cles: interneurons in the brain and spinal cord to motor neurons to muscles.

Thinking Critically

1. A nerve signal is most like the burning of a dynamite fuse. (The same drop of water that goes in one end of the pipe comes out the other. The same is true for a car on a highway. With the nerve sig-nal or burning fuse, no substance is carried from one end to the other; there is only a change—a membrane voltage change or a flame. Unlike nerve signal transmission, the burning of a fuse can occur only once.)
2. The results show the cumulative effect of all incoming signals on neuron D. Comparing experiments #1 and #2, the more nerve sig-nals D receives from C, the more it sends; C is excitatory. Because neuron A is not varied here, its action is unknown; it may be ei-ther excitatory or mildly inhibitory. Comparing experiments #2 and #3, neuron B must be a strongly inhibitory neurotransmitter, because when B is transmitting, D stops.

1. *Some issues and questions to consider:* Could the diagnosis be wrong? What are the chances of a "miraculous" recovery? Could medical personnel be held responsible for murder or manslaughter? Might the resources needed to keep a comatose patient alive be better used elsewhere? What about the financial and emotional burden on the family? When the U.S. Supreme Court dealt with a similar situation, they asked whether the patient had made her wishes known. Should a family be considered to represent the wishes of a comatose member?

2. *Some issues and questions to consider:* What is the role of alcohol in crime? What are its effects on families and in the workplace? In what ways is the individual responsible for alcohol abuse? The family? Society? Who is affected by alcohol abuse? How effective are treatment and punishment in curbing alcohol abuse? Who pays for alcohol abuse and consequent treatment or punishment? Why do people use alcohol? Is it possible to enjoy alcohol without abusing it?

Chapter 29 Answers

Testing Your Knowledge

Multiple Choice: **1.** d (He could hear the tuning fork against his skull, so the cochlea, nerve, and brain are OK. Apparently sounds are not being transmitted to the cochlea; therefore, the bones are the problem.) **2.** b **3.** d **4.** a **5.** b **6.** e

True/False: **1.** F (Receptors gradually become less sensitive.) **2.** T **3.** T **4.** F (It senses the scent of its home stream.) **5.** T **6.** F (The semicircular canals sense movement; the saccule and utricle sense body position.) **7.** F (Hair cells are mechanoreceptors.) **8.** F (hearing)

Describing, Comparing, and Explaining

1. Scent molecules enter the nasal cavity, dissolve in mucus, and bind to protein molecules on cilia of certain sensory receptor cells. This changes the permeability of the receptor cells, allowing ions to enter and altering the membrane potential. This change in receptor potential causes the receptor cells to release extra neurotransmitter, triggering action potentials in sensory neurons. An increased rate of action potentials signals the brain that the receptor cells detect a scent stimulus.

2. Sound waves strike the eardrum, which moves the bones of the middle ear. The bones vibrate fluid in the cochlea, which moves the basilar membrane. Hair cells on the basilar membrane move against the overlying tissue, causing permeability change and receptor potentials in the hair cells. The hair cells stimulate sensory neurons, which transmit action potentials to the brain. Louder sounds move hair cells more, generating a greater frequency of action potentials. Different pitches affect different parts of the basilar membrane; different hair cells transmit to different parts of the brain.

3. Sensation is the detection of stimuli (light) by the photoreceptors of the retina and transmission of action potentials to the brain. Perception is the interpretation of these nerve impulses—sorting out the patterns of light and dark, and determining their meaning.

Thinking Critically

1. You might expect the eyes of an owl to be adapted for collecting more light. They are large, relative to the size of the bird, with large pupils that let in more light. Their retinas have high concentrations of rods, which function in dim light.

2. Paired sensory receptors enable an animal to compare the intensity of the stimulus on either side and determine the direction the stimulus is coming from. Also, a comparison of slightly different images seen by the eyes enables the brain to perceive depth and distance.

3. Do the turtles hear the surf? Plug the ears of some turtles and not others. If turtles without earplugs head for the water and turtles with ear plugs get lost, they probably hear the ocean. Or do they smell the water? Plug their nostrils, etc.

Science, Technology, and Society

1. *Some issues and questions to consider:* The sound is loud enough to impair hearing, but how long an exposure is necessary for this to occur? Does exposure have to occur all at once, or is damage cumulative? Who is responsible, concert promoters or listeners? Should there be regulations regarding sound exposure at concerts (as there are for job-related noise)? Are young people mature and aware enough to heed such warnings?

2. *Some issues and questions to consider:* Who is questioning the benefits and safety of radial keratotomy? Could there be a conflict of interest? How much risk is there associated with the procedure? Might the operation simply be ineffective, or could vision be made worse? Are patients aware of the risk? How inconvenient is nearsightedness to the patient? Are the beneficial effects of the surgery permanent? Are the benefits worth the risk?

Chapter 30 Answers

Testing Your Knowledge

Multiple Choice: **1.** c **2.** e **3.** a (Water supports aquatic animals, reducing the effects of gravity.) **4.** d **5.** c **6.** d **7.** e **8.** d **9.** a (Each neuron controls a smaller number of muscle fibers.)

True/False: **1.** F (It is a ball-and-socket joint.) **2.** F (Gravity is the main impediment for land animals.) **3.** T **4.** T **5.** F (It is part of the appendicular skeleton.) **6.** F (ATP provides this energy.) **7.** T **8.** T **9.** T **10.** F (Tendons connect muscles to bones.) **11.** F (Blood is manufactured in red bone marrow.) **12.** T

Describing, Comparing, and Explaining

1. The bird's wings are airfoils, with convex upper surfaces and flat or concave lower surfaces. As the wings beat, air passing over them travels farther than air beneath. Air molecules above the wings are more spread out, lowering pressure. Higher pressure beneath the wings pushes them up.

2. Advantages of an exoskeleton include strength, good protection for the body, and flexibility at joints. The major disadvantage is that the exoskeleton must be shed periodically as the insect grows, leaving the insect temporarily weak and vulnerable.

3. Action potentials from the brain travel down the spinal cord and along a motor neuron to the muscle. The neuron releases a neurotransmitter, which triggers action potentials in a muscle fiber membrane. These action potentials initiate the release of calcium from the ER of the cell. Calcium enables myosin heads of the thick filaments to bind with actin of the thin filaments. ATP provides energy for the movement of myosin heads, which causes the thick and thin filaments to slide along one another, shortening the muscle fiber. The shortening of muscle fibers pulls on bones, bending the arm. If more motor units are activated, the contraction is stronger.

4. The human skeleton has a larger skull than the baboon, balanced on top of (instead of in front of) the backbone. The human backbone is S-shaped, not arched as the baboon's is. The human pelvic girdle is shorter, rounder, and oriented more vertically. The

human hand is adapted for gripping, and our feet are adapted for support of the entire body and bipedal locomotion.

Thinking Critically

1. Chemical A would work better, because acetylcholine triggers contraction. Blocking it would prevent contraction. B would actually increase contraction, because Ca^{2+} signals filaments to interact and slide.
2. Circular muscles in the earthworm body wall decrease the diameter of each segment, squeezing internal fluid and lengthening the segment. Longitudinal muscles shorten and fatten each segment. Different parts of the earthworm can lengthen while others shorten, producing a crawling motion. The whole roundworm body moves at once, because of a lack of segmentation. The body can only shorten or bend, not lengthen, because of a lack of circular muscles. Roundworms simply thrash from side to side.
3. ATP causes the myosin heads of the thick filaments to detach from the thin filaments (Figure 30.9B, Step 1). If there is no ATP present, the myosin heads remain attached to the thin filaments, and the muscle fiber remains fixed in position.

Science, Technology, and Society

1. *Some issues and questions to consider:* Are the places where you live, work, or attend class accessible to a person in a wheelchair? Would you have trouble with doors, stairs, drinking fountains, toilet facilities, and eating facilities? What kinds of transportation would be available to you, and how convenient would they be? What activities would you have to forego? How might your disability alter your relationships with your friends and family? How well would you manage on your own?
2. *Some issues and questions to consider:* The major negative consequence is that the bones will become completely hardened before they have grown to their full size. This means that the steroid user will be shorter than he or she would otherwise have been. Do steroid users know this? Do they think the effects of steroids are temporary or reversible? Why is such emphasis placed on appearance and athletic prowess? Will steroid users later regret their use? If steroid use were not against the rules, would it be acceptable?
3. *Some issues and questions to consider:* Do young people look that far ahead? What does it take to get them to think about their later years? Who will they believe? Will scare tactics work? Or are good health habits simply their own responsibility?

Chapter 31 Answers

Testing Your Knowledge

Multiple Choice: **1.** e (Xylem forms just inside cambium; old xylem is pushed inside.) **2.** b **3.** e **4.** c
True/False: **1.** F (Meristems are the growth points.) **2.** T **3.** T **4.** T **5.** T **6.** F (Perennial plants live for many years.) **7.** F (The root cap protects the tip of the growing root.)
Matching: **1.** f **2.** b **3.** e **4.** a **5.** c **6.** d

Describing, Comparing, and Explaining

1. Bees deposit the pollen on a carpel, and a pollen tube grows to the ovary at the base of the carpel. Sperm travel down the pollen tube and fertilize egg cells in ovules. The ovules grow into seeds, and the ovary grows into the flesh of the fruit (actually, in an apple, just the core). As the seeds mature, the fruit ripens and falls (or is picked).
2. Bulbs, root sprouts, and runners are all examples of vegetative reproduction. Vegetative reproduction is less wasteful and costly than sexual reproduction and less hazardous for young plants.

Thinking Critically

1. Xylem lies toward the inside of the stem, phloem toward the outside. As the vascular tissue bends to enter the leaf, this arrangement leaves xylem on top and phloem on the bottom.
2. Modern methods of plant breeding and propagation have increased crop yields but have decreased genetic variability, so plants become more vulnerable to epidemics. Primitive varieties of crop plants could contribute to gene banks and be used for breeding new strains.
3. Tomato: Fruit (ripened ovary). Onion: Bulb (underground shoot). Celery stalk: Leaf stalk. Peanut: Seed (ovule). Strawberry: Fruit (ripened ovary). Lettuce: Leaf blades. Artichoke: Terminal bud. Beet: Root.

Science, Technology, and Society

Some issues and questions to consider: Why are the forests being cut, when preserving them is more beneficial in the long run? Why can't the developing nations develop new forest products themselves? Should developing and developed nations share in the profits from the forest? What incentive is there for a company to develop a new product if they can't keep the profits? What about companies paying an exploration fee for exclusive rights to take plants from a certain area? What about a contract for the developing country to get a percentage of the profits from new products? Are there reasons to preserve the forest, other than profits and beneficial new products?

Chapter 32 Answers

Testing Your Knowledge

Multiple Choice: **1.** d **2.** d **3.** e **4.** b
True/False: **1.** F (The guard cells close the stomata.) **2.** F (transpiration-cohesion-adhesion) **3.** T **4.** T **5.** F (Pressure flow moves sugar from leaves to roots.) **6.** F (Carbon, oxygen, nitrogen, and hydrogen are macronutrients.) **7.** T

Describing, Comparing, and Explaining

1. If plant starts to dry out, K^+ is pumped out of guard cells. Water follows by osmosis, guard cells become flaccid, and stomata close. This prevents wilting, but keeps leaves from taking in carbon dioxide, needed for photosynthesis.
2. Soil bacteria break down organic matter, releasing nitrogen. They also fix nitrogen from the air, and they convert ammonium to nitrate, which plants are more able to absorb.
3. Fine rock and clay particles hold water and nutrients. Humus slowly releases nutrients, has air spaces, and holds water. Soil organisms aerate the soil and recycle organic matter.

Thinking Critically

1. The hydrogen ions in acid rain displace positively charged nutrient ions from negatively charged clay particles.
2. As nitrogen builds up in their environment, nitrogen-fixing bacteria slow their rate of nitrogen fixation. This is because the bacterial genes that code for nitrogen-fixing enzymes are switched off.
3. Air in the xylem tubes interferes with the cohesion of water molecules that helps pull them up the stem. Without water, the flower wilts.

Science, Technology, and Society

1. *Some issues and questions to consider:* How were the farmers assigned or sold "rights" to the water? How is the price established when a farmer buys or sells water rights? Is there enough water for everyone who "owns" it? What kinds of crops are the farmers

growing using the water? What will the water be used for in the city? Are there other users with no rights, such as wildlife? Is there any effort being made to curb urban growth and conserve water? Should millions of people be living in what is essentially a desert? What are the reasons for farming desert land?

2 *Some issues and questions to consider:* What kinds of changes are being made in food plants via genetic engineering? How do these changes compare with changes in crop plants produced by traditional breeding techniques? How might these genetic changes affect a person who eats an altered fruit or vegetable? What possible environmental problems might result from genetically engineered plants? What kinds of safeguards and regulations protect people and the environment from possible hazards?

3. *Some issues and questions to consider:* What kinds of changes in crops might make organic farming easier? Could an upturn in organic farming result in incentives to breed and market new crops? Might the introduction of genetically engineered strains lead to genetic uniformity and increased succeptibility to pests? Could this lead to a dependence on pesticides and inorganic fertilizers? Might engineered crops become weeds? How will organic farmers control them?

Chapter 33 Answers

Testing Your Knowledge

Multiple Choice: **1.** b **2.** c (A short-day plant requires a long night. The 13-hour night of answer C is the only uninterrupted night longer than 12 hours.) **3.** b **4.** a **5.** d **6.** b

Matching: **1.** g **2.** e **3.** f **4.** a **5.** b **6.** c **7.** h **8.** d

True/False: **1.** F (Sleep movements continue.) **2.** F (Night length controls these responses.) **3.** T **4.** F (Location of the biological clock is unknown.) **5.** T **6.** T **7.** T

Describing, Comparing, and Explaining

1. Fruit produces ethylene gas, which triggers ripening and aging of the fruit. Ventilation prevents buildup of ethylene and delays its effects.

2. The terminal bud produces auxins, which counter the effects of cytokinins from the roots and inhibit the growth of axillary buds. If the terminal bud is removed, the cytokinins predominate, and lateral growth occurs at the axillary buds.

3. The ratio of abscisic acid (ABA) to gibberellins determines whether buds or seeds will remain dormant. A higher concentration of ABA causes seeds and buds to remain dormant. A higher concentration of gibberellins causes seeds to germinate and buds to grow.

Thinking Critically

1. The red wavelengths in the room lights quickly convert the phytochrome in the mums to the P_{fr} form, which inhibits flowering in a long-night plant. The mums will not flower unless Jon can rig up some far-red lights. Exposure to a burst of far-red light would convert the phytochrome to the P_r form, allowing flowering to occur.

2. She could grow some dandelions under constant light and note whether the sleep movements continue.

3. He could remove leaves at different stages of being eaten to see how long it took for changes to occur in nearby leaves. He could capture the "hormone" in an agar block, as in the phototropism experiments in Module 33.1, and apply it to an undamaged plant. Or he could block "hormone" movement out of a damaged leaf or into a nearby leaf.

4. On Earth, it is not possible to study the effects of phototropism without gravitropism also affecting the plant. In space, phototropism could be studied in a gravity-free environment, eliminating possible effects of gravitropism. For example, you could compare plants' response to light on Earth and in a gravity-free environment. Does a plant need to know which way is up to bend toward light? It is also possible that circadian rhythms are affected by Earth's rotation (aside from the day-night cycle). It would be possible to investigate these effects in space, free of Earth's rotation. You could grow plants in constant light on Earth and in space to see if there is any difference in their timekeeping ability.

5. This is probably due to a change in the ratio of auxin to ethylene. Growing leaves produce auxin, but photosynthesis drops in the leaves on shaded branches, so they cannot grow as fast. As their auxin production drops relative to ethylene, the ethylene causes aging and drying of the branches, and they eventually break off. Branches that can't "pay their own way" are shed.

Science, Technology, and Society

1. *Some issues and questions to consider:* Is the hormone safe for human consumption? What are its effects in the environment? Could its production produce impurities or wastes that might be harmful? What kinds of tests need to be done to demonstrate its safety? How much does it cost to make and use? Are the benefits worth the costs and risks? Are there reasons for using an artificial chemical on food simply to improve its appearance?

2. *Some issues and questions to consider:* How dangerous are the chemicals? How are the chemicals tested, and what criteria are used to decide if they are safe or unsafe? Who carries out the tests and evaluates the results? How do the chemicals affect the price, availability, and safety of food? How will continuing or discontinuing their use affect the farmer and the consumer? Should society control exposure to artificial chemicals (approve them, ban them), or are individuals responsible for their own consumption of these products? Is information on growth conditions and processing of foods readily available?

3. *Some issues and questions to consider:* Will this cause us to use even more herbicides on food crops? What are the possible health and environmental effects? Might the engineered crop plants interbreed with weeds, producing resistant pests?

Chapter 34 Answers

Testing Your Knowledge

Multiple Choice: **1.** a **2.** e **3.** b **4.** d

True/False: **1.** F (A population consists of such individuals.) **2.** F (It looks marshy because of poor drainage and slow evaporation.) **3.** T **4.** F (Phytoplankton are the producers.) **5.** F (Oxygen availibility limits aquatic organisms.) **6.** T

Matching: **1.** f **2.** c **3.** e **4.** h **5.** b **6.** d **7.** a **8.** g

Describing, Comparing, and Explaining

1. At the source, nutrient content is generally low; the current is stronger; there is usually less sediment, lower temperature, and higher dissolved O_2 concentration; and food is from attached algae or organic material from land. At the mouth, the nutrient content is generally higher, and the current slower; there is usually more sediment, higher temperature, and lower dissolved O_2 concentration; most food is from phytoplankton.

2. Depends on student's choice.

3. The sun's rays strike the equator directly. Near the equator, heated

air rises and then cools, and moisture condenses and falls as rain. Dry air descends and spreads out 30 degrees north and south of the equator, forming deserts.

Thinking Critically
1. Apparently, unrelated animals adapted in similar ways to similar environments, temperate grassland and savanna.
2. An estuary is where fresh water merges with seawater. Warm, nutrient-rich water from a river mingles with seawater. Shallow water allows light to penetrate to the bottom. The mix of warmth, nutrients, and light makes an estuary very productive; lots of food is available for young organisms, in a relatively sheltered coastal environment.

Science, Technology, and Society
Some issues and questions to consider: What reasons are there for protecting areas like this? Is a public refuge or preserve the only way to protect species and habitats? Is there some way to protect organisms without confiscating land? Does protection take precedence over property rights? Would the prairie have been better off without the publicity? Should the public be able to condemn land to protect it? Should there be state or federal laws to protect endangered plants on private land? Should there be laws to protect endangered habitats, like this patch of prairie?

Chapter 35 Answers

Testing Your Knowledge
Multiple Choice: **1.** e **2.** c **3.** c (Seedlings would tend to grow only in gaps with sufficient light, spacing the trees fairly evenly.) **4.** a **5.** c
True/False: **1.** T **2.** F (They exhibit a clumped pattern.) **3.** T **4.** T

Describing, Comparing, and Explaining
1. Exponential growth is accelerating, unlimited population growth described by the equation $G = rN$. It might occur briefly when there are no factors slowing growth. Logistic growth occurs when density-dependent factors slow growth as population density approaches carrying capacity. It is described by $G = rN (K - N)/K$
2. Carrying capacity depends on both the species and the environment, because different environments have different capacities to support different organisms. For example, a meadow could support more rabbits than horses. An irrigated meadow could support more horses than a meadow without irrigation.
3. Survivorship is the fraction of individuals in a given age interval that survive to the next interval. It is a measure of the probability of surviving at any given age. A survivorship curve shows the fraction of individuals in a population surviving at each age interval during the life span. Almost all oysters die young, with a few living a full life span. Few humans die young; most live out a full life span and die of old age. Squirrels suffer constant mortality during the life span, with an equal chance of surviving at all ages.

Thinking Critically
1. Equilibrial populations tend to live in fairly stable environments, held near carrying capacity by density-dependent limiting factors. They reproduce later than opportunistic species, and have fewer offspring. Human intrusion is density-independent, and the lower reproductive rate of equilibrial species makes it hard for them to recover from this kind of disruption.
2. There would be about 15 beavers in the sixth generation. The beaver population grows exponentially. At a rate of 1.1 offspring/parent/generation, there would be 3.3 beavers in the sixth

generation. This is exponential growth too, but the population grows more slowly. Numbers do not have to double for exponential growth to occur.
3. Using the formula for the mark-recapture method, there are (185 × 208)/35 = 1099 fish in the lake. Density = 1099/120 = 9 fish/hectare in the lake as a whole. Their clumped dispersion increases their density in the reed beds. Density in the reed beds is 1099/30 = 37 fish/hectare.

Science, Technology, and Society
1. *Some issues and questions to consider:* How does population growth in developing countries relate to food supply, pollution, and use of natural resources such as fossil fuels? How are these things affected by population growth in developed countries? Which of these factors is most critical to our survival? Are they pressured more by the growth of developing or developed countries? What will happen as developing countries become more developed? Will it be possible for everyone to live at the level of the developed world?
2. *Some issues and questions to consider:* Will lack of funding undermine other aspects of population control, such as education and contraception? Could this lead to greater population growth and even more abortions? Should we apply the same standard to ourselves and other countries? Is the U.S. obliged to help other countries solve their population problems in their own ways? Or should the U.S. call the tune if it is paying the bills?

Chapter 36 Answers

Testing Your Knowledge
Multiple Choice: **1.** c **2.** c **3.** e (The bat and moth have adapted to each other.) **4.** d

Describing, Comparing, and Explaining
1. These animals are secondary or tertiary consumers, at the top of the energy pyramid. Stepwise energy loss means not much energy is left for them; thus, they are rare.
2. Cutting forests reduces transpiration, and irrigation depletes groundwater reserves, both disrupting the water cycle. Burning fossil fuels adds CO_2 to the air, which may lead to global warming. Nitrogen and phosphates in fertilizers may pollute water.

Thinking Critically
1. It looks as if the kangaroo rat is a keystone predator in the desert. By preferring one plant for food, it kept this plant from outcompeting others. Removing the rats reduced prey diversity. (The situation is similar to the sea star example described in Module 36.4.)
2. For 10,000 kilocalories in leaves, 1000 kilocalories end up in *Pieris*, 100 in *Apanteles,* 10 in ichneumons, and 1 calorie in chalcids. There is a 10% energy transfer from each trophic level to the next.
3. There are many possible answers to this question; refer to Module 36.14, and be creative. *Some questions to consider:* What are the main reservoirs of carbon? What chemical form does the carbon take in each reservoir? What processes move it from one reservoir to another (and convert it from one form to another)? How can carbon travel long distances (from land to sea, for example)? How long might a carbon atom stay in one place?

Science, Technology, and Society
1. *Some issues and questions to consider:* What relationships might exist in the mussels' native habitat that are altered in the Great Lakes? What are possible predators? Competitors? Parasites? How might

the mussel compete with Great Lakes organisms? Might competitive exclusion occur? What happens to displaced competitors? Might the Great Lakes species adapt in some way? Might the mussels adapt? Could possible solutions present problems of their own?

2. *Some issues and questions to consider:* What was the original role of the wolves in the ecology of Yellowstone? How might the ecological communities in the park have changed as a result of their removal? How might human activities in the park and surrounding areas have changed since then? Would reintroducing wolves return the park to a more natural state? Is this important? Why? How might reintroducing wolves alter the current status quo in the park? Do the ranchers have reason to fear the reintroduction of wolves? Who has to bear the losses if the wolves interfere with ranching? How might they be compensated? How much are we willing to pay to return wolves to their natural habitat?

Chapter 37 Answers

Testing Your Knowledge

Multiple Choice: **1.** d (Innovation, or reasoning, is most often seen in primates. Most animals do not exhibit this kind of problem-solving ability.) **2.** c **3.** c (Such clocks allow them to compensate for Earth's rotation and the resulting change in the sun's position in the sky.) **4.** a
True/False: **1.** F (Altruism is taking such a risk.) **2.** F (A search image helps animals find food.) **3.** T **4.** T **5.** F (Biological clocks can run slow or fast.)

Describing, Comparing, and Explaining

1. Main advantage: Flies do not live long. Innate behaviors can be performed the first time without learning, enabling flies to find food, mates, etc., without practice. Main disadvantage: Innate behaviors are rigid; flies cannot learn to adapt to specific situations.

2. A dominance hierarchy minimizes energy wasted in fighting, maximizes efficiency in hunting, and ensures that some in the pack will get adequate resources. A territory ensures adequate resources for the pack.

3. Courtship behaviors reduce aggression between potential mates and confirm their species, sex, and physical condition. Environmental changes such as rainfall, temperature, day length, and presence of females probably lead frogs to start calling, so these would be the proximate causes. The ultimate cause relates to evolution. Fitness (reproductive success) is enhanced for frogs that engage in courtship behaviors.

4. A bird is genetically programmed to imprint on the song of its own species during a critical period. The tendency to listen to and repeat the song is innate, but the song itself is learned.

Thinking Critically

1. You could pair desirable behaviors (sitting and rolling over) with rewards (a dog biscuit or "Good dog!") and undesirable behaviors (digging) with punishment ("Bad dog!"). The dog would learn to perform behaviors associated with rewards and not to perform behaviors associated with punishment.

2. If the helper cannot obtain its own territory or mate, it may help its relatives. Since they share some of its genes, the bird is indirectly enhancing its own fitness by helping its relatives raise their young. This is kin selection. Alternatively, the bird may help because the favor may be repaid later. This would be reciprocal altruism. The scientists were looking for kin selection.

3. Place substitute eggs and chicks in nests at different times to see whether the birds imprint on them during a critical period or reject

them after imprinting on their own eggs or chicks. Western gulls probably imprint on both their eggs and young, because they nest so close together that eggs or chicks might get mixed up. Herring gulls nest farther apart, so their eggs are unlikely to get mixed up. They probably imprint only on chicks. Kittiwakes probably imprint on neither, because their nest sites provide little chance of a mix-up.

Science, Technology, and Society

1. Identical twins are genetically the same, so any differences between them are due to environment. This enables researchers to sort out the effects of "nature" and "nurture" on human behavior. The data suggest that many aspects of human behavior are inborn. Some people find these studies frightening because they seem to leave less room for free will and self-improvement than we would like.

2. A particular species of animal may have features that make it best suited to answer a specific biological question. Squids, for example, have a giant nerve fiber that made possible the discovery of how all nerve cells function. Animal experiments are important in medical research. Many vaccines that protect humans against deadly diseases have been developed through animal experiments. Animals also benefit as vaccines are developed against their own brands of pathogens. Researchers point out that the number of animals used in research is a small fraction of those killed as strays by animal shelters and a miniscule fraction of those killed for human food. They also maintain that modern laboratory facilities are models for responsible and considerate treatment of animals. What are some medical treatments or products that have undergone testing in animals? Have you benefited from any of them? Are there alternatives to using animals in experiments? Would alternatives put humans at risk? Are all kinds of animal experiments equally valuable? In your opinion, what kinds of experiments are acceptable and what kinds are unnecessary? What kinds of treatment are humane and inhumane?

Chapter 38 Answers

Testing Your Knowledge

Multiple Choice: **1.** d **2.** b **3.** c **4.** b **5.** c **6.** b **7.** a (The U.S. uses more energy each year than all these other areas combined.)
True/False: **1.** T **2.** F (The lemur and loris are prosimians.) **3.** T **4.** F (They were probably scavengers.) **5.** F (We know of no significant human biological changes since that time.) **6.** T **7.** T **8.** F (Humans and chimps are thought to have evolved from a common ancestor.) **9.** F (CFCs from air conditioners and refrigerators are the major culprits.) **10.** F (They may have driven many large animals to extinction.)

Describing, Comparing, and Explaining

1. Several primate characteristics related to our hands make it easy for us to make and use tools—mobile digits, opposable fingers and thumb, and great sensitivity of touch. Primates also have forward-facing eyes, which enhances depth perception and eye-hand coordination, and a relatively large brain.

2. Chimpanzees make and use simple tools, raid other social groups of their own species, can learn sign language and may use symbolic communication in the wild, and seem self-aware and able to use some form of reasoning to solve problems.

3. Our intelligence and culture—accumulated and transmitted knowledge, beliefs, arts, and products—have enabled us to overcome our physical limitations and alter the environment to fit our needs and desires.

4. The australopithecines roamed the African savanna, probably in small groups. They may have subsisted on nuts and seeds, birds' eggs, and whatever animals they could catch or scavenge from larger predators. They lived mostly on the ground but would climb trees to pick fruits, escape predators, or steal carcasses left by leopards and other large cats.

5. Carbon dioxide and several other gases in the atmosphere absorb infrared radiation and thus slow the escape of heat from Earth. This is called the greenhouse effect. The greenhouse effect is beneficial to life on Earth; without CO_2 in the atmosphere, the temperature at the surface of Earth would be only about $-18°C$—too cold for most life to exist.

6. Fossil-fuel consumption, industry, and agriculture are increasing the quantity of greenhouse gases—such as CO_2, methane, and nitrous oxide—in the atmosphere. This could trap more heat and raise atmospheric temperatures 2–5 degrees Celsius over the next century. Logging and clearing of forests for farming contribute to global warming by reducing the uptake of CO_2 by plants (and adding CO_2 to the air when trees are burned). Global warming could shift patterns of precipitation, turning farmland into deserts. It could also cause polar ice caps to melt, raising ocean levels and flooding coastal areas. Global warming is an international problem; air and climate do not recognize international boundaries. Greenhouse gases produced by industrialized nations add to those resulting from deforestation in less-developed tropical nations. Cooperation and commitment to a less-consumptive lifestyle will be needed to slow global warming.

Thinking Critically

1. Most anthropologists think that humans and apes diverged from a common apelike ancestor 4–5 million years ago. Primate fossils 3–10 million years old might help us understand how humans first evolved.

2. Bipedalism would have allowed our ancestors to see farther and run faster in the more open savanna environment, enabling them to find food and escape predators more easily. Bipedalism also freed the hands to carry objects and use tools. Some anthropologists believe that tool users with larger brains would have been favored by natural selection.

3. If global warming causes CO_2 to bubble out of the ocean into the air, the additional CO_2 in the atmosphere could compound the problem, trapping even more heat and making global warming worse. On the other hand, if the ocean can hold a large amount of CO_2, perhaps it can absorb some of the excess CO_2 we are adding to the atmosphere, and thus slow global warming.

4. These birds might be affected by pesticides while in their wintering grounds in Central and South America, where such chemicals may still be in use. The birds are also affected by deforestation throughout their range.

Science, Technology, and Society

1. Current counts indicate there are about 1.5 million species of living things. Assume 80% of all living things (not just plants and animals) live in tropical rain forests. This means there are $1.5 \times 0.8 = 1.2$ million species there. If half the species survive, this means 0.6 million species will go extinct in 100 years, or 0.6 million/100 = 6000 per year. This means that 6000/365 = 16 species will disappear per day, or almost one per hour. If there are 30 million species on Earth, 24 million live in the tropics, and 12 million will disappear in the next century. This is 120,000 per year, 329 per day, or 14 per hour.

2. *Some issues and questions to consider:* How does our use of fossil fuels affect the environment? What about oil spills? Disruption of wildlife habitat for construction of oil fields, pipelines, etc.? Burning of fossil fuels and possible climate change and flooding from global warming? Pollution of lakes and destruction of property by acid rain? Health effects of polluted air on humans? How are we paying for these "side effects" of fossil-fuel use? In taxes? In health insurance premiums? Do we pay a nonfinancial price in terms of poorer health and quality of life? Could oil companies be required to pick up the tab for environmental effects of fossil-fuel use? Could these costs be covered by an oil tax? How would this change the price of oil? How would a change in the price of oil change our pattern of energy use, our lifestyle, and our environment?

3. *Some issues and questions to consider:* Does the current evidence constitute "proof" that global warming is occurring? Why is it difficult to prove conclusively that human activities have an effect on global warming? How much information do we need before we act? What are the advantages and disadvantages of acting now versus waiting and gathering more data? Which of the following would be more serious, and why? Acting now to curb global warming, and later finding out that we were wrong, that global warming is not a threat? Or failing to take action now, and later finding out that global warming is a real and serious problem?

4. *Some issues and questions to consider:* How do population growth, resource consumption, pollution, and reduction in biodiversity relate to sustainability? How do poverty, economic growth and development, and political issues relate to sustainability? Why might developed and developing nations take different views of a sustainable society? What would life be like in a sustainable society? Have any steps toward sustainability been taken in your community? What are the obstacles to sustainability in your community? What steps have you taken toward a sustainable lifestyle? How old will you be in 2030? What do you think life will be like then?

Photograph Credits

Frontispiece: © Art Wolfe.

Contents: p. xv: © Frans Lanting/Minden Pictures. p. xviii: © John Sohlden/Visuals Unlimited. p. xix: © David Scharf. p. xx: © Blair Seitz/Photo Researchers, Inc. p. xxi: © Frans Lanting/Minden Pictures. p. xxii: © Manfred Kage/Peter Arnold. p. xxiii: © J. Collins/Photo Researchers, Inc. p. xxiv: © Malcolm Bolton/Photo Researchers, Inc. p. xxv: © Art Wolfe. p. xxvi: © Norman Owen Tomalin/Bruce Coleman, Inc. p. xxvii: © Runk/Schoenberger/Grant Heilman Photography, Inc. p. xxviii: © L. West/Photo Researchers, Inc. p. xxix: © J.P. Varin/Jacana/Photo Researchers, Inc.

Chapter 1: Chapter opener: © Raphael Gaillarde/Gamma Liaison. 1.0: © Raphael Gaillarde/Gamma Liaison. 1.1A: © Christine L. Case. 1.1B: © M.I. Walker/Photo Researchers, Inc. 1.1C: © David Matherly/Visuals Unlimited. 1.1D: © Leonard Lee Rue III/Earth Scenes. 1.1E: © Michael and Patricia Fogden. 1.2A1: © Frans Lanting/Minden Pictures. 1.2A2: © Kjell B. Sandved/Photo Researchers. 1.2A3: © Kjell B. Sandved/Visuals Unlimited. 1.2B: © American Museum of Natural History. 1.3A: Neg. No. 326795. Photo Courtesy Department of Library Services/American Museum of Natural History. 1.3C1: © Frans Lanting/Minden 1991. 1.3C2: © Fred Felleman/Allstock. 1.5A: © Kevin Scully. 1.5B: Roberto Roca/ Nature Conservancy. 1.6A: Roberto Roca/ Nature Conservancy. 1.6B: © Richard Wagner, UCSF Graphics. 1.8: Roberto Roca/ Nature Conservancy. 1.9: © CNRI/Science Photo Library/Photo Researchers, Inc. 1.10A: © Alex Kerstitch. 1.10B: Richard Wagner/UCSF Graphics. 1.11A: © M. Abbey/Visuals Unlimited. 1.11B: © Don Fawcett/Science Source/Photo Researchers, Inc. 1.12A: © Gregory Dimijian, M.D./Photo Researchers, Inc. 1.12B: © P. Henry/Jacana/Photo Researchers, Inc. 1.14A: *The New York Times,* August 16, 1993. 1.14B1: © Kevin Schafer/Martha Hill. 1.14B2: © G. Prance/Visuals Unlimited.

Chapter 2: Chapter opener: © Branson Reynolds. 2.0B: © Robert Fletterick/UCSF. 2.1B: © Dr. Jeremy Burgess/Science Photo Library/Photo Researchers, Inc. 2.1C: © Manfred Kage/Peter Arnold, Inc. 2.1D: © John Shaw/Tom Stack & Associates. 2.2B: Courtesy of James Pushnik, California State Univ., Chico. 2.5A & B: © Howard Sochurek. 2.10B: © Kevin Schafer. 2.11: © G.I. Bernard/Animals Animals. 2.12: © Al Tielemans/Duomo 1989. 2.16: © 1991 Maresa Pryor/Earth Scenes.

Chapter 3: Chapter opener: © Franklin Wagner/Image Bank. 3.0: Vollrath & Edmunds, *Nature* 340:305–317. 3.4A: © 1991 Scott Camazine/Photo Researchers, Inc. 3.6: © Alain McGlaughlin. 3.7A: © Biophoto Associates/Photo Researchers, Inc. 3.7B: © L.M. Beidler, Florida State Univ. 3.7C: © Biophoto Associates/Photo Researchers, Inc. 3.8A: © Frans Lanting/Minden Pictures. 3.10: Gerard Vandystadt/Photo Researchers, Inc. 3.11: © Fran Heyl & Associates. 3.14B: Courtesy of the Graphics Systems Research Group, IBM U.K. Scientific Centre. 3.19b: Courtesy of Linus Pauling. 3.19B: © Renee Fadiman.

Chapter 4: Chapter opener: Courtesy Jan Hinsch, Leica, Inc. 4.1A1: © Glenn Hoffman. 4.1A2, B2, C2: Courtesy of Carl Zeiss, Inc. 4.1B1,C1: W.L. Dentler, Univ. of Kansas/Biological Photo Service. 4.9: Dr. Barry F. King, Univ. of California, Davis, School of Medicine/Biological Photo Service. 4.10: G.T. Cole, Univ. of Texas, Austin/Biological Photo Service. 4.11: R. Rodewald, Univ. of Virginia/Biological Photo Service. 4.13A: W.P. Wergin, courtesy of E.H. Newcomb, Univ. of Wisconsin, Madison. 4.13B: © Roland Birke/Peter Arnold, Inc. 4.15: Courtesy of W.P. Wergin and E.H. Newcomb, Univ. of Wisconsin/Biological Photo Service. 4.16: Courtesy of Dr. Nicolae Simionescu. 4.17A: Courtesy of Drs. J.V. Small and G. Rinnerthaler. 4.17B: Courtesy of Dr. John Heuser, Washington School of Medicine, St. Louis. 4.18: W.L. Dentler, Univ. of Kansas/Biological Photo Service. 4.21: © CNRI/Science Photo Library/Photo Researchers.

Chapter 5: Chapter opener: © Keith Kent/Peter Arnold, Inc. 5.0: © James E. Lloyd, 1977. 5.1, 5.2A: © Alain McGlaughlin. 5.2B: © Jerry Berndt/Stock Boston. 5.10A: Courtesy of David Robertson. 5.10B: © Joseph Hoffman, Yale Univ. School of Medicine. 5.19C1: © M. Abbey/Visuals Unlimited. 5.19C2: © D.W. Fawcett/Science Source/Photo Researchers, Inc. 5.19C3: M.M. Perry and A.B. Gilbert, *J. Cell Sci.* 39(1979): 257. Copyright 1979 The Company of Biologists Ltd.

Chapter 6: Chapter opener: © Eric Schabtach and Ira Herskowitz. 6.0: © David Madison 1991. 6.1: © David Madison 1990. 6.6: © Stephen Frisch. 6.15C: Obester Winery, Half Moon Bay, CA. Photo by Kevin Schafer. 6.18: Jessie Cohen, National Zoological Park, Smithsonian Institution.

Chapter 7: Chapter opener: © Larry Lefever/Grant Heilman Photography, Inc. 7.0 : © John Sohlden/Visuals Unlimited, 1987. 7.1: © Larry Ulrich. 7.1: David J. Wrobel, Monterey Bay Aquarium/Biological Photo Service. 7.1D: © Frederick D. Atwood. 7.2 : © M. Eichelberger/Visuals Unlimited. Courtesy of W.P. Wergin and E.H. Newcomb, Univ. of Wisconsin/Biological Photo Service. 7.3A: © Runk/Schoenberger/Grant Heilman. 7.7: © Chris-

tine L. Case. 7.13A: © Tom and Pat Leeson.

Chapter 8: Chapter opener: © Dr. V.D. Vacquer/Univ. of California, San Diego. 8.0: © Daniel W. Gotshall/Visuals Unlimited. 8.1A: © Biophoto Associates/Photo Researchers, Inc. 8.1B: Courtesy of The Inouye and Yaneshiro Families. 8.3B: © Lee D. Simon/Photo Researchers, Inc. 8.4: © Andrew S. Bajer/Univ. of Oregon. 8.5A: © Biophoto Associates/Photo Researchers. 8.7: © Ed Reschke. 8.8: Courtesy of Richard Macintosh. 8.9A: © David M. Phillips. 8.9B: Micrograph by B.A. Palevitz. Courtesy of E.H. Newcomb, Univ. of Wisconsin. 8.11A: © Biophoto Associates/Photo Researchers, Inc. 8.11C: © 1988 Custom Medical Stock Photography. 8.12A: Carolina Biological Supply. 8.12B: © Ed Reschke. 8.12C: © Biophoto Associates/Science Source/Photo Researchers, Inc. 8.18B: Courtesy of Stan Short/Jackson Laboratory. 8.19A: © Cabisco/Visuals Unlimited. Page 147: Carolina Biological Supply.

Chapter 9: Chapter opener: © Hans Reinhard/Bruce Coleman Photography. 9.2A: Bettmann Archives. 9.5B: Science Education Resources Pty. Ltd., Blackburn, Australia. 9.8.1: © Marc Romanelli/The Image Bank. 9.8.2: © Eliane Sulle/The Image Bank. 9.8.3&4: © Ogust/Image Works. 9.8.5: © Eliane Sulle/The Image Bank. 9.8.6: Anthony Loveday, Benjamin/Cummings. 9.8.7: © Ogust/The Image Works. 9.8.8–10: Anthony Loveday, Benjamin/Cummings. 9.12: © Bill Longcore/Photo Researchers, Inc. 9.17A: From *Thomas Hunt Morgan: The Man and His Science,* Garland Allan (Princeton University Press, 1978). Photo by Dr. Tore Mohr, Fredrikstad, Norway. 9.18.1: Merlin G. Butler, Cytogenetics Laboratory, Vanderbilt Univ. 9.19A1: © Jean Claude Revy/PhotoTake, NYC. 9.19A2: © Carolina Biological Supply/PhotoTake, NYC.

Chapter 10: Chapter opener: Harold Fisher, Univ. of Rhode Island and Robley Williams, Univ. of California, Berkeley. 10.2D: © Ken Eward/Science Source/Photo Researchers, Inc. 10.3A: Cold Spring Harbor Archives. 10.3B: From J.D. Watson, *The Double Helix,* Atheneum, New York, 1968, p. 215. © 1968 by J.D. Watson. Courtesy of Cold Spring Harbor Laboratory Library Archives. 10.3D3: Richard Wagner, UCSF Graphics. 10.5B: © Christine L. Case. 10.8: From *Prog. Nucleic Acid Res. and Mol. Biology III,* 35 (1964) Academic Press Photo by D.M. Prescott, Ph.D., University of Colorado Medical School. 10.9C: © M.A. Rould, J.J. Perona, P. Vogt, and T.A. Steitz, *Science* 246 (1 December 1989): cover. Copyright 1989 by the American Association for the Advancement of Science. 10.17A: © Visuals Unlimited. 10.19: © Lennart Nilsson, Boehringer Ingelheim International.

Chapter 11: Chapter opener: © David Scharf. 11.4: © Ed Reschke. 11.7A2: Courtesy of R.D. Kornberg. 11.7A3: G.F Bahr, Armed Forces Institute of Pathology/Biological Photo Service. 11.7B: © Hans Reinhard/Bruce Coleman, Inc. 11.8A: P. Bryant, Univ. of California, Irvine/Biological Photo Service. 11.8B: Courtesy of Dr. Jose Mariano Amabis, University of Sao Paulo. 11.13A: © E.R. Turner, Indiana Univ.. 11.18A,B: UPI/Bettmann Newsphotos. 11.18C: Reuters/Bettmann.

Chapter 12: Chapter opener: © Dr. Dennis Kunkel/PhotoTake, NYC. Chapter opener (inset): © Nita Winter. 12.0: Courtesy of UCSF School of Medicine. 12.2C: Courtesy of Huntington Potter and David Dressler/Harvard Medical School. 12.3A: © Associated Press/World Wide Photos. 12.3B: Alain McGlaughlin, Benjamin/Cummings. 12.12: Courtesy of Genentech. 12.13: © 1987 David J. Cross. 12.15: Courtesy of Dr. R.L. Brinster, School of Veterinary Medicine, Univ. of Pennsylvania. 12.16: Courtesy of Univ. of California, San Francisco.

Chapter 13: Chapter opener: From the Collection of the Vineyard Museum/Dukes County Historical Society. 13.1: Victor Eroschenko, Benjamin/Cummings. 13.2: © Department of Clinical Cytogenetics, Adder Brookes Hospital, Cambridge/Science Photo Library/Photo Researchers, Inc. 13.3A: © CNRI/Science Photo Library/Photo Researchers, Inc. 13.3B: © Richard Hutchings/Photo Researchers, Inc. 13.8B: © Dick Zimmerman/Shooting Star. 13.9A: Courtesy of School of Optometry/Univ. of California, Berkeley. 13.9B: © Culver Pictures, Inc. 13.10: © Will and Deni McIntyre/Photo Researchers, Inc. 13.11C: © Blair Seitz/Photo Researchers, Inc. 13.11D: © Howard Sochurek. 13.13A, B: Courtesy of Dr. Nancy Wexler, Columbia Univ. 13.16A: © Stephen Ferry/The Gamma Liaison Network. 13.16B: © Courtesy of Cellmark Diagnostics. 13.18: Applied Biosystems, Foster City. 13.19A: Robin Heyden, Benjamin/Cummings. 13.19B: Courtesy of Dr. Nancy Wexler, Columbia Univ. 13.19C: © 1993 Davis Freeman.

Chapter 14: Chapter opener: © Joe McDonald/Animals Animals. 14.1A: © Frans Lanting/Minden Pictures. 14.1B: Larry Burrows, Life Magazine © Time Inc. 14.3A: © John Colwell/Grant Heilman Photography, Inc. 14.3B1: © Gerard Lacz/Peter Arnold. 14.3B2&3: © Ron Kimball. 14.3B4&5: © Gerard Lacz/Peter

Pat Leeson. 14.3C4: © D. Robert Franz/The Wildlife Collection. 14.3C5: © Joe McDonald/Tom Stack and Associates. 14.4A: OSF/Animals Animals. 14.4B: © Breck P. Kent/Animals Animals. 14.5: USAF , NOAA/NESDIS at Univ. of Colorado, CIRES/National Snow and Ice Data Center. 14.7A: © Leonard Lee Rue III/Photo Researchers, Inc. 14.10B: © Frank S. Balthis. 14.12: © R. Andrew Odum/Peter Arnold, Inc. 14.16: © Gunter Ziesler/Peter Arnold, Inc. 14.17: © S. Rhia/Custom Medical Stock Photography.

Chapter 15: Chapter opener: © E. R. Degginger/Animals Animals. 15.0: Robert E. Hynes © National Geographic Society. 15.1A1: Don and Pat Valenti/Tom Stack & Associates. 15.1A2: © John Shaw/Tom Stack & Associates. 15.1B1: © John Eastcott/Yva Momatiuk/Woodfin Camp & Associates. 15.1B2: © Sven-Olof Lindblad/Photo Researchers, Inc. 15.1B3: © Nita Winter. 15.1B4: © George Holton/Photo Researchers, Inc. 15.3B: © Robert Lee/Photo Researchers, Inc. 15.3C: © John Eastcott/The Image Works. 15.4: © Larry Ulrich/DRK Photo. 15.4A: © John Shaw/Bruce Coleman, Inc. 15.4B: © M.P.L. Fogden/Bruce Coleman. 15.5A: © David Cavagnaro. 15.6: Courtesy of the Univ. of Amsterdam. 15.7A: © Marge Lawson. 15.8C1: © T. Kitchin/Tom Stack and Associates. 15.8C2: © Bob McKeever 1981/Tom Stack & Associates. 15.8C3: © Mike Andrews/Animals Animals. 15.9: Annalisa Kraft, Benjamin/Cummings.

Chapter 16: Chapter opener: Donna Braginetz, courtesy of Natural History Magazine. 16.1A: © Doug Lee. 16.1B: © Margo Crabtree. 16.1C1: © Manfred Kage/Peter Arnold, Inc. 16.1C2: © Chip Clark. 16.1D: © Tom Till. 16.1E: © Walter Hodge/Peter Arnold, Inc. 16.2: © Tom Bean 1986/DRK Photo. 16.5C: © Zig Leszczynski/Animals Animals. 16.6A: © 1992 Kevin Schafer. 16.6B: © John Meehan/Photo Researchers, Inc. 16.8A: © Jim Strawser/Grant Heilman. 16.8: © Richard Schiell/Earth Scenes. 16.9A: © Jane Burton/Bruce Coleman. 16.9C: © The Walt Disney Company. 16.13: © Tom McHugh/Photo Researchers, Inc.

Chapter 17: Chapter opener: © Jonathan Blair 1990/Woodfin Camp and Associates. 17.0: S.M. Awramik, Univ. of California, Santa Barbara/Biological Photo Service. 17.1A: © Sigurgier Jonasson. 17.1B,D: S.M. Awramik, Univ. of California, Santa Barbara/Biological Photo Service. 17.3: © Roger Ressmeyer/Starlight. 17.6B: Sidney Fox, Univ. of Miami/Biological Photo Service. 17.7: Dr. Tony Brain/David Parker/Science Photo Library/Photo Researchers, Inc. 17.8: © Christine L. Case. 17.9: © David M. Phillips/Visuals Unlimited. 17.10A: J.R. Waaland, Univ. of Washington/Biological Photo Service. 17.10B: © Christine L. Case. 17.12: © Larry Ulrich/DRK Photo. 17.13A: © Lee D. Simon/Science Source/ Photo Researchers, Inc. 17.13B: S. Abraham and E.H. Beachey, VA Medical Center, Memphis. 17.13C: © H.S. Pankratz, T.C. Beaman/Biological Photo Service. 17.13D: © David Scharf/Peter Arnold. 17.14.1: © Frederick D. Atwood. 17.14.2: © Michael Gabridge/Visuals Unlimited. 17.15A: © Christine L. Case. 17.15B1: © L. West/Bruce Coleman. 17.15B2: Centers for Disease Control. 17.16A: The Bettmann Archive. 17.17: Douglas Munnecke/Biological Photo Service. 17.18A: Courtesy of Exxon Corporation. 17.18B: Courtesy of Susan Barnett. 17.21: © Roland Birke/Peter Arnold. 17.22A: © Peter Parks/Animals Animals. 17.22B: John Mansfield/Univ. of Wisconsin. 17.22C: © Masamichi Aikawa. 17.22D1: © Roland Birke/Peter Arnold, Inc. 17.22D2: © Eric Grave/Science Source/Photo Researchers, Inc. 17.23.1: Courtesy of Matt Springer, Stanford Univ. 17.23.2: Courtesy of Robert Kay, Cambridge Univ. 17.23.3: Robert Kay, Cambridge Univ. © The Company of Biologists, Ltd. 17.24.1: © E.R. Degginger/Earth Scenes. 17.24.2: © George Loun/Visuals Unlimited. 17.25A: © Biophoto Associates/Photo Researchers, Inc. 17.25B: © Manfred Kage/Peter Arnold, Inc. 17.25C1: From W.J. Snell, "Mating in *Chlamydomonas*," J. Cell Biol. 68(1976)48–69. 17.25C2: © Manfred Kage/Peter Arnold. 17.26A: © Lawrence Naylor/Photo Researchers, Inc. 17.26B: © Gary Robinson/Visuals Unlimited. 17.26C: © D.P. Wilson/Eric and David Hosking/Photo Researchers, Inc.

Chapter 18: Chapter opener: © Thomas Hovland/Grant Heilman Photography. 18.0: © J.M. Graham/Univ. of Florida. 18.1B: © Rod Planck/Tom Stack & Associates. 18.2A: © L.E. Graham. 18.2B: © Chip Clark 1988. 18.3B: © Stephen Kraseman/DRK photos. 18.3C1: © Glenn Oliver/Visuals Unlimited. 18.3C2: © Walter H. Hodge/Peter Arnold, Inc. 18.5.2: © Glenn Oliver/Visuals Unlimited. 18.6.2: © Milton Rand/Tom Stack and Associates. 18.7: Field Museum of Natural History, Neg# 75400c, Chicago. 18.9.1: © Dr. William M. Harrow/Photo Researchers, Inc. 18.10A: © Pat O'Hara/DRK Photo. 18.10B: © Lara Hartley. 18.12A: © M.E. Warren/Photo Researchers, Inc. 18.14A: © Jane Burton/Bruce Coleman, Inc. 18.14B: © 1993 Lara Hartley. 18.14C: © Dwight R. Kuhn. 18.15A: © D. Wilder. 18.15B: © Bob and Clara Calhoun/Bruce Coleman. 18.15C: © Merlin D. Tuttle/Bat Conservation International. 18.16A: © John D. Cunningham/Visuals Unlimited. 18.16B: N. Allin and G.L. Barron, Univ. of

Guelph/Biological Photo Service. 18.16C: Steven D. Aust. 18.17B: © Kjell B. Sandved/Visuals Unlimited. 18.17C: © Kerry T. Givens/Tom Stack & Associates. 18.17D: © James W. Richardson/Visuals Unlimited. 18.19A: © Robert Lee/Photo Researchers. 18.19C: © Mark Newman/Tom Stack & Associates. 18.20A: © John D. Cunningham/Visuals Unlimited. 18.20B: © Leonard Lee Rue III/Photo Researchers. 18.21: © Mark Wicklaus/Western Michigan Univ.

Chapter 19: Chapter Opener: © Norbert Wu. 19.0: © Chris Huss/Wildlife Collection. 19.1: Anthony Mercieca/Photo Researchers. 19.3A: C.R. Wyttenbach, Univ. of Kansas/Biological Photo Service. 19.4A: © Gwen Fidler/Comstock. 19.4B: © Claudia Mills/Friday Harbor Labs. 19.6B: Centers for Disease Control. 19.6C: © Drs. Kessel and Shih/Peter Arnold. 19.8A, B: © L.S. Stepanowicz/Photo Researchers, Inc. 19.9B: © Robert and Linda Mitchell. 19.9C: © 1988 Chris Huss. 19.9D: H.W. Pratt/Biological Photo Service. 19.9E: © Carl Roessler/Animals Animals. 19.9F: J.W. Porter, University of Georgia/Biological Photo Service. 19.10B: © John Gerlach/Tom Stack & Associates. 19.10C: © Michael Neveux. 19.11A: © Jeff Goodman/NHPA. 19.11B: © Kjell B. Sandved. 19.11C: © Robert and Linda Mitchell. 19.12B: © Zig Leszczynski/Animals Animals. 19.12C1: © Robert and Linda Mitchell. 19.12C2: © Paul Skelcher/Rainbow. 19.12C3: © David Scharf. 19.12D: © Tom Stack/Tom Stack and Associates. 19.12E: © Robert and Linda Mitchell. 19.14B: © Gary Milburn/Tom Stack. 19.14C: © David Wrobel. 19.15A: Ralph Buchsbaum. 19.15B: © Robert Brons/Biological Photo Service. 19.17A: Tom McHugh, Steinhart Aquarium/Photo Researchers. 19.18A: Steinhart Aquarium/Photo Researchers, Inc. 19.18B2: © Fritz Prenzel/Animals Animals. 19.18B1: © Richard Ellis/Photo Researchers. 19.19A: © 1990 Suzanne L. and Joseph T. Collins/Photo Researchers, Inc. 19.19B: © Hans Pfletschinger/Peter Arnold. 19.20A: © Robert and Linda Mitchell. 19.20B: © 1988 C. Allan Morgan. 19.20C: The Natural History Museum, London. 19.21A: John P. O'Neill, *Science* 295(5 February 1993): cover. Copyright 1993 by the American Association for the Advancement of Science. 19.21B: © B.K. Wheeler/VIREO. 19.21C: © David Dare Parker/Auscape International. 19.22A: © Jean-Paul Ferrero/Auscape International. 19.22B: © E.R. Degginger/Animals Animals. 19.22C: © Mitch Reardon/Photo Researchers, Inc. 19.24: © C. Seghers/Photo Researchers, Inc.

Chapter 20: Chapter opener: © G.L. Kooyman/Animals Animals. 20.1: © Janice Sheldon. 20.7: © Ed Reschke. 20.10A: © Mehau Kulyk/Science Photo Library/Photo Researchers, Inc. 20.10A2: © Howard Sochurek. 20.10B&C: © Howard Sochurek. 20.10D: Courtesy of M.E. Raichle. 20.11C: © Science Photo Library/Photo Researchers, Inc. 20.12A: © Frans Lanting/Minden Pictures.

Chapter 21: Chapter opener: Richard Schlecht © National Geographic Society. 21.0: © C. Allan Morgan/Peter Arnold, Inc. 21.1A: © Nigel Dennis/NHPA. 21.1B: © Jett Britnell/DRK Photos. 21.1C: © Tom Eisner. 21.1D: © Hans Pfletschinger/Peter Arnold, Inc. 21.1E: Wide World Photos, Inc. 21.12B: © Mary Mitchell. 21.14B: © William Thompson.

Chapter 22: Chapter opener: © 1991 Jim Brandenburg/Minden Pictures. 22.0: © Dennis and Bo Hansen/Ric Ergenbright Photography. 22.6C: © CNRI/Photo Researchers Inc. 22.7: © Martin Rotker. 22.8B3: © Professor Dr. Hans Rainer Dunker, Justus Liebig Univ., Giessen.

Chapter 23: Chapter opener: © Kennan Ward 1986/DRK. 23.0: © Malcolm Boulton/Photo Researchers, Inc. 23.1A: © Ed Reschke. 23.8B: © Mitch Kezar/PhotoTake NYC. 23.12A: © D.W. Fawcett/Photo Researchers, Inc. 23.14: © David M. Phillips/Visuals Unlimited. 23.15.1–5: Victor Eroschenko, Benjamin/Cummings. 23.16: © Lennart Nilsson/Boehringer Ingelheim International GmbH/*The Body Victorious*, Delacorte Press. 23.17: © Teri Gilman 1992.

Chapter 24: Chapter opener: © Stephen Dunn/Allsport. 24.1A: © Lennart Nilsson/Boehringer Ingelheim International GmbH. 24.5D: © Bruce Iverson/Photo Researchers. 24.10A: Courtesy of A.J. Olson, Scripps Clinic and Research Foundation. 24.12B: Custom Medical Stock Photography. 24.14: © Lennart Nilsson, Boehringer Ingelheim International GmbH.

Chapter 25: Chapter opener: © Marty Snyderman 1992. 25.0: © Art Wolfe. 25.2A: © Robert and Linda Mitchell. 25.3: © E. Lyons/Bruce Coleman. 25.4: © Merlin D. Tuttle/Bat Conservation International. 25.6: © David Madison 1986. 25.7A: © Charles Preiter/Visuals Unlimited. 25.7B: © John Crowe.

Chapter 26: Chapter opener: © Michael P. Gadomski/Photo Researchers, Inc. 26.6A: © Ivan Polunin/Bruce Coleman, Inc. 26.11: AP/Wide World Photos. 26.12: © 1989 Kennan Ward.

Chapter 27: Chapter opener: © David Crews, Photo modification by Joseph Maas, Paradigm[3]. 27.1A: D. Wrobel. 27.1B: Jim Solliday/Biological Photo Service. 27.1C: © Hans Pfletschinger/Peter Arnold, Inc. 27.1D: © Dwight Kuhn. 27.2B: C. Edelman/La Vilette/Photo Researchers, Inc. 27.8: Anthony Loveday, Ben-co/Visuals Unlimited. 27.12C: Thomas Poole, SUNY Health Science Center. 27.12D: © Hans Pfletschinger/Peter Arnold, Inc. 27.13B: Dr. Jean-Paul Thiery, Reproduced from *J. of Cell Bio.* 1983, Vol. 96, pp. 462–473. Rockefeller Univ. Press. 27.17A-D: © Lennart Nilsson/*A Child is Born*, Dell Publishing. 27.17E: © 1992 Keith/Custom Medical Stock Photography. 27.19: © Andy Walker, Midland Fertility Services/Science Photography Library/Photo Researchers, Inc.

Chapter 28: Chapter opener: © Norbert Wu. 28.0.2: A.L. Hodgkin and R.D. Keynes, *J. Phys.* 131(1956):615. 28.8: Dr. E.R. Lewis, ERL-EECS, Univ. of California, Berkeley. 28.18A: Courtesy of Roger Sperry. 28.19A: Custom Medical Stock Photography. 28.21: © Daniel Gotshall/Visuals Unlimited.

Chapter 29: Chapter opener: © Michio Hoshino/Minden Pictures. 29.0: © Keith Gunnar/Bruce Coleman, Inc. 29.3C1: Oxford Scientific Films/Animals Animals. 29.3C2: Dr. R.A. Steinbrecht. 29.3D: © Joe McDonald/Animals Animals. 29.4B: © John L. Pontier/Animals Animals. 29.4B2: © Syd Greenberg/Photo Researchers, Inc. 29.9 D1: Courtesy Dr. Andrew Forge, University College London Medical School. 29.9D: Courtesy Dr. Andrew Forge, Univ. College London Medical School.

Chapter 30: Chapter opener: © Stephen Dalton/NHPA. 30.1A: © Marty Snyderman. 30.1B: © John Cancalosi/Tom Stack & Associates. 30.1C: © Gerard Lacz/NHPA. 30.1E: © Stephen J. Krasemann/DRK Photo. 30.2A: © Dwight Kuhn. 30.2B: © Tony Florio/Photo Researchers, Inc. 30.2C: © E.R. Degginger/Animals Animals. 30.2D1: © Marty Snyderman. 30.2D2: © David Wrobel. 30.6A: © Carolina Biological Supply/PhotoTake, NYC. 30.8: Courtesy of Clara Franzini-Armstrong. 30.11A: © Matthew Mulbry. 30.11B: © Bruce Frisch. 30.12A: © Al Tielemans 1990/Duomo. 30.12B: © Norman Owen Tomalin/Bruce Coleman, Inc.

Chapter 31: Chapter opener: National Park Service. 31.0: © Tom McHugh/Photo Researchers, Inc. 31.1A: Courtesy of J. Thorsch. 31.1B: Univ. of California, Santa Barbara, Photography Department. 31.4C top: © Larry Mellichamp/Visuals Unlimited. 31.4C bottom: © Kevin Schafer. 31.5B, C2: © Nels Lersten, Iowa State University. 31.5D2,4, © Bruce Iverson. 31.5E2: George J. Wilder/Visuals Unlimited. 31.5F.2: © Randy Moore/Visuals Unlimited. 31.6B: © Ed Reschke. 31.6C, D2, 31.7C: © Ed Reschke. 31.8B1: © Runk/Schoenberger/Grant Heilman Photography, Inc. 31.12A: W. H. Hodge/Peter Arnold. 31.12A: W. H. Hodge Peter Arnold. 31.12C: Anthony Loveday, Benjamin/Cummings. 31.13A: © Barry Runk/Grant Heilman. 31.13B: © Runk/Schoenberger/Grant Heilman. 31.14A: © Kevin Schafer. 31.14B: © Frank Balthis. 31.14C: © Galen Rowell, Mountain Light. 31.14D: J.N.A. Lott, McMaster Univ./Biological Photo Service. 31.15A: Runk Schoenberger/Grant Heilman. 31.15B: Plantek/Photo Researchers, Inc.

Chapter 32: Chapter opener: © Jane Grushow/Grant Heilman. 32.0: The Bettmann Archive. 32.1B: © Renee Lynn/Photo Researchers, Inc. 32.2A: © Runk/Schoenberger/Grant Heilman. 32.5A: Biophoto Associates/Photo Researchers, Inc. 32.5C: © M.H. Zimmermann. 32.6B: © Grant Heilman/Grant Heilman Photography. 32.7A, C, D: Courtesy of James Pushnik, California State Univ., Chico. 32.7B: © Holt Studios/Earth Scenes. 32.8A: Steven C. Wilson/Entheos. 32.9A: © Runk/Schoenberger/Grant Heilman. 32.9B: © Grant Heilman/Grant Heilman Photography. 32.10: Courtesy of Stephen Moore. 32.11: © R.L. Peterson, Univ. of Guelph/Biological Photo Service. 32.12A: © Kevin Schafer. 32.12B: © Jim Strawser/Grant Heilman. 32.12C: © Robert and Linda Mitchell. 32.12D: © Jeff Lepore/Photo Researchers, Inc. 32.14A: © Breck P. Kent/Earth Scenes. 32.14B: From E.H. Newcomb and S.R. Tandon, Univ. of Wisconsin, Madison/Biological Photo Service. 32.15A: Louisiana State Univ. Public Relations. 32.16: © John C. Sanford/Cornell Univ.

Chapter 33: Chapter opener: © Jeanne White/National Audubon Society/Photo Researchers, Inc. 33.0: © E.R. Degginger/Photo Researchers, Inc. 33.1A: © John Kaprielian/Photo Researchers, Inc. 33.3A: © David Newman/Visuals Unlimited. 33.4: © Walter Chandoha. 33.5A: © Tugio Sasaki, Institute for Agricultural Research, Japan. 33.5B: Fred Jensen/Univ. of California, Davis. 33.6: © Larry Ulrich. 33 7A: © Runk/Schoenberger/Grant Heilman, Inc. 33.7B: © Ed Reschke. 33.8: © John Colwell/Grant Heilman Photography, Inc. 33.9A: © Michael Evans, Ohio State University. 33.9B: © Peter G. Aitken/Photo Researchers, Inc. 33.10: © Frank B. Salisbury. 33.13.1: Robin Heyden. 33.13.2: Univ. of Connecticut.

Chapter 34: Chapter opener: © National Geographic Society/Emory Kristof. 34.0A: © J. Edmond/Visuals Unlimited. 34.0B: © Dudley Foster/Woods Hole Oceanographic Institute. 34.1: © Carl Wirsen/Woods Hole Oceanographic Institute. 34.2A: Courtesy of NASA. 34.2B: © Stephen J. Krasemann/Photo Researchers, Inc. 34.3: The Bettmann Archive. 34.4: © Arthur C. Smith III/Grant Heilman Photography. 34.5: © Erwin and Peggy Bauer/Natural Selection.

34.7A: © M.E. Warre/Photo Researchers, Inc. 34.7B: © Anne Wetheim Rosenfeld. 34.7D: © Douglas Faulkner/Photo Researchers, Inc. 34.8A: © Steve Solum/Bruce Coleman, Inc. 34.8B: © Wendell Metzen/Bruce Coleman, Inc. 34.10: © Peter Ward/Bruce Coleman. 34.11: Alain McLaughlin, Benjamin/Cummings. 34.12: © Jonathon Scott/Planet Earth Pictures. 34.13: © Charlie Ott/Photo Researchers, Inc. 34.14: © Doug Wechsler/Earth Scenes/Animals Animals. 34.15: © Tom McHugh/Photo Researchers, Inc. 34.16: © John Shaw. 34.17: © Lars Egede-Nissan/BPS. 34.18: © J. Warden/Superstock, Inc.

Chapter 35: Chapter opener: © Maslowski/Visuals Unlimited. 35.2A: Jerry Wolff. 35.2B: © Kirtley Perkins 1983/Visuals Unlimited. 35.2C1: © Robert P. Carr/Bruce Coleman, Inc. 35.2C2: © Breck R. Kent/Earth Scenes/Animals Animals. 35.3B: © Erwin and Peggy Bauer/Natural Selection. 35.4B: © Runk/Schoenberger/Grant Heilman. 35.5: Alan Carey/Photo Researchers, Inc. 35.6A: John Endler, Univ. of Calif., Santa Barbara. 35.6B: © R. Calentine/Visuals Unlimited. 35.6C: © Max and Bea Hunn/Visuals Unlimited. 35.8C1: © Jan Halaska/Photo Researchers, Inc. 35.8C.2: © Wesley Bocxe/Photo Researchers, Inc. 35.9A: © Alain Evrard/Photo Researchers, Inc. 35.9E: © Arvind Garg. 35.10.1: © Cynthia Johnson/TIME Magazine. 35.10.2: © Breton Littlehales.

Chapter 36: Chapter opener: © Y. Sato. 36.0.1: © John Serrao/Visuals Unlimited. 36.0.2: © Norm Thomas/Photo Researchers, Inc. 36.0.3: © E.R. Degginger/Photo Researchers, Inc. 36.1: © Marcia W. Griffen/Animals Animals. 36.3A: © L.E. Gilbert, Univ. of Texas, Austin/Biological Photo Service. 36.3B: © Kevin Schafer/Tom Stack & Associates. 36.3C: © C. Allan Morgan/Peter Arnold, Inc. 36.3D1: © J. Alcock/Visuals Unlimited. 36.3D2: © L. West/Photo Researchers, Inc. 36.3E1: © E.S. Ross. 36.3E2: © Runk/Schoenberger/Grant Heilman. 36.4A: Dr. Gregory A. Antipa. 36.4B: © William E. Townsend/Photo Researchers, Inc. 36.5A: Australian News and Information Bureau. 36.5B: © Robert and Linda Mitchell. 36.6A: © Mitchell Bleier/Peter Arnold, Inc. 36.6B: Tom Bean. 36.17A: © John D. Cunningham/Visuals Unlimited. 36.17B: Furnished by Robert Pierce/U.S. Department of Agriculture, Northeastern Experimental Forest Station, Durham, NH. 36.18: D.W. Schindler, *Science* 184 (24 May 1974): 897.

Chapter 37: Chapter opener: © Frans Lanting/Minden Pictures. 37.0: Alan Rabinowitz, New York Zoological Society. 37.3B2: © Ian Wyllie/Survival Anglia. 37.3B3: © Michael Leach/OSF/Animals Animals. 37.5A: Thomas McAvoy, Life Magazine. 37.5B2: © Calvin Larsen/Photo Researchers, Inc. 37.6A: © David Weintraub 1986/Photo Researchers,Inc. 37.6B1: © Harry Engels/Animals Animals. 37.6B2: © 1992: Stephen J. Krasemann/Photo Researchers, Inc. 37.7B: F. DeWaal/Primate Center Library. 37.9A: © Stouffer/Animals Animals. 37.9B: © Thomas Ives. 37.10, 37.11B: © François Gohier/Photo Researchers, Inc. 37.11C: © Jonathan Blair/Woodfin Camp. 37.12A: © Kenneth Mallory/New England Aquarium. 37.12B: © John Cancalosi/Peter Arnold, Inc. 37.12C: © Runk/Schoenberger/Grant Heilman Photography. 37.12D1: © Joe McDonald/Animals Animals. 37.12D2: © Kevin Schafer/Peter Arnold, Inc. 37.12E: © Joe McDonald/Animals Animals. 37.14: © Gordon Wiltsie.37.15: Renee Lynn/Benjamin/Cummings. 37.16A: © Ken Regan/Camera 5. 37.16B: © Warren and Genny Garst/Tom Stack & Associates. 37.17A: © Bill Bachman/Photo Researchers, Inc. 37.17B: © Jean-Michel/Jacana/Photo Researchers, Inc. 37.17C © Bruce Davidson/Animals Animals. 37.18A: © Peter Roberts. 37.18B: © Anthony Mercieca/Photo Researchers, Inc. 37.19A: © J.P. Varin/Jacana/Photo Researchers, Inc. 37.19B: © Kenneth Lorenzen, Univ. of California, Davis. 37.20A: © Frans Lanting/Minden Pictures. 37.20B: © 1990 Richard R. Hansen/Photo Researchers, Inc. 37.21: © Micha Bar'Am/Magnum. 37.22: © Ivan Polunin/Bruce Coleman.

Chapter 38: Chapter opener: © 1993 Frans Lanting/Minden Pictures. 38.0: © Frans Lanting/Minden Pictures. 38.1: © Martin Wendler/Peter Arnold, Inc. 38.2: © E.H. Rao/Photo Researchers, Inc. 38.2B1: © Norman Owen Tomalin/Bruce Coleman, Inc. 38.2B2: © Wardene Weisser/Bruce Coleman, Inc. 38.2C: © Chicago Zoological Park/Tom McHugh/Photo Researchers, Inc. 38.3A: © E.R. Degginger/Animals Animals. 38.3B: © Ulrich Nebelsiek/Peter Arnold, Inc. 38.3C: © Nancy Adams E.P.I./Tom Stack & Associates. 38.3D: © Ken Regan/Camera 5. 38.5A: © John Reader 1982. 38.5B: © John Gurche. 38.7: © Tom McHugh/Photo Researchers, Inc. 38.9: © Anthony Bannister/Earth Scenes. 38.10: © Elliot Erwitt/Magnum Photos, Inc. 38.11.1: © Mary Evans Picture Library/Photo Researchers, Inc. 38.11.2: Courtesy of NCR. 38.12A: © Al Grillo/Alaska Stock Images. 38.14A: © David Hiser/Photographers Aspen. 38.14B: © Richard Shiell/Earth Scenes. 38.15A: © Zig Leszczynski/Animals Animals. 38.15B: © Stoddart/Katz Pix/Woodfin Camp & Associates, Inc. 38.16: © Art Wolfe.

APPENDIX 5 Illustration and Text Credits

The artists who contributed directly to *Biology: Concepts and Connections* are listed on the copyright page at the front of the book. Illustrations from Neil Campbell's *Biology* (© 1987, 1990, and 1993 The Benjamin/Cummings Publishing Company) that are used in *Biology: Concepts and Connections* are the work of Barbara Cousins, Chris Carothers, Raychel Ciemma, Pamela Drury-Wattenmaker, Cecile Duray-Bito, Janet Hayes, Darwen and Valley Hennings, Georg Klatt, Linda McVay, Kenneth Miller, Fran Milner, Elizabeth Morales-Denney, Carla Simmons, Carol Verbeeck, or John and Judy Waller.

The following figures and tables are adapted from Neil Campbell, *Biology*, 3rd ed. (Redwood City, CA: Benjamin/Cummings, 1993). © 1993 The Benjamin/Cummings Publishing Company: Figures 2.2A, 2.4A, 2.7A, 2.8, 2.9, 2.10A, 2.13, 2.14, 2.17A, 3.1, 3.2, 3.4B, 3.4C, 3.5, 3.9, 3.14A, Table 4.1, Figures 4.2, 4.3, 4.4, 4.5A, 4.5B, 4.10, 4.11, 4.14, 4.15, 4.16, 4.17B, 4.18A, 4.19A, 4.19B, Table 4.20, Figures 5.4C, 5.5B, 5.6, 5.8, 5.11B, 5.12, 5.13A-D, 5.14A, 5.14B, 5.15A, 5.15B, 5.16, 5.17, 5.19A, 5.19B, 5.21, 6.7A, 6.8, 6.11B, 6.14, 6.16, 7.2, 7.5, 7.6A, 7.7A-C, 7.8, 7.9, 7.10, 7.12A-C, 7.14, 8.3A, 8.6, 8.9B, 8.10, 8.11B, 8.14, 8.17, 9.2C, 9.2D, 9.3A, 9.3B, 9.5A, 9.5B, 9.7, 9.10A, 9.14, 9.16A, 9.16B, 9.17B, 9.18A-D, 9.19B-D, 10.1B, 10.1C, 10.2B, 10.3D, 10.5A, 10.6, 10.7A, 10.9A, 10.9B, 10.10A-C, 10.11B, 10.12, 10.14A, 10.14B, 10.15B, 10.16, 10.17C, 10.18A, 10.18B, 11.2A, 11.2B, 11.5A, 11.5B, 11.8C, 11.9A, 11.14, 12.2A, 12.2B, 12.5, 12.6, 12.7, 12.8, 12.14, 13.2, 13.3C, Table 13.5, Figures 13.4A, 13.4B, 13.6A, 13.6B, 13.11A, 13.11B, 13.14A-D, 13.17, 15.2, 15.6A. 15.8A, 15.8B, Table 16.3, Figures 16.5B, 16.9B, 16.10A, 16.10B, 16.14A, 16.15A, 16.16, 17.1C, 17.3B, 17.5, 17.6A, 17.6C, 17.11, 17.20A, 17.20B, 18.1A, 18.3A, 18.19B, 19.3C, 19.3D, 19.4C, 19.5, 19.6A, 19.6B, 19.7, 19.9A, 19.10A, 19.12A, 19.12E, 19.13A, 19.14A, 19.16B, 19.17C, 19.18A, 20.1, 20.4, 20.5, 20.6, 20.11A, 20.11B, 20.12B, 21.3A, 21.3B, 21.4, 21.6, 21.7, 21.8, 21.10B, 21.10C, 21.11, 21.12C, 21.16, 22.2, 22.3, 22.4, 22.5B, 22.8A, 22.8B, 22.11A, 22.11B, 23.2A, 23.2B, 23.3A, 23.3B, 23.4C, 23.6, 23.7, 23.9A, 23.9B, 23.10, 23.11B, 23.23B, 23.13, 23.16A, 24.2, 24.5A-C, 24.6, 24.7, 24.8A, 24.8B, 24.11, 24.12, 24.13B, 24.13C, 25.5A, 25.5B, 25.9A, 25.11, 26.3A, 26.4A, 26.4B, 26.5A, 26.5B, 26.6B, 26.7, 26.8, 26.10, 27.3B, 27.4A-C, 27.5, 27.9B, 27.12A, 27.12C, 27.13A, 27.14, 27.16A-F, 28.2, 28.4A, 28.6, 28.7, 28.8, 28.11A-C, 28.12, 28.13, 28.14, 28.15B, 28.17, 29.3A, 29.5, 29.6, 29.9A-C, 29.10, 29.12A, 29.12C, 30.3C, 30.7, 30.8, 30.9B, 30.10A, 30.10B, 31.2, 31.3A, 31.4B, 31.5D-F, 31.6A, 31.6D, 31.7A-C, 31.9, 31.10, 31.11A, 31.11B, 31.13, 32.1A, 32.2B, 32.3, 32.4, 32.6A, 32.8A-C, 32.13, 33.1B-D, Table 33.2, Figures 33.11, 33.12A, 33.12B, 34.6A-E, 34.7C, 34.9, 35.2B, 35.3B, 35.5, 35.7, Table 35.7, 35.8B, 35.9B-D, 36.2A, 36.2B, 36.8, 36.13, 36.15, 36.16, 37.2, 37.3A, 37.7A, 37.9A, 37.19B-D.

The following figures are adapted from Neil Campbell, *Biology*, 2nd ed. (Redwood City, CA: Benjamin/Cummings, 1990). © 1990 The Benjamin/Cummings Publishing Company: Figures 8.19B, 9.11, 9.13, 10.4B, 10.15A, 10.17B, 11.9B, 15.3A, 16.15B, 17.27, 20.2, 24.13A, 25.1, 31.8A, 33.3B, 36.7, 36.10.

The following figures are adapted from Neil Campbell, *Biology*, 1st Edition (Menlo Park, CA: Benjamin/Cummings, 1987). © 1987 The Benjamin/Cummings Publishing Company: Figures 5.18, 15.5A, 15.5B.

Several other Benjamin/Cummings books are sources of multiple figures adapted for *Biology: Concepts and Connections*. The original artists for these figures include Martha Blake and Charles Hoffman, as well as several of the artists for Neil Campbell's *Biology*.

The following figures are adapted from Lawrence G. Mitchell, John A. Mutchmor, and Warren D. Dolphin, *Zoology* (Menlo Park, CA: Benjamin/Cummings, 1988). © 1988 The Benjamin/Cummings Publishing Company: Figures 2.6, 19.1B, 19.15, 19.16, 19.20A, 20.0, 20.1, 21.10B, 21.12A, 25.5B, 25.8, 28 Opener, 29.12B, 30.1, 30.2E, 37.1.

The following figures are adapted from Elaine M. Marieb, *Human Anatomy and Physiology*, 2nd ed. (Redwood City, CA: Benjamin/Cummings, 1992). © 1992 The Benjamin/Cummings Publishing Company: Figures 4.8A, 20.9A–J, 21.5, 22.10A, 22.12, 24.4, 24.9, 26.8, 27.2A, 27.2C, 27.3A, 27.18B, 29.7A, 29.7B, 30.3A, 30.5, 30.6B, 30.9A, 30.10B.

The following figures are adapted from Elaine M. Marieb, *Human Anatomy and Physiology*, 1st ed. (Redwood City, CA: Benjamin/Cummings, 1989). © 1989 The Benjamin/Cummings Publishing Company: Figures 8.12B, 25.12, 29.4C.

The following figures are adapted from Gerard J. Tortora, Berdell R. Funke, and Christine L. Case, *Microbiology: An Introduction*, 4th ed. (Redwood City, CA: Benjamin/Cummings Publishing, 1992). © 1992 The Benjamin/Cummings Publishing Company: Figures 10.1A, 12.4, 12.8, 17.16B, 17.17, 24.10.

The following figures are adapted from C. K. Mathews and K. E. van Holde, *Biochemistry* (Redwood City, CA: Benjamin/Cummings, 1990). © 1990 The Benjamin/Cummings Publishing Company: Figures 6.12, 6.13, 8.8, 11.7A.

Other artists and sources of illustrations in *Biology: Concepts and Connections* are listed below, along with sources for tables and text quotations.

Figure 1.14A: Reuter News Service, reprinted in *The New York Times*, August 16, 1993, Science Section B7.

Figures 3.15C and D: Adapted from C. C. F. Blake et al., *Journal of Molecular Biology 88(1974):1–12.* **Module 3.19,** Talking About Science: Adapted from an interview in Neil Campbell, *Biology,* 1st ed. (Menlo Park, CA: Benjamin/Cummings, 1987).

Figure 4.18B: From B. Alberts, D. Bray, J. Lewis, M. Raff, K. Roberts, and J. D. Watson, *Molecular Biology of the Cell,* 2nd ed. (New York: Garland, 1989).

Table 6.3: Data from C. M. Taylor and G. M. McLeod, *Rose's Laboratory Handbook for Dietetics,* 5th ed. (New York: Macmillan, 1949), p. 18; J. V. G. A. Durnin and R. Passmore, 1967. *Energy and Protein Requirements,* in *FAO/WHO Technical Report No. 522,* 1973; W. D. McArdle, F. I. Katch, and V. L. Katch. *Exercise Physiology* (Philadelphia: Lea & Feibiger, 1981); R. Passmore and J. V. G. A. Durnin, Physiological Reviews 35 (1955): 801–840.

Figure 9.12: From D. Suzuki, A. Griffith, J. Miller, and R. Lewontin, *Introduction to Genetic Analysis,* 4th ed.(New York: W. H. Freeman, 1989). Copyright © 1989 by W.H. Freeman and Company. Reprinted by permission.

Chapter 10 Opening Figure: p. 171: Adapted from James D. Watson, N. H. Hopkins, J. W. Roberts, J. A. Steitz, and A. M. Weiner, *Molecular Biology of the Gene, 4th ed.* (Menlo Park, CA: Benjamin/Cummings, 1987). © 1987 James D. Watson.

Figure 11.5B: Adapted from W.M. Becker, *The World of the Cell* (Redwood City, CA: Benjamin/Cummings, 1986), p. 592. **Table 11.17:** Data from L. A. G. Ries, B. F. Hankey, and B. K. Edwards, eds., *Cancer Statistics Review 1973–1987,* NIH Publication 90-2789 (Bethesda, MD: National Institutes of Health, 1990).

Figure 13.1: ©Michelin. Adapted from the *Green Guide to Washington, DC.* (Greenville, SC: Michelin Travel Publications, 1991). **Figure 13.7:** Reprinted by permission of the publishers from *Everyone Here Spoke Sign Language,* by Nora Ellen Groce (Cambridge, MA: Harvard University Press). © 1985 by Nora Ellen Groce. **Module 13.13,** Talking About Science: Adapted from an interview in Neil Campbell, *Biology,* 2nd ed. (Redwood City, CA: Benjamin/Cummings, 1990). **Module 13.19:** Text quotation from James Watson, Science, April 6, 1990. Copyright © 1990 by the American Association for the Advancement of Science. Text quotation from Leroy Hood, *Science News,* January 21, 1989. Science Service, Inc.

Module 15.9, Talking About Science: Adapted from interview in Neil Campbell, *Biology,* 3rd ed. (Redwood City, CA: Benjamin/Cummings, 1993).

Figure 16.5D: From W. K. Purves and G. H. Orians, *Life: The Science of Biology,* 2nd ed. (New York: Sinauer Associates, 1987), p. 1180.

Figure 16.7: Art adapted from "What Caused the Mass Extinction?" © October 1990, by Scientific American. All Rights Reserved. **Figure 16.11:** Adapted from Biruta Hansen, © 1986 *Discover* Magazine.

Module 17.3, Talking About Science: Adapted from an interview in Neil Campbell, *Biology,* 2nd ed. (Redwood City, CA: Benjamin/Cummings, 1990).

Table 21.15: "Weighty Issues/Evaluating Diets." Chicago: American Dietetic Association, 1987. C. L. Rock and A. M. Coulston, "Weight-Control Approaches: A Review by the California Dietetic Association," *Journal of the American Dietetic Association* 88(1988):44–48. **Table 21.17:** Data from RDA Subcommittee, *Recommended Dietary Allowances,* Washington, DC: National Academy Press, 1989; M. E. Shils and V.R. Young, *Modern Nutrition in Health and Disease* (Philadelphia: Lea & Feibiger, 1988). **Table 21.18:** Data from (1) M. E. Shils, "Magnesium," in M. E. Shils and V. R. Young, eds., *Modern Nutrition in Health and Disease* (Philadelphia: Lea & Feibiger, 1988); (2) V. F. Fairbanks and E. Beutler, "Iron" [same as (1)]; (3) N. W. Solomons, "Zinc and Copper" [same as (1)]; (4) RDA Subcommittee, *Recommended Dietary Allowances* (Washington, DC: National Academy Press; 1989); (5) E. J. Underwood, *Trace Elements in Human and Animal Nutrition* (New York: Academic Press, 1977).

Figure 22.10B: Copyright © Irving Geis.

Chapter 24 Opening essay: Excerpt from Magic Johnson with Ray Johnson, "I'll Deal With It." Reprinted courtesy of *Sports Illustrated* from the November 18, 1991 issue. Copyright © 1991, Time, Inc. All Rights Reserved.

Figure 26.9: Data from C.J. Byrne, D.F. Saxton, et al., *Laboratory Tests: Implication for Nursing Care,* 2nd ed. (Menlo Park, CA: Addison-Wesley, 1986). © 1986 Addison-Wesley Publishing Company.

Figure 27.3D: Adapted from Robert Crooks and Karla Baur, *Our Sexuality, 5th ed.* (Redwood City, CA: Benjamin/Cummings, 1993), Figure 5.9, © 1993 The Benjamin/Cummings Publishing Company. **Table 27.8:** Data from R. Hatcher et al., *Contraceptive Technology, 1990-1992* (New York: Irvington, 1990), p. 134.

Figure 28.7: Adapted from Wayne M. Becker and David W. Deamer, *The World of the Cell,* 2nd ed.(Redwood City, CA: Benjamin/Cummings, 1991), Figure 20.16. © 1991 The Benjamin/Cummings Publishing Company.

Table 28.9: Adapted from H. R. Roberts and J. J. Barone, *Food Technology,* 37 (9):32-39. Copyright © 1983 Institute of Food Technologists. **Figure 28.15A:** Adapted from A. S. Romer and P. S. Parsons, *The Vertebrate Body, 6th ed.* (Philadelphia: Saunders, 1986), p. 569. © 1986 by Saunders College Publishing. **Figure 28.18:** Excerpt from Roger W. Sperry, "Left-Brain, Right-Brain," *Saturday Review,* August 9, 1975, pp. 30–33.

Figure 30.3B: Baboon skeleton from Fleagle, J. G., *Primate Adaptations and Evolution* (San Diego: Academic Press), 1988.

Module 34.11, Talking About Science: Adapted from an interview in Neil Campbell, *Biology,* 3rd ed. (Redwood City, CA: Benjamin/Cummings, 1993).

Figure 35 Opening Figure, p.679: From Cleveland Hickman, *Zoology, 8th ed.*(St. Louis, MO: Mosby, 1988), p. 557. **Table 35.7:** Data from F. Black and H. D. Skipper, Jr., *Life Insurance,* 12th ed. (Englewood Cliffs, NJ: Prentice-Hall, 1988). **Figure 35.8A and B:** Data from United Nations, *The Sex and Age Distribution of Populations. The 1990 Revision of the United Nations Global Population Estimates and Projections.* (New York: United Nations, 1991), p. 368.

Figure 36.9: From Robert Leo Smith, *Ecology and Field Biology, 4th ed.* (New York: HarperCollins, 1990) © 1990 by Robert Leo Smith. Reprinted by permission of HarperCollins College Publishers. **Figure 36.17C:** From G. E. Likens et al., "Effects of Forest Cutting and Herbicide Treatment on Nutrient Budgets in the Hubbard Brook Watershed Ecosystem," *Ecological Monographs* (1970), 40. Copyright © 1966 by Ecological Society of America. Reprinted by permission. **Figure 36.19:** From W. K. Purves and G. H. Orians, *Life: The Science of Biology,* 3rd ed. (New York: W.H. Freeman, 1992), p. 1140. © 1992 by Sinauer and W.H. Freeman and Co. Reprinted with permission of W.H. Freeman and Co.

Figure 37.3B: Adapted from N. B. Davies and M. Brooke, "Coevolution of the Cuckoo and Its Hosts," *Scientific American,* January 1991, pp. 94-95. © 1991 by Scientific American, Inc. All rights reserved. **Figure 37.5B:** Courtesy of Masakazu Konishi. **Figure 37.11C:** Adapted from T. H. Waterman, ed., *Animal Navigation* p. 109. Copyright © 1989 by Scientific American Books. Reprinted with permission of W.H. Freeman and Company. **Module 37.16,** Talking About Science: Adapted from an interview in Neil Campbell, *Biology,* 3rd ed. (Redwood City, CA: Benjamin/Cummings, 1993). **Figure 37.18A:** From J. McIntyre, *The Common Loon, Spirit of Northern Lakes* (Minneapolis, MN: University of Minnesota Press, 1988.) © 1988 University of Minnesota Press. Reprinted by permission.

Chapter 38 Opening Essay: Text quotation from David Quammen, "A Murder in Madagascar," *Audubon,* January 1991. **Table 38.12:** Data from Population Reference Bureau (Washington, DC, 1992); *World Resources 1992-1993 Annual Report,* United Nations Environment Program and United Nations Development Program. **Figure 38.12B:** From G. Tyler Miller, *Living in the Environment,* 2nd ed. (Belmont, CA: Wadsworth, 1979), p. 87. © 1979 Wadsworth Publishing Company. All Rights Reserved. **Module 38.13:** Text quotation from Wallace Broecker, in Norman Myers "The Heat Is On." *Greenpeace* 14 (3), May/June 1989. Text quotation from P. D. Jones and T. M. L. Wigley, "Global Warming Trends," Scientific American, August 1990. **Figure 38.13C:** Data from P. D. Jones and T. M. L. Wigley, "Global Warming Trends," *Scientific American,* August 1990.

GLOSSARY

A

abiotic (A-by-OT-ik)   Pertaining to the nonliving components of an ecosystem, such as air, water, or temperature.

abscisic acid (ABA) (ab-SIS-ik)   A plant hormone that inhibits cell division and promotes dormancy; interacts with gibberellins in regulating seed germination.

absorption   The uptake of small nutrient molecules by an organism's own body; the third main stage of food processing, following digestion.

accommodation   The automatic changes made by the eye as it focuses on near objects.

acetylcholine (uh-SEE-tul-KOLE-een)   A nitrogen-containing neurotransmitter; among other effects, it slows the heart and makes skeletal muscles contract.

acetyl CoA (ASS-uh-teel)   Acetyl coenzyme A; a high-energy organic molecule that fuels the Krebs cycle. For each molecule of glucose that enters glycolysis, two acetyl fragments from two acetyl CoA molecules enter the Krebs cycle.

achondroplasia (a-KON-druh-PLAY-zhuh)   A form of human dwarfism caused by a single dominant allele; the homozygous condition is lethal.

acid   A substance that increases the hydrogen ion concentration in a solution.

acid chyme (kime)   The acidic mixture of food and gastric juice in the vertebrate stomach.

acid precipitation   Rain, snow, sleet, hail, drizzle, etc., with a pH below 5.6; can damage or destroy organisms by acidifying lakes, streams, and possibly land habitats.

acrosome (AK-ruh-soam)   A membrane-bounded sac at the tip of a sperm; contains enzymes that help the sperm penetrate an egg.

actinomycete (ak-TIN-oh-MY-seat)   One of a group of eubacteria characterized by a mass of branching cell chains called filaments.

action potential   A change in the voltage across the plasma membrane of a neuron; a nerve signal. Action potentials are self-generating.

activator   A protein that switches on a gene or group of genes.

active immunity   The type of immunity achieved by exposure to antigens, which stimulate the body to produce antibodies or immune cells.

active site   A small part of an enzyme molecule that attaches to a substrate molecule by means of weak chemical bonds; typically, a pocket or groove on the enzyme's three-dimensional surface.

active transport   The movement of a substance across a biological membrane against its concentration gradient, aided by specific transport proteins and requiring cellular work.

adaptive radiation   The emergence of numerous species from a common ancestor introduced to new and diverse environments.

adhesion (ad-HE-zhun) The tendency of different kinds of molecules to cling to one another.

adipose tissue (AD-ih-POSE)   A type of connective tissue whose cells contain fat.

adrenal cortex (uh-DREEN-ul KOR-tex)   The outer portion of an adrenal gland, controlled by ACTH from the anterior pituitary; secretes hormones called glucocorticoids and mineralocorticoids.

adrenal gland (uh-DREEN-ul)   One of a pair of endocrine glands located adjacent to a kidney in mammals; composed of an outer cortex and a central medulla.

adrenal medulla (uh-DREEN-ul med-UL-uh)   The central portion of an adrenal gland; controlled by nerve signals; secretes the fight-or-flight hormones epinephrine and norepinephrine.

adrenocorticotropic hormone (ACTH) (uh-DREEN-oh-KOR-tik-oh-TRO-pik)   A protein hormone secreted by the anterior pituitary; stimulates the adrenal cortex to secrete glucocorticoids.

aerobic (air-OH-bik)   Containing oxygen; an organism, environment, or cellular process that requires oxygen.

aestivation (ES-tih-VAY-shun)   An animal's state of reduced activity (torpor) during periods of high environmental temperatures and reduced food and water supplies.

age structure   The proportion of individuals in different age groups in a population.

aggregate fruit   A fruit that develops from a flower with many carpels.

agonistic behavior (AG-un-NIS-tik) Confrontational behavior involving a contest waged by threats, displays, or actual combat, which settles disputes over limited resources, such as food or mates.

AIDS   Acquired immune deficiency syndrome; the late stages of HIV infection; characterized by a reduced number of T cells; usually results in death caused by other diseases.

alcoholic fermentation   A metabolic process performed by yeasts and some bacteria in anaerobic environments. Following glycolysis, enzymes remove CO_2 from pyruvic acid and reduce it, forming ethyl alcohol; simultaneously, NADH is oxidized, thus recharging the cell with a supply of NAD^+ to keep glycolysis working.

alga (AL-guh) (plural, *algae*)   One of a great variety of protists, most of which are unicellular or colonial photosynthetic autotrophs with chloroplasts containing the pigment chlorophyll *a*; heterotrophic and multicellular protists closely related to unicellular autotrophs are also regarded as algae.

alimentary canal (AL-ih-MEN-teh-ree)   A digestive tract consisting of a tube running between a mouth and an anus.

allantois (uh-LAN-toe-iss)   An extraembryonic membrane that develops from the yolk sac; helps dispose of the embryo's nitrogenous wastes; forms part of the umbilical cord in mammals.

allopatric speciation (AL-oh-PAT-rik)   The formation of a new species as a result of an ancestral population's becoming isolated by a geographical barrier.

allele (uh-LEEL)   An alternative form of a gene.

allergen (AL-er-jen)   An antigen that causes an allergy.

allergy   A disorder of the immune system caused by an abnormal sensitivity to antigens; triggered by histamines released from mast cells.

alpha helix (AL-fuh HE-lix)   The spiral shape resulting from the coiling of a polypeptide in a protein's secondary structure.

alternation of generations   A life cycle in which there is both a multicellular diploid form, the sporophyte, and a multicellular haploid form, the gametophyte; a characteristic of plants and multicellular green algae.

altruism (AL-troo-IZ-em)   Behavior involving self-sacrifice.

alveolus (al-VEE-oh-lus) (plural, *alveoli*)   One of millions of tiny sacs within the vertebrate lungs where gas exchange occurs.

amine (uh-MEEN)   An organic compound with one or more amino groups.

amino acid (uh-MEE-noh)   An organic molecule containing a carboxyl group and an amino group, serves as the monomer of proteins.

amino acid sequencing   A process that determines the sequence of amino acids in a polypeptide; used to compare protein structure.

amino group   In an organic molecule, a functional group consisting of a nitrogen atom bonded to two hydrogen atoms.

amniocentesis (AM-nee-oh-sen-TEE-sis)   A technique for diagnosing genetic defects while a fetus is in the uterus; a sample of amiotic fluid, obtained via a needle inserted into the amnion, is analyzed for telltale chemicals and defective fetal cells.

amnion (AM-nee-on)   The extraembryonic membrane that encloses the fluid-filled amniotic sac containing the embryo.

amniotic egg (AM-nee-ah-tik)   A shelled egg in which an embryo develops within a fluid-filled amniotic sac and is nourished by yolk; produced by reptiles, birds, and egg-laying mammals, it enables them to complete their life cycles on dry land.

amoeba (uh-ME-ba)   A type of protist characterized by great flexibility and the presence of pseudopodia.

amoebocyte (uh-ME-buh-SITE)   An amoebalike cell that moves by pseudopodia, found in most animals; depending on the species, may digest and distribute food, dispose of wastes, make skeleton, fight infections, and change into other cell types.

anabolic steroid (an-uh-BAH-lic STEHRoid)   A synthetic variant of the male hormone testosterone.

anaerobic (an-air-OH-bik)   Lacking oxygen; an organism, environment, or cellular process that lacks oxygen and may be poisoned by it.

anaphase (AN-uh-faze)   The third stage of mitosis. Anaphase begins when the centromeres of duplicated chromosomes divide and sister chromatids separate from each other; anaphase ends when a complete set of daughter chromosomes are located at each of the two poles of the cell.

anaphylactic shock (AN-uh-feh-LAK-tik)   A potentially fatal allergic reaction caused by extreme sensitivity to an allergen; includes an abrupt dilation of blood vessels and a sharp drop in blood pressure.

anchoring junction   A junction that rivets tissue cells to each other or to an extracellular matrix and allows materials to pass from cell to cell.

androgen (AN-druh-jen)   A steroid sex hormone secreted by the gonads; promotes development and maintenance of the male reproductive system and male body features.

anemia (uh-NEE-me-ah)   Having an abnormally low amount of hemoglobin or a low number of red blood cells, such that the body cells do not receive enough oxygen.

angiosperm (AN-jee-oh-spurm)   A flowering plant, which forms seeds inside a protective chamber called an ovary.

animal behavior   What an animal does in interacting with its environment.

Annelida (uh-NEL-ih-duh)   The phylum that contains the segmented worms, or annelids; characterized by uniform segmentation; includes earthworms, polychaetes, and leeches.

annual   A plant that completes its life cycle in a single year or growing season.

anterior   Pertaining to the front, or head, of a bilaterally symmetrical animal.

anterior pituitary   An endocrine gland adjacent to the the hypothalamus and the posterior pituitary; synthesizes several hormones, including some that control the activity of other endocrine glands.

anther   A sac in which pollen grains develop, situated at the tip of a flower's stamen.

anthropoid (AN-thruh-poid)   A member of a primate group made up of the apes (gibbons, orangutan, gorilla, chimpanzee, and bonobo), monkeys, and humans.

anthropomorphism (AN-thruh-puh-MOR-fiz-em)   The tendency to ascribe human feelings, motivation, or conscious awareness to other animals.

antibody (AN-tih-BOD-ee)   A protein dissolved in blood plasma that attaches to a specific kind of antigen and helps to counter its effects.

anticodon (AN-tee-KO-dahn)   On a tRNA molecule, a specific sequence of three nucleotides that is complementary to a codon triplet on mRNA.

antidiuretic hormone (ADH) (AN-tee-DIE-yoo-RET-ik)   A hormone made by the hypothalamus and secreted by the posterior pituitary; promotes retention of water by the kidneys.

antigen (AN-tih-jen)   A foreign (nonself) molecule that elicits an immune response.

antigen-presenting cell (APC)   One of a family of white blood cells (e.g., a macrophage) that ingests a foreign substance or a microbe and attaches antigenic portions of the ingested material to its own surface, thereby displaying the antigens to a helper T cell.

antihistamine (AN-tee-HISS-tuh-meen)   A drug that

interferes with the action of histamine, providing temporary relief from an allergic reaction.

anus (A-nus)   The opening from a digestive tract through which undigested waste is expelled.

aorta (a-OR-tuh)   An artery that conveys blood directly from the heart to other arteries.

aphotic zone (a-FOE-tik)   The region of an aquatic ecosystem beneath the photic zone, where light does not penetrate enough for photosynthesis to take place.

apical dominance (APP-ih-kul)   In a plant, the hormonal inhibition of axillary buds by a terminal bud.

apical meristem (APP-ih-kul MAIR-ih-stem)   A meristem at the tip of a plant root, or in the terminal and axillary bud of a shoot.

apicomplexan (APP-ee-kum-PLEX-un)   One of a group of parasitic protozoans, some of which cause human diseases.

appendicular skeleton (APP-en-DIK-yoo-ler)   Components of the skeletal system that support the fins of a fish or the arms and legs of a land vertebrate; cartilages and bones of the shoulder girdle, pelvic girdle, and the forelimbs and hind limbs.

appendix (uh-PEN-dix)   A small, fingerlike extension of the vertebrate cecum; contains a mass of white blood cells that contribute to immunity.

aqueous humor (AY-kwee-us HYOO-mer)   Plasmalike liquid in the space between the lens and the cornea in the vertebrate eye; helps maintain the shape of the eye, supplies nutrients and oxygen to its tissues, and disposes of its wastes.

archaebacteria (ARK-ee-bak-TEER-ee-uh)   A lineage of prokaryotes represented today by a few groups of bacteria that inhabit extreme environments. Some biologists place archaebacteria in their own kingdom, separate from all other bacteria.

archenteron (ar-KEN-teh-RON)   The endoderm-lined cavity formed during gastrulation; the digestive cavity of a gastrula.

artery   A vessel that carries blood away from the heart to other parts of the body.

arteriole (ar-TEHR-ee-ole)   A vessel that conveys blood between an artery and a capillary bed.

arthritis (ar-THRY-tiss)   A skeletal disorder characterized by inflamed joints and deterioration of cartilage between bones.

Arthropoda (ar-THROP-uh-duh)   The most diverse phylum in the animal kingdom; includes the horseshoe crab, arachnids (e.g., spiders, ticks, scorpions, and mites), crustaceans (e.g., crayfish, lobsters, crabs, barnacles), millipedes, centipedes, and insects. Arthropods are characterized by a chitinous exoskeleton, molting, jointed appendages, and a body formed of distinct groups of segments.

artificial pacemaker   An electronic device surgically implanted near the AV node in the heart; emits electrical signals that trigger normal heart muscle contractions.

artificial selection   Selective breeding of domesticated plants and animals to promote the occurrence of desirable inherited traits in offspring.

asexual reproduction   The creation of offspring by a single parent, without the participation of sperm and egg.

association   Learning that a particular stimulus or response is linked with a reward or punishment.

astigmatism (uh-STIG-muh-TIZ-em)   Blurred vision caused by a misshapen lens or cornea.

atom   The smallest unit of matter that retains the properties of an element.

atomic number   The number of protons in each atom of a particular element.

atomic weight   The approximate total mass of an atom; given as a whole number, the atomic weight approximately equals the mass number.

ATP synthase   A cluster of several different enzymes (membrane proteins) in the mitochondrial cristae. ATP synthases function in chemiosmosis with adjacent electron transport chains, using the energy of a hydrogen-ion concentration gradient to make ATP. ATP synthases also provide a port through which hydrogen ions can diffuse into the matrix of a mitochondrion.

atrium (A-tree-um) (plural, *atria*)   A heart chamber that receives blood from the veins.

auditory canal   Part of the vertebrate outer ear that channels sound waves from the pinna or outer body surface to the eardrum.

australopithecines (aw-STRAY-loh-PITH-eh-sins)   The first hominids; scavenger-gatherer-hunters who lived on African savannas between about 4 million years ago and 1.5 million years ago.

autoimmune disease (AW-toe-ih-MYOON)   An immunological disorder in which the immune system turns against the body's own molecules.

autonomic nervous system (AW-tuh-NAHM-ik)   A subdivision of the motor nervous system of vertebrates that regulates the internal environment; made up of sympathetic and parasympathetic subdivisions.

autosome (AW-tuh-soam)   A chromosome whose genes are not directly involved in determining the sex of an organism.

autotroph (AW-tuh-trofe)   An organism that makes its own food, thereby sustaining itself without eating other organisms or their molecules. Plants, algae, and photosynthetic bacteria are autotrophs.

auxin (AWK-sin)   One of a family of plant hormones having a variety of effects; chiefly promotes the growth and development of shoots.

AV (atrioventricular) node (A-tree-oh-ven-TRICK-yoo-ler)   A signal relay point in the heart wall between the right atrium and right ventricle. Signals from the AV node ensure the atria will empty before the ventricles contract.

axial skeleton (AK-see-ul)   Components of the skeletal system that support the central trunk of the body. In a vertebrate, the skull, backbone, and rib cage make up the axial skeleton.

axillary bud (AK-seh-LAIR-ee)   An embryonic shoot present in the angle formed by a leaf and stem.

axon (AK-son)   A neuron fiber that conducts signals to another neuron or to an effector.

B

bacillus (buh-SIL-us) (plural, *bacilli*)   A rod-shaped bacterium.

backbone   A series of segmental units called vertebrae, present in all vertebrates.

bacterial cell wall   A fairly rigid, chemically complex wall that protects the bacterial cell and helps maintain its shape.

bacterial flagellum   A long surface projection that propels a bacterial cell through its liquid environment; totally different from the flagella on eukaryotic cells.

bacteriophage (bak-TEER-ee-oh-FAJE)   A virus that infects bacteria. Also called phage.

bark   All the tissues external to the vascular cambium in a plant that is growing in thickness. Bark is made up of secondary phloem, cork cambium, and cork.

Barr body   A highly compacted *X* chromosome, of which almost all the genes are inactive; an inactive *X* chromosome that lies along the inside of the nuclear envelope in cells of female mammals.

barrier method   A method of contraception that prevents egg and sperm from meeting.

Bartholin's glands (BAR-toe-linz)   Glands near the vaginal opening of a human female that secrete lubricating fluid during sexual arousal.

basal body (BAY-sul)   A eukaryotic cell organelle consisting of a 9 + 0 arrangement of microtubules; may organize the microtubule assembly of a cilium or flagellum; structurally identical to a centriole.

basal ganglia (GANG-lee-uh)   Clusters of nerve cell bodies located deep within the cerebrum; important in motor coodination.

basal metabolic rate (BMR)   The number of kilocalories a resting animal requires to fuel its essential body processes for a given time.

base   A substance that decreases the hydrogen ion concentration in a solution.

Batesian mimicry (BAYTZ-ee-un MIM-ih-kree)   A type of mimicry in which a species that a predator can eat looks like a different species that is poisonous or otherwise harmful to the predator.

B cell   A type of lymphocyte that develops in the bone marrow and later produces antibodies.

behavioral biology   The scientific study of behavior.

behavioral ecology   The scientific search for ultimate causes of behavior.

behavioral isolation   A type of prezygotic barrier between species. Two species remain isolated because individuals of neither species are sexually attracted to individuals of the other species.

benign tumor (bih-NINE TWO-mur)   An abnormal mass of cells that remains at its original site in the body.

benthic zone   A seafloor, or the bottom of a freshwater lake, pond, river, or stream.

biennial (by-EN-ee-ul)   A plant that completes its life cycle in two years.

bilateral symmetry   An arrangement of body parts such that an organism can be divided equally by a single cut passing longitudinally through it. A bilaterally symmetrical organism has mirror-image right and left sides.

bile   A solution of bile salts secreted by the liver, which emulsifies fats and aids in their digestion.

binary fission (BY-neh-ree FIH-zhun)   A means of asexual reproduction in which a parent divides into two individuals of about equal size.

binomial (by-NOME-ee-ul)   A two-part, Latinized name of a species; for example, *Homo sapiens*.

biodiversity   The variety of species that make up a community; refers to species richness (the total number of different species) and to the relative abundance of the different species.

biogeography   The geographical distribution of species.

biological clock   An internal timekeeper that controls an organism's circadian rhythms; marks time with or without environmental cues but requires daily signals from the environment to remain tuned to a period of about 24 hours.

biological community   An assemblage of all the organisms living together and potentially interacting in a particular area; characterized by its biodiversity, prevalent form of vegetation, stability, and trophic structure.

biological magnification   The accumulation of persistent chemicals in the living tissues of consumers in food chains.

biological species   A population or group of populations whose members have the potential in nature to interbreed and produce fertile offspring; a particular kind of organism. Members of a species possess similar inherited characteristics and have the ability to interbreed.

biology   Life science, or the scientific study of life.

biomass   The amount, or mass, of organic material in an ecosystem.

biome (BY-ome)   A terrestrial biological community, largely determined by climate, usually classified according to the predominant vegetation, and characterized by adaptations of its organisms.

biosphere   The global ecosystem; that portion of Earth that is alive; all of life and where it lives.

biotechnology   The use of living organisms (often microbes) to perform useful tasks; may involve recombinant DNA.

biotic (by-OT-ik)   Pertaining to the living components (the organisms) of a biological community.

blastocoel (BLAS-toe-SEAL)   A central, fluid-filled cavity in a blastula.

blastocyst (BLAS-toe-sist)   A mammalian embryo (equivalent to a blastula), made up of a hollow ball of cells that results from cleavage and that implants in the endometrium.

blastopore (BLAS-toe-pore)   A small indentation on one side of a blastula where cells that will form endoderm and mesoderm leave the surface and move inward.

blastula (BLAS-tyoo-luh)   An embryonic stage that marks the end of cleavage during animal development; a hollow ball of cells in many species.

blind spot   The place on the retina of the vertebrate eye where the optic nerve passes through the eyeball and where there are no photoreceptor cells.

blood   A type of connective tissue with a fluid matrix called plasma in which blood cells are suspended.

blood pressure   The force that blood exerts against the walls of blood vessels.

body cavity   A fluid-containing space between the digestive tract and the body wall.

bone   A type of connective tissue, consisting of living cells held in a rigid matrix of collagen fibers embedded in calcium salts.

bottleneck effect   Genetic drift resulting from a drastic reduction in population size.

Bowman's capsule   A cup-shaped swelling at the receiving end of a nephron; collects the filtrate from the blood.

breathing   The alternation of inhalation with exhalation, supplying a lung or gill with O_2-rich air or water and expelling CO_2-rich air or water.

breathing control center   A brain center that directs the activity of organs involved in breathing.

bronchus (BRONK-us) (plural, *bronchi*)   One of a pair of breathing tubes that branch from the trachea into the lungs.

bronchiole (BRONK-ee-ole)   A thin breathing tube that branches from a bronchus within a lung.

brown alga   One of a group of marine, multicellular, autotrophic protists, the most common and largest type of seaweed. Brown algae include the kelps.

bryophyte (BRYE-eh-FITE)   One of a group of plants that lack xylem and phloem; a nonvascular plant. Bryophytes include mosses and their close relatives.

bud   In a plant, an undeveloped shoot covered with modified leaves called scales that protect underlying embryonic tissues.

budding   An asexual means of reproduction. A new individual developed from an outgrowth of a parent splits off and lives independently.

buffer   A chemical substance that resists changes in pH by accepting H^+ ions from or donating H^+ ions to solutions.

bulbourethral gland (BUL-bo-yoo-REE-thrul)   One of a pair of glands near the base of the penis in the human male that secrete fluid that lubricates and neutralizes acids in the urethra during sexual arousal.

C

C_3 plant   A plant that fixes carbon from CO_2 directly into the three-carbon compound 3-PGA in the Calvin cycle.

C_4 plant   A plant that can fix carbon from CO_2 into a four-carbon compound. The enzyme involved has a very high attraction for CO_2 and can fix carbon when CO_2 levels are low in the leaf. The four-carbon compound passes the fixed carbon to the Calvin cycle, maintaining sugar production and preventing photorespiration when leaf stomata are closed. Keeping stomata closed during the day is an adaptation for conserving water.

calcitonin (KAL-sih-TONE-in)   A peptide hormone secreted by the thyroid gland; lowers the blood calcium level.

Calvin cycle   The second of two stages of photosynthesis, the Calvin cycle is a cyclic series of chemical reactions that occur in the stroma of the chloroplast, using the carbon in CO_2 and the ATP and NADPH produced by the light reactions to make the energy-rich sugar molecule, G3P.

cAMP   Cyclic adenosine monophosphate; a molecule synthesized from ATP, which serves as a second messenger, mediating the effect of many steroid hormones in target cells.

CAM plant   A plant whose stomata are open and that fixes carbon only at night. Carbon is fixed into a four-carbon compound, and the carbon is passed to the Calvin cycle during the day. Keeping stomata open only at night is an adaptation for conserving water.

cancer cell   A cell that is not subject to normal control mechanisms and that will divide continuously, often killing the organism if unchecked.

capillary (KAP-ih-lair-ee)   A microscopic blood vessel that conveys blood between an artery and a vein or between an arteriole and a venule; enables the exchange of nutrients and dissolved gases between the blood and interstitial fluid.

capillary bed   A network of capillaries that infiltrate every organ and tissue in the body.

capsule   A sticky layer that surrounds the bacterial cell wall, protects the cell surface, and sometimes helps glue the cell to surfaces.

carbohydrate (KAR-bo-HI-drate)   A class of biological molecules consisting of simple single-monomer sugars (monosaccharides), disaccharides, or other multi-unit sugars (polysaccharides).

carbon fixation   The incorporation of carbon from atmospheric CO_2 into the carbon in organic compounds. During photosynthesis in a C_3 plant, carbon is fixed into a three-carbon sugar, as it enters the Calvin cycle. In C_4 and CAM plants, carbon is fixed into a four-carbon sugar.

carbon skeleton   The chain of carbon atoms that forms the structural backbone of an organic molecule.

carbonyl group (KAR-beh-nil)   In an organic molecule, a functional group consisting of a carbon atom linked by a double bond to an oxygen atom.

carboxyl group (kar-BOK-sul)   In an organic molecule, a functional group consisting of an oxygen atom double-bonded to a carbon atom that is also bonded to a hydroxyl group.

carboxylic acid   An organic compound containing a carboxyl group.

carcinogen (kar-SIN-eh-jen)   A cancer-causing agent, such as X-rays, ultraviolet light, tobacco, and many chemicals.

carcinoma (KAR-sih-NO-muh)   Cancer that originates in the coverings of the body, such as skin or the linings of the intestinal tract.

cardiac cycle (KAR-dee-ak)   The alternating contractions and relaxations of the heart.

cardiac muscle   Striated muscle that forms the contractile tissue of the heart.

cardiac output   The volume of blood per minute that the left ventricle pumps into the aorta.

cardiovascular system (KAR-dee-oh-VAS-kyuh-ler)   A closed circulatory system with a heart and branching network of arteries, capillaries, and veins.

carnivore   (1) An animal that eats other animals. (2) An organism that eats heterotrophs.

carpel (KAR-pul)   The female part of a flower, consisting of a stalk with an ovary at the base and a stigma, which traps pollen, at the tip.

carrier   An individual that is heterozygous for an inherited trait (often a disorder) but who does not show symptoms of that trait.

carrying capacity   In a population, the number of individuals that an environment can maintain.

cartilage (KAR-til-ij)   A type of connective tissue, consisting of living cells embedded in a rubbery matrix with collagenous fibers.

Casparian strip (kas-PAIR-ee-un)   A waxy barrier in the walls of endodermal cells in a root; prevents water and ions from entering the xylem without crossing one or more cell membranes.

cecum (SEE-kum) (plural, *ceca*)   A blind outpocket of a hollow organ such as an intestine.

cell   A basic unit of living matter separated from its environment by a membrane; the fundamental structural unit of life.

cell body   The part of a cell, such as a neuron, that houses the nucleus.

cell cycle   An orderly sequence of events (including interphase and the mitotic phase) from the time a cell divides to form two daughter cells to the time those daughter cells divide again.

cell division   The reproduction of cells.

cell junction   A structure that connects tissue cells to one another.

cell-mediated immunity   Components of the immune system mediated by T cells.

cell theory   The principle that all life is composed of cells and that all cells come from other cells.

cellular differentiation   Specialization in the structure and function of cells; one of the processes that transforms cells into tissues, and cells resulting from a zygote into an adult organism.

cellular metabolism (meh-TAB-ul-izm)   The sum total of the chemical activities of cells.

cellular respiration   The aerobic harvesting of energy from food molecules; the energy-releasing chemical breakdown of food molecules, such as glucose, and the storage of potential energy in a form that cells can use to perform work; involves glycolysis, the Krebs cycle, the electron transport chain, and chemiosmosis.

cellular slime mold   A type of protist that has unicellular amoeboid cells and multicellular reproductive bodies in its life cycle.

cellulose (SEL-yuh-lohs)   A large polysaccharide composed of many glucose monomers linked into cablelike fibrils that provide structural support in plant cell walls.

central canal   A hollow space in the spinal cord, containing cerebrospinal fluid and continuous with the ventricles of the brain.

central nervous system (CNS)   The integration and command center of the nervous system; the brain and, in vertebrates, the spinal cord.

central vacuole (VAK-yoo-ohl)   A membrane-enclosed sac occupying most of the interior of a mature plant cell; has diverse roles in reproduction, growth, and development.

centralization   (1) The presence of a central nervous system (CNS), distinct from a peripheral nervous system. (2) The evolution of a nervous system with central control.

centriole (SEN-tree-ole)   A structure in an animal cell, composed of cylinders of microtubules arranged in a 9 + 0 pattern. An animal cell usually has a pair of centrioles, which are involved in cell division.

centromere (SEN-troh-mere)   The region of a chromosome where two sister chromatids are joined and where spindle microtubules attach to the chromosome during mitosis and meiosis. The centromere divides at the onset of anaphase during mitosis and anaphase II of meiosis.

cephalization (SEF-uh-le-ZAY-shun)   (1) The concentration of a nervous system at the anterior end. (2) The evolution of a "head" end with sensory structures and a brain.

cerebellum (SEH-reh-BELL-um)   Part of the vertebrate hindbrain; mainly a planning center that interacts closely with the cerebrum in coordinating body movement.

cerebral cortex (seh-REE-brul KOR-tex)   A folded sheet of gray matter forming the surface of the cerebrum. In humans, it contains integrating centers for higher brain functions such as reasoning, speech, language, and imagination.

cerebral hemisphere   The right or left half of the vertebrate cerebrum.

cerebrospinal fluid (seh-REE-bro-SPINE-ul)   Blood-derived fluid that surrounds, protects from infection, nourishes, and cushions the brain and spinal cord.

cerebrum (seh-REE-brum)   The largest and most sophisticated vertebrate brain center, made up of right and left cerebral hemispheres.

cervix (SUR-viks)   The neck of the uterus, which opens into the vagina.

chaparral (CHAP-uh-RAL)   A biome dominated by spiny evergreen shrubs adapted to periodic drought and fires; found where cold ocean currents circulate offshore, creating mild, rainy winters and long, hot, dry summers.

chemical bond   An attraction between two atoms resulting from a sharing of outer-shell electrons or the presence of opposite charges on the atoms; the bonded atoms gain complete outer electron shells.

chemical cycling   The circular passage of materials within an ecosystem.

chemical element   A substance that cannot be broken down to other substances by ordinary chemical means; scientists recognize 92 chemical elements occurring in nature.

chemical energy   Energy stored in the chemical bonds in molecules.

chemical reaction   A process leading to chemical changes in matter; involves the making or breaking of chemical bonds.

chemiosmosis (KEM-ee-oz-MOH-sis)   The production of ATP using the energy of hydrogen-ion gradients across membranes to phosphorylate ADP; powers most ATP synthesis in cells.

chemoheterotroph (KEM-oh-HET-er-oh-TROFE)   An organism that obtains its energy and carbon skeletons by consuming organic molecules.

chemoreceptor (KEM-oh-reh-SEP-ter)   A sensory receptor that detects chemical changes within the body or a specific kind of   molecule in the external environment.

chiasma (kye-AZ-muh) (plural, *chiasmata*)   The microscopically visible site where crossing over has occurred between chromatids of homologous chromosomes during prophase I of meiosis.

chloroplast (KLOR-uh-plast)   An organelle found in plants and photosynthetic protists. Enclosed by two concentric membranes, a chloroplast absorbs sunlight and uses it to power the synthesis of organic food molecules (sugars).

choanocyte (ko-AN-eh-site)   A flagellated feeding cell found in sponges. Also called a collar cell, it has a collarlike ring that traps food particles around the base of its flagellum.

Chordata (kore-DAY-tuh)   The phylum of the chordates; characterized by a dorsal, hollow nerve cord, a notochord, gill structures, and a post-anal tail; includes lancelets, tunicates, and vertebrates.

chorion (KOR-ee-ON)   The outermost extraembryonic membrane, which becomes the mammalian embryo's part of the placenta.

chorionic villi sampling (KOR-ee-on-ik VILL-eye)   A technique for diagnosing genetic defects while the fetus is in the uterus. A small sample of the fetal portion of the placenta is removed and analyzed.

chorionic villus (KOR-ee-on-ik VILL-us)   An outgrowth of the chorion, containing embryonic blood vessels. As part of the placenta, chorionic villi absorb nutrients and oxygen from, and pass wastes into, the mother's bloodstream.

choroid (KOR-oid)   A thin, pigmented layer in the vertebrate eye, surrounded by the sclera. The iris is part of the choroid.

chromatin (KRO-muh-tin)   Diffuse, very long, coiled fibers of DNA with proteins attached; the form taken by the chromosomes when a eukaryotic cell is not dividing.

chromosome (KRO-muh-soam)   A threadlike, gene-carrying structure found in the nucleus of all eukaryotic cells and most visible during mitosis and meiosis. Chromosomes consist of DNA and protein.

chromosome theory of inheritance   A basic principle in biology stating that genes are located on chromosomes and that the behavior of chromosomes during meiosis accounts for inheritance patterns.

ciliate (SIL-ee-it)   A type of protozoan that moves by means of cilia.

cilium (SIL-ee-um) (plural, *cilia*)   A short appendage that propels protists through the water and moves fluids across the surface of many tissue cells in animals. In common with eukaryotic flagella, cilia have a 9 + 2 arrangement of microtubules covered by the cell's plasma membrane.

circadian rhythm (sur-KAY-dee-un)   A biological cycle of about 24 hours that persists in an organism even in the absence of external cues; a pattern of activity that is repeated daily.

circulatory system   The organ system that transports materials such as nutrients, O_2, and hormones to body cells and transports CO_2 and other wastes from body cells.

cladistic taxonomy (kluh-DIS-tik tak-SAHN-eh-mee)   An approach to classifying organisms entirely according to when branches arise in a phylogenetic lineage.

class   In classification, the taxonomic category above the order and below the phylum.

classical conditioning   Learning that an arbitrary stimulus is associated with an important signal.

classical evolutionary taxonomy   An approach to classifying organisms that takes into account the apparent degree of divergence between phylogenetic lineages.

cleavage (KLEE-vij)   (1) In animal development, the succession of rapid cell divisions without growth, converting the zygote into a ball of cells. (2) Cytokinesis in animal cells and in some protists, characterized by pinching of the plasma membrane.

cline   A gradation in an inherited trait along a geographical continuum; variation in a population's phenotypic features that parallels an environmental gradient.

clitoris (KLIT-er-is)   An organ in the female that engorges with blood and becomes erect during sexual arousal.

clonal selection (KLONE-ul)   The formation of a lineage of genetically identical cells that recognize and attack the specific antigen that stimulated their formation. Clonal selection is the mechanism that underlies the immune system's specificity and memory of antigens.

closed circulatory system   A circulatory system in which blood is confined in vessels and remains distinct from the interstitial fluid.

Cnidaria (nigh-DARE-ee-uh)   The phylum that contains the hydras, jellyfishes, sea anemones, corals, and related animals characterized by cnidocytes, radial symmetry, a gastrovascular cavity, polyps, and medusae.

cnidocyte (NIDE-eh-site)   A specialized cell for which the phylum Cnidaria is named; consists of a capsule containing a fine coiled thread, which, when discharged, functions in defense and prey capture.

coccus (KOK-us) (plural, *cocci*)   A spherical bacterium.

cochlea (KOK-lee-uh)   A coiled tube in the inner ear of birds and mammals that contains the hearing organ.

codominant   Pertaining to two or more alleles that are expressed in a heterozygote.

codon (KO-dahn)   A three-nucleotide sequence in mRNA that specifies a particular amino acid or termination signal and functions as the basic unit of the genetic code.

codon recognition (KO-dahn)   During translation, the pairing of an anticodon on a tRNA molecule carrying an amino acid with the complementary mRNA codon in the A site of a ribosome.

coelom (SEE-loam)   A body cavity completely lined with mesoderm.

coenzyme (ko-EN-zime)   An organic substance (usually a vitamin or a compound synthesized from a vitamin) that acts as a cofactor, helping an enzyme catalyze a metabolic reaction.

coevolution   Evolutionary change, in which adaptations in one species act as a selective force on a second species, inducing adaptations that in turn act as a selective force on the first species; mutual influence on the evolution of two different interacting species.

cofactor   A nonprotein substance (such as a copper, iron, or zinc atom, or an organic molecule) that helps an enzyme catalyze a metabolic reaction. *See* coenzyme.

cognition   Conscious thinking, awareness of self, judgment, and the use of language.

cohesion (ko-HE-zhun)   The tendency of molecules of the same kind to stick together.

collecting duct   A tube in the vertebrate kidney that concentrates urine while conveying it to the renal pelvis.

collenchyma cell (ko-LEN-kim-uh)   In plants, a cell with a thick primary wall and no secondary wall, functioning mainly in supporting growing parts.

colon (KO-lun)   *See* large intestine.

commensalism (keh-MEN-sul-izm)   A symbiotic relationship in which one partner benefits without significantly affecting the other.

communicating junction   A channel between adjacent tissue cells through which water and other small molecules pass freely.

community   *See* biological community.

competitive exclusion principle   The concept that populations of two species cannot coexist in a community if their niches are identical. Using resources more efficiently and having a reproductive advantage, one of the populations will eventually outcompete and eliminate the other.

competitive inhibitor   A substance that reduces the activity of an enzyme by entering the enzyme's active site in place of the substrate; a competitive inhibitor's structure mimics that of the enzyme's substrate.

complement   A family of nonspecific defensive, antimicrobial blood proteins; may amplify the inflammatory response and the immune response.

compound   A substance containing two or more elements in a fixed ratio; for example, table salt (NaCl) consists of one atom of the element sodium (Na) for every atom of chlorine (Cl).

compound eye   The photoreceptor in many invertebrates; made up of many tiny light detectors, each of which detects light from a tiny portion of the field of view.

computed tomography (CT) (tuh-MOG-ruh-fee)   A technology that uses a computer to create X-ray images of illuminated sections through the body.

concentration gradient   A regular change (incline or decline) in the intensity or density of a chemical substance. Cells often maintain concentration gradients of H^+ ions across their membranes. When a gradient exists, the ions or other chemical substances involved tend to move from where they are more concentrated to where they are less concentrated.

cone   (1) In vertebrates, a photoreceptor cell in the retina, stimulated by bright light and enabling color vision. (2) In conifers, a reproductive structure bearing pollen or ovules.

conifer   A gymnosperm that produces cones.

conjugation (KON-jeh-GAY-shun)   The union (mating) of two bacterial cells or protist cells and the transfer of DNA between the two cells.

connective tissue   Tissue consisting of cells held in a nonliving substance called a matrix; the cells produce the matrix.

continental drift   A change in the position of continents resulting from the incessant slow movement (floating) of the plates of Earth's crust on the underlying molten mantle. It has caused continents to periodically fuse and break apart throughout geological history.

contraception   The prevention of pregnancy.

controlled experiment   A component of the scientific method whereby a scientist carries out two parallel tests, an experimental test and a control test. The experimental test differs from the control by one factor, the variable.

convergent evolution   Adaptive change resulting in nonhomologous (analogous) similarities among organisms. Species from different evolutionary lineages come to resemble each other (evolve analogous structures) as a result of living in very similar environments.

coralline alga (KOR-uh-lin AL-guh)   A type of red alga with walls encrusted with hard, limy deposits; often important in reef building.

cork   The outermost protective layer of a plant's bark, produced by the cork cambium.

cork cambium (KAM-bee-um)   Meristematic tissue that produces cork cells during secondary growth of a plant.

cornea (KOR-nee-uh)   The transparent frontal portion of the sclera, which admits light into the vertebrate eye.

coronary artery (KORE-uh-nair-ee)   The large blood vessel that conveys blood from the aorta to the tissues of the heart.

corpus callosum (KOR-pus kuh-LO-sum)   The thick band of nerve fibers that connect the right and left cerebral hemispheres in placental mammals.

corpus luteum (KOR-pus LOO-tee-um)   A cluster of endocrine tissue that develops from an ovarian follicle after ovulation; secretes progesterone and estrogen during pregnancy.

cortex (1) In plants, the ground tissue system of a root, made up mostly of parenchyma cells, which store food and absorb minerals that have passed through the epidermis. (2) In vertebrates, the outer portion of the kidney and of the adrenal gland.

cotyledon (KOT-ih-LEED-un) The first leaf that appears on an embryo of a flowering plant; a seed leaf. Monocot embryos have one cotyledon; dicot embryos have two.

countercurrent exchange Transfer of a substance from a fluid or volume of air moving in one direction to another fluid or volume of air moving in the opposite direction.

countercurrent heat exchanger Parallel blood vessels that convey warm and cold blood in opposite directions, maximizing heat transfer to the cold blood.

covalent bond (ko-VALE-ent) An attraction between atoms that share one or more pairs of outer-shell electrons; symbolized by a single line between the atoms.

cranial nerve (KRAY-nee-ul) In the peripheral nervous system of vertebrates, a nerve that carries signals to or from the brain.

crista (KRIS-tuh) (plural, *cristae*) A fold of the inner membrane of a mitochondrion. Enzyme molecules embedded in cristae make ATP.

critical period In animal behavior, the specific time during which imprinting takes place.

crop A pouchlike organ in a digestive tract where food is softened and may be stored temporarily.

cross *See* hybridization.

cross-fertilization The fusion of sperm and egg derived from two different individuals.

crossing over The exchange of segments between chromatids of homologous chromosomes during synapsis in prophase I of meiosis; also, the exchange of segments between DNA molecules in bacteria.

cuticle (KYOO-tih-kul) (1) In plants, a waxy coating on the surface of stems and leaves that helps retain water. (2) In animals, a tough, nonliving outer layer of the skin.

cyanobacteria (sy-AN-oh-bak-TEER-ee-uh) Photosynthetic, oxygen-producing bacteria, formerly called blue-green algae.

cyclic electron flow In the light reactions of photosynthesis, the circular route that electrons take, passing from a photosystem through an electron transport chain and back to the photosystem. ATP is generated in the process.

cystic fibrosis (SIS-tik fy-BRO-sis) A genetic disease that occurs in people with two copies of a recessive allele; characterized by an excessive secretion of mucus and consequent vulnerability to infection; fatal if untreated.

cytokinesis (SY-toh-kih-NEE-sis) The division of the cytoplasm to form two separate daughter cells. Cytokinesis usually occurs together with telophase of mitosis, and the two processes make up the mitotic (M) phase of the cell cycle.

cytokinin (SY-toh-KINE-in) One of a family of plant hormones that promotes cell division, retards aging of flowers and fruits, and may interact antagonistically with auxins in regulating plant growth and development.

cytoplasm (SY-toh-PLAZ-em) Everything inside a cell between the plasma membrane and the nucleus; consists of a semifluid medium and organelles.

cytoskeleton A meshwork of fine fibers that provides structural support for the eukaryotic cell.

cytotoxic T cell (SY-toh-TOK-sik) A type of lymphocyte that attacks body cells infected with pathogens.

D

decomposition The breakdown of organic materials into inorganic ones.

dehydration synthesis (dee-hy-DRAY-shun SIN-thuh-sis) A chemical process in which a polymer forms as monomers are linked by the removal of water. One molecule of water is removed for each pair of monomers linked. Also called condensation.

dehydrogenase (DEE-hy-DRAH-jen-ace) An enzyme that catalyzes a chemical reaction during which one or more hydrogen atoms are removed from a molecule.

deletion The loss of a nucleotide from a gene by mutation; a mutational deficiency in a chromosome resulting from loss of a fragment through breakage.

denaturation (dee-NAY-chur-A-shun) A process in which a protein unravels and loses its specific conformation; can be caused by changes in pH, salt concentration, or environmental temperature.

dendrite (DEN-drite) A neuron fiber; in a motor neuron, one of several short, branched projections that convey nerve signals toward the cell body.

density-dependent factor A population-limiting factor whose effects depend on population density.

density-dependent inhibition The arrest of cell division that occurs when cells grown in a laboratory vessel touch one another; also, the general decrease in the rate of cell division as a cell population in a laboratory vessel becomes more dense.

density-independent factor A population-limiting factor whose occurrence and effects are independent of population density.

deoxyribonucleic acid (DNA) (DEE-ok-see-ry-boh-noo-KLAY-ik) The genetic material that organisms inherit from their parents; a double-stranded helical macromolecule consisting of nucleotide monomers with deoxyribose sugar and the nitrogenous bases adenine (A), cytosine (C), guanine (G), and thymine (T). *See* gene.

derived character A feature found in members of a lineage but not found in ancestors of the lineage.

desert A biome characterized by organisms adapted to sparse rainfall (less than 30 cm per year).

desertification The conversion of other kinds of biomes to deserts.

determinate growth Limited growth; termination of growth after reaching a certain size, as in most animals.

detritivore (deh-TRITE-ih-vore) An organism that derives its energy from organic wastes and dead organisms.

detritus (deh-TRITE-us) Nonliving matter.

diabetes mellitus (DY-uh-BEET-is meh-LITE-es) A human hormonal disease in which body cells cannot absorb enough glucose from the blood and become energy-starved; body fats and proteins are then consumed for their energy. Insulin-dependent diabetes results when the pancreas does not produce insulin. Noninsulin-dependent diabetes results when body cells fail to respond to insulin.

dialysis (dy-AL-ih-sis) Separation and disposal of metabolic wastes from the blood by mechanical means; a means of performing the functions of the kidneys.

diaphragm (DY-ah-fram) The sheet of muscle separating the chest cavity from the abdominal cavity in mammals; its contraction expands the chest cavity, and its relaxation reduces it.

diastole (dy-ASS-tuh-lee) The stage of the heart cycle in which the heart muscle is relaxed, allowing the chambers to fill with blood.

diatom (DY-eh-tom) A unicellular photosynthetic alga with a unique, glassy cell wall containing silica.

dicot (DY-kot) A flowering plant whose embryo has two seed leaves, or cotyledons.

differentiation *See* cellular differentiation.

diffusion The tendency of particles of any kind to spread out spontaneously from where they are more concentrated to where they are less concentrated; the tendency of molecular order to become disordered.

digestion The mechanical and chemical breakdown of food into molecules small enough for the body to absorb; the second main stage of food processing, following ingestion.

digestive system The organ system that ingests food, breaks it down into smaller chemical units, and absorbs the nutrient molecules.

dihybrid cross (DIE-HI-brid) An experimental mating of individuals in which the inheritance of two traits is tracked.

dikaryon (die-KAIR-ee-on) A fungal mycelium that possesses two separate haploid nuclei per cell.

dikaryotic phase (die-KAIR-ee-ot-ik) A series of stages in the life cycle of many fungi in which cells contain two nuclei.

dinoflagellate (DINE-uh-FLAJ-uh-lit) A unicellular photosynthetic alga with two flagella situated in perpendicular grooves in cellulose plates covering the cell.

diploid cell (DIP-loid) In the life cycle of an organism that reproduces sexually, a cell containing two homologous sets of chromosomes, one set inherited from each parent; a $2n$ cell.

directional selection Natural selection that acts against the relatively rare individuals at one end of a phenotypic range.

disaccharide (di-SAK-uh-ride) A double-unit sugar molecule, consisting of two monosaccharides linked by dehydration synthesis.

dispersion pattern The manner in which individuals in a population are spaced within their area. Three types of dispersion patterns are clumped (individuals are aggregated in patches), uniform (individuals are evenly distributed), and random (patternless, unpredictable distribution).

distal tubule The portion of a nephron that helps refine filtrate and empties it into a collecting duct.

diversifying selection Natural selection that favors extreme over intermediate phenotypes.

diversity *See* biodiversity.

DNA-DNA hybridization (HY-brid-uh-ZAY-shun) Comparison of the DNAs of different species by measuring the extent of hydrogen bonding between single strands of the DNA.

DNA fingerprint An individual's unique collection of DNA restriction fragments, detected by electrophoresis.

DNA ligase (LY-gase) An enzyme, essential for DNA replication, that catalyzes the covalent bonding of adjacent DNA nucleotides; used in genetic engineering to paste a specific piece of DNA containing a gene of interest into a bacterial plasmid.

DNA polymerase (pul-IM-ur-ase) An enzyme that assembles DNA polynucleotides.

DNA sequencing Determination of the nucleotide sequences of segments of DNA; the most direct and precise method of determining whether two species share an ancestor.

doldrums (DOLE-drums) An area of calm or very light winds near the equator, caused by rising, warm air.

dominance hierarchy The ranking of individuals one above the other by social interactions; usually maintained by agonistic behavior.

dominant allele In a heterozygote, the allele that is fully expressed in the phenotype.

dorsal Pertaining to the back of a bilaterally symmetrical animal.

double bond A type of covalent bond in which two atoms share two pairs of electrons; symbolized by a pair of lines between the bonded atoms.

double fertilization The formation of both a zygote and a cell with a triploid nucleus. The cell with the triploid nucleus develops into the endosperm of a flowering plant.

double helix (HE-lix) The structure of DNA, referring to its two adjacent polynucleotides spirally wound around each other.

Down syndrome A human genetic disease resulting from having an extra chromosome 21; characterized by heart and respiratory defects, and varying degrees of mental retardation.

Duchenne muscular dystrophy (di-SHEN DIS-truh-fee) A human genetic disease caused by a sex-linked recessive allele; characterized by progressive weakening and a loss of muscle tissue.

duodenum (doo-AH-duh-num) The first portion of the vertebrate small intestine after the stomach, where acid chyme from the stomach is mixed with bile and digestive enzymes.

duplication Repetition of part of a chromosome resulting from fusion with a fragment from a homologous chromosome; can result from an error in meiosis or from mutagens.

dynein arm (DY-neen) A protein extension from a microtubule doublet in a cilium or flagellum; involved in energy conversions that drive the bending of cilia and flagella.

E

eardrum A connective tissue sheet separating the outer ear from the middle ear that vibrates when stimulated by sound waves and passes the waves to the middle ear.

Echinodermata (ee-KIGH-noh-DER-ma-tuh) The phylum of echinoderms, such as sea stars, sea urchins, and sand dollars; characterized by a rough or spiny skin, a water vascular system, an endoskeleton, and radial symmetry in adults.

ecological succession Transition in the species composition of a community, such as what occurs after a flood, fire, or volcanic eruption.

ecology The scientific study of how organisms interact with their environments.

ecosystem (EE-koh-sis-tem) All the organisms in a given area along with the abiotic factors with which they interact; a biological community and its physical environment.

ectoderm (EK-toh-durm) The outer layer of three embryonic cell layers in a gastrula; forms the skin of the gastrula and gives rise to the epidermis and nervous system in the adult.

ectopic pregnancy (ek-TOP-ik) The implantation and development of an embryo outside the uterus.

ectotherm (EK-toh-thurm) An animal that warms itself mainly by absorbing heat from its surroundings.

effector cell A muscle cell or gland cell that carries out a command from the nervous system; in the immune system, a T cell or B cell specialized to defend against a specific antigen.

ejaculation (eh-JACK-yoo-LAY-shun) Discharge of semen from the penis.

electroencephalogram (EEG) (eh-LEK-troh-en-SEF-el-o-gram) A graph that shows the patterns of electrical activity in the brain during arousal and sleep.

electromagnetic energy Solar energy, or radiation, which travels in space as rhythmic waves and can be measured in photons.

electromagnetic receptor A sensory receptor that detects energy of different wavelengths, such as electricity, magnetism, and light.

electron A particle with a single negative electric charge; one or more electrons orbit the nucleus of an atom.

electron carrier A molecule that conveys electrons; one of several membrane proteins in electron transport chains in cells. Electron carriers shuttle electrons during the redox reactions that release energy used to make ATP.

electron microscope (EM) An instrument that focuses an electron beam through, or onto the surface of, a specimen.

An electron microscope achieves a thousandfold greater resolving power than a light microscope; the most powerful EM can distinguish objects as small as 0.2 nanometers.

electron shell An energy level at which an electron orbits the nucleus of an atom.

electron transport chain A sequence of electron-carrier molecules (membrane proteins) that shuttle electrons during the redox reactions that release energy used to make ATP; located in the cristae of mitochondria.

electronegativity The tendency for an atom to pull electrons toward itself.

electrophoresis (eh-LEK-tro-for-EE-sis) A technique for separating and purifying macromolecules. A mixture of molecules is placed on a gel between a positively charged electrode and a negatively charged one; negative charges on the molecules are attracted to the positive electrode, and the molecules migrate toward that electrode; the molecules separate in the gel according to their size and rate of migration.

elimination The passage of undigested wastes out of the digestive tract; the fourth and final stage of food processing.

embryo (EM-bree-oh) A developing stage of a multicellular organism; in humans, the stage in the development of offspring from the first division of the zygote until body structures begin to appear.

embryonic disc (EM-bree-on-ik) An early mammalian embryo consisting of a two-layered plate of cells; one layer will form ectoderm and mesoderm, the other will form endoderm.

embryonic induction During development, the influence of one group of cells on another group of cells.

embryo sac The female gametophyte, contained in the ovule of a flowering plant.

emergent property A functional feature that results from the precise organization and interaction of component parts.

endergonic reaction (EN-der-gahn-ik) An energy-requiring chemical reaction that yields products with more potential energy than the reactants. The amount of energy stored in the products equals the difference between the potential energy in the reactants and that in the products.

endocrine gland (EN-deh-krin) A ductless gland that synthesizes hormone molecules and secretes them directly into the bloodstream.

endocrine system The organ system consisting of ductless glands that secrete chemical signals called hormones and the molecular receptors on or in target cells that respond to the hormones; cooperates with the nervous system in regulating body functions and maintaining homeostasis.

endocytosis (END-oh-sigh-TOE-sis) The movement of materials into the cytoplasm of a cell via membranous vesicles or vacuoles.

endoderm (EN-doh-durm) The innermost of three embryonic cell layers in a gastrula; forms the archenteron in the gastrula, and gives rise to the innermost linings of the digestive tract and other hollow organs in the adult.

endodermis The innermost layer (a one-cell-thick cylinder) of the cortex of a plant root; forms a selective barrier determining which substances pass from the cortex into the vascular tissue.

endomembrane system A network of membranous organelles that partition the cytoplasm of eukaryotic cells into functional compartments. Some of the organelles are structurally connected to each other, whereas others are structurally separate but functionally connected by the traffic of membranous vesicles between them.

endometrium (EN-doh-MEE-tree-um) The inner lining of the uterus in mammals, which is richly supplied with blood vessels that provide the maternal part of the placenta and nourish the developing embryo.

endorphin (en-DOR-fin) A pain-inhibiting hormone produced by the brain and anterior pituitary; also serves as a neurotransmitter.

endoskeleton (EN-doh-SKEL-eh-ton) A hard skeleton located within the soft tissues of an animal; includes spicules of sponges, the hard plates of echinoderms, and the cartilage and bony skeletons of many vertebrates.

endosperm A nutrient-rich structure formed by the union of a sperm cell with two polar nuclei during double fertilization; provides nourishment to the developing embryo in angiosperm seeds.

endospore A thick-coated, protective cell produced within a bacterial cell exposed to harsh conditions.

endosymbiotic hypothesis (EN-doh-SIM-by-OT-ik) A hypothesis on the origin of the eukaryotic cell, maintaining that the forerunners of eukaryotic cells were symbiotic associations between small prokaryotic cells living inside larger prokaryotes.

endotherm An animal that derives most of its body heat from its own metabolism.

endotoxin A poisonous component of the cell walls of certain bacteria.

energy The capacity to perform work, or to move matter in a direction it would not move if left alone.

energy coupling In cellular metabolism, the use of energy released from exergonic reactions to drive essential endergonic reactions.

energy flow The passage of energy through the components of an ecosystem.

energy of activation The amount of energy that reactants must absorb before a chemical reaction will start; abbreviated E_A.

entropy (EN-truh-pee) A measure of disorder; one form of disorder is heat, which is random molecular motion.

enzyme (EN-zime) A protein that serves as a chemical catalyst, changing the rate of a chemical reaction without itself being changed into a different molecule in the process

epidermis (EP-ih-DER-mis) (1) In plants, the tissue system forming the protective outer covering of leaves, young stems, and young roots. (2) In animals, the living layer or layers of cells forming the protective covering, or outer skin.

epididymis (EP-ih-DID-eh-mis) A long coiled tube into which sperm pass from the testis and are stored until mature or until ejaculation.

epinephrine (EP-ih-NEF-rin) An amine hormone secreted by the adrenal medulla; prepares body organs for fight or flight; also serves as a neurotransmitter. Also called adrenaline.

epithelial tissue (EP-ih-THEEL-ee-ul) A sheet of tightly packed cells lining organs and cavities. Also called epithelium.

epithelium (EP-ih-THEEL-ee-um) (plural, *epithelia*) Epithelial tissue.

equilibrial life history A life history characterized by the production of relatively few offspring that receive parental care, have a relatively high rate of survival, and that mature slowly. Populations with equilibrial life histories tend to have relatively stable numbers.

erythrocyte (eh-RITH-ruh-site) A red blood cell.

essential amino acid An amino acid that an organism cannot itself synthesize and must obtain in food. Nine of the 20 kinds of amino acids that humans need to make proteins are essential amino acids.

esophagus (eh-SOF-eh-gus) The channel through which food passes in a digestive tract; usually receives food from the pharynx.

estrogen (ES-truh-jen) A steroid hormone secreted by the gonads; maintains the female reproductive system and promotes development of female body features.

estuary (ES-choo-air-ee) An area where fresh water merges with seawater.

ethylene A gas that functions as a hormone in plants; promotes aging processes, such as fruit ripening and leaf drop.

eubacteria (YOO-bak-TEER-ee-uh) The lineage of prokaryotes that includes all bacteria living today except archaebacteria.

eugenics (yoo-JEN-iks) A socially rejected practice, among humans, of attempting to eliminate genetic disorders and "undesirable" inherited traits by selective breeding.

eukaryotic cell (yoo-KAIR-ee-OT-ik) A type of cell that has a membrane-enclosed nucleus and other membrane-enclosed organelles. All organisms except bacteria are composed of eukaryotic cells.

Eustachian tube (yoo-STAY-shun) An air passage between the middle ear and throat of vertebrates, serving to equalize air pressure on either side of the eardrum.

eutrophication (YOO-trofe-ih-KAY-shun) An increase in productivity of an aquatic ecosystem.

evolution All the changes that transform life on Earth, leading to the diversity of organisms. Evolution results from genetic change and consequent phenotypic change in a population or species over generations.

excretion (ek-SKREE-shun) The disposal of nitrogen-containing metabolic wastes.

excretory system (EK-skruh-tor-ee) The organ system that disposes of nitrogen-containing waste products of cellular metabolism.

exergonic reaction (EK-ser-gahn-nik) An energy-releasing chemical reaction in which the reactants contain more potential energy than the products. The reaction releases an amount of energy equal to the difference in potential energy between the reactants and the products.

exocytosis (EK-soh-sigh-TOE-sis) The movement of materials out of the cytoplasm of a cell via membranous vesicles or vacuoles.

exon (EK-sahn) In eukaryotes, a coding portion of a gene.

exoskeleton A hard, external skeleton that protects an animal and provides points of attachment for muscles.

exotoxin A poisonous protein secreted by bacteria.

exponential growth model A mathematical description of idealized, unregulated population growth.

external fertilization The fusion of gametes that parents have discharged into the environment.

extractive reserve A conservation area where people are permitted to harvest products of the natural environment at a rate that does not destroy the ecosystem.

extraembryonic membranes (EX-truh-EM-bree-AHN-ik) Four membranes (the yolk sac, amnion, chorion, and allantois) that form a life-support system for the developing embryo of a reptile, bird, or mammal.

eye cup The simplest type of photoreceptor; a cluster of photoreceptor cells shaded by a cuplike cluster of pigmented cells; detects light intensity and direction.

F

F_1 generation The offspring of two parental (P generation) individuals; F_1 stands for first filial.

F_2 generation The offspring of the F_1 generation; F_2 stands for second filial.

facilitated diffusion The passage of substances across a biological membrane down their concentration gradients, aided by specific transport proteins.

facultative anaerobe (FAK-ul-tay-tiv AN-uh-robe) An organism that makes ATP by aerobic respiration if oxygen is present, but that switches to fermentation when oxygen is absent.

family In classification, the taxonomic category above the genus and below the order.

farsightedness An inability to focus on close objects; occurs when the eyeball is shorter than normal and the focal point of the lens is behind the retina. Also called hyperopia.

fat A large lipid molecule made from an alcohol called glycerol and three fatty acids; a triglyceride. Most fats function as energy-storage molecules.

feces (FEE-seez) The waste products of digestion.

fertilization The union of the nucleus of a sperm cell with the nucleus of an egg cell. Fertilization produces a zygote, which is 2*n*, from two haploid gametes.

fertilization membrane A barrier that forms around an egg seconds after fertilization, preventing additional sperm from penetrating the egg.

fetal-alcohol syndrome (FAS) Birth defects caused by alcohol passed from a mother's bloodstream into the fetal bloodstream via the placenta.

fetoscopy (fee-TOS-cup-ee) A technique for examining a fetus for anatomical deformities. A needle-thin tube containing a viewing scope is inserted into the uterus, giving a direct view of the fetus.

fetus (FEE-tus) A developing human from about eight weeks to birth; has all the major structures of an adult.

F factor A piece of DNA that can exist as a plasmid; carries genes for making sex-pili and other structures needed for conjugation, as well as a site where DNA replication can start; F stands for fertility.

fiber (1) In plants, a long, slender sclerenchyma cell that usually occurs in a bundle. (2) In animals, an elongate, supportive thread in the matrix of connective tissue; (3) an extension of a neuron; (4) a muscle cell.

fibrin (FY-brin) The activated form of the blood-clotting protein fibrinogen, which aggregates into threads that form the fabric of a blood clot.

fibrinogen (FY-brin-uh-jen) The plasma protein that is activated to form a clot when a blood vessel is injured.

fibrous connective tissue A connective tissue with a matrix of densely packed parallel bundles of collagen fibers; forms tendons, which attach muscles to bones, and ligaments, which join bones together.

filter feeder Animal that actively screens small organisms or food particles from the water and then ingests them.

filtrate Fluid extracted by the excretory system from the blood or body cavity. The excretory system produces urine from the filtrate after extracting valuable solutes from it and concentrating it.

filtration In a kidney, the extraction of water and small solutes, including metabolic wastes, from the blood by the nephrons.

first law of thermodynamics The natural law that the total amount of energy in the universe is constant and that energy can be transferred and transformed, but it can never be destroyed; also called the principle of energy conservation.

fission A means of asexual reproduction in which a parent separates into two or more individuals of about equal size.

fitness The contribution that an individual makes to the gene pool of the next generation, relative to the contribution of other individuals in the population.

fixed action pattern (FAP) A genetically programmed, virtually unchangeable behavioral sequence.

flagellate (FLAJ-uh-lit) A protozoan that moves by means of one or more flagella.

flagellum (fluh-JEL-um) (plural, *flagella*) A long appendage that propels protists through the water and moves fluids across the surface of many tissue cells in animals. A cell may have one or more flagella. Like cilia, flagella have a 9 + 2 arrangement of microtubules covered by the cell's plasma membrane.

flower In an angiosperm, a short stem with four sets of modified leaves, bearing structures that function in sexual reproduction.

fluid feeder An organism that eats nutrient-rich fluids from another living organism.

fluid mosaic A description of membrane structure, depicting a cellular membrane as a mosaic of diverse protein molecules embedded in a fluid bilayer made of phospholipid molecules.

fluorescence A glow produced when the excited electrons of an illuminated molecule fall back to their original energy level.

follicle (FOL-ih-kul) A cluster of cells that surround, protect, and nourish a developing egg cell in the ovary; also secretes estrogen.

follicle-stimulating hormone (FSH) A protein hormone secreted by the anterior pituitary; stimulates the production of eggs by the ovaries and sperm by the testes.

food chain A sequence of food transfers from producers through several levels of consumers in an ecosystem.

food-conducting cells A specialized, living plant cell with thin primary walls; arranged end-to-end, they collectively form phloem tissue. Also called sieve-tube member.

food web A network of interconnecting food chains.

forebrain One of three ancestral and embryonic regions of the vertebrate brain; develops into the thalamus, hypothalamus, and cerebrum.

fossil A preserved remnant or impression of an extinct organism.

fossil fuel An energy deposit formed from the remains of extinct organisms.

fossil record The chronicle of evolution over millions of years of geological time engraved in the order in which fossils appear in rock strata.

founder effect Random change in the gene pool that occurs in a small colony.

fovea (FO-vee-uh) An eye's center of focus and the place on the retina where photoreceptors are highly concentrated.

fragmentation A mechanism of asexual reproduction in which a single parent breaks into parts that regenerate into whole new individuals.

fruit A ripened, thickened ovary of a flower, which protects dormant seeds and aids in their dispersal.

fruiting body A stage in an organism's life cycle that functions only in reproduction; for example, a mushroom is a fruiting body of many fungi.

functional group An assemblage of atoms that form the chemically reactive part of an organic molecule.

fungus (plural, *fungi*) A heterotrophic eukaryote that digests its food externally and absorbs the resulting small nutrient molecules. Most fungi consist of a netlike mass of filaments called hyphae. Molds, mushrooms, and yeasts are examples of fungi.

G

gallbladder Organ that stores bile and releases it as needed into the small intestine.

gametangium (GAM-eh-TANJ-ee-um) (plural, *gametangia*) A reproductive organ that houses and protects the gametes of a plant.

gamete (GAM-eet) A sex cell; a haploid egg or sperm; union of an egg and a sperm during fertilization, produces a zygote.

gametic isolation (guh-MEE-tik) A type of prezygotic barrier between species. The species remain isolated because male and female gametes of the different species cannot fuse, or they die before they unite.

gametophyte (guh-MEE-tuh-FITE) The multicellular haploid form in the life cycle of organisms undergoing alternation of generations; mitotically produces haploid gametes that unite and grow into the sporophyte generation.

ganglion (GANG-lee-un) (plural, *ganglia*) A cluster of nerve cell bodies belonging to the axons and dendrites making up nerves.

gas exchange *See* respiration.

gastric gland A tubular structure in the vertebrate stomach that secretes gastric juice.

gastric juice A digestive fluid of mucus, pepsinogen, and strong acid secreted by the stomach.

gastric ulcer An open sore in the lining of the stomach, resulting when pepsin and hydrochloric acid destroy the lining tissues faster than they can regenerate.

gastrin A digestive hormone that stimulates the secretion of gastric juice.

gastrovascular cavity (GAS-tro-VAS-kyul-ler) A digestive compartment with a single opening, the mouth; may function in circulation, body support, waste disposal, and gas exchange, as well as digestion.

gastrula (GAS-troo-luh) The embryonic stage resulting from gastrulation in animal development. Most animals have a gastrula made up of three layers of cells: the ectoderm, endoderm, and mesoderm.

gastrulation (GAS-troo-LAY-shun) The phase of embryonic development that follows cleavage and transforms the blastula into a gastrula. Gastrulation adds more cells to the embryo and sorts the cells into two distinct cell layers.

gene A discrete unit of hereditary information consisting of a specific nucleotide sequence in DNA (or RNA, in some viruses); located on the chromosomes in eukaryotes.

gene cloning The production of multiple copies of a gene.

gene expression The process whereby genetic information flows from genes to proteins; the flow of genetic information from the genotype to the phenotype.

gene flow The gain or loss of alleles from a population by movement of individuals or gametes into or out of the population.

gene pool All the genes in a population at any one time.

genetic drift A change in the gene pool of a population due to chance.

genetic markers (1) Alleles tracked in genetic studies. (2) Specific sections of DNA that earmark a particular allele. The sections contain specific restriction sites (points where restriction enzymes cut the DNA), which occur only in DNA that contains the allele.

genetic recombination The production, by crossing over, of chromosomes with gene combinations different from those in the chromosomes before crossing over.

genetics The science of heredity.

genome (JEN-nome) A complete (haploid) set of an organism's genes; an organism's genetic material.

genomic library (jen-OME-ik) A set of DNA segments from an organism's genome; each segment is carried by a plasmid or phage.

genotype (JEEN-oh-tipe) The genetic makeup of an organism.

genus (JEE-nus) (plural, *genera*) In classification, the taxonomic category above the species and below the family; the first part of a species' binomial; for example, *Homo*.

germinate (JER-mih-NATE) To begin to grow.

gestation (je-STAY-shun) The state of carrying developing young within the female reproductive tract.

gibberellin (JIB-er-EL-in) One of a family of plant hormones that trigger the germination of seeds and interact with auxins in regulating growth and fruit development.

gill An extension of the body surface of an animal, specialized for gas exchange and/or suspension feeding.

gizzard A pouchlike organ in a digestive tract, where food is mechanically ground.

glans The rounded, highly sensitive head of the clitoris in females and penis in males.

glomerulus (glum-AIR-yoo-lus) (plural, *glomeruli*) Part of a nephron consisting of the capillaries that are surrounded by Bowman's capsule. Together, a glomerulus and Bowman's capsule produce the filtrate from the blood.

glucagon (GLUKE-uh-gahn) A peptide hormone secreted by islet cells in the pancreas; raises the level of glucose in the blood.

glucocorticoid (GLUKE-oh-KOR-tih-koid) One of a family of steroid hormones secreted by the adrenal cortex; increases the level of glucose in the blood; helps maintain the body's response to long-term stress.

glycogen (GLY-kuh-jen) A complex, extensively branched polysaccharide of many glucose monomers; serves as an energy-storage molecule in liver and muscle cells.

glycolysis (gly-KOL-eh-sis) The multistep chemical breakdown of a molecule of glucose into two molecules of pyruvic acid; the first stage of cellular respiration in all organisms; occurs in the cytoplasmic fluid.

glycoprotein (gly-koh-PRO-teen) A macromolecule consisting of one or more polypeptides linked to short chains of sugars.

Golgi apparatus (GOLE-jee) An organelle in eukaryotic cells consisting of stacks of membranes that modify, store, and ship products of the endoplasmic reticulum.

gonad A sex organ in an animal; an ovary or a testis.

gradualist model The view that evolution occurs as a result of populations becoming isolated from common ancestral stock and gradually becoming genetically unique as they are adapted by natural selection to their local environments; Darwin's view of the origin of species.

granum (GRAN-um) (plural, *grana*) A stack of hollow disks formed of thylakoid membranes in a chloroplast. Grana are the sites where chlorophyll traps solar energy and converts it to chemical energy during the light reactions of photosynthesis.

gravitropism (greh-VIH-tro-PIZ-em) A plant's growth response to gravity.

gray matter The area of the vertebrate brain and spinal cord that consists mainly of nerve cell bodies.

green alga One of a group of photosynthetic protists that includes unicellular, colonial, and multicellular species. Green algae are plantlike in having biflagellated cells (gametes in colonial and multicellular species), chloroplasts with chlorophyll *a*, cellulose cell walls, and starch.

greenhouse effect The warming of the atmosphere caused by carbon dioxide, which absorbs infrared radiation and slows its escape from the Earth's surface.

ground tissue system A tissue of mostly parenchyma cells that makes up the bulk of a young plant and is continuous throughout its body. The ground tissue system fills the space between the epidermis and the vascular tissue system.

growth hormone (GH) A protein hormone secreted by the anterior pituitary; promotes development and growth and stimulates metabolism.

guard cell A specialized epidermal cell in plants that regulates the size of a stoma, allowing gas exchange between the surrounding air and the photosynthetic cells in the leaf.

gymnosperm (JIM-noh-spurm) A naked-seed plant; its seed is said to be naked because it is not enclosed in a fruit.

H

habitat A place where an organism lives.

habitat isolation A type of prezygotic barrier between species; the species remain isolated because they breed in different habitats.

habituation Learning not to respond to unimportant stimuli.

hair cell A type of mechanoreceptor that detects sound waves and other forms of movement in air or water.

haploid cell (HAP-loid) In the life cycle of an organism that reproduces sexually, a cell containing a single set of chromosomes; an *n* cell.

Hardy-Weinberg principle The concept that the shuffling of genes that occurs during sexual reproduction, by itself, cannot change the overall genetic makeup of a population.

heart attack Death of cardiac muscle cells and a failure of the heart to deliver enough blood to the body.

heartwood In the center of trees, the darkened, older layers of secondary xylem made up of cells that no longer transport water and are clogged with resins.

heat The amount of energy resulting from the movement of molecules in a body of matter. Heat is energy in its most random form.

helper T cell A type of lymphocyte that helps activate

other types of T cells and may help stimulate B cells to produce antibodies.

hemoglobin (HE-moh-glo-bin)   An iron-containing protein in red blood cells that reversibly binds oxygen and conveys it to body tissues.

hemophilia (HE-muh-FIL-ee-uh)   A human genetic disease caused by a sex-linked recessive allele, characterized by excessive bleeding following injury.

hepatic portal vessel   A large blood vessel that conveys blood from capillaries surrounding the intestine directly to the liver.

herbivore   (1) An organism that eats only plants. (2) An organism that eats only autotrophs (plants, algae, and autotrophic bacteria).

hermaphroditic (her-MAF-ruh-DIT-ik)   Pertaining to an individual animal that has both female and male gonads and produces eggs and sperm.

hermaphroditism (her-MAF-ruh-di-tizm)   A condition in which an individual has both female and male gonads and functions as both a male and female in sexual reproduction by producing both sperm and eggs.

heterotroph (HET-ur-oh-TROFE)   An organism that cannot make its own organic food molecules and must obtain them by consuming other organisms or their organic products; a consumer or a decomposer in a food chain.

heterozygote advantage (HET-ur-oh-ZY-gote)   Greater reproductive success of heterozygous individuals compared to homozygotes; tends to preserve variation in gene pools.

heterozygous (HET-ur-oh-ZY-gus)   Having different alleles for a given trait.

hibernation   Long-term torpor during cold weather; an animal's metabolic rate is reduced, and it survives on energy stored in body fat.

high-density lipoprotein (HDL)   A cholesterol carrier in the blood, made up of cholesterol and other lipids surrounded by a single layer of phospholipids in which proteins are embedded. An HDL carries less cholesterol than a related lipoprotein, LDL, and may be correlated with a decreased risk of blood vessel blockage.

hindbrain   One of three ancestral and embryonic regions of the vertebrate brain; develops into the medulla oblongata, pons, and cerebellum.

histamine (HISS-tuh-meen)   A chemical alarm signal released by injured cells that causes blood vessels to dilate during an inflammatory response.

histone   A small protein molecule associated with DNA and important in DNA packing in the eukaryotic chromosome.

HIV   Human immunodeficiency virus; the retrovirus that attacks the human immune system and causes AIDS.

homeobox (HOME-ee-OH-box)   A specific sequence of DNA that regulates patterns of differentiation during the development of an organism.

homeostasis (HOME-ee-oh-STAY-sis)   The steady state of body functioning; a steady state characterized by a dynamic interplay between outside forces that tend to change an organism's internal environment and the internal control mechanisms that oppose such changes.

homeotic gene (HOME-ee-OT-ik)   A master control gene that controls the overall body plan of a plant or animal by controlling the developmental fate of groups of cells.

hominid (HAHM-ih-nid)   A species on the human branch of the evolutionary tree; a member of the family Hominidae, including *Homo sapiens* and our ancestors.

homologous chromosomes (hoh-MOL-uh-gus)   The two chromosomes that make up a matched pair. Homologous chromosomes are of the same length, centromere position, and staining pattern and possess genes for the same traits at corresponding loci. One homologous chromosome is inherited from the organism's father, the other from the mother.

homologous structures   Structures that are similar in different species because the species have a common ancestry.

homozygous (HO-moh-ZY-gus)   Having identical alleles for a given trait.

hormone   A regulatory chemical signal synthesized by an endocrine gland, secreted into the blood, and carried in the blood to target cells that respond to the signal.

human chorionic gonadotropin (HCG) (KOR-ee-ahn-ik goh-NAD-uh-TROPE-in)   A hormone secreted by the chorion; maintains the corpus luteum of the ovary during the first three months of pregnancy.

humoral immunity   Components of the immune system mediated by B cells.

humus (HYOO-mus)   Decomposing organic material found in topsoil.

Huntington's disease   A human genetic disease caused by a dominant allele; characterized by uncontrollable body movements, and degeneration of the nervous system; usually fatal 10–20 years after symptoms begin.

hybrid   The offspring of parents of two different species or of two different varieties of one species; the offspring of two parents that differ in one or more inherited traits; an individual that is heterozygous for one or more pair of genes.

hybrid breakdown   A type of postzygotic barrier between species; the species remain isolated because the offspring of hybrids are weak or infertile.

hybrid inviability   A type of postzygotic barrier between species; the species remain isolated because hybrid zygotes do not develop or hybrids do not become sexually mature.

hybridization   The cross-fertilization of two different varieties of an organism or of two different species. Also called cross.

hybrid sterility   A type of postzygotic barrier between species; the species remain isolated because hybrids fail to produce functional gametes.

hydrocarbon   A chemical compound composed only of the elements carbon and hydrogen.

hydrogen bond   A type of weak chemical bond formed

when the partially positive hydrogen atom participating in a polar covalent bond in one molecule is attracted to the partially negative atom participating in a polar covalent bond in another molecule.

hydrolysis (hi-DROL-ih-sis)   A chemical process in which molecules are broken down as water is added to the bonds linking the monomers composing them; an essential part of digestion.

hydrophilic (HI-druh-FIL-ik)   Pertaining to molecules that are soluble in water.

hydrophobic (HI-druh-FO-bik)   Pertaining to molecules that do not mix with water.

hydrostatic skeleton (HI-druh-STAT-ik)   A skeletal system composed of fluid held under pressure in a closed body compartment; the main skeleton of most cnidarians, flatworms, nematodes, and annelids.

hydroxyl group (hi-DROK-sul)   In an organic molecule, a functional group consisting of a hydrogen atom bonded to an oxygen atom.

hymen   A thin membrane that partly covers the vaginal opening in the human female; ruptured by sexual intercourse or other vigorous activity.

hypercholesterolemia (HI-per-ko-LES-tur-ah-LEEM-ee-uh)   An inherited human disease characterized by an excessively high level of cholesterol in the blood.

hypertension   Abnormally high blood pressure; a persistent blood pressure of 140/90 or higher.

hypertonic solution   In comparing two solutions, the one with the greater concentration of solutes.

hyperventilating   Taking several deep breaths so rapidly that the CO_2 level in the blood is reduced, causing the breathing control centers to temporarily shut down breathing movements.

hypha (HI-fuh) (plural, *hyphae*)   One of many filaments making up the body of a fungus.

hypoglycemia (HI-po-gly-SEE-mee-uh)   An abnormally low level of glucose in the blood; results when the pancreas secretes too much insulin into the blood.

hypothalamus (HI-po-THAL-uh-mus)   The master control center of the endocrine system, located at the ventral portion of the vertebrate forebrain. The hypothalamus functions in maintaining homeostasis, especially in coordinating the endocrine and nervous systems; secretes hormones of the posterior pituitary and releasing hormones that regulate the anterior pituitary.

hypothesis (hi-POTH-uh-sis) (plural, *hypotheses*)   An educated guess that a scientist proposes as a tentative explanation for a specific phenomenon that has been observed.

hypotonic solution   In comparing two solutions, the one with the lesser concentration of solutes.

I

imitation   Learning by watching and mimicking the actions of others.

immune system   The organ system that protects the body by recognizing and attacking specific kinds of pathogens and cancer cells.

immunity   Resistance to specific body invaders.

immunodeficiency disease   An immunological disorder in which the immune system lacks one or more components, allowing the body to become susceptible to infectious agents that would ordinarily not be pathogenic.

imprinting   Learning that is limited to a specific critical period in an animal's life and that is reversible.

incomplete dominance   A type of inheritance in which F_1 hybrids have an appearance that is intermediate between the phenotypes of the parental varieties.

indeterminate growth   Unlimited growth; continuing to grow throughout life, as in most plants.

indoleacetic acid (IAA)   A naturally occurring auxin, a plant hormone.

inferior vena cava (VEE-nuh KAY-vuh)   A large vein returning O_2-poor blood to the heart from the lower, or posterior, part of the body.

inflammatory response   A nonspecific body defense; redness, heat, and swelling in damaged tissues. Inflammation is caused by a release of histamine and other chemical alarm signals, which trigger increased blood flow and a local increase in white blood cells and fluid leakage from the blood.

ingestion   The act of eating; the first main stage of food processing.

inhibiting hormone   A hormone, secreted by the hypothalamus; inhibiting hormones make the anterior pituitary stop secreting hormones.

innate behavior   Genetically programmed, virtually unchangeable behavior.

inner ear   One of three main regions of the vertebrate ear; includes the cochlea, organ of Corti, and semicircular canals.

innovation   The ability to perform a correct or appropriate behavior on the first attempt without prior experience; reasoning.

insulin   A protein hormone secreted by islet cells in the pancreas; lowers blood glucose level.

integration   The interpretation of sensory signals within neural processing centers of the central nervous system.

integumentary system (in-TEG-yoo-MEN-ter-ee)   The organ system consisting of the skin and its derivatives, such as hair and nails in mammals; helps protect the body from drying out, mechanical injury, and infection.

interferon (IN-ter-FEER-on)   A nonspecific defensive protein produced by virus-infected cells and capable of helping other cells attack viruses.

intermediate   One of the compounds that form between the initial reactant in a metabolic pathway, such as glucose in glycolysis, and the final product, such as pyruvic acid in glycolysis.

intermediate filament   An intermediate-sized fiber that is one of the three main kinds of fibers making up the cytoskeleton of eukaryotic cells; ropelike, made of fibrous proteins.

internal fertilization   The fusion of gametes within the female's body.

interneuron (IN-ter-NOOR-on)   A nerve cell, entirely within the central nervous system, that integrates sensory signals and may relay command signals to motor neurons.

internode   The portion of a plant stem between two nodes.

interphase   The period in the life cycle of a cell when the cell is not dividing. During interphase, a cell's metabolic activity is very high, its chromosomes and many cell parts are duplicated, and the cell may increase in size. Interphase lasts for 90% of the total time of the cell cycle.

interspecific competition   A contest between individuals of two populations that require a limited resource; may inhibit population growth and help structure communities.

interstitial fluid (IN-ter-STISH-yul)   An aqueous solution that surrounds body cells, and through which materials pass back and forth between the blood and the body tissues.

intertidal zone (IN-ter-TIDE-ul)   A shallow zone where the waters of an estuary or ocean meet land.

intestine   The region of a digestive tract between the gizzard or stomach and the anus, where chemical digestion and nutrient absorption usually occur.

intrinsic rate of increase   An organism's inherent capacity to reproduce.

intron (IN-trahn)   In eukaryotes, a noncoding portion of a gene, or a part of a gene that is not expressed.

inversion   A change in a chromosome resulting from reattachment in a reverse direction of a chromosome fragment to the original chromosome. Mutagens and errors during meiosis can cause inversions.

invertebrate   An animal that lacks a backbone.

in vitro fertilization (VEE-tro)   Uniting sperm and egg in a laboratory container, followed by the placement of a resulting early embryo in the mother's uterus.

ion (EYE-on)   An atom or molecule that has gained or lost one or more electrons, thus acquiring an electric charge.

ionic bond (eye-ON-ik)   An attraction between two ions with opposite electric charges; the electrical attraction of the opposite charges holds the ions together.

iris   The colored part of the vertebrate eye, formed by the anterior portion of the choroid.

islet cells (EYE-lit)   Clusters of endocrine cells in the pancreas that produce insulin and glucagon.

isomers (EYE-sum-ers)   Organic compounds with the same molecular formula but different structures and, therefore, different properties.

isotonic solution (eye-soh-TAHN-ik)   In a comparison of two solutions, both have equal concentrations of solutes.

isotope (EYE-so-tope)   A variant form of an atom; isotopes of an element have the same number of protons and electrons but different numbers of neutrons.

K

karyotype (KAIR-ee-uh-tipe)   A method of organizing the chromosomes of a cell in relation to number, size, and type.

kelp   A giant brown alga, up to 100 m long, that forms extensive undersea forests.

keystone predator   A predator species that reduces the density of the strongest competitors in a community, thereby helping maintain species diversity.

kilocalorie (kcal)   1000 calories; a quantity of heat equal to 1000 calories; used to measure the energy content of food.

kinesis (kih-NEE-sis)   Random movement in response to a stimulus.

kinetic energy (kih-NET-ik)   Energy that is actually doing work; the energy of a mass of matter that is moving. Moving matter performs work by transferring its motion to other matter, such as leg muscles pushing bicycle pedals.

kinetochore (kih-NET-oh-core)   A specialized protein structure at the centromere region on a sister chromatid. Spindle microtubules attach to the kinetochore during mitosis and meiosis.

kingdom   In classification, the broadest taxonomic category (above the phylum or division).

kingdom Animalia (AN-eh-mal-ee-uh)   The taxonomic group that contains the animals.

kingdom Fungi (FUN-jie)   The taxonomic group that contains the fungi.

kingdom Monera (moh-NAIR-uh)   The taxonomic group that contains the bacteria, or prokaryotes.

kingdom Plantae (PLANT-ay)   The taxonomic group that contains the plants.

kingdom Protista (pro-TIS-tuh)   The taxonomic group that contains the unicellular eukaryotes called protists.

kin selection   The concept that altruism evolves because it increases the number of copies of a gene common to a genetically related group of organisms; a hypothesis about the ultimate cause of altruism.

Koch's postulates   A set of diagnostic criteria used to determine which infectious agent causes a disease.

Krebs cycle   The metabolic cycle that is fueled by acetyl coA formed after glycolysis in cellular respiration; chemical reactions in the Krebs cycle complete the metabolic breakdown of glucose molecules to carbon dioxide; occurs in the matrix of mitochondria and supplies most of the NADH molecules that carry energy to the electron transport chain.

L

labia majora (LAY-bee-uh mu-JOR-uh)   A pair of outer thickened folds of skin that protect the female genital region.

labia minora (LAY-bee-uh mi-NOR-uh)   A pair of inner folds of skin, bordering and protecting the female genital region.

labor   A series of strong, rhythmic contractions of the uterus that expel a baby out of the uterus and vagina during childbirth.

lactic acid fermentation   A metabolic process performed by muscle cells without oxygen and by some bacteria in anaerobic environments. Following glycolysis, enzymes reduce pyruvic acid, forming lactic acid; simultaneously, NADH is oxidized, thus recharging the cell's supply of NAD$^+$ to keep glycolysis working.

large intestine   The tubular portion of the vertebrate alimentary tract between the small intestine and the anus; functions mainly in water absorption and formation of feces. Also called colon.

larva (plural, *larvae*)   An immature individual that is structurally and often ecologically very different from an adult.

larynx (LAIR-inks)   The voicebox, containing the vocal cords.

lateral   Pertaining to the side of a bilaterally symmetrical animal.

lateral line system   A row of sensory organs along each side of a fish's body. Sensitive to changes in water pressure, it enables a fish to detect minor vibrations in the water.

leaf   The main site of photosynthesis in a plant; consists of a flattened blade and a stalk (petiole) that joins the leaf to the stem.

learning   A behavioral change resulting from experience.

lens   The light-focusing structure in an eye; focuses light rays onto the retina.

leukemia (loo-KEE-me-ah)   A type of cancer of the blood-forming tissues; characterized by an excessive production of white blood cells and an abnormally high number of them in the blood; cancer of the bone marrow cells that produce leukocytes.

leukocyte (LOO-kuh-site)   A white blood cell.

lichen (LY-ken)   A mutualistic association between a fungus and an alga or between a fungus and a cyanobacterium.

life cycle   The entire sequence of stages in the life of an organism, from the adults of one generation to the adults of the next.

life history   The series of events from birth through reproduction to death.

life table   A listing of survivals and deaths in a population in a particular time period and predictions of how long, on average, an individual of a given age will live.

light microscope (LM)   An optical instrument with lenses that refract (bend) visible light to magnify images and project them into a viewer's eye or onto photographic film.

light reactions   The first of two stages in photosynthesis, the light reactions are the steps that absorb solar energy and convert it into chemical energy in the form of ATP and NADPH. The light reactions power the sugar-producing Calvin cycle, but produce no sugar themselves.

limbic system (LIM-bik)   A functional unit of several integrating and relay centers located deep in the human forebrain. The limbic system interacts with the cerebral cortex in creating emotions and storing memories.

linked genes   Genes located close enough together on a chromosome to be inherited together.

lipid   An organic compound consisting mainly of carbon and hydrogen atoms linked by nonpolar convalent bonds, and therefore mostly hydrophobic. Lipids include

fats, waxes, phospholipids, and steroids that are insoluble in water.

liver   The largest organ in the vertebrate body. The liver performs diverse functions such as producing bile, preparing nitrogenous wastes for disposal, and detoxifying poisonous chemicals in the blood.

lobe-finned fish   An extinct bony fish with saclike lungs and strong, muscular fins supported by bones. Lobefins were capable of spending some time on land and may have been ancestral to the earliest amphibians.

local regulator   A chemical messenger that is secreted into the interstitial fluid and causes changes in cells very near the point of secretion. Prostaglandins and neurotransmitters are local regulators.

locomotion   Active movement from place to place.

locus (plural, *loci*)   The particular site where a gene is found on a chromosome. Homologous chromosomes have corresponding loci.

logistic growth model   A mathematical description of idealized population growth that is restricted by limiting factors.

long-day plant   A plant that flowers in late spring or early summer when day length is increasing.

loop of Henle (HEN-lee)   The portion of a nephron, in the vertebrate kidney, that helps concentrate the filtrate while conveying it between a proximal tubule and a distal tubule.

loose connective tissue   Connective tissue whose matrix is a loose weave of fibers; holds other tissues and organs in place.

low blood pressure   A systolic blood pressure reading persistently below 100 mm Hg.

low-density lipoprotein (LDL)   A cholesterol carrier in the blood, made up of cholesterol and other lipids surrounded by a single layer of phospholipids in which proteins are embedded. An LDL carries more cholesterol than a related lipoprotein, HDL, and high LDL levels in the blood correlate with a tendency to develop blocked blood vessels and heart disease.

lung   An internal sac, lined with moist epithelium, where gases are exchanged between inhaled air and the blood.

luteinizing hormone (LH) (LOO-tee-in-EYE-zing)   A protein hormone secreted by the anterior pituitary; stimulates ovulation in females and androgen production in males.

Lyme disease   A debilitating human disease caused by the bacterium *Borrelia burgdorferi*; characterized at first by a red rash at the site of tick bite and, if not treated, by heart disease, arthritis, and nervous disorders.

lymph   A fluid similar to interstitial fluid circulated in the lymphatic system.

lymphatic system (lim-FAT-ik)   The organ system through which lymph circulates; includes lymph vessels, lymph nodes, and the spleen. The lymphatic system helps remove toxins and pathogens from the blood and interstitial fluid, and returns fluid and solutes from the interstitial fluid to the circulatory system.

lymphocyte (LIM-fuh-site)   A white blood cell that

produces the immune response, found mostly in the lymphatic system.

lymphoma (lim-FOH-muh)   Cancer of the tissues that form white blood cells.

lysogenic cycle (lie-so-JEN-ik)   A type of viral replication cycle in which the viral genome becomes incorporated into the bacterial host chromosome as a prophage; new phage is not produced, and the host cell is not killed or lysed.

lysosomal storage disease (LYE-suh-SOAM-ul)   A serious hereditary disorder resulting from the absence of one or more of the hydrolytic enzymes found in lysosomes.

lysosome (LYE-suh-soam)   A digestive organelle in eukaryotic cells; contains hydrolytic enzymes that digest the cell's food and wastes.

lytic cycle (LIT-ik)   A type of viral replication cycle resulting in the release of new viruses by lysis (breaking open) of the host cell.

M

macroevolution   The main events in the evolutionary history of life on Earth.

macromolecule   A giant molecule in a living organism: a protein, carbohydrate, or nucleic acid.

macronutrient   A chemical substance that an organism must obtain in relatively large amounts, compared to the small amounts of micronutrients required.

macrophage (MAK-roh-faje)   A large, amoeboid, phagocytic white blood cell that develops from a monocyte.

magnetic resonance imaging (MRI)   Imaging technology that uses magnetism and radio waves to induce hydrogen nuclei in water molecules to emit faint radio signals. A computer creates images of the body from the radio signals.

magnification   An increase in the apparent size of an object.

malignant tumor   An abnormal tissue mass that can spread into neighboring tissue and to other parts of the body.

mantle   In a mollusk, the outgrowth of the body surface that drapes over the animal. The mantle produces the shell and forms the mantle cavity.

marsupial (mar-SOO-pee-ul)   A pouched mammal, such as a kangaroo, opossum, or koala. Marsupials give birth to embryonic offspring that complete development while housed in a pouch and attached to nipples on the mother's abdomen.

marsupium (mar-SOO-pee-um) (plural, *marsupia*)   The external pouch on the abdomen of a female marsupial.

mass number   The sum of the number of protons and neutrons in an atom's nucleus.

mast cell   A vertebrate body cell that produces histamine and other molecules that trigger the inflammatory response.

mating type   A group of sexually compatible individuals in a population. Certain species of prokaryotes, protists, and fungi have mating types.

matter   Anything that occupies space and has mass.

mechanical isolation   A type of prezygotic barrier between species; the species remain isolated because structural differences between them prevent fertilization.

mechanoreceptor (meh-KAN-oh-reh-SEP-ter)   A sensory receptor that detects physical deformations in the environment, associated with pressure, touch, stretch, motion, and sound.

medulla oblongata (meh-DUL-uh OB-long-GAH-tuh)   Part of the vertebrate hindbrain continuous with the spinal cord. The medulla passes data between the spinal cord and forebrain; also controls autonomic, homeostatic functions, including breathing, heart and blood vessel activity, swallowing, digesting, and vomiting.

medusa (meh-DOO-suh) (plural, *medusae*)   One of two types of cnidarian body forms; an umbrellalike body form. Also called a jellyfish.

megareserve   An extensive region of land that includes one or more areas undisturbed by humans, surrounded by a buffer zone where disturbances are not extensive and do not threaten the undisturbed area.

meiosis (my-OH-sis)   In a sexually reproducing organism, the division of a single diploid nucleus into four haploid daughter nuclei. Meiosis, and cytokinesis accompanying it, produce haploid gametes from diploid cells in the reproductive organs of the parents.

memory cell   One of a clone of long-lived lymphocytes formed during the primary immune response. A memory cell remains in a lymph node until activated by exposure to the same antigen that triggered its formation. When activated, a memory cell forms a large clone that mounts the secondary immune response.

menstrual cycle (MEN-stroo-ul)   The hormonally synchronized cyclic buildup and breakdown of the endometrium of some primates, including humans. Also called uterine cycle.

menstruation (MEN-stroo-AY-shun)   Uterine bleeding resulting from shedding of the endometrium during a menstrual cycle.

meristem (MAIR-eh-STEM)   Plant tissue consisting of undifferentiated cells that divide and generate new cells and tissues.

mesoderm (MEZ-oh-durm)   The middle layer of the three embryonic cell layers in a gastrula; gives rise to muscles, bones, dermis of the skin, and to most other organs in the adult.

mesophyll (MEZ-oh-fil)   The ground tissue of a leaf; the main site of photosynthesis.

messenger RNA (mRNA)   The type of ribonucleic acid that encodes genetic information from DNA and conveys it to ribosomes, where the information is translated into amino acid sequences.

metamorphosis (MET-uh-MOR-fuh-sis)   The transformation of a larva into an adult.

metaphase (MET-uh-faze)   The second stage of mitosis. During metaphase, all the cell's duplicated chromosomes are lined up at an imaginary plane equidistant between the poles of the mitotic spindle.

metastasis (meh-TAS-reh-sis)   The spread of cancer cells beyond their original site.

microevolution A change in a population's gene pool over a succession of generations; evolutionary changes in species over relatively brief periods of geological time.

microfilament The thinnest of the three main kinds of fibers making up the cytoskeleton of the eukaryotic cell; a solid, helical rod usually composed of the globular protein actin. Microfilaments help some cells change shape and move by assembling at one end while disassembling at the other.

micrograph A photograph taken through a microscope.

micronutrient An element that an organism needs in very small amounts and that functions as a component or cofactor of enzymes.

microtubule The thickest of the three main kinds of fibers making up the cytoskeleton of eukaryotic cells; a straight, hollow tube made of globular proteins called tubulins. Microtubules form the basis of structure and movement of cilia and flagella.

microtubule organizing center A specialized place in the cell where microtubules of the mitotic spindle begin to form.

microvillus (plural, *microvilli*) A microscopic projection on the surface of a cell. Microvilli increase a cell's surface area.

midbrain One of three ancestral and embryonic regions of the vertebrate brain; develops into sensory integrating and relay centers that send sensory information to the cerebrum.

middle ear One of three main regions of the vertebrate ear; a chamber containing three small bones (the hammer, anvil, and stirrup), which convey vibrations from the eardrum to the oval window.

mineral In nutrition, the chemical elements, other than carbon, hydrogen, oxygen, and nitrogen, that an organism requires for proper body functioning.

mineralocorticoid (min-er-AL-oh-KOR-te-koid) One of a family of steroid hormones secreted by the adrenal cortex. Mineralocorticoids help maintain salt and water homeostasis and may increase blood pressure in response to long-term stress.

mitochondrial matrix (MY-toh-KON-dree-ul MAY-triks) The fluid contained within the inner membrane of a mitochondrion.

mitochondrion (MY-toh-KON-dree-on) (plural, *mitochondria*) An organelle in the eukaryotic cell where cellular respiration occurs. Enclosed by two concentric membranes, it is where ATP is made.

mitosis (MY-toh-sis) The division of a single nucleus into two genetically identical daughter nuclei. Mitosis and cytokinesis make up the mitotic (M) phase of the cell cycle.

mitotic spindle A spindle-shaped structure formed of microtubules and associated proteins that is involved in the movements of chromosomes during mitosis and meiosis.

modern synthesis A comprehensive theory of evolution that incorporates genetics and includes most of Charles Darwin's ideas, focusing on populations as the fundamental units of evolution.

molecular biology The study of the molecular basis of genes and gene expression; molecular genetics.

molecule Two or more atoms held together by chemical bonds.

Mollusca (mol-LUS-kuh) The phylum that contains the mollusks; characterized by a muscular foot, mantle, mantle cavity, and radula; includes gastropods (snails and slugs), bivalves (clams, oysters, scallops), and cephalopods (squids and octopuses).

molting In arthropods, the process of shedding an old exoskeleton and secreting a new, larger one.

monoclonal antibody (MON-oh-KLONE-ul) An antibody secreted by a clone of cells and, consequently, specific for the one antigen that triggered the clone.

monocot (MON-uh-kot) A flowering plant whose embryos have a single seed leaf, or cotyledon.

monoculture The cultivation of a single plant variety in a large land area.

monocyte (MAHN-uh-site) A nonspecific defensive, phagocytic white blood cell that can engulf bacteria and viruses in infected tissue; has a large oval or horseshoe-shaped nucleus.

monoecious (mon-EE-shus) Having individuals that produce both sperm and eggs (usually refers to plants).

monohybrid cross An experimental mating of individuals in which the inheritance of a single trait is tracked.

monomer (MON-uh-mer) A chemical subunit that serves as a building block of a polymer.

monosaccharide (MON-oh-SAK-uh-ride) The smallest sugar molecule; a single-unit sugar. Monosaccharides are the building blocks of more complex sugars and polysaccharides.

monotreme (MON-uh-treem) An egg-laying mammal, such as the duck-billed platypus.

morphs Two or more different kinds of individuals or forms of a phenotypic trait in a population.

motor division The part of the vertebrate peripheral nervous system made up of motor neurons.

motor neuron (NOOR-on) A nerve cell that transmits command signals from the central nervous system to muscles or glands.

motor output The conduction of signals from a processing center in a central nervous system to effector cells.

motor unit A motor neuron and all the muscle fibers it controls.

mouth An opening through which food is taken into an animal's body.

mucous membrane (MYOO-kus) Smooth, moist epithelium that lines the digestive tract and air tubes leading to the lungs.

Müllerian mimicry (myoo-LER-ee-un MIM-ih-kree) A mutual mimicry by two species, both of which are poisonous or otherwise harmful to a predator.

multiple fruit A fruit that develops from a group of separate flowers tightly clustered together.

muscle tissue Tissue made of bundles of long cells called muscle fibers.

muscular system The organ system made up of all the skeletal muscles in an animal's body.

mutagen (MYOO-tuh-jen) A chemical or physical agent that interacts with DNA and causes a mutation.

mutagenesis (MYOO-tuh-jen-ih-sis) The creation of a mutation.

mutation (myoo-TAY-shun) A change in the nucleotide sequence of DNA; the ultimate source of genetic diversity.

mutualism (MYOO-choo-ul-iz-um) A symbiotic relationship in which both partners benefit.

mycelium (my-SEEL-ee-um) (plural, *mycelia*) The densely branched network of hyphae in a fungus.

mycorrhiza (MY-ko-RY-za) (plural, *mycorrhizae*) A mutualistic association of plant roots and fungi.

myelin sheath (MY-eh-lin) A series of cells, each wound around, and thus insulating, the axon of a nerve cell in vertebrates. Each pair of cells in the sheath is separated by a space called a node of Ranvier.

myofibril (MY-oh-FY-brul) A contractile thread in a muscle cell (fiber) made up of many sarcomeres. Longitudinal bundles of myofibrils make up a muscle fiber.

N

NAD⁺ Nicotinamide adenine dinucleotide; a coenzyme present in all cells that assists enzymes by conveying electrons (actually hydrogen atoms) during the redox reactions of metabolism; the plus sign indicates that the molecule is oxidized and tends to pick up hydrogen atoms.

natural killer cell A nonspecific defensive cell that attacks cancer cells and infected body cells, especially those harboring viruses.

natural selection Differential success in reproduction by different phenotypes resulting from interactions with the environment. Evolution occurs when natural selection produces changes in the relative frequencies of alleles in a population's gene pool.

nearsightedness An inability to focus on distant objects; occurs when the eyeball is longer than normal and the lens focuses distant objects in front of the retina. Also called myopia.

negative feedback Inhibition of a chemical reaction, metabolic pathway, or hormone-secreting gland by the products of the reaction, pathway, or gland. As the concentration of the products builds up, the product molecules themselves inhibit the chemical reaction, metabolic pathway, or secretion by the gland; a primary mechanism of homeostasis.

Nematoda (NEM-eh-TOAD-uh) The phylum that contains the roundworms, or nematodes; characterized by a pseudocoelom, a cylindrical, wormlike body form, and a tough cuticle.

neoteny (nee-OT-eh-nee) The retention of juvenile body features in an adult.

nephron The tubular excretory unit and its associated blood vessels of the vertebrate kidney; extracts filtrate from the blood and refines it into urine.

nerve A ropelike bundle of neuron fibers (axons and dendrites) tightly wrapped in connective tissue.

nerve cord An elongate bundle of nerves, usually extending longitudinally from the brain or anterior ganglia. One or more nerve cords and the brain make up the central nervous system in many animals.

nerve net A weblike system of neurons; characteristic of radially symmetrical animals such as *Hydra*.

nervous system The organ system that forms a communication and coordination network among all parts of an animal's body.

nervous tissue Tissue made up of neurons and supportive cells.

neural tube (NOOR-ul) An embryonic cylinder that develops from ectoderm after gastrulation; gives rise to the brain and spinal cord.

neuromuscular junction A synapse between an axon of a motor neuron and a muscle fiber.

neuron (NOOR-on) A nerve cell; the fundamental structural and functional unit of the nervous system; specialized for carrying signals from one location in the body to another.

neurosecretory cell (NOOR-oh-SEEK-reh-tor-ee) A nerve cell that synthesizes hormones and secretes them into the blood, as well as conducting nerve signals.

neurotransmitter A chemical messenger that carries information from a transmitting neuron to a receiving cell, either another neuron or an effector cell.

neutral liquid Pure water or an aqueous solution that is neither acidic nor basic and has a pH of 7.

neutral variation Genetic variation that provides no apparent selective advantage for some individuals over others.

neutron An electrically neutral particle (having no electric charge), found in the nucleus of an atom.

neutrophil (NOO-truh-fil) A nonspecific defensive, phagocytic white blood cell that can engulf bacteria and viruses in infected tissue; has a multilobed nucleus.

niche (nich) The sum total of a population's use of the biotic and abiotic resources of its environment; the role a population plays in its environment.

nitrogen fixation The conversion of atmospheric nitrogen (N_2) into nitrogen compounds (NH_4^+, NO_3) that plants can absorb and use.

nitrogenous base (nigh-TRAH-jen-us) An organic base that contains the element nitrogen.

node The point of attachment of a leaf on a stem.

node of Ranvier (RAHN-vee-ay) An unmyelinated region on a myelinated axon of a nerve cell, where signal transmission occurs.

nodule On a plant, a swelling that houses nitrogen-fixing bacteria.

noncompetitive inhibitor A substance that impedes the activity of an enzyme without entering an active site; by binding to the outside of the active site, a noncompetitive inhibitor changes the shape of the enzyme so that the active site no longer functions; often involved in negative feedback control mechanisms.

noncyclic electron flow   In the light reactions of photosynthesis, the noncircular route that electrons take, passing from one photosystem through an electron transport chain to a second photosystem and ending up in NADPH. ATP is generated, water is split, and O_2 is liberated in the process.

nondisjunction   An accident of meiosis or mitosis in which both members of a pair of homologous chromosomes or both sister chromatids fail to separate.

nonpolar covalent bond   An attraction between atoms that share one or more pairs of electrons equally because the atoms have similar electronegativity.

nonrandom mating   The selection of mates other than by chance.

nonself molecule   A foreign antigen; a molecule that is not part of an organism's body.

nonsteroid hormone   A regulatory chemical synthesized from amino acids that acts on target cells via a second messenger such as cAMP. Amine hormones, peptide hormones, and protein hormones are nonsteroids.

norepinephrine (NOR-ep-ih-NEF-rin)   An amine hormone produced by the adrenal medulla; prepares body organs for fight or flight; also serves as a neurotransmitter. Also called noradrenaline.

notochord (NO-tuh-kord)   A flexible, cartilagelike, longitudinal rod located between the digestive tract and nerve cord in chordate animals; present only in embryos in many species.

nuclear envelope   A double membrane perforated with pores, enclosing the nucleus and separating it from the rest of the eukaryotic cell.

nucleic acid (noo-KLAY-ik)   A polymer consisting of many nucleotide monomers; serves as a blueprint for proteins and, through the actions of proteins, for all cellular activities. The two types are DNA and RNA.

nucleoid region (NOO-klee-oid)   The region in a prokaryotic cell consisting of a concentrated mass of DNA.

nucleolus (noo-KLEE-uh-lus)   A combination of DNA, RNA, and proteins within the nucleus of a eukaryotic cell; the site where ribosome subunits are made.

nucleosome (NOO-klee-uh-soam)   The beadlike unit of DNA packaging in a eukaryotic cell; consists of DNA wound around a protein core made up of eight histone molecules.

nucleotide (NOO-klee-oh-tide)   An organic monomer consisting of a five-carbon sugar covalently bonded to a nitrogenous base and a phosphate group. Nucleotides are the building blocks of nucleic acids.

nucleus (plural, *nuclei*)   (1) An atom's central core, containing protons and neutrons. (2) The genetic control center of a eukaryotic cell.

O

ocean current   A riverlike flow pattern in the ocean.

old-growth forest   An ancient forest that has never been seriously disturbed by humans. Trees dominating an old-growth forest may be thousands of years old.

omnivore   (1) An animal that eats plants and animals. (2) An organism that eats autotrophs and heterotrophs.

oncogene (ONK-oh-jean)   A cancer-causing gene.

oogenesis (OH-uh-JEN-eh-sis)   The formation of ova.

open circulatory system   A circulatory system in which blood is pumped through open-ended vessels and out among the body cells. In an animal with an open circulatory system, blood and interstitial fluid are one and the same.

operator   In DNA, a sequence of nucleotides that determines whether RNA polymerase can attach to the promoter and move along the genes.

operculum (oh-PUR-kyoo-lum) (plural, *opercula*)   A protective flap on each side of a fish's head that covers a chamber housing the gills.

operon (OP-ur-on)   A unit of genetic function common in prokaryotes; a cluster of genes with related functions, along with a promoter and an operator.

opportunistic life history   A life history characterized by the production of many offspring, few of which survive; surviving offspring mature and reproduce quickly. Populations with opportunistic life histories tend to expand rapidly when environmental conditions are favorable and may die off abruptly when conditions change.

optimal foraging   Feeding behavior that provides maximal energy gain with minimal energy expense and minimal time spent in searching for and eating food.

order   In classification, the taxonomic category above the family and below the class.

organ   A structure consisting of several tissues adapted as a group to perform specific functions.

organ of Corti (KOR-tee)   The hearing organ in birds and mammals, located within the cochlea.

organelle (or-gan-EL)   A structure with a specialized function within a cell.

organic compound   A chemical compound containing the element carbon.

organism   An individual living thing, such as a bacterium, fungus, protist, plant, or animal.

organ system   A group of organs that work together in performing vital body functions.

orgasm   Rhythmic contractions of the reproductive structures, accompanied by extreme pleasure, at the peak of sexual excitement in both sexes; includes ejaculation by the male.

orientation behavior   Directed movement resulting in an organism's placing itself in a specific position in reference to an environmental cue.

osmoconformer (OZ-moh-con-FORM-er)   An organism whose body fluids have a solute concentration equal to that of its surroundings. Osmoconformers do not have a net gain or loss of water by osmosis.

osmoregulation   Control of the gain and loss of water and dissolved solutes in an organism.

osmoregulator   An organism whose body fluids have a

solute concentration different from that of their environment and that must use energy in controlling water loss or gain.

osmosis (oz-MOH-sis)   The movement of water across a selectively permeable membrane.

osteoporosis (OS-tee-oh-puh-RO-sis)   A skeletal disorder characterized by thinning, porous, and easily broken bones; most common among women after menopause and probably correlated with low estrogen levels.

outer ear   One of three main regions of the ear in reptiles, birds, and mammals; made up of the auditory canal and, in many birds and mammals, the pinna.

oval window   In the vertebrate ear, a membrane-covered gap in the skull bone, through which sound waves pass from the middle ear into the inner ear.

ovarian cycle (oh-VAIR-ee-un)   Hormonally synchronized cyclic events in the mammalian ovary, culminating in ovulation.

ovary   (1) In flowering plants, the basal portion of a carpel in which the egg-containing ovules develop. (2) In animals, the female gonad, which produces egg cells and reproductive hormones.

oviduct (OH-vih-dukt)   The tube that conveys egg cells away from the ovary.

ovulation (AH-vyoo-LAY-shun)   The release of an egg cell from an ovarian follicle.

ovule (OH-vyool)   A reproductive structure in a seed plant; contains the female gametophyte and the developing egg. An ovule develops into a seed.

ovum (OH-vum) (plural, *ova*)   An unfertilized egg, or female gamete.

oxidation   The loss of electrons from a substance involved in a redox reaction; always accompanies reduction.

oxytocin (OK-sih-TOE-sin)   A peptide hormone made by the hypothalamus and secreted by the posterior pituitary; stimulates contraction of the uterus and mammary gland cells.

P

pacemaker   The SA (sinoatrial) node; a specialized region of cardiac muscle that maintains the heart's pumping rhythm by setting the rate at which the heart contracts.

paleontologist (PAY-lee-un-TOL-uh-jist)   A scientist who studies extinct life and its evolution, as recorded in the fossil record.

pancreas (PAN-kree-us)   A gland with dual functions: The nonendocrine portion secretes digestive enzymes and an alkaline solution into the small intestine via a duct; the endocrine portion secretes the hormones insulin and glucagon directly into the blood.

Pangaea (pan-JEE-uh)   The supercontinent consisting of all the major landmasses of Earth fused together. Continental drift formed Pangaea near the end of the Paleozoic era.

parasitism (PAIR-eh-SIH-tiz-em)   A symbiotic relationship in which the parasite, a type of predator, lives within or on the surface of a host, from which it derives its food.

parasympathetic division   One of two sets of neurons in the autonomic nervous system; generally promotes body activities that gain and conserve energy, such as digestion and reduced heart rate.

parathyroid glands (PAIR-uh THY-roid)   Four endocrine glands embedded in the surface of the thyroid gland; secrete parathyroid hormone.

parathyroid hormone (PTH)   A peptide hormone secreted by the parathyroid glands; raises blood calcium level.

parenchyma cell (pur-EN-kim-uh)   In plants, a relatively unspecialized cell with a thin primary wall and no secondary wall; functions in photosynthesis, food storage, and aerobic respiration, and may differentiate into other cell types.

passive immunity   Temporary immunity obtained by acquiring ready-made antibodies or immune cells; lasts only a few weeks or months because the immune system has not been stimulated by antigens.

passive transport   The diffusion of a substance across a biological membrane, without the cell's performing any work.

pathogen   A disease-causing organism.

pattern formation   During development, the emergence of major body parts.

pedigree   A family tree representing the occurrence of heritable traits in parents and offspring across as many generations as possible.

pelagic zone (pel-AJ-ik)   The region of an ocean occupied by seawater.

penis   The male organ of copulation and urination.

peptide bond   The covalent linkage between two amino acid units, formed by dehydration synthesis.

peptide bond formation   During translation, the formation of a peptide bond as an amino acid is added to a growing polypeptide chain on a ribosome.

peptidoglycan (PEP-tid-oh-GLY-kan)   A polymer of complex sugars cross-linked by short polypeptides.

perception   The brain's meaningful interpretation, or conscious understanding, of sensory information.

perennial (peh-REN-ee-ul)   A plant that can live for many years.

perforin (PUR-fer-in)   A protein secreted by a cytotoxic T cell, which lyses (ruptures) a target by perforating its membrane.

peripheral nervous system (PNS)   The network of nerves and ganglia carrying signals into and out of the central nervous system.

peristalsis (PER-ih-STAHL-sis)   Rhythmic waves of contraction of smooth muscles. Peristalsis propels food through a digestive tract and also enables many animals, such as earthworms, to crawl.

permafrost   Continuously frozen ground found in the tundra.

petal   A modified leaf of a flowering plant. Petals are the often colorful parts of a flower that advertise it to insects and other pollinators.

P generation The parent individuals from which offspring are derived in studies of inheritance. P stands for parental.

phage (FAJE) *See* bacteriophage.

phagocyte (FAG-geh-site) A white blood cell (e.g., a neutrophil or a monocyte) that engulfs bacteria, foreign proteins, and the remains of dead body cells.

phagocytosis (FAG-oh-sigh-TOE-sis) Cellular eating; a type of endocytosis whereby a cell engulfs macromolecules, other cells, or particles into its cytoplasm.

pharynx (FAIR-inks) The organ in a digestive tract that receives food from the oral cavity; in terrestrial vertebrates, the throat region where the air and food passages cross.

phenotype (FEE-noh-tipe) The expressed traits of an organism.

phenylketonuria (PKU) (FEN-ul-KEE-tuh-NOOR-ee-uh) A human genetic disease caused by a recessive allele; characterized by an inability to properly break down the amino acid phenylalanine, an accumulation of toxins in the blood, and mental retardation.

pheromone (FAIR-uh-mone) A perfumelike chemical signal that functions in communication between individual organisms, influencing behavior and body functions much like a hormone does.

phloem (FLO-um) The portion of a plant's vascular system that conveys phloem sap throughout a plant; made up of sieve-tube members.

phloem sap The solution of sugars, other nutrients, and hormones conveyed throughout a plant via phloem.

phosphate group (FOS-fate) A functional group containing the element phosphorus covalently bonded to four oxygen atoms. Phosphate groups are important in energy transfer and as components of nucleic acids.

phospholipid (FOS-fo-LIP-id) A type of lipid molecule containing the element phosphorus and two fatty acids. Phospholipids make up the inner bilayer of biological membranes.

phosphorylation (fos-for-uh-LAY-shun) The transfer of a phosphate group, usually from ATP, to a molecule. Nearly all cellular work depends on ATP energizing other molecules by phosphorylation.

photic zone (FOE-tik) The region of an aquatic ecosystem into which light penetrates and where photosynthesis occurs.

photon (FOE-tahn) A quantum, or specific amount, of light energy. The shorter the wavelength of light, the greater the energy of a photon.

photoperiod The length of a day relative to the length of a night.

photophosphorylation (FOE-toe-fos-for-uh-LAY-shun) The production of ATP by chemiosmosis during the light reactions of photosynthesis. ATP is made by adding a phosphate group to ADP using energy harvested from sunlight.

photopsin (foe-TOP-sin) One of a family of visual pigments in the cones of the vertebrate eye that absorb bright, colored light.

photoreceptor A type of electromagnetic receptor that detects light.

photorespiration In a plant cell, the breakdown of a two-carbon compound produced by the Calvin cycle. The Calvin cycle produces the two-carbon compound, instead of its usual three-carbon product G3P, when leaf cells fix O_2, instead of CO_2. Photorespiration produces no sugar molecules or ATP.

photosynthesis (FOE-toe-SIN-thuh-sis) The process by which plants, autotrophic protists, and some bacteria use light energy to make sugars and other organic food molecules from carbon dioxide and water. Photosynthesis is the food-making process upon which most organisms depend.

photosynthetic autotroph (FOE-toe-sin-THET-ik AW-toe-trofe) An organism that uses light energy to make food molecules.

photosystem A light-harvesting "antenna" unit of the chloroplast's thylakoid membrane.

phototropism (foh-TOT-treh-PIZ-em) Growth of a plant shoot toward or away from light.

pH scale A measure of the relative acidity of a solution, ranging in value from 0 (most acidic) to 14 (most basic); pH stands for potential hydrogen and refers to the concentration of hydrogen ions (H^+).

phylum (FYE-lum) (plural, *phyla*) In classification, the taxonomic category above the class and below the kingdom; members of a phylum all have a similar general body plan.

phylogenetic tree (FYE-lo-jeh-NET-ik) A diagram that traces the likely evolutionary relationships among organisms.

phylogeny (fye-LAHJ-en-ee) The evolutionary history of a species or group of related species.

phytochrome (FYE-teh-krome) A colored protein in plants that contains a special set of atoms that absorbs light.

phytoplankton (FYE-toh-PLANK-tun) Algae and cyanobacteria that drift passively in aquatic environments.

pili (PIL-eye) (singular, *pilus*) Short projections on the surface of bacterial cells that help bacteria attach to surfaces; specialized sex pili are used in conjugation to hold the mating cells together.

pineal gland (PIN-ee-ul) An outgrowth of the vertebrate brain that secretes the hormone melatonin. Melatonin coordinates daily and seasonal body activities, such as reproductive activity, with environmental light conditions.

pinna (PIN-uh) The flaplike part of the outer ear, projecting from the body surface of many birds and mammals; collects sound waves and channels them to the auditory canal.

pinocytosis (PIN-oh-sigh-TOE-sis) Cellular drinking; a type of endocytosis whereby the cell takes fluid and dissolved solutes into small membranous vesicles.

pith Part of the ground tissue system of a dicot plant. Pith fills the center of a stem and may store food.

placenta The organ that provides nutrients and oxygen

to the embryo and helps dispose of its metabolic wastes; formed of the embryo's chorion and the mother's endometrial blood vessels.

placental mammal   A mammal whose young complete their embryonic development in the uterus, nourished via the mother's blood vessels in the placenta.

plasma   The liquid matrix of the blood in which the blood cells are suspended.

plasma cell   An antibody-secreting B cell; an effector B cell.

plasma membrane   The thin, living border of a cell that sets it off from the outside environment and acts as a selective barrier to the passage of ions and molecules between the cell and the environment.

plasmid   In a bacterial cell, a small ring of DNA separate from the bacterial chromosome.

plasmodesma (PLAZ-moh-DEZ-muh) (plural, *plasmodesmata*)   An open channel in a plant cell wall, through which strands of cytoplasm connect from adjacent walls.

plasmodial slime mold (plaz-MO-dee-ul)   A type of protist that has amoeboid cells, flagellated cells, and an amoeboid syncytial feeding stage in its life cycle.

platelet   A piece of cytoplasm from a large cell in the bone marrow of a mammal; a blood-clotting element.

plate tectonics   Forces within planet Earth that cause movements of the crust, resulting in continental drift, volcanoes, and earthquakes.

Platyhelminthes (PLAT-ee-hel-MIN-theez)   The phylum that contains the flatworms, the bilateral animals with a thin, flat body form, gastrovascular cavity or no digestive system, and no body cavity; the free-living flatworms, flukes, and tapeworms.

pleated sheet   The folded arrangement of a polypeptide in a protein's secondary structure.

pleiotropy (PLEE-ih-tro-pee)   The ability of a single gene to have multiple phenotypic effects.

polar covalent bond   An attraction between atoms that share electrons unequally because the atoms differ in electronegativity. The shared electrons are pulled closer to the more electronegative atom, making it partially negative and the other atom partially positive.

polar molecule   A molecule containing polar covalent bonds.

pollen grain   In a seed plant, the male gametophyte that develops within the anthers of stamens.

pollination   In seed plants, the delivery, by wind or animals, of pollen from the male parts of a plant to the stigma of a carpel on the female.

polygenic inheritance (POL-ee-JEN-ik)   The additive effect of two or more gene loci on a single phenotypic trait.

polymer (POL-uh-mer)   A large molecule consisting of many identical or similar molecular units, called monomers, strung together.

polymerase chain reaction (PCR) (puh-LIM-eh-rase)   A techique used to obtain many copies of DNA molecules. A small amount of DNA mixed with the enzyme DNA polymerase, DNA nucleotides, and other chemicals replicates repeatedly in a test tube.

polymorphic   Having several or many different forms; pertaining to a population in which two or more morphs are present in readily noticeable frequencies.

polynucleotide (POL-ee-NOO-klee-oh-tide)   A polymer made up of many nucleotides covalently bonded together.

polyp (POL-ip)   One of two types of cnidarian body forms; a columnar, hydra-like body.

polypeptide   A chain of amino acids linked by peptide bonds. The amino acid at one end of the chain has a free amino group, and the one at the other end has a free carboxyl group.

polyploid (POL-eh-ploid)   Having more than two complete sets of chromosomes.

polysaccharide (POL-ee-SAK-uh-ride)   A carbohydrate polymer consisting of hundreds to thousands of monosaccharides linked by dehydration synthesis.

pons   Part of the vertebrate hindbrain that functions with the medulla oblongata in passing data between the spinal cord and forebrain and in controlling autonomic, homeostatic functions.

population   A localized group of interacting individuals belonging to one species.

population density   The number of individuals of a species in a unit area or volume.

population dynamics   Changes in population size resulting from various forces that control and regulate populations over time.

population genetics   The study of genetic changes in populations; the science of microevolutionary changes in populations.

population-limiting factor   An environmental factor that restricts population growth.

Porifera (por-IF-er-uh)   The phylum that contains the sponges, characterized by choanocytes, a porous body wall, and no true tissues.

positive feedback   A control mechanism in which a change in some variable triggers mechanisms that amplify change.

positive-ion exchange   A mechanism whereby root hairs take up certain positively charged inorganic ions. H+ ions released from root hairs replace positively charged nutrient ions that adhere to clay particles in the soil; the root hairs then absorb the free nutrient ions.

positron-emission tomography (PET)   Imaging technology that uses radioactively labeled biological molecules, such as glucose, to obtain information about metabolic processes at specific locations in the body. The labeled molecules are injected into the bloodstream, and a PET scan for radioactive emissions determines which tissues have taken up the molecules.

post-anal tail (POST-ANE-ul)   A tail situated behind the anus; a diagnostic feature of chordate animals.

posterior   Pertaining to the rear, or tail, of a bilaterally symmetrical animal.

posterior pituitary   An extension of the hypothalamus

composed of nervous tissue that secretes hormones made in the hypothalamus; a temporary storage site for hypothalamic hormones.

postzygotic barrier (POAST-zy-GOT-ik)   A reproductive barrier that prevents zygotes from developing into fertile adults.

potential energy   Stored energy; the capacity to perform work that matter possesses because of its location or arrangement. Water behind a dam or chemical bonds both possess potential energy.

preadaptation   A structure that has evolved in one environmental context and later becomes adapted for a different function in a different environmental context.

predation   An interaction between species in which one species, the predator, eats the other, the prey.

predator   A consumer in a biological community.

pregnancy   The condition of carrying developing young within the female reproductive tract.

prepuce (PRE-pyoos)   Foreskin; a fold of skin covering the head of the clitoris and penis.

pressure-flow mechanism   The process whereby phloem sap is transported in a vascular plant; caused by higher pressure at a sugar source than at a sugar sink. Cells at a sugar source load sugar into phloem, and water follows osmotically, increasing the water pressure within the phloem; cells at a sugar sink extract sugar from the phloem, and water follows, decreasing the pressure within the phloem.

prevailing winds   The major air movements over Earth's surface.

prey   An organism eaten by a predator.

prezygotic barrier (PREE-zy-GOT-ik)   A reproductive barrier that prevents mating between species or hinders fertilization if mating between species is attempted.

primary consumer   An organism that eats only autotrophs.

primary growth   Growth in the length of a plant root or shoot produced by an apical meristem.

primary immune response   The initial immune response to an antigen, which appears after a lag of several days.

primary oocyte (OH-eh-SITE)   A diploid cell, in prophase I of meiosis, that can be hormonally triggered to develop into an ovum.

primary phloem (FLO-um)   Phloem tissue formed by cells of the vascular cylinder in a young plant.

primary productivity   The rate at which an ecosystem's producers convert solar energy to chemical energy.

primary spermatocyte (spur-MAT-eh-SITE)   A diploid cell in the testis, which undergoes meiosis I.

primary structure   The first level of protein structure; the specific sequence of amino acids making up a polypeptide chain.

primary xylem (ZY-lum)   Xylem tissue formed by cells of the vascular cylinder in a young plant.

primitive character   A feature found in members of a lineage and also in the ancestors of the lineage; an ancestral trait.

principle of independent assortment   A general rule in inheritance that when gametes form during meiosis, each pair of alleles for a particular trait segregate independently; also known as Mendel's second law of inheritance.

principle of segregation   A general rule in inheritance that individuals have two alleles for each gene; and that when gametes form by meiosis, the two alleles separate, and each resulting gamete ends up with only one allele of each gene; also known as Mendel's first law of inheritance.

producer   An autotroph (a plant, alga, or autotrophic bacterium).

product   Material that results from a chemical reaction.

progesterone (pro-JES-teh-roan)   A steroid hormone secreted by the corpus luteum of the ovary; maintains the uterine lining during pregnancy.

progestin (pro-JES-tin)   One of a family of steroid hormones, including progesterone, produced by the mammalian ovary; progestins prepare the uterus for pregnancy.

prokaryotic cell (pro-KAIR-ee-OT-ik)   A type of cell lacking a membrane-enclosed nucleus and other membrane-enclosed organelles; found only in the kingdom Monera; a bacterial cell.

prolactin (PRL) (pro-LAK-tin)   A protein hormone secreted by the anterior pituitary; stimulates milk production in mammals.

promoter   A specific nucleotide sequence in DNA, located at the start of a gene, that instructs RNA polymerase where to start transcribing RNA.

prophage (PRO-faje)   Phage DNA that inserts by genetic recombination into a specific site on a bacterial chromosome.

prophase   The first stage of mitosis. During prophase, duplicated chromosomes condense from chromatin, and the mitotic spindle forms and begins moving the chromosomes toward the center of the cell.

prosimian (pro-SIM-ee-un)   A member of the primate group comprised of lorises, bushbabies, tarsiers, and lemurs.

prostaglandin (PROS-tuh-GLAN-din)   One of a large family of local regulators secreted by virtually all tissues and serving a wide variety of regulatory functions.

prostate gland (PROS-tate)   A gland in human males that secretes an acid-neutralizing component of semen.

protein   A biological polymer (macromolecule) made of one or more chains of amino acid monomers.

protist (PRO-tist)   A member of the kingdom Protista.

proton   A particle with a single positive electric charge, found in the nucleus of an atom.

proto-oncogene (PRO-toe-ONK-oh-jean)   A normal gene that can be converted to a cancer-causing gene.

protozoan (PRO-teh-ZO-un)   A protist that lives primarily by ingesting food; a heterotrophic, animal-like protist.

proximal tubule   The portion of a nephron immediately downstream from Bowman's capsule in the vertebrate kidney; conveys and helps refine filtrate.

proximate cause   In behavioral biology, the immediate explanation for an organism's behavior; the interactions of an organism with the environment or the particular environmental stimuli that trigger a behavioral response in the organism.

pseudocoelom (SOO-doe-SEE-loam)   A body cavity that is in direct contact with the wall of the digestive tract.

pseudopodium (SOO-doe-PODE-ee-um) (plural, *pseudopodia*)   A temporary extension of an amoeboid cell. Pseudopodia function in moving cells and engulfing food.

pulmonary artery   A large blood vessel that conveys blood from the heart to a lung.

pulmonary circuit   One of two main blood circuits in terrestrial vertebrates; conveys blood between the heart and the lungs.

pulmonary vein   A blood vessel that conveys blood from a lung to the heart.

pulse   The rhythmic stretching of the arteries caused by the pressure of blood forced through the arteries by contractions of the ventricles during systole.

punctuated equilibrium   The view that speciation occurs in spurts that punctuate long periods of little change.

Punnett square   A diagram used in the study of inheritance to show the results of random fertilization.

pupil   The opening in the iris, which admits light into the interior of the vertebrate eye; muscles in the iris regulate its size.

purine (PYOOR-een)   One of two families of nitrogenous bases found in nucleotides. Adenine (A) and guanine (G) are purines.

pyloric sphincter (PIE-lore-ik SFINK-ter)   In the vertebrate digestive tract, a muscular ring that regulates the passage of food out of the stomach and into the small intestine.

pyrimidine (pir-IM-ih-deen)   One of two families of nitrogenous bases found in nucleotides. Cytosine (C), thymine (T), and uracil (U) are pyrimidines.

Q

quantitative trait   A genetic trait, such as human skin color and height, that varies in a population along a continuum.

quaternary consumer (KWA-tur-ner-ee)   An organism that eats tertiary consumers.

quaternary structure   The fourth level of protein structure; the shape resulting from the association of two or more polypeptide subunits.

R

radial symmetry   An arrangement of the body parts of an organism like pieces of a pie around an imaginary central axis. Any slice passing longitudinally through a radially symmetrical organism's central axis divides it into mirror-image halves.

radioactive dating   A method using half-lives of radioactive isotopes to determine the age of fossils and rocks.

radioactive isotope   An isotope whose nucleus decays spontaneously, giving off particles and energy.

radula (RAD-yoo-luh)   A toothed, rasping organ used to scrape up or shred food; found in many mollusks.

reabsorption   In a kidney, reclamation of water and valuable solutes from the filtrate.

reactant   A starting material in a chemical reaction.

reaction center   In a photosystem in a chloroplast, the pair of chlorophyll *a* molecules that trigger the light reactions of photosynthesis. The reaction center donates an electron excited by light energy to the primary electron acceptor molecule of the photosystem.

reading frame   The specific grouping of codons (nucleotide triplets) in DNA.

receptor   On or in a cell, a specific protein molecule whose shape fits that of a specific molecular messenger, such as a hormone.

receptor-mediated endocytosis (END-oh-sigh-TOE-sis)   The movement of specific molecules into a cell by the inward budding of membranous vesicles. The vesicles contain proteins with receptor sites specific to the molecules being taken in.

receptor potential   The electrical signal produced by sensory transduction.

recessive allele   In a heterozygous individual, the allele that has no noticeable effect on the phenotype.

reciprocal altruism (AL-troo-IZ-em)   In animal behavior, a selfless act repaid at a later time by the beneficiary or by another member of the beneficiary's social system.

recombinant DNA   A DNA molecule carrying a new combination of genes.

recombinant DNA technology   Techniques for synthesizing recombinant DNA in vitro and transferring it into cells, where it can be replicated and may be expressed; also known as genetic engineering.

Recommended Dietary Allowance (RDA)   Recommendations for daily nutrient intake established by the U.S. National Academy of Sciences.

rectum   The terminal portion of the large intestine where the feces are stored until they are eliminated.

red alga   One of a group of marine, multicellular, autotrophic protists, which includes the reef-building coralline algae.

red blood cell   A blood cell containing hemoglobin, which transports O_2. Red blood cells are nucleated except in mammals. Also called erythrocyte.

red bone marrow   Blood-forming tissue found in bone cavities.

red-green color blindness   A category of common, sex-linked human disorders involving several genes on the *X* chromosome; characterized by a malfunction of light-sensitive cells in the eyes; affects mostly males but also heterozygous females.

redox (REE-doks)   Short for oxidation-reduction; chemical reactions in which electrons are lost from one substance (oxidation) and added to another (reduction). Oxidation and reduction always occur together.

reduction   The gain of electrons by a substance involved in a redox reaction; always accompanies oxidation.

reflex An automatic reaction to a stimulus.

regeneration The regrowth of body parts from pieces of an organism.

regulatory gene A gene that codes for a protein, such as a repressor, that controls the transcription of a gene or group of genes.

releasing hormone A hormone, secreted by the hypothalamus, that makes the anterior pituitary secrete hormones.

REM sleep Rapid eye movement sleep; a period of sleep when the brain is highly active, brainwaves are fairly rapid and regular, and the eyes move rapidly under the closed eyelids. Most dreams occur during REM sleep.

repressor A protein that blocks the expression of a gene or operon.

reproduction The creation of new individuals from existing ones.

reproductive barrier A biological feature of a species that prevents it from interbreeding with other species even when populations of the two species live together.

reproductive system The organ system that functions in perpetuating the species.

resolving power A measure of the clarity of an image; the ability of an optical instrument to show two objects as separate.

respiration (1) Gas exchange, or breathing; the exchange of gases between an organism and its environment. An aerobic organism takes up O_2 and gives off CO_2. (2) Cellular respiration; the aerobic harvest of energy from food molecules by cells.

respiratory surface The moist part of an animal where gas exchange with the environment occurs.

respiratory system The organ system that functions in exchanging gases with the environment; it supplies the blood with O_2 and disposes of CO_2.

resting potential The voltage across the plasma membrane of a resting neuron.

restriction enzyme A bacterial enzyme that cuts up foreign DNA, thus protecting bacteria against intruding DNA from phages and other organisms. Restriction enzymes are used in recombinant DNA technology to cut pieces of DNA with genes of interest out of chromosomes.

restriction fragments Pieces of DNA cut up by a restriction enzyme; used in RFLP (restriction fragment length polymorphism) analyses.

retina (RET-ih-nuh) The light-sensitive layer in an eye; made up of photoreceptor cells and sensory neurons.

retrovirus An RNA virus that reproduces by means of a DNA molecule; an important class of cancer-causing viruses. Retroviruses transcribe their RNA into DNA and then insert the DNA into a cellular chromosome.

reverse transcriptase (tran-SKRIP-tase) An enzyme, encoded by some RNA viruses, that catalyzes the synthesis of DNA on an RNA template.

RFLPs Restriction fragment length polymorphisms ("riflips"); the differences in homologous DNA sequences that are reflected in different lengths of restriction fragments produced when DNA is cut apart with restriction enzymes.

rhizome (RY-zoam) A horizontal stem that grows below the ground.

rhodopsin (ro-DOP-sin) Visual pigment, in the rods of the vertebrate eye. Rhodopsin absorbs dim light.

ribonucleic acid (RNA) (ry-boh-noo-KLAY-ik) The type of nucleic acid that consists of nucleotide monomers with a ribose sugar and the nitrogenous bases adenine (A), cytosine (C), guanine (G), and uracil (U); usually single-stranded; functions in protein synthesis and as the genome of some viruses.

ribosomal RNA (rRNA) (RY-bah-so-mul) The type of ribonucleic acid that, together with proteins, makes up ribosomes; the most abundant type of RNA.

ribosome (RY-beh-soam) A cell organelle consisting of two subunits (constructed in the nucleolus) and functioning as the site of protein synthesis in the cytoplasm.

RNA editing Changes in the sequence of coding nucleotides in an mRNA molecule before it is translated; may include nucleotide substitutions, insertions, and deletions.

RNA polymerase (puh-LIM-eh-rase) An enzyme that links together the growing chain of RNA nucleotides during transcription.

RNA splicing The removal of introns and joining of exons in RNA, forming an mRNA molecule with a continuous coding sequence; occurs before mRNA leaves the nucleus.

rod A photoreceptor cell in the vertebrate retina, enabling vision in dim light.

root cap A cone of cells at the tip of a plant root that protects the root's apical meristem.

root hair An outgrowth of an epidermal cell on a root, which increases the root's absorptive surface area.

root pressure The upward push of xylem sap in a vascular plant, caused by active pumping of minerals into the xylem by root cells.

root system All of a plant's roots that anchor it in the soil, absorb and transport minerals and water, and store food.

rough endoplasmic reticulum (reh-TIK-yuh-lum) A network of interconnected tubular membranes in a eukaryotic cell's cytoplasm. Rough ER membranes are studded with ribosomes that make membrane proteins and secretory proteins. The rough ER constructs membrane from phospholipids and proteins.

R plasmid A type of bacterial plasmid that carries genes for enzymes that destroy antibiotics. R plasmids protect bacteria from the lethal effects of antibiotics such as penicillin and tetracycline.

rule of addition A law of probability stating that the chance of an event's occurring in two or more alternative ways is the sum of the separate probabilities of the alternative ways.

rule of multiplication A law of probability stating that the chance of two or more independent events' occurring simultaneously is the product of the separate probabilities of each event.

ruminant mammal (ROO-min-ent) A mammal with a four-chambered stomach housing microorganisms that can digest cellulose; examples are cattle, deer, and sheep.

S

SA (sinoatrial) node (SY-noh-A-tree-ul) The pacemaker of the heart.

saccule (SAK-yool) A fluid-filled inner ear chamber containing hair cells that detect the position of the head relative to gravity.

sapwood Light-colored, water-conducting secondary xylem in a tree.

sarcoma (sar-KO-muh) Cancer of the supportive tissues, such as bone, cartilage, and muscle.

sarcomere (SAR-ko-meer) The fundamental unit of muscle contraction, composed of thin filaments of actin and thick filaments of myosin; the region between two Z lines in a myofibril.

saturated Pertaining to fats and fatty acids whose hydrocarbon chains contain the maximum number of hydrogens and therefore have no double covalent bonds. Saturated fats and fatty acids solidify at room temperature.

savanna A biome dominated by grasses and scattered trees.

scanning electron microscope (SEM) A microscope that uses an electron beam to study the detailed architecture of a cell or other specimen surfaces.

scientific method The logical process used by scientists to answer questions about nature. The method's key elements are observations, questions, hypotheses, predictions, and tests.

sclera (SKLER-uh) A layer of connective tissue forming the outer surface of the vertebrate eye. The cornea is the frontal part of the sclera.

sclereid (SKLER-id) In plants, a very hard, dead sclerenchyma cell found in nutshells and seed coats; a stone cell.

sclerenchyma cell (skler-EN-kim-uh) In plants, a supportive cell with rigid secondary walls hardened with lignin.

scrotum A pouch of skin outside the abdomen that houses a testis; functions in cooling sperm thus keeping them viable.

search image Concentration by an animal on a particular kind of food that is abundant.

seasonal migration The regular movement of animals from one place to another at particular times of the year.

secondary consumer An organism that eats primary consumers.

secondary growth An increase in a plant's girth, involving cell division in the vascular cambium and the cork cambium.

secondary immune response The immune response elicited when an animal is exposed to an antigen a second time; a more rapid, stronger, and longer lasting response than the primary immune response.

secondary oocyte (OH-eh-SITE) A haploid cell resulting from meiosis I in oogenesis, which will become an ovum after meiosis II.

secondary phloem (FLO-um) Phloem tissue produced by the vascular cambium during secondary growth of a plant.

secondary spermatocyte (spur-MAT-eh-SITE) A haploid cell resulting from meiosis I in spermatogenesis, which will become a sperm cell after meiosis II.

secondary structure The second level of protein structure; the regular patterns of coils or folds of a polypeptide chain.

secondary xylem (ZY-lum) Xylem tissue produced by the vascular cambium during secondary growth of a plant.

second law of thermodynamics The natural law that energy conversions reduce the order of the universe. Energy changes are always accompanied by entropy, an increase in disorder, such as heat.

second messenger A molecule such as cAMP that in response to a hormone triggers a change in enzymatic activity in the hormone's target cell.

secretion (1) The discharge of molecules synthesized by a cell. (2) In the vertebrate kidney, the discharge of wastes from the blood into the filtrate from the nephron tubules.

seed A plant embryo packaged with a food supply within a protective covering.

seed coat A tough outer covering of a seed, formed from the outer coat of an ovule; in a flowering plant, it encloses and protects the embryo and endosperm.

seed dormancy Temporary suspension of growth and development of a seed.

segmentation Subdivision along the length of an animal body into a series of repeated parts called segments.

selective permeability (PUR-mee-uh-BIL-ih-tee) A property of biological membranes that allows some substances to cross more easily than others and blocks the passage of other substances altogether.

self-fertilization The fusion of sperm and egg that are produced by the same individual organism.

self protein A protein on the surface of an antigen-presenting cell, which holds a foreign antigen and displays it to helper T cells. Each individual has a unique set of self proteins that serve as molecular markers for the body. Lymphocytes do not attack self proteins unless the proteins are displaying foreign antigens; therefore, self proteins mark normal body cells as off-limits to the immune system.

semen (SEE-men) The sperm-containing fluid that is ejaculated during orgasm.

semicircular canals Fluid-filled channels in the inner ear that detect changes in the head's rate of movement.

seminal vesicle (SEM-ih-nul VES-ih-cul) A gland in males that secretes a fluid (a component of semen) that lubricates and nourishes sperm.

seminiferous tubule (SEM-ih-NIF-er-us) A coiled sperm-producing tube in a testis.

sensation A feeling, or general awareness, of sensory stimuli.

sensory adaptation The tendency of sensory neurons to become less sensitive when they are stimulated repeatedly.

sensory division The part of the vertebrate peripheral nervous system (PNS) made up of sensory neurons.

sensory input The conduction of signals from sensory receptors to processing centers in the central nervous system.

sensory neuron (NOOR-on) A nerve cell that receives information from sensory receptors and transmits signals toward the central nervous system.

sensory transduction Conversion of a stimulus signal to an electrical signal by a sensory receptor cell; stimulus detection.

sepal (SEE-pul) A modified leaf of a flowering plant. A whorl of sepals encloses and protects the flower bud before it opens.

sex chromosome A chromosome that determines whether an individual is male or female.

sex-linked gene A gene located on a sex chromosome.

sexually transmitted disease (STD) A contagious disease, such as AIDS, gonorrhea, and syphilis, spread by sexual contact.

sexual reproduction The creation of offspring by the fusion of two haploid sex cells (gametes), forming a diploid zygote.

shoot system All of a plant's stems, leaves, and reproductive structures.

short-day plant A plant that flowers in late summer, fall, or winter, when day length is shortening.

sieve plate A pore in the end wall of a sieve-tube member through which phloem sap flows.

sieve-tube member A food-conducting cell in a plant; chains of sieve-tube members make up phloem tubes.

sign stimulus A stimulus that triggers a fixed action pattern.

simple fruit A fruit that develops from a flower with a single carpel and ovary.

single-lens eye A photoreceptor containing one lens that focuses light on the retina, producing a single, uninterrupted image.

sister chromatid (KRO-muh-tid) One of the two identical parts of a duplicated chromosome in a eukaryotic cell. Sister chromatids consist of exact copies of a long, coiled DNA molecule with associated proteins. Sister chromatids are joined at the centromere of a duplicated chromosome.

skeletal muscle Striated muscle attached to the skeleton. Contraction of striated muscles provides voluntary movements of the body.

skeletal system The organ system that provides body support and protects body organs such as the brain, heart, and lungs.

sliding filament model The model that explains the relationship between the structure of a sarcomere and its function, which is contraction; the explanation of how muscles contract. A sarcomere contracts when it consumes energy and its thin (actin) filaments slide across its thick (myosin) filaments.

small intestine In the vertebrate alimentary canal, the tubular organ between the stomach and large intestine; functions mainly in chemical digestion and absorption of nutrients.

smooth endoplasmic reticulum A network of interconnected tubular membranes in a eukaryotic cell's cytoplasm. Smooth ER lacks ribosomes. Enzymes embedded in the smooth ER membrane function in the synthesis of molecules, especially lipids.

smooth muscle Muscle made up of cells without striations, found in the walls of organs such as the digestive tract, urinary bladder, and arteries.

social behavior Any kind of interaction between two or more animals, usually of the same species.

sociobiology The study of the biological basis of social behavior.

sodium–potassium (Na⁺–K⁺) pump A membrane transport mechanism made of proteins that move sodium out of, and potassium into, the cell against their concentration gradients.

soil horizon A distinct layer of soil.

solute (SOL-yoot) A substance that is dissolved in a solution.

solution A fluid mixture of two or more substances, consisting of a dissolving agent, the solvent, and a substance that is dissolved, the solute.

solvent The dissolving agent in a solution. Water is the most versatile solvent known.

somatic cell (so-MAT-ik) Any cell in a multicellular organism except a sperm or egg cell.

somatic nervous system The division of the motor nervous system of vertebrates composed of neurons that carry signals to skeletal muscles.

somite (SO-mite) A block of mesoderm in a chordate embryo. Somites give rise to vertebrae and other segmental structures.

speciation (SPEE-shee-A-shun) The origin of new species.

species *See* biological species.

sperm A male gamete.

spermatogenesis (spur-MAT-oh-JEN-eh-sis) The formation of sperm cells.

sperm nucleus The nucleus of a sperm cell; fuses with an egg cell nucleus during fertilization.

spinal cord The dorsal, hollow nerve cord in vertebrates; lies within the backbone and, with the brain, makes up the central nervous system.

spinal nerve In the peripheral nervous system of vertebrates, a nerve that carries signals to or from the spinal cord.

sporangium (spuh-RANJ-ee-um) (plural, *sporangia*) A bulbous structure at the tips of some branches.

spore (1) In plants and algae, a haploid cell that can develop into a multicellular individual without fusing with another cell. (2) In prokaryotes, protists, and fungi, any of a variety of resistant life cycle stages capable of surviving unfavorable environmental conditions.

sporophyte (SPOR-uh-FITE)   The multicellular diploid form in the life cycle of organisms undergoing alternation of generations; results from a union of gametes and meiotically produces haploid spores that grow into the gametophyte generation.

stabilizing selection   Natural selection that favors intermediate variants by acting against extreme phenotypes.

stamen (STAY-men)   A pollen-producing male reproductive part of a flower, consisting of a stalk, and bearing an anther.

starch   A storage polysaccharide found in the roots of plants and certain other cells; consists of only glucose monomers.

start codon (KO-dahn)   The specific three-nucleotide sequence to which an initiator tRNA binds, starting translation of genetic information.

stem   Part of a plant's shoot system that supports the leaves and reproductive structures.

stem cell   A type of cell in the bone marrow that gives rise to all types of blood cells.

steroid (STEHR-oid)   A type of lipid whose carbon skeleton is bent to form four fused rings: three 6-sided rings and one 5-sided ring; examples are cholesterol, testosterone, and estrogen.

steroid hormone   A regulatory chemical consisting of a lipid made from cholesterol; activates transcription of specific genes in target cells.

stigma (STIG-muh) (plural, *stigmata*)   The sticky tip of a flower's carpel, which traps pollen grains.

stimulus (plural, *stimuli*)   A factor that triggers sensory transduction and nerve signals.

stoma (STO-muh) (plural, *stomata*)   A pore surrounded by guard cells in the epidermis of a leaf. When stomata are open, CO_2 enters a leaf, and water and O_2 exit. A plant conserves water when its stomata are closed.

stomach   A pouchlike organ in a digestive tract, which grinds and churns food and may store it temporarily.

stop codon (KO-dahn)   On mRNA, one of three triplets (UAG, UAA, UGA) that signal genetic translation to stop.

stretch receptor   A type of mechanoreceptor sensitive to changes in muscle length; detects the position of body parts.

strict anaerobe   An organism, such as many bacteria that live deep in the soil, that can survive only where there is no O_2.

stroma (STRO-muh)   A thick fluid enclosed by the inner membrane of a chloroplast. Sugars are made in the stroma by the enzymes of the Calvin cycle.

stromatolite (stro-MAT-ul-lite)   Rock formed of layered, fossilized bacterial mats.

substrate   (1) A specific substance (reactant) on which an enzyme acts. Each enzyme recognizes only the specific substrate or substrates of the reaction it catalyzes.   (2) A surface in or on which an organism lives.

substrate feeder   An organism that lives in or on its food source and eats its way through the food.

substrate-level phosphorylation   Formation of ATP occurring when an enzyme transfers a phosphate group from an organic molecule (one of the intermediates in glycolysis and the Krebs cycle) to ADP.

sugar sink   A location in a plant where sugar is stored or consumed.

sugar source   A location in a plant where sugar is produced, either by photosynthesis or by breaking down stored starch.

sugar-phosphate backbone   A repeating pattern of a sugar component of a nucleotide, covalently bonded to a phosphate group of another nucleotide, covalently bonded to a sugar, and so on; the structure of polynucleotides in nucleic acids.

summation (suh-MAY-shun)   The overall effect of all the information a neuron receives at a particular instant.

superior vena cava (VEE-nuh KAY-vuh)   A large vein that returns O_2-poor blood to the heart from the upper body and head.

supporting cell   In the nervous system, a cell that may protect, insulate, reinforce, and assist neurons.

suppressor T cell   A type of lymphocyte that inhibits B cells and other T cells. Suppressor T cells terminate immune activities after an infection has been eliminated.

surface tension   A measure of how difficult it is to stretch or break the surface of a liquid.

suspension feeder   An animal that traps small organisms or food particles in a film of mucus on its gills or other surface projections; the food is then swept into the mouth and ingested.

swim bladder   A gas-filled internal sac that helps bony fish maintain buoyancy.

symbiotic relationship (SIM-by-OT-ik)   An ecological relationship between organisms of two or more different species that live in direct contact; a close association between two or more species. Also called symbiosis.

sympathetic division   One of two sets of neurons in the autonomic nervous system; generally prepares the body for energy-consuming activities, such as fighting or flight.

sympatric speciation (sim-PAT-trik)   The formation of a new species as a result of a genetic change that produces a reproductive barrier between the changed population (mutants) and the parent population.

synapse (SIN-aps)   A junction between two neurons. Electrical or chemical signals are relayed from one neuron to another at a synapse.

synapsis (seh-NAP-sis)   The pairing of duplicated homologous chromosomes, forming a tetrad, during prophase I of meiosis. During synapsis, chromatids of homologous chromosomes sometimes exchange segments by crossing over.

synaptic cleft (seh-NAP-tik)   A narrow gap separating a synaptic knob of a transmitting neuron from a receiving neuron.

synaptic knob   A relay point at the tip of a transmitting neuron's axon, where signals are relayed to another neuron or to an effector.

syncytium (sin-SIH-shum) (plural, *syncytia*) A single mass of cytoplasm undivided by membranes and containing many nuclei.

systematics The scientific study of biological diversity and its classification.

systemic circuit One of two main blood circuits in terrestrial vertebrates; conveys blood between the heart and the rest of the body.

systole (SIS-tuh-lee) The contraction stage of the heart cycle when the heart chambers actively pump blood.

T

T$_3$ *See* triiodothyronine.

T$_4$ *See* thyroxine.

taiga (TIE-guh) A biome dominated by conifer trees; a band of coniferous forest extending across North America and Eurasia south of the arctic tundra; also found just below alpine tundra on mountainsides in temperate zones.

target cell A cell that responds to a regulatory signal, such as a hormone.

taxon (TAK-sahn) (plural, *taxa*) A proper name, such as phylum Chordata, class Mammalia, or *Homo sapiens*, in the taxonomic hierarchy used to classify organisms.

taxis (TAK-sis) (plural, *taxes*) Virtually automatic orientation toward or away from a stimulus.

taxonomy (tak-SAHN-eh-mee) The branch of biology concerned with identifying, naming, and classification of species.

T cell A type of lymphocyte responsible for cell-mediated immunity. T cells differentiate under the influence of the thymus.

telophase (TEL-uh-faze) The fourth and final stage of mitosis. During telophase, daughter nuclei form at the two poles of a cell. Telophase usually occurs together with cytokinesis.

temperate deciduous forest (deh-SIJ-oo-us) A biome dominated by large trees that lose their leaves seasonally; located mostly between latitudes 35° and 50° where rainfall is relatively heavy and evenly distributed throughout the year.

temperate grassland A treeless, temperate-zone biome dominated by various species of grasses.

temperate zones Latitudes between the tropics and the Arctic Circle in the north and the Antarctic Circle in the south; regions with milder climates than the tropics or polar regions.

temperature A measure of the intensity of heat, reflecting the average kinetic energy or speed of molecules.

temporal isolation A type of prezygotic barrier between species; the species remain isolated because they breed at different times.

tendon Fibrous connective tissue connecting a muscle to a bone.

terminal bud Embryonic tissue at the tip of a shoot, made up of developing leaves and a compact series of nodes and internodes.

terminator A special sequence of nitrogenous bases that marks the end of a gene; it signals RNA polymerase to release the RNA molecule, which then departs from the gene.

territory An area that an individual or individuals defend and from which other members of the same species are usually excluded.

tertiary consumer (TER-she-air-ee) An organism that eats secondary consumers.

tertiary structure The third level of protein structure; the overall, three-dimensional shape of a polypeptide in a protein.

testcross The mating between an individual of unknown genotype for a particular trait and an individual that is homozygous recessive for that same trait.

testis (plural, *testes*) A male gonad in an animal; produces sperm and reproductive hormones.

testosterone (tes-TOS-teh-roan) An androgen hormone that stimulates an embryo to develop into a male and promotes male body features.

tetrad A paired set of homologous chromosomes, each composed of two sister chromatids. Tetrads form during prophase I of meiosis.

thalamus (THAL-uh-mus) An integrating and relay center of the vertebrate forebrain; sorts and relays selected information to specific areas in the cerebral cortex.

theory A widely accepted explanatory idea that is broad in scope and supported by a large body of evidence.

thermodynamics (THER-moh-dy-NAM-iks) The study of energy transformations that occur in a collection of matter.

thermoreceptor A sensor that detects heat or cold.

thermoregulation The maintenance of internal temperature within a tolerable range.

thick filament A bundle of parallel strands of the protein myosin in a muscle fiber.

thigmotropism (thig-MOH-treh-PIZ-em) Growth movement of a plant in response to touch.

thin filament Two strands of the protein actin coiled together with regulatory protein in a muscle fiber.

threshold potential The minimum change in a membrane's voltage that must occur to generate a nerve signal.

thylakoid (THY-luh-koid) A disklike sac formed by the inner membrane of the chloroplast. Thylakoid membranes contain chlorophyll and the enzymes of the light reactions of photosynthesis. A stack of thylakoids is called a granum.

thymus gland (THY-mus) An endocrine gland in the neck region of mammals that is active in establishing the immune system; secretes several hormones that promote development and differentiation of T cells.

thyroid gland (THY-roid) An endocrine gland that secretes thyroxine (T$_4$), triiodothyronine (T$_3$), and calcitonin.

thyroid-stimulating hormone (TSH) A protein hormone secreted by the anterior pituitary; stimulates the thyroid gland to secrete its hormones.

thyroxine (T$_4$) (thy-ROK-sin) An amine hormone secreted by the thyroid gland; stimulates metabolism in virtually all body tissues.

tight junction A junction that binds tissue cells together and forms a leak-proof sheet.

Ti plasmid A bacterial plasmid that induces tumors in plants cells that it infects; often used as a vector to introduce new genes into plant cells. Ti stands for tumor-inducing.

tissue A cooperative unit of many similar cells that perform a specific function.

tissue system One of the three components (epidermis, vascular tissue, and ground tissue) continuous throughout a vascular plant and making up its roots and shoots.

topsoil The uppermost layer (horizon) of soil, subject to extensive weathering.

torpor (TOR-per) A state of reduced activity by an endotherm. Torpor reduces energy consumption because the metabolic rate, body temperature, heart rate, and breathing rate decrease.

trace element An element that is essential for the survival of an organism but only in minute quantities.

trachea (TRAY-kee-uh) The windpipe; the portion of the respiratory tube between the larynx and the bronchi.

tracheae (TRAY-kee-ee) Tiny tubes that branch throughout an insect's body, enabling gas exchange between outside air and body cells.

tracheid (TRAY-kee-id) A tapered, porous, water-conducting and supportive cell in plants. Chains of tracheids or vessel elements make up the water-conducting, supportive tubes in xylem.

trade winds The dominant winds of the tropics, created by dry air descending and moving toward the equator from higher latitudes.

transcription The transfer of genetic information from DNA into an RNA molecule.

transduction (1) The transfer of bacterial genes from one bacterial cell to another by a phage. (2) *See* sensory transduction.

transfer RNA (tRNA) A type of ribonucleic acid. Each tRNA molecule carries a specific anticodon, picks up a specific amino acid, and conveys the amino acid to the appropriate codon on mRNA.

transformation The incorporation of new genes into a cell from DNA that the cell takes up from the fluid around it.

transgenic organism An organism that contains genes from another species.

translation The transfer of genetic information from an mRNA molecule into a polypeptide, involving a change of "language" from nucleic acids to amino acids.

translocation (1) During protein synthesis, the movement of a tRNA molecule carrying a growing polypeptide chain from the A site to the P site on a ribosome. (2) A change in a chromosome resulting from a chromosomal fragment attaching to a nonhomologous chromosome; can occur as a result of an error in meiosis or from mutagenesis.

transmission electron microscope (TEM) A microscope that uses an electron beam to study the internal structure of thinly sectioned specimens.

transpiration The evaporative loss of water from a plant.

transpiration-adhesion-cohesion mechanism An explanation of how xylem sap ascends from a plant's roots to its leaves. Xylem is moved upward by a pull exerted by transpiration; the pull is relayed downward along a string of water molecules held together by cohesion and helped upward by adhesion.

transport vesicle A tiny sac containing molecules produced by a cell. The vesicle buds from the endoplasmic reticulum and is released to the outside of the cell from the plasma membrane.

transposon (trans-POZE-ahn) A transposable genetic element, or "jumping gene"; a segment of DNA that can move from one site to another and serve as an agent of genetic change.

trial-and-error learning Learning to associate a particular behavioral act with a reward or punishment.

triglyceride (try-GLIS-eh-ride) A fat.

triiodothyronine (T₃) (TRY-eye-ode-doe-THY-ruh-neen) An amine hormone secreted by the thyroid gland; stimulates metabolism in virtually all body tissues.

trimester In human development, one of three 3-month-long periods of pregnancy.

triplet code Genetic information encoded in DNA and RNA as a linear set of three-nucleotide-long sequences that specify the amino acids for polypeptide chains; the basis of the flow of genetic information from gene to protein; the genetic code.

trophic structure (TRO-fik) The feeding relationships in an ecosystem; determines the route of energy flow and the pattern of chemical cycling in an ecosystem.

trophoblast (TROFE-uh-blast) The outer portion of a blastocyst. Cells of the trophoblast secrete enzymes that enable the blastocyst to implant in the endometrium of the uterus.

tropical forest A biome located near the equator where temperatures are warm, days are 11–12 hours long year-round, and where rainfall determines the dominant vegetation.

tropical rain forest A tropical forest where rainfall exceeds 250 cm per year; the most complex and diverse of all biological communities.

tropics Latitudes between 23.5° north and south.

tropism (TRO-piz-em) A growth response that changes a plant's shape or makes it grow toward or away from a stimulus.

true-breeding variety Organisms for which sexual reproduction produces offspring with inherited traits identical to those of the parents.

tubal ligation The removal of a section of each oviduct to prevent eggs from reaching the uterus, as a means of contraception in a female.

tuber An enlargement at the end of a rhizome, in which food is stored.

tumor An abnormal mass of cells that forms within otherwise normal tissue.

tumor-suppressor gene   A gene whose products inhibit cell division, thereby preventing uncontrolled cell growth.

tundra   A biome at the northernmost limits of plant growth and at high altitudes, characterized by dwarf woody shrubs, grasses, mosses, and lichens.

U

ultimate cause   In behavioral biology, the evolutionary explanation for an organism's behavior.

ultrasound imaging   A technique for examining a fetus for anatomical deformities. High-frequency sound waves echoing off the fetus are used to produce an image of the fetus.

unsaturated   Pertaining to fats and fatty acids whose hydrocarbon chains lack the maximum number of hydrogen atoms and therefore have one or more double covalent bonds. Unsaturated fats and fatty acids do not solidify at room temperature.

ureter (yoo-REE-ter)   A duct that conveys urine from the kidney to the urinary bladder.

urethra (yoo-REE-thruh)   A duct that conveys urine from the urinary bladder to the outside. In the male, the urethra also conveys semen out of the body during ejaculation.

urinary bladder   An organ that collects urine from the kidneys and periodically empties during urination.

urine   Wastes produced by the excretory system; the product of filtration, reabsorption, and secretion by a kidney.

uterus (YOO-ter-us)   In the female reproductive system, an organ where eggs may be fertilized and the development of young occurs; the womb, or site of pregnancy, in a mammal.

utricle (YOO-trih-kul)   A fluid-filled inner ear chamber containing hair cells that detect the position of the head relative to gravity.

V

vaccination (VAK-sih-NAY-shun)   A procedure that presents the immune system with a harmless variant or derivative of a pathogen, thereby stimulating the immune system to mount defenses against the pathogen.

vaccine (VAK-seen)   A harmless variant or derivative of a pathogen used to stimulate a host organism's immune system to fight the pathogen.

vacuole (VAK-yoo-ohl)   A membrane-enclosed sac that is part of the endomembrane system of a eukaryotic cell; has diverse functions.

vagina (vu-JI-nuh)   Part of the female reproductive system between the uterus and the outside opening; the birth canal in mammals; also accommodates the male's penis and receives sperm during copulation.

vascular bundle (VAS-kyu-ler)   A strand of vascular tissues (both xylem and phloem) in a plant stem.

vascular cambium (VAS-kyu-ler KAM-bee-um)   During secondary growth of a plant, the cylinder of meristematic cells, surrounding the xylem and pith, which produces secondary xylem and phloem.

vascular plant   A plant with xylem and phloem.

vascular system   A network of narrow tubes that extends throughout a plant; made up of xylem, which channels water and minerals upward from the roots, and phloem, which channels sugars produced in the leaves throughout the plant.

vascular tissue system   The continuum of xylem and phloem throughout the plant body.

vas deferens (vas DEF-er-enz) (plural, *vasa deferentia*)   Part of the male reproductive system that conveys sperm away from the testis; the sperm duct; in humans, the tube that conveys sperm between the epididymis and the common duct that leads to the urethra.

vasectomy (veh-SEK-tum-ee)   Surgical removal of a section of the two sperm ducts (vasa deferentia) to prevent sperm from reaching the urethra; a means of contraception in the male.

vector   In molecular biology, a piece of DNA, usually a plasmid or a viral genome, that can move genes from one cell to another.

vegetative reproduction   Asexual reproduction by a plant.

vein   (1) In plants, a vascular bundle in a leaf, composed of xylem and phloem. (2) In animals, a vessel that returns blood to the heart.

ventilation   A mechanism that provides contact between an animal's respiratory surface and the air or water to which it is exposed. Contact between a respiratory surface and air or water enables gas exchange to occur.

ventral   Pertaining to the undersurface, or bottom, of a bilaterally symmetrical animal.

ventricle (VEN-trih-kul)   (1) A heart chamber that pumps blood out of a heart. (2) A space in the vertebrate brain, filled with cerebrospinal fluid.

venule (VEN-yool)   A vessel that conveys blood between a capillary bed and a vein.

vertebra (VER-teh-bruh) (plural, *vertebrae)*   One of a series of segmented units making up the backbone of a vertebrate animal.

vertebrate (VER-teh-BRATE)   A chordate animal with a backbone; includes agnathans, cartilagnous fishes, bony fishes, amphibians, reptiles, birds, and mammals.

vessel element   A short, open-ended, water-conducting and supportive cell in plants. Chains of vessel elements or tracheids make up the water-conducting, supportive tubes in xylem.

villus (VIL-us) (plural, *villi)*   (1) A fingerlike projection of the inner surface of the small intestine. (2) A fingerlike projection of the chorion of the mammalian placenta. Villi increase the surface area of these organs.

visual acuity   The ability of the eye to distinguish fine detail.

vital capacity   The maximum volume of air that a respiratory system can inhale and exhale.

vitamin   An organic nutrient that an organism requires in very small quantities. Vitamins generally function as coenzymes.

vitreous humor (VIH-tree-us HYOO-mer)   A jellylike substance filling the space behind the lens in the vertebrate eye; helps maintain the shape of the eye.

vocal cord   One of a pair of stringlike tissues in the larynx. Air rushing past the tensed vocal cords makes them vibrate, producing sounds.

W

water-conducting cells   Specialized, dead plant cells with lignin-containing secondary walls, arranged end-to-end, forming xylem tissue. *See* tracheid; vessel element.

water vascular system   In echinoderms, a radially arranged system of water-filled canals that branch into extensions called tube feet. The system provides movement and circulates water, facilitating gas exchange and waste disposal.

wavelength   The distance between crests of waves, such as those of the electromagnetic spectrum.

wax   A type of lipid molecule consisting of one fatty acid linked to an alcohol; functions as a waterproof coating on many biological surfaces, such as apples and other fruits.

westerlies   The prevailing winds in temperate zones, blowing from west to east.

wetland   An ecosystem intermediate between an aquatic one and a terrestrial one. Wetland soil is saturated with water permanently or periodically.

white blood cell   A blood cell that can fight infections and kill cancer cells. Also called leukocyte.

white matter   An area in the brain and spinal cord consisting of myelinated axons and dendrites.

wood ray   A column of parenchyma cells radiating from the center of a log and transporting water to its outer living tissues.

wood   Secondary xylem of a plant. *See* heartwood; sapwood.

X

xylem (ZY-lum)   The nonliving portion of a plant's vascular system that provides support and conveys xylem sap from the roots to the rest of the plant. Xylem is made up of vessel elements and/or tracheids.

xylem sap   The solution of inorganic ions conveyed in xylem from a plant's roots to its shoots.

Y

yellow bone marrow   Fat tissue found in bone cavities.

yolk plug   A cluster of endodermal cells at the surface of an amphibian gastrula. The yolk plug marks the position of the blastopore and the site of the future anus.

yolk sac   An extraembryonic membrane that develops from endoderm; produces the embryo's first blood cells, germ cells, and gives rise to the allantois.

Z

zooplankton (ZO-oh-PLANK-tun)   Animals that drift in aquatic environments.

zygote (ZY-goat)   The fertilized egg, which is diploid, that results from the union of a sperm cell nucleus and an egg cell nucleus.

NOTE: A *t* following a page number indicates tabular material and an *f* following a page number indicates an illustration.

A

ABA. *See* Abscisic acid
Abiotic factors in ecosystem, 660, 662
  organisms' adaptation to, 663
ABO blood groups, multiple alleles determining, 159
Abscisic acid, 644*t*, 647
Absorption, nutrient, 413
  in small intestine, 420–421
Absorptive feeders, 412
Accessory glands, digestive, 416
Accommodation, 573
Acetylcholine, 552
Acetyl CoA (acetyl coenzyme A), in Krebs cycle, 98, 99*f*
Achondroplasia, 237
Acid chyme, 419
Acid precipitation, 28, 753
Acids, 27
Acquired characteristics, inheritance of, 256
Acquired immune deficiency syndrome (AIDS), 191, 465, 482, 526*t*
Acrosome, 528
ACTH, 503*t*, 505, 511
Actin, 64, 592–593
Actinomycetes, 324–325
Action potential, 549
  neurotransmitters and, 552
  propagation of, 550
Activators, in operon, 197
Active immunity, 468
Active site, 77
Active transport, 84
Adam's apple. *See* Larynx
Adaptation, sensory, 569
Adaptive change, natural selection and, 268
Adaptive radiation, 283
Addition, rule of, 156
Adenine, 175
Adenosine triphosphate. *See* ATP
ADH. *See* Antidiuretic hormone
Adhesion, in transpiration, 626, 627
Adipose tissue, 399
Adrenal glands, 503*t*, 510–511
Adrenaline. *See* Epinephrine
Adrenocorticotropic hormone (ACTH, adrenocorticotropin), 503*t*, 505, 511
Aerobic environment, yeasts in, 89
Age structure, population growth and, 688
Agglutination, 476
Aggregate fruits, 617
Aging, genetic control of, 210
Agnathans, 382
Agonistic behavior, 732
Agriculture
  plant hormones used in, 649
  polyploid plants in, 285
  protein content of crops and, 637
  recombinant DNA technology used in, 224–225
  as second stage of culture, 751

vegetative reproduction and, 620
AIDS, 191, 465, 482, 526*t*
Air, plant nutrients in, 624
Air movements, global, 665
Air pollution, cigarette smoke and, 439
Albatross, wandering, 395, 407*f*
Albinism, 237*t*
Alcohol
  as depressant, 553
  fetus affected by maternal ingestion of, 537
Alcoholic fermentation, 103
Alcohols, 35
Aldehydes, 35
Algae, 334–337
  brown, 336
  coralline, 336
  green, 335, 337
    plants evolving from, 343
    plants differentiated from, 342–343
  red, 336
Alimentary canal, 414, 415*f*. *See also* Digestive system
  human, 416
Alkali, 27
Allantois, 515, 537
Alleles, 152
  dominant, 152
  on homologous chromosomes, 153
  multiple, 159
  recessive, 152
Allergens, 481
Allergies, 481
Alpha helix, 44
Alpha interferon, genetic engineering in production of, 222*t*
Alternation of generations, 336*f*, 337, 346–347, 614–615
Altruism, 738
Alveolus (alveoli), 438*f*, 439
Alzheimer's disease, 237*t*
Amines, 35
Amino acids, 35, 42
  as cellular fuel, 104
  essential, 426
  peptide bonds linking, 43
  sequence of in primary protein structure, 44, 45*f*
  synthesis of, 317
Amino acid sequencing, in systematics, 306–307
Amino group, 35, 42
Ammonia, 491, 710
Amnesia, 563
Amniocentesis, 240
Amnion, 515, 537
Amniotic egg, 385, 515
Amoebas, 11*f*, 332
  reproduction in, 128
Amoebocytes, 366
Amphetamine, 553
Amphibians, 384, 388
Amygdala, in memory, 563
Anabolic steroids, 41
Anaerobes, facultative, 103
Anaerobic environment, yeasts in, 89
Anaerobic respiration, fermentation as, 103

Analogous structures, 305
Anaphase, 133
Anaphase I, 140, 142
Anaphylactic shock, 481
Anchorage, cell division affected by, 136
Anchoring junctions, 66
Androgens, 503*t*, 512
Angina pectoris, 454
Angiosperms (flowering plants), 345, 350–354, 603
  animal interaction affecting evolution of, 354
  life cycle of, 352–353, 614–620
  reproduction in, flower and, 351, 614
Animal behavior. *See* Behavior
Animal cells, 56–57
  cytokinesis in, 135
  junctions between, 66
  plant cells differentiated from, 57
  surfaces of, 66, 67
Animal fats, 40. *See also* Fats
Animal husbandry, recombinant DNA technology used in, 225
Animalia (animal kingdom), 2, 3, 310. *See also* Animals
  origins of, 365
Animal Minds (book), 725
Animal phyla, 388. *See also specific type*
Animals, 3. *See also specific type*
  bilateral symmetry in, 368–369
  body cavity of, 370
  defenses of against predators, 698–699
  definition of, 364
  diversity of, evolution of, 363–392
  exchange with environment and, 406–407
  life cycles of, 364
  migration of, 728–729
    salmon, 567
  movement of, 583–598
    evolution of diverse means of, 584–585
    muscles and, 592–594
    navigation and, 728–729
    robotics and, 595
    skeletons and, 586–591
  Phylogenetic tree, 389
  structure and function of, unifying concepts of, 395–409
Annelida (annelids), 375, 388
Annuals, 610
Ants, 379, 583, 701
Antagonistic hormones
  calcitonin and PTH, 507
  insulin and glucagon, 508
Antenna, photosystem, 115
Antennae, chemoreceptors on, 571
Anterior, 368
Anterior pituitary gland, 503*t*, 504, 505
Anther, 351, 614
Anthropoids, 745
Anthropomorphism, 725
Antibiotics
  as enzyme inhibitors, 78
  production by actinomycetes, 325
  production by fungi, 360
  resistance to, R plasmids and, 215
Antibody, 468, 472, 475

Hydrocarbons, 34
Hydrochloric acid, in digestion, 418, 419f
Hydrogen, electron shell of, 22f
Hydrogen bonds, 24
   in ice, 26
   in water, 24
      cohesiveness and, 25
      temperature moderation and, 25
Hydrogen carriers, in redox reactions, 92–93
Hydrolysis, 36
Hydrophilic compounds, 35
Hydrophobic compounds, lipids as, 40
Hydrostatic skeleton, 586
Hydrothermal vents, 659
Hydroxyl group, 35
Hymen, 519
Hymenoptera, 379
Hypercholesterolemia, 237t
   genetics of, 158–159
   membrane defects and, 85
Hyperopia, 574
Hypertension, 456–457
Hyperthyroidism, 506
Hypertonic solutions, 82
Hyperventilation, 441
Hypha (hyphae), 356
Hypoglycemia, 509
Hypothalamus, 503t, 504, 504–505, 559
Hypothesis, scientific, 6
Hypothyroidism, 506
Hypotonic solutions, 82

I

I. *See* Iodine
IAA. *See* Indoleacetic acid
Ice, hydrogen bonds in, 26
Ice-minus bacterium, 224
Iguana, marine, 254–255
IL-2. *See* Interleukin 2
Imaging, 404–405
Imitation, 722t, 724
Immune system, 402–403, 465–483. *See also*
      Immunity
   AIDS affecting, 191, 465, 482
   cancer and, 479
   self recognition and, 480
   as specific defense mechanism, 468
   stress affecting, 481
Immunity (immune response), 468. *See also*
      Immune system
   active, 468
   AIDS affecting, 191, 465, 482
   allergies and, 481
   cell-mediated, 468–469
      T cells in, 478–479
   failure of, 480–481
   humoral, 468
      antibodies in, 475
      B cells in, 474–475
   passive, 468
   primary, 473
   secondary, 473
   self recognition and, 480
   stress affecting, 481
Immunization, 468
Immunodeficiency diseases, 480. *See also* AIDS

Immunological memory, 468, 473
Immunosuppression, 480
Imprinting, 722t, 722–723
Incest, innate resistance to, 739
Incomplete dominance, 158–159
Independent assortment, Mendel's principle
      of, 154–155
   linked genes and, 163
   rules of probability and, 156
Indeterminate growth, 610
Indoleacetic acid, plant growth affected by,
      644–645
Industrial Revolution, 751
Industrial waste, bacteria in cleanup of, 328–329
Infection
   lymphatic system in, 471
   nonspecific defenses against, 466
Inferior vena cava, 451
Inflammatory response, 467
Influenza virus, 190
Ingenhousz, J., 109
Ingestion, in food processing, 413
Ingestive feeders, 412
Inhalation, 440
Inheritance, 149–169
   cellular basis of, 127–147
   chromosome theory of, 162–163
   crossing over in, 144–145, 164
   human traits and, 157
   independent assortment in, 154–155
   linked genes and, 163
   Mendelian, 150–151
      chromosome behavior and, 162–163
      probability and, 156
   patterns of, 149–169
   polygenic, 161
   probability and, 156
   segregation in, 153
   sex-linked, 166–167
   X-linked, 167
Inhibiting hormones, 504
Inhibitors, 78
   competitive, 78
   noncompetitive, 78
Initiation
   of transcription, 181
   of translation, 184, 185f
Initiation codon, 184
Innate behavior, 720–721
   social, culture and, 740
Inner ear, 576
   balance receptors in, 578
Innovation, 722t, 725
Insecticides, as enzyme inhibitors, 78
Insects, 378–379, 388
   chemoreceptors on antennae of, 571
   exoskeleton in, 586–587
Insulin, 503t, 508–509
   for diabetes, 509
   formation of active form, 204
   genetic engineering in production of, 222t,
      223
Integration, 546
   central, 580
   interneurons in, 547
Integumentary system (skin), 403

as nonspecific defense, 466
   sensory receptors in, 570
Intercellular junctions. *See* Cell junctions
Intercourse, sexual, for internal fertilization,
      517. *See also* Reproduction
Interferons, 466
   genetic engineering in production of, 222t
Interleukin 2, genetic engineering in produc-
      tion of, 222t
Intermediate filaments, 64
Intermediates, in glycolysis, 96
Internal clocks. *See* Biological clocks
Internal environment, control of, 485–497
Internal fertilization, 517
Internal membranes, 56–57
Interneurons, 547
Internodes, of plant stem, 604
Interphase, 131, 132, 142
Interphase I, 140, 142
Interspecific competition, 697
Interstitial fluid, 406
Intertidal zone, 666
Intestine, 414
   large, 422
   small, 420–421
Intrauterine devices, 527
Intrinsic rate of increase, 682
Introns, 203
Inversions, 234
Invertebrates, 381
   nervous systems in, 554
   single-lens eye in, 572
In vitro fertilization, 541
Iodine, 428t
Ion channels, neurotransmitters affecting, 552
Ionic bonds, 22–23
Iris, 573
Iron, 428t
   in plant nutrition, 630
Irrigation, 633
Islands, evolutionary principles demonstrated
      on, 283
Islet cells, insulin produced by, 508
Isobutane, 34
Isomers, 34
Isotonic solution, 82
Isotopes, 20
   radioactive, 20, 21
IUDs. *See* Intrauterine devices

J

Jaguars, 717
Jaosolo, B., 743
Jaw, hinged, 382
Jellyfishes, 367, 388
Joints
   ball-and-socket, 589
   hinge, 589
   pivot, 589
Jumping genes, 216

K

K. *See* Potassium
Karyotype, 230, 231f
Kcal. *See* Kilocalories
Kelp, 336

Sinoatrial node, 453
Sister chromatids, 130, 132, 133, 139
Skeletal muscle, 400, 592
Skeletal systems (skeletons), 403, 586–591
   appendicular, 588
   axial, 588
   development of, 590–591
   disorders of, 589
   human, 588–589
   hydrostatic, 586
   interaction with muscles and, 591
Skin, 403
   as nonspecific defense, 466
   sensory receptors in, 570
Skull, 382
Skunks, 276–277
Sleep, 562
Sleep cycle, 562
Sliding filament model, of muscle contraction, 592–593
Slime molds
   cellular, 333
   plasmodial, 334
Slugs, 372–373, 388
Small intestine, 420
Smell, sense of, 570–571, 579
   in salmon, 567
Smoking, 439
   cancer and, 209, 439
Smooth endoplasmic reticulum, 59
   functions of, 67
Smooth muscle, 400
Snails, 372–373, 388
Snakes, 385, 446–447
Snomax, genetic engineering in production of, 222t
Social behavior, 731–740
   agonistic, 732
   altruism, 738
   culture and, 740
   dominance hierarchies and, 732, 733
   evolutionary basis of, 739
   mating, 734–735
   signalling in, 736–737
   territorial, 734
Sociobiology: The New Synthesis (book), 739
Sociobiology, 739
Sodium, 428t
Sodium channels (sodium gates), 550, 552
Sodium-potassium pump, 548
Soil, 632
   biosphere affected by, 662
   conservation of, 633
   horizons (layers) of, 632
   plant nutrients in, 624
Solar energy, biosphere affected by, 662
Solute concentrations, osmosis affected by, 82
Solution, 26
Solvent, 26
   water as, 26
Somatic cells, 130
   mutations in, cancer caused by, 208
Somatic nervous system, 555
Somites, 533
Song imprinting, 723
Sound, hair cells in detection of, 570, 576–577

Speciation, 279. *See also* Evolution
   geographical isolation and, 280
   on islands, 283
   sympatric, 284
Species
   in binomial, 304
   biological, 278–279
   in community, diversity and, 696
   competition among, community structure and, 697
   definition of, 263, 278–279
   development of. *See* Speciation
   endangered, reduced variation in, 270–271
   geographical isolation and, 283
   loss of, biodiversity crisis and, 757
   origin of, 277–289
   polyploid, 284
   populations differentiated from, 279
   reproductive barriers separating, 280–281
Sperm, 516, 520, 521, 528f
   formation of, 522–523
Spermatocytes
   primary, 523
   secondary, 523
Spermatogenesis, 522–523
Spermicides, 527
Sperry, R., 561
S phase, 131
Sphincters, precapillary, capillary blood flow controlled by, 457
Spiders, 33, 377, 388
Spinal cord, 554, 557
   central canal of, 555
Spinal nerves, 555
Spindle microtubules, 132–133f, 34
Spine, 554
Spirilla, 321
Spirochetes, 321
Sponges (Porifera), 366, 388
   endoskeleton in, 587
Spontaneous generation, 316
Sporangium (sporangia), 343
Spore, 343
Sporophyte generation, 336f, 337, 346, 347, 614
Squids, 373, 388
   nervous system of, 545
Stability, community, 696
Stabilizing selection, 273
Stamens, 351, 614
Staphylococcus aureus, 326
Starch, 39
   as cellular fuel, 104
Starlings, 679
STDs. *See* Sexually transmitted diseases
Stem cells, for blood cell disorder therapy, 462
Stems, plant, 604
   gibberellins affecting growth of, 646–647
   modifications of, 605
   secondary growth of, 612–613
Steppes, 674
Sterility, hybrid, as reproductive barrier, 281
Sterilization, 527
Steroid hormones, 501
Steroids, 41
   anabolic, 41
Stigma, 351, 614

Stimulants, 553
Stimulus, 548
   intensity of, receptor potential strength and, 568, 569
   movement in response to, 728
   sensory receptor cell detection of, 568–569
   sign, 721
Stimulus-response integration, 580
Stimulus-response learning, 724
Stirrup, 576
Stomach, 414, 418–419
Stomata, plant leaf, 111, 342
   opening and closing of in transpiration, 627
Stop codon, 184
Storage proteins, 42
Streams, 668
Streptococcal infection, rheumatic fever after, 480
Stress
   adrenal glands and, 510–511
   immune system affected by, 481
   plant, abscisic acid and, 647
Stretch receptors, 570
Striated muscle. *See* Skeletal muscle
Stroma, chloroplast, 63, 111
   Calvin cycle occurring in, 113
Stromatolites, 313, 315
Structural proteins, 42
Sturtevant, A.H., 165
Structure-function relationship, 396–409
Substrate feeders, 412
Substrate-level phosphorylation, in ATP generation, 94
Substrates, 77
Sucrose, 38
Sugar-phosphate backbone, 174
Sugars, 38, 39. *See also specific type*
   phloem in transport of, 628–629
Sugar sink, 628
Sugar source, 628
Sulfur, 428t
Summation, 552
Sundew plant, 635
Superior vena cava, 451
Supporting cells, 546
Suppressor T cells, 478
Surface tension of water, 25
Surface-to-volume relationship, cell size and, 55
Surrogate motherhood, 541
"Survival of the fittest," 272
Survivorship curves, 687
Suspension feeders, 412
Swallowing reflex, 417
Swamp forests, 348
Sweating, water loss and, 490
Swim bladder, 383
Swimming, locomotion by, 584
Symbiotic relationships, 330, 700–701
Symmetry
   bilateral, 368, 388
   radial, 367, 388
Sympathetic division, of autonomic nervous system, 556–557
Sympatric speciation, 284
Synapses, 551